Polymer Technology Dictionary

Tony Whelan MSc, Consultant

*Formerly Director
London Polymer Consultants Ltd
New Southgate, London, UK*

CHAPMAN & HALL

London · Glasgow · New York · Tokyo · Melbourne · Madras

Published by Chapman & Hall, 2–6 Boundary Row, London SE1 8HN, UK

Chapman & Hall, 2–6 Boundary Row, London SE1 8HN, UK

Blackie Academic & Professional, Wester Cleddens Road, Bishopbriggs, Glasgow G64 2NZ, UK

Chapman & Hall Inc., One Penn Plaza, 41st Floor, New York NY10119, USA

Chapman & Hall Japan, Thomson Publishing Japan, Hirakawacho Nemoto Building, 6F, 1-7-11 Hirakawa-cho, Chiyoda-ku, Tokyo 102, Japan

Chapman & Hall Australia, Thomas Nelson Australia, 102 Dodds Street, South Melbourne, Victoria 3205, Australia

Chapman & Hall India, R. Seshadri, 32 Second Main Road, CIT East, Madras 600 035, India

First edition 1994

© 1994 Chapman & Hall

Typeset in 9/10pt Times by Variorum Publishing Limited, Rugby
Printed in Great Britain by
St Edmundsbury Press Limited, Bury St Edmunds, Suffolk

ISBN 0 412 58180 9

Apart from any fair dealing for the purposes of research or private study, or criticism or review, as permitted under the UK Copyright Designs and Patents Act, 1988, this publication may not be reproduced, stored, or transmitted, in any form or by any means, without the prior permission in writing of the publishers, or in the case of reprographic reproduction only in accordance with the terms of the licences issued by the Copyright Licensing Agency in the UK, or in accordance with the terms of licences issued by the appropriate Reproduction Rights Organization outside the UK. Enquiries concerning reproduction outside the terms stated here should be sent to the publishers at the London address printed on this page.

The publisher makes no representation, express or implied, with regard to the accuracy of the information contained in this book and cannot accept any legal responsibility or liability for any errors or omissions that may be made.

A catalogue record for this book is available from the British Library

Library of Congress Cataloging-in-Publication data
Whelan, Tony.
  Polymer technology dictionary / Tony Whelan. — 1st ed.
    p. cm.
  ISBN 0 412 58180 9
    1. Plastics — Dictionaries.   2. Polymers — Dictionaries.   I. Title.
TP1110.W45   1993
668.9'03 — dc20   93-18950
                                                                    CIP

∞ Printed on acid-free text paper, manufactured in accordance with ANSI/NISO Z39.48-1992 (Permanence of Paper).

Polymer Technology Dictionary

CONTENTS

Preface . . . . *vii*

Notes to reader . . . . *viii*

Dictionary
    Greek 1
    Numeric 2
    Alphabetic 7

Tables
    1a. Standard (based on ISO and ASTM) abbreviations of plastics   491
    1b. Standard (based on ISO and ASTM) abbreviations of rubbers   492
    2a. Letters used to modify abbreviations for plastics (ISO and ASTM) 493
    2b. Commonly-used letters used to modify abbreviations for plastics (i.e. in addition to Table 2a) 493
    2c. Symbols used for fillers and/or reinforcing materials . . . 493
    3. Some commonly-used abbreviations and trade names of plastics and thermoplastic elastomers 494
    4. Some abbreviations and names of plastics and elastomers . 496
    5a. Some trade names/trade marks, abbreviations and suppliers of polymers and polymer compounds, sorted by alphabetical order of trade name 504
    5b. Some trade names/trade marks, abbreviations and suppliers of polymers and polymer compounds, sorted by alphabetical order of abbreviation 518
    6. Drying conditions for injection moulding materials . . . 532
    7. Heat contents of some moulding materials 533
    8. Shrinkage values 534
    9. Relative densities of some compounding ingredients and other materials 535
    10. Plastics identification chart 538
    11. Suggested temperatures and loads for MFR tests . . . 539
    12. Moisture content limit for good injection mouldings . . . 539
    13. Suggested temperature settings for high shear rate rheometry . 539
    14. Moldflow data for PA 6 540
    15. Carbon black classification 540

Appendices
    A. SI units - advice on use 541
    B. SI prefixes 541
    C. Unit conversion 542
    D. Temperature conversion 554
    E. Relative atomic masses (atomic weights) 555
    F. The Greek alphabet 555

PREFACE

For many years I have been actively involved in the testing, selection and processing of plastics materials. More recently I have specialized in teaching the subject of polymer technology to people employed in industry. As a result of this experience I realized that there was a need for a book, similar to Mark Alger's *Polymer Science Dictionary*, devoted to the technology of those polymers known as plastics and as rubbers. This is justified by the number of industries including adhesives, coatings, fibres, paints, plastics and rubbers which are based on polymer science and technology and the vast range of different industries which use plastics and rubbery materials. People who are extensively employed with the selection and use of plastics and rubbers, yet have limited training and experience of such materials should find this book particularly useful.

Rather than simply providing a short definition for each term, I have adopted the approach used by Mark Alger in the complementary volume, *Polymer Science Dictionary*, which is to provide an explanation of what many of the terms mean together with necessary background information. Of course this means that many of the entries are longer than is usual in a conventional dictionary. There are also many terms included which are not specific to polymer technology but are valuable in this type of handbook. These include many entries on measurements and units which arise as a result of the scientific and technological communities using different systems as well as some polymer science terms.

The information presented in this book is intended to give the reader a general overview of polymeric materials and their additives together with information on the processing, testing and properties of certain polymeric materials.

> The data presented in the *Polymer Technology Dictionary* does not imply any legally binding assurance of certain properties, ease of processing or suitability for a given purpose. The book is based on my knowledge and experience and represents my personal opinions. To the best of my knowledge the information is accurate; however I do not assume any responsibility whatsoever for the accuracy and completeness of such information. I strongly recommend that the user should seek out and adhere to manufacturers' and/or suppliers' detailed information and current instructions for the handling of each grade or type of material. Any determination of the suitability of a material for any use contemplated by the user, and the manner of processing is the sole responsibility of the user. There are many factors which affect the processing and properties of polymers, this book does not relieve the user from carrying out tests and experiments in order to satisfy themselves that a material is suitable for their chosen application.
>
> It is not suggested or guaranteed that any hazards outlined in this publication are the only ones which exist and before using any equipment, processing technique or material mentioned here, the user is responsible for ensuring that health and safety standards are met. Remember there is a need to reduce human exposure to many organic materials to the lowest possible limits, in view of possible long-term adverse affects.
>
> I am not responsible for ensuring that proprietary rights are not infringed or relevant legislation is observed. The book also includes some names which are, or are asserted to be proprietary trade names or trademarks. The use of such proprietary trade names or trademarks does not apply for legal purposes of a non-proprietary or general significance, nor is any other judgement implied concerning their legal status.

Whilst every effort has been made to check the accuracy of the information contained in these volumes, no material should ever be selected and specified for a component or product on a paper exercise alone. The purpose of these volumes is to provide enough information for a short list of candidates for testing and to reduce the number of fruitless tests. No liability can be accepted for loss or damage resulting from the use of information contained herein. Thanks are due to my many friends and colleagues throughout the polymer industry for their useful help and advice.

Tony Whelan
London 1993

NOTES TO READER

The text was sorted electronically and although this gives the order that one would expect, there are some peculiarities which are worth mentioning.

Look up any entry according to the first letter or number given. For example, 1-butene and 1-butylene should be looked up under 1 (one) in the numerical entry section. Greek letters also have their own separate entries and a Greek alphabet is listed at the end of the dictionary.

Lower case letters are placed before upper case letters so that the letter a is filed before A. If either the lower case letter or the upper case letter is part of a word, or phrase, which contains numbers or symbols then these numbers or symbols influence how the entry is filed. All such numbers and symbols are given priority over letters of the alphabet: symbols include spaces, hyphens, oblique strokes (/), periods (full stops), and commas: these are sorted in the order listed.

The entries for butene and butylene would be filed as follows but please bear in mind than many other entries appear in between these entries.

1-butene
1 butene
but-1-ene
butene
butene-1
butene-type material
butene type material
butylene.

Italicized wording that appears in entries throughout the dictionary indicates words or phrases which are cross-referenced to alternative entries for the reader's information.

Greek entries

α
An abbreviation used for the *degree of degradation*.

α anomaly
The original term used for the *glass transition temperature* (T_g).

α cellulose
An abbreviation used for *alpha-cellulose*.

α hardness value
A measure of the hardness of a material. See *Rockwell hardness*.

α olefin
One of a series of unsaturated hydrocarbons which are *olefins* substituted on the α carbon atom. See *alkenes*. Very important *monomers* with the formula $CH_2 = CHR$ where R is an alkyl or cycloalkyl group. Where R is CH_3 then propylene is obtained. Where R is C_2H_5 then butylene is obtained. Where R is $CH_2CH(CH_3)_2$ then methyl pentene is obtained. See *poly-(α olefins)*.

α particle
A helium nucleus which contains two protons and two neutrons: has a double positive electric charge.

α Rockwell hardness value
A measure of the hardness of a material. See *Rockwell hardness*.

α sulphur
See *rhombic sulphur*.

α-iso rubber
See *iso-rubber*.

α-methyl styrene
A vinyl monomer which has a boiling point of approximately 165°C and which is sometimes used as a replacement for styrene to make plastics materials (see *petroleum resins*). May be represented as $CH_2 = C.Me.\phi$ where Me is the methyl group and ϕ is a benzene ring (both are joined to the same carbon atom). This monomer may be polymerized to poly-α-methyl styrene. See *alpha methyl styrene*.

α-pinene resin
A *pinene resin* made from α-*pinene*.

α-transition
See *glass transition temperature*.

α-trihydrate
See *aluminium hydrate* and *aluminium trihydrate*.

β cellulose
An abbreviation used for *beta-cellulose*.

β naphthol
Also known as naphthalen-2-ol. May be represented as $C_{10}H_7OH$. This material has a boiling point of 285°C, a melting point of 122°C and a relative density (RD) of 1·22. A white solid material used, for example, as an antioxidant for rubbers. See *naphthol*.

β particles
High velocity electrons emitted from nuclei in radioactive decay.

β ray gauge
See *beta ray gauge*.

β-iso rubber
See *iso-rubber*.

β-methylacrylic acid
See *crotonic acid*.

β-naphthylamine
See *phenyl-β-naphthylamine* (in which it occurs as a contaminant).

β-oxynaphthoic acid
Abbreviations used for this material are BON or BONA. β-oxynaphthoic acid is also known as 3-hydroxy-2-naphthoic acid. It is a coupling agent used in *organic pigment manufacture*.

β-pinene resin
A *pinene resin* made from β-*pinene*.

β-thionaphthol
See *naphthyl-β-mercaptan*.

β-transition
See *secondary transition*.

γ
This symbol is widely used to indicate elastic *shear strain*. Occasionally *shear rate* is denoted by γ without the dot above it. This practice should not be encouraged because of the possibility of confusion with elastic shear strain. Use $\dot{\gamma}$ which is the Greek letter gamma with a superimposed dot: called gamma dot. $\dot{\gamma}_w$ is used as an abbreviation for the true shear rate at the wall of a die or capillary. The dot above the gamma denotes a first derivative with respect to time. $\dot{\gamma}_{w,a}$ is used as an abbreviation for the *apparent wall shear rate* at the wall of a die or capillary. It equals $4Q/R^3$ where Q is the output rate and R is the die radius.

γ-aminopropyltriethoxysilane
A *coupling agent* often used for *epoxide resins*.

γ cellulose
An abbreviation used for *gamma-cellulose*.

γ gauge
See *gamma ray gauge*.

γ transition
See *glass transition*.

γ-chloropropylene oxide
See *epichlorhydrin*.

γ-methacryloxypropyltrimethoxysilane
A *coupling agent* often used for *unsaturated polyester resin systems*. See *methacrylatosilane*.

γ-rays
A form of *high energy radiation*. See *radiation - effect of*.

γ-transition
See *secondary transition*.

δ
An abbreviation used for *solubility parameter*.

Δ
The upper case Greek letter delta is used to indicate a change or difference. For example:

ΔH = delta H is the heat change in a chemical reaction: the difference between two different values of enthalpy;

ΔP = delta P the difference between two different values of pressure (P) or the pressure drop in rheological studies;

ΔT = delta T the difference between two different values of temperature. A change of 1°C is approximately equal to a change of 1·8°F. °C/°F is approximately equal to 1·8.

ε
An abbreviation used for *strain* and for *dielectric constant*.

ε-caprolactam
See *caprolactam*.

η
An abbreviation used for *efficiency of reinforcement*. Also used as an abbreviation for the *viscosity of a solution* (see *relative viscosity*). η is sometimes used to denote the coefficient of viscosity of a melt (see *apparent viscosity*).
  η_0 = the viscosity of a pure solvent (see *relative viscosity*).
  η_{rel} = *relative viscosity*.
  η_{sp} = *specific viscosity*.
  $[\eta]$ = *limiting viscosity number*.

λ
An abbreviation used for *thermal conductivity*.

μ
Used to denote the coefficient of viscosity of a *Newtonian material*. μ_a is used as an abbreviation for *apparent viscosity*.

μ in
An abbreviation used for *micro-inch*.

μ sulphur
An abbreviation used for insoluble *sulphur*.

ρ
An abbreviation used for:
  density;
  Reynold's number: and,
  volume resistivity.

σ
An abbreviation used for *stress*.
  σ_B = *flexural strength*.
  σ_v = *volume resistivity*.

Σ_t
An abbreviation used for *tensile strain*.

Σ_B
An abbreviation used for *elongation at break*.

τ
The Greek letter tau used to represent *shear stress*.
  τ_w = *shear stress* at a capillary wall
  $\tau_{w\ true}$ = true shear stress at a capillary wall.
See *capillary rheometer*.

Ω
The abbreviation used for ohm (resistance).
  Ω.cm = ohm.centimetre.
  Ω.m = ohm.meter. See *volume resistivity*.

ω-aminoenanthic acid
Also known as ω-aminoheptanoic acid. The monomer for *nylon 7*.

ω-aminoheptanoic acid
See *ω-aminoenanthic acid*.

ω-aminoundecanoic acid
See *aminoundecanoic acid*.

Numeric entries

0° nylon belt
The fabric belt of a *steel-braced* radial tyre. This secures the tread against deforming (squirming) out of shape under load and minimizes tyre fatigue, for example, during prolonged high speed driving.

°B
An abbreviation used for degrees *Baumé*. See *Baumé scale*.

°C
An abbreviation used for degrees Celsius or centigrade. See *centigrade scale*.

°R
An abbreviation used for degrees *Réamur*.

1/2S d/2
See *method S2 d2*.

1-butene
See *butene*.

1-butylene
See *butene*.

1-chloro-2,3-epoxy propane
See *epichlorhydrin*.

1-hydropentafluoro propylene
See *hydropentafluoro propylene*.

1,1-bis-(4-hydroxyphenyl) cyclohexane
See *bisphenol Z*.

1,1,3-tris-(4-hydroxy-2-methyl-5-t-butylphenyl) butane
A *phenylalkane antioxidant*. See *phenolic antioxidant*.

1,2 BR
An abbreviation used for 1,2-polybutadiene. See *polybutadiene rubber*.

1,2-benzenedicarboxylic acid
See *o-phthalic acid*.

1,2-butadiene
See *butadiene* and *styrene-butadiene rubber*.

1,2-diaminoethane
See *ethylene diamine*.

1,2-dichloroethane
Also known as dichloroethane or DCE.

1,2-dihydroxy-2,2,4-trimethylquinoline, polymerized
An abbreviation used for this type of material is TMQ. This is a strongly discolouring antioxidant (see *staining antioxidants*). See *dihydroquinoline derivatives* and *ketone-amine condensates*.

1,2-diphenylethene
See *stilbene*.

1,2-polybutadiene
An abbreviation used for this type of material is 1,2 BR. See *butadiene rubber*.

1,2-polymerization
See *styrene-butadiene rubber* and *randomizing agent*.

1,2-propylene glycol
See *propylene glycol*.

1,2-propylene glycol mono-laurate
See *propylene glycol mono-laurate*.

1,2-propylene glycol mono-oleate
See *propylene glycol mono-oleate*.

1,2-propylene mono-stearate
See *propylene glycol mono-stearate*.

1,3-benzenedicarboxylic acid
See *isophthalic acid*.

1,3-diene
A monomer which contains two double bonds, that is, the main chain may be represented as C=C–C=C: such materials are also *vinyl monomers* but are usually considered separately as 1,3-dienes.

1,3-dihydroxybenzene
See *resorcinol*.

1,3-diphenyl-2-thiourea
See *thiocarbanilide*.

1,3-pentadiene
See *piperylene*.

1,3,5-triamino-2,4,6-triazine
See *melamine*.

1,4-benzenedicarboxylic acid
See *terephthalic acid*.

1,4 BDO
An abbreviation used for *1,4 butane diol*.

1,4 butane diol
An abbreviation used for this type of material is 1,4 BDO. A *diol* which may be used as a *chain extender*.

1,4-cyclohexanedimethanol
See *cyclohexanedimethanol*.

1,4-cyclohexylene glycol
See *cyclohexanedimethanol*.

1,4-cyclohexylenedimethylene terephthalate/isophthalate
See *poly-(1,4-cyclohexylenedimethylene terephthalate-co-isophthalate)*.

1,4-diazabicyclo-2,2,2-octane
Also known as diaminobicyclooctane or as, triethylene diamine. A tertiary amine which is an often used component of a *polyether foam catalyst system*. Often referred to as DABCO. DABCO catalyses both urethane and urea formation and when used with an organometallic compound there is a synergistic effect on urethane formation. This means that, for example, most *reaction injection moulding (RIM) formulations* use mixed catalysts. A polyether foam could use stannous octoate, dimethylethanolamine and 1,4-diazabicyclo-2,2,2-octane as a catalyst system. The stannous octoate may be replaced by dibutyl tin dilaurate if there is a danger of hydrolysis in water-containing blends. See *flexible polyurethane foam*.

1,4-dihydroxy benzene
See *hydroquinone*.

1,4-epoxy butane
See *tetrahydrofuran*.

1,4-polybutadiene rubber
See *butadiene rubber*.

1,5-naphthalene-diisocyanate
See *naphthalene-1,5-diisocyanate*.

1,5 PDO
An abbreviation used for *1,5 pentane diol*.

1,5 pentane diol
An abbreviation used for this type of material is 1,5 PDO. A *diol* which may be used as a *chain extender*.

1,6-diaminohexane
See *hexamethylene diamine*.

1,6 HDO
An abbreviation used for *1,6 hexane diol*.

1,6-hexamethylene diisocyanate
See *hexamethylene diisocyanate*.

1,6-hexanediamine
See *hexamethylene diamine*.

1,6 hexane diol
An abbreviation used for this type of material is 1,6 HDO. A *diol* which may be used as a *chain extender*.

2-hydroxybutanedioic acid
See *malic acid*.

2:4:5-T
An abbreviation used for *2:4:5-trichlorophenoxyacetic acid*.

2:4:5-trichlorophenoxyacetic acid
An abbreviation used for this material is 2:4:5-T. A hormone used in low concentrations to stimulate *latex yield*. In high concentrations it will kill rubber trees.

2-(2'-hydroxy-5'-methylphenyl)-benzotriazole
A *benzotriazole* derived from *2,2'-hydroxy-phenyl-benzotriazole* and used as an *ultraviolet absorber*.

2-(3'-tertiary-butyl-2'-hydroxy-5'-methylphenyl)-5-chlorobenzotriazole
A *benzotriazole* derived from *2,2'-hydroxy-phenyl-benzotriazole* and used as an *ultraviolet absorber*.

2-(4-morpholinyl-mercapto)-benzthiazole
See *N-oxydiethylbenzothiazolesulphenamide*.

2-benzothiazole-dithio-N-morpholine
An abbreviation used for this material is BDTM. A vulcanization *accelerator*.

2-chloro-1,3-butadiene
See *chloroprene*.

2-chloroethane phosphonic acid
A *yield stimulant* used in *natural rubber production*. The active ingredient is ethylene.

2-chloroethanol
See *ethylene chlorohydrin*.

2-ethylidenebicyclo-(2,2,1)-5-heptene
See *ethylidene norbornene*.

2-hydroxy benzophenone derivatives
The derivatives of 2-hydroxy benzophenone (for example, 2-hydroxy-4-methoxy-benzophenone) are widely used as *ultraviolet stabilisers*. The ortho placing of the hydroxyl group, relative to the carbonyl group, allows tautomeric shifts to occur which absorb energy and help to make the incoming radiation harmless. By varying the organic substituent (the X substituent) the wavelength of the UV radiation which is absorbed may be controlled: X is an alkyl or alkoxy group. By varying this group, the wavelength which causes decomposition, or degradation, of a particular polymer can be absorbed. See *substituted benzophenone*.

2-hydroxy-4-methoxy-benzophenone
A ultraviolet stabilizer. See *substituted benzophenone*.

2-hydroxy-4-methoxy-5-sulpho-benzophenone
A ultraviolet stabilizer. See *substituted benzophenone*.

2-hydroxy-4-octoxy-benzophenone
A ultraviolet stabilizer. See *substituted benzophenone*.

2-mercaptobenzthiazole
See *mercaptobenzthiazole*.

2-mercaptoimidazoline
See *ethylene thiourea*.

2-mercaptonapthalene
See *naphthyl-β-mercaptan*.

2-methyl-1-propanol
See *isobutyl alcohol*.

2-methyl-1,3-butadiene
See *isoprene*.

2-methylbuta-1,3-diene
See *isoprene*.

2-methylpropan-1-ol
See *isobutyl alcohol*.

2-naphthalene mercaptan
See *naphthyl-β-mercaptan*.

2-naphthalene-thiol
See *naphthyl-β-mercaptan*.

2-nitro biphenyl
See *o-nitrobiphenyl*.

2-pentenoic acid
This acid is, for example, liberated when copolymers of *polyhydroxybutyrate* are heated. That is, from a HB-HV copolymer.

2-phenylindole
A *metal-free organic*, heat stabilizer used with *polyvinyl chloride (PVC)*: it is used as a co-stabilizer, in non-toxic bottle and pipe applications, with a *calcium/zinc stabilizer*, an epoxy compound and a phosphite chelator.

2-propanone
See *acetone*.

2,2'-dibenzthiazyl disulphide
An additive used in conjunction with *N-oxydiethylbenzothiazolesulphenamide (NOBS)* to give an *accelerator system* with less of a delayed action, and less processing safety, than NOBS. See *benzthiazyl disulphide*.

2,2'-dihydroxy diethylamine
See *diethanolamine*.

2,2'-dihydroxy-4-methoxy-benzophenone
A tetra-substituted benzophenone. An ultraviolet stabilizer. See *substituted benzophenone*.

2,2'-dihydroxy-4,4'-dimethoxy-5-sulpho-benzophenone
A tetra-substituted benzophenone. An ultraviolet stabilizer. See *substituted benzophenone*.

2,2'-hydroxy-phenyl-benzotriazole
The organic materials known as *benzotriazoles* are derived from this material. Included in this group of *ultraviolet absorbers* and/or *ultraviolet stabilizers*, are 2-(2'-hydroxy-5'-methylphenyl)-benzotriazole and 2-(3'-tertiary-butyl-2'-hydroxy-5'-methylphenyl)-5-chlorobenzotriazole.

2,2'-iminodiethanol
See *diethanolamine*.

2,2'-thiobis-(4-methyl-6-tert-butylphenol)
Also known as 2,2'-thiobismethyl butyl phenol. A light-coloured, *phenylsulphide antioxidant* with a melting point of about 85°C and which is used synergistically with *carbon black*. Useful in rubbers, polystyrene and polyolefins. See *phenolic antioxidant*.

2,2'-thiobismethyl butyl phenol
See *2,2'-thiobis-(4-methyl-6-tert-butylphenol)*.

2,2-azobisisobutyronitrile
See *azobisisobutyronitrile*.

2,2-bis-(hydroxymethyl)-1,3-propanediol
See *pentaerythritol*.

2,2-bis-4-hydroxyphenyl propane
See *bisphenol A*.

2,2-dimethylpropane-1,3-diol
See *neopentylene glycol*.

2,2,4-trimethylpentane-1,3-diol di-isobutyrate
Also known as Texanol isobutyrate. The di-isobutyrate of 2,2,4-trimethyl-1,3-pentanediol. An abbreviation used for this material is TXIB (also used for 3,3,5-trimethylpentane-1,4-diol di-isobutyrate). A *plasticizer* for PVC which is used for non-stain flooring and in *plastisol applications*, for example, where low and consistent plastisol viscosity is required.

2,3-dimethyl-1,3-butadiene
See *dimethyl butadiene*.

2,3,7,8 TCDD
An abbreviation used for *dioxin*. See *dioxins*.

2,3,7,8-tetrachlorodibenzo-para-dioxin
An abbreviation used for this type of material is 2,3,7,8 TCDD or TCDD. See *dioxin*.

2,4-tolylene diisocyanate
Also known as toluene-2,4-diisocyanate. An isomer of *tolylene diisocyanate*.

2,4,5-triketoimidazolidine
See *polyparabanic acid*.

2,5-dimercapto-1,3,4-thiadiazole
At approximately 1 phr this material is used as a curative for *acrylic rubber*.

2,6-di-t-butyl-4-methylphenol
Also known as 4-methyl-2,6-di-t-butyl phenol or as, 2,6-di-tertiarybutyl-p-cresol. A *hindered phenol*. An *antioxidant*.

2,6-dimethyl phenol
See *2,6-xylenol*.

2,6-ditertiarybutyl-p-cresol
See *2,6-di-t-butyl-4-methylphenol*.

2,6-tolylene diisocyanate
Also known as toluene-2,6-diisocyanate. An isomer of *tolylene diisocyanate*.

2,6-xylenol
Also known as 2,6-dimethyl phenol. This isomer of *xylenol* has a melting point of approximately 49°C and a boiling point of 212°C. It is used to make the monomer for *polypropylene oxide-type materials*.

3-hydroxy-2-naphthoic acid
See *β-oxynaphthoic acid*.

3-hydroxybutyrate
An abbreviation used for this material is HB. See *polyhydroxybutyrate*.

3-hydroxypentanoate
Also known as 3-hydroxyvalerate or HV. See *polyhydroxybutyrate*.

3-hydroxyvalerate
Also known as *3-hydroxypentanoate* or *HV*. See *polyhydroxybutyrate*.

3-methoxy-4-hydroxybenzaldehyde
See *vanillin*.

3,3'-dimethyl-4,4'-diphenyl diisocyanate
Also known as bitolyl diisocyamate. An abbreviation used for this material is TODI. This off-white, solid material has a melting point of 70°C and a relative density (RD or SG) of 1.20. May be used as the isocyanate component to make *polyurethane rubbers*. See *tolylene diisocyanate*.

3,3',4,4' benzophenonetetracarboxylic dianhydride
An abbreviation used for this material is BTDA: used as a monomer to prepare some *polyimides*.

3,3,5-trimethylpentane-1,3-diol di-isobutyrate
An abbreviation used for this material is TMIB. The di-isobutyrate of 3,3,5-trimethyl-1,3-pentanediol. A low-staining *plasticizer* for *polyvinyl chloride (PVC)*. See *2,2,4-trimethylpentane-1,3-diol di-isobutyrate*.

3,3,5-trimethylpentane-1,4-diol di-isobutyrate
An abbreviation used for this material is TXIB. The diisobutyrate of 3,3,5-trimethyl-1,4-pentanediol. A *plasticizer* for *polyvinyl chloride (PVC)*. See *2,2,4-trimethylpentane-1,3-diol di-isobutyrate*.

3,5-dimethyl phenol
See *3,5-xylenol*.

3,5-xylenol
Also known as 3,5-dimethyl phenol. This isomer of *xylenol* has a melting point of approximately 63°C and a boiling point of 225°C. As it has the three reactive positions available (two ortho and one para), it is used to make *phenol-formaldehyde-type materials* with improved chemical resistance for coating purposes: it is more oil-soluble than a phenol-based PF resin.

4.4'-methylenedianiline
See *4,4'-diamino diphenyl methane*.

4-decyloxy-2-hydroxy-benzophenone
An ultraviolet stabilizer. See *substituted benzophenone*.

4-hydroxy-3-methoxy-benzaldehyde
See *vanillin*.

4-isopropyl aminodiphenylamine
An additive used in rubber formulations as an *antioxidant* and as an *antiozonant*. A brownish/purplish solid material with a melting point of approximately 70°C and a relative density of 1.17.

4-methyl-2,6-di-t-butylphenol
See *2,6-di-t-butyl-4-methylphenol*.

4-nitroperfluorobutyric acid
See *nitrosoperfluorobutyric acid*.

4-tert-butyl phenol
See *tert-butyl phenol*.

4-tert-butyl-1,2-dihydroxybenzene
See *p-tert-butyl catechol*.

4,4'bis-(dimethylamino)-benzophenone
Also known as Michler's ketone. A commonly used aromatic ketone for *photoresist* purposes. A *photosensitizer* which improves the photo-response of *polyvinyl cinnamate*.

4,4'-butylidene-bis-(6-t-butyl-m-cresol)
A bisphenol-type of *antioxidant*. A non-staining, *chain breaking antioxidant* widely used in rubbers and plastics materials.

4,4'-diamino diphenyl methane
An abbreviation used is DAPM. Also known as diamino diphenyl methane or as, 4.4'-methylene aniline or as, 4.4'-methylenedianiline or as, p,p'-diaminodiphenylmethane or as, bis-4-(aminophenylmethane). This material is produced from aniline and formaldehyde and has a melting point of approximately 92°C. It has been used as an *antioxidant, accelerator* and *activator* for rubbers although it causes brown staining and decreases processing safety. Used as a curing agent for *epoxide resins* and as a monomer for the synthesis of *polyamide* and *polyimide-type materials*.

4,4'-dicyclohexylmethane diisocyanate
See *4,4'-diphenyl methane diisocyanate*.

4,4'-dihydroxy-diphenyl methane
See *bisphenol F*.

4,4'-diphenyl methane diisocyanate
Also known as diphenylmethane di-isocyanate or as 4,4'-dicyclohexylmethane diisocyanate or as, di-p-isocyanate phenyl methane or as, methylene bis-(4-phenyl isocyanate) or as, methylene-bis-(4,4'-phenyl-isocyanate). An abbreviation used for this material is MDI. Obtained by the reaction of formaldehyde with aniline. The unpurified reaction product, which is a mixture of isomers and polynuclear amines, may be phosgenated to give polymeric MDI - this is an isocyanate used in the manufacture of *rigid polyurethane foams*. If the reaction product is purified, to give the pure 4,4' isomer then after reaction with phosgene, pure 4,4'-diphenylmethane diisocyanate is the result. Pure MDI is used to make polyurethane rubbers.

4,4'-dithiomorpholine
An abbreviation used for this material is DTDM. An example of a *sulphur donor vulcanization system* which is not an *accelerator* in that formulation.

4,4'-thiobis-(6-t-butyl-o-cresol)
Also known as 4,4'-thiobis-(6-tert-butyl-2-cresol). A *phenylsulphide antioxidant* with a melting point of about 125°C and a specific gravity of 1.08. A non-staining antioxidant for rubbers. See *phenolic antioxidant*.

4,4'-thiobis-(6-t-butyl-m-cresol)
Also known as 4,4'-thiobis-(6-tert-butyl-3-cresol) or as, bis-(2-methyl-4-hydroxy-5-t-butylphenyl)sulphide. A *phenylsulphide antioxidant* with a melting point of about 145°C and a specific gravity of 1.08. A non-staining antioxidant for rubbers. Used synergistically with *carbon black* in rubbers and polyolefins. See *phenolic antioxidant*.

4,4'-thiobis-(6-t-butyl-3-cresol)
See *4,4'-thiobis-(6-t-butyl-m-cresol)*.

4,4'-thiobis-(6-tert-butyl-3-cresol)
See *4,4'-thiobis-(6-t-butyl-m-cresol)*.

4,4'-thiobis-(6-tert-butyl-2-cresol)
See *4,4'-thiobis-(6-t-butyl-o-cresol)*.

4,4'-dithiomorpholine
See *morpholine disulphide*.

4,4'-oxybis-(benzenesulphonylhydrazide)
Also known as p,p'-oxybisbenzene sulphonyl hydrazide. An abbreviation used for this material is OB. This white crystalline powder has a decomposition temperature over the range 130 to 160°C, a melting point of about 140°C and a relative density (RD) of 1.56. A *blowing agent* used, for example, for rubbers: decomposes to give nitrogen and water. Can be used in conjunction with *sodium bicarbonate* to give more even cell structures.

6-ethoxy-2,2,4-trimethyl-1,2-dihydroxyquinoline
An abbreviation used is ETMQ. This is a strongly discolouring, liquid antioxidant (see *staining antioxidant*). See *dihydroquinoline derivatives* and *ketone-amine condensates*.

6-hexanolactam
See *caprolactam*.

6PPD
An abbreviation used for N-1,3-dimethylbutyl-N'-phenyl-p-phenylenediamine. See *antiozonant* and *antioxidant*.

9-octadecenoic acid
See *oleic acid*.

11-aminoundecanoic acid
See *aminoundecanoic acid*.

12-hydroxy-cis-9-octadecanoic acid
See *ricinoleic acid*.

24M4 test
Refers to a carbon black test. Structure levels in *carbon black*, may be determined by an *oil adsorption method* or by, the *DBP adsorption method*. These adsorption methods measure the *structure* before the black is added to the compound and before the structure is degraded by shear. To simulate what happens during shearing, the black is compressed by a load of 24,000 psi (170 MPa): the DBP value is measured before and after compression. The DBP number of the compressed sample may be reported in $cm^3/100$ g.

65:35 TDI
Also referred to as 65/35 TDI. An abbreviation used for the mixture of 2,4-tolylene diisocyanate with 2,6-tolylene diisocyanate. 65 parts of the first isomer to 35 parts of the second isomer. See *tolylene diisocyanate*.

80:20 TDI
Also referred to as 80/20 TDI. An abbreviation used for the mixture of 2,4-tolylene diisocyanate with 2,6-tolylene diisocyanate. 80 parts of the first isomer to 20 parts of the second isomer. See *tolylene diisocyanate*.

94 V-0
See *Underwriters laboratory UL 94 vertical burning test*.

94 V-1
See *Underwriters laboratory UL 94 vertical burning test*.

94 V-2
See *Underwriters laboratory UL 94 vertical burning test*.

100% modulus
An abbreviation used for this term is M_{100}. This is the stress needed to produce an elongation of 100 per cent in a rubbery material. Not a true modulus but the *tensile stress* at 100% extension.

200% modulus
An abbreviation used for this term is M_{200}. This is the stress needed to produce an elongation of 200 per cent in a rubbery material. Not a true modulus but the *tensile stress* at 200% extension.

300% modulus
An abbreviation used for this term is M_{300}. This is the stress needed to produce an elongation of 300 per cent in a rubbery material. Not a true modulus but the *tensile stress* at 300% extension. For *carbon black* loaded compounds, 300% modulus is primarily a function of carbon black structure. Increased loadings of reinforcing and semi-reinforcing blacks has a very significant effect on increasing 300% modulus. The carbon black particle size appears to have a secondary effect on modulus although in *butyl rubber* small particle size contributes to higher modulus at high loadings.

A

a
An abbreviation used for a *Mark-Houwink constant*. The letter a is also sometimes used for *acceleration* and for *atto*. See *prefixes - SI*.

A
This letter is used as an abbreviation for:
abrasion - when used in connection with *carbon black*:
adipate - see, for example, *dialphanyl adipate*;
acetyl;
alkyl - see, for example, *alkyl ricinoleate*;
work;
atactic;
ampere;
area - for example, see *volume resistivity*;
asbestos; and,
amorphous;

A/B conversion
See *method A to B conversion*.

A calender
A *calender* named after the letter A. The rolls are arranged in the shape of the upper case letter A, that is, in the shape of a triangle.

A/CPE/S
An abbreviation used for *acrylonitrile-chlorinated polyethylene-styrene copolymer*.

A/EPDM/S
An abbreviation used for *acrylonitrile-ethylene/propylene-styrene rubber*.

A/F
An abbreviation used for *across the flats*.

A plate
Part of an *injection mould* which is attached to the fixed platen by the top clamping plate: the *sprue bush* passes through this part of the mould. See *standard mould set*.

A type H-NBR
See *fully saturated nitrile rubber*.

A type screw
A type, or size, of *screw* used in *injection moulding*. Has the lowest *shot capacity* and the highest *injection pressure*. See *screw size*.

Å
An abbreviation used for *Ångström unit*.

Å.U.
An abbreviation used for *Ångström unit*.

A-B-A
See *linear triblock polymer*.

a-c
An abbreviation used for alternating current.

A-glass
A type of soda-lime glass which is relatively cheap. Also known as alkali glass or as, window glass. Has relatively poor chemical resistance to, for example, hydrolysis and a refractive index which cannot be easily matched to that of *unsaturated polyester resins* - this means that sheeting of good clarity cannot be made. The percent composition by weight is approximately SiO_2 72·0%, Al_2O_3 0·6%, CaO 10·0%, MgO 2·5%, Na_2O 14·2% and B_2O_3 8·0%. Not widely used as fibrous reinforcement. See *E-glass* and *S-glass*.

a-PP
An abbreviation used for *atactic polypropylene*.

A-stage
An early stage in the preparation of certain thermosetting resins; the material is still soluble in certain liquids and may be liquid or, capable of becoming liquid upon heating. Such resins are sometimes referred to as 'resols'. (See also *B-stage* and *C-stage*).

AA
An abbreviation used for *adipic acid*.

AAS
An abbreviation used for *acrylate-acrylonitrile-styrene*. See *acrylate-styrene-acrylonitrile*.

AATCC
An abbreviation used for American Association of Textile Chemists and Colorists.

ab
A prefix added to the words used for electrical *units* so as to give the names of the corresponding units in the electro-magnetic system, for example, abampere and abvolt.

AB block copolymer
An alternative name for a diblock *copolymer* such as *styrene-butadiene (SB) copolymer*.

ABA
An abbreviation used for a triblock copolymer such as SBS. See *linear triblock polymer*.

abbreviations - filled engineering plastics
For individual materials, a modified version of the abbreviation used for the base polymer is often employed. For example, G stands for glass and F stands for fibre i.e. GF stands for glass fibre. GF may be seen before or after the abbreviation for the base polymer. That is, GFABS, or ABS-GF, or ABS GF. If a number appears before the GF then it usually means the percentage concentration by weight of the glass fibre.

As a class of materials, filled engineering plastics are also known as filled engineering thermoplastics materials or as, reinforced thermoplastics (RTP) or as, reinforced thermoplastics materials (RETP or REP). When the thermoplastics material contains a fibrous filler (e.g. glass) the term fibre reinforced thermoplastic (FRTP) is sometimes used. As most composite materials contain short glass fibres the phrase *short fibre reinforced plastic (SFRP)* is also encountered. If the composite material contains longer glass fibres, the phrase *long fibre reinforced plastic (LFRP)* is sometimes used.

abbreviations - plastics copolymers
It is common practice, for materials based on more than one monomer, to use the names of the monomers to identify, or name, the polymer. The names of the monomers may be separated by:
a space or,
by commas followed by a space or,
by an oblique stroke or slash (/) or,
by hyphens or,
by a combination of any of the above.
To save on space, and to give consistency, the hyphenated system has been used in this publication that is, the names of the monomers have been separated by hyphens. However, if one of the starting materials is a polymer then, if it is a copolymer or terpolymer, the names of the monomer units for that polymeric starting material are separated by an oblique stroke. For example, AES is acrylonitrile-ethylene/propylene-styrene rubber.

It should be noted that sometimes the same abbreviation, i.e. the same set of letters, is used by different authors and/or authorities, for different materials. For example, EAM is sometimes used for both ethylene-vinyl acetate rubber and for ethylene-methyl acrylate rubber.

It is recommended (ISO) that an oblique stroke / be placed between the two monomer abbreviations, for example, E/P for an ethylene propylene copolymer: the oblique strokes may be omitted when common usage so dictates according to ISO 1043-1:1987 (E). The major ingredient/monomer is usually mentioned first and the other polymer is often only mentioned if it is above a certain percentage, for example, 5%. (Some, so-called homopolymers are in actual fact copolymers but the second monomer is only present in minor amounts).

Copolymers may also be abbreviated in a different way, for example, to P(HB-co-HV) where the P stands for 'poly' and the letters in brackets refer to the monomer units: the 'co' indicates a copolymer.

See, for example, **table 1** and ISO 1043.

abbreviations - plastics homopolymers

The abbreviations used for plastics homopolymers take the form of a short string of capital letters; each capital letter refers to a part of the common name. If the plastics material begins with 'poly' then the first letter is P: the other letter(s) are derived from the monomer unit. Names for homopolymers, such as polystyrene and polyethylene, are thus shortened to PS and PE respectively (see **table 1** and ISO 1043). Source-base abbreviations are nearly always used to describe polymer materials which are homopolymers.

abbreviations - plastics mixtures

When mixtures are made from two or more polymers (blends or alloys), they are commonly represented by the abbreviations used for the individual materials but each abbreviation is separated by an oblique stroke, for example, a mixture of styrene acrylonitrile copolymer and ethylene-vinyl acetate copolymer would be represented as SAN/EVA. However, In ISO 1043 it is suggested that the symbols for the basic polymers be separated by a plus (+) sign and that the symbols be placed in parentheses. For example, a mixture of *polymethyl methacrylate* and *acrylonitrile-butadiene-styrene* should be represented as (PMMA + ABS). That is, a mixture of poly(methyl methacrylate) and acrylonitrile/butadiene/styrene should be represented as (PMMA + ABS).

abbreviations - rubbery materials

Rubbery materials, and compositions based upon them, are commonly referred to by a number of capital letters which refer to the monomers on which the polymer is based. The standard recommended practice (D1418-72A) issued by the American Society for Testing Materials (ASTM) is the most widely used for the nomenclature of rubbers and lattices. ISO 1629 'Schedule for symbols for plastics and rubbers' (BS 3502 Part 2) is in many ways similar. Such documents recommend that the rubbers be grouped and coded into a number of classes (designated by capital letters) according to the chemical composition of the main polymer chain. For example, the classes are:

M rubbers having a saturated chain of the polymethylene type;
N rubbers having nitrogen in the main polymer chain;
O rubbers having oxygen in the main polymer chain;
Q rubbers having silicon and oxygen in the main polymer chain;
R rubbers having an unsaturated carbon chain, for example natural rubber and synthetic rubbers derived at least partly from diolefins;
T rubbers having sulphur in the main polymer chain: rubbers having sulphur, carbon and oxygen in the main polymer chain;
U rubbers having carbon, oxygen and nitrogen in the main polymer chain; and,
Z rubbers having nitrogen and phosphorous in the main polymer chain.

Of these various types, the 'R' and 'M' classes are the most commercially important. For example, the 'M' class includes:

CM chloropolyethylene (chlorinated polyethylene);
CSM chlorosulphonylpolyethylene (chlorosulphonated polyethylene);
EPDM terpolymer of ethylene, propylene and a diene with the residual unsaturated portion of the diene in the side chain;
EPM copolymers of ethylene and propylene; and,
IM polyisobutylene.

The 'R' class is defined by inserting the name of the monomer(s) before the word 'rubber' from which it was prepared (except for natural rubber). The letter immediately preceding the letter R signifies the diolefin from which the rubber was prepared (except for natural rubber). Any letter(s) preceding the diolefin letter signifies the comonomer(s). Commonly encountered members of this class are:

BR butadiene rubber (BR is also known as polybutadiene);
CR chloroprene rubber (CR is also known as polychloroprene;
IIR isobutene-isoprene rubber (butyl rubber).
IR isoprene synthetic (IR is also known as cis-polyisoprene);
NBR nitrile-butadiene rubber (NBR is also known as acrylonitrile butadiene rubber or as, nitrile butadiene rubber);
NR natural rubber; and,
SBR styrene-butadiene rubber.

abbreviations - standard and non-standard

Many standards organizations [for example, the American Society for Testing and Materials (ASTM), the British Standards Institution (BSI) and the International Standards Organisation (ISO)] issue standards which specify what letters shall be used for abbreviations. See ASTM standard D1600-86 (that is standard number D1600 revised/published in 1986) and called 'Standard abbreviations of terms related to plastics. Also see ISO 1043.

Both standard and non-standard abbreviations are widely used in the polymer industry: more than one abbreviation may therefore be used for the same material. For example, the *thermoplastic elastomer*, known as polyether ester elastomer, may be referred to as PEEL or, as COPE (from copolyester) or, as TEEE (thermoplastic elastomer ether ester) or, as YBPO (an American suggestion) or as TPE-E. Some abbreviations are shown in **table 1**.

aberration

A deviation from the normal or typical: usually used to mean the blurring or distortion of an image by a lens or mirror. In optics means that light rays cannot be brought to a sharp focus. See *chromatic aberration*.

abhesion

The loss in adhesion caused by, for example, a surface treatment or coating.

ABIBN

An abbreviation used for *azobisisobutyronitrile*.

abietic acid

See *rosin*.

ablation

Dissipation of heat caused by atmospheric friction.

ablative polymer compound

A substance usually used to protect against heat caused by atmospheric friction, for example, between a space craft and the Earth's atmosphere on re-entry. The material may burn away, which protects as the burning is endothermic and/or it may form a protective, insulating char.

ABR

An abbreviation used for acrylic ester butadiene rubber. See *acrylic rubber*.

abrasion resistance

The resistance of a substance to wear, for example, the resistance of a rubber component to wearing away by contact with a moving abrasive surface.

abrasive
A substance used for rubbing, or grinding, down surfaces, for example, emery.

abrasive water jetting
Water-jet cutting in which abrasive particles are incorporated in the jet. See *water jet cutting*.

abs
An abbreviation used for *absolute*.

ABS
An abbreviation used for *acrylonitrile-butadiene-styrene*.

ABS copolymer
See *acrylonitrile-butadiene-styrene*.

ABS terpolymer
See *acrylonitrile-butadiene-styrene*.

abscissa
Usually used in the sense of 'the abscissa of a point': one of a set of numbers that determines the location of a point, for example, on a graph. It is the distance from the (vertical) y axis on a graph as measured along a line parallel to the x axis.

absolute
An abbreviation used for this term is abs. Means that something is not relative but is independent or fixed: for example, the *absolute zero of temperature*, as distinct from zero on an arbitrary scale such as the Celsius/centigrade temperature scale.

absolute alcohol
Ethanol which contains greater than 99%, by weight, of pure ethanol.

absolute electrical units
See *units*.

absolute humidity
The amount of water vapour in the atmosphere: measured in terms of the number of kilograms, or grams, of water vapour in one cubic meter of air.

absolute pressure
The *pressure* measured against zero pressure. The sum of atmospheric and *gauge pressure*.

absolute system of units
Any system of units which uses the least possible number of fundamental units. A system of units in which the scales for most quantities are derived from a small number of fundamental quantities. For example, mass, time, length, temperature and, sometimes, an electrical quantity.

absolute temperature
Also known as thermodynamic temperature and has the symbol K (kelvin). The thermodynamic temperature is a measure of the thermal energy of random motion of particles of a system in thermal equilibrium. The triple point of water is defined as being 273.16 kelvins. The zero is at -273.16 degrees Celsius (absolute zero) and the magnitude of the unit of thermodynamic temperature is the same as the degree on the Celsius scale. Absolute temperature is approximately obtained by adding 273 to the Celsius/centigrade figure.

absolute zero
The lowest temperature that is theoretically possible and the zero of thermodynamic temperature. 0 K = $-273.15°C$, that is $-459.67°F$.

absorptance
The ratio of the luminous flux absorbed by a body to the flux falling on that body: this term has replaced absorptivity.

absorptiometer
A device used to measure the structure of *carbon black* by measuring the amount of *dibutyl phthalate* needed to fill the voids in a 100 g sample of the black.

absorption
The uniform penetration of one substance into another across either a liquid or solid surface. See *adsorption*.

absorption coefficient
The ratio of, the sound energy absorbed at a boundary to the sound energy falling on the boundary.

absorption - radiation
The absorption of radiation is accompanied by a rise in temperature: dull black surfaces absorb the most incident radiation and brightly polished surfaces absorb the least. Dull black surfaces are also the best radiators of energy and brightly polished surfaces radiate the least.

absorption spectra
A spectrum in which lines are missing due to absorption of that particular wavelength by passage through a medium such as an atmosphere.

ac
An abbreviation used for alternating current.

AC drive
An *electric screw drive*.

ACC
An abbreviation used for the *Automotive Composites Association* - a USA based organization.

accelerated ageing
A test procedure which attempts to simulate the effects of ageing on a compound or on a component: such testing may be done at elevated temperatures. See *artificial weathering*.

accelerated sulphur system
A *vulcanization system* for unsaturated rubbers. A vulcanizing system which is based on, for example, the *vulcanizing agent (sulphur)* an *accelerator (MBT)*, an accelerator activator (*zinc oxide*) and a fatty acid (*stearic acid*). The use of *accelerators* permits the uses of lower sulphur levels and gives more *efficient vulcanization (EV)*. Zinc oxide reacts with the accelerator to form a zinc salt which, in turn forms a perthiosalt (sometimes called a perthioaccelerator): this in turn reacts with the rubber to form cross-links. The modern tendency is to use lower sulphur levels, to use *delayed action accelerators*, to use accelerator combinations (see *synergism*) and to use *pre-vulcanization inhibitors*.

Accelerated sulphur systems are used to vulcanize the major commercial rubbers, for example, *natural rubber* and *synthetic rubbers* such as *styrene-butadiene-styrene rubber*, *synthetic polyisoprene rubber*, *polybutadiene rubber*, *butyl rubber*, *nitrile rubber* and *ethylene-propylene-diene monomer (EPDM) rubber*. This is because the systems are relatively cheap, are easily used on standard equipment, their use gives compounding flexibility and good properties to the vulcanizates. The toxicity hazards are, in general, known and understood.

acceleration
Abbreviation used is the letter a and sometimes b. The rate of change of velocity or speed. Measured in ms^{-2} or $ft\ s^{-2}$.

acceleration - due to gravity
Abbreviation used g. The rate of change of velocity or speed (acceleration) of a body falling in a vacuum. Measured in ms^{-2} or $ft\ s^{-2}$. Average value $9.80665\ ms^{-2}$ or $32.174\ ft\ s^{-2}$.

accelerator
A chemical added to speed up a reaction or to reduce the temperature at which a chemical reaction occurs. Usually associated with *cross linking (crosslinking)*. For example, with the cross linking (vulcanization) of rubber compounds: such materials may be called vulcanization accelerators. Both *inorganic* and *organic accelerators* are known. Organic accelerators are now the most important although the inorganic accelerators, used for the improved crosslinking of rubber compounds, were discovered first.

For example, the effect of lead oxide on the curing of rubber derived from the use of such oxides as colorants for rubberized fabrics. The effect of organic compounds on the curing of rubber derived from the attempted use of ammonia as a *blowing agent*.

Accelerators used, in the cross linking of rubber compounds have been classified as slow, medium-fast, fast, semi-ultra fast and ultra-fast (very rapid accelerators are called ultraaccelerators). Accelerators may also be divided into direct or delayed action types. Accelerators used include *dithiocarbamates, guanidines, sulphenamides, thiazoles, thiuram disulphides* and *xanthates*. By combining different accelerators, a wide range of vulcanization, and end-use or physical property, behaviour can be obtained.

Accelerators improve the efficiency of vulcanization (as they promote the formation of *mono-* and *di-sulphide crosslinks* rather than intra-molecular reactions). By their use, at say 1 phr, the amount of sulphur required is considerably reduced, for example, from 10 phr to 3 phr. By using a *delayed action accelerator*, it is possible to obtain rapid and efficient vulcanization at high moulding temperatures without *scorching*, at the lower, shaping or mixing temperatures: more than one accelerator may be used to obtain the desired characteristics.

Rubber accelerators are usually used in conjunction with an *activator*, for example, with a mixture of *zinc oxide* and *stearic acid*. This is because accelerators may only function efficiently in the presence of an *activator*. Natural rubber vulcanizes more quickly than the synthetic rubbers and so needs less of the more expensive accelerator system.

Basic materials such as lime (calcium oxide) or magnesium oxide are also used in novolak-based, phenolic moulding powders as accelerators. The use of accelerators is also common in the glass reinforced plastics (GRP) industry where *unsaturated polyester resins* are set by, for example, the action of a *catalyst* (an *initiator* which is often a peroxide) and an accelerator: such a material is *cobalt naphthenate*. By the use of systems, such as *methyl ethyl ketone peroxide* and *cobalt naphthenate*, curing can be effected at room temperature. See *accelerator deactivation*.

accelerator activator
A material, or combination of materials which activates an *accelerator* or makes it more efficient. For example, *zinc oxide* and a fatty acid such as *stearic acid*.

accelerator combinations
See *synergism*.

accelerator condensation products
A type of delayed action accelerator used in rubber formulations. Compounds from which the accelerator, and in some cases also the activator, are released when the rubber compound is exposed to high temperatures. Such materials are however, safe to use at processing temperatures. See *accelerator*.

accelerator deactivation
With some fillers (for example, *silica*) the *accelerator* efficiency is reduced because of surface bonding. Cure may be retarded and, in order to maintain a given cure rate, additional accelerator is required (compared to black formulations). Glycols can also be used to coat the filler before the accelerator is added as glycols are preferentially absorbed. However, the use of glycols for this purpose has diminished: silane *coupling agents* are now preferred.

accelerator mixtures of
In order to obtain desirable properties in both the compound and the vulcanizate, mixtures of *accelerators* are often used in the rubber industry. For example, the triangular accelerator system suggested by the Vanderbilt Laboratory uses *mercaptobenzthiazole, benzthiazyl disulphide* and *tetramethyl thiuram disulphide*. The use of the first two ingredients gives the compound a wide vulcanization range and good ageing properties. Benzthiazyl disulphide also controls *scorch tendency* while the tetramethyl thiuram disulphide ensures uniform cure.

acceptable quality level
An abbreviation used for this term is AQL. Means that a pre-determined level of defects will be accepted.

accumulator
A device used to store hydraulic fluid under pressure or, a part of the processing equipment (for example, in *blow moulding*) where melt is stored until required.

accumulator extrusion blow moulding
Also known as accumulator blow moulding. An abbreviation used is AEBM. Technique used to stop excessive parison sag when producing large blow mouldings by *extrusion blow moulding*. The melt is stored until required, in an *accumulator head*, and then it is ejected rapidly from that storage head - usually by means of a ram or piston.

accumulator head
Used in *extrusion blow moulding* to store melt until required: fits onto the discharge end of the *extruder*. Used to stop excessive parison sag when producing large blow mouldings: the melt is stored until required, in the accumulator head, when it is ejected rapidly - usually by means of a ram or piston. See *accumulator extrusion blow moulding*.

accurate
Refers to the closeness of a measured value to the true or accepted value. For most experimental measurements a true value is not known so the accuracy cannot really be determined. The accuracy of a measurement may be expressed in terms of the error or of the percentage error.

acetal
A thermoplastics material based on polymers having a predominance of *acetal groups*, or linkages, in the main chain. An abbreviation used for this type of material is POM. Also known as acetal plastics or as, polyoxymethylene or as, polyacetal or as, polyformaldehyde or as acetal homopolymer (POM-H) or as, acetal copolymer (POM-CO sometimes POM-C or POM-K). Best not to use the abbreviation POM-C for acetal copolymer as 'C' is also more generally used for chlorination; use instead POM-CO.

Acetals are classed as engineering thermoplastics. Acetals are hard, tough and resilient; they exhibit good creep resistance and dimensional stability. The good impact resistance is maintained at temperatures as low as −50°C. Of the two types, homopolymers have the highest tensile strength, flexural strength, fatigue resistance and hardness. Copolymers have the better thermal stability, alkali and hot water resistance and processability. Both types are crystalline and are naturally white with low moisture absorption. For outdoor use, a black-filled grade, or a UV stabilized grade, is essential. Electrical insulation properties are good but it is difficult to get materials with UL VO ratings (limiting oxygen index or LOI is approximately 15%).

Acetals can withstand fairly high temperatures for long periods of time. For example, the copolymer can withstand 80 to 100°C/176 to 212°F in air (about 20°C/36°F lower in water) for years and short term exposure to 140°C/284°F for several hours: Exposure to high temperatures causes embrittlement: at 120°C/248°F the copolymer is seriously embrittled after 3 months. Homopolymers have a slightly higher temperature resistance: the heat resistance of both is raised by the addition of glass fibre (GF). For example, it is raised by approximately 5°C by the addition of 25% glass fibre (GF).

POM commonly competes with PA (*nylon*) 66 or 6. Such polyamide materials have better resistance to abrasion and impact but acetals have better fatigue endurance, creep and water resistance and are stiffer. Both nylons and acetals have low coefficients of friction: the bearing properties of POM are improved by blending with PTFE, silicones or phosphor bronze. Polycarbonate has better creep resistance than POM. Special grades of acetals are offered to suit the applications which this material has created. For example, copolymer grades are offered as basic grades, grades with improved sliding properties, reinforced grades (e.g with glass fibre) and high impact grades. [Super tough (five times tougher than normal) may be produced by alloying with an elastomer]. The relatively poor processing characteristics of homopolymers have been improved (for example, Delrin II) by improved stabilization: in other respects it is similar to the original material (Delrin I) and is offered at the same price.

POM is resistant to stress cracking, biological attack and solvents. The copolymer has better chemical resistance, for example, it has good fuel resistance and will withstand fuels containing ethanol and up to 20% methanol. Acetals dissolve only in hot (over 70°C/158°F) solvents - such as chlorophenols and benzyl alcohol. Swollen by long-term immersion in water and solvents. POM-H is not resistant to acids and bases: copolymer is resistant to bases and will for example, withstand 50%, hot sodium hydroxide. POM is attacked by dilute mineral acids, oxidizing agents and fairly strong organic acids. Very rapid attack by concentrated nitric acid. Restricted range of pH use - more for homopolymers than copolymers. Very few solvents are known for POM; one is hexafluoracetone sesquihydrate. It is used as an adhesive but it is expensive and toxic. When it is required to sterilize POM, it is better to use superheated steam or ethylene oxide rather than high levels of gamma radiation.

The density is approximately $1\cdot41$-$1\cdot43$ gcm^{-3} (solid, non-filled material). As the natural colour of the material is translucent white, then a wide colour range is possible: it i sold in both compounded colours and as natural material for colouring on the injection moulding machine by techniques such as dry-colouring, masterbatching and liquid colouring.

POM is widely used as a light engineering materials and is often processed by *injection moulding* into components such as gears and cams. However, various impurities and contaminants have an adverse effect on the thermal stability of acetal (particularly the homopolymer). For instance, when changing from a halogen-containing polymer (*polyvinyl chloride* or a chlorine-based flame retardant) or other acid generating polymers, purge with HDPE (preferably of MFR value of 1-2) first, as rapid degradation and even explosions may occur. Degradation can even result from the use of the wrong pigment, lubricant or traces of glass filled nylon (GF PA); therefore, thorough purging of the cylinder assembly is essential so as to ensure that degradation is kept to a minimum.

Acetal is used where the components are required to be tough, resilient, stiff and to have good dimensional stability; very good snap-fits are possible. Components have good creep resistance and resistance to tensile and flexural fatigue stress. The natural material has a very attractive, white colour and is pleasing to the touch. Such properties, together with the low coefficient of friction, help explain the use of acetals in gears, bearings and conveyor components (such as load bearing hooks). In bearing applications, the rolling contact wear for acetals is less than that for PA 66. Lubricating the acetal bearings with oil, for example, with Shell Vitrea Oil 100, improves bearing performance.

There is more copolymer (POM-CO) used than homopolymer (POM-H) as it is generally considered to be easier to process. Its high resistance to boiling water, low water absorption and resistance to detergents, explains why it is used for the injection moulding of electric kettles. Because of their stability under hot water conditions, acetals are used in plumbing, for example, as valve and pump housings. Acetals have also been used for the manufacture of bicycle frames and wheels.

acetal copolymer
See *acetal*. An abbreviation used for such materials is POM-CO.

acetal groups
Chemical groups which, for example, form the basis of polyacetals, or of acetal plastics, and consist of -CH$_2$-O- groups. See *acetal*.

acetal homopolymer
An abbreviation used for such materials is POM-H. See *acetal*.

acetal linkages
See *acetal groups*.

acetal plastics
See *acetal*.

acetal thermoplastics
See *acetal*.

acetals
See *acetal*.

acetate
Sometimes used as an abbreviation for the plastics material, cellulose acetate: also used for the fibres made from *regenerated cellulose fibres* (rayon). See *secondary cellulose acetate* and *cellulosic plastics*.

acetate fibres
Fibres made from cellulose acetate in which at least 74% of the hydroxyl groups on the original cellulose, but not more than 92%, have been acetylated. See *acetylation* and *triacetate fibres*.

acetic acid
Also known as ethanoic acid. May be represented as CH$_3$COOH. A colourless liquid with a strong pungent smell of vinegar. This material has a relative density of $1\cdot06$, a melting point of approximately 17°C and a boiling point of 118°C. Made from butane or ethanol. Used, for example, to make *cellulose acetate*: acts as a coagulant for *natural rubber* (2% solution).

acetic anhydride
Also known as ethanoic anhydride and is the anhydride of *acetic acid*. May be represented as (CH$_3$CO)$_2$O. A colourless liquid with a strong pungent smell. This material has a boiling point of 140°C. Used, for example, to make *cellulose acetate*.

acetonated diphenyl amines
An abbreviation used is ADPA. See *substituted diphenyl amines*.

acetone
Also known as dimethyl ketone (DMK) or, ketopropane or as, 2-propanone or as, propan-2-one, or as propanone. It is a colourless flammable liquid with a pleasant smell and is used, for example, as a solvent. This material has a relative density (RD or SG) of 0.79, a melting point of −94°C and a boiling point of 56.3°C. It is miscible with water in all proportions. It is a solvent for many natural polymers and resins, for example, for *cellulose acetate*. May be used to determine acetone soluble sulphur in rubber compounds and uncured material in phenolics. Also used to remove plasticizers, antioxidants etc. from rubber compounds before identification by further solvent extraction with high boiling point solvents and subsequent infra-red analysis. It is a good solvent for uncured *nitrile butadiene rubber (NBR)* but a poor solvent for uncured *chloroprene rubber (CR)*, *butyl rubber (IIR)*, *natural rubber (NR)*, *styrene-butadiene rubber (SBR)* and *thiokol rubber (T)*. See *acetone extraction*.

acetone extraction
The extraction of a material with acetone. Used, for example, to determine the degree of cure of phenolic resins as acetone extracts the uncured resin. Acetone will also extract non-rubber constituents from unvulcanized rubber compounds.

From vulcanizates, it will extract mineral oils, resins, sulphur and waxes.

acetone-diphenylamine condensate
An abbreviation used is ADPA. A diphenylamine derivative used as an *antioxidant*. See *ketone-amine condensates*.

acetyl-n-butyl citrate
A *citrate plasticizer*.

acetylation
The process used to introduce an acetyl group into an organic molecule such as *cellulose*. See *cellulose acetate*.

acetylcellulose
See *cellulose acetate*.

acetylene black
A high structure, *carbon black* made by the high temperature thermal decomposition of acetylene. Used, for example, in dry cell batteries because of the very high electrical conductivity possible from this material. See *carbon black*.

acid acceptor
A material which accepts or neutralises an acid. An acid acceptor is used in chlorine containing rubbers where it is used to accept hydrochloric acid. Red lead (Pb_3O_4) and magnesium oxide (MgO) are the standard acid acceptors. Dibasic lead phosphite, dibasic lead phthalate, basic lead silicate, litharge (PbO), magnesium stearate, potassium stearate and sodium stearate are all examples of acid acceptors.

acetylene trichloride
See *trichloroethylene*.

acetyltriethyl citrate
A *citrate plasticizer*.

achromatic lens
A lens corrected for *chromatic aberration* so that false colour is reduced. Obtained by combining elements or components made of different types of glass. See *chromatic aberration*.

acicular
A needle-like shape. See *pigment*.

acid
A substance which liberates hydrogen ions when in solution: it reacts with a base to produce a salt plus water only and turns litmus red.

acid casein
Solid casein obtained from skim milk by acidifying to a pH of 4·6. This material is used to make fibres. See *casein*.

acid coagulation
The process of obtaining a solid material from *latex* using an acid: the preferred method of *coagulation*. For example, acetic or formic acid is added to *natural rubber latex*. The amount used depends upon the concentration of the latex. The latex is diluted to approximately a 12% solids content and about 10 parts of 0·5% formic acid is added to make the latex coagulate. Coagulation occurs at an iso-electric point of between 4·8 and 5·1, that is, coagulation occurs when the pH falls below approximately 5. Immediately after coagulation the latex is further processed before decomposition, or degradation, by bacteria occurs. The coagulum may be processed to give *air dried sheet* or *ribbed smoked sheet*. See *natural rubber*.

acid dye
An anionic dye which is usually applied from an acid solution: such a dye has an affinity for wool.

acid number
Also known as acid value or as AV. A measure of the concentration of the carboxyl end-groups in a polymer, for example, a *polyester*: obtained by measuring the amount (in milligrams) of potassium hydroxide (KOH) which will react with one gram of the polymer.

acid pigments
A class or type of *organic pigment*. Acid and basic pigments give strong, bright pigments which give colours ranging from blue, red, violet and green. Generally such systems are not very lightfast or bleed-resistant and are therefore used where long life is not a requirement.

acid reclamation processes
Used to obtain *reclaimed rubber* from textile-containing material. Reclaimed rubber is obtained from un-wanted vulcanized material by heating with say, 10 to 25% sulphuric acid at 95°C for about 6 h. This removes textiles and may also dissolve mineral fillers. Washing with water and alkalis follows: the material is then steam heated at approximately 200°C. See *reclaiming processes - rubber*.

acid value
See *acid number*.

acid-type dye
A *dye* in which one or more of the aromatic groups is stabilized with an acid group, for example, with the sodium salt of sulphonic acid (see *azo acid dye*). For example, anthraquinone dyes are used to make acid-type dyes.

ACM
An abbreviation used for *acrylic rubber*.

acoustic hood
A box that, for example, covers a machine (for example, a printer or granulator) and contains some of the noise made.

acoustical testing
See *surface bonding*.

acrawax
A group of synthetic waxes which are used, for example, to prevent the adhesion of polymer compounds to processing equipment.

acre
A unit of square measure or of area equal to 4,840 square yards or 4,046·86 square metres. Thought to be based on the area which could be ploughed by a team of oxen in a day. May be obtained by considering an area which is one chain wide by 10 chains (one furlong) long.

acrolein
Also known as propenal or as, acrylaldehyde. An aldehyde with the formula C_2H_3CHO and with a boiling point of 53°C. Used to make *acrylic plastics*.

across the flats
An abbreviation used for this term is A/F. A method of measuring nut sizes and bolt heads. Nuts and bolt heads are measured by the distance between the jaws, or flats, of the spanner needed to turn them and quoted as, for example, ⅞" A/F.

across the sheet
A term used in *calendering* and *extrusion*. For example, it means perpendicular to the calendering direction.

acrylaldehyde
See *acrolein*.

acrylamate reaction injection moulding
See *polyacrylamate reaction injection moulding*.

acrylate modified styrene acrylonitrile
See *acrylate-styrene-acrylonitrile*.

acrylate rubber
See *acrylic rubber*.

acrylate-acrylonitrile-styrene
See *acrylate-styrene-acrylonitrile*.

acrylate-butadiene rubber
An abbreviation used for this type of material is ABR. See *acrylic rubber*.

acrylate-styrene-acrylonitrile

An abbreviation used for this type of material is ASA. Also known as acrylate styrene acrylonitrile (AAS) or as, acrylic ester: sometimes, confusingly, referred to as styrene acrylonitrile copolymer.

The commercial success of *acrylonitrile-butadiene-styrene (ABS)* led to the development of many other multiphase materials. ABS consists of a rubbery phase dispersed in a glassy, or plastics, phase. Unfortunately the rubber used (polybutadiene) is not particularly light resistant, as it is prone to oxidation, and so ABS plastics are not naturally weather resistant. If the polybutadiene is replaced with a rubber that contains no main chain, double bonds (i.e the unsaturation is removed) then a more heat and light resistant material results. When an acrylic ester rubber (or elastomer) is used then ASA results.

ASA has good resistance to yellowing and to light; its light resistance can be further improved by the use of UV stabilisers. However, the best light resistance is, as usual, possessed by black grades. More resistant to UV, outdoor exposure, heat, yellowing and ESC than ABS; in general however, the properties are similar to those of ABS. Low temperature, impact resistance is not as good as ABS but is retained after weathering. The electrical insulation properties are similar to those of standard PS. Blending ASA with PC raises the *heat distortion temperature (HDT)*: the Vicat softening point (can be raised by 30°C/86°F) and the impact strength can also be increased.

This type of material is resistant to saturated hydrocarbons, oils (animal, vegetable and mineral), water, aqueous solutions of salts, and dilute acids and alkalis. It is not resistant to aromatic and chlorinated hydrocarbons; esters, ethers, ketones and various chlorinated hydrocarbons, for example, methylene chloride, ethylene chloride and trichloroethylene. Also not resistant to concentrated inorganic acids. Solvents such as 2-butanone, methyl ethyl ketone, dichloroethylene, and cyclohexane may be used to join ASA to itself or to ABS.

ASA has a density of approximately $1.07\ \text{gcm}^{-3}$. The natural colour of this material is an opaque, yellowish white but it is not often seen as this; usually seen as bright, glossy injection mouldings. It is best to dry all ASA for 2-4 hours at 80°C to 85°C before *injection moulding* into components such as telephones and exterior trim for automobiles/cars. ASA is used for items such as telephones as the material can have good gloss, high impact strength and ease of flow. Used for exterior trim for cars because of its resistance to impact, heat deformation and to graying (caused by exposure to UV and to hot water) when ABS is used. ASA is also used for the engine covers of mopeds and for the rear lamp mouldings for heavy vehicles. It also has been used to make windsurfer boards, garden furniture, traffic signs, electric fan components, solar panel covers, wash basins and for fittings which are resistant to hot water

Mouldings made from ASA pick up very little dust during manufacture or storage due to advantageous antistatic properties. However, dirty mouldings can be easily cleaned with soap, or detergents, as the material has good *environmental stress cracking resistance* (better than ABS). An important property of ASA is the ability to withstand processing without colour changes occurring. Thin walled components, when moulded in ABS, may be subject to colour change during storage (yellowing); the colour change can take place weeks or months after being moulded, as is caused for example, through excessive barrel residence times. Whenever this problem arises, it is suggested that the ABS material is replaced with ASA.

acrylate-styrene-acrylonitrile blend

An abbreviation used for this type of material is ASA blend. A *blend* or alloy based on the thermoplastics material *acrylate-styrene-acrylonitrile (ASA)*. For example, blending ASA with *polycarbonate (PC)* raises the heat distortion temperature (HDT): the Vicat softening point (the VSP can be raised by 30°C/86°F) and the impact strength can also be increased.

acrylic

See *polymethyl methacrylate*.

acrylic acid

Also known as propenoic acid. A material with the formula $CH_2=CHCOOH$. For example, used to make polyacrylic acid and thermosetting acrylic coatings. See *methyl acrylate*.

acrylic acid ester rubber

See *acrylic rubber*.

acrylic acid terpolymer NBR

See *carboxylated nitrile rubber*.

acrylic elastomer

See *acrylic rubber*.

acrylic ester

See *acrylate-styrene-acrylonitrile*.

acrylic ester-acrylonitrile copolymer rubber

See *acrylic rubber*.

acrylic ester-butadiene rubber

See *acrylic rubber*.

acrylic fibre

A manufactured, polymeric fibre composed of at least 85% by weight of acrylonitrile units $\{-(CH_2\text{-}CHCN)\text{-}\}$.

acrylic monomer

A *monomer* used to make an *acrylic polymer*, for example, acrylic acid or a substituted acrylic acid. That is, $CH_2=CXCOOY$: where

X = Y = H it is acrylic acid;
X = H and Y = CH_3 it is methyl acrylate;
X = CH_3 and Y = CH_3 it is methyl methacrylate;
X = H and Y = CN it is acrylonitrile; and
X = H and Y = C_2H_5 it is ethyl acrylate;

Such monomers readily polymerize so that a wide range of homopolymers and copolymers is possible. See *acrylic plastics* and *acrylic rubber*.

acrylic plastics

Plastics based on polymers made with acrylic acid or, a structural derivative of acrylic acid. See, for example, *polymethyl methacrylate*.

acrylic polymer

See *acrylic plastics* and *acrylic rubber*.

acrylic rubber

Also known as acrylate rubber or as, acrylic acid ester rubber or as, acrylic elastomer or as, polyacrylic elastomer or as, polyacrylate elastomer or as, polyacrylate rubber. An abbreviation used for this type of material is ACM: this abbreviation designates copolymers of ethyl, or other, acrylates and a small amount of a *cure site monomer* such as vinyl chloroacetate.

Acrylic rubbers are rubbery polymers, based on *acrylic monomers*, which contain polar groups to impart hydrocarbon resistance but which do not contain reactive carbon-to-carbon double bonds. ACM rubbers have good oil resistance and also good high temperature resistance: for this reason they are used for shaft seals in automotive applications. Ethylene-vinyl acetate (EAM) rubber gives better low temperature flexibility but has greater swelling in oil.

ACM rubbers are, for example, cured using a system which typically based on a curative, an accelerator, an activator and a retarder. An example is:

curative, 2,5-dimercapto-1,3,4-thiadiazole at approximately 1 phr;

accelerator, *tetrabutyl thiuram disulphide* at approximately 2·5 phr;

activator, such as lead stearate at approximately 1 phr; and, retarder, such as *zinc stearate* at approximately 1 phr.

Acrylic rubbers are included in the category of *speciality rubber* because of their good oil resistance and their resistance to high temperatures.

acrylonitrile

An abbreviation used for the material is AN. Also known as propenonitrile or as vinyl cyanide. An unsaturated monomer with the formula CH_2:CHCN and with a boiling point of 78°C. An important feedstock for the polymer industry and used to make *acrylic plastics*, fibres and synthetic rubbers. When acrylonitrile is used to make a copolymer rubber, then in the name for that rubber, acrylonitrile is often shortened to nitrile.

acrylonitrile/butadiene copolymer
See *nitrile rubber*.

acrylonitrile/butadiene rubber
See *nitrile rubber*.

acrylonitrile ethylene propylene styrene
See *acrylonitrile-ethylene/propylene-styrene rubber*.

acrylonitrile ethylene propylene styrene polymer
See *acrylonitrile-ethylene/propylene-styrene rubber*.

acrylonitrile-butadiene-styrene

An abbreviation used for this type of material is ABS. Also known as acrylonitrile butadiene styrene or as, styrene/copolymer blends or as, ABS copolymers or as, ABS terpolymers. Sometimes referred to as poly-(1-butenylene-g-1-phenylethylene-co-1-cyanoethylene).

The thermoplastics materials known as ABS were originally made by blending a lightly cross-linked *nitrile rubber (NBR)* into a *styrene-acrylonitrile (SAN)* copolymer. Now such materials are commonly made by polymerizing styrene and acrylonitrile onto polybutadiene (contained in a polybutadiene latex): the resultant grafted polybutadiene phase (the rubbery phase) is then melt compounded with styrene acrylonitrile (SAN) which is the plastics phase. Also added at this stage are additives such as stabilizers, lubricants and colorants. The SAN phase comprises more than 70% of the total composition. The grafted polybutadiene phase, which has a high rubber content, may also be used as an impact modifier for other plastics such as *polyvinyl chloride (PVC)*.

By varying the monomer ratios, the way in which they are combined, the size (and the amount) of the rubber particles, the cross-link density of the rubber particles, and the molecular weight of the SAN, it is possible to produce a wide range of materials which differ in their impact strength, ease of flow, colour, etc. In general, as the molecular weight of the SAN is increased, the strength and rigidity of the ABS increases: as the rubber content increases, the strength, hardness, heat resistance and rigidity of the ABS decreases.

Broadly speaking, this material can be divided into injection moulding grades and extrusion grades. In turn, each of these two major divisions can be sub-divided into medium, high and very high impact grades: can also have other grades such as high heat, plating and flame retardant grades. Injection moulding grades have much lower viscosities than extrusion grades: achieved by using lower molecular weight SAN and/or plasticizers or lubricants.

In general however, ABS is a hard tough material with good resistance to impact even at low temperatures. It has low water absorption and is a good electrical insulator: electrical properties are unaffected by changes in humidity. It is normally available in transparent or opaque colours and the resultant mouldings can have a high gloss and are dimensionally stable: this material gives good reproduction of the mould surface. The surface is resistant to scuffing but the material has poor weathering properties. Superior heat resistance and impact strength compared to *high impact polystyrene (HIPS)*: higher flexural modulus than PP. Not so notch sensitive as PC and PA.

When *injection moulding* the highest gloss levels are achieved with moderate melt temperatures, high mould temperatures, fast filling speeds and moderate levels of packing pressure. For electroplating applications, the melt and mould temperatures should be as high as possible; the fill speed should be slow and the packing pressure moderate. To minimize the warping of injection moulded components, mould as for electroplating but use high mould filling speeds.

This type of material is more resistant to organic chemicals (e.g. carbon tetrachloride) than PS. Resistant to staining and to alkalis, acids (not concentrated oxidizing acids), salts, oils and fats; also resistant to the majority of alcohols and hydrocarbons. Concentrated phosphoric and hydrochloric acid have little effect. Most products have good *environmental stress cracking (ESC) resistance*. ABS is not resistant to aromatic and chlorinated hydrocarbons; esters, ethers, ketones and various chlorinated hydrocarbons, for example, methylene chloride, ethylene chloride and trichloroethylene. Methyl ethyl ketone and methylene chloride may be used to solvent weld ABS.

The density of the material is approximately 1·07 gcm^{-3} (solid, non-filled material). Wide colour range possible as the natural base colour is ivory or white, depending upon the type of polymerization process used, the ingredients used etc. The ratio of A:B:S is approximately 20:30:50. These materials are hygroscopic and will absorb 0·2 to 0·35% water in 24 hours at room temperature: drying is therefore often necessary before processing.

ABS materials can have useful properties from -40 to 100°C. For example, it has a high notched impact strength (can reach 11·5 ft lbs per inch of notch), a *heat distortion temperature* that can be 100°C, good stiffness, excellent processability, a high gloss appearance and moderate cost: may be successfully electroplated. Widely used in the automotive industry e.g. for radiator grilles, mirror housings, wheel covers, air inlets and instrument panels. For office machine housings, grades are available which are colour-fast and fire retardant.

ABS materials can be tailored to emphasize a property (e.g. heat resistance, impact strength etc.) so as to satisfy specific customer requirements. In the automotive industry (a major market for ABS), there is tremendous interest in low gloss grades (i.e. matt finish materials) as these do not require post-painting.

ABS is processed by injection moulding (grilles and mirror housings), extrusion (pipe and sheet) and by thermoforming (boats and caravan components). ABS is widely used for pipes and pipe fittings as solvent welding is possible: refrigeration is another major market for ABS. The appliance market uses ABS for power tools, vacuum cleaners, fans and kitchen appliances. A major growth area is in the business machine and consumer electronics sector where this material is used in computer housings, word processors, copying machines and telephones. Components which give electromagnetic shielding (EMI), may be made by the incorporation of conductive materials such as carbon fibre, graphite fibre or aluminum flake into the base material. The use of ABS in load bearing applications continues to grow as its good dimensional stability and low creep properties are appreciated. Transparent grades of ABS offer direct competition to PC where moderate impact properties and clarity is required. Such materials have replaced PC for the use of fizzy-drink machine covers.

By blending ABS with other plastics it is possible to extend the range of use of ABS-type materials. See *acrylonitrile-butadiene-styrene blend*.

acrylonitrile-butadiene-styrene alloy
See *acrylonitrile-butadiene-styrene blend*.

acrylonitrile-butadiene-styrene blend
A blend or *alloy* of *acrylonitrile-butadiene-styrene (ABS)* with another polymer(s) - usually another thermoplastics material but could be a rubber and/or a *thermoplastic elastomer*.

By blending ABS with other polymers, it is possible to extend the range of use of ABS-type materials. For example, the use of polycarbonate (PC) or styrene-maleic anhydride (SMA) improves the *heat distortion temperature* while the use of polyvinyl chloride (PVC) improves flame retardancy: these alloys have higher viscosities than standard ABS. For example, PC/ABS blends are used where the heat resistance of ABS is not good enough, for example, for hair driers, irons and coffee makers.

During polymerization, if part or all of the styrene is replaced by a monomer such as α-methylstyrene (AMS), then a high heat grade results as the heat resistance is improved: such high heat grades have a low melt viscosity at a reasonable cost. Clear ABS grades may be made by using methyl methacrylate (MMA) as a fourth monomer; this improves transparency as it helps to match the refractive index of the other monomers. Grades are now available offering up to approximately 80% light transmission and a haze level of 10% with other properties similar to those of medium impact, standard ABS materials.

By the incorporation of chlorinated polyethylene into SAN, ABS-type materials known as ACS result. These have better flame retardancy, heat resistance, weatherability and resistance to dust deposition than ABS but also have poorer processing stability. Olefin modified SAN results from the incorporation of olefin elastomers into SAN: similar properties to ABS but have better weathering properties.

acrylonitrile-butadiene-styrene copolymer
See *acrylonitrile-butadiene-styrene*.

acrylonitrile-butadiene rubber
See *nitrile rubber*.

acrylonitrile-chlorinated polyethylene-styrene copolymer
Also known as acrylonitrile-styrene-chlorinated polyethylene. An abbreviation used for this material is ACS or A/CPE/S. A terpolymer, which is a thermoplastics material, made by polymerizing *acrylonitrile* and *styrene* in the presence of *chlorinated polyethylene*. Similar in many ways to *acrylonitrile-butadiene-styrene* but has better weathering properties and fire resistance.

acrylonitrile-chloroprene rubber
An abbreviation used for this type of material is NCR. Also known as chloroprene-acrylonitrile copolymer or rubber. See *chloroprene rubber*.

acrylonitrile-ethylene/propylene-styrene rubber
Also known as acrylonitrile-styrene/EPR rubber or as, acrylonitrile-styrene/EPR elastomer. An abbreviation used for this material is AES. That is, *styrene-acrylonitrile copolymer* blended or grafted onto an ethylene-propylene rubber (EPR) and/or an ethylene-propylene-diene monomer rubber (EPDM). A terpolymer, which is a thermoplastics material and which is sometimes referred to as acrylonitrile ethylene propylene styrene polymer (A/EPDM/S). Similar in many ways to *acrylonitrile-butadiene-styrene* but has better weathering properties.

acrylonitrile-isoprene rubber
See *nitrile-isoprene rubber*. An abbreviation used for this type of material is NIR.

acrylonitrile-methyl methacrylate
An abbreviation used is AMMA. A thermoplastics material which is a *copolymer* of *acrylonitrile* and *methyl methacrylate*.

acrylonitrile-styrene-acrylate copolymer
See *acrylate-styrene-acrylonitrile*.

acrylonitrile-styrene/EPR elastomer
See *acrylonitrile-ethylene/propylene-styrene rubber*.

acrylonitrile-styrene/EPR rubber
See *acrylonitrile-ethylene/propylene-styrene rubber*.

acrylonitrile-styrene-chlorinated polyethylene
See *acrylonitrile- chlorinated polyethylene-styrene copolymer*.

ACS
An abbreviation used for *acrylonitrile-chlorinated polyethylene-styrene copolymer*. An abbreviation used for the American Chemical Society.

activate
See *activator*.

activated
When used in connection with a *filler*, usually means that the filler has been treated (heated or coated) so as to make it more absorbent or, to improve dispersion or, to improve adhesion between the polymer and the filler (substrate).

activated alumina
Alumina which has been *activated*. Prepared from various hydrated aluminas by controlled heating so as to remove the water of constitution. Made from *aluminium trihydrate* that has been dehydrated so as to leave a very porous absorbent structure. Used, for example, for drying purposes. See *alumina*.

activated calcium carbonate
A *precipitated calcium carbonate* which has been treated with *stearic acid* or a stearate so as to aid dispersion. This material has a relative density (RD or SG) of 2·6.

activated carbon
Charcoal which has been treated (for example, heated) to improve its absorbent characteristics. Used in, for example, waste recovery systems.

activated charcoal
See *activated carbon*.

activation
See *sensitisation*.

activator
A chemical or compound added to a formulation in order to improve the rate of cross-linking A chemical or compound added to activate or 'kick' a reaction. Activators are usually associated with cross-linking where such chemicals are used with an *accelerator* in order to realise the full potential of the accelerator. For example, for diene rubbers, the activator is a mixture of a metal oxide (usually *zinc oxide*) and a fatty acid (usually *stearic acid*). Where good transparency is required a fatty acid salt (*zinc stearate*) is used in place of the acid.

activator mixture
One of the mix ingredients used in *flexible polyurethane (PU)* foam production by the *slab-stock process*. Consists of water, a catalyst (for example, dimethylcyclohexylamine), an emulsifier (for example, sulphonated castor oil), a structure modifier (*silicone oil*) and paraffin oil (controls pore size and minimises foam splitting).

active copper compounds
See *copper*.

active environment
An environment which promotes *environmental stress cracking*. The active environment depends upon the particular polymer, but commonly includes detergent, oils, greases and fats. Gases or vapours may also cause failure.

activity
A term sometimes used in place of 'reinforcing' in connection with fillers for rubbers. Higher activity means higher reinforcing action, for example, with carbon black: inactive means no reinforcing action. Medium activity means that the reinforcing action lies in between no reinforcing action and a good reinforcing action. See *carbon black*.

actual temperature
That temperature which is measured as opposed to that which is set.

actuator
A device which induces movement. A component of a hydraulic, or a pneumatic, circuit and which provides movement when required: a motor cylinder. For example, an actuator may be attached to an injection moulding mould to move the splits or side cores.

acyl
A univalent organic group with the general formula RCO-. May be considered as being derived from a carboxylic acid RCOOH. See *diacyl peroxide*.

acyl anhydride
An acid anhydride which contains the *acyl group*.

acyl halide
An acid halide which contains the *acyl group*.

acyl peroxide
A *peroxide* which contains the *acyl group*. See *diacyl peroxide*.

adapter
Also spelt adaptor. Connects an extruder to the die and funnels the melt into the die; used to attach the dies to the machine. May also change the direction of melt flow as there are angled, crosshead and offset adapters.

adapter - pipe fitting
Also spelt adaptor. A pipe fitting which is fitted into the inlet and/or the outlet holes of a system so as to effect operative compatibility.

adaptive control
A control system which automatically changes settings in response to changes in machine performance so as to try and produce products of the specified quality. A machine is said to have adaptive control if it 'adapts' itself to meet a change so as to improve performance, for example, in *injection moulding*.

ADC
An abbreviation used for *azodicarbonamide*. The same initials may also be used for ammonium diethyl dithiocarbonate.

added sulphur
One of the ways the *sulphur* in rubber compounds, on analysis, may be classified. Elemental sulphur: sulphur which is added to achieve vulcanization. See *sulphur analysis*.

addition cross-linked silicone polymer
A *room temperature vulcanizing silicone rubber system*. There are two main categories of this type of material, which are *condensation cross-linked silicone polymers* and addition cross-linked silicone polymers. This type of material requires absolute cleanliness during processing as many substances will interfere with the cure. See *two-pack systems* - room temperature vulcanizing silicone rubbers.

addition polymerization
Polymerization in which monomers are linked together without the splitting off of water or other simple molecules.

additive
A substance which is added to a polymer to modify one or more properties.

Most rubber and plastics products are composed of a mixture of several materials in addition to the basic polymer. Such additives are used, for example, to enhance the appearance of the final product, to allow product production by mass production methods and to extend service life. The chief groups of materials used for such purposes are *plasticizers*, *anti-ageing additives* (*heat stabilizers*, *antioxidants* and *ultraviolet stabilizers*) *colorants*, *lubricants*, *flame retardants* and *blowing agents*. Not every plastics material, uses all of the additives listed: some formulations are relatively simple while others, noticeably those based on *polyvinyl chloride (PVC)*, are very complex.

With many materials, variations of basic formulas are available with additives to provide, for example, improved heat resistance or, weatherability. Some formulations offer improved impact strength while others, which contain fillers, are used where the products require greater flexural strength and heat distortion temperature. Processing and performance modifiers can be added; these include, for example, in the case of thermoplastics materials, antistatic and nucleating agents: such additives may form part of a *masterbatch system*. The properties of, for example, plastics products may therefore be dramatically changed not only by the processing conditions employed but by the use of additives. However, to get the best results from additives, they must be very well dispersed within the basic polymeric material: this is why *melt mixing* is very important to the polymer industry.

The additives may be dispersed, at the end-use level, within the polymer so as to form a compound which is, for example, sold by the raw material supplier. Alternatively, a *masterbatch* may be produced and subsequently incorporated by either the supplier or by the processor. Rubbers are normally compounded *in-house* whereas thermoset moulding powders are supplied already compounded. There is a tendency to supply thermoplastics as *natural materials* which are then, for example, coloured on the machine by a *masterbatch*. It is most important that any additive, is added at a definite, pre-selected ratio so that, for example, the flow properties of the resultant blend, or compound, are consistent.

additive flame retardant
A type of flame retardant which is added to the polymer system but which does not enter into a chemical reaction with the polymer so as to become part of the chemical structure of the polymer. See *flame retardant* and *aluminium trihydrate*.

additives
The collective name for the class of materials which are used to modify the properties or processing of polymers. See *additive*.

adhesion promoter
A material which will improve the adhesion between a polymer and substrate. For example, *m-phenylene diamine* has been used to improve the adhesion of rubber to tyre cords.

adhesives
The collective name for the class of non-metallic materials which can join solids by adhesion and cohesion. That is, they can surface bond to the substrates and the adhesive then has sufficient internal strength to withstand use.

adiabatic extrusion
Extrusion performed without the addition of external heat. After the extruder has been warmed up, there is no external heat supplied to the machine as the heat necessary to plasticize (plasticise) the material comes from conversion of the drive energy.

adiabatic machine
A machine, for example, a plastics *extruder* or *injection moulding machine*, which is designed to operate without using external heaters.

adiabatically
Usually applied to heating of plastics materials and means that the material is heated by working. See *adiabatic extrusion*.

adipate
The reaction product of adipic acid and an alcohol. An *adipic acid ester*. The alcohol may be linear or branched and/or mixed alcohols may be used. A wide variety of adipates are made for use as plasticizers for materials such as *polyvinyl*

chloride (PVC). In plasticizer terminology, the letter A is used to stand for this term. See, for example, *dioctyl adipate*, *di-iso-decyl adipate* and *aliphatic diester*.

adipates
The collective name given to those materials which contain the *adipate group* and which are used as *plasticizers*.

adipic acid
A linear carboxylic acid which is derived from cyclohexane by oxidation. This material has a melting point of 153°C. May be represented as $HOOH(CH_2)_4COOH$. A precursor for PA 66 and its copolymers: also used to make *plasticizers*. See *adipate*.

adipic acid ester
The reaction product of *adipic acid* and an alcohol and which is commonly called an *adipate*.

admer
An adhesive material which is mixed into a base polymer in order to promote adhesion to another polymer. For example, an admer may be mixed into PE in order to promote adhesion to PA.

ADPA
An abbreviation used for *acetone-diphenylamine condensation product*.

ADS
An abbreviation used for *air dried sheet*.

adsorbed layer thickness method
See *t method*.

adsorption
Absorption where only the surface is involved in the absorption process. The material absorbed (the adsorbate) is present as a film on the surface of the adsorbant.

AEB
An abbreviation used for average extent of burning in ASTM D 635. A *horizontal combustion test*.

AEBM
An abbreviation used for *accumulator extrusion blow moulding*.

AECO
An abbreviation used for *allyl-group-containing-epichlorhydrin rubber*. See *epichlorhydrin rubber*.

AEM
An abbreviation used for *ethylene-methyl acrylate rubber*.

aeration
Means that air, or gas, bubbles are entrained/accumulated in a liquid, for example, in the hydraulic fluid. In this case, the hydraulic components will operate erratically because of the compressibility of the trapped air.

aerogel
See *silica* and *silica gel*.

aerosil
A highly *active white filler* used to give tear and rupture resistance to vulcanizates. See *fumed silica* and see *pyrogenic process*.

aerosol container
Also known as an aerosol. A container, usually metal, which contains ingredients under pressure: such ingredients are dispensed as a spray when a button on the top of the container is depressed. See *hydrochlorofluorocarbon*, *butane* and *pentane*: also see *chlorofluorocarbon*.

aerosol propellant
The material(s) which pushes the ingredients from an *aerosol container*. Until comparatively recently, *chlorofluorocarbons (CFC)*, for example, CFC 11 and CFC 12, were widely used as aerosol propellants now, many aerosols are CFC free. *Propane* and *butane* have replaced CFC in many cases: under pressure in the aerosol they are liquids but vaporise in the spray. Also like CFC, they maintain the same spray pressure until almost all the ingredients have been expelled. They are considered to be non-toxic for many applications. However, they are not suitable for many medical applications and they are flammable.

AES
An abbreviation used for *acrylonitrile-ethylene/propylene-styrene rubber*.

AFMU
An abbreviation used for *nitroso rubber*.

AFNOR
An abbreviation used for Association Française de Normalisation. See *standards organizations*.

after-bake
See *postcure*.

aftermixer
Part of a *reaction injection moulding (RIM) system* used to improve mixing and homogeneity and used after the *mixhead*. For example, a flow diverter runner is kinked or stepped along its length to improve mixing - this type of mixing device traps less gas than a harp aftermixer.

ageing
Sometimes spelt aging. The effect on materials of exposure to an environment (for example, heat and light) for an interval of time: the process of exposing materials to an environment for an interval of time.

agglomerate
An association of primary particles which can be ruptured if sufficient shear is applied. Often means pigment agglomerate.

aggregate
See *carbon black - structure*.

aging
See *ageing*.

AGS acid
Mixed acids used to prepare low temperature plasticizers for PVC. A mixture of adipic, glutaric and succinic acids. See *plasticizer*.

Ah
An abbreviation used for *ampere hour*.

AH antioxidant
A *chain breaking antioxidant* which is often depicted as AH. The H represents an active hydrogen atom which is available for reaction with a propagating free radical (R·). Secondary aryl amines are the most important type of AH antioxidant. See *oxidation*.

AIBN
An abbreviation used for *azobisisobutyronitrile*.

AH salt
See *nylon salt*.

air assist
Term used in moulding to describe an ejection technique. Ejection after moulding is usually done using pins but sometimes air is introduced into the mould. That is, air assistance is also used.

air bag
See *curing bag*.

air blast
An ejection technique used during, for example, *injection moulding* of a *thermosetting plastic*, to remove loose flash. A blast of compressed air applied across the surface of the mould as the product is ejected.

air dried sheet
An abbreviation used for this material is ADS. One of the forms in which *natural rubber* is supplied. A pale (light coloured) form of natural rubber. The *coagulum* produced from *natural rubber latex*, is extruded and milled while being thoroughly washed to remove impurities. Sheets are produced which, after being air-dried for several days, are pressed into thicker sheets. The product has virtually no odour and is a light brown colour. It has a high degree of purity and gives vulcanizates of good mechanical properties. See *extra white crepe* and *specially prepared rubbers*.

air gap
The vertical distance between the die lips and the nip, for example, in *extrusion lamination*.

air gauging
Also called pneumatic gauging. An inspection method used, for example, for the measurement of thickness by determining the movement of a surface: can provide feed-back signals for use in a closed loop system.

Compressed air, at constant pressure, will flow through a jet at a constant rate. If a surface is moved towards the jet then the flow will be obstructed and the flow of air will be decreased. By observing the change in the air flow, for example, with a variable area flowmeter or with a pressure gauge, the movement of the surface can be detected. As the jet does not contact the work, it is a non-contact gauge which is not susceptible to wear and which does not damage the surface finish of the work. For the gauging of rough or porous surfaces, a contact gauging element is used: the contact gauging element has a stylus and movement of the stylus changes the air flow.

air inhibited resin
A resin, for example, an *unsaturated polyester resin*, the surface cure of which is inhibited by air.

air mile
See *nautical mile*.

air purge
See *air shot*.

air ring
A cooling ring. Used, for example, in the production of *lay flat film* so as to cool the film as it leaves the die lips. Mounted on the *extrusion die* and often fed with air from an electrically driven fan. See *dual lip air ring*.

air shot
Term usually associated with *injection moulding* and means that the contents of the cylinder, or barrel, have been *purged* into the air for inspection or for *melt temperature measurement*.

An air shot technique is often used for melt temperature measurement but take care as an accident can easily occur. Wear safety glasses/face shield and heat resistant gloves. Purge the plastics material onto a heat resistant, non-metallic board, remove from the nozzle area and immediately insert a thermocouple into the melt: recharge the injection cylinder by rotating the screw. Shoot the new charge from the barrel, place on the board and insert the now hot thermocouple into the new shot: stir and note the highest temperature. The melt temperature is the highest temperature noted. Ensure that there is no unauthorized person in the vicinity of the machine and ensure that everyone involved appreciates the dangers.

air side
The side of a seal which faces outwards, or toward atmosphere, as opposed to the fluid being sealed.

air slip forming
A *thermoforming process* which is based on *plug assist forming*. Air is introduced into the space between the plug and the sheet, as the plug moves to distort the sheet, so as to reduce sheet marking.

air trapping
An *injection moulding* fault caused by air becoming trapped in the cavity. Both air trapping and mould fouling can be minimised by using, for example, a *vacuum injection mould*. See *bumping* and *breathing*.

air-assisted ejection
Air can be used in *injection moulding* in order to break the vacuum that can exist within certain moulds and/or, in order to cause component or sprue ejection. Very flexible materials are more successfully ejected by compressed air either completely (as is the case for most small-draw mouldings) or partially (as is the case for large-area deep-draw articles). Many moulds incorporate both mechanical and air-assisted ejection. In these cases an air valve in the mould acts as a vacuum breaker, thus avoiding suck-back, and mechanical ejection completes the operation.

air-dried
See *air dried sheet*.

air-inhibition of cure
Retardation or inhibition of cure in the presence of air. See *surfacing agent* and *air inhibited resin*.

air-knife coating
A *spread coating method* in which the substrate is coated by an element which uses a jet of air in place of a *doctor blade*.

airflow switch
A switch fitted on the exit side of the cooling circuit of an electric motor which prevents motor overheating if the air flow fails; useful in a powder environment.

AISI
An abbreviation used for the USA-based organization called the *American Iron & Steel Institute* which, for example, gives steel an AISI coding. For example, AISI 420 is a hardenable *corrosion resistant steel*.

alcaligenes eutropus
A bacterium used to produce the biopolymer *polyhydroxybutyrate*.

aldehyde-amine accelerator
Reaction products of an aldehyde and an amine which function as an *accelerator* for a rubber compound. Not as important as other accelerators. Included in this class are materials such as *formaldehyde-ethylene diamine condensate*, *formaldehyde-p-toluidine condensate*, *butyraldehyde-aniline condensate*, *heptaldehyde-aniline condensate*, *hexamethylene tetramine* and *tricrotonylidine tetramine*.

aldehyde-condensing agent
A chemical grouping, or chemical, which produces a cross-link in freshly tapped *natural rubber latex* by reacting with carbonyl groups in the rubber molecule. Such materials react with, for example, aldehyde groups which otherwise would cause cross-linking of natural rubber. Examples of such chemicals include hydroxylamine hydrochloride and/or hydrazine hydrate. See *constant viscosity rubber* and *storage hardening*.

aldehyde-condensing group
A chemical group which produces a cross-link in freshly tapped *natural rubber latex* by reacting with an *aldehyde-condensing agent*. See *storage hardening*.

aldehyde-group cross-linking
See *storage hardening*.

alicyclic
Means that an organic compound is mainly aliphatic but also contains saturated, carbon-ring structures.

aliphatic
Concerned with organic compounds which do not contain ring structures. See *aliphatic hydrocarbon*.

aliphatic diesters
Usually refers to a group of plasticizers which are noted for their good *low temperature properties*. Such desirable properties are often accompanied by high levels of volatility and oil extraction. See *adipate*, *azelate* and *sebacate*.

aliphatic hydrocarbon
An organic compound which contains only carbon and hydrogen. Based on, for example, *alkanes*, *alkenes*, *alkynes* and their derivatives.

aliphatic petroleum resins
See *petroleum resins*.

aliphatic polyamide
A *polyamide* which contains aliphatic groups as linkages. See *linear aliphatic polyamide* and *nylon*.

aliphatic polyimide
A *polyimide* which contains aliphatic groups as linkages. Such materials can have softening points below 150°C.

alkali cellulose
That which is formed when caustic soda acts on *cellulose*. Cellulose is turned into alkali cellulose by the action of warm caustic soda in, for example, *viscose rayon production*.

alkali glass
See *A-glass*.

alkali metals
Group 1A of the Periodic table contains the elements lithium, sodium, potassium, rubidium, caesium and francium.

alkali reclamation processes
Used to obtain *reclaimed rubber* from un-wanted vulcanized material by heating with say, 4 to 10% caustic soda solution, and oils, at approximately 185°C for up to 24 h. The use of such a procedure removes textiles. Washing with water and drying follows. Useful for *natural rubber compounds* but not for synthetic compounds as cyclization often results.

alkali refining
An *oil treatment*. The *oil* is treated with alkali so as to lower acidity, to make it less reactive and to improve its colour.

alkaline earth metal sulphate filler
A sulphate based on an alkaline earth metal. Such materials are used as fillers. For example, see *barium sulphate* and *anhydrous calcium sulphate*.

alkaline earth metals
Group 11A of the Periodic table contains the elements beryllium, magnesium, calcium, strontium, barium and radium.

alkane
An organic saturated material, that is, it contains no double bonds and has the general formula C_nH_{2n+2}. Also known as a paraffin or a member of the methane series.

alkenamer
See *polyalkenamer rubber*.

alkene
An organic material which contains one or more double bonds, that is two carbon atoms are linked by $C=C$. It is unsaturated. Also known as an olefin or as an olefine. See *conjugated* and *diene*.

alkenes
A series of unsaturated hydrocarbons which are more commonly called *olefins*.

alkoxymethylation
The chemical process whereby groups such as $-CH_2OCH_3$ and $-(CH_2O)_2CH_3$ are grafted onto a polyamide to give a material which is softer and tougher than the parent *nylon* and of a lower melting point. See *methylmethoxy nylon*.

alkyd
Many years ago the term alkyd was applied to ester-type materials; the 'al' came from the alcohol and the 'kyd' from the acid. See *granular polyester moulding compound* and *polyester moulding compound*. See *alkyd resin*.

alkyd moulding compound
An abbreviation used for this type of material is AMC. See *granular polyester moulding compound* and *polyester moulding compound*.

alkyd plastics
Plastics based on *alkyd resin*.

alkyd resin
May also be referred to as an alkyd or as a polyester alkyd. The term usually means a polyester resin based on, for example, glycerol and phthalic anhydride: such resins are sometimes known as glyptals. A polyester convertible into a cross-linked form; for this, the resin requires a reactant of functionality higher than two or, it must have double bonds.

The number of cross-links are reduced if *fatty acids* are incorporated as these react with some of the hydroxyl groups on the glycerol. If the fatty acids are unsaturated, the alkyd resin becomes more soluble and will harden in air. Such oil modified resins, modified with *a drying oil*, are used to make surface coatings and to extend rubbers. The use of an unsaturated acid anhydride (for example, *maleic anhydride*) also helps in the drying properties.

alkyl benzyl phthalate
A group of *plasticizers* for *polyvinyl chloride (PVC)* which is typified by *butyl benzyl phthalate (BBP)*. This type of material has good solvent action at low processing temperatures and good processing properties: this is an advantage when highly filled compounds (for example, flooring compounds) are being processed. Gives compounds which resist staining and fungal attack. These plasticizers tend not to migrate into the flooring adhesive (which could cause adhesive softening).

alkyl diaryl phosphate
A phosphate which contains one alkyl and two aryl groups. See *mixed alkyl aryl phosphate*.

alkyl ricinoleate
An organic compound, which contains an *alkyl group*, and is prepared from *ricinoleic acid*. An ester: esters of ricinoleic acid which are used as, for example, *plasticizers*.

alkyl substituted polyphenylene oxide
See *polyphenylene oxide modified*.

alkyl sulphonic acid ester
See *alkylsulphonic acid ester*.

alkyl-aryl p-phenylene diamine
An additive, an *antiozonant*, which protects a rubber compound against dynamic ozone attack. When used in conjunction with waxes then the system will protect against static ozone attack. Such materials include N-isopropyl-N'-phenyl-p-phenylenediamine (IPPD) and N-1,3-dimethylbutyl-N'-phenyl-p-phenylenediamine (6PPD). See *antiozonant* and *antioxidant*.

alkylated diphenylamines
See *substituted diphenylamines*.

alkylated phenols
An abbreviation used for alkylated phenols is APH. Additives used in rubber compounds to protect a component against the effects of *ageing*. Monophenol derivatives which are classed as non-discolouring *antioxidants*.

alkylation
See *N-alkylation*.

alkylene sulphide rubber
An abbreviation used for this material is ASR. These are sulphur containing polymers which are of very limited commercial value. Such materials may be based on homopolymers of polypropylene sulphide or, on copolymers of

polypropylene sulphide and ethylene sulphide. Can also have alternating ether/thioether polymers: the basic formula of such a polymer could be $H(OCH_2CH_2SCH_2CH_2)_nOH$. Commercial materials have additional pendant unsaturation as thioalkyl glycidylether units are incorporated. Vulcanized by zinc-free, sulphur polyamide systems and by sulphur with thiazole-type accelerators. Claimed to have good heat and solvent resistance with a low glass transition temperature of approximately $-65°C$.

alkylmagnesium halide.
See *Grignard reagent*.

alkylsulphonic acid ester
An abbreviation used for this type of material is ASE. Also known as alkyl sulphonic acid ester or as, n-alkylsulphonate or as, alkylsulfonic acid ester. A sulphonic acid mono-ester which is probably better known by the trade name/trademark Mesamoll. This material is used as a *plasticizer* and is a primary plasticizer similar to a general purpose *phthalate*.

alkyne
An organic material which contains one or more triple bonds. That is, it is unsaturated. The simplest members of this series have the general formula C_nH_{2n-2} and the series may be known as the ethyne or acetylene series.

all skin fibre
Viscose fibre in which the coagulation and stretching are controlled to give a more highly oriented, uniform structure. See *viscose rayon*.

all-electric machine
A type of *injection moulding machine* which uses electric motors to obtain machine movements, moulding and clamping pressures. The machine does not contain a hydraulic system. With such machines it is now possible to produce mouldings under clean room conditions.

all-warp fabric
Means that the fabric mainly consists of cords running in one direction only (the warp) and these are held together by a few, widely spaced threads which run at right angles to the warp cords. These widely spaced strands (the welt) keep the warp (the tyre cords) at the correct spacing relative to each other during handling and processing. (Steel is produced in weftless form from a creel feeding direct into the *calender*). Such a precaution ensures that the tyre cords do not cross and touch one another when the tyre is in use, as if this happened they could either saw through one another of generate heat through friction. See *tyre-cord line*.

allene ladder polymer
See *ladder polymer*.

allophanate group
A chemical grouping which may be represented as -NHCO.N.COO-. The reaction of a *urethane group* and an *isocyanate* results in the formation of such a group in *polyurethane manufacture*, for example, R·NHCO·N·COO·R' (where R and R' are organic groups).

alloy
Originally a composition, or blend, based on two or more metals: the term can cover compositions which contain a non-metal, for example, iron carbon alloys See *ferroalloys*. In the plastics industry the term is used for compositions, or blends, which are based on two or more polymers the properties of which are significantly better than would be expected from a simple blend. Such an improvement may result from chemical bonding or grafting between two immiscible polymers. See *blend nomenclature*.

alloy steels
See *ferroalloys*.

allyl chloroformate
See *poly-(thiocarbonyl fluoride)*.

allyl glycidyl ether terpolymer or rubber
See *epichlorhydrin rubber*.

allyl group
A group based on allyl alcohol: this is $CH_2=CH-CH_2OH$. See *allyl plastics*.

allyl moulding materials
A common name for this type of thermosetting material is allyl moulding material. An abbreviation used for this type of material is DAP and/or DAIP. Also known as allyls or as, allylics or as, allyl moulding compounds or as, DAP.

If the monomer *diallyl phthalate* (or *diallyl isophthalate*) is heated than it will react or polymerize: the solid reaction product (an allyl resin) can be blended with fillers, peroxide catalysts, etc., to give moulding powders which are usually referred to as a DAP (or DAIP) moulding material or compound. Because the materials are esters, i.e. products of an acid and alcohol, they are sometimes referred to as 'alkyds' (see GPMC) but this practice is not recommended. Use of the initials DAP and DAIP is reserved for the moulding compounds which have the same names as their monomers, i.e. diallyl phthalate and diallyl isophthalate.

DAP is sometimes referred to as ortho diallyl phthalate and DAIP materials are sometimes referred to as meta diallyl phthalate. Both types of resin may be blended with a variety of fillers so as to give a range of grades, some of which have very high impact strengths: achieved by using high concentrations of long glass fibre (GF). Easier processing grades are obtained if the GF concentration and/or length is reduced: can replace some or all of the GF with mineral fillers. Compounds with lower densities, or specific gravities, result if the GF is replaced by an organic-based fibre, for example, nylon or terylene.

DAP/DAIP compounds are suitable for continuous use in the temperature range 160–180°C/320 to 356°F. They exhibit excellent electrical insulating properties which are retained under severe environmental conditions (high temperature and high humidity) and the dimensional stability is exceptional - partly due to negligible after-shrinkage. DAP (allyls) and alkyd moulding materials have, broadly speaking, similar properties; end-users and processors would employ similar processing conditions for both materials. They are both supplied as dry granular solids but DAIP materials are more expensive; they do however, retain their excellent electrical properties even under severely humid conditions and also resist biological attack.

The natural colour of the base compound is water-white and with the common fillers used, a light, beige coloured material is obtained. This enables a wide range of colours to be produced, e.g. reds, blues, whites, etc.).

Can be compression, transfer or injection moulded. For optimum properties, post-bake the mouldings in an air circulating oven at 150–190°C/302–374°F, for periods of 2–16 h. These materials are resistant to a wide range of chemicals, e.g. 30% sulphuric acid, 10% nitric acid, 25% ammonium hydroxide, gasoline/petrol, paraffin, lubricating oil, diesel, hydraulic brake fluid, acetone and, chlorinated hydrocarbons (such as trichloroethane and carbon tetrachloride; also resistant to conditions of high humidity. Alkyds can be hydrolysed or saponified, DAP's cannot. Can withstand continuous exposure to 160°C/320°F and DAIP can withstand 180°C/356°F. These figures are not quite as good as epoxy, or epoxide, material which may be formulated to withstand 220°C/428°F.

These materials are not resistant to sodium hydroxide: some grades have poor resistance to 10% solutions. 10% oxalic acid can attack some grades. The diallyl phthalate materials (DAP) are better than PF under conditions of dry and wet heat. Allyl materials are superior to PF in their tracking resistance and in the colour range available. They show a lower shrinkage on cure than PF but, cracking around inserts can occur due to the lower elasticity of this type of material.

They are high density materials which have densities in the range 1·73–2·08 gcm^{-3} (the actual density will largely depend on the filler type and concentration). Both DAP and DAIP, are often used to produce very complex components which are moulded to very high accuracy as they are used in very demanding applications where consistency is important. For example, demanding electrical/electronic applications where, retention of electrical and mechanical properties under extreme conditions, is of paramount importance, e.g. connectors, potentiometer housings, switches, relays, circuit breakers and coil bobbins. Widely used in military applications where absolute reliability is demanded. DAP may be seen on very delicate components having wall sections as low as 0·25 mm/0·01 in.

allyl plastics
A *plastics* material made from an *allyl resin*. See *allyl moulding materials*.

allyl resin
A *resin* made by polymerization of chemical compounds containing the *allyl group*. See *allyl moulding materials*.

allyl-group-containing-epichlorhydrin rubber
See *epichlorhydrin rubber*.

allylics
See *allyl moulding materials*.

allyls
See *allyl moulding materials*.

along the sheet
A term used in *calendering* and *extrusion*. For example, it means parallel to the calendering direction. See *machine direction*.

alpha hardness value
A measure of the hardness of a material. See *Rockwell hardness*.

alpha methyl styrene
This term is used to stand for both a monomer and for the polymer produced from that monomer. An abbreviation used for this material is AMS. See *poly-α-methylstyrene*.

alpha sulphur
See *rhombic sulphur*.

alpha-cellulose
Also known as α-cellulose. A pure form of *cellulose* which has a high molecular weight and a bright, white colour. It results when impurities (for example, lignin and hemicelluloses) are removed from *cotton linters* by an alkali treatment under pressure (dilute sodium hydroxide at 150°C) followed by bleaching (sodium hypochlorite). If wood is selected as the raw material (for example, for rayon production) then, α-cellulose may be defined as that which does not dissolve in an 18% (approximately) solution of sodium hydroxide after 30 minutes at 20°C.

alphanumeric
A character that can be a numeral or a letter of the alphabet. Also means in computer language, a type of field within a record that can hold letters or numbers.

alternating ether/thioether polymer
See *alkylene sulphide rubber*.

alum
Potassium aluminium sulphate $K_2SO_4Al_2(SO_4)_3 24H_2O$. A white powder with a relative density of 1·75 which may be used to *coagulate rubber latex*.

alumina
An example of a *whisker-forming material*. See *aluminium oxide*.

alumina trihydrate
See *aluminium trihydrate*.

alumina trihydroxide
May be represented as $Al(OH)_3$. An ore which may be referred to gibbsite or as hydragillite: crystallized alumina trihydroxide is the principle ingredient of tropical bauxite which is used to prepare *aluminium trihydrate*.

aluminium
This element (Al) is, with boron, in Group 111A of the periodic table. It has a relative density (RD or SG) of 2·7 and is made from *aluminium oxide*. It is a lustrous white metal which melts at 660°C and which boils at approximately 2,400°C. A malleable, ductile material which is a good conductor of heat and electricity. Because of the presence of a thin film of inert oxide the material is corrosion resistant and is not attacked by water or steam. Can remove this film chemically, with mercury or mercuric chloride solution, and this removal will allow rapid corrosion. Aluminium has been used as a filler for thermosetting materials - in order to improve thermal and electrical conductivity. Has also been used to coat other fillers (for example, glass spheres) for the same reasons: the electrical conductivity of some systems can be very good.

A major use of this metal, in the plastics industry, is for the manufacture of moulds - particularly when the mould is not required to withstand high pressures. Such applications include, for example, *blow moulding* and *powder moulding*. The light weight and improved thermal conductivity, compared to steel, are a great attraction. However, this metal is not so wear resistant as steel. Cast aluminium is used for moulds and has a relative density (RD or SG) of approximately 2·9. See *aluminium mould*.

aluminium alkyl
A compound of aluminium and an organic alkyl group. An example is aluminium triethyl which is used with titanium trichloride to give a *Ziegler-Natta catalyst system*.

aluminium containers
A container made from *aluminium*. Now used to make light weight, two-piece cans for carbonated soft drinks (CSD). Also widely used to make foil laminates based, for example, on PE-coated paper with an interlayer of aluminium (Al) foil. Such foil laminates are fabricated into a range of products which include cartons for liquid packaging and, which are used, for example, for the packaging of milk and fruit juices.

aluminium flakes
A *flake pigment* based on *aluminium*. Metallic colours are produced by the incorporation of flakes of aluminium or of copper or of bronze. As aluminium flake poses a fire hazard, it is often used as a paste with plasticizer.

aluminium hydrate
See *aluminium hydroxide* and *aluminium trihydrate*. The use of Greek letters in the naming of alumina hydrates was introduced by the Aluminium Company of America (Alcoa) in approximately 1930 and the letters are associated with those hydrates that occur most abundantly in nature.

aluminium hydroxide
May be represented as $Al(OH)_3$. May be referred to as α-trihydrate or as, aluminium trihydrate or as, hydrated alumina. All these names are misleading as the materials do not contain water of hydration but only hydroxide and hydroxy-oxide groups.

aluminium mould
A mould made from aluminium or an alloy of aluminium. Such moulds are used because of their good heat conductivity: the moulds can be manufactured from rolled semi-finished goods or from castings. Wrought aluminium plate has been used to produce relatively cheap injection moulds which because of their lightness are easier to handle than steel moulds. When this material is cast, the cooling system may be incorporated at the casting stage. However, the material is

much softer than steel and, in general, the surface is not as good; because of such factors this material is used to construct moulds where the number of shots is between 10,000 and 50,000 (although 100,000 components have been made).

aluminium oxide
Also known as alumina. Used, for example, to make aluminium and for cement manufacture. Occurs naturally as emery and as corundum. Aluminium oxide has been used as a filler with thermosetting plastics as the product is hard and has a low coefficient of expansion. Aluminium oxide has been used to obtain high modulus fibres as the modulus of alumina is, for example, 55×16^6 psi. The fibre material has a relative density (RD or SG) of 3.9. The melting point of alumina is 2015°C.

aluminium silicate
Also known as hydrated aluminium silicate. This fine white material has a relative density (RD or SG) of 2.60. Often used in a hydrated precipitated form. A white powder which may be used as a filler in rubber compounds but is relatively expensive. Imparts hardness, wear and tear resistance for light coloured goods particularly in the footwear industry. Not so active or reinforcing as *calcium silicate*. May slightly decrease vulcanization rate. More expensive than mixtures of *silica* and *china clay* which may be used to replace the silicate in rubber compounds.

aluminium stearate
This white solid material has a melting point of 160 to 170°C/320 to 338°F. Used as a *mould lubricant* and as an *internal lubricant*, for example, for high temperatures curing, rubber compounds. Its use, slightly retards cure. Used extensively as a thickener for lubricants as it forms gels with hydrocarbon oils. Helps to form matt and semi-matt finishes with paints: can give textile waterproofing.

A *lubricant* used with *nylon materials*, for example, PA 66. Tumble coating the granules before moulding with approximately 0.1%, eases ejection and also reduces ejection forces (for both nucleated and un-nucleated grades) and so, minimizes ejection problems.

aluminium triethyl
See *Ziegler-Natta catalyst system.*

aluminium trihydrate
Also known as α-trihydrate, aluminium hydrate, alumina trihydrate and hydrated alumina. Produced by the action of sodium hydroxide on tropical bauxite (natural hydrated aluminium oxide or hydrated oxide of aluminium. (See *alumina trihydroxide*.)

Aluminium trihydrate (ATH) is an inexpensive, white, *flame retardant filler*: it is an *additive flame retardant* which functions as a filler, a smoke suppressant and as a flame retardant. It has a relative density (RD or SG) of 2.4, a Mohs hardness of 3 and an index of refraction of 1.57. It is used as a flame retardant/smoke suppressant with both thermoplastics and with thermosetting materials: such as, for example, with polyester moulding compounds (PMC). Compounds based on this filler may be translucent.

When used with thermoplastics, the addition of large amounts of this material is, however, required in order to achieve flame retardancy. Such large additions can change the properties of the basic plastics material (for example, the flow properties) and introduce processing problems. For example, because of aluminium trihydrate decomposition, or degradation, at processing temperatures above 200°C/392°F. When used with PMC it also changes the flow properties but the change is acceptable or desirable. By replacing approximately half of the filler in *sponge rubber*, with ATH, a more fire retardant sponge results.

ATH functions as a flame retardant by absorbing the initial heat of combustion of the polymer, by decomposing endothermically and by evolving steam. Available in a range of particle sizes and surface treatments.

aluminizing
The coating of a substrate with a layer of aluminium. For example, by coating *aramid fibres* with aluminium, the heat resistance of the fibres is raised significantly.

AMC
An abbreviation used for alkyd moulding compound. See *granular polyester moulding compound* and *polyester moulding compound.*

American Institute of Petroleum
See *American Petroleum Institute.*

American Iron and Steel Institute
An abbreviation used for this USA-based organization is AISI. The AISI issues codings for *steel* which are widely used in the plastics industry: their designation is the same as the Society of Automotive Engineers. The prefix used for mould steels is P.

American Petroleum Institute
An abbreviation used for this USA-based organization is API. This organization issues standards or specifications, for example, for drill pipe. Copies of their publications can usually be obtained from the national standards office of a particular country, for example, from the *British Standards Institution (BSI)* in the UK.

American Society for Testing and Materials
An abbreviation used for this USA-based organization is ASTM. A *standards organization* responsible for issuing standards or specifications. Copies of their publications can usually be obtained from the national standards office of a particular country, for example, from the *British Standards Institution (BSI)* in the UK. See *abbreviations.*

AMFI
An abbreviation used for automatic melt flow indexer. See *flow rate.*

amide group
A group represented as -NHCO- or as (-NH-CO-). May also be represented as -(CO-NH)- Because this linkage is the fundamental group in the *nylon group of plastics materials*, such materials are often known as *polyamides.*

amidex fibre
A manufactured fibre composed of at least 50% by weight of one or more esters of a mono-hydric alcohol and *acrylic acid.*

amine
An organic compound formed by replacing one or more of the hydrogen atoms of *ammonia* with organic hydrocarbon groups (R). Can have primary amines RNH_2, secondary amines R_2NH, and tertiary amines R_3N. See *aromatic amine* and *aryl amine.*

Amines are used to achieve the curing or setting of *epoxides* and as *antioxidants* in rubber compounds (the early antioxidants developed for rubbers were, for example, aromatic amines). Amines are also used to treat clays so as to, for example, improve dispersion and to improve antioxidant efficiency. One amine, octadecylamine, may be used to minimise dust formation in asbestos. See *amine antioxidant.*

amine antioxidant
An additive which is used to prevent oxygen attack on, for example, diene rubbers and polyolefins. In rubber technology amine antioxidants are classed as staining antioxidants. May be sub-divided into *phenylnaphthylamines, ketone-amine condensates, substituted diphenylamines* and *substituted p-phenylene diamines.*

amino group
May be represented as $-NH_2$. See *amine.*

amino plastics
See *aminoplastics*.

amino resin
A resin made by polycondensation of a compound containing amino groups, such as urea or melamine, with an aldehyde, such as formaldehyde, or an aldehyde-yielding material: gives UF and MF plastics materials, respectively. See *aminoplastic*.

amino-formaldehyde resin
A resin based on the reaction between a material which contains amine groups, or amide groups, and *formaldehyde*. Used to make, for example, *urea-formaldehyde moulding materials* and *melamine-formaldehyde moulding materials*.

aminocrotonate
An ester of aminocrotonic acid used as a *metal-free organic*, heat stabilizer with *unplasticized polyvinyl chloride (UPVC)* it is used, in non-toxic bottle applications, with a co-stabilizer such as an *epoxy compound*.

aminoplast
See *aminoplastic*.

aminoplastic
Also called amino plastic or, aminoplast or, aminoresin. Plastics based on *amino resins*. An aminoplastic is a plastics material formed by the reaction between materials containing *amine groups*, or amide groups, and aldehydes. For example, between *urea* and *formaldehyde*. See *melamine-formaldehyde* and *urea-formaldehyde*.

aminopolymer
See *urea-formaldehyde* and *melamine-formaldehyde*.

aminoundecanoic acid
Also known as 11-aminoundecanoic acid or as, ω-aminoundecanoic acid. The monomer for *nylon 11* derived from *castor oil*.

AMMA
An abbreviation used for *acrylonitrile-methyl methacrylate*.

ammonia
A gas which is very soluble in water and which has a pungent irritating odour. Has the formula NH_3. When added to water there is partial combination to give ammonium hydroxide (NH_4OH): dissociation of this to ammonium ions and hydroxyl ions gives an alkaline solution. Obtained commercially from atmospheric nitrogen and widely used as a refrigerant and to make plastics materials. Also used to preserve rubber latex. Ammonia and *chlorofluorocarbons (CFC)* are widely used as refrigerant gases. About half of the world's industrial refrigerant is still based on ammonia. Now considered by some to be an acceptable alternative to CFC.

ammonium bicarbonate
This material has a relative density (RD or SG) of 1·58 and may be represented as NH_4HCO_3. Used a *blowing agent* for rubber compounds. Decomposes at approximately 65°C.

ammonium carbonate
Also known as hartshorn salt. May be represented as $(NH_4)_2CO_3$. This solid material has a relative density (RD or SG) of 1·59 and decomposes rapidly above 70°C/158°F. It is used as a *blowing agent* for the manufacture of sponge rubbers usually in conjunction with *sodium bicarbonate*: on its own the pore size may be unacceptably large. Try 10 parts of sodium bicarbonate to 1 part of ammonium carbonate.

ammonium caseinate
A soluble form of *casein*.

amorphous
The term amorphous usually refers to a plastics material which is not crystalline and which is therefore usually transparent, for example, *polyvinyl chloride (PVC)*. See *amorphous thermoplastics materials*.

amorphous fluoropolymer
This type of thermoplastics material (for example, Teflon AF DuPont). has many of the properties associated with the semi-crystalline, thermoplastics fluoropolymers. It is however, non-slippery and is optically clear. This material has the lowest dielectric constant of known plastics, good creep resistance and has excellent resistance to chemical attack. The mechanical properties are retained up to the end-use temperatures of approximately 300°C/572°F. Can be melt processed on equipment used for thermoplastics fluoropolymers: can also be cast into thin films and pin-hole free coatings as it is soluble in selected perfluorocarbon solvents. Suggested uses include release films and coatings, lens covers for microwaves and radar devices and in fibre optics.

amorphous silica
Silica which lacks a crystalline or semi-crystalline structure. Most forms of silica are considered amorphous but this is often because of the extreme fineness of the crystal structures. See *cryptocrystalline* and *Illinois silica*. Silica gel, *precipitated* and *fumed silica* are regarded as amorphous forms of silica which may be considered as being condensed polymers of silicic acid. However, *diatomaceous earth* is a micro-crystalline form.

amorphous thermoplastics material
An amorphous, *thermoplastics material* is usually a hard, clear, rigid material with a low shrinkage and a low impact strength: such a material is *polystyrene*.

amp
An abbreviation used for *ampere*.

ampere
The basic SI unit of electric current which has the symbol A. Defined as the current that when flowing through two, very long, parallel wires of negligible cross-section, separated by one metre in a vacuum, results in a force between the wires of 2×10^{-7} newtons per metre of length. See *Système International d'Unité*.

ampere-hour
An abbreviation used for this term is Ah. A practical unit of electric charge: the charge flowing through a conductor in one hour when the electric current has a value of one *ampere*.

ampere-turns
A unit of magnetomotive force: the number of turns in a coil times the current (amperes) flowing through that coil.

amphibole
A type of *asbestos*.

amplitude
The height of the crest of a wave; its maximum displacement from zero.

AMS
An abbreviation used for *alpha methyl styrene*.

amt
An abbreviation used for amount.

amu
An abbreviation used for *atomic mass unit*.

amylum
See *starch*.

AN
An abbreviation used for *acrylonitrile*.

analogue setting
Means that a pointer is set on a scale. For example, when an analogue temperature controller is set, a pointer is rotated until it points at the desired scale value.

analogue signal
An AC or DC signal (voltage or current) that represents continuously variable quantities such as pressure or speed.

analysis of flow capillary rheometer
See *capillary rheometer*.

anatase titanium dioxide
Also known as anatase. One of the crystalline forms of *titanium dioxide*. *Rutile titanium dioxide*, because of its higher index of refraction (2·76) has better *hiding power*. Although slightly whiter than the rutile form, the anatase form is also not recommended for outdoor use as it blocks out less ultra-violet light than the rutile form.

anchor
See *sprue puller*.

anchor platen
One of the platens of an *injection moulding machine*: the platen at the (clamping) end of the machine and which accepts the thrust from the locking system. See *fixed platen* and *moving platen*.

AND-gate
A logic circuit with two or more input channels: an output pulse is only delivered if there is a pulse on all input channels.

angle of braid
The acute angle formed between any strand of a braid and a line which is parallel to the axis of a hose.

angle of helix
Also known as angle of lay or reinforcement angle. The acute angle formed between any strand of a helical reinforcement and a line which is parallel to the axis of a hose.

angle of lay
See *angle of helix*.

anglehead adapter
Also spelt anglehead adaptor. An adaptor which changes the direction of melt flow, that is, it swings the melt through an angle of, for example, 60°.

Ångström unit
A unit of length which has the abbreviations Å or Å.U. It is equal to 0·1 nanometres or 10^{-10} metres.

anhydrous calcium sulphate
An *alkaline earth metal sulphate filler* with the formula $CaSO_4$. Available as a particulate filler and as a fibrous filler. The particulate filler may have a particle size in the region of 7 μm. The fibrous filler is a single crystal fibre of diameter 2 μm and of average length 60 μm: an *asbestos replacement*.

anhydrous sodium/potassium/aluminium silicate
See *nepheline syenite*.

aniline
Also known as aminobenzene or as, phenylamine. May be represented as $C_6H_5NH_2$. This material has a relative density (RD or SG) of 1·02, a boiling point of 184°C and a melting point of -6°C. Aniline is used to prepare aniline dyes. Was used as an organic *accelerator* (red oil) but is now considered to be too slow. Forms the basis of a powerful class of *antioxidants* (staining antioxidants). May be used to form a polymer with formaldehyde, that is, *aniline formaldehyde*.

aniline formaldehyde
Also known as aniline resin. A group of thermoplastics materials produced from aniline and formaldehyde in aqueous acid media and then subsequently polymerized. This type of material has good electrical properties but is difficult to mould because of partial cross-linking. Of limited commercial importance.

aniline p
An abbreviation for *aniline point*.

aniline point
An indication of the aromatic content of an *oil* and which is determined by measuring the lowest temperature at which the oil is miscible with an equal volume of aniline. Measured in °C. See *ASTM oils*.

anilox roller
An engraved stainless steel roller which holds ink in the recesses of the design (in cells) and acts as a metering device to the *stereo* in *flexographic printing*.

animal oil
See *oil*.

anionic dye
See *acid dye*.

anionic exchanger
An *ionic polymer* whose matrix contains bound cations (for example, $-CH_2N^+(CH_3)_3$) pendant to the main chain.

anionic ionic polymer
An *ionic polymer* whose matrix contains bound anions ($-SO_3^-$ or, $-CO_2^-$ or, $-PO_3^{2-}$) pendant to the main chain. Most ionic polymers are anionic.

anionic surface active agents
Surface active agents are classified as being anionic surface active agents or as, *cationic surface active agents* or as, *non-ionic surface active agents*. Anionic surface active agents are typified by *soaps* in which the active groups are negatively-charged ions.

anisotropic behaviour
Having different properties in different directions. The properties of many polymeric products differ if measured in different directions. That is, they are anisotropic. For example, the tensile strength is greater in the direction of orientation. Impact strength is also affected by such *frozen-in orientation*. In an *Izod impact test*, where the sample has been injection moulded with the gate at one end of the sample, the molecules will be roughly aligned with the long axis of the sample. Thus to snap the sample, in a standard Izod test, would require fracture across the elongated molecules thus giving a higher impact strength than would be recorded with un-oriented samples. On the other hand, if impact strength was being measured by dropping a weight onto a flat plate, lower impact strengths will be recorded with more oriented mouldings. This is because fracture can occur more easily parallel to the direction of orientation since this largely requires fracture between, rather than across, molecules.

anisotropy
Having different properties in different directions: with thermoplastics materials, this often arises because of *frozen-in orientation*. See *melt processing orientation introduction*.

ANM
An abbreviation used for acrylic ester acrylonitrile copolymers. An abbreviation used for ethyl, or other acrylate, ester and acrylonitrile copolymers.

annealed compression moulded sheet
A flat sheet which has been produced by *compression moulding* and then annealed by, for example, heating the sheet to a temperature above the melting point of the polymer and then slow cooling at a controlled rate of, say, 5°C/hour. This removes the previous thermal history and results in sheet with very low internal stress.

annealing
The heating of a product to condition and/or to relieve stresses. For example, if *nylon 6* components of high heat stability are required, anneal in a non-oxidizing oil, for example, for 20 minutes at 150°C/302°F.

annular area
A ring shaped area. Such an area is the effective area of the rod-side of a cylinder piston: the cross-sectional area of the rod must be subtracted from the piston area.

annular die
A *die* which contains an annulus and used, for example, for tube extrusion: such a gap is formed by a torpedo which forms the inner surface of the *tube*.

anode
The positive (+) electrode of an electrical device (or in electrolysis) to which electrons, or negative ions, are attracted. Since in *dielectric heating* the anode collects the electrons after they have done their work, it must absorb their kinetic energy which is then dissipated, for example, by forced cooling.

ANSI
An abbreviation used for American National Standards Institute.

ante-chamber
A shaped recess in an injection mould which allows the *injection nozzle* to be positioned close to the cavity.

anthracene
An aromatic chemical which is obtained, for example, from *anthracene oil*. Anthracene may be used, in turn, used to prepare *phthalic anhydride*.

anthracene oil
The fraction, obtained during the distillation of *coal tar*, and which distils at from 270 to 400°C is called anthracene oil and from this may be obtained *anthracene*.

anthranyl-9-mercaptan
A *peptiser*. A material which accelerates the oxidative breakdown of rubbers and of vulcanizates. Can be used, for example, to reclaim vulcanized rubber, so as to form reclaim, by steam heating the finely ground rubber in a caustic soda solution for several hours.

anthraquinone
An abbreviation used for this material is AQ. Derived from anthracene. A tri-cyclic hydrocarbon which may be represented as $C_6H_4(CO)_2C_6H_4$. Used to make *anthraquinone dyes*.

anthraquinone dyes
A class or type of organic *dye* which is based on *anthraquinone*. Such dyes have better heat resistance and weather resistance than *azo dyes* but are more expensive. Used in acrylics, for example, solvent red 111 is used for the red lenses on cars. Such dyes are also compatible with polystyrene, cellulosics and phenolics. See *anthraquinone pigments*.

anthraquinone pigments
A class or type of *organic pigment* which give transparent colours ranging from yellow, red to blue. Relatively high cost systems which are lightfast. Heat stability may be suspect above 180°C/356°F and such colorants can have poor chemical resistance, in some circumstances, leading to colour loss. See *anthraquinone dyes*.

anthraquinones
The term usually refers to a class of *organic pigments*. See *anthraquinone pigments*.

anti-ageing additive
An *additive* used to minimize the effects of degradation. Such additives are widely used with polymeric materials as heat and light will otherwise cause the properties of such materials to change: to change either in use or, during processing. Heat and light cause chemical reactions to occur within the material, or on the surface, and of the many chemical reactions possible, oxidation, ozone attack, dehydrochlorination and ultra-violet (UV) attack are the most common. Which type of chemical attack occurs depends, in the first instance, on the type of polymer and the particular circumstances.

For example, as polyolefins (PO) readily degrade by oxidation, antioxidants are widely associated with polyolefins such as polyethylene (PE) and polypropylene (PP). PVC degrades more readily by dehydrochlorination and so additives which restrict this type of attack are more widely used in PVC technology. Ultra-violet (UV) attack is common with most polymeric materials and so, UV stabilizers are usual in products which are to be used outdoors. Ageing by ozone is peculiar to the rubbery class of materials and is combated and controlled by the use of anti-ozonants.

See *antioxidant*, *heat stabilizer*, *ultra-violet (UV) stabilizer* and *anti-ozonant*.

anti-ageing additives
The collective name for the class of materials which hinder degradation. See *anti-ageing additive*.

anti-ageing chemicals
Chemicals added to hinder degradation, for example, by oxidation. See *anti-ageing additive*.

anti-aging additive
See *anti-ageing additive*.

anti-block additive
See *antiblock agent* and *matting agent*.

anti-extrusion ring
A ring assembled with a seal and which prevents the sealing material being extruded/forced into the clearance space.

anti-plasticization
A term used in *polyvinyl chloride (PVC) technology*. At low levels of addition, a *plasticizer* often embrittles a PVC compound. This effect is sometimes called anti-plasticization. Anti-plasticization is exhibited with other polymers, for example, with *polycarbonates*. See *critical concentration*.

antiblock agent
An additive used to stop blocking, or sticking, of two surfaces and usually associated with film production; such an additive may be dispersed in the resin before extrusion or, dusted on after extrusion. See *diatomaceous earth*.

aniline resin
See *aniline formaldehyde*.

anticoagulant
An additive used to stop the premature coagulation of rubber latex. For example, *natural rubber latex* is only naturally stable for a few hours after *tapping*. Coagulation may be prevented by retarding bacterial and enzyme action. Examples of coagulants include *ammonia*, *sodium sulphite*, *sodium carbonate*, and *formaldehyde*.

antidegradant
See *stabilizer*.

antimony (III) sulphide
See *antimony sulphide*.

antimony (V) sulphide
See *antimony pentasulphide*.

antimony oxide
See *antimony trioxide*.

antimony pentasulphide
Also known as antimony sulphide or as, antimony (V) sulphide or as, antimony-V-sulphide or as golden sulphur (antimony (III) sulphide my also be referred to as an antimony sulphide). May be represented as Sb_2S_5. Antimony (V) sulphide is a orange-yellow solid sometimes used as a pigment: it is an *inorganic accelerator* with a relative density of 4.12.

antimony pentoxide
A *flame retardant*. See *antimony trioxide* which is more widely used.

antimony sulphide
Also known as antimony (III) sulphide or as, antimony trisulphide or as, stibnite. May be represented as Sb_2S_3 (antimony (V) sulphide may also be referred to as a antimony sulphide).

Antimony (III) sulphide is a crystalline solid, used as a pigment, with a melting point of approximately 550°C. May be black or red. Without free sulphur this material has a relative density (RD or SG) of 3·6. It is an inorganic vulcanization *accelerator* and is used, for example, for colouring ebonite.

antimony tetrasulphide
A golden yellow material with a relative density of 2·6 to 3·0 and sometimes used as a pigment.

antimony trioxide
May also be known as antimony oxide. Obtained from *antimony sulphide* by controlled oxidation. This material has a relative density (RD or SG) of 5·4. It is used to obtain flame retardant polymeric materials: used in conjunction with halogenated materials, for example, pentabromophenol. Functions synergistically with such materials.

antimony trisulphide.
See *antimony sulphide*.

antioxidant
A material, or additive, which helps to prevent oxidation of a polymer. As *polyolefins (PO)* readily degrade by oxidation, antioxidants are widely associated with polyolefins (such as *polyethylene (PE)* and *polypropylene (PP)*) although the early antioxidants were developed for rubbers. Such materials were aromatic amines and may now be classed as *staining antioxidants*. Antioxidants may be grouped, or classified, in various ways, for example, into *chain breaking antioxidants* and into *preventive antioxidants*. Various *synergistic combinations* are widely used as are *chelating agents (metal deactivators)*.

For thermoplastics materials such as *polyethylene (PE)*, *polypropylene (PP)*, *acrylonitrile-butadiene-styrene* and *high impact polystyrene*, the use level ranges between 0·05 to 1 phr. As the usage of polyolefins grows, so the usage of antioxidants will grow. See *oxidation*.

antiozonant
An additive which is used to prevent ozone attack. Ageing by ozone is peculiar to the rubbery class of materials and is combatted and controlled by the use of antiozonants. May provide a barrier to the attack of ozone by forming a surface film of wax or, they may react with ozone or the products of ozone attack (peroxides or ozonides). Such additives may prevent the formation of ozonides or retard the decomposition, or degradation, of ozonides: the decomposition, or degradation, is a chain reaction. Substituted p-phenylene diamines, such as the *alkyl-aryl p-phenylene diamines*, are the most effective. The protection given by such materials is improved when they are used in conjunction with waxes. See *alkyl-aryl p-phenylene diamine*.

antiqueing
See *scumbing*.

antistatic agent
Also known as an antistat. An antistatic agent is an additive added to a polymer, often a thermoplastics material, in order to reduce static problems; such problems could be caused by dust attraction or by static electricity interfering with product handling, for example, the handling of film. Examples of antistatic agents include quaternary ammonium compounds for polystyrene and polyethylene glycol alkyl esters for polyethylene. When dispersed in a polymeric compound, the antistat blooms to a surface (e.g. of a film) and forms a conductive layer which disperses the build-up of static charge.

The use of such agents may not improve the finish of the components directly but they do allow the component to look attractive for a longer period, simply because, for example, they reduce dust attraction which is unsightly.

antistatic compound
A compound with electrical insulation properties in between those of a conductor and insulator. To be classed as antistatic, a polymer compound has to have a *volume resistivity* in the range 10^2 to 10^8 ohm.cm.

A_0
The original cross-sectional area of a *tensile strength specimen*.

APE
An abbreviation used for *aromatic polyester carbonate*.

apex strip
See *bead filler*.

APH
An abbreviation used for alkylated phenols.

APHA
An abbreviation used for American Public Health Association.

API
An abbreviation used for American Petroleum Institute.

APME
An abbreviation used for *Association of Plastics Manufacturers in Europe*.

APO treatment
A textile treatment using *tris-(1-aziridinyl) phosphine oxide*. Such a treatment can however result in fabric yellowing.

apothecaries' measure
A measurement system used in dispensing, for example, for the measurement of liquids. In the UK, 20 minims = 1 fluid scruple, 3 fluid scruples = 1 fluid drachm, 8 fluid drachms = 1 fluid ounce, 20 fluid ounces = 1 pint and 8 pints = 1 imperial gallon (277·274 cubic inches). Also 1 drop = 1 grain, 60 drops = 1 drachm, 1 drachm = 1 teaspoonful, 2 drachms = 1 dessertspoonful, 4 drachms = 1 tablespoonful, 2 ounces = 1 wineglassful and 3 ounces = 1 teacupful. In the US, 60 minims = 1 fluid dram, 8 fluid drams = 1 fluid ounce, 16 fluid ounces = 1 pint and 8 pints = 1 US gallon (231 cubic inches).

apothecaries' weight
A system of weights used in dispensing, for example, for the measurement of drugs. In the UK, 20 grains = 1 scruple, 3 scruple = 1 drachm, 8 drachms = 1 ounce and 12 ounces = 1 pound. The grain, ounce and pound are the same as those used in troy weight. In the US, 20 grains = 1 scruple, 3 scruple = 1 dram, 480 grains = 1 ounce and 12 ounces = 1 pound.

APP
An abbreviation used for atactic polypropylene. See *polypropylene*.

apparent
That which is observed but without correction.

apparent crown increase
See *roll crossing*.

apparent density
The weight in air of a unit volume of material. See *density*.

apparent flow curve
Obtained by plotting the *shear stress* at the capillary wall (τ_w) against the apparent *shear rate* at the capillary wall ($\dot{\gamma}_{w,a}$). See *flow curve*.

apparent powder density
Commonly used to assess the powder density of a *thermosetting moulding powder*. A powder (115 ml) flows from a funnel into a measuring cylinder (100 ml), from a specified height (say 25 mm). After removing the excess with a straight edge, the cylinder is weighed: two measurements are made and the

results expressed in g/ml. Specified in, for example, ISO, BS and ASTM standards. The apparent powder density may be used to calculate *bulk factor*.

apparent shear rate
An abbreviation used for the apparent shear rate at the wall of a die is $\dot{\gamma}_{w,a}$. The expression for the shear rate at the wall of a tube or die is very ungainly, or cumbersome, and in practice it is often simplified to give $\dot{\gamma}_{w,a} = 4Q/\pi R^3$. Where $\dot{\gamma}_{w,a}$ is the apparent (or uncorrected) shear rate at the wall of a die.

For Newtonian fluids, the apparent (or uncorrected) shear rate at the wall of a die ($\dot{\gamma}_{w,a}$) is equal to the true shear rate ($\dot{\gamma}_w$). However, with non-Newtonian fluids, such as polymer melts, the two are not equal but are related by the equation:

$$\dot{\gamma}_w = [(3n' + 1)/4n']\dot{\gamma}_{w,a}. \text{ Where}$$
$$n' = [d\log(R\Delta P/2L)]/[d\log(4Q/\pi R^3)].$$

However, there is a unique relationship between shear stress and $\dot{\gamma}_{w,a}$ and this means that this simplified relationship may be used for scale-up work.

apparent viscosity
Sometimes referred to as structure viscosity. May be defined as the ratio between *shear stress* and *shear rate*, over a narrow range, for a non-Newtonian fluid such as a plastics melt. The apparent viscosity at a given *shear rate* is obtained by dividing the *shear stress* at the wall by the apparent wall shear rate. That is, apparent viscosity is obtained by dividing τ_w (the shear stress) by $\dot{\gamma}_{w,a}$ (the apparent wall shear rate). An abbreviation used for this term is η or μ_a. This is the apparent viscosity at a particular shear rate. In practice the word 'apparent' is often omitted and the symbol μ is used. This practice should not be encouraged as the symbol μ is strictly used to denote the coefficient of viscosity of a *Newtonian material* i.e. a material whose viscosity is independent of shear rate.

If apparent viscosity is defined as the shear stress at the wall/true shear rate at the wall then, in this case the *Rabinowitsch correction* will need to be invoked.

apparent wall shear rate
See *apparent shear rate*.

appearance surface
The surface which will be seen in use and which therefore has to be of a specified standard.

applesauce
Rough, wavy appearance of an extrudate; also referred to as orange peel, shark skin and flow patterns.

applicator
A device for applying energy to a product in *high frequency* or *microwave heating*.

appraisal
Inspection of the results of performance after completion.

AQL
An abbreviation used for *acceptable quality level*.

aquaplaning
A vehicle steering and driving defect. A loss of direct road contact caused by a *tyre* riding on a layer of water and caused, for example, by insufficient *tread depth*.

aramid
An *aromatic polyamide*. A long chain synthetic *polyamide* in which at least 85% of the amide groups are attached directly to two aromatic rings. Because of the intractable nature of this type of material it is only usually seen as a fibre. See *aramid fibre*.

aramid fibre
A manufactured, polymeric *aramid* fibre: high modulus, high heat resistant materials which burn only with difficulty and do not burn like more conventional polyamides (*nylon*). The first commercial aramid fibre was probably poly-(m-phenyleneisophthalamide) which is better known as Nomex (DuPont). Other aramid fibres include those based on poly-(p-benzamide) and poly-(p-phenyleneterephthalamide) which is better known by the name Kevlar (DuPont). Kevlar-type materials may be used for composite reinforcement (has a higher heat resistance than Nomex) and/or for tyre reinforcement. Short aramid fibres are also used to reinforce engineering thermoplastics materials: in such cases the fibres may be kinked, rather than straight, and this gives more isotropic mouldings. As the fibres are also softer than glass, abrasion problems are reduced. By coating the fibres with aluminium, the heat resistance of the fibres is raised significantly.

arc resistance testing
Tests performed to determine the resistance of a polymer compound to electric arcs such as appear in high voltage switch-gear and in circuit breakers. See, for example, DIN 53484 which uses a carbon arc formed between two carbon rods. See *resistance to tracking*.

arc tracking
The formation of a conducting track across a polymer surface caused by an electric arc. See *arc resistance testing*.

Ardichvili's equation.
An equation used to calculate the *roll separating force* generated during *calendering*. Ardichvili's equation offers the advantage of relative simplicity and gives a good basis for comparative tests as information obtained on one calender may be used to calculate pressures on another. The roll separating force $F = 2\mu Vrw(1/h_0 - 1/H)$ where:

μ = viscosity (the material is assumed to be Newtonian)
V = velocity of roll surface (rolls are assumed to have equal speed and no slip is assumed between the material and the roll)
r = radius of rolls (the rolls are assumed to be of equal diameter)
w = effective length of the roll
h_o = roll separation at the nip
H = height of material in front of the nip.

are
A unit of measurement in the *metric system* of measurement. A metric surface measure which is 100 square *metres* or 119·6 square *yards*. An are is 1/100 of an hectare.

arithmetical mean deviation
Also referred to as centre-line average and abbreviated to R_a. The arithmetical average value of the departure of a profile measurement above and below a reference line along the sampling length. A measure of *roughness* which is expressed in μm or μ in. The larger the number, the rougher the surface and the larger the *roughness grade number*.

Arizona Parking Lot Test
An *automotive fogging test*.

armouring
A protective covering, for example, a protective covering over a hose applied to prevent mechanical damage.

aromatic amine
An *amine* which contains *aromatic groups*, for example, p-phenylene diamine derivatives which are widely used to protect rubber compounds against *oxidation*.

aromatic diazo compound
The basis of diazo pigments and which are prepared from aromatic primary *amines* by *diazotization*.

aromatic group
An organic chemical group which, for example, contains an unsaturated ring structure derived from *benzene*.

aromatic halogen flame retardant
A *flame retardant* which contains a *halogen* (for example, *bromine*) and which is based on an aromatic organic material. For example, tribromotoluene.

aromatic oil
See *petroleum oil plasticizers*.

aromatic petroleum resins
See *petroleum resins*.

aromatic polyamide
An *aromatic polymer*: a polyamide which is based on main-chain, aromatic groups. Many different types of *polyamide (PA)* are possible but, for various reasons, those based on aliphatic chemicals are preferred. In general, an aromatic PA will take longer to prepare, will colour easily during polymerization and will decompose before melting. However, because they contain main chain aromatic groups, they possess better heat resistance and stiffness than an aliphatic polyamide. See *aramid* and *polyaryl amide*.

aromatic polyamide-hydrazide
See *polyamide-hydrazide polymer*.

aromatic polyester
A polyester derived from monomers in which all the hydroxyl and carboxyl groups are linked directly to aromatic nuclei. A material based on linked ring structures: a fully aromatic, *saturated polyester*. See, for example, *polyarylate*.

aromatic polyester carbonate
An abbreviation used for this type of material is APE. A *polycarbonate* with a higher heat resistance than a *bisphenol A polycarbonate*: the heat resistance can approach 192°C dependant on the percentage of iso-phthalic acid used during polymer manufacture. Such a material is also known as a *polyarylate*.

aromatic polyester liquid crystal polymer
An abbreviation used for this type of material is aromatic polyester LCP. See *liquid crystal polymer*.

aromatic polyether ketone
See *polyether ether ketone*.

aromatic polyimide
An aromatic polyimide polymer which contains no aliphatic groups. See *polyimide*.

aromatic polymer
A polymer containing aromatic rings in the main chain. Often based on fused benzene rings. May be partially aromatic (for example, *polyethylene terephthalate*) or completely aromatic (for example, polyphenylene). Polymers which contain aromatic groups (for example, p-phenylene) have desirable properties such as *high heat distortion temperatures (HDT)*, thermal stability and stiffness. The presence of rings raises the resistance to thermal degradation as the thermal energy is dissipated along the aromatic chain.

Processing of aromatic polymers is however difficult because of the high processing temperatures required and because of the high pressures needed to obtain the desired flow rate in melt processes such as *injection moulding*. Easier processing is obtained if the aromatic groups are spaced with other groups or linkages, for example, spacing them with aliphatic groups or segments (provided by, for example, adipic acid sections). This gives a semi-aromatic polymer which is easier to process than the fully aromatic structure. Because of the stiff polymer chains in aromatic polymers, and the high set-up temperatures, it is easy to get high moulded-in strains with this type of material. The use of high mould temperatures, low packing pressures and subsequent annealing help to reduce this and give more stable components. Be careful that over-packing does not occur because of the use of large runners and gates (necessary because of the high melt viscosity).

Heterocyclic polymers are also sometimes classed as aromatic polymers. Can also have mixed systems based on, for example, a repeat unit of a benzene ring linked via a heterocyclic, five membered ring.

Aromatic polymers often have excellent heat stability but are often *intractable* and so may be produced from low molecular weight intermediates by, for example, solution casting and sold in finished form. For example, in film form and used for electrical insulation. See *polyphenylene-1,3,4-oxadiazole* and *polyphenylene*.

aromatic polysulphone
See *sulphone polymers*.

aromatic processing oil
A type of *processing oil*. There are three primary groups of processing oil. The three groups are aromatic, naphthenic and paraffinic processing oils.

aromatic sulphide polymer
See *polyphenylene sulphide*.

aromatic sulphone
See *sulphone polymers*.

ARP
An abbreviation used for *asbestos reinforced plastics*. See *fibre reinforced plastics*.

artificial bone
See *calcium hydroxyapatite*.

artificial cotton
A term sometimes used for *high wet modulus modal*. See *viscose rayon - high tenacity fibre*.

artificial silk
A term formerly used for *rayon*. See *viscose*.

artificial weathering
The degradation of polymers caused by exposure to an artificial climate. Exposure in a laboratory to simulated weather conditions. The weathering resistance may be assessed, for example, by exposure of standard samples to selected light and to artificial rain in a test apparatus under specified conditions: such conditions may include the temperature, the intensity of the light, the wavelength of the light, the orientation of the sample to the light etc.

The changes may be cyclic, involving changes in temperature, relative humidity, radiant energy, and any other elements found in the atmosphere in various, but usually specified, geographical areas. The laboratory exposure conditions are usually intensified beyond those encountered in actual outdoor exposure in an attempt to achieve an accelerated effect (accelerated ageing). See *weathering*.

aryl
An organic, univalent radical obtained by abstracting a hydrogen atom from an *aromatic hydrocarbon*.

aryl amine
An *amine* obtained by substituting a hydrogen atom, or hydrogen atoms, of *ammonia* with an aromatic organic group. See *secondary aryl amine* and *aromatic amine*.

ASA
An abbreviation used for *acrylate-styrene-acrylonitrile*.

asbestos
A generic name for a group of minerals (natural hydrated silicates) which can be made to yield fine mineral fibres by mechanical working. There are two major types of asbestos, serpentine and amphiboles. Serpentine is also known as *chrysotile* and it is this type of asbestos which yields fibres which are soft, flexible and silky and therefore used to make fabrics. There are five different types of amphiboles but, in general, fibres from such materials are harsh, stiff and relatively brittle. Asbestos, from hornblende, has a relative density (RD or SG) of 2·7 to 3·6. The serpentine-type of asbestos has a relative density (RD or SG) of 2·3 to 2·8.

At one time asbestos materials were widely used as reinforcements for plastics materials (for example, for phenolics) but now, because of toxicity problems caused by the fineness of the fibres, the use of such fillers is declining. Dust formation in asbestos may be minimized by adding very fine PTFE, by treatment with octadecylamine and/or by pelletizing.

Asbestos replacement is difficult because any material which could replace asbestos in all its applications would probably constitute a similar health hazard - it is the fine fibrous, or fibrillar, nature of this mineral which is responsible for the desirable characteristics and also for the health hazard.

A major replacement material is *glass fibre (GF)* which can have a reproducible fibre diameter of approximately 10 μm - which is well outside the respirable range. The heat resistance of GF fibres can be raised by coatings. For example, by coating with vermiculite, GF fibres retain useful tensile strength up to 750°C which is approximately 250°C above the temperatures at which GF normally melts. However, in general, asbestos fibres give products which are stiffer, more fire resistant and which have a lower coefficient of expansion. Another replacement material is fibrous *anhydrous calcium sulphate* which is a single crystal fibre of diameter 2 μm and of average length 60 μm. See *asbestos cement composite* and *asbestos polymer composite*.

asbestos cement composite
Asbestos-cement products (for example, Sindanyo) may be in the form of rods, tubes or plates. A mechanically strong and heat resistant material used in plate-form, for example, to insulate an *injection mould* from the *platens*. Pin-fibrillated polypropylene fibre has been used to replace *asbestos* in cement-based composites.

asbestos polymer laminate
A laminate based on *asbestos* impregnated with *phenolic*, *melamine* or *silicone resins* (for example, Silumite) and used as an insulator in the electrical industry.

asbestos reinforced thermoplastic
An abbreviation used for this type of material is ARP. A moulding compound based on *chrysotile fibre* and on, for example, *polyvinyl chloride (PVC)*. Such a composite is used for the manufacture of road signs.

asbestos reinforced thermoset
An abbreviation used for this type of material is ARP. A moulding compound, based on *chrysotile fibre*, used in the manufacture of bearings.

ASE
An abbreviation used for *alkylsulphonic acid ester*.

asepsis
The state of being free from pathogenic organisms.

aseptic
Of or relating to asepsis.

ash content
The amount of inorganic material in a polymer compound. May be measured by strongly heating, for example, to 750°C a known weight of the material in air. Ash content = weight of ash × 100/initial weight of material. See, for example, ASTM D 2584.

aspect ratio
An abbreviation used for this term is H/S. The ratio of the *section height (H)* to the *section width (S)* of a *tyre*.

asphaltic bitumen
This term is sometimes used as another name for *bitumen*.

ASR
An abbreviation used for *alkylene sulphide rubber*.

Assam rubber tree
See *Ficus elastica*.

assembly time
See *closed assembly time* and *open assembly time*.

assignable
Capable of being specified.

assignable cause
An undesirable *variation* from the manufacturing requirements which is due to error, for example, mechanical or human error. Such errors included incorrect settings, faulty materials, faulty machines and incorrect specifications.

assignable variations
Intentional changes made, for example, to machine settings. By making deliberate changes to machine settings, the output (the product or mouldings) of a machine will be changed. An easy way of showing this is to weigh the components: changes in weight can be correlated, or assigned, to deliberate changes in machine settings.

Association Française de Normalisation
An abbreviation used for this French organization is AFNOR. A *standards organization* responsible for issuing standards or specifications. Copies of their publications can usually be obtained from the national standards office of a particular country, for example, from the *British Standards Institution (BSI)* in the UK. See *standards organizations*.

Association of European Isocyanate Producers
An abbreviation used for this organization is ISOPA. This non-profit making organization, founded by some major chemical companies, operates as a sector group under the auspices of the European Chemical Industry Federation (CEFIC).

Association of Plastics Manufacturers in Europe
An abbreviation used for this European organization, based in Brussels, is APME.

assy
A shortened form of the word 'assembly'.

ASTM
An abbreviation used for the American Society for Testing and Materials.

ASTM ignition test
See *ignition test ASTM*.

ASTM oils
Oils which have different aniline points, standardized by the *ASTM* and which may be used to assess the swelling resistance of rubbery materials.

ASTM standard motor fuels
Fuels standardized by the *ASTM* and which may be used to assess the swelling resistance of rubbery materials. May

be based on iso-octane and on mixtures of iso-octane and toluene.

at
An abbreviation used for *atmosphere*.

atactic poly-(propylene)
See *atactic polypropylene*.

atactic polymer
A polymer in which there is no regular, repeating structure. The repeat units, along the polymer chain, do not have the same configuration: that is, the groups of atoms do not have the same orientation in space. See *isotactic polymer, syndiotactic polymer* and *α olefin*.

atactic polypropene
See *atactic polypropylene*.

atactic polypropylene
Also known as atactic polypropene or as, atactic poly-(propylene). An abbreviation used for this type of material is APP: PP-A and a-PP are also used. A type of *polypropylene* in which there is no regular, repeating structure: there is a random distribution of the configuration of the repeat units. Produced as a by-product in isotactic PP manufacture and separated by solvent extraction with hexane. A weak rubbery material which is used in conjunction with asphalt as a road surfacing material.

ATB
An abbreviation used for average time of burning in ASTM D 635. A *horizontal combustion test*.

atm
An abbreviation used for *atmosphere*.

atmosphere
In chemical terms it means the gaseous envelope or medium in which a reaction is occurring. In astronomy, it means the gaseous mantle which surrounds a star or planet. It is also the term used to describe a unit of pressure which is the normal pressure of air at sea level: equivalent to approximately 101·33 Pa or 760 mmHg or 14·72 psi. An abbreviation used for this unit is, at or atm.

atmosphere Earth
The gaseous mantle which surrounds the planet Earth has been divided into several layers or strata. These are, the troposphere, the stratosphere, the ionosphere and the exosphere. The atmosphere stretches upwards for hundreds of miles but most of its mass is concentrated in the bottom four or five miles, that is, in the troposphere.

atmosphere inert
Usually means without oxygen and is obtained by flushing the system with an inert gas such as nitrogen.

atmospheric pressure
That *pressure* exerted by the atmosphere: this varies from location to location. At sea level it is equivalent to approximately 101·33 Pa or 760 mmHg or 14·72 psi.

atom
The smallest unit of an element which retains the character of that element. Atoms are composed of sub-atomic particles such as electrons, protons and neutrons: the number and arrangement of such constituent particles determine the element. Atoms join together to form a molecule or molecules.

atomic mass
The mass of an isotope of an element measured in *atomic mass units*.

atomic mass unit
Also known as a dalton. A unit used to express the mass of an individual isotope of an element and which is based on the isotope of carbon-12. It is 1/12 of the mass of an atom of carbon-12: approximately equal to $1·7 \times 10^{-24}$ g.

atomic number
The number of protons in the nucleus of an atom of an element. Also known as the proton number and given the symbol Z. See *atomic mass unit*.

atomic orbital
Sometimes referred to as AO or as orbital and is the region surrounding an atomic nucleus inhabited by an electron. The density of the AO, at any point, is proportional to finding the electron at that point.

atomic weight
Average mass of the atoms of a specimen of an element and measured in *atomic mass units*. See *relative atomic mass*.

attapulgite clay
A type of clay which is similar to sepiolite. Both are crystalline or paracrystalline clays with chain-like structures.

attenuation
The conversion of energy into heat when electromagnetic energy passes through a medium.

attenuation device
A short tunnel, of restricted aperture, placed at the inlet and outlet of a continuous *high frequency* or *microwave heating oven*. Such a device minimizes energy losses.

atto
An abbreviation used for this term is a. See *prefixes SI*.

attribute
A property or characteristic.

attributes characteristic
A *characteristic* which is either present or absent and which is measured by counting the frequency of occurrence. For example, the number of short mouldings in a batch.

ATV tyres
An abbreviation used for all terrain vehicle tyres.

AU
An abbreviation used for a *polyester polyurethane rubber*.
  AU-I = isocyanate crosslinkable.
  AU-P = peroxide crosslinkable. See *polyurethane rubber*.

auger regranulator
A type of *regranulator* widely used in *injection moulding* and which is fed with a wide-bladed screw or auger. Regranulation of feed systems can be carried out at the moulding press by means of an auger regranulator assembly. The entrance section of the regranulator is positioned adjacent to the chute of the moulding machine into which, the feed system falls when ejected from the mould. Once the material has been regranulated, it is then automatically blended at the required ratio with virgin material and fed directly back into the moulding machine hopper. This type of process has two distinct advantages over a separate system; it minimises the risk of contamination and the need to redry hygroscopic materials however, this type of assembly does take up invaluable floor space around the moulding machine.

auguring
Term used in injection moulding and means that the screw pushes itself backwards too quickly when the screw is rotated (so as to cause material feeding on an in-line screw machine). Some *back pressure* may be needed simply to stop the screw pushing itself (auguring) too easily out of the cylinder, or barrel; for example, 5 bar or, 0·5 MNm^{-2} or, 73 psi may be needed.

auto-coagulation
Spontaneous or natural coagulation of *latex*. See *tapping*.

autoclave
A heating device used, for example, to vulcanize rubber components. A large steam heated vessel.

autoclave moulding
See *bag moulding*.

autoflex die
A *sheet die* in which the lip opening is controlled by the expansion of thermal bolts which, in turn, are responding to thickness sensors.

automatic control
Machine operating mode in which the equipment continues production until a fault develops.

automatic hide turning device
A mixing aid which cuts the *hide*, or gelled sheet, on a *two-roll mill* at pre-set intervals to approximately half the roll width and tumbles it back into the *nip*.

automatic melt flow indexer
An abbreviation used for this type of test equipment is AMFI. A *melt flow index apparatus* which performs the test cycle automatically: such equipment stores, and manipulates, the test data using a built-in microprocessor. A conventional *melt flow indexer* applies the load to the material by means of a weight: an AMFI may apply a constant load to the piston via a microprocessor-controlled drive.

automatic mould changing
The changing of a mould, for example, an injection mould, without significant human intervention. Time-saving on this job means less machine down time, greater productivity and quicker availability of the machine setter for other work. The mould changer comprises of, for example, a tool carriage, common manifold fittings (to accommodate all tool/water and electrical connections), hydraulic mould clamps and a feeding and locating mechanism. The new tool is loaded onto the carriage and preheated. At a signal from the control unit the clamp is released and the old mould is pulled out onto the vacant position on the carriage. A new mould is then moved into position between the press platens, aligned and locked into position by automatic hydraulic clamps. All manifold connections are also made automatically.

automatic operation
Usually refers a production process, such as injection moulding, and means that moulding will continue until a specified fault develops.

automatic profile control system
A system which automatically adjusts itself so as to control the size of an emerging profile, for example, *lay flat film*. The die has a number of heating/cooling units around its circumference so that melt viscosity can be varied, to give the required output at that point, without adjusting the die gap. The thickness is measured and the temperatures adjusted accordingly. Improves bubble stability and output.

automatic quality control
Quality control which ie performed automatically, for example, by the production machine. Because of the power of the microprocessor, used for the control system, it is now relatively easy to incorporate features so that verification of product quality is possible during the manufacture of each item. Verification of product quality at the point of manufacture can be documented in *statistical quality control (SQC) records*.

For example, to monitor on-line, an *injection moulding machine* must be equipped with appropriate transducers and measuring equipment. For example, with a melt thermocouple (in the machine nozzle), pressure transducers (in both the mould and the hydraulic line) and a linear displacement transducer (to measure linear screw movement). The output from these transducers is monitored and displayed as an *average chart* on a visual display unit (VDU). During production therefore, the data is automatically gathered and displayed on the VDU. Production drifts, or trends, can therefore be easily spotted and, out-of-specification products rejected completely or, diverted, for inspection.

automatic screen changer
A unit which removes one screen pack from the melt stream path in, for example, extrusion and replace it with another pack when, for example, the pressure drop, as measured by a pressure sensor, reaches a preset value.

automatic seal
Term applied to seals that are pressure energized: the term is specifically used to classify certain types of flexible lipseals.

automotive fogging
The build-up of an oily layer on automobile (car) windows. Plasticizer loss from *polyvinyl chloride (PVC)* leathercloth and crash-pads can contribute to this problem. The use of *trimellitate* and *terephthalate plasticizers* helps prevent car window fogging which may result from the use of unsuitably plasticized PVC upholstery.

Samples of the PVC are exposed to high temperatures (for example, 60, 75 and 90°C) for specified times and the reflectance of an exposed surface is measured: fogging reduces reflectance. *Low fogging stabilizers* are available.

Automotive Composites Association
An abbreviation used for this organization id ACC. A USA based organization (Chrysler, Ford and General Motors) whose aims are to develop cost effective structural polymer composites.

autooxidation process oxidation
See *oxidation mechanism*.

autosynergistic
A single material which behaves as a *synergistic combination*. For example, p-phenylene diamine *antioxidants*.

auxiliary inhibitor unsaturated polyester resin
Such inhibitors include resorcinols and phenols. See *unsaturated polyester resin*.

auxochrome
A colour strengthening group in a *dye*, for example, an amine group.

av
An abbreviation used for average. See *mean average*.

AV
An abbreviation for acid value. See *acid number*.

avdp
An abbreviation used for *avoirdupois*.

average
A measure of central tendency i.e. usually that which is obtained by totalling the value of items in a set and dividing by the number of individuals in a set (this is the mean average).

average chart
A *process control chart* on which average values are plotted. Such a chart has five lines, the average value and two sets of tramlines. The upper set of tramlines has two lines called the

upper warning limit (UWL) and the upper action limit (UAL): these limit line correspond to probability points on the normal distribution curve of, say, 95% and 99·9%. The lower set of tramlines has two lines called the lower warning limit (LWL) and the lower action limit (LAL): these limit line correspond to probability points on the normal distribution curve of, say, 95% and 99·9%. See *production capability*.

average molecular weight
See *molecular weight average*.

Avogadro's number
Also known as Avogadro's constant. The number of atoms or molecules in one mole of a substance: $6.022\ 57 \times 10^{23}$ mol^{-1}.

avoirdupois
An abbreviation used for this term is avdp. See *avoirdupois weight*.

avoirdupois weight
A system of weights introduced for general, or everyday, use and based on the pound. In the UK, 16 drams = 1 ounce, 16 ounces = 1 pound, 14 pounds = 1 stone, 28 pounds = 1 quarter, 4 quarters = 1 hundredweight and 20 hundredweight made one ton. In the US, 100 pounds = 1 hundredweight. One pound avoirdupois contains 7,000 grains. See *US Customary Measure* and *UK system of units*.

AW
An abbreviation sometimes used for atomic weight.

awl vent
A deliberate penetration of the *sidewall rubber* of a tubeless *tyre* used to prevent air pressure build-up in the carcass.

axial clearance
The clearance which exists between the sealing element and the inside face of a cover.

Ayem
A trade name/trademark for a *low melting point metallic alloy* based on zinc. See *zinc-based alloy*.

AZDN
An abbreviation used for *azobisisobutyronitrile*.

azelaic acid
A linear carboxylic acid which is derived from oleic acid by ozonolysis. A precursor for PA 69: also used to make *plasticizers*. May be represented as $CH_3(CH_2)_7CH=CH(CH_2)_3COOH$.

azelaic acid ester
The reaction product of *azelaic acid* and an alcohol and which is commonly called an *azelate*.

azelate
An *azelaic acid ester*. The reaction product of azelaic acid and an alcohol is commonly called an azelate. The alcohol may be linear or branched and/or mixed alcohols may be used. A variety of azelates are made for use as plasticizers for materials such as *polyvinyl chloride (PVC)*. See, for example, *dioctyl azelate* and *plasticizer*.

azelates
The collective name for the class of materials which contain the *azelate* group and which are a group of plasticizers.

azeotropic mixture
A constant boiling mixture: a mixture of two or more liquids which has a specific boiling point at a certain pressure.

azide
A compound which contains the univalent azide group -N$_3$ or ion N$_3^-$. A bisazide compound contains two such groups and a monoazide compound contains one group. An azide is used to make an azide-type photo-sensitizer with, for example, *cyclized rubber*. Used with, for example, aqueous base-soluble polymers (such as novolak resins and poly-4-hydroxy styrene) to make a *photoresist*. The azide can be tuned, by altering the molecular structure, so that it responds to a particular frequency. In turn, for example, the novolak resin responds to a particular frequency.

azine dye
A class or type of organic *dye* which includes nigrosine. Used to produce jet black in, for example, *acrylonitrile-butadiene-styrene* materials.

azlon fibre
A manufactured, polymeric fibre composed of a regenerated, naturally-occurring protein. See *casein*.

azo acid dye
A type of colorant which is based on an *azo dye*. One or more of the aromatic groups is stabilized with an acid group, for example, the sodium salt of sulphonic acid. Such a dye is very compatible with, for example, phenolic plastics materials.

azo compound
A compound which contains an *azo group* attached to two carbon atoms: such materials are widely used as *blowing agents*: they are chemical blowing agents. These organic materials can be structured so that they decompose over a fairly narrow temperature range, for example, at the melt processing temperatures of a particular polymer. See *diazonium salts*.

azo dyes
A class or type of organic dye which is of relatively low cost. Contains the azo group -N=N- and this group is often sandwiched between aromatic groups. Can be used for plastics materials such as phenolics, polymethyl methacrylate, polystyrene and unplasticized polyvinyl chloride. See *acid azo dye*.

azo group
Two nitrogen atoms linked by a double bond. May be represented as -N = N-. See *azo compound*.

azo pigment
A class or type of *organic pigment*, based on an *azo dye*, and which is of relatively low cost. Bright semi-transparent colours, for example, reds are common but can also have yellow. Such pigments have good bleed resistance but are susceptible to chemical attack which can cause reversion of the pigment to an *azo dye*. Mono-azo (*monazo*) and *diazo pigments* are of two types, *pigment dyes* and *precipitated azo pigments*.

azo-tipped polystyrene
See *comb-grafted natural rubber*.

azobisformamide
See *azodicarbonamide*.

azobisisobutyronitrile
Also known as 2,2'-azobisisobutyronitrile or as azoisobutyronitrile. Abbreviations used include ABIBN and AIBN and AZDN. An *azo compound* commonly used as an *azo initiator* for free radical polymerization.

azocyclohexyl nitrile
A colourless solid *azo compound* used as a *blowing agent*. The decomposition, or degradation, temperature is approximately 114°C.

azodicarbonamide
An abbreviation used for this material is ADC. Also known as azobisformamide. A pale yellow solid *azo compound* used as a *blowing agent*. The decomposition, or degradation, temperature of 190 to 230°C/374 to 446°F can be reduced by the use of a metal salt which is called a *kicker*.

azodicarboxylate-functional polystyrene
See *comb-grafted natural rubber*.

azodicarboxylic acid diamide
A solid *azo compound* used as a *blowing agent*. The decomposition, or degradation, temperature is approximately 140°C.

azodicarboxylic acid diethyl ester
A solid *azo compound* used as a *blowing agent*. The decomposition, or degradation, temperature is approximately 108°C.

azohexahydrobenzonitrile
A solid *azo compound* used as a *blowing agent*. The decomposition, or degradation, temperature is approximately 103°C.

azo initiator
An *azo compound* which can be decomposed to give free radicals. R–N=N–R decomposes to give 2R· plus N_2. The free radicals (R·) can then be used to initiate polymerization. See *azobisisobutyronitrile*.

azoisobutyronitrile
See *azobisisobutyronitrile*.

B

b
An abbreviation sometimes used for:
acceleration (more usual to use a); and,
width (of a test specimen). See *flexural strength*.

B
This letter is used as abbreviation for:
benzyl - in *plasticizer abbreviations*. For example, *butyl benzyl phthalate (BBP)*;
block copolymer;
bromo - in *plasticizer abbreviations*. For example, *tri-(2,3-dibromopropyl) phosphate (TDBP)*;
butoxy - in *plasticizer abbreviations*. For example, *tri-(2-butoxyethyl) phosphate (TBEP)*;
butyl - in *plasticizer abbreviations*. For example, *butyl benzyl phthalate (BBP)*;
degrees Baumé. See *Baumé scale*.
filler - If a plastics material contains a filler then this letter may be used to show the presence of boron or, beads or, spheres or, balls.

B plate
Part of an *injection mould*. Attached to the moving platen by a clamping plate: the *ejector pins* usually pass through this part of the mould. See *standard mould set*.

B type H-NBR
See *partially saturated nitrile rubber*.

B type screw
A type, or size, of *screw* in *injection moulding*. Has intermediate *shot capacity* and *injection pressure*.

B.th.u.
An abbreviation used for *British thermal unit*.

B-stage
An intermediate stage in the reaction of certain thermosetting resins in which the material swells when in contact with certain liquids (for example, acetone or monomer) and softens when heated, but may not entirely dissolve or fuse. The resin in an uncured thermosetting moulding compound is usually in this stage and is sometimes referred to as a resitol. (See also *A-stage* and *C-stage*).

Ba-Cd stabilizers
An abbreviation used for *barium/cadmium stabilizer*.

BAA
An abbreviation used for *butyraldehyde-aniline (condensate)*.

back connected
A hydraulic term which means that pipe connections are on normally un-exposed surfaces of hydraulic equipment.

back flow valve
See *back-flow stop valve*.

back injection moulding
Also known as textile back moulding and as the decor mould process. An *injection moulding process* used to mould a thermoplastics material onto a substrate, for example, onto a substrate such as a fabric or textile. Used to produce components, such as shelves, for the automotive industry. For example, based on *polypropylene (PP)* and tufted velour.

back plate
A support plate in a mould: the rear plate in which there is provision for the attachment of the mould half to the *platen*.

back pressure
Term associated with *injection moulding* and with in-line screw machines. It is the pressure that the screw must generate, and exceed, before it can move back. Back pressure is generated by rotating the screw against the restriction of the plastics material which is contained in the cylinder, or barrel. Back pressures may reach 250 bar or, 25 MNm^{-2} or, 3600 psi. The use of such high back pressures can improve colour dispersion and material melting, but it is paid for as it increases the screw retraction time, reduces fibre lengths in filled systems and imposes stress on the injection moulding machine. Keep as low as possible and in any event do not exceed 20% of the machines (maximum rated), injection moulding pressure. See *back pressure programming*.

back pressure hydraulic
A hydraulic term which means that there is pressure on the discharge side of a load.

back pressure programming
Term associated with *injection moulding* and with screw machines. Means that it is possible to program, or alter, the *back pressure*. This is done so as to compensate, for example, for the effective reduction in screw length which occurs during plasticization; such a reduction means less heat input and therefore a drop in temperature. However, on many machines it is difficult to sensibly set the machine as there is no easy way of measuring the effects of the changes.

back rinding
See *backrinding*.

back roll
Also known as the rear roll. The back roll of a *two roll mill* is the one furthest away from the operator.

back-flow stop valve
Also known as a ring check valve or as a screw valve. A non-return valve on the tip of, for example, the *screw* in an injection moulding machine and which prevents shot loss during injection. To stop the injection pressure being transmitted to the screw's thrust bearings, on a *rubber injection moulding machine*, it is important that there is a *back-flow valve* between the plunger and the screw.

backpressure
See *back pressure*.

backrinding
Also called back rinding or, suck back or, rind-back or flashback (flash back) and retracted spew. A moulding fault associ-

ated with rubber compounds. The surface of the component is torn or gouged at the mould parting line. A rubber moulding defect experienced when the moulding pressure is suddenly released and the product tears at the spew line.

bacteria
The plural of bacterium. A very large group of micro-organisms which are, for example, concerned with fermentation and polymer manufacture. A given species of bacteria will reproducibly synthesize polymers of a fixed molecular weight under specified conditions. See *biopolymer* and *polyhydroxybutyrate*.

baffle
A device, a plate, mounted in a reservoir and which separates the pump inlet from the pump return lines. May also mean an obstruction used to change flow direction in a *high speed mixer*. May also mean a metal strip, or plug, fitted in the coolant circuit of a mould so as to restrict flow to a prescribed path.

bag moulding
This term covers autoclave moulding, pressure bag moulding and vacuum bag moulding. A moulding process for reinforced plastics (*GRP moulding*) in which the impregnated reinforcement is consolidated against the mould by means of a pressure difference applied via a rubber bag or blanket: usually by means of air (but may be steam, water or vacuum). The pressure difference may be generated by means of a vacuum or compressed air thus giving rise to the different names.

Bagley end correction
Also known as the Bagley correction and as the Bagley entrance correction - it is of greater practical importance than the *Rabinowitsch correction*. A correction applied to rheology data so as to counter viscous and elastic melt effects at the die entrance. It is thought that with polymers, approximately 95% of the entrance effects in polymer melts are due to chain uncoiling and subsequent storage of elastic energy. the application of such a correction has the effect of lengthening the die (capillary).The Bagley correction may be measured by using a range of dies which ideally have the same radius (R) but different lengths (L). The radius R of the die may be measured and n (the correction) is found experimentally. This is done by using dies of differing length, but constant diameter, and studying the flow behaviour for a specific polymer. Pressure is plotted against shear rate for each die. For a specific shear rate, the pressure required for each of the dies may be obtained and used to produce plots of L/R. The correction, the Bagley end correction, is obtained from the negative intercept of the base line.

Establishing the Bagley correction can be lengthy and so various alternatives have been developed. See *knife edged die procedure* and *Couette-Hagenbach method*.

Bagley entrance correction
See *Bagley end correction*.

bake-on mould release
A mould release agent based on, for example, a *silicone polymer*, applied as a liquid to a metal surface and cured in-situ by heating to about 215°C. See *silicone mould release agent*.

Bakelite
A trade name/trademark often used for *phenolic moulding materials* or the products made from such materials.

balance tray
A tray placed under the mould (for example, an injection mould) and onto which the mouldings fall after being ejected; if the moulding is of the correct weight then *injection moulding* continues and the components are accepted.

balanced runner mould
A multi-cavity *injection mould* in which the flow path length between each cavity and the *sprue* is of the same length. It is relatively easy to design a balanced runner lay-out when the number of cavities equals an exponential power of 2, i.e. 2, 4, 8, 16, 32, etc. The feed system in a balanced runner mould provides each cavity with the best chance of feeding simultaneously and undergoing the same packing pressures. Its use therefore should result in components which are similar. See *runner lay-out*.

balata
A natural polymer of isoprene which is used to make high quality, golf ball covers. The raw material is a latex and is obtained by *tapping* the tropical tree known as mimusops balata. The latex contains approximately 50% of a trans-polyisoprene structured material and the remainder is largely resinous material (approximately 35%). See *natural rubber*, *gutta-percha* and *hydrogenated natural rubber*.

bale
A form of supply, for example, for *natural rubber*.

bale cutter
A machine used to reduce the size of rubber bales. See *rubber - raw material form*.

ball or pin impression method
An *environmental stress cracking test*. It is suggested that an *annealed compression moulded sheet* is used as such sheet has a low internal stress level. Strip specimens are cut from this compression moulded sheet or the complete product utilised. Alternatively, if the reduction in tensile or flexural strength is to be measured then standard tensile or flexural specimens are used. See, for example, ISO 4600 (1981).

2·8 mm holes are drilled centrally into the specimens and then reamed out to 3 mm. After storage under standard conditions for 24 h the diameters of 5 holes, selected at random, are measured to 0·005 mm. Oversize balls or pins are then pressed into the holes. One hour after insertion of the pins the specimens are immersed in the liquid medium for 20 h (short test) at a specified temperature. On removal from the liquid they are dried on blotting paper and allowed to stand for three hours before stress cracking is evaluated. Either, the oversize at which cracks are visible and the time for their appearance are recorded or, the oversize found at which a 5% reduction in either the tensile or flexural force occurs (5% compared to when there in no pin inserted into the holes).

ballast
The material used to load a *weight-loaded accumulator*.

ballotini
See *glass ballotini*.

Ballotini
See *glass ballotini*.

balls of glass
See *glass beads*.

BAMA
An abbreviation used for the British Aerosol Manufacturers Association.

bambooing
A periodic extrusion defect which gives the product the appearance of bamboo. See *melt fracture*.

Banbury mixer
One of the most well known mixers used by the polymer industry. An *internal mixer* particularly widely used in the rubber industry: a *batch mixer* which has a fixed capacity and which is often referred to as a Banbury.

Banbury-Lancaster process
A reclamation process for rubber compounds which uses an *internal mixer*. See *reclaiming processes*.

band
That which forms around the roll of a *two-roll mill* and which is cut and worked so as to achieve mixing. When the band is cut fully across, the mixed material may be removed from the mill in the form of a sheet or hide.

In tyre technology a band is also known as pocket. A pre-assembly of some of the plies of the *tyre carcass*.

banded spherulite
See *ringed spherulite*.

bank
The roll of material in a *nip*, for example, in the *feed nip*.

bank marks
Irregular areas of surface roughness and associated with the *calendering* of *polyvinyl chloride (PVC)*. Product marking caused by, for example, poor thermal homogeneity of the PVC.

bar
A unit of pressure widely used on, for example, *injection moulding machines*. A pressure of 10^6 dynes per square centimetre or 10^5 pascals or, 0·986 923 atmospheres.

bar colorant
A *masterbatch* supplied in the form of bars or strips and which is chipped, metered and blended with thermoplastics material on the processing equipment, for example, an *extruder*.

bar-bell
See *dumb-bell*.

Barber-Coleman Impressor
A spring based instrument used for *hardness measurements* and which is readily portable. See *Barcol hardness*.

Barcol hardness
The test reading or value obtained by the use of a Barcol hardness tester. A measure of *hardness* obtained using the Barber-Coleman Impressor. There are three models: one for soft metals and the harder plastics, another for softer plastics and the third is for extremely soft plastics and rubbers.

The instrument is placed in contact with the flat, well supported specimen and then using hand pressure only the hardened steel, truncated cone indentor is forced into the material. Barcol hardness is quoted together with the model number used.

Barcol hardness is used, for example, to test the state of cure of an *unsaturated polyester resin*: the higher the reading the more complete is the cure.

barite
Naturally occurring *barium sulphate* is a mineral known as barite or as baryte. See *barium sulphate*.

barium
An alkaline earth metal (Ba) which is far too reactive to occur naturally. It ignites spontaneously in moist air. Barium sulphate is used as a filler in, for example, rubber compounds. Barium compounds are also used as *heat stabilizers* for *polyvinyl chloride (PVC)* - usually in conjunction with cadmium compounds.

barium/cadmium/lead stabilizer
A polymer stabilizer based on the metals barium, cadmium and lead which is used as a *heat stabilizer*. Compounds of such metals are used as, for example, heat stabilizers for *polyvinyl chloride (PVC)* because of the good stabilizing performance of such a *mixed metal stabilizer*: a *heavy metal stabilizer* used in, for example, the extrusion of *unplasticized polyvinyl chloride* because of the ease of processing and the subsequent weather resistance.

barium/cadmium salt
See *barium/cadmium stabilizer*.

barium/cadmium stabilizer
May also be known as cadmium/barium stabilizer. May also be known as a barium/cadmium salt and is a type of metal soap stabilizer. They are *heat stabilizers* used with *polyvinyl chloride (PVC)* and are classed as synergistic mixtures. For example, a mixture of cadmium octoate and barium ricinoleate gives more protection to PVC than a similar quantity of either salt: the cadmium salt gives good initial colour while the barium salt extends this period of good colour retention. To reduce *plate out problems*, liquid systems, based on phenates, may be used. However, solid soaps with high cadmium content are widely used in unplasticized *polyvinyl chloride (PVC) formulations*.

barium/cadmium stabilizer package
A total stabilizer system designed to be added in one lot or package. Could include, for example, barium/cadmium phenate (2·5 phr as a primary stabilizer), epoxidized oil (4 phr as a synergistic heat stabilizer), stearic acid (0·5 phr to improve clarity by improving compatibility), trisnonyl phenyl phosphite (1 phr as an antioxidant) and zinc octoate (0·5 phr as a synergistic stabilizer to improve colour). See *metal complex stabilizers*.

barium/calcium/zinc stabilizer
A polymer stabilizer based on the metals barium, calcium and zinc which is used as a *heat stabilizer*. Compounds of such metals are used as, for example, heat stabilizers for *polyvinyl chloride (PVC)* because of the relatively good stabilizing performance of such *a mixed metal stabilizer*. Such a stabilizer is often combined with a *metal-free organic stabilizer*.

barium chromate
May be represented as $BaCrO_4$. Also known as barium yellow or as, baryta yellow. Yellow crystalline material used as a yellow *pigment*: has a relative density of 4·5.

barium ricinoleate
A metal soap used as a *heat stabilizer* for *polyvinyl chloride (PVC)*. May be used a component of a synergistic mixture. For example, a mixture of cadmium octoate and barium ricinoleate gives more protection to PVC than a similar quantity of either salt. See *barium/cadmium stabilizer*.

barium stearate
May be represented as $(C_{17}H_{35}COO)_2Ba$. White crystalline powder used, for example, as a dusting agent for rubber compounds and as a *stabilizer/lubricant* in *polyvinyl chloride (PVC) technology*.

barium sulphate
An *alkaline earth metal sulphate filler* with the formula $BaSO_4$. Naturally occurring barium sulphate is a mineral known as barite (baryte): referred to as barytes or as, barytes white or as, heavy spar or as, schwerspat. The synthetic material is called *blanc fixe* and is a finer, purer material obtained by precipitation.

The natural material has a relative density (RD or SG) of 4·4 to 4·6. Used as an inert, white, high gravity filler. Dry ground material has a maximum particle size of approximately 60 μm while wet ground material has a maximum particle size of approximately 20 μm. Barium sulphate has little pigmenting power and a colour which ranges from white to pale brown. Does not effect cure rate in rubber compounds. Used in acid resistant rubber compounds and to obtain X-ray opaque materials.

barium yellow
See *barium chromate*.

barium/zinc stabilizer
A polymer stabilizer based on the metals barium and zinc which is used as a *heat stabilizer*. Compounds of such metals are used as a *heat stabilizer* for *polyvinyl chloride (PVC)* because of the relatively good stabilizing performance of such *a mixed metal stabilizer*. Such a stabilizer is often combined with a *metal-free organic stabilizer*.

bark specks
Impurities in *natural rubber* caused by the presence of bark.

barrel
A hollow chamber in which the *screw* (or screws) operates. If one screw is used the barrel has a cylindrical cross-section: if two screws are used, the barrel has a figure of eight (8) cross-section. It is the screw and the cylinder, or barrel, which interact to convey, melt and generate pressure on a polymeric material within either an *injection moulding machine* or an *extruder*. It is essential that this is done in a controlled way as the output must be uniformly plasticized material, of constant composition, at the required, controllable rate. To achieve this, the barrel must be made very accurately; the total out-of-alignment error, after all machining, must be less than one half of the screw/barrel clearance. A barrel is rated in terms of its diameter, for example, in inches or millimetres.

barrel construction
Because the pressures generated within either an injection moulding machine or an extruder can reach very high values, the machine cylinder, or *barrel*, must be made to withstand these high pressures. A barrel is made form thick walled, alloy steel tube or pipe and is usually designed to operate at up to 25,000 psi or 175 MNm^{-2} in the case of an *injection moulding machine*; the minimum burst pressure is approximately twice this, for example, 50,000 psi or 350 MNm^{-2}. The feed throat is cut through the barrel wall and the size of the feed opening is approximately the same size as the barrel diameter. At the other end of the barrel, provision is made to attach the *die* or *nozzle* e.g. by means of a clamping nut or end-cap.

barrel diameter rheometer
An abbreviation used for this term is D. Measured in, for example, cm. See *capillary rheometer - measuring* $\dot{\gamma}_{w,a}$ and τ_w.

barrel finishing
See *tumble finishing*.

barrel heating for thermoplastics
Most injection moulding machines and extruders for thermoplastics have electrically heated barrels. The barrel is usually heated by means of resistance coils or cuffs, which are strapped or bolted around the cylinder, or barrel. Upon demand, initiated by a thermocouple, electricity is passed through the resistance wire, inside the coil, and the resistance to flow causes the temperature to rise. If a *PID controller* is used then, when the set point is approached the power is progressively reduced and then finally turned off completely at the set point. For a given machine, the actual cylinder settings to achieve a desired melt temperature, will depend on, for example, the *screw* rotational speed, the pressure within the system and the throughput.

barrel measure
A measure of capacity or volume the size of which depends upon the country and the usage. In the UK, it is usually 36 gallons (Imperial) which is 0·1636 cubic metres or 163·6 litres. One barrel of oil is 42 gallons (US) or 0·159 cubic metres or 159 litres. Abbreviations used for barrel are bbl or bl. To convert from barrels (US) to cubic metres multiply by 0·1590.

barrel polishing
See *tumble polishing*.

barrel residence time
The average time for which a material is contained with the cylinder, or barrel. This is important because, for example, the rate of decomposition of plastics, is dependent on both temperature and time. For example, a plastic may be degraded by a short exposure to a high temperature or, by a longer exposure to a lower temperature. How long the plastic is in the cylinder is therefore important.

For injection moulding, the actual residence time may be determined experimentally by measuring the time taken for coloured plastic to pass through the cylinder. May be calculated very roughly by the following formula for in-line screw, injection moulding machines;

$$\frac{\text{rated capacity of injection cylinder (g)} \times \text{cycle time (s)}}{\text{shot weight (g)} \times 30.}$$

barrel temperature settings
The temperatures set on the control instruments so as to achieve a desired melt temperature. It is the *melt temperature* which is important and any cylinder temperatures quoted (in the literature) are only guidelines. Where there is no experience of processing a particular grade of material, then start with the lowest settings recommended. Usually the first zone temperature is set at the lowest value as this helps prevent premature melting and bridging of the material (resin) in the feed throat, The temperatures of the other zones then gradually increase until the nozzle/die is reached. In injection moulding, the nozzle temperatures is often slightly lower than the front barrel temperatures so as to prevent *drooling*. In *extrusion*, the die temperatures are often slightly higher so as to put a higher gloss on the product.

barrel wear
An increase in the operating clearance of the screw/barrel system and/or wear of, for example, the barrel in the *hopper* region. In *extruders*, and in *injection moulding machines*, the *screw* and barrel assembly operate in a very aggressive environment which can cause severe wear problems; each year the environment gets worse as material modification becomes more common with, for example, abrasive fillers (such as *glass*) being added to more plastics materials.

To improve the wear resistance of the cylinder or barrel, it may be modified, or lined. Modification is by, for example, *nitriding* or *ion implantation* but these treatments are not as good as *lining*. As it is easier to replace a *screw* than a barrel then, the barrel must be harder than the screw. See *bimetallic barrel*.

barrelling
Also known as tumbling. A finishing process for mouldings made from relatively brittle materials which are flashed. Used, for example, for *thermosetting plastics*. The mouldings plus, for example, wooden pegs, are placed in an octagonally-shaped barrel which is slowly rotated (at approximately 30 rpm). See *cryogenic tumbling*.

barrier design screw
See *barrier screw*.

barrier layer
A separate layer of material, usually in an extrudate or a blow moulding, which is there to stop, or hinder, the passage of another material, for example, a gas. The barrier layer may be a layer of another plastics material, a metal layer, or a dispersed plate-like filler (such as *mica*). In *blow moulding*, a barrier layer is a layer of thermoplastics material (or materials) incorporated into a package or bottle in order to hinder the passage of a flavour, an odour or a gas in or out of the package. See *coextrusion*.

barrier properties
A package is said to have barrier properties if it hinders the passage of a flavour, an odour or a gas in or out of the package. Permeation of gases and liquids, through the plastics container, can be one of the major factors which determine the shelf life of the product. The transmission rates of oxygen, carbon dioxide and flavour constituents are usually of the greatest interest. Permeation is a function of materials, design and processing method. See *permeability to gases* and *liquids*.

barrier screw
A barrier design screw is a *two start screw*; as the resin melts it is transferred to the other flight. That is, the screw employs the melt pool and solid bed separation principle which gives improved output per rpm and a lowering of melt temperature. Mixing sections may also use this principle, for example, the Egan mixing section uses barrier flights. See *screw mixing sections*.

baryta yellow
See *barium chromate*.

baryte
Naturally occurring *barium sulphate*.

barytes
Naturally occurring *barium sulphate*.

barytes white
See *barium sulphate*.

BAS
An abbreviation used for *British Antarctic Survey*.

base coat
A lacquer or varnish whose application forms part of the production process for the *vacuum metallization* of plastics. For example, injection mouldings are coated with a base coat so as to prevent out-gassing, improve metal adhesion, improve surface finish and minimise top coat solvent attack of the plastics moulding. A clear high gloss polymeric coating which may be based on either a *thermoplastics material* or a *thermosetting material*, for example, an alkyd resin.

base colour
The natural colour of a material. Due to variations in the base colour of polymers, the amount of colorant addition will vary and will be dependent upon the colour shade and the intensity required.

basic dye
A cationic dye which combines with acids: such a dye has an affinity for wool but is, for example, used for dyeing *acrylic fibre*.

basic lead carbonate
Also known as cerussa, flake lead, flake white, lead white and white lead. May be represented as $2PbCO_3 \cdot Pb(OH)_2$ - from lead (II) carbonate hydroxide. Has a relative density of 6·64 and an index of refraction of 2·01. Used as a *heat stabilizer* for *polyvinyl chloride (PVC)* at a level of from 2 to 8 phr. It decomposes to give carbon dioxide, which results in gassing, and this limits its uses. Also prone to sulphur staining and like all lead compounds it is regarded as toxic. See *lead stabilizer* and *lead carbonate*.

basic lead chromate
See *chrome red*.

basic lead silicate
Also known as lead silicate or as, basic silicate white lead. A composite pigment with a core of silica surrounded by a white pigment: because of this has a relative density which is less than that of *basic lead carbonate*. Used as an *acid acceptor* for chlorine containing rubbers. This white powdered material has a relative density (RD or SG) of approximately 5·8. An activator for natural rubber and synthetic rubbers such as nitrile rubber and styrene-butadiene rubber. An *acid acceptor* for chlorine containing rubbers.

basic magnesium carbonate
May be represented as $MgCO_3 \cdot Mg(OH)_2 \cdot 3H_2O$ or $3MgCO_3 \cdot Mg(OH)_2 \cdot 3H_2O$ See *magnesium carbonate*.

basic pigment
A class or type of *organic pigment*.

basic sulphate white lead
See *basic white lead*.

basic unit
The starting point of a system of measurement. See *fundamental unit* and *Système International d'Unité*.

basic white lead
Also known as basic sulphate white lead (BSWL) or as, basic lead sulphate. May be represented as $PbO \cdot PbSO_4$. The concentration of PbO is much less than that of the $PbSO_4$ (about one half to one quarter). Has a relative density of approximately 5·9. Used as a *heat stabilizer* for *polyvinyl chloride (PVC)*.

basic zinc carbonate
Also known as zinc carbonate. Occurs naturally as hydrozincite, calamine and as Smithsonite. An *activator*, *filler* and *pigment* for rubbers. Has a variable composition but is a white amorphous powder: the relative density (RD or SG) is approximately 4·4.

batch
A load for a particular piece of equipment.

batch mixer
A mixing machine used to prepare polymer compounds. A machine which produces a specified amount of compound after the mixing time has lapsed, for example, two litres after 5 minutes mixing. One of the most well known mixers used by the polymer industry is an *internal mixer* known as the *Banbury mixer*. A batch mixer has a fixed volumetric capacity.

Baumé scale
A scale of degrees used for the *relative density (rd or sg) of liquids*. Abbreviation used B or °B or Bé. Degrees Baumé = $144 \cdot 3$ (rd $-$ 1)/rd. For liquids lighter than water, degrees Baumé = $140 - 130$/rd.

BBP
An abbreviation used for *butyl benzyl phthalate*.

BCA
An abbreviation used for *dibutoxyethyl adipate*.

BBS
An abbreviation used for *blow* and *barrier system*.

BCHP
An abbreviation used for *butyl cyclohexyl phthalate*.

BCPC
An abbreviation used for *bisphenol chloral polycarbonate*.

BCS
An abbreviation used for the UK-based organization called the British Composites Society.

BDS
An abbreviation used for *butadiene-styrene block copolymer*.

BDTM
An abbreviation used for *2-benzothiazole-dithio-N-morpholine*.

Be Cu casting
An abbreviation used for *beryllium-copper casting*. See *hot hobbing*.

Bé
See *Baumé scale*.

Be-Cu
An abbreviation used for *beryllium-copper*.

Be-Ni
An abbreviation used for *beryllium-nickel*.

bead
A *tyre component*: that part which is shaped to fit the rim. It has a core of several inextensible strands with the plies wrapped around the core.

bead core
A *tyre component*: that part used to make the *bead* and which, for example, consists of several rubber covered wires in the form of a hoop.

bead filler
Also known as an apex strip. A *tyre component*: a strip of hard rubber which is placed on top of the *bead core* so as to prevent void formation by the plies of fabric which are anchored around the bead core.

bead polymerization
See *suspension polymerization*.

bead toe
A *tyre component*: that part of the *bead* which is opposite the *bead heel*.

bead heel
A *tyre component*: that part of the *bead* which fills the angle formed by the junction of the rim flange and the rim.

bead wrapper
A rubberized tape fabric used to prevent distortion of the *bead core* during tyre building, handling and curing.

becquerel
An SI derived unit which has a special symbol, that is Bq. It is the SI derived unit of radioactivity and is the activity of a radionuclide that decays at an average rate of one spontaneous nuclear transition per second. See *Système International d'Unité*.

beeswax
May also be known as cera alba or as, cera flava. This material has a relative density (RD or SG) of 0·96 and a melting point of approximately 63°C. It is a mixture of esters based on high molecular weight acids (for example, palmitic acid) and high molecular weight alcohols (for example, myricyl alcohol). It is a pale yellow, non-toxic, amorphous wax: can be used, for example, in rubber technology to improve the gloss on rubber extrudates. May be applied to the surface in emulsion form: when so applied may give some protection against ozone cracking.

behaviour on heating plastics
Thermoplastics soften on heating whereas thermosetting plastics (thermosets) do not - once the temperature reaches a certain point they decompose. An amorphous, thermoplastics material will soften over a wider temperature range than a semi-crystalline, thermoplastics material; these have sharper melting points.

Behre quick plastimeter
A *plastimeter*. A test apparatus used to measure the extrusion speed of rubber compounds. See *plasticity*.

bel
Ten decibels.

Bell Telephone Laboratories test method
Also called the Bell test or the bent strip method. An *environmental stress cracking test*. See, for example, ASTM D 1693-66). It is suggested that an *annealed compression moulded sheet* is used as such sheet has a low internal stress level.

Ten rectangular specimens, each 38 × 19 mm, are blanked out of the compression moulded sheet to give a specimen blank: each blank is then conditioned prior to nicking. by, for example, heating in steam or water at 100°C for 1 h followed by up to a day at room temperature. Each specimen has a cut 19 mm long inserted centrally and parallel to the length, that is, it is nicked. (Cut dimensions depend on the grade of polyethylene being tested.) The specimens are bent to a U-shape, through an angle of 135°, inserted into a length of brass channel and then immersed in an alkyl-aryl polyethylene glycol detergent (Igepal CO-630) in a large test tube which is stoppered and placed in a temperature controlled bath at 50°C. The proportion of the total number of specimens that crack in a given time, for example 24 hours, is obtained or, an F_{50} value is obtained.

Bell test
See *Bell Telephone Laboratories test method*.

belt
A flexible band, which passes around pulleys, and which is used to transmit motion, power or materials. A *tyre component*: that layer of material underneath the *tread* which restricts the *carcass* in a circumferential direction.

Bemelman's process
A reclamation process for vulcanized rubber which utilizes the fact that textiles are charred by heating with naphthalene. The cut pieces of rubber are heated with 2% naphthalene in a dry atmosphere of carbon dioxide and ammonia for about 2 h at elevated temperatures. See *reclaiming processes*.

bending test
See *flexural properties*.

bent strip method
See *Bell Telephone Laboratories test method*.

bentonite clay
Often referred to as bentonite. Naturally occurring aluminium silicate with a relative density (RD or SG) of 2·50. A clay-like material similar to *diatomaceous earth*. Used as an thixotropic additive in, for example, *unsaturated polyester resins* to prevent the resin dripping, or running, when the resin is applied to a vertical surface. Also used as a stabilizer/filler for *latex foam*.

benzene
Also known as benzol. A six-membered ring compound which may be represented as C_6H_6. This material has a relative density (RD or SG) of 0·88 and a boiling point of approximately 80°C. A colourless, liquid, *aromatic hydrocarbon* used, for example, as a solvent but this application is limited by its toxicity. It is a good solvent for uncured *nitrile butadiene rubber* (low acrylonitrile NBR), *chloroprene rubber* (CR), *butyl rubber (IIR)*, *natural rubber* (NR), *styrene-butadiene rubber* (SBR) and *thiokol rubber (T)*. This chemical causes a large amount of swelling, or gel formation, of uncured high acrylonitrile NBR. Will dissolve *polystyrene* and its copolymers, some acrylics and *polyvinyl acetate*. This material is important to the polymer industry as it is also used as an intermediate in the manufacture of, for example, styrene and phenol. Prepared by distillation from petroleum.

benzene chloride
See *chlorobenzene*.

benzene sulphonyl hydrazide
Also known as benzenesulphonylhydrazide or as, benzenesulphonhydrazide. May be represented as $C_6H_5SO_2NHNH_2$. An abbreviation used for this material is BSH. A gray/white powder with a relative density (RD or SG) of 1·41. Used as a *blowing agent*, for example, in the preparation of microcellular rubber as it yields nitrogen on heating. Has a decomposition, or degradation, temperature of 120 to 140°C but this temperature is lowered in the presence of bases. Add last to the mix and use, for example, dibenzthiazyl disulphide as an accelerator.

benzene-1,3-dicarboxylic acid
See *isophthalic acid*.

benzene-1,3-diol
See *resorcinol*.

benzene-1,3-disulphonyl hydrazide
May be represented as $NH_2NHSO_2C_6H_5SO_2NHNH_2$. A *blowing agent* with a decomposition, or degradation, temperature of approximately 120°C.

benzene-1,4-diamine
See *para-phenylene diamine*.

benzenecarboxylic acid
See *benzoic acid*.

benzenesulphonhydrazide
See *benzene sulphonyl hydrazide*.

benzidine yellows
Prepared from dichlorbenzidine. See *diazo pigments*.

benzidines
See *diazo pigments*.

benzoic acid
Also known as benzenecarboxylic acid. May be represented as C_6H_5COOH. This white, crystalline material has a relative density (RD or SG) of 1·27 and a melting point of 121°C. Used at an approximate concentration of 3 to 6 phr so as to obtain an increase in the hardness of rubber vulcanizates. In black rubber mixes it softens the unvulcanized stock and acts as a weak *retarder*. Retards vulcanization when used with mercapto-type *accelerators*. It is also used as a food preservative.

benzol
See *benzene*.

benzophenone
May be represented as $(C_6H_5)_2CO$. Also known as diphenyl ketone or as, benzoylbenzene or as, diphenylmethanone. A material which can promote photo-degradation in some polymers and yet, forms the basis of many *ultraviolet stabilizers* and/or *ultraviolet absorbers*.

benzothiazole
An alternative term for *thiazole*. See *accelerator*.

benzothiazole sulphenamide accelerator
A benzothiazole sulphenamide accelerator is derived from *mercaptobenzothiazole (MBT)* by bonding an *amine* to the mercapto sulphur atom. The accelerator becomes active as the amine is removed on heating and *vulcanization* proceeds rapidly after the retarded start as the base activates the MBT. The *thiazole accelerators* can be sub-divided into *mercapto accelerators* and *benzothiazole sulphenamide accelerators*.

benzothiazole sulphenamide accelerators
The collective name for the class of materials which are used as rubber *accelerators* (see *benzothiazole sulphenamide accelerator*, *mercaptobenzothiazole (MBT)* and *benzthiazyl disulphide*).

Processing safety is highest with N,N-dicyclohexylbenzothiazole-2-sulphenamide (DCBS): it then decrease from N-morpholinothiobenzothiazole-2-sulphenamide (MBS) to N-t-butylbenzothiazole-2-sulphenamide (TBBS) to N-cyclohexylbenzothiazole-2-sulphenamide (CBS). All sulphenamides give high cross-link densities: higher than mercapto accelerators.

benzothiazyl sulphenamide accelerators
A class of materials which are used as *accelerators* for rubbers. See *sulphenamides* and *benzothiazole sulphenamide accelerators*.

benzotriazoles
Also known as hydroxy-benzotriazoles. Ultraviolet absorbing compounds widely used to improve the *ultraviolet (UV)* resistance of polymers. Benzotriazoles are derived from 2,2'-hydroxy-phenyl-benzotriazole. Included in this group are 2-(2'-hydroxy-5'-methylphenyl)-benzotriazole and 2-(3'-tertiary-butyl-2'-hydroxy-5'-methylphenyl)-5-chlorobenzotriazole.

benzoyl peroxide
See *dibenzoyl peroxide*.

benzoyl superoxide
See *dibenzoyl peroxide*.

benzoylbenzene
See *benzophenone*.

benzyl alcohol
This material has a relative density (RD or SG) of 1·04 and a boiling point of 205°C. A high boiling point solvent for *cellulose acetate*, *cellulose ethers*, *natural rubber* and *polystyrene*.

benzthiazyl-disulphide
Also known as dibenzothiazyldisulphide or dibenzothiazyl disulphide or as, 2,2'-dibenzthiazyl disulphide or as bis-(2,2'-benzothiazolyl) disulphide. An abbreviation used for this material is MBTS. Derived from *mercaptobenzothiazole (MBT)* by oxidation: contains two sulphur atoms (-S-S-) obtained by joining two molecules of MBT and eliminating hydrogen (from the SH group). This material has a relative density (RD or SG) of 1·5 and a melting point of approximately 175°C. A white to pale yellow powder with little or no smell, An *accelerator* for natural rubber with a similar action to MBT but with a slower rate of cure - particularly below 140°C. A delayed action accelerator. Lower tendency to scorch than MBT. Like MBT it is non-discolouring and non-staining, acts as an antioxidant and has a long curing range. Frequently employed in synergistic mixtures with secondary accelerators (for example, with DPG, TMT and ZDC). Functions as a retarder for polychloroprene (below 1%).

benzyl 2-ethylhexyl adipate
See *benzyl octyl adipate*.

benzyl butyl phthalate
See *butyl benzyl phthalate*.

benzyl octyl adipate
An abbreviation used for this type of material is BOA. Also known as benzyl 2-ethylhexyl adipate. A low temperature *plasticizer* for *polyvinyl chloride*.

bergmehl
See *diatomaceous earth*.

Berlin blue
See *ferric ferrocyanide*.

beryl
Beryllium silicate: the ore for *beryllium*. Beryllium oxide occurs in beryl.

beryllium
A hard, greyish white metal which occurs in Group 11A of the Periodic table: it is one of the alkaline earth metals. When

pure it is malleable and ductile: beryllium and its compounds are poisonous. It is used to make corrosion resistant alloys and for fibre manufacture. This material has a relative density (RD or SG) of 1·84 and a melting point of 1,280°C. Obtained from *beryl*. See *specific modulus*.

beryllium nickel
See *beryllium-nickel*.

beryllium oxide
Has been used to make fibres which have a relative density (RD or SG) of approximately 1·8 and a melting point of 2,550°C. See *specific modulus*.

beryllium-copper
An abbreviation used for this type of material is Be-Cu or Be Cu. An alloy of copper which contains approximately 2·75% beryllium and 0·5% cobalt. A light, corrosion resistant alloy used for mould manufacture when heat removal is a problem during moulding. It has a thermal conductivity which is roughly three times that of steel (if the thermal conductivity of tool steel is taken as 1 then the figure for stainless steel would be 0·4, for cast-zinc alloy 2·8, for beryllium-copper 3, and for aluminium 4). A beryllium-copper insert may form part of the mould if that part of the moulding is not cooling quickly enough. Beryllium-copper is also used in moulds which are cooled by refrigerated water as this medium can result in condensation on the surface of the mould. Such condensation can cause rusting of a steel form but it does not corrode Be Cu. Another advantage of beryllium-copper is that it may be pressure-cast (*hot hobbed*) to the shape required (i.e. machining may be avoided) and the resultant moulds have a harder surface than aluminium moulds. They therefore have a longer working life than aluminium moulds but are usually more expensive. With a hardness on the Rockwell C scale of approximately 35, beryllium-copper can be used for runs of approximately 50,0000 shots, unless unusually abrasive materials are being moulded.

When beryllium-copper is used to make moulds, tests should be performed to ensure that nothing in the compound will attack the material. For example, ammonia, formed by using azodicarbonamide as a *blowing agent*, will attack unprotected beryllium-copper. Nickel is a suitable protective system. Alternatively a non-contaminating *blowing agent* should be used. When *natural rubber* compounds are moulded, using beryllium-copper, bonding problems may arise unless the cores and cavities are chromium-plated. An alternative solution is to use a castable metal which has similar casting properties to beryllium-copper but which does not suffer from the same bonding problems. For example, an *iron-based casting alloy*.

beryllium-nickel
An abbreviation used for this type of material is Be-Ni or Be Ni. A higher hardness material than *beryllium-copper* - it has a hardness of up to 50 Rockwell C. This type of material is therefore more suitable for processing highly abrasive materials or for use on longer running jobs than those used for beryllium-copper.

BET method
An abbreviation used for *Brunauer, Emmett* and *Teller method*. See *nitrogen absorption method*.

BET value
Obtained from a *nitrogen absorption method*. Defines the surface area of a filler in terms of m^2/g nitrogen. The higher the value, the more interaction there is between the filler and the rubber and the more the reinforcement effect.

beta ray gauge
β-ray gauge or β ray gauge. A thickness measuring device which relies on the fact that, for a given formulation, β-ray absorption, or *particle absorption*, is directly proportional to mass per unit area. As the composition is fixed, then the output can be set to read product thickness in continuous processes such as *calendering* and *extrusion*.

In this system a low energy radioactive source, which emits beta particles, is mounted close to the moving sheet and the transmitted radiation is detected by an ionisation chamber. The amount of particle transmission varies with the thickness and composition of the material through which it passes. If the composition is constant therefore the degree of ionisation and hence the conductivity of the chamber will vary, in effect, with the thickness. These changes in the conductivity can be converted to give a visual read-out of thickness or weight per unit area and/or used to actuate calender roll adjusting motors which move the bearing blocks and therefore adjust roll separation.

The read-out gives the *gravimetric thickness value*, that is, the notional average thickness. For an un-embossed sheet, the gravimetric thickness is the same as the geometric thickness. For an embossed sheet, the gravimetric thickness is not the same as the geometric thickness and is more relevant.

beta-naphthylamine
See *phenylnaphthylamines*.

beta-ray gauge
See *beta ray gauge*.

BeV
An abbreviation used for one thousand million electron-volts. May also be referred to as GeV: that is *giga electron-volts*.

BFS
An abbreviation used for *blow, fill* and *seal*.

BHT
An abbreviation for *butylated hydroxytoluene*.

bias
The angle at which a textile material is cut with respect to the running edge of the fabric.

bias angle
The acute angle formed between the cutting line and the cords during the production of *tyre cord*, fabric plies. The acute angle formed between the cutting line and the warp, of a reinforcing fabric, of a hose.

bias-ply tyre
See *cross-ply tyre*.

biaxial orientation
Directional orientation obtained in a product by simultaneous stretching in two directions. It is important in film manufacture, piping, bottles, and other hollow containers, as it enhances the hoop strength and fracture resistance of the product. A process for the improvement of tube or sheet properties by stretching the material in two directions at right angles. For example, along and across the extrusion direction. See *injection blow moulding*.

biaxially oriented film
A film produced by stretching: the original film is stretched in two directions at right angles so as to increase *orientation* but reduce the thickness.

bicarbonate of soda
See *sodium bicarbonate*.

bicomponent fibre
Also known as conjugate fibre. A fibre which consists of two polymers: the two polymers may either lie side-by-side or a core-and-shell type of structure may be involved.

biconstituent fibre
A fibre which contains two polymers: one of the two polymers is in short fibril form and is dispersed in a matrix of the second polymer.

bicyclo-(2,2,1)-heptene-2
See *norbornene*.

bicyclopentadiene
See *dicyclopentadiene*.

biddiblack
A cheap, black mineral filler of relatively coarse particle size: produced in Bideford, UK.

BIDP
An abbreviation used for *butyl isodecyl phthalate*.

BIIR
An abbreviation used for a *halogenated butyl rubber*.

billion
Now commonly accepted as being one thousand million (an American billion). In the UK it was a million times a million.

billow forming
A variation on *drape forming*. A *thermoforming process* which uses a *male mould* and which offers a way of producing components of relatively deep draw cheaply: such components can have uniform wall thickness distribution. The sheet is heated by an infra-red heater (mounted approximately 150 mm above the sheet) while being held in a clamping frame. At *forming temperature*, the heater is removed and air is introduced into the space between the sheet and the mould so as to make the sheet 'billow' up. The sheet is then moved relative to the mould, for example, the *male mould* is driven upwards and passes through the clamping frame and into the bubble of material. A vacuum is applied to the space between the sheet and the male mould: this completes the forming operation. The product is then cooled by mould contact, or by blowing air across the forming, and ejected. Such a process is popular as the machine and mould costs are low and the process is simple to operate.

bimetallic barrel
Also known as a bimetallic cylinder. A *barrel* which is based on two different metallic materials and in which one metallic material surrounds the other. The inner layer, in contact with the *melt*, possesses wear and/or corrosion resistance. For example, the substrate may be a good quality tensile steel (AISI 4140 - formerly known as EN 19) which resists crushing. The 1.5 mm thick *lining* may be based on a nickel matrix reinforced with, for example, particles of iron boride or of *tungsten carbide*.

bimodal distribution
A frequency distribution wherein individual observations cluster around two peaks: a frequency distribution which is, in effect, a mixture of two normal distributions.

bin cure
The slow curing of rubber compounds during storage.

bin curing
Also known as setting up. The premature curing of a rubber compound while in storage.

binder
A term used in the coating industry and which means the resin or polymeric/plastics material. See *dusting oil*.

binder resin
When used in the *reinforced plastic industry*, the term refers to the resin, which holds together the reinforcement.

binomial distribution
A distribution of data, derived from *attributes inspection*, wherein individual observations cluster towards zero on the scale in a heavily skewed manner.

biocide
A compound added to prevent decomposition, or degradation, due to the action of natural organisms such as bacteria and fungi. Such compounds are used, for example, on rubber plantations. See *biodegradation*.

biodegradable
Capable of being degraded, or decomposed, by micro-organisms.

biodegradation
Decomposition, or degradation, due to the action of natural organisms such as bacteria and fungi. Many natural polymers (for example, *cellulose*) are biodegradable whereas many synthetic polymers of high molecular weight are resistant. Polyesters, which contain ester groups in the main chain, are not resistant neither are polyester polyurethanes.

biopolymer
A polymer produced by biosynthesis. A material produced by *biopolymerization*. The best known biopolymer, and the one most commercially used, is *natural rubber*: another is *cellulose*. Biopolymers may be homopolymers or copolymers. See *polyhydroxybutyrate*.

biopolymerization
The process whereby a *biopolymer* is produced. See *polyhydroxybutyrate*.

bioproduction
Term used to indicate that a polymer is produced by the action of a *bacterium*. See *polyhydroxybutyrate*.

biosynthesis
The synthesis of complex materials, such as polymers, from simple materials by living organisms. See *natural rubber* and *polyhydroxybutyrate*.

biotite
See *glimmer*.

bipolymer
Another term for *copolymer*.

bis-(2,2'-benzothiazolyl) disulphide
See *benzthiazyl-disulphide (MBTS)*.

bis-(2,2,6,6,-tetramethyl-4-piperidyl) sebacate
A *hindered piperidine*.

bis-dibromopropylether of tetrabromobisphenol A
An *organo-bromine flame retardant*. Such a bromine-containing compound tends to be more powerful than the equivalent chlorine-containing compound: often used in conjunction with *antimony trioxide*. For example, *polypropylene 94%, bis-dibromopropylether of tetrabromobisphenol A 4% and antimony trioxide 2%*.

bis-(diethylene glycol monoethylether) phthalate
See *dicarbitol phthalate*.

bis-(diethylthiophosphoryl) trisulphide
An abbreviation used for this material is ETPT. An example of a *sulphur donor vulcanization system* which is not an *accelerator* in that formulation.

bis-(dimethylbenzyl) carbonate
A *plasticizer* for *polyvinyl chloride (PVC)* which has low volatility and high permanence: has good flame retardancy and heat stability. It is compatible with polystyrene (PS), polyvinyl butyral (PVB) and styrene-butadiene rubber (SBR). It is incompatible with cellulose acetate (CA).

bis-(dimethylbenzyl) ether
A plasticizer for polyvinyl chloride (PVC) and its copolymers. Also for ethyl cellulose and for many synthetic rubbers. Has good water resistance, heat stability and can give good electrical insulation properties.

bis-(ethylene glycol monobutylether) adipate
See *dibutoxyethyl adipate*.

bis-4-(aminophenylmethane)
See *4,4'-diamino diphenyl methane*.

bis-[2-hydroxy-5-methyl-3(1-methylcyclohexyl)-phenyl]methane
A *phenylalkane antioxidant*. See *phenolic antioxidant*.

bisazide compound
A compound which contains two *azide* groups. For example, 3,3'-diazidediphenyl sulphone.

biscuit
A slug, or charge, of material fed to a mould, for example, in gramophone record moulding. Biscuit is sometimes used to denote the total output of a multi-impression *flash mould*: that is, the mouldings and the web of *flash* which connects them (see *lift*).

bismuth oxychloride flakes
See *flake pigments*.

bisphenol
An abbreviation used for this type of aromatic structure is BP. See, for example, *bisphenol A*.

bisphenol A
An abbreviation used for this material is BPA. An aromatic material which is also known as 4,4'-dihydroxy-diphenyl propane or as, diphenylol propane or as, 2,2-bis-4-hydroxyphenyl propane. Consists of two phenol groups linked to a central carbon atom which also carries two methyl groups and which may be represented as $C(CH_3)_2(\phi_2)_2$: where ϕ is the phenol group. Used, for example, to make *polycarbonate materials* and as a curing agent - see *diglycidyl ether of bisphenol A*.

bisphenol A polycarbonate
An abbreviation used for this type of material is PC or, PC-BPA or BPA-PC. The most common type of *polycarbonate*. This type of material combines good transparency (light transmission approximately 90%) excellent impact strength, high heat resistance (Vicat B of approximately 148°C), good electrical insulation properties, reasonable ease of processing and a favourable price (compared to other heat resistant thermoplastics materials). Where higher heat resistance is required, then an *aromatic polyester carbonate (APE)* may be considered: the heat resistance can approach 192°C dependant on the percentage of iso-phthalic acid during polymer manufacture. However, such materials have high melt viscosities, for example, 1,600 Pa·s at 360°C and at 1000 s^{-1}. A bisphenol A polycarbonate could have a melt viscosity of, for example, 120 Pa·s at 360°C and at 1000 s^{-1}.

bisphenol AF
May be considered a derivative of *bisphenol A* which consists of two phenol groups linked to a central carbon atom which also carries two fluoromethyl groups and which may be represented as $C(CF_3)_2(\phi_2)_2$: where ϕ is the phenol group. Used, for example, as a curing agent - see *fluororubber*.

bisphenol C
An abbreviation used for this material is BPC. An aromatic material which consists of two methyl-substituted phenol groups linked to a central carbon atom which also carries two methyl groups and which may be represented as $C(CH_3)_2(\phi_2)_2$: where ϕ is the methyl-substituted phenol group. Used, for example, to make *polycarbonate materials* of high heat resistance and with good hydrolysis resistance.

bisphenol chloral polycarbonate
An abbreviation used for this material is BCPC. This material has been mixed with *polymethyl methacrylate (PMMA)*, with which it is miscible, to give blends which have low gas permeability coefficients.

bisphenol F
An abbreviation used for this material is BPF. Also known as 4,4'-dihydroxy-diphenyl methane. An aromatic material which consists of two phenol groups linked to a central carbon atom which also carries two hydrogen atoms and which may be represented as $CH_2(\phi_2)_2$: where ϕ is the phenol group. Used, for example, to make *polycarbonate materials* of high heat resistance.

bisphenol of the phenol alkane type
See *phenylalkane* and *bridge hindered phenol*.

bisphenol of the phenolic sulphide type
A *bridge hindered phenol*. See *phenolic antioxidant* and *phenylsulphide*.

bisphenol S
May be considered a derivative of *bisphenol A* which consists of two phenol groups linked to a central sulphur atom which also carries two oxygen groups and which may be represented as $S(O)_2(\phi_2)_2$: where ϕ is the phenol group. Used, for example, as a curing agent with a *fluororubber*.

bisphenol Z
An abbreviation used for this material is BPZ. Also known as 1,1-bis(4-hydroxyphenyl) cyclohexane. An aromatic material which consists of two phenol groups linked to another central aromatic group (a saturated aromatic ring or Ar) and which may be represented as $Ar(\phi_2)_2$: where ϕ is the phenol group. Used, for example, to make *polycarbonate (PC)* materials of high heat resistance. This type of PC material has been used for electrical insulation film.

bite
See *nip*.

bitolyl diisocyamate
See *3,3'-dimethyl-4,4'-diphenyl diisocyanate*.

bitumen
Asphalt is sometimes used as another name for bitumen in the USA. In the UK, asphalt is a mixture of bitumen with particulate minerals. The base material (bitumen) can be a black, sticky solid (or semi-solid) and is a complex mixture of hydrocarbons. It occurs naturally but is also made synthetically: it is the crude oil residues left after distillation. It has a relative density (RD or SG) of approximately 1·04. Natural bitumen softens over the temperature range 130 to 160°C. Because of its colour it is only used in compounds where colour is not important. Used as a *plasticizer* and processing aid. For example, it may be used in rubber cable compounds as it aids extrusion and improves electrical insulation properties. Also used in heavily loaded rubber compounds so as to make them easier to handle during mill mixing.

biuret group
A chemical group which may be represented as -NHCO·N·CONH-. The reaction of a *urea group* and an isocyanate results in the formation of such a group in *polyurethane manufacture*, for example, R·NHCO·N·CONH·R' (where R and R' are organic groups).

black
See *carbon black*.

black box
A phrase used to describe a device whose method of working is ill-defined or, understood poorly if, at all.

black pigments
The most widely used black pigment is *carbon black*. Others include *antimony trisulphide*, *bone black*, *vegetable black*, *graphite* (natural graphite is called plumbago) *iron oxide blacks* and *nigrosine*. See *shade*.

black value
See *S-value*.

blacklead
See *graphite*.

blade ejector
A rectangular *ejector element* used, for example, for the ejection of thin components from an *injection mould*.

blanc fixe
The type of *barium sulphate*, known as blanc fixe, has a relative density (RD or SG) of approximately 4·25. It is finer, purer form of barium sulphate which is obtained by converting the ore to the soluble chloride: sodium sulphate is then added to precipitate the barium sulphate. Also obtained as a by-product in hydrogen peroxide manufacture.

blank
A component produced by *blanking*. See *specimen blank*.

blanket
A continuous flexible band, for example, used in *knife-on-blanket coating*.

blanking
A process used to produce blanks (or shapes) by a cutting, pressing operation. See *specimen blank*.

blast finishing
A finishing process for mouldings, usually thermoset mouldings. The mouldings are blasted with an air stream which contains fine particles, for example, nut shell particles which remove the *flash*.

bleaching agent
A chemical that improves the whiteness of a substance or material, for example, a textile by removing the natural colour or greyness.

bleed
A controlled removal of a small amount of fluid. See *bleed-off*.

bleed nipple
A key operated valve through which air or fluid can be removed from a system.

bleed out
See *exudation*.

bleed-off
A hydraulic term which means that a controllable portion of the pump delivery is diverted to the reservoir.

bleeding
Migration of a colorant into another material.

blend
A simple mix: a mixture produced as a result of a blending operation and which is fed to another mixing machine. In the plastics industry, a blend is usually associated with the production of mixtures by relatively simple, low temperature processes such as tumbling. The product of such an operation is probably now best referred to as a *preblend* as the tendency is to use the term 'blend' to describe a melt-mixed *compound* of two different plastics, for example, two different thermoplastics.

blend manufacture of
The term blend is now being used in the plastics industry to describe a multi-component system which contains two or more different, high polymer systems. Current commercial interest is on making thermoplastic blends as such materials can still, for example, be injection moulded. The major reasons for blending thermoplastics materials are to improve impact strength and/or to improve processability. Can have *miscible blends* and *immiscible blends*.

The major methods of making blends of materials, based on thermoplastics materials, are melt mixing methods or techniques. Mechanical mixing with or without concurrent chemical reaction(s) using, for example, a *twin screw extruder*, is commercially very important.

blend nomenclature
When mixtures, produced by *melt mixing*, are made from two or more polymers (see *alloy*), it is suggested that the symbols for the starting polymers be separated by a plus sign, and the whole be placed in parentheses i.e. (A + B). If styrene acrylonitrile copolymer or, SAN is blended with the copolymer made from ethylene and vinyl acetate or, E/VAC then it would be represented as (SAN + E/VAC) in the 'standard' system i.e by ISO 1043-1 1987 (E). More commonly however, it would be referred to as SAN/EVA or, as SAN/EVAC. The major ingredient is usually mentioned first and the other polymer is often only mentioned if it is above a certain percentage, for example, 5%. If below this figure it is not often mentioned. In the case of both copolymers and blends, it would seem reasonable to indicate the percentage by weight of each ingredient, for example, 70:30, but this information is not often not, readily available. See *abbreviations*.

blend rubber-like
See *thermoplastic rubber-like material*.

blend with vulcanized rubber
Usually means a blend of a *thermoplastics material* with a rubber. See *dynamic vulcanization*.

blender
A machine used to produce a *blend* or a *preblend*.

blending
The procedure used to produce a mixture - a *blend* or *preblend*. In the plastics industry, the term blend is associated with the production of mixtures (by relatively simple, low temperature processes) and with the production of melt mixtures of materials. The major reasons for melt blending *thermoplastics materials* are usually to improve impact strength and/or to improve processability. Because of their ease of manufacture, using (for example, *twin-screw compounding extruders*) there is a lot of interest in blends of plastics or, in blends of plastics with elastomers: either may be modified with fillers or with *glass fibre*. See *blend* and *alloy*.

blister
An imperfection, a rounded elevation on the surface of a plastic, with boundaries that may be more or less sharply defined, somewhat resembling in shape a blister on the human skin.

blister ring
See *shear ring*.

blk
An abbreviation used in rubber technology for black. See *carbon black*.

block copolymer
Also known as a block polymer. A *copolymer* in which the repeating units, based on one monomer type, occur in long sequences.

block copolymer thermoplastic elastomer
A *block copolymer* which is a *thermoplastic elastomer*. The use of block copolymers has yielded materials such as *styrene-*

butadiene block copolymer (SBS) and *polyether ester (PEEL)* - two very important *thermoplastic elastomers*.

Many SBS type-materials have the structure of a central, elastomer (E) block linked to two plastics (P) blocks (i.e. P-E-P): other copolymers may be represented as -(P-E)-X- where 'X' is a link group. Another type of block copolymer has the repeating structure $(P-E)_n$: where P is a hard, plastics block (either polyurethane, polyester or polyamide): these materials have broad, molecular weight distributions of both the individual blocks and the final polymer.

block crumb rubber
See *block rubber*.

block polyether ester
See *polyether ester*.

block polyetherester
See *polyether ester*.

block polymer
See *block copolymer*.

block press
A moulding machine which consolidates piles of sheets into a large block: after cooling the sheet is transferred to a *slicing machine*. Thick sheets produced from dough on *sheeting rolls* is pressed in a block press. See *cellulosics*.

block rubber
Also known as *block crumb rubber*. Natural rubber in block or bale form: made from *comminuted rubber* which has been compressed into a block or bale. See *Heveacrumb process*.

blocked diphenylmethane diisocyanate
See *urethane cross-linking*.

blocking
The tendency of sheets or films to stick together under light pressure making separation by normal, as opposed to sliding forces, difficult. Blocking is particularly troublesome with thin, flexible films. It can be controlled by altering the surface so that, for example, microscopic protrusions are formed on the surface and these stop the film surfaces coming into contact with each other. Fine particle size, inorganic materials, such as *kieselguhr*, act in this way for *polyethylene*. The addition of some of the additive may reduce blocking while the addition of even more will give a more matt finish (see *matting agent*).

blocking agent
An additive added to a polymeric compound so as to reduce *blocking*. See *diatomaceous earth*.

blocking group
See *protecting group*.

blooming
Term used when a colorant migrates to the surface of a product and forms a dusty layer; *dyes* are more susceptible than *pigments* to *blooming*.

blow and barrier system
The fluorination process used to improve the permeability resistance of *polyethylene (PE) containers* to, for example, automotive fuels. The inside of the container is exposed to fluorine and this causes hydrogen atoms on the PE molecule to be replaced with fluorine. In this way, polar groups are introduced into the thermoplastics material and it is the introduction of such groups which makes the plastics material less attractive to the hydrocarbon (non-polar) fuel.

blow, fill and seal
An abbreviation used for this process is BFS. A *blow moulding process* in which the container is produced (blown), filled and then sealed in one sequence.

blow moulded container
A container produced by *blow moulding*. Such a container is based on a *thermoplastics* material and most commonly used for the packaging of liquids. Polyethylene (PE) plastics materials made the plastics bottle acceptable when, in the 1950s, 'squeeze' bottles were first used for washing up liquid (detergent). Such bottles offered lightness in weight, good impact strength and ease of dispensing (by squeezing). Now a wide range of plastics materials, and plastics materials combinations, are used in an effort to overcome two of the major disadvantages of PE: these are opacity and/or permeability. Because of developments in materials, material combinations and equipment, plastics containers (produced by *blow moulding*) are now firmly established as they offer an attractive, energy-efficient alternative to traditional materials. Such containers are available in a wide range of sizes, textures, shapes and colours. The containers may range from clear to opaque and their appearance may be enhanced by a range of decorating techniques. The containers can be made efficient in their use of space as complex shapes can be readily produced, for example, fuel tanks for cars. In many cases, plastics are chosen because of their strength and good resistance to breakage.

CONTAINER IDENTIFICATION Because of concern about the environment, some countries, or states within countries, have passed laws which require plastics bottles to carry codes which identify the material(s) of manufacture (this is done to protect the environment against litter). Such codes may be based on those issued by the Plastics Bottle Institution (PBI of the USA) and are an aid to sorting before recycling. About 20% of all bottles in the USA are recycled - HDPE and PET account for approximately 75% of all plastics bottles. Curbside sorting and collection is usual. Therefore a scheme, or system, for identifying, collecting and recycling, for example, PET bottles, needs to be operated by suppliers and users: in this case the reclaimed material is used for fibres. Concerns about the recycling of plastics have inhibited usage of plastics materials in can-type applications.

CONTAINER LEAKAGE Leakage is controlled by the design and fit of the closure and, by the bottle neck finish. If the neck is, in effect, compression moulded, during *blow moulding*, then this usually leads to tighter neck tolerances and less leakage. The thread type and the number of threads per unit of length and the length of the thread all influence leakage. Buttress threads are often used as they pull the closure down uniformly: if the closure contains a soft, resilient material, as a gasket, then this will also help to prevent leakage.

CONTAINER STERILIZATION The use of plastics materials for bottles, instead of glass, can mean a loss of the traditional heat sterilization step in many cases: this is simply because many plastics materials have too low a heat distortion temperature (HDT). When sterile products are being packed then, the bottles must be sterile to receive sterile contents. The bottle could be flushed with hydrogen peroxide and sterile water but this is relatively expensive. Another, cheaper approach is to blow the bottle with sterile air and surround the machine with a sterile air curtain. The bottle neck is sealed before storage. The filling station has a purified air zone in which the necks are cropped before the bottles are filled and capped.

CONTAINERS FOR LIQUIDS Most containers for liquids are bottles and most bottles are used to contain drinks, for example, water, beer and fruit squash: other liquids include sauces and salad dressings. Materials for the containment of liquids include *glass, tin-plate, aluminium* and *plastics*.

The primary objective in the design and development of containers for such food products, is to maximize the performance to cost ratio. Each product has certain performance criteria, which it must satisfy. These are usually a combination of containment, protection, marketing and end-use functions. The choice of materials of construction, structural design and moulding method can be major factors in deciding how the package performs the functions necessary for the product being contained.

IMPROVING PERMEABILITY Most thermoplastics materials are permeable to gases and vapours. This means that a material, for example, a plastics material, will allow diffusion. In turn, this means that ordinary plastic bottles (based on just one untreated material) will lose gas, from fizzy drinks such as lemonade because they are not impermeable to the gas. The plastics container can be made less permeable by, for example, *fluorination*, *coextrusion* and *biaxial orientation*. See *barrier properties*.

IMPROVING PRINTABILITY Many of the containers produced by blow moulding are made from polyolefins. The surface of products made from polyolefins is not very receptive to printing inks. One way of improving the printability is to modify the surface, make it more polar, so as to make it more receptive to inks; techniques used include *flaming* and *corona discharge*.

blow moulder
A machine used to produce products by *blow moulding*.

blow moulding
A process used to produce hollow enclosed components. A shaping technique used for *thermoplastics materials* which uses compressed air to make a hot material conform to the shape of a surrounding mould. Blow moulding is commonly used to produce bottles, or containers, from thermoplastics materials such as *polyvinyl chloride (PVC)* and *high density polyethylene (HDPE)*. There are two main blow moulding processes and these are *injection blow moulding (IBM)* and *extrusion blow moulding (EBM)*.

blow up ratio
Also known as blow ratio. The ratio of the diameter of the bubble to the die diameter in blown film extrusion (see *tubular film process*). In *blow moulding*, it refers to the ratio of the product diameter to the parison diameter: the larger the ratio, the more easily will the container distort (because of reversion) on hot filling.

blowing
An *oil treatment*. The *oil* is heated while air is blown through the oil. This impairs brushability, and often, colour and colour retention: it does improve wetting, flow, gloss and drying properties.

blowing agent
A material, an additive, which brings about a reduction in density by the introduction of gas cells. A great many polymeric products are based on cellular materials: that is, on materials which are filled with gas cells. Either a gas (usually nitrogen) is added to the processing equipment or, a chemical compound (a blowing agent) is used to generate the gas when required. Widely used blowing agents, particularly for plastics materials, are *azo compounds*. These organic materials can be structured so that they decompose over a fairly narrow temperature range, for example, at the melt processing temperatures of a particular thermoplastics material.

Sodium carbonate is still widely used to blow rubber compounds as are ammonium carbonate and bicarbonate. Dinitrosopentamethylene tetramine is used for *microcellular products* such as shoe soling. Stearic acid often aids the decomposition of such rubber blowing agents.

Any blowing agent must decompose in a consistent manner over a temperature range which suits the polymer and the gases liberated must not attack, or be allowed to attack, the processing equipment or the personnel.

blowing stick
A *mandrel* used in *injection blow moulding*.

blown film extrusion
A process for making plastics film by extruding a tube and blowing it up to several times the extruded diameter: also known as *tubular film extrusion* or as, lay-flat film extrusion.

blue
A *primary* colour. A wavelength in the visible spectrum with a wavelength of approximately 470 nm.

blue circle rubber
A *technically classified rubber* which is fast curing. See *circle rubber*.

blue pigments
Phthalocyanines are the major blue *organic pigments*. Inorganic pigments are often based on cobalt: other blue pigments are based on elements such as aluminium, silicon, tin, zinc, or titanium in addition to cobalt. See *ultramarine blue*, *manganese blue*, *Prussian blue*, *cobalt blue* and *titanium blue*.

BMC
An abbreviation used for *bulk moulding compound* (a *polyester moulding compound* or *PMC*).

BMC TP
An abbreviation used for bulk moulding compound - thermoplastics. See *thermoplastic bulk moulding compound*.

BOA
An abbreviation used for *benzyl octyl adipate*.

Board of Trade unit
Also known as the BOT unit. A kilowatt-hour.

bodied adhesive
See *solvent cement*.

boiling point
The temperature at which a liquid boils freely under the external pressure. Boiling points are usually quoted, unless otherwise stated, as being those obtained when the pressure is 760 mm of mercury.

bolls
See *cotton*.

bolometer
A sensitive instrument used to measure heat radiation.

bolster
Also called a chase. Part of a mould: an outer frame used to retain the *punch* or *cavity components*. The insert/bolster assembly constitutes a mould plate.

BON
An abbreviation used for β-*oxynaphthoic acid*.

BONA
An abbreviation used for β-*oxynaphthoic acid*.

bonded fibre fabric
A fabric structure consisting of one or more webs, or masses, held together with a bonding agent or by fusion.

bonding agent
An additive or a surface treatment which improves adhesion between, for example, a rubber and the fibres of a synthetic fabric. Adhesion may be improved by coating the fibre with a bonding agent by a pre-treatment process and/or, incorporating the bonding agent into the rubber.

In the latter process the rubber may be compounded with resorcinol (a formaldehyde donor) and silica: this type of compound will adhere to clean untreated rayon and nylon fabrics. Coating the fibre with a bonding agent is still the most popular bonding method and such systems are usually based upon a mixture of polymers dissolved or suspended in water: they are preferred as they avoid the need for an expensive solvent (which must be extracted and recovered) and present a low fire risk. See *resorcinol-formaldehyde-latex dip*, *brass coating* and *coupling agent*.

bone black
A black pigment: a weaker black pigment than *carbon black*. Produced when bones are calcined and contains calcium phosphate and *calcium carbonate*.

booster
An auxiliary or secondary *accelerator*. See *hydraulic intensifier*.

booted
Computer terminology and which means the process of loading a computer with software.

BOP
An abbreviation used for *butyl octyl phthalate*.

BOPP
An abbreviation used for biaxial orientated polypropylene. See *biaxial orientation* and *polypropylene*.

borate glass
A glassy material which is obtained, for example, from boric acid: an amorphous polymer of boric oxide.

boron
Boron (B) is an element which occurs, with aluminium in Group 111A of the Periodic table. Has a high melting point of 2,300°C and a low relative density (RD). This element is used to harden steel: also used to make continuous fibre for use in reinforced plastics materials as it has a high tensile strength (3·4 GPa), a Young's modulus of approximately 310 GPa and a relative density of approximately 2·6. Such fibres are, however, expensive to produce (by *chemical vapour deposition*): *carbon fibres* are cheaper and can give similar properties in plastics composites. See *specific modulus*.

boron fibre
A *fibre* based on *boron*. To protect the fibres against chemical attack, such fibres have been protected with silicon carbide or silicon nitride.

boron nitride
A polymer of empirical formula BN. Obtained, for example, from boric acid: the boric acid is heated with urea under a nitrogen atmosphere. A network-type of polymer like *graphite*.

boron trifluoride
A Lewis acid catalyst which will, for example, polymerize isobutylene to give polyisobutylene. See *butyl rubber*.

borosilicate glass
Obtained, for example, from boric acid, silica and a metal oxide such as sodium oxide. Such glasses have better thermal shock resistance than *soda-lime glasses* as they have lower coefficients of thermal expansion. A type of glass which is noted for its high resistance to thermal shock and to alkalis. Such materials are relatively easily drawn into fibres and are such are used in composites. The best known example of a borosilicate glass is often referred to as Pyrex. See *E-glass* and *C-glass*.

boss
A significantly raised area on the surface of a moulding.

BOT unit
See *Board of Trade unit*.

bottom
The maximum injection point: the furthest point to which a screw can be safely pushed into a barrel. See *screw cushion*.

bottom ejection
Ejection from the lower part of the mould. Compression moulds may have both top and bottom ejection so as to ensure automatic operation.

bottom nip
The last nip of a *calender*: the gauging nip. See *inclined Z calender*.

bottom plate
Part of a mould: a steel plate which is bolted to the lower part of a mould and usually used to secure the mould to the *platen*. The plate may be cored for heating or cooling.

bounce
See *hysteresis* and *resilience*.

bouncing putty
A rubber compound which has the soft, pliable consistency of putty and which will bounce if dropped: the same material will fracture on impact if struck quickly. The addition of *zinc stearate* will reduce the tendency to fracture at high impact rates. Made by heating a *polydimethylsiloxane* with boric oxide and ferric chloride: a borosilicone which contains approximately three -Si-O-B- links for every 100 silicon atoms.

bound antioxidant
An antioxidant which is chemically bonded to the polymer/black so as to reduce volatility and to reduce loss in use. See *network-bound antioxidant*.

bound rubber
That rubber, in a rubber compound, which is attached to a filler and which is not soluble in the usual rubber solvents. Associated with *carbon black* and in particular with reinforcing blacks. Forms a network structure with the black.

bound sulphur
One of the ways the *sulphur* in rubber compounds, on analysis, amy be classified. That sulphur which is chemically bound, or chemically attached, to the rubber. See *sulphur analysis*.

Bourdon gauge
An instrument used to measure pressure by measuring the amount of deflection imparted to a partly-flattened, curved tube by the pressure.

bowl
Another name for *roll*. See *calender*.

bp
An abbreviation used for boiling point.

BP
An abbreviation used for *bisphenol*. For example:
BPA = *bisphenol A*;
BPC = *bisphenol C*;
BPF = *bisphenol F*; and,
BPZ = *bisphenol Z*;

BPA-PC
An abbreviation used for a *polycarbonate* based on *bisphenol A*.

BPF
An abbreviation used for the British Plastics Federation.

Bq
An abbreviation used for *becquerel*.

Br Imp
An abbreviation used for British Imperial. See *Imperial*.

BR
An abbreviation used for *butadiene rubber*.

BR-A
An abbreviation formerly used for *nitrile rubber*.

Brabender
In plastics technology usually means the torque measuring device known as the *Brabender plastograph*.

Brabender Aquameter
An instrument used to measure the *moisture contents* of polymers. This type of instrument is available in 3 versions. The basic instrument for moisture contents above 2%, the L version for levels of less than 1% moisture and the K version for materials with surface moisture only. The L version is of most interest for polymers.

In this instrument a known weight of polymer is heated and brought into intimate contact with calcium carbide in a sealed reaction chamber. The moisture driven out of the polymer reacts with calcium carbide producing acetylene gas which increases the pressure in the reaction vessel. The pressure generated by the acetylene, which is directly proportional to the amount of water in the polymer, is registered on a high precision pressure gauge. The gauge is calibrated directly in percentage water and reads from 0 to 1·0% for a sample weight of 20 g. The scale is sub divided to read to 0·01 g. For very accurate low moisture levels It would be advisable to calibrate the instrument prior to use.

Brabender plastograph
A torque measuring device which, for example, may be equipped with a small extruder as an attachment (a range of other attachments are available). Used to make checks on the incoming raw material: for example, the extruder is fitted with a rod die and the machine is set at specified barrel temperatures so as to give an appropriate melt temperature. The extrusion behaviour (output, die swell and melt temperature) is measured over a range of screw speeds and graphs are plotted of output against screw speed, melt temperature against screw speed and die swell against screw speed.

One of the big problems is maintaining the *set temperatures* as any alteration in screw speed alters the extrusion cylinder temperatures: the *melt* temperature often alters significantly as the screw speed is changed. As a range of screw speeds are employed, the shear rates employed can become quite high and so, this type of equipment can be regarded as a *high shear rate device*.

braided hose
A hose in which the reinforcing material has been applied by braiding.

brake valve
A hydraulic valve used, for example, to prevent excessive pressure build-up when decelerating or stopping a load.

branched chain alcohol
An alcohol which contains one or more carbon atoms joined at a branch in the main chain. Used to make, for example, plasticizers, that is, to make *branched chain phthalates* (which are phthalic acid esters). Synthesis of branched chain alcohols, with up to 10 carbon atoms, became widespread in the 1950s and these relatively cheap materials were esterified with *phthalic anhydride* to make plasticizers.

branched chain phthalate
A *phthalate* made from a *branched chain alcohol(s)* and used as a *plasticizer*. Branched chain phthalates are phthalic acid esters: the name for such materials commonly contains the prefix 'iso'. Compared to phthalates from linear (normal) alcohols, branched chain phthalates give plasticized *polyvinyl chloride (PVC) compounds* with worse low temperature properties but with better oil extraction resistance: other properties are similar.

branched chain trimellitate plasticizer
A *plasticizer* based on trimellitic anhydride and a branched-chain alcohol. See *trimellitate plasticizer*.

brass
Alloys of copper and zinc. This type of material has a relative density (RD or SG) of approximately 8·4 to 8·7. It is used to improve the bond between rubber and metal, for example, between rubber and steel. See *brass coating*.

brass coating
Used to improve the bond between rubber and metal, for example, steel. Such a bond is stable to heat and solvents. Used to coat *steel tyre cord*. Brass coating is useful in, for example, *injection moulding* processes where the high pressures involved could displace chemical bonding agents.

breakdown strength
See *dielectric strength*.

breaker
A *tyre component*: a layer of *ply material* between the *tread* and the *belt*.

breaker plate
A disc or plate which has a number of holes drilled in the flow (extrusion) direction which permit the plastics material to pass or, to flow. It is usually used to support screens (gauzes) and this assembly, is placed between the end of the *extruder barrel* and the die holder so as to remove contamination or, to improve mixing.

breaker strip
A component used to make a *breaker*.

breaking strain
An abbreviation used for this term is γ. It equals 6D h/L, where D is the deflection at midspan, b is the width of the specimen and L the span width. See *flexural strength*.

breaking strength
See *flexural strength*.

breather
A component of a hydraulic circuit which allows air movement so as to maintain *atmospheric pressure*.

breathing
The opening of a mould, during the moulding of a *thermosetting material*, at an early stage in the cure cycle so as to vent the mould. For example, during injection moulding of a thermosetting plastic, just before the mould fills, injection is momentarily stopped and the mould is slightly parted or bumped. After final mould closing, injection is completed under second-stage pressure. See *compression moulding*.

bridge press
High frequency welding equipment consists of a press, a generator and a materials handling system. A bridge press is used for the welding of components with large surface areas (for example, greater than 1 m^2) as this gives more accurate alignment of the press components than the more common *C-type press*. The platen, which carries the electrodes, slides on *tie rods*: both up-stroking and down-stroking presses are available and the force generated may reach 100 tonnes.

bridged hindered phenol
Also known as a *polynuclear phenol*. A *phenolic antioxidant*. There are three main classes (i) *phenylalkanes*, (ii) *phenylsulphides*, and (iii) *polyphenols*.

bridged screw
A screw which will not transport material because an arch, or bridge, of material has formed in the hopper; often removed by *rodding*.

bridging agent
See *vulcanizing agent*.

brilliant green
The sulphonated derivative of *malachite green*.

Brinell hardness
A measure of *hardness*: it is measured with a hardened steel ball of diameter D which is forced into the material under test with a force F for 15 seconds. After removal of the indentor the diameter d of the indentation is measured in two directions at right angles.
The Brinell hardness number $HB = 2F/\pi D\{D - \sqrt{(D^2 - d^2)}\}$.

The indentors used are normally 1·25 or 10 mm in diameter and the loads are selected to give F/D^2 ratios of 30, 10, 5 or 1. The test is not suitable for most plastics materials because of the recovery which takes place after removal of the load and the poorly defined impression.

Brinell hardness number
An abbreviation used for this term is HB. The ratio of the load in kilograms (in a *Brinell hardness* test) to the area of the depression in square millimetres.

BRITE
An abbreviation used for the European organization called Basic Research in Industrial Technologies for Europe.

British Antarctic Survey
An abbreviation used for this organization is BAS. Scientists working for this organization noted the depletion of *stratospheric ozone* in a patch or hole during the 1970s. There is now concern that ozone depletion may no longer be confined to polar areas but will occur over inhabited areas where the increased levels of *ultraviolet light* will cause problems, for example, high levels of blindness.

British Electrotechnical Committee
A standards committee which is responsible for UK participation in *CENELEC*. Participation in CENELEC is via the British Electrotechnical Committee, the Electrotechnical Standards Council of the *British Standards Institution*.

British Engineering system of units
See *FPS gravitational*.

British imperial liquid and dry measure
See *UK system of units*.

British Standard
An abbreviation used for this term is BS. British Standards are issued by the *British Standards Institution*. BS references without a prefix are to British Standards in the general series. Some British Standards use prefixes: the meaning of the prefixes are as follows:

A	Aerospace series;
AU	Automobile series;
CECC	BS CECC;
CP	Codes of practice;
DD	Drafts for Development;
MA	Marine series;
PD	Published document;
PP	Education publications; and
QC	BS QC publications.

British Standard 5750
An abbreviation used for this term is BS 5750. A *quality assurance standard* entitled 'Quality systems' and which is equivalent to ISO 9000 and to EN 29000. See *BS 5750: 'Quality systems'*.

British Standard softness number
A number which indicates the softness of a material. The higher the number the softer the material. An abbreviation used for this term is BS softness number.

A flat sheet or disc is used which is between 8 and 10 mm thick, conditioned at the test temperature and humidity. The time of conditioning is important and, for example, it is recommended to be seven days for plasticised *polyvinyl chloride (PVC) compounds* whose hardness is known to change over several days after moulding.

The test is performed using a *durometer* which measures the increase in depth of penetration of a steel ball, 2·5 mm in diameter, into a flexible material, when the force on the ball is increased from 0·30 N to 5·70 N. The minor or contact load of 0·30 N is applied for 5 seconds and then the load is increased to 5·7 N. The reading on the dial is noted when 5·7 N has been applied for 30 seconds. Four readings are taken at different points on the surface none of which is close to the edge of the specimen and the average is expressed as the BS softness number.

British Standards Institution
An abbreviation used for this UK-based organization is BSI. A *standards organization* responsible for issuing standards or specifications. This independent national organization prepares British Standards (BS) and is the UK member of the International Organization for Standardization (ISO). UK sponsor of the British National Committee of the International Electrotechnical Commission. BSI also provides, for example, information and certification for *BS 5750*. Information on BSI publications is given in the *British Standards Institution catalogue* (the BSI catalogue).

British Standards Institution catalogue
An abbreviation used for this term is BSI catalogue. Information on British Standards Institution (BSI) publications is given in the BSI catalogue. This catalogue now lists over 10,000 British Standards and their ISO equivalents: such standards are used in all industries and technologies. There is an index which can be used to locate, for example, standards which are concerned with thermoplastics materials.

British Standards Institution Kitemark
More commonly known as the BSI Kitemark or as the Kitemark. The Kitemark is the British Standards Institution's (BSI's) registered certification trade mark which is licensed for use if the BSI is satisfied that products are produced to national or international standards. The presence of a Kitemark on a product indicates to the customer that a standard of quality beyond that required by legislation is being achieved by the product. Any product for which a standard, or specification, exists with precise quantitative criteria can carry a Kitemark. Companies that already produce goods to a recognized standard, under a BS 5750 quality system, can obtain a Kitemark relatively easily. Independent tests on both raw materials, compounds and products are required followed by independent sampling and testing so as to ensure that quality of production is being maintained. This product certification mark carries BSI's assurance that products comply with national or international standards. See *compliance* and *British Standards Institution Safety Mark*.

British Standards Institution registered firm
More commonly known as BSI registered firm or as a registered firm. A firm whose quality system has been assessed by the British Standards Institution (BSI) against *BS 5750*. Such companies are entitled to display the registered firm symbol on their letterheads and publicity material. The registered firm symbol indicates that the firm is subjected to ongoing surveillance so as to ensure that the prescribed quality system is maintained to the desired standard.

British Standards Institution registered stockist
More commonly known as BSI registered stockist or as a registered stockist. Registered stockist purchases products, components and materials from quality assured sources, that is from *BSI registered firms*, and maintains the quality of those goods while they are under their control: this ensures the continuity of the quality chain through to the purchaser. The registered stockist system is designed to assure purchasers that supplies obtained from stockists maintain their original quality and conformity with specification.

British Standards Institution Safety Mark
An abbreviation used for this term is BSI Safety Mark. In addition to the *British Standards Institution Kitemark* there is the BSI safety mark which can be applied for when the standards cover specific safety requirements.

British thermal unit
An abbreviation used for this unit is Btu or B.T U. and/or B.th.u. The amount of heat required to raise the temperature of one pound of water by 1°F. It is equivalent to 1055·06 *joules*.

brittleness temperature
See *cold flex temperature*.

brittleness temperature test
See *low temperature flexibility* and *cold flex temperature*.

broad MWD
An abbreviation used for broad molecular weight distribution.

brominated butyl rubber
See *halogenated butyl rubber*.

brominated natural rubber
The reaction product of the halogen, bromine and natural rubber. See *chlorinated natural rubber*.

bromine
The elements of Group V11b of the Periodic table are known as halogens and consist of fluorine, chlorine, bromine, iodine and astatine. Bromine does not occur naturally as it is too reactive: it occurs as bromides, for example, in sea water from which it is extracted by treatment with chlorine. It is a reactive element and is a dark red liquid (Br_2) which is very soluble in water (bromine water). Melts at $-7°C$ and boils at 59°C. Bromine compounds are used, for example, as *flame retardants*. All halogens, except iodine, should be treated with great care as they are either highly poisonous and/or very irritating (iodine vapour is irritating and poisonous).

bromine flame retardant
A *flame retardant* which contains *bromine*, for example, an organo-bromine compound. Such bromine-containing compounds tend to be more powerful than chlorine-containing compounds: often used in conjunction with *antimony trioxide*.

bromine hexadecyltrimethylammonia
See *cetyl trimethylammonium bromide absorption*.

bromo-isobutene-isoprene rubber
See *halogenated butyl rubber*.

bromobutyl rubber
See *halogenated butyl rubber*.

bromotetrafluorobutene
A cure-site monomer introduced into a *fluororubber* so as to ease vulcanization.

bronze
Metallic alloys of, for example, copper and tin or of, copper and aluminium (aluminium bronze). Bronze has been used to fill nylon materials so as to produce a bearing material which can be injection moulded.

bronze flakes
A *flake pigment* based on *bronze*. Metallic colours are produced by the incorporation of aluminium flakes or of, copper flakes or of bronze flakes. Copper flakes and bronze flakes may tarnish during processing unless coated or protected with, for example, a transparent resin.

bronze powders
Metallic colours are produced by the incorporation of bronze powders which contain zinc: as the zinc content increases the colour becomes progressively less reddish. Bronze may tarnish during processing unless coated or protected with, for example, a transparent resin. See *flake pigments*.

Brookfield viscometer
A *rotational viscometer*. See *concentric cylinder viscometer*.

brought forward
Term used in *injection moulding* and means that the *injection unit* is brought into contact with the *sprue bush*.

brown crepe
See *estate brown crepe*.

brown factice
May be referred to as sulphur (dark) factice or as, dark factice or as, hot-type factice. See *factice* and *sulphurized oil*.

brown iron oxide
Iron (III) oxide. May be represented as Fe_2O_3. A red-brown material used as a *pigment*. This material has a relative density (RD or SG) of 5·15 and a melting point of 1,565°C.

brown pigments
Such pigments are often based on iron but may also contain aluminium, chromium, titanium and/or zinc. See *iron oxide pigments*. Inorganic pigments which do not contain iron may be based on aluminium, antimony, chromium, manganese, tin and/or zinc.

Brunswick green
A green pigment which is also known as lead chrome green. Obtained from *chrome yellow* and *Prussian blue*. Not very stable to sunlight and to chemicals (unstable to alkalis) but has good hiding power.

brown types
Term used in the *natural rubber industry* for *inferior grades of rubber*. See *scrap*.

BRP
An abbreviation used for boron reinforced plastics. See *fibre reinforced plastics* and *boron*.

Brunauer, Emmett and Teller method
A *nitrogen absorption method* used to determine filler surface area.

BS
An abbreviation used for *British Standard*.

BS softness number
An abbreviation used for *British Standard softness number*.

BS 5750
An abbreviation used for *British Standard 5750* which is entitled 'Quality systems'. See *BS 5750: 'Quality systems'*.

BS 5750: 'Quality systems'
An abbreviation used for *British Standard 5750* which is entitled 'Quality systems'. This standard is in several parts and most of these parts have an *ISO* or *EN* equivalent. This standard is applicable to manufacturing and to service industries and shows how an effective quality system can be established and maintained. By prescribing a course of action for the prevention of errors, rather than their correction, the approach developed ensures that the product or service for sale is right first time.

BS 5750: Part 0:1987 An abbreviation used for part 0 of *British Standard 5750*, issued in 1987 and entitled 'Principal concepts and applications'.

Section 0.1:1987 is entitled 'Guide to selection and use' and is identical with ISO 9000-1987 and with EN 29000.

Section 0.2:1987 is entitled 'Guide to quality management and quality system elements' and is identical with ISO 9004-1987 and with EN 29004-1987.

BS 5750: Part 1:1987 An abbreviation used for part 1 of *British Standard 5750*, issued in 1987 and entitled 'Specification for design/development, production, installation and servicing'. This *British Standard* is identical with ISO 9001-1987 and with EN 29001-1987.

BS 5750: Part 2:1987 An abbreviation used for part 2 of *British Standard 5750*, issued in 1987 and entitled 'Specification for production and installation'. This *British Standard* is identical with ISO 9002-1987 and with EN 29002-1987.

BS 5750: Part 3:1987 An abbreviation used for part 3 of *British Standard 5750*, issued in 1987 and entitled 'Specification for final inspection and test'. This *British Standard* is identical with ISO 9003-1987 and with EN 29003-1987.

BS 5750: Part 4:1990 An abbreviation used for part 4 of *British Standard 5750*, issued in 1990 and entitled 'Guide to the use of BS 5750 Pt 1 1987, BS 5750 Pt 2 1987 and to BS 5750 Pt 3 1987'.

BS 5750: Part 8:1991 An abbreviation used for part 8 of *British Standard 5750*, issued in 1991 and entitled 'Guide to quality management and quality systems elements for services'. This *British Standard* is identical with ISO 9004-2.

BS 5750: Part 13:1991 An abbreviation used for part 13 of *British Standard 5750*, issued in 1991 and entitled 'Guide to the application of BS 5750 Pt 1 1987 to the development, supply and maintenance of software'. This *British Standard* is identical with ISO 9000-3.

BS CECC

A standard issued by the *British Standards Institution (BSI)*. If the standard has 5 digits after BS CECC then it is a *CECC publication* adopted without change by BSI. If the standard has 8 digits after BS CECC then it is a British Standard harmonized with CECC. See *British Standard*.

BS EN

A *British Standard* implantation of the English language versions of European standards (ENs). The *British Standards Institution (BSI)* has an obligation to publish all ENs and to withdraw any conflicting British Standards (BSs) or parts of BSs. This has led to a series of BS ENs which use the EN number.

BSH

An abbreviation used for *benzene sulphonyl hydrazide*.

BSI

An abbreviation used for *British Standards Institution*.

BSI Handbook 22:1992

See *Quality Assurance Handbook 22*.

BSI Kitemark

See *British Standards Institution Kitemark*.

BSI registered firm

See *British Standards Institution registered firm*.

BSI registered stockist

See *British Standards Institution registered stockist*.

BSI Safety Mark

See *British Standards Institution Safety Mark*.

BS ISO/IEC

A *British Standard* implantation which comprises the *ISO* or *IEC text*, without any national deviation, and which has a front and back cover indicating the UK committee responsible. The date referenced in the identifier is the date of the international standard.

BS QC

A standard issued by the *British Standards Institution (BSI)*. A harmonized system of quality assessment for electronic components. The identity of a British Standard (BS) with an *IEC QC publication*, is indicated by the presence on the front cover of the number of the QC (quality control) publication preceded by BS. Many such publications are adopted without change in content or format as *British Standards*.

BSWL

An abbreviation used for basic sulphate white lead. See *basic white lead*.

BTDA

An abbreviation used for 3,3',4,4' *benzophenonetetracarboxylic dianhydride*.

Btu

An abbreviation used for *British thermal unit*.

BTU

An abbreviation used for *British thermal unit*.

bubble blow ratio

Diameter of bubble: diameter of die. See *tubular film process*.

bubble process

Another name for blown film extrusion. See *tubular film process*.

bubbles of glass

See *glass microspheres*.

bulging

Distortion of a blow moulded product caused, for example, by using too high a product temperature during hot filling.

bulk density

Bulk density is the weight per unit volume of a material including voids which are inherent as tested. May be determined by a pyknometer method. Three specimens which may be granules or fragments of moulded articles are used: each must be of minimum weight of 1 g. An empty, dry pyknometer is weighed (m_1), then the dry pyknometer plus sample is weighed (m_2). After degassing the pyknometer is placed in a thermostatically-controlled bath and filled with distilled water to the limits (m_3). The pyknometer is emptied and filled with distilled water to the limits (m_4). Relative density is ($m_2 - m_1$) $\times$ d/($m_3 - m_4$) $\times$ ($m_2 - m_1$). See, for example, ASTM D 792-66.

bulk factor

The ratio of the volume of a given mass of moulding material to its volume in the moulded form. The bulk factor is also equal to the ratio of the density of the material to its apparent density in the unmoulded form.

bulk moulding compound

An abbreviation used for this type of material is BMC. Originally, BMC was different from *dough moulding compound* (DMC) in that it was formulated to give mouldings of improved quality and finish.

Resins (for example, isophthalic resins give better hot strength and stability) are chemically thickened with magnesium oxide so as to reduce filler segregation and to improve surface finish by reducing porosity. Large shrinkage reductions are obtained by adding a thermoplastic powder such as *polystyrene*. Nowadays the distinction between DMC and BMC has become blurred and often, the terms are used interchangeably although, strictly speaking, they are different.

bulk polymerization
See *mass polymerization*.

bulk thermoplastics materials
See *commodity thermoplastics materials*.

bulked yarns
Yarns that have been treated so as to increase their volume, for example, by crimping of the yarns. See *viscose rayon*.

bulking station
The place where *field latex* is taken where it may be preserved with, for example, *ammonia*.

bumped off
Removal of an undercut component by utilizing the mould opening movement during *ejection* from the mould: unscrewing devices are not used.

bumping
The rapid opening and closing of a mould, during the early stages of cure, so as to reduce air trapping. See *breathing*.

bun
The long, continuous block of *flexible polyurethane (PU) foam* produced, by a *slab-stock process*, and later cut to the required size.

buna rubber
An abbreviation used for this term is buna. The term 'buna' is derived from butadiene and from natrium (sodium). A *synthetic rubber*, first used in the 1920's and made by the sodium-catalyzed polymerization of butadiene. Later the term covered styrene butadiene copolymers (Buna S) and acrylonitrile butadiene copolymers (Buna N).

burn cleaning
The cleaning of metal machine parts by pyrolytic decomposition, or degradation, in a nitrogen atmosphere at temperatures in he region of 550°C/1022°F.

burn mark
A region of burnt or degraded material.

burning
An *injection moulding* fault which occurs as a result of compressing trapped gases (for example, air): suitable *mould venting* will alleviate this type of problem and will also help to improve the strength of any weld.

burning behaviour tests
Tests used, for example, to assess how fast a material will burn once ignition has occurred: the amount and temperature of evolved gases may also be measured. Laboratory tests are commonly used to assess burning behaviour because they are relatively easy to perform and can give reproducible results. However, it must be emphasised that the results of such tests must not be used to predict how a material will perform in a real fire situation. In spite of this limitation there are two tests whose results are widely quoted and these are *limiting oxygen index (LOI)* and the *Underwriters Laboratory UL 94 vertical burning test*.

burning drops
See *lighter test*.

burning speed
See *Underwriters Laboratory horizontal burning test*.

burning test
See *burning behaviour tests*, *Underwriters laboratory UL 94 vertical burning test*, *needle burning test* and *lighter test*.

bush
Sometimes called a bushing. A part of a machine, for example, part of a *two-roll mill*, which is there to protect the mill from excessive loads: it does this by failing if the system is over-loaded. The replaceable sleeve in which a guide pin, or dowel, slides.

bush bearing
Also known as a smooth bush bearing or as, a sleeve bearing or as, a phosphor-bronze bush bearing. A type of bearing for a *calender*.

bushel
An abbreviation used for this unit is bu or bsh. A unit of measurement (for dry measurements) and equal to 4 pecks. An Imperial bushel is 2,219·36 cubic inches and a US bushel is 2,150·42 cubic inches (35·24 litres). See *US Customary Measure* and *UK system of units*.

bushing
A tubing die section that forms the outside diameter of the tube. Also another name for bush. See *glass fibre manufacture*.

Buss Kno-Kneader
A single-screw continuous compounder widely used in the *polyvinyl chloride (PVC)* industry. Each turn of the screw is accompanied by a back-and forth movement of the special screw. It is special in that there are three gaps running the full length of the screw: into these gaps kneading pins, or teeth, protrude from the *barrel*.

but-1-ene
See *butene*.

butadiene
Also known as buta-1,3-diene or as butadiene-1,3. At 0°C this material has a relative density (RD or SG) of 0·65 and a boiling point of −2·6°C. Obtained, for example, by petroleum cracking: a very important monomer for the preparation of several *synthetic rubbers*, for example, *styrene-butadiene rubber* and *nitrile rubber*. See *diene polymerization*.

butadiene rubber
Also known as cis–polybutadiene or as, cis-1,4-polybutadiene rubber or as, 1,4-polybutadiene or as, polybutadiene rubber or as, poly-(1-butenylene). An abbreviation used for this material is BR: sometimes PB or PBD is used.

Because of an asymmetric carbon atom it is possible, by *diene polymerization*, to produce both *atactic* and *stereoregular polymers* from *butadiene*: the stereoregular polymers are the ones preferred for rubbers and are, for example, prepared in solution using an alkyl-lithium or, a coordination catalyst (for example, a *Ziegler-Natta catalyst*).

There are four stereoregular forms of polybutadiene rubber, cis-1,4-polybutadiene, trans-1,4-polybutadiene, isotactic 1,2-polybutadiene, and syndiotactic 1,2-polybutadiene. Commercial materials are not 100% stereoregular and so different combinations of the *above* four forms are produced. Isotactic 1,2-polybutadiene and syndiotactic 1,2-polybutadiene are not rubbery materials: it is the cis material which is preferred for rubber use as the trans material has a high melting point which inhibits rubberiness. Commercial materials may be classified as high cis-1,4-polybutadiene (approximately 96%), medium-cis (92%) and low-cis (approximately 40%). The higher the cis-1,4 content the lower will be the *glass transition temperature* of a vulcanizate and the better will be the low temperature properties. The higher the 1,2 content, the higher will be the glass transition temperature but the wet traction properties improves while the abrasion resistance is worse. A coordination catalyst will yield higher levels of the cis-1,4-polybutadiene while the alkyllithium catalyst yields a BR with a higher 1,2-polybutadiene content.

BR is the only rubber which has better resilience than natural rubber: BR is a *general purpose rubber*. Like other butadiene polymers, it has poor gum strengths (see *styrene-butadiene rubber* or SBR) and so it is essential to compound this polymer with reinforcing fillers. By using reinforcing blacks it is possible to produce mouldings which have excellent abrasion resistance. Another noteworthy feature of this polymer is its ability to accept large quantities of oil and black - such additions help, of course, to reduce compound costs. Both BR and SBR resist breakdown by mechanical shear. Another peculiarity of BR is that, because of its chemical structure, it requires less sulphur than other diene rubbers to achieve an optimum crosslink density.

Because of this material's reasonable resistance to ozone, and its good oxidation and chemical stability, it has been adopted as a replacement for *natural rubber* for some applications, e.g. engine mounts. However, the major use of BR is as a tyre material, particulary in blends with NR and SBR. Blends are used because of the poor mixing behaviour on *two-roll mills*, poor tensile and tear strength and poor wet traction properties of BR vulcanizates. Sulphenamides are used as primary *accelerators* and thiurams as secondary accelerators.

butadiene rubber coordination catalyst based
Abbreviations used for this type of material put the chemical abbreviation for the catalyst before that for butadiene rubber (BR), for example, Ti-BR. Such butadiene rubbers are prepared in solution using a coordination catalyst (for example, a *Ziegler-Natta catalyst*) based on a transition element such as cobalt, nickel, neodymium or titanium.

Co-BR = butadiene rubber based on a cobalt catalyst. A high-cis BR.

Ni-BR = butadiene rubber based on a nickel catalyst. A high-cis BR.

Ti-BR = butadiene rubber - titanium catalyst based. A medium-cis BR.

Nd-BR = butadiene rubber based on a neodymium catalyst. Nd-Br grades are claimed to have improved green strength, tack, abrasion resistance, better groove cracking resistance, and lower heat build up in tyres.

butadiene rubber lithium catalyst based
An abbreviation used for this material is Li-BR. A *butadiene rubber (BR)* which is prepared in solution using an alkylithium catalyst. This type of BR usually exhibits a narrow molecular weight distribution and the polymer exhibits considerable cold flow compared to one prepared with a coordination catalyst (such as titanium). A low-cis BR.

butadiene-1,3
See *butadiene*.

butadiene-acrylonitrile copolymer
See *nitrile rubber*.

butadiene-acrylonitrile copolymer diene rubber
See *nitrile rubber*.

butadiene-acrylonitrile copolymer rubber
See *nitrile rubber*.

butadiene-acrylonitrile rubber
See *nitrile rubber*.

butadiene-styrene block copolymer
See *styrene-butadiene block copolymer*.

butadiene-styrene copolymer
See *styrene-butadiene-styrene block copolymer* and *styrene-butadiene rubber*.

butadiene-vinyl pyridine copolymer
See *vinyl pyridine rubber*.

butan-2-ol
See *secondary butyl alcohol*.

butan-2-one
An alternative name for *methyl ethyl ketone*.

butane
A saturated, straight chain, hydrocarbon of formula C_4H_{10}. A member of the alkane series. See *CFC replacement* and *pentane*.

butane lighter test
See *lighter test*.

butanol
Also known as *butyl alcohol*.

butene
An *olefin*: a monomer for *butylene plastics*. Also known as butylene or as, 1-butylene or as 1-butene or as, but-1-ene. May be represented as $CH_2 = CHC_2H_5$. See *polybutylene*.

butene-type material
See *linear low density polyethylene*.

butenedioic acid
See *maleic acid*.

buttress
Tyre terminology: the lower part of the shoulder area of a *tyre*.

butyl
See *butyl rubber*.

butyl 2-ethylhexyl phthalate
See *butyl octyl phthalate*.

butyl acetate
This material has a boiling point of approximately 127°C. It is a good solvent for cellulose nitrate (CN), polyvinyl acetate (PVAC), polyvinyl chloride (PVC) and polystyrene (PS). It is a good solvent for uncured chloroprene (CR), natural (NR) and styrene butadiene (SBR) rubbers. It is a poor solvent for uncured thiokol (T) rubbers. It is of intermediate solvating power for uncured nitrile rubbers (NBR).

butyl acetyl ricinoleate
A *plasticizer* for *cellulose esters* and *polyvinyl chloride*. This material has a relative density (RD or SG) of 0·93.

butyl alcohol
Also known as butanol. Used as a solvent for natural resins and for aminoplastic resins. This material has a relative density (RD or SG) of approximately 0·8.

butyl benzthiazyl sulphenamide
Also known as N-t-butylbenzothiazyl-2-sulphenamide and as N-tert-butyl-2-benzothiazyl sulphenamide or as, N-t-butyl-benzothiazole-2-sulphenamide. An abbreviation used for this material is TBBS. This yellow solid material has a boiling point of greater than 105°C and a relative density (RD) of 1·29. A delayed action *accelerator*. See *benzothiazyl sulphenamide accelerators*.

butyl benzyl phthalate
Also known as benzyl butyl phthalate and often abbreviated to BBP. This material has a relative density (RD or SG) of 1·10 and a boiling point of 370°C. It is a plasticizer for *polyvinyl chloride (PVC)*. This type of material has good solvent action at low processing temperatures and good processing properties: this is an advantage when highly filled compounds (for example, flooring compounds) are being processed. It resists staining, migration and water absorption. Useful for expanded products because of the low processing temperatures which minimizes premature expansion of the *blowing agent*. Used, for example, for expanded fabric coatings, vinyl flooring and slush moulding. It is compatible with cellulose acetate butyrate (CAB), cellulose nitrate (CN), ethyl

cellulose (EC), polystyrene (PS), polyvinyl acetate (PVAC) and polyvinyl butyral (PVB). It has limited compatibility with cellulose acetate (CA) and polymethyl methacrylate (PMMA).

butyl cyclohexyl phthalate
An abbreviation used for this material is BCHP. A *plasticizer* for polyvinyl chloride. See *phthalate*.

butyl halogenated product
See *halogenated butyl rubber*.

butyl isodecyl phthalate
An abbreviation used for this material is BIDP. A *plasticizer* for polyvinyl chloride. See *phthalate*.

butyl nonyl phthalate
An abbreviation used for this material is BNP. A *plasticizer* for polyvinyl chloride. See *phthalate*.

butyl octyl phthalate
Abbreviation used BOP. Also known as butyl 2-ethylhexyl phthalate. It is a plasticizer for *polyvinyl chloride (PVC)* which has low volatility and gives fast compound fusion at processing temperatures. It resists staining. Used, for example, for garden hose, strippable plastisols, vinyl flooring and slush moulding. It is compatible with cellulose nitrate (CN), ethyl cellulose (EC) and polystyrene (PS). It has limited compatibility with cellulose acetate (CA), polyvinyl butyral (PVB) and polymethyl methacrylate (PMMA). It is incompatible with cellulose acetate (CA) and polyvinyl acetate (PVAC). See *phthalate*.

butyl oleate
Also known as n-butyl oleate. A *plasticizer* for cellulose acetate, other cellulose esters and ethers and for polyvinyl chloride. Also for thermosetting materials such as alkyd resins and for phenolics.

butyl rubber
Also known as isobutylene-isoprene rubber which gives the abbreviation IIR. Also known as isobutylene-isoprene copolymer or as, isobutene-isoprene rubber or as, isobutene-isoprene copolymer or as, poly-(1,1-dimethylethylene-co-1-methyl-1-butenylene) or as, poly-(isobutene-co-isoprene) or as, poly-(1,1-dimethylethylene-co-1-methyl-1-butenylene).

This material has been classified as a *general purpose rubber*. Butyl rubber is a copolymer of isobutylene and isoprene: a copolymer is made because polyisobutylene will not vulcanise on its own. The amount of isoprene is relatively small (e.g. up to 3%) and, for example, by varying the amount of this monomer, a range of grades can be produced. Increasing the isoprene content decreases the resistance to ageing but increases the rate of cure. The isoprene is added in order to make the final polymer capable of being vulcanised by sulphur but it is found that if butyl is mixed with other unsaturated rubbers, then the sulphur will react preferentially with the added rubber and thus upset the cure reactions. For this reason, butyl rubber is incompatible with many other rubbers, for example, with *natural rubber (NR)*, styrene-butadiene rubber (SBR), nitrile rubber (NBR) and with chloroprene rubber (CR).

Butyl rubber is noted for its high impermeability to gases, its excellent ageing resistance and good chemical resistance. Because of these characteristics, IIR is used for inner tubes and for curing bags and bladders. Despite its good chemical resistance, butyl rubber is not classified as a solvent-resistant polymer, although it is better than NR and SBR in this respect. The material is also noted for being relatively slow curing and mouldings produced from this polymer exhibit low resilience (e.g. at room temperature a butyl ball will not rebound very far when dropped). See *halogenated butyl rubber*.

butyl stearate
Also known as n-butyl stearate. A secondary *plasticizer* for cellulose esters, ethers and for rubbers. Acts as a processing aid as it is also a *lubricant*. Also used as a mould release agent.

butyl xanthogen disulphide
See *dibutyl xanthogen disulphide*.

butyl-2-ethylhexyl phthalate
See *butyl octyl phthalate*.

butylated hydroxytoluene
An abbreviation used is BHT. Also known as 2,6-di-t-butyl-4-methylphenol. Used as an *antioxidant*. A simple hindered phenol.

butylene
See *butene*.

butylene plastics
Plastics materials based on the polymerization of *butene* (butylene) or, on the copolymerization of butene with one or more unsaturated compounds and in which the butene is in the greatest amount by weight. See *polybutylene*.

butyloxy radical
This may be represented as $(CH_3)_3C-O\cdot$ and is formed, for example, by the decomposition of di-t-butyl peroxide by *homolytic decomposition*.

butyraldehyde-aniline condensate
An abbreviation used for this type of material is BAA. An *aldehyde-amine compound*: such a compound function as an extremely rapid *accelerator* in a rubber compound and gives high cross-link densities.

by-pass
A secondary passage for fluid flow in hydraulic equipment.

C

c
An abbreviation used for *centi*. See *prefixes - SI*. The same letter is also used for cubic and for cycles.

c/s
An abbreviation used for cycles per second. See *hertz*.

C
This letter is used as an abbreviation for;
    capryl. For example, dicapryl phthalate (DCP). See *phthalate*.
    carbon;
    cellulose;
    channel (*carbon black*);
    chlorinated;
    chloro or chloride;
    citrate in *plasticizer* abbreviations. For example, tributyl o-acetyl citrate (TBAC);
    controlled - as in *controlled rheology polypropylene*;
    coulomb;
    cyclo;
    cresyl;
    crystalline;
    dithiocarbamate as in *zinc diethyl dithiocarbamate (ZDC)*;
    and filler. If a plastics material contains a filler then the letter may be used to show the presence of carbon or, chips, or, cuttings.

C_4 LLDPE
A *linear* low density polyethylene material which uses butene as a co-monomer.

C₈ LLDPE
A *linear* low density polyethylene material which uses octene as a co-monomer.

C scale
The scale of *Rockwell hardness* used for metals such as *steel*.

C type H-NBR
See *partially saturated nitrile rubber*.

C type screw
A *screw* used in injection moulding. Has the highest shot capacity but gives the lowest injection pressure. See *screw size - injection moulding*.

C-glass
A type of borosilicate glass which is not as widely used for fibre manufacture, for use in *glass reinforced plastics*, as *E-glass*: however, it has superior resistance to corrosion by *acids* and *alkalis*. The percent composition by weight is approximately SiO_2 62·0 to 65·0%, Al_2O_3 1·0%, CaO 6·0%, Na_2O/K_2O 12 to 18%, B_2O_3 3 to 4%.

C-scan
See *ultrasonic C-scan evaluation*.

C-stage resin
An insoluble and infusible, cross-linked resin: an insoluble and infusible, cross-linked *phenol-formaldehyde resin*. See *A-stage* and *B-stage resin*.

C-type press
High frequency welding equipment consists of a press, a generator and a materials handling system. Welding is commonly performed using a C-type press or a *bridge press*. High frequency welding is most commonly performed using a C-type press which is coupled to a generator with an output of up to approximately 6 kW. The press is based on a C frame structure, that is, the front and sides are open for easy insertion of the work. Once the work has been positioned a foot pedal is depressed and this causes the electrode assembly to pressed against the work (with a pressure of up to approximately 10 kg/sq cm of weld area) for a preset time.

CA
An abbreviation used for cellulose acetate. See *cellulosics*.

CAB
An abbreviation used for cellulose acetate-butyrate. See *cellulosics*.

cable cord belt
A *V-belt* in which the tension member consists of a single layer of cord.

cable-covering process
See *extrusion wire-covering process*.

CAD/CAM
An abbreviation used for computer-aided design and computer-aided manufacture.

cadmate cure system
A cadmium based curing system for rubbers which gives improved heat resistance because of the short length of the resultant cross-links. Based on, for example, cadmium diethylthiocarbamate (1·5 phr), cadmium oxide (2 to 5 phr), magnesium oxide (5 phr), MBTS, (1·5 phr) and sulphur (<1 phr). See *curing systems* and *efficiency parameter*.

cadmate efficient vulcanization system
See *cadmate cure system* and *efficiency parameter*.

cadmium
Group 11b of the Periodic table consists of the elements zinc, cadmium and mercury. A *transition element (Cd)* which is a soft, white, lustrous metal with a melting point of 321°C and a boiling point of 767°C: the relative density is 8·6. Some cadmium compounds are used, for example, as *heat stabilizers* for *polyvinyl chloride (PVC)* (usually in association with barium compounds). Cadmium compounds are also used as pigments - see *cadmium pigments*. There is great concern about the toxicity of cadmium compounds.

cadmium/barium stabilizer
See *barium/cadmium stabilizer* and *heavy metal stabilizer*.

cadmium diethylthiocarbamate
An *accelerator* for rubber compounds. See *cadmate cure system* and *dithiocarbamates*.

cadmium/lead stabilizer
A polymer stabilizer based on the metals cadmium and lead which is used as a *heat stabilizer*. Compounds of such metals are used as, for example, heat stabilizers for *polyvinyl chloride (PVC)* because of the good stabilizing performance of such a *mixed metal stabilizer*: a *heavy metal stabilizer* used in, for example, the extrusion of *unplasticized polyvinyl chloride*.

cadmium lithopones
Cadmium pigments which have been precipitated onto *barium sulphate* so as to reduce costs. See *cadmium red*.

cadmium octoate
A metal soap used as a *heat stabilizer* for polyvinyl chloride (PVC). May be used as a component of a synergistic mixture. See *stabilizer* and *metal soap stabilizer*.

cadmium oxide
This material has a relative density (RD or SG) of 8·20. Used in rubber technology to give vulcanizates of high heat resistance in, for example, *natural rubber compounds*. Cadmium oxide is also used with PA 66 so as to stop the plate-out of oxidation products on the surfaces of *nylon 66 injection mouldings*. As with all cadmium compounds, there is great concern about the toxicity of such a material and restrictions on use. See *cadmate cure system*.

cadmium pentamethylene dithiocarbamate
A fast *accelerator* for rubbers with a slight delayed action. This material has a relative density (RD or SG) of 1·82, a melting point of approximately 240°C and a formula of $[(CH_2)_5N·CS·S]_2Cd$. See *dithiocarbamates*.

cadmium pigments
A class or type of *inorganic pigment* based on the element cadmium: based on calcined cadmium sulphides and selenides in various proportions. Some grades may also be known as *chemically pure cadmiums*. The sulphide is orange and incorporation of the selenide changes the colour towards red. Relatively, high priced pigments which give bright colours in the yellow, orange, red and maroon range. Heat resistance is good, for example, up to 330°C and to over 500°C for short term exposures. For use at temperatures above 290°C/554°F a cadmium yellow is more heat resistant than chrome yellow. Such pigments can be lightfast particularly in dark shades. Cadmium pigments can be sensitive to a combination of light and moisture (as found in outdoor usage). However, in general, good chemical and light resistance - particularly for reds. As with all cadmium compounds, there is great concern about the toxicity of these materials and restrictions on use. Cadmium red has a relative density (RD or SG) of 4·4 to 5·4, cadmium yellow 4·1 to 4·6 and cadmium sulphide 4·4.

cadmium red
Also known as cadmium selenide lithopone. A group of red pigments which are resistant to heat, light, acids and alkalis. See *cadmium pigments*.

cadmium ricinoleate
A white powder used as a *stabilizer/lubricant* for polyvinyl chloride. This material has a relative density (RD or SG) of 1·11 and a melting point of 105°C.

cadmium selenide
A red powder (CdSe) with a relative density (RD or SG) of 5·8 and a melting point of 1,350°C. A red pigment which is resistant to heat, light, acids and alkalis. See *cadmium pigments*.

cadmium selenide lithopone
Cadmium pigments which have been precipitated onto *barium sulphate* so as to reduce costs. See *cadmium red*.

cadmium sulphide
See *cadmium yellow* and *cadmium pigments*.

cadmium yellow
Also known as cadmium sulphide (CdS) or as, jaune brilliant or as, capsebon. A yellow *cadmium pigment*.

cadmium/zinc stabilizer
A polymer stabilizer based on the metals cadmium and zinc which is used as a *heat stabilizer*. Compounds of such metals are used as, for example, heat stabilizers for *polyvinyl chloride (PVC)* because of the good stabilizing performance of such a *mixed metal stabilizer*: a *heavy metal stabilizer* used in PVC foams.

cadmiums
The collective name for the class of materials which contain the element cadmium and which are used as pigments. See *cadmium pigments*.

CAF
An abbreviation used for *compressed asbestos fibre*.

cake
A package of *continuous filament yarn* produced during the *spinning* of *viscose rayon*.

cal
An abbreviation used for *calorie*.

calamine
A naturally occurring form of zinc carbonate. See *basic zinc carbonate*.

calandrette line
See *calendrette line*.

calcination
Strong heating of a material, for example, to 750°C in air. See *calcined clay*.

calcined clay
A *clay* which has been *calcined* so as to improve electrical resistivity and to improve whiteness.

calcite
A hard, compact crystalline variety of *limestone*.

calcium
An alkaline earth metal which is too reactive to occur naturally. It is a hard gray metal with a melting point of approximately 851°C, a boiling point of 1,480°C and a relative density of 1·55. Its compounds are widely used as fillers: see, for example, *calcium carbonate* and *whiting*. Calcium compounds are also used as non-toxic *heat stabilizers* for *polyvinyl chloride (PVC)*.

calcium/aluminium/zinc stabilizer
A polymer stabilizer based on the metals calcium, aluminium and zinc which is used as a *heat stabilizer*. Compounds of such metals are used as, for example, heat stabilizers for *polyvinyl chloride (PVC)* because of the relatively good stabilizing performance of such a *mixed metal stabilizer*. Such a stabilizer is often combined with a *metal-free organic stabilizer*.

calcium/antimony stabilizer
A polymer stabilizer based on the metals calcium and antimony which is used as a *heat stabilizer*. Compounds of such metals are used as, for example, heat stabilizers for *polyvinyl chloride (PVC)* because of the relatively good stabilizing performance of such a *mixed metal stabilizer*. Such a stabilizer is often combined with a *metal-free organic stabilizer*.

calcium carbonate
May be represented as $CaCO_3$. Another name is carbonic acid lime. Widely used as a *filler* in both the rubber and plastics industries as the material is cheap and white. May be obtained by grinding chalk to give *whiting*: such products are relatively coarse and are non-reinforcing fillers. The particles obtained from ground limestone have a particle size in the range of 0·5 to 30 μm. (Much finer particles, of better colour, may be obtained - see *precipitated calcium carbonate* and *activated calcium carbonate*.) Such fillers are used in *polyvinyl chloride (PVC)* compounds and in *glass reinforced plastics (GRP)*. In rubber technology, it has been found that a more reinforcing filler is obtained if the filler is coated with a carboxylated unsaturated polymer. Calcium carbonate (activated and precipitated) has a relative density (RD or SG) of approximately 2·6. Calcium carbonate (whiting) has a relative density (RD or SG) of 2·70.

calcium chloride
A white deliquescent material with a melting point of approximately 770°C. A coagulant for *latex*.

calcium hydroxide
Also known as caustic lime and slaked lime This material has a relative density (RD or SG) of 2·28. See *lime*.

calcium hydroxyapatite
Also known as hydroxyapatite. A calcium/phosphorus material which may be represented as $Ca_{10}(PO_4)_2(OH)_2$. This material is one of the major constituents of bone. Particulate calcium hydroxyapatite can be used to produce composites which are of interest as replacements for artificial bone. Particulate calcium hydroxyapatite (used for bone china) can be obtained from calcined bone or, made synthetically. The synthetic material is used for artificial bone and in such a case it is thought that the structure and size of the particles should be the same as that found in the natural material.

CALCIUM HYDROXYAPATITE FILLED POLYETHYLENE Calcium hydroxyapatite filled *polyethylene* (of a molecular weight of approximately 400,000) is used as a material for use as a bone substitute as it is biocompatible. The volume fraction of calcium hydroxyapatite may reach 0·5 as the composite is still tough.

CALCIUM HYDROXYAPATITE FILLED POLYHYDROXYBUTYRATE Calcium hydroxyapatite filled *polyhydroxybutyrate (PHB)* is being evaluated as a material for use as a bone substitute as it is biocompatible and biodegradable.

calcium metasilicate
See *wollastonite*.

calcium oxide
May be represented as CaO. Also known as quicklime. A white amorphous powder made by heating *calcium carbonate* and with a melting point of approximately 2,600°C. Reacts vigorously with water (slaking) to form slaked lime or $CaOH)_2$. Calcium oxide may be used to reduce porosity in rubber compounds. An intermediate used to produce *precipitated calcium carbonate*. See *lime*.

calcium silicate

The natural material is derived from *wollastonite*. Synthetic calcium silicate is made from *diatomaceous silica* and *lime* to give colloidal particles of large surface area and with large amounts of trapped air. It is also known as hydrated calcium silicate or as hydrous calcium silicate. This synthetic material has a relative density (RD or SG) of 2·26 to 2·40 and may be represented as $CaO \cdot 3SiO_2 \cdot xH_2O$. Often used in the hydrated precipitated form: a fine white powder which may be classed as a semi-reinforcing filler in rubber compounds. May slightly decrease vulcanization rate: the use of diphenylguanidene may offset this. Calcium silicate is more expensive than mixtures of *silica* and *china clay* which may be used to replace the silicate in compounds. At high loadings calcium silicate can still give soft vulcanizates.

calcium soap

A calcium salt of an acid such as stearic, palmitic and/or oleic acid. These materials are used to assist dispersion of fillers in rubber compounds, for example, in *natural rubber compounds*. In light coloured compounds, calcium soaps of saturated *fatty acids* are preferred. See *calcium stearate*.

calcium stearate

A *calcium soap* which may be represented as $(C_{17}H_{35}COO)_2Ca$. Often used in the form of a white, finely divided powder as a mould lubricant, in rubber compression moulding, as it has a melting point of approximately 150 to 155°C/302 to 320°F. Also used to improve flow in both extrusion and injection moulding of rubbers: especially useful for white compounds. Used as a heat stabilizer and processing aid in *polyvinyl chloride* (PVC) technology. This material has a relative density (RD or SG) of 1·04.

calcium sulphate

See *anhydrous calcium sulphate*.

calcium sulphate hexahydrate

See *plaster of Paris*.

calcium/zinc stabilizer

A polymer stabilizer based on the metals calcium and zinc which is used as a *heat stabilizer*. Compounds of such metals are used as, for example, heat stabilizers for *polyvinyl chloride* (PVC) because of the relatively good stabilizing performance of such *mixed metal stabilizers* which are also considered to be non-toxic. Such a stabilizer is often combined with a *metal-free organic stabilizer*.

calender

The term calender may refer specifically to the machine used to produce film or sheet or it may be more loosely applied to the whole calendering layout: in which case it would therefore include the feed devices, haul-off etc. To avoid confusion the term *calendering* train (or calendering line) is sometimes used when the total system is being discussed.

A calender is a machine based on rolls which is used to produce film or sheet at very high speeds and, usually, of high quality; a hot compound is turned into sheet by being passed through a series of nips formed by sets of rolls.

calender configurations

See *calender types*.

calender drive

That which rotates the rolls on a calender. Rolls are commonly driven by variable speed DC motors and at low values of fed-in power the efficiency of such motors drops. Over-dimensioning of calender motors, which may be deemed necessary if a variable production is required, is undesirable if large electrical losses and decreased accuracy of rotational speed are to be avoided. The motor speed controller must be capable of maintaining the rolls at constant speed.

Not every roll of a *calender* requires the same drive energy. For example, with an *inverted L calender* it would be expected that the largest torque would be measured on roll number 2, followed by roll no. 3, followed by roll no. 4 and then roll no. 1. See *single motor drive* and *unit drive*.

calender feed

That which is fed to a *calender*: the output from a mixer. For example, a *continuous mixer* is commonly used as the primary mixer as this type of mixer has a relatively short dwell time. Two roll mills are used after the mixer as in many cases the output from such continuous mixers is not suitable as a calender feedstock. A *wig-wag* is often used to distribute the melt mixed material across the feed nip.

calender forces

See *forces - calendering*.

calender grain

The difference in properties between the machine direction and the cross-wise direction and which is caused by *orientation*.

calender identification

The rolls may be arranged in different ways or configurations and each configuration is named after letters of the alphabet. Such letters may be obtained by drawing a side view of the roll arrangements and joining the centres of the rolls. Can have, for example, in the case of a four roll calender an *L calender*, an *inverted L calender*, a *Z calender* and an *inverted Z calender*.

calender nip settings

If a four roll calender for *polyvinyl chloride* (PVC) is considered then, for example, if the calender is required to produce 0·25 mm PVC sheet (0·010") then, for running, the top nip may be set at 0·75 mm (0·030"): the second nip is set at 0·50 mm (0·020") and the third nip is set at 0·25 mm (0·010").

calender - number of rolls

Machines are made with 2, 3, 4 or 5 rolls respectively with 3 or 4 roll machines being the most popular. As the number of rolls is increased so does the cost and complexity of the machine: however, more uniform gauge sheet, freer from porosity results. The actual configuration chosen depends upon many factors. See *calender - roll configuration*.

calender operation

See *four roll calender operation*.

calender - roll adhesion

Hot *polyvinyl chloride (PVC)* will, unless grossly over-lubricated, adhere to the hottest roll of a *nip-forming pair* and/or, to the fastest running roll. Surface finish is very important and the material will remain on, or transfer to, the roll with the dullest (matt) surface finish irrespective of speed or temperature differences.

calender - roll configuration

The actual configuration, or arrangement, of the rolls depends upon many factors including type of feed, ease of feed, nip accessibility, arc of contact required for the material, running speed, labour requirements, initial costs, running costs, accuracy required etc. See *calendering*.

calender - roll finish

Both highly polished rolls and matt finish rolls are used. The finish affects not only the finish of the sheet but also how the material behaves during processing. See *calender - roll adhesion*.

calender - roll heating

Rolls can be drilled or bored and steam or hot water circulated through the rolls to give heating or cooling as required. For slow running machines, the roll is often centrally bored whereas for fast running machines, *drilled rolls* are commonly employed. This is because once a production *calender* is at operating temperature, and material is passing through the nips, then considerable heat can be generated. Heat removal is then required if the line speed of the calender is to be maintained without product deterioration or changes in roll temperatures and therefore in dimensions (with nitrile and neoprene feedstocks heat generation can be particularly troublesome). Therefore, rolls are commonly heated with high pressure hot water (HPHW) using drilled rolls. A *heated oil circulating system* could also be used. No matter which system is fitted, slow heating and cooling must be employed if thermal shock is to be avoided. For example, a temperature change of 1°C/min is recommended. See *two-roll mill*.

calender - roll number
Calender roll identification is usually by number: the number increases in value as the *feed-path* is traced through the *stack* from the feed nip. The roll closest to the feed device, or which is offset from the *superimposed stack*, is called roll 1. See *calendering*.

calender - roll rotation
See *single motor drive* and *unit drive*.

calender - roll speeds
On a four roll calender, roll speeds normally increase from roll 1 to 4 as it is found that materials tend to stick to the faster moving roll of a pair. This is provided that each roll has the same finish. If one roll has a matt finish then the material will stick to this roll: this is presumably because of the increased area in contact with the material.

calender rolls - materials of construction
Calender rolls are normally constructed from chilled cast iron although steel rolls can be made. The various layers within the calender bowl should be of uniform thickness and centrally positioned in relation to the roll axis as otherwise faults will develop on heating. A rough surface could develop on heating cast iron because this material contains different materials (cementite and perlite) and these have different coefficients of expansion. Because of such effects some rolls are hot ground.

The surface finish of the bowls is extremely important as the sheet can only be as good as the rolls with which it is in contact. Hard chromium plated bowls are sometimes used and it is essential that the chromium plating is tightly bound to the substrate as if flaking occurs only limited repairs are possible. See *calender - roll adhesion*.

calender rolls - sizes
Rolls on production calenders can be supplied with a diameter (body width) of from say 400 mm (16 in) up to 915 mm (36 in): the total length of the 915 mm (36 in) roll would be of the order of 6500 mm (256 in) and could have an overall weight of approximately 17 tonnes. The roll body dimensional inaccuracy would probably not exceed 0·0002 mm (8×10^{-6} in), the concentricity at 180°C up to 0·005 mm (2×10^{-4} in) and deviation from parallel along the shaft axis of up to 0·002 mm (8×10^{-5} in). These tolerances being offered so that the sheet thickness is consistent.

calender side frames
The frames used to support, or carry, the *calender rolls*. May also be called gables. Such frames are very sturdy and are of cast iron tubular construction.

calender type
See *calender identification*.

calendered film
Film produced by a *calendering process*.

calendered sheet
Sheet produced by *calendering*.

calendered sheet cooling
The cooling that is necessary (to minimize *blocking*) is commonly accomplished by passing the sheets over a series of water cooled rolls or cans. A number of rolls are used as it is found that the rate of cooling can then be controlled: this is important for certain products, for example, unplasticized *polyvinyl chloride (PVC)* sheet. See *cooling can*.

calendered sheet - thickness measurement and control
The *nips* and the haul-off speed are constantly monitored and adjusted during running so as to produce sheet of the required dimensions. Before any adjustment can be made it is necessary to know, and to know very precisely, the thickness and the thickness distribution both across and along the sheet.

At slow calendering speeds simple measuring systems can be used for example, micrometers and spring loaded rollers. At high speeds such systems become impractical or unreliable and non-contact systems based on *radiation gauges* have become popular. Of such systems the *beta ray gauge* is the most well known.

calendered sheet trimming
In *calendering*, the edge of the sheet may not be of the same thickness, as the rest of the sheet and sheet width may also vary. Edges are therefore removed and *reclaimed* by either *cold trimming* or *hot trimming*.

calenderette
A term sometimes used to denote a machine based on a series of rolls. Used as the haul off in sheet extrusion or used to produce sheet, for example, from a thermoplastics material in powder form. Not of such heavy construction as a calender.

calendering
A process in which a thermally softened compound is converted into a sheet form by the action of a *calender*. The process is used to produce sheet which may consist of (i) a single layer of material, (ii) a layer of material reinforced with fabric or, (iii) layers or plies of material bonded together.

Multi-layer construction is possible by laminating a series of flexible substrates together, that is, using a calender to generate the conditions necessary to achieve bonding. Decorative and/or durable compositions are possible by bonding the calendered sheet to metallic or wooden substrates in a post calendering operation. The range of products possible may be widened still further by using different polymers in separate layers or differing polymer compounds and/or cellular materials. Calendered sheet forms the feed stock for other processes, for example, welding, thermoforming and moulding.

The rubber industry was initially responsible for using the calender for the production of sheet based on polymer compounds and many rubber factories possess machines which produce sheet suitable for say, hose or boot manufacture. It is however in plastics factories that one finds the most spectacular machines, that is, those capable of producing *polyvinyl chloride (PVC)* sheet with extraordinary accuracy, for example, to within ±2·5 μm (0·000 1 in) at high speeds (say 150 m/min or 488 ft/min).

A calendering line, or train, suitable for the production of such PVC film or sheet is an extremely expensive proposition. Not only is the calender itself expensive but it has associated with it a series of other items which must be matched to the calender. Matched that is in terms of both quantity requirements and of quality.

In terms of tonnage, two processes dominate thermoplastics film and sheet production: these two major processes are calendering and *extrusion*. Broadly speaking extrusion is used to produce the thinner sheet, based on PE and PP, and calendering is used to produce the thicker sheet which is based on PVC. Approximately 25% of all PVC passes through a calender and as PVC is a *commodity thermoplastics material*, calendering is an important process. Product thickness over the range 0·030–0·6 μm are common (0·001 to 0·025 in). If thicker sheets are required then they may be obtained by plying or lamination and thinner materials may be obtained by stretching.

calendering - coating
See *tyre-cord line*.

calendering - control of
Gauges and/or sensors provide raw data which must be processed if control actions and/or information is required. Digital computers are economically attractive and give very compact control systems which can handle large amounts of data extremely quickly so that better quality sheet results.

On a 3 roll calender, producing plain unsupported sheeting, the sheet may be scanned by a single transmission *beta ray gauge A* 1500 mm (60 inch) wide sheet may be scanned in 6 seconds and 20 measurements may be taken during that time. The weight per unit area may then be calculated for an individual *zone* and the sheet forming gap may be adjusted, if necessary, automatically. More than one type of gauge may be fitted to the machine (see *sheet measurement - dual gauge system*).

Once the computer has initiated a control action then it can monitor the results by reading the installed transducers. For example, linear variable differential transducers (LVDT) would measure the roll positions and high resolution, television width sensors could measure the width of the material continuously. Sheet weight per unit area is monitored by means of *radiation gauges*. For coating, the object is to maintain the thickness to the specified target value (on both sides of say a tyre cord fabric) and across its width.

Such a system can give target weights and coating thicknesses of a high order of accuracy both along and across the sheet. High quality sheet can be produced consistently and reliably and the control strategy can be tuned or modified to suit the system under consideration. For example roll eccentricity can be allowed for in the control scheme.

Measured variables and set-points may be displayed on video screens in analogue and/or digital form and gauge standardisation may be performed automatically at pre-set intervals or by push button initiation. Either the operator or management can communicate easily with the system, via a keyboard, and the information displayed can be copied with an associated printer.

calendering - extrusion differences
Because of the way the sheet is formed by the calender roll nips, calendered sheeting differs from extruded sheeting in some respects. The calender is not required to de-aerate, homogenise and fuse the *polyvinyl chloride (PVC)* composition: it accepts hot, fused PVC which it forms into sheet by surface contact with the rolls.

On a four roll calender, with three nips, the surface is re-worked twice after it has been formed in the feed nip: the core of the sheet is not substantially re-worked. Differential cooling of the sheet and core sets up stresses in the final sheet which can, for example, lead to wrinkling (because of stress relaxation) in the final product. If the feed is non-uniform, with regard to temperature and distribution across the bank, then these factors will also cause stresses and subsequent stress relaxation in service.

calendering - flexible PVC sheet production
The *polyvinyl chloride (PVC)* polymer is, for example, first blended with *plasticiser* in a *ribbon blender* and then melt mixed in an *internal mixer*, for example, a Banbury mixer: at this stage the other ingredients of the mix are added, *stabiliser*, *lubricant*, *pigment* etc. After mixing the compound is discharged onto a *two-roll mill* for storage and then passed to a *strainer extruder*. The strip output from this machine is passed under a metal detector and is then fed to a *four roll calender*. The sheet so produced may be *hot trimmed*, removed from the calender, embossed if required, cooled, measured and wound up into rolls of appropriate size. If the product is not of the quality required, for example, not of the required thickness, then roll adjustments are made during running.

calendering forces
See *roll separating forces - calendering*.

calendering line
Also known as a calendering train - see *calender*.

calendering material - requirements of
Any calendering compound must be formulated so that its elasticity is not too high as otherwise the amount of swelling (see *die swell*) that occurs as the material emerges from between the rolls could be excessive. The material must have a reasonable melt strength and must not adhere too strongly to the rolls: these properties are necessary if the material is to be handled successfully, for example, peeled smoothly and consistently from the rolls.

If the compound is going to be stretched into thinner sheet then a high melt strength is required. The compound must be capable of forming a sheet of the required thickness, at the speed desired, without causing radical changes in material properties or in machine operating characteristics. For example, if the calender is run fast, the *shear heat* generated within the compound could cause thermal degradation of a *polyvinyl chloride (PVC)* material if the compound is inadequately stabilized and/or if the heat is not removed quickly. Under certain conditions a deposit may build up on the rolls (plate-out) which then effects the machines operating characteristics and the appearance of the product thus obtained.

Melt viscosity should not be markedly temperature dependent over a convenient processing range. It should be capable of being controlled or adjusted to suit the product, for example, a high viscosity is required if thick sheet is being manufactured from a rubber compound.

However the calender must not be considered in isolation, for example, to minimise gauge variations, and to obtain a consistently high surface finish, the material must be fed to the calender at a constant rate: it must be homogenous with regard to temperature and to composition.

calendering materials
Polymer compositions based on *polyvinyl chloride (PVC homopolymer* or *copolymer)* and on rubber (*natural* and *synthetic*) are the major materials processed by *calendering*.

PVC is used to make clear and opaque sheet, which may be used in upholstery, license holders, wallets and toys: by appropriate formulation the hardness may vary from rigid to very flexible. Very thin sheet is produced by calendering and stretching: in such cases thicker sheet is first produced and this is then stretched at stretch ratios of up to 8:1. The products are used in adhesive and recording tape. PVC as a flooring material is now well established and such flooring may be continuous or in tile form. Both are produced by calendering and attractive patterns may be obtained by printing: the printed layers may be protected by a thin, clear layer (or top coat of PVC) which has a very good surface finish: the assembly may be backed by a cellular or heavily filled layer so as to obtain the required thickness.

Bridging the gap between rubbers and plastics are *thermoplastic elastomers* and such materials form useful calender feedstocks as they are relatively easy to handle by calendering. Other plastics such as *acrylonitrile butadiene styrene (ABS)* and *polypropylene* are also calendered into high quality sheeting.

The major application of rubber is in tyres and in such an application the rubber is reinforced with fabric to give a durable, resilient composition which, so far, is unequalled in terms of price and performance. Other uses for rubber coated fabric is in boots and belting. Moulding blanks may be prepared by guillotining a sheet which has been prepared on a fairly simple calender, for example, a two-roll calender.

The processes known as *frictioning* and *skin coating* both produce supported products i.e. the rubber is supported or reinforced by fabric. Rubber sheet may be produced over the thickness range 0.4 to 1.5 mm (0.015 to 0.062 in) although thicker sheets are possible. If thick, homogeneous sheet is required (i.e. free of air bubbles) it is customary to ply up a number of thin sheets to obtain the necessary thickness. For products such as moulding blanks where air bubbles can be tolerated, 10 mm (0.4 in) thick sheet can be calendered directly.

calendering mixing system
See *mixing system - calendering*.

calendering - rubber compounds
In rubber technology the term calendering embraces a number of activities that make use of the calendering machine:
   sheet production, where the rubber sheet produced is rolled into a liner (a fabric sheet);
   frictioning, where a quantity of rubber is forcibly rubbed into the interstices of a fabric; and,
   skin coating or topping, where a thin sheet of rubber is rolled out and applied to the surface of a previously prepared fabric.

calendering thickness variations
See *thickness variations - calendering*.

calendering train
See *calendering line*.

calendrette line
Also spelt calandrette line and sometimes calanderette line. A sheet production facility which consists of an *extruder*, often a *planetary extruder*, which feeds a relatively simple *calender*. For example, to obtain a uniform feed bank, the extruder output may be in the form of a number of spaced-out strands. The spaced-strand feed is used with a three-roll *stack* to give sheet at up to approximately 1.5 m width: the output may be 750 kg/h.

calibrating blow pin
A blow pin used in *extrusion blow moulding*. As well as introducing the blowing air, this pin also shapes, or calibrates, an internal diameter on an opening in a blow moulding. If the mould closes around the pin then it is referred to as 'clamping stroke calibration'. If the pin, or mandrel, enters the closed mould, then this is known as 'mandrel stroke calibration'.

calorie
An abbreviation used for this unit is cal. The amount of energy required to change the temperature of 1 gram of water by 1°C over the temperature range 14.5 to 15.5°C. This 15° calorie is equal to 4.1855 joules. The International Calorie is equal to 4.1868 joules. See *specific heat*.

calorie - large
A kilogram calorie or 1,000 *calories*.

calorie$_{it}$
An International Calorie or International Table Calorie. This is equal to 4.1868 joules. See *calorie*.

cam
A hardened, steel member fitted to a mould plate so as to actuate, for example, splits and side cores.

camber correction
See *roll separating forces - compensating for*.

cambering.
See *roll cambering*.

camelback
A shaped extrudate produced for rubber *retreading*.

camphor
Obtained originally from camphor trees by distillation of the wood of such trees: that is obtained from camphor oil. Now obtained from turpentine oil. A translucent, whitish material with a pleasant smell. This material has a melting point of 178°C, a boiling point of 204°C and a relative density (RD or SG) of 1.0. Best known as a *plasticizer* for cellulose nitrate.

can
An abbreviation used for *cooling can* or *roll*.

can-type containers
See *tinplate containers*.

candela
The SI basic unit of luminous intensity with the abbreviation cd. Sometimes referred to as a new candle. Defined as the luminous intensity of 1/600,000 of a square metre of a black body at the temperature of solidification of platinum. One square centimetre of such a black body at this temperature (2,046 K) would have a brightness equal to 60 candela. See *Système International d'Unité*.

candle test
A form of the *limiting oxygen index (LOI)* test where a stick of the solid polymer (the test specimen) is burnt like a candle. In the LOI test, a stick of the solid polymer (the test specimen) is burnt vertically (like a candle) and the concentration of oxygen, in an oxygen/nitrogen mixture, which will just sustain burning is measured. The test yields a number or value of, say, 27. This value is taken, for solid polymers, as being the self extinguishing limit for a candle test as candle-like burning does not take account of convective heating. For flame retardancy purposes, the higher the number, the better.

candlepower
A measure of luminous intensity. The illuminating capacity of a light as compared with that given by a standard candle. See *candela*.

caoutchouc
The Peruvian name for crude *natural rubber*. Obtained by heating natural rubber latex. The sticky wax-like substance so obtained was used, for example, for water-proofing.

CAP
An abbreviation used for computer aided production. CAP is also used as an abbreviation used for cellulose acetate-propionate. See *cellulosics*.

capability index
See *process capability index*.

capacitance
The property of a system which enables it to store electrical charge. When a potential difference of one volt appears between the plates of a capacitor, charged by one coulomb of electricity, then the capacitor has a capacitance of one *farad*. That is, a capacitance of one farad requires a charge of one coulomb to raise its potential by one volt. See *Système International d'Unité*.

capacitive heating
See *high frequency heating*.

capacitive sensor
A location or thickness sensor, for example, which uses the presence of the plastics material as a dielectric in a circuit so as to indicate (measure) thickness: also used to measure the proximity of two articles to each other.

capillary die
A rod die with a small bore or diameter: often used in rheological studies. See *capillary rheometer*.

capillary rheometer
A test instrument used to study the rheological properties of polymer melts by forcing the melt through a *capillary die*. This type of *rheometer* is of greatest interest to the plastics technologist as it provides data in the shear rate range used in *injection moulding* and *extrusion*. The data so obtained, may in turn be used, for example, to size dies or runner systems. A capillary rheometer, when used in the processing shear rate range experienced during melt processing, may be called a high shear rate rheometer

capillary rheometer - analysis of flow
The analysis of *shear flow* in a *capillary rheometer* that is employed to obtain flow data, makes some important assumptions. Of these the most important are that:

(i) the flow is isothermal;
(ii) there is no slip on the tube wall;
(iii) the melt is incompressible;
(iv) the flow pattern is the same right down the tube (die); and,
(v) dissipation of energy at the die entrance, or due to chain uncoiling, is negligible.

If the assumptions made, so as to obtain flow data, are assumed to be valid then it may be shown that:

τ_w = the shear stress at the wall of the die = $PR/2L$, and,

$$\dot{\gamma}_w = \frac{(3n' + 1) \times 4Q}{4n' \times \pi R^3} = \text{the shear rate,}$$

where, $n' = \dfrac{d\log (RP/2L)}{d\log (4Q/\pi R^3)}$.

τ_w = the shear stress at the wall of the die,
P = the measured pressure,
R = the die radius,
L = the die length,
$\dot{\gamma}_w$ = the shear rate at the wall of the die, and,
Q = the volumetric output rate.

capillary rheometer - measuring $\dot{\gamma}_{w,a}$ and τ_w
Let the ram speed of the rheometer = V (measured in cm/s) and the barrel diameter = D (measured in cm). Then the *volumetric flow rate* $Q = \pi V D^3/4$ (Eq 1). This is the same in both the barrel and the capillary. Now the *apparent shear rate* at wall is given by $\dot{\gamma}_{w,a} = 4Q/\pi R^3$ (Eq 2). Where R is the radius of the capillary in cm. Substituting equation 1 into equation 2 and simplifying, $\dot{\gamma}_{w,a} = VD^3/R^3$ s^{-1} is obtained.

If the barrel diameter is 9·525 mm then, the formula becomes $\dot{\gamma}_{w,a} = 90 \cdot 726V/R^3$ s^{-1} (when the capillary radius is in mm and the ram speed is in mm/s). Since the shear stress at the wall is given by $\tau_w = PR/2L$ then, for a 20:1 die, $\tau_w = P/40$ (where P is the recorded pressure at the entrance to the capillary).

By plotting τ_w against $\dot{\gamma}_{w,a}$ a flow curve may be obtained. See *apparent viscosity*.

capillary tube method
A method used to obtain an accurate measurement of the melting point of a crystalline polymer such as a *polyamide (PA)*. A thin strip of PA is cut with a microtome and placed inside a capillary tube. The tube plus contents are then slowly heated inside a heating apparatus until the sharp edges of the specimen become rounded. See, for example, ASTM D 2117.

capped end
Also known as a sealed end. A *rubber hose end* in which the wall section has been sealed with rubber.

capped polyol
Also known as a tipped polyol. A *polyol* which has been made more reactive. For example, to increase the reactivity of a *polyoxypropylene polyol*, the materials may be capped or tipped with ethylene oxide.

capping
See *end capping*.

caproester
A type of polyester *thermoplastic polyurethane*, based on polycaprolactone, which is noted for its hydrolysis resistance.

caprolactam
Also known as ϵ-caprolactam or as 6-hexanolactam. The monomer for *nylon 6*.

caprolactam reaction injection mouding
See *nylon 6 reaction injection moulding*.

capsebon
See *cadmium yellow* and *cadmium pigments*.

capstan
A large drum device used to pull extrudates, such as wire or cable, by wrapping the extrudate around the drum so as to provide sufficient friction to get a nonslip drive.

car window fogging
See *automotive fogging*.

carat
In Troy weight, 3·17 *grains* make 1 carat (0·2053 g). A metric carat is equivalent to 200 milligrams.

carbamates
A group of rubber *accelerators*.

carbamic acid
May be represented as $NH_2 \cdot COOH$. The esters of this material are known as *urethanes*. A carbamic acid, an unstable compound of the type $R \cdot NH \cdot COOH$ where R is an organic group, is formed in *polyurethane manufacture* by the reaction of water and an isocyanate.

carbamide
See *urea*.

carbolic acid
See *phenol*.

carbon
This element (C) occurs in Group 1VB of the Periodic table along with silicon, germanium, tin and lead. It occurs naturally as pure forms as graphite and diamond and, in an impure form, as coal: *carbon black* is a very important filler for the polymer industry. It is an infusible, unreactive element which however, forms an incredible number of compounds. It is an essential constituent of all life forms and most commercial polymers are based on very high molecular weight carbon compounds. This element is remarkable for its ability to form long chain compounds.

carbon arc resistance
See *arc resistance testing*.

carbon arc test
A carbon arc is sometimes used for light/heat testing but see *arc resistance testing*.

carbon black
A fine powdered form of polycrystalline carbon with a *graphite type of structure*. Each small piece of graphite-type material, associates with other pieces so as to form basically spherical particles. The outer layers of these spherical par-

ticles have a high degree of order with the small pieces, which make up the particle, lying tangentially to the surface. Particle sizes may be from 10 to 250 nm in furnace blacks and from 120 to 500 nm in thermal blacks.

Particle size may be determined by electron microscopy: particle surface area may be determined by gas or liquid absorption methods, for example, by *cetyl trimethylammonium bromide (CTAB) absorption, nitrogen absorption,* or by *iodine absorption*. Liquid phase adsorption methods are used for product control and specification: for example, iodine absorption (absorption of iodine from potassium iodide solution) is the standard ASTM method. However, CTAB absorption is in some ways better as this method is not so affected by factors which limit the iodine absorption. Factors such as surface oxidation and absorbed hydrocarbons.

The major use for carbon black is in rubber compounds as its use significantly improves properties such as abrasion resistance and tensile strength. The particles, of most blacks used for rubber compounding, are non-porous and so there may not be a great deal of difference between absorption methods which distinguish between internal and external surfaces of the particles. For example, the *t method* and the *BET method* may give similar results.

Carbon black consists mainly of carbon (at least 95% and often >99%) with some hydrogen and oxygen: these elements are present, for example, as surface carboxylic groups, phenolic hydroxyl groups and quininoid structures.

Carbon black may be made by various methods and the method of manufacture may be used to classify the black (see *carbon black - classification*). The most important type of black for the rubber industry is furnace black (sometimes called oil furnace black) which is made by the incomplete combustion of liquid feedstocks from petroleum fractionation. Of the many grades possible, grades such as high abrasion furnace (HAF), general purpose furnace (GPF) and fast extrusion furnace (FEF) are probably the ones most widely used by the rubber industry.

Carbon black is also used to pigment and stabilize plastics materials, for example, polyolefins. A medium colour black (MCF) of particle size 17 nm and with a *BET value* of approximately 210 m^2g, is usually used for such applications. Only a small percentage of black is used but the outdoor weathering properties are improved tremendously. Much higher concentrations of black (approximately 50%) may be added to thermoplastics if the plastics material is cross-linked. Carbon black has a relative density (RD or SG) of from 1·81 to 2·04. (Diamond is 3·52 and graphite is 2·27.)

carbon black - classification
Carbon black may be made by various methods and the method of manufacture may be used to classify the black. The most important types of black are furnace blacks, thermal blacks, channel blacks, acetylene blacks, and lamp black. The furnace blacks account for approximately 95% of the market while the thermal blacks, which give coarser materials, account for approximately 4% of the market. Structure is low in thermal blacks but can be high in furnace carbon blacks: see *carbon black - structure*. If the other ingredients of the mix are ignored then, the properties of a black filled, cured rubber compound will largely depend upon the particle size of the black, the particle size distribution of the black, the *structure* and the amount of *black* present.

One system of classification uses the method of manufacture to classify the black. From furnace blacks comes F, from thermal blacks comes T and from channel blacks comes C. As the presence of carbon black significantly improves properties such as abrasion resistance (A) then this system of classification used the method of manufacture and the effect on the properties to classify the black. For example, high abrasion furnace black became HAF. Other letters are used, for example, HS for high structure, LS for low structure, FF for fine furnace, SC for super conducting, HP for hard processing and LC for low colour. Some of the older names are shown in **table 15** (this table is not complete as there are many more carbon blacks).

To improve on this confusing system, the ASTM introduced a system which uses a letter followed by three numbers. The letter is N for normal cure rate and S for slow cure rate. The first of the three numbers indicates the particle size of the carbon black. This system of classification uses the *iodine absorption* (a measure of surface area) and *structure* to classify the *black*.

Generally the finer the particle size, the greater is the amount of reinforcement. In general, the *structure* affects the processability of the compound with high structures giving lower nerve and stiffer unvulcanized compounds (as compared to lower structure carbon blacks).

carbon black - structure
The individual primary particles of a filler may combine to form aggregates or structures. With *carbon black*, the primary particles are fused into structures like clusters or chains - the primary structures. It is these which are referred to as 'structure' (see also *24M4 test*). Structure is low in thermal blacks but high in furnace blacks.

The number of particles in each structure is important but so is the way the particles are grouped or clustered as such groupings will affect the rubber/filler interaction. An open aggregate is one in which the aggregate has a high bulk while a clustered aggregate is more compact. More open-structured blacks are available which give, for example, improved abrasion resistance. Such improved technology blacks, or new technology blacks, are obtained because of modifications to the production process: modifications such as improved reactor and burner designs. Such improvements give lower surface area blacks with fewer large aggregates. The aggregates in these blacks are smaller and more uniform in size with more open structures. This type of structure gives more *bound rubber* which in turn results in better strength and road wearing qualities in tyres.

Structure also affects the processability of the compound with high structures giving ease of incorporation, high compound viscosity and low die swell: the cured compound will have high modulus, high hardness and good wear resistance. That is, lower nerve and stiffer unvulcanized compounds will result as compared to the use of low structure (LS) carbon blacks. LS carbon blacks give low dynamic heat build up, high strengths (tensile and tear) and good crack growth resistance. Structure levels may be determined by an *oil adsorption method* or by the *DBP adsorption method*. (Oil extension potential of a compound increases with increasing structure.)

carbon black - effect on 300% modulus
For carbon black loaded compounds, 300% modulus is primarily a function of carbon black structure and loading: with compounds containing the highest structure blacks having the highest modulus. Increased loadings of reinforcing and semi-reinforcing blacks has a very significant effect on increasing 300% modulus. The carbon black particle size appears to have a secondary effect on modulus although in *butyl rubber* small particle size contributes to higher modulus at high loadings.

carbon black - pigments
A class or type of *organic pigment*. The most important black, organic pigment. Used because of low cost, blackness (jetness) and good colouring strength. Can be finely dispersed in polymers and may make the compound conductive. Another major reason for their use is because of the excellent light stability which *carbon black* can give to polymer compounds.

carbon dioxide content
The amount of carbon dioxide (CO_2) measured in g/l of a carbonated drink, for example, 8 g/l of carbon dioxide (CO_2) is possible when PET bottles are used.

carbon disulphide
A highly inflammable solvent. This material has a relative density (RD or SG) of 1·26 and a boiling point of approximately 47°C. It is a good solvent for uncured *chloroprene rubber (CR)*, *nitrile butadiene rubber* (low acrylonitrile NBR), *natural rubber (NR)* and, *styrene-butadiene rubber (SBR)*. This chemical causes some swelling of uncured high acrylonitrile NBR and thiokol rubbers. Will dissolve cellulose ethers and is used in the manufacture of viscose rayon: will swell plastics such as polyethylene (PE), *polyvinyl chloride (PVC)* and *polymethyl methacrylate (PMMA)*.

carbon fibre
An abbreviation used for this type of material is CF. Carbon fibre is also known as graphite fibre and can be obtained as black, silk-like threads with a diameter in the range of approximately 7 microns. Such materials are used to make composites from both thermoplastics and from thermosets. Long carbon fibres are usually used to make thermoset composites (often based on *epoxide resins*) and short fibres are used to make, for example, composites based on thermoplastics materials. Both types of material may be referred to as a carbon fibre reinforced plastic (CFRP or CF RP). The materials based on thermosets are often processed as *sheet moulding compounds* or as, *dough moulding compounds*. The materials based on *thermoplastics materials* are often processed by *injection moulding* although, for example, dough moulding offer advantages such as fibre length retention.

Continuous lengths of carbon fibre are preferred, as the final carbon fibre, can be woven into tape, cloth or chopped into shorter lengths: the chopped material may be used to make, for example, non-woven fabrics. By fibre alignment in a *preform*, a high performance, carbon fibre sheet moulding compound may be produced.

CF was originally produced from a *cellulose rayon fibre* but now, it is often produced from continuous *polyacrylonitrile (PAN) fibre*: by varying the heat treatment a range of fibres are produced, for example, to give high modulus (HM CF), high strength (HS CF), and high strain carbon fibre. Lower cost fibres, with a higher modulus but lower strength, may also be produced from pitch - see *mesophase pitch-based carbon fibre*. Carbon fibre may also be produced by *chemical vapour deposition* of hydrocarbons. To improve the bonding between CF and the polymer, *surface treatments* are often employed during the fibre production process.

CF is thermally conductive and electrically conductive along the fibre direction: the high electrical conductivity of such materials can result in corrosion of metal alloys if such alloys are joined by mechanical fastening to the CF material. See *fibre reinforced plastics* and *high performance sheet moulding compound*.

carbon fibre dough moulding compound
An abbreviation used for this type of material is CF DMC or DMC CF. A *dough moulding compound* which contains some, or all, *carbon fibre*. See *fibre reinforced plastics* and *high performance sheet moulding compound*.

carbon fibre reinforced plastic
An abbreviation used for this type of material is CFRP or CF RP. A composite material: either a thermoplastics material or a thermosetting material filled, or reinforced, with *carbon fibre*: may also include other fillers, for example, *glass fibre*. See *fibre reinforced plastics* and *high performance sheet moulding compound*.

carbon fibre reinforced thermoplastics material
An abbreviation used for this type of material is CFRTP or CF RTP. A composite material, based on a *thermoplastics* material filled, or reinforced, with *carbon fibre*: may also include other fillers, for example, *glass fibre*. See *fibre reinforced plastics* and *high performance sheet moulding compound*.

carbon fibre sheet moulding compound
An abbreviation used for this type of material is CF SMC or SMC CF. A *sheet moulding compound* which contains some, or all, *carbon fibre*. See *fibre reinforced plastics* and *high performance sheet moulding compound*.

carbon tetrachloride
A highly toxic solvent. This material has a relative density (RD or SG) of 1·63 and a boiling point of approximately 77°C. It is a good solvent for uncured *chloroprene rubber (CR)*, *butyl rubber*, *natural rubber (NR)* and, *styrene-butadiene rubber (SBR)*. It is a poor solvent for uncured high acrylonitrile NBR and *thiokol rubbers*. This chemical causes some swelling of uncured low acrylonitrile NBR. Will dissolve natural resins and *polystyrene* but its use is best avoided for toxicity reasons. Miscible with most organic solvents but immiscible with water.

carbonated soft drinks
An abbreviation used for this term is CSD. See *blow moulding* and *injection blow moulding*.

carbonic acid lime
See *calcium carbonate*.

carbonization
Also called coking. An *injection moulding* fault experienced with, for example, PBT where black specks in the mouldings indicate that degradation, or decomposition, has taken place. For example, within the channels of the *hot runner manifold*.

carbonization of coal
The heating of coal in the absence of air at temperatures of the order of 500 to 1,300°C. *Coal tar*, *coal gas*, *gas carbon*, *coke* and *ammoniacal liquor* are produced.

carbonyl chloride
See *phosgene*.

carbonyl group
May be represented as = CO.

carbonyl-group cross-linking
See *storage hardening*.

carbonylation reaction
See *OXO process*.

carborundum
See *silicon carbide*.

carbowax
A group of polyethylene glycols which may be represented as $HO(CH_2CH_2O)_xH$. They are of comparatively low molecular weight (below 20,000) and of low melting point (below 65°C). Used, for example, as mould lubricants and release agents.

carboxy-nitroso rubber
An abbreviation used for this type of material is AFMU: CNR is also used. Also known as carboxynitroso rubber. A terpolymer formed from trifluoronitrosomethane, tetrafluoroethylene and approximately 1% of nitrosoperfluorobutyric acid. May be vulcanized by using metal oxides or organometallic compounds, for example, chromium trifluoroacetate. Silica is used as filler and the products have excellent flame and chemical resistance. The use of a third monomer, nitrosoperfluorobutyric acid, gives easier curing than *nitroso rubber*.

carboxylated chloroprene rubber
An abbreviation used for this type of material is XCR. Also known as carboxylic-chloroprene rubber. See *carboxylated rubber*.

carboxylated latex
A *latex* based on a neutralized *ionomer*. Such a latex is used for dipping and coating applications, for example, for carpet backing, woollen treatments, adhesives and glove manufacture. See, for example, *carboxylated polybutadiene ionomer* and *mixed vulcanization*.

carboxylated nitrile rubber
May also be referred to as acrylic acid terpolymer NBR or as, carboxylic-nitrile rubber. An abbreviation used for this type of material is X-NBR or XNBR. A terpolymer of butadiene, acrylonitrile and a diene monomer with carboxylic acid groups (based on, for example, acrylic acid or methacrylic acid) which has been cross-linked ionically using zinc oxide. The ionically cross-linked compounds, exhibit higher abrasion strength, tensile strength and tear resistance but poorer compression set and scorch resistance than *nitrile rubber*.

The non-crosslinked material is very reactive as it can be *vulcanized* not only by metal oxides, such as *zinc oxide*, but also by *sulphur*. Because of filler interaction, such a material can be reinforced with non-black fillers.

carboxylated polybutadiene ionomer
An *ionomer* based on butadiene, acrylonitrile, styrene and, say, methacrylic acid (up to approximately 6% acid units) which when neutralized with, for example, zinc oxide, undergoes ionic cross-linking so as to produce a material which is effectively a vulcanizate. This is a carboxylated polybutadiene rubber: an *elastomeric ionomer*. The product neutralized with zinc oxide may be referred to as a zinc vulcanizate. A product neutralized with sodium hydroxide may be referred to as a sodium vulcanizate. The vulcanizates are high tensile strength materials which have poor compression set, high stress relaxation and poor high temperature resistance. Mainly used as carboxylated polybutadiene latex and used for dipping and coating applications, for example, for carpet backing, woollen treatments, adhesives and glove manufacture.

carboxylated polybutadiene latex
A *latex* based on an *ionomer*. See *carboxylated polybutadiene ionomer*.

carboxylated polybutadiene rubber
The neutralized product of *carboxylated polybutadiene ionomer*. In general, the vulcanizates are high tensile strength materials which have a poor compression set, high stress relaxation and poor high temperature resistance.

carboxylated rubber
Also known as a carboxylated elastomer. A rubber which contains a carboxyl comonomer, for example, acrylic or methacrylic acid, and which may be vulcanized with polyvalent metal oxides such as zinc oxide. The vulcanizates can have higher strengths than reinforced sulphur vulcanized non-carboxylated rubbers but the cross-links can be thermally labile. Because of *scorch problems*, carboxylated rubbers are often used in latex form as binders for non-woven fabric. See *carboxylated nitrile rubber*.

carboxylated styrene-butadiene rubber
An abbreviation used for this type of material is XSBR. Also known as carboxylic-styrene butadiene rubber. See *carboxylated rubber*.

carboxylic-butadiene rubber
See *carboxylated polybutadiene rubber*.

carboxylic-nitrile rubber
See *carboxylated nitrile rubber*.

carboxylic-styrene butadiene rubber
See *carboxylated styrene-butadiene rubber*.

carburising
A process for introducing carbon into the structure of steel, for example, *mild steel*. The components are heated and dropped into *carburising powder*. This gives a hard wear-resistant surface, a *case hardened surface* (which is useful when parts must slide or move over one another) and a tough core. Carburising depth may be 0.6 to 1 mm. Guide pins and bushes are treated in this way as the components must move one relative to the other.

carburising powder
A powder used in *carburising*. A carbon-containing compound, for example, containing 20% of a metallic carbonate and charcoal.

carcass
Also known as a casing. The rubber bonded structure of the *tyre* which is integral with the *bead*.

carding
A textile process used to bring order to an entangled mass of fibres: the entangled mass of fibres are worked between two surfaces (such as rollers) which are covered with sharp points.

carnauba wax
A hard wax which is obtained from the carnauba palm. This material has a relative density (RD or SG) of 0.99 and a melting point of 80 to 85°C. In the polymer industry it is used as a mould release. Used, for example, to ease mould sticking with fluoroelastomer rubbers (FKM).

carob seed flour
Also known as locust bean flour. Obtained from the fruits of seratonia sillqua and used a diluent and thickening agent for latex. See *creaming*.

carriage
Part of an *injection moulding machine*: the injection unit assembly which can be moved so as to make contact with the *sprue bush*. The term is also used for that part of the side core assembly, or side cavity assembly, which provides for the guiding and operating functions of the design.

carrier
Also known as a vehicle. That which is used to carry an additive. See *masterbatch*.

carrier resin
A material used as the base for a masterbatch. A low melting point, thermoplastic material is used as the base for a universal *masterbatch*.

carrot
See *sprue* and *sub-sprue*.

cartridge
A term used in *hydraulics* and which refers to the replaceable element of a filter for fluids or, the term refers to the pumping unit from a vane pump - the rotor, ring, vanes and one or more side plates. A component which can be removed and replaced easily without being fully dismantled.

cartridge heater
An electric *resistance heater* which is inserted into a hole in the workpiece: usually shaped in the form of a circular rod.

cartridge valve
A term used in *hydraulics* and which refers to a valve which performs directional, pressure or flow control functions. The valve is mounted in a *manifold* and often only has two positions, open or closed (or 'on' or 'off'). Because of this, hydraulic systems based on such valves may be referred to as *digital hydraulics*.

Such valves have brought big improvements to hydraulically powered machines as they operate faster (about five times faster), allow higher flows and close off more quickly than spool valves. Such valves are smaller and as they fit inside a manifold (about ¾ of the valve is inside a block) then, oil leaks are reduced. This makes for a tidier, safer moulding shop and saves on contamination. Such valves do not usually make for a closed loop system. However, their use results in very good reproducibility from shot-to-shot and from run-to-run. A moulding machine based on such valves, should be capable of producing mouldings whose weight varies by less than 0·2%.

CAS/CAP
An abbreviation used for computer aided setting/computer aided production.

cascade
Part of a *mixing* or *blending machine*. May consist of a series of flaps or baffles which divide and re-combine the granular materials as they fall through the cascade of a *mixing hopper*. The blended material is then deposited in the hopper of the moulding or extrusion machine. Melt mixing is then provided by the screw of the moulding or extrusion machine.

cascade control
A form of control in which the output of one controller is used to control, or feed, another.

cascade extruder
A type of *twin screw machine* in which, in effect, two extruders are connected so that the melt output from the first feeds the second. The two screws are connected by a passageway which can be vented so that devolatization is possible. This type of machine may also be classified as a vented extruder. It offers the advantage that each screw may be driven at a separate speed: valves may be used to optimize the output from each stage.

cascade extrusion
Extrusion performed with a *cascade extruder*.

case hardened
A steel which has been hardened by the introduction of carbon into the surface structure (of a low carbon steel). The wear resistance of, for example, mild steel can be significantly improved by *hardening* by, for example, *carburising*.

case hardening
Usually associated with metal technology where the term refers to the production of a hard surface on a softer core, for example, on a steel core pin. Can also be used to describe the appearance of *smoked sheet* which contains a central, thin white stripe: caused through incorrect drying, for example, at too high temperatures.

case hardening steel
A low carbon steel which when made into a component will harden on the outside when *case hardened* by, for example, *carburising*. A chrome-manganese steel (AISI-type P2) is a commonly used case hardening steel as it can be hobbed and also used for the manufacture of relatively large injection moulds.

case strapping
A highly oriented strip of extruded material: such strips are usually based on polyethylene or polypropylene. See *fibre tape*.

casein
A regenerated protein obtained from skimmed milk: caseinogen is converted to insoluble casein. For example, solid casein is obtained by acidifying to a pH of 4·6 (to give acid casein) or by the addition of rennin (an enzyme) - this product is sometimes called paracasein or rennet casein.

A white to yellowish powder which has been used as a stabilizer/thickener for latex. A latex/casein mixture may be used to improve the adhesion between rubber and fabric in the tyre and belting industries. SG of casein is approximately 1·25 to 1·30. Soluble forms of casein include ammonium caseinate and sodium caseinate. Casein has been used to make *fibres* and *plastics materials*. See *casein-formaldehyde*.

casein-formaldehyde
A plastics material prepared from *casein* and *formaldehyde*. An abbreviation used for this material is CF (CF has also been used for cresol-formaldehyde). Casein-formaldehyde plastics materials are usually produced by the *dry process*. The moist, powdered, *rennet casein* is extruded by means of a short screw extruder into, for example, rods. If required the material can be cut into button blanks or, shaped by further pressing into sheets. The product is then hardened with *formaldehyde* (formolising). At one time this material was widely used to make buttons as it is an attractive horn-like material. The *formolising process* can however take a very long time, for example, months for thick sections

caseinate
A derivative of *casein*.

caseinogen
See *casein*.

casing
See *carcass*.

cast acrylic sheet
See *cast polymethyl methacrylate sheet*.

cast film
A film made by depositing a layer of a thermoplastics material, either molten, in solution or, in dispersion, onto a surface and then solidifying and removing the film from the surface.

cast iron
Pig iron which has been remelted, mixed with *steel scrap* and cast into cool moulds: similar physical properties to *pig iron*.

cast nylon 6
Produced from Σ-caprolactam by *monomer casting*. Large products can be relatively easily made from this material, for example, propellers for ships.

cast PC
An abbreviation used for cast *polymer concrete*.

cast polymethyl methacrylate
An abbreviation used for this type of material is cast PMMA. Polymethyl methacrylate produced from *methyl methacrylate* by *monomer casting* within a mould. To reduce the amount of heat evolved during *polymerization*, some polymer is dissolved in the monomer and the casting temperature is regulated during polymerization. Sheet, rod and tube amy be produced in this way.

Sheet is, for example, produced from MMA by *monomer casting* in a glass plate assembly called a cell. As the monomer is cast against glass, a high gloss finish is obtained.

cast polyurethane elastomer
Also known as reaction moulded polyurethane elastomer or as, cast PU elastomer. A polyurethane elastomer made from a *prepolymer* and a *diol* or a *diamine*: the prepolymer is made from a *polyether polyol* or a *polyester polyol* and a *diisocyanate*. The ingredients are mixed at about 90°C, poured into moulds and then curing is completed in an oven. Such products are noted for their wear resistance and solvent resistance. Used, for example, to make printing rollers and to make cast tyres. See *reaction injection moulding*.

cast sheet
Sheet produced from a *monomer* by *monomer casting* within a mould. See *cast polymethyl methacrylate sheet*.

casting
The process used to produce *cast film* or to produce mouldings.

In *extrusion* technology refers to a flat film process whereby a film is cast into water or, against a water cooled roll.

Casting is a low pressure moulding process which usually employs a liquid feedstock (based on a monomer or on a monomer/polymer mixture) which is poured, or cast, into an appropriate mould. This liquid feedstock sets to the shape of the mould after the mould is filled as, for example, hardeners or catalysts, were previously incorporated into the liquid feedstock. Usually associated with thermosetting plastics (such as *epoxy resins* and *unsaturated polyester resins*) although both rubbery components and thermoplastics components can be produced by casting. As there is little, or no shaping pressure, the mould costs can be extremely low. See, for example, *liquid silicone rubber*.

castor oil
Derived from ricinus communis. A *dual purpose vegetable oil*. Castor oil is obtained from castor beans and consists mainly of glyceryl ricinoleate. First pressing castor oil is the medical grade and is obtained by cold mechanical crushing of the seeds. Commercial castor oil is obtained from the residue by extraction with solvents. The oil is used as a plasticizer/extender in cellulose derivatives and as a non-drying oil in alkyd resins.

A major use of this oil is as a drying oil. This oil, dehydrated castor oil, is formed by the action of dehydrating agents on the oil which introduce additional double bonds by removing water. Dehydrated castor oil has better drying properties than *linseed oil*: its water and alkali resistance is almost as good as *tung oil*.

Polyamides are derived from castor oil (see *nylon*). It is also used to produce *comminuted rubber* as the oil is incompatible with *natural rubber*. 0·3% of the oil may be added to *latex* before *coagulation* and removed by washing after coagulation. Spraying this oil onto coagulated rubber will also assist comminution. This material has a relative density (RD or SG) of 0·96.

catalyst
A catalyst is a substance which alters the rate at which a chemical reaction occurs but which is itself unchanged at the end of a reaction. In plastics technology, for example, the term may be used in connection with room temperature curing systems such as those used with *unsaturated polyester resin*. Unsaturated polyester resins are set by, for example, the action of a catalyst (a peroxide) and an accelerator. By the use of systems, such as *methyl ethyl ketone peroxide* and *cobalt naphthenate*, curing can be effected at room temperature: the catalyst is not however unchanged at the end of the reaction. See *copper*.

catalyzed polyester resin
A resin mix, based on *unsaturated polyester resin*, which has had the hardening ingredients added (*accelerator* and *catalyst*) and which will therefore set or cure.

cationic dye
See *basic dye*.

cationic exchanger
An *ionic polymer* whose matrix contains bound anions ($-SO_3^-$ or, $-CO_2^-$ or, $-PO_3^{2-}$) pendant to the main chain. Most ionic polymers are anionic ionic polymers.

cationic ionic polymer
An *ionic polymer* whose matrix contains bound cations (for example, $-CH_2N^+(CH_3)_3$) pendant to the main chain. Most ionic polymers are however *anionic*.

cationic surface active agents
Surface active agents are classified as being *anionic surface active agents* or as, cationic surface active agents or as, *non-ionic surface active agents*. Cationic surface active agents are typified by quaternary ammonium compounds in which the active groups are positively-charged quaternary ammonium ions.

cats' eyes
See *fish eyes*.

cauprene
A rubbery material obtained by heating polybutadiene dibromide with zinc dust.

caustic lime
See *calcium hydroxide*.

caveat
The name given to the hard rubber discovered by Nelson Goodyear. Based on a mix of *natural rubber* with 25 phr of sulphur and 75 phr of magnesia which is heated for approximately 6 h at 150°C.

cavitation
A term used in *hydraulics* and which means that the fluid has boiled inside the pump as a result of an inlet restriction. The pressure has been reduced to the vapour pressure.

cavity
The space in a mould which is used to shape the polymer compound: the shaped space, the female space, which gives a moulding outside dimensions.

cavity dimension
See *cavity sizing*.

cavity insert
A metal block which carries the *cavity*. The insert is carried by a *bolster* so as to form the cavity plate.

cavity plate
A metal plate, or block, which contains the *cavity*. The cavity may be sunk directly into the metal (integer cavity) or, it may be sunk into a *cavity insert* which, in turn, is carried by a *bolster*.

cavity pressure
This is the pressure that the plastics material exerts inside the mould cavity. This pressure tries to open the mould and also tries to distort the mould. In, for example, *injection moulding* it may be sensed by means of a transducer and the signal so produced used to actuate the switch from first stage pressure to second stage, holding pressure. This process is called *cavity pressure control* or *CPC*.

cavity pressure changeover control
More usually known as *cavity pressure control* or as *CPC control*.

cavity pressure control
Also known as cavity pressure changeover control but more usually referred to as CPC or as, CPC control. Term used in *injection moulding* to indicate that the final mould filling part of the moulding cycle is pressure controlled (by means of a pressure or of a force transducer). There are two main types of CPC control and these are classified according to transducer location: (i) within the injection mould and where the transducer is in direct contact with the melt and (ii) outside the injection mould.

However, in general, of the various *VPT options* currently available, CPC appears to be the most suitable as control is exercised from where it is required, that is, from a signal generated from within the mould: the pressure in the mould controls, for example, moulding shrinkage and component weight.

CPC has not made a great impression on the injection moulding industry because of the amount of mould modification required, setting difficulties, transducer calibration problems, ease of transducer damage and the high cost of transducer repair. Many moulders also believe that there has been a significant improvement in machine design and operation so that, there is no longer a need for such sophistication. However, even the accuracy of a modern machine, can be significantly improved by the use of a *VPT transfer* initiated by a pressure transducer.

cavity pressure control - direct sensing
A *CPC system* which uses a *direct pressure transducer*. In general, direct pressure sensing is best for *cavity pressure control* and should be selected wherever possible. However, the transducers are more expensive than the indirect type as they must be built to withstand melt temperatures and still give consistent readings over long production periods: a witness mark will be present on the component and in some cases, this will be objectionable.

cavity pressure control - indirect sensing
A *CPC system* which uses an indirect *pressure transducer*. Such an indirect sensor is used where there are space limitations within the mould, or hardened steel cavities have been used. The transducer is located in the mould ejector plate and force is transferred to it, by one of the ejector pins: the use of pins can sometimes cause errors as the pins can stick or bend.

The injection mould is modified so that it will accept the transducer. An accurately machined slot is made in the ejector plate behind an ejector pin in one of the mould cavities. The transducer is then carefully pushed into the slot and, connected to the controller which, in turn, is connected to the electrical switch which controls the change from high to low pressure.

During use, the pressure measuring system should be checked periodically so as to ensure that the preset values are being obtained. This often means simply using the built-in shunt calibration system. When this circuit is energized, the display should read 80% of full scale.

cavity pressure control - multi-cavity mould
With a balanced, multi-cavity mould, it is usually best to fit the transducer in a typical cavity - if they are all identical. If one cavity is causing the most problems then locate the transducer in that cavity. With an unbalanced, multi-cavity, mould it is best to measure the pressure within a cavity (or the runner system) located 50% to 75% along the melt flow path. If the filling pattern is unknown then the pressure should be measured near the sprue or within the machine nozzle.

cavity pressure control - setting
Select a critical, or peak, cavity pressure at which to change from *filling pressure* to *hold pressure*. This pressure is selected by observing what *critical cavity pressure* is associated with an acceptable moulding. A *switch pressure*, is then set on the controller and when this pressure is reached (during mould filling) the switch is made from high to low pressure. The highest cavity pressure is noted and if different from the desired cavity pressure then the switch pressure is adjusted. Some adjustment is demanded because the injection moulding machine is unable to provide instantaneous switchover as the pressure rise in the cavity is so rapid.

cavity pressure control - single cavity mould
Pressure is highest in the gate region and therefore locating the transducer in the gate region would therefore seem an obvious choice. However, such a location can often lead to the formation of *pressure spikes* and an erratic signal can be the result. For example, if one of the teeth of the spikes is occasionally higher than the switching pressure then, premature switching may be the result. For monitoring purposes, a gate location is good but it is not necessarily good for control purposes. Locating the pressure transducer away from the gate, towards the centre of the component, is generally the best location point.

cavity pressure control - transducer calibration
Ensure that the equipment has been calibrated to suit the size of ejector pin being used if an indirect type of transducer is being used. This is because the indirect type of transducer is a force transducer. Pressure (P) is force (F) per unit area (A). So, if the pressure is required then the area of pin in contact with the melt must be known. The force on the ejector pin equals the area of the ejector pin (A) multiplied by the cavity pressure. That is, $F = P \times A$. The area (A) = the diameter $(D)^2/4$. Or, $0.7855 \times D \times D$. The direct type of pressure transducer is a true pressure transducer and so such calculations, to establish which range of transducer to use, are not necessary.

cavity register
Matching, angled faces on a mould which give mould alignment when the mould is closed.

cavity sizing
May also be referred to as cavity dimensioning. The calculation of a cavity size so as to allow for *shrinkage*.

The *cavity dimensions* for a given mould (for example, an injection mould) can be calculated by using $D_c = D_p + D_p S + D_p S^2$. Where D_c is the cavity dimension, D_p is the dimensions of the moulding and S is the linear shrinkage. A simple step by step approach is recommended when specifying mould cavity dimensions. This approach is as follows:

A. List and number each dimension of the product;
B. Decide on the type and grade of thermoplastics material;
C. Obtain the shrinkage range for the material;
D. Calculate the maximum and minimum shrinkage values for each dimension of the product;
E. Determine the cavity size for each product dimension;
F. Specify the mould making tolerance for each cavity/core dimension.

Quoted *shrinkage values* should only be taken as a guide, for cavity sizing purposes, as in the case of injection moulding, for example, part thickness, cavity pressure and the time for which that pressure is applied, all markedly influence moulding shrinkage. Part geometry, and changes in flow path length, cause pressure differences in the mould and these in turn result in different shrinkage values in different directions. If the shrinkage is non-uniform then warping of the component may result.

CB
An abbreviation used for *carbon black*.

CBS
An abbreviation used for cyclohexylbenzothiazyl sulphenamide (N-cyclohexylbenzothiazyl-2-sulphenamide). See *accelerator*.

cc
An abbreviation used for *cubic centimetre*. See *Prefixes - SI*.

CC
An abbreviation used for conductive channel (black). See *carbon black - classification*.

cd
An abbreviation used for *candela*. The term cd/m^2 is used as an abbreviation for candela per square metre.

Cd-Ba
An abbreviation used for cadmium-barium. See *barium/cadmium stabilizers*.

CDP
An abbreviation used for *cresol diphenyl phosphate*. A triaryl phosphate.

CECC
In English, this abbreviation stands for the Electronics Components Committee of *CENELEC*. CECC is also a prefix used in British Standards to indicate *BS CECC*. CECC and IECQ are certification systems for electronics components in Europe and world wide. See *British Standard*.

CEFIC
An abbreviation used for *European Chemical Industry Federation*.

CEI
An abbreviation used for the organization *Commission Electrotechnique Internationale*. The French language version of International Electrotechnical Commission.

CEI hot wire test - see *hot wire test*.
CEI needle burner test - see *needle burner test*.
CEI tracking resistance - see *resistance to tracking*.

cell
A small cavity surrounded partially or completely by walls. Cells are introduced into polymers in order to make cellular materials.

cell - closed
A *cell* totally enclosed by its walls and hence not interconnecting with other cells.

cell nucleating agent
An additive used in cellular polymers to control cell size. An example is sodium bicarbonate.

cell - open
A *cell* not totally enclosed by its walls and hence interconnecting with other cells.

cell stabilizer
An additive used in cellular *polyurethane polymers* to control and stabilize the expanding material: also assists in additive dispersion. An example is a silicone-polyether block copolymer.

cellular
When applied to polymeric products this term means that the density of the material has been reduced by expansion. Means that the polymer contains gas bubbles. The term covers *expanded*, *foam* and *sponge*.

cellular rubber
Means that the rubber component polymer contains bubbles and/or is porous. Cellular rubber, with closed cells, is produced by exposing a rubber compound to a pressure of approximately 200 atmospheres of nitrogen. After vulcanization, the pressure is released and the dissolved nitrogen expands the compound and leaves a closed cell structure. A similar process has been used to produce cellular polyethylene - in this case the structure is open cell.

cellular wire-covering process
An *extrusion wire-covering process* which covers the wire with a *cellular polymer*.

cellulose
This naturally occurring material is the chief constituent of the cell walls of plants and is an essential constituent of wood, cotton, paper etc. The major commercial sources of this natural high polymer are cotton and wood: cellulose is very abundant in nature. For chemical purposes, *cotton linters* and *wood pulp* are the major sources.

Cellulose is a polysacharide of fibrous form which consists of linked rings. It is made more soluble by molecular weight reduction and by producing derivatives via the hydroxyl groups (three per ring). Such derivatives include *cellulose esters* (for example, cellulose acetate and cellulose acetate butyrate) cellulose ethers (for example, ethyl cellulose) fibres (regenerated cellulose fibres such as *rayon*) and film (cellophane-type materials). See *fibre*, *ginning saw-gin*, and *alpha-cellulose*.

cellulose acetate fibres
See *acetate fibres*.

cellulose acetate rayon
See *secondary cellulose acetate*.

cellulose plastics
See *cellulosic plastics*.

cellulose rayon carbon fibre
A *carbon fibre* produced from *rayon*. A continuous tow of the highly drawn textile fibre is carbonized while being held under tension to give *high strength (HS) carbon fibre*. If the material is stretched while being carbonized then the modulus of the product is increased giving high modulus carbon fibre. This high temperature treatment causes further orientation of the crystallites but is difficult and expensive to do. To minimize the use of very high temperature stretching, many types of carbon fibre are now *polyacrylonitrile carbon fibres* or *mesophase pitch-based carbon fibre*.

cellulose triacetate
Also called primary acetate or triacetate. Cellulose which has a degree of substitution of approximately 3, that is, the three hydroxyl groups on the *cellulose* have been reacted: more usually the degree of substitution is about 2·5. A fibre forming material which may be, for example, dry spun from dichlorethane solution. This material has a higher softening point than *secondary cellulose acetate*.

cellulose xanthate
See *sodium xanthate cellulose*.

cellulosic plastics
This term covers both cellulose esters and cellulose ethers. Cellulose esters are, commercially, the most important and include both inorganic and organic esters. It is the organic esters which are injection moulded (the inorganic one is cellulose nitrate (CN) which is never *injection moulded* as it is too dangerous). Organic esters include cellulose acetate (CA) which is also known as cellulose diacetate or, as secondary acetate or, as acetylcellulose or, as acetylcelluloid or, as cellulose ethanoate. Cellulose acetate butyrate (CAB) is also known as cellulose ethanoate butanoate. Cellulose acetate propionate (CAP) is known as cellulose ethanoate propanoate or, mistakenly as cellulose propionate (CP) or as cellulose propanoate.

Cellulose is a naturally occurring high polymer which cannot be melt processed (moulded) unless it is modified by a combination of the following: (i) chemical modification (esterification or etherification), (ii) molecular weight reduction and, (iii) plasticisation. The most commercially important route is by esterification, that is, making cellulose esters (e.g. CA) or, mixed esters (e.g. CAB).

Cellulose acetate (CA), is made when cellulose is esterified with acetic acid: it is the secondary acetate which forms the basis of CA injection moulding materials. The secondary acetate is dry blended, and then melt compounded with plasticizers (such as diethyl phthalate), extenders (such as castor oil), stabilisers (such as phenyl salicylate) and colouring systems: fillers are rarely used as the material is often selected for its clarity, toughness and attractive appearance. CA is the most commercially important cellulose derivative but, as with all cellulose plastics its usage is now relatively small.

Cellulose acetate butyrate (CAB) and cellulose acetate propionate (CAP) are mixed esters: these materials are prepared by methods similar to those used for making CA. That is, by reacting pretreated cellulose with appropriate esterification mixtures which are based on the acids and/or, the acid anhydride(s). In general, as these materials contain longer side chains (and therefore more hydrocarbon), they are not so water absorbent as CA; also have a lower density, a lower heat distortion temperature (HDT), are softer and of slightly easier flow. Permeability to water vapour is lower than CA and electrical properties similar. They are slightly lower in tensile strength than CA but have higher elongations at break and slightly higher impact strength. The propionates are intermediate between the acetates and the butyrates but have high flexural strength and impact strength: it does not have the rancid odour associated with CAB (this odour is worse at higher temperatures). CAB is however, more widely used. This material is used where a material which is tougher and more weather resistant than CA is required. It is specially formulated for outdoor use.

Alkaline cellulose is reacted with ethyl chloride to form the cellulose ether, ethyl cellulose (EC). This material is more alkali resistant than the cellulose esters but not as acid resistant. Not as clear as the esters but still available in many transparent, translucent and opaque colours.

In general, cellulosics are hard, stiff materials which can be compounded with plasticizers to improve their flow, or ease of processing, and to improve the flexibility of finished mouldings. They are extremely tough plastics and are reasonable electrical insulators but have high gas permeability and medium water vapour transmission rates.

By varying the ester, the type and the amount of *plasticiser*, many different grades of cellulosic plastics are made which differ in their flow behaviour (see *flow temperature*). Shrinkage is of the order of 0·003 to 0·010 mm/mm (0·3 to 1·0%) and post-moulding shrinkage is negligible. The addition of glass fibre will usually reduce the shrinkages values by approximately half

Cellulosics are resistant to aliphatic hydrocarbons (such as hexane), fatty oils and mineral oils. The esters are resistant to aliphatic hydrocarbons, petrol, detergents, oils and greases. CA has the best resistance to aromatic and chlorinated hydrocarbons (carbon tetrachloride resistant): such resistance is good when cold but deteriorates at elevated temperatures. CAB has excellent resistance to perspiration and EC is alkali resistant.

The esters are relatively un-affected by weak acids and bases (may be swollen on long contact) but are not resistant to strong acids and bases. Attacked by esters of the low molecular weight alcohols and by low molecular weight ketones and alcohols. The relatively high water absorption of these materials can cause dimensional changes: CA absorbs about twice as much as the others. These materials will dissolve in solvents such as acetone and trichloromethane, but will not dissolve in solvents such as carbon tetrachloride. CA is soluble in acetone and dioxane. The cellulose ether, ethyl cellulose, is more alkali resistant than the cellulose esters but not as acid resistant.

CA has the highest density while CAB and CAP are generally tougher and easier to process. CA is 1·26 to 1·30, CAB is 1·15 to 1·21, CAP is 1·18 to 1·23 and EC 1·09 to 1·17 gcm^{-3}.

The heat distortion temperatures (HDT) of the cellulosic materials can overlap but the highest HDT is available from CAP: this is followed by CAB and CA then by EC. Impact strength is roughly in the same order.

These materials will absorb large amounts of water. CA will absorb up to 6% (usually 4·5%) water in 24 hours, CAB up to 2·2%, CAP up to 2·8% and EC up to 2%. This means that drying is often necessary, before melt processing, but do not use drying temperatures of above 90°C and do not overdry. For example, dry in a desiccant drier for 1-2 hours at 85°C/122°F.

In general cellulosics are hard, tough and can be glass clear (not EC): they are available in beautiful colours. Because of their relatively high water absorption, and the dangers of plasticiser migration, very precise mouldings are not often produced in cellulosics. The type and amount of plasticiser affect the fire resistance and the food contact use.

CA is often selected for its clarity, toughness and attractive appearance. CA materials have good lightfastness and can transmit up to 90% of light; the refractive index ranges between 1·49 and 1·51. Such materials have a good gloss and are pleasant to handle as they are pleasing to the touch. They exhibit a self polishing effect in use and electrostatic build up is quickly dispersed (because of their water absorption). Stresses around inserts rapidly decay as the material creeps. Available in food contact grades. Used in toys, tool handles, writing equipment, transparent rigid containers, shields, lenses and electrical appliance housings. Because of the wide colour range available, it is particularly suitable for use in the world of fashion: used to make spectacle frames, buckles, earrings, hairdressing and toilet articles. Special effects such as wood grains, tortoiseshell, onyx, jade, and other minerals, can be relatively easily, injection moulded in CA. CA is also available in self extinguishing grades for crystal, transparent and opaque colours.

CAB is used for toys, pen and pencil barrels, decorative plaques, tool handles and machine guards. Because of its toughness and sweat resistance the material is used in steering wheels and suitcase handles. EVA-modified materials available, which are more temperature and weather resistant than plasticised compounds.

CAP is used in lighting fixtures, safety goggles, flash cubes and brush handles Can be even more weather and solvent (for example, benzene) resistant than CAB: used in high quality sunglasses, spectacle frames and ski goggles. Infrared absorbing grades available which are used for protective goggles and screens. EVA-modified materials available, which are still transparent, hydrolysis, impact and craze resistant, which have better strength, stiffness and creep resistance: they do not suffer from the same plasticiser loss problems as the parent materials.

Ethyl cellulose (EC) is made in heat resistant formulations, high impact formulations and in food contact grades. This material is flexible and tough - particularly at low temperatures. It is used in flashlight cases and for extinguisher components; also used where alkali resistance is required.

cellulosics
See *cellulosic plastics*.

Celsius scale
Originally known as the Centigrade scale but renamed in 1948 to honour A. Celsius who originated the scale in the 18th century. The scale is based on the freezing and boiling points of water with 100 degrees separating the two points. Each degree is known as a °C. This means that temperatures below the freezing point of water are negative which is objectionable from the point of view of a consideration of the meaning of *temperature*.

Celsius temperature
A temperature reading on the *Celsius scale*.

cement
See *solvent cement*.

CEN
An abbreviation used for the *Comité Européen de Normalisation*.

CEN standards
Test specifications or standards issued by the *Comité Européen de Normalisation*.

CENELEC
An English language translation for this abbreviation is the European Organization for Electrotechnical Standardization. See *Comité Européen de Normalisation*.

cenospheres
Absorbent spheres made from the fusion of smaller primary particles during polymer manufacture. Associated with the production of *polyvinyl chloride (PVC)* by *emulsion polymerization*: such particles absorb plasticizer very rapidly unlike the particles usually produced by emulsion polymerization.

centare
One square *metre*. 10·764 square feet.

centi
A prefix which means one hundredth. For example, a;

centigram is 0·01 gram;
centilitre is 0·01 litre;
centimetre is 0·01 metre.
centipoise is 0·01 poise. See *prefixes - SI*;

centigrade scale
A temperature scale now known as the *Celsius scale*.

centimetre
A non-SI unit which is permitted within the SI system. The abbreviation for centimetre is cm. It is 0·01 m or 0·3937 in. See *Système International d'Unité*.

centimetre-gram-second system
Also known as the CGS system or as the cgs system. A system of measurement which uses the *centimetre (cm)* as the basic unit of length, the basic unit of mass is the *gram (g)* and the basic unit of time is the *second (s)*. The derived unit of force is the dyne (dyn).

centipoise
An abbreviation used for this unit is cP: it is 0·01 *poise* or 10^{-3} newton second per square metre.

centistokes
An abbreviation used for this unit is cSt: 0·01 stokes. 1 cSt is $10^{-6} m^2 s^{-1}$.

central feed machine
A hydraulic *injection moulding machine* which does not contain its own power pack: the necessary hydraulic fluid is drawn from an external hydraulic line.

central limit theory
A statistical concept which means that the distribution of averages tends to follow the normal distribution curve and used, for example, to ensure that a normal distribution curve is followed when performing a sampling run. The output should be divided into small lots so that averages (of say, four or five individual measurements) are used for *quality control purposes*.

centrally drilled roll
A roll which has one fluid circulation channel centrally drilled along its longitudinal axis. The steam, or steam/water mixture, is introduced into the roll via a spray pipe which lies in the central channel. Used to heat a *two roll mill*. See *peripheral drilled rolls*.

centre plate
One of the three plates of a *three-plate mould*: the central plate which is linked to the two other main plates.

centre-core winder
A winding, or *reeling*, system for flexible sheet in which the centre of the core, on which the sheet is wound, is driven. With centre-core winding it is possible to reel under constant tension if hydraulic motors are used (see *surface winder*).

centre-line-average
An abbreviation used for this term is CLA. See *arithmetical mean deviation*.

centrifugal casting
A *casting process* used to form hollow cylindrical components by high speed rotation of a tubular mould. A monomer, a prepolymer or a polymer dispersion may be used as the polymer feedstock: this is fused by, for example, heating during rotation.

centrifugal moulding
A *casting process* used to form hollow cylindrical components by high speed rotation of a tubular mould. A thermoplastics powder is used as the polymer feedstock: this is fused by heating during rotation and subsequent cooling.

centrifuging
A method of producing *concentrated latex*: may be called the Utermark process. The particles of rubber in *natural rubber latex* will separate, or *cream*, on standing as they have a lower relative density (RD) than water. The particles of rubber have a RD of 0·91 and the *serum* has a RD of 1·02. By centrifuging, continuous separation is possible into two fractions. The cream fraction has a rubber concentration of approximately 60% and the skim fraction has a concentration of approximately 5% rubber: this may be recovered by coagulation. See *concentrated latex*.

ceramic-matrix composite
An abbreviation used for this type of material is CMC. A *composite material* in which the continuous phase is based on a ceramic: reinforcement is provided by *fibres* or by *whiskers*.

ceresin wax
Also called hard paraffin or purified mineral wax. A paraffin which contains alicyclic structures is sometimes also called ceresin or ceresine: an isoparaffin which contains *alicyclic structures* is sometimes called isoceresin or isoceresine. They are roughly similar and difficult to separate. A hard brittle wax: a *paraffin wax* with a melting point of approximately 70 to 100°C, a relative density (RD or SG) of approximately 0·93: it is used as a substitute for beeswax. May be obtained from *ozocerite* by refining with sulphuric acid. Such materials are sometimes used as *ozone protection waxes in rubber compounds*.

ceresine
An alternative way of spelling ceresin. See *ceresin wax*.

cerussa
See *basic lead carbonate*.

cetyl trimethylammonium bromide absorption
An abbreviation used for this term is CTAB absorption. A liquid absorption method, using cetyl trimethylammonium bromide (CTAB), which assesses the filler area (*carbon black*) which is available to the rubber. CTAB may also be referred to as bromine hexadecyltrimethylammonia.

This method and the *BET value* may rank or assess carbon blacks in the same order as the blacks used for rubbers are relatively non-porous. CTAB absorption is in some ways better than iodine absorption as this method is not so affected by factors which limit the iodine absorption method. Factors such as surface oxidation and absorbed hydrocarbons.

CF
An abbreviation used for conductive furnace (black). See *carbon black - classification*. CF is also used as an abbreviation for *casein formaldehyde*, *cresol formaldehyde* and for *carbon fibre*.

CF DMC = carbon fibre dough moulding compound;
CF HS = high strength carbon fibre;
CF HM = high modulus carbon fibre;
CFRTP = carbon fibre reinforced thermoplastics material;
CFRP = carbon fibre reinforced plastics; and
CF SMC = carbon fibre sheet moulding compound. See *fibre reinforced plastics* and *high performance sheet moulding compound*.

CFC
An abbreviation used for chlorofluorocarbon. For example:

CFC 11 = chlorofluorocarbon 11 = Freon 11;
CFC 12 = chlorofluorocarbon 12 = Freon 12;
CFC free = chlorofluorocarbon free (see *aerosol*);
CFC reclamation = chlorofluorocarbon reclamation; and,
CFC replacement = chlorofluorocarbon replacement.

See *aerosol*, *chlorofluorocarbon* and *reclamation*.

cfm
An abbreviation used for cubic foot per minute.

CFM
An abbreviation used for *polychlorotrifluorethylene rubber*. An abbreviation for *continuous filament mat*.

CFRP
An abbreviation used for *carbon fibre reinforced plastics*. See *fibre reinforced plastics*.

cgs system
An abbreviation used for *centimetre-gram-second system*.

cgs electromagnetic system = centimetre-gram-second electromagnetic system. See *units - cgs electromagnetic*.
cgs electrostatic system = centimetre-gram-second electrostatic system. See *units - cgs electrostatic*.

CGS system
An abbreviation used for *centimetre-gram-second system of measurement*.

CH
Letters used in *plasticizer* abbreviations for *cyclohexyl*. For example, butyl cyclohexyl phthalate (BCHP). See *phthalate*.

chafer
A *tyre component*: that outer part of the bead which protects against rim chafing.

chain
A unit of measurement which contains 100 links and which is 22 yards. See *UK system of units*.

chain breaking antioxidant
An additive, an *antioxidant*, which functions by interrupting the oxidative chain reaction which is causing decomposition, or degradation. Chain breaking antioxidants are amines and phenols (hindered phenols such as 2,6-di-t-butyl-4-methylphenol). The most important type of antioxidant of this class is an *AH antioxidant*. See *oxidation - mechanism*.

chain extender
A material which causes *chain extension*. For example, in *polyurethane* technology, a *diol* (such as *ethylene glycol*) may be considered as a diol chain extender.

chain extension
Increasing the molecular weight of a polymer by, for example, lengthening the molecular chain. See *prepolymer process*.

chain propagation
See *propagation*.

chain structure
The structure of a chain polymer. Long chain polymers are made from monomers by *polymerization*. For example, *polyethylene (PE)* is made from the monomer *ethylene*. Ethylene is a simple, low molecular weight (mass) material which has a molecular weight of 28. When the gaseous ethylene is joined together (polymerized) to give a long chain structure, the molecular weight is increased dramatically. If a thousand ethylene molecules were joined together, to make, for example, a film grade of PE, the molecular weight would be 28,000. However, not all the chains would be of the same size: some would be larger than the average value and some would be smaller than the average. Some, for example, may only contain 20 ethylene units and others may contain a hundred thousand ethylene units. What is also important is how the ethylene units are joined together: for example, if two ethylene units add onto another growing chain at the same point, then chain branching will be the result. This means that polyethylene is not a simple, long chain material but one which contains a large number of short and long side branches. These branches effect how the material flows and how the molecules pack together (crystallize) on cooling.

chalking
A product defect which is caused by surface degradation of the plastics material and which exposes the pigment; can be confused with *blooming*.

chamber
A term used in *hydraulics* and which refers to a compartment within a hydraulic unit.

change of phase
The conversion of a material from one of the physical states into another. For example, he conversion of a material from a liquid to a gas.

change of state
See *change of phase*.

channel black
See *carbon black*.

channel blow
A technique used in *blow moulding* in which a sacrificial channel is used to direct the air flow.

Chardonnet silk
Fibres made from cellulose nitrate solutions: the first commercial artificial fibre. See *rayon*.

charge
A term used in *hydraulics* and which refers, for example, to the filling of an *accumulator* with fluid under pressure. May also be called supercharge.

charge pressure
The pressure at which fluid is forced into a hydraulic system. See *charge*.

Charpy impact test
A *pendulum impact test - Charpy*.

Charpy test specimen
A specimen used in a Charpy impact test. See *pendulum impact test - Charpy*.

chase
See *bolster*.

chattering
Also called double spew. A rubber moulding defect experienced with components of thick cross-section: the outside cures before the inside has completed its thermal expansion and on mould opening the component has a split, or gouged out, appearance.

CHC
An abbreviation once used for epichlorhydrin copolymer rubber. See *epichlorhydrin rubber*.

CHDM
An abbreviation used for *cyclohexanedimethanol*.

check ring
A non-return valve on the tip of the screw (a *check valve*) which prevents melt loss during injection moulding as the ring seals against a seat on the end of the screw.

check valve
A term used in *hydraulics* and which refers to a valve which allows flow in one direction only. A valve used on the tip of a *screw* to prevent melt or gas loss (see *check ring*). For example, on an *injection moulding machine* a common type of check valve is a *ring check valve*.

cheek plate
Part of a *two-roll mill*: two are mounted above each roll of the mill, one at each end and the distance between them is capable of being adjusted. They are so shaped, and overlapped, that they contain the mix/compound between themselves and the rolls. The effective width of the rolls is controlled by the setting of the cheek plates.

cheese
A cylindrical package of *yarn*.

chelating agent
A chelator, or *metal deactivator*, is an additive which will sequester, or lock up, unwanted metal ions. Some metal ions can promote decomposition, or degradation, of polymers, for example, copper promotes the decomposition, or degradation, of polypropylene. See *antioxidant*.

chelator
A *metal deactivator*. See *chelating agent*.

chemical blowing agent
See *blowing agent*.

chemical bonding agent
Usually used in connection with fillers or with fibres. The chemical bonding agent is used in an effort to improve the properties of the composite. Such coupling agents contain a chemical group which will react with a group on the filler/fibre surface and another group which will react with, or dissolve in, the polymer. See *glass fibre*.

chemical recycling
Thermal degradation of polymers into a lower molecular weight product, for example, an oil or grease which may be suitable for use as a refinery feedstock. A major problem is halogen contamination of such a feedstock (from the use of *polyvinyl chloride*). See *degradative extrusion*.

chemical resistant silicone rubbers
See *oil resistant silicone rubbers*.

chemical rubber
A term sometimes used for *chlorosulphonated polyethylene*.

chemical unsaturation
See *unsaturated compound*.

chemical vapour deposition
An abbreviation used for this term is CVD. A deposition process used to produce thin films and fibres from gases or vapours. For example, synthetic diamond in *film* and *filament* form are grown from hydrocarbon vapour. Small area surfaces can be coated directly and thin, self-supporting film can be produced. Typical of applications to benefit from the ability to produce very hard, scratch resistant coatings are optical systems (lenses, windows and prisms), cutting tools and semiconductor devices. The surfaces of cutting tools have been coated with layers which may be up to 10 microns thick.

Carbon fibre may be produced by, for example, chemical vapour deposition of hydrocarbons (methane, benzene or naphthalene) onto a filament produced by catalytic growth around a sub-microscopic iron particle. Pyrolytically deposited carbon fibre is the most graphitizable fibre known and when heat treated to 3,000°C the electrical properties approach those of *highly oriented pyrolytic graphite*. Fibres of up to 50 mm in length have been made.

chemically cross-linked polyethylene
A *cross-linked polyethylene* produced by the action of a chemical additive (for example, a *peroxide* on polyethylene), which produces free radicals. See *cross-linked polyethylene*.

chemically crumbed rubber
See *Heveacrumb process*.

chemically foamed, polymeric material
A cellular, polymeric material in which the cells are formed by gases generated from thermal decomposition of a chemical *blowing agent*.

chemically loaded, molecular sieve
See *molecular sieve accelerator*.

chemically pure cadmiums
An abbreviation used for this group of material is CP. A group of inorganic pigments which are based on calcined cadmium sulphides and selenides in various proportions. Because of their good temperature resistance, such materials are still used for engineering thermoplastics materials. See *cadmium lithopones*.

chemically softened natural rubber
See *softened rubber*.

chemically-bound sulphur
Sulphur which reacted with rubber so as to cause *vulcanization*.

chert
A form of *silica* which is white or off-white (for example, tan or gray) and which is probably biochemical in origin (see *diatomaceous earth*). Flint, jasper, fossiliferous chert, oolite chert, novaculite, porcelanite and tripoli are all forms of chert. Tripoli is, for example, used as a filler.

cheval vapeur
Equivalent to 0·736 kilowatts or 735·499 watts. Equivalent to one *pferde-stärke (PS)* or one metric horsepower.

chill roll
The relatively cold roll against which a plastics film is cast after being extruded from a slot die; the chill roll temperature may be relatively high, for example, 80°C/175°F. See *chill roll casting*.

chill roll casting
A *flat film process* for the production of *plastics film*. An *extrusion process* in which a thermoplastics film is extruded from a slot die and fed directly onto the chill roll which is cold compared to the melt temperature. It is important that the extruded web does not tear on stretching: that is, it is capable of high *draw down*. One phenomenon associated with chill roll casting is that of *neck-in*. It is found that more elastic melts, which can maintain a tension in the extrusion direction, are less liable to neck-in.

When a water-bath is used for cooling, carry-over of water can be a problem, and hence the chill-roll process, which involves no direct water contact with the film, is usually preferred commercially.

chill roll process
See *chill roll casting*.

chiller

A refrigeration system: part of an *indirect cooling system* used to keep a mould at temperatures below ambient. Many moulds are operated using chilled cooling water or water/anti-freeze mixtures in an effort to improve productivity or to remedy poor mould designs. Within such chillers there is a mechanical refrigeration unit which cools the circulating fluid by passing it through a heat exchanger. The heat exchanger is of the shell-and-tube type, and a liquid (formed from a compressed cooled gas) is circulated through the tubes whilst the mould circulation fluid is passed through the shell. Heat is exchanged and as a result cold fluid is returned to the mould and the liquid returns to the gaseous state; the gas is then compressed and cooled so that the heat exchange may continue.

Water/glycol (anti-freeze) mixtures are often preferred for low temperature work. When chillers are employed it is often found that the lowest temperature possible from the unit cannot be used because water condenses on the mould surfaces when the mould is open - either when the mould is open for normal ejection or when the machine has been stopped so that the mould may be cleared of an obstruction.

china clay

An alternative name is kaolin. Hydrated aluminium silicates of formula $Al_2O_3 2SiO_2 2H_2O$. This material has a relative density (RD or SG) of 2·5. A very widely used filler particularly in the rubber industry for general mechanical components. Fine particle grades can be reinforcing in rubber compounds. This filler can absorb accelerators, for example, MBS and MBTS, and due allowance for this, should be made when formulating with this filler. The finest white grades are also used as fillers in plastics materials such as *polyvinyl chloride (PVC)*. In liquid plastics materials, such as *unsaturated polyesters* and *epoxides*, china clay may be used to control viscosity.

Chinese blue

See *Prussian blue*.

Chinese white

See *zinc oxide*.

Chinese wood oil

See *tung oil*.

chisel gate

A type of *submarine gate* in which the shape of the gate resembles the edge of a chisel. It is sometimes used on large components where automatic separation from the runner system is required.

chlorendic acid

An abbreviation used is HET acid. Also known as hexachloro-endomethylene-tetrahydrophthalic acid. Used as a fire retardant comonomer for *unsaturated polyesters* because of the high chlorine content. However, the resulting resins can have inferior heat and light resistance compared to more conventional materials.

chlorendic anhydride

An abbreviation used is HET anhydride. Also known as hexachloro-endomethylene-tetrahydrophthalic anhydride. Used as a fire retardant comonomer for *unsaturated polyesters* because of the high chlorine content. However, the resulting resins can have inferior heat and light resistance compared to more conventional materials.

chlorethene

See *vinyl chloride*.

chlorinated biphenyl

See *chlorinated diphenyl*.

chlorinated butyl rubber

A *halogenated butyl rubber*. This material may contain approximately 1% of chlorine. Vulcanizes more readily than non-chlorinated *butyl rubber* and with a wider range of materials, for example, with *zinc oxide*.

chlorinated diphenyl

A mixture of chlorinated diphenyl and terphenyl compounds which can be solids or liquids. This type of material has a relative density (RD or SG) of from 1·2 to 1·7: the RD depends on, for example, the chlorine content. This may range from 20 to 60% chlorine. The higher the chlorine content the better is the compatibility with *polyvinyl chloride (PVC)*. Used as *plasticizers* for PVC as such materials impart fire resistance. Also used as plasticizers for *cellulose esters* and for rubbers.

chlorinated hydrocarbon

A *hydrocarbon* which contains chlorine as part of its molecular structure. Common examples include carbon tetrachloride, chlorobenzene, chloroform and trichloroethylene.

chlorinated hydrocarbon wax

Also known as chlorinated paraffin or as chloroparaffin. A chlorinated paraffin wax can be either liquid or solid and can contain from 40 to 70% chlorine: the higher the chlorine content, the more solid is the material and the better is the compatibility with PVC. Used as softeners/plasticizers for plastics material, such as *polyvinyl chloride (PVC)*, as such materials impart fire resistance. Also used with rubbers to give flame and weatherproof impregnations: in such cases, use in conjunction with *antimony trioxide*.

chlorinated natural rubber

Also known as rubber chloride or as, chlorinated rubber. The reaction product of the halogen, *chlorine* and *natural rubber*. During chlorination, substitution and cyclization reactions occur until finally about 65% chlorine may be incorporated. A pale cream thermoplastics material which is non-inflammable and chemically resistant. This material has a relative density (RD or SG) of approximately 1·64. Has been used for chemical coatings and for flame resistant coatings where it is applied from solution. Such materials have been largely replaced by *polychloroprene-based materials*.

chlorinated paraffin wax

See *chlorinated hydrocarbon wax*.

chlorinated poly(vinyl chloride)

See *chlorinated polyvinyl chloride*.

chlorinated poly(vinyl chloride) plastics

An abbreviation used for this type of material is CPVC or PVC-C. Plastics based on chlorinated poly(vinyl chloride) in which the chlorinated poly(vinyl chloride) is in the greatest amount by weight. See *chlorinated polyvinyl chloride*.

chlorinated polyether

Also known as oxetane polymer or as, poly-(3,3-bis-(chloromethyl)oxacyclobutane or as, poly-(3,3-bis-(chloromethyl)-oxatane (a trade name is Penton). Prepared from 3,3-bis-(chloromethyl)oxacyclobutane. This thermoplastics material is known for its very good chemical and solvent resistance and, for this reason, has been used as a protective coating material in, for example, the chemical industry.

chlorinated polyethylene

Polyethylene (PE) which has been reacted with chlorine: both low density and high density PE have been used. Also known as chloro-polyethylene. An abbreviation used for this material is CM: sometimes CPE or PE-C is used. This material has a relative density (RD or SG) of approximately 1·06 at 25% chlorine content: the RD depends on the chlorine content.

The introduction of chlorine atoms initially makes the PE rubbery as it destroys crystallinity if this introduction is random, for example, if chlorination is done in solution. A chlorine content of approximately 25% gives a rubbery polymer which on precipitation may be obtained in a powder form. This type of material may be compounded with *carbon black*, oil, a heat stabiliser (for example, organometallic stabiliser) and a curing system (for example, a peroxide) to give cured rubbers possess which possess good resistance to oil, heat, flame, ozone and to weathering. The presence of chlorine give fire resistance but causes problems when the rubber does burn. Compared to *chloroprene rubber (CR)* this material is cheaper and because of this, consumption of CR is decreasing.

Thermoplastic chlorinated polyethylene is used as an additive for other polymers, for example, it is used as an *impact modifier* for polyvinyl chloride (PVC). The amount of chlorine introduced is chosen so as to give a material which is semi-compatible with the PVC.

chlorinated polypropylene
Polypropylene (PP) which has been reacted with chlorine: this type of material has good resistance to heat, light, and to chemicals. One suggested use is as a paint carrier or vehicle; another is as a component of printing inks.

chlorinated polyvinyl chloride
An abbreviation used for this type of material is CPVC or, PVC-C. Also known as chlorinated poly(vinyl chloride). The plastics material is based on *polyvinyl chloride (PVC)* which has been post-chlorinated (reacted with chlorine) so as to give a chlorine content of approximately 67%. Post chlorination is used to improve the heat resistance of the PVC: the softening temperature is increased to approximately 100°C. This type of material has been suggested for use as a piping material which is exposed to hot water. See *chlorinated poly(vinyl chloride) plastics*.

chlorinated rubber
The reaction product of the halogen, *chlorine* and a *rubber*. See *chlorinated natural rubber*. Chlorinated rubbers are included in the category of *speciality rubber* because of their good oil and temperature resistance.

chlorination
The process whereby chlorine is introduced into a polymer structure. Polymers may be reacted with chlorine, for example, so as to improve the fire resistance. The introduction of chlorine will also reduce crystallinity and will probably increase the density. See, for example, *chlorinated polyethylene*. The introduction of chlorine into a rubber usually means that the rubber vulcanizes more readily than the non-chlorinated material and with a wider range of materials. See *chlorinated natural rubber*.

chlorine
The elements of Group V11b of the Periodic table are known as halogens and consist of fluorine, chlorine, bromine, iodine and astatine. Chlorine does not occur naturally as it is too reactive: it occurs as chlorides, for example, as sodium chloride from which it is extracted by electrolysis. It is a very reactive element and is a greenish-yellow diatomic gas (Cl_2) which is slightly deeper in colour than fluorine and with a similar odour. Melts at -102°C and boils at -34°C. Used, for example, in the plastics industry to make *polyvinyl chloride (PVC)* and in the rubber industry to make polychloroprene. Chlorine compounds are also used as *flame retardants*. All halogens, except iodine, should be treated with great care as they are either highly poisonous and/or very irritating (iodine vapour is irritating and poisonous). See *chlorination* and *chlorine monoxide*.

chlorine flame retardant
A *flame retardant* which contains *chlorine*, for example, an organo-chloro compound. Bromine-containing compounds tend to be more powerful than chlorine-containing compounds: often used in conjunction with *antimony trioxide*.

chlorine monoxide
A brown gas of formula Cl_2O. This material has a boiling point of 3°C and a melting point of -12°C. An atmospheric pollutant. See *stratospheric chlorine*.

chlorine-containing rubber
A rubber which contains the element chlorine as a part of the rubber molecule. The chlorine may be incorporated into the monomer (*chloroprene rubber*) or put in after polymerization, for example, *chlorinated polyethylene*. The presence of this element gives fire resistance but causes problems when the rubber does burn. See *chlorinated rubber*.

chloro-isobutene-isoprene rubber
See *halogenated butyl rubber*.

chloro-polyethylene
See *chlorinated polyethylene*.

chlorobenzene
Also known as benzene chloride or as, monochlorobenzene or as, phenyl chloride. May be represented as C_6H_5Cl. This material has a relative density (RD or SG) of 1.10 and a boiling point of approximately 132°C. It is a good solvent for uncured *chloroprene rubber (CR)*, *nitrile-butadiene rubber (NBR)*, *natural rubber (NR)* and *styrene-butadiene rubber (SBR)*. This chemical causes some swelling of *butyl rubber (IIR)* and *thiokol (T)* rubbers.

chlorobutyl rubber
See *halogenated butyl rubber*.

chloroethylene
See *ethylene chloride*.

chlorofluorocarbon
An abbreviation used for this type of material is CFC. A carbon-based compound which contains chlorine and fluorine atoms. The first, commercially-produced CFC was Freon 12 which is dichlorodifluoromethane. It appeared to be the perfect replacement for ammonia, the toxic and volatile refrigerant in use when this CFC was developed (developed by Thomas Midgley in 1930). Because of the non-toxic, non-flammable nature of such materials, and their stability, they were developed for other applications. CFCs, such as Freon 11 and Freon 12, are used in three main areas (a) refrigeration and air conditioning (b) blowing of *polyurethane foams* (particularly rigid foams) and, (c) as *aerosol propellants*. CFCs are also used as cleaners in the electronics industry (see *terpenes*).

Now known that *stratospheric ozone* is depleted by *stratospheric chlorine* which depends on, for example, CFC emissions. Chlorofluorocarbons are *greenhouse gases* which account for approximately 25% of the global warming effect. Freon 11 is given a *global warming* potential of 1. Because of the dangers posed by CFC use, there is great commercial interest in replacing such materials with substances which have less *ozone depletion potential*. See, for example, *hydrochlorofluorocarbon* and *pentane*. See *reclamation*.

chlorofluorocarbon plastics
Plastics based on polymers made with monomers composed of chlorine, fluorine and carbon only. See for example, *polychlorotrifluoroethylene*.

chlorofluorohydrocarbon plastics
Plastics based on polymers made with monomers composed of chlorine, fluorine, hydrogen and carbon only. See *polychlorotrifluoroethylene*.

chloroform
A chlorinated hydrocarbon which is also known as trichlormethane and which has the formula of $CHCl_3$. A colourless, sweet-smelling liquid which is a good solvent for a wide range of polymers but which is very toxic. This material has a relative density (RD or SG) of 1·5 and a boiling point of approximately 61°C. It is a good solvent for uncured *nitrile butadiene rubber (NBR) chloroprene rubber (CR), natural rubber (NR), styrene-butadiene rubber (SBR)* and *thiokol rubber (T)*. Of more limited use as a solvent for *butyl rubber (IIR)*. Will also dissolve *cellulose esters* and *polymethyl methacrylate*. The starting material for the production of *polytetrafluoroethylene*.

chloroform extract
That which is extracted by chloroform from a polymer compound. For example, after extraction with acetone, rubber is extracted with chloroform to detect the presence of bituminous-type materials

chloromethyl oxirane
See *epichlorhydrin*.

chloromethyl oxirane copolymer rubber
An abbreviation used for this type of material is ECO. See *epichlorhydrin rubber*.

chloromethyl oxirane rubber
An abbreviation used for this type of material is CO. See *epichlorhydrin rubber*.

chloromethyl oxirane terpolymer rubber
An abbreviation used for this type of material is AECO. See *epichlorhydrin rubber*.

chloroparaffin
See *chlorinated hydrocarbon wax*.

chloropolyethylene
An abbreviation used for this type of material is CM. See *chlorinated polyethylene*.

chloroprene
Also known as 2-chloro-1,3-butadiene. May be made by the addition of hydrochloric acid to vinyl acetylene. The monomer for *chloroprene rubber* by *diene polymerization*.

chloroprene acrylonitrile rubber
Also known as chloroprene acrylonitrile copolymer. An abbreviation used for this type of material is NCR. A *chloroprene copolymer rubber*.

chloroprene copolymer rubber
A rubber based on a copolymer of *chloroprene*. For example, with acrylonitrile or with, sulphur or with, 2,3-dichloro-1,3-butadiene. Such rubbers are produced in order to reduce crystallization and/or, to improve low temperature properties and/or, to reduce gel levels when homopolymers of *chloroprene* are made. If the copolymer is capable of being peptized then it may be referred to as a peptizable chloroprene copolymer rubber, for example, chloroprene sulphur rubber.

Both chloroprene homopolymer rubbers and chloroprene copolymer rubbers are made. Compared to a homopolymer it is found that, in general, a peptizable copolymer will give better tear strength, flex cracking resistance, hot strength, better adhesion to other rubber surfaces, higher modulus and higher hardness. The homopolymer has better tensile strength, lower compression set and better heat resistance.

chloroprene dichlorobutadiene rubber
A *chloroprene copolymer rubber* which incorporates 2,3-dichloro-1,3-butadiene.

chloroprene rubber
This is better known as neoprene, or as neoprene rubber. May also be known as polychloroprene or as polychloroprene rubber or as, polychlorobutadiene or as, poly-(2-chloro-1,3-butadiene) or as, poly-(1-chloro-1-butenylene). An abbreviation used for this type of material is CR (from chloroprene rubber). This material (CR) was the world's first commercial synthetic organic rubber and is probably better known by the trade name/trademark, Neoprene.

Most CR is predominantly trans-1,4-polychloroprene: this is obtained by emulsion polymerization using free radical polymerization. Lowering the polymerization temperatures increases the proportion of the trans isomer. Low polymerization temperatures are used to produce grades of CR useful in adhesives: such a quick grab CR has a high cohesive strength without curing.

Both chloroprene homopolymer rubbers and *chloroprene copolymer rubbers* are made. In general, mouldings made from this type of material are highly resistant to oxidation and can be compounded (e.g. with *carbon black* and *antioxidants*) so that they exhibit good long-term outdoor weathering properties. The material is also resistant to ozone cracking and CR mouldings have a high level of resistance to flex-cracking. Mouldings also exhibit good resistance to aliphatic (paraffinic) oils and solvents: because the material contains a large amount of chlorine in its structure, it has very good flame resistance. However, mouldings made from this material lose their flexibility relatively quickly as the temperature is lowered and it is also found that the raw polymer, and compounds based on this polymer, harden during storage due to crystallisation.

The polymer can be softened by mastication but the rate of change is slower than that found with *natural rubber*. The material has a tendency to stick to metal surfaces (this is characteristic of chlorinated rubbers) and in general compounds based on this polymer tend to be scorchy, particularly those based on a *peptizable copolymer*, during processing.

Vulcanization is based on a non-sulphur system, for example, for a sulphur containing copolymer it is based on *zinc oxide* and *magnesium oxide*. For a homopolymer, vulcanization is based on *zinc oxide* (5 phr), *magnesium oxide* (4 phr), *accelerator* (1 phr of ethylene thiourea) and *retarder* (0·75 phr of TMTD). Antioxidants and antiozonants are also used to improve vulcanizate performance.

Consumption of chloroprene rubber is decreasing because of the availability of cheaper chlorine-containing rubbers (for example, *chlorinated polyethylene*) and also because it contains chlorine. The presence of this element gives fire resistance but causes problems when the rubber does burn. Typical applications include joint seals, bridge bearings, hose, belting, cable sheathing and tarpaulins.

chloroprene sulphur rubber
Also known as chloroprene sulphur copolymer. An abbreviation suggested for this type of material is SCR. A *chloroprene copolymer rubber*.

chlorosilane
A material which is based on a central silicon atom to which is attached hydrogen atom(s) or organic group(s) and chlorine group(s). Used in the manufacture of polyorganosiloxanes or silicones. See *dimethyl dichlorosilane* and *silicone rubber*.

chlorosulphonated polyethylene
Also known as chlorosulphonated polyethylene rubber or as, chlorosulphonylpolyethylene. Sometimes referred to as chemical rubber but probably better known by the Du Pont trade name/trademark, Hypalon. An abbreviation used for this type of material is CSM. A rubbery material produced by reacting polyethylene (PE) with chlorine and with sulphur dioxide in the presence of light. Both linear and branched PE may be used and typical polymers contain approximately 1% sulphur and up to 43% chlorine. Both -Cl and -SO_2Cl groups are introduced into the PE chains: the introduction of chlorine

groups destroys crystallinity and the introduction of reactive sulphonyl chloride groups provides sites suitable for vulcanization by a variety of systems. For example, with a metal oxide and an organic accelerator; or with a metal oxide, a polyfunctional alcohol and an organic accelerator; or with an epoxide resin and an organic accelerator. What is used depends on the property which it is desired to optimize. In general, the cured products have good resistance to ozone, oxygen, heat, weathering, fire and oil. Chlorosulphonated rubbers are included in the category of *speciality rubbers* because of their good oil and temperature resistance.

chlorosulphonated polyethylene rubber
See *chlorosulphonated polyethylene*.

chlorosulphonylpolyethylene
See *chlorosulphonated polyethylene*.

chlorotrifluoroethylene
An abbreviation used for this monomer is CTFE. May be represented as $CF_2 = CFCL$. Such a material is used to make fluoropolymers - see, for example, *fluororubber* and *polychlorotrifluoroethylene*.

chlorotrifluoroethylene-ethylene alternating copolymer
See *chlorotrifluoroethylene-ethylene copolymer*.

chlorotrifluoroethylene-ethylene copolymer
Also known as chlorotrifluoroethylene-ethylene alternating copolymer or, poly-(chlorotrifluoroethylene-co-ethylene). An abbreviation used for this material is ECTFE. A 1:1 alternating copolymer of chlorotrifluoroethylene and ethylene which is a thermoplastics material with good chemical, creep and impact resistance. The electrical insulation properties are good and the material is fire resistant.

choke
A term used in *hydraulics* and which refers to a restriction, or a reduction in cross-sectional area, in a line which is relatively large with respect to its diameter.

choke bar
Part of an *extruder assembly*. That portion of a sheet die which is used to selectively restrict the flow so as to even out variations in sheet thickness; it forces the material out to the die extremities and restricts passage through the centre of the die.

choke plate
Part of an extruder assembly. A single hole unit which is used between the end of the extruder barrel and the die holder to produce a controlled pressure drop in the melt.

chopped fibre composite
See *short fibre composite*.

chopped glass fibre
Also known as chopped glass or as, chopped strand glass fibre. Relatively short *glass fibres* which are produced by chopping continuous fibres: such continuous fibres are based on filaments of approximate diameter 10 to 20 μm bound with an adhesive. For the *reation injection moulding (RIM) process*, the chopped fibre length may be 1·6 mm and is comparatively uniform. When dispersed in the polyol, the adhesive dissolves and the filaments separate to give a high viscosity, shear thinning suspension. Because of the high viscosity which results on fibre addition, approximately 15% by weight is the maximum that can be used in RIM moulding.

chopped strand mat
An abbreviation used is CSM. A form of glass reinforcement widely used in the *hand lay-up process*. CSM consists of *chopped strands* approximately 50 mm/2 in in length bound together by a resinous binder: a random fibre mat.

chopper gun
The spray gun used in *spray up*. See *spray gun*.

CHR
An abbreviation used for allyl-group-containing epichlorhydrin rubber. See *epichlorhydrin rubber*.

chroma
The purity of a colour: the extent to which a colour departs from white. Also referred to as saturation.

chromatic aberration
A lens defect whereby false colour is produced by use of the lens. When ordinary light (*polychromatic light*) passes through an uncorrected lens then the shorter wavelengths will be more strongly bent or refracted than the longer wavelengths. For example, violet rays will be brought to focus close to the lens: each colour, or wavelength, will be brought to focus at slightly different points so producing a series of differing images. Characterised by prismatic colouring at the edges of, and within, the image. By using a compound lens, made of differing glasses, this defect can be minimised.

chrome finish
See *methacrylatochromic chloride*.

chrome green
A class or type of inorganic pigment which is based on mixtures of *chrome yellow* and *Prussian blue (iron blue)*. Attacked by alkalis (even by calcium carbonate), stained by sulphur and considered to be toxic. See *chromium oxide*.

chrome oxide
See *chromium oxide*.

chrome red
Also known as basic lead chromate. May be represented as $PbCrO_4 \cdot PbO$. A red pigment.

chrome reds
The collective name for a class or type of *inorganic pigment* based on *lead chromate*.

chrome yellow
Also known as *lead chromate*. May be represented as $PbCrO_4$. A yellow pigment. A class or type of inorganic pigment which can be obtained in several shades of yellow. See *chrome yellows*.

chrome yellows
A class or type of *inorganic pigment* based on *lead chromate* and available in different shades of yellow-orange. By blending lead chromate with lead sulphate can obtain lighter colours such as primrose and lemon shades. By combining lead molybdate, with the chromate and sulphate, then molybdate orange is obtained.

Such pigments are relatively cheap, bright and have good *hiding power*, however, the chemical resistance can be poor and they are sensitive to staining by hydrogen sulphide. Toxicity considerations rule them out for many applications. Heat resistance can be improved by encapsulation with *silica*. At temperatures above 220°C/428°F, chrome yellow will discolour unless surface treated. For use at temperatures above 290°C/554°F a cadmium yellow is more heat resistant.

chromic oxide
See *chromium oxide*.

chromium
A *transition element (Cr)* which is a very hard, lustrous gray/white metal with a high melting point (1,857°C) and boiling point (2,672°C): the relative density is 7·2. Does not occur naturally. Noted for its excellent corrosion resistance: used to make steel alloys which are noted for their hardness, strength and corrosion resistance. Many stainless steels contain

between 12 and 18% of chromium. The metal is also used to plate both steel and material materials: see *acrylonitrile-butadiene-styrene* or *ABS*.

chromium oxide
Also known as chrome oxide and as chromic oxide (see *chrome green*). An oxide of chromium which may be represented as Cr_2O_3. A dull green material used for the production of weatherable green and yellow-green colours: reflects infra-red light. Grades free from rubber poisons are available. This material has a relative density (RD or SG) of from 3·5 to approximately 5·21. The hydrated form (Guignet's green) may be represented as $Cr_2O_3 2H_2O$: it is a more brilliant green but dehydrates at 500°C. See *chromium oxide greens*.

chromium oxide greens
A class or type of *inorganic pigment* based on *chromium oxide*. Relatively, high priced pigments which have low colouring strength and opacity. Such pigments have good heat stability at high temperatures, are chemically inert and are lightfast: make good camouflage colours. See *chromium oxide* and *chrome green*.

chromium trifluoroacetate
An organometallic compound used, for example, to *vulcanize carboxy-nitroso rubber*.

chromophore
A colour giving group such as the azo group. Chromophores are colour-forming groups. See *colouring of polymers*.

chrysotile
Also known as white asbestos: a hydrated magnesium silicate with a fibrous structure. The fibre diameter is of the order of 0·01 to 1 μm and the length is 1 to 40 μm. Retains all of its strength to at least 400°C. Chrysotile fibre is the basic raw material for the *asbestos textile industry*. Most modern friction lining materials are moulded from compounds which contain short chrysotile fibres (<70%), fillers and synthetic polymers to give for example, asbestos-reinforced thermoset bearings. The friction and wear characteristics of *asbestos*, and its decomposition product forsterite, influence the braking efficiency and service life of the products with high operating temperatures promoting higher friction.

CI
An abbreviation used for *colour index*.

cinnamate resin
A *resin* which contains the cinnamate group. See, for example, *polyvinyl cinnamate* and *photo-resist*.

Cinpres process
A *gas injection process* for injection moulding. In such processes, a gas is injected into the melt in order to form pressurised voids in thick sections and so give smooth surfaces with, for example, high shrinkage materials such as *polypropylene (PP)*. By the use of such processes it is possible to reduce the clamp force requirements significantly, for example, by approximately a third.

circle rubber
A *technically classified rubber*. A *natural rubber* whose cure properties have been assessed or graded by cure rate testing on the ACS test compound. Blue circle is fast curing, yellow circle is medium curing and red circle is slow curing.

circuit
A term used in *hydraulics* and which refers to an arrangement of components and inter-connecting pipes which together perform a specific function within the hydraulic system. The complete system of internal *flow-ways* within a mould. In filament winding, a circuit is the winding produced by a single revolution of the mandrel or form.

circular flash gate
See *ring gate*.

circular mil
An abbreviation used for this term is cmil. A unit of area used, for example, for measuring the cross-section of wires. It is the area of a circle having a diameter of one *mil*.

circulatory flow
A type of flow which is common for non-Newtonian fluids: the flow lines cross the flow direction. See *secondary flow*.

cis form
See *cis-trans isomerism*.

cis-1,2-ethylene dicarbonic acid
See *maleic acid*.

cis-1,4-butadiene
See *butadiene rubber* and *styrene-butadiene rubber*.

cis-1,4-polybutadiene rubber
See *butadiene rubber*.

cis-1,4-polyisoprene
See *cis-polyisoprene rubber*.

cis-9-octadecenoic acid
See *oleic acid*.

cis-configuration
See *cis-trans isomerism*.

cis-octadec-9-enoic acid
See *oleic acid*.

cis-polybutadiene rubber
A rubbery polymer produced by *stereoregular polymerization*. See *butadiene rubber*.

cis-polyisoprene rubber
A rubbery polymer produced by *stereoregular polymerization*. Also known as cis-1,4-polyisoprene or as *isoprene rubber*.

cis-trans isomerism
Isomerism which results as a result of functional groups being differently positioned with respect to a central ring or with respect to a double bond. For example, in diene rubbers such isomerism results as a result of the positioning of the chain portions on either side of the double bond. If the two are on the same side, then the isomer is referred to as a cis-configuration. If the two are on opposite sides of the double bond, then the isomer is referred to as a trans-configuration. Rotation to the other configuration is prevented by the double bond. See, for example, *trans-polyisoprene*.

CISPR
An abbreviation used for the *International Special Committee on Radio Interference*.

citrate plasticizer
The reaction product of citric acid and an alcohol: a high molecular weight ester, for example, tributyl o-acetyl citrate. Such materials are usually *primary plasticizers* with good low temperature properties and are used in medical and food applications because of their low toxicity. Their resistance to extraction, particularly to soapy water extraction, is not very good.

CLA
An abbreviation used for *centre-line-average*.

clamp force regulation
See *locking force*.

clamp stroke
Also known as opening stroke. This is quoted in either inches or millimetres for an *injection moulding machine*. It is the maximum distance over which the *moving platen* can be made

to move and if the *mould thickness* is subtracted from this figure, the answer indicates what space is available for moulding ejection.

clamping force
The amount of force, usually measured in tonnes or in kiloNewtons, used to hold a mould closed and so oppose the opening force. In blow moulding, a clamping force of 1·25 × the blow pressure × the projected area, is suggested. See *locking force*.

clamping line pressure
The pressure in the hydraulic line, or pipe, which is connected to the hydraulic actuator and which causes the component parts of the mould to be held together during, for example, mould filling in *injection moulding*.

clamping plate
See *stationary plate*.

clamping pressure
That pressure which is obtained by dividing the *clamping force* by the projected area. The pressure needed to hold the component parts of the mould together during, for example, mould filling in *injection moulding* and needed to oppose the injection pressure.

The maximum available pressure should not be automatically used but should be estimated from a consideration of the *projected area* of the moulding and the injection moulding pressure. For most injection moulded components, it is approximately 2 tons per square inch (2 tsi) or 31 megaNewtons per square meter (31 MNm^{-2}). This is however, a low figure and should only be treated as a very rough rule of thumb because, for example, once the moulding has any depth then side wall forces must also be considered.

clamping stroke calibration
See *calibrating blow pin*.

clamping system
That which applies the *clamping/locking force* on a moulding machine. With injection moulding machines, *direct lock machines* (direct ram) give, in general, more consistent clamping forces, and are more maintenance free, than *toggle machines*. Over the size range of approximately 100 to 800 tonne, toggle machines are often preferred because of ease of manufacture and cost considerations.

clarifying agent
Also known as a clearing agent. An additive which improves the clarity of a polymeric material or product. For example, *trixylyl phosphate*, is used for *casein-formaldehyde* as it can double the light transmittance through a thick sample.

Clash and Berg test
See *cold flex temperature*.

classical ladder polymer
See *ladder polymer*.

classification of tests
See *tests*.

clays
Finely divided materials formed by the weathering of silicate rocks. The product, a complex hydrated alumino-silicate, is usually coloured by impurities such as iron oxide: *china clay* is a purer clay and is white. See *calcined clay*.

cleaning
Part of the production process for *electroplating*, and *vacuum metallization*, of plastics in which injection mouldings are cleaned (for example, with alkaline cleaners) so as to remove substances which would interfere with the chemical etching stage (see *surface conditioning*) and subsequent adhesion. To avoid plating problems it is best to minimise moulding contamination before this stage is reached.

cleaning compound
See *purge compound*.

clean out piston
Part of a *reaction injection moulding (RIM) machine*: that part of a reaction injection moulding *mixhead* which removes the mixed reactants from the mixhead.

clearance bush seal
A type of *rotary shaft seal*.

clearing agent
See *clarifying agent*.

cling film
A soft pliable film which can either cling, or stick, to itself or to other objects and which is widely used in packaging; this property is shown by thin, soft films based on soft *polyvinyl chloride (PVC)* or *ethylene-vinyl acetate (EVA)*.

clippings
See *cuttings*.

clock gauge
See *dial test indicator*.

clone
A group of plants which originated from the same individual by asexual production from buds or cuttings. All members of the group are genetically identical to each other and to the parent. See *natural rubber*. The yield of rubber has been increased by breeding high yielding clones which may yield about 2,500 kg of *natural rubber* per hectare per year.

closed assembly time
The time between the assembly of a coated surface and the application of pressure in *adhesive bonding*.

closed centre valve
A *hydraulic valve* in which all ports are isolated (blocked or closed) when the spool is in the neutral, or central, position.

closed circuit
An electrical circuit in which there is a complete path for the flow of electrical current. A term used in *hydraulics* and which refers to a piping arrangement in which the fluid from the pump returns to the pump inlet after passing through the system.

closed centre circuit
A *circuit* in which flow is blocked when the valve is in neutral and pressure at the pump outlet is maintained at the maximum pressure control setting. See *closed centre valve*.

closed loop
Control terminology used when the control system checks to see if its commands have been obeyed and then enforces them if they have not. See *closed loop system*.

closed loop, dimension control system
A system which feeds back information on the product dimensions and then adjusts parameters, such as extrusion line speed. to correct for dimension shifts.

closed loop system
A system in which the output is continuously monitored and compared with what is required. The *error signal* is used to control the output of the loop so as to obtain what is required or set. That is, corrective action is applied.

In *injection moulding*, it is easier, and cheaper, to close the loop at the hydraulic valve: the oil flow rate through, or across, the valve is monitored and controlled. On larger machines it may be worth monitoring and controlling screw speed but where the stroke is relatively small (e.g. <100 mm) then, valve monitoring is often preferred.

Many users of *statistical process control (SPC) systems* do not like existing *closed loop control* as what is available at present is not considered good enough: the changes that a closed loop system can introduce, may cause more problems and variations than it solves. What is needed is the application of an artificial intelligence to the machine so that the effect of any changes can be predicted and assessed; the only changes allowed, would then be sensible ones.

closed mould processes
Moulding techniques used in the reinforced plastics industry to reduce *styrene emissions* and to improve on traditional *GRP moulding* such as *hand lay up*. Two mould halves are used: the most common processes are cold press moulding, *vacuum assisted resin injection (VARI)*, *resin transfer moulding (RTM)* and *structural resin injection moulding (SRIM)*.

closed-cell cellular plastics
Cellular plastics in which almost all the cells are non-interconnecting.

clouds
Dull or opaque patches in sheet rubber.

CLTE
An abbreviation used for coefficient of thermal expansion.

clustered aggregate
See *carbon black - structure*.

cm
The abbreviation for *centimetre*.
   cm^2 = square centimetre; and,
   cm^3 = cubic centimetre. See *prefixes - SI*.

CM
An abbreviation used for *chlorinated polyethylene*.

CMC
An abbreviation used for carboxymethylcellulose. See *ion exchange resin*.

CMC
An abbreviation used for *ceramic-matrix composite*.

cmil
An abbreviation used for *circular mil*.

cmpd
An abbreviation used for compound.

CN
An abbreviation used for cellulose nitrate. See *cellulosics*.

CNC
An abbreviation used for *computer numerical control*.

CNG
An abbreviation used for compressed natural gas.

CNR
An abbreviation used for *carboxy-nitroso rubber*. CNR is also used as an abbreviation for *technically specified (natural) rubber* from China.

co
An abbreviation used for *copolymer*. See *abbreviations - plastics copolymers*.

Co-BR
An abbreviation used for *butadiene rubber* based on a cobalt catalyst. See *butadiene rubber*.

co-spun rove
A *yarn* which contains more than one type of fibre.

co-woven fabric
Also called a hybrid fabric. A fabric reinforcement made from different materials. Each fibre is processed and then combined (woven) to produce a hybrid fabric in which, for example, the reinforcement fibres (carbon fibre) alternate in the warp and the weft. See *hybridized reinforcement fabrics*.

coacervate
See *polysalt*.

coacervating agent
A chemical added to *latex* to cause coacervation or *coagulation*, for example, acetic or formic acid. See *natural rubber*.

coacervation
Also called *coagulation*: obtained when a *coacervating agent* is added to latex.

coagulant
A material which causes coagulation of *latex*. For example, *acetic acid* or *calcium chloride*.

coagulating bath
A place where setting or hardening occurs. For example, in *viscose rayon production*, the chemicals in the coagulating bath (sulphuric acid, sodium sulphate and zinc sulphate) turn the sodium xanthate cellulose solution back into solid cellulose. That is, *cellulose* is regenerated.

coagulation
The process of obtaining a solid material from latex. *Acid coagulation* is usually preferred. Acetic or formic acid is added, for example, to *natural rubber latex* to make the latex coagulate or curdle. In this way, solid rubber may be obtained from the latex. The first stage of coagulation is *coalescence*. See, for example, *natural rubber*.

coagulum
That which is obtained by *coagulation* of, for example, *natural rubber latex*. In this case, a sponge-like rubber mass is obtained.

coal gas
This material is produced by carbonizing coal: that is, by heating coal in the absence of air at temperatures of the order of 500 to 1,300°C. See *carbonization* of coal and *coal tar*.

coal tar
This material is produced by the *carbonization* of coal. Coal tar can be the source of an incredible number of organic chemicals although its importance in recent years has decreased because of the growth of the petrochemical industry. It is a black viscous liquid which is a mixture of over two hundred chemicals - many of them are aromatic. May be distilled into, for example, five main fractions. The fraction which distils at up to 170°C is called light oil and from this may be obtained *benzene, toluene, xylenes* and *pyridine*. The fraction which distils at from 170 to 230°C is called middle oil and from this may be obtained *phenol, cresols* and *naphthalene*. The fraction which distils at from 230 to 270°C is called heavy oil and from this may be obtained creosote. The fraction which distils at from 270 to 400°C is called anthracene oil and from this may be obtained anthracene. The residue, approximately 55%, is *coal tar pitch*. Neutralised coal tar is also used to protect the cut surfaces of Hevea trees during latex tapping.

coal tar pitch
This material is produced during the distillation of *coal tar*. This material has a relative density (RD or SG) of approximately 1·2 and was, at one time, used as a processing aid and as a compound stiffener for rubber.

coalescence
The first stage of coagulation is coalescence: for example, *natural rubber particles* combine to from larger particles.

coated fabric
A *fabric* which has a thin layer of material applied by a *coating process*: usually the fabric remains flexible after coating.

For example, rubber-coated fabric is produced by *frictioning*. The fabric used may be based on natural or synthetic materials and examples of each are cotton and rayon respectively.

Coated fabrics are used for example to make the carcass plies of a tyre (which give the tyre its strength) and to make the tread bracing components. Such components are used in radial or belted bias tyres to increase the modulus in the tread area and thus reduce tread pattern movement and distortion.

coating
A thin layer of material applied by a *coating process*. See, for example, *calendering*. In extrusion, coating is the process of applying a molten polymer web to a moving substrate.

coating - blow moulding
Coating with a relatively impermeable material is a widely used method of reducing the permeability of a *blow moulding*. Often the coating is *polyvinylidene dichloride (PVDC)* although other materials could be used - glass is one such material; high molecular weight, crosslinked PE (put on by plasma polymerization) is another.

Polyethylene terephthalate (PET) bottles are coated with PVDC to improve their barrier properties. This is because a key factor, in the usage of any polymer as a packaging material, is the shelf life that can be achieved for the contents of the package. Coating of PET bottles is usually carried out immediately after the bottle is formed and is done by either dip coating, flow coating or spray coating (spray coating gives the highest coating rate). The coating thickness is roughly 10 microns: such a coating thickness would approximately double the shelf life of the products.

coating process
The process of applying a thin layer of material to a substrate. See, for example, *calendering*.

cobalt
A *transition element (Co)* which is a very hard metal resembling iron in appearance. It is ferromagnetic with a high melting point (1,480°C) and boiling point (2,870°C): the relative density is 8·9. It is used to make steel alloys which retain their hardness at high temperatures: used, for example, to make cutting tools and permanent magnets.

cobalt aluminate
A blue inorganic material with a relative density of approximately 3·5 and which is used as a pigment. See *cobalt aluminate blues*.

cobalt aluminate blues
A class or type of *inorganic pigment*. Blue pigments prepared from cobalt oxide and alumina. May be represented as $CoO.Al_2O_3$ but the composition varies: if chromium is added then the colour moves towards green (a more greenish-blue results). Relatively, high priced pigments which have low colouring strength and opacity. Such pigments have good heat stability at high temperatures, are chemically inert and are usually lightfast.

cobalt blue
See *cobalt aluminate blues*.

cobalt chrome aluminate
A bluish-green inorganic material with a relative density of approximately 4·1 and which is used as a pigment. See *cobalt aluminate blues*.

cobalt green
See *Riemann's green*.

cobalt naphthenate
A metal salt, a soap-like material, derived from *naphthenic acid*. An *accelerator* used in styrene solution to promote room temperature curing of *unsaturated polyester resins*: used with, for example, methyl ethyl ketone peroxide. The cobalt concentration in the salt, is commonly 1%.

cobalt octoate
A metal salt used in *silicone resin production* to achieve cross-linking so as to give a cross-linked *polyorganosiloxane*. An *accelerator* used in dimethyl phthalate solution to promote room temperature curing of *unsaturated polyester resins*: used because the solutions are stable and give little colour to laminates.

cobweb whisker
A *whisker* of very small diameter, for example, approximately 20 nm.

coconut oil
An *edible vegetable oil* derived from cocos nucifera. A non-drying oil obtained by crushing *copra* and known as copra oil. The triglycerides of this oil contain large amounts of lauric and myristic acid residues. The oil is used to make soap, margarine and cooking fats: sometimes used in, for example, *alkyd resins* so as to give good colour retention.

coefficient of friction
In the case of *sliding friction*, this is given by the ratio of the tangential force which is required to maintain motion (without acceleration) to the normal force at the contact surface. If F_k is the friction when steady sliding is obtained then the kinetic coefficient of friction is given by F_k/R. Where R is the normal force at the contact surface. In the case of *static friction*, it is given by the ratio of the friction found when the body is just on the point of moving to the normal force at the contact surface. If F_s is the friction then the static coefficient of friction is given by F_s/R. Where R is the normal force at the contact surface.

coefficient of viscosity
Another way of saying *viscosity*. See *kinematic viscosity*.

coex
An abbreviation used for *coextrusion* or for, *coextruded*.

coextruded
Means that a product consists of more than one layer: such a product is produced by *coextrusion*.

coextrusion
The extrusion process whereby two, or more, melt streams are combined in the die so as to make an extrusion of two, or more, layers of plastics materials; one of these layers is often a barrier layer based on say, *polyvinyl alcohol (PVAL)*. By laminating two or more layers of different polymers together a product may be produced which has properties far superior to that obtained when only one polymer is used.

Such products are used in packaging where the use of two or more polymers produces a plastics product with superior gas diffusion resistance. In the *blow moulding* field it is possible to combine co-extrusion with biaxial orientation and so produce strong, light weight bottles which give a long storage life to the products.

coextrusion blow moulding
An *extrusion blow moulding process* used in the production of mouldings with more than one layer of plastics material: such a *coextrusion* process is used to improve barrier properties.

coextrusion capping
The extrusion process used to produce a plastic product which is topped or capped with another plastics material.

coherent units
Units based on basic units from which all derived units may be obtained by multiplication or division without the introduction of numerical factors. See *Système International d'Unité*.

cohesive energy density
The energy of vaporization per molar volume of a material. The square root of the cohesive energy density of a material is the *solubility parameter*.

coinage metals
The three metals in Group 1B of the periodic table (copper, silver and gold) are sometimes known as the coinage metals.

coking
See *carbonization*.

cold bend temperature
The lowest temperature noted during a bending test at which cracking does not occur: a specified piece of material is wound around a mandrel of a specified size.

cold creeping flow
Flow which occurs when the melt is being sheared when it is relatively cold. See *VPT setting* and *orientation*.

cold cure
Also called cold vulcanization. A process or technique used to vulcanize rubber compounds at room temperatures. Such processes were rendered obsolete by the development of *ultra-accelerators*. Sulphur chloride may be used in solution (carbon disulphide or benzene) to cold cure, thin-walled articles made from *natural rubber* at room temperatures (Parkes process).

cold cured foam
A *flexible polyurethane (PU) foam* produced by the *cold moulding process*. Also known as high resilience foam, or HR foam, as such foams have the best elasticity of PU foams.

cold curing - unsaturated polyester resin
The hardening, or curing, of a resin composition using room (ambient) temperatures, for example, 25°C. Cold curing can be achieved using a so-called catalyst (such as *methyl ethyl ketone peroxide*) and an accelerator (such as *cobalt naphthenate*). For fast cure and a long pot life (for example, body fillers for boats and cars) a tertiary amine and benzoyl peroxide are used.

cold drawing
Plastic deformation at relatively low temperatures. For example, with some plastics materials a neck is formed during a tensile test. if the test is continued, then elongation occurs by movement of the neck along the sample.

cold elastomer
See *cold rubber*.

cold flex temperature
The temperature measured in a *cold flex temperature test*. For example, in the Clash and Berg test it is that temperature at which, under a specified torque, the test specimen is deflected through an arc of 200°.

cold flex temperature test
A brittleness temperature test used to assess the low temperature properties of plasticizers or of plasticized compounds. A rectangular strip of *plasticized PVC* (64 × 6·4 × 1·27 mm thick) is cooled in a bath of methylated spirits and solid carbon dioxide. The temperature is raised at a rate of 2°C/minute and the strip is twisted using a specified torque: initially the strip is twisted by more than 400°. A graph is plotted of temperature against time and the temperature for a deflection of 200° is read off the graph and called the *cold flex temperature*. See *low temperature flexibility*.

cold flow
Usually refers to the flow of a thermoplastics material which occurs at less than the optimum melt temperature.

cold full machine
Term usually used in extrusion or injection moulding and means that the plasticization cylinder is full of polymer compound which is below processing temperatures. A cold, full machine sometimes happens through a power failure or, the barrel, could be left full deliberately. For example, when processing an easily-oxidized material (say a *polyolefin*) the machine may be left with material in the barrel so as to prevent oxidation. See *warming up*.

cold hardening
The hardening of a rubber or elastomer which occurs when the material is stored at low temperatures: usually due to crystallisation.

cold moulding
A type of *compression moulding* in which the moulding is formed at room temperature and subsequently baked, or set, at elevated temperatures.

cold moulding process
A *direct foam moulding process* used to produce *direct moulded foam*. The cold moulding process requires high molecular weight, reactive *polyether polyols* with a high proportion of primary hydroxyl groups - higher than that used for the *hot moulding process*. Isocyanates with a functionality greater than 2 are also required. Cold cured foams have good flexibility and give comfortable seating whereas hot cured foam gives better damping of vehicle vibrations. See *flexible polyurethane foam*.

cold press moulding
A *closed mould process* which is also known as coldpress. An abbreviation used is CPM. This process is used for polyester moulding compounds (PMC): the PMC is pressed between unheated matched moulds under low pressures (70 to 280 kN/m^2 or 10 to 40 psi).

Two unheated mould halves are used and these are mounted in a press: the tool is loaded with a fibrous filler (glass), catalyzed resin is then poured into the female half and the mould is closed. This forces the mix to fill the cavity: after *curing* the component is removed. A relatively cheap moulding method which can use *soft tooling*.

Resins are formulated to have a short pot life and a very short cure time, with a controlled exotherm of about 120°C. Reinforcement is usually *glass* but sisal and hessian have been used. Continuous strand mat is popular as it gives good draping and resin flow. This process fits between *hand lay-up (HLU)* and hot press moulding. Products have smooth surfaces both sides and are more uniform than HLU: lower labour costs than HLU. Lower capital costs than both HLU and hot press moulding.

cold rolling
Cold deformation achieved by passing a sheet or film of material through rolls set at low temperatures, for example, below the melting point/softening point of a particular plastics material. Cold rolling will improve the ductility of brittle *polyhydroxybutyrate (PHB)* film and sheet as spherulitic cracks are healed by this process. *Self seeding* will also improve ductility of PHB. Such cold rolling will also improve the properties of polycarbonate (PC).

cold rubber
Rubber, for example, emulsion styrene-butadiene rubber (E-SBR), produced at comparatively low polymerization temperatures. Initially E-SBR was polymerized at approximately 30°C: now, E-SBR is produced at approximately 5°C. See *hot rubber*.

cold runner mould - thermoplastics material
An *injection mould* in which the *runner* is ejected with the components. If cold runner moulds are used in preference to

runnerless moulds, for economic reasons, then hot sprue bushings should be incorporated in the mould so as to reduce the material content within the feed system and to allow faster cycle times to be achieved.

cold runner mould - thermosetting material
Also called a warm runner mould and used to eliminate the production of waste sprues and runners during the injection moulding of *thermosetting plastics* and *rubbers*. A *three-plate mould construction* is required so as to contain a manifold system which distributes the plasticised material to each cavity. The manifold is maintained at a temperature intermediate between that of the barrel and that of the mould. For example, when rubber is being moulded by this technique the feed system is commonly held at a temperature between 75 and 90°C by oil circulation.

cold slug
A part of an *injection moulding* charge that is cooler than the bulk of the moulding charge, for example, that which is contained inside the *cold slug well* of an injection mould.

cold slug well
A depression, or recess, immediately opposite the *sprue* and which contains the *sprue-puller pin*. Its function is to accept the first part of the shot, i.e. that part of the shot which was located at the tip of the nozzle and which could therefore be colder than the bulk of the shot in injection moulding.

The importance of a cold slug well is often overlooked; its inclusion prevents semi-solid material from entering the injection mould cavity. The presence of such material considerably affects the surface finish and strength of the product.

cold trimming
A *trimming system* for sheet where the edges are removed after the *calendered sheet* has been cooled: the edge trim is reeled, granulated, blended in a definite ratio with virgin material and then mixed in for example, an *internal mixer*. For flexible *polyvinyl chloride (PVC)*, such cold trimming is done by using a weighted, circular knife which bears against a hardened steel roller. This is driven by continuously variable DC motors and the contact pressure may be varied to suit the material being cut. For more rigid materials, pairs of knives with edges bearing against each other are found more effective.

cold vulcanization
See *cold cure*.

cold vulcanizing agent
See *sulphur chloride*.

cold water absorption
A measure of how much cold water a material will absorb under specified conditions and over a certain time. See *water absorption*.

cold-setting adhesive
An *adhesive* which cures at relatively low temperatures, for example, below 30°C.

cold-type factice
See *white factice*.

coldpress
See *cold press moulding*.

collapse
Inadvertent densification of cellular material during manufacture resulting from breakdown of cell structure.

collapsible core mould
Part of an *undercut mould* used to put an internal *undercut* into a moulded component. The collapsible core consists of two main parts: a bush which moves segmentally inwards during mould opening and a core with a conical top. The core pushes the segments into the closed position during mould closing and after moulding the undercut is released by the core being moved back.

colloidal silica
Solid, fine particle-size *silica*, whose particle size is less than that of carbon black and synthesized by a *precipitation method*. The term also refers to stable solutions of amorphous silica (silica sols) in water. Particle size 3 to 100 nm. See *synthetic silica*.

colloidal sulphur
An allotropic form of *sulphur* of very fine particle size: mainly used in *latex technology*.

Colmonoy
A trade name/trade mark for a family of hard, corrosion resistant alloys. This type of material is based on nickel. Used, for example, to surface screws which operate in aggressive environments. Has been used to surface worn nitrided screws: used for *extruders* which produce *lay-flat film*. Has a hardness in the region of 56 Rockwell C. Not as hard as *nitriding* but sufficiently hard to give a new lease of life to a worn component.

colophony
See *rosin*.

colorant
An additive used to impart colour. Generally speaking, colorants may be divided into two major types and these are *dyes* and *pigments*. The choice of a colorant is influenced by the heat stability of the colorant, the lightfastness, the weatherability, the migration resistance, the chemical resistance, the toxicity that it imparts and the economics of use.

Most polymeric components are coloured by *mass colouring techniques*. The colorant system may be dispersed, by melt mixing, in a large volume of the polymer to form a *fully compounded material* which is then sold as being ready for use. Alternatively a *masterbatch* may be produced and incorporated by the processor or, the colouring system may be added to the polymer and then dispersed throughout the polymer by the processor. The object must be uniform dispersion of the colorant without agglomeration, streaking or gel formation so as to achieve the most efficient use of the colorant system.

Fully compounded material gives consistently good results but unfortunately it is expensive, particularly when only a relatively small lot is required. The approximate relative costs of colouring thermoplastics would be fully compounded material 100, *dilute masterbatch 30, concentrated masterbatch 20, liquid colouring 20* and *dry colour 15*.

It is possible to colour only the skin in order to save on colorant, colorant costs and compounding costs: this may be done by, for example, painting, sandwich moulding and extrusion. Surface dyeing is sometimes performed with polar materials such as polyamides (PA), for example, PA fibres. Theoretically it is possible to introduce colour-forming groups (chromophores) into the polymer but this is not commercially important. See *liquid colour* and *masterbatch colour*.

colorimeter
A machine used to measure colour in terms of the three *primary* colours. More precise measurements, or characterisation of colour, may be made by means of a *spectrophotometer*.

colour
The sensation obtained when *light* of a certain wavelength reaches the retina of the eye. Light has the characteristics of *hue, saturation* and, for a colouring system, *lightness*. Most people can distinguish a large numbers of colours, for example, 10 million. The colour of a pigment is related to the wavelengths of the incident and reflected light. Black pigments absorb throughout the entire wavelengths and white pigments reflect all wavelengths.

colour - change of
With thermoplastics materials the first sign of degradation is often a slight change of colour during melt processing. This slight change is important as in many applications, the material was used, or selected, because of the colour (or the lack of colour for a transparent material) possible. Any colour change cannot usually be tolerated: a *stabilizer* may be classed as good, simply because it delays colour formation and not necessarily because it stops degradation.

colour composition
A colour system may be made up by mixing the three *primary* colours and the intensity of such colours may be lightened or darkened by the addition of white or black (see *shade* and *tint*). Longest wavelength in the visible spectrum is red at approximately 700 nm: shortest wavelength in the visible spectrum is blue at 400 nm. Green has a wavelength of about 530 nm.

colour compound
Also called *fully compounded material*.

colour concentrate
A *masterbatch*.

colour index
An abbreviation used is CI. An international coding system which groups *colorants* into groups or families and gives a colorant a number, for example, 244 for red.

colour-dust contamination
Contamination caused by coloured dust drifting in the atmosphere. See *dry colouring*.

colouring of polymers
See *colorant*.

colour match
An acceptable matching or reproduction of an original colour. The degree of exactness of the match is usually limited by price and end-quality. Ideally the two compounds should match in terms of hue, chroma, shade etc.

columns
Cylindrical metal rods which, for example, form part of an *outrigger system* on a mould.

colza oil
See *rape seed oil*.

COM
An abbreviation used for mechanical comminution (process). See *comminution process*.

comb graft copolymer
See *comb polymer*.

comb polymer
Also known as a comb graft copolymer. A type of graft *copolymer* in which there is a main polymer chain from which regularly protrude, approximately uniform branches of another polymer.

comb-grafted natural rubber
A *polymer modified natural rubber*. A *thermoplastic natural rubber* in which natural rubber (NR) has been modified by, for example, grafting *polystyrene side chains* onto the NR molecule so as to give a comb-grafted NR. A reactive *prepolymer*, *azo-tipped polystyrene* (azodicarboxylate-functional polystyrene) of a molecular weight of about 8,000, is subjected to high shear in the presence of NR in an *internal mixer*: temperatures greater than the softening point of the PS are necessary, for example, 10 m at 95°C. The products are tough flexible materials which dissolve in the solvents for PS and for polyisoprene. See *natural rubber - thermoplastic blends*.

comb-grafting
The process used to produce a *comb-grafted polymer*, for example, *comb-grafted natural rubber*.

combing
A textile process used to bring order to a mass of fibres: the mass of fibres are straightened by a comb which also removes shorter fibres and impurities.

combustion speed
See *Underwriters Laboratory horizontal burning test*.

Comité Européen de Normalisation
An abbreviation used for this EEC organisation is CEN. An English translation is the European Committee for Standardization. Also known as the European Organization for Standardization. This organization was founded in 1961 and comprised the national standards bodies of EEC countries, EFTA countries and Spain. This organization prepares *European Standards (EN standards)* that are published without variation of text as national standards in countries approving them. The European Committee for Electrotechnical Standardization (CENELEC), which is the electrotechnical counterpart of CEN, was founded in 1973: CENELEC prepares European Standards (EN standards) for identical publication nationally.

Comité Européen de Normalisation members
An abbreviation used for this term is CEN member. Such members are the national standards bodies of Austria, Belgium, Denmark, Finland, France, Germany, Greece, Ireland, Italy, Netherlands, Norway, Portugal, Spain, Sweden, Switzerland, and the United Kingdom (UK).

command signal
A term used in *hydraulics* and which is also called input signal. An external signal which represents a new position or velocity for a servo-valve.

comminute
To reduce to small pieces or flakes. See *comminuted rubber*.

comminuted rubber
Natural rubber which is baled crumb rubber. See *Heveacrumb process*.

comminution process
A technique or process used to reduce *natural rubber coagulum* to *crumb rubber* by the use of dicing machines, rotary cutters and/or hammer mills. See *Heveacrumb process*.

Commission Electrotechnique Internationale
An abbreviation used for this organization is CEI. The French language version of *International Electrotechnical Commission*.

commodity thermoplastics materials
A group of materials: the major (large tonnage) materials such as the *polyolefins*, *polyvinyl chloride plastics* and *styrene-based plastics*.

Because of their ease of manufacture from readily available monomers, their low cost and the versatility of the materials, polyolefin usage has grown dramatically in recent years. Both homopolymers and copolymers are available in this family as are rubbers and thermoplastics materials. The members of the polyolefin plastics family include *low density polyethylene (LDPE)*, *high density polyethylene (HDPE)* and *polypropylene (PP)*. The term *polyethylene (PE)* therefore covers a range of materials.

Polyvinyl chloride (PVC) plastics covers both homopolymers and copolymers both of which may be used without plasticizers thus giving *unplasticized grades (UPVC)* and *plasticized grades (PPVC)*. The term styrene-based plastics also covers a range of materials, for example, the homopolymer *polystyrene (PS)*, the copolymer *styrene-acrylonitrile (SAN)*, the rubber-toughened material called *high impact polystyrene (HIPS)* and *acrylonitrile-butadiene-styrene (ABS)*.

The importance of bulk thermoplastics cannot be over-emphasized; approximately 70% of all plastics used fall into this category.

common names - plastics materials
The names of most thermoplastics begin with 'poly' and then this term, which means 'many' is followed by the old fashioned name for the monomer from which the plastic is derived i.e. the name is source-based. Because of this practice names such as polystyrene and polyethylene, for *homopolymers*, are common. When the plastics material has more than one word in the name, parentheses, or brackets, may be put around the words so that poly(vinyl chloride) results. However this practice is not universal and so the same term without the brackets is also used i.e. polyvinyl chloride. Source-based nomenclature is not however, universally used and so names such as 'acetals' and 'cellulosics' are also encountered. To add to the confusion, many plastics are known by more than one name, for example, an *acetal* may be known as polyformaldehyde or as, polyoxymethylene. See *abbreviations* and **table 1**.

common tests
A test which is widely used. The most common type of tests performed on plastics materials are tests such as impact strength determination and tensile testing. Tensile testing and hardness testing are most commonly performed for rubbers and their compounds. It is for reasons of speed, economy and convenience, that tests such as tensile strength measurement are performed.

comparative tracking index
An abbreviation used for this term is cti. See *resistance to tracking*.

compatibilization
In thermoplastics blend technology this term means making a commercially useful material from an incompatible system. See *immiscible blend*. The aim is to reduce the inter-facial tension gradient so that coalescence of the dispersed phases is avoided and adhesion is improved. There are two approaches (i) addition of a pre-formed *compatibilzing agent* and, (ii) *reactive compatilization*.

compatibilizer
See *compatibilizing agent*.

compatibilizing agent
Also known as a compatibilizer or as, a pre-formed compatibilizing agent. An additive: a material used to improve the properties of polymer blends or alloys. A compatibilizing agent is an additive which is compatible with *immiscible polymers*. Such a material may be, for example, a block copolymer. For example, the two ends of the *block copolymer* may be each soluble in a different polymer. ABA-type copolymers are used for this purpose. See *reactive compatilization*.

compatible blend
A commercially useful *blend*, or system, based on an *immiscible mixture* of two or more thermoplastics materials. See *incompatible blend* and *alloy*.

compatible plasticizer
A *plasticizer* which can be used as the sole plasticizer and which will not exude from a material in use.

compensator control
A term used in *hydraulics* and which refers to a type of displacement control which is used for variable displacement pumps and motors. When the system pressure exceeds the set pressure then the displacement is altered.

complex material
See *composite material*.

compliance
Complying to a standard. A standard number on a product is a manufacturers claim of compliance with a standard. It does not indicate assessment by an independent body as does, for example, a BSI Kitemark. See *British Standards Institution Kitemark*.

compo
An abbreviation sometimes used for compound. Also used to describe the brown crepe-type rubbers produced from *natural rubber*. See *compo crepe*.

compo crepe
One of the forms in which *natural rubber* is supplied. An inferior grade of rubber obtained from lumps, tree scraps, smoked sheet cuttings etc.

composite
A shaped product made from a *composite material*.

composite material
A complex material: sometimes simply referred to as a *composite*. A combination of two or more materials each of which retains its identity in the finished component. The most common examples are based on *glass fibre* and resins although thermoplastics composites are receiving considerable commercial interest. See, for example, *thermoplastic bulk moulding compound* and *specific modulus*.

composite moulding
The production of *injection mouldings* where a part of the moulding need not be produced during the moulding cycle. Many finished components consist of different materials which have been combined during the moulding process. The term includes both *insert moulding* and *outsert moulding*.

composition density
The density of a composite material. The addition of inorganic fillers and fibres usually increases the density of polymer mixtures, as such materials usually have a density greater than that of the polymer. To calculate the density of a polymer composition divide the total mass of polymer composition by the total volume. For example, if 100 g of UP (density 1·28 g/cc) is mixed with 50 g of glass (density 2·55 g/cc) then the compound density will equal:

$$100 + 50 / (100/1·28 + 50/2·55) = 1·52 \text{ g/cc}.$$

This assumes that all the air spaces, or voids, in the mixture are filled.

compound
The intimately mixed material which results when the polymer (rubber or plastic) and the compounding ingredients are melt mixed, or compounded, in a *batch* or *continuous mixer*.

compound blending
Once ingredients have been selected they are weighed to a preset formulation: it is common practice in the thermoplastics industry to *preblend* the ingredients together before they are fed to the melt processing equipment, for example, to an *injection moulding machine*. This pre-blending operation may be performed, for example, on the injection moulding machine or it may be performed as a separate operation.

HAND BLENDING OPERATIONS. For laboratory work, or for the small scale production of thermoplastics compounds, a hand shaken polyethylene bag, provides a useful blender. The ingredients are carefully weighed into a clean polyethylene bag which is then slowly inflated from a clean, dry air supply. By twisting the end of the bag, it may be sealed so that the shaking operation can commence. It is best to use a large bag as it is important not to fill the bag more than half full; mixing will take approximately 5 minutes. The big advantage of this method is that it is cheap, quick and there is little risk of contamination.

MECHANICAL BLENDING. Rotating blenders, which are usually based on either drums or conical containers, are widely employed in the thermoplastics industry to blend granules with other granules or with additives. A system based on a steel drum is the simplest that can be imagined. A steel drum is partially filled (not more than two-thirds full) with the required ingredients and then the sealed drum is slowly rotated, at say 25 rpm by an electric motor, until a uniform blend is obtained: this takes approximately 25 minutes. Simply rolling the drum, for example, along the workshop floor, will not produce the required mixing action as there is no reason for the ingredients to become distributed along the length of the drum. A folding, spreading action is required and this is most easily achieved by rotating the drum end-over-end and at a slight angle. Therefore, as the drum turns the ingredients are folded upon themselves when they fall into the corners' of the drum and are spread (or tumbled apart) as the drum rotation is continued. Such a system is relatively cheap as the drum may be one that is already available. The material may have been supplied in the drum and in such a case, the chance of contamination is reduced. However, this system is not as effective as a *conical blender*.

compound lens
See *chromatic aberration*.

compound room
See *weighing-up room*.

compounded material
In plastics technology, this term means material which contains all additives, that is *fully componded material*. Because of the ease with which colour may be added at, for example, the injection moulding machine, there is increasing use of *masterbatches* in conjunction with *natural material*. However, compounding still remains the most accurate colouring technique and gives the most precise, and reproducible, colour. It also gives the best density of colour and is the most suitable for small runs. Most commodity plastics are coloured, for example, on the injection moulding machine whereas most engineering plastics are sold already coloured i.e. fully compounded.

In rubber technology the term means material which contains all additives except the *sulphur* and/or, it means material which contains all additives including sulphur.

compounder
A machine used for melt mixing. For example, a *compounding extruder*.

compounding
The process used to produce a *compound*. The preparation of a compound usually involves melt mixing: that is, mixing when the base polymer is soft and pliable. Such mixing uses both *dispersion mixing* and *distributive mixing*.

compounding extruder
An extruder which is operated so as to give good homogeneity and/or which has been designed to give good homogeneity or *good dispersion*. For example, the mixing efficiency of a *twin screw extruder* can be increased by incorporating mixing elements part-way along the extrusion screws. These may take various forms (e.g. reversed screw flights, kneading discs. pins etc.) and with some machines their length, number and form may be easily changed when required. The material is heat-softened, in an extrusion section, passed into the mixing section and then into another twin screw section: this process may be repeated several times e.g. in devolatilizing machines or, in machines used to carry out chemical reactions.

compressed asbestos fibre
An abbreviation used for this material is CAF. This material is mainly used for the manufacture of joints and gaskets. It is made from high grade *asbestos fibre* well opened and intimately bonded with selected polymers so as to give a material which can withstand 600°C.

compressibility
A term used in *hydraulics* and which refers to a change in the volume of unit volume of fluid when subjected to a unit change in *pressure*.

compression mould
A mould used in *compression moulding*. See *flash mould, semi-positive mould* and *positive mould*.

compression moulded sheet
A sheet produced by *compression moulding* and used, for example, for testing purposes. Such sheets are widely used as they are relatively easily and cheaply prepared using a *frame mould*. The test samples are cut from the compression moulded sheet after moulding: as only one or two sheets are required, the relatively long production cycle with *thermoplastics* is acceptable. Sheet size is commonly $150 \times 150 \times 1.5$ mm ($6 \times 6 \times 0.62$ in).

PLASTICISED POLYVINYL CHLORIDE (PPVC)
A blank which is slightly smaller than the cavity is cut from a milled sheet which is thicker (for example, by about 0.5 mm) than the centre sheet thickness and placed in the cold mould. The mould assembly is then placed in a steam heated, and water-cooled, compression moulding press at 150°C: contact pressure only is applied. After 10 minutes a force of approximately 20 tonnes is applied while the mould is cooled under pressure. Once cold, shaped cutters are then used to punch the samples from the cooled sheet.

POLYOLEFIN SHEETS
Insert soft aluminium sheets (of 0.25 mm thickness) between the centre plate and the outer plates of the *frame mould*. This will allow the moulded sheet material to shrink uniformly so that a void-free sheet is produced. The temperatures employed will be approximately 150°C for LDPE but about 175°C for *polypropylene (PP)*: an electric press will be needed for PP. In other respects the procedure is as for PVC.

RUBBER SHEETS
A blank which is slightly smaller than the cavity is cut from a milled sheet which is thicker (for example, by about 0.5 mm) than the centre sheet thickness and placed in the preheated mould. The mould assembly is then placed in a heated compression moulding press at 150°C and the moulding force is applied immediately and left applied until the setting reaction is complete: no cooling period is, of course, necessary. Often a cure time of 20 minutes is used: after the cure time has lapsed the sheet is removed from the hot mould. Once cold, shaped cutters are then used to punch the samples from the cooled sheet or, the test samples are cut from the sheet using a knife or scissors.

THERMOSET SHEETS
When moulding thermoset sheets then the moulding force is applied immediately and left applied until the setting reaction is complete: no cooling period is, of course, necessary. The test samples may be cut oversize from the thermoset sheet and then filed to size against hardened steel templates.

compression moulding
A moulding technique in which a heat softened material is shaped by pressure applied to a *compression mould*.

THERMOSETS
This moulding process was the first high pressure moulding process used by the plastics industry and when automated it is capable of giving high speed production of attractive components. This moulding technique is normally restricted to thermosetting materials (rubbers and plastics) as when it is employed for thermoplastics very long cycles normally result

unless special equipment is used, for example, as in the production of vinyl gramophone records.

A measured amount of the thermoset compound is placed in the cavity of a heated mould, for example, set to 150°C. The mould is attached to the platens of a hydraulic press so that rapid closing, followed by pressure application to the mould, may be achieved. When heat and pressure are applied to the material it flows and fills the mould cavity: excess material (known as flash) escapes from the mould. The heat applied causes cross-linking thus hardening the material. As a result of this hardening process the moulding sets into the shape of the cavity and can be removed after a predetermined time, e.g. 3 minutes. This time can be reduced if the material is preheated (for example, to 70°C) before being placed in the mould.

When the moulding is removed from the press it still has attached to it a thin web of flash and this must be removed, by buffing or sanding, before the moulding can be used. Despite the flash produced the amount of waste produced by this process is relatively low as, for example, there is no feed system to be re-used as in injection moulding.

Thermosetting plastics mouldings with a high gloss can be produced by compression moulding, at high rates, if fine powders are used and if the powder is preheated before being placed in the mould. This is because the moulding powder is in contact with a hot polished-metal surface continuously and is not required to flow very far.

The process has been used for large and small mouldings which do not require extremely close tolerances and which do not involve the use of delicate inserts. Such inserts would be disturbed when the material starts to flow. In the rubber industry the process has been used to make car mats, seals, gaskets, etc. In the plastics industry the process is used for materials such as aminoplastics which are moulded into products such as bottle caps, light fittings, tableware and switch plates. Because of the forces involved, and because there are two main parts to the mould, tool costs are fairly high. The mould must be made of tool steel and it is the machining of this material which makes the mould expensive.

THERMOPLASTICS
When compression moulding a thermoplastics material, the product must be set to shape by cooling the mould before it can be ejected: this can take a long time unless the equipment is specifically designed for this process. Even if the equipment is so designed (as for gramophone records) then the process will be energy inefficient because of the rapid heating and cooling required if the operation is to be a commercial success.

Compression moulding of thermoplastics is therefore only usually performed when the number of mouldings is relatively small and/or when there are specific advantages in doing so. For example, it is found that moulded sheets have more uniform properties when produced by compression moulded as compared to injection moulding. That is, there is a reduction of *anisotropy* and this feature is useful in the production of test samples. It was found many years ago that good reproduction of the mould surface can be obtained and this is particularly important in gramophone record production. In order to get strong stiff composites based on engineering thermoplastics materials there is now considerable commercial interest in thermoplastics composites: such composites may be compression moulded so as to retain fibre integrity.

compression moulding press
A press used for *compression moulding*: a compression moulding machine. Could be a *mechanical press* but, more commonly, a *hydraulic press*.

compression moulding shrinkage
The *shrinkage* experienced as a result of *compression moulding*. Because of the lower mould temperatures employed, in compression moulding, *shrinkage* may be lower than that found in *injection moulding*: it will also be more uniform.

compression ratio
Screw terminology. The ratio of the volume of one flight of a screw at the feed end of a machine to the volume of one flight at the discharge end. The channel depth often changes along the length of a screw used in plastics processing; it is deepest under the hopper and shallowest at the screw tip. This means that the screw has a compression ratio and it is there to compensate for the effective reduction in volume that occurs on melting the plastics granules. The compression ratio is related to channel depth so if the depth is 0·373" under the hopper and 0·125" at the tip, then the compression ratio is said to be 3:1. Typically general purpose screws have a compression ratio of approximately 2·0/2·5:1 and a *length to diameter ratio (L:D)* of about 20:1.

compression set
The set, or change in dimensions which remains after a *compressive strain* has been removed from a test sample for a certain time, for example, 30 minutes.

compression zone
In blow moulding, it is that part of a mould which is also called a dam. The dam forces material back into the pinch area, on mould closing, and thus leads to localized thickening in this weak area. A compression zone is also present on many extruder screws, towards the discharge end, where its presence results in melt pressure generation. See *compression ratio*.

compression-less extruder screw
An extrusion screw which does not have a compression zone: a *constant depth screw*. See *vario screw* and *zero compression screw*.

compressive strength
The ability of a material to withstand a compressive stress. A strength test performed in compression. The stress needed to cause failure of a material in compression. The strength determined from the load and the test piece dimensions in a compression test. See *crushing strength*.

computer aided design
An abbreviation used for this term is CAD. See *computer-aided design*.

computer aided selection of materials
Selection of a material using a computer. Many systems are available for this purpose, for example, from the suppliers of thermoplastics materials. Often such a system is based on the user answering pre-selected questions - the answer guides the user to the next question, and so on, and then to the material.

computer control
A mode of machine operation where, for example, the *extruder* and the extrusion line, are under the control of a process computer, or microprocessor.

computer numerical control machine
An abbreviation used for this term is CNC machine. For example, a numerically controlled milling machine which is under the control of a computer. A program is prepared which specifies the shape of the component and the cutting conditions; for checking purposes the computer may draw on a plotter the shape to be cut and the cutting path. After checking, the program is converted into machine-tool language by the computer and fed into the CNC machine. The bolster plates, followed by the cavity shape in soft metal, are then automatically machined using optimum cutting rates. When CNC machines are fitted with the correct tooling, and when used with the correct lubricants, steels with a hardness of over 60 Rockwell C can be machined. CNC is more flexible than NC as once NC tapes are cut they are awkward to modify.

With CNC the tape is simply the output of an easily edited program; dimensions and instructions are keyed into the computer or fed in from tape.

computer-aided design
An abbreviation used for this term is CAD and, as the term implies, a computer is used by the mould designer to shape model the proposed component or moulding, translate that shape into an efficient mould design and produce fully annotated engineering drawings. If while the CAD system is being used for design purposes a numerical-control (NC) tape is also generated, the system is known as CAD/CAM. See *stereolithography*.

computer-aided drafting
The use of a computer to provide a fast and accurate method of generating fully annotated engineering drawings without the need for a drawing board. A computer-driven drafting table will produce blueprint drawings which are fully dimensioned and which contain appropriate instructions and legends. Such systems are very useful when repetitive or routine work is being handled. For example, if a multi-cavity mould is being designed, the designer need only complete one cavity; the computer program will then reproduce this drawing in any selected position. Standard sections or profiles may also be stored so that they may be recalled as and when required.

con
An abbreviation used for *constantan*. See *thermocouple*.

concave tread tyre
This type of *cross-ply tyre* is also known as a depressed-crown contour moulded tyre. When un-inflated the tread shape is concave; when inflated the tread shape flattens and this puts more uniform pressure on the *contact patch*. This results in improved road grip and reduced wear. However, the tyre must be at the correct inflation pressure to realise the benefits.

concentrate
A *masterbatch*.

concentrated latex
Latex whose rubber content has been increased. For example, as tapped, natural rubber latex may have a rubber content of <40%; this may be increased to approximately 60 to 65% by *creaming*, *centrifuging*, *electrodecanting* and *evaporation*.

concentrated masterbatch
See *universal masterbatch*.

concentric cylinder viscometer
A *rotational viscometer* in which the shearing force between two concentric cylinders is measured. The inner concentric cylinder is rotated at a constant speed and the resistance to movement is measured. Commercial examples of such machines include the Brookfield viscometer and the Ferranti viscometer.

condensate
The hot water which is formed when steam cools. The condensate must be frequently removed from a steam heated system if successful heating is to be obtained. This is most commonly done by the use of a *steam trap* but the prevention of condensation by good insulation must never be forgotten.

condensation
The change of vapour into liquid, for example, the conversion of steam to water (steam is colourless what is often termed steam is in fact tiny drops of water). In chemistry, the term condensation means a reaction between molecules which eliminates water or some other simple substance. The term polycondensation is sometimes used when a polymer results: used, for example, to make polyamides. See *condensation polymerization* and *condensate*.

condensation cross-linked silicone polymer
A *silicone* polymer. Silanol end groups in branched polymers are condensed to give a cross-linked structure in a reaction catalyzed by a base. See *multi-functional organosilicon cross-linking agent*.

condensation polymer
A *step-growth polymer* produced by *step-growth polymerization*. A polymer made by *condensation polymerization*. A synthetic polymer in which the long chain structure was built up by the chemical reaction which occurred between reactive, or functional, groups.

condensation polymerization
Also known as polycondensation. Polymerization in which monomers are linked together with the splitting off (elimination) of water or other simple molecules to give a *condensation polymer*.

condition
Sometimes used to refer to the test condition in *flow rate testing*. For example, condition 190/2·16, formerly known as condition E, means that the specified temperature is 190°C and the specified weight is 2·16 kg.

condition E
See *condition* and *melt flow rate*.

conditioning
Subjecting a test specimen to a controlled environment. See *test sample conditioning*.

conditioning tank
A vessel used to store, and/or condition components, on, for example, a *reaction injection moulding* machine. A conditioning tank accepts material from a *supply tank* and is used to control temperature and degree of dispersion of a reactant by low pressure recirculation through that vessel. During the recirculation process, an inert gas may be injected into the liquid reactant so as to compensate for shrinkage in the final product: the introduction of the gas may be referred to as nucleation.

conducting compound
A compound which is an electrical conductor. To be classed as conducting, a polymer compound has to have a volume resistivity less than 100 ohm.cm. For insulation purposes the high resistivity of plastics is an advantage but, in some cases, it can be a serious disadvantage as it results in a high static charge, build-up: this in turn can result in dust pick-up and/or spark generation. The established way of improving conductivity is by adding a conductive filler such as, a high structure, *carbon black*. The addition of lubricants can minimize the generation of static while the addition of some semi-incompatible liquids can cause static to leak away. See *antistatic agent*.

cone calorimeter
A *fire testing* apparatus which was originally developed for the measurement of heat release of materials. The sample of material is heated on a load cell by a cone-shaped heater. The combustion products are drawn away, at a controlled rate, through an exhaust system where solid material is collected by a soot collection filter: the smoke density may be measured by a laser detection beam and the temperatures of the evolved smoke and gases measured. Samples of gas may be removed for chemical analysis and identification.

conical blender
A mixing device used for blending or pre-blending of solid thermoplastics materials with their additives. This type of blender uses a cone-shaped container to improve the *folding, spreading action* which is desired. Equipment for mixing small batches, e.g. 10 kg (22 lbs) is formed from two circular cones

which are clamped across their base after the ingredients have been added. Such a construction means that the interior is readily available for cleaning. Conical blenders used for large batches are usually fabricated so that they are of one-piece construction and access is by means of hatches located at each end of the double-cone: machines are available which will mix 1 tonne of material when *compound blending*.

conjugate fibre
Also known as *bicomponent fibre*.

conjugated
Means that a chemical compound has two double bonds separated by a single bond. See *diene*.

consistency
A term sometimes used in rubber technology in place of *thixotropy*.

consistency index
An abbreviation used for this term is K. See *power law equation*.

constant depth screw
Also known as a parallel screw. A screw in which the channel depth is constant along its length i.e. the root of the screw is of the same constant diameter. Such a design is sometimes used in *blow moulding* machines as such a design minimizes heat generation.

constant viscosity natural rubber
An abbreviation used for this term is CVNR. See *constant viscosity rubber*.

constant viscosity rubber
An abbreviation used is CV rubber. Also known as constant viscosity natural rubber (CVNR) or as viscosity stabilised natural rubber. Natural rubber that has been chemically treated so as to prevent unwanted hardening reactions (cross-linking reactions) occurring on storage. Freshly tapped *latex* is treated with an *aldehyde-condensing agent*, for example, with 0·15% of hydroxylamine hydrochloride which reacts with aldehyde groups which otherwise would cause cross-linking of the natural rubber. The use of constant viscosity natural rubber (CVNR) can substantially reduce, and in some cases eliminate, the energy intensive mastication process. See *storage hardening*.

constantan
An abbreviation used for this term is con. An alloy of copper and nickel and which contains from 10 to 55% of nickel. The electrical resistance of such alloys does not vary with temperatures. Used to make thermocouples: the *thermocouple* is based on, for example, the use of iron (Fe) and of constantan.

constitutional unit
See *repeating unit*.

contact adhesive
An *adhesive* which bonds under slight pressure when two surfaces, coated with the adhesive, are brought into contact.

contact batching
See *surface winder*.

contact heating
Term used in *thermoforming* and means that the sheet is heated through contact with a heated platen, or sole plate, so as to achieve rapid heating rates. When used with biaxially oriented material, such heating can reduce initial changes in sheet thickness. Air applied through the *female mould*, blows the sheet up against the heating platen. See *pressure assisted thermoforming*.

contact moulding process
See *hand lay-up process*.

contact patch
That part of the *tyre tread* in contact with the road and which, for example, produces the cornering force when the tyre is distorted by steering. The contact patch then runs at a small angle to the direction of travel of the wheel: this angle is known as the *slip angle*. A smooth tyre would give the best grip on a dry road, as the contact patch would be large, but would be virtually useless on a wet road as there would be limited road contact.

contact pressure
A term used in *compression moulding*: a low pressure is applied to the mould. The pressure is only just sufficient to keep the mould in contact with the platens.

contact pressure moulding
A method of moulding, or laminating, in which a low pressure, usually less than 70 kPa (10 psi), is used.

contact ultrasonic welding
An *ultrasonic welding* technique for thermoplastics which is sometimes called near-field welding. This ultrasonic welding technique is used for joining film and sheet: the interface is less than 6 mm/0·25" from the point of ultrasonic contact.

contact viewing
See *see-through clarity*.

contact welding
A welding process used for thermoplastics materials and which relies on the use of a *resistance heater*. A tip or wheel is heated and pressed against the plastic surface which is in contact with another plastic substrate. Both continuous contact welding and discontinuous contact welding is performed. See, for example, *impulse welding* and *hot plate welding*.

container barrier properties
See *barrier properties*.

contamination
A term used in *hydraulics* and which refers to any foreign material that hinders the performance of a hydraulic system.

continuous cure
See *continuous vulcanization process*.

continuous fibre reinforced prepreg
A *prepreg* containing continuous fibres. May also be known as a continuous fibre reinforced towpreg if the fibres (for example, glass or graphite) are 'towed' through, for example, a solution of a thermoplastics material. The solvent (for example, methylene chloride for polyetherimide and polyarylsulphone) is removed by heating and then the fibre assembly passes through heated consolidation dies to produce the continuous strip of prepreg (for example, 152 mm wide by 0·2 mm thick): such a prepreg may contain 60% fibre by volume. That is, *pultrusion* is used to produce the prepreg from thermoplastics coated fibres. This composite material is then fabricated to shape by, for example, *compression moulding* which consolidates the stacked plies of the prepreg.

The fibre type and *orientation* has a tremendous effect on properties. The temperatures used depend upon the polymer. For example, press temperatures of 345°C are suggested for PEI and PAS: PEEK prepreg may require 370°C. Moulding pressures of 2,000 kPa are suggested: the pressure may need to be applied or held for 15 minutes before cooling. Release the pressure at the end of the cooling stage.

continuous fibre reinforced towpreg
See *continuous fibre reinforced prepreg*.

continuous filament mat
Also known as continuous strand mat or as, swirl mat. An abbreviation used for this material is CFM. Glass is drawn through bushings and the resultant fibre is deposited on a

belt evenly and randomly so as to produce a layer of mat (CFM consists of randomly orientated, looped filaments). As the belt progresses the mat is treated with a *binder*, heated, calendered (where necessary), rolled and packed. The use of such a mat gives products of good strength in *polyester moulding compositions*: it offers very little resistance to resin flow and is quickly impregnated in processes such as *resin transfer moulding*. Continuous filaments provide better and more consistent mechanical properties than mats made from *chopped strands*.

continuous filament yarn
There are two main types of *yarn*; continuous filament yarn and *spun yarn*. Synthetic continuous filament yarns are the most common type used in *composites*. They are based on parallel, continuous filaments of uniform cross-sectional area. Commonly based on *glass* but see *specific modulus*.

continuous mixer
A mixing machine used to prepare polymer compounds and which produces a continuous flow of compound as long as it is fed with the ingredients of the mix: such machines are usually based on *extruders*. See *batch mixer*.

continuous rotary curing machine
See *rotocure*.

continuous roving moulding compound
A wound *polyester moulding compound*.

continuous screen changer
A straining system fitted to an *extruder*: a continuous strip of mesh is drawn across the melt stream so that it is not necessary to stop the extruder when the screen becomes blocked. Such devices may be actuated by a *pressure transducer* mounted within the die.

continuous strand mat
See *continuous filament mat*.

continuous vulcanisation
An abbreviation used for this term is CV. Vulcanization which is done on a continuous basis rather than a batch basis. For continuous vulcanisation of extruded products, the extrudate is passed continuously through a heated bath. Two main types of bath are used, a *molten salt bath* and *fluid bed continuous vulcanisation*. See *rotocure*.

continuously impregnated compound
See *polyester moulding compound*.

contouring
See *roll cambering*.

contra-rotation
To rotate in opposite directions. See *twin screw extruder*.

control loop
The signal circuit that provides feedback information for *closed loop*, process control.

control system
A device used to regulate the function of a unit. The instruments, and power controlling units, which are used to hold machine temperatures, pressure, rate, and other production parameters, to the set values.

Control of Substances Hazardous to Health
The Control of Substances Hazardous to Health (COSHH) regulations came into force in the UK on October 1st 1989. Such regulations seek to safeguard the health of workers. For example, in the glass reinforced plastics (GRP) industry where styrene emissions, during open mould curing, are known to introduce dangers to health. COSHH regulations spell out very clearly that it is now the responsibility of employers to make their personnel aware of the nature of hazards/risks and what precautions are needed.

controllability
A term used in *hydraulics* and which refers to the finest adjustable increment possible for a *pump*.

contacting rotary shaft seal
A type of *rotary shaft seal*.

controlled rheology polypropylene
A type of *polypropylene (PP)* which has a narrow molecular weight distribution: this gives shrinkage uniformity. The high molecular weight fraction can be removed from the PP by the use of peroxides.

controlled rotational casting
A rotational casting process used to insulate pipe with *rigid polyurethane (PU) foam*. Streams of reacting PU are sprayed over a rotating pipe or, jets circle a pipe spraying the pipe with streams of reacting PU.

controller
A controller is a discrete, or dedicated, instrument used to control, for example, temperature, in the production operation. Usually named after that which it is controlling. Can have, for example, a temperature controller, a speed controller, a pressure controller etc.

convoluted hose
A hose which has regular annular bellows-like corrugations: such a hose may be made by moulding or by extrusion.

cooler
A term used in *hydraulics* and which refers to a heat exchanger which removes excessive heat from the hydraulic system.

cooler-blender
A cooled mixing device which is, for example, used to cool the output from a *high speed mixer*. Rapid cooling is achieved by using a large (about twice the capacity of the high-speed mixer), water-cooled vessel equipped with a slowly rotating sweep e.g. at 30 rpm. After the temperature has fallen to an acceptable level the dry blend is conveyed to a storage silo so that any electrostatic charges can be dissipated. Such charges are introduced during the high-speed mixing operations and unless they are removed, e.g. by storage in a well-earthed silo, they can interfere with material handling and/or feeding.

cooling can
A cooling roll used in, for example, *calendering*. The cooling cans are usually of double skin construction with forced water circulation between the inner and the outer skin. Such a construction minimises problems of balancing and can give smoother rotation than that found for a hollow drum filled with water. As such drums are commonly 0·6 m (2 ft) in diameter balancing is an important consideration.

Cooling rolls are carried in roller bearings and may be driven by DC motors which transmit the drive through reduction gear-boxes. Such motors may be linked to the main calender drive so that if the speed of the calender is altered the cooling roll speed follows suit.

cooling channel
A channel or coil through which a fluid is circulated so as to remove heat, or to control the temperatures, during processing. The majority of moulds are designed with insufficient cooling channels and therefore require a longer period of time to dissipate the heat than should ideally be required. This in turn, directly effects the overall cycle time.

For example, when processing thermoplastic materials, the heat which is carried into the *injection mould* is removed by (a) conduction into the machine, (b) radiation, and, (c) a fluid circulated through the mould. In most cases (c) is the most important and is usually achieved by water circulation although water/glycol mixtures and oil are also used. For a given material, the rate at which this heat is removed from the mould is dependent upon the number of cooling channels, the length of cooling channels, the diameter of the cooling channels and the rate at which fluid is circulated. It is essential to ensure that the right weight/volume of fluid is circulated, at the correct rate, for each part of the mould.

It is a good rule-of-thumb guide to commence with a cooling channel diameter of 10 mm (0.4 in), to locate each channel at approximately one and a half diameters from the surface of the mould cavity and separated by three diameters pitch. The cooling circuits should be positioned symmetrically so that each cavity, in a multi-cavity mould, receives the same treatment. If possible, each cavity should be treated as a separate cooling circuit and should be equipped with its own inlet, outlet, flow-control valve and a temperature measuring point. Even on a single-impression mould it will be much better to employ a number of short independent cooling circuits with each circuit being capable of independent control. What this means is that parallel cooling is very often preferred to series cooling.

cooling fixtures
Holding devices or jigs (sometimes fitted with air or water cooling) used for holding and setting a formed shape.

cooling fluid
Liquid circulated through a system in order to remove heat, for example, circulated through the *cooling channels* of an *injection mould*. Such fluids may be at a relatively high temperature in order to obtain the required properties in the product.

cooling roll
See *cooling can*.

cooling table
A support for a cooling extrudate which is often a roller conveyor. See *extrusion*.

coordination
Means that a group of controllers are connected together so that they may all be changed at the same time from a single point: often applied to drive systems, for example, the speeds of the doser, the extruder and the haul-off may all be ramped up or down simultaneously.

coordination polymer
A polymer which consists of groups (for example, *phosphinate* groups) linked together with metal atoms.

COPE
An abbreviation used for copolyetherester. See *polyether ester*.

copolycarbonate
A *polycarbonate copolymer*.

copolyester
A synthetic *copolymer* which contains ester-type linkages and which is based on more than one type of acid monomer and/or, on more than one type of alcohol-type monomer. A term sometimes used to describe a type of *polyethylene terephthalate (PET)* polymer, for example, Kodar PETG. A copolymer of irregular structure which is based on *terephthalic acid*, *isophthalic acid* and the glycol known as cyclohexanedimethanol. An *amorphous thermoplastics material* (the partial use of isophthalic acid retards crystallinity) which can give high clarity mouldings and formings. This type of thermoplastic polyester can tolerate up to 0.02% moisture and has a relatively wide, processing temperature range: it should still be dried before *melt processing*. Although screws suggested for *polyolefin-type materials* can be used for extrusion, best results have been obtained with *barrier-type screws*. The molecular weight of the material used for extrusion is of the order of 26,000 (number average molecular weight). See *glycol modified polyethylene terephthalate*.

copolyether ester
See *polyether ester*.

copolyetherester
See *polyether ester*.

copolymer
Sometimes called a biopolymer. A polymer made by polymerizing two monomers together, for example, *ethylene-vinyl acetate (EVA)*: the second monomer is added to improve properties such as adhesion or impact strength.

Most synthetic copolymers are random copolymers: that is, the repeat units occur along the chain in a purely random manner. With an alternating copolymer, the repeat units are randomly separated by the other unit. Block copolymers have long sequences of one unit and then long sequences of the other unit. A diblock copolymer contains two blocks: each contains one type of repeat unit. A triblock copolymer contains three blocks: two of the blocks (usually the end blocks) contains one type of repeat unit while the central block contains the other type of repeat unit. A tapered block copolymer, has long sequences of one type of repeat unit which gradually contain more and more of the other repeat unit. A graft copolymer, has chains of one polymer grafted onto chains of another polymer. Such synthetic polymers do not have the precise, regular structure of natural materials.

To locate a copolymer look for the name of the major monomer. For example, for copolymers of α-methylstyrene and styrene, see *styrene-α-methyl styrene copolymers*. See *abbreviations*.

copolymer repulsion effect
Also known as the copolymer effect. Used to make a useful *blend* based on two polymers which are immiscible. An immiscible A/B system can be compatibilized by copolymerizing B, or A, with a third monomer C so as to form an A/BC system. Repulsive inter-action between the B and C components exceeds that between A and B, or between A and C, and makes the system miscible for the appropriate B/C composition ratio.

copolymerization
The technique used to produce a *copolymer*, i.e. polymerizing two monomers together.

copper
This element (Cu) has a relative density of 8.93 and occurs in Group 1B of the periodic table. The three metals in Group 1B of the periodic table (copper, silver and gold) are sometimes known as the coinage metals. It is a soft, lustrous metal which is renowned for its high electrical and thermal conductivity. It melts at 1,083°C.

Even minute traces of copper (ppm range) can cause the catalytic decomposition, or degradation, of polymers such as *natural rubber* or *polypropylene*. All copper compounds are not equally active or, have the same catalytic effect: can have active copper compounds and passive copper compounds. The strongest catalytic effect is found with compounds based on fatty acids (for example, stearates and oleates) and with other rubber soluble materials. The least active are inorganic compounds while complex copper compounds (for example, the phthalocyanine colour pigments) are inactive or passive. Ingredients may be added to, for example, rubber compounds in order to inhibit the effect of the copper. See *inhibitor* and *oxidation*.

copper arsenite
See *Scheel's green*.

copper diethyl dithiocarbamate
A dark brown *accelerator*, for example, for styrene-butadiene rubber. An abbreviation used is CUDC. This material has a relative density (RD or SG) of 1·70 and a melting point of 196°C. May be represented as $[(C_6H_5)_2N \cdot CS \cdot S]_2Cu$. See *dithiocarbamates*.

copper dithiocarbamates
See *dithiocarbamates*.

copper flakes
A *flake pigment* based on copper. Metallic colours are produced by the incorporation of *aluminium flakes* or of, *copper flakes* or of, *bronze flakes*. Copper flakes and bronze flakes may tarnish during processing unless coated or protected.

copper inhibitor
A material added to polymer compounds in order to render the copper inactive. See *copper*.

copper phthalocyanine blue
The major *phthalocyanine pigment*: a blue pigment which consists of a copper atom linked to 4 isoindole rings via 4 nitrogen atoms.

copra
Obtained by drying the meat or flesh of the coconut. When copra is crushed an oil is obtained. See *coconut oil*.

copra oil
See *coconut oil*.

cord
A tightly twisted thread which may be made from both natural and synthetic materials. Used as a reinforcement in, for example, for tyres.

core
In *extrusion*, a core is a circular tube on which flexible sheet is wound, for example, a cardboard core. In moulding, it is the solid part of the mould which gives the inside shape to the moulding i.e. the protruding or male part.

core insert
A metal block which incorporates the *core*. The insert is fitted into a *bolster* to form the *core plate*.

core pin
The centre unit, of a tubing die, used to form the inner wall of the extrudate: also known as a *torpedo*. Core pins are also used to form holes in *injection moulded components*.

core plate
A plate, or metal block, which incorporates the *core*. The core may be formed from a solid piece of metal (integer core) or, it may of multi-part construction (an insert bolster).

core shaft
The rear portion of a rotating core.

core shaft plate
A steel plate located behind the *gear plate* in an *unscrewing mould*.

cored screw
A *screw* which has been bored with an axial hole so that fluid may be passed along the screw for temperature regulation. See *extrusion*.

cork
A natural material obtained from the bark of the evergreen tree called a cork oak. When ground, cork (powdered cork) is used a filler in rubber compounds as its incorporation gives the compound a high degree of resilience and compressibility. Used in flooring, gaskets and tiles. The relative density (RD or SG) of cork can range from 0·4 to 1·4.

cornering force
The force which the *contact patch* of a *tyre* exerts when it is distorted by steering: the force which pushes the vehicle away from the straight-ahead path.

corona discharge
A corona is a luminous discharge, around a conductor, caused by the voltage exceeding a critical value. Such a discharges is used as a technique for modification of a surface so as to make it more receptive to inks by, for example, oxidation; it is used to treat *polyolefins*.

corotation
See *twin screw extruder*.

corrosion resistant steel
A *steel* which will resist corrosion because of its chemical composition. Such steels are usually supplied relatively soft and are hardened after mould manufacture: this gives a steel with a hardness in the region of $R_C 52$. Such a steel is AISI 420 which is magnetic and can be oil quenched: another is X 36 CrMo 17 which contains 17% chromium and 1% molybdenum. See *stainless steel*.

corundum
Aluminium oxide. This common mineral is noted for its hardness (on the Mohs scale it is 9): it is used as an abrasive. The relative density (RD or SG) is 4·0.

COSHH
An abbreviation used in the UK for *Control of Substances Hazardous to Health*.

cost per unit volume
See *density*.

cotton
A naturally occurring material which is a major source of *cellulose* for the chemical industry. Harvested from an annual plant, a shrub, which grows in sub-tropical regions. The cotton lint is obtained from the seed pods or bolls: the fibres are attached at their bases to hard black seeds from which they must be removed by *ginning* (see *alpha-cellulose*). Derivatives of cotton for the polymer industry, include *cellulose esters* (for example, cellulose acetate and cellulose acetate butyrate) cellulose ethers (for example, ethyl cellulose) fibres (*regenerated cellulose fibres*) and film (*cellophane*). The relative density (RD or SG) of cotton, and of cotton flock, is 1·45.

Fabrics made from *cotton* have several desirable characteristics. For example, unlike wool it is not as liable to shrink nor to attack by moths. It is stronger wet than dry and will therefore withstand repeated washing. Very thin fabrics are possible which are cool, pleasing to the touch and of an attractive finish. However, cotton is easily creased but this disadvantage can be minimized by resin treatments using amino resins. Urea resins are widely used to impart crease resistance and 'drip dry' qualities. The fabric is passed through aqueous solutions of hydroxy methyl ureas and the resin is hardened by heating (130 to 160°C/266 to 320°F) in the presence of metal salt catalysts. See *rayon*.

cotton gin
See *saw-gin*.

cotton linters
The short fibre removed from *cotton seeds* after the long fibres (used in textiles) have been removed by *ginning*.

cottonseed oil
An edible vegetable oil derived from gossypium hirsutum or gossypium barbadense. Cotton fibres are attached at their bases to hard black seeds from which they must be removed (see *ginning*). A brown-yellow oil is obtained from these seeds: the relative density (RD or SG) of this oil is 0·92.

Couette-Hagenbach method
An alternative to the *Bagley correction* and used for correcting for end effects. This is a 'difference' method in which two dies of different lengths L_1 and L_2 are used and for each of which the output Q for a series of pressure drops ΔP are obtained. Since for each die at a given shear rate the entrance effect is the same, the pressure drop and length of the shorter die can be subtracted from those of the longer die to give an estimate of the pressure drop $\Delta P_1 - \Delta P_2$ for a hypothetical die of length $L_1 - L_2$. From this the true shear stress at wall may thus be obtained from $\tau_w = (\Delta P_1 - \Delta P_2)/2(L_1 - L_2)$.

While this approach does correct for the end effects it gives no idea of the magnitude of the error which could be of interest in scale-up.

coulomb
An SI derived unit which has a special symbol, that is, C. It is the unit of electric charge. It is the quantity of electricity transported by one ampere in one second. That is, it is 1 ampere second. See *Système International d'Unité*.

coumarone
Also known as benzofuran. See *coumarone-indene resin*.

coumarone resin
A resin based on the acid polymerization of *coumarone*: more usually it is a resin based on a mixture of coumarone and indene as these two materials have similar boiling points. See *coumarone-indene resin*.

coumarone-indene resin
Also known as coumarone resin or as, indene-coumarone resin or as indene resin. An abbreviation used is CIR. Coumarone and indene are obtained when *coal tar naphtha* is fractionated at approximately 170°C. A copolymer is obtained if a mixture of these two materials is treated, for example, with concentrated sulphuric acid at temperatures below 0°C. Dark coloured materials (ranging from sticky liquids to solids) which are compatible with a wide range of other materials, for example, with paints, plastics (PVC) and rubbers. Used as, for example, softeners in rubbers. The relative density (RD or SG) of this type of material is approximately 1·11.

counter ions
The ions which neutralise the charges of the bound ion in an ionic polymer. The counter ions (counterions) may be univalent, divalent, trivalent or polyvalent.

counter rotation
See *twin screw extruder*.

counter-draft
See *reverse taper*.

counterbalance valve
A term used in *hydraulics* and which refers to an automatic control valve which prevents a load from over-riding an *actuator*.

coupling
A chemical reaction used in *organic pigment manufacture*. Azo pigments are formed by *diazotization* (of a primary aromatic amine) and coupling (with say, 3-hydroxy-2-naphthoic acid or BON) so as to achieve the desired high molecular weight, structure. See *diazonium salts*.

coupling agent
A material which couples, or bonds, a *filler* to a resin matrix and so improves properties such as modulus. See *silane coupling agent*.

cover
An external protective covering of polymeric material over a hose applied, for example, to prevent mechanical damage.

cP
An abbreviation used for *centipoise*.

CP
An abbreviation used for *cellulosic plastics*. CP may also be used for cellulose propionate. An abbreviation used for *chemically pure (cadmiums)*.

CPA
An abbreviation used for the Composites Processors Association - a UK based organization.

CPC
An abbreviation used for *cavity pressure control*.

CPD
An abbreviation for *cadmium pentamethylene dithiocarbamate*. See *dithiocarbamates*.

CPE
An abbreviation used for *chlorinated polyethylene*.

CPET
An abbreviation used for crystalline *polyethylene terephthalate*.

CPI
An abbreviation used for cyclized polyisoprene. See *cyclized rubber/azide-type photo-sensitizer*.

CPM
An abbreviation used for *cold press moulding*.

CPU
Abbreviation for central processing unit; part of a computer.

CR
An abbreviation used for *controlled rheology (polypropylene)*. CR is also used as an abbreviation for *chloroprene rubber* and for *cyclized rubber*.

Cr-Mn steel
A chrome-manganese steel. For example, AISI-type *P2* is a commonly used *case hardening steel*. See *mould steel*.

Cr-Mn-Mo steel
A chromium-manganese-molybdenum steel. See *pre-hardened steel*.

Cr-Mn-Mo-S steel
A chromium-manganese-molybdenum-sulphur steel. See *pre-hardened steel*.

cracker
A type of *two-roll mill*. A mill of heavy construction with deeply corrugated rolls.

cracking
The production of simple chemicals from complex ones by heating or, the appearance of fine cracks, or crazes, in a product.

cracking pressure
A term used in *hydraulics* and which refers to the pressure at which a valve begins to open or pass fluid.

crammer feeder
A hopper unit which forces plastic material into the feed throat of an *extruder*.

crash cooling
Fast cooling. The barrel of an *injection moulding machine* used for rubbers, and for thermosetting plastics, may be fitted with cooling circuits so that, when activated, significant compound curing does not occur.

crater
A small, shallow surface imperfection.

crazing
Apparent fine cracks at, or under, the surface of a plastics component; see *cracking*. Such hairline cracks may be generated within an *unsaturated polyester resin*, on curing, and are caused by stresses introduced during cure.

crazing effect
Also called weather skin and elephant skin. A defect that can occur with non-black rubber components and which is caused by weathering and light ageing. See *oxidation*.

cream
That which is obtained as a result of *creaming*. See *concentrated latex*. Rubber latex will separate, or cream, on standing as the particles have a lower relative density than water. The particles of rubber in natural rubber latex have a relative density of 0·91 and the *serum* has a relative density of 1·02.

creaming
A continuous method of producing *concentrated latex*. Natural rubber latex will *cream* on standing because of density differences: creaming can be accelerated by adding small quantities of colloids such as *sodium alginate* and/or *carob seed flour*. In 4 to 6 days, after addition of colloids, latex will separate into the cream which contains approximately 60% rubber and into a serum which contains little rubber. Creaming is sometimes referred to as Traube's process.

creep
Deformation over a long period of time under mechanical stress. The time-dependent part of strain resulting from stress. If a load is applied to a plastics component under specified conditions of temperature and humidity a deformation or strain will result. Creep is defined as the total strain which is time dependent resulting from the applied stress. The stress may be applied by a *tensile* or a *compressive load*. Creep is a time dependent deflection.

creep fracture
See *static fatigue*.

creep modulus/time curve
A way of expressing the results of *creep testing*. A modified form of the *isometric stress-log time curve* obtained by plotting creep *modulus* against log time, A family of tensile creep modulus curves, as a function of strain and log time, may be obtained. The creep modulus may be substituted into classical elastic design formula.

creep rupture
Failure induced in a ductile material as a result of embrittlement caused by ageing under an applied load. See *creep*.

creep test
A long term test used to measure *creep*.

creep testing
Testing performed to determine the effects of *creep*. Short term tests, such as *tensile* or *flexural tests*, are not capable of giving information which could be used, for example, in the design of a continuously stressed component in a particular environment. This is because the conditions of test may not reflect the way in which the component is to be loaded.

Creep tests are often performed by weight loading a plastics or rubber sample maintained at a specified temperature. periodic measurements of deflection or extension are taken, possibly over several years. Such tests indicate that quite small loads applied for long periods can cause significant deformations of plastics materials. However, such deformation is not necessarily permanent. Because of the industrial relevance, creep tests are commonly performed on *engineering thermoplastics*.

A *dumb-bell specimen* is chosen which has a very long parallel or waisted section, of approximately 80 mm, so that the small deformations which results on loading may be measured accurately. The material is tested at a specific temperature by applying a weight by way of a lever system. Because even quite large loads will produce only small extensions a very sensitive and accurate extensometer is required. Extensometers based on the optical lever principle have been widely used, but those based on the moiré-fringe principle and on very sensitive displacement transducers are more easily adapted to automatic data recording and processing by computer analysis. Care is needed to ensure that no electrical or thermal drift occurs in the strain measuring device.

For convenience, the specimens are often machined from *tensile test dumb-bells* or actual mouldings. Sample conditioning is extremely important and may involve regulation of test temperature and/or humidity for several months so as to ensure that the specimens are in equilibrium with their surroundings. For example, in the case of nylon 66 such storage will allow water take-up to occur as well as allowing post-moulding crystallization to develop.

creep-curve
A way of expressing the results of *creep testing*. The *strain* resulting from a constant stress is plotted against time. By using different loads a family of curves may be assembled, but in order to get a more convenient scale, strain is plotted against log-time. Such an assembly of curves provides a relationship between stress, strain and time under specified conditions of temperature and humidity.

crepe rubber
One of the forms in which *natural rubber* is supplied. After *coagulation water* is excluded by compression and sheets of rubbers are formed by rolling and drying or smoking. See *pale crepe* and *extra white crepe*.

crepeing battery
A production unit of say, three *two-roll mills* in which *natural rubber* is thoroughly washed so as to remove impurities and to produce *pale crepe*.

cresol
An aromatic material which may be known as hydroxytoluene or as, methylphenol. The three isomers, ortho, meta and para occur in cresylic acid: it is the meta isomer which is reacted with formaldehyde to make *cresol-formaldehyde resins*.

cresol diphenyl phosphate
See *cresyl diphenyl phosphate*.

cresol-based resole
A *resole* which uses cresol as a starting material rather than phenol. Gives better electrical insulation properties than phenol-based resoles.

cresol-formaldehyde
An abbreviation used for this type of material is CF (see also casein formaldehyde). A polymer formed by condensing cresol with *formaldehyde*; the isomer used is m-cresol. Thermosetting polymers which are sometimes used in place of, or with *phenol-formaldehyde*, to give improved chemical resistance. Such resins are sometimes known as cresols.

cresols
See *cresol-formaldehyde* and *methylphenols*.

cresyl diphenyl phosphate
Also known as cresol diphenyl phosphate or as, tolyl diphenyl phosphate. An abbreviation used for this material is DPCP or DPCF or CDP. A *triaryl phosphate*. A phosphoric acid derivative. This material has a relative density (RD or SG) of 1·21, a melting point of $-40°C$ and a boiling point of approximately $390°C$. May be represented as $(CH_3C_6H_4O)(C_6H_5O)_2PO$.

A low cost, flame retardant plasticizer with slightly superior low temperature properties than *tricresyl phosphate*. It is a plasticizer for *polyvinyl chloride (PVC)* used, for example, for *plastisols*. It is compatible with cellulose acetate butyrate (CAB), cellulose nitrate (CN), ethyl cellulose (EC), polymethyl methacrylate (PMMA), polystyrene (PS), polyvinyl acetate (PVAC) and polyvinyl butyral (PVB). It has limited compatibility with cellulose acetate (CA). Used as a softener and extender with styrene-butadiene rubber and with nitrile rubber: does not affect vulcanization.

CRF
An abbreviation for channel replacement furnace (black). See *carbon black*.

critical cavity pressure
The pressure at which, ideally, the change from filling to hold pressure is made when *cavity pressure control* is being employed. The pressure associated with an acceptable moulding. See *switch pressure*.

critical concentration
In *polyvinyl chloride* (PVC) technology, the term means the concentration above which a *plasticizer* exhibits normal behaviour. At low levels of addition, a plasticizer often embrittles a PVC compound. This effect is sometimes called anti-plasticization. Critical concentration is inversely proportional to plasticization efficiency. The critical concentration is high for TCP; higher than for DOP which in turn is higher than for DOS.

critical linear extrusion rate
An extrusion rate which if exceeded will cause extrusion defects. See *sharkskin* and *melt fracture*.

critical point
The point on the shear rate-shear stress diagram (flow curve) where *melt fracture* occurs.

critical shear rate
A *shear rate* rate which if exceeded will cause extrusion defects. See *sharkskin* and *melt fracture*.

critical speed
An abbreviation used for this term is u_c. See *Reynold's number*.

critical temperature
The temperature above which a gas cannot be liquefied by the use of changes in temperature alone. In rubber technology it is the temperature below which vulcanization no longer occurs. The temperature below which an *accelerator* is inactive.

critical velocity
The velocity at which the flow of a liquid changes from streamlined to turbulent. See *Reynold's number*.

CRMB
An abbreviation used for cyclized rubber masterbatch. See *cyclized rubber*.

crocidolite
A type of *asbestos* which yields a stiff and brittle fibre.

crocking
The staining of a surface caused by contact with a pigmented rubber component. See *blooming*.

cross axis adjustment
See *roll crossing*.

cross barrier mixer
See *dispersive mixing section*.

cross breaking strength
Also known as maximum surface stress in bending or as, transverse strength or as, modulus of rupture. The ability of a material to withstand a bending stress. The stress needed to cause failure of a material, in beam-form, in bending.

cross contamination
See *dry colouring*.

cross laminate
A laminate in which some of the layers of material are oriented approximately at right angles to other layers; used to make, for example, strong film for packaging.

cross-axis roll adjustment
See *roll crossing*.

cross-dyeing
A dyeing process in which one type of fibre in a mixture is dyed and another is not dyed.

cross-link
A chemical bond which links polymer molecules together; cross-linking increases, for example, the molecular weight and the viscosity. See *natural rubber*.

cross-link density
See *degree of cross-linking*.

cross-linkable
A polymer which is capable of being cross-linked.

cross-linked
A material which contains cross-links: that is, the long polymeric chains are bound together by chemical bonds. A material which is cross-linked may also be called cured or, set or, vulcanized. See *cross-linked polymer*.

cross-linked plastics material
A plastics material which contains cross-links. Prepared, for example, from a *thermosetting polymer*. See *thermosetting plastics material*.

cross-linked polyethylene
A *cross-linked plastics material*. Also known as vulcanized polyethylene. An abbreviation used for this type of material is PE-X. Other abbreviations include XPE or X-LPE or VPE. The introduction of *cross-links* reduces molecular packing and therefore crystallization and gives a material with lower *modulus* and hardness. At temperatures above the melting point the material still has the cross-linked, network structure so retains some strength.

The PE may be cured (cross-linked) with radiation or chemically, for example, with a peroxide system such as cumyl peroxide: copolymers with vinyl acetate are preferred for peroxide cured XPE as are grades which contain *carbon black*. Cross-linked polyethylene may also be produced by grafting: a trialkylvinylsilane is grafted onto the PE, in the presence of a peroxide. The PE may be cross-linked when the trialkylvinylsilane is hydrolysed: a siloxane cross-link is formed by hydrolysis of the alkoxy groups. Grafting and extrusion may be performed on the same extruder and the extrudate cross-linked on standing in water. This type of process has been used for pipe and for wire-covering (Sioplas process).

Cross-linked polyethylene is also available as a cellular material and is used in the automotive industry for sound deadening purposes in the form of mats.

cross-linked polyethylene foam
A cellular material usually based on *low density polyethylene* which may involve chemical cross-linking (Furukawa process and the Hitachi process) or, radiation cross-linking (Sekisui process). See *cross-linked polyethylene*.

cross-linked polymer
A polymer in which the plastic molecules are linked or tied together; crosslinking increases the molecular weight and the viscosity.

cross-linked polyorganosiloxane
See *silicone resin*.

cross-linking
Also known as crosslinking or curing. In rubber technology more commonly referred to as *vulcanization* and in thermoset technology as curing or as setting. The process whereby *cross-links* are introduced into a polymeric system (see *cross-linking additives*). Many gels, seen in extrusion processing of thermoplastics materials, are based on crosslinked plastics material.

cross-linking additive
An additive added to a polymer in order to achieve cross-linking. Many polymers can be *cross-linked* after shaping (usually during processing) so that either a vulcanized rubber (an elastomer) or a thermosetting plastics material results. Two well known cross-linking systems are *sulphur* (used with diene rubbers) and *peroxides* (used with some rubbers and polyolefins). See *accelerator* and *activator*.

cross-ply tyre
Also known as a crossply tyre or as, a bias-ply tyre. A tyre in which the *plies* of the textile reinforcement are at 45° to the direction of movement and alternate plies cross each other at an acute angle: the carcass cord runs diagonally from bead-to-bead. Although such tyres give a comfortable ride most automotive tyres are now *radial-ply tyres* as, for example, such tyres last very much longer.

crossflow
Flow in an extrusion die at right angles to the primary flow direction.

crosshead adaptor
In *crosshead extrusion*, the adaptor (also spelt adapter) swings the melt through, for example, 90°.

crosshead extrusion
An extrusion process wherein the extrudate comes out of the machine at right angles to the barrel axis: used in *wire covering* and in some tube lines.

crosslink
See *cross-link*.

crosslinked
See *cross-linked*.

crosslinked polyethylene
See *cross-linked polyethylene*.

crosslinking
See *cross-linking*.

crossply tyre
See *cross-ply tyre*.

crosswise laminate
A *laminate* in which the layers of *anisotropic reinforcement* are arranged at right angles to one another.

crotonic acid
Also known as β-methylacrylic acid. A softener for *synthetic rubbers*. This material is obtained when *polyhydroxybutyrate* is heated at melt processing temperatures.

crown
See *main head*.

crown increase
See *roll crossing*.

crown - of a roll
The increased thickness of a *roll* at its centre. See *roll separating forces*.

crown - of a tyre
The road contacting area between the shoulders of a *tyre*.

crowned bun
A *bun* which is of increased thickness in the centre: a domed block. See *slab-stock process*.

crowning
See *roll cambering*.

crows feet
Also known as pine trees. A defect found in *calendering*. This defect is caused by the stock folding as it enters the nip. Normally the folded material fuses together but if it is cold, V-shaped marks (resembling crows feet) result.

CRT
An abbreviation for cathode ray tube which is a more precise name for the screen of an electronic display.

crude MDI
An abbreviation used for crude *4,4'-diphenyl methane diisocyanate*.

crude oil
Also known as *mineral oil* and as petroleum.

crumb form
See *block crumb rubber*.

crumb rubber
See *block crumb rubber*.

crushing strength
A strength test performed in compression, on a cylinder, which equals the load at fracture (W) divided by the cross-sectional area (A). See, for example, ASTM D 695-68T. See *compressive strength*.

cryogenic fragmentation
A recycling process used, for example, for waste tyres. The worn or discarded tyres are shredded to a chip form and then cooled to approximately $-80°C$ using liquid nitrogen. A hammer mill is then used to pound the chips: as a result of this milling action, the chips separate into steel, textile and rubber fragments which may then be material and size classified.

cryogenic tumbling
Also known as cryogenic barrelling. The *barrelling* of mouldings at temperatures which are low enough to make the flash brittle, for example, of rubber mouldings. Commercially, liquid nitrogen may be used to reduce the temperatures.

cryoscopy
A molecular weight determination method which gives the number average molecular weight. See *molecular weight determination*.

cryptocrystalline
A term used to describe the micro-crystallinity sometimes found in some grades of, for example, silica. Means that the crystallinity is so fine that it cannot be detected even by microscopic examination. See *silica*.

crystal polystyrene
See *polystyrene*.

crystalline
Usually refers to a plastics material which contain crystalline areas or zones of crystallinity. See *crystalline thermoplastics materials*.

crystalline plastics material
A thermoplastics material: a *crystalline thermoplastics material*.

crystalline polyvinyl chloride
An abbreviation used for this type of material is crystalline PVC. Polyvinyl chloride (PVC) with increased crystallinity compared to conventional PVC. Crystalline *polyvinyl chloride* may be produced by low temperature, free radical polymerization using gamma radiation. Compared to conventional PVC, crystalline PVC has greater heat, solvent, and creep resistance: the brittleness of this type of material may be offset by the use of thermoplastic *chlorinated polyethylene*.

crystalline silica
A form of silica obtained by, for example, crushing, pulverizing and purifying quartzite. The chemical inertness, and purity (>99.6% silica) make it a useful reinforcing filler for *silicone rubber*.

crystalline thermoplastics material
A *crystalline plastics material*: a *thermoplastics material* which contains amorphous material and so may also be known as a semi-crystalline, thermoplastics material. Such plastics are usually tougher, softer, but can have a higher heat distortion temperature, than an *amorphous, thermoplastics material*: such plastics are also translucent, or opaque, have a high shrinkage and a high specific heat. The best known example of a semi-crystalline, thermoplastics material is the plastics material known as *polyethylene (PE)*.

crystallinity
That amount of a polymer which is ordered or, which consists of crystalline zones or areas.

CS
See *casein formaldehyde*.

CSD
An abbreviation used for carbonated soft drinks. See *blow moulding*.

CSM
An abbreviation used for *chlorosulphonated polyethylene*. An abbreviation used for *chopped strand mat*.

CSP
An abbreviation used for *chlorosulphonated polyethylene*.

CSR
An abbreviation used for standard China rubber. A *technically specified rubber*.

cSt
An abbreviation used for centistokes.

CTA
An abbreviation used for cellulose triacetate. See *cellulose*.

CTAB absorption
An abbreviation used for *cetyl trimethylammonium bromide absorption*.

cti
An abbreviation used for comparative tracking index. See *resistance to tracking*.

cu
An abbreviation used for cubic.

cubic measure
Measurement of volume in cubic units. Units of area and volume are derived from unit of length (l).
  cubic centimetre. 1 cm^3 = 1 cc = 0.061 023 7 in^3. The term millilitre (ml) is sometimes used interchangeably with cubic centimetre even though the *litre* is not used for very accurate scientific measurements.
  cubic foot. 1 ft^3 = 1/27 yd^3 = 1,728 in^3 = 0.0283168 m^3. One cubic foot of water weighs approximately 28.317 kg or 62.428 lbs.
  cubic inch. 1 in^3 = 1/1,728 ft^3 = 16.387064 cm^3.
  cubic metre. 1 m^3 = 35.3147 ft^3. See *stere*. One cubic metre of water weighs 1,000 kg or 1 tonne (approximately 1 ton).
  cubic yard. 1 yd^3 = 27 ft^3 = 0.764 555 m^3. One cubic yard of water weighs approximately 764.555 kg or 1685.553 lbs.
See *UK system of units* and *US customary measure*.

CUDC
An abbreviation used for *copper diethyl dithiocarbamate*.

cuff heater
Another term for *heater band*.

cull
See *transfer moulding*.

CUMD
An abbreviation used for *copper dimethyl dithiocarbamate*.

cup flow test
A test used to assess the ease of flow of a thermoset moulding compound. This test uses a standardized cup mould, operated under specified temperature conditions and a specified amount of moulding material. The moulding material is placed in the cup-shaped cavity and the press is closed at a specified speed. The instant that the *line pressure gauge* shows a pressure reading, a clock is started: it is stopped when the *flash* stops moving. The cup flow time is the difference between the two readings measured in seconds.

cup flow time
The result of a *cup flow test*.

cup lump
A form of *natural rubber* which is sometimes referred to as cuplump and which is produced by *auto-coagulation* of *natural rubber latex*. A field coagulum: the coagulated material which accumulates in the *tapping cup* after the last collection. See *Hevea Brasiliensis*. Ammonia, formaldehyde or sodium sulphite may be added to the tapping cup to prevent premature coagulation.

cuprammonium process
Used to produce *cuprammonium rayon*. Alpha cellulose (α cellulose) is dissolved in cuprammonium liquor at low temperatures, filtered, spun and then acid hardened. See *rayon*.

cuprammonium rayon
Rayon produced by the *cuprammonium process*, gives the finest, silk-like yarn of the regenerated *cellulose fibres* known as *rayon*. Also known as cupro. See *fibres*.

cupro
See *cuprammonium rayon*.

curatives
Additives used in conjunction with polymers and which are used to cause setting, *cross-linking* or *vulcanization*.

curdle
Another name for *coagulation*. Acetic or formic acid is added to *natural rubber latex* to make the latex curdle or coagulate.

cure
To *cross-link* or *vulcanize* a system. The changing of a soluble, fusible polymer to an insoluble, infusible polymer. To change the properties of a polymeric system into a more stable, usable condition by the use of heat, radiation, or reaction with chemical additives: cure may also be accomplished, for example, by removal of solvent or by crosslinking. The conditions under which *cross-linking* or *vulcanization* occurred.

cure cycle
The schedule of time periods, at specified conditions, to which a reacting thermosetting material is subjected to reach a specified property level.

cure index
A term used in rubber technology. The torque interval $t_{35} - t_5$ is known as the cure index when a *Mooney viscometer* is used to measure cure behaviour.

cure retardation
See *accelerator* and *accelerator deactivation*.

cure site monomer
See *cure-site monomer*.

cure time
The period of time for which a reacting thermosetting material is exposed to specific conditions, so as to reach a specified property level. See *steam heating*.

cure-site monomer
A monomer which when incorporated in the polymer facilitates vulcanization. See *fluororubber*.

cured
See *cross-linked*.

curie
A unit used to measure the activity of a radioactive material and defined as being that quantity of a radioactive isotope which decays at the rate of 3.7×10^{10} disintegrations per second. To convert from curie to *becquerel* multiply by 3.7×10^{10}.

curing bag
An inflatable bag made of a heat resistant rubber (for example, *butyl*) and used during tyre manufacture to exert moulding, or shaping, pressure.

curing bladder
See *curing diaphragm*.

curing diaphragm
Also known as a curing bladder. A flexible component of a tyre press which fulfils the same function as a *curing bag*.

curing systems
See *cross-linking additives*.

curing tube
A thin-walled *curing diaphragm*.

curtain coating
A coating method in which the substrate is passed through a stream of a fluid composition which falls from a slit: the excess may be removed by, for example, a *doctor blade*.

cushion
A term used in *hydraulics* and which refers to a device inside an *actuator* which provides controlled deceleration. A device which restricts fluid flow at the outlet of a cylinder thus slowing piston rod motion.

cushion-type tyre
A narrow *tyre*: a tyre with an *aspect ratio* of approximately 95%.

cut off units
Devices such as saws, shears, flying knives, and other devices and which are used for cutting extrusions to predetermined length.

cut-layers
As applied to laminated plastics; a condition of the surface of machined, or ground, rods and tubes and of sanded sheets in which cut edges of the surface layer, or lower laminations, are revealed.

cut-off
See *flash land*.

cut-off line
See *parting line*.

cuttings
Also known as clippings. Cuttings of *natural rubber sheet* which contain air bubbles or impurities.

CV
An abbreviation used for continuous vulcanization. An abbreviation used for *constant viscosity (rubber)*.

CV natural rubber
An abbreviation used for *constant viscosity natural rubber*.

CVD
An abbreviation used for *chemical vapour deposition*.
  CVD CF = a *carbon fibre* produced by *chemical vapour deposition*.

CVNR
An abbreviation used for *constant viscosity natural rubber*.

cwt
An abbreviation used for hundredweight. See *UK system of units* and *US customary measure*.

cyanate
Also called a fulminate. A salt or ester of *cyanic acid*. A trifunctional ring, called a cyanurate ring, can be used to build up highly cross-linked polycyanurate structures via *reaction injection moulding*: this type of polymer is less brittle than the *poly-isocyanurate*. See *cyanuric acid*.

cyanic acid
An isomer of *isocyanic acid* which may be called fulminic acid: forms salts called cyanates or fulminates. The acid may be represented as HOCN.

cyanuramide
See *melamine*.

cyanuric acid
Also known as tricyanic acid. A trimer of *cyanic acid* which consists of a six-membered ring with alternating -NH- and -CO- groups.

cycle
The repeating sequence of events which constitute a moulding operation.

cycle counter
A device which totals the number of cycles that a machine has performed.

cycle time
The total time necessary to complete the sequence of events which are used to produce one moulding, for example, an injection moulding or, one shot (a set of injection mouldings).

cyclic
Means that a substance contains ring structures.

cyclic amide
See *lactam*.

cyclic ester
See *lactone*.

cyclic ether group
See *epoxide group*.

cyclic oil
This material has a relative density (RD or SG) of 0·92. Used as a general purpose plasticizer for natural and synthetic rubbers. Of light colour and of little odour: does not affect colour of rubber compounds or the rate of cure. Acts as a processing aid as, for example, it improves flow during moulding and extrusion. Can reduce mixing time. Up to 20 phr may be used, especially in heavily loaded compounds. Available in emulsion form for use with rubber latex.

cyclic oil emulsion
This material has a *cyclic oil* content of approximately 60%. Used with latex (for example, used to make balloons and fine gloves) as it improves extensibility without decreasing strength. Up to 20 parts by weight of the emulsion may be used, for example, with a 60% latex.

cyclic ozonides
Formed between ozone and diene rubbers. See *ozone-induced degradation*.

cyclic thioether group
Similar to an *epoxide group* but sulphur is used in place of oxygen. Compounds which contain a cyclic thioether group may be used to make polymers by ring opening polymerization. See, for example, *propylene sulphide*.

cyclization
The introduction of *cyclic groups*: changing a polymeric material by introducing cyclic ring structures. See *cyclized rubber*.

cyclized natural rubber
See *cyclized rubber*.

cyclized polyisoprene rubber
See *cyclized rubber*.

cyclized rubber
Also known as cyclorubber and as isomerized rubber. The reaction product of, for example, *natural rubber (NR)* and an acid-type material, for example, sulphuric acid or chlorostannic acid: these cause the development of cyclic ring structures. The rubber is usually *natural rubber* although another diene rubber (for example, synthetic polyisoprene) may be used. When cyclized with an acid the products may be called thermoprenes. Rubber which has been cyclized with a hydrogen halide may be called Pliolite or Pliofilm: see *hydrohalogenated natural rubber*. Thermal cyclization is also possible.

As a result of the reaction, with the acid, internal cyclization occurs and the product has the same empirical formula as NR, that is, $(C_5H_8)_n$. However, the unsaturation may be reduced to about 57% of the original unsaturation: the *isoprene units* of the rubber molecule react to give a saturated six-membered ring and the rubber becomes a partial *ladder polymer*. The rubber loses elasticity and there is an increase in density as the reaction proceeds.

The term, cyclized rubber, refers to a range of thermoplastics materials whose properties range from tough to brittle and which are light coloured. Because of the chemical resistance of cyclized rubber it was used, for example, for printing inks: because of hardness improvements, without a significant increase in density, it was used for reinforcing shoe-soling made from natural rubber.

Cyclized rubber is sunlight resistant, chemical resistant and water resistant. Soluble in aromatic and aliphatic hydrocarbons: not compatible with alkyds. Acid and alkali resistant: capable of sulphur vulcanization and of being plasticized (for example, with ester-type plasticizers). When mixed with natural or synthetic rubbers, the products may be formulated to be hard, tough and light coloured and was used to make soles, heels and golf ball covers. This type of material, when based on natural rubber, has a relative density (RD or SG) of 0·99 and a softening point of approximately 90 to 120°C.

cyclized rubber/azide-type photo-sensitizer
A negative working photoresist system. A bisazide/cyclized rubber resist based on a cyclized, or partially cyclized, polyisoprene (CPI) rubber (*cis-polyisoprene*). Partial cyclization is employed so as to raise the glass transition temperatures (T_g) of the rubber and improve hardness: the partially cyclized material has good adhesion and film forming properties. In such a system, cross-linking of the linear polymer backbone occurs by the light induced decomposition of a *photosensitizer* so as to generate active species (photo-crosslinking). A good solvent is used to remove un-exposed rubber compound. Used where high resolution is not required as the solvents used to remove the un-exposed rubber compound may cause image distortion.

cyclized rubber masterbatch
An abbreviation used is CRMB. Consists of, for example, a 50:50 mixture of *cyclized rubber* and *natural rubber*.

cyclo-hexylamine
An *amine* used, for example, to produce *cyclohexylbenzothiazyl sulphenamide*.

cyclohexadiene-1,4-dione
See *p-benzoquinone*.

cyclohexane
May also be known as hydrobenzene and as hexamethylene. A ring compound derived from benzene. May be represented as C_6H_{12}. This material has a relative density (RD or SG) of 0·78 and a boiling point of approximately 81°C. A solvent for many rubbers but not for *nitrile rubber*. Used, for example, for adjusting the viscosity of rubber cements.

cyclohexanedimethanol
An abbreviation used for this type of material is CHDM. A glycol which is also known as 1,4-cyclohexanedimethanol or as, 1,4-cyclohexylene glycol. A symmetrical *glycol* which is based on a saturated six-membered ring and which may be represented as $HO-CH_2-Ar-CH_2-OH$ where Ar is $(CH)_2(CH_2)_4$. When reacted with dimethyl terephthalate will give a polymer with a higher melting point than *polyethylene terephthalate (PET)*. See *copolyester*.

cyclohexanol
May also be known as hexalin or as, hexahydrophenol. A ring compound derived from benzene. May be represented as $C_6H_{11}OH$. This material has a relative density (RD or SG) of 0·97 and a boiling point of approximately 161°C. A solvent for many rubbers but not for *nitrile rubber*. A solvent for cellulose ethers, low molecular weight silicones and phenol-formaldehyde resins. Used with epoxy resins to improve surface finish.

cyclohexanone
This material has a relative density (RD or SG) of 0·94 and a boiling point of approximately 157°C. A solvent for many plastics and some rubbers. For example, for *cellulosics, natural resins, polyvinyl chloride (PVC), polystyrene (PS)* and *phenol-formaldehyde resins (PF)*. It is a good solvent for uncured *chloroprene rubber (CR), nitrile butadiene rubber (low acrylonitrile NBR), natural rubber (NR), styrene-butadiene rubber (SBR)* and *thiokol rubber (T)*. It is a poor solvent for uncured high acrylonitrile NBR. This chemical causes a large amount of swelling, or gel formation, of uncured IIR rubbers.

cyclohexylbenzothiazyl sulphenamide
An abbreviation used for this material is CBS. This material is also known as N-cyclohexylbenzothiazyl-2-sulphenamide or as, N-cyclohexylbenzothiazylsulphenamide or as, N-cyclohexylbenzothiazole-2-sulphenamide or as, N-cyclohexyl-2-benzothiazole sulphenamide. The product of the reaction between cyclo-hexylamine and *mercaptobenzthiazole*. A delayed action *accelerator* with good processing safety but which is very active at high temperatures. Zinc oxide and stearic acid are necessary in the mix. This pale grey material has a melting point of approximately 97°C and a relative density (RD) of 1·30.

cyclohexylene glycol
See *cyclohexanedimethanol*.

cyclorubber
See *cyclized rubber*.

cylinder
A term used in *hydraulics* and which refers to a linear *actuator* which converts fluid power into linear motion and force. Consists of a piston, and piston rod, operating inside a cylindrical bore. See *linear actuator*. The word 'cylinder' is often used for *barrel* in, for example, injection moulding.

cylinder temperature
The temperature of a cylinder or barrel. Should differentiate between *set temperature* and the actual or *measured temperature*.

D

d
An abbreviation used for *deci*. See *prefixes - SI*.

D
This letter is used as an abbreviation for:
 barrel diameter of, for example, a *capillary rheometer*;
 decyl - in plasticizer abbreviations. For example, butyl isodecyl phthalate (BIDP);
 density;
 derivative - in control terminology;
 di - in plasticizer abbreviations. For example, *dicapryl phthalate (DCP)*;
 dithiocarbamate - as in *zinc ethyl phenyl dithiocarbamate*;
 filler - this letter may be used to show the presence of powder in plastics compositions.

d-c
An abbreviation used for direct current.

D-glucose
See *glucose*.

D-shaped ejector pin
An *ejector pin* which has been modified so that the ejection surface is semi-spherical in cross-section (shaped like the letter D).

$D_{79}A$
An abbreviation used for *dialphanyl adipate*.

$D_{79}P$
An abbreviation used for *dialphanyl phthalate*.

$D_{79}S$
An abbreviation used for *dialphanyl sebacate*.

$D_{79}Z$
An abbreviation used for *dialphanyl azelate*.

da
An abbreviation used for *deca*. See *prefixes - SI*.

$DA_{79}A$
An abbreviation used for *dialphanyl adipate*.

$DA_{79}P$
An abbreviation used for *dialphanyl phthalate*.

$DA_{79}S$
An abbreviation used for *dialphanyl sebacate*.

$DA_{79}Z$
An abbreviation used for *dialphanyl azelate*.

DAA
An abbreviation used for *dialphanyl adipate*.

DABCO
An abbreviation used for *1,4-diazabicyclo-2,2,2-octane*.

DAIP
An abbreviation used for *diallyl isophthalate*. See *allyl moulding materials*.

Dakes' machine
A machine used for the continuous production of *latex foam*. A rotor mixes the latex and air, under pressure, within a cylindrical chamber: on leaving the chamber the mixture froths or foams before being set.

dalton
See *atomic mass unit*.

dam gate
A *gate* used in *reaction injection moulding* (RIM) which is similar to a *flash gate* used in thermoplastics injection moulding.

damping
The decrease in the amplitude of an oscillation, or wave motion, with time: damping in rubber components results from *hysteresis*.

damping factor
See *loss tangent*.

dancer roll
A roll used, for example, on a sheet extrusion line to control the sheet (line) tension and which compensates for line speed variations.

DAP
An abbreviation used for *dialphanyl phthalate* and for *allyl moulding materials*.

DAPM
An abbreviation used for *4.4'-diamino diphenyl methane*.

daraf
The reciprocal of the *farad*.

dark factice
Also known as sulphur factice. See *brown factice*.

dart
The striker used in the determination of film impact strength. See *falling weight impact strength (test)*.

DAS
An abbreviation used for *dialphanyl sebacate*.

day
A practical unit of time which may however, be defined in different ways. Usually taken to mean the time taken for the Earth to make one complete revolution around its axis.

daylight
The working distance between two rolls or platens.

DAZ
An abbreviation used for *dialphanyl azelate*.

dB
An abbreviation used for *decibel*.

DB
Letters used in *plasticizer terminology* for dibenzoate, for example, for *dipropylene glycol dibenzoate* (DPDB).

DBA
An abbreviation used for *dibenzyl amine*. An abbreviation used for *dibutyl adipate*.

DBEA
An abbreviation used for *dibutoxyethyl adipate*.

DBIM
An abbreviation used for *direct blend injection moulding*.

DBP
An abbreviation used for *dibutyl phthalate*.

DBP adsorption method
An abbreviation used for *dibutyl phthalate adsorption method*.

DBP adsorption value
See *dibutyl phthalate adsorption method*.

DBS
An abbreviation used for *dibutyl sebacate*.

dc
An abbreviation used for direct current.

DC drive
An abbreviation used for *direct current drive*.

DCBS
An abbreviation used for *N,N-dicyclohexylbenzothiazole-2-sulphenamide*.

DCHP
An abbreviation used for *dicyclohexyl phthalate*.

DCL
An abbreviation used for *digital* and *closed loop*. The same letters are used for decorative continuous laminates.

DCP
An abbreviation used for *dicapryl phthalate*.

DCP
An abbreviation used for *dicyclopentadiene*. See *polydicyclopentadiene reaction injection moulding*.

DDA
An abbreviation used for *didecyl adipate*.

DDL
An abbreviation used for *direct digital link*.

DDP
An abbreviation used for *didecyl phthalate*.

DE
Letters used in *plasticizer* abbreviations for diethyl. For example, diethyl phthalate (DEP).

de-vent
A term used in *hydraulics* and which refers to the action of closing the vent connection of a *pressure control valve* thus allowing the valve to operate at the adjusted pressure setting.

deactivation
To make something less active. See *accelerator deactivation*.

dead fold
The ability of a film or foil to be creased without heat and to retain that crease; shown by metal foils.

dead-band
That part of a temperature controller range in which the control is not affecting the temperature: the power is off and the temperature is allowed to drift. A term used in *hydraulics* and which refers to a region of no-response where an error signal will not cause a corresponding change to occur.

dead-time
A term used in *hydraulics* and which refers to the time delay between two related actions.

deadband
See *dead-band*.

deadtime
See *dead-time*.

deca
A prefix which means ten. For example:
   decagram = 10 grams or 0.01 kilogram;
   decalitre = 10 litres;
   decameter = 10 metres; and,
   decastere - 10 cubic metres = 13,079 cubic yards.
Sometimes spelt deka. Abbreviation da. See *prefixes - SI*.

decabromodiphenyl ether
An *organo-bromine flame retardant*. Such a bromine-containing compound tends to be more powerful than the equivalent chlorine-containing compound: often used in conjunction with *antimony trioxide*. For example, *polycarbonate 93%, decabromodiphenyl ether 5% and antimony trioxide 2%*.

decahydronaphthalene
Also known as dekalin. This material has a relative density (RD or SG) of 0.88 and a boiling point of approximately 190°C. Produced when naphthalene is hydrogenated. A solvent for natural resins and for hydrocarbon rubbers. At elevated temperatures it will dissolve polyethylenes.

decanedoic acid
See *sebacic acid*.

deci
A prefix which means one tenth. For example:
   deciare is 0.1 are:
   decibel is 0.1 bel. An abbreviation used for this term is dB.
   decigram is 0.1 g;
   decilitre is 0.1 litre;
   decimetre = 0.1 metre:
   decistere is 0.1 stere;
   decitex is 0.1 tex.
Abbreviation d. See *prefixes - SI*.

deckle rod
A metal rod, or insert, used to close off the end of a flat film die so as to allow the extrusion of product of narrower width. Two such rods are used: one at each end of the *die*.

decompression
Melt pressure reduction. For example, the action of reducing melt pressure by pulling the screw back after it has finished rotating in an *injection moulding machine*: also known as suckback. Decompression is also a term used in *hydraulics* and which refers to the slow release of fluid under pressure so as cause the fluid pressure to fall.

decompression zone
A zone or region in, for example, a *vented barrel* where the melt is decompressed so as to allow the release of volatile matter (gas). Such a decompression zone is used, for example, to remove volatile matter and/or to dry material.

decor mould process
See *back injection moulding*.

decorative continuous laminate
An abbreviation used for this type of material is DCL. A laminate in sheet form which is usually based on thermosetting resins. See *melamine formaldehyde*.

decorative laminate
Also known as a plastics laminate or as a decorative laminate. A *laminated plastics sheet* which has a decorative pattern moulded into the surface. A laminate in sheet form which is usually based on thermosetting resins. See *melamine formaldehyde*.

decorative rib
A decorative, circumferential raised pattern on the *sidewall*.

decyl
This term is commonly seen in terms used to describe a *plasticizer*: it means that the side chain contains 10 carbon atoms.

dedicated
A machine or system which is exclusively reserved for, or can only do, one particular job or function. For example, a dedicated screw is a *screw* which has been designed to suit one type of material. Common examples include screws for *nylon (PA 66)* and for *polyvinyl chloride (PVC)*.

deflection temperature under flexural load
See *heat distortion temperature*.

deflection temperature under load
An abbreviation used for this term is DTUL. See *heat distortion temperature*.

deflocculation agent
An additive to a system which breaks down agglomerates or which prevents agglomerates from forming.

Defo plastimeter
Also known as a Defo meter. With this machine the amount of load needed to give a constant change in thickness of a test piece, when contained between parallel plates, is measured. The weight in grams needed to compress a test sample 10 mm in height and 10 mm in diameter to 4 mm: the sample is pre-warmed to 80°C and the load is applied for 30 s. The weight in grams is called the Defo hardness or DH.

degating
The removal of the *feed* from an injection moulding, for example, by utilizing the mould opening movement using a submarine gate (the edges of the submarine gate need to be kept sharp so as to ensure effective degating). The removal of the feed system from the moulded component by manual means, for example, by cropping, cutting by means of scissors or scalpel or, by simple twisting.

DEGB
An abbreviation used for *diethylene glycol dibenzoate*.

degC
An abbreviation for *degrees Celsius*. Avoids the problem of writing the degree sign above the line of print. See *Système International d'Unité*.

degF
An abbreviation for degrees Fahrenheit. See *Fahrenheit scale*.

deglazing
See *surface conditioning*.

degradable polymer
A polymer which will degrade in specific circumstances. See *degradant* and *polyhydroxybutyrate*.

degradant
An additive. A material added to cause product decomposition, or degradation and usually associated with the plastics industry.

Generally speaking, polymeric components are noted for their long life however, once the component has fulfilled the function required then it often becomes a nuisance and the once valuable component becomes classed as litter: this problem is most serious for plastics materials used in packaging. One way of overcoming this problem is to lay the plastics material open to decomposition, or degradation. this may be done by, for example, incorporating a degradant, for example, a *filler* such as starch into the plastics material. Another route is to use a biodegradable plastics material such as *polyhydroxybutyrate (PHB)* either on its own or in combination with other plastics. Another way is to incorporate a UV absorber into the material so that the UV energy produces chemical entities which reduce the molecular weight of the plastics material.

degradation
A deleterious change in the chemical structure, physical properties, or appearance of a polymeric material.

degradative extrusion
Plastics materials are either, for example, reclaimed and used on their own, reclaimed and used as additives for other similar plastics compounds or, cracked into petrochemical raw materials (*chemical recycling*). Degradative extrusion is an extrusion process which deliberately sets out to induce degradation (so as to produce petrochemical raw materials from a thermoplastics material), by exposing that material to high temperatures, shear and a degrading atmosphere (for example, oxygen or water vapour). Often the output from the extruder, which may be a cascade system, is in the form of an oil, or of a semi-solid material, which is then used as a source of energy or as a raw material. Prior to being transferred to the second stage of the cascade, the melt (heated to say 300°C) may be filtered (to remove fillers) dehalogenated (so as to remove chlorine from *PVC*) by applying a vacuum to the partially decomposed melt. Further degradation occurs in the second stage which operates at say, 500°C. See *reclamation*.

degree
A sub-division of a temperature scale or a unit used to measure angle: a circle has 360 degrees. See *Système International d'Unité*.

degree of cross-linking
An abbreviation used is q. The fraction of repeat units in a polymer molecule which are *cross-linked*. Also known as density of cross-linking and cross-link density.

degree of crystallinity
The amount of crystallinity in a semi-crystalline, thermoplastics material. Usually expressed as a percentage: very dependent on the method used for measurement.

degree of degradation
An abbreviation used is α. The fraction of breakable bonds that have been broken in random scission decomposition, or degradation.

degree of polymerization
An abbreviation used is DP. The number of repeat units in a polymer molecule. A measure of molecular weight as the molecular weight of the polymer molecule is $DP \times M_m$. Where M_m is the *molecular weight of the repeat unit*.

degree of substitution
An abbreviation used for this term is DS. The average number of groups substituted. See *secondary cellulose acetate*.

degrees absolute
See *degrees Kelvin*.

degrees Baumé
See *Baumé scale*.

degrees Celsius
A non-SI unit which is permitted within the SI system. Degrees Celsius were formerly known as *degrees centigrade* but the more widely used degrees centigrade, were changed to honour the originator of the scale. That is, the 18th century astronomer Anders Celsius. An SI unit which has a special symbol, that is °C. See *Système International d'Unité*.

degrees centigrade
Based on the freezing and boiling points of water with the difference divided into 100 parts or degrees. Now known as *degrees Celsius* and commonly referred to as °C. It follows that temperatures below those of the freezing point of water are negative and this can be misleading. Ideally *temperature* should be measured on a scale which increase from *absolute zero*. See *degrees Kelvin*.

degrees Kelvin
Represented as K. Defined as being 1/273.16 of the temperature difference between absolute zero and the triple point of water. One degree Kelvin is equal in magnitude to one degree Celsius but at any temperature the value will be different as the scales are defined differently. See *Kelvin* and *Système International d'Unité*.

degrees Réamur
See *Réamur scale*.

degrees Twaddell
See *Twaddell scale*.

DEHP
An abbreviation used for diethylhexyl phthalate, See *dioctyl phthalate*.

dehalogenation
The removal of a halogen from a material, for example, the removal of chlorine from a material which contains *polyvinyl chloride (PVC)*. See *degradative extrusion*.

dehydrated castor oil
A fast *drying oil* obtained by heating castor oil to approximately 250°C. Contains more unsaturation than *castor oil* and is non-yellowing and water resistant. Used in *alkyd resins*.

dehydrochlorination
The loss of hydrochloric acid (HCl). Polyvinyl chloride (PVC) degrades more readily by dehydrochlorination, than by oxidation, and so additives which restrict this type of attack are more widely used in PVC technology.

deka
An alternative way of spelling *deca*.

dekalin
See *decahydronaphthalene*.

Dekur
An abbreviation used for *Deutsch Kunstofferecycling* or *German Plastics Recycling*.

delamination
The separation of a product into layers.

delayed action accelerator
An additive which permits rapid and efficient vulcanization at the high moulding temperatures without allowing *scorching*, at the lower shaping, or mixing temperatures. Such accelerators are typified by the *sulphenamides*. The desired characteristics are obtained by chemical modification of an *accelerator* so that the active group is blocked until it is freed by being heated to high temperatures. See *cyclohexylbenzothiazyl sulphenamide*.

delivery
A term used in *hydraulics* and which refers to the output of fluid from a pump, or another component in a hydraulic system. Usually measured in gallons per minute or litres per second.

delivery hose
See *softwall hose*.

delta
The upper case, Greek letter delta Δ is used to show a change or difference, for example, between two different values. Read as 'a change in' or, 'a change of'. For example:

delta H = ΔH = the heat change in a chemical reaction. The difference between two different values of enthalpy.

delta P = ΔP = the difference between two different values of pressure.

delta T = ΔT = the difference between two different values of temperature. A change of 1°C is equal to a change of approximately 1.8°F. °C/°F is equal to 1.8.

demould time
That time, usually measured in seconds, from the end of mould filling to the beginning of component ejection. See *reaction injection moulding process*.

demoulding
The removal of a component from a mould.

den
An abbreviation used for *denier*.

denesting agent
An *additive* which helps formed blisters, or containers, based on, for example, a thin sheet of *copolyester* to separate during filling or handling: an *antiblock agent*.

denier
An abbreviation used for this term is den. A unit of weight used for textiles. Used to measure the fineness or coarseness of fibres. The weight in grams of 9,000 metres of the fibre or yarn. To convert to kg/m multiply by 1.111×10^{-7}. To convert to *tex* multiply by 0.111.

density
Defined as mass per unit volume. Has the units of mass and of volume. When expressed in grams per cubic centimetre (g/cc) it is numerically equal to *relative density* (specific gravity). Nowadays it is often given in kg.m^{-3} (kilogram per cubic metre) although grams per cubic centimetre is more common (g.cc^{-1} is the same as g/cc and gcm^{-3}). It is suggested, for ease of understanding, that when density is discussed, that the units used should be Mg/m^3 rather than kg/m^3. This gives values which have the same numerical values as the well established g/cm^3 values or SG values. 1 Mg/m^3 = 1 g/cm^3 = 1,000 kg/m^3. See *composition density*, *apparent density* and *bulk density*.

density - apparent
See *apparent density*.

density - cost per unit volume
If a material has a low density it can be a tremendous advantage as materials are usually bought by weight and sold by volume. Supposing 10 cm^3 of a material is required to produce a particular item. If a material with a density of 1.4 g/cm^3 is used then $1.4 \times 10 = 14$ g of material will be needed for each item and 1000 divided by 14 = 71 items could be made from 1 kilogramme. However, if the density of the material were 0.9 g/cm^3 then only 9 g of material per item would be needed and it would be possible to produce 1000 divided by 9 = 111 items from one kilo. This means that the cost per unit volume is often more useful than the cost per pound of material. To obtain the unit volume (specific volume) of a substance, divide the figure one by the density value.

density - filled material
If a polymeric material is filled, then the density will usually be higher than that of the unfilled material: this is because most inorganic fillers, the most common fillers, have a relatively high density. (To calculate the density of a polymer composition divide the total mass of the polymer composition by the total volume.) If a weighed sample of the material is completely burnt then, the inorganic *filler* content may be easily estimated. See *composition density*.

density - gases
The densities of gases is much lower than that of solids and liquids: gases have densities 1/1,000 that of solids and liquids. At room temperatures, the density of hydrogen is approximately 0.000 084 g/cm^3. At room temperatures, the density of carbon dioxide is approximately 0.001 8 g/cm^3.

density gradient column
A device used for measuring *density*. The density of such a column gradually changes from top to bottom and so a small piece of a material dropped into it will come to rest at the point where its density matches that of the column. Glass floats of known density are used to calibrate the column. See *flotation method*.

density - measurement of
Very often the absolute density of a thermoplastics material is not required; what is required is an approximate value so that, for example, the separation of unidentified lots of material can be performed. May be obtained by seeing if the material sinks or floats in a limited range of liquids. These may include water and saturated magnesium chloride; the former has a density of 1 gcm^{-3} and the latter has a density of 1.34 gcm^{-3}. If an accurate measurement of density is required then this may be performed by a *flotation* or a displacement method. For foamed plastics, then weigh a known volume: for example, a cube of the material is cut and weighed. The density is weight over volume. See *relative density*.

density of cross-linking
See *degree of cross-linking*.

density - optical
A measure of refractive index difference between materials. A material is said to have a greater optical density than another if its refractive index, for a particular wavelength of light, is also greater.

density - solids
The densities of solids, as measured on the cgs scale, differs by a factor of greater than 100. Balsa is 0.16, oak 0.71, water 1.00, magnesium 1.74, sodium 2.32, aluminium 2.70, iron 7.90 and lead is 11.3 g/cm^3. The materials osmium and iridium are the most dense with an approximate value of 22.6 g/cm^3.

The density of a particular polymer is a function of the mass of the individual atoms in the molecules and the way that the molecules are packed. If a polymer is based on just carbon and hydrogen then its density will be relatively low, for example, below 1 g/cm^3. Inclusion of other atoms, for example, chlorine or fluorine will increase the density and, for a given molecular formula, an increase in crystallinity will also increase the density.

deodorant
A deodorant or odorant is an additive to a polymer compound used to remove an odour, rather than mask it: examples used for *polyvinyl chloride (PVC)* include perborates and persulphates.

DEP
An abbreviation used for *diethyl phthalate*.

depolymerization
A reduction in molecular weight; the opposite of *polymerization*.

depolymerized rubber
Also known as liquid rubber. Natural rubber which has been oxidatively degraded by being heated in an *internal mixer* at about 120°C together with a *peptizer*. A pourable liquid results which may be compounded, in a Z-blade mixer, vacuum vented and then poured into moulds. When *vulcanized* the products have relatively poor physical properties because of the low molecular weight of the polymer. See *fluid rubber*.

deproteinized natural rubber
An abbreviation used for this term is DPNR or DP-NR. Also known as deproteinated natural rubber or as, low nitrogen natural rubber (LN-NR). This material is prepared by diluting *natural rubber latex* and reacting it with an enzyme that removes the proteins from the rubber. The resultant material gives vulcanizates of low creep, low stress relaxation and reproducible modulus: it is used for engineering components.

depth
Screw terminology meaning the perpendicular distance from the top of the *thread* to the *root*.

derivative
Referred to as 'D'; this term is added to a temperature controller in order to prevent overshooting on start up.

derived unit
Units of measurement are of two kinds: *fundamental units* (basic units) and derived units. For example, the meter and the kilogram are fundamental units of the metric system whereas the litre and the hectare are derived units. See *Système International d'Unité*.

desiccant
A substance which absorbs water, for example, anhydrous calcium chloride.

desiccant drier
A type of hot air drier for thermoplastics materials in which the moisture is removed from the air by means of a *desiccant*. A popular method of desiccant drying, passes the air over/through a *molecular sieve desiccant* which removes the water from the air: this lowers the dew point of the air to, for example, −40°C/−40°F. The hot, dry air is then used to dry the plastics material.

desiccation
The removal of moisture. See *desiccant drier*.

DET
An abbreviation used for *diethylene triamine*.

DETDA
An abbreviation used for *diethyl toluene diamine*.

detent
A term used in *hydraulics* and which refers to the device used to lock the valve spool in a desired position.

Deutsch Kunstofferecycling
An abbreviation used for this German based organization is Dekur. German Plastics Recycling. An organization set up by Verwertungsgesellschaft für gebrauchte Kunstoffe and Duale System Deutschland (DSD) to coordinate the recycling of plastics packaging wastes deposited by consumers in the yellow household collection bins issued by the DSD.

Deutsches Institut Für Normung
An abbreviation used for this organization is DIN. See *standards organisations*.

devolatilizing extruder
An *extruder* designed, or operated, so that volatiles or vapours, are removed from the polymer when it is in melt form. See *vented extruder*.

devulcanization
The removal of *chemically-bound sulphur* from vulcanized rubber. See *reclaiming processes*.

dextrimized starch
A size: a surface treatment applied to *glass fibre* so as to improve the resistance of the fibres to damage caused by handling.

dextrose
See *glucose*.

DFM
An abbreviation used for directed fibre preform. See *preform*.

DGEBA
An abbreviation used for *diglycidyl ether of bisphenol A*.

DH
An abbreviation used for Defo hardness. See *Defo plastimeter*.

DHP
An abbreviation used for *diheptyl phthalate*.

DHX
Letters used in *plasticizer* abbreviations for dihexyl (see *H*). For example, dihexyl phthalate (DHXP). See *phthalate*.

DHXP
An abbreviation used for dihexyl phthalate. See *phthalate*.

di-(2-ethylhexyl)
An abbreviation used for this prefix is DEH. Compounds which begin with this prefix are listed under *dioctyl* entry. For example, for di-(2-ethylhexyl) phthalate see *dioctyl phthalate*.

di-(3,3,5-trimethylhexyl) phthalate
More commonly known as *dinonyl phthalate*.

di-(3,3,5-trimethylhexyl) sebacate
See *dinonyl sebacate*.

di-(benzene carbonyl) peroxide
See *dibenzoyl peroxide*.

di-(n-octyl n-decyl) phthalate
See *octyl decyl phthalate*.

di-1-methylheptyl phthalate
See *dicapryl phthalate*.

di-2-ethylhexyl
See di-(2-ethylhexyl). An abbreviation used for this prefix is DEH. Compounds which begin with this prefix are listed under *dioctyl* entry. For example, for di-2-ethylhexyl phthalate see *dioctyl phthalate*.

di-C(6-8-10) phthalate
See *linear phthalates* and *plasticizer*.

di-C(7-9) phthalate
See *dialphanyl phthalate* and *plasticizer*.

di-C(7-9-11) phthalate
See *linear phthalates* and *plasticizer*.

di-ester
See *diester*.

di-heptyl
See *diheptyl*.

di-hexyl
See *dihexyl*.

di-iso-
An abbreviation used for this prefix is DI. This prefix is most commonly seen in terms used to describe *plasticizers*, for example, di-iso-butyl phthalate (DIBP). Such compounds are esters. The term di-iso- means that the side chain is branched and that two such chains are combined with a central group.

di-iso-butyl
An abbreviation used for this prefix is DIB. This prefix is most commonly seen in terms used to describe *plasticizers*, for example, di-iso-butyl phthalate (DIBP). Such compounds are esters. The term di-iso-butyl means that the side chain is branched and contains 4 carbon atoms (from iso-butyl alcohol) and that two such chains are combined with a central group (-OOC-ϕ-COO- where ϕ is based on a benzene ring and is C_6H_4). The side chain is $(CH_3)_2CHCH_2$-. When this is combined with the central group -OOC-ϕ-COO- then the following is obtained $(CH_3)_2CH-CH_2-OOC-\phi-COO-CH_2-CH(CH_3)_2$.

di-iso-butyl adipate
An abbreviation used is DIBA. DIBA has been used as a low temperature *plasticizer* for *polyvinyl chloride (PVC)* and for some rubbers, for example, for butyl, chloroprene, natural and styrene-butadiene rubbers. Has a boiling point of approximately 282°C but lacks permanence in polymer compositions. Not compatible with cellulose acetate but is compatible with polystyrene and ethyl cellulose.

di-iso-butyl phthalate
An abbreviation used for this material is DIBP. A *plasticizer* for polyvinyl chloride. See *phthalate*.

di-iso-decyl
An abbreviation used for this prefix is DID. This prefix is most commonly seen in terms used to describe *plasticizers*, for example, di-iso-decyl phthalate (DIDP). Such compounds are esters. The term di-iso-decyl means that the side chain is branched, contains 10 carbon atoms (from the alcohol) and that two such chains are combined with a central group (-OOC-ϕ-COO- where ϕ is based on a benzene ring and is C_6H_4). The side chain is $C_{10}H_{21}$-. When this is combined with the central group -OOC-ϕ-COO- then the following is obtained $C_{10}H_{21}-OOC-\phi-COO-C_{10}H_{21}$.

di-iso-decyl adipate
Abbreviation used DIDA. Obtained by reacting decanols (mixed isomers) with adipic acid. Gives good *low temperature* properties with, for example, *polyvinyl chloride (PVC)* and its copolymers and has good permanence. See *di-iso-decyl*.

di-iso-decyl phthalate
An abbreviation used is DIDP. Also known as didecyl phthalate (DDP). Obtained by reacting decanols (mixed isomers) with phthalic anhydride. This material is based on branched alcohols and is a low volatility plasticizer for, for example, *polyvinyl chloride (PVC)* and its copolymers. It is a *low efficiency plasticizer*. See *di-iso-decyl*.

di-iso-heptyl
An abbreviation used for this prefix is DIH. This prefix is most commonly seen in terms used to describe *plasticizers*, for example, di-iso-heptyl phthalate (DIHP). Such compounds are esters. The term di-iso-heptyl means that the side chain is branched, contains 7 carbon atoms (from the alcohol) and that two such chains are combined with a central group (-OOC-ϕ-COO- where ϕ is based on a benzene ring and is C_6H_4). The side chain is C_7H_{15}-. When this is combined with the central group -OOC-ϕ-COO- then the following is obtained $C_7H_{15}-OOC-\phi-COO-C_7H_{15}$.

di-iso-heptyl phthalate
An abbreviation used for this material is DIHP. A *plasticizer* for *polyvinyl chloride*. See *phthalate*.

di-iso-hexyl phthalate
An abbreviation used for this material is DIHXP. A *plasticizer* for *polyvinyl chloride*. See *phthalate*.

di-iso-nonyl
An abbreviation used for this prefix is DIN. This prefix is most commonly seen in terms used to describe *plasticizers*, for example, *di-iso-nonyl phthalate (DINP)*. Such compounds are esters. The term di-iso-nonyl means that the side chain is branched, contains 9 carbon atoms (from the alcohol) and that two such chains are combined with a central group (-OOC-ϕ-COO- where ϕ is based on a benzene ring and is C_6H_4). The side chain is C_9H_{19}-. When this is combined with the central group -OOC-ϕ-COO- then the following is obtained C_9H_{19}-OOC-ϕ-COO-C_9H_{19}.

di-iso-nonyl adipate
An abbreviation used for this type of material is DINA. A *low temperature plasticizer*.

di-iso-nonyl phthalate
Abbreviation used DINP. Obtained by reacting nonanols (mixed isomers) with *phthalic anhydride*. This material is based on branched alcohols and is a *plasticizer* for, for example, *polyvinyl chloride (PVC)* and its copolymers. Similar in action to *dinonyl phthalate*. The use of DINP, together with a phosphate plasticizer (in place of a linear plasticizer such as DOP in PVC) will give compounds with greater resistance to heat ageing and flame resistance.

di-iso-octyl
An abbreviation used for this prefix is DIO. This prefix is most commonly seen in terms used to describe *plasticizers*, for example, *di-iso-octyl phthalate (DIOP)*. Such compounds are esters. The term di-iso-octyl means that the side chain is branched, contains 8 carbon atoms (from the alcohol) and that two such chains are combined with a central group (-OOC-ϕ-COO- where ϕ is based on a benzene ring and is C_6H_4). The side chain is C_8H_{17}-. When this is combined with the central group -OOC-ϕ-COO- then the following is obtained C_8H_{17}-OOC-ϕ-COO-C_8H_{17}.

di-iso-octyl adipate
Abbreviation used DIOA. This material has a relative density (RD or SG) of 0.93. A *low temperature plasticizer for*, for example, *polyvinyl chloride (PVC)* and its copolymers which is obtained by reacting octanols (mixed isomers from the Oxo process) with *adipic acid*.

di-iso-octyl azelate
Abbreviation used DIOZ. A *low temperature plasticizer* used, for example, with polyvinyl chloride (PVC) and its copolymers. Obtained by reacting octanols (mixed isomers from the Oxo process) with *azelaic acid*. It is compatible with cellulose acetate (CA), cellulose acetate butyrate (CAB), cellulose nitrate (CN) and ethyl cellulose (EC). It has limited compatibility with polyvinyl acetate (PVAC) and is incompatible with polymethyl methacrylate (PMMA), polystyrene (PS) and polyvinyl butyral (PVB).

di-iso-octyl fumarate
An abbreviation used is DIOF. An unsaturated *diester* which may be used to prepare copolymers.

di-iso-octyl maleate
An abbreviation used is DIOM. An unsaturated *diester* which may be used to prepare copolymers.

di-iso-octyl phthalate
Abbreviation used DIOP. Also known as di-2-ethylhexyl isophthalate. The ester obtained by reacting orthophthalic acid with isooctanol (mixed isomers from the *Oxo* process). This material has a relative density (RD or SG) of 0.98. Phthalates which contain 8 carbon atoms are the most commercially important *plasticizers*. Included in this category are *dialphanyl phthalate (DAP)*, *dioctyl phthalate (DOP)* and *di-iso-octyl phthalate (DIOP)*. DOP is the yardstick by which the other plasticizers are judged but for economic reasons DIOP or DAP may be preferred. (Compared to DAP, DIOP has less odour and is more permanent but DAP has better heat stability). Similar to DOP but has more permanence. It is compatible with cellulose acetate butyrate (CAB), cellulose nitrate (CN), ethyl cellulose (EC), polymethyl methacrylate (PMMA) and polystyrene (PS). It has limited compatibility with polyvinyl butyral (PVB). It is incompatible with cellulose acetate (CA) and polyvinyl acetate (PVAC).

di-iso-octyl sebacate
Abbreviation used DIOS. A *low temperature plasticizer* for polyvinyl chloride (PVC) which has low volatility and resists water extraction. Used interchangeably with di-octyl sebacate (DOS) but DIOS is preferred for plastisol use. It is compatible with cellulose nitrate (CN), ethyl cellulose (EC), polymethyl methacrylate (PMMA) and polystyrene. It has limited compatibility with cellulose acetate butyrate (CAB), polyvinyl acetate (PVAC) and polyvinyl butyral (PVB). It is incompatible with cellulose acetate (CA).

di-iso-pentyl phthalate
An abbreviation used for this type of material is DIPP. See *plasticizer*.

di-iso-propyl
This prefix is most commonly seen in terms used to describe *accelerators* for rubber compounds. The term di-iso-propyl means that the side chain is branched, contains 3 carbon atoms (from the alcohol) and that two such chains are combined with a central group. The side chain is $(CH_3)_2CH$-.

di-iso-tri-decyl phthalate
See *di-tri-decyl phthalate*.

di-isobutyrate
Also spelt diisobutyrate.

di-isobutyrate of 2,2,4-trimethyl-1,3-pentanediol
See *2,2,4-trimethylpentane-1,3-diol di-isobutyrate*.

di-isobutyrate of 3,3,5-trimethyl-1,4-pentanediol
See *3,3,5-trimethylpentane-1,4-diol di-isobutyrate*.

di-isocyanate
Also spelt diisocyanate. A compound which contains two *isocyanate groups*.

di-n-
This prefix is most commonly seen in terms used to describe *platicizers*. However, in *ISO terminology*, the letter n is not used in plasticizer abbreviations to indicate normal. See ISO 1043 Part 3 Sept 1986. For example:

di-n-butyl adipate = *dibutyl adipate*;
di-n-butyl phthalate = *dibutyl phthalate*;
di-n-decyl adipate = *didecyl adipate*;
di-n-decyl phthalate = *didecyl phthalate*;
di-n-hexyl azelate = *dihexyl azelate*;
di-n-octyl adipate = *dioctyl adipate*;
di-n-octyl phthalate = *dioctyl phthalate*;
di-n-octyl-n-decyl phthalate = *octyl decyl phthalate*;
di-n-tridecyl phthalate = *di-tri-decyl phthalate*.

However, the absence of the letter N, in terms used to describe *plasticizers*, should not be automatically taken to mean that the ester is based on a normal (straight chain) alcohol. It

di-naphthyl-p-phenylene diamine
An abbreviation used is DNPPD. Also known as N,N'-di-naphthyl-p-phenylene diamine. This grey, solid material has a relative density (RD or SG) of 1·20 and a melting point of approximately 233°C. A chain breaking *antioxidant* and also an *activator* for mercapto-type *accelerators*. Because of the risk posed by impurity β-naphthylamine the use of this material has declined.

di-nitroso-pentamethylene tetramine
Also known as dinitrosopentamethylenetetramine. Abbreviation used DNO and DNPT. A chemical *blowing agent* which decomposes over the range 160 to 200°C and which is used in, for example, the production of microcellular rubber.

di-nonyl
See *dinonyl*. For example, for di-nonyl sebacate see *dinonyl sebacate*.

di-o-tolylguanidine
Abbreviation used DOTG. This material has a relative density (RD or SG) of 1·19 and a melting point of approximately 170°C. It is a slow/medium speed *accelerator* used for the *sulphur curing of rubbers*. Slightly faster curing than *diphenyl guanidine (DPG)*. This material, like DPG, is rarely used alone but is used in *synergistic combinations* with primary accelerators such as the mercapto accelerators. It is a secondary accelerator. See *guanidine*.

di-octyl
See *dioctyl*.

di-p-isocyanate phenyl methane
See *4,4'-diphenyl methane diisocyanate*.

di-sec-octyl phthalate
See *dicapryl phthalate*.

di-sulphide cross-link
A cross-link based on two sulphur atoms. A link formed by two sulphur atoms, joined together, and which cross-links molecules together. That is, -S-S-.

The introduction of mono-sulphide cross-links and of di-sulphide cross-links, during *vulcanization*, improves vulcanizate properties as opposed, for example, to the formation of polysulphide cross-links which do not. To obtain more *efficient vulcanization*, use small amounts of sulphur (0·5 to 2 phr) and relatively large amounts of *accelerator* (3 to 6 phr). Even lower *efficiency parameter values (E)* may be obtained by using a thiuram efficient vulcanization system or, a *cadmate efficient vulcanization system*.

di-t-butyl peroxide
One of the best known *dialkyl peroxides*. Also known as di-tert.butylperoxide. The formula may be represented as $(CH_3)_3C$-O-O-$(CH_3)_3$ which on *homolytic decomposition* gives the tert butyloxy radical $(CH_3)_3C$-O·.

di-tert butylperoxide
See *di-t-butyl peroxide*.

di-tri-decyl phthalate
Also known as di-iso-tri-decyl phthalate or as, ditridecyl phthalate. An abbreviation used for this material is DITDP or DTDP or DITP. (DTDP is used as di-n-tridecyl phthalate and is not used as a plasticizer). This material has a relative density (RD or SG) of 0·95. It is a *plasticizer* for *polyvinyl chloride (PVC)* and for vinyl chloride (VC) copolymers which is used for high temperature cable insulation because of the good high temperature properties its use imparts. It is a low volatility plasticizer which is used, for example, for automotive upholstery as it gives compounds which retain their flexibility for a long time, which resist plasticizer extraction by hot, soapy water and which resist fogging. It is a mixed phthalate ester made from alcohols which contain 13 carbon atoms (a mixture of tetramethyl nonanol isomers) and which, in turn, are made by the *Oxo process*. It is compatible with cellulose acetate butyrate (CAB), cellulose nitrate (CN), ethyl cellulose (EC), polystyrene (PS), polyvinyl acetate (PVAC) and polyvinyl butyral (PVB). It has limited compatibility with polymethyl methacrylate (PMMA). It is incompatible with cellulose acetate (CA).

dia
An abbreviation used for diameter.

diacyl peroxide
Also known as acyl peroxide. An organic peroxide which contains two organic *acyl groups* (*alkyl* and/or *aryl*) connected via a peroxy radical. That is RCO-O-O-OCR. Benzoyl peroxide is the best known diacyl peroxide: another is lauryl peroxide. Such materials are widely used as free radical polymerization initiators as they may be decomposed, or degraded, by heat or UV light to give organic free radicals. See *organic peroxide*.

dial test indicator
An abbreviation used for this term is DTI. Also known a dial gauge or as, a clock gauge. A measuring device used to measure relatively small movements or displacements. The displacement of a plunger or stylus is shown on a circular dial by means of a rotating pointer.

The amount of tie bar extension can be measured by mounting a DTI on a stand so that the end of the spring loaded plunger makes contact with the end (vertical) face of the *tie bar*. The stand may be designed to accommodate 2 or 4 DTI's (i.e. one for each tie bar), so that individual tie bar extensions can be measured simultaneously. Such measurements should be recorded on the mould setting/record sheet. See *locking force*, *plunger type dial test indicator* and *lever type dial test indicator*.

dialkyl (C_7-C_9)
See the entries under *dialphanyl*.

dialkyl peroxide
An organic peroxide which contains two alkyl (R) groups connected via a peroxy radical. That is R-O-O-R. Dicumyl peroxide is the best known dialkyl peroxide: another is di-t-butyl peroxide. Such materials are relatively stable to heat and are therefore used for the high temperature production of organic free radicals. See *organic peroxide*.

dialkyl phthalates
Refers to a group of plasticizers which are based on phthalic anhydride and an aliphatic alcohol. As the ester which results has two side groups (alkyl groups) it is referred to as dialkyl. See *plasticizer* and *phthalate*.

diallyl isophthalate moulding material
An abbreviation used for this type of material is DAIP. More expensive than a similar diallyl phthalate moulding material (DAP) but has better heat resistance: they can withstand continuous operating temperatures of 180°C/356°F (DAP is approximately 160°C/320° F). The isophthalate-based materials are therefore recommended for higher temperature applications. See *allyl moulding materials*.

diallyl phthalate
An abbreviation used for this material is DAP. An ester which consists of two side chains (CH_2=CH-CH_2-) combined with a central group (-OOC-φ-COO- where φ is based on a benzene ring and is C_6H_4). When these are combined with the

central group -OOC-ϕ-COO- then the following is obtained CH$_2$=CH-CH$_2$-OOC-ϕ-COO-CH$_2$-CH=CH$_2$.

This liquid material is used to make the thermosetting plastics material which is also known as DAP. The monomer DAP, on which the thermosetting plastics material is based, may also be used as a polymerizable plasticizer for cellulose esters and *polyvinyl chloride (PVC)*. Has also been used with *unsaturated polyester resins*, in place of styrene, to give finished, GRP products with greater heat resistance than styrene-based materials. See *allyl moulding materials*.

diallyl phthalate moulding material
See *allyl moulding materials*.

diallyl phthalate plastics
An abbreviation used for this material is DAP. Thermosetting plastics materials based on the monomer *diallyl phthalate*. See *allyl moulding materials*.

dialphanol
See the entries under *dialphanyl*.

dialphanyl
An abbreviation used for this prefix is DA. This prefix is most commonly seen in prefixes used to describe *plasticizers*. Such compounds are esters which are (or were) based on a mixture of C7, C8 and C9 branched chain, oxo-alchols. That is, CH$_3$(CH$_2$)$_6$OH to CH$_3$(CH$_2$)$_8$OH isomers reacted with, for example, phthalic anhydride. It is thought that the term alphanyl is derived from the ICI trade name of Alphanol 79 - that used for the C7 to C9 alcohols.

dialphanyl adipate
Also known as dialkyl (C$_7$-C$_9$) adipate or as, dialphanol adipate or as, heptylnonyl adipate. An abbreviation used is DA$_{79}$A or D$_{79}$A or DAA. A *low temperature plasticizer*. See *dialphanyl phthalate*.

dialphanyl azelate
Also known as dialkyl (C$_7$-C$_9$) azelate or as, dialphanol azelate or as, heptylnonyl azelate. An abbreviation used is DA$_{79}$Z or D$_{79}$Z or DAZ. A *low temperature plasticizer*. See *dialphanyl phthalate*.

dialphanyl phthalate
This material is also known as dialkyl (C$_7$-C$_9$) phthalate or as, dialphanol phthalate or as, as heptylnonyl phthalate. An abbreviation commonly used for this material is DAP but this abbreviation is also used for *diallyl phthalate*: DA$_{79}$P is therefore suggested. Sometimes D$_{79}$P is used. This *plasticizer* is the phthalate ester of C7 and C9 oxo-alcohols. That is, CH$_3$(CH$_2$)$_6$OH and CH$_3$(CH$_2$)$_8$OH isomers reacted with phthalic anhydride.

This material has a relative density (RD or SG) of 1·00. It is a relatively cheap plasticizer for *polyvinyl chloride (PVC)* which has good solvent action at comparatively low processing temperatures and has good heat stability. It is used as a general purpose plasticizer for PVC and for polar rubbers such as *nitrile* and *chloroprene* rubbers. On average the alcohols on which it is based contain 8 carbon atoms and so it is similar to *dioctyl phthalate* (DOP).

Phthalates which contain 8 carbon atoms are the most commercially important plasticizers. Included in this category are dialphanyl phthalate (DAP), dioctyl phthalate (DOP) and di-iso-octyl phthalate (DIOP). DOP is the yardstick by which the other plasticizers are judged but for economic reasons DIOP or DAP may be preferred. DIOP has less odour and is more permanent but DAP has better heat stability.

dialphanyl sebacate
Also known as dialphanol sebacate or as, heptylnonyl sebacate or as, dialkyl (C$_7$-C$_9$) sebacate. An abbreviation used is DA$_{79}$S or D$_{79}$S or DAS. A *low temperature plasticizer*. See *dialphanyl phthalate*.

dialphanyl undecyl phthalate
See *heptyl nonyl undecyl phthalate*.

diamine
A chemical compound which is a base and which is based on ammonia: two of the three hydrogen atoms are replaced by organic groups (R). That is, R$_2$NH. Such materials are used as, for example, *accelerators*. The simplest example of a diamine is dimethyl amine. See *amine*.

diaminobicyclooctane
See *1,4-diazabicyclo-2,2,2-octane*.

diamino diphenyl methane
See *4.4'-diamino diphenyl methane*.

diamond
A pure, or very nearly pure, form of carbon which is highly crystalline. Noted for its extreme hardness and brilliance. Diamond is amongst the most thermally conductive materials known and is transparent to the spectrum of electromagnetic radiation from the *ultra-violet (UV)* through to the *infra-red*.

dianisidines
A class of *organic pigments*. See *diazo pigments*.

diaphragm gate
Also known as a disc gate. A *gate* used in *injection moulding* for the moulding of, for example, components which must contain holes, e.g. telephone hand-sets. The runner feeds into a thick, disc section, which almost completely covers the hole, and then the material flows through a thin restricted slit (which extends completely around the circumference of the disc) into the cavity. After moulding, removal of the sprue and the circular disc creates the hole which is required. Provided that the mould is adequately vented, *air-trapping* and *weld line formation* is not usually a problem when this type of gate is employed.

diatomaceous earth
This material is also known as diatomite or as, Fuller's earth or as, fossil flour or as, infusorial earth or diatomic earth. This material has a relative density (RD or SG) of 2·15. A natural micro-crystalline, or *cryptocrystalline*, form of *silica* based on the remains of diatoms. Most *silica* is amorphous, but because of crystallinity which has occurred with age, diatomaceous earth is thought to be a micro-crystalline form.

diatomic earth
See *diatomaceous earth*.

diatomite earth
See *diatomaceous earth*.

diazo compound
An organic compound which contains two nitrogen atoms (which may form an azo link, -N=N-) but only one nitrogen atom is attached to a carbon atom. For example, benzenediazonium chloride is C$_6$H$_5$N$^+$≡NCl$^-$. Such compounds form *diazonium salts*. Aromatic diazo compounds form, by azo coupling, azo compounds which are used as colorants.

diazo condensation pigments
Organic *diazo pigments* which are produced in two main steps or stages. The first is *diazotization* to produce two *azo dyes* and then these two dyes are combined, or coupled, into one high molecular weight, *diazo pigment*. A very wide range of high molecular weight compounds can be produced in this way. Used in place of *inorganic pigments* such as those based on cadmium and lead because of toxicity considerations.

diazo pigments
These pigments contain two *azo groups* and are also known as benzidines. For example, benzidine yellows are prepared from dichlorobenzidine. If the chlorine atoms on this material

are replaced with methoxy groups then the diazo pigments are called dianisidines. See *pigment*.

diazonium group
The diazonium cation ($-N^+ \equiv N$) which is formed by *diazotization* of an aromatic primary *amine*.

diazonium salt
An ionic compound which contains the diazonium group or cation bound electrovalently to an acid anion, for example, Cl^-. For example, $C_6H_5N^+ \equiv NCl^-$. Prepared from aromatic primary *amines* by *diazotization*. Such salts are used in the manufacture of *organic pigments* as they will, for example, couple with aromatic amines and phenols to produce azo compounds: a wide range of colours is possible from such compounds.

diazotization
The process used to prepare aromatic diazo compounds (the basis of *diazo pigments*). A salt of an aromatic primary *amine* (RNH_2) is reacted with nitrous acid at 0 to 5°C: this converts the $-NH_2$ group into the *diazonium group* $-N^+ \equiv N$. See *organic pigments*.

DIBA
An abbreviation used for *diisobutyl adipate*.

dibasic lead phosphite
A *lead stabilizer* or a basic lead stabilizer for *polyvinyl chloride* (PVC). Dibasic lead phosphite is used as a heat stabilizer for PVC: as it absorbs ultra-violet light, it also helps to protect the system against light. However, it is more expensive than other lead stabilizers, for example, the carbonate or the sulphate. Dibasic lead phosphite is also used as an acid acceptor in chlorine containing rubbers where it is used to accept hydrochloric acid.

dibasic lead phthalate
A comparatively expensive, *lead stabilizer* for *polyvinyl chloride* (PVC) which is suited for high temperature applications, for example, in electrical insulation applications (wire coating) as it has a low reactivity with *plasticizers*. Also used as a heat stabilizer in gramophone records (because of the low noise) and in expanded PVC formulations (where *azodicarbonamide* is used as a blowing agent). In this application the stabilizer acts as a *kicker*. Dibasic lead phthalate is also used as an acid acceptor in chlorine containing rubbers where it is used to accept hydrochloric acid.

dibenzo-1,4-pyran
See *xanthene*.

dibenzothiazole disulphide
See *benzthiazyl-disulphide*.

dibenzothiazyl disulphide
See *benzthiazyl-disulphide*.

dibenzothiazyldisulphide
See *benzthiazyl-disulphide*.

dibenzoyl peroxide
An *organic peroxide* which may also be known as benzoyl peroxide or as di-(benzene carbonyl) peroxide. That is RCO-O-O-OCR where R is C_6H_5. This material has a melting point of approximately 104°C. Decomposed into organic free radicals either by heat or by ultra-violet light. When pure and dry it will decompose readily so is best handled as a damp powder or as a paste (with *plasticizer*) or dispersed on an inert solid filler. See *diacyl peroxide*.

dibenzoyl-p-quinone dioxime
Also known as p,p'-dibenzoyl-p-quinone dioxime or as, p-quinone dioxime benzoyl ester. An abbreviation used is DPQD. A solid, brown powder which is used to give fast sulphur-free cures particularly for butyl rubber (where it is used in combination with lead tetroxide). More processing safety than p-quinone dioxime and has a more pronounced delayed action.

dibenzthiazyl disulphide
See *benzthiazyl-disulphide*.

dibenzyl amine
An abbreviation used is DBA. This material has a relative density (RD or SG) of 1·03 and is a pale yellow to brown liquid. An amine activator for rubber compounds and for rubber latex when used with, for example, *accelerators* such as zinc butyl xanthate and butyl xanthogen disulphide. With these ingredients it forms *ultrafast accelerators*. Use zinc oxide and sulphur but not fatty acids as these retard.

dibenzyl ether
This material has a relative density (RD or SG) of 1·02 and a boiling point of approximately 296°C. It is a good solvent for uncured *thiokol rubber* (T). It is a poor solvent for uncured *butyl rubber*. This chemical is an intermediate solvent for uncured *chloroprene rubber (CR)*, *nitrile rubber*, *natural rubber (NR)* and, *styrene-butadiene rubber (SBR)*. Used as a softener, to improve the elasticity of some butadiene-based rubbers.

dibenzyl sebacate
An abbreviation used is DBS. A *plasticizer* for *polyvinyl chloride* (PVC) and for *synthetic rubbers* which is known for the good low temperature properties which its use imparts.

diblock
Polymeric materials in which two blocks of two different materials are connected together. See *styrene block copolymer*.

diblock copolymer
See *copolymer* and *styrene block copolymer*.

DIBP
An abbreviation used for *di-iso-butyl phthalate*.

dibutoxyethyl adipate
Abbreviation used BCA or DBEA. Also known as dibutyl-cellosolve adipate and bis-ethylene glycol monobutylether adipate. A low temperature *plasticizer*. This material has a relative density (RD or SG) of 0·99, a melting point of -34°C and a boiling point of approximately 200°C (at 4 mm).

dibutyl adipate
Sometimes known as di-n-butyl adipate. An abbreviation used for this material is DBA (sometimes DNBA is seen but see *di-n-*). A liquid *plasticizer*. See *phthalate*.

dibutyl amine
An *activator* for some *accelerators*. This material has a relative density (RD or SG) of 0·75, a melting point of -67°C and a boiling point of approximately 167°C.

dibutyl ammonium oleate
An *activator* for *thiazoles* and *thiurams* which gives good processing safety. Also functions as a processing aid and *softener*. Reduces the need for stearic acid and improves filler dispersion in rubber compounds. This yellow/brown liquid has a relative density (RD or SG) of 0·87.

dibutyl phthalate
Abbreviation used DBP. Sometimes known as di-n-butyl phthalate or as, phthalic acid dibutyl ester but can also have di-iso-butyl phthalate. A liquid plasticizer used, for example, for cellulose nitrate but too volatile for general use with other polymers. When used with *polyvinyl chloride* (PVC), this material gives compounds which have high water extractability - especially in thin sections. This material has a relative density (RD or SG) of 1·05 and a boiling point of approximately 339°C. See *dibutyl phthalate adsorption method*.

dibutyl phthalate adsorption method
An abbreviation used for this term is DBP adsorption method. Structure levels in *carbon black*, may be determined by an *oil adsorption method* or by, the DBP adsorption method. The amount of *dibutyl phthalate* (DBP) which is just sufficient to form an oil or paste is determined using an adsorptometer or a plastograph. The void volume between the particles is filled with the liquid and the plastograph shows a torque maximum when all the voids have been filled and the mixture becomes a paste. The DBP adsorption value is expressed in ml of DBP per 100 g of filler, for example, 15 ml DBP per 100 g of carbon black. (The DBP number of the sample may be reported in $cm^3/100$ g.) The larger the oil or DBP value is, at constant black surface, the stronger are the forces between the *carbon black particles*. See also *24M4 test*.

dibutyl sebacate
Abbreviation used DBS. This material has a relative density (RD or SG) of 0·94 and a boiling point of approximately 345°C. As it is prepared from *sebacic acid* (a diacid) it is a diester. May be represented as $CH_3(CH_2)_3OOC(CH_2)_8COO(CH_2)_3CH_3$. It is best known as a very good, *low temperature plasticizer* for *polyvinyl chloride* (PVC). However, industrial use of such traditional low temperature plasticizers is now reduced because of the availability of cheaper materials, for example, the *nylonates*.

This material was sanctioned by the FDA for food contact: resists oil and solvent extraction. It is compatible with cellulose nitrate (CN), ethyl cellulose (EC), polymethyl methacrylate (PMMA) and polystyrene (PS). It has limited compatibility with cellulose acetate butyrate (CAB), polyvinyl acetate (PVAC) and polyvinyl butyral (PVB). It is incompatible with cellulose acetate.

dibutyl tin diacetate
Also known as dibutyl-Sn-diacetate. A catalyst used to help set, or cure, *liquid silicone resins*.

dibutyl tin dilaurate
Also known as dibutyl-Sn-dilaurate. For example, used as a catalyst for the preparation of *flexible polyurethane foams* and which has a melting point of approximately 23°C. Used in place of *stannous octoate* where greater resistance to hydrolysis is required. Also used to help set, or cure, *liquid silicone resins* and as a *heat stabilizer* for *polyvinyl chloride* (PVC). See *dibutyl tin stabilizer*.

dibutyl tin maleate
Also known as dibutyl-Sn-maleate. A catalyst used to help set, or cure, *liquid silicone resins*.

dibutyl tin stabilizer
An *organotin* compound widely used as, for example, a *heat stabilizer* for *polyvinyl chloride (PVC)* where transparency is required. The general formula for such materials is $(C_4H_9)_2SnX_2$ where X is an organic group - such as the laurate group (see *dibutyl tin dilaurate*).

dibutyl xanthogen disulphide
Also known as butyl xanthogen disulphide An *accelerator* with a relative density of approximately 1·16. See *dibenzyl amine*.

dibutyl-p-phenylene diamine
Also known as N,N'-disec-butyl-p-phenylene. An *antiozonant* and *antioxidant* for *natural* and *synthetic rubbers*.

dibutyl-Sn
See the entries under *dibutyl tin*.

dibutylcellosolve adipate
See *dibutoxyethyl adipate*.

dicapryl phthalate
Abbreviation used DCP. Also known as di-1-methylheptyl phthalate and as di-sec-octyl phthalate. This material has a relative density (RD or SG) of 0·97, a melting point of −60°C and a boiling point of approximately 225°C (15 mm). A *plasticizer* used, for example, with *polyvinyl chloride* (PVC). Similar to dioctyl phthalate but gives better weathering and a low initial viscosity: for this reason is used in *plastisols*. It is compatible with cellulose acetate butyrate (CAB), cellulose nitrate (CN), ethyl cellulose and polyvinyl butyral (PVB). It has limited compatibility with cellulose acetate (CA) polystyrene and polyvinyl acetate (PVAC). It is incompatible with cellulose acetate (CA), cellulose acetate butyrate (CAB), cellulose nitrate (CN), ethyl cellulose (EC), polymethyl methacrylate (PMMA), polystyrene (PS), polyvinyl acetate (PVAC) and polyvinyl butyral (PVB). Softener for natural and synthetic rubbers or elastomers.

dicarbitol phthalate
Also known as diethoxyethyoxyethyl phthalate, or as, diethyldiglycol phthalate or as, bis-diethylene glycol monoethylether phthalate. A liquid *plasticizer* for cellulose esters and ethers, for *polyvinyl chloride* (PVC) and its copolymers and for some synthetic rubbers.

dichlorodifluoromethane
Also known as *Freon 12*. See *chlorofluorocarbon*.

dichloroethane
Also known as 1,2-dichloroethane or DCE. See *ethylene chloride*.

dichloromethane
See *methylene chloride*.

dicumyl peroxide
Probably the best known *dialkyl peroxide*. Has a melting point of 42°C. Useful as a high temperature, free radical initiator for *unsaturated polyesters*, *rubbers*, *polyolefins* and *silicones* at temperatures greater than 100°C. Sometimes referred to as dicup. Gives vulcanizates of good heat resistance.

dicup
See *dicumyl peroxide*.

dicy
See *dicyandiamide*.

dicyandiamide
Also known as dicy. This solid material has a melting point of approximately 110°C. Used as, for example, a high temperature, curing agent for *epoxide resins*. Gives a long pot life at room temperatures but causes curing to occur at temperatures greater than approximately 145°C. When heated above its melting point, melamine is formed. Can be used to make *dicyandiamide-formaldehyde resin*.

dicyandiamide-formaldehyde resin
A polymer formed when dicyandiamide is reacted with formaldehyde to give comparatively low molecular weight polymers. These may be used to improve the wet strength of paper and to improve the dyeability of *cotton fabrics*.

dicyclohexyl phthalate
An abbreviation used is DCHP. This material is used as a plasticizer for *polyvinyl chloride* (PVC): it has low water solubility. This material has a relative density (RD or SG) of 1·15, a melting point of approximately 59°C and a boiling point of approximately 212°C (at 5 mm). It is also compatible with, for example, cellulose nitrate (CN), ethyl cellulose (EC) and polystyrene (PS).

dicyclopentadiene
Also known as bicyclopentadiene or as, endo-4,7-methylene-4,7,8,9-tetrahydroindene. Sometimes abbreviated to DCP. This aromatic material has a melting point of 32°C and a boiling point of 170°C and is based on cyclopentadiene. Sometimes used in *reaction injection moulding* (RIM) where the highly strained norbornene ring is probably opened first but the less strained cyclopentene ring also opens to form cross-linked structures. That is, *polydicyclopentadiene* is formed in *reaction injection moulding* (RIM).

dicyclopentadiene resin
A petroleum resin based on dicyclopentadiene: a low molecular weight resin obtained by thermal polymerization of petroleum fractions which contain large portions of *dicyclopentadiene*. The softening point of such a material can reach 175°C if the molecular weight is sufficiently high. See *polydicyclopentadiene reaction injection moulding*.

DIDA
An abbreviation used for *di-iso-decyl adipate*.

didecyl adipate
An abbreviation used is DDA. A *plasticizer* for *polyvinyl chloride (PVC)* and its copolymers which is noted for its permanence and the good *low temperature properties* which it imparts.

didecyl phthalate
An abbreviation used for this material is DDP. A *plasticizer* for polyvinyl chloride. See *phthalate* and *di-iso-decyl phthalate*.

DIDP
An abbreviation used for *di-iso-decyl phthalate (diisodecyl phthalate)*.

die
The assembly, located at the end of an *extruder*, which contains an orifice used to shape a plastics melt. The word die is sometimes used in place of mould.

die air vent
A passage in a pipe or profile *die* used to permit the passage of air into the interior of hollow extrusion.

die cart
A trolley which supports the *die* and holds it at the correct height for machine connection.

die entry angle
Angle of convergence of melt entering the extrusion *die*.

die entry region
In an extruder, this corresponds to the point where melt moves into the die parallel portion of the die. See *melt fracture*.

die gap
The distance, or separation, of the die lips.

die land
The land is the straight, or parallel, section through an extrusion die at the *die lips*.

die length
The effective length of the die is often taken as being the *die land*. An abbreviation used for the effective length of the die in, for example, shear flow is L. See *capillary rheometer*.

die lips
That part of the *die*, which forms the product, and which is located at the die exit.

die parallel
That part or portion of the *die lips*, where the die walls or lands are parallel to each other.

die plate
A term sometimes used in place of mould half or plate. For example, in a *two plate mould*, the moving head die plate is attached to the moving platen and the fixed head die plate is attached to the fixed platen.

die radius
The effective radius of the *die* is often taken as being the radius in the *die land region*. An abbreviation used for this term in, for example, *shear flow* is R. See *capillary rheometer*.

die spider
A legged unit used to support the torpedo i.e. that die section in the melt stream which forms the interior of a hollow section.

die swell
The swelling of an extrudate on emergence from a *die*: the extrudate swells so that its cross-section, immediately as it leaves the die, is greater than that of the die orifice. For a capillary die, the ratio extrudate diameter/die diameter is known variously as the die swell ratio, swelling ratio or as puff-up ratio. For a slit die, the relevant ratio is thickness of extrudate:depth of slit.

Die swell occurs because, as the melt is sheared through the die, the molecules become extended (with the greatest orientation near the die wall). On emergence from the die, the molecules tend to coil up and thus contract in the flow direction and expand in directions perpendicular to the flow. If an extrudate is cut at the die face, it will be noticed that the leading edge of the extrudate is convex thus indicating greatest contraction in the flow direction nearest to the wall where the shear has been greatest.

die swell - factors affecting
Experimental work has shown that die swell:
i) increases with increase in extrusion rate, or more specifically shear rate, up to a critical shear rate - see *melt fracture*;
ii) decreases with increase in temperature at a given shear rate or extrusion rate;
iii) is little affected by temperature at a fixed shear stress;
iv) decreases with increased length of die parallel at a fixed shear rate;
v) is somewhat greater through a slit die than through a capillary die and also increases more rapidly with increasing shear rate; and that,
vi) die swell increases with an increase in the ratio reservoir diameter:capillary diameter (although little affected at ratios above 10:1).

die swell - compensating solid extrudates
It is common practice to compensate solid extrudates for *die swell* by stretching, or drawing down, the extrudate so that it can just pass through a sizing die. It should be noted that *drawing down* will cause *molecular orientation*. The method is somewhat limited where the solid extrudate has a varying section thickness since shear rates, and hence die swell, will be higher at the thinner sections. Such thinner sections may also have a shorter die parallel, to ensure that extrusion rates are constant across the cross-section, and this will alter the die swell even further.

die swell - compensating hollow extrudates
In the case of pipe and tubing, compensation for *die swell* is complicated by the fact that the extrudate is usually inflated to the dimensions of a sizing die. In this case it may be assumed that on emergence from the die that, the wall thickness will expand appropriate to the die swell for the *shear rate* used but that, the wall thickness will be reduced proportionally to the amount of inflation given by the ratio of diameter of sizing die:external diameter of die.

die swell - diameter determination

If the extrudate is smooth and circular, the measurements may be made using a projection microscope: a laser micrometer allows very accurate measurements of both the extrudate and of the die. If the sample is not round and smooth (for example, due to collapse of an unvulcanized rubber extrudate or if, the extrusion exhibits *sharkskin* or *melt fracture*) the best procedure is to weigh (M) a known length (L) of the extrudate and from a knowledge of the density (σ) calculate the extrudate diameter (D) by the formula $D^2 = 4M/\pi\sigma L$. For amorphous materials, it is satisfactory to use the density of the solid polymer. In the case of crystalline polymers however the density will depend on the rates of cooling from the die and for accurate measurement the density should be obtained from the extruded sample.

die swell - measurement of

Usually measured on samples of circular cross-section that have been extruded downwards from a capillary *rheometer* fitted with a rod die. Accurate measurement, and interpretation of die swell measurements, requires care. The weight of the extrudate is liable to cause drawing down and the softness of the extrudate allows distortion. To ensure that a circular cross-section is obtained from the die, the extrudate must be given time to freeze while suspended since it can distort if it is allowed to lie on a surface before it has hardened. Drawing down could be avoided by extruding into a bath containing a liquid of the same density as the polymer and which is placed immediately below the die. See *draw down*.

die swell ratio
See *die swell*.

die wall shear rate
See *apparent shear rate*.

die-face pelleter

A cutting, or chopping, device used to produce granules from a thermoplastics material after extrusion by *die-face pelletizing*.

die-face pelletizing

The production of granules from a thermoplastics material using a *die-face pelleter* and an *extruder*. A method of producing thermoplastics granules after compounding. As the material emerges from the die, in rod form, a rotating knife cuts the rods into pellets of a regular size which are then cooled. See *polyvinyl chloride*.

dielectric

A term used to describe an electric insulator, that is, a material with a resistivity greater than $10^{10}\Omega.$cm. Only dielectric materials can sustain a potential difference when placed in a high frequency or microwave frequency field and can therefore be heated by *dielectric heating* (unlike good conductors). Not all dielectric materials have the same ability to being heated in a high frequency field. See *loss factor*.

dielectric breakdown

The permanent loss of insulating properties of a *dielectric material* caused by the application of high voltages and subsequent changes to the polymer: often the polymer between the electrodes becomes burnt. See *dielectric strength*.

dielectric constant

Also known as relative permittivity or as, specific inductive capacity or as, permittivity. A measure of how well a material will store an electrical charge. When the material is going to be used as an insulator then a low dielectric constant is needed; when the material is going to be used in condenser applications then a high dielectric constant is needed. It is the ratio of the capacity of a condenser made with a plastics material over the capacity of an identical condenser made with air as the dielectric. As this is a ratio it has no dimensions. Often given the symbol ϵ or ϵ_r and measured using a Schering bridge using a sheet of material (40 × 40 × 3 mm) which is made the *dielectric* of a condenser.

A measure of the ability of the material to store electric charge through *polarization*. One of the properties used to assess a material as an electrical insulator: for such a purpose, a knowledge of the dielectric constant over a range of temperatures is needed. Both temperature and frequency affect dielectric constant. The dielectric constant is frequently equal to the square of the refractive index and both may be calculated from a knowledge of the chemical bonds present. See, for example, ASTM D 150.

dielectric heating

The heating of a *dielectric material* by the application of a *high frequency*, alternating electric field which causes *polarization*. A heating process which uses the output from a high frequency generator. Operation is at radio frequencies: when the frequency is below 300 MHz this gives *high frequency heating* and when the frequency is above 300 MHz this gives *microwave heating*.

The application of a high frequency field causes a rapid and comparatively uniform temperature rise within the material. For every material there is an optimum frequency and ideally this frequency should be used. However, high frequency oscillators of the type used for high frequency and microwave heating cause radio interference and as a result there are restrictions on the frequencies which can be used. The permitted wavelengths may differ from country to country. For example, the *Industrial, Scientific* and *Medical bands* in the UK are as follows.

Frequency MHz	Wavelength m	Tolerance ±MHz	Application
13·56	22·1	0·07	High frequency heating
27·12	11·0	0·163	High frequency heating
915	0·33	13	Microwave heating
2450	0·12	50	Microwave heating.

The electrodes for dielectric heating take different forms dependent upon the frequency of the field. In the lower range (below 300 MHz), the applicator may take the form of, for example, a parallel plate capacitor or of a concentric tubular capacitor. To generate the frequencies a conventional valve oscillator (a thermionic device known as a triode valve) is used. At microwave frequencies, conventional valves cannot generate sufficient power and a thermionic device known as a magnetron is used together with tubes or waveguides in place of wires.

Dielectric heating is used for drying, preheating and curing. The amount of power (P) absorbed by a material is given by $P = 2\pi f V^2 \epsilon_0 \epsilon_r \tan \delta$. Where P = power density in watts per cubic metre: f = the applied frequency in hertz: V = the voltage gradient across the material in volts per metre: ϵ_0 = the dielectric permittivity of free space (8·85 farads per metre): ϵ_r = the dielectric constant of the material and δ is the loss tangent in radians.

dielectric loss index
See *loss factor*.

dielectric loss tangent
See *loss tangent*.

dielectric material
An insulating material. See *dielectric*.

dielectric strength
Sometimes called electric strength. Could be called insulation strength as it is a measure of how well a material, a *dielectric*, can

withstand a voltage. The resistance to *dielectric breakdown* and specified as the voltage at which dielectric breakdown occurs divided by the specimen thickness. An abbreviation used for this term is E: it is also known as breakdown strength.

It is defined as the voltage difference per unit of thickness which will cause catastrophic failure of the dielectric: breakdown occurs when there is a sudden flow of current through the material. It is very dependent upon thickness so that, for example, the quoted dielectric strength for a 0·001 in film (in volts/mil) is often twice that for a 0·005 in film of the same material (25·4 V/mil = 1 Kv/mm). For some materials, increasing humidity decreases the results; the result also decreases rapidly with increasing A.C. frequency.

Test specimen thickness is usually >3 mm and the size is 100 × 100 mm. An alternating current of 50 Hz is applied to the material and the voltage is steadily increased until flash-over occurs in between 10 and 15 s. If V_f is the flash-over voltage and th the specimen thickness then $E = V_f/th$ in kV/mm. See, for example, ASTM D 149.

diene
An *alkene* which contains two double bonds. See *conjugated* and *diene polymerization*.

diene polymerization
Polymerization of a material with two double bonds. Usually means polymerization of a *conjugated 1,3-diene monomer*. For example, polymerization of butadiene, isoprene or chloroprene. Polymerization of a monomer which may be represented as $CH_2=CX-CH=CH_2$. Where $X = H$ = butadiene; where $X = CH_3$ = isoprene and, $X = Cl$ = chloroprene.

Within such monomers there are two double bonds and one of these bonds remains after polymerization. Because of this remaining double bond, the carbon-to-carbon bond angles and the presence of an asymmetric carbon atom, it is possible, by diene polymerization, to produce *stereoregular polymers* from, for example, butadiene. If this is represented as $CH_2=CH-CH=CH_2$ or $C1=C2-C3=C4$, then there are four stereoregular forms of polybutadiene rubber, cis-1,4-polybutadiene, trans-1,4-polybutadiene, isotactic 1,2-polybutadiene, and syndiotactic 1,2-polybutadiene (polymerization across the 1,2 bond gives similar products to polymerization across the 3,4 double bond).

With *isoprene* then, if this is represented as $CH_2=C(CH_3)-CH=CH_2$ or $C1=C\#2-C3=C4$, then there are six stereoregular forms of polyisoprene rubber, cis-1,4-polyisoprene, trans-1,4-polyisoprene, isotactic 1,2-polyisoprene, syndiotactic 1,2-polyisoprene, isotactic 3,4-polyisoprene and syndiotactic 3,4-polyisoprene. Because of the C atom, identified as C#, polymerization across the 1,2 bond gives different products to polymerization across the 3,4 double bond.

Chloroprene may, in theory, take up the same steric forms as isoprene.

diene rubber
A rubber formed from a *diene monomer* or from diene monomers. Can have both homopolymers and copolymers. An example of a homopolymer is *butadiene rubber* and that of a copolymer is *styrene-butadiene rubber*.

diene rubber degradation
The degradation of a *diene rubber*. Most of the degradation of diene rubbers (*natural rubber* and *synthetic diene rubbers*) is due either to oxygen attack or to ozone attack. See, for example, *oxidation*.

diester
A material which contains two ester groups or linkages. An organic ester prepared from a *polyol* (for example, *propylene glycol*) and an organic acid (for example, *ricinoleic acid*) and in which two of the hydroxyl groups were used in the esterification reaction. See, for example, *propylene glycol diricinoleate*. When the term is used in connection with *polyvinyl chloride (PVC)* and with *plasticizers*, it is usually taken to mean *aliphatic diesters*.

diethanolamine
Also known as 2,2'-dihydroxy diethylamine or as, 2,2'-iminodiethanol. A colourless liquid with a relative density of 1·09 and a boiling point of approximately 270°C. A strong base which neutralises the acidity of clays. Used as an *activator* in rubber compounds.

diethoxyethyoxyethyl phthalate
See *dicarbitol phthalate*.

diethyl ether
See *ether*.

diethyl hexyl
See *di-2-ethylhexyl*.

diethyl phthalate
An abbreviation used is DEP. Sometimes referred to as phthalic acid diethyl ester. A low molecular weight *phthalate plasticizer*. This material has a relative density (RD or SG) of 1·12. because of its volatility, this material is seldom used as a *plasticizer* with *polyvinyl chloride* (PVC). Sometimes used in conjunction with *dibutyl phthalate* to plasticize cellulose acetate. A plasticizer for *natural rubber* and for *synthetic rubbers*: does not affect vulcanization.

diethyl toluene diamine
An abbreviation used for this material is DETDA. An *amine* used to achieve cross-linking in *polyurethane foam* technology, for example, in *reinforced reaction injection moulding (RRIM)*. When using powders, for example, reclaimed material, in RRIM processes the DETDA can become adsorbed by the powder: the adsorption can be minimized by using three separate streams fed to the mixing head. One is the powder-charged polyol, the second is the isocyanate and the third is the DETDA which can contain catalysts and internal mould release agents.

diethyldiglycol phthalate
See *dicarbitol phthalate*.

diethylene glycol
An abbreviation used for this material is DEG. This material has a relative density of 1·12 and a boiling point of approximately 245°C. May be represented as $[HOCH_2CH_2OCH_2CH_2OH]$. This material is an *activator* in silicate-filled, rubber compounds: also acts as a *plasticizer* for *natural* and *synthetic rubbers*. When used with highly filled rubber compounds, it improves the water resistance of such compounds. Sometimes used for the preparation of relatively flexible, *unsaturated polyester resins*. When reacted with acids, the products (esters) are used as softeners and plasticizers. Either one, or both, hydroxyl groups may be reacted. An example where one hydroxyl group is reacted is *diethylene glycol monolaurate*. An example where both are reacted is *diethylene glycol dibenzoate*. DEG is a solvent for uncured *phenol-formaldehyde* and *urea-formaldehyde resins*.

diethylene glycol dibenzoate
A diester of *diethylene glycol*. Abbreviation used DEGB. This material has a relative density (RD or SG) of 1·18, a melting point of 16 to 28°C and a boiling point of approximately 235°C (5 mm). This material is a plasticizer for *polyvinyl chloride* (PVC) and its copolymers.

diethylene glycol monolaurate
A mono-ester of *diethylene glycol*. This oily liquid material has a relative density (RD or SG) of 0·95 and a melting point of approximately 10 to 18°C. It is used as a plasticizer for *natural* and *synthetic rubbers*.

diethylene oximide
See *morpholine*.

diethylene triamine
An abbreviation used for this material is DET. This colourless liquid material has a relative density (RD or SG) of 0·96, a melting point of −39°C and a boiling point of approximately 207°C. It is an activator in emulsion polymerization and is a curing agent for *epoxide resins*. Gives fast curing even at room temperatures and reasonable properties to the cured material.

differential area
The difference between two areas.

differential cylinder
A *cylinder* in which the two opposed piston areas are unequal.

differential pressure
The effective result of two pressures acting on separate areas.

differential scanning calorimetry method
An abbreviation used for this term is DSC method. In the DSC method, of measuring *specific heat*, a linear increase in temperature is maintained in two cells, one of which usually contains the test specimen in an aluminium sample pan, the other, a similar but empty sample pan. The difference in power input (heat flow rate) needed to maintain the same temperature in both cells as the temperature is raised (or lowered) is recorded as a function of temperature. If the equipment is first calibrated using, for example, synthetic sapphire whose specific heat is well documented, the specific heat of an unknown can be found from: $C_{ps} = C_{pc} \times (m_c/m_s) \times \{(D_s - D_e)/D(_c - D_e)\}$.

Where,
C_{ps} = specific heat of the specimen;
C_{pc} = specific heat of the calibrant, for example, synthetic sapphire:
m_c = mass of calibrant (for example, synthetic sapphire);
m_s = mass of specimen;
D_s = instrument output with specimen in one sample pan;
D_c = instrument output with calibrant in one sample pan; and,
D_e = instrument output with both sample pans empty.

However, this method of calculation has now been largely superseded as modern equipment has sophisticated data storage facilities and computer programmes which enable specific heats over a wide range of temperature to be calculated quickly and efficiently to accuracies of ±2 per cent. See *enthalpy data* and **table 7**.

diffusion
Diffusion (spreading out) occurs as a result of natural processes which tend to equal out the concentration of a given species in a particular environment. Diffusion through a polymer occurs by the small molecules, for example, gases and vapours, passing through voids, and other gaps, in the structure of the plastics material. Diffusion therefore depends on the size of the gaps and, on the size of the small molecules. See *permeability*.

diffusion printing
See *printing techniques*.

digital and closed loop
An abbreviation used for this term is DCL. A system which uses microprocessor-based circuitry, mounted within a *servo-valve*, which is capable of responding to *digital command signals*.

digital hydraulics
A system whereby oil flow rates and pressures are set digitally and the values are regulated or controlled, in discrete steps, by means of *cartridge valves*. Widely used on *injection moulding machines* as such systems are very reliable and give reproducible results.

digital setting
The setting of a parameter to a precise numerical value; usually done by means of thumbwheels or, a key-pad. Widely used on, for example, *injection moulding machines*.

digital signal
A voltage or current which only has two distinct values.

diglycidyl ether of bisphenol A
An abbreviation used for this material is DGEBA. This material is prepared by reacting an excess of *epichlorhydrin* with *bisphenol A*. It is the basis of many *epoxy resins*: such resins are cured most commonly with amines and anhydrides.

DIH
An abbreviation used in plasticizer terminology for *di-iso-heptyl*. For example, *di-iso-heptyl phthalate*.

diheptyl phthalate
An abbreviation used for this type of material is DHP. A *phthalate plasticizer*. See *plasticizer*.

dihexyl
This prefix is most commonly seen in terms used to describe *ester plasticizers*, for example, dihexyl adipate. The term dihexyl means that the chain contains 6 carbon atoms and that two such chains are combined with a central group which may be represented as -OOC-ϕ-COO- (where ϕ is based on a benzene ring and is C_6H_4). The total is C_6H_{13}-OOC-ϕ-COO- C_6H_{13}. An abbreviation used for dihexyl is DH or DHX.

dihexyl azelate
An abbreviation used for this type of material is DHXZ. Similar to *dioctyl azeleate* in the properties it imparts. It is compatible with cellulose acetate (CA), cellulose acetate butyrate (CAB), cellulose nitrate (CN) and ethyl cellulose (EC). It has limited compatibility with polyvinyl acetate (PVAC) and is incompatible with polymethyl methacrylate (PMMA), polystyrene (PS) and polyvinyl butyral (PVB).

dihexyl phthalate
An abbreviation used for this material is DHXP. A *plasticizer* for polyvinyl chloride. See *phthalate*.

DIHP
An abbreviation used for *di-iso-heptyl phthalate*.

DIHX
Letters used in *plasticizer* abbreviations for di-iso-hexyl. For example, *di-iso-hexyl phthalate (DIHXP)*. See *phthalate*.

DIHXP
An abbreviation used for *di-iso-hexyl phthalate*.

dihydric
A material is dihydric if it contains two hydroxyl groups, for example, a *diol*. See *diethylene glycol*.

dihydroquinoline derivatives
Additives used as *antioxidants*. For example, polymerized 1,2-dihydroxy-2,2,4-trimethylquinoline (TMQ). May be referred to

as 1,2-dihydroxy-2,2,4-trimethylquinoline, polymerized. Also in this category is 6-ethoxy-2,2,4-trimethyl-1,2-dihydroxyquinoline (ETMQ). These are strongly discolouring antioxidants. See *staining antioxidants*.

diiso-
Entries which begin with this prefix are listed under di-iso. For example, see *di-iso-butyl*.

dilatancy
Also known as *shear thickening*. A type of *non-Newtonian flow behaviour* in which the *apparent viscosity* increases as the *shear rate* increases. See *shear-induced crystallization*.

dilauryl thiodipropionate
An abbreviation used for this material is DLTDP. An example of a *thioester antioxidant*. A white powder with a melting point of approximately 40°C which is used, for example, as a *stabilizer* for *polyolefins*.

dilute masterbatch
A *masterbatch* which uses the parent polymer (usually a *thermoplastics material*) as the carrier for the colouring system. Such masterbatches may contain up to 20% of the colouring system and would probably be prepared by extrusion: supplied in chip or granule form. Up to 5%, of this dilute masterbatch may need to be added to natural polymer for colouring purposes. Such systems can give very good results at comparatively low costs and are clean and easy to handle. However, to get the best results, it may be necessary to use a special mixing screw for their preparation and it will be probably necessary to use a separate dilute masterbatch for each type of polymer. Where a large number of different materials are being run, it will be necessary to exercise strict control if mistakes are to be avoided. See *universal masterbatch*.

dilute solution viscometry
See *dilute solution viscosity*.

dilute solution viscosity
A molecular weight determination method which gives the solution viscosity, average molecular weight.

dimensional stability
The ability of a component to resist a particular environment for a specified time. Often assessed in the case of plastics components by exposing the specimen to oven heating.

dimer
Two monomer units linked together.

dimethyl acetamide
An abbreviation used for this material is DMAC. This material is used as a solvent for, for example, aromatic polyamides, polyimides and polybenzimidazoles.

dimethyl amine
The simplest example of a diamine is dimethyl amine. A *diamine* which may be represented as $(CH_3)_2NH$. This gaseous material has a relative density (RD or SG) of 0·68 and a boiling point of approximately 7°C. It is an *accelerator* for rubber compounds. See *dithiocarbamates*.

dimethyl benzene
See *xylene*.

dimethyl butadiene
Also known as 2,3-dimethyl-1,3-butadiene or as, dimethylbutadiene. A *substituted butadiene*. A *methyl substituted butadiene*: methyl groups are substituted for hydrogen atoms on the butadiene molecule. See *methyl rubber*.

dimethyl dichlorosilane
A *chlorosilane* which has two methyl groups (CH_3) and two chlorine atoms (Cl) attached to each silicone atom. That is, $(CH_3)_2SiCl_2$. Hydrolysis gives, for example, the linear *polyorganosiloxane* known as *polydimethylsiloxane*. Very pure material is required for the monomer for *silicone rubber*.

dimethyl diphenyl thiuram disulphide
An abbreviation used for this material is MPTD. This material has a relative density (RD or SG) of 1·33 and a melting point of approximately 175°C. Dimethyl diphenyl thiuram disulphide is a secondary *accelerator* used with, for example, tetramethyl diphenyl thiuram disulphide as the combination has good processing safety, gives fast curing at low temperatures and the vulcanizates have good ageing properties.

dimethyl formamide
An abbreviation used for this material is DMF. May be represented as $HCON(CH_3)_2$. This material has a relative density (RD or SG) of 0·95, a melting point of −61°C and a boiling point of approximately 153°C. Miscible with water, ethanol, ether, ketones and benzene. Will dissolve solvent resistant materials such as *polyvinyl chloride* and *polyacrylonitrile*.

dimethyl ketone
See *acetone*.

dimethyl phenol
See *xylenol*.

dimethyl phthalate
An abbreviation used for this material is DMP. This oily, colourless material has a melting point of −1°C and a boiling point of approximately 284°C. The SG is 1·19. It is a diester prepared from phthalic anhydride and methyl alcohol but is too volatile for general use as a plasticizer: used in conjunction with *diethyl phthalate* as a *plasticizer* for *cellulose acetate*.

dimethyl silicone
See *polydimethylsiloxane*.

dimethyl silicone elastomer
See *polydimethylsiloxane* and *silicone rubber*.

dimethyl silicone gum
The un-vulcanized polymer on which *silicone rubbers* are based. Such a gum may be turned into a cross-linked elastomer by heating with an organic *peroxide*, for example, benzoyl peroxide.

dimethyl sulphoxide
An abbreviation used for this material is DMSO. A solvent for, for example, *polyacrylonitrile* and *polyvinyl acetate*. Will also dissolve *cellulose nitrate*, *cellulose acetate* and materials containing *sulphur*.

dimethylbutadiene
See *dimethyl butadiene*.

dimethylbutadiene elastomer
See *methyl rubber*.

dimethylbutadiene polymer
See *methyl rubber*.

dimethylbutadiene rubber
See *methyl rubber*.

dimethylcyclohexylamine
An amine catalyst used for *polyester foams*. See *flexible polyurethane foam*.

dimethyldichlorosilane
See *dimethyl dichlorosilane*.

dimethylethanolamine
A component of a flexible *polyether foam catalyst system*. A polyether foam could use stannous octoate, dimethylethanolamine and 1,4-diazabicyclo-2,2,2-octane as a catalyst system. The stannous octoate may be replaced by dibutyl tin dilaurate if there is a danger of hydrolysis in water-containing blends.

dimethylsilicone elastomer
See *dimethylsilicone rubber*.

dimethylsilicone rubber
A *silicone rubber* which contains methyl groups. An abbreviation used for this type of material is MQ. See *polydimethylsiloxane*.

dimethylol ethylene urea
An abbreviation used for this material is DMEU. This material is used as crease-resistant treatment for cellulose fabrics. When heated, to approximately 130°C, it forms a urea-formaldehyde polymer which is also bonded to the cellulose. DMEU is formed by the reaction of formaldehyde with ethylene urea.

dimyrstyl thiodipropionate
An example of a *thioester antioxidant*.

DIN
An abbreviation used for Deutches Institut Für Normung. Usually refers to a German test method or standard issued by the Deutscher Normenausschuss. See *standards organizations*.

DINA
An abbreviation used for *di-iso-nonyl adipate*.

dinitrosopentamethylenetetramine.
See *di-nitroso-pentamethylene tetramine*.

dinonyl
Also spelt di-nonyl and sometimes referred to as di-(3,3,5-trimethylhexyl). An abbreviation used for this prefix is DN. This prefix is most commonly seen in terms used to describe *plasticizers*, for example, *dinonyl phthalate (DNP)*. Such compounds are esters. The term di-nonyl means that the side chain is substantially linear (but branched at the end) contains 9 carbon atoms (from the alcohol) and that two such chains are combined with a central group (-OOC-ϕ-COO- where ϕ is based on a benzene ring and is C_6H_4). The side chain is C_9H_{19}-. When this is combined with the central group -OOC-ϕ-COO- then the following is obtained C_9H_{19}-OOC-ϕ-COO-C_9H_{19}. The side chain arrangement is $CH_{33}CCH_2CH(CH_2)CH_2CH_2COO$-.

dinonyl phthalate
Abbreviation used DNP. Sometimes referred to as di-(3,3,5-trimethylhexyl) phthalate or as, di-nonyl phthalate. This material has a relative density of 0.97. It is a *low efficiency plasticizer* and has a lower plasticizing efficiency than other *general purpose plasticizers* (such as DAP, DIOP and DOP) and this means that more of the DNP can be incorporated to give compounds of the same hardness. The gelation rate with PVC is less than that of DAP, DIOP and DOP. If used, the difficulty of incorporating a larger quantity of lower solvating plasticizer should not be forgotten. DNP is also useful if low viscosity plastisols are required.

dinonyl sebacate
Abbreviation used DNS. Sometimes referred to as di-(3,3,5-trimethylhexyl) sebacate or as, di-nonyl sebacate. A *low temperature plasticizer*.

DINP
An abbreviation used for *di-iso-nonyl phthalate*.

dinuclear phenols
Structures based on phenols which contain two aromatic nuclei. Formed, for example, during *resole manufacture*. See *polynuclear phenols*.

DIOA
An abbreviation used for *di-iso-octyl adipate*.

dioctyl
This prefix is most commonly seen in terms used to describe *plasticizers*, for example, the ester *dioctyl phthalate*. The prefix di-octyl, or dioctyl, is **usually** another way of writing di-2-ethylhexyl (as in di-2-ethylhexyl phthalate). However, the absence of the letter N, in terms used to describe plasticizers, should not be automatically taken to mean that the ester is based on a normal (straight chain) alcohol. It is common practice to use abbreviations without N for esters which may be based on branched chain alcohols.

The term di-2-ethylhexyl (commonly expressed as dioctyl) means that the side chain contains a branch of 8 carbon atoms (from the alcohol, 2-ethylhexanol) and that two such chains are combined with a central group (-OOC-ϕ-COO- where ϕ is based on a benzene ring and *later* is C_6H_4). The side chain is C_8H_{17}-. When this is combined with the central group -OOC-ϕ-COO- then the following is obtained C_8H_{17}-OOC-ϕ-COO-C_8H_{17}. The side chain arrangement is $CH_3CH_2CH_2CH_2CH(C_2H_5)CH_2$- (from the 2-ethylhexanol). If a straight chain alcohol is used then the side chain arrangement is $CH_3(CH_2)_7$-.

dioctyl adipate
An abbreviation used for this material is DOA. Also known as di-2-ethylhexyl adipate. This material has a relative density of 0.93. A high efficiency *plasticizer* for *polyvinyl chloride* (PVC) which imparts good low temperature flexibility. Used, for example, in film and in plastisols for low temperature use. Said to be suitable for food contact use. It is compatible with *ethyl cellulose* (EC) and *polystyrene* (PS). It has limited compatibility with *cellulose acetate butyrate* (CAB), *cellulose nitrate* (CN), *polymethyl methacrylate* (PMMA) and *polyvinyl butyral* (PVB). It is incompatible with *cellulose acetate* (CA) and *polyvinyl acetate* (PVAC). See *low temperature plasticizer*.

dioctyl azelate
An abbreviation used for this material is DOZ. Also known as di-2-ethylhexyl azelate. DOZ is used as a *plasticizer*. Very good low temperature properties and high compatibility with *polyvinyl chloride* (PVC). Has low volatility and low water extraction in compounds. Imparts good handle and drape characteristics to film and sheet. It is compatible with *cellulose acetate butyrate* (CAB), *cellulose nitrate* (CN), *ethyl cellulose* (EC), *polystyrene* (PS) and *polyvinyl acetate* (PVAC). It has limited compatibility with *cellulose acetate* (CA). It is incompatible with *polymethyl methacrylate* (PMMA) and *polyvinyl butyral* (PVB). See *low temperature plasticizer*.

dioctyl isophthalate
Also known as di-ethylhexyl isophthalate. An abbreviation used is DOIP. This material is used as a *plasticizer*. This material is based on isophthalic acid - *di-iso-octyl phthalate* is based on *phthalic anhydride*. The use of the meta isomer of phthalic acid, compared to the ortho isomer, results in compounds which have lower extraction (by water, oils and solvents), lower volatility and good mar resistance. It is a plasticizer for *polyvinyl chloride* (PVC) which is used for applications such as handbags, furniture upholstery and automotive upholstery. Gives plastisols of comparatively low viscosity. It is compatible with *cellulose nitrate* (CN), *ethyl cellulose* (EC), *polystyrene* (PS) and *polyvinyl butyral* (PVB). It has limited compatibility with *cellulose acetate butyrate* (CAB) and *polymethyl methacrylate* (PMMA). It is incompatible with *cellulose acetate* (CA) and *polyvinyl acetate* (PVAC).

dioctyl maleate
An abbreviation used for this material is DOM. A *plasticizer*.

dioctyl phthalate
An abbreviation used for this material is DOP. As this material may be known as di-2-ethylhexyl phthalate, or diethylhexyl

phthalate, then the abbreviation DEHP is sometimes used. DOP is used as a *plasticizer*. This material has a relative density (RD or SG) of 0·99 and a boiling point of approximately 370°C. Two side chains are joined to a benzene ring (ortho position as phthalic anhydride is used as the starting material) and each is represented as -COOCH$_2$CH(CH$_2$CH$_3$)(CH$_2$)$_3$CH$_3$.

Phthalates which contain 8 carbon atoms, in each side chain, are the most commercially important plasticizers. Included in this category are *dialphanyl phthalate* (DAP), dioctyl phthalate (DOP) and *di-iso-octyl phthalate* (DIOP). DOP is the yardstick by which the other plasticizers are judged but for economic reasons DIOP or DAP may be preferred. DOP is similar to DIOP but has less permanence. Has reasonable oil and water extraction resistance when used in *polyvinyl chloride* (PVC), moderate volatility and moderate low temperature properties.

DOP is compatible with *cellulose acetate butyrate* (CAB), *cellulose nitrate* (CN), *ethyl cellulose* (EC), *polymethyl methacrylate* (PMMA) and *polystyrene* (PS). It has limited compatibility with *polyvinyl butyral* (PVB). It is incompatible with *cellulose acetate* (CA) and *polyvinyl acetate* (PVAC). Also used in rubbers, for example, for electrical insulation compounds.

dioctyl sebacate
Also known as di-2-ethylhexyl sebacate. An abbreviation used for this material is DOS. This material has a relative density (RD or SG) of 1·50, a melting point of approximately −55°C and a boiling point of approximately 249°C. This material is best known as a *low temperature plasticizer* for *polyvinyl chloride* (PVC). Industrial use of such traditonal materials is now reduced because of the availability of cheaper materials, for example, *nylonates*. Used interchangeably with di-iso-octyl sebacate (DIOS) but DIOS is preferred for plastisol use.

DOS is a low temperature plasticizer for polyvinyl chloride (PVC) which has low volatility and resists water extraction. It is compatible with *cellulose nitrate* (CN), *ethyl cellulose* (EC), *polymethyl methacrylate* (PMMA) and *polystyrene*. It has limited compatibility with *cellulose acetate butyrate* (CAB), *polyvinyl acetate* (PVAC) and *polyvinyl butyral* (PVB). It is incompatible with *cellulose acetate* (CA). Used for automobile seat covers, coated fabrics and low temperature insulation. Also used for rubber compounds.

dioctyl terephthalate
Also known as di-2-ethylhexyl terephthalate or as di-(2-ethylhexyl) terephthalate. An abbreviation used is DOTP. A *phthalate ester plasticizer* which is based on terephthalic acid and octanols. It is roughly similar to the corresponding o-phthalate ester (prepared from phthalic anhydride) but is best compared with an o-phthalate with one more carbon atom: terephthalates are much less volatile than o-phthalates. Has good fogging resistance which is useful, for example, for *polyvinyl chloride* (PVC) *compounds* which are to be used in automotive interiors. The use of terephthalates helps prevent car window fogging which may result from the use of plasticized PVC upholstery. See *di-octyl isophthalate*.

dioctyl tin stabilizer
A group of organic compounds which contain tin (Sn) and which are used, for example, as heat stabilizers for *polyvinyl chloride* (PVC) particularly where transparency is required. (An organotin may also be known as a *dibutyl tin stabilizer*). The general formula for the dioctyl compounds is (C$_8$H$_{17}$)$_2$SnX$_2$ where X is an organic group - such as the laurate group (see *dibutyl tin dilaurate*). Similar to the dibutyl tin stabilizers but of lower toxicity and of lower heat stabilizing power on an equal weight basis (as they contain less tin). See *heat stabilizer*.

DIOF
An abbreviation used for *di-iso-octyl fumarate*.

diol
A diol is a dihydric alcohol derived from an aliphatic hydrocarbon by substituting two hydroxyl groups for two hydrogen atoms. Diols are also known as glycols.

diol chain extender
A *diol* which is used as a *chain extender*, for example, *ethylene glycol*. Other diol chain extenders include 1,4 butane diol (1,4 BDO), 1.5 pentane diol (1,5 PDO) and 1,6 hexane diol (1,6 HD-O).

DIOM
An abbreviation used for *di-iso-octyl maleate*.

DIOS
An abbreviation used for *di-iso-octyl sebacate*.

dioxane
Also known as dioxan. This material has a relative density (RD or SG) of 1·04, a melting point of approximately 12°C and a boiling point of approximately 101°C. A cyclic ether which is a colourless, flammable liquid: it is a toxic solvent which is miscible with water in all proportions. Used as a solvent and a dehydrating agent. It is a solvent for *polyvinyl chloride* (PVC), *cellulose acetate* (CA), *cellulose acetate butyrate* (CAB), *cellulose nitrate* (CN), *ethyl cellulose* (EC), *polystyrene* (PS), *polyvinyl acetate* (PVAC) *natural rubber* (NR) *natural resins*, *unsaturated polyesters* and *epoxide resins*.

dioxin
An aromatic chlorinated hydrocarbon which is very toxic. The chlorine containing compound 2,3,7,8-tetrachlorodibenzo-para-dioxin (2,3,7,8 TCDD or TCDD). See *dioxins*.

dioxins
Also known as polychlorinated dibenzo-para-dioxins (PCDDs). A family of aromatic chlorinated hydrocarbons which are very toxic: similar to polychlorinated dibenzofurans (PCDFs) and to polychlorinated biphenyls (PCBs). Dioxins are toxic chemicals which may be produced during monomer manufacture, polymer manufacture and polymer incineration. For example, of *polyvinyl chloride*.

DIOZ
An abbreviation for *di-iso-octyl azelate*.

dip coating
A coating method in which the substrate is passed through a bath containing a fluid composition: the excess is removed by, for example, a *doctor blade*.

Dip coating is used in blow moulding to cover bottles with, for example, *polyvinylidene dichloride* (PVDC) so as to improve permeability. Immediately after the moulding operation, the bottles are dipped into the PVDC latex, drained for approximately 30 s and, oven dried for 2 to 3 minutes at 65 to 70°C: output rates of up to 5000 bottles per hour can be achieved.

dip moulding
Term associated with *polyvinyl chloride* (PVC) *plastisol moulding*. A PVC *plastisol*, or *paste*, is used to make relatively soft mouldings. A *male mould*, or former, is heated and dipped into the PVC paste. The coated former is then removed, inverted and heated in an oven until the paste gels at say 160°C: if necessary the former is dipped several times to build up the thickness. After the final fusion, the coated former is cooled and the product removed. Used to make gloves and boots.

dip process
Also known as the reclaimator process. A *reclamation process* for vulcanized rubber which achieves the required softening by working the vulcanized rubber compound in a modified extruder. The material is subjected to high temperatures (200°C) while being sheared by a screw which develops pressure.

dipentamethylene thiuram disulphide
An abbreviation used for this material is DPTD and/or PTD. This yellow solid material has a relative density (RD or SG) of approximately 1·39 and a melting point of approximately 112°C. An *activator* for *thiazoles*. When used at approximately a 4% level can give *sulphurless cures* as this material has about 10% sulphur available. Such cured materials have good ageing properties and good heat resistance. See *accelerator*.

dipentamethylene thiuram monosulphide
An abbreviation used for this material is DPTM and/or PTM. This yellow, solid material has a relative density (RD or SG) of approximately 1·39 and a melting point of approximately 100°C. An *activator* for *thiazoles*. A very safe *accelerator* with a pronounced delayed action.

dipentamethylene thiuram tetrasulphide
An abbreviation used for this material is DPTT. This greyish solid material has a relative density (RD or SG) of approximately 1·44 and a melting point of approximately 114°C. An *ultra-fast accelerator*. An example of a *sulphur donor vulcanization system* which is an *accelerator* as well as being a sulphur donor. This material has about 25% sulphur available for vulcanization. The cured materials have good ageing properties and good heat resistance if extra sulphur is not added.

diphenyl amine
An abbreviation used for this material is DPA. See *substituted diphenyl amines*.

diphenyl amine acetone condensation products
Ketone amine antioxidants. Antioxidants for rubber compounds. See *antioxidants* and *ketone amine condensates*.

diphenyl cresyl phosphate
See *cresol diphenyl phosphate*.

diphenyl ethylene diamine
A *diamine* which is a very good *antioxidant* for rubbers. This material has a relative density (RD or SG) of 1·14 to 1·21, a melting point of 60 to 75°C and a boiling point of approximately 228°C.

diphenyl guanidine
Abbreviation used DPG. This white crystalline material has a relative density (RD or SG) of approximately 1·19 and a melting point of approximately 146°C. It is a slow/medium speed *accelerator* used for the sulphur curing of rubbers. Slightly slower curing than di-o-tolylguanidine (DOTG). This material, like DOTG, is rarely used alone but is used in synergistic combinations with primary accelerators such as the mercapto accelerators. It is a secondary accelerator. Vulcanization temperature is approximately 135 to 160°C.

diphenyl guanidine absorption
See *diphenyl guanidine number*.

diphenyl guanidine acetate
An abbreviation used is DPGA. An *accelerator*: a booster for thiazole accelerators with a delayed action.

diphenyl guanidine number
An abbreviation used for this term is DPG number. Diphenyl guanidine absorption is a measure of filler influence on vulcanization rate: as absorption increases, so does retardation of vulcanization.

diphenyl guanidine oxalate
An abbreviation used is DPGO. An *accelerator*: a booster for thiazole accelerators with a delayed action.

diphenyl guanidine phthalate
An abbreviation used is DPGP. An *accelerator*: a booster for thiazole accelerators with a delayed action. This white solid material has a relative density (RD or SG) of approximately 1·21 and a melting point of approximately 178°C. It is a slow accelerator when used alone but does give improved safety, compared to pure guanidines, when used as a booster for the sulphur curing of rubbers.

diphenyl ketone
See *benzophenone*.

diphenyl octyl phosphate
Also known as octyl diphenyl phosphate. An abbreviation used for this material is DPOP or ODP or DPOF. A *mixed alkyl aryl phosphate*.

diphenyl phthalate
An abbreviation used for this material is DPP. A solid, *phthalate ester plasticizer* with a high boiling point of approximately 405°C and of low water solubility. A *plasticizer* for *polyvinyl chloride* (PVC), *ethyl cellulose, cellulose nitrate* and for *polystyrene*.

diphenyl thiourea
See *thiocarbanilide*.

diphenyl urea
See *thiocarbanilide*.

diphenyl-p-phenylene diamine
An abbreviation used is DPPD. This grey solid material has a relative density (RD or SG) of approximately 1·21 and a melting point of approximately 145°C. An *antioxidant*, for example, for rubbers which may be represented as C_6H_5-NH-C_6H_5-NH-C_6H_5.

diphenylmethane di-isocyanate
See *4,4'-diphenylmethane diisocyanate*.

diphenylmethanone
See *benzophenone*.

diphenylol propane
See *bisphenol A*.

diphenylolpropane
See *bisphenol A*.

diphenylpicryl hydrazyl
A *radical scavenger* which is a stable free radical. Also known as DPPH. An additive used, in small quantities, to prevent gel formation during rubber mixing.

diphenylthiourea
A *urea derivative* used as a *heat stabilizer*: a *metal-free organic, heat stabilizer*.

diphenylurea
A *urea derivative* used as a *heat stabilizer*: a *metal-free organic, heat stabilizer*.

DIPP
An abbreviation used for *di-iso-pentyl phthalate*.

dipropylene glycol dibenzoate
An abbreviation used for this material is DPDB. This material is a *plasticizer* for *polyvinyl chloride (PVC)* and also for rubbers. This material has a relative density (RD or SG) of 1·13, a melting point of from −12 to −35° and a boiling point of approximately 230°C (at 5 mm).

direct action accelerator
See *accelerator*.

direct blend injection moulding
An abbreviation used for this term is DBIM. An *injection moulding process* used to produce a *rubber modified material*, for example, *rubber modified polypropylene*. The individual materials (i.e. rubber and PP) are fed to the injection moulding machine. Good mixing within the injection moulding machine is essential if useful products are to be obtained.

direct clamp machine
See *direct lock machine*.

direct cooling system
Also known as an open cooling system: a mould heating/cooling system used when the maximum temperature of use is below approximately 85°C.

One extremely popular system basically consists of an electric immersion heater and an electric pump. This pump circulates water over the immersion heater and then through the mould. The temperature of the water is continuously measured and when the set-point temperature is reached the electric heater is turned off; if the temperature exceeds another slightly higher set-point then a solenoid valve is opened and some of the over-heated water is dumped. Mains cold water is introduced into the system to compensate for this loss. Such units are popular because they are relatively compact and inexpensive (compared to an *indirect system cooling system*).

direct current drive
An abbreviation used for this term is DC drive. Standard, electric-motor drive systems for dosers, extruders and pullers: used as they give good speed consistency.

direct digital link
An abbreviation used for this term is DDL. A term used in *hydraulics* and which refers to the digital feedback system as used with servovalves. This allows the valve to communicate with a microprocessor-based controller which is capable of issuing *digital command signals*. Valve position details are, in turn, relayed to the controller and an *error signal* is issued.

direct dye
A dye used to colour, for example, natural fabrics by immersion in an aqueous solution of the *dye*.

direct foam moulding process
A moulding process used to produce *direct moulded foam*. See *flexible polyurethane foam*.

direct hydraulic
A term used in the injection moulding industry and means that an injection moulding machine is a 'direct hydraulic-type' of machine. See *direct lock machine*.

direct hydraulic drive
A term associated with *injection moulding* and which means that the *screw* is rotated by means of an hydraulic motor: the motor is coupled directly to the screw without using a gearbox.

direct hydraulic lock
System for closing and clamping a mould which uses a ram or cylinder acting directly against the platen. See *direct lock machine*.

direct lithography
A *lithographic printing technique*.

direct lock machine
Also known as a direct thrust machine or as a direct clamp machine or as a direct hydraulic-type machine. Term used in the injection moulding industry to describe a machine which has a certain type of *clamping system*. That is, a large area ram, usually powered hydraulically, is used to obtain the clamping force. To save on the usage of hydraulic fluid, and to speed up cycle times, *jack rams* may be used.

direct moulded foam
A foamed product produced in the shape required by reaction in a mould using a *direct foam moulding process*. See *flexible polyurethane foam*.

direct pressure measurement transducer
Sometimes known as a direct pressure transducer. A *pressure transducer* which is in direct contact with the melt. A true pressure transducer and so calculations, to establish which range of transducer to use in conjunction with a selected pin, are not necessary as for indirect sensing. Used in injection moulding to achieve *cavity pressure control*. The tip of the transducer is in direct contact with the melt.

direct pressure sensing
This uses a *direct pressure measurement transducer* in injection moulding to achieve *cavity pressure control*.

direct thrust machine
See *direct lock machine*.

direct-adhesive process
See *resorcinol-formaldehyde-silica system*.

directional valve
A term used in *hydraulics* and which refers to a valve which directs fluid flow, or prevents fluid flow, to selected ports or channels. That is, it directs fluid to the selected part of the circuit.

disc gate
See *diaphragm gate*.

discontinuously coated fabric
A *plastic coated fabric* which has areas of the substrate left un-coated.

disilane
A silicon-based material which consists of two *silane units* joined together. The formula of such a material may be represented as $H_3Si-SiH_3$. See *trilsilane*.

disiloxane
A *siloxane* which has the structure $H_3Si-O-SiH_3$: if some of the hydrogen atoms are replaced by organic groups then the compound should be referred to as an organosiloxane, for example, as an organodisiloxane.

dispersed rubber
A synthetic dispersion similar to latex: difficult to produce but can be made from reclaim. See *reclaiming of rubber*.

dispersion
The uniform distribution of the particles of compounding ingredients.

dispersion coating
A plastics material is dispersed in a liquid which is then applied to a film and the liquid is evaporated to leave a coating; this type of processing is used, for example, when the coating resin cannot be melt processed or dissolved.

dispersion mixing
The breaking down of additives into smaller, primary particles usually by a shearing-type of action. Once the additives have been dispersed to the required degree throughout the body of the compound then a uniform distribution, of the well mixed ingredients, must be obtained within the compound. See *distributive mixing*.

dispersion of fillers
See *filler dispersion*.

dispersion of sulphur
See *sulphur dispersion*.

dispersive and distributive mixing
Ideally the melt mixing system, for example, the extrusion system, should destroy, or reduce, additive agglomerates such as, for example, pigment agglomerates. That is, it should give dispersive mixing. Once the solid agglomerates have been destroyed then uniform distribution (random spatial orien-

tation) of the dispersed additive is required: this is known as distributive mixing. The two mixing processes are not totally separated. For example, when dispersive mixing is performed there will always be some distributive mixing but the converse does not follow. Could have, for example, a uniform distribution of unbroken pigment particles. If, in a subsequent process, the stresses acting on the agglomerates became too high then, the agglomerates would break up and colour streaks would result. See *mixing section*.

dispersive mixing section
A *mixing section* designed for *dispersive mixing*. A large number of dispersive mixing sections are now available. They may be classified into four main groups (i) shear or blister rings, (ii) fluted mixers (such as the Egan, Maddock or Zorro mixing sections), (iii) cross barrier mixers (such as the EVK and the straight cross-channel barrier mixing sections) and, (iv) planetary gear (PG) extruder mixers.

Planetary gear extruder (PGE) mixers are probably the best all round, single screw machines. However, it should be possible to improve dispersion mixing on an unmodified extruder by the use of relatively simple dispersion mixing sections, for example, by the use of *fluted mixing sections* such as the Maddock mixing section.

displacement
A term used in *hydraulics* and which refers to the quantity of fluid which passes through a pump or *actuator* in a single revolution or stroke. Could also mean the quantity of fluid which is passed by a pump or *actuator* in a given time.

displacement method
Used for measuring relative *density*. Weigh a sample of the polymer in air (W_A) and then in water (W_W). Then *relative density* is: $W_A - W_W/W_A$. This method is usually used for materials with a relative density greater than that of water. See *flotation method* and *density gradient column*.

dissipation factor
For most dielectrics (insulators) power factor and dissipation factor mean the same and are a measure of how much power is converted to heat; this conversion is obviously undesirable in an insulator and so the power factor should be as low as possible. See *loss factor*.

distearyl thiodipropionate
An example of a *thioester antioxidant* which has a melting point of approximately 60°C. An abbreviation used is DSTDP. A *preventive antioxidant* which decomposes peroxides: used particularly in polyolefins and in combination with a *hindered phenol*.

distearylpentaerythritol diphosphite
A *phosphite antioxidant*. A *preventive antioxidant*.

distortion
An unwanted change of shape. For distortion of extrudates see *melt fracture* and *sharkskin*.

distributive mixing
Once additives has been dispersed to the required degree throughout the body of the compound (see *dispersion mixing*) then a uniform distribution, of the well mixed ingredients, must be obtained within the whole mass of the compound. It is this part of the mixing operation which is called distributive mixing as the ingredients are being randomly distributed in space.

distributive mixing sections
A *mixing section* designed for *distributive mixing*. A large number of distributive mixing sections are now available. They may be classified into four main groups (i) slotted flight mixers, (ii) pin mixers, (iii) cavity mixers and, (iv) variable depth mixers. Slotted flight mixers are probably the best distributive mixing sections: a *Dulmage mixing section* (Dow) is probably the best known.

disubstituted p-phenylene diamines
See *substituted p-phenylene diamines*.

disulphide bridge
A link formed by two sulphur atoms, joined together, and which join molecules together. That is, -S-S-. See *di-sulphide cross-links*.

disulphur dichloride
See *sulphur chloride*.

DITD
An abbreviation used in *plasticizer terminology* for *di-iso-tri decyl*. For example, di-iso-tri-decyl phthalate. See *di-tri-decyl phthalate*.

DITDP
An abbreviation used for di-iso-tri-decyl phthalate. See *di-tri-decyl phthalate*.

dither
A term used in *hydraulics* and which refers to a low amplitude AC signal which is applied to a torque motor or solenoid, in addition to the input signal, to improve resolution. Used, for example, to minimize the effects of *hysteresis* and *deadband*.

dithio dimorpholine
See *morpholine disulphide*.

dithiocarbamate
A compound which contains the structure $>$N-C(=S)-S-. The *dithiocarbamates* are *accelerators* for rubbers.

dithiocarbamates
Dithiocarbamates are *accelerators* (ultra-fast accelerators) for rubbers and which are salts of unstable or hypothetical disubstituted dithiocarbamic acids. Secondary aliphatic *amines* react with carbon disulphide to produce dithiocarbamic acids. See *dimethylamine*.

The compounds may be alkali salts, amino salts or heavy metal salts. Zinc dithiocarbamates are the best known, and the most powerful, but can also have bismuth, cadmium, copper and lead dithiocarbamates. Such dithiocarbamates can be very powerful *accelerators* and tend to be used, for this reason, in latex technology and for use in synergistic combinations) with slower curing rubbers such as EPDM and IIR. Nickel dithiocarbamates can act as anti-ageing additives, for example, for CR and CSM. Dithiocarbamates may be oxidized to the *thiuram disulphides*.

dithiodimorpholine
See *morpholine disulphide*.

dithiophosphates
A group of rubber *accelerators* useful for low temperature curing, for example, at 100°C: such vulcanizates do not cause brown staining in the presence of copper.

DITP
An abbreviation used for *di-tri-decyl phthalate*.

ditridecyl phthalate
See *di-tri-decyl phthalate*.

diundecyl phthalate
An abbreviation used is DUP. A low volatility *plasticizer* used, for example, with *polyvinyl chloride* (PVC) and capable of giving compounds with reasonable *low temperature properties*. The brittleness temperature of an 80 durometer compound with dibutyl phthalate (DBP) could be 0°C; the brittleness temperature of an 80 durometer compound with

diundecyl phthalate (DUP) could be −40°C and that of an *aliphatic diester* could be as low as −70°C. Diundecyl phthalate may be used in blends with a *trimellitate plasticizer*. See *undecyl dodecyl phthalate*.

divinyl benzene
An abbreviation used is DVB. A *divinyl monomer* which is used to introduce cross-linking in polymers. For example, into *polystyrene* (PS) so as to obtain PS with a higher softening point. DVB is also used to make *ion exchange resins*.

divinyl monomer
A monomer which contains two vinyl groups. A divinyl monomer is used to introduce cross-linking in polymers. For example, *divinyl benzene* is used with *polystyrene (PS)* so as to obtain PS with a higher softening point. Divinyl monomers may also be used to reduce solubility and to improve hardness.

divinyl polymerization
The polymerization of a *divinyl monomer*.

dixylyl disulphide
A dark brown oil used during the *reclaiming of rubber*.

DLTDP
See *dilauryl thiodipropionate*.

DMAC
An abbreviation used for *dimethyl acetamide*.

DMBS
An abbreviation used for *N-dimethyl benzthiazyl sulphenamide*.

DMC
An abbreviation used for *dough moulding compound*: a *polyester moulding compound* or PMC.
  DMC CF = *carbon fibre dough moulding compound*.
  DMC regrind = dough moulding compound regrind.

DMEU
An abbreviation used for *dimethylol ethylene urea*.

DMF
An abbreviation used for *dimethyl formamide*.

DMHPPD
An abbreviation used for *N,N'-di-(3-methyl heptyl)-p- phenylene diamine*.

DMP
An abbreviation used for *dimethyl phthalate*.

DMS
An abbreviation used for dimethyl sebacate. See *plasticizer*.

DMSO
An abbreviation used for *dimethyl sulphoxide*.

DMT
An abbreviation used for the dimethyl ester of *terephthalic acid*. See *polyethylene terephthalate*.

DN
This prefix is most commonly seen in terms used to describe *plasticizers*: it is used for di-nonyl, dinaphthyl and for di-nitroso. In ISO plasticizer terminology no letter is used to indicate normal (n or N is not used) so the abbreviation DN is not used for di-normal. For example:

  DNBA   = di-n-butyl adipate but see *dibutyl phthalate*;
  DNDP   = di-n-decyl phthalate. but see *didecyl phthalate*;
  DNHZ   = di-n-hexyl azelate but see *dihexyl azelate*;
  DNOA   = di-n-octyl adipate but see *dioctyl adipate*.
  DNODP = di-n-octyl n-decyl phthalate but see *octyl decyl phthalate*; and,
  DNOP   = di-n-octyl phthalate but see *dioctyl phthalate*.

The absence of the letter N, in terms used to describe *plasticizers*, should not be automatically taken to mean that the ester is based on a normal (straight chain) alcohol. It is common practice to use abbreviations without N for esters which may be based on branched chain alcohols. See *dioctyl* and *dioctyl phthalate*.

DNA
An abbreviation used for Deutscher Normenausschuss; the organisation responsible for issuing *DIN standards*.

DNO
An abbreviation used for *di-nitroso-pentamethylene tetramine*.

DNP
An abbreviation used for *dinonyl phthalate*.

DNPD
An abbreviation used for *di-naphthyl-p-phenylene diamine*.

DNPPD
An abbreviation used for *di-naphthyl-p-phenylene diamine*.

DNPT
An abbreviation used for *di-nitroso-pentamethylene tetramine*.

DNS
An abbreviation used for *dinonyl sebacate*.

D_o
An abbreviation used for *overall diameter*.

DO
An abbreviation used for *diol*.

DOA
An abbreviation used for *dioctyl adipate* or *di-2-ethylhexyl adipate*. See *dioctyl*.

doctor blade
Also called a doctor knife. A bar or blade used in coating so as to remove excess material, spread the coating and give uniform coating. For example, to obtain the desired coating weight in a *spreading process*, the treated substrate is drawn below the knife which removes the excess polymer solution.

doctor knife
See *doctor blade*.

dodecanelactam
Also known as laurolactam. The monomer for *nylon 12* which is derived from butadiene.

dodecanoic acid
Alternative name for *lauric acid*.

DOIP
An abbreviation used for *dioctyl isophthalate* or *di-2-ethyl-hexyl isophthalate*. See *dioctyl*.

dolomite
Also known as pearl spar. A type of *limestone* which contains about 45% magnesium carbonate: if the magnesium carbonate content is lower than this, then the mineral may be known as a magnesian limestone. This off-white, solid material has a relative density (RD or SG) of 2·34. It is a naturally occurring carbonate of magnesium and calcium ($MgCO_3CaCO_3$) which occurs in vast amounts, for example, in Scandinavia. Used, for example, as a filler with unsaturated polyester resins: the filler is produced by grinding and so may be known as ground dolomite.

dolly
Also known as a *pig*. A small roll of material cut from the warming mill and used to renew the *feed bank* in rubber calendering.

DOM
An abbreviation used for *dioctyl maleate* or *di-2-ethylhexyl maleate*.

domain crosslink
A heat-fugitive cross-link which differ from traditional, chemical crosslinks in that the domain crosslinks are thermally reversible. See *domain theory*.

domain theory
A theory advanced to explain the rubbery behaviour of a *styrene block copolymer*. In the bulk state, the *polystyrene (PS)* short-end segments agglomerate and, at temperatures below the glass transition temperature (T_g), these agglomerates (the domains) act as strong, multifunctional tie-points. That is, although the segments are chemically connected they are incompatible and this results in the formation of PS domains, or clusters, in a continuous rubbery phase. These domains act as crosslinks, and/or reinforcing fillers, so that the rubbery polybutadiene (BR) phase exhibits high elasticity just like a conventional, vulcanised elastomer. On heating, the PS domains are destroyed so that shaping in a mould is possible, that is, the material behaves as a thermoplastic at 'high' temperatures and as a rubber at 'low' temperatures. Such 'domain crosslinks' differ from traditional, chemical crosslinks in that the domain crosslinks are thermally reversible.

dome
In reinforced plastics terminology, an end of a filament-wound, cylindrical container.

domed block
See *slab-stock process*.

DOP
An abbreviation used for *dioctyl phthalate* or *di-2-ethylhexyl phthalate*. See *dioctyl*.

dope
A thick solution of resin, plus additives such as colorants, in a solvent.

DOS
An abbreviation used for *dioctyl sebacate* or *di-2-ethylhexyl sebacate*. See *dioctyl*.

dose fed
See *starve fed*.

doser
A device which meters/doses the resin, and/or additives, into the extrusion cylinder or barrel.

DOTG
An abbreviation used for *di-o-tolylguanidine*.

DOTP
An abbreviation used for *dioctyl terephthalate* or *di-2-ethylhexyl terephthalate*. See *dioctyl*.

double bank calendering
A *calendering process* used in rubber technology where the product is required to be of better quality (good finish or narrower thickness distribution) than that possible from the *single bank* technique. Two nips are used with the first acting as a feed to the second. Gauging takes place at the second nip but this is only possible because it is being fed with thermally homogeneous material: a *pencil-bank* is present at the second nip. As the sheet width is virtually the same on all rolls then the nip sizes are virtually the same in both nips.

Butyl rubbers and butadiene acrylonitrile rubbers tend to stick to a cooler roll and for this reason the following temperature range profile would be used in double bank calendering of these materials.

	°C	°F
Top roll	90 to 105	194 to 221
Centre Roll	70 to 90	158 to 194
Bottom roll	90 to 110	194 to 230

double breathing
A moulding technique used to allow the escape of gases. The mould is partly opened twice. See *breathing*.

double parallel screw
Also know as a stepped screw or as a nylon screw. A screw which two parallel sections (feed and metering) connected by a tapered transition zone: such a tapered transition zone is usually very short for *nylon extrusion*.

double spew
See *chattering*.

double strand polymer
See *ladder polymer*.

double-acting
A term used in *hydraulics* and which refers to a component, such as a *cylinder*, which is capable of operating in two directions.

double-acting cylinder
A *cylinder*, which is capable of operating in two directions.

double-cone blender
A mixing device used for blending or pre-blending of solid thermoplastics materials with their additives. See *compound blending*.

dough
A very thick solution of resin, plus additives such as colorants, in a solvent: the material is semi-solid and is prepared in a dough-mixer. See *cellulosics*.

dough moulding compound
An abbreviation used for this type of material is DMC. Also known as premix; a *polyester moulding compound* (PMC). See *bulk moulding compound* and *thick moulding compound*.

In its simplest form, DMC consists of an *unsaturated polyester resin* (dissolved in styrene), chopped glass (approximately 6 mm, or 0·23 in, fibre length), mineral filler (to reduce shrinkage and costs, particulate fillers such as calcite or dolomite, are used), mould release agent and catalyst. These are thoroughly mixed together to a dough-like consistency in, for example, a Z blade mixer. The glass is added last and care is taken to avoid breaking down the fibres (by either reducing the length or, by causing filamentization). The curing catalyst is a heat activated, organic peroxide such as t-butyl perbenzoate; zinc stearate is a commonly used *lubricant*. The resin is normally a styrenated, unsaturated polyester (UP) but the monomer (*styrene*) may be replaced by others, such as *diallyl phthalate*. The compounds may be formulated to maximize a particular property, for example, flame resistance, impact strength, etc.

For the economic production of high strength, glossy products from these putty-like, polyester moulding compounds, compression, transfer or injection moulding must be employed.

DMC is available as an extruded rope or in 'dough' form. Ropes enable moulded charges to be measured by length so as to eliminate weighing. The dough-form is supplied as 'cheeses' and each cheese is wrapped in film and supplied inside a box, or keg, so as to minimize styrene loss. Nylon or cellophane film is often used instead of PE to prevent styrene evaporation. Avoid skin contact with this type of material and only use this type of material in a well ventilated area.

DMC is offered in a range of opaque colours which, on moulding, can give a glossy attractive finish. The surface is hard but not as hard as PF. DMC is usually seen as glossy, pastel-coloured mouldings. The natural colour is beige but this depends on the fillers used.

DMC is widely processed by *injection moulding* but take care to avoid fibre degradation. On screw machines, doughs are fed into the barrel from a *stuffer box*; such forced charging can lead to *fibre degradation*. DMC is an easy flow material (BMC is much thicker) and so low pressures may be used.

DMC is resistant to outdoor exposure: on prolonged exposure (several years) there may be some loss of colour and gloss but, mechanical properties are usually retained. Mouldings can withstand high temperatures (up to 160°C/320°F) for long periods and 200°C/392°F, for short term exposure. DMC resists water, alcohols, aliphatic hydrocarbons, detergents, greases and oils. DMC is not resistant to ketones and chlorinated hydrocarbons; the resistance to aromatic hydrocarbons, acids and alkalis is also not very good and worsens as the temperature is increased.

DMC is heavily loaded with filler at a filler to resin ratio of around 2:1 and the fibre content can range from 10-25%. Densities are therefore high, for example, 1.7 gcm^{-3}. DMC is a highly filled, reinforced material which has a high flexural modulus and a low coefficient of thermal expansion - similar to metals. DMC is used where precision, strength, attractive appearance and resistance to corrosion and high temperatures, at an economic price, are required.

Used in microwave cookware and mid-temperature range oven cooking. Before use, bake at the highest temperature envisaged, say 180°C/356°F, for 15 minutes so as to remove traces of volatiles (the human palate is very sensitive to traces of styrene). DMC is commonly seen in and around the home, as sandwich toaster components, shrouds for domestic irons, etc. It is also used in office machinery for machine covers: because of its good impact resistance and reduced flammability rating - it can have a rating of *UL 94-V0*. DMC has excellent dimensional stability and good water resistance. It is now being used in automotive applications, such as headlight reflectors. DMC is also widely used in the electrical industry, for instance for meter boxes and lids.

dough moulding compound regrind
An abbreviation used for this type of material is DMC regrind or DMC-R. The ground material obtained when a component made from a *dough moulding compound* is ground or reduced in size. See *particle recycling*.

dough - rubber
A rubber compound swollen by a solvent.

dough-filter
A machine used to prepare a purified *dough*. A hydraulic ram forces the dough through filter cloth. See *cellulosics*.

dough-mixer
A machine used to prepare a *dough*. A *kneader* with a tightly fitting lid. See *cellulosics* and *hardening rolls*.

Dow mixing section
See *Dulmage mixing section*.

dowel pin
Also known as a *guide pin* or as a leader pin. Guide pins are used to ensure that a mould closes correctly.

downstream equipment
The auxiliary units used on, for example, an *extrusion line*, after the die, and which are used to cool, shape, control and, if necessary, cut the extrudate.

downstroke press
A press in which the direction of force application is from above: from its open position, the *moving mould* half moves downwards against the *fixed mould half*.

DOZ
An abbreviation used for *dioctyl azelate* or *di-2-ethylhexyl azelate*. See *dioctyl*.

dp
An abbreviation used for dew point and for *degree of polymerization*.

DP
An abbreviation used for *degree of polymerization*.

DPA
An abbreviation used for diphenyl amine. See *substituted diphenyl amines*.

DPCF
An abbreviation used for diphenyl cresyl phosphate. See *cresyl diphenyl phosphate*.

DPCP
An abbreviation used for diphenyl cresyl phosphate. See *cresyl diphenyl phosphate*.

DPDB
An abbreviation used for *dipropylene glycol dibenzoate*.

DPG
An abbreviation used for *diphenyl guanidine*.

DPG number
An abbreviation used for *diphenyl guanidine number*.

DPGA
An abbreviation used for *diphenyl guanidine acetate*.

DPGO
An abbreviation used for *diphenyl guanidine oxalate*.

DPGP
An abbreviation used for *diphenyl guanidine phthalate*.

DPNR
An abbreviation used for *deproteinized natural rubber*.

DPOF
An abbreviation used for *diphenyl octyl phosphate*.

DPOP
An abbreviation used for *diphenyl octyl phosphate*.

DPP
An abbreviation used for *diphenyl phthalate*.

DPPD
An abbreviation used for *diphenyl-p-phenylene diamine*.

DPPH
An abbreviation used for *diphenyl picryl hydrazyl*.

DPQD
An abbreviation used for *dibenzoyl-p-quinone dioxime*.

DPR
An abbreviation used for *depolymerized rubber*.

DPTD
An abbreviation used for *dipentamethylene thiuram disulphide*.

DPTH
An abbreviation used for *thiocarbanilide*.

DPTM
An abbreviation used for *dipentamethylene thiuram monosulphide*.

DPTT
An abbreviation used for *dipentamethylene thiuram tetrasulphide*.

DPTU
An abbreviation used for *thiocarbanilide*.

dr
An abbreviation used for *dram*.

drab reclaim
A type of reclaimed rubber.

drachm
A UK unit of volumetric measurement in *apothecaries' weight*.

draft
Usually refers to an angle: the relieve angle on the side of a mould and which is there to ease separation of the moulding from the mould. The taper on a mould which allows for easy extraction of the moulding.

draft angle
The angle of the taper from the vertical i.e. from the direction of mould opening.

drag flow
A type of flow that predominates in single screw *extrusion*; the melt is dragged towards the die by the interaction between the *screw* and the *barrel*. The polymer must wet both the screw and barrel to produce drag flow effects.

dragging
Also referred to as picking up. Mould surface damage often caused by mis-alignment of moving mould components and subsequent scoring.

drain
A term used in *hydraulics* and which refers to the channel through which fluid returns to the reservoir under low *pressure*.

dram
An abbreviation used for this term is dr. A unit of measurement in Avoirdupois weight and in US Apothecaries' measure. See *UK system of units* and *US Customary Measure*.

drape forming
A male forming technique. A *thermoforming process* which uses a *male mould* and which offers a way of producing components of relatively deep draw cheaply. The sheet is heated by an infra-red heater (mounted approximately 150 mm above the sheet) while being held in a clamping frame. At *forming temperature*, the heater is removed and the sheet is moved relative to the mould, for example, the *male mould* is driven upwards and passes through the clamping frame. A vacuum is applied to the space between the sheet and the male mould: this completes the forming operation. The product is then cooled by mould contact, or by blowing air across the forming, and ejected. Such a forming can have produce relatively strong components, for example, the corners of box-type components will be thick because the material which makes the corners is chilled by contact with the rising mould. The *wall thickness distribution* may be improved by *prestretching* the sheet - see *billow forming*.

draw down
Also spelt drawdown. The stretching of a product by pulling it away faster than the natural production rate. Usually associated with extrusion processes where the extrudate may be subject to extensive stretching after leaving the die, for example, in the manufacture of film. The stretching of an extrudate so as to produce a product of the correct size. In extrusion coating the resin is drawn down from the die gap thickness to the coating thickness. Drawing down will cause *molecular orientation*.

draw ratio
The die gap thickness divided by the thickness of the final plastics layer; the degree to which the extrudate has been drawn in a particular direction.

draw resonance
A form of drawdown failure characterised by the wandering or wavering of the edge of a molten film or, by the loss of thickness uniformity across the width of a coating.

drawdown
See *draw down*.

drawing
The stretching of an extrudate to produce orientation. See *Godet roll* and *draw down*.

drawing down
See *draw down*.

DRC
An abbreviation used for *dry rubber content*.

dressing
See *size*.

drex
The mass in grams of 10,000 metres of *yarn*.

dribble
Material lost from a machine by *drooling*, for example, from an injection moulding machine nozzle.

dribbling
See *drooling*.

drilled roll
A *roll* with a number of holes, or channels, located close to the roll surface. For calendering, such rolls are best as the length of path from the heating medium to the product is short and this means better heat exchange and more accurate temperature control. Such accurate temperature control is important as surface temperature variations of ±0·05°C may cause surface irregularities of approximately 0·3 μm (0·000 01 in).

drooling
Sometimes called dribbling. Means an undesirable loss of melt from a machine, for example, from an *injection moulding machine*.

droop
See *offset*.

drop
A unit of measurement in *apothecaries' measure*.

drop method
A simple method used to obtain heat content data (see *enthalpy*). A mass of heated polymer is purged from a machine, for example, an injection moulding machine, into a paper envelope and then quickly dropped into a well-lagged container of water, whose subsequent temperature rise is monitored until it reaches a maximum. Then, if no heat losses are assumed to have occurred, the heat transferred from the polymer equals the heat gained by the water and container or,

$$M_p S_p (T_1 - T_2) = M_w S_w (T_2 - T_0) + M_c S_c (T_2 - T_0).$$

Where:

M_p = mass of polymer (kg);
S_p = specific heat of polymer (J/kgK);
M_w = mass of water (kg);
S_w = specific heat of water (J/kgK);
M_c = Mass of container (kg);
S_c = specific heat of container (J/kgK);
T_1 = initial polymer temperature (K);
T_0 = initial water temperature (K) and that of the container;
T_2 = observed water temperature.

Although heat losses can be minimised by lagging the container, a more accurate result can be obtained by finding and applying a cooling correction so as to obtain that temperature which would have been reached if there had been no heat losses. See **table 7**.

drop test
Known also as an impact test: a destructive test on overall product quality which is, for example, performed by filling a *blow moulded* container with liquid and then dropping the capped, or uncapped, container vertically onto a hard surface. so as to establish the maximum drop height it will withstand without fracture. Such a test may also be done at low temperatures.

drug room
A term formerly used in the rubber industry for a compound, or *weighing-up*, room.

drum blender
A mixing device used for blending or pre-blending of solid thermoplastics materials with their additives. See *compound blending*.

dry blend
Also spelt dry-blend. A dry compound prepared without heat fluxing or the addition of solvent; usually used in *polyvinyl chloride* (PVC) technology where it is also known as a powder blend or as a premix. Prepared either in a *high speed mixer* or in *ribbon blender*. In some respects, high speed mixer units do not have the same flexibility of operation, as a *ribbon blender*, as once a mix is made it must be dumped from the mixer.

For example, to produce an unplasticized PVC (UPVC) dry blend the PVC polymer is added together with other ingredients, except the *lubricant*, into the chamber of a *high speed mixer*. This mixture is blended at high speed until a temperature of between 110 to 125°C is reached. Within this temperature range, a lubricant and/or impact modifier may be added, then the resultant mix is dumped into a cooler/blender. A large cooler/blender is used to cool the mix, to improve the batch-to-batch consistency and to also reduce the electrostatic charges on the dry blend. The dry blend may then be used directly or pelletized by using an extruder. Pelletizing improves dispersion but adds to the cost and reduces the thermal stability of the material.

To improve the batch-to-batch consistency of a dry blend, the additives used in the dry blend (e.g. stabilizers, lubricants, processing aids, impact modifiers and fillers) can now be incorporated in *one shot* sachets. Because of the dust problem which is associated with UPVC dry blends, additives are now also being supplied in a granular, flake or spaghetti form. Although the dispersion characteristics of the latter forms are not as good as when the additive is a powder, they are preferred by the processors because of safety considerations.

Use of PVC dry blends is increasing within the injection moulding industry. This is because, for example, different types of PVC compounds can be readily developed so as to suit the end-application of the component.

dry colour
Dry, solid *pigment* or *dye* in either solid or chip form. Dry colour may be supplied in the form of pre-weighed sachets or in bulk: it is pre-blended with the polymer before introduction to the throat of, for example, an *injection moulding machine*.

dry colouring
A method of colouring polymers, particularly thermoplastics materials. One of the cheapest ways of colouring plastics materials is by the use of *dry colour*. A very wide range of thermoplastics may be coloured in this way at a comparatively low cost and with a relatively small colour inventory. The colorants, and any other additives, should be thoroughly pre-blended with the polymer before introduction to the throat of the processing machine (for example, to an injection moulding machine) or, alternatively, the dry colour may be automatically metered and blended at the machine. It can be difficult to meter the colour accurately and this can give rise to the development of inconsistent colours. If a chip form of the colorant is used, then colour-dust contamination may be minimised: such dust-free grades also flow more easily than the original powder form and this suits automatic metering equipment better. If the material is pre-mixed, before feeding to the machine, then colour contamination can occur as it is difficult to clean the mixers thoroughly. No matter which system of *blending* is used, thorough blending is necessary before the material reaches the screw.

dry cycling speed
The amount of time needed for a moulding machine to do one complete cycle when it is operated dry, that is, for example, without any material in the barrel.

dry printing
See *hot foil marking*.

dry process
A process used to produce *casein-formaldehyde*. The *rennet casein* is finely ground and blended with water in a dough mixer: it is also mixed with *pigments* and a *clarifying agent*. When mixed the material should still be a free-flowing powder which can be consolidated by *extrusion*.

dry rubber content
An abbreviation used for this term is DRC. The total solids content of latex. Obtained by *coagulation, sheeting* and *drying of latex* at 70°C.

dry spinning
A fibre-forming process and which typically uses a solution of a fibre-forming polymer in a solvent. The *fibres* are formed by forcing the solution through a spinneret: the fibres are then hardened by evaporation of the solvent. See *acetate fibre*.

dry-blend
See *dry blend*.

dry-ground
Usually refers to a method of filler preparation, for example, dry-ground limestone. If the mineral is ground dry, as compared to being wet-ground, then the product is of coarser particle size and more un-even size distribution than the wet-ground material but cheaper. See *limestone*.

dry-spot
An imperfection in reinforced plastics, an area of incomplete surface film where the reinforcement has not been wetted with resin.

drying
Usually refers to the action of removing water from a material prior to melt processing. Most plastics materials are supplied dry and ready for use although some are 'wet' and must be dried before being melt processed by, for example, injection moulded. Many plastics, and particularly engineering plastics, absorb water (they are hygroscopic) and if they have absorbed water, they must be dried before use.

If drying is required, then dry in either a hot air oven, in a desiccant drier or, in a vacuum drier. The last two methods are the more efficient as they are quicker and reduce the water content to lower values. They take approximately half the time required for a hot air drier and, for some materials, they are the only way of removing the moisture.

Hygroscopic polymers such as *nylon* and *polycarbonate* may be kept dry (by the use of, for example, hopper dryers) during processing: other polymers, such as *acetals* rarely need drying if the material is stored correctly. Compounding, particularly with carbon black, may increase the tendency to absorb moisture.

While most moulders have drying facilities they rarely bother to check either the moisture content of the material as supplied, after storage or during processing. The most usual drying procedure is to dry for a set number of hours at a particular temperature. For example 6 h at 130°C for PET. This may not always be satisfactory as many factors affect the drying process. For example, the higher the initial moisture content the longer the time to reach the required limit. Therefore there is a real need to measure *moisture content*.

drying agent
An additive used to absorb traces of moisture in a rubber mix and/or, used to measure *moisture content*. Such an additive may be based on calcium oxide or phosphorous pentoxide.

drying oil
An oil which hardens or sets on exposure to air. Films of such oils harden because of the high degree of unsaturation present in the fatty acid residues of the triglycerides of which they are made. See *linseed oil* and *tung oil*.

drying to constant weight
Heating a material until no further weight loss is experienced. See *oven drying assessment*.

DS
An abbreviation used for *degree of substitution*.

DSC method
An abbreviation used for *differential scanning calorimetry method*.

DSD
An abbreviation used for *Duale System Deutschland*.

DSTDP
An abbreviation used for *distearyl thiodipropionate*.

DTBC
An abbreviation used for *N,N'-dithiobishexahydro-2,4-azepinone*.

DTBP
An abbreviation used for *di-tert-butyl peroxide* and/or for *di-tert-butyl-p-cresol*.

DTDM
An abbreviation used for *morpholine disulphide*.

DTDP
An abbreviation used for *ditridecyl phthalate*.

DTI
An abbreviation used for *dial test indicator*.

dual durometer extrusion
Extruding a shape with a soft and a hard material, usually based on *polyvinyl chloride (PVC) materials*; a form of *co-extrusion*.

dual lip air ring
An *air ring* which has two sets of lips so that two streams of air can be forced over *lay flat film* so as to improve the cooling rate.

dual purpose vegetable oil
An oil which can be used as an *edible vegetable oil* or as an industrial vegetable oil. Examples of such oils include perilla, oiticica, castor and tung oil. See *vegetable oil*.

Duale System Deutschland
An abbreviation used for this German organization is DSD. A privately organized, packaging collection system: the organization picks up waste, sorted material from private homes. To assist in this task 22 manufacturers 170 processors formed the *Verwertungsgesellschaft für gebrauchte Kunstoffe*. The collected plastics material is sorted, for example, into bottles, thick films, cups, expanded polystyrene and into a mixed polymer fraction. Typically the sorted waste is shredded, washed and then separated by flotation (density difference or fractionation) processes: the relatively pure material (for example, 99% pure) is then dried and compounded before re-sale. See *Deutsch Kunstofferecycling*.

ductile fracture
Also known as tough fracture.

Dulmage mixing section
May be referred to as a Dow mixing section. A *distributive mixing section*. It has several, shallow-angled, multi-flighted sections separated by tangential grooves. The section is formed by machining, over approximately 3 D length, several flights and then making three cuts completely across (around) the flights. Because these grooves do not overlap, the barrel surface is not completely wiped by the rotating screw. See *Saxton mixing section*.

dumb-bell shaped
Having the shape of a *dumb-bell*. See *gauge section*.

Dunlop pendulum
An instrument used to measure *resilience*.

Dunlop tripsometer
An instrument used to measure *resilience*.

DUP
An abbreviation used for *diundecyl phthalate*.

Du Pont 903 moisture evolution analyser
This instrument measures *moisture contents* of polymers down to 10 parts per million (ppm) and even 1 ppm with special care. A known weight of sample is placed in the sample oven and heated to a preselected temperature for a chosen time. The moisture driven off is carried by dry nitrogen gas into an electrolytic cell where it is absorbed by phosphorous pentoxide (P_2O_5), a very efficient drying agent, supported on fine platinum wires. In the cell the current required to hydrolyse the water into hydrogen and oxygen is measured. The current is integrated for the time the sample is heated so that the total current, for hydrolyzing all the released water, is found. As the current needed is directly proportional to the moisture present the digital read-out of the instrument is scaled to read the amount of water directly.

Du Pont mixing section
See *Saxton mixing section*.

duraluminium
This material has a relative density (RD or SG) of 2·8. It is an alloy of aluminium and copper (approximately 4%): also contains small amounts of magnesium, manganese and silicon. This light, hard alloy is sometimes used to make moulds, for example, for *blow moulding*.

duromer
This term is sometimes seen in books which have been translated into English and means a highly cross-linked material: may mean an *ebonite type* of material or, a *thermosetting plastics* material. Such a material is also referred to as a thermoelast.

durometer
A device for measuring the *hardness*, or resistance to indentation, of a compound.

durometer hardness
The test result obtained when *hardness* is measured using a *durometer*.

dusting
The action of applying a *dusting agent* to rubber surfaces.

dusting agent
A powder which is put onto a sticky material in order to prevent adhesion. A substance which reduces the surface tack of unvulcanized rubber.

dusting oil
Also called a binder as it is used to bind a colorant to a thermoplastics material. Used, for example, with *polybutylene terephthalate* so as to prevent separation of colorant and polymer when using dry colorants during tumble mixing: a typical binder is based on a paraffin oil plus a glycol ester.

DV
An abbreviation used for dynamic vulcanization or for *dynamically vulcanized*.

DVB
An abbreviation used for *divinyl benzene*.

DVNR
An abbreviation used for *dynamically vulcanized natural rubber*.

dwell injection moulding
Packing of extra material into the injection mould so as to compensate for shrinkage. See *dwell under pressure*.

dwell period
See *dwell time*.

dwell pressure
See *dwell under pressure* and *hold pressure*.

dwell thermosetting material
A pause in the application of pressure, during the moulding of a *thermosetting material*, at an early stage in the cure cycle so as to vent the mould. See *compression moulding*.

dwell time
Term used in *injection moulding*: the time for which a *hold pressure* is applied. Also known as holding period, holding phase, holding time, hold period, dwell time, dwell period or packing time. See *dwell under pressure*.

dwell under pressure
Also known as dwell pressure or as, hold pressure or as, second stage pressure or as, packing pressure. The pressure applied to a moulding cavity, during the *dwell time*, so as to compensate for *shrinkage* by pushing in additional material. During the injection moulding operation, on an in-line screw machine, the screw is held in the forward position after the initial mould filling so as to give a dwell under pressure.

The final mould filling part of the moulding cycle may be done under a different pressure to that used for initial mould filling (so as to avoid over-packing). Usually this second stage pressure is lower than the first stage (but not for some semi-crystalline, thermoplastics materials). When moulding some semi-crystalline, thermoplastics materials, for example *nylon* and *acetal*, then the use of second stage pressure may not be required as abrupt changes in pressure can cause undesirable changes in crystalline structure.

After the screw has penetrated a pre-set distance into the barrel then the pressure is changed to the second stage pressure by means of a micro-switch and striker or, a transducer inside the mould initiates the pressure switch-over. This drop is usually set to occur just as the melt fills the mould. The *screw cushion* should be adjusted to give the minimum consistent with good pressure transmission, for example, 0·12 in/3 mm on small machines is suggested.

Apart from the energy savings, when a lower pressure is used, other benefits result, for example, lower moulding weight, more consistent moulding weight and lower stress levels in the mouldings (which results in parts with superior properties). See *cavity pressure control*.

dye
A soluble colouring system which gives a transparent colour (most colouring systems used for polymers are now based on *pigments*). Included in the category of dyes are *anthraquinone*, *azine*, *azo* and *xanthene-type materials*.

dyn
An abbreviation for *dyne*.

dyn/cm = dyne per centimetre. To convert to Newtons per meter multiply by 1×10^{-3}.

dyn cm^{-2} = dyne per square centimetre. Units used to measure *shear stress* in the CGS system. 1 dyn cm^{-2} = 0·002 088 lbf ft^{-2} = 0·1 Nm^{-2} = 0·000 014 5 lbf in^{-2} or psi.

dyn s cm^{-2} = dyne-second per square centimetre, that is, a *poise*. Units used to measure *viscosity* in the *CGS system*.

dynamic rotary shaft seal
See *rotary shaft seal*.

dynamic viscosity
An alternative name for the coefficient of viscosity of a fluid. An alternative name for the viscosity of a fluid. The usual symbol is μ but η is sometimes used. In the c.g.s system, the units are dyn s cm^{-2} and such units may be referred to as *poise*.

dynamic vulcanization
An abbreviation used for this term is DV. A process used to produce a *rubber modified material*, for example, based on *polypropylene* (PP) and *ethylene-propylene rubber* (EPDM), in which the rubber is cross-linked. Such compounds may be produced by intensively melt mixing the PP with the rubber: melt mixing may be performed, for example in an internal mixer, at a temperature greater than the melting point of the PP. After the melt mixing stage, vulcanizing agents are then added: the mixing is continued until the mixing torque peaks. Shortly after this peak, the mix is dumped, cooled and granulated.

To avoid damaging the PP, peroxides are not used as the curing system: one based on sulphur may be used. For example, a suggested formulation is EPDM 100, PP 50 phr, zinc oxide 5 phr, stearic acid 1 phr, sulphur 2 phr, tetramethylthiuram disulphide (TMTD) 1 phr and mercaptobenzothiazole 0·5 phr. As with traditional, rubber compounds, large quantities of oil and black may be added to such materials, for example, 100 phr of each. Fillers may also be used. By ringing the changes on the types, and quantities, of materials used, it is possible to produce a very wide range of compounds.

Within the melt-mixed compound, a two phase structure should result in which the rubber is in the form of small (1 to 2 micron) particles which are embedded in a continuous phase of PP. The rubber particles are cured or cross-linked but if enough plastic (PP) is present, the compound can still be easily processed by techniques used for thermoplastics materials. See *dynamically vulcanized polyolefin rubber*.

Compared to compounds which contain non-crosslinked rubber, the dynamically vulcanized (DV) materials have substantially improved compression set, permanent set resiliency and high temperature shape retention. They are a very rapidly growing sector of the *thermoplastic elastomer industry* and are, for example, widely used in the automotive industry.

dynamically crosslinked polyolefin elastomer
See *dynamically vulcanized polyolefin rubber*.

dynamically crosslinked polyolefin rubber
See *dynamically vulcanized polyolefin rubber*.

dynamically vulcanized natural rubber
An abbreviation used for this type of material is DVNR. See *dynamic vulcanization*.

dynamically vulcanized polyolefin rubber
Such materials may be referred to as TPO-XL or, as a TPV. TPO-XL stands for thermoplastic polyolefin which incorporates cross-linked rubber and TPV stands for *thermoplastic vulcanizate*. This type of material may also be referred to as an *elastomeric alloy thermoplastic vulcanizates* (EA-TPV): they are two phase systems in which a crosslinked rubber phase is dispersed in a continuous plastics phase, which is usually a polyolefin. Considerable improvements in properties result, compared to *rubber modified polypropylene*, if cross-linking is introduced into the rubber particles by *dynamic vulcanization*. If there is enough thermoplastics material present, the compound can still be easily processed by thermoplastic techniques. In DV-type materials, the particulate rubber domains are fully vulcanized and so the materials have good hot oil resistance, compression set resistance at elevated temperatures and high temperature utility. A factor contributing to the high temperature properties, is that the rubber particles, in finished products, interact to form a network of touching, loosely bound together aggregates: these aggregates are destroyed on heating so as to allow melt processing.

The rubber cannot be extracted, by more than 3%, by cyclohexane at 23°C (cross-link density is at least 7×10^{-5} moles/ml of elastomer as determined by swelling). This means that such materials are resistant to a wide variety of oils, solvents and chemicals. (Where excellent oil resistance is required then use a material which contains nitrile rubber.) Highly polar fluids, such as alcohols, ketones, glycols, esters and aqueous solutions of acids, salts and bases have little effect. DV materials will swell in aromatic solvents, halogenated organic solvents and hot petroleum oils. Harder grades are more heat and oil resistant. A DV material of hardness 64 Shore A, would have an SG of approximately 0.97 while a 50 Shore D material would have an SG of 0.94. DV materials have a pronounced, rubbery feel: the softer grades feel like a piece of vulcanized rubber.

The major application of these materials has also been in the automotive industry. Comparative lightness (density 0.94 to 0.97 gcm^{-3}), low final component cost and good chemical resistance are the major factors responsible for the continuous growth in consumption of TPO-XL type materials (DV). Such materials are a very rapidly growing sector of the *thermoplastic elastomer industry* and are, for example, widely used in the automotive industry for convoluted bellows, steering wheels and mirror frames. In the engine compartment, such materials can be used where moderately good resistance to oils, lubricating greases and fuels is sufficient. Some grades will tolerate continuous service temperatures of 125°C/257°F: short term exposure to 140°C/284°F, can be tolerated. At the present time, this type of system gives the most commercially successful, rubber-like materials from *blends* of plastics and rubbers.

dyne
Sometimes abbreviated to dyn. The unit of force in the *centimetre-gram-second system of measurement*. That force which will impart to a mass of one gram an acceleration of one centimetre per second per second. To convert to Newtons multiply by 1×10^{-5}. A dyne-second per square centimetre is a *poise*.

E

e
An abbreviation used for *strain*.

E
This letter is used as an abbreviation for:
  dielectric strength;
  elasticity modulus - see *Young's modulus*;
  emulsion - see, for example, *butadiene rubber*;
  energy;
  epoxidized - for example, *epoxidized soya bean oil (ESO)*;
  ethyl - in *plasticizer* abbreviations. For example, diethyl phthalate (DEP);
  ethylene - in polymer abbreviations;
  exa - see *prefixes - SI*; and,
  Moore's efficiency parameter. See *efficiency parameter*.

E mark - tyre
If a *tyre* carries the letter E, this shows that it is made to EEC specifications and will meet what is shown in the *tyre numbering*.

E/P
Term sometimes used for *ethylene propylene rubber*.

E-BR
An abbreviation used for emulsion *butadiene rubber*.

E-glass
A type of glass which was originally used for its good electrical insulation properties but now widely used as the fibrous material in *glass reinforced plastics*. Used, for example, in *dough moulding compounds* and in *sheet moulding compounds*. A borosilicate glass with good resistance to hydrolysis and with a refractive index which can be matched to that of *unsaturated polyester resins* - this means that sheeting of good clarity can be made (this is not the case for A-glass). The percent composition by weight is approximately SiO_2 54.3%, Al_2O_3 15.0%, CaO 17.3%, MgO 4.7%, Na_2O/K_2O 0.6% and B_2O_3 8.0%.

E*
An abbreviation used for the *energy causing breakage* in a *falling weight impact strength*.

EA
An abbreviation used for *elastomeric alloy*.
  EA-MPR = *elastomeric alloy melt processable rubber*.
  EA-TPV = *elastomeric alloy thermoplastic vulcanizate*.
See *thermoplastic elastomer*.

EAA
An abbreviation used for *ethylene-acrylic acid*.

EAM
See *ethylene-vinyl acetate copolymer*.

earth wax
See *ozocerite*.

ease of flow
A term which is concerned with melt viscosity. Thermoplastics materials at processing temperatures, differ widely in their viscosity, or ease of flow: some materials, for example, nylon 66 (PA 66), are very easy flowing whereas others, for example, *unplasticized polyvinyl chloride (UPVC)* are very stiff flowing. Each material is also available in a range of grades each of which also has a different flow behaviour. The position is made even more complicated by the fact that the flow properties of plastics are non-Newtonian and so there is not a linear relationship between pressure and flow. What all this means is that the flow properties cannot be represented meaningfully by one figure and so flow testing over a range of conditions is often performed: the results may be presented in the form of a *flow curve*. However, because of the expense involved in doing such testing, simple tests, for example *melt flow rate*, are still very widely used. See *cup flow test*.

ease of flow - extrusion testing
Checks may be made on incoming raw materials, thermoplastics materials, by using a small, single-screw extruder as a *rheometer*. A die is specified and the machine is set at specified temperatures. The extrusion behaviour is measured over a range of screw speeds and graphs are plotted of output against screw speed, temperature against screw speed and die swell against screw speed. In this way the flow behaviour is characterized. One of the big problems is maintaining the set temperatures as, any alteration in screw speed alters the extrusion cylinder temperatures and the melt temperature.

ease of flow - injection moulding assessment
The flow tendency, or ease of flow, of a thermoplastics material may be assessed by, for example, *melt flow rate*, *high shear rheometry* and *spiral flow testing*. However, such tests require special equipment to be set aside and so, these tests are not commonly employed in the moulding production industry. It is obviously easier to perform the flow test during the moulding operation.

An injection moulding machine may be used to assess or study the flow behaviour of plastics materials by, for example, measuring the *flow path: wall thickness ratio*, the *spiral flow length*, the *flow tab length* or the *minimum moulding pressure*.

easy flow grade
A grade of a polymer which is more easy flowing than the normal or standard grades of polymer. Easy flow grades of thermoplastics materials are available which permit the production of thin-walled, high-gloss products by *injection moulding*. For example, in the case of polystyrene (PS), such grades are based on a comparatively low molecular weight polymer which contains an internal lubricant such as butyl stearate or liquid paraffin. Small, regularly-sized pellets are produced by *extrusion*, which are then coated with an external lubricant such as zinc stearate or a wax: such pellets feed into an injection moulding machine quickly, melt easily and flow comparatively easily.

EBO
An abbreviation used for *ethylene bis-oleamide*.

ebonite
The material may also be known as hard rubber or as, vulcanite or as, whalebone caoutchouc. A hard, black ebony-like material obtained when an unsaturated rubber is heated with relatively large amounts of sulphur (25 to 40 phr). Rubber: sulphur ratios may be 68:32 and long curing times are typical. The formation of ebonite is exothermic which can cause problems in thick article production (also because of the evolution of hydrogen sulphide gas). To reduce the exotherm and shrinkage, ebonite dust may be used as a filler.

At one time ebonite was, for example, widely used for battery boxes and for smoker's pipe stems: this is because it is relatively chemically inert and water resistant: it also has good electrical insulation properties. Above the glass transition temperature (T_g), which may range from 35 to 90°C, it behaves as a thermoplastics material: this means that the heat resistance is not outstanding. It is better for compounds which contain high sulphur contents and which have been vulcanized for very long times. In many applications, ebonite has been replaced by cheaper thermoplastics components. Ebonite-type materials may be prepared from unsaturated rubbers such as *natural rubber, synthetic polyisoprene, butadiene rubber, styrene-butadiene rubber* and *nitrile rubber*. Cannot be formed from butyl rubber of from thiokol-type rubbers. Ebonite-type materials can also be made from *depolymerized rubber*.

ebonite dust
This material has a relative density (RD or SG) of 1·15 to 1·2. Finely ground ebonite dust is used as a filler for *ebonite*.

It acts as a processing aid (reduces shrinkage) but will also effect other final properties.

EBS
An abbreviation used for *ethylene bis-stearamide*.

ebulliometry
A *molecular weight determination method* which gives the number average molecular weight.

EC
An abbreviation used for ethyl cellulose. See *cellulose ether*.

ECH
An abbreviation used for *epichlorhydrin*.

ECO
An abbreviation used for epichlorhydrin (copolymer) rubber. See *epichlorhydrin rubber*.

ECTFE
An abbreviation used for *chlorotrifluoroethylene-ethylene copolymer*.

ECU
An abbreviation used for electrochemical unit.

ECVM
An abbreviation used for *European Council of Vinyl Manufacturers*.

edge bead
The build up of thickness at the edges of a extruded film and which is caused by *neck in*. The cross section of an edge bead is tear-drop shaped and tapers into the film; internal reduction rods may be used to minimise edge beads.

An edge bead can also be an increased thickness of sheet commonly caused by *hot trimming*. An edge bead can give problems in *reeling* because the thickness will steadily increase as the sheet is wound, for example, this could cause stretching.

edge defects
Defects on the edge of a test sample, for example, cuts or notches. The presence of such defects could cause premature sample failures during *testing*.

edge tear
Partial tearing, or ripping, of a film along its edge and which is also known as 'nip in'. The line speed and the extruder output rate, at which edge tear occurs, determines the minimum coating thickness.

edge trim
That which is removed from the edge of a sheet. For example, in *calendering*, the trim is reeled, granulated, blended in a definite ratio with virgin material and mixed in, for example, an *internal mixer*. See *trimming*.

EDM
An abbreviation used for *electro-discharge machining*.

EDTA
An abbreviation for the sodium salt of *ethylene diamine tetra-acetic acid*.

EEA
An abbreviation used for *ethylene-ethyl acrylate copolymer*.

EEC
An abbreviation used for the European Economic Community. See *Comité Européen de Normalisation*.

effective screw length
The axial distance as measured from the front edge of the feed opening to the forward end of the *flighted length* of a *screw*.

effective tie bar length
The length of *tie bar* between the tie bar nuts. It is this length, on a *toggle lock machine* which controls the amount of *locking force* generated.

efficiency of reinforcement
Also known as efficiency factor for reinforcement or as, fibre efficiency factor or as, Krenckel's efficiency factor of reinforcement. An abbreviation used is η. The ratio of the amount of fibre contributing to reinforcement to the amount not contributing. Usually refers to the Young's modulus when the comparison is made with a continuous fibre reinforced material in which all the fibres lie parallel to the direction of stressing.

efficiency parameter
May also be known as Moore's efficiency parameter or as E. This parameter accounts for the sulphur atoms which are ineffective in forming *efficient cross-links*. The larger the value of E, the larger is the number of sulphur atoms bound per cross-link site. Normal, or conventional, vulcanization systems will give an E value of approximately 7 to 12: this will give vulcanizates which suffer from poor heat stability, reversion and permanent set. Reducing E to approximately 3, by the use of a *thiuram efficient vulcanization system* or, to 1 by using a *cadmate* efficient vulcanization system, will improve such properties.

efficiency proportion
A term used in *polyvinyl chloride* (PVC) *technology* which may be defined as the number of parts of plasticizer required per 100 parts of PVC to give a specified property. For example, to give a modulus of 1100 lbf/in^{-2} at 100% elongation the following number of parts of plasticizer must be added. Dinonyl phthalate 70, di-iso-octyl phthalate 64, dioctyl phthalate 63 and dialphanyl phthalate 59. See *plasticizer efficiency*.

efficient cross-links
See *efficient vulcanization*.

efficient vulcanization
An abbreviation used for this term is EV. Sulphur vulcanization in which the sulphur is efficiently used or, used to form *mono-sulphide cross-links* and *di-sulphide cross-links*. Such cross-links improve vulcanizate properties as opposed, for example, to the formation of polysulphide cross-links which do not. To obtain more efficient vulcanization, use small amounts of sulphur (0.5 to 2 phr) and relatively large amounts of *accelerator* (3 to 6 phr). Even lower *efficiency parameter* values (E) may be obtained by using a thiuram efficient vulcanization system or, a cadmate efficient vulcanization system.

efficient vulcanization system
A *vulcanization system* which uses the sulphur in an efficient manner. See *efficient vulcanization*.

EG
An abbreviation used for *ethylene glycol*.

Egan mixing section
A *fluted mixer*: a *dispersive mixing section* used in extrusion mixing.

ejection
The removal of a moulded component, or of moulded components, from the mould which produced them: most commonly done by means of an *ejector pin*. When an ejection system is fitted, to an injection mould, it is usually only fitted in one half of the mould. Provision must therefore be made to retain the moulding on the appropriate mould half; this is usually the *moving-mould half*. For example, as the mould opens the *spray* is held on the moving-mould half by the *sprue-puller pin*.

ejection system
That which ejects components from a mould. Most ejection systems are mechanical ejection systems. The part configuration and the plastics material being moulded affect the choice of ejection system. See *pin ejection, ejector sleeve, stripper ring, stripper blade* and *stripper plate*.

ejection tie bar
A bar which connects the *ejector plate* to the *ejector frame*.

ejector bar
Part of the *ejector plate assembly*: a metal bar which transmits the ejection force to the *ejection element*.

ejector blade
Also called a knockout blade. A system of ejection fitted to, for example, an *injection mould* and used to eject by working against a part edge or against a moulded-in rib; however, the machined slot required to house such a blade can be expensive to produce and difficult to maintain.

ejector element
That part of the ejector system (for example, an *ejector pin*) which applies the ejection force to the moulding.

ejector frame
A frame which is attached to the ejection plunger of a press and which secures the *ejection tie bars*.

ejector grid
That part of the mould assembly which supports the mould plate and which provides a space for the *ejector plate system* to operate.

ejector pin
A rod, or pin, which moves when the mould opens and, in doing so, ejects the moulding from the mould.

ejector plate
Part of the *ejector plate assembly*: a metal plate which transmits the ejection force to the *ejection element*.

ejector plate assembly
Part of a mould: that part which locates and secures the *ejector elements* (pins) and allows them to move in unison. A metal assembly which transmits the ejection force to the ejection element and then returns the ejection assembly after ejection. Consists of an *ejector plate, retaining plate* and an *ejector rod*.

ejector plate return pin
Also called a surface pin. Part of a mould: a pin which returns the ejector assembly during mould closing.

ejector rod
Part of the *ejector plate assembly*: a circular metal rod which actuates and guides the *ejector plate assembly*.

ejector rod bush
A hardened steel *bush* fitted in the back plate of a mould and through which, the *ejector rod* operates.

ejector sleeve
Also called a knockout sleeve or as, a sleeve ejector. A system of ejection fitted to, for example, an *injection mould* and used to eject a cylindrical part. The tubular pin, or sleeve, may be moved over a central core so as to cause component ejection.

ejector system
The whole system which is responsible for removing mouldings from the mould; may be actuated mechanically, hydraulically, pneumatically or electrically.

EL
An abbreviation used for *electro-luminescent display*.

elastic effects
The application of shear during processing operations tends to straighten out polymer molecules: when the shearing process ceases the molecules, providing they are still molten, tend to coil up again. If cooling occurs rapidly after shear, the re-coiling may not be complete. Such uncoiling/re-coiling processes can give rise to a number of effects which, because they are related to the uncoiling/re-coiling in vulcanized rubber, are often referred to as elastic effects. The most important elastic effects are *die swell*, *melt fracture*, *sharkskin*, *frozen-in orientation* and *draw down*.

elastic limit
The greatest *stress* which a material is capable of carrying in a *tensile test*, without any permanent strain remaining after the *strain* has been completely released.

elastic melt extruder
An *extruder* which utilises the *Weissenberg effect*. Basically the machine contains two discs and one disc rotates against the other. The thermoplastics material is introduced into the gap at the edge and turned into a polymer melt: this melt then moves towards the centre of the disc. As it moves towards the centre the nip, or gap, becomes narrower so as to compensate for the volume reduction which occurs on melting, for example, one disc is convex. The melt then passes through a hole in the centre of the disc towards the die. To improve melt uniformity, a short *screw* (a low L:D ratio screw) is placed between the discs and the die. An elastic melt extruder is a very compact machine as the discs can be mounted vertically.

elastic memory
Term used to describe the recovery properties of plastics melts i.e. the melt behaves elastically as it will recover, after being stretched, by an applied load. See *elastic effects*.

elastic modulus
Also known as stiffness constant. See *modulus*.

elastic modulus for uniaxial extension
See *modulus*.

elastic modulus in flexure
Sometimes called elastic modulus in bend. An abbreviation used for this term is E which equals $L^3/(4bh^3)$ times F/Y. Where F/Y is the slope of the initial linear load-deflection curve, b is the width of the specimen, d is the thickness of the specimen and L the span width. See *modulus*.

elastic turbulence
See *melt fracture*.

elastic yield
Elastic deformation. The deformation which disappears when the deforming force is removed.

elasticity
The ability of a material to regain its original shape after removal of the force which caused deformation.

elasticity modulus
See *Young's modulus*.

elasticoviscous fluid
A fluid which exhibits predominantly viscous flow but which also exhibit partial elastic recovery after deformation. See *die swell*.

elastomer
This word is sometimes used in place of *rubber* but in this publication it will refer to a *rubbery material* which has a loose cross-linked structure. A *rubber compound* whose chains cannot readily slip past each other because, for example, the polymer has been *vulcanized*. Such a material can be extended, or stretched, by approximately 100% and on removal of the retracting force it rapidly springs back to its original dimensions. That is, it snaps back. See *thermoplastic elastomer*.

elastomer modified
A grade of a thermoplastics material which has been modified with a *rubber* or with an elastomer. For example:

elastomer modified polybutylene terephthalate. A type of *thermoplastic elastomer* obtained by melt mixing *polybutylene terephthalate (PBT)* with a synthetic *acrylic rubber*. Such a material is used to make car spoilers - painted on line and stoved at temperatures of up to 140°C/284°F;

elastomer modified polypropylene. A type of *thermoplastic elastomer* obtained by melt mixing *polypropylene* with an *ethylene-propylene rubber*. See *rubber modified polypropylene*.

elastomeric alloy
An abbreviation used for this type of material is EA. A melt mixed compound of a plastics material with a crosslinked rubber, A combination of two or more polymers which has significantly better properties than what would be expected from a simple blend. The two major types of elastomeric alloy are classed as *thermoplastic vulcanizates* (TPV or TPO-XL) and *melt processable rubbers (MPR)*. See *dynamically crosslinked polyolefin rubber*.

elastomeric alloy melt processable rubber
An abbreviation used for this type of material is EA-MPR. This type of material has also been called a halogenated polyolefin alloy thermoplastic rubber or a thermoplastic chloroolefin elastomer.

Single phase combinations of two, or more, polymer systems. For example, a mix of a highly plasticized combination of a chlorinated polyolefin, an ethylene vinyl acetate copolymer and an acrylic ester rubber is thought to give the EA-MPR known as Alcryn (DuPont). Such materials often have lower processing temperatures than EA-TPVs and, because of this, may be processed on relatively simple, rubber-type equipment. That is, for example, with comparatively short L:D ratio screws and at low temperatures: plastics processing equipment is however preferred. Vulcanization is not required but in many ways treat such materials as though *unplasticized polyvinyl chloride* (UPVC) was being handled. Keep *melt temperatures* below 180°C, use corrosion-resistant materials and avoid melt mixing with *acetal*. See *thermoplastic vulcanizate*.

elastomeric alloy thermoplastic vulcanizate
An abbreviation used for this material is EA-TPV. Elastomeric alloy thermoplastic vulcanizates are also referred to as TPE-OXL or TPO-XL. Such materials are two phase systems in which a crosslinked rubber phase is dispersed in a continuous plastics phase, which is usually a polyolefin such as *polypropylene (PP)*. The use of a crosslinked rubber improves the chemical resistance and, improves physical properties such as compression set and retention of properties at elevated temperatures. When the crosslinked rubber is an ethylene propylene rubber then, the material may be referred to as an EPDM-TPV. When the crosslinked rubber is a nitrile rubber then, the material may be referred to as an NBR-TPV. See *dynamic vulcanization*.

elastomeric ionomer
An *ionomer* which yields a *thermoplastic elastomer*. For example, *carboxylated polybutadiene ionomer*.

elastomeric material
A compound which is based on a *rubber* (see *elastomer*) and other mix ingredients, and which has been cross-linked or *vulcanized* after being shaped or moulded.

elastomeric polyamide
See *polyether block amide*.

elastothiomer
See *polysulphide rubber*.

electric discharge machining
See *electro-discharge machining*.

electric insulator
A material with a resistivity greater than $10^{10}\Omega.cm$: sometimes referred to as a *dielectric*. See *dielectric strength*.

electric polarization
See *polarization*.

electric screw drive
A term associated with *injection moulding* which means that the *screw* is rotated by means of an electric motor. For example, a pole-reversing, three-phase electric servo-motor via a gearbox. On a conventional *injection moulding* machine, the screw is rotated by means of an hydraulic motor. For machines with a large throughput of material, the use of an electric motor (for example, an AC servo motor via a gearbox) is claimed to save on energy usage. See *screw drive* and *hydraulic screw drive*.

electric strength
See *dielectric strength*.

electrical properties
Usually means those properties which are concerned with electrical insulation. One of the earliest uses for plastics was for electrical insulation: they were chosen because they could be easily shaped into non-conductive components by processes such as moulding and extrusion.

The effectiveness of the electrical insulation that plastics and rubbers provide depends on the chemical structure of the polymer, process residues, impurities and the level and type of any additives in the compound. Thus it is possible to produce both plastic and rubber compounds which are classed as semi-conducting (antistatic) or even electrically conducting by the incorporation of suitable additives, for example, *carbon black*. A loading of 30% of a special carbon black in ethylene vinyl acetate can reduce the *volume resistivity* to 10 ohm.cm. Two other terms widely used when discussing the electrical insulation behaviour of polymers are *surface resistivity* and *insulation resistance*. Also see *resistivity*, *dielectric strength* and *dielectric constant*.

electrical resistance
See *ohmic resistance*.

electrical resistance element
A heating element which relies on the resistance to flow of an electric current through a wire element based, for example, on a nickel-chrome alloy. Heating elements form the basis of many indirect heating processes. There are two principal types of resistance heaters and these are resistance elements wound on a refractory former, and mineral-insulated metal-sheathed resistance elements.

Resistance elements wound on a refractory former are used in ovens and radiant infra-red heaters: for protection the element may be mounted in a quartz tube. The element may be woven in a mesh with a heat-resistant fibre, for example, *asbestos* or *quartz*, to form a flexible heater.

Mineral-insulated metal-sheathed resistance elements insulate the heating element from the protective case with a compressed layer of an inorganic insulator, for example, magnesium oxide. They are available in various forms. For example, in the form of a *heater band* (a cuff heater), a *strip heater* and a *cartridge heater*. More than one element type may be used on a *mould* or *die* in an effort to get the most uniform temperature distribution.

electro-discharge machining
An abbreviation used for this term is EDM. This technique is also known as electric discharge machining or as, spark erosion or as, burning and, as these names suggest, the metal is removed by generating a spark between the workpiece and an electrode (carbon or copper). The electrode is made in the reverse shape of the part required. Both the workpiece and the electrode are immersed in a fluid (temperature-controlled) which cools the workpiece and flushes away the eroding material. The workpiece is first roughly shaped under water, using a roughing electrode, and then finished with finishing electrodes under oil. Using this process it is possible to machine blind holes, curved holes, sharp angles, etc. The process is not restricted to the shaping of steel but can be used for other materials, such as *tungsten carbide*.

The electrodes for such EDM equipment may be produced on wire erosion machines. These use a continuous wire electrode which cuts through the workpiece just as wire may be used to cut cheese. The wire electrode may be used to cut profile dies and to make long cuts; tapers of up to 10° may be produced.

Computer numerical control (CNC) may be applied to EDM, controlling such parameters as table motion and gap conditions. Up to five axes may be run simultaneously on a CNC machine. The five axes include X, Y, Z and U - the last, a rotating and cutting action on the quill; the fifth axis can be used for such operations as rotating or indexing the workpiece.

electro-hydraulic
A component of a hydraulic system which converts an electrical signal into a hydraulic signal. See *electro-hydraulic servo-valve*.

electro-hydraulic servo-valve
A component of a hydraulic system: a directional type of valve which converts an electrical signal into a hydraulic output. Used, for example, to alter the *injection speed* during a shot in *injection moulding*.

electro-luminescent display
An abbreviation used for this term is EL. Part of the computer control system of, for example, an *injection moulding* machine. The display system of, for example, an *intelligent terminal*.

electro-magnetic interference
See *electromagnetic interference*.

electro-rheological fluid
Also known as a smart fluid. An abbreviation used for this type of material is ER. A fluid which when exposed to an electric field thickens, or gels, to a semi-plastic state: the change of state is proportional to the strength of the electric field so the thickening effect can be regulated. On removal of the field, the material changes back to its original form.

electrode
A metal plate, or formed shape, used to conduct electricity. Two electrodes make the capacitor of a *high frequency* welding machine and between which is created the alternating electric field necessary to cause *polarization*.

electrodecanting
A continuous method of producing *concentrated latex* which is also called electrodialysis. The latex is concentrated because the negatively charged latex particles migrate in an electric field. A permeable membrane prevents the rubber particles reaching, and coagulating, on the anode: the lower density *cream* is separated by decanting.

electrodialysis
See *electrodecanting*.

electroforming
Deposition of metal, for example, nickel, from solutions using electroplating. By electroplating it is possible to build up, for example, hard nickel to a thickness of 5 mm (0·2 in). With a hardness of <48 Rockwell C and the virtual non-corrodibility of nickel, inserts made by electroforming have a life expectancy comparable to the best tool steels. Since there is no heat involved in the process, the reproduction of the original master is precise, although allowances must still be made for *shrinkage*. Excellent surface reproduction is obtained but the metal build-up is slow and this makes the process somewhat more expensive than casting.

electrohydraulic
See *electro-hydraulic*.

electroless plating
Also known as electroless metal deposition. A metallization technique whereby a metal such as copper or nickel is deposited by chemical means on the surface of a plastic component before *electroplating*. This is necessary as plastics are not normally electrically conductive and cannot therefore be made the cathode in a plating bath. Can have electroless copper plating and electroless nickel plating. See *electroplating of plastics*.

electrolyte
A compound which conducts an electric current and which is simultaneously decomposed by that current: the current is carried by ions. Usually electrolytes are liquids or molten solids (acids, bases or solids) although the use of a solid electrolyte has several advantages for the construction of a solid battery (a lithium rechargeable battery). For example, lack of electrolyte leakage, long shelf life and a wide temperature range of use. See *polyethylene oxide-salt complex*.

electromagnetic interference
An abbreviation used for this term is EMI. The modification of a signal, in say a machine, caused by a magnetic field which is being radiated from another device or from a conductor. Many natural and man-made sources radiate random, broad range frequency, electromagnetic radiation in an uncontrolled manner and must be guarded against over the range of, say, 30 to 1,000 MHz.

Such electromagnetic shielding can be obtained with a thermoplastics material with about 13 percent stainless steel fibres. By plating high strength *microspheres* it is possible to obtain an electrically conducting filler, capable of withstanding moulding, and which gives light weight components suitable for use, for example, as business machine housings and electronics packaging (as EMI shielding is obtained). Aluminum coated, glass fibres or aluminum flake should give the same effect at a lower cost than competing metal fibres: faster cooling should also result from the uses of such fillers.

EMI shielding is determined by the attenuation of a high frequency signal transmitted through a test sample. The shielding effect (α) is expressed in decibels (dB) by:
$\alpha = 20 \log_{10} (E_b/E_a)$. Where E_b is the field intensity (V/m) without shielding and E_a is the field intensity (V/m) with shielding.

electromagnetic radiation
Radiation which consists of electromagnetic waves. See *electromagnetic interference*.

electromagnetic shielding
See *electromagnetic interference*.

electromagnetic spectrum
That which encompasses all electromagnetic radiation: that is, from the longest to the shortest. From audio waves (10 Hz) to radio waves, to microwaves, to infra-red, visible, ultra-violet, X and gamma-rays (10^{18} Hz). The whole range of frequencies over which radiation is propagated. Visible light is only a very small part of this spectrum.

electromagnetic units
See *units*.

electron beam
A form of *high energy radiation*. Such a beam is used, for example, in place of *ultraviolet radiation* or visible light because of the finer definition possible in electronic *photoresist* applications.

electron charge polymer
See *oxidation-reduction polymer*.

electron microscopy surface area
An abbreviation used for this term is EM surface area. A measure of the surface area of fillers as measured by electron microscopy.

An electron microscope may be used to measure the particle diameter of a number (for example, 1,200 to 1,500) of individual particles of a filler such as *carbon black* (under a magnification of up to 75,000 times). From such data, particle size distribution curves may be plotted and the average particle size determined. From the average particle size determination, the total surface area (the EM surface area) may be calculated if the particles are assumed to be spherical. Such an assumption is incorrect as, for example, the filler particles may be porous. See *nitrogen absorption* and *BET value*.

electron-volt
An abbreviation used for this unit is eV. A unit of *energy*. The increase in energy when the potential of an electron is raised by one *volt*. 1 eV is equal to 1.60×10^{19} J.

electroplating
The deposition of a metal by electrolysis: the component to be plated is made the cathode in a plating bath which contains a solution of the metal required on the component. When electroplating plastics, it is usual to use more than one metal so as to get the appearance required together with reasonable permanence in use. After *electroless plating*, layers of copper, nickel and chromium may be applied using techniques developed for metals. In order to eliminate problems caused through copper corrosion, just nickel and chromium are also used. See *electroplating of plastics* and *metallizing*.

electroplating of plastics
The coating of plastics components by *electroplating*. In order to obtain the best results (good finish and adhesion) mouldings which contain low levels of strain, and which are of excellent finish, must be used. The components must be etched and surface treated so as to obtain a coating of metal by *electroless plating*. The thermoplastics material associated with electroplating is *acrylonitrile-butadiene-styrene (ABS)* and for this material the main steps are *injection moulding* of the components, cleaning, surface conditioning, sensitisation, electroless plating and electroplating.

Such plated plastics components are widely used in the automotive industry for mirrors and grilles: also used for plumbing and bathroom fixtures, such as, taps and levers. The components replace zinc-based die castings. See *metallizing*.

electrostatic fluid bed coating
A coating method which uses a *fluid bed*. This is formed by inserting electrodes through the bed: when the electrodes are energised with an electrostatic charge a cloud of powder is formed, for example, *polyethylene* powder. When an earthed metal object is passed through the cloud it is coated uniformly with plastics material. After immersion coating, the coated metal is heated to complete fusion before being cooled.

electrostatic spray coating
A coating method which uses an electrostatic spray gun and a plastics material in powder form. The plastics material in powder form is passed through a spray gun. An electrode is located at the end of the spray gun so that when the electrode is energised, with an electrostatic charge, a cloud of powder is formed, for example, *polyethylene* powder. When the earthed metal object is passed through, or near, the cloud it is coated uniformly with plastics material. After coating, the coated metal is heated to complete fusion.

electrostatic unit
See *units*.

element
A material which cannot be divided into chemically simpler substances. Each part of the element has the same atomic number. There are 106 such elements of which 92 occur naturally: the lightest of the elements is hydrogen and the heaviest, of these natural elements, is uranium. Most commercial polymers are based on the element *carbon*.

elephant foot
The thickened base of a grafted tree. See *Hevea brasiliensis*.

elephant skin
See *crazing effect* and *oxidation*.

ELO
An abbreviation used for *epoxidized linseed oil*.

elongation
In *tensile testing*, the increase in length between bench, or *gauge*, marks on a *dumb-bell*: the extension is expressed as a percentage of the original distance between the marks.

elongation at break
The percentage elongation at break (Σ_B) is usually expressed as a percentage of the original length and this, therefore, may be expressed as longer length (l_1) minus original gauge length (l_o) divided by original gauge length (l_o) times 100.

elongational flow
Also known as tensile flow or stretching flow. See *tensile viscosity*.

elongational viscosity
See *tensile viscosity*.

EM surface area
An abbreviation used for *electron microscopy surface area*.

embedding
A *casting process* used to encase components in a polymer by pouring a monomer, a prepolymer or a polymer dispersion over the component while that component is contained in a mould. The polymer is cured by, for example, heating and the encased component is then removed.

embedding layer
The layer of polymeric material in which a reinforcing helix is embedded: see *smooth bore hose*.

embossing
A process for producing a contoured surface on a substrate. A sheet-shaping technique which uses a *male* and *female mould* to shape heated sheet. An alternative is *hydroforming*.

embossing roll
A steel roll used to emboss a pattern on *calendered sheet*. Consists of an engraved metal roll which is cooled with chilled water. See *embossing unit*.

embossing unit
An ancillary unit used, for example, to emboss a pattern on *calendered sheet*. Consists of an engraved metal roll (which is cooled with chilled water), a rubber covered steel roller (cooled by water circulation), a contact roll running in a water bath (which acts as a cooling roll for the rubber-covered roll and a take-off roll. With such a system thermal degradation of the rubber-to-metal bond on the roll is minimised. The nip pressure exerted by the embossing roll can be controlled hydraulically and pressures of up to 175 kNm^{-1} of working face (1,000 psi) can be employed. The rubber roll is usually driven and again the speed can be varied. Once the take-off speed at the embossing unit has been determined then it may be related to the main *calender drive* so that if calender speed is altered the speed of such an ancillary unit is also altered. If frequent pattern changes are required then the steel embossing rolls could be mounted on a turret arrangement so that the roll pattern can be easily changed.

emergency stop
A machine stop which when operated, stops a machine very quickly. See *safety stop*.

emery
This material has a relative density of 3·7 to 4·0. An abrasive filler sometimes used to make polishing or grinding wheels from rubber compounds. See *aluminium oxide*.

emf
An abbreviation used for *electromotive force*.

EMI
An abbreviation used for *electromagnetic interference*.

emu
An abbreviation used for *electromagnetic unit*.

emulsifying agent
Also known as an emulsifier. A material, for example, a detergent, which helps to form or stabilize an *emulsion*.

emulsion
A colloidal solution in which the dispersed phase consists of minute droplets of liquid.

emulsion paint
A very popular type of paint which is water-based. Highly pigmented, plasticized lattices are used, for example, based on *polyvinyl acetate*.

emulsion polymerization
A very popular method of producing synthetic polymers as the heat liberated on *polymerization* is transferred to the water in which the reaction occurs. The monomer is suspended in the water by the action of a surfactant and some of the monomer is held within surfactant micelles. A water soluble initiator is used to initiate the reaction while the mixture is stirred: by the use of active initiators, 'cold' polymerization is possible. It is thought that polymerization commences in the micelles the number of which control the molecular weight of the final polymer. The growing polymer chains are fed with more monomer from the dispersed droplets of monomer. The latex which results may be used as it is or, the water may be removed by, for example, spray drying. The resultant solid polymer often contains large amounts of surfactant which can impair electrical and optical properties. See *polyvinyl chloride* where PVC-E means that the polymer was made by emulsion polymerization.

emulsion synthetic rubber
An abbreviation used for this material is E-SR. A *synthetic rubber* produced by *emulsion polymerization*. For example:

emulsion butadiene rubber = E-BR; and,
emulsion styrene butadiene rubber = E-SBR.

EN
An abbreviation used for Europäishe Norme, Norme Européene or *European Standard*.

ENB
An abbreviation used for *ethylidene norbornene*.

ENCAF
An abbreviation used for *ethyl N-phenylcarbanoylazoformate*.

encapsulation
A process used for applying a polymer coating to components: the polymer is applied by techniques such as brushing, dipping, spraying or moulding.

enchained ionic polymer
See *ionene* and *ionic polymer*.

end capping
Also known as capping. A technique used to produce a more stable polymer by reacting, or changing, unstable end groups into more stable end groups. Used, for example, to produce more thermally stable *acetal polymers*.

end effects
See *flow*.

end group
The atom or group at the end of a polymer chain. See *end capping*.

end group analysis
Also known as end group assay. The determination of the number of groups at the end of a polymer chain. The concentration of end groups in moles per gram of polymer. See *end capping*.

end reinforcement
Reinforcement built into the end of a hose so as to provide additional strength: reinforcement applied to the end of a hose so as to provide additional strength.

end to end distance
The distance separating the two ends of a polymer chain. An abbreviation used for this term is r.

endo-4,7-methylene-4,7,8,9-tetrahydroindene
See *dicyclopentadiene*.

ends
The number of strands in a *roving*.

energy
The capacity for doing work. An abbreviation used for this term is E. The derived SI unit is the *joule*. See *power*.

energy attenuation
See *attenuation*.

energy causing breakage
An abbreviation used for the energy causing breakage, in an *instrumented falling weight impact strength test*, is E*. It has been found that a reasonably good approximation of the energy causing breakage (E*) can be obtained by multiplying the impact speed (V_0) by the area under the force-time graph up to the point where failure began if the speed of impact (V_0) remains virtually unchanged during the impact.

energy output
See *power*.

energy penetration
See *penetration depth*.

engineering plastics
An abbreviation used for these types of material is EP. See *engineering thermoplastics*.

engineering plastics blends
Usually means *engineering thermoplastics blends*.

engineering plastics material
An abbreviation used for this type of material is EP. See *engineering thermoplastics*.

engineering strain
The *strain* obtained by using the original sample dimensions is sometimes referred to as 'nominal' or as 'engineering' strain.

engineering stress
The *stress* obtained by using the original sample dimensions is sometimes referred to as 'nominal' or as, 'engineering' stress. See *nominal stress*.

engineering thermoplastic
An abbreviation used for this type of material is EP or, ETP. Engineering plastics are a group of polymers which offer a combination of some of the following, high strength, stiffness, toughness, resistance to wear, chemical attack and heat. The major materials in this group are the *polyamides, acetals, polycarbonates, thermoplastic polyesters* and *modified polyphenylene oxide*. See *engineering thermoplastics blends*.

engineering thermoplastics blend
An engineering-type of thermoplastics material which is a melt mixed, blend of materials. May be obtained by blending *engineering thermoplastics* together or, by blending engineering thermoplastics with *commodity thermoplastics materials* and/or with *thermoplastic elastomers*. Such blends are often classed as *alloys* and offer a similar spectrum of properties to *engineering thermoplastics*.

enhancement factor
The ratio of the modulus of a filled material to that of the unfilled material at low strain values.

enlarged end
A hose end in which the internal diameter at the end is larger than that of the main body of the hose.

ENM
An abbreviation suggested for *hydrogenated nitrile rubber*. Hydrogenation removes the double bonds and for this reason the abbreviation ENM has been suggested as the polymer could have been derived from *ethylene* and *acrylonitrile*.

ENR
An abbreviation used for *epoxidized natural rubber*.

ENR 10 = natural rubber (NR) which contains 10 mole percent of epoxide groups.
ENR 20 = NR which contains 20 mole % of epoxide groups.
ENR 30 = NR which contains 30 mole % of epoxide groups.
ENR 50 = NR which contains 50 mole % of epoxide groups. This material is not compatible with *natural rubber*.

ENs
The plural of *EN (Europäishe Norme)*. In English, this means *European Standards*.

enthalpy
Also known as heat content. The amount of heat energy that a system contains. H is used as an abbreviation. ΔH is used as an abbreviation for the enthalpy change in a system. See *heat* and **table 7**.

entrance effect correction
See *Bagley end correction* and *Couette Hagenbach method*.

entropy
An abbreviation used for this term is S. The entropy of a system is a measure of its disorder.

Entwicklungsgesellschaft für die Wiederverwertung von Kunstoffen
An abbreviation used for this German-based organization is EWvK. An English translation is Plastics Recycling Development. A joint venture between BASF, Bayer and Hoechst which coordinates and develops uses for recycled materials.

environmental shrinkage
See *post-moulding shrinkage*.

environmental stress cracking
If a material is stressed below its yield point in air it may crack after a long period. However, if it is simultaneously stressed and exposed to an active environment, the time for cracking or crazing may be dramatically reduced. This type of failure is referred to as environmental stress cracking (ESC). Sometimes called stress corrosion cracking.

The active environment depends upon the particular polymer, but commonly includes detergents, oils, greases and fats. Gases or vapours may also cause failure. The closer the stress is to the yield stress in air the faster the cracking or crazing will take place. The environment alone often has no detrimental effect on the unstressed material. The stress may be internal, that is, built-in as a result of processing.

Because internal stress combined with a suitable stress crack agent can cause ESC, it is possible to use the phenomenon to check the level of internal stress in mouldings and extrusions. Simple immersion tests are often used and are commonly performed on specimens cut from *annealed, compression moulded sheet*. The specimens are subjected to a known level of stress or strain and placed in an active environment (such as a detergent). The time for cracks to appear is often used to quantify the measurement.

Testing for resistance to ESC is complicated by many factors such as the level of internal stress in the specimens, the aggressiveness of the stress crack agent, the test temperature, the time of contact with the particular environment and specimen dimensions. As a result many tests have been devised but few standardised. See *Bell Telephone Laboratories test method, Landers test, ball* or *pin impression method* and *International Electrotechnical Commission*.

There are standardised test methods for assessing the ESC resistance of finished products such as bottles, pipes and cables. They do, of course, evaluate not only the material but also the quality of the processing and design. Most of these tests involve exposing the product to a specific stress crack agent, often at elevated temperature, and examining for cracking after a specified time. Cables are usually bent to a specified radius and pipes pressurised in an attempt to simulate possible service conditions.

ESC is associated with *polyethylene* (PE). Low molecular weight polyethylenes are more prone to stress cracking than high molecular weight materials of the same density. This is true whether they are subjected to the Bell test (ASTM D 1693-66) or the Landers test (ASTM D 2552-66T). For polymers of the same molecular weight but different densities the position is less clear. The higher density material usually performs better in the *Landers test* but not necessarily in the *Bell test* as this is a constant strain test and the level of stress imparted is proportional to material rigidity. Homopolymers with narrow molecular weight distributions are also known to be more ESC resistant, presumably because of the absence of the low molecular weight fractions. Copolymerization with small amounts, usually less than 5 per cent, of ethylene vinyl acetate is also known to improve the ESC resistance of polyethylene.

EOT
A rubber which has sulphur, carbon and oxygen in the main polymer chain. A rubber with polysulphide linkages in which, for example, the polysulphide linkages are separated by organic groups (R groups) such as $-CH_2-CH_2-O-CH_2-O-CH_2-CH_2-$ and other R groups. For example, by $-CH_2-CH_2-$.

EP
An abbreviation used for *epoxy*: also used for engineering plastics (material) or for, *engineering thermoplastics material*.

EP(D)M
An abbreviation used for *ethylene-propylene diene monomer rubber* and for *ethylene-propylene rubber*.

EP(D)M blend
An abbreviation used for polypropylene/ethylene propylene diene monomer blend: an abbreviation used for polypropylene/ethylene propylene rubber blend. See *rubber modified polypropylene*.

EPC
An abbreviation used for *easy processing channel* (black). S300. See *carbon black*.

EPDM
An abbreviation used for *ethylene-propylene diene monomer rubber*. See *ethylene-propylene rubber*.

EPDM-TPV
An abbreviation used for a *thermoplastic vulcanizate* based on an *ethylene propylene diene monomer rubber*.

epichlorhydrin
Also known as epichlorhydrine or as, 1-chloro-2,3-epoxy propane or as, chloromethyl oxirane or as, γ-chloropropylene oxide. An abbreviation used for this material is ECH. This liquid material has a relative density (RD or SG) of 1·18, a melting point of −26°C and a boiling point of approximately 116°C. This material contains an *epoxy group* and when reacted with a diol, this material gives a diglycidyl ether - the basis of epoxy resin production. ECH may be polymerized, by ring opening polymerization, to give the homopolymer, *epichlorhydrin rubber* (CO from chloromethyl oxirane). Ethylene oxide is copolymerized with epichlorhydrin to give epichlorhydrin copolymer rubbers (ECO from ethylene - chloromethyl oxirane). A terpolymer is made from allyl glycidyl ether, ethylene oxide and epichlorhydrin to give epichlorhydrin terpolymer rubbers (AECO from allyl - ethylene - chloromethyl oxirane).

epichlorhydrin copolymer rubbers
The collective name for the class of materials which are rubbery copolymers of *epichlorhydrin*. Ethylene oxide is copolymerized with *epichlorhydrin* to give epichlorhydrin copolymer rubbers. See *epichlorhydrin rubber*.

epichlorhydrin homopolymer rubber
See *epichlorhydrin rubber*.

epichlorhydrin rubber
Abbreviations used for this type of material include AECO or CO or ECO or CHR or CHC or ETER: epichlorhydrin (homopolymer) rubber is CO or EC: epichlorhydrin copolymer rubber is ECO or CHC: allyl-group-containing epichlorhydrin rubber or, epichlorhydrin-ethylene oxide-allyl glycidyl-ether terpolymer or rubber is AECO or ETER or ETE. See *epichlorhydrin*.

Epichlorhydrin rubbers are included in the category of *speciality rubber* because of their ozone resistance and because of the oil and fuel resistance of CO and ECO types. Such materials also have low gas permeability and resist burning. The homopolymer has a relatively high brittle point but this may be reduced by plasticization (using an ester-type *plasticizer*) and/or by copolymerization. The 1:1 copolymer does not resist fire as well as the homopolymer, is more permeable but has a lower brittle point and a lower density.

Carbon black is widely used as a filler, for example, an FEF black. Such polymers are vulcanized through the chlorine atoms with diamines, ureas and thioureas, for example, ethylene thiourea and lead oxide. The terpolymer is sulphur vulcanizable and is more resistant to *sour gas*.

epichlorhydrin rubbers
The collective name for the class of materials which are rubbers and which are based on *epichlorhydrin*.

epichlorhydrin-ethylene oxide copolymer
See *epichlorhydrin rubber*.

epichlorhydrin-ethylene oxide-allyl glycidyl ether rubber
A terpolymer. See *epichlorhydrin rubber*.

epichlorhydrin-ethylene oxide-allyl glycidyl ether terpolymer
See *epichlorhydrin rubber*.

epichlorhydrine
See *epichlorhydrin*.

EPM
An abbreviation used for ethylene-propylene monomer. See *ethylene-propylene rubber*.

epoxidation
The introduction of the epoxy group. See *epoxidized oil* and *epoxidized natural rubber*.

epoxide
Usually means *epoxide resin*. See *epoxide moulding compounds*.

epoxide equivalent
The weight of an *epoxy resin*, in grams, which contains one equivalent of epoxide groups.

epoxide moulding compounds
Also known as epoxides and as epoxies. An abbreviation used for this type of material is EMC; EP is also used.

Epoxide moulding compounds are based on *epoxy resin* blended with fillers (e.g. *glass* and *mineral fillers*), a mould release agent (e.g. *zinc stearate*) and, a hardening system. There are two principal classes of hardeners, or curing systems, and these are amines (e.g. diaminodiphenylmethane - DDM) or, acid anhydrides (e.g. phthalic anhydride). Chain extension and cross-linking, which produce the thermoset structure, proceed via addition-type reactions so that volatile by-products are not produced and low shrinkages on curing, or hardening, are obtained. The polyether structure, of the cured resin, resists many forms of chemical attack and the mouldings are therefore relatively inert. This type of structure means that mouldings can also withstand relatively high temperatures: rigidity decreases only slightly with rising temperatures, up to 150°C/302°F, and the mouldings exhibit good creep resistance at elevated temperatures. Cured materials have excellent electrical properties even under extreme climatic conditions.

Epoxy novolaks are made in a similar way to bisphenol A resins but are based on novolak resins having a multiplication of phenolic groups instead of bisphenol A-type groups. Such resins when cured with an aromatic amine, have the best chemical resistance of EMC materials.

An advantage of EMC is that its properties are unaffected by the length of cure time, provided a minimum is exceeded. Components possessing thick and thin wall sections are uniformly cured throughout as the thin sections do not become overcured while waiting for the thicker sections to cure: this cannot be said for other thermosets.

Available as granular solids which are compression, injection and transfer moulding into products with a limited colour range if amine hardeners are used: those cured with anhydrides, may be produced in light colours (white, beige, grey, etc.). Long glass fibre (GF) filled grades are mainly compression and transfer moulded so as to avoid *fibre degradation*. Components can be moulded to very close tolerances as a result of the low variation in mould shrinkage and the absence of *post moulding shrinkage*.

Cured materials are resistant to a wide range of aqueous solutions and solvents. Being cross-linked the resin will not dissolve without decomposition, but will be swollen by liquids of similar solubility parameter to the cured resin. Chemical resistance is dependent on both the hardener and the resin as these, will determine the linkages formed, e.g. acidic hardeners form ester groups which will be less resistant to alkalis but the cured material has superior electrical properties. Such materials have good resistance to aqueous acids and salts and, to aliphatic hydrocarbons. Amine cured materials have, in addition, better alkali, inorganic salt and, organic chemical resistance. EMC is not resistant to polar organic solvents (e.g. methyl ethyl ketone, tetrahydrofuran and industrial methylated spirits. Alkalis attack acid-cured materials whereas amine-cured compounds have good resistance.

Such materials are usually filled with inorganic fillers (e.g. silicas, metal silicates, glass fibre, etc.) and so the density of EMC is often 2 gcm^{-3}. No obvious sign of shrinkage and mouldings can be produced with a very high order of flatness and a good surface finish (resembles machined metal in this respect). Some mouldings have incredible wall thicknesses, e.g. 20 mm/0.8 in.

The most important applications of epoxy moulding compounds have been in electrical engineering. This is because of the high tracking resistance, arc resistance and dielectric strength of epoxy components - even under extremely varying climatic conditions. Because of their good chemical resistance, epoxy compounds are used in chemical plant applications. Their outstanding dimensional stability, and lack of internal stresses, allows EMCs to be used where the moulded components contain metallic inserts (e.g. automotive distributor caps). Typical uses vary from spark plug connectors, thermostat control mechanisms, pump assemblies and circuit breaker assemblies.

epoxide reaction injection moulding
An abbreviation used for this term is EP RIM or epoxy RIM. Also known as epoxy reaction injection moulding. A *reaction injection moulding* (RIM) process which produces highly cross-linked products: the process of producing such a moulding.

Such *epoxide systems* are usually based on *bisphenol A resins*, *liquid aliphatic diamines*, for example, di-(cyclohexylamino) methane and a catalyst such as calcium nitrate. The temperature of the reactants is typically 60°C, mould temperatures are of the order of 120°C while demould times may be 90 s: the adiabatic exotherm temperature for a typical formulation without filler can be 150°C. As the reaction may be slow at mixing temperatures, the use of an impingement mixer may not be necessary. Strict stoichiometric balance is often not necessary for epoxy RIM.

Epoxide mouldings offer high modulus, solvent resistance and high heat distortion temperatures (HDT): brittleness can be a problem with such reaction injection mouldings as can thermal oxidation during moulding.

epoxide resin
See *epoxy resin*.

epoxide sheet moulding compound
A *sheet moulding compound* based upon an epoxide resin. See *high performance sheet moulding compound*.

epoxides
See *epoxide moulding compounds*.

epoxidized linseed oil
An abbreviation used for this material is ELO. A *plasticizer* for polyvinyl chloride. See *epoxidized oil* and *linseed oil*.

epoxidized materials
Materials which have had their structure modified by the introduction of the epoxy group. See *epoxidized oil* and *epoxidized natural rubber*.

epoxidized natural rubber
An abbreviation used for this material is ENR. A modified *natural rubber (NR)*. Natural rubber latex is epoxidized with *formic acid* and *hydrogen peroxide*: this reaction is cheap and easy to perform on the trialkylethylenic double bonds of NR.

Best done at low acid concentrations and at moderate temperatures if the production of furanized NR is to be avoided.

The introduction of epoxide groups raises the *glass transition temperature* of *natural rubber*, makes the product less permeable to air and improves the oil resistance: this is because of the increase in polarity. Can have various grades, for example, ENR 10 contains 10 mole percent of epoxide groups, ENR 20 contains 20 mole percent of epoxide groups and ENR 30 contains 30 mole percent of epoxide groups. ENR 50 is comparable in air permeability properties to *butyl rubber* and has the oil resistance, at room temperatures, of a medium *nitrile rubber*. The modified NR can be cured with the normal *sulphur vulcanization systems* although *semi-efficient* and *efficient curing systems* give the best heat ageing to the compounds. There is significant reinforcing action with *silica fillers* without the use of *coupling agents*. For example, tyre-tread compounds based on ENR 25 and silica, give good wet grip and low rolling resistance. This type of material may be a useful additive for polar plastics materials such as *polyvinyl chloride* (PVC).

epoxidized oil
An oil which contain the *epoxy* or *oxirane ring*. Because of the good light stability of epoxidized oils, such materials are used as *heat stabilizers/plasticizers* in plastics materials such as *polyvinyl chloride*. Also used as stabilizers and protective agents in materials such as *chlorinated polyethylene*.

epoxidized soya bean oil
Also known as epoxidized soybean oil. This material has a relative density of 0.99. An abbreviation used is ESO or ESBO. Because of the good light stability of *epoxidized oils*, such materials are used as *plasticizers* in plastics materials such as *polyvinyl chloride*. See *soya bean oil*.

epoxidized soybean oil
See *epoxidized soya bean oil*.

epoxies
See *epoxide moulding compounds*.

epoxy
See *epoxy plastics* and *epoxy resin*.

epoxy compound
An organic compound which contains the *epoxy ring* and which is used, for example, as a *synergistic co-stabilizer* with a *phosphite chelator* in a *polyvinyl chloride* (PVC) formulation.

epoxy novolak
A type of *epoxy resin*. Made in a similar way to *bisphenol A* epoxide resins but are based on novolak resins having a multiplication of phenolic groups instead of bisphenol A-type groups. Such resins when cured with an aromatic amine, have the best chemical resistance of EP materials. As the cross-link density is high, cured materials have high heat distortion temperatures. See *epoxide moulding compounds*.

epoxy number
The gram equivalent of epoxide oxygen per 100 g of an *epoxy resin*.

epoxy plastics
Thermoplastic or thermosetting plastics containing ether or hydroxyalkyl repeating units (or both) resulting from the ring-opening reactions of lower molecular weight polyfunctional oxirane resins (or compounds) with catalysts or, with various polyfunctional acidic or basic co-reactants. See *epoxy resin* and see *epoxide moulding compounds*.

epoxy resin
Also known as epoxide or, epoxy resin or, epoxide resin or, ethoxyline resin. An abbreviation used for this type of material is EP. Epoxy resin is used to describe both the low molecular weight oligomer prepolymer (which contains two or more *epoxide* groups per molecule) and the cross-linked material prepared from such an oligomer. For example, if epichlorohydrin is reacted with *bisphenol A* in the presence of an alkali, then syrups (or brittle resins) of low molecular weight are produced. Such, low molecular weight intermediates have a backbone structure of a polyether with hydroxy groups (-OH) along its length and, a reactive glycidyl ether group (epoxy group) at each end: that is, they are complex diglycidyl ethers (diglycidyl ethers of bisphenol A or DGEBA). Most commercial polymers are based on DGEBA.

Such materials are then turned into thermosetting structures by cross-linking or hardening using a hardening system. There are two principal classes of hardeners, or curing systems, and these are either *amines* (e.g. diaminodiphenylmethane - DDM) or, acid anhydrides (e.g. *phthalic anhydride*). Chain extension and cross-linking, which produce the thermoset structure, proceed via addition-type reactions so that volatile by-products are not produced and low shrinkages on curing, or hardening, are obtained.

Epoxies are probably best known as good adhesives. Good bonding with such materials is achieved, partly because of the low curing shrinkage but also because of the polar hydroxy group; this helps initial wetting on polar substances. Hence the resins stick well to glass, stone, metal, etc.

The polyether structure, of the cured resin, resists many forms of chemical attack and cured products are therefore relatively inert. This type of structure means that the cured material can also withstand relatively high temperatures: cured materials can have excellent electrical properties even under extreme climatic conditions and are tough. See *epoxide moulding compounds*.

epoxy ring
A three membered ring which consists of an oxygen atom linked to two carbon atoms. Epoxidation of, for example, *natural rubber* enhances hydrocarbon resistance and increases the glass transition temperature: this is because of the increase in polarity. See *epoxy resin* and *epoxide moulding compounds*.

epoxyethane
See *ethylene oxide*.

EPR
An abbreviation used for *ethylene-propylene rubber*.

EPROM
An abbreviation used for erasable, programmable, read only memory; a data storage device which forms part of a computer system.

EPS
An abbreviation used for *expanded polystyrene*.

equilibrate
To settle down. To allow machine settings, for example, temperatures, to come to equilibrium or to their natural values.

equilibrium centrifugation
See *sedimentation equilibrium method*.

ER
An abbreviation used for *electro-rheological fluid*.

ERCOM
A German based co-operative organization engaged in the re-processing of automotive components made from fibre reinforced composites.

erg
A unit of work in the *centimetre-gram-second system* of measurement. The work done by a force of 1 dyne acting through a distance of one centimetre. One erg is 10^{-7} joules.

error
May be defined as the difference between a measured value and the true (or most probable) value. The error is stated as a positive value even if the difference is negative.

error detection
A feature used to ensure that transmitted information is not distorted during transmission from one device to another.

error signal
The sum of a command signal and a feedback signal. See *closed loop system*.

erythrene
A former name for *butadiene*.

ESBO
See *epoxidized soya bean oil*.

ESC
An abbreviation used for *environmental stress cracking*.

ESCR
An abbreviation used for *environmental stress crack resistance*.

Esemar
See *standard Malaysian rubber (SMR)*.

ESO
An abbreviation used for *epoxidized soya bean oil*.

essential oil
The volatile aromatic oil found in the stems and leaves of plants. A highly-volatile, odoriferous oil isolated from a single botanical species and which bears the name from the plant from which it was obtained. See *fixed oil*.

estate
A *natural rubber plantation* whose size is greater than approximately 40 ha. In Malaysia such a plantation would yield about 1,500 kg of natural rubber per hectare per year.

estate brown crepe
One of the forms in which *natural rubber* is supplied. Obtained, for example, from *cup lump* and *cleaned tree bark scrap*.

ester
An organic compound (corresponding to an inorganic salt) which is obtained by the reaction of an organic acid with an alcohol. Many esters are pleasant smelling materials which are used for flavouring. In the polymer industry, they are used, for example, as *plasticizers*. Many *fats* and *oils* are esters.

ester exchange reaction
See *ester inter-change reaction*.

ester gum
Also known as a rosin gum. Ester gums are the esters produced by the reaction of rosin with an alcohol such as *glycerol*. Such esters are harder than the original material and so give harder varnishes.

ester inter-change reaction
A chemical reaction whereby one organic radical (R_1) on an *ester* is exchanged with another organic radical (R_2) from an alcohol. Used to prepare *polycarbonates*. If this reaction is performed during processing, then it may result in the compatibilization of an otherwise incompatible system.

ester TPU
An abbreviation used for a polyester *thermoplastic polyurethane*.

esterification
The formation of an *ester*.

esu
An abbreviation used for electrostatic unit.

ETA extraction
Extraction using a mixture of 70% by volume of *ethyl alcohol* and 30% by volume of toluene so as to determine the amount of vulcanizable rubber.

etching
See *surface conditioning*.

ethanal
An alternative way of saying acetaldehyde.

ethanol
See *ethyl alcohol*.

ethanoyl
The ethanoyl group is also known as the acetyl group and has the formula CH_3CO-.

ethene
An alternative way of saying *ethylene*. For example, for ethene-vinyl ethanoate copolymer, see *ethylene-vinyl acetate copolymer*.

ether
Also called diethyl ether and ethoxy ethane. A highly inflammable solvent. This material has a relative density (RD or SG) of 0.72 and a boiling point of approximately 35°C. It is a good solvent for uncured *natural rubber (NR)* and *styrene-butadiene rubber (SBR)*. It is a poor solvent for uncured *thiokol rubber (T)* and *nitrile rubber*. This chemical causes some swelling of uncured *chloroprene rubber (CR)* and *butyl rubber (IIR)*. When mixed with a small amount of ethanol, it is a useful solvent for *cellulose nitrate* and *polyvinyl acetate (PVAC)*.

ether group
May be represented as -O-. See *polyether imide*.

ether TPU
An abbreviation used for a polyether *thermoplastic polyurethane*.

ether-group containing rubber
The *abbreviations* used for such materials contain O.

etherified
The reaction of a polymer with an alcohol. For example, *urea-formaldehyde polymers* may be reacted with an alcohol (for example, n-butanol) to give resins which are more soluble in organic solvents than the base polymer: used with *alkyd resins* to make stoving enamels.

ethoxy ethane
See *ether*.

ethoxylin resin
See *epoxy resin*.

ethoxyline resin
See *epoxy resin*.

ethyl
The ethyl group is an alkyl group which has the formula CH_3CH_2- or C_2H_5-.

ethyl acetate
Also known as ethyl ethanoate. It is an ester which has the formula $CH_3COOC_2H_5$. This material has a relative density (RD or SG) of 0.9 and a boiling point of approximately 77°C. It is a good solvent for *cellulose nitrate*, *cellulose-acetate butyrate*, *ethyl cellulose*, *polyvinyl acetate* and *polystyrene*. It will cause swelling of *polyvinyl chloride*, *polymethyl methacrylate* and *polyethylene*. Will dissolve cellulose acetate when mixed with ethanol. It is not a good solvent for uncured rubbers. For example, it is a poor solvent for uncured SBR and T rubbers

ethyl alcohol

but causes some swelling of uncured IIR and NR rubbers. it is of intermediate solvating power for uncured CR and NBR rubbers.

ethyl alcohol
Also called ethanol. A flammable solvent with the formula CH_3CH_2OH. This material has a relative density (RD or SG) of 0·78 and a boiling point of approximately 78°C. It is a useful solvent for varnishes as it dissolves, for example, *phenolformaldehyde* and *natural resins*.

ethyl benzene
Made from *benzene* and *ethylene*. The precursor for *styrene*.

ethyl carbamate
See *urethane*.

ethyl cellulose
An abbreviation used for this material is EC. See *cellulose ether*.

ethyl ethanoate
See *ethyl acetate*.

ethyl methyl ketone
See *methyl ethyl ketone*.

ethyl N-phenylcarbanoylazoformate-modified natural rubber
An abbreviation used for this material is ENCAF-modified natural rubber (NR). The use of ethyl N-phenylcarbanoylazoformate (ENCAF) modifies *natural rubber* by introducing highly polar groups during, for example, mill mixing or when the rubber is in *latex* form. The product is a modified natural rubber which has reduced permeability and reduced solvent swelling. This material is however, difficult to prepare and relatively expensive: the compression set is unacceptably high and the material has poor thermal ageing properties. Modification of *natural rubber* with p-nitroso-ethyl N-phenylcarbanoylazoformate gives a modified natural rubber with even better resistance to solvent swelling.

ethylene
Also known as ethene: a member of the *alkene series of hydrocarbons*. A very important *monomer* with the formula C_2H_4. An *unsaturated material* which is usually a gas as it has a boiling point of approximately −104°C. See *polyethylene*.

ethylene acrylate copolymer
See *ethylene-methyl acrylate rubber*.

ethylene acrylate rubber
See *ethylene-methyl acrylate rubber*.

ethylene bis-oleamide
An abbreviation used for this type of material is EBO. A *lubricant*.

ethylene bis-stearamide
An abbreviation used for this type of material is EBS. A *lubricant* with a relative density of 0·97.

ethylene bridge
A chemical group which links two molecules together. For example, if a polymer contains methyl side groups, then cross-linking may occur through dehydrogenation and the formation of an ethylene bridge $-CH_2-CH_2-$.

ethylene chloride
Also known as ethylene dichloride or as, dichlorethane (DCE), or as, 1,2-dichlorethane or as ,chloroethylene. A colourless liquid with a formula of CH_2Cl-CH_2Cl. This material has a relative density (RD or SG) of 1·26 a melting point of −36°C and a boiling point of approximately 84°C. A solvent for bitumens, resins and rubbers. Used to produce polysulphide rubbers.

ethylene chlorohydrin
Also known as 2-chloroethanol. A colourless liquid with a formula of CH_2Cl-CH_2OH. This material has a relative density (RD or SG) of 1·20 a melting point of −67°C and a boiling point of approximately 128°C. A solvent for *natural rubber* and *synthetic rubbers*.

ethylene diamine
Also known as 1,2-diaminoethane. A colourless, strongly alkaline liquid with a formula of $NH_2CH_2CH_2NH_2$. This material has a relative density (RD or SG) of 0·89, a melting point of 9°C and a boiling point of approximately 117°C. Used a stabilizer for *natural rubber latex* and as an antigelation agent for rubber cements.

ethylene diamine carbamate
A white solid material with a relative density (RD or SG) of 1·37 and a melting point of approximately 150°C. Used as a vulcanizing agent for *fluororubbers*.

ethylene diamine tetra-acetic acid
An abbreviation used for this type of material is EDTA. It is the sodium salt of this solid acid which is used as a stabilizer for *natural rubber latex* (as a substitute for *ammonia*): a copper and manganese inhibitor for rubber.

ethylene dichloride
See *ethylene chloride*.

ethylene glycol
An abbreviation used for this material is EG. Also known as ethane-1,2-diol. This material has a relative density (RD or SG) of 1·12 and a boiling point of approximately 197°C. A diol with the formula $HOCH_2CH_2OH$. Used to prepare polyester plastics, polyester prepolymers and polyester plasticizers.

ethylene ionomer
An *ionomer* which contains ethylene units.

ethylene oxide
Also known as epoxyethane or as, oxirane, or as, oxiran. A colourless, flammable gas made by the oxidation of ethylene and which may be represented as CH_2CH_2O: a ring compound. Used to make, for example, ethylene glycol and to make epichlorhydrin copolymer rubbers (the ethylene oxide is copolymerized with *epichlorhydrin*. See *polyethylene oxide*.

ethylene oxide-chloromethyloxiran
An abbreviation used for this type of material is ECO. An epichlorhydrin copolymer rubber. See *epichlorhydrin rubber*.

ethylene plastics
Plastics based on polymers of ethylene, or copolymers of ethylene, the ethylene being in present in the greatest amount by weight. See *polyethylene*.

ethylene propylene elastomer
See *ethylene-propylene rubber*.

ethylene propylene rubber
See *ethylene-propylene rubber*.

ethylene propylene rubber polypropylene blend
An alternative way of saying polypropylene/ethylene propylene rubber blend. See *rubber modified polypropylene*.

ethylene stimulation
See *stimulation tapping*.

ethylene sulphide
This organosulphide compound may be represented as CH_2CH_2S. Used to make, for example, *alkylene sulphide rubbers*.

ethylene thiourea
Also known as 2-mercaptoimidazoline. An abbreviation used is ETU. A toxic, vulcanizing agent, for example, for polychloroprene. See *chloroprene rubber*.

ethylene vinyl acetate copolymer
See *ethylene-vinyl acetate copolymer* and *ethylene-vinyl acetate rubber*.

ethylene-acrylic acid copolymer
An abbreviation used for this material is EAA. Also known as ethylene-acrylic acid. Random or block copolymers of ethylene and acrylic acid are available which contain up to approximately 20% acid units. Such plastics materials are noted for their good adhesive properties, particularly to metals such as aluminium. Because of this feature, and their toughness, such materials are used as a component of laminates for packaging.

ethylene-acrylic elastomer
See *ethylene-methyl acrylate rubber*.

ethylene-ethyl acrylate copolymer
An abbreviation used for this type of material is EEA. A copolymer based on *ethylene* and *ethyl acrylate* and which is similar to the plastics copolymers based on ethylene and vinyl acetate (EVA) but with better heat and abrasion resistance: EVA type materials are cheaper, tougher and of greater clarity. See *ethylene-vinyl acetate copolymer*.

ethylene-methyl acrylate rubber
Also known as ethylene-acrylate copolymer or as, ethylene-acrylate rubber or as, ethylene-acrylic elastomer or as, ethylene-methyl acrylate terpolymer rubber. May be better known by the trade name/trade mark, Vamac (Du Pont). An abbreviation used for this type of material is AEM or EAM. A terpolymer of ethylene, methyl acrylate and a third, *cure-site monomer* which contains a pendant carboxylic group. A terpolymer rubber which is used when *chloroprene rubber* fails to meet high temperature resistance requirements. When compounded with reinforcing fillers, for example, SRF carbon black, cured materials (cured with primary diamines or peroxides) exhibit good resistance to heat, ageing, weathering, ozone and to aliphatic hydrocarbons. Has moderate oil resistance and is used in the automotive industry for seals, O-rings and constant-velocity joints.

ethylene-methyl acrylate terpolymer rubber
See *ethylene-methyl acrylate rubber*.

ethylene-propylene copolymer
See *ethylene-propylene rubber*.

ethylene-propylene diene monomer
See *ethylene-propylene rubber*.

ethylene-propylene elastomer
See *ethylene-propylene rubber*.

ethylene-propylene monomer
See *ethylene-propylene rubber*.

ethylene-propylene rubber
This type of material has been classified as a *general purpose rubber*. These materials are sometimes referred to as polyolefin rubbers and are produced in two main types; the saturated copolymers (which are referred to as EPM or EPR) and the terpolymers (which are referred to as EPDM). Where one does not wish to differentiate between the two types of rubber, then occasionally the abbreviation EP(D)M is used.

A rubber based on *ethylene copolymerized* with *propylene* which is an amorphous, rubbery material if the ethylene content is high: commercial polymers may contain 30 to 50% of ethylene units. Such a copolymer may be referred to as poly(methylene-co-propylene). Usually, all types of this rubber have a propylene content of between 25 and 55% by weight: such a copolymer can only be vulcanised with an organic peroxide. In order to obtain a sulphur vulcanisable material, a small percentage of a third monomer (3 to 10% of a diene monomer, for example, ethylidene norbornene) is incorporated so as to make a terpolymer; this terpolymer may be referred to as EPDM.

As these polymers have very poor gum strengths, it is necessary to compound them with reinforcing fillers. In some ways these materials resemble *butyl rubber*, i.e. they have very good chemical stability and are noted for their good ageing resistance. They also have a high resistance to breakdown during mechanical operations which generate shear. However, they are easier processing than butyl rubbers and have better resilience at ordinary temperatures. A major application for both types of ethylene-propylene rubbers is in blends with thermoplastics materials (such as polypropylene); such blends are widely used to produce items such as car bumpers. See *rubber modified polypropylene*.

ethylene-propylene terpolymer
A terpolymer based on *ethylene*, *propylene* and, *ethylidene norbornene*. A terpolymer is used in order to obtain a sulphur vulcanisable material: approximately 3 to 10% of a diene monomer (ethylidene norbornene) is incorporated. This terpolymer may be referred to as EPDM. See *ethylene-propylene rubber*.

ethylene-propylene-ethylidenenorbornyl sulphonate
See *sulphonated ethylene-propylene-diene monomer rubber*.

ethylene-tetrafluoroethylene copolymer
An abbreviation used for this material is ETFE. See *tetrafluoroethylene-ethylene copolymer*.

ethylene-vinyl acetate copolymer
Also simply known as ethylene-vinyl acetate. An abbreviation commonly used for this type of material is EVA. Polymerizing vinyl acetate (VA) with ethylene, disrupts the crystal structures that are present in polyethylene (PE) and eventually gives an amorphous material. By varying the percentage of vinyl acetate (VA) in the composition, polymers with significantly different properties are produced. As the percentage of VA is increased the transparency and flexibility of the copolymers increases. Broadly speaking, there are three different types of EVA copolymer, which differ in the vinyl acetate (VA) content and the way the materials are used.

The EVA copolymer which is based on a low proportion of VA (approximately up to 4%) may be referred to as vinyl acetate modified polyethylene. It is a copolymer and is processed as a thermoplastics material - just like *low density polyethylene*. It has some of the properties of a low density polyethylene but increased gloss (useful for film), softness and flexibility. Generally considered as a non-toxic material.

The EVA copolymer which is based on a medium proportion of VA (approximately 4 to 30%) is referred to as *thermoplastic ethylene-vinyl acetate copolymer* and is a *thermoplastic elastomer material*. It is not vulcanized but has some of the properties of a rubber or of *plasticized polyvinyl chloride* - particularly at the higher end of the range. May be filled and both filled and unfilled materials have good low temperature properties and are tough. The materials with approximately 11% VA are used as hot melt adhesives.

The EVA copolymer which is based on a high proportion of VA (greater than 40%) is referred to as *ethylene-vinyl acetate rubber*: this is compounded and vulcanized like a traditional rubber. See *ethylene-ethyl acrylate copolymer*.

ethylene-vinyl acetate copolymers
The collective name for the class of materials which contain vinyl acetate as a comonomer. See *ethylene-vinyl acetate copolymer*.

ethylene-vinyl acetate rubber
An abbreviation used for this material is EVM or EAM or EVA. Also known as EVA rubber or as, vulcanizable ethylene-vinyl acetate rubber or as, vulcanizable EVA rubber. An *ethylene-vinyl acetate copolymer* which is based on a high proportion of vinyl acetate (greater than, for example, 40%).

Can be cross-linked by peroxides, for example, by *di-t-butyl peroxide* together triallyl cycanurate as a co-agent: may also be cured by high energy radiation. Antioxidants can interfere with cure and are therefore often omitted. The acidity associated with some fillers, for example, china clay, also interferes with cure.

EVA rubbers are used for seals, for example, in automotive applications as they have good heat resistance (better than EPDM-type materials). Acrylic rubbers (ACM rubbers) are also used for automotive applications as they have good oil resistance and also good high temperature resistance: for this reason they are used for shaft seals. Ethylene-vinyl acetate (EAM) rubbers gives better low temperature flexibility but have greater swelling in oil.

ethylene-vinyl alcohol copolymer
Also simply known as ethylene-vinyl alcohol. An abbreviation used for this material is EVAL: also known as EVOH and/or EVAl. A *copolymer* which is based on *ethylene* and on vinyl alcohol and which is obtained by the hydrolysis of *ethylene-vinyl acetate copolymers*. A water-sensitive copolymer which is noted for its low permeability and used as a component of laminates and/or, of *blow mouldings*.

ethylidene norbornene
An abbreviation used for this type of material is ENB. Also known as ethylidenenorbornene or as, 2-ethylidenebicyclo-(2,2,1)-5-heptene. A monomer used to prepare an *ethylene-propylene terpolymer*.

ethylidenenorbornene
See *ethylidene norbornene*.

ETMQ
An abbreviation used for *6-ethoxy-2,2,4-trimethyl-1,2- dihydroxyquinoline*.

ETPT
An abbreviation used for *bis(diethylthiophosphoryl) trisulphide*.

ETS
An abbreviation used for *European Telecommunication Standards*.

ETSI
An abbreviation used for *European Telecommunications Standards Institute*.

ETU
An abbreviation used for *ethylene thiourea*.

EU
An abbreviation used for *polyether polyurethane rubber*.

EU polyurethane rubber
A type of *polyurethane rubber* which is based on polyols containing ether groups.

Euromap
The European Committee of Machinery Manufacturers for the Plastics and Rubber Industries. This organization issues recommendations on machine design and use and which are available from the national association of a particular country, for example, the BPF in the UK.

Europäishe Norme
See *European Standard*.

European Chemical Industry Federation
An abbreviation used for this organization is CEFIC. See *Association of European Isocyanate Producers*.

European Committee for Electrotechnical Standardization
See *European Organization for Electrotechnical Standardization*.

European Committee for Standardization
An English translation of *Comité Européen de Normalisation*.

European Council of Vinyl Manufacturers
An abbreviation used for this European organization, based in Brussels, is ECVM.

European Organization for Electrotechnical Standardization
An abbreviation used for this organization is CENELEC. Also known as the European Committee for Electrotechnical Standardization. An organization which issues *European Standards*, for example, in the field of electrical engineering (wires, electrical appliances etc). UK participation in CENELEC is via the British Electrotechnical Committee, the Electrotechnical Standards Council of the *British Standards Institution*.

European Standard
The English translation of Europäishe Norme (Norme Européene). An abbreviation used for this term is EN. A standard issued by, for example, the *Comité Européen de Normalisation (CEN)*: such a standard is prefixed by EN, for example, EN 29002-1987. Such standards are adopted on the principle of consensus amongst member countries and adopted by a weighted majority. It is mandatory for all CEN/CENELEC members to implement an EN as a national standard and to withdraw any of their own national standards that conflict with it regardless of the way they voted.

European Standards (ENs) are usually available from, for example, the standards organization of a particular country: in the UK this is the *British Standards Institution (BSI)*.

The official languages of an EN may be English, French or German. However, a version in another language has the same status as the official versions of an EN if it is made by translation under the responsibility of a CEN member into its own language and submitted to CEN Central Secretariat.

European Telecommunication Standards
An abbreviation used for this term is ETS. Standards produced by the *European Telecommunications Standards Institute* (ETSI).

European Telecommunications Standards Institute
An abbreviation used for this organization is ETSI. This organization produces European Telecommunication Standards (ETS).

eutectic salt mixture
A solid solution of two or more materials which has the lowest freezing point of all possible mixtures of those materials. For the *liquid curing method*, a eutectic salt mixture is used which may be based on 53% potassium nitrate, 40% sodium nitrite and 7% sodium nitrate. Because of the high temperatures used (up to 240°C), and the good heat transfer characteristics, rapid vulcanization times are obtained.

eV
An abbreviation used for *electron-volt*.

EV
An abbreviation used for *efficient vulcanization*.

EVA
An abbreviation used for *ethylene-vinyl acetate copolymer*.

EVA rubber
See *ethylene-vinyl acetate rubber*.

EVAc
An abbreviation sometimes used for *ethylene-vinyl acetate copolymer*.

evaporation
A method of producing *concentrated latex* by removing water either by heating or, by heating and evaporation under vacuum. The final concentration can reach approximately 66% rubber and the *concentrated latex* contains all the original *serum constituents*, for example, alkalis and soaps. Such a process may also be referred to as the Revertex process.

EVK
A cross barrier mixer. See *dispersive mixing section*.

EWvK
An abbreviation used for *Entwicklungsgesellschaft für die Wiederverwertung von Kunstoffen* (Plastics Recycling Development).

exa
An *SI prefix* which means 10^{18} and which has the abbreviation E.

exact
An adjective used in connection with measurements and with units. Exact is used for quantities that have an infinite number of significant figures. Exact quantities are either decided arbitrarily or are obtained by counting objects individually, that is one by one. No measured quantities other than those measured by simple counting can be exact. See *accurate*.

excitation
The process of raising an electron to a higher orbital. That is, raising it to a higher energy state.

excited state
An atom is said to be excited when an electron has been raised to a higher orbital than the one it occupies in the ground state.

exfoliate
The expansion of minerals, like *vermiculite*, when heated. For example, exfoliated vermiculite is expanded vermiculite.

exosphere
The gaseous mantle, the atmosphere, which surrounds the planet Earth has been divided into several layers or strata. The exosphere is the outermost one, that is, the one above the *ionosphere* and which gradually merges into space.

exotherm curve
A plot of temperature versus time during resin cure. Used to establish the *peak exotherm*, the *gel time* and the *cure time*.

exothermic heat
Heat liberated from a curing resin system. Heat liberated during polymerization.

expandable plastic
A plastic in a form capable of being made cellular by thermal, chemical, or mechanical means.

expanded
An alternative word to cellular but which often implies that solid starting materials were used. See *expanded polystyrene* and *expanded rubber*.

expanded coated fabric
A *plastic coated fabric* which contains one or more layers of a cellular structure: often based on *polyvinyl chloride* compositions.

expanded plastics
See *cellular plastics*.

expanded polyethylene
A cellular plastics material based on *polyethylene* (PE): a cellular plastics product produced from solid starting materials. A chemical *blowing agent* is incorporated into the PE mix and during, for example, *extrusion*, this blowing agent decomposes to release a gas which causes subsequent expansion. Used for cable insulation as such products have a very low *dielectric constant* (approximately 1·45). Expanded polyethylene, in sheet form, has also been made by dissolving a gas (for example, nitrogen) in PE sheets in a pressurized, heated autoclave: the gas dissolves in the PE and on subsequent processing, expansion occurs. See *expanded polystyrene*.

expanded polymer
An alternative to cellular polymer: the term often implies that solid starting materials were used. See *expanded polystyrene* and *expanded rubber*.

expanded polystyrene
Also known as foamed polystyrene. An abbreviation used for this material is EPS: may also be known as PS-E or XPS or PS-X.

A cellular plastics material based on polystyrene (PS): a cellular plastics product produced from solid starting materials, often by a two-stage process. A physical *blowing agent*, such as pentane, is incorporated into the PS during, for example, suspension polymerization. The output is in the form of fine beads which contain trapped pentane.

When these beads are heated, then the pentane (a blowing agent) changes state to give gas which causes expansion of the heat-softened PS beads: the beads are heated in steam and/or water at approximately 100°C to give pre-expanded beads (puffed-up, or expanded, beads). At this stage the beads are delicate as on cooling the pentane condenses and/or is lost and a partial vacuum develops inside the beads. The beads are therefore stored, for approximately 6 h, before being processed into blocks or mouldings. The pre-expanded beads are packed into a mould and heated by steam: fusion then occurs inside the steam-heated mould. It is found that a two-stage process gives products of very low density (for example, 0·016 Mg/m^3.

Expanded polystyrene mouldings are widely used for packaging as the products are cushioned against impact partly because of the soft, resilient nature of the EPS but also because the moulding can be shaped to suit the product. Used for heat insulation, for example, of the roofs of buildings where it is laid in place before asphalt-based compositions are applied. Higher density foams are made from PS by extrusion and/or by injection moulding. See *structural foam*.

expanded rubber
A cellular rubber: a cellular rubber product produced from solid starting materials and which possesses a closed cell structure. A chemical blowing agent is incorporated into a high viscosity rubber mix and during, for example, moulding this blowing agent decomposes to release a gas which causes subsequent expansion. See *expanded polystyrene*.

exponential decay
A type of decay in which the rate of decay is proportional to the quantity of material present.

extended nozzle
A long *nozzle* which projects deep into the injection mould plate so as to permit a reduction in *sprue length*. The *sprue* may be dispensed with entirely by using a long nozzle which extends into the base of the component.

Such nozzles are suitable for single-impression work or, in the form of a manifold nozzle, for multi-impression moulding. One type of extended nozzle, used for single-impression work, has a small flat machined on the nozzle end. This small flat forms part of the base of the mould and as the mould is kept cool provision must be made to keep this mould tip hot. This may be done by fitting a heater to the nozzle tip or by making the nozzle from a material of high thermal conductivity (for example, *beryllium-copper*). The contact area between the nozzle and an enlarged water-cooled *sprue bush* (necessary to take up the pressure of the nozzle) is kept as small as possible.

extended rubber
See *oil extended rubber*.

extender
An additive. Both solid and liquid materials are sometimes used as extenders although the term is often restricted to liquid materials. Extenders are sometimes referred to as 'liquid fillers' and are commonly associated with *plasticized polyvinyl chloride* (PPVC) where there use, reduces compound costs as the extender is of relatively low cost. The chlorinated waxes used in this way also act as *flame retardants*. With PPVC compounds, the extenders used are relatively incompatible, whereas in rubber compounds the extenders used are relatively compatible. See *plasticizer*.

extension
The addition of an extender: the addition of a large proportion (approximately 50%) of a rubber *processing oil* to a compound. See *elongation* and *lake*.

extensional flow
A type of flow which occurs when a melt is being pulled or extruded: occurs, for example, during drawing from a die.

extensometer
A device used to measure extensions in, for example, *tensile strength/elongation determinations*.

external coating
A coating applied to an external surface. With some *blow moulded* containers, a coating or layer, may be present as either an inside layer, an outside layer, or both. Outside coating may be used to protect the contents against ingress of, for example, oxygen or water. Such a coating may result in *scalping*.

external lubricant
An additive. A *lubricant* added to a compound in order to prevent the compound sticking to the processing equipment. *Stearic acid* is an example of a lubricant which is added to, for example, *polyvinyl chloride* (PVC) in order to prevent the compound sticking to the processing equipment. The excessive use of an external lubricant can cause *plate out*. See *internal lubricant*.

external mould release
A *mould release* which is applied to the surface of a mould. Mould release can be eased by the use of an *internal mould release* and/or an *external mould release*.

external plasticization
The addition of a *plasticizer* in order to ease processing and/or to alter compound properties. See *internal plasticization*.

external plasticizer
An additive. A *plasticizer* added to a compound in order ease processing and/or to alter compound properties. Most plasticizers are liquids and are commonly *phthalates*.

external softening
Process used to increase the softness of a polymer by the addition of a *softener*.

external stabilizer
An *additive* which is used to minimize the effects of degradation caused by, for example, heat and/or light. See *stabilizer*.

external undercut
See *undercut*.

externally corrugated hose
A hose which contains a reinforcing helix and in which the outer cover has been formed into corrugations between the helix. Can have *rough bore hose*, *smooth bore hose*, and *semi-embedded hose*.

extra white crepe
One of the forms in which *natural rubber* is supplied. Produced in the same way as *pale crepe* but, before *coagulation* the latex is first treated with *sodium bisulphite* and *xylyl mercaptan* to reduce colour. See *white* and *pale crepes*.

extractable sulphur
One of the ways the *sulphur* in rubber compounds, on analysis, may be classified. That sulphur which is not chemically bound, or chemically attached, to the rubber and which may be removed by solvents such as acetone or acetone/chloroform. Includes elemental sulphur and sulphur contained in soluble compounds.

extrudate
The shaped material exiting from the *extrusion die*.

extrudate collapse
An *extrusion fault*: the extrudate collapses when being hauled-off. In the case of *natural rubber* may be minimized by the use of *superior processing rubber*.

extrudate diameter/die diameter
A ratio known as the *die swell ratio*, *swelling ratio* or as *puff-up ratio*.

extruded film
Film produced by an *extrusion process*.

extruded foam
An extruded product produced when the plastic mass is expanded by the formation of gas cells.

extruded HIPS sheet
An abbreviation used for *high impact polystyrene sheet* which has been produced by *extrusion*. This material is used in *thermoforming processes*.

extruded net
A net produced by extrusion and whose production did not involve a weaving process. Polymers in net form are commonly used in the packaging industry to package fruit. By utilising the *extrusion process*, nets may be produced at relatively low cost as a weaving process is eliminated. The polymer, usually based on polyethylene, may be extruded through slots cut in a circular die (see *tubular film process*). By rotating the outside of the die, relative to the centre, during haul-off a circular net may be produced. By, for example, adjusting the speed of rotation and the haul-off speed a wide range of patterns may be produced. If the slots are large, then heavy gauge nets may be produced which are suitable for gardening applications.

extruded profile
A continuous shape produced by extrusion. Profiles of open or hollow section can be extruded through specially-designed profile dies; each shape presents its own problems of die design and post-extrusion handling. Profile dies can be difficult and expensive to make: without experience, lengthy processes of trial and error may be required to establish both die design and extrusion conditions, for example, to prevent preferential flow in the thicker sections. For this reason, a design change may necessitate a new die.

Such dies can be satisfactory for small sections but a simpler, more reliable, technique is more desirable for larger section, e.g. for diffusers for fluorescent lighting fittings. This technique involves post-forming sections extruded from tube dies. Compared with profile dies, large tube dies can be made at low cost and their symmetry facilitates analysis and uniform flow. If desired, the inner lip of the die can have a reeded surface which imparts a pattern to the extrudate.

The emerging reeded tube is slit by a knife mounted in the die face, and the slit tube passes to a shaping and cooling jig. The shaping is done by alternate inner and outer formers with polished forming surfaces: cooling is accomplished by internal and external perforated copper rings through which is directed a gentle flow of air. The haul-off can be of the rubber caterpillar-band type, and sections can be cut off by a travelling saw. Trial and error may be necessary to establish conditions and former design but, adjustments to formers are made more quickly and cheaply than to a profile die.

extruder
The machine used for *extrusion*. Most machines are screw extruders and use a *screw*, or screws, to transport and heat the material through the *barrel*: a screw/barrel assembly functions as a wiped surface heat exchanger and can also be used for melt mixing or compounding.

extruder barrel
The main section of the extruder in which the extruder screw turns; also known as the extrusion cylinder or as the *barrel*.

extruder classification
Extruders may be classified by three figures, for example, 1-60-24. The first number states how many screws the machine has, the second number specifies the screw diameter in millimetres (mm) and the third number specifies the effective screw length as a multiple of the screw diameter. In the example given therefore, the machine is a single screw machine which has a screw of diameter 60 mm and a length of 24 diameters. (L:D ratio of 24:1).

extruder feedback control system
A method of controlling extruder output by the adjustment of speed or back pressure so as to maintain a constant output rate.

extruder screw
The unit used to propel material through an *extruder* and to generate pressure on the melt. See *screw*.

extruder - use for reclamation
See *dip process* and *reclaiming processes*.

extrusion
The production of semi-finished or finished components using an *extruder*. Although it is possible to extrude thermosetting plastics materials, using ram extruders, most of the plastics that are extruded are *thermoplastics materials*. This is because such materials are more widely used than *thermosetting plastics* and also because they are easier to handle in a heated machine, for example, they do not set or harden inside the machine.

The term 'heated machine' was used because most extrusion processes are melt extrusion processes. That is, the thermoplastics material is turned into a soft, pliable material (the melt) by heating the material inside a *barrel*: it is most commonly conveyed through the barrel by means of a *screw*. This heating process is performed inside a machine called an extruder and the product, the extrudate, is then formed by forcing the heated material through an orifice called a *die*.

Once the extrudate leaves the die, it can either be set to the shape produced or, its shape may be altered and then it may be set to shape. The equipment which handles the extrudate is called the post extrusion equipment or the haul-off and in terms of size it is far larger than the extrusion machine itself. One reason for this, is the slow rate of cooling of thermoplastics materials. See *extrusion - products of*.

extrusion behaviour
See *extrusion testing*.

extrusion blow moulding
An abbreviation used for this term is EBM. A *blow moulding process* in which a parison, or tube, is produced by extrusion. This is the major blow moulding process as very fast production is possible using equipment which is relatively cheap.

An *extruder* is used to form the thermoplastics material into a tube: this emerges in a downwards direction from the shaping orifice or die. Usually the still hot material is then blown into the shape of a mould: a mould closes around the tube, or parison, and this closing action causes the tube to be sealed or pinched at one end. The product is then blown into the shape of the mould with compressed air (air pressures of up 7 atmospheres are usually used).

As one moulding is cooling, another mould may be made ready to receive a fresh parison. The split moulds can be mounted on a turntable below the extruder die, or, alternatively, the melt can be delivered to two or more static moulds by arranging for melt to flow, in turn, to different dies connected to the same extruder. All extrusion-blown mouldings involve welds of the open-ended parison, and as these are potential sources of weakness care must be taken in the design of the mould to ensure the formation of strong welds.

In some processes (form-and-fill machines) a liquid product is injected into the bottle during the cooling stage as this may simplify production and increase the output (cooling) rate if the liquid is cold. For example, milk at 4°C has been packaged in this way.

Most thermoplastics can be extrusion blow moulded although the process is associated with polyethylene (PE) and with polyvinyl chloride. The products of blow moulding are nearly always bottles or containers which are used for food packaging as EBM proves to be an economic and fast route to such products. Such containers should either allow the product to be seen (through the container) or they should be attractively decorated so as to improve sales.

Blow moulding is not, however, the only way in which plastics bottles can be made: bottles have been made by *thermoforming*, followed by a welding operation. EBM is also used for the production of large containers and barrels, and in this area the technique competes against processes such as *rotational moulding* which uses as its starting material, PE in powder form.

extrusion coating
The application of a polymer to a substrate using an *extruder*. See *overcoating*.

extrusion compounding
Melt compounding performed with an *extruder*.

extrusion cylinder temperatures
The temperatures of the cylinder or barrel of an extruder. Should differentiate between *set temperature* and the actual or *measured temperature*.

extrusion degradation
See *degradative extrusion*.

extrusion laminating
The application of a polymer to a substrate using an *extruder*. See *overcoating* and *extrusion-lamination process*.

extrusion lamination
See *overcoating* and *extrusion-lamination process*.

extrusion line
The whole extrusion system, for example, materials handling equipment, the *extruder* and the *post extrusion equipment*.

extrusion materials
Most materials used in *extrusion* processes are *thermoplastics materials*. Polypropylene is already well established as an *extrusion material*: it is now one of the major materials used

for film or sheet: the other major plastics materials used for sheet are *polyethylene (PE)* and *polyvinyl chloride (PVC)*. These materials are all thermoplastics and it is the comparative ease with which thermoplastics can be converted into useful products at high speeds which has led to a large increase in their usage in recent years.

extrusion of sheet
The production of sheet by extrusion. See *sheet extrusion*.

extrusion of tube
The production of tube by extrusion. See *thermoplastics tube*.

extrusion plastimeter
A test apparatus used to measure *plasticity*. Such machines measure the time taken for a test sample to flow through a die under, for example, constant pressure. See *plastimeter*.

extrusion - products of
The products of extrusion include the following.
1. Feedstock for other processes - melt mixing using a compounding extruder gives good dispersion and a continuous output: this output can be chopped into regularly-sized pellets, or blanks, which often form the feed to moulding machines.
2. Thermoplastics film - this is often welded into bags or sacks and may be used to produce *fibrillated fibre*.
3. Thermoplastics sheet - used for cladding and as the feedstock for thermoforming machines.
4. Plastics pipe - used for conveying potable water, for drains, gas conveying etc.
5. Plastic tubing - used for garden hose, in automobiles, in laboratories etc.
6. Plastic insulated wire and cable - used for conveying electricity (for example, in the home and in automobiles) and for communications etc.
7. Profiles - for curtain tracks, window and door frames, home siding and gaskets.
8. Filaments - used for brushes, rope and twine.
9. Nets - used for gardening and for packaging.
10. Plastics coated paper and metal - used in the packaging industry.

extrusion testing
This term usually means a *flow test*: a high shear rate flow test. Used to check incoming raw material by using a small, single-screw extruder as a rheometer. Usually a rod die is specified and the machine is set at specified temperatures. The extrusion behaviour is measured over a range of screw speeds and graphs are plotted of output against screw speed, temperature against screw speed and die swell against screw speed. One of the big problems is maintaining the set temperatures as, any alteration in screw speed alters the extrusion cylinder temperatures and the melt temperature. As a range of screw speeds are employed, the shear rates employed can become quite high.

extrusion valve
An adjustable restriction in the melt stream which is used to control back pressure in an *extruder*.

extrusion wire-covering process
A coating process used to apply a polymer compound to wire or cable: most commonly used for *thermoplastics* materials. Wire, or a bunched cable core, is passed through a torpedo which is mounted in a *crosshead* and which directs the conductor centrally into the die orifice; here the conductor comes into contact with the softened plastics material and the coated wire is drawn away from the die under such conditions as to give a predetermined insulation thickness.

The insulated conductor then passes through one or more cooling baths to a haul-off or capstan maintained at constant speed and then on to a wind-up drum. Cooling must be gradual - if the insulation is cooled too rapidly the outer layers will freeze and contract, resulting in voids between the insulation and the conductor.

For most materials the position of the tip of the torpedo relative to the die parallel is very important because it controls the tightness of the covering of the wire. Some machines use a vacuum system to control the tightness of covering. An accumulator is inserted between the capstan and the wind-up arrangements so as to allow completed reels to be changed without interfering with the extrusion operation.

Cellular coatings on wire can be extruded directly from a compound containing a *blowing agent*. Extrusion conditions should be such that the blowing agent decomposes on the last few flights of the screw and the melt issues from the die in an unblown state and expands to its final dimensions as it leaves the die. Premature expansion leads to a product of rough external finish.

extrusion-lamination process
The application of materials such as polyethylene or polypropylene to paper, fabrics, metal foils and various other flexible substrates by the extrusion-lamination process is a rapid and economical method of producing laminates for packaging. The plastics material is extruded in the form of a *flat film* through a slot die and the molten film is cast into a nip-roll assembly formed by a water-cooled steel casting roll and a rubber-covered pressure roll. The substrate to be coated is fed continuously from an unwind position over the rubber pressure roll into the nip, which presses the molten film on to it so as to form a laminate. As the laminate runs round the steel roll the plastic layer is cooled sufficiently so that it can be taken off by the wind-up mechanism. The thickness of the applied coating is controlled by the extruder output rate and the speed of the substrate.

exudation
Also called bleed out or sweat out. The migration of liquids to a surface.

F

f
An abbreviation used for *fento*. See *prefixes - SI*.

F
This letter is used as an abbreviation for:
farad;
force - which causes failure, for example, *tensile strength at break*;
fibre;
filler;
flexible;
fluid/liquid state - in polymer abbreviations;
fluorinated;
fluorine;
furnace (black). See *carbon black - classification*.
fumarate; and,
phosphate - for example, *trioctyl phosphate (TOP)*.

F calender
A *calender* named after the letter F. See *inverted L calender*.

F_{50} value
The time for 50% of the specimens to fail a specified test, for example, a *falling weight impact strength test*. The time for 50% failure is often found by plotting the results on log-normal probability paper. For example, in *impact testing* this figure is the energy of the striker which caused 50 per cent failure of a large number of sheet specimens. See *Bell Telephone Laboratories test method*.

fabric
A cloth made from fibres: the cloth may be made from the fibres by weaving, knitting or felting. Woven fabrics are most important to the reinforced plastics industry where they are used to make *composites*. Such fabrics usually have uniform weight per unit area. See *yarn*.

fabric bonding agent
See *bonding agent*.

fabric coating machine
A machine used to produce a *coated fabric*. See *four-roll calender*.

fabric liner
See *liner*.

fabric - treatment of
The treatment of a *fabric* before it is used in a process such as *calendering*. When used for rubber coating cotton fabrics need no pre-treatment, other than drying, before use. Synthetic fabrics, such as rayon and nylon, are normally pre-treated in order to improve the adhesion between the rubber and the fibres. This is necessary as such materials may be relatively smooth, with few fibre ends which can become embedded in the rubber, and possess a low specific adhesion for the polymer, that is, intermolecular attraction between the fibre and the polymer is low. In the absence of a *bonding agent*, adhesion between the polymer and the synthetic fibre is therefore low.

fabric-braced radial
A *radial tyre* which has a multi-ply textile belt beneath the tread.

fabricating - thermoplastics
The manufacture of plastic products from moulded parts (rods, tubes, sheeting, extrusions), and other forms by appropriate operations such as punching, cutting, drilling, and tapping; also includes the fastening of plastic parts together or, to other parts by mechanical devices, adhesives, heat sealing, or other means.

face cutting
The cutting of extruded rods into pellets at the *die* face. See *underwater pelletizing*.

face seal
A type of *rotary shaft seal*.

factice
Factice is the name given to the polymerized products of the reaction between *sulphur* (or sulphur monochloride) and various oils, for example, *rape seed oil* or *linseed oil*. Used as a rubber substitute or additive so as to reduce nerve and improve processing. Two main classes, white and brown. Brown factice is obtained from the reaction between sulphur (10 to 30%) and oil at elevated temperatures: it has a relative density (RD or SG) of 1·0 to 1·1. White factice has a relative density (RD or SG) of 1·06 to 1·15 and is obtained from the reaction between sulphur monochloride (about 15%) and oil at low temperatures. Factice may be extended with *mineral oils*.

White factice is an ingredient of eraser mixings: in such mixes lime may be used to speed up the rate of cure. White factice may also be used in black mixings which have a tendency to scorch, for example, based on furnace blacks.

White factice may be referred to as sulphur chloride (white) factice or as, cold-type factice. Brown factice may be referred to as sulphur (dark) factice or as, dark factice or as, hot-type factice. See *sulphurized oil*.

Fahrenheit degree
Denoted as °F. Equal to 5/9°C. See *Fahrenheit scale*.

Fahrenheit scale
A temperature scale in which the melting point of ice is taken as 32°F and the boiling point of water is taken as 212°F.

falling weight impact strength
An abbreviation used for this term is FWIS. The energy which on average causes 50% of test specimens to fail in a *falling weight impact strength test*. A falling weight impact strength (FWIS) test is possibly the most useful type of impact strength determination as the test is usually performed on actual products or, on samples cut from them. End-use service performance can be easily simulated. There are two main types of test employed and these have been called conventional falling weight and instrumented falling weight.

falling weight impact strength test
CONVENTIONAL. An abbreviation used for this term is FWIS test - conventional. In this test, widely used for testing film, sheet and pipe, the sample is struck by a weight attached to a hemispherical striker or dart which falls from a constant height. By testing a large number of specimens the energy which on average causes 50% of them to fall is determined. This is called the falling weight impact strength (FWIS).

As film is most commonly tested by this technique, the method employed for testing such materials will be outlined. The BS method requires that 14 m (45 ft) of film be available and that the film width is 260 mm; so that at least 60 successive but not overlapping test portions are obtained. Specimen preparation is usually by extrusion and it is usual to leave the film in the form of a strip.

A vacuum is used to clamp the film taut and then a dart is dropped onto it from a specified height. If the film is punctured then this is recorded as a failure and the procedure is repeated for a range of dart masses. Sets of ten test pieces are used with at least six different dart masses. The percentage of failures for each dart mass is recorded.

The results are quoted as an M_{50} value where this figure is the mass, in grams, of the dart that would be expected to break 50% of a large number of specimens. For sheet the F_{50} value is quoted and this figure is the energy of the striker which caused 50% failure, i.e. mass of dart (kg) $\times$ acceleration due to gravity (m^{-2}) $\times$ height of fall (m). In this case the result is given in Joules (J).

A graph is drawn on probability paper of dart mass against the percentage of test pieces punctured for each set of ten test pieces. The best fit, straight line is drawn through the points and the M_{50} value read off. This is reported as the falling weight impact strength of the film.

INSTRUMENTED. Also known as instrumented falling weight. This test is not commonly encountered as the equipment is expensive. This test is used to test mouldings, film or sheet. A dart, or striker, is used which is so heavy that it will break the specimen easily and without a significant decrease in striker speed. A transducer is mounted behind the nose of the striker and the information that this generates may be used to calculate the energy to break the specimen and measure the maximum force. Only a small number of components or test specimens are required for this test.

The test is normally performed at room temperature and the signal that the transducer generates as it breaks the sample is amplified, recorded and/or displayed. To get more consistent results, and to stop the sample wrapping around the nose of the striker, a striker with a flat face is sometimes used. The speed of the striker is measured photo-electrically. It is relatively easy to perform this test with only a few specimens over a range of temperatures so as to obtain information about ductile to brittle transitions. To show the effect of processing on properties, the results obtained by testing the actual components may be compared against results obtained when compression moulded specimens of the same thickness, are used. Compression moulded samples are assumed to be free of orientation, that is, *isotropic*.

It has been found that a reasonably good approximation of the energy causing breakage (E^*) can be obtained by multiplying the impact speed (V_0) by the area under the force-time graph up to the point where failure began if the speed of impact (V_0) remains virtually unchanged during the impact. That is $E^* = V_0 \int F dt$.

If the dart speed does change during the time it is forcing its way through the specimen, then a correction must be applied to E^*. The corrected value of the energy to cause breakage (E) is then calculated from $E = E^* (1 - E^*/4E_0)$ where E_0 is the kinetic energy of the dart at the moment of impact.

falling weight impact test
See *falling weight impact strength test*.

false colour
See *chromatic aberration*.

family mould
A type of *injection mould* which is used to produce a number of different components at the same time. For example, family moulds are commonly used to mould self-assembly construction kits. Despite the widespread use of this type of mould, in general it is not good practice to have cavities of different sizes fed from the same feed system as problems may be experienced in production and during the end-use of the components. This is because the cavities usually fill at different rates thus resulting in over-packing for the cavities which fill first; this in turn can result in component warpage.

However, the production of associated mouldings from the same mould does confer certain advantages as all the components needed to make the final article are kept together. The handling, storage and moving around of components can add 35% to the cost of the moulding. Family moulds have been successfully run when the runners have been made deliberately uneven so that each of the differently sized cavities fills at the same time.

fan gate
A type of side *gate* used in *injection moulding*. A wide gate which is formed by opening or 'fanning' the *runner* out so that it blends into the component being moulded. The gate cross-section is still relatively small, or restricted, so that finishing is still comparatively easy. The fan gate helps to spread the material into the cavity and can thus help to minimise weld lines; it is used on components having flat thin sections, for example, boxes and lids. To minimise pressure drops the land length of the gate should be kept as small as possible providing that adequate strength of the mould metal in this area can be maintained.

far-field welding
See *remote ultrasonic welding*.

farad
An SI derived unit which has a special symbol, that is F. It is the SI derived unit of capacitance and is equivalent to one coulomb per volt. A capacitance of one farad requires a charge of one coulomb to raise its potential by one volt. See *Système International d'Unité*.

fat
Glycerides which are solid at room temperatures. See *oil*.

fatigue
The weakening of an elastomer, or of a rubber component, due to repeated distortion.

fatigue strength
The ability of a material to withstand vibration before failure occurs. A load, which varies periodically, is applied to the test specimen and the number of cycles withstood by the test specimen is determined. The *stress* is chosen to simulate working conditions, for example, sinusoidal or step-wise.

fatty acid
A monobasic organic acid with the formula RCOOH. Such an acid is *stearic acid*.

fault finding
A procedure used to determine the origin of a defect. Such a procedure should be logical and systematic if a fault, or faults, are to be eliminated. This seven-point procedure is based upon one suggested by John Bown in his book 'Injection Moulding of Plastic Components' published by McGraw Hill 1979.

Faults should be clearly described and all the possible causes should be examined. The effect of the fault should be taken into account and when the cause has been identified the necessary steps should be taken to eliminate it and to prevent its recurrence.

(1) Name the fault. Ensure that the terms used to name a fault are unambiguous and are known to all concerned. This may seem obvious but some faults are given more than one name - even in the same establishment a variety of names may be used. Decide which of the names will be used and ensure that the terms selected are known to all in the establishment.

(2) Describe the fault. Describe all common faults in the simplest possible terms without ascribing any possible cause.

(3) Find the cause of the fault. This may be a lengthy process since it requires consideration of material, machine, mould and process.

(a) Material. Check that the correct grade and type of material is being used, examine for contamination and make sure the material complies with the manufacturer's specification. Ensure that the material is dry and that the correct level of regrind is being used. Observe the effect of regrind addition on component properties and on processing characteristics (compared to the virgin material).

(b) Machine. Check that the machine appears to be functioning correctly and that the machine is set correctly. Check that the thermocouples are not loose in their mounting holes and are of the correct type. Ensure that the hopper block, the barrel and the hydraulic oil are at the correct temperature. Ensure that the screw cushion is set correctly and that it is being held in production. Ensure that the screw rotates for the correct time and that the rotational speed is as specified.

(c) Mould. Make sure that the mould is properly set, that all parts are at the correct temperature and that all the parts function smoothly and correctly. Ensure that ejection is consistent and that the mould open time is as specified.

(d) Process. Check that pressures, temperatures, rates, and times are set as recommended by the setting sheet and/or by the supplier of the material. Check the speed and regularity of the moulding cycle and that the times of opening and closing the safety gate are as set and are consistent.

(4) Determine the effect of the fault. If the fault renders the component unusable or unsaleable it must be rectified. If it is only of minor significance then, it may be unnecessary to try to eliminate it entirely. However, clearance to continue production with the minor fault must be obtained in writing.

(5) Determine where the responsibility for the fault lies. This may only be of academic interest, but if the fault recurs the operator - as well as the material, machine, mould and process - needs to be checked.

(6) Take immediate action to avoid the fault. If this is not done then the reasons for continuing production with the fault should be entered on the record sheet.

(7) Take steps to prevent a recurrence of the fault. Make full records of the conditions used when the fault was present and of the conditions used when the fault was eliminated. Note any repairs and alterations which were made to the mould or the machine and any variations in type, grade or quality of material. If rework (i.e. re-ground material) is used, note the proportion used and the quality. Keep labelled samples of what is being produced - both with and without the fault.

FB
An abbreviation used for load at break. See *flexural properties*.

fc
An abbreviation used for *foot-candle*.

Fe Con
A type of *thermocouple* which is based on the use of iron (Fe) and of constantan (Con) wires. Sometimes referred to as fe con.

FED
An abbreviation used for *formaldehyde-ethylene diamine condensate*.

Federal Motor Vehicle Safety Standard
An abbreviation used for this term is FMVSS. A rate of flame spread test used to assess the burning behaviour of materials used for vehicle interiors. For example, for *plasticized polyvinyl chloride* (PPVC) sheeting.

feed
Usually refers to what is fed to a piece of processing equipment, for example, to an *injection moulding machine*. It should be noted that the feed to machines involved in processing thermoplastics (such as injection moulding machines) is very often a mixture of virgin material, regranulated material (regrind) and colorant (often contained in a solid masterbatch). Such materials must be kept clean and dry; a regular ratio of the materials must also be used if machine operation and part consistency (such as surface appearance) is to be maintained. See *feed form* and *raw material feed*.

feed bank
The bank of material in the *feed nip*. A small roll of material a *pig*) is cut from the warming mill and used to renew the feed bank in rubber *calendering*.

feed form
Processing equipment, for example, an injection moulding machine, can be fed with plastics (resins) or compounds in various forms: the feed may be fine powder, regranulated material or, pellets. If the material is available in more than one feed form, then feeding problems will probably be encountered if a mixture of feed forms is used.

In terms of feeding efficiency, spherical granules (of approximately 3 mm/0.125 in diameter) are the most efficient and fine powder is usually the worst. Regranulated material can be almost as bad as fine powder; cube cut granules are better and *lace cut granules* are better still. Because of the feeding differences, machines must be fed with a consistent raw material mix. This particularly applies to *masterbatch mixes*.

The use of such a mixture can lead to significant cost savings as a compounding step may be eliminated. Most commonly, *masterbatches* are only used to impart colour to the finished product. The use of such a mixture can sometimes cause problems. The usual problem is one of colour shade differences between different machines; another is separation of the masterbatch from the plastics material in the hopper.

feed nip
The *nip* which is fed with material, for example, on a *calender*.

feed plate
The first plate in an underfeed mould design and which often contains the sprue and runner system.

feed port
The entry hole in the *barrel* into which the material is fed.

feed setting
A moulding machine setting. For example, the distance over which the screw rotates on an *in-line screw machine*.

feed system
That which feeds the mould with polymer melt in, for example, an injection mould or a transfer mould; consists of the *sprue*, the *runners* and the *gates*.

The sprue should be as short as possible, of full round cross-section and of conical shape with an included angle of approximately 5°. Runners should also be of full round cross-section and should incorporate cold slug wells. The runners should be connected to the sprue and gates via a rounded radius. Gates should also be connected to the moulding by a rounded radius; short lands and smooth cross-sectional variations are essential.

Runners, like sprues, are usually short in length and generous in diameter as this reduces pressure loss and permits the application of adequate follow up pressure. However, if they are made too large then excessively long cycles and large material losses result. If they are made too small then the large amount of pressure which is lost is transferred into heat. This heat will show up in the regions where the material is being sheared the most, that is, in the gate regions. Such local temperature rises can be very high and can lead to severe material degradation.

The feed system, for example the gate, should be sized so that the maximum *shear rate* is less than 20,000 s^{-1} in the case of a heat sensitive material such as *unplasticized polyvinyl chloride* (UPVC). Maximum shear rates are quoted for other materials, however in general, the gate should be sized so that the maximum *shear rate* is less than 50,000 s^{-1}. The gate diameter should not be, for example, less than half the component thickness - gates should be large to stop premature freezing and excessive shearing i.e. 0.8 to 0.9 t where t is the wall thickness of the component. Taper all vertical surfaces of the feed system in order to get satisfactory, balanced and easy ejection. Generous tapers and rounded corners, help to ease ejection. Construct the mould from a wear resistant steel (high chrome, high carbon tool steel) and design the moulds so that areas prone to wear can be replaced in service. Tungsten-carbide, shaped by electrodischarge machining (EDM), has been used for the gate sections.

In multi-component moulds, the individual cavities should be made using techniques which give good repeatability so that cavities are as identical as possible and replacement does not cause problems. For such multi-impression moulds, a balanced runner design is usually preferred; generous radii should be incorporated at each runner junction so as to prevent unnecessary pressure drops. When selecting the gating position, ensure that there is a continuous flow of material at all points of the flow path with no stagnation at sharp corners or, at chips and scratches in any metal surface. Avoid regions of compression and decompression and keep the flow path very streamlined.

feed throat
That which contains the *feed port* in the *barrel* for the material. To prevent the material from fusing in this region, the feed throat temperature is often regulated, for example, by water cooling.

feed zone
Term applied, for example, to injection moulding and to extrusion and refers to the barrel *zone* adjacent to the hopper.

This term is also used to describe one of the zones of a screw. That is, the first zone: the one which conveys the material from the *feed port* or *hopper*, through the feed zone of the barrel.

feed zone - cooling
Term applied, for example, to injection moulding and to extrusion and means that, the zone immediately under the hopper is cooled. Until recently, the feed zone on, for example, an injection moulding machine, was simply kept cool by circulating water through channels in the feed throat: if this was not done then, the hopper could become 'bridged' i.e. blocked. Now, for the processing of, for example, *engineering plastics*, the temperature of the feed zone is measured and controlled at a relatively high level so as to improve plasticization and/or, cycle times. May be done by only allowing water flow when the temperature has reached a pre-set value (a solenoid valve in the water supply is opened) or, a fluid at the preset temperature is circulated through the feed throat.

feed-path
Also called feedpath and may also be referred to as sheet path. The route taken by the polymer compound through the *calender*.

feedback
A signal or device which monitors the action of, for example, a hydraulic component. See *feedback signal*.

feedback signal
The output signal from a *feedback element*. See *closed loop system*.

feedpath
See *feed-path*.

feet of water
A unit of pressure with the abbreviation ft water and which is equal to 0·0295 atmospheres.

FEF
An abbreviation used for fast extrusion furnace (black) or, fast extruding furnace (black). N550. See *carbon black*.

female forming
Forming, or *thermoforming*, which is performed using a *female mould*. See *vacuum forming*.

female half
That part of a mould which has a negative, or hollow, shape and into which the *male half* fits. That part of the mould which forms the outside of a container.

female mould
A mould which has a hollow shape. For example, a mould into which sheet is drawn in *thermoforming*: air is removed from between the heated sheet and the mould via small holes or slots machined into the mould. This gives a *wall thickness distribution* which is different to that obtained from a *male mould*. In female forming it is only the sheet over the cavity which is available to the component: the sheet which contacts the mould first, as the sheet is drawn down, chills and cannot then be moved. This means that any corners, thin excessively as the corner is progressively formed. See *vacuum forming*, *drape forming* and *hand lay-up*.

fento
An SI prefix which means 10^{-15} and which has the abbreviation f. See *prefixes SI*.

FEP
An abbreviation used for *fluorinated ethylene-propylene copolymer*.

FEP fluorocarbon
See *fluorinated ethylene propylene copolymer*.

fermentation
A change in organic materials brought about by the action of living organisms. Usually means alcoholic fermentation but now fermentation is being used to produce polymers. See *polyhydroxybutyrate*.

ferni
A unit of length equivalent to 10^{-13} centimetres.

Ferranti viscometer
A *rotational viscometer*. See *concentric cylinder viscometer*.

ferric
Iron (Fe) in its +3 oxidation state. So, ferric oxide is sometimes referred to as iron III oxide.

ferric ferrocyanide
Also known as Berlin blue or as, Chinese blue or as, mineral blue or as, Paris blue or as, Prussian blue. May be represented as $Fe_4[Fe(CN_6)]_3$. A blue pigment.

ferric oxide
May be represented as Fe_2O_3. A brown to reddish material used as a pigment and sometimes referred to as iron III oxide. Occurs naturally as haematite. See *iron oxide*.

ferro
A prefix for iron (Fe) often seen in combination with the name of another metallic element so as to indicate a ferroalloy: such materials are ferroalloys or *alloy steels*. For example:

ferroaluminium - an *alloy* of *aluminium* and *iron* which contains up to approximately 80% aluminium;

ferrotungsten - an alloy of tungsten and iron.

ferrous
Iron (Fe) in its +2 oxidation state. So, ferrous oxide is sometimes referred to as iron II oxide. See *iron oxide*.

ferrous metal detector
A *metal detector* used to detect the presence of iron-based materials in the feed fed to a machine such as a *calender*.

FF
An abbreviation used for fine furnace (black). N440. See *carbon black*.

FFA
An abbreviation used for the USA-based organization, Fiberglass Fabrication Association.

FFKM
An abbreviation used for *perfluorinated elastomer*. See *fluororubber*.

FFS
An abbreviation used for *form-fill-seal*.

FGD
An abbreviation used for flue gas desulphurization.

Fi-Fo policy
An abbreviation used for *first in - first out policy*. Sometimes referred to as FI-FO or fi-fo.

Fiberglass Fabrication Association
An abbreviation used for this USA-based organization is FFA.

fibre
A material whose length is very much greater than its diameter or breadth: the diameter is of the order of 10 to 50 μm. Successful textile fibres are flexible and resilient so that they can, for example, be knitted or woven into fabrics. Fibres are classed as being natural or synthetic: both are often organic

polymers although metal fibres and inorganic polymer fibres are used. In, for example, the plastics industry, an inorganic polymer fibre (glass fibre) is most widely used.

The addition of fibres to improve the properties of materials has been known for many centuries. However, once the concentration of fibres reaches approximately 2% then it is difficult to add fibres to a matrix: above 2% the matrix is added to the fibres. Even so, it is difficult to incorporate more than 50% (by volume) and still retain ease of shaping during component manufacture. See *fibres - classification*, *fibres - natural* and *fibres - man-made*.

fibre bloom
The appearance of fibres through the surface of a *GRP laminate*. Minimized by using a thick *gel-coat*, for example, 0.5 mm thick.

fibre coupling
See *filler coupling*.

fibre degradation
This term usually means reduction of fibre length caused by *shear* during, for example, *injection moulding* of *fibre reinforced thermoplastics material*. To minimise fibre degradation avoid passing the material through a machine where it will be sheared by, for example, a screw (see *thermoplastic bulk moulding compound*). If injection moulding is used, then use a *reverse temperature profile* on the barrel, a *low work screw*, a *free-flow*, sliding check ring and an adequately sized feed system. Avoid using high back pressures, keep the screw rotational speed as low as is possible and use a large gate.

If a *dough moulding compound* is processed using a screw *injection moulding machine*, each turn of the screw cuts off one layer of compacted material. Fibre degradation occurs by filamentization of the fibre bundle, and by the breakdown of filaments into elements of lower aspect ratio. Fibre degradation decreases, with an increase in barrel temperature and with an increase in gate size or diameter.

Degradation in the feed system, particularly when passing through the gate, is very serious. To minimize fibre degradation, *injection-compression moulding* may be used. Injection moulds must be designed to minimize fibre damage and orientation, for example, use large gates and runners and avoid abrupt changes in direction. It must be accepted that there will be fibre degradation through the feed system and that the mouldings will be *anisotropic*.

fibre efficiency factor
See *efficiency of reinforcement*.

fibre from film
See *fibrillated film fibre*.

fibre - glass
See *glass fibre*.

fibre mat reinforced reaction injection moulding
See *structural reaction injection moulding*.

fibre optics
The use of glass fibres for the transmission of information using light.

fibre orientation - textiles
See *Godet wheel* and *viscose rayon*.

fibre positioning
Ideally the fibres of a composite should be capable of being positioned so that they can give the optimum level of reinforcement. An appropriate number of reinforcing fibres must be orientated to counteract the stresses which the product will encounter during its useful life. This is because, for example, a thermoset matrix is weak and is incapable of carrying significant loads alone. See *specific modulus*.

fibre reinforced plastic
An abbreviation used for this type of material is FRP. See *fibre reinforced thermoplastics material* and *glass reinforced plastic*.

fibre reinforced thermoplastics material
Obtained by the addition of a fibre, which is often glass, to a thermoplastics material. Such a reinforced thermoplastics material may be referred to as a fibre reinforced thermoplastic (FRTP) or as a *short fibre reinforced plastic (SFRP)* as the fibre length is small. As fillers are often used with engineering thermoplastics then RETP, standing for 'reinforced engineering thermoplastic' is also seen. Mixtures of fillers and fibres are often used with thermoplastics materials in order to obtain the desired balance of properties, for example, stiffness without excessive orientation during *injection moulding*.

As the fibre/filler content of such composites increases, surface gloss decreases but, usually, stiffness and *heat distortion temperature (HDT)* increases. Because of the high specific gravity of the fibres/fillers used (often between 2 and 5), and the difference in the refractive index of the filler and the polymer, filled thermoplastics are usually opaque and feel heavy. The stiffness of such materials can be quite remarkable and the surface is usually satin-smooth with a hard, mar resistant finish. Mould surfaces may therefore be textured, or vapour blasted, as gloss finishes often give no advantage. However, these materials still have a comparatively low modulus. Such a low value is obtained because the amount of fibre added is relatively small and because, the fibre length is very low, for example, 0.5 mm/0.02 in. See *specific modulus* and *filled engineering thermoplastics*.

fibre reinforcement
See *glass fibre reinforcement* and *reinforcement*.

fibre show
Strands or bundles of fibres (fibres) not covered by resin which are at, or above, the surface of a reinforced plastic.

fibre strengthening
If a material has a fibrous reinforcement incorporated then the composite is fibre-strengthened if the elastic deformation is reduced by the fibres carrying a large part of the load. See *specific modulus*.

fibres - classification
Fibres may be divided into natural fibres and into man-made fibres. These, in turn, may be sub-divided in various ways. Natural fibres may be divided into those of vegetable origin, those of animal origin and into those of mineral origin. Man-made fibres may be divided into those made from synthetic polymers and into those made from natural polymers. Fibres may also be divided in other ways: for example, into mineral fibres, inorganic fibres, high temperature fibres, ceramic fibres etc.

fibres - man-made
Man-made fibres may be divided into those made from *synthetic polymers* and into those made from natural polymers. Those made from synthetic polymers include *polyamide* (PA) fibres, *polyester* (for example, PET) fibres, *polyacrylonitrile* (PAN) fibres, *polyolefin* (PE and PP) fibres and *polyurethane* fibres. Those made from natural polymers include *cellulose fibres, cellulose ester fibres, protein fibres, alginate fibres, natural rubber fibres* and *silica* and *silicate fibres*.

fibres - natural
Natural fibres include *cotton, wool* and *silk* although there are many others. Natural fibres may be divided into those of vegetable origin, those of animal origin and into those of mineral origin. Fibres of vegetable origin include bast fibres (such as flax and jute) the leaf fibres (such as sisal and manila) and the

seed fibres (such as *cotton*). Those of animal origin include wool and those of mineral origin include *asbestos*.

fibres - other than glass
Glass fibres (GF) are favoured for composite materials as they are relatively cheap, inert and give reasonable properties. However, their is considerable commercial interest in using fibres other than glass in composite materials in order to extend the range of use of such composite materials. Fibres from materials such as boron (B), aluminium oxide (Al_2O_3) and carbon (graphite fibres) are used. Of these 'exotic' fibrous materials, *carbon fibre (CF)* is the one most commonly used - either on its own or, in blends with glass fibre GF) as *specific modulus* is improved.

fibres - synthetic
A fibre made from a synthetic polymer: this category includes *polyamide* (PA) fibres, *polyester* (for example, PET) fibres, *polyacrylonitrile* (PAN) fibres, *polyolefin* (PE and PP) fibres and *polyurethane* fibres. *Glass* may also be considered a synthetic, inorganic polymer.

fibril
A small diameter fibre. In crystallography, the term is taken as meaning the fibre-like aggregation of crystal units seen, for example, in *spherulites*.

fibrillated film fibre
Also referred to as oriented film fibre or as, film fibre. Based on a highly oriented strip, or tape, of extruded material and usually based on *polyethylene* or *polypropylene*. After production of *film tape*, fibre may be produced by, for example, twisting the highly oriented tape which then breaks down into a mass of fine fibrils.

fibrous filler
A *filler* based on a fibrous form of a material, for example, *glass fibre* or *carbon fibre*. Fibrous calcium silicate is a low cost, fibre-like *filler* which has been suggested as an alternative to *mica* and to *wollastonite*. For fibrous calcium sulphate, see *anhydrous calcium sulphate*.

field coagulum
The solid, rubbery material obtained when *natural rubber latex* coagulates spontaneously, for example, in the *tapping cup* by auto-coagulation. Such material may be converted to *crumb form* by various size reduction devices. See *cup lump* and *tree lace*.

field latex
Freshly tapped *natural rubber latex*. This is taken to a central factory, or bulking station, where it is sieved and blended so as to improve the uniformity. It may be preserved with, for example, ammonia: sodium sulphite and formaldehyde are also used. It is either concentrated, by removing part of the water, to give *latex concentrate* or, it is deliberately *coagulated* to give solid dry rubber.

fifteen degree calorie
See *calorie*.

Fikentscher K-value
See *K-value*.

Fikentscher number
See *K-value*.

filament
A long continuous fibre such as silk: a fibre of indefinite length. Most filaments are used as *yarn* although, if thick enough, monofilament is used for some purposes.

filament count
The number of filaments in a *yarn*.

filament tow
A collection of many (several thousand) *filaments*.

filament winding
Process used to produce tubes and pipes form plastics materials and a long-fibre reinforcement. Traditionally based on a thermosetting plastic, such as *unsaturated polyester resin (UP)*, and on glass fibre: now other resins are being used and other reinforcements, for example, *carbon fibre*.

When a thermosetting plastic is used, such as UP, the reinforcement is impregnated with catalyzed resin and then wound onto a mandrel in a predetermined pattern and under controlled tension. The covered mandrel is then rotated while being heated until the resin hardens. After cooling the composite tube may be withdrawn from the mandrel, using a hydraulic jack, or it may be left in place. If a thermoplastics material is used then a hot gas torch may be used to melt the resin: a pressure roll may be used to consolidate the assembly: this allows the production of concave shapes. See *filament winding - two axis winding*.

filament winding - five axis winding
Also known as five axis filament winding. A *filament winding process* which gives greater control over the way the reinforcement is laid down. In conventional *filament winding*, the two axis are the rotation of the mandrel and the carriage traversing motion (along the X axis). In addition, with five axis winding, can have control over the feed eye rotation and over the cross feed (horizontal and vertical).

filament winding - two axis winding
Conventional *filament winding*. The two axis are the rotation of the mandrel and the carriage traversing motion (along the X axis). In practice, this limits the winding angle of the fibre to within the range 30 to 75°.

filled engineering plastics
A filled engineering plastic is also known as a filled engineering thermoplastics material or as, reinforced thermoplastic (RTP) or as, reinforced engineering thermoplastics material (RETP or REP). When the thermoplastics material contains a fibrous filler (e.g. glass) then the term fibre reinforced thermoplastic (FRTP) is sometimes used. As most composite materials contain short glass fibres the phrase *short fibre reinforced plastic (SFRP)* is also encountered. If the composite material contains longer glass fibres, the phrase *long fibre reinforced plastic (LFRP)* is sometimes used.

Materials modification, such as with *fibres* or with *fillers*, is extensively adopted with thermosetting plastics and now, to an increasing extent, with engineering thermoplastics; this is done so as to obtain a desirable combination of properties. It is not done simply to save money as often a component, made from a filled compound, is the same price as one made from the unfilled plastics material. This is because of the high density of most fillers and because of the high compounding costs. Many of the fillers used are fibrous fillers as the use of such materials improves properties such as *modulus*: a commonly used fibrous filler is glass (GF). By the use of such fillers it is possible to lift, or move, a plastics material from one category to another. In the case of *polypropylene (PP)*, a commodity plastics material, it can be changed into an engineering plastics material by materials enhancement.

In general the viscosity of filled thermoplastics is higher than that of the base polymer and the shrinkage is much reduced but there may be a warping problem caused by fibre orientation and *anisotropy*. Glass spheres are used to achieve a more uniform reduction in shrinkage (both across and along the flow direction).

The stress-cracking tendency of some materials is reduced by GF addition while the creep resistance of filled materials is better than that of the parent polymer. Low temperature impact resistance is sometimes better than that of the parent plastics material. Heat distortion temperature and stiffness is often markedly better: elongation at break and gloss are usually reduced. For a given semi-crystalline, thermoplastics material, the heat resistance may be improved by increasing the amount of crystallinity and/or by adding fillers. Such fillers compensate for the softening effect of the amorphous phase when the plastics material is above its glass transition temperature (T_g). The effect of GF is most marked for those plastics which have a crystalline content of about 50%; for such materials, adding glass fibre (GF) can improve the heat resistance by approximately 100°C. Adding GF to an amorphous, thermoplastics material usually only has a small effect on the heat resistance; it may only be improved by approximately 10°C. This is because the heat resistance of such materials is controlled by their glass transition temperature (T_g).

Filled materials are not resistant to those substances which attack the base polymer. In FRTP the interface between resin and glass can be attacked by alkalis. Use correct moulding conditions (such as high temperature) so that there is no glass at the surface which will draw alkali into the moulding.

As mineral fillers have lower specific heats than polymers, filler addition reduces heat removal requirements by lowering the heat content. Filler addition also usually gives a more rigid material which may be ejected at higher temperatures; filled compounds may also set up more quickly as the filler can act as a nucleating agent in crystalline polymers. Significantly faster cycling should therefore result for FRTP compared to the base polymer.

The largest areas in which such materials have been used are, for example, electrical-electronic, automotive and household/consumer uses. Nylons account for half of this total: approximately one third of all the *nylon* (PA) used is reinforced with *glass*. The addition of *glass*, stiffens the material but may not impart toughness. Impact strength at low temperatures is usually improved by fibre addition but this benefit may not be retained at higher temperatures. Even with added filler the modulus of plastics materials is lower than that of metal but, if high stiffness is a requirement, it can usually be gained by making the composite into a thicker component based on a cellular material. Impact strength can also be improved by the addition of rubbers: fire resistance can be improved by the use of additives such as *aluminum trihydrate*.

The use of fillers in conjunction, with engineering thermoplastics, is therefore helping to overcome some of the basic disadvantages of plastics and is allowing such composite materials to penetrate the huge metal replacement market. For example, as die casting replacements where intricate shapes may be produced economically from corrosion-resistant materials whose properties can be tailored to suit an application.

filled reaction injection moulding

An abbreviation used for this term is filled RIM. A *reaction injection moulding*, based on for example, a *polyurethane*, which has been filled with, for example, an inorganic filler (such as *flake glass*) before being moulded: the process of producing such a moulding.

The amount of filler in such systems is often comparatively low as the reactant viscosity is too high for pumping if high filler loadings are used. For example, for *flake glass* approximately 20% by weight in the polyol is the maximum that can be used. This loading gives an optimum between modulus improvement, coefficient of thermal expansion improvement and impact strength reduction. In filled formulations, *chopped glass*, *hammer milled glass*, *flake glass*, *mica* and *wollastonite* have all been used to improve properties such as tensile strength and modulus. The addition of such fillers reduces the exotherm during *reaction injection moulding* (RIM). In such formulations the dried fillers are often only added to the polyol as the fillers may catalyze unwanted isocyanate reactions.

Because of the high viscosity of filled reactants, and the need for continuous recirculation, moving cavity or concentric screw pumps are used on the *reaction injection moulding* (RIM) machines.

filler

An additive. A material added to a polymer in order to reduce compound cost and/or, to improve processing behaviour and/or, to modify product properties. In recent years, improving the fire resistance of plastics has become extremely important but, ideally, such improvements must not detract from other properties. When large amounts of filler are used, then this reduction of properties can be difficult to overcome.

Fillers are usually solid materials either of fine particle size (*particulate fillers* such as china clay) or they are fibrous fillers (*fibres*). Fibrous fillers are often based on *glass fibre*. Fillers may also be classified as organic or as, inorganic.

Natural organic fillers, such as *woodflour*, were amongst the first fillers used by the plastics industry. It was used in materials such as *phenol-formaldehyde (PF)* in order to control blistering, to reduce shrinkage and to improve impact strength: they also reduced compound costs. Although the structure of such a material is an advantage, it is an absorbent fibrous material, the fact that it is inflammable, and also absorbs water, can be a serious disadvantage in service. *Inorganic fillers* are therefore preferred in most applications: the use of such fillers commonly improves heat resistance and stiffness but does little, if anything, for impact strength. For most applications a pure, white filler of fine particle size is preferred: this saves on colorant costs. See *non-reinforcing filler* and *reinforcing filler*.

filler coupling

The bonding, or coupling of a filler to a polymer. The level of use of most *inorganic fillers*, in a *thermoplastics material*, is restricted by the severe reduction in properties, such as impact strength and gloss, which results if more than a few percent of filler are added. (More filler may be added to a *rubber* because in some ways the rubber acts as a liquid). Such a fall-off in properties may be lessened if the filler is coupled to the polymer. Such coupling, to an inert surface, is often achieved by the use of a *coupling agent*.

Silanes are the most commonly used systems for inorganic fillers based on *silicates*. The use of a silane coupling agent with a silicate filler can give an immediate improvement in properties such as stiffness and also helps the component to maintain this improvement when it is subjected to a wet or humid atmosphere. Unfortunately there is no one *coupling agent* suitable for all polymers and all fillers.

Some of the newer types of coupling agent can be added directly to the plastic, thus saving filler pretreatment: neo-alkoxy titanates and zirconates are available in pellet form and can be added like a colour masterbatch. A major problem with coupling agents is realizing the expected property improvements. Much is promised but many people find that the improvement does not happen in practice. For this reason other, more tolerant, systems are sometimes used: such systems include maleated polypropylene and epoxidized polybutylene.

filler dispersion

It is important to ensure that a particulate filler is uniformly and finely dispersed in the compound. Usually assessed by the effect on compound properties. May be assessed by examination of a prepared, or of a fracture, surface under a microscope. See *Roninger method*.

filler - grades of

It is usual to offer fillers in a range of grades. Such grades may differ in particle size and size distribution, particle shape and porosity, the chemical nature of the surface and the presence of impurities. Altering each of these has a tremendous effect on composite properties. Altering particle size and size distribution also gives a way of reaching high, or low, filler loadings; altering particle shape and porosity offers ways of achieving reinforcement (as in the case of *talc* and *polypropylene*). The chemical nature of the surface is of tremendous importance as a filler cannot be used to best advantage in a polymer unless there is good adhesion between them. In particular the filler particle/polymer interface will not be stress bearing and will therefore provide a point of mechanical weakness: *filler coupling* is therefore necessary.

filler polymer
A type of *polyvinyl chloride* (PVC) polymer used to make a *rigisol*. A PVC polymer with a low *plasticizer* absorption and a high packing density. Such a polymer is usually of relatively large particle size and when mixed with the smaller paste polymer increases particle packing so as to make more plasticizer available for particle lubrication. This means that the plasticizer concentration, for a given viscosity, can be reduced.

filler rod
A rod of *thermoplastics material* used in *hot gas welding*. The rod is thermally softened with a stream of heated gas from a welding torch and pressed into a groove located between the thermoplastics sheets which are to be joined so as to provide a joint-filling material. The rod is of the same type of plastics material as the sheets.

filler strip
See *flipper*.

filler surface
The amount of filler surface in contact with the polymer. The amount of filler surface in contact with the polymer is very important as if there is a large surface area, and a small particle size, reinforcement is probable See *BET surface* and *EM surface*.

filler - effect on appearance
It is commonly found that the incorporation of a *filler* nearly always worsens, or roughens, the surface finish. The larger the particle size, the greater is the effect: this is why small particle size fillers are preferred by the polymer industry. Surface treating such fillers before use, for example, with a *lubricant*, improves dispersion, hinders filler agglomeration and gives mouldings of a better surface finish.

Fibre filled thermoplastics materials nearly always have a relatively poor surface finish: even though abrasion during processing reduces the filler length to below 1 mm, the injection mouldings produced have a matt finish. In such a case the use of an expensive highly-polished mould would be a waste and it would be better to use a fine textured mould finish, for example, a leather-grain finish. A wide range of such finishes are now available which differ in roughness and pattern: their use is popular in the automotive industry where, for safety reasons, matt finishes are preferred as they reduce the amount of un-wanted reflected light.

filling polyurethane foam
A *semi-flexible polyurethane foam* behind a suitable facing, for example, behind a *thermoforming*.

filling pressure
The pressure required during mould filling, for example, the pressure required to maintain a mould filling speed in *injection moulding*.

film
A term applied to thin sheet and usually taken to apply to thicknesses of 0·010 in/0·25 mm or below.

film fibre
See *fibrillated film fibre*.

film gate
See *flash gate*.

film impact strength
See *falling weight impact strength (test) - conventional*.

film production processes - plastics
In terms of tonnage, two processes dominate thermoplastics *film* and *sheet* production: these two major processes are *calendering* and *extrusion*. Broadly speaking extrusion is used to produce the thinner sheet, based usually on *polyethylene* and *polypropylene*, and calendering is used to produce the thicker sheet which is often based on *polyvinyl chloride*.

film tape
Narrow tape produced from *film*. Such products are usually based on *polyethylene* or *polypropylene* (PP) and the products may be used as string, ropes or woven into products such as sacks or carpet backing.

The polymer may be extruded on a single screw machine into a flat sheet using the *chill roll process*. The flat sheet may then be slit into tapes by either stationary or rotating knives. Godet rolls are then used to stretch the tape whilst it is heated in a hot air oven: in the production of film tape the width of the tape may decrease to $\frac{1}{3}$ of its original slit width as a result of this stretching process. Annealing may then be performed on the tape before it is wound into packages, or cheeses, suitable for weaving. See *fibrillated film fibre*.

filter
A component of a hydraulic circuit which removes solid particles from the hydraulic fluid. An essential but often overlooked component of a hydraulic circuit.

filter pack
See *screen pack*.

fine ceramic fibre
A *fibre* of small diameter which is produced by, for example, the pyrolysis of a polycarbosilane precursor to give a *silicon carbide fibre* whose diameter is approximately 15 μm.

fines
A plastics material (a resin) which is in a fine powder form: fines are, for example, produced when a plastics material is chopped into granules and in this case are undesirable.

finger seal
A type of *rotary shaft seal*.

finish
A treatment which is applied to *glass fibres* in order to improve adhesion between the resin and the glass.

finished component testing
See *testing of finished components*.

finishing
Treatments applied to the output of a machine in order to improve the value of a product. Finishing may included machining, deflashing, and decoration.

finite element analysis
A mathematical procedure used to find flow effects as well as stress effects by dividing the analyzed space into discrete 'finite' elements which are related to each other by constitutional equations. The method uses computer techniques to analyze stresses and flows in the absence of exact solutions.

fire
Burning or combustion: a chemical reaction which is accompanied by heat, light and the evolution of volatiles and *smoke*. A fire can be broken down into elements each, or all of which, should be capable of being assessed by a fire test. See *fire testing*.

fire behaviour tests
See *fire testing*.

fire retardant
See *flame retardant*.

fire testing
A fire can be broken down into elements each, or all of which, should be capable of being assessed by a fire test. For example:
  i)   the temperature or heat load required to promote initial ignition;
  ii) the temperature or heat load required to sustain that initial ignition;
  iii) the speed at which combustion penetrates the component;
  iv) the contribution made to the fire by the combustion;
  v)   the quantity of combustion products released as the fire progresses;
  vi) the speed of release of those combustion products as the fire progresses;
  vii) the density of those combustion products;
  viii) the opacity of those combustion products;
  ix) the irritancy of those combustion products; and,
  x)   the toxicity of those combustion products.

Ideally the fire test should also compare the fire performance of materials and how they would behave in a true, or real, fire where, for example, there is interaction of materials and differences of ambient temperature and humidity. Because of the difficulties involved in assessing fire performance, a number of test procedures are used to assess the listed elements. See, for example, *limiting oxygen index* and *thermal decomposition product testing*. It is now thought that fire is, in general, less life threatening than the accompanying *smoke*. See *burning behaviour tests*, *smoke* and *toxic gas generation*.

Firestone extrusion plastimeter
A *plastimeter*. A test apparatus used to measure the extrusion speed of rubber compounds and to evaluate rod, or thread, extrusion properties. See *plasticity*.

first stage
See *prepolymer process*.

first stage pressure
That pressure needed, during injection moulding, in order to maintain the desired mould filling speed. Once the mould is full then this high pressure may not be necessary, or even desirable. In many cases, a high, first stage pressure may therefore be followed by a lower, second stage pressure. When moulding some semi-crystalline, thermoplastics materials, for example nylon and acetal then the use of second stage pressure may not be desirable as abrupt changes in pressure can cause undesirable changes in crystalline structure. See *cavity pressure control*.

first surface metallization
Part of the production process for *vacuum metallization*, of plastics in which the upper, or outer surface, of injection mouldings, after treatment with a *base coat*, are coated with metal in a vacuum chamber.

first-in, first-out policy
An abbreviation used for this term is Fi-Fo policy. A *stock control policy*.

Fischer melting method
See *Karl Fischer method*.

Fischer's yellow
See *potassium cobalt nitrite*.

fish eyes
Also known as cats' eyes or nibs. Defects in an extrudate (usually *film*) which look like the eyes of a fish and which are caused by small particles of unmelted resin; small globular masses which have not blended completely into the surrounding material. For example, such defects may be due to the presence of high molecular weight material or gel. Also associated with the *calendering* of *polyvinyl chloride* (PVC) where they are seen as hard un-dispersed polymer particles in the final product.

fit and finish testing
Most tests performed on mouldings, to judge fitness for use, are of the 'fit and finish' type; that is, if the appearance of the moulding is acceptable and, if it fits into the finished assembly, then it is satisfactory. See *injection moulding* and *testing*.

fitness-for-purpose testing
Tests performed in an effort to decide if a product is fit for an intended purpose. The need to carry out adequate fitness-for-purpose and quality control testing has been increased greatly in recent years by legislation on liability for product performance. New legislation is making the risk of punitive damage claims for product liability a very real possibility which manufacturers and suppliers must face and guard against; particularly by ensuring that their materials and products are not in any way defective. Hence the need for adequate testing.

fitting line
A moulded line on a *tyre* which serve as a fitting guide.

five axis filament winding
See filament winding - five axis winding

five roll calender
Also known as a five bowl calender. A *calender* with five main rolls. That is, with four nips. The rolls may be arranged in different ways or configurations. Can have, for example, an *L calender* or an *inverted L calender*. Such a machine is rather unusual as most calenders are three roll calenders or four roll calenders. Such a machine may be used for the production of very thin sheeting to tight tolerances or, used where the extra nip is utilized for laminating.

fixed head die plate
The *die plate* which is attached to the fixed platen.

fixed mould half
One of the halves of an injection mould and which is attached to the *fixed platen*. This mould half accepts the thrust from the locking system via the *moving mould half* and which is mounted on the *moving platen*.

fixed oil
An *oil* obtained from an oilseed: an oil of animal origin. A fixed oil is different from an *essential oil*. A fixed oil is a triglyceride: a *fatty acid ester* of glycerol. The properties of such oils are largely determined by the properties of the *fatty acid* component as this comprises approximately 95% of the triglyceride molecule. If there is only one fatty acid residue (only one fatty acid used) then the product is a simple triglyceride: if there is more than one fatty acid residue (more than one fatty acid used) then the product is a mixed triglyceride.

fixed plate
Part of a mould: a steel plate which is bolted permanently to a press. The plate may be cored for heating or cooling.

fixed platen
Also known as the stationary platen. One of the platens of an *injection moulding machine*: one of the platens located in the centre of the machine and on which the *fixed mould half* is mounted. This *platen* moves very little and accepts the thrust from the locking system via the *moving platen*.

fixed point
An accurately reproducible point on a temperature scale, for example, the ice point of water.

fixed safety guard
A machine guard which is fixed, or bolted, over a hazardous area, for example, over the *nip* of two rolls. See *movable interlocked safety guard*, *safety bar*, *safety stop* and *plugged braking*. Training and a responsible attitude towards safe working are very important safety considerations as undue reliance should not be put upon guards.

fixed-gap spread coating
A *spread coating method* in which there is a fixed gap between the moving substrate and the spreading device (for example, a *doctor blade*).

F_k
An abbreviation used for the friction when steady sliding is obtained. See *kinetic coefficient of friction*.

F_k/R
An abbreviation used for the *kinetic coefficient of friction*.

FKM
An ASTM designation/abbreviation used for a type of *fluororubber*.

fl
An abbreviation used for *foot-lambert*.

fl oz
An abbreviation used for fluid ounce.

flake glass
Also known as glass flake. Relatively small pieces of *glass* which are produced by hammering glass bubbles and passing the pieces through a metal screen, for example, the sieve opening size may be 0.4 mm (1/64"). For the *reaction injection moulding* (RIM) process, the flake size may nominally be a maximum of 0.4 mm (1/64") but is often considerably smaller and the size distribution is very non-uniform. Not so regular in shape as *mica* particles but of lower density and can give higher, puncture impact resistance at lower temperatures. Such a platy filler is used where warpage is a problem in *reaction injection moulding* (RIM): approximately 20% by weight in the polyol is the maximum that can be used.

flake lead
See *basic lead carbonate*.

flake pigment
A class or type of *inorganic pigment* which may be used to produce a range of metallic, pearlescent and iridescent colours. Pearlescent and iridescent colours may be obtained with flakes of *basic lead carbonate* or of, bismuth oxychloride or of, titanium dioxide coated mica. Such plate-like pigments have a high refractive index. See *metallic colours*.

flake white
See *basic lead carbonate*.

flame resistance
Resistance of polymer compounds to the effects of *fire*. See *flame retardant*.

flame retardant
May mean that a system resists burning but can also refer to a type of additive that is, to a material capable of increasing *fire resistance*. An additive: a material which reduces the tendency to burn of a polymer compound. Such a material usually contains one, or more, of the elements antimony, boron, halogen or phosphorus. Such a material may also act as a *smoke suppressant*.

Most polymers burn as they are organic materials; that is, they are carbon-based. It is natural for a carbon-based material to combine with atmospheric oxygen so as to form carbon monoxide and carbon dioxide. Now, halogens, such as chlorine and bromine, form compounds which are naturally flame retardant and by the use of such elements, and their compounds, polymers can be made flame retardant.

In the case of *polyvinyl chloride* (PVC), the halogen is 'built-in' into the plastics material: with other plastics the halogen must be added in the form of a halogen-containing compound such as tribromotoluene: the efficiency of such compounds is often improved by the use of *antimony trioxide* or *antimony pentoxide*. Unfortunately the use of halogens can introduce smoke and fume problems and so, other flame retardants may be preferred: examples of others include *aluminium trihydrate*, *aluminium oxide* and *phosphates*. Phosphorus has been used with, for example, PA 66 so as improve flame retardancy. Aluminum trihydrate (ATH) functions as a flame retardant and as a *smoke suppressant*.

To get useful flame retardant properties, the addition of large amounts of ATH is, however, required and such large additions can change the properties of the basic plastics material. For example, the flow properties are changed - the compound becomes stiffer flowing as more flame retardant filler is added. As with other fillers, the surface finish also worsens and processing problems can be introduced. For example, because of ATH decomposition, or degradation, at processing temperatures (in the region of 200°C) processing may be very difficult at temperatures approaching 200°C.

Other commercial system may be based on, for example, zinc stannate or, zinc hydroxystannate or, magnesium hydroxide or, magnesium carbonate or, magnesium-zinc complexes. Combination systems are also available, for example, with a *phosphate ester plasticizer*.

flame retardant - action mechanisms
Flame retardants appear to function by one, or a combination, of the following mechanisms (i) interference with the flame propagation mechanism, (ii) dilution of the air supply by the production of incombustible gases, (iii) endothermic absorption of heat and (iv) the formation of an impervious fire-resistant coating or char.

flame retardant plasticizer
A plasticizer which improves the fire resistance of a plasticized compound. Tricresyl phosphate and cresol diphenyl phosphate are both flame retardant plasticizers for *polyvinyl chloride*.

flame retarder
See flame retardant.

flaming
A technique used for the modification of a surface (a polyolefin surface) so as to make that surface more receptive to inks. The surface is briefly wiped with a gas flame so as to cause surface oxidation which introduces polar groups. The introduction of polar groups into the *polyolefin* structure, improves the wetting by the ink and the bond between the ink and the plastics surface.

flaming mode
A way of smoke testing in which the sample is ignited. See *National Bureau of Standards smoke chamber*.

flaming test
See *National Bureau of Standards smoke chamber*.

flammability
The term flammability is now used in place of inflammability. A material which is flammable is easily set on fire and is combustible.

The flammability of plastics in rod form may be determined by, for example, ISO R 1210-1970. A horizontal bar or rod of length >80 mm, width 10 to 15 mm and thickness 3 to

5 mm is tested as delivered. The end of a horizontal rod is exposed to a specified flame for 60 s. After the flame has been withdrawn the specimen is classed as category 1 if it does not burn; it is classed as category 2 if it partly burns for less than 15 s and as category 3 if it burns for more than 15 s or is completely burnt. See *horizontal combustion test*.

flammability tests
A test used to assess *flammability*. The tests used may be divided into *ignition tests* and into *burning tests*. See *burning behaviour test* and *fire testing*.

flap
A ring of rubber used to prevent the base of the *inner liner* chafing in service.

flash
Excess, unwanted material around the edge of an extrusion or, a moulding.

flash back
See *back-rinding*.

flash gate
A type of side *gate* used in *injection moulding*. This type of gate is also known as a film gate and it is usually a slit which can extend along the complete side of the component. Initially, the gate thickness would be uniform but if it was found that the mould was not filling with the same rate at all points then metal would be removed from the gate at the slow-filling areas. For example, the gate land length could be reduced. Such gates work best with a full round runner system and with the gate carried in each mould half. By feeding the material in this way, that is, along the *parting line* of the mould, distortion and non-uniform surface shrinkage are minimised. It will be found that more uniform filling will be obtained if the runner extends beyond the gate, e.g. by about 6 mm (0.25 in). These gates have been used for large flat components, as their use minimises distortion. Also used for fibre-filled materials and for metallic-coloured materials - as this type of gate helps to eliminate unsightly flow or weld lines.

flash groove
Also called a spew groove. A groove used to allow *flash* to escape from a cavity.

flash ignition temperature
A term used in the *ignition test* (ASTM). The lowest initial temperature of air passing around the test specimen at which a sufficient amount of combustible gas is evolved to be ignited by a pilot flame.

flash land
Also called a cut-off or a shut-off. The *land* of a mould across which excess materials escape. Such a land is commonly 6 mm/0.25 in wide for a *flash mould*.

flash line
Also called a spew line. A raised line on a moulded surface: the line which shows where the *flash* escaped from the cavity.

flash mould
A type of mould which is designed to allow for the easy escape of excess material; the excess material (approximately 5%) escapes across the *flash land*. Such a mould is widely used in *compression moulding* of *thermosetting plastics* as it is relatively cheap and can give products of acceptable quality, for example, for relatively simple mouldings of shallow depth.

flash pocket
A part of the mould, for example, outside the pinch area in a blow mould, which is designed to accept excess material: it is usually 1.5 × the parison thickness for each mould half.

flash-back
See *backrinding*.

flash-over voltage
An abbreviation used for this term is V_f. See *dielectric strength*.

flat bark crepe
An inferior grade of *natural rubber*.

flat curing
A property of a rubber compound obtained from a curing curve. The value of a particular property, for example, *tensile strength*, is plotted against time of vulcanization and if the curve shows a levelling off (a plateau effect) the compound is said to be flat curing.

flat film
Film produced by the *flat film process*.

flat film process
Sometimes referred to as the cast film process. In the extrusion of flat film, the melt is extruded from a straight slot die and cooled either by quenching in water or by casting onto highly polished water-cooled chill rolls, both systems involving rapid cooling of the film. The cooled film is wound up on take-off rolls.

In the casting process high temperatures may be used because of the fast cooling possible and therefore much higher output rates can be achieved than in the *tubular film process*. The slot die is usually centre-fed and is fitted with accurately controlled heaters along its length. When a water-bath is used for cooling, carry-over of water can be a problem, and hence the chill-roll process, which involves no direct water contact with the film, is usually preferred. The rolls must be highly polished because the film surface is an exact reproduction of the roll surface.

flat temperature profile
A temperature profile which has the same value for each *zone*.

flat-band method
A *tyre building procedure* employed on a *tyre-building machine*.

flatting additive
See *matting agent*.

flatting agent
See *matting agent*.

flax
A natural *fibre*: a bast fibre which is obtained from a plant and which is mainly *cellulose*.

flaxseed oil
See *linseed oil*.

flecking
Also called fleck marking. Associated with the *calendering* of *polyvinyl chloride* (PVC). Product marking caused by, for example, excessive *external lubricant* and/or *under-gelation of the PVC*.

flexibilizer
A term sometimes used in the rubber industry for *plasticizer*.

flexible polyester polyurethane foam
This type of *flexible polyurethane foam* is formed by the reaction of an isocyanate (for example, *tolylene diisocyanate*) with a polyester polyol to give a polyester foam.

flexible polyether polyurethane foam
This type of *flexible polyurethane foam* is formed by the reaction of an isocyanate (for example, *tolylene diisocyanate*) with a polyether polyol to give a polyether foam.

flexible polyurethane foam
Also known as flexible PU foam and as soft polyurethane foam. A cellular material, or expanded material, based on

a *polyurethane* and which is soft, flexible and resilient: the material usually has an open cell structure.

The material is formed by the reaction of an isocyanate (for example, *tolylene diisocyanate*) with a linear, high molecular weight:
- polyester polyol to give a polyester foam;
- polyether polyol to give a polyether foam; or with a,
- natural polyol, for example, *castor oil*.

The foam structure is generated by the reaction of water with the isocyanate (produces carbon dioxide and urea cross-links) and/or, by the inclusion of a *blowing agent*, for example, a *chlorofluorocarbon*. The density of the final foam depends upon the amount of gas evolved and it may be reduced, for example, by increasing the amount of water and isocyanate used.

Complex catalyst mixtures may be needed to obtain the desired balance of properties at a specified rate: must obtain the correct balance of chain extension, chain branching and gas generation. A polyether foam could use stannous octoate, dimethylethanolamine and 1,4-diazabicyclo-2,2,2-octane: the stannous octoate may be replaced by dibutyl tin dilaurate if there is a danger of hydrolysis in water-containing blends. The catalysts used for polyester foams are much simpler as polyesters contain primary hydroxyl groups and tin compounds are not required - use, for example, dimethylcyclohexylamine.

Flexible polyurethane foam production may be obtained by the *one-shot process*, the *prepolymer process* and the *quasi-prepolymer process*. The polyether foams are used for upholstery applications as the foams are resilient, elastic and show little compression set. Polyester foams have better solvent resistance, tensile strength and load bearing properties: such foams are used in textile lamination and footwear applications. The most common method of producing flexible PU foam is by a *slab-stock process*. Flexible polyurethane foams have good ageing and oxidation resistance: they are less flammable than rubber foams.

flexible PU foam
See *flexible polyurethane foam*.

flexible slab-stock foam
Flexible polyurethane (PU) foam produced by a *slab-stock process*.

flexographic printing
A *printing technique* in which the ink is transferred to the substrate, for example, a film, from a surface which carries the required design in relief. A high speed printing process (up to 5 m/s) in which the ink is transferred to the film by means of a *stereo* which has a raised legend. The stereo is mounted on a plate cylinder with adhesive and ink is transferred to the stereo from an *anilox roller*: this is fed from an ink bath or fountain via a rubber inking roller. Flexographic inks are based on solutions of pigmented resins (for example, polyamides or cellulose nitrate) in alcoholic solvents or, on acrylics in alkali and water.

flexogravure printing
See *printing techniques*.

flexomer
A low modulus (low stiffness) plastics material: can be as soft and as flexible as *plasticized polyvinyl chloride* (PPVC), without plasticizer addition. Some *copolymers* and *thermoplastic elastomers* can be considered as flexomers.

flexural modulus
The modulus determined from the load and the test piece dimensions in a flexural test. See *flexural properties*.

flexural properties
Flexural properties are usually measured in order to obtain a measure of stiffness or rigidity. Often measured by applying a stress at the centre of a rectangular bar which is supported at two other points: that is, three-point loading is applied in a bending test. During the test the force applied and the deflection which results is measured and the test may be easily performed on a *universal testing machine*.

The specimens used for flexural or bending tests are simple rectangular bars the length of which must be at least 20 times the thickness. The width and depth can vary but the preferred dimensions are width 10 mm and depth 4 mm. The result of bending the specimens in a *three point bending jig* (on which the span or distance between the two outer supports is set 15 to 17 times the thickness) is a load/deflection curve or graph. From this curve *flexural strength* and *modulus* are determined.

If the specimen is brittle and breaks at very low strain (less than 0.05) then, the breaking or flexural strength can be calculated from $\sigma_B = 3 F_B L/2 b h^2$. Where σ_B is the flexural strength, F_B = load at break, L = span width, b and h are width and thickness of the specimen respectively. If the sample does not break but simply bends then the stress measured when the sample deflection reaches 1.5 times the sample thickness is used. The breaking strain is often given the symbol γ and equals $6D h/L$, where D is the deflection at midspan. The elastic modulus in flexure, symbol E, can be calculated from $E = L/4 b h^3$ times F/Y. Where F/Y is the slope of the initial linear load-deflection curve. As for tensile testing, five specimens should be tested and the arithmetic mean and standard deviation of the five individual results reported. Where there is *anisotropy*, five specimens in two directions at right angles should be tested.

For thermoplastics, specimen preparation is most commonly performed by injection moulding where an end-gated bar is used. For thermosets, such as phenol-formaldehyde, compression moulding is normally employed. For unsaturated polyesters and epoxides, hand lay-up or spray lay-up techniques are used to produce larger sheets from which the test specimens are cut by sawing and filing. All samples should be conditioned at the test temperature and appropriate humidity for a standard time, for example, 24 h.

flexural properties - typical results (at 23°C)

	UPVC	LDPE	PA 66[a]	PC
Flexural strength				
(at rupture or yield)				
10^3 psi	10–16	N/A	6.1	13.5
MPa	69–110	N/A	42	93
Flexural modulus				
10^3 psi	300–500	8–60	185	340
MPa	2070–3450	55–413	1270	2340

[a] As conditioned to equilibrium with 50 per cent relative humidity
[b] Plastics tested to ASTM D 790.

Note that as with tensile testing, the results obtained will depend upon the testing conditions, the method of sample preparation and the grade of polymer employed.

flexural test
See *flexural properties*.

flexural strength
The strength determined from the load and the test piece dimensions in a flexural test. Flexural strength is also known as breaking strength. See *flexural properties*.

flight
The space enclosed by the *screw thread* and the surface of the screw root, in one complete revolution, or turn, of the screw.

flighted length
The axial length of the flighted portion of a *screw*. See *effective screw length*.

flipper
Also known as a filler strip. A *tyre component*: a narrow strip of rubberized fabric, which enfolds the *bead* components, and used to produce a gradual change in stiffness from bead to *sidewall*.

floating cavity plate
A steel plate which carries the *cavity* in an underfeed-type of injection mould. See *centre plate*.

floating plate
See *centre plate*.

floating platen
A *platen* of a multi-platen press which is located between the fixed platen and the moving platen: the use of this type of platen allows the creation of extra daylights. As the press is closed, the moving platen contacts the floating platen and the two then move together.

floating plug
An *extrusion* aid used to avoid the need for sealing extruded tube. The internal floating plug may be attached by a hook arrangement to the *torpedo*: the extruded tube passes over the plug.

floating reclaim
A type of *reclaimed rubber*.

floating-knife coating
A *spread coating* method in which the substrate is held against a *doctor blade* by substrate tension.

flocculate
To form an agglomerate of particles (e.g. clusters) which are held together by relatively weak forces.

flocculation
The coagulation of smaller particles into particles of larger size, for example, in rubber technology the first stage of coagulation.

flood feeding
A feeding technique for thermoplastics processing equipment whereby the hopper of the machine is filled. For example, with a *single screw extruder* the hopper is filled and the screw takes what material it wants: such a scheme of flood feeding is often not possible with a *twin screw extruder*.

flooded head
A term used in *hydraulics* and which refers to a pump inlet port which is subjected to pressure from a head of fluid.

flop
The change in *shade* between two different angles of viewing and which is often exhibited by *metallic pigments*.

flotation method
An accurate density measurement used, for example, with *polyethylene* (PE) and which uses a water/alcohol mixture. Small pieces of the PE are placed in a beaker which also contains a water/alcohol mixture. The density of the water/alcohol mixture should be approximately 0.9 gcm^{-3} (obtained by mixing 35 ml of methylated spirits with 25 ml of distilled water at 23°C). Additional water may then be slowly added to the beaker. When the density of the liquid is the same as the density of the polymer, then the pieces of PE will remain suspended. It is important to add the additional water slowly and to keep the solution well stirred, these precautions ensure that the liquid density is uniform and that the heat of mixing does not cause the temperature to rise to an unacceptable level. The density of the water/alcohol mixture may then be determined by a weighing bottle technique. See *density gradient column*.

flow behaviour index
The flow index of the power law equation in the *power law equation*. An abbreviation used is n.

flow coating
A coating method in which a solution of a polymer is poured over a substrate. For example, in *blow moulding* this process is sometimes used to coat *polyethylene terephthalate* (PET) bottles with *polyvinylidene dichloride (PVDC)* so as to reduce permeability. Immediately after moulding the PVDC is poured over the bottles in a thick stream, drained for approximately 30 s and then, dried for 2 to 3 minutes at 65 to 70°C. Up to 5,000 bottles an hour may be so coated.

flow control
A term used in *hydraulics* and which refers to a device which regulates the rate at which fluid flows in a circuit.

flow control valve
Hydraulic valve used to throttle, or regulate, the amount of oil fed to a cylinder or actuator and so control speed.

flow curve
Sometimes referred to as a rheogram. A flow curve is a plot of *shear stress* (τ_w) against *apparent shear rate* at the wall ($\dot{\gamma}_{w,a}$). This is done because the expression, for the *shear rate* at the wall of a die, is rather cumbersome and in practice it is just as useful to use the much simpler expression $\dot{\gamma}_{w,a} = 4Q/\pi R^3$. It may be shown that, as with the true shear rate, there is a unique relationship (i.e. just one relationship) between shear stress and $\dot{\gamma}_{w,a}$: this simplified expression may be used for scale up work.

Polymer melts are often *pseudoplastic*. There is not therefore, a linear relationship between pressure and flow. What all this means is that the flow properties cannot be represented meaningfully by one figure and so flow testing over a range of conditions is required. The information so obtained may be presented in the form of a series of flow curves - each one representing a particular temperature. The flow curves provided by most laboratories are generally obtained by plotting τ_w against $\dot{\gamma}_{w,a}$. By plotting shear stress in Nm^{-2} against shear rate in s^{-1}, a viscosity in Nsm^{-2} may be obtained by reading from the flow curve. Conversion from shear rate to shear stress is obtained for plastic melts by reading from a *flow curve*. This is done because, the equations used to describe the flow behaviour of *pseudoplastic fluids* are imprecise.

The wall of the die was initially specified: however, this is simply a place at which the shear stress and shear rate may be calculated from a knowledge of flow rates and pressures. The relationship between τ and $\dot{\gamma}$ will however be independent of the position in the flow system and indeed will be equally applicable to flow in other geometries, e.g. slits. For this reason it is quite correct to plot τ against $\dot{\gamma}$ without printing the subscripts. Occasionally shear rate is denoted by γ without the dot above it. This practice should not be encouraged since this symbol is widely used to indicate elastic shear strain.

The information necessary to generate a flow curve may be obtained from a variety of types of machines - these include cone and plate viscometers, co-axial cylinder viscometers and *capillary rheometers*.

flow data
Shear flow data obtained from, for example, a *capillary rheometer*. Such information is available in several materials data bases, for example, in the ones maintained by Moldflow (Europe) Ltd. The data is used to help predict how an injection mould will fill. For this purpose the effect of changing temperature at a constant *shear rate* is needed together with values which show, for example, the effect of changing *shear rate* at a constant temperature (see *flow curve*).

The figures in **table 14** were obtained when samples of *nylon 6 (PA 6)*, produced by Akzo, were tested. (Such results are typically obtained using a die with high L:D ratio, for example, 20:1.) These figures clearly show that the three, injection moulding grades tested, have very different viscosities, with the first grade having the lowest viscosity that is, it is the easiest flowing grade. It would be used where mould filling is difficult, or where long flow lengths are involved. The viscosity of all three materials falls as the shear rate is increased, that is, mould filling becomes easier. Raising the melt temperature, while keeping the shear rate constant, reduces the amount of injection pressure required to maintain a certain rate of flow.

flow direction
The direction in which the *melt* is flowing.

flow diverter runner
See *aftermixer*.

flow - energy absorption
It is wrongly assumed that all of the energy involved in forcing the material through a die is absorbed between the ends of the tube. Some energy is used up in feeding, or funnelling, molten polymer molecules into the die: that is, from the barrel or reservoir into the die parallel. Some of this energy absorption is due to the extra shearing that occurs just above the die; much of it is believed to be due to the energy involved in uncoiling the polymer chains as they are suddenly subjected to an increase in shear. This extra energy input is equivalent to an increase in the effective length of the die. See *Bagley correction*.

flow index
The flow behaviour index in the *power law equation*. An abbreviation used for this term is n.

flow leaders
Regions of an *injection mould* designed to offer an easy flow path to the polymer melt: often regions of slightly increased wall thickness. By the use of such regions it is possible to arrange for all parts of a mould to fill at the same time and so save over-packing.

flow line
A line on a moulded surface which shows the flow direction in the cavity.

flow mark
A mark on, or in, the surface of a moulding which shows the flow path of the material. Such a mark is particularly noticeable in areas where the stream of material divided during moulding.

flow moulding
A method of *injection moulding* used to obtain large shot weights from relatively small, injection moulding machines; this usually involves rotating the screw during injection and during the *dwell time*. May also be called intrusion.

flow orientation
The *orientation* caused by melt flow: the partial alignment of the polymer molecules.

flow path:wall thickness ratio
The result of a *flow test* which is expressed as a ratio: a way of indicating the ease of flow of a plastics material. A high shear rate flow test. The test is performed on an *injection moulding machine* under specified conditions and using 'typical moulds'. If the ratio is quoted as being 150:1 then this means that if the wall thickness of the moulding is 1 mm/0.04 in then, the maximum length of flow possible will be approximately 150 mm/6 in. Because the amount of flow possible is dependent upon wall thickness, flow ratios may be quoted for a range of wall thicknesses, for example, 1, 2 and 3 mm/0.04, 0.08 and 0.12 in.

flow promoter
An additive: a material which eases flow. See *flow testing*.

flow properties representation
See *flow curve*.

flow rate
A term used in *hydraulics* and which refers to the volume of flow which occurs in a unit of time. A term used to describe a low shear rate flow test in which a heated plastics material (for example, polyethylene or PE) is forced through a die by a weight (see **table 11**). The die length and diameter are specified as are the temperature, load and piston position in the cylinder, or barrel (the position as the timed measurement is being made). The amount of plastic extruded in 10 minutes is measured and reported as the flow rate (FR). FR = m × 600/t where t is the cut-off interval expressed in seconds (s). The test is more usually referred to as MFR or MFI; the terms mean the same and stand for melt flow rate and melt flow index respectively.

The results are reported, if the test was done according to ASTM D 1238, as, for example, FR-190/2.16 = 2.3. This means that the temperature was 190°C and a load of 2.16 kg was used. That is, the condition used for the test was Condition 190/2.16. Under these conditions, 2.3 g of the plastics material was extruded in 10 minutes. If more plastics material had been extruded in the 10 minute period, then the material would be a more easy flowing material. If less plastics material had been extruded in the 10 minute period, then the material would be a less easy flowing (or stiffer flowing) material. See *method A* and *method B*.

flow rate - equipment details
The details quoted here are those specified by ASTM Method D 1238-86. Cylinder size, length (L) 162 mm/6.375 in: diameter (D) 50.8 mm/2 in. Cylinder bore D 9.5504 mm/0.376 in. Die L 8.0000 mm/0.315 in: D 2.0955 mm/0.0825 in. Piston land L 6.35 mm/0.250 in: D 9.4742 mm/0.3730 in. The metal surfaces must be very well finished, for example, so as to produce 12 rms or better in accordance with ANSI B 46.1. The set temperatures must be capable of being held to within ±0.2°C and the combined weight of the piston and load must be within ±0.5% of the specified load.

flow rate - increasing shear rate
Flow rate (FR) testing is a low shear rate test, for example, the shear rate may only be $1\ s^{-1}$. Larger weights than 2.16 kg may be used and different temperatures may also be used: what is used depends upon the material and upon the grade of material (see **table 11**). For *unplasticized polyvinyl chloride* (UPVC) a weight of 20 kg/44.1 lb may be employed and the temperature suggested in ASTM D 3364 is 175°C/347°F. When the test is run with such a high load, or with the even higher load of 21.6 kg/47.62 lb, then it may be referred to as the high load melt index or the HLMI.

flow rate ratio
An abbreviation used for this term is FRR. Flow rate ratio is obtained by dividing the *flow rate* obtained when a large load was used by the flow rate obtained when a smaller load was used. The load ratio is general 10:1 and on some machines may be measured automatically during a single test run.

This modified *MFR* test is done as polymer viscosity varies as a function of *shear rate*: the usual single point determination by a *melt indexer* does not therefore fully characterize material flow. FRR is measured so as to obtain more useful information, from a *melt flow indexer*. For example, in blow moulding the sag behaviour of the parison is best measured by a low shear flow while the flow through the die lips is best measured by a higher shear flow. The load ratio is general 10:1.

flow rate - report
The flow rate (FR) is reported, if the test was done according to ASTM D 1238, as the rate of extrusion in grams in 10 minutes, for example, FR-190/2·16. The results are sometimes shown as MFR (190, 2·16) = 2·3, that is melt flow rate is reported. They could be shown as, MFR (190, 21·2) = 2·3. This means that the temperature was 190°C and a load of 21·2 Newtons was used. It is therefore important to specify in any report, or table, the test procedure used for testing, the nature and physical form of the material fed to the cylinder, the temperature and load used, details of any material conditioning (for example, drying), the procedure used (for example, *Method A* or *B*) and any unusual behaviour of the plastics material seen during the test.

flow rate - standards and methods
Flow rate testing is governed by various international standards, for example, ASTM Method D 1238-86 and ISO R1133. Such standards specify orifice size, melt temperature, heat chamber size and piston tip diameter and the method of conducting the melt index test: the object being to obtain consistent results from different melt indexers. Two basic methods have been developed for melt index testing, *Method A* and *Method B*. Method A is the traditional manual testing method whereas Method B uses electronic sensing of plunger displacement and calculates the flow data from such measurements. Once set up, Method B is simpler to run and more accurate for routine testing.

flow rate tester
The test equipment used to conduct a *flow rate test*. May be referred to as an imposed pressure rheometer or as a shear stress rheometer.

flow rate testing
An assessment of material viscosity. Of all the *flow tests* used by the plastics industry, flow rate testing (sometimes called melt flow rate, or melt flow index) is most widely used. It is traditionally associated with the testing of *polyethylene materials* for quality control purposes, for example, to determine lot-to-lot consistency of material batches. Also used to test other materials, to determine barrel residence times within plastics processing equipment or, to assess regrind content within materials or mouldings. It is a low shear rate test although the shear rate can be increased by using larger weights. See *flow rate*.

flow rates
Results obtained from a *flow rate* test.

flow tab
A thin, graduated channel which is often located at the end of the *runner* system. During *injection moulding* the length of the flow tab is noted. If the flow tab length alters, then this is due to either a change in machine settings, operating procedures or material properties. Used to assess a materials flow behaviour during injection moulding: a high shear rate flow test.

Measurement of *flow tab length* is performed on an injection moulding machine under the production conditions and using the production mould. As material is forced into the cavity, then the flow tab is also partially filled at the same time. After ejection, with the component, the length of the flow tab is measured. The result is expressed, for example, as a certain length produced under the specified production conditions and is entered on the production records. The test is not a straight forward rheological test as hot material is flowing into a mould which is at a different temperature to the melt. This test is used to assess the ease of flow of thermoplastics materials.

flow tab length
The length of material in a *flow tab*.

flow temperature
A flow test result used, for example, for *cellulosic plastics*. By varying the ester, the type and the amount of plasticiser, many different grades of *cellulosic plastics* are made which differ in their flow behaviour. Cellulosics may be classified in accordance to ASTM D-1562 whereby a flow temperature is determined. A standard apparatus is used to determine at what temperature a 1" rod will be extruded in 2 minutes under a pressure of 1500 psi/10·3 MNm^{-2} using a predried cylinder of material as the initial feed. The higher the temperature (to get this flow length) the stiffer, or the harder, is the flow. Cellulosics may be divided into various categories for example, S_2, MS, H and H_2. The flow temperature of each is S_2 130°C/266°F, MS is 140°C/284°F, H is 155°C/311°F and H_2 is 160°C/320°F. The hard flow types, of the cellulose ester materials, have similar flow properties to PS. Hard flows (for example, H_2) have high hardness, stiffness and the highest tensile strength. As the flow becomes easier, or softer, these properties decrease and the materials become more impact resistant. CAB is available in easier flows than CAP.

flow tendency
See *ease of flow*.

flow test
A test performed to assess or study the flow behaviour of polymer materials or compounds. Purpose-built machines may be used for this purpose and/or production machines may be used. Examples of purpose-built machines include *plastimeters* and *melt indexers*: examples of production machines include *injection moulding machines* and *extruders*. See *flow path*: wall thickness ratio, *spiral flow length*, *flow tab length* and *minimum moulding pressure*.

flow testing - imposed rate
See *imposed rate type rheometer*.

flow testing - moisture content
With many plastics materials, the level of water/moisture in the material fed to the processing equipment must be kept below very small values. For example, the material fed to an injection moulding machine must have a moisture level of below 0·2%. This is usually to prevent the production of mouldings with a poor surface finish; however, the water can act as a flow promoter. This means that if the flow properties of a production material are being assessed then the sample used, for flow testing, must have the same water content as the production material.

flow tests - high shear rate
Flow tests which are performed under conditions of relatively high shear. The shear rate used is typical of that found in processes such as *extrusion* and *injection moulding* and is often above, for example, 100 s^{-1} and can reach 100,000 s^{-1}. Such testing is often done in a *capillary rheometer*.

flow tests - importance of
Because most methods of shaping plastics are *melt processes*, the measurement of melt flow properties is extremely important. A large number of tests have been devised to measure such properties but, in general, such tests may be grouped into low shear rate flow tests and high shear rate flow tests.

flow tests - low shear rate
Flow tests which are performed under conditions of relatively low shear. The shear rate is not typical of that found in processes such as *extrusion* and *injection moulding* and is often below, for example, 10 s^{-1}. See *flow rate*.

flow - thermoset moulding material
The flow, and cure rate, of a moulding material is determined by the amount of resin, the degree of condensation of the resin

and the residual volatile content, for example, the moisture content of the compound. If the resin is only lightly condensed during manufacture then, the resulting moulding material will flow easily but will require comparatively long cure times: the mouldings will exhibit high shrinkage and will not exhibit the best mechanical properties. Increasing the degree of condensation will worsen the flow but improve the properties of the mouldings. The use of approximately 1% of a plasticizer, or flow promoter such as dibutyl phthalate, may be incorporated in some low shrinkage grades to get adequate flow. The ease of flow of a thermoset moulding compound is often assessed by the *cup flow test*.

flow time
The time taken for a polymer solution, or of a solvent, to flow through the capillary of a viscometer. See *solution viscometry*.

flow velocity
A term used in *hydraulics* and which refers to the linear speed of fluid flow at a specified point in a hydraulic circuit.

flow-way
A waterway. A hole or channel in an *injection mould* through which fluid is circulated so as to maintain a desired temperature.

flowers of sulphur
An allotropic form of *sulphur*: mainly used in rubber technology as insoluble sulphur. It is practically insoluble in carbon disulphide. Fine powdered *sulphur* produced by the condensation of sulphur during the distillation of crude sulphur. Fine crystals of sulphur which revert to the normal form on standing.

flowers of zinc
See *zinc oxide*.

flue gas desulphurization
An abbreviation used for this term is FGD. Flue gas desulphurization systems are used to clean flue gases from, for example, the sulphur dioxide emitted by coal-fired power plants. This is a major market for *glass reinforced plastics* materials: such materials are used because of the relative ease of fabricating large mouldings which have good corrosion resistance.

fluid
A term used in *hydraulics* and which refers to a liquid which is specially formulated, and used, to transmit power in a hydraulic system.

fluid bed
A bed or bath which looks and acts like a liquid but which is based on dry material. Glass *ballotini* is commonly used for vulcanization. When the bed is aerated, through a porous base, this gives a 'fluid' which is chemically inert, non-wetting and non-boiling. Air is commonly used for *fluid bed coating* purposes (with plastics) whereas steam may be used for *continuous vulcanization*. The pressure drop across the fine mesh or gauze, of the base, is of the order of 1 psi or 7 kNm^{-2}. Both *horizontal* and *vertical fluidised beds* are used.

fluid bed coating
A coating method which uses a *fluid bed*, for example, based on *polyethylene* (PE) powder and used for coating heated metal objects with plastics materials. After immersion in the bed, the coated metal may be heated to complete fusion. To improve coating uniformity the bed may be stirred or shaken. Epoxy resins are applied in this way for electrical insulation purposes. See *electrostatic fluid bed coating*.

fluid bed vulcanization
A *continuous vulcanisation method* which utilises a *steam-heated fluid bed*. The extruded rubber product is drawn through the heated bed and is covered by a fine rain of glass beads; the profile floats through the bed and there may be no need for guides, rollers etc, to touch the vulcanizate.

fluid drachm
A unit of measurement in *apothecaries' measure*.

fluid flow rate - mould cooling
See *mould cooling*.

fluid measure
See *apothecaries' measure*.

fluid natural rubber
See *fluid rubber*.

fluid ounce
An abbreviation used for this term is fl oz. A unit of measurement in Apothecaries' measure. $\frac{1}{20}$ of a pint in the UK and $\frac{1}{16}$ of a pint in the US. To convert from fluid oz (US) to m^3 multiply by $2 \cdot 957 \times 10^{-5}$. See *US Customary Measure* and *UK system of units*.

fluid rubber
A *natural rubber* composition based on natural rubber together with a large amount of mineral *oil*. A low melting point jelly which can be poured at relatively low temperatures and then vulcanized. See *depolymerized rubber*.

fluid scruple
A unit of measurement in *apothecaries' measure*.

fluidics
Also known as fluid logic. The study, design and application of the use of jets of fluid to carry out logic and amplification functions.

fluidity
The reciprocal of viscosity. In the cgs system, the unit is the rhe.

fluorescence
The emission of radiation of a different wavelength to that which was absorbed. When the source of radiation is removed, fluorescence ceases (unlike phosphorescence). Fluorescence and *phosphorescence* are examples of luminescence.

fluorinated elastomer
See *fluorosilicone rubber*.

fluorinated ethylene propylene copolymer
An abbreviation used for this type of material is FEP. It is also known as tetrafluoroethylene - hexafluoropropylene copolymer or as, FEP fluorocarbon or simply as fluorinated ethylene propylene. Sometimes referred to poly-(tetrafluoroethylene-co-hexafluoropropylene).

A random copolymer based on tetrafluoroethylene and hexafluoropropylene. A thermoplastics material which was developed to give the desirable properties of *polytetrafluoroethylene (PTFE)* and the ease of processing of a thermoplastics material.

PTFE and FEP are similar in their excellent dielectric properties, chemical inertness, toughness at low temperature, low coefficient of friction, anti-stick properties and weatherability. These properties are a result of the strong carbon-fluorine, primary valence bond (which lends stability to the resin), the relative size of the carbon and fluorine atoms (which permits compact molecular chain formation), the high molecular weight of the resins (which discourages solubility and promotes toughness), and their relatively weak secondary valence forces (which dictate the degree of attraction between molecules).

The property differences between PTFE and FEP lie mainly in their upper service temperature limitations. PTFE can be used continuously at 260°C/500°F while FEP has an upper limit of 205°C/401°F for continuous service. Other, more minor, differences include a higher modulus of elasticity for FEP at low temperatures, lower permeability to most liquids and gases for FEP, a slightly lower coefficient of friction for PTFE and a somewhat higher dissipation factor for FEP at very high frequencies (0·0012 versus 0·0004 at 109 cps and room temperature). Mechanical properties are similar to those of PTFE but FEP has better impact strength and wear resistance; fatigue resistance is inferior. Both have excellent weathering resistance. The physical properties of FEP are not outstanding but are maintained over a wide temperature range - useful properties maintained from $-250°C/-418°F$ to $200°C/392°F$. Resists ignition and flame spread.

The processing characteristics of FEP and PTFE are quite different. FEP resin has a melt viscosity of between 104 and 105 poises, which permits it to be processed on conventional thermoplastic equipment. However the material has a high melting point and a high melt viscosity. If both PTFE and FEP are suitable for a particular application then it must be decided whether the components are better made by specialized PTFE processing techniques or conventional processing using FEP. Either method may be followed by supplementary finishing operations: factors such as shape, size, number of components and tolerances must be taken into account. It is difficult to injection mould components with wall thickness below 1 mm (0·040 in). Complex parts, requiring multiple machining operations on PTFE shapes, can be cheaper in FEP at a production level of a few thousand mouldings.

As with other thermoplastics, fillers are added in order to improve, for example, wear resistance and resistance to plastic deformation. But many fillers cause processing problems. Glass fibres and graphite have been successfully used but filled grades are not marketed by Dupont.

FEP is almost completely inert to most chemicals; small weight increases by halogenated organic solvents (less affected than PTFE as FEP is less permeable). High molecular weight grades have the best stress-crack resistance but the worst flow. Stress-cracking is rarely observed with PTFE and FEP but can occur at high temperatures in the presence of some halogenated hydrocarbons. FEP is not resistant to alkali metals and halogens in some circumstances. Chemically attacked by molten alkali metals, gaseous fluorine, chlorine trifluoride and oxygen difluoride. This material has a high density (similar to PTFE) of 2·12 to 2·17 gcm^{-3} (solid, non-filled material): components made from natural FEP are quite transparent with a bluish tint. As the natural colour of the material is transparent then a wide colour range is possible; this includes both transparent and opaque colours.

Molten FEP adheres well to other materials; has been used to join PTFE components to metal. This type of material is used where heat, fire and/or chemical resistance is required together with the ability to be melt processed into complex shapes; the low friction and good anti-stick properties are also useful attributes. Used to make bellows, laboratory ware, filters, gaskets, seals, bearings, valves etc. To promote component life avoid the use of sharp corners as these concentrate stress - use large radii: never below 0·5 mm (0·020 in).

fluorinated rubbers
The collective name for the class of rubbery materials which contain a fluorine group. See *fluorosilicone rubber*.

fluorination
The introduction of *fluorine* into a thermoplastics material such as a *polyethylene (PE) material*. Sometimes done with plastics materials so as to reduce the permeability of PE blow mouldings used for fuel tanks. The term 'blow and barrier system' refers to the fluorination process used to improve the permeability resistance to PE containers to, for example, automotive fuels.

The inside of the container is exposed to fluorine and this causes hydrogen atoms on the PE molecule to be replaced with fluorine. In this way, polar groups are introduced into the thermoplastics material and it is the introduction of such groups which makes the plastics material less attractive to the hydrocarbon (non-polar) fuel.

fluorine
The elements of Group V11b of the Periodic table are known as halogens and consist of fluorine, chlorine, bromine, iodine and astatine. Fluorine does not occur naturally as it is too reactive: it is the most reactive element known. It is a greenish-yellow diatomic gas (F_2) which is slightly paler in colour than chlorine and with a similar odour. Melts at $-220°C$ and boils at $-188°C$. Has a low critical temperature of approximately $-129°C$ which means that liquid fluorine cannot be stored at room temperatures. Used, for example, in the blow moulding industry to reduce the permeability of blow mouldings used for fuel tanks. All halogens, except iodine, should be treated with great care as they are either highly poisonous and/or very irritating (even iodine vapour is irritating and poisonous).

fluorine-containing rubber
A rubber which contains the element fluorine as a part of the rubber molecule. The presence of this element gives heat and fuel resistance but causes problems when the rubber does burn. See *fluororubber* and *fluorosilicone rubber*.

fluoro-silicone elastomer
See *fluorosilicone rubber* and *silicone rubber*.

fluorocarbon
An organic compounds in which some, or all, of the hydrogen atoms have been replaced by *fluorine*.

fluorocarbon plastic
See *fluoroplastic*.

fluorohydrocarbon plastics
Plastics based on polymers made with monomers composed of fluorine, hydrogen and carbon only.

fluoronitrosorubber
See *nitroso rubber* and *fluororubber*.

fluorophosphazene rubber
See *polyphosphazene* and *fluororubber*.

fluorophosphonitrilc polymer
See *polyfluorophosphazene rubber* and *fluororubber*.

fluoroplastic
A plastic based on polymers made from monomers containing one or more atoms of fluorine, or copolymers of such monomers with other monomers, the fluorine-containing monomer(s) being in greatest amount by mass. See, for example, *fluorinated ethylene propylene copolymer* and *polytetrafluoroethylene*.

fluoropolymer
A polymer whose repeating units contain *fluorine*: both *fluoroplastics* and *fluororubbers* are available. There are several fluoropolymers that can be processed by injection moulding, including *fluorinated ethylene-propylene (FEP)*, *ethylene tetrafluoroethylene (ETFE)*, *ethylene-chlorotrifluoroethylene (ECTFE)*, *chlorotrifluoroethylene (CTFE)*, *perfluoroalkoxy tetrafluoroethylene (PFA)* and *polyvinylidene fluoride (PVDF)*. In general, these materials have excellent chemical and heat resistance, they are good electrical insulators and have high impact strength.

fluoropolyphosphazene polymer
See *polyphosphazene* and *fluororubber*.

fluoropolyphosphazene rubber
See *polyphosphazene* and *fluororubber*.

fluororubber

A *synthetic rubber* which contains fluorine as part of its molecular structure. Fluororubbers have excellent heat, burning, chemical, weathering and solvent resistance: they have excellent resistance to concentrated acids.

One of the first commercially successful fluororubbers was the copolymer based on *vinylidene fluoride* and *chlorotrifluoroethylene* (designated CFM): another was the copolymer based on vinylidene fluoride and *hexafluoropropylene*. Copolymers are used as they have a relatively high fluorine content and a low *glass transition temperatures*: the chain symmetry of the PVDF is destroyed. Terpolymers are now extensively produced which also contain *tetrafluoroethylene*. Such materials may be cured by an amine-based system or by a bisphenol-based system: both require a metal oxide to be present and both generate water on curing. This can lead to processing problems, for example, moulding splitting and for this reason a fourth monomer may be used so as to achieve peroxide cross-linking. That is, a tetrapolymer (based on for example, bromotetrafluorobutene) is used.

An ASTM designation/abbreviation used for fluororubbers based on vinylidene fluoride and either hexafluoropropylene or, hydropentafluoro propylene is FKM. An ISO designation or abbreviation used for such fluororubbers is FPM. The FKM rubbers dominate the fluoroelastomer market. About half of the FKM rubbers used are terpolymers with PTFE: as these have a slightly higher fluorine content they have better heat and oil resistance. The copolymers can however, exhibit excellent compression set resistance.

FKM materials are usually prepared by free radical emulsion polymerization and the products of the compounded, cured materials are used as O-rings and gaskets for the aerospace industry. Such fluororubbers have excellent heat resistance and chemical resistance. For example, to aliphatic hydrocarbons, aromatic hydrocarbons, chlorinated solvents and to petroleum fluids: ketones, monoesters and ethers cause swelling. Steam and hot water resistance is poor unless litharge is used as part of the cure system together with a capped diamine. Low molecular weight fluororubber may be used as a *plasticizer* or processing aid for a fluororubber. FKM rubbers can also be cured with *bisphenol AF* and a metal oxide, for example, magnesium oxide (a *hydrofluoric acid acceptor*). Such compounds have good processing safety but the vulcanizates are not so resistant to some fuels as amine-cured systems. Best vulcanizate properties are obtained if fillers (<30% *carbon black*) are incorporated and if the products are post-cured, for example, for 24 h at 230°C.

Polymers based on *perfluoro(methyl vinyl ether)* and tetrafluoroethylene have no C-H groups and have excellent thermal stability as a result but cannot be cross-linked. A cross-linking site is provided by the introduction of a *cure-site monomer* so as to make a terpolymer. For example, an aromatic perfluorovinyl ether may be used and cross-linking is achieved by an amine-based system via the para-fluorine atom on the benzene ring. Such fluororubbers may be known as perfluorinated elastomer (FFKM or PFE) and have a level of heat and chemical resistance better than other elastomers: they are however very expensive and difficult to process.

Other types of fluororubber include *polyphosphazene, tetrafluoroethylene-propylene copolymer, fluorosilicone, carboxynitroso rubber, perfluoroalkylenetriazine* and *poly-(thiocarbonyl fluoride)*.

fluorosilicone

Also known as fluorosilicone elastomer or as, fluorinated rubber or as, fluoro-silicone rubber or elastomer or as, silicone rubber containing fluoro, vinyl and methyl groups. An abbreviation used for this type of material is FVMQ. See *silicone rubber*.

A highly specialized *fluororubber*. A polymer based on alternating silicon and oxygen atoms in the main chain: the silicon atoms has two organic groups (R & R') attached. May also be known as a *polysiloxane*. The most common polymer is *polydimethylsiloxane* where R = R' = CH_3. If some of the R groups are trifluoropropyl groups then fluorosilicones result: these materials are copolymers which usually contain vinyl groups as well as a fluoropropyl side group. Vulcanizates have excellent heat resistance, fuel resistance and possess low temperature flexibility.

fluorotriazine elastomer
See *perfluoroalkylenetriazine*.

flushing
A process used to replace one material with another. For example, during pigment manufacture, an organic liquid (*plasticizer*) is added to a wet pigment in order to displace the water: this may then be easily removed by decanting.

flushing compound
See *purge compound*.

fluted barrier mixing element
A *dispersive mixing section*: see *Maddock mixing section*.

fluted barrier section
A *dispersive mixing section*: see *Maddock mixing section*.

fluted mixer
A *dispersive mixing section*: a region of a screw which is there to improve mixing. Grooves or flutes machined into a screw torpedo and which lie parallel to the long axis of the screw. Examples include the Egan, Maddock and Zorro mixing sections.

fluted mixing section
See *fluted mixer*.

flux density
See *tesla*.

FMVE
An abbreviation used for *perfluoro-(methyl vinyl ether)*.

FMVSS
An abbreviation used for the Federal Motor Vehicle Safety Standard.

foam
Strictly speaking, a suspension of a gas in a liquid. In polymer technology, the term is usually applied to a cellular material which was produced from a liquid, For example, *foamed rubber* is produced from *latex*.

foam stabiliser
A material used, for example, to stabilize the rising foam in a *one-shot process*. See *silicone-polyether block copolymer*.

foamed plastics
See *cellular plastics*.

foamed polymer
A cellular material produced from a liquid, For example, *foamed rubber* is produced from *latex*.

foamed polystyrene
See *expanded polystyrene*.

foamed rubber
A cellular, or porous, material produced from a vulcanizable latex: the vulcanizable latex is aerated, gelled and cured. See *foam*.

FoE
An abbreviation used for *Friends of the Earth*.

fogging
The increase in *haze* of a film due to condensation. See *automotive fogging*.

fogging temperatures
The temperatures used in a fogging test such as the Arizona Parking Lot Test. See *automotive fogging*.

foil
Term sometimes applied to very thin, but very stiff, sheet.

foil laminate
A laminate which contains a layer of metal foil. Based, for example, on polyethylene-coated paper with an interlayer of aluminium (Al) foil and most commonly produced by *extrusion*. Such foil laminates are fabricated into a range of products which include cartons for liquid packaging and which are used, for example, for the packaging of milk and fruit juices.

folding, spreading action
Term used in *compound blending* to describe what is required during a pre-blending operation. The mixture is combined, separated, re-combined etc.: this action is needed, for example, for the blending, or pre-blending, of thermoplastics materials with their additives. See *compound blending*.

foot
An abbreviation used for this unit is ft. A unit of length being first defined in England as being equivalent to 12 *inches*. One foot = 1/3 of a yard and is equivalent to 30·48 cm or 304·8 mm. See *UK system of units* and *US Customary Measure*.

foot-candle
A unit of illumination with the abbreviation fc and which is equal to one *lumen* per square foot. See *lux*.

foot-lambert
A unit of luminance with the abbreviation fl and which is equal to the luminance of a uniform diffuser emitting one *lumen* per square foot.

foot-pound
A unit of *work*. The work done by a force of one pound acting through a distance of one foot.

foot-pound per inch of notch
An abbreviation used for this term is ft lb/in of notch. A way of expressing the results obtained from an *Izod impact test*. The actual test values are divided by the width of the specimen used in the test.

foot-pound-second
A system of measurement which uses the foot (ft) as the basic unit of length, the basic unit of mass is the pound (lb) and the basic unit of time is the second (s). The derived unit of force is the poundal (pdl). This system is sometimes known as FPS absolute or Imperial (FPS absolute). See also *FPS gravitational*.

foot-poundal
A unit of *work* in the *foot-pound-second system*. The work done by a force of one poundal acting through a distance of one foot.

force
That which produces motion or changes in motion. Measured in *newtons*, *dynes* or *poundals*. If the abbreviation F is used for force then F = m a. Where m is mass (kg) and a is the acceleration (ms^{-2}) then F is in newtons.

force - of a mould
See *punch*.

force transducer
A *transducer* which measures force. See *cavity pressure control*.

force-time graph
A graph of force (in Newtons) versus time (in milliseconds). This may be displayed on an oscilloscope and/or recorded for transmission to a computer: this enables computation of areas under the force-time graph, and other calculations, to be made quickly without errors in a *falling weight impact strength test*.

forced charging
Forcing a material into a barrel or mould. For example, the loading of a *dough moulding compound* into an injection moulding machine, is done by forced charging from a stuffing box.

form and fill
A packaging process in which the container is formed and the contents inserted, on line; used in, for example *blow moulding*, to get high speed production because the rapid cooling which results when cold liquid is injected into the bottles.

form pin
A circular, hardened steel pin which is used for the moulding of an internal undercut.

form-fill-seal
Also known as FFS: a packaging process in which a film is formed into a container, filled and then sealed.

formable sheet
See *postforming sheet*.

formaldehyde
Also known as methanal. May be represented as HCHO. An abbreviation used is F when the formaldehyde is used to make polymers, for example, PF. Formaldehyde is a gas (boiling point of approximately −21°C) with an irritating pungent smell. It is the simplest aldehyde and is made by the oxidation of methyl alcohol. Used to make both thermosetting plastics materials and thermoplastics materials. For example, *phenol formaldehyde (PF)* and *acetals (POM)*. See *formalin*.

formaldehyde donor
A substance which is capable of giving *formaldehyde* during processing, for example, *hexamethylene tetramine*.

formaldehyde-based plastics - processing of
When processing any plastics material which is based on formaldehyde, care must be taken and the fumes evolved treated as harmful: good ventilation of the work-place is essential. Formaldehyde is toxic and carcinogenic, and so free formaldehyde is the major cause for concern when dealing with the use of aminoplastics which come into contact with food. The toxicity of free melamine and of free urea is low in comparison with that of formaldehyde. However, as melamine is slightly soluble in water and does have a diuretic effect on animals, care should be exercised when dealing with foodstuffs. Standards exist, for example, BS 2782, which specify how free formaldehyde is measured in *melamine-formaldehyde (MF) resins*.

formaldehyde-ethylene diamine condensate
An abbreviation used is FED. An *aldehyde-amine compound*: such compounds function as *accelerators* in rubber compounds.

formaldehyde-p-toluidine condensate
An abbreviation used is FPT. An aldehyde-amine compound which functions as an *accelerator* in rubber compounds.

formalin
A solution of *formaldehyde* (for example, 40%, in water) which also contains methyl alcohol as a stabilizer. 30% formaldehyde solution has a relative density of 1·09.

formed blister
A shape produced by *thermoforming*: see *denesting agent*.

formic acid
Also known as methanoic acid. A coagulating agent for *natural rubber*. May be represented as HCOOH. To coagulate one kilo of *natural rubber* approximately 5 g of concentrated acid is needed.

forming
A process in which the shape of plastic pieces such as sheets, rods, or tubes is changed to a desired shape or configuration In plastics technology, the term does not usually include such operations as moulding, casting, or extrusion, in which shapes or pieces are made from moulding materials or liquids. See *thermoforming*.

forming box
A cooled *die* extension which is used for sizing and cooling extruded products such as rod: melt is pumped into the forming box and solid rod slowly emerges.

forming temperature
The temperature (or temperature range) at which a forming process is performed. For example, during thermoforming, extruded *high impact polystyrene* (HIPS) sheet is therefore usually heated by an infra-red heater (mounted approximately 150 mm above the sheet) while being held in a clamping frame. The forming temperature is not usually measured but the sheet is judged to be at the correct temperature when the sheet tightens up after initially slackening on heating. This time is measured and used to set the heating (forming) time.

forming time
See *forming temperature*.

formolising
The hardening of a material, for example, *casein-formaldehyde*, with *formaldehyde*.

forsterite
A decomposition product of *asbestos*: a non-fibrous silicate. See *chrysotile fibre*.

fossil flour
See *diatomaceous earth*.

four bowl calender
See *four roll calender*.

four colour printing process
The most widely used type of *process printing*.

four nip calender
A five roll *calender*.

four roll calender
Also known as a four bowl calender. A *calender* with four main rolls. That is, with three nips (a three nip calender). The rolls may be arranged in different ways or configurations. Can have an *L calender*, an *inverted L calender*, a *Z calender* and an *inverted Z calender*. The most popular type of calender for long runs in the polymer industry: used, for example, for tyre cord production and for flexible *polyvinyl chloride (PVC)* production. Fewer rolls may not give sheet of the finish and accuracy required whereas a larger number of rolls increases the cost and machine complexity.

four roll calender - plastics usage
The four rolls form three nips; the first a feed pass, the second a metering pass and the third a sheet formation, gauging and finishing pass. Plasticized *polyvinyl chloride* (PVC), from the *strainer extruder* is fed to the *feed nip*, for example, the top, off-set nip of an inverted L calender. If the calender is required to produce 0·25 mm sheet (0·010") then, for running, the top nip may be set at 0·75 mm (0·030"): the second nip is set at 0·50 mm (0·020") and the third nip is set at 0·25 mm (0·010").

As most PVC formulations adhere to the hottest roll (compounds containing calcium stearate appear to be an exception) the temperature gradient is adjusted so that the final roll (roll 4) is the hottest at say 170°C: the other rolls decrease in 5°C increments so that roll 1 is at 155°C. Heating of the rolls is commonly achieved using high pressure hot water (HPHW).

The temperatures quoted are not necessarily the material temperatures as, at high speeds, large amounts of frictional heat can be generated. On the system mentioned temperatures of 200°C could be reached in the last nip at high running speeds with certain formulations. By running the rolls at differing speeds the amount of shear generated, and therefore the frictional heat generated, in the 4 nips can be altered. Roll speeds normally increase from roll 1 to 4 as it is found that materials tend to stick to the faster moving roll of a pair. This is provided that each roll has the same finish. If one roll has a matt finish then the material will stick to this roll: this is presumably because of the increased area in contact with the material. See *calendering*.

four roll calender - rubber technology usage
A great deal of rubber sheeting is produced by *single* and *double bank calendering methods*. However with *synthetic rubbers* even two banks may not be enough if very accurate sheet at high output rates is required. For this reason four roll machines are used: such machines are usually of the *inclined Z* or of the *inverted L type*. Apart from the production of synthetic sheeting such 4 roll machines may also be used to:

(i) give sheet of increased thickness by plying; and,
(ii) coat fabric, both sides, in one pass

Such fabric coating machines are widely used in the tyre industry and for this application the inclined Z calender may be used. Sheeting is produced in the first and last nip and these two sheets are brought to the centre nip and simultaneously laid on the fabric so as to give a fabric coated on both sides in one operation. See *calendering*.

four way valve
A directional valve which has four separate flow paths through the valve.

four-nucleus bisphenol A polycarbonate
A heat resistant *polycarbonate*: an aromatic polycarbonate with a high heat resistance but which is relatively brittle compared to a *polycarbonate* based on *bisphenol A*.

fp
An abbreviation used for freezing point.

FPM
An ISO designation/abbreviation used for a *fluororubber*.

fps
An abbreviation used for feet per second. See *fps system* and *foot-pound-second*.

fps system
The *foot-pound-second system of measurement*.

FPS gravitational
A system of measurement which is sometimes known as the British Engineering system of units or as, the British Engineering system. The basic units in this system are those of length (the *foot*), time (the *second*), and force (the *pound*). The unit of mass is the *slug*. Also see *foot-pound-second*.

FPT
An abbreviation used for *formaldehyde-p-toluidine condensate*.

FR
An abbreviation for *flow rate*. FR is also used as an abbreviation for *flame retardant* and/or *fire resistant*.

fracture area
A term used in, for example, impact testing: obtained by multiplying the width of the specimen by the depth behind the notch. See *pendulum impact test*.

frame mould
Sometimes called a picture frame mould or a simple frame mould. A simple mould used to produce compression moulded sheets for testing purposes. Such a compression mould consists of three pieces:

1. one top sheet (e.g. 250 × 250 × 6 mm);
2. one similarly sized centre sheet of an appropriate thickness (e.g. 1·4 mm and with a hole of the required size in the centre (e.g. 150 × 150 mm); and,
3. one bottom sheet (e.g. 250 × 250 × 6 mm).

The top and bottom sheets should be of good surface finish and free from obvious defects such as scratches. See *compression moulding*.

free extrusion
Extrusion into a cooling unit without the use of forming fixtures or sizing devices.

free radical
A very reactive chemical grouping which, because of its reactivity, cannot normally exist on its own but which exists for short periods of time during chemical reactions. Usually depicted as R·. For example, free radicals are generated by degradation of a *free radical initiator* during a polymerization reaction, for example, by thermal degradation of a *peroxide*.

free radical initiator
A material which generates *free radicals*. For example, free radicals are generated by degradation of a free radical initiator (R-R) during a polymerization reaction, for example, by thermal degradation of a *peroxide* to give R·.

free radical oxidation
See *oxidation*.

free radical polymerization
Also known as radical polymerization. The most common type of chain polymerization and one in which the active centres are free radicals. These free radicals are usually generated by degradation of a *free radical initiator*, for example, thermal degradation of a *peroxide*.

free radical scavenger
See *radical scavenger*.

free radical trap
See *radical scavenger*.

free sulphur
One of the ways the *sulphur* in rubber compounds, on analysis, may be classified. That elemental sulphur which is not chemically bound, or chemically attached, to the rubber. The amount of sulphur which can be extracted from a cured product by *acetone*. See *sulphur analysis*.

free-flow, sliding check ring
A *check ring* which is designed not to obstruct flow and so puts minimum obstruction in the way of the melt stream. See *fibre length*.

freeze dried concentrate
A *masterbatch* based upon a wax (melting point of approximately 93°C/200°F. The wax is the carrier for the colorant system. As the wax is a solid at room temperatures more of this type of *masterbatch* can be added than is possible for liquid colour. Can incorporate up to 4%.

freeze grinding
The removal of unwanted material, such as flash, from a formed product by making the unwanted material brittle by lowering the temperature: the unwanted material is then removed by tumbling or grinding. Associated with soft flexible materials such as rubber mouldings or with soft plastics materials. Liquid air or nitrogen may be used as the cooling medium.

freeze line
The place on a *blown film* bubble where a semi-crystalline, thermoplastics material, usually a *polyolefin material*, starts to crystallize as it cools.

French chalk
See *talc*.

French ochre
A yellow pigment which is a mixture of clay and yellow hydrated iron oxide. Has a relative density of approximately 2·9. Light resistant but not heat resistant: changes colour, to red, at approximately 180°C/356°F.

French polish
A solution of *shellac* in methylated spirits. Used to seal a porous surface in *plug manufacture*. The sealed surface is then polished with a wax polish and then treated with a release agent which allows the *unsaturated polyester (UP) resin* to wet out on the plug surface.

Freon
A DuPont trade name/trademark for a group of halogenated hydrocarbons: a *chlorofluorocarbon* or *CFC*. Individual members are identified by numbers after the word Freon. The first, commercially-produced CFC was Freon 12 which is dichlorodifluoromethane. It appeared to be the perfect replacement for ammonia, the toxic and volatile refrigerant in use when this CFC was developed (developed by Thomas Midgley in 1930). Freon 12 may also be known as CFC 12.

Freon 11 is trichlorofluoromethane. Freon 12 is dichlorodifluoromethane. Freon 22 is chlorodifluoromethane. Freon 113 is trichlorotrifluoroethane. Freon 114 is dichlorotetrafluoroethane.

Freon 11 and Freon 12 are used in three main areas (a) refrigeration and air conditioning (b) blowing of *polyurethane foams* (particularly rigid foams) and, (c) as *aerosol propellants*. Now known that *stratospheric ozone* is depleted by *stratospheric chlorine* which partly depends on CFC emissions.

frequency distribution
To define a frequency distribution it is necessary to know: (a) its location or level, which is usually expressed by the average and, (b) the spread or dispersion, which is usually expressed by the standard deviation (σ or SD). See *normal distribution curve*.

freshening
Restoration of *tack* by solvent wiping.

friable bale
See *technically specified rubber*.

friction
The resistance to movement between two bodies in contact.

friction lining materials
Most modern friction lining materials and pads are moulded from compounds which contain short *chrysotile fibres* (<70%), *fillers* and *synthetic polymers*.

friction ratio
The ratio of the speed of rotation (the rpm) of two similarly sized rolls. For example, on a *two-roll mill* the front roll may rotate at 18 rpm and the rear roll may rotate at 20 rpm: in such a case the friction ratio is 1:1·11.

friction - stick-slip
The friction process in which *static* and *sliding friction* alternates. See *coefficient of friction*.

friction welding
A *welding technique* used for *thermoplastics materials* in which the necessary heat is generated by friction. See *spin welding*, *ultrasonic welding* and *oscillating plate welding*.

frictional heating
Heat generated by friction. For example, heat generated within the *extruder* by turning the *screw* at high speeds.

frictioned fabric
A rubber coated fabric which has been produced by *frictioning*.

frictioning
A *calendering process* used in the rubber industry to force rubber into the interstices of a fabric. May be performed on a 3-roll *calender* by adjusting the top nip so as to give the required thickness of material on the centre roll. With the bottom nip open the fabric is located in the bottom nip and then the bottom bowl is raised until the fabric bites against the stock. The fabric is then moved forward and a small *bank* forms in front of the bottom nip. A film of stock remains stuck on the centre roll during the frictioning operation and is replenished from the first nip.

With frictioning the aim is to get maximum adhesion and impregnation of the fabric. This is done by passing the cloth through the nip slowly and rotating the rubber-covered centre roll quickly against it under pressure. Rubber is therefore forced into the fabric because of the pressure and the difference in speed between the rubber and the fabric. A *friction ratio* of up to 2 may be employed.

Although pressure is required for good impregnation it must not be too high as otherwise fabric damage will result. The cloth must not be too tightly woven or contain too much size. It must be dry, free from fluff and strong enough to withstand the frictioning process. The selvedge should be removed before processing.

The stock should be soft and tacky and such a consistency is achieved by using softeners such as *pine tar* and working at high temperatures. For this reason an appropriate curing system must be used. Reclaimed rubber is also used but drying fillers such as magnesium carbonate and silica should not be used. Although *whiting* is the most widely used filler it is not as good as *titanium dioxide* or *zinc oxide* for this applications as these fillers promote tackiness in frictioning stocks. For the same reason natural rubber is the most common polymer used to make such stocks.

If both sides of a cloth need to be frictioned then this is commonly done by making two passes through the same calender. Two calenders could be used of course if the output required justified such a large expense.

Friends of the Earth
An abbreviation used for this organization is FoE. An environmental protection group or league.

front cavity plate
See *moving plate*.

front connected
A term used in *hydraulics* and which refers to a system where the pipe connections are on exposed surfaces of hydraulic components.

front plate
Part of an injection mould: a steel plate which is used to attach the fixed *mould half* to the *fixed platen*.

front roll
The front roll of a *two roll mill* is the one nearest the operator.

front slip angle
See *slip angle*.

frosting
A surface dulling of rubber vulcanizates, for example, caused by the action of *ozone* in regions of high humidity.

frozen-in orientation
Orientation associated with thermoplastics materials and which arises because of distortion of polymer molecules, from a random coil configuration, by the application of external stresses: the *orientation* is then set by rapid cooling. External stresses are applied during extrusion, moulding and other shaping operations. In most processing operations it is generally desirable to 'set' the polymer as soon as possible after it has been shaped, for example by cooling in a water bath after extrusion through a die. In such circumstances, the polymer molecules may not have time to coil up completely before the melt, in effect, freezes.

Such frozen-in orientation is greater when the melt has been subjected to higher stresses and reduced intervals between shearing and setting of the melt. Such conditions would be obtained when using low melt temperatures and low temperatures after shaping (for example, low injection mould temperatures or low extrusion cooling bath temperatures).

frozen in strain
See *frozen-in orientation* and *melt processing*.

frozen prepreg systems
Another term used to describe *low temperature moulding epoxy prepreg systems*.

frozen rubber
Raw *natural rubber* which has been stored at temperatures below 10°C/50°F.

FRP
An abbreviation used for *fibre reinforced plastic*.

FRR
An abbreviation used for *flow rate ratio*.

FRTP
An abbreviation used for *fibre reinforced thermoplastics*.

F_s/R
An abbreviation used for *static coefficient of friction*.

FSD
An abbreviation used for *full scale deflection*.

FSi
See *FVMQ* and *silicone rubber*.

ft
An abbreviation used for *foot* and/or for *feet*. For example:

ft H_2O = feet of water;
ft water = feet of water;
ft-lbf = foot pound-force; and,
ft ib/in of notch = foot-pound per inch of notch.

FT
An abbreviation used for fine thermal (black). N880. See *carbon black*.

full flow
A term used in *hydraulics* and which means that the *fluid* must pass through the filter element.

full machine
Term usually used in extrusion or injection moulding and means that the plasticization cylinder is full of polymer compound/resin. See *warming up*.

full spiral method
See *tapping methods*.

fuller's earth
See *diatomaceous earth*.

fully aromatic
An aromatic polymer which contains no aliphatic groups. For fully aromatic polyester, see *aromatic polyester*. For fully aromatic polyimide, see *polyimide*.

fully compounded material
Also called a colour compound. A compound which contains the ingredients required and which has been melt compounded. Term used to describe a thermoplastics compound fed to, for example, an *injection moulding machine*: nothing else is added. For many years, fully compounded material was widely used in the plastics industry as there was no real alternative until the advent of the *in-line screw machine*. The advent of such machines allows the use of a *masterbatch*. Despite the disadvantages, some moulders still prefer fully compounded material because, for example, the raw material supplier may have recommended the use of a particular grade for a specified application. This guarantee, or recommendation implies shared responsibility if anything goes wrong, e.g. through a toxicity claim. High colorant loadings may dictate the use of fully compounded material.

The injection moulding of fully compounded material gives consistently good results but unfortunately it is expensive, particularly when only a relatively small lot is required. For example, a moulder who only wishes to purchase a small lot may find it difficult to locate a supplier who will supply the fully compounded material at an economic price. There is also a tendency for the size of minimum orders to be increased by the raw material suppliers and for the range of fully compounded materials to be decreased. Many processors are finding it economically attractive to change over to *in-house colouring*: the *natural material* is then coloured at the machine using a *masterbatch*.

fully hardening steel
Also known as through hardening steel. A *mould steel* with a relatively high carbon content which may be hardened by a relatively simple heat treatment: components made from such a steel may be corrected at a later stage as the *steel* is hard throughout and not just at the surface. However, there is a risk of cracking as the steel does not have a tough core (see *case hardened steel*). A popular grade is X 45 Cr-Mn-Mo 4 steel which is a very tough chromium-manganese-molybdenum steel.

fully positive mould
See *positive mould*.

fully round runner
A runner with a circular cross-section. A fully round runner gives the lowest pressure losses but it is expensive to machine as it must be cut in each half of the mould and cannot be used in every type of mould; for example, in two-plate moulds where a sliding action occurs across the parting line. Where the fully round runner is expensive or inconvenient to use then the *trapezoidal runner* is used as it is almost as efficient in terms of pressure losses.

fully saturated nitrile rubber
Nitrile rubber with no unsaturation. This type of material is only cross-linked with peroxides but gives the best heat and hot oil resistance: sometimes known as peroxide-cured A type. See *hydrogenated nitrile rubber*.

fully-embedded hose
See *smooth bore hose*.

fulminate
See *cyanate*.

fumarate units
See *maleic anhydride*.

fumaric acid
See *maleic acid*.

fumed silica
Also known as fume silica or as, aerosil or as, pyrogenic silica or as, thermal silica. Fine *8 silica*, whose particle size is less than that of precipitated silica and of carbon black, and synthesized by pyrogenic processes. Primary particle size may be 15 μm: precipitated silica may be 15 to 20 μm. May be produced by burning silicon tetrachloride, with natural gas, at approximately 1,000°C. Fumed silica is a very fine silica used, for example, with silicone rubbers as it gives better heat resistance than *precipitated silica*. May be too expensive for general use. See *pyrogenic process*.

functional group
See *organofunctional silane*.

fundamental physical constants
Also known as fundamental constants or as universal constants. Constants which occur naturally and which are often interdependent so that a precise knowledge of a few allows the establishment of the others. Examples include the electronic charge and the Avogadro number.

fundamental unit
Units of measurement are of two kinds: fundamental units (basic units) and *derived units*. For example, the meter and the kilogram are fundamental units of the metric system whereas the litre and the hectare are derived units.

furan plastics
Plastics based on *furan resins*.

furan resin
A resin in which the furan ring is an integral part of the polymer chain and represents the greatest amount by mass.

furanized natural rubber
An abbreviation used for this material is furanized NR. The reaction product obtained if high acid concentrations and high temperatures are used during the production of *epoxidized natural rubber*. The NR chain is modified so that it contains a five-membered ring which contains -O-. Such materials give limp, leathery vulcanizates and are of little technological importance.

furnace black
The most common type of *carbon black*.

Furukawa process
A process used to produce *cross-linked polyethylene foam* by chemical cross-linking.

fusible core
A core or form made from a low melting point metal or alloy: such a core is used in a moulding process to produce components which have hollow sections. See *fusible core process*.

fusible core mould
Part of an *undercut mould* used to put an internal *undercut* into a moulded component. By using a *fusible core*, it is possible to produce mouldings with a very complex inner section. For example, a manifold may be produced from a thermosetting material.

fusible core process
A moulding process which uses a *fusible core*. Such a core is used in a moulding process to produce components which have hollow sections. After moulding around the core, the moulding plus the core is heated and the metal is removed by pouring so that the metal may be used again. Components produced by such a process include engine manifolds and sports racquets. Injection moulding of *short fibre reinforced thermoplastics* and compression moulding using *bulk moulding compound*, have been used.

FUSION

fusion
The formation of a hard fused system. See *fusion time*.

fusion time
The time from catalyst addition to the formation of a hard fused system. When, for example, an *unsaturated polyester resin* is set, or hardened, this is done by the addition of a *catalyst* system. Initially, the original liquid system forms a solid *gel*: then the resin hardens further. May be assessed by the resistance to penetration of a glass rod. May be arbitrarily taken as the time when the rod cannot be pushed into the resin system. See *gelation time*.

FVMQ silicone rubbers
A fluorosilicone rubber based on *polydimethylsiloxane* where some of the methyl groups have been replaced with fluoro and vinyl groups. FVMQ-types of silicone rubbers are used where high resistance to chemical attack (fuel, oil and solvent) is required. See *polyorganosiloxane*.

FWIS
An abbreviation used for *falling weight impact strength*.

FZ
A rubber which has nitrogen and phosphorous in the main polymer chain. The -P=N- chain has flouroalkoxy groups on the P atoms or aryloxy groups on the P atoms: the aryloxy groups are phenoxy and substituted phenoxy groups. See *phosphonitrilic polymer*.

G

g
This letter is used as an abbreviation for *gram*. Also used for the *acceleration* due to gravity.

G
This letter is used as an abbreviation for:
 filler. For example, used If a plastics material contains glass or, ground (filler);
 giga. See *prefixes - SI*; and,
 shear modulus;

g-mol wt
An abbreviation used for gram molecular weight.

gables
See *calender side frames*.

gal
An abbreviation sometimes used for *gallon* and for gallons. See *UK system of units* and *US Customary Measure*.

gallon
An abbreviation used for this term is gal.
 gallon - Imperial. Defined as the volume occupied by 10 *pounds* of distilled water under specified conditions. Contains eight pints and is equal to 4·546 090 litres or 277·419 432 cubic inches.
 gallon - US dry. This is equal to 4·404 884 litres.
 gallon - US liquid. Defined as being 231 cubic inches. Contains eight pints and is equal to 3·785 412 litres.
 An imperial gallon and a US gallon are not the same size. One imperial gallon equals 1·200 949 US gallons.
 One US gallon is 0·833 674 imperial gallons.

gals
An abbreviation sometimes (wrongly) used for gallons. See *gallon*.

galvanized
Iron which is protected by a layer of zinc.

gamma dot
The Greek letter gamma with a dot above it, is used for *shear rate*. That is, $\dot{\gamma}$.

gamma ray gauge
May be referred to as a γ gauge. A measuring device used in *calendering*. A gamma gauge responds primarily to wire weight but is slightly affected by polymer (for example, rubber) weight.

Garvey die
An extrusion die used in rubber technology to assess the extrusion behaviour of a rubber compound: a die designed to show the typical faults that can occur in a mix with poor extrusion characteristics. The die has two sharp corners and an acute-angled wedge section: the longest length is 9/16"/14·288 mm and the widest cross-section is 1/4"/6·350 mm. Mixes are rated by measuring surface characteristics, sharpness of extruded angle and *die swell*. This figures are reported separately or combined to give a single figure of merit.

gas
A gas is a state of matter which is differentiated from the solid and liquid states by very low density and viscosity.

gas injection moulding
An *injection moulding process* which injects gas together with the melt. A *semi-crystalline thermoplastics material*, such as *polypropylene* (PP), has a relatively high *shrinkage* and this means that it difficult to mould thick-sectioned components with a good surface finish.
 A technique that has been developed to overcome this problem relies on the injection of gas (for example, nitrogen) during the moulding operation. The gas may be injected into the flowing melt stream via the *injection nozzle* and, if this done, it is found that the gas stays in discrete pockets in the thicker sections of the moulding. As the moulding cools, the trapped gas exerts a pressure which pushes the thermoplastics material against the walls of the injection mould thus significantly reducing shrink marks. Thick-sectioned components, such as bathroom fittings have been produced in this way and the surface quality is such that the gas-containing, injection mouldings may be electroplated so as to give metal replacements from PP.

gas lighter test
See *lighter test*.

gas oil
Diesel oil.

gas permeability
This is usually independent of thickness and is often expressed with respect to the pressure difference per unit thickness. There will be a difference between two gases, for example, between nitrogen (N_2) and carbon dioxide (CO_2). See *vapour permeability* and *permeability*.

gas phase polymerization
A chain polymerization process in which the monomer is in the gaseous state. The monomer is fed to a solid polymerization catalyst where the polymerization occurs at the solid surface. Used to produce *polyolefins*.

gas reinforced injection moulding
See *gas injection moulding*.

gas testing
Toxic gas testing is still in its infancy as can be judged by the fact that few standards exist. It is a very complex subject as the amount and type of gas emitted depends not only on the polymer used but on associated additives and a particular fire situation. The methods that are used include mass spectra, gas chromatography and infra red analysis. See *smoke testing*.

gasoline
The name given to petrol in, for example, North America.

gassing
A *moulding fault*. Air trapped in the mould, for example, during *injection moulding*, and which on mould filling becomes very hot and so causes marks and burns on the injection mouldings.

gate
A valve, for example, the slide at the base of the hopper used to stop and start the flow of material or, the same term is used for part of the die. A hinged gate may be used to attach the adapter to the extruder. In injection moulding, the term means the restriction within the feed system which connects the *runner* to the cavity. A narrow restriction in the runner system. As the gate size is much smaller than the runner, the moulding may be easily separated from the feed system at this point when desired. It is recommended that initially the gates be made deliberately small so that they may be opened up after moulding trials.

Many different types, or styles, of gate are used within the feed system. If the gate seriously interferes with material flow then it may be known as a restricted gate; conversely where there is no serious flow obstruction, the gate may be known as an unrestricted gate. See, for example, *tab gate*, *sprue gate* and *pin-point gate*.

gate land
The length of the gate as measured from the *runner* to the cavity. In general, the *gate land length* for edge, pin or film type gates should not be greater than 0·75 mm (0·030 in) in length.

gate location
The *gate* (of an injection mould) should be located so that easy and therefore economical separation of the component from the *runner* system is obtained. To prevent *jetting*, the gate should be located so that the incoming plastics material impinges against an obstruction within a short distance, for example, within 6 mm (the use of *programmed injection* can, however, eliminate this requirement).

As far as possible the material should enter the cavity at the point where the wall thickness of the component is greatest; as the gate area is commonly a highly stressed area the gate should not be located in a position which will be exposed to stress in service. Aesthetic reasons may also restrict gate location areas to hidden or masked parts of the moulding. If the flowing melt stream is divided, e.g. by a core, then when the melt recombines a *weld line* and/or *burning* may result: mould venting will alleviate burning and will also help to improve the strength of the weld.

If possible, the gate should be located so that, if a weld has to be produced, the divided melt is recombined as quickly as possible. That is, the weld is located as close to the gate as possible. Where part of a component is required to have the best 'see-through' properties then that part should not be located in a region remote from the gate. Some components, e.g. large mouldings, are made via more than one gate (for example, four) as this shortens the flow length and results in a lower clamping pressure requirement. However, unless there are very good reasons for their use, multiple gates should not be used.

gate shear rate
If the *gate* (of an injection mould) is considered to be either a hole or a slit then it is relatively easy to calculate the shear rate for a particular mould and machine; whether such formulae are appropriate for high-speed mould filling is debatable. To calculate the shear rate which exists in a particular gate, one needs to know the rate of mould filling and the dimensions of the gate.

The Newtonian shear rate for a circular orifice is given by $\dot{\gamma} = 4Q/\pi r^3$ where $\dot{\gamma}$ is the shear rate (s^{-1}), Q is the volumetric flow rate ($cm^3 s^{-1}$); and r is the radius of the orifice (cm). For example, if the mould is filled in 0·03 s with 8 g of material (which has a density of unity) through a gate of diameter 1·6 mm then the shear rate will be: $(4 \times 8/0·03)/3·142 \times 0·08^3$. Such shear rates are possible but are not typical of those normally found in injection moulding. High shear rates may cause, for example, severe polymer degradation to occur with some materials - for such materials a minimum size of gate may be recommended by the materials manufacturer. In general, the *shear rate* in injection moulding is kept below 50,000 s^{-1}.

The shear rate in a rectangular gate (or slit gate) is $\dot{\gamma} = 6Q/Lt^2$ where $\dot{\gamma}$ is the shear rate (s^{-1}), Q is the volumetric flow rate ($cm^3 s^{-1}$), L is the slit length (cm) and t is the slit thickness in cm (not the land length).

gate size
Because of a lack of suitable information (for example, material temperature, viscosity and rate of fill) gate sizes are not normally calculated but are usually obtained by experiment. The gate of an injection mould is, therefore, usually made deliberately small (e.g. 50% of part thickness) and then metal is removed, after sampling, in order to achieve the desired filling rate and/or pattern. This is because it is easier to remove metal than it is to put it back. One of the biggest dangers with this approach is that the final gate size will be rather small and this will result in mouldings of poor surface finish and non-uniform shrinkage.

If the thickness of the moulding at the gate is t, then the gate thickness will range from 0·5 to 0·9 t; it will be 0·5 t for low-viscosity melts such as those found with nylon, and it will be 0·9 t for a high viscosity material such as *unplasticized polyvinyl chloride (UPVC)*. The width of a rectangular gate is usually 0·75–1·5 t. Gate length (gate land length) is usually kept as short as possible and is often dictated by the strength of the mould metal; it is commonly approximately 1 mm (0·04 in) in length.

With a multi-cavity mould, the short shots produced during the moulding trials should be used to indicate which gates need modification or balancing. This may be done by increasing the cross-sectional area of the gate or by decreasing the land length of the gate. The former alters both shear rate and pressure drop; the latter only reduces pressure drop and has the virtue of being more accurate. For example, a gate may only require a very slight adjustment to enable the cavity to fill, and on a land length of 0·06 in (1·5 mm) removal of 0·002 in (0·05 mm) of metal will give a reduction in pressure drop of about 3%. Removal of the same amount of metal on the diameter of a round gate will alter the pressure drop by about 12%.

gate vestige
A blemish seen on an injection moulding: a mark which shows gate location, for example, when using a hot runner mould or nozzle. To prevent gate vestige, or a short fibre attached to the gate, it is best to use hot runner nozzles that incorporate a shut-off valve, or needle, and very accurate temperature control.

gauge
An instrument on which the measured value is displayed, in analogue form, on a dial. Gauge is often used as a measure of film thickness (100 gauge is 1 mil or 0·001 in).

gauge bands
Also known as piston rings. Regions of increased thickness on a roll of film; produced through winding thicker film in the same place or, on the same part of the reel each revolution. May be eliminated by, for example, rotating the *die*.

gauge length
The original length of the *gauge section*.

gauge pressure
The pressure read on a gauge and which does not take account of the pressure exerted by the atmosphere. That is, the pressure in excess of the atmosphere. See *absolute pressure*.

gauge section
That portion of a test sample which is actually tested or deformed. For example, the sample region in the waist region of a *dumb-bell*, over which deformation occurs, is the gauge section.

gauge thickness
Gauge is often used as a measure of film thickness (100 gauge is 1 mil or 0·001 in).

gauge variation
Variations in product thickness. See, for example, *calendering*.

gauging
Measurement. See *in-process gauging* and *post-process gauging*.

gauging nip
The final nip of a *calender*.

gauss
The unit of magnetic flux density in the *cgs system*. One gauss is equal to 10^{-4} tesla.

gaussian chain
A randomly jointed chain used as a model for a polymer chain or molecule; the statistical distribution of chain end-to-end distances is a Gaussian, or normal, distribution.

gaussian distribution
A normal distribution.

gear plate
Part of an *injection mould*. A steel plate machined to accommodate the sun and planet gear system of an *unscrewing mould*.

gear pump
A type of pump which gives movement to a liquid material by propelling it through the system by the action of two intermeshing gears. See *gear pump extruder*.

gear pump extruder
An extruder which uses intermeshing gears to pump the plastic or resin. Used, for example, to produce fibres from polymers which are low viscosity liquids at processing temperatures, for example, polyamides. Such a device is also fitted to a single-screw extruder in order to give a steadier pressure, and therefore a smoother output, from the die.

gel
A semi-solid system consisting of a network of solid aggregates in which liquid is held: a cross-linked polymer system, insoluble in solvents, but which may be swollen by solvents. In extrusion usually means the lumps seen in film and which result, for example, from resin degradation or crosslinking.

gel cellophane
A film of regenerated *cellulose* which after being produced is kept wet.

gel coat
Also spelt gelcoat or gel-coat. Also known as a surface coat. A protective surface layer of resin: a protective coating. Associated with *unsaturated polyester laminates* where the gel coat is applied to the mould surface and allowed to harden before bulk, reinforced layers are applied by hand or by spray. Traditionally associated with *hand lay up moulding* but also used with *resin transfer moulding*. The gel coat stops water penetrating between the fibre and the resin in the reinforced layer.

gel particles
Cross-linked polymer in, for example, a thermoplastics resin or system. Such particles are sometimes seen in *polyethylene* where, on *lay flat extrusion*, they form elastic particles in the film.

gel point
The onset of *gelation*.

gel rubber
The benzene insoluble portion of rubber.

gel time
The time taken to form a *gel* from a liquid polymer system. The time, taken from the addition of the curing system, for a liquid resin to change to a *gel*. In the case of an *unsaturated polyester resin* may be assessed by the resistance to penetration of a glass rod. The time, measured in for example, minutes, from the addition of the catalyst system to the setting of the resin to a soft gel which can just be easily penetrated by a glass rod. See *exotherm curve*.

gel-coat
See *gel coat*.

gel-type ion exchange resin
A type of *ion exchange resin* with a dense internal structure. The first type of ion exchange resin to be commercially introduced was gel-type polystyrene resin. See *styrene ionomer*.

gelatin
Also known as gelatine. A protein obtained from collagen by heating with water. Water soluble and used, for example, as an adhesive and a textile size. A solution of this material will set or *gel*.

gelatine
See *gelatin*.

gelation
The formation of a *gel* from a liquid polymer system. A term usually used in *polyvinyl chloride* (PVC) technology to denote that the PVC particles will gel, or fuse, under the combined effects of heat and pressure. The term is also used to describe the setting (or hardening) of a PVC plastisol on heating. The term is also used to describe the setting (or hardening) of a liquid resin system after it has been catalyzed. When, for example, an *unsaturated polyester resin* sets, or hardens, the original liquid system initially forms a solid gel which then hardens further. See *gel time*.

gelation time
See *gel time*.

gelcoat
See *gel coat*.

gelled
A term used to describe a system which has set or fused, for example, *polyvinyl chloride* (PVC). Gelled PVC results when PVC particles have fused, or gelled, under the combined influence of heat and pressure.

gelling agent
A material which causes gelation of *latex*. Also known as a heat sensitizing agent. See *latex processes*.

General Motors heat test
See *heat sag*.

general purpose
An abbreviation used is GP. For example, general purpose furnace (black) is GPF (see *carbon black*).

general purpose natural rubber
Also known as GPNR or as, or as NR-GP or as, general purpose NR. A *natural rubber grade* made by combining

latex and *field coagulum*: the ratio may be 60:40. A viscosity stabilized blend of latex grade rubber and field coagulum material. A large volume grade of *natural rubber* with consistent processing properties.

general purpose phenol-formaldehyde
General purpose PF. The filler, *woodflour*, gives reasonable properties at an acceptable cost and so woodflour-filled PF is regarded as a general purpose material. See *phenolic moulding material*.

general purpose plasticizer
A widely used *plasticizer*, for example, a *phthalate plasticiser* in which there are approximately 8 carbon atoms in the side groups.

general purpose polystyrene
An abbreviation used for this material is GPPS. See *polystyrene*.

general purpose rubber
Also known as GP rubber. A widely used rubber. For example, one which is widely used in the tyre industry. Included in this category are *butadiene rubber*, *butyl rubber*, *natural rubber*, *polyisoprene rubber* and *styrene-butadiene rubber*.

general purpose silicone rubber
A VMQ-type material may be considered as a general purpose *silicone rubber*.

generator
A device for converting mains frequency energy to *high frequency* or *microwave frequency energy*.

geographical mile
See *nautical mile*.

German Machinery Plant Manufacturers Association
An abbreviation used for this German-based organization is VDMA.

GeV
An abbreviation used for one thousand million electron-volts: that is *giga electron-volts*. May also be referred to as BeV in North America.

gf den
An abbreviation used for gram-force per denier.

GFT
An abbreviation used for *glass fibre test*.

giga
An SI prefix which means 10^9. Abbreviation used G. See *prefixes - SI*.

gilbert
The unit of magnetomotive force, in electromagnetic units, in the *cgs system*. It is equal to $10/4\pi$ ampere-turns. To go from gilberts to amperes multiply by 0·7958.

gill
A unit of measurement which is ¼ of a pint. See *UK system of units* and *US Customary Measure*.

gilsonite
Asphalt with a melting point of approximately 110 to 160°C and a relative density of 1·15. Used, for example, as an extender and tackifier in rubber technology.

Giorgi system
The complete *MKS system of measurement*. Combines the units from mechanics with the usual electrical units. That is, the units of length (the metre), mass (the kilogram), time (the second), and force (the newton) are combined with the volt, the ampere, the ohm, the farad and the coulomb. Also included are the weber/m² for flux density and the henry for inductance.

glacial acid
Pure *acetic acid*.

glass
An inorganic product of fusion which has cooled to a rigid condition without crystallizing. Silica sand (*silica*) and limestone (*calcium carbonate*) are the basic ingredients from which the inorganic, amorphous polymer known as glass is manufactured. The other ingredients used depend upon the end-properties required in the finished glass but include aluminium oxide, boric oxide, sodium carbonate, potassium carbonate and magnesium oxide. The use of such materials will, for example, cause a reduction of softening point and so facilitate processing. In the bulk state, at room temperatures, glasses are stiff and hard but because of the absence of surface defects, useful fibres can be produced. *Borosilicate glass* is commonly used to make the *glass fibre* products used by the plastics industry.

glass balloons
See *glass microspheres*.

glass ballotini
See *glass beads*.

glass balls
See *glass beads*.

glass beads
Also known as glass balls or as, balls of glass. Small particle size, solid spheres of glass available in a range of particle sizes. Has the appearance of a fine white filler: used as a filler in plastics materials to reduce, for example, shrinkage. Used in, for example, *unsaturated polyester resins* and sometimes in *thermoplastics materials*. If surface treated can act as a reinforcing filler in some systems. Glass spheres of approximate diameter 0·1 mm to 0·25 mm. Used, for example, in *fluid bed vulcanization*.

glass bubbles
See *glass microspheres*.

glass cloth
Also known as a glass fabric. A cloth made from *glass fibres*: the cloth is usually made from the fibres by weaving. Woven glass fabrics are most important to the reinforced plastics industry where they are used to make *composites*. Such fabrics usually have uniform weight per unit area: this may be used to classify the cloth along with the type of weave. See *yarn*.

glass container
A container, such as a bottle made from *glass*. Glass has been used for such a long time that its disadvantages have become accepted: the disadvantages are brittleness and heaviness. The advantages of this extraordinary material include clarity, attractive colours, chemical inertness, rigidity, hardness, impermeability and temperature resistance. It is derived from cheap raw materials which are readily available in most parts of the world. (Precision, mass produced bottles, made in a split mould, were first patented in 1821).

glass fabric
See *glass cloth*.

glass fibre
A synthetic *fibre* made from *glass*. Synthetic fibres are easily made from *glass*. The ingredients are charged into a furnace where they are fused at high temperatures to form molten glass. Fibres can be made directly from this or, the glass may be formed into marbles, cooled, inspected, sorted, re-heated and then turned into fibres. Continuous filament strands are formed by allowing the molten glass to flow through small holes in the baseplate of a platinum melter (a bushing). As the fibres form they are drawn away at high speed and

approximately 50 fibres are combined to form a strand. A lubricating size, designed to suit the polymer of the future composite, is applied so as to minimise fibre breakage and to improve resin to glass adhesion. The strand may be converted into *chopped strand mat* (CSM), *rovings* or *chopped strands* and used in *glass reinforced plastics moulding*.

Glass fibres (GF) are one of the plastics industries favourite fillers as glass is relatively cheap, is inert and gives reasonable properties in many plastics materials. The use of glass fibre often improves, for example, heat resistance and stiffness of a *thermoplastics material*. A rough rule of thumb is that if a thermoplastics material has a large difference between the heat distortion temperatures measured at the two common loads (264 and 66 psi) then it is worth reinforcing so as to improve the heat resistance.

Because of the differences in density between inorganic materials, such as glass, and polymers it is found that large amounts, by weight, of inorganic fillers must be added in order to obtain significant improvements and/or, cost savings. The density differences can cause problems in metering and mixing while the abrasive nature of such inorganic materials introduces wear problems. Anisotropy of fibres causes differences to arise in components and, fibre attrition (or breakdown) can cause serious problems, for example, warping of components and a loss of strength.

glass fibre test
An abbreviation used for this term is GFT. This moisture content test is used for assessing the moisture content of granules of glass-reinforced grades of polycarbonate. A glass container with a capacity of about $2-3$ cm^3 is filled with granules and then heated to 310–320°C for 3 minutes in an oil bath. The granules are examined after brief cooling. For safe processing they must be glass-like and not whitish and porous after heating.

glass finish
A material applied to the surface of *glass fibres*, used to reinforce plastics, and intended to improve the physical properties of such reinforced plastics over that obtained using glass reinforcement, without finish.

glass flake
See *flake glass*.

glass mat
A random arrangement of *glass fibres* which are lightly bonded together: the bonding is just sufficient to allow handling. Such mats are usually sold on the basis of weight per unit area.

glass mat (reinforced) thermoplastics (material)
An abbreviation used for the term, glass mat thermoplastics, is GMT. GMT is however, taken to refer to a glass mat reinforced thermoplastics material. A composite material based on a thermoplastics material reinforced with a glass mat and supplied in sheet form. Cut sections from the sheet are shaped by *compression moulding* into strong, stiff composite mouldings. See *glass reinforced thermoplastic composite*.

glass mat thermoplastics
An abbreviation used for this term is GMT. See *glass mat (reinforced) thermoplastics (material)*.

glass microspheres
Also known as glass bubbles or as, bubbles of glass or as glass balloons. Small particle size, hollow spheres of glass available in a range of particle sizes. Such a material has the appearance of a fine white filler: used as a filler in plastics materials to reduce the density and to produce *syntactic foams*. Used in, for example, *unsaturated polyester resins*: care is need during mixing to get good wetting without breaking the spheres. Glass microspheres, can be introduced dry into the resin stream of a *spray gun* so as to minimise filler mixing problems and produce light-weight composites.

glass reinforced nylon
A *polyamide (PA)* which contains glass as a *filler*. Often the *glass* is in the from of short glass fibres. Such compounds are often *injection moulded*. When injection moulding glass reinforced PA, the gate must be positioned so that the orientation of the fibres is in the direction of the longest length of the component. This tends to produce mouldings having lower shrinkage values and less distortion. Glass reinforced grades may require higher clamping pressures (compared to non-reinforced grade), for example, 5 to 7 tsi/77 to 108 MNm^{-2}: this is because of the higher injection speeds needed to produce acceptable components. Glass reinforced grades cause high wear and this can be minimized by using ion implantation on screws and barrels or, by using screws with Stellite flights. Glass reinforced, impact-modified grades of PA 6 are available which are good electrical insulators and which can also withstand reasonably high temperatures without distortion - useful for power tool housings.

glass reinforced plastics
An abbreviation used is GRP. The term usually refers to room temperature curing systems such as *unsaturated polyester resins* and *glass fibre* or, *epoxide resins* and *glass fibre*. See *glass reinforced plastic moulding*.

glass reinforced plastics moulding
An abbreviation used for this term is GRP moulding. The moulding processes whereby glass reinforced plastics mouldings are produced: a moulding or product which consists of a plastics material in which is embedded glass fibre. Boats are the most widely known example of this type of product and now quite large craft (i.e. greater than 40 m in length) are made using GRP techniques. As the plastics materials used for such products are often based on polyester resins (an *unsaturated polyester resin*) the initials GRP may also be taken to stand for glass reinforced polyester.

Such resins are solutions of an *unsaturated polyester resin* dissolved in styrene which can be set by an appropriate hardening system, for example, based on *methyl ethyl ketone peroxide* and *cobalt naphthenate* (also called a catalyst system). Once the resins have been catalyzed they will then set to shape at room temperature without eliminating any volatile by-products. As the materials are liquids they conform to the shape of the mould without the necessity of applying pressure: as no volatile by-products are eliminated during the setting operation, pressure is again not required. Moulds are therefore relatively cheap and this means that the production of large components can be economically attractive. See *hand lay-up*.

GRP laminates may be coated or protected, with a *gel-coat* which is, for example, 0·5 mm thick: such gel coats protect against weathering and *fibre bloom*. Such a gel coat is incorporated during the moulding process: if damaged in use, the gel coat can sometimes be successfully repaired although this is difficult to do well and is time-consuming.

The part may also be gel-coated during moulding, sanded subsequently and then painted using, for example, acrylic paints: this allows the production of large structures from separate components and gives a uniform colour to the assembly. Alternatively *in-mould coatings* may be used.

glass reinforced polyester moulding
See *glass reinforced plastics moulding*.

glass reinforced polyurethane moulding
The production of a moulding (reinforced with *glass fibre*) from a *polyurethane* material or, a moulding (reinforced with glass fibre) which is based on a polyurethane polymer. See *reaction injection moulding* and *structural reaction injection moulding*.

glass reinforced thermoplastic composite
A thermoplastics-based composite in which most of the fibrous reinforcement is based on glass fibres. For example, if the fibres are 50 mm in length, but not less than 12 mm in length, then the composite may be called a *long glass fibre thermoplastic composite*.

glass reinforced thermoplastic polyurethane
See *reinforced thermoplastic polyurethane*.

glass rovings
Rovings made from *glass*. See *roving*.

glass rubber transition
See *glass transition*.

glass rubber transition temperature
See *glass transition temperature*.

glass transition
Also known as the glass rubber transition or as, the rubber glass transition or as, γ transition or as, α anomaly or as, second order transition or as, vitrification. The change which occurs on heating or cooling a polymer. That is, a polymer becomes glass-like, or brittle, on cooling. See *glass transition temperature*.

glass transition temperature
Also known as T_g or as, the glass-rubber transition temperature or as, second order transition temperature. The temperature at which a polymer becomes glass-like on cooling or, the temperature at which a polymer becomes rubber-like on heating. Not usually a sharp, precisely defined point but more a temperature range. See *glass transition*.

glass-ceramics
Materials which are chemically similar to *glass* but which are crystalline. Based on, for example, lithium and magnesium alumino-silicates and have, for example, very small coefficients of thermal expansion and isotropic mechanical properties. Components are made by casting then re-heating to above the glass transition temperature (T_g) so as to induce crystallisation.

glassy state
Means that a polymer has formed a glass-like structure and is relatively brittle. May be formed, for example, in rubbers, by cooling the material to below the *glass transition temperature* (approximately 20°C below T_g). In the glassy state, the rubber will behave as a brittle solid and may be fractured by a sharp blow. Useful in processes such as *freeze grinding*.

glaze
Usually associated with pottery where a glass-like material is used to improve the surface finish. The surface finish of some plastics materials has also been improved by glazing. For example, the surface finish of *high impact polystyrene* sheet is improved by heating - preferably in an inert atmosphere.

glazing
See *glaze*.

glimmer
Also known as biotite or as, muscovite. A silicate material with a relative density of approximately 2.8. Used as a dusting agent for rubbers.

gloss
A measure of the ability of a material to reflect light. An optical feature of all materials whether transparent or opaque. When a beam of light strikes a surface the amount reflected will depend not only on the type of surface finish and refractive index of the material but also on the angle of incidence and the angle of viewing of the reflected beam used in the measurement. Because of this many standard methods of measuring gloss exist, varying mostly in the angle of incidence and reflection used.

In many instances tests are made on specimens cut from finished products. No further preparation is usually necessary unless test temperature and humidity are known to affect the result (when careful conditioning prior to testing will be needed). Specimens made for comparison purposes by injection and/or compression moulding need to be made very carefully as the result can be affected by many factors. The following are important in *injection moulding*: mould temperature, the rate of mould filling and mould finish.

glucose
Also known as dextrose and grape-sugar. May be represented as $C_6H_{12}O_6$ and is a six-membered ring. A colourless, crystalline solid with a melting point of approximately 145°C. The naturally occurring form is dextrorotatory or d-glucose. *Cellulose* consists of long chains of glucose units - up to 10,000 glucose units. Glucose may be regarded as the monomer for cellulose.

glyceride
An ester of *glycerol* and a *fatty acid*. The fatty acids of glycerides are long chain saturated, and unsaturated, organic acids: the metal salts od these acids are soaps. Glycerides which are solid at room temperatures are often referred to as fats. Glycerides which are liquid at room temperatures are often referred to as oils. See *triglyceride* and *oil*.

glycerin
See *glycerol*.

glycerine
See *glycerol*.

glycerine derivatives
Also known as *glyceryl derivatives*.

glycerine monostearate
See *glyceryl monostearate*.

glycerol
Also known as glycerin or as, glycerine or as, propane-1,2,3-triol or as, 1,2,3-propane-triol. May be represented as CH_2OH-$CHOH$-CH_2OH. A thick, syrupy liquid with a melting point of approximately 18°C, a boiling point of 290°C and a relative density of 1.27. A colourless, liquid triol which is water soluble. Used as a processing aid in rubber extrusion and in rubber reclaiming. See *glyceryl derivatives*.

glyceryl derivatives
Also known as glycerine derivatives. The glyceryl derivatives of organic acids are widely used in the polymer industry. For example, *glyceryl monostearate*, is used as a softener in the rubber industry and as a lubricant in the plastics industry. One, two or three hydroxyl groups, of *glycerol*, may be used to make derivatives. For example, can have the mono-acetate, the di-acetate and the tri-acetate.

glyceryl monostearate
Also known as glycerine monostearate or as, monostearine. An abbreviation used is GMS. A white, waxy solid with a melting point of approximately 58°C and a relative density of 0.97. A *glyceryl derivative* used as a softener in the rubber industry. Used as a *lubricant* in the plastics industry, particularly for *unplasticized polyvinyl chloride* (UPVC).

glycidyl
The prefix used for an epoxy containing group which is commonly found in *epoxy resins*.

glycidyl ester
The epoxy containing group which is formed by the reaction of *epichlorhydrin* and a carboxylic acid.

glycidyl ether
The group which is formed by the reaction of *epichlorhydrin* and an alcohol. Glycidyl ethers are commonly used in *epoxy resins*: such materials are diglycidyl ethers.

glycidylamine epoxy.
See *tetraglycidyl methylene dianiline epoxy resin*.

glycol modified PET
An abbreviation used for *glycol modified polyethylene terephthalate (PET)*. See *copolyester*.

glycol modified polyethylene terephthalate
An abbreviation used for this type of material is glycol modified PET. A *copolyester* which may be made by adding a second glycol (cyclohexanedimethanol or CHDM) during polymer manufacture. For example, *terephthalic acid* may be reacted with two glycols (ethylene glycol and cyclohexanedimethanol). See *poly-(1,4-cyclohexylenedimethylene terephthalate-co-isophthalate*.

glycolysis
A procedure for re-using, or recycling the *polyether polyols* in *flexible polyurethane foam*. Foam flakes are digested in a glycol mixture at moderate temperatures (approximately 200°C) and pressures: excess glycol is removed and *polyether polyol* added so as to give glycolysis polyol which is re-used. See *particle recycling process*.

glyptal
A type of alkyd resin which is formed from *glycerol* and *phthalic anhydride* and modified by the use of a *fatty acid*. See *granular polyester moulding compound* and *polyester moulding compound*.

GMS
An abbreviation used for *glyceryl monostearate*.

GMT
An abbreviation used for glass mat thermoplastics. See *glass mat* (reinforced) *thermoplastics* (material).

gmw
An abbreviation used for gram molecular weight.

Godet unit
A multiple roll, or wheel, drive unit which is usually used on monofilament and strip production; the extrudate is drawn along by the friction between the extrudate and the rolls. Such a unit is commonly used to orientate monofilament. The fibre is wrapped around the wheel and drawn away from the spinneret at a speed which will give the required fibre diameter. By the use of a second, on-line wheel, rotating faster than the first, the fibre can be stretched to give controlled orientation. See *viscose rayon*.

gold blocking
See *hot foil marking*.

gold rubber
Rubber sheet which has a gold-like appearance. Produced from *latex sheet* which is painted with an aluminium powder/rubber solution and then vulcanized with sulphur chloride.

gold stamping
See *hot foil marking*.

golden sulphur
See *antimony pentasulphide*.

goldenrod
See *solidago*.

good dispersion
Dispersion which, for example, results in the absence of obvious visible defects, e.g. caused by poor *additive* dispersion. To get the best results from additives, they must be very well dispersed within the basic polymeric material. Melt mixing using, for example, a *compounding extruder*, is very important to the polymer industry as it gives good dispersion. It also gives a continuous output which can be in the form of rods: this output can then be chopped into regularly-sized pellets and these often form the feed to moulding machines. However, in some cases a lower level of dispersion is tolerated to save on costs and to give operational flexibility (see *masterbatches*).

Goodrich plastimeter
A parallel plate type of *plastimeter*.

Goodyear pendulum
An instrument used to measure *resilience*.

GOST standards
Standards used in the USSR.

Gough-Joule effect
Also known as the Gough effect or as, the Joule-Gough effect. The increase of modulus in strained rubber which occurs when the temperature is raised.

GP
An abbreviation used for general purpose, as in general purpose polystyrene (GPPS). Also used for *granular polyester*.

GP rubbers
An abbreviation used for *general purpose rubbers*.

GPF
An abbreviation used for general purpose furnace (black). N660. See *carbon black*.

gpm
An abbreviation used for gallons per minute. To go from gallons per minute to cubic metres per second, multiply by 0·2271.

GPMC
An abbreviation used for *granular polyester moulding compound*. A *polyester moulding compound* or *PMC*.

GPO
An abbreviation used for propylene oxide (copolymer) rubber. See *propylene oxide rubber*.

GPPS
An abbreviation used for general purpose *polystyrene*.

gr
An abbreviation used for *grain*.

GR
An abbreviation once used for Government Rubber. For example:

- **GR-A** is Government Rubber - Acrylonitrile (see *nitrile rubber*);
- **GR-I** is Government Rubber - Isobutylene (see *butyl rubber*);
- **GR-M** is Government Rubber - Monovinylacetylene (see *chloroprene rubber*);
- **GR-P** is Government Rubber - Polysulphide (see *thiokol rubber*);
- **GR-S** is Government Rubber - Styrene (see *styrene-butadiene rubber*);
- **GR-S-X** is Government Rubber - Styrene (test or experimental). See *styrene-butadiene rubber*.

grade
A material type: a modification, or variation, of a base material, for example, a *thermoplastics material* is often sold in a wide range of grades or types. Once a thermoplastics material has been selected for a particular application, then a high viscosity grade should be selected if the components are to be subjected to severe mechanical stresses; this is because the high viscosity grades usually have the highest molecular weight and exhibit the best mechanical properties. However, in some cases, for example, in *injection moulding*, this advice cannot be followed as unacceptable levels of *frozen-in strains* result. Easy flow grades are preferred for filling thin walled sections or for use where very smooth surfaces are specified.

graft copolymer
A type of *copolymer* which is formed, for example, by polymerizing a monomer in the presence of a formed polymer. Some of the new polymer formed, grafts, or joins, onto the existing polymer chain.

grafted latex
See *polymethyl methacrylate grafted natural rubber*.

grafted natural rubber
A modified *natural rubber (NR)*. See *polymethyl methacrylate grafted natural rubber*.

grafting
See *graft copolymer*.

grain
A unit of weight with the abbreviation gr. In many measurement systems, a grain is the smallest unit. Originally, for example, determined by the weight of a specified number of grains of wheat. In the United Kingdom and United States systems, the grain is identical. Has the same value in apothecaries', avoirdupois and Troy systems. The equivalent value is 0·064 798 9 grams.

gram
One thousandth of a *kilogram*. Abbreviated to g. Also spelt gramme.

gram cal
An abbreviation for gram calorie. See *calorie*.

gram weight
A unit of force in the cgs system. The pull of gravity on a *gram mass*.

gramme
An alternative way of spelling *gram*.

gramme cal
An abbreviation for gramme calorie. See *calorie*.

granular moulding compound
An abbreviation used for this type of material is GMC. See *granular polyester moulding compound* and *polyester moulding compound*.

granular polyester compound
See *granular polyester moulding compound* and *polyester moulding compound*.

granular polyester moulding compound
An abbreviation used for this type of material is GPMC. Also known as a granular polyester compound or as, granular moulding compound (GMC) or as, nodular moulding compound (NMC) or as granular polyester or as, alkyd or as, a polyester alkyd or as an alkyd moulding compound (AMC). A *polyester moulding compound*.

This type of material is based on an *unsaturated polyester resin (UP)*. Although UP resins are normally made from phthalic anhydride, maleic anhydride and a glycol such as propylene glycol, many other starting materials may be used. These may be chosen to give a solid resin after reaction: such a resin may then be blended with monomers, other than *styrene*, to give solid systems. When a non-volatile monomer (such as *diallyl phthalate* or *diallyl isophthalate*) is used, the moulding composition which results after fillers, catalysts, etc. have been added, is sometimes known as an alkyd, a polyester alkyd or, an alkyd moulding compound (AMC). However, the term alkyd is not nowadays widely used; the term granular polyester moulding compound is preferred. Use of the initials PMC, standing for pelletized moulding compound, can lead to confusion as the same initials are now being used as a generic term for all moulding compounds based on unsaturated polyesters for example, *dough moulding compound* or DMC) and for powder mould coating.

A wide range of GPMC grades is possible as even if the resin and monomer are kept constant, the filler type, size and concentration may be varied. Like DMC, this type of material usually contains both mineral and fibrous fillers - the fibrous filler is usually glass fibre (GF).

Although GPMC can be compression and transfer moulded successfully it is usually injection moulded and, in general, the surface finish obtained is far better than can be achieved from, for example, *dough moulding compound (DMC)*. Unlike DMC, GPMC can be moulded into complex shapes without causing resin/filler separation in deep sections, such as ribs and bosses. Not as strong as DMC (the GF is shorter) but has the advantage of being a dry granular solid which looks, and handles, like pellets of a thermoplastic material. Because the material is pelleted, there is no need for an expensive *stuffer unit*.

Moulded components are resistant to high temperatures - GPMC can withstand up to 220°C/428°F, whereas DMC would not be recommended above 160°C/320°F. The chemical resistance is similar to DMC. It is not resistant to boiling water, more affected than *allyl plastics* on exposure to humidity or boiling water. As with DMC, acidic environments can cause problems due to the widespread usage of calcium carbonate fillers in this type of material. The moulded material can have a high density - around 2 g/cm^3. This type of material is available in a wide range of colours, has good heat resistance and resists burning - oxygen index is approximately 50% and the material is rated as UL 94 V-0.

GPMC materials are used for the manufacture of automotive accessories such as distributor caps, ignition coil caps and lamp sockets. Also used for domestic appliance applications such as iron handles, toaster or grill housings and switches for washing machines and dishwashers. The good mechanical stability and excellent insulating qualities, are properties which allow the successful use in components such as connector boxes for telecommunication systems and casings for explosion proof switches. A major development is in the use of GPMC for microwave cookware as the materials have good transparency to microwaves and can withstand cooking conditions. As these materials can withstand up to 220°C/428°F, microwave trays can be hot filled, frozen and reheated without apparent damage. DMC dishes are glossier but must be post-baked to remove traces of styrene.

Allyl and GPMC materials have, broadly speaking, similar properties, use, and employ similar processing conditions. Both are dry, granular solids. GPMC materials are, however, weaker mechanically than allyl materials, have a lower resistance to cracking around inserts, do not maintain their electrical properties so well under extreme conditions, but are cheaper.

granulated rubber
See *crumb rubber*.

granulator
A machine which reduces plastics components to pieces of a smaller size: such a machine is used to grind the plastics com-

ponents for re-use and does not necessarily produce regular *granules*.

granules
A raw material feed form; the output from a *granulator* and which is often a rough, or coarse, cut.

grape-sugar
See *glucose*.

graphite
Also known as plumbago or as, blacklead. A crystalline form of *carbon* which has good lubricating properties. Because of the structure, this material is also a good conductor of heat and electricity. See *carbon fibre*.

graphite reinforced thermoplastic prepreg
A *prepreg* containing continuous *carbon fibres*. See *continuous fibre reinforced prepreg*.

gravimetric thickness
May be represented as t and obtained by weighing a sample of known area and then determining the density. Obtained by dividing the mass by the area and the density. Given by a *beta ray gauge*.

gravitational system of units
One which uses the weight of a standard body as the basic force unit. Usually associated with the *FPS gravitational* (imperial or British Engineering system of units).

gravity
A shortened form of specific gravity. See also *acceleration*.

gravure printing
A *printing technique* in which the ink is transferred to the substrate, for example, a film, from an engraved surface, that is, one on which the design has been recessed. The image is etched (recessed) in the form of small cells on the surface of a metal cylinder, for example, photographically. The tiny cells are of different depths so that differing amounts of ink are held: the surplus is removed by a doctor blade. The ink-covered gravure roller is then pressed against the film which is backed by an impression cylinder. The inks are based on resins and solvents, such as toluene, and the ink dries by evaporation of the solvent with heat often being used. Ink viscosity is of the order of 2 poise. Although the gravure cylinders are expensive to produce this process is suited to the printing of non-porous substrates, for example, coated cellulose films.

gray
An SI derived unit which has a special symbol, that is Gy. It is the SI derived unit of absorbed dose of ionizing radiation. It is the energy in joules absorbed by one kilogram of irradiated material (Jkg^{-1}). The gray is equivalent to 100 rads. See *Système International d'Unité*.

greasy sheet
A fault in raw, natural rubber sheet caused through insufficient washing at the production stage.

green
In rubber technology means raw or uncured.

green book
A publication issued by the Rubber Manufacturers Association Inc of New York and entitled International Standards of Quality and Packing for Natural Rubber.

green compound
An un-cured or un-vulcanized compound.

green dot
A green recycling dot, or circle, carried by packaging material intended for recycling in Germany. See *Duale System Deutschland*.

green pigments
Inorganic green pigments are often based on cobalt: other green pigments are based on elements which include elements such as aluminium, antimony, chromium, magnesium, nickel and titanium in addition to cobalt. Such green pigments are brighter than the chromium oxides. See also *Brunswick green, chrome green, chromium oxide, hydrated chromium oxide, Malachite green, permanent green, titanium green* and *pigment green* (examples of *inorganic pigments*). Phthalocyanine green is an *organic pigment*.

green strength
Strength before curing of a rubber compound. The tensile strength of an un-cured or un-vulcanized compound.

green tyre
An un-cured or un-vulcanized tyre produced on a *tyre-building machine*, from the separate tyre components, and vulcanized in a *tyre press*.

Grignard process
A technique used, for example, for the synthesis of chlorosilanes by reacting a *Grignard reagent* with a silicon compound in an ether solution.

Grignard reagent
An alkylmagnesium halide. A chemical compound which has the formula RMgX: X is usually chlorine if R is alkyl and X is usually bromine if R is aryl. Used in the *Grignard process*.

gross ton
A long ton of 2,240 pounds.

ground
A term used in *hydraulics* and which refers to a point of zero reference in an electrical circuit: circuit voltages are compared to that point.

ground chalk
See *whiting* and *calcium carbonate*.

ground dolomite
See *dolomite*.

ground limestone
See *limestone*.

ground state
The lowest, and the most stable, state of an atom.

grown tyre
A *tyre* which has expanded or grown due to service conditions.

grown tyre overall diameter
The maximum overall diameter of an inflated *tyre* with an additional tolerance for growth in service.

grown tyre overall width
The maximum overall width of an inflated *tyre* with an additional tolerance for growth in service.

GRP
An abbreviation used for glass reinforced plastics and/or for glass reinforced polyester. See *glass reinforced plastic moulding*.

GRP laminate
An abbreviation used for *glass reinforced plastics laminate*.

GRP moulding
See *glass reinforced plastic moulding*.

guanidine
Also known as iminourea. May be represented as $HN:C(NH_2)_2$. A white crystalline solid with a melting point of approximately 50°C and which is strongly basic. Used to produce, for example, *accelerators* for rubbers. See *guanidines*.

guanidines
Chemical compounds with the general formula of (ArNH)$_2$C=NR where Ar is an aromatic ring and R is an alkyl group, a phenyl group or a hydrogen atom. In general, guanidines are slow/medium speed *accelerators* used for the sulphur curing of rubbers. They are rarely used alone but used in synergistic combinations with primary accelerators such as the mercapto accelerators: they are secondary accelerators. See *guanidine*.

guard
Usually means *machine guard*.

guarding
See *machine guard*.

guide bush
A hardened steel *bush* which accommodates a *guide pillar*.

guide pillar
A hardened steel circular pillar which is used to locate two mould halves.

guide pin
Also known as a leader pin or as, a *dowel pin*.

guide strip
A flat steel bar attached to a mould plate and which is used to guide, for example, side cores or side cavities.

guide tip
Also known as a guider tip. Part of a *torpedo*; that part through which wire passes and emerges into the melt.

guider tip
See *guide tip*.

Guignet's green
See *hydrated chromium oxide*.

gum elastic
See *natural rubber*.

gum space
The distance between the guide tip of a *torpedo* and the *die*.

gummi
A German term for rubber.

gummi elasticum
A former term for rubber. See *natural rubber*.

gummi optimum
A former term for rubber. See *natural rubber*.

gun
See *spray gun*.

gusset
The folded-in portion of flattened tubular film or, a piece of material used to give additional size or strength to a component.

gutta-percha
Also known as gutta. A natural polymer of isoprene. At one time used for cable insulation but now, not often seen. The raw material is a high viscosity latex and is obtained from the leaves of the tropical tree known as *Palquium oblongifolium*. The latex contains approximately 70% of a trans-polyisoprene structure and approximately 11% resin: the remainder is largely water and some dirt. Rapidly degraded by air, light and ozone and therefore often stored underwater. At room temperatures it is a crystalline material which resembles *high density polyethylene (HDPE)*. Before the advent of PE it was used for hydrofluoric acid, storage bottles. See *natural rubber*, *balata* and *hydrogenated natural rubber*.

Gy
An abbreviation used for *gray*.

gypsum
A very common mineral: natural hydrated calcium sulphate. An *alkaline earth metal sulphate* with the formula CaSO$_4$2H$_2$O. Used to make *plaster of Paris*. Hydrated calcium sulphate loses water of crystallisation when heated to form plaster of Paris.

H

h
An abbreviation used for:
   hecto; and for,
   hour.

H
This letter is used as an abbreviation for:
   enthalpy;
   high;
   homopolymer;
   henry;
   hexa;
   hexyl;
   hydro;
   section height (of a *tyre*); and,
   whisker-type filler. That is, it is used to modify polymer abbreviations when a compound contains such a filler;

H/C
An abbreviation used for *Heveacrumb*.

H/S
An abbreviation used for *aspect ratio*.

h-h
An abbreviation used for head-to-head. See *positional isomerism*.

H-NBR
An abbreviation used for *hydrogenated nitrile rubber*. See *nitrile rubber*.

h-t
An abbreviation used for head-to-tail. See *positional isomerism*.

ha
An abbreviation used for *hectare*.

HA
An abbreviation used for *heptaldehyde-aniline condensate*.

haematite
See *ferric oxide*.

HAF
An abbreviation used for high abrasion furnace (black). N326.

HAF-HS	= high abrasion furnace - high structure (black).
HAF-HS (NT)	= high abrasion furnace - high structure (new technology black).
HAF-LS	= high abrasion furnace (black) - low structure. See *carbon black*.

hairline cracks
See *crazing*.

half spiral alternating daily
A *tapping method* for *natural rubber*.

half spiral method
A *tapping method* for natural rubber. The most frequently used *tapping method* for obtaining *natural rubber latex*.

halide catalyst
See *Ziegler-Natta catalyst*.

halobutyl rubber
See *halogenated butyl rubber*.

halogen
The electronegative elements of Group VIIb of the Periodic table are known as halogens and consist of *fluorine, chlorine, bromine, iodine* and astatine.

halogen flame retardant
A *flame retardant* which contains a *halogen*, for example, bromine.

halogen-containing rubber
A rubber which contains a *halogen* element, for example, chlorine, as a part of the rubber molecule. The halogen may be incorporated into the monomer (*chloroprene rubber*) or put in after polymerization, for example, *chlorinated polyethylene*. The presence of such an element gives fire resistance but causes problems when the polymer does burn.

halogenated alkyl phosphate
Flame retardant plasticizers which have poor compatibility with *polyvinyl chloride*. Examples include tri(monochloroproyl) phosphate and tri(2-chloroethyl) phosphate. See *plasticizer*.

halogenated butyl rubber
Also called halogenated isobutene-isoprene rubber or, halogenated isobutylene-isoprene rubber or, halobutyl rubber or, brominated butyl rubber or, chlorinated butyl rubber. Abbreviations used for this type of material are BIIR or CIIR. XIIR has also been used but the use of X can lead to confusion as, in ISO terminology, X indicates the carboxylic group.

Butyl rubber is slow curing because of the low degree of unsaturation: the speed of curing is improved by halogenation, for example, with chlorine or with bromine: this give chlorobutyl rubber and bromobutyl rubber respectively. The halogen adds to the isoprene units and as there are less 3% of these, the amount of halogen addition is relatively small.

The rubbers may be cured with peroxides and *zinc oxide*: they are more compatible with other rubbers, have better adhesion and heat resistance than butyl rubber. Such materials have improved resistance to gas permeability, compared to butyl rubber, and are therefore used for the inside liners of tyres. Other properties are similar but the materials are more water sensitive and expensive.

Bromobutyl rubber is more expensive than chlorobutyl rubber but is faster curing, has better flex cracking resistance and gives a higher state of cure. In many respects CIIR and BIIR are formulated and processed as *butyl rubber*.

halogenated isobutene-isoprene rubber
See *halogenated butyl rubber*.

halogenated isobutylene-isoprene rubber
See *halogenated butyl rubber*.

halogenated natural rubber
The reaction product of a halogen and *natural rubber*. The reaction product of, for example, chlorine and *natural rubber* is used for chemical and heat resistant coatings. See *chlorinated natural rubber*.

halogenated polyolefin alloy thermoplastic rubber
See *elastomeric alloy melt processable rubber*.

halogenated polyolefin thermoplastic rubber
See *elastomeric alloy melt processable rubber*.

halogenation
The chemical process whereby a *halogen* is introduced into an organic compound. See, for example, *halogenated butyl rubber*.

halon
A compound that has in its structure, atoms of chlorine, bromine and fluorine. Halons are used in fire-fighting: have a high *ozone depleting potential* or *ODP*. Halon 1301 has a value of 10·5. Classed as *Group 11* by the *Montreal Protocol* and so are due for phasing out of commercial use in the near future as their high ODP values mean that they constitute a significant threat to atmospheric *ozone*.

HALS
An abbreviation used for *hindered amine light stabiliser*.

hammer milled glass
Also known as milled glass or as, milled fibre. Relatively short *glass fibres* which are produced by hammering glass filaments against a metal screen so as to cause size reduction. Such filaments have an approximate diameter of 10 to 20 μm: the milled fibres are not held or bound together. For the *reaction injection moulding* (RIM) process, the fibre length may nominally be 1·6 mm ($^1/_{16}$") but is often considerably shorter, for example, 0·2 mm, and the size distribution is very non-uniform. When dispersed in the polyol, the filaments give a shear thinning suspension whose viscosity is lower than that of a *chopped glass fibre* suspension (because of the lower aspect ratio). Because of the reduced viscosity which results on fibre addition, approximately 40% by weight, in the polyol, can be used in RIM moulding. This is approximately 25% in the final part.

hand blending
The manual blending of a compound, for example, before mixing. See *compound blending*.

hand lay-up
An abbreviation used for this term is HLU. Also called the contact moulding process. A labour intensive, moulding technique widely used to produce components from *unsaturated polyester (UP) resins* and *glass fibre (GF) reinforcement*, for example, *chopped strand mat (CSM)*. As there is no shaping pressure involved, the mould costs are low and very large mouldings may be readily produced (see *glass reinforced plastic moulding*). Although the process appears simple, care must be taken during manufacture to avoid close contact with both the resin and the reinforcement: work in a well-ventilated area, wear a mask, gloves and goggles and after work, wash very thoroughly.

Only one mould is finally needed for component production and this may be either a male mould or a female mould. Which one is used usually depends on the location of the smooth finish which mould contact gives. Most commonly this is required on the outside of the moulding, for example, a boat, and so a female mould is used for production. However, a male mould (a *plug*) is used to produce the (production) female mould.

A gel-coat is first applied to the treated plug surface and allowed to partially harden. Once the gel-coat is touch dry, further catalyzed resin is painted over the surface, or over part of the surface in the case of a large moulding. Cut and shaped CSM is then applied to the resin-painted surface and more resin applied to the CSM: further layers of resin and CSM may be applied to build up the thickness. Once the resin has penetrated the mat, the assembly is consolidated using a ribbed metal roller. If required stiffening, in the form of blocks or ribs of wood, may be incorporated into the structure. The resin to glass ratio is approximately 1:3.

The assembly is then allowed to partially harden and at a suitable stage the over-lapping edges are cut off. After further hardening, for example, for 24 h the moulding is removed from the mould. The surface of the moulding is then washed, to remove the release agent and inspected for blemishes. If necessary the surface is repaired with a stopping compound, based on a filled UP resin, sanded and polished. The female mould so produced is then used to make the male mouldings which have the required glossy, outside surface.

Such a moulding process is comparatively simple and the mould and equipment costs, even for large mouldings, can be low. Very large complex shapes can therefore be manufactured using relatively unskilled labour. The process is, however, slow, messy, and could be dangerous (due to the styrene fumes). See *resin transfer moulding* and *spray-up*.

hand operation
A method of machine operation. The operator initiates each part of the moulding cycle by, for example, pushing buttons so as to achieve the right sequence of events or operations.

hang up
A problem associated with irregular, or uneven, flow through a *die* or *nozzle*; may be caused by dirt, uneven temperature or by poor die design (for example, by a lack of streamlining).

hanging bar ejection
A method of moulding ejection which relies on the *ejector mechanism* being actuated by *tie bars* which, in turn, are hanging from the head of the press.

hansa yellows
Yellow pigments which are *azo pigments*, or monoazo pigments, and which are based on diazotized, aromatic amines. Such pigments have good brightness, colour strength and chemical resistance but, in general are not used in plastics because of their tendency to bloom and bleed: heat and light resistance can be poor. *Permanent yellow* is an exception.

HAO
An abbreviation used for *higher alpha olefin*.

hard block
Part of a copolymer: that part of the copolymer structure which is relatively stiff and often inter-molecularly hydrogen bonded. Such hard blocks often act as thermo-labile crosslinks in a *thermoplastic elastomer*. See, for example, *polyurethane block copolymer*.

hard grade
The terms hard and soft usually refer to appropriate grades of materials such as *thermoplastic elastomers*. For example, with *polyether ester elastomer* if the hardness is 34-47 Shore D then the material is referred to as a soft grade: if the hardness is 55-72 Shore D then the material is referred to as a hard grade.

hard material alloy steel
A hard material alloy produced by powder metallurgy: a *steel* which contains a very high *titanium carbide content* (approximately 33%). As supplied this type of material is relatively soft and easily machined: after machining the components can be hardened by heating, in a vacuum furnace, to give a hardness in the region of $R_C 70$. Such a material is used in regions where high abrasion is expected.

hard natural rubber/polyolefin blend
A *blend* with a low proportion of natural rubber, for example, below 50% at say, 15%. See *natural rubber/polyolefin blend*.

hard paraffin wax
A wax in which the *alkane chain* is of relatively high molecular weight and contains between approximately 40 and 70 carbon atoms: such a wax will have a relatively high melting point (of up to 105°C). See *ceresin wax*.

hard rubber
See *ebonite*.

hard rubber dust
See *ebonite dust*.

hard thermoplastics/elastomer blend
A category of *thermoplastic elastomer*, a blend, which may be referred to as a rubber modified material. Such blends can be roughly divided into two categories. These are, those in which the plastics material has a low level of crystallinity and which derive their rigidity from a base plastics material which has a relatively high glass transition temperatures (T_g), for example, *polyvinyl chloride* (PVC). The second class is based on thermoplastics materials which have a high level of crystallinity, for example, polypropylene (PP) or its copolymers.

PVC, and its copolymers, are most often converted to TPE-type materials by melt mixing them with *plasticizers* which are compatible with the plastics material. Both polymeric and low molecular weight plasticizers are used with PVC to give elastomeric compounds. With the semi-crystalline, thermoplastics materials, it is necessary to mix them with rubbers/elastomers, with which they are not molecularly compatible, so as to form the required softer, elastomeric materials. This is because plasticizers are not compatible with a semi-crystalline, thermoplastics material. Usually *ethylene-propylene rubber* is used to give a *rubber modified polypropylene*. See *dynamically crosslinked polyolefin elastomer*.

hardened
A term applied to *steel* when *martensite* is formed on cooling. See *case hardened*.

hardener
An additive which initiates hardening or setting. See *initiator* and *catalyst*.

hardening rolls
A pair of warmed rolls used to reduce the solvent content of *dough*: the dough is then passed to *sheeting rolls*. See *cellulosics*.

hardening time - unsaturated polyester resin
The time measured in minutes, for example, from the setting of the resin (*gelation*) to the point when the resin is hard enough for the component to be removed from the mould.

hardness
Commonly means the resistance to indentation or penetration by a blunt indentor when loaded with a spring: both the indentor and the spring have precisely fixed characteristics. For example, hardness 0 may mean a penetration depth of 2.54 mm and a spring loading of 56 g: hardness 100 may mean a penetration depth of zero mm and a spring loading of 822 g.

Because of the ways in which hardness is usually measured it is not a fundamental property of a material. The techniques used rely on scratching or indentation and these involve compression, tension and shearing forces which may or may not penetrate the surface, and depend on the condition of the surface. Many tests and scales of hardness have evolved over the years. See *Moh hardness* and *hardness indentation test methods*.

hardness degrees
See *International Rubber Hardness Degrees*.

hardness indentation test method
A test used to assess *hardness*. These tests fall into two categories. Those in which:

(1) the indentor and load are chosen so that a largely non-reversible deformation is produced. These tests are mainly used for metals when it is possible to satisfactorily measure the dimensions of the indentation after removal of the load; and

(2) the loaded indentor remains in contact with the test material and the depth of penetration is measured after a fixed time. These tests are typically used for rubbers and

hardwall hose
A hose which contains a concentric reinforcing helix of, for example, wire.

hartshorn salt
See *ammonium carbonate*.

haul off
That system which pulls the extrudate away from the *die*; usually a series of rolls or, a caterpillar haul off. See *post extrusion equipment*.

hazards
Hazards indicate the potential of a substance to cause harm.

Hayes plasticity
A *flow test*. This test uses a Macklow Smith plastometer (rheometer) and is a low shear rate test. The test is commonly used to assess, and compare, the flow properties of *plasticized polyvinyl chloride* (PPVC). It is similar to a *flow rate test* in that a piston is used to force the heated material from a barrel and die assembly. It differs from the melt flow test in that the piston is heated and the piston is driven at a constant, pre-selected speed: the pressure generated is displayed on a gauge or chart. A range of testing speeds may be employed so as to further characterize the PVC compound.

haze
The amount of light scattered in a forward direction, either by surface irregularities and/or inhomogeneities within the material. The latter can be for example, dust, filler particles or *crystalline regions* of different refractive index from the rest of the material. When the percentage of haze is greater than 30% the material is considered to be translucent rather than transparent. It is possible to distinguish between haze due to surface roughness from that due to inhomogeneities by coating a small area of the specimen on both sides with a liquid of the same refractive index as the polymer, e.g. 1·49 for polypropylene. If the transmission in this wetted area improves then the haze was mainly due to surface imperfections.

After a *light transmittance measurement*, and with the specimen still in position, the integrating sphere hazemeter is set so that all the available light reaches the photocell. The galvanometer reading is increased from its value below 100 to read 100 exactly by using the sensitivity setting of the galvanometer. The sphere is then moved so that all the light except that scattered by the haze passes into the light trap. The scattered light or haze then registers on the photocell which outputs the value directly on the galvanometer.

HB
A rating in the *Underwriters Laboratory horizontal burning test*. For a thickness of between 3·05 mm and 12·7 mm the material is rated as HB if the combustion speed is less than 38·1 mm/min. HB is also used as an abbreviation for *Brinell hardness number*.

HB-HV copolymer
See *polyhydroxybutyrate*.

HCC
An abbreviation used for *high colour channel black*.

HCFC
An abbreviation used for *hydrochlorofluorocarbon*.

HDI
An abbreviation used for *hexamethylene diisocyanate*.

HDPE
An abbreviation used for *high density polyethylene*.

HDT
An abbreviation used for *heat distortion temperature* or heat deflection temperature.

HDT 264 psi = heat distortion temperature at 264 psi.
HDT 66 psi = heat distortion temperature at 66 psi.
HDT at 120°C and 0·45 MPa = heat distortion temperature at a heating rate of 120°C/h and using a load of 0·45 MPa (66 psi).
HDT at 120°C and 1·8 MPa = heat distortion temperature at a heating rate of 120°C/h and using a load of 1·8 MPa (264 psi).

head
A term used in *hydraulics* and which refers to the vertical distance between a selected point in the hydraulic system and the fluid surface in the reservoir.

head correction
A correction applied to rheological data or measurements, so as to compensate for pressure losses in the reservoir or barrel. If the length of melt in a reservoir of radius R_b is L_b and the length and radius of the capillary are L_c and R_c respectively then, for a true power law fluid the ratio of the pressure drop in the reservoir or barrel (ΔP_b) to the pressure drop in the capillary or die (ΔP_c) is given by $\Delta P_b/\Delta P_c = (L_b/L_c)(R_c/R_b)$.

In the case of the measurement of *melt flow rate*, most standards now specify that measurements must be made with the piston within fairly closely defined limits above the die so that in all measurements the amounts of material in the reservoir are similar.

head effect
See *Bagley end correction*.

head pressure
The pressure measured inside the *die head*. See *thrust load*.

head-and-gate end
The *die end* of an *extruder*.

head-to-head
An abbreviation used for term is h-h. See *positional isomerism*.

head-to-tail
An abbreviation used for term is h-t. See *positional isomerism*.

Health and Safety Executive
An abbreviation used for this UK-based organization is HSE.

heat
Energy that is transferred from one body or system to another as a result of a temperature difference.

heat capacity
The amount of heat energy required to change the temperature of a body, or of a substance, by 1°C or 1K. The units of heat capacity are cal/°C or J/K. See *specific heat capacity*.

heat conductivity
See *thermal conductivity*.

heat content
The amount of *heat energy* that a system contains. The amount of heat that is contained in the melt per hour, can be calculated if the output rate, the specific heat and the melt temperature is known. This is because mass, multiplied by the *specific heat*, multiplied by the temperature difference equals heat content or *enthalpy*.

In injection moulding technology, the temperature difference between a typical processing temperature, and a typical mould temperature are used to calculate the amount of heat that must be removed. The temperature difference is obtained by subtracting the mould temperature from the melt temperature (see **table 7**). By multiplying the total number of shots produced per hour by the shot weight, the mass may be calculated. As specific heat changes with temperature, an average specific heat, over the temperature range from melt to room temperature, is used.

It is often found that only approximately half of this heat must be removed from the component by the mould as the cool surface layers of the moulding provide sufficient rigidity so that ejection can be obtained even though the temperature of the inside of the moulding may still be very hot.

heat curable silicone rubber
See *high temperature vulcanizing silicone rubber*.

heat distortion temperature
Also called the heat deflection temperature. Should be called heat deflection under flexural load or, the heat deflection under load (HDUL) but is often abbreviated to HDT (which may stand for heat distortion temperature or heat deflection temperature).

In test specifications it is called deflection temperature under flexural load and two flexural loads are quoted. A plastic bar (for example, $110 \times 10 \times 4$ mm which is $4.4 \times 0.4 \times 0.16$ inches) is subjected to three-point bending by a load which produces a maximum stress at its mid-point of either 1.8 MPa or, 0.48 MPa (264 or 66 psi) while being heated. For example, if the breadth (b) is 4 mm, the depth (d) is 10 mm and the stress required is 1.8 MPa then, the load (F) in Newtons which has to be applied to the mid-point (if the span (L) is 100 mm) is equal to $2Pbd^2/3L$. That is, $2 \times 1.8 \times 100/ 3 \times 100$ which equals 4.8 N. This load is applied and the temperature raised at 120°C per hour; when the 10 mm thick bar deflects by 0.32 mm (approximately 12 thou or, 0.012") that temperature, in °C, is recorded and called the HDT. It is also quoted in °F.

heat deflection under flexural load
See *heat distortion temperature*.

heat deflection under load
An abbreviation used for this term is HDUL. See *heat distortion temperature*.

heat energy
See *heat*.

heat exchanger
A term used in *hydraulics* and which refers to a device which transfer energy from one medium to another medium. For example, from the hydraulic fluid to water.

heat history
The total amount of heat received by a compound.

heat input
See *heat content*.

heat lines
Closely spaced lines running in the *machine direction* and associated with the *calendering* of PVC. Product marking caused by, for example, poor release of the sheet from the roll aggravated by over-heating of the PVC.

heat mark
Extremely shallow depression or groove in the surface of a plastics component visible because of a sharply defined rim or a roughened surface (see *sink-mark*).

heat removal
See *heat content*.

heat resistance
Usually means the heat resistance of a thermoplastics material as measured by the effect of a rising temperature on a test specimen. The two most widely used standardized tests for measuring heat resistance are the *Vicat softening point (VSP)* and the *heat distortion temperature (HDT)*. The HDT is more widely used in the USA; VSP is more widely quoted in Europe.

The Vicat softening temperatures (VST) of amorphous, thermoplastics material are usually quite close to their glass transition temperature (T_g) when 10 N and 50°C are used; these conditions do not approximate to the melting points (T_m) for semi-crystalline, thermoplastics materials. Best to use, for this purpose, the HDT at 120°C per hour.

	PMMA	PVC	PP
VST at 50°C and 10 N	114	85	134
HDT at 120°C and 1.8 MPa	97	64	67
HDT at 120°C and 0.45 MPa	106	70	127
Glass transition temperature	105	80	-23
Melting point	—	—	170

The numbers quoted are in °C.

For a given semi-crystalline, thermoplastics material the heat resistance may be improved by increasing the amount of crystallinity and/or by adding fillers. Such fillers compensate for the softening effect of the amorphous phase when the plastics material is above its glass transition temperature (T_g). The effect of GF is most marked for those plastics which have a crystalline content of about 50%; for such materials, adding glass fibre (GF) can improve the heat resistance by approximately 100°C. Adding GF to an amorphous, thermoplastics material usually only has a small effect on the heat resistance; may only be improved by approximately 10°C. This is because the heat resistance of such materials is controlled by their *glass transition temperature* (T_g). See *unplasticized polyvinyl chloride*.

heat resistant polycarbonate
See *aromatic polycarbonate* and also see, for example, *tetramethylcyclohexane bisphenol polycarbonate*.

heat - rubber
In the rubber industry, a heat is a number of products which have been vulcanized at the same time, for example, in a press or autoclave.

heat sag
A measure of resistance to elevated temperatures for plastics materials, for example, a *reaction injection moulding* (RIM). A General Motors heat test which is expressed in millimetres (mm): the deflection of a 3.2 mm thick bar, overhung for 1 h at 121°C or for, 0.5 h at 163°C.

heat sealing
A *welding technique* used for *thermoplastics materials* in thin sheet form: the surfaces to be joined are heated to *melt temperature* and pressure is applied: the assembly is then cooled under pressure. See *impulse welding*.

heat sensitizer
An additive which when added to *latex compounds* makes such compounds heat sensitive. For example, ammonium salts give room temperature, stable compounds but if the latex compound is heated to approximately 50°C then, rapid coagulation occurs.

heat sensitizing
The process whereby *latex* is made heat sensitive.

heat sensitizing agent
See *gelling agent*.

heat softening
Thermal softening of a material so as to allow processing. For example, to extrude a thermoplastics material, the material must first be softened so that it can be shaped; this softening is most commonly done by heat. This step may also be called plasticisation, plastication, plasticization and thermal softening.

heat softening point
The temperature at which a flat-ended needle penetrates a specified distance into a heated plastics material. See *Vicat softening point*.

heat stabilizer
An additive which improves the resistance of a polymer system to decomposition, or degradation, caused by elevated temperatures. Associated with *polyvinyl chloride* (PVC) technology and particularly with unplasticized PVC (UPVC) technology. PVC degrades readily by dehydrochlorination and so additives which restrict this type of attack are widely used with PVC. Such additives may be grouped, or classified, in various ways, for example, into *lead compounds*, *metal soaps* (for example, barium/cadmium salts and zinc/calcium salts), *organophosphites*, *organotins*, and *epoxidized oils*.

heat traced
A term used in *reaction injection moulding* (RIM) technology and which means that a machine is heated/insulated so that it can operate at elevated temperatures so that the processing of materials which are solid at room temperatures is possible. For example, caprolactam for *nylon 6 reaction injection moulding*.

heat treatment
See *thermoset heat treatment*.

heat welding
See *welding*.

heat-fugitive cross-link
A bond which becomes ineffective at elevated temperatures. See *styrene block copolymer*.

heated oil circulating system
A heating system used, for example, for heating *calender rolls*: the pressures involved with this type of system are much lower than those used for *high pressure hot water*, for example, the pressure may only be 0.35 MNm^{-2} (50 psi). To compensate for the lower specific heat of this heat exchange medium it must be circulated faster than *high pressure hot water*. See *drilled rolls*.

heated platen
A *platen* at elevated temperatures which is usually used to heat materials indirectly, that is, via the mould: platens are heated by either steam or electricity. A platen can be used to heat materials directly in some processes. See *contact heating*.

heated probe
Part of an *injection mould* used, for example, in an *insulated runner mould* and which fits into the centre of the nozzle; nozzle temperature is accurately controlled by using a probe which contains both a heating coil and a thermocouple. Some probes are now available which are capable of being operated at two temperatures; the temperature at the tip of the probe is reduced when mould filling is completed so that the plastics material hardens and this plug of material helps to minimise *drooling*.

heated sprue bush
A *sprue bush* which incorporates a heater element so as to maintain the material at an elevated temperature.

heater band
An *electrical heating element* in the form of a sleeve or cuff. Most commonly such heater bands are resistance elements and are clamped around the barrel of the machine: the power to the band is regulated by a *thermocouple-actuated instrument*.

heating
Raising the temperature of a machine to operating temperatures (see *heating up*) and supplying heat once the machine is at operating temperatures.

Once a machine, for example, an *injection moulding machine* or a *calender*, is operational then the heat necessary to plasticize the material can come, either from the heating system or, from the effort needed to propel the material through the system. Often such machines operate *adiabatically*.

For example, once an *in-line screw*, injection moulding machine is operational then the heat necessary to plasticize the material is often derived from the effort needed to turn the screw. (The amount of heat needed can be calculated if the specific heat of the plastic is known and if, the difference between the processing temperature and the input temperature is also known). It is now thought best to adjust the *screw rotational speed* so that the screw is rotating for as long as is possible as this will minimize *melt temperature over-ride*.

If the screw is rotating for approximately the whole of the cooling cycle this will minimize the amount of frictional heat generated and decrease temperature overshoot. When operated in this way the barrel heaters will simply put in enough heat to bring the temperature up to the set point and overshoot should not occur. This is important as most moulding machines have no means of removing excess heat, that is, they do not have a cooling system on the barrel.

heating time
See *forming temperature*.

heating up
Often means the raising of a machine to operating temperatures. See *warming up* and *heating*.

heavy metal stabilizer
A polymer stabilizer, for example, a *heat stabilizer*, based on a metal such as cadmium and/or lead: the metal may be combined with an organic group such as stearate or octoate. Compounds of such metals are used as, for example, heat stabilizers for *polyvinyl chloride* (PVC) because of the good stabilizing performance of such mixed metal stabilizers. There is now growing concern about the use of such toxic metals and various alternatives are available. Could use a *metal-free organic* and/or a stabilizer based on another metal or mixtures of metals. For example, could use a *calcium/zinc stabilizer*, *barium/zinc stabilizer*, *barium/calcium/zinc stabilizer*, *strontium/zinc stabilizer*, *strontium/zinc/tin stabilizer*, *calcium/antimony stabilizer*, *calcium/aluminium/zinc stabilizer*, and *magnesium/zinc stabilizer*.

heavy oil
The fraction, obtained during the distillation of *coal tar*, and which distils at from 230 to 270°C is called heavy oil. From this may be obtained creosote.

heavy spar
See *barium sulphate*.

HEC
An abbreviation used for *hydroxyethyl cellulose* (a *cellulose ether*).

hectare
A measure of area which is 10,000 *square metres* or 2.471 05 acres. See *are*.

hecto
A prefix which means one hundred. An abbreviation used for this term is h. See *prefixes - SI*. For example:
hectogram = one hundred grams or 0.1 kilogram;
hectolitre = 100 litres; and,
hectometre = 100 metres;

Hefner candle
A measure of light intensity which is abbreviated to HK. See *candela* (cd). 1 cd = 1.1 HK.

height of fall
An abbreviation used for this term is H_f. See *resilience*.

height of rebound
An abbreviation used for this term is H_r. See *resilience*.

helical distortion
An extrusion defect. See *melt fracture*.

helix angle
The angle between the *screw thread* and the transverse plane of the *screw*.

Henecke machine
A machine used to produce *flexible polyurethane foam* by the *slab-stock process*.

Henecke process
A *slab-stock process*.

henry
An SI derived unit which has a special symbol, that is H. It is the SI derived unit of inductance (self and mutual). One henry is equivalent to the inductance of a closed circuit in which an electromotive force of one volt is produced by a current which varies at the rate of one ampere per second. See *Système International d'Unité*.

hepta
A prefix meaning seven. For example, heptane is the seventh member of the *alkane series*: that is, it is a paraffin.

heptaldehyde-aniline condensation product
The condensation product of heptaldehyde and aniline is abbreviated to HA. Also known as heptene base. This condensation product is an ultra-accelerator. See *accelerator*.

heptane
Also known as n-heptane. A member of the *alkane series* (a paraffin): the seventh member. This material has a relative density (RD or SG) of 0·68 and a boiling point of 98·4°C. Sometimes used as a solvent, for example, for rubbers (with the exception of nitrile rubber).

heptene base
See *heptaldehyde-aniline condensation product*.

heptyl
The use of this prefix indicates that the compound contains seven carbon atoms. See *dialphanyl phthalate*.

heptyl diphenylamine
A brown liquid with a relative density of approximately 0·97 and which is used as an *antioxidant* for rubbers - particularly for neoprene-type rubbers.

heptyl nonyl undecyl phthalate
An abbreviation used for this type of material is HNUP. Also referred to dialphanyl undecyl phthalate. A higher *phthalate plasticizer*. A *plasticizer* based on alcohols with a range of carbon atoms, for example, from 7 to 11.

heptylnonyl
More commonly referred to as *dialphanyl*. For example:

   heptylnonyl adipate = *dialphanyl adipate*;
   heptylnonyl azelate = *dialphanyl azelate*; and,
   heptylnonyl sebacate = *dialphanyl sebacate*.

hertz
An SI derived unit which has a special symbol, that is Hz. It is the SI derived unit of frequency. One repetition of a regular occurrence in one second. Has the dimensions of s^{-1}. See *Système International d'Unité*.

HET acid
An abbreviation for hexachloro-endomethylene-tetrahydrophthalic acid. See *chlorendic acid*.

HET anhydride
An abbreviation for hexachloro-endomethylene-tetrahydrophthalic anhydride. See *chlorendic anhydride*.

hetero-polymerization
See *copolymerization*.

heterocyclic compound
An organic ring compound which contains elements other than carbon in the ring structure.

heterogeneous composition
Of uneven, or non-uniform, composition.

heterolytic decomposition
With peroxides, such as *di-t-butyl peroxide*, decomposition can occur homolytically into peroxide radicals (*homolytic decomposition*) or heterolytically into ions. For the *peroxide cross-linking of rubbers*, homolytic decomposition is preferred.

heterolytic fission
See *heterolytic decomposition*.

heteropolymer
A polymer containing more than one type of repeat unit. See *copolymer*.

Heusler's alloys
Ferromagnetic alloys that are unusual in that they are not based on iron, cobalt or nickel but are based on copper, manganese and aluminium.

hevea
See *Hevea brasiliensis*.

hevea latex
Latex derived from *natural rubber*. See *Hevea brasiliensis*.

Hevea brasiliensis
Also known as hevea. Rubber latex can be obtained from a variety of plants and trees but the most satisfactory one for most purposes is the species of tree known as Hevea brasiliensis. This tree originally came from the Amazon forests but is now cultivated extensively in other regions of Equatorial climate (see *plantation rubber*). This is because the tree needs a fertile soil, high steady temperatures (approximately 27°C/80°F) and high uniform rainfall. The tree has leaves which resemble those of the elm and when the tree is cut, it exudes a milky white fluid which is called *latex* or *rubber latex*. See *tapping*. There are approximately 20 sub-species of this tree and the average yield obtained is about 2,000 kg/ha per year. See *natural rubber*.

Hevea rubber
See *natural rubber*.

heveacrumb
See *Heveacrumb process*.

Heveacrumb process
An abbreviation used for this process is H/C process. A technique or process used to produce *block crumb rubber* from *natural rubber coagulum*. Castor oil (approximately 0·3 phr) is added to rubber *latex* or to wet crumb so as to give chemically crumbed rubber. The oil promotes crumbling of the *coagulum* as it passes through the *crepeing rolls*: the rubber is subsequently pressed into flat blocks or bales. The amount of oil may be reduced if *zinc stearate* is also used.

Heveaplus MG rubber
A modified *natural rubber (NR)*. A grafted natural rubber which contains various percentages of grafted methyl methacrylate (MMA). For example, MG 30 contains 30% of *polymethyl methacrylate*, MG 40 contains 40% of *polymethyl methacrylate* and MG 49 contains 49% of *polymethyl methacrylate*. See *polymethyl methacrylate grafted natural rubber*.

HEX
An abbreviation for *hexachloro-cyclo-pentadiene*.

hexa
A prefix meaning six. An abbreviation used for *hexamethylene tetramine*.

HEXA
An abbreviation used for *hexamethylene tetramine*.

hexabromocyclododecane
An *organo-bromine flame retardant*. Such a bromine-containing compound tends to be more powerful than the equivalent chlorine-containing compound: often used in conjunction with *antimony trioxide*. For example, *polystyrene* 96%, *hexabromocyclododecane* 3% and *antimony trioxide* 1%.

hexachloro-cyclo-pentadiene
An abbreviation used for this material is HEX. Used, for example, to prepare fire retardant monomers. See *chlorendic acid* and *chlorendic anhydride*. May also be used in place of sulphur to cross-link rubbers.

hexachloro-endomethylene-tetrahydrophthalic acid
See *chlorendic acid*.

hexachlorobenzene
A solid material with a relative density of 2·0, a melting point of 227°C and a boiling point of 326°C. May be represented as C_6Cl_6. Used as a diluent for rubbers.

hexadecanoic acid
See *palmitic acid*.

hexafluoropropylene
An abbreviation used for this monomer is HFP. May be represented as $CF_2 = CFCF_3$. Such a material is used to make fluoropolymers - see, for example, *fluororubber*.

hexahydrophenol
See *cyclohexanol*.

hexahydrophthalic anhydride
A curing agent for *epoxide resins* which gives light coloured components with high heat resistance and good electrical properties.

hexahydropyrazine
See *piperazine*.

hexalin
See *cyclohexanol*.

hexamethyl phosphoric acid triamide
An abbreviation used is HPT. A clear liquid which has been used as an *ultra-violet light stabilizer* in, for example, vinyl polymers.

hexamethylene
See *cyclohexane*.

hexamethylene diamine
Also known as 1,6-diaminohexane or as, 1,6-hexanediamine. An abbreviation used for this material is HMD. A monomer for the production of *polyamides (PA)*. For example, for PA 6, PA 66 and PAA 6.

hexamethylene diamine carbamate
Also known as HMDA carbamate and as HMDAC. A *vulcanizing agent* for a *fluororubber*.

hexamethylene diammonium adipate
Also known as *nylon 66 salt*.

hexamethylene diammonium sebacate
Also known as *nylon 610 salt*.

hexamethylene diisocyanate
Also known as 1,6-hexamethylene diisocyanate or as, 1,6-hexane diisocyanate. An abbreviation used for this material is HDI. An isocyanate used to prepare polyurethanes. More toxic than *MDI* and *TDI* and therefore not so widely used. Used, for example, to make polyurethane fibres.

hexamethylene tetramine
Also known as hexa or as, HEXA or as, hexamine. Abbreviations used are HMT and HMTA. A white crystalline solid which is prepared from ammonia and formaldehyde $(CH_2)_6N_4$. Has a relative density of 1·3 and sublimes at 270°C without melting. A major use is for the curing of *phenolics*, for example, for the curing of novolak resins: also used as a secondary accelerator in rubbers, for example, to boost the cure of mercapto and thiuram *accelerators*.

hexane
A member of the alkane series (a paraffin) and which is also known as n-hexane. This material has a relative density (RD or SG) of 0·66 and a boiling point of 69°C. It is a good solvent for uncured *butyl rubber (IIR)*, *natural rubber (NR)* and, *styrene-butadiene rubber (SBR)*. It is a poor solvent for uncured high acrylonitrile *nitrile rubber (NBR)* and *thiokol rubber (T)*. This chemical causes some swelling of uncured low acrylonitrile NBR and chloroprene (CR).

hexyl group
A univalent radical with 5 isomers. May be represented as C_6H_{13}- and abbreviated to HX (or H). The term hexyl is usually seen in connection with terms used to describe *plasticizers*, for example, dihexyl phthalate (DHXP).

hexyl octyl decyl adipate
An abbreviation used for this type of material is HXODA. A higher *adipate plasticizer*. A *plasticizer* based on alcohols with a range of carbon atoms, for example, from 6 to 10.

hexyl octyl decyl phthalate
An abbreviation used for this type of material is HXODP. A higher *phthalate plasticizer*. A *plasticizer* based on alcohols with a range of carbon atoms, for example, from 6 to 10.

H_f
An abbreviation used for height of fall. See *resilience*.

HF
An abbreviation used for high frequency. For example:

    HF heating   = *high frequency heating*;
    HF preheating = *high frequency preheating*; and
    HF welding   = *high frequency welding*.

HF acceptor
An abbreviation used for *hydrofluoric acid acceptor*.

HFC
An abbreviation used for *hydrofluorocarbon*.

HFP
An abbreviation used for *hexafluoropropylene*.

HI
An abbreviation used for high impact. For example:

    HI-PMMA = *high impact polymethyl methacrylate*; and,
    HIPS   = *high impact polystyrene*.

HIC
An abbreviation used for household and industrial chemicals. Such terminology is used, for example, to describe certain grades of HDPE which are designed for use in such markets.

hide
That which is obtained from a *two-roll mill* by cutting the *band*. The mixed material may be removed from the mill in the form of a sheet: this is called a hide (presumably because, in appearance, it resembles the hide of an animal).

hiding power
A measure of the ability of a colouring systems to impart opacity. Related to the ability of the pigment to scatter light. Inversely proportional to the particle size of the pigment but is also dependent on the index of refraction.

high abrasion furnace black
An abbreviation used for this material is HAF. See *carbon black*.

high carbon nylons
A term used for *polyamide materials* such as *nylon 11* or, *nylon 12*.

high cis BR
An abbreviation used for high cis-polybutadiene rubber. See *polybutadiene rubber*.

high cis-polybutadiene rubber
Also known as high cis BR. See *polybutadiene rubber*.

high colour channel black
An abbreviation used for this material is HCC. A channel black which has high tinting capacity. See *carbon black*.

high density polyethylene
An abbreviation used for this type of material is HDPE. Sometimes referred to as high density polyethene or as, polyethylene - high density (PE-HD). Other names used for this material include high density poly(methylene), low pressure polyethylene, linear high density polyethylene. Ultra high molecular weight, high density polyethylene is abbreviated to UHMW-HDPE.

Under conditions of comparatively low temperature and pressure, ethylene can be polymerized to give a plastics material (HDPE) which is substantially free from branching. This is achieved by using stereospecific catalysts, for example, of the Ziegler-Natta type, which direct the incoming monomer and make the polymer chain grow in a very ordered way: the resultant material is a semi-crystalline, thermoplastics material. Because of its regularity, and lack of chain branching, this material has a higher level of crystallinity than low density polyethylene (LDPE). This increase gives materials which have a higher density, rigidity, tensile strength, hardness, heat distortion temperature, chemical resistance, viscosity and resistance to permeability; however the impact strength is lower. HDPE has good dynamic fatigue resistance but not as good as polypropylene (PP); some living hinge effect. As with all polyolefins, the weathering resistance is satisfactory but can be improved, for example, by the addition of *carbon black*. Compared to PP homopolymer, HDPE has better resistance to low temperature impact and to oxidation. PP has a higher Vicat softening point, better resistance to flexing, a higher hardness, a higher tensile and elongation and will also cycle faster in injection moulding: the gloss on mouldings can be similar.

The water vapour permeability of this type of polyethylene (PE), as with all types of PE, is low. PE is permeable to gases and vapours but linear low density polyethylene (LLDPE) and HDPE are less permeable to gases and vapours than LDPE. Permeability for organic vapours is least for alcohols and then increases from acids to aldehydes and ketones, to esters, ethers, hydrocarbons and halogenated hydrocarbons. (Permeability decreases with density). Some grades of HDPE, high molecular weight grades, are accepted as being suitable for containers for oil and petrol: they have been used for fuel tanks. In some cases the formed containers have been chemically modified, by *fluorination* or *sulphonation*, so as to make the material almost impermeable to fuels. HDPE will resist deformation by boiling water.

The properties of HDPE are significantly controlled by its *melt flow index*, density and molecular weight distribution values. The effects of changes in molecular weight, density and molecular weight distribution (MWD) are as follows.

Property	Effect of increasing:		
	MFR	Density	Breadth of MWD
Pseudoplasticity			Increases
Tensile strength at yield		Increases	
Tensile strength at break	Decreases	Increases	
Elongation at break	Decreases	Decreases	
Impact strength	Decreases	Slight Decrease	Decreases
Modulus		Increases	
Transparency		Decreases	
Long term load bearing		Increases	
ESC resistance	Decreases	Decreases	
Softening temperature		Increases	
Melt strength	Decreases		Increases
Melt elasticity	Decreases		Increases
Melt fracture tendency	Decreases		Decreases
Gloss			Decreases

HDPE, with a density of 0.940 to 0.965 g.cm^{-3}, is sometimes known as a Type 3 PE. The grades which have a density below 0.96 are produced by using a second monomer at low levels (<1%). Strictly speaking they are therefore copolymers of PE with another olefin, for example, with butene-1 or, with hexene-1. The use of the second monomer, reduces the density by introducing short, side chain branching. Such materials may be known as MDPE. HDPE/MDPE is widely used in the blow moulding industry: approximately 55% of all plastics used in blow moulding are HDPE/MDPE materials.

There is no known solvent at room temperature for HDPE but it is soluble in hydrocarbons and aromatic hydrocarbons at temperatures above 60°C/140°F. Such materials will also cause swelling at room temperature as this plastic is permeable to such materials. Also swollen by white spirit and carbon tetrachloride; in this respect HDPE is better than PP. Theoretically less resistant to oxidation than LDPE but in practice they are similar; could be due to catalyst residues.

PE is associated with *environmental stress cracking*. Minimize the environmental stress cracking effect by reducing the residual stresses in the moulding, by careful component design and by using the lowest MFR grade at any particular density level. The light resistance is satisfactory; the cheapest way of improving this is by black incorporation. Carbon black filled HDPE compounds are now being used for applications where very long term weatherability is required. Has limited resistance to oxygen at elevated temperatures; antioxidants are used for protection. Unless so protected the electrical properties will suffer. Resists aromatic and chlorinated hydrocarbons more than LDPE.

The natural colour of the material is a milky white and so a wide colour range is possible; transparent mouldings are not possible. This material may not be joined to itself using solvents as there is no solvent at room temperature. Because of its inert, 'non-stick' surface it is difficult to bond using adhesives; some success with contact or hot melt adhesives. If the surface is made polar, for example by using a flame or an electrical discharge, then this material may be bonded to metals using epoxides or nitrile-phenolic adhesives; such treatments also improve printability.

This material is strong and stiff, even at low temperatures. For these reasons it is used to produce components which are required to have a reasonable impact strength at low temperatures, e.g. bottle crates, containers used in refrigeration, fish boxes, storage vessels for paint, adhesives etc. It is also used for containers where its rigidity and resistance to gas permeability is an asset when used for food storage boxes. Widely used for household goods, e.g. buckets, bowls, kitchenware; although it has a higher density than PP the cost difference in money/kilo still enables HDPE to compete with PP for various applications. Other uses for HDPE include dustbins, over-caps for the aerosol spray cans, security seal caps for medicine bottles, bases for *polyethylene terephthalate* (PET) bottles and toys.

A major use for HDPE is in blow moulding applications. Large containers or drums (for example, of 180 litres/45 gallons capacity) toys and car components (fuel tanks and spoilers) are typical applications. HDPE is used in such applications because of its high impact strength and, because it can retain its properties at low temperatures, for example, $-18°$ C. To obtain the required properties in the finished product (for example, strength and creep resistance), it is often necessary to use high molecular weight material - known as HMW HDPE or, as HDPE HMW.

Injection moulding grades of HDPE are manufactured having a narrow molecular weight distribution compared to grades used for *extrusion* and *blow moulding*. The molecular weight of an injection moulding grade, or of a grade for small blow mouldings, may be 75,000. For *extrusion blow moulding*, a commercial HDPE will have a molecular weight in the range of 50,000 to 250,000. High molecular weight (HMW) material will be higher than this, say 250,000 to 1 million. Ultra high molecular weight materials (known as UHMW HDPE or HDPE-UHMW) will have molecular weights greater than 1 million; it has tremendous properties but moulding is very difficult.

high elasticity

A property displayed, or possessed by some materials, for example, by lightly cross-linked, *natural rubber*. This type of material may be easily stretched by large amounts (several hundred percent) and when the stretching force is removed, it will retract, or spring back, very quickly.

high energy radiation

This term covers X-rays, γ-rays, proton beams, electron beams and neutron beams. Exposure of polymer to high energy *radiation* causes localized formation of high concentrations of *free radicals* which may combine to form cross-links or they may cause degradation by chain scission.

high frequency heating

Also known as radio frequency heating or as, capacitive heating: a *dielectric heating process*. Sometimes abbreviated to RF heating or to HF heating. A heating process which uses the output from a high frequency generator: a heating process which uses radio waves with a frequency less than 300 MHz. Generators are oscillators (a triode valve) and RF voltages and currents are carried by conductors insulated from the ground: the electric fields are created by electrodes as in electrostatics and unlike *microwave heating*. In a high frequency heater, polar molecules, such as water, interact with the high frequency field and generate heat as a result of *polarization*.

high frequency heating processes

Heating processes which uses electronically generated power at high frequencies; includes, for example, *dielectric heating*, induction heating and ultrasonic heating. Dielectric heating includes *high frequency heating* and *microwave heating*.

high frequency preheating

Also known as HF preheating or HF heating. A *dielectric heating process* which is usually associated with the heating of *thermosetting materials* as the water which such materials contain, interacts with the high frequency field and as a result heat is generated. Extremely rapid, and uniform preheating (before, for example, *compression moulding*) is possible.

A uniformly shaped charge of the material is placed between the plates and usually a small air gap is left between them as this simplifies continuous operation and also allows for the escape of volatiles. An alternating electric field (frequency of say 80 MHz) is applied to the two parallel metal plates so as to cause *polarization* and a subsequent rapid temperature rise.

The use of high preheat temperatures (up to approximately 150°C/302°F) and high mould temperatures can reduce cycle times by about 50%. If the material is preheated to such high temperatures, then the equipment must be powerful enough to reach the desired temperature in less than 30 s so as to avoid moulding problems. Preheating, when applied to compression moulding, improves appearance, electrical properties, dimensional stability and output. See *high frequency heating processes*.

high frequency welding

Also known as HF welding. May also be referred to as radio frequency welding (RF welding or sealing) or as, high frequency sealing. A welding process which uses the output from a high frequency generator at, for example, 27.12 MHz. A strip electrode (a tool rule) is commonly used and is pressed (by a pneumatic ram) against the two sheets to be welded both during the heating and cooling periods. A high frequency field is momentarily applied, to generate heat by *polarization*, and after cooling the electrode is removed from the weld.

This process is a very rapid welding technique and heating times may be less than 1 s. To speed up the output water-cooled electrodes may be needed: welding is most commonly performed using a *C-type press* or a *bridge press* for a short pre-selected time.

Usually associated with the welding of *plasticized polyvinyl chloride* (PPVC) film or sheet; this process cannot be used to weld all plastics but is peculiarly suited to PPVC as this type of compound is a high loss material which interacts with the high frequency field. Other materials which can be HF welded include nylon 6, nylon 66, polyvinylidene chloride (PVDC), heavily plasticized cellulosics (for example, cellulose acetate and cellulose acetate butyrate), and thermoplastic polyurethane. The frequency used depends upon the material and the size of the job, for example, 70 MHz for nylon and 13.56 MHz for large weld masses. See *high frequency heating processes*.

high gum compound

A rubber compound which contains rubber, curatives and the minimum of filler. Sometimes called a pure gum compound.

high impact polymethyl methacrylate

Also known as high impact methacrylate. An impact resistant polymethyl methacrylate polymer made, for example, by the incorporation of a rubbery material into a *polymethyl methacrylate polymer*. This gives a high impact polymethyl methacrylate which is sometimes abbreviated to HI-PMMA or PMMA-HI.

high impact polystyrene

An abbreviation used for this type of material is HIPS and/or TPS for toughened polystyrene. Also known as impact polystyrene (IPS) or as, rubber toughened polystyrene or as, toughened polystyrene. This material is made by polymerising styrene in the presence of a rubber. For example, *polybutadiene* (BR) is dissolved in styrene and on polymerisation (caused by heat and catalysts) the rubber comes out of solution to form fine rubber particles suspended in the plastic phase; these rubber particles are chemically grafted to the *polystyrene* (PS) matrix. This grafting gives a material with a much higher impact strength than PS but with lower tensile strength, stiffness, hardness and Vicat softening point; the transparency of PS is however lost by the incorporation of the rubber and the gloss of HIPS is often lower.

However, high gloss-high impact grades are now available thus providing the required balance between impact strength and gloss. These grades give a brilliant finish to mouldings thus meeting the needs of many domestic products. The gloss ratings are comparable to, and in some cases better than, *acrylonitrile-butadiene-styrene*.

Normally the resistance to light of HIPS is poor and the material has poorer resistance to chemicals than PS. As the rubber content is increased, so is the impact strength but the strength, stiffness, hardness and Vicat softening point, falls. Some high impact grades may be made from high-cis polybutadiene and the material contains approximately 9% rubber.

This type of material is resistant to aqueous solutions of salts, acids and alkalis (if strong oxidising agents are absent). Not resistant to aromatic and chlorinated hydrocarbons; esters, ethers, aldehydes and, ketones. The natural shade is an off-white but the material is not often seen that colour as it is usually pigmented, e.g. white, pale grey, etc. The density is similar to that of PS, i.e. 1.05 g/cm^{-3}.

This material is used where a certain impact resistance is required and where the relatively poor chemical resistance is not important or, can be turned to good advantage, for example, to solvent weld injection moulded components together. Scale model kits are a good example of this type of application. Also used for household appliances, toys, disposable cups, appliance housings etc. As a result of its cheaper raw material price the high gloss-high impact grades of HIPS are now being used for components which were previously made from ABS. Typical products include portable television and radio housings, computer hardware cabinets, clocks, vacuum cleaner housings and vacuum cleaner accessories. A large outlet for this material is in the form of extruded sheet: such sheet is commonly thermoformed into products such as refrigerator liners.

high load melt index
An abbreviation used for this term is HLMI. A *melt flow rate test* obtained from *high load melt index testing*.

high load melt index testing
An abbreviation used for this term is HLMI testing. Used in unplasticized *polyvinyl chloride (UPVC) technology* where the *melt flow index test* is performed with a large load, for example, of 21.6 kg/47.62 lb and at 190°C. See *flow rate*.

high modulus carbon fibre
An abbreviation used for this type of material is HM CF or CF HM. A *carbon fibre* which has been produced so as to optimize the *modulus strength* of the fibre. HM CF is also sometimes known as type II carbon fibre.

high modulus polyurethane reaction injection moulding
An abbreviation used for this term is HM PU RIM. A polyurethane reaction injection moulding which has been formulated so as to give relatively high values of modulus or stiffness. See *filled polyurethane reaction injection moulding*.

high molecular weight
An abbreviation used for this term is HMW.

high nitrile NBR
A *nitrile rubber* with a high acrylonitrile content, for example, 45%.

high performance sheet moulding compound
Also called structural sheet moulding compound (SMC). An SMC which has better strength properties than general purpose *sheet moulding compound*. Such improvements in strength, and other properties, result from high glass fibre concentrations, for example, 65% and the use of high performance, resin matrices.

Vinyl ester resins are used where, for example, very good chemical resistance is required. Epoxy/*epoxide* resins offer good chemical resistance and good retention of properties up to approximately 150°C/302°F. Mouldings have very low porosity and can be moulded in sections as thin as 0.5 mm/ 0.02 in and as thick as 150 mm/6 in. *Phenolic* resins are used because of their good flammability characteristics together with low smoke and reduced levels of toxic by-products. *Polyimides* are of interest because they give mouldings which can be used at temperatures greater than 300°C/572°F.

Mouldings made from such materials are moving into applications requiring mechanical strength, temperature resistance and dimensional stability rivalling, or even surpassing, that of metals. The easy production of complex shapes by *compression moulding* also opens metal markets, for example, a wide variety of fairings, covers and access doors are being used in aircraft. Glass is the most common reinforcement but epoxy, vinyl ester and polyimide SMC with *carbon fibre* are now available. Typically, structural SMC contains 50 to 70% by weight, 40 to 50% by volume, of randomly orientated, chopped fibres in the resin matrix.

high performance SMC
An abbreviation used for *high performance sheet moulding compound*.

high polymer
A polymer of high molecular weight: that is, with a large number of repeat units. The final product consists of many identical, repeat units. Because the final molecular weight, or mass, is so large the material may also be referred to as a macromolecule.

high pressure
An abbreviation used for this term is HP.

high pressure forming
See *pressure assisted thermoforming*.

high pressure hot water
An abbreviation used for this term is HPHW. Heated water which because it is under pressure can reach relatively high temperatures. When high pressure hot water is used for the heating of *calender* rolls, temperatures of up to 200°C/392°F may be required and the rolls controlled to approximately 1°C/1.8°F (when thin UPVC is being calendered). The pressurized water can both heat the rolls and remove the heat produced at high production speeds. See *peripheral drilled rolls*.

To get the operating temperature required with HPHW on a high speed PVC *calender*, the system may need to be under a pressure of 3.5 MNm^{-2} (500 psi) and for this reason some calenders are equipped with a lower pressure, *heated oil* circulating system. Pressures involved with this system are much lower e.g. 0.35 MNm^{-2} (50 psi) but to compensate for the lower specific heat of this heat exchange medium it must be circulated faster than water.

high pressure impingement mixing
An abbreviation used for this term is HPIM. See *reaction injection moulding*.

high pressure moulding processes
Moulding processes which employ greater than a few atmospheres shaping pressure are classed as high pressure moulding processes. Included are *injection moulding*, *transfer moulding* and *compression moulding*. As the shaping pressures are high, mould and equipment costs are also high. This in turn means that large numbers of components must be produced to make such processes economically attractive.

high pressure polyethylene
See *low density polyethylene*.

high pressure steam
Steam at high pressures which can generate relatively high temperatures. Used in the polymer industry, for example, for

high resilience foam
An abbreviation used for this type of material is HR foam: such foams have the best elasticity of *polyurethane foams*. See *cold cured foam*.

high rubber content masterbatch
A *masterbatch* which contains a large proportion of rubber dispersed in a thermoplastics material and used, for example, to produce *rubber modified polypropylene*. This is because there is a big difference in viscosity between a rubber and the plastics material which can lead to mixing problems as even the lowest molecular weight rubber has a higher viscosity than any grade of polypropylene (PP). If only a small amount of rubber addition is being considered, for example, to reduce cavitation during injection moulding, then it may be best to use such a masterbatch - particularly if mixing is done on the *injection moulding machine*. At low rubber addition levels, the resultant blends process like PP.

high shear device
A piece of equipment in which the shear rates experienced by the material are similar to those found in *extrusion* or *injection moulding*. See *flow test*.

high shear rate flow test
A *flow test* performed on a *high shear rate rheometer*.

high shear rate rheometer
Machine used to study *high shear rate rheometry*. A *capillary rheometer*, when used in the processing *shear rate* range, may be called a high shear rate rheometer.

high shear rate rheometry
Flow testing: high shear rate flow testing. Such flow testing is commonly done by forcing the heat-softened, plastics material through a *capillary rheometer* die, using a ram, at a known ram speed and material temperature. The pressure opposing flow is measured or, the force needed to maintain the specified flow rate is measured. The ram speed is then changed and the new force/pressure (needed to maintain this speed), is measured and recorded. The ram speed is then changed, the force recorded etc. This procedure is repeated at different cylinder, or barrel, temperatures. For each ram speed a force/pressure is recorded. Knowing the barrel dimensions and the ram speed, the volumetric flow rate through the die can be calculated. In turn the shear stress and the shear rate may then be estimated and used to construct *flow curves*. Characterization of a material's shear flow properties is therefore done, over a temperature range, using an imposed rate, laboratory capillary rheometer (LCR).

The temperatures listed in **table 13** are those suggested for injection moulding grades of those particular materials: the most useful data is obtained at those temperatures used in production and so, **table 13** is given for guidance only. The shear stress and shear rate conditions used for the test, should also closely approximate those used in the production process so that, for example, the flow properties are measured at several shear rates which are representative of those used at different points in the production process.

high shear rate testing
See *high shear rate rheometry*.

high speed mixer
A mixing device used for blending or pre-blending of solid thermoplastics materials, for example, *unplasticized polyvinyl chloride* (UPVC), with additives. Such units are basically cylindrical in shape and usually, with respect to the mixing tank itself, the height is approximately equal to the diameter. The rotor, or mixing impeller, is situated as near to the base of the mixing tank as possible.

The machine is driven from an electric motor which, on the smaller units, is usually a variable speed design. On most production-size units it is usually a two-speed motor which drives, through a pulley and vee-belt arrangement, the central shaft which enters through the base of the mixing tank. Most of the available machines are similar in basic design although there are small variations, these usually being associated with the form of the mixing rotor or impeller. This impeller is shaped rather like a ship's propeller but usually it has a relatively large number of blades, e.g. 4 or 6. Driving the impeller through the mix not only mixes the ingredients but causes a rapid temperature rise. This rise in temperature, and the associated turbulence, is used to cause rapid plasticizer absorption: the output is often called a dry-blend or a premix.

Premixes, for calendering, can be manufactured using a high speed mixer: thought that such units do not have the same flexibility of operation, as a *ribbon blender*, as once a mix is made it must be dumped from the mixer.

high speed mixer - operation of
High speed mixers may be operated in many different ways and the following procedure is only one of the many variations which could be imagined even for a simple *unplasticized polyvinyl chloride* (UPVC) formulation. The impeller is set in motion at a relatively low speed and ingredients such as polymer, *filler* and *impact modifier* are added - the *lubricant* is added at a later stage. The speed is then increased and as a result of the friction generated, and the heat supplied via the heated outer jacket, the temperature of the ingredients rapidly rises (on small mixers, impeller speeds of 3500 rpm are sometimes used).

While the temperature is rising, blending is occurring as the impeller causes the material to be thrown outwards and upwards: it then falls back into the central vortex which has been created. The *baffle* breaks up streamlined flow and directs the material into the working area.

Material temperature is sensed (by means of a thermocouple inserted through the baffle) and when it has exceeded the melting temperature of the lubricant (e.g. 100°C), this compounding ingredient is added. This melts and coats the particles. The blend is then dumped into a *cooler-blender* when a temperature of 130°C has been reached. The lubricant is added late in the mixing cycle so as to ensure that the initial heat build-up is rapid, for example, within 8 minutes. By cooling the material a more consistent feed to the injection moulding machine is obtained as otherwise, unless the hopper was heated, a variable feedstock temperature would result. Rapid cooling is usual so as to minimise the risk of thermal degradation.

high speed tensile test
A *tensile test* performed at high speed. See *pendulum impact test*.

high strain carbon fibre
A *carbon fibre* which has been produced so as to optimize the *breaking strain* of the fibre.

high strength carbon fibre
An abbreviation used for this type of material is HS CF or CF HS. A *carbon fibre* which has been produced so as to optimize the *tensile strength* of the fibre. HS CF is sometimes also known as type I carbon fibre.

high structure carbon black strength
See *carbon black*.

high styrene resin
A *styrene-butadiene rubber* which contains a relatively high levels of styrene, for example, 40% or more. For general purpose use, the level of styrene is 23·5%. High styrene levels give harder compounds which give, for example, good road grip in tyres.

high temperature
An abbreviation used for this term is HT.

high temperature conditioning
Also known as post curing. Used for *thermoplastic polyurethane* as this post moulding treatment permits stress relaxation and allows crystalline regions to stabilize.

high temperature reaction injection moulding
An abbreviation used for this term is HT RIM. See *nylon 6 reaction injection moulding*.

high temperature thermoplastics fibre
An abbreviation used for this material is HT thermoplastics fibre. A fibrous material based on a heat resistant thermoplastics material, for example, made from polyether ether ketone. To get supple, drapeable fabrics (with the desired stiffness) these fibres may be combined with carbon fibre. See *hybridized reinforcement fibre*.

high temperature vulcanizing silicone rubber
Also called heat curable silicone rubber. An abbreviation used for this type of material is HTV silicone rubber. Gum-like materials (*VMQ* and *PVMQ types*) which are processed by traditional rubber processing techniques but cross-linked by *peroxides*. Silicone rubbers can be classified according to the technology employed for their processing into high temperature vulcanizing silicone rubber, *room temperature vulcanizing silicone rubber* and *liquid silicone rubber*.

high tenacity fibre
See *viscose rayon*.

high tenacity rayon
See *viscose rayon*.

high wet modulus modal
Also known as HWM modal. Sometimes called artificial cotton. See *viscose rayon*.

high yielding clone
A tree which gives a higher than average yield of *natural rubber*. Such a *clone* may yield about 2,500 kg of natural rubber per hectare per year.

high-speed mixer
See *high speed mixer*.

high-vinylidene copolymer
See *polyvinylidene chloride*.

higher alpha olefin - polyethylene
An abbreviation used for this type of material is HAO-PE. Such materials are co- or terpolymers which contain up to 10% of octene, 4-tetramethylpentene-1 or, propylene. VLDPE, means *very low density polyethylene*, and such materials can have densities in the region of 0·88 to 0·91 g/cm^{-3}. Such very light HAO grades are hardly crystalline and are rubber-like materials. They can be highly filled and used like other thermoplastics materials or, they can be used to enhance certain properties (for example, crack resistance) of other polyolefins.

highly oriented pyrolytic graphite
An abbreviation used for this type of material is HOPG. A form of graphite which is a highly ordered structure. The thermal conductivity of highly oriented pyrolytic graphite can be six times that of *copper* at room temperatures.

highly polar plasticizer
A *plasticizer* which contains a large number, or percentage, of polarisable groups so that there is strong interaction between the polar polymer (for example, *polyvinyl chloride*) and the polar plasticizer (for example, a *phosphate plasticizer*).

highly saturated nitrile rubber
An abbreviation used for the term highly saturated nitrile is HSN: however HSN is taken to mean highly saturated nitrile rubber. See *nitrile rubber* and *hydrogenated nitrile rubber*.

hindered amine light stabiliser
An abbreviation used for this term is HALS. A photo-stabilizer used, for example, with *polypropylene* to improve the light resistance of that type of material. Such stabilizers are based on hindered piperidine materials and act in a similar manner to that of an *antioxidant*. See *stabilizer*.

hindered phenol
An additive which is used to prevent oxygen attack on, for example, diene rubbers and polyolefins (PO), by acting as a chain breaker. Also known as a mononuclear phenol. A phenol-type material, usually an *antioxidant*, which has bulky side-groups attached in the ortho position on the benzene ring: such groups hinder, or restrict, chemical reactions. The bulky side-groups (for example, t-butyl groups) thus prevent the phenol-type material from being oxidized. Hindered phenols such as *2,6-di-t-butyl-4-methylphenol* are often used in non-toxic PO formulations as such materials have a low level of staining.

hindered piperidine
A piperidine compound which is used, for example, as a *hindered amine light stabiliser*. For example, bis-(2,2,6,6,-tetramethyl-4-piperidyl) sebacate.

HIPS
An abbreviation used for *high impact polystyrene*.

Hitachi process
A process used to produce *cross-linked polyethylene foam* by chemical cross-linking.

HK
An abbreviation used for *Hefner candle*. HK is also used as an abbreviation used for Knoop hardness number.

HLMI testing
An abbreviation used for *high load melt index testing*.

HLU
An abbreviation used for *hand lay-up*.

HMD
An abbreviation used for *hexamethylene diamine*.

HM CF
An abbreviation used for *high modulus carbon fibre*.

HM PU RIM
An abbreviation used for *high modulus polyurethane reaction injection moulding*.

HMDA carbamate
An abbreviation used for *hexamethylene diamine carbamate*.

HMDAC
An abbreviation used for *hexamethylene diamine carbamate*.

HMF
An abbreviation used for high modulus furnace (black). N601. See *carbon black*.

HMFI
An abbreviation used in the UK for Her Majesty's Factory Inspector.

HMT
An abbreviation used for *hexamethylene tetramine*.

HMTA
An abbreviation used for *hexamethylene tetramine*.

HNUP
An abbreviation used for *heptyl nonyl undecyl phthalate*.

hobbing
The cold forming, or coining, of a desired impression into a block (blank) of prepared hobbing steel using a hardened steel master (hob) of the desired impression, thus creating an

exact negative cavity with a fine surface finish. The surface on the master hob is the finish that is transferred to the cavity; sharp internal corners should be avoided and the sides of the hob should be tapered, e.g. by 1° per side.

In the USA the hob is most often made from the *AISI steels* A-2 and A-6 (both of which are air hardening) and the cavity steel is made from AISI P-4 and P-5. The cavity blank size is approximately twice the hob dimensions and all surfaces should be true; a good polish is necessary on the surface to be hobbed.

The blank is retained within a soft steel ring and the steel of the cavity blank is then displaced upwards as the lubricated hob is pushed into the blank. The pressure needed depends on the detail required and on the projected area of the hob - typical pressures are of the order of 1,500 MNm^{-2} (217,000 psi). After hobbing the cavities are annealed and cleaned.

Hoekstra plastimeter
A parallel plate type of *plastimeter*.

hold period
See *dwell time*.

hold pressure
Term used in *injection moulding*: that pressure which is needed to compensate for shrinkage. See *dwell under pressure*.

holding period
See *dwell time*.

holding phase
See *dwell time*.

holding time
See *dwell time*.

hole
Term often associated with the depletion of *stratospheric ozone* in a patch or hole over the Antarctic. See *British Antarctic Survey*.

hollow enclosed components
See *blow moulding* and *rotational moulding*.

hollows
Term used to describe the products of *blow moulding*, for example, toys and ducting.

homoatomic polymer
See *homopolymer*.

homochain polymer
See *homopolymer*.

homogeneous material
A material which has a uniform dispersion of additives.

homologues
See *polymer homologues*.

homolytic decomposition
An even splitting, or decomposition of a material. For example, with peroxides, such as *di-t-butyl peroxide*, decomposition can occur homolytically into peroxide radicals or heterolytically into ions. For the peroxide cross-linking of rubbers, homolytic decomposition is preferred. That is, for example, $(CH_3)C-O-O-(CH_3)$ which on homolytic decomposition gives the tertiary butyloxy radical $(CH_3)C-O\cdot$.

homopolymer
Also known as a homochain polymer or as a, homoatomic polymer or as an, isochain polymer or as a, unichain polymer. A polymer based on one monomer: a polymer which contains only one type of repeat unit in the main chain.

Hooke's law
This states that *stress* is proportional to *strain*. Young's modulus is the constant of proportionality. Hooke's law is no longer obeyed beyond the *proportional limit*.

HOPG
An abbreviation used for *highly oriented pyrolytic graphite*.

Hopkinson's process
A process used to produce blocks of *natural rubber* which contains all the serum constituents: white flakes of rubber are produced by *spray drying* ammonia stabilized latex. The material is compressed into blocks.

hopper
The material holding unit which is attached to the barrel at the feed port and used to hold, and then feed, the plastics material into the *extruder* or moulding machine.

Material usage (for a *hygroscopic material*) should be calculated so as to determine how much material should be loaded into the hopper and thus prevent storage in that hopper for longer than 1 hour: ideally a heated hopper assembly should be used.

Consider the example of an *injection moulding machine* which is producing 6 components with a cycle time of 24 s. The weight of each component and the feed system is 14 g and 12 g respectively and, therefore, the total shot weight is $(6 \times 14) + 12 = 96$ g. The material consumption (Q) in kg/h is total shot weight $\times$ 3600/1000 $\times$ cycle time (s). This equals $96 \times 3600 / 1000 \times 24 = 14.4$ kg/h. Therefore the hopper should be filled with 14.4 kg, the level noted accordingly, and it should not be topped up beyond this point.

hopper block
That part of an injection moulding machine or of an extruder, which carries the feed hopper. To ensure that the material freely enters the feed section of the *screw*, cooling water is passed around the *hopper throat assembly*.

The flow rate and temperature of the cooling water is important as these variables affect how the material is *plasticized* within the barrel. Usually, the water passing around the hopper throat is from the mains supply and is therefore subject to a seasonal temperature variation. This variation can affect the melt temperature and cause product variation. In order to overcome this problem the use of an indirect, temperature controlled system (similar to a mould temperature controller) should be installed with the flow rate and temperature of coolant standardized for each machine. Alternatively fit a solenoid actuated valve into the coolant supply: this should only permit water flow when the temperature exceeds a preset value.

hopper slide
Also known as the *throat gate*.

hopper throat
The base of the hopper where, for example, material flows onto the *screw* in an *injection moulding machine*.

horizontal combustion test
See *Underwriters Laboratory horizontal burning test*; also see ASTM D 635 and ISO R 1210–1970.

horizontal flash mould
A type of *flash mould* in which the mating surfaces (the flash land) are at right angles to the direction of mould opening. The flash escapes in a horizontal direction.

horizontal flash semi-positive compression mould
A *semi-positive compression mould* in which the *flash* initially escapes horizontally. When using horizontal flash, semi-positive type moulds, for fibrous filled materials such as *dough moulding compound* (DMC), particular attention should be given to the width of the lands used. As DMC cures very quickly,

problems can arise when material is trapped and cures on the land surface; this prevents further flow of material across the land and subsequently restricts the amount of pressure that can be applied to the material in the cavity. The land should therefore be approximately 1·6–5 mm wide or 0·06–0·2 in.

horizontal fluidised bed
A *fluid bed* through which, for example, an extruded profile travels horizontally with respect to the ground.

horizontal locking and horizontal injection
A type of *injection moulding machine* arrangement. The horizontal locking and horizontal injection arrangement is the most common as it gives a convenient working height and allows the components to fall from the mould under the influence of gravity.

horizontal locking and vertical injection
A type of *injection moulding machine* arrangement in which the injection unit is above the clamping unit and injection is usually into the top parting line of the mould. This gives a compact machine: pneumatic machines are often of this type.

horizontal machine
The term usually refers to a design of *injection moulding machine*: with such a machine the injection unit is parallel to the floor. This is the most common type, or layout of injection moulding machine.

horizontal semi-positive mould
See *horizontal flash semi-positive compression mould*.

horn
Also called a sonotrode. That which transmits the ultrasonic vibrations through, and into, the components to be joined in *ultrasonic welding*.

hornblende
See *asbestos*.

horsepower
An abbreviation used for this unit is hp. A unit for measuring the rate of work: a unit of power measurement. Abbreviation used h.p. or hp. Equivalent to 550 foot-pounds per second or 745·7 watts. Similarly a metric horsepower is a unit for measuring the rate of work: sometimes known as cheval vapeur or pferde-stärke (PS) or pferde-kraft. Equivalent to 0·736 kilowatts or 735·499 watts. One metric horsepower is the rate of doing work of 75 m kg/s which is equivalent to 550 foot-pounds per second.

hot air ignition furnace testing
See *ignition test*.

hot cured foam
A *flexible polyurethane foam* produced by the *hot moulding process*.

hot curing - unsaturated polyester resin
The hardening, or curing, of a unsaturated polyester resin composition using elevated temperatures, for example, 140°C. Hot curing can be achieved using a so-called catalyst such as *benzoyl peroxide*.

hot edge trim
That which is removed from *calendered sheet* by *hot trimming*: the trim is usually re-cycled.

hot fillable
A plastics container which is capable of withstanding the temperatures of that which is to be packaged (which is at an elevated temperature).

hot foil marking
A dry process used to apply a coating of metal to a plastics substrate, by transferring the metal, usually aluminium, from a *metallized foil* to the substrate using a heated die. Also known as hot stamping, gold stamping, gold blocking, product marking, hot-press marking and dry printing. See *metallizing*.

hot gas welding
A welding technique used for *thermoplastics materials*, for example, *polyethylene* (PE) and *unplasticized polyvinyl chloride* (UPVC), in which the necessary heat is generated by a stream of hot gas (air or nitrogen). The gas is heated in an electrically-heated, welding torch and used to heat both the edges of the sheets and a *filler rod*. The thermally softened rod is pressed into a groove located between the thermoplastics sheets which are to be joined so as to provide a joint-filling material. The rod is of the same type of plastics material as the sheets: usually the sheets to be joined are of the same material.

hot hobbing
Also known as pressure-casting. A process used to produce a pressure casting from a *hob*. In one process the steel model of the required component is covered with molten *beryllium-copper* (e.g. at 1,300°C). Pressure is then applied to the cooling metal, by means of a plunger, until the alloy has hardened; a post-casting hardening stage will probably also be needed. Provided that allowance has been made for the shrinkage of beryllium-copper, accurately sized mould components can be produced and once a master has been made it may be used for any number of cavities. The cooling system may be incorporated at the casting stage and finely detailed mould components may be produced in this way: the pressure cast beryllium-copper cavities are supplied to the mould maker annealed and hardened (approximately R_C38 which is soft enough to machine but hard enough for a long life).

hot manifold block
See *hot runner plate*.

hot melt adhesive
An adhesive, based on a thermoplastics material, which relies on the adhesive powers of hot thermoplastic compounds. The hot adhesive (in melt form) is applied to the substrate, the other component is then held in position until the polymer sets by cooling. Such adhesives are easy to apply and set rapidly but can suffer from creep. May be based on a variety of polymers, for example, *polyamides*.

hot mixing process
An internal mixing process used for rubber compounding which uses relatively high mixing temperatures which are achieved by using, in turn, relatively high rotor speeds in an *internal mixer*.

hot moulding process
A *direct foam moulding process* used to produce *direct moulded foam*. Can have a hot moulding process and a cold moulding process. The hot moulding process requires reactive *polyether polyols* with a high proportion of primary hydroxyl groups - higher than that used for *slab-stock production*. The reactive polyether polyol is reacted with the *tolylene diisocyanate* (TDI) and most of the foaming is brought about by the water/isocyanate reaction, that is, the foam is water blown. Cold cured foams have good flexibility and give comfortable seating whereas hot cured foam gives better damping of vehicle vibrations. See *flexible polyurethane foam*.

hot plate welding
Sometimes called hot shoe welding or contact welding. A welding technique used for products based on *thermoplastics materials*, for example, for *polyethylene* (PE) pipe. An electrically-heated plate is used and the cut surface of the pipe are pressed against the heated plate: after a predetermined time, the two pipes are withdrawn and pressed together while

hot press moulding
This process is used for polyester moulded products which are required in large quantities: either a *preform* or a *polyester moulding compound (PMC)* is used.

hot press moulding - preform
Also known as preform moulding. A known quantity of the catalyzed resin and the reinforcement, in the form of a *preform*, is placed between heated matched moulds (heated to 100 to 170°C). Comparatively low pressures (50 bar, 700 psi or 4,900 kN/m^2) are then applied for the specified cycle time, for example, 3 minutes. Generally the pressures required are lower than those used for a *polyester moulding compound* based on *glass*. This process is a matched die process which gives products which have smooth surfaces both sides and a uniform thickness: lower labour costs than *hand lay-up (HLU)* and a higher output rate than *cold press moulding (CPM)*. Higher capital costs than both hand lay-up and cold press moulding.

hot rubber
Rubber, for example, emulsion styrene-butadiene rubber (E-SBR), produced at comparatively high polymerization temperatures. Initially E-SBR was polymerized at approximately 30°C: now, E-SBR is produced at approximately 5°C to give the so-called *cold rubber*.

hot runner manifold
Also called a hot runner plate or, a hot manifold block. Part of a *hot runner mould* which is made from steel and whose temperature is capable of being accurately maintained at an elevated value, e.g. by cartridge heaters and thermocouple-actuated instruments. The melt passes through channels in the hot manifold block and is fed into the cavities via internally heated nozzles. Unless problems with drooling or blocking are to be experienced it is very important that the temperature control, particularly in the nozzle region, should be very accurate.

hot runner mould
This term is applied to an *injection mould*, used for thermoplastics materials, in which the runner system is kept at an elevated temperature by means of cartridge heaters: this stops the runner system from setting. The mould has a *hot runner manifold* which is insulated from that part which carries the cavities and it is also separated from the rest of the mould by insulating air gaps and/or insulated blocks. The mould must be constructed to take account of the differential expansion which occurs as a result of operating parts of the mould at different temperatures. Despite the obvious advantages of such a *runnerless mould* it must be appreciated that there are some major disadvantages. For example, the moulds are relatively complicated and require expensive temperature-control equipment. (The solution to this problem should become easier as *microprocessor-controlled machines* become standard as such machines can have a large number of temperature-control circuits incorporated into them, e.g. 20.) The moulds are more difficult to maintain and to operate, and colour/material changes can be difficult.

hot runner plate
See *hot runner manifold*.

hot shoe welding
See *hot plate welding*.

hot stamping
See *hot foil marking*.

hot stretching
The stretching of a material so as to introduce *orientation*. For example, extruded *polystyrene* is stretched at approximately 130°C so as to give a material which is stronger in the stretching direction but which has a lower *heat distortion temperature* (HDT) - the low HDT arises because of the inherent instability of oriented molecules. See *biaxial orientation*.

hot tip mould
See *insulated hot-tip mould*.

hot trimming
A *trimming system* for sheet where the edges are removed before the *calendered sheet* cools: on many systems this is done by mounting two circular knives against the final roll. Pressure is exerted on the knives by means of weighted levers and the width of trim is readily adjusted. Marking of the rolls is minimised by making the knife blades out of a plastics material (such as *polyamide*) and with such a system the hot edge trim can be immediately conveyed back to the mixing system. However it is commonly found that the edge of the sheet is beaded by hot trimming and this can give problems in *reeling* because the thickness will steadily increase as the sheet is wound: this could cause stretching.

hot water absorption
A measure of how much hot water a material will absorb under specified conditions and over a certain time. See *water absorption*.

hot wire test
Sometimes referred to as the *CEI hot wire test*. This test is intended to simulate the temperature stresses which might be caused through over-heating, for example, incandescent components or over-loaded resistors: it also evaluates the risk of ignition. A hot wire at a specified temperature (for example, 650, 750, 850 or 950°C) is applied to the horizontal test specimen with a force of 1 N and for a period of 30 s: the wire or the specimen are moved towards one another over a distance of 7 mm. The time between application and ignition is noted together with the time between the start and the end of burning and the maximum height of the flame.

hot-press marking
See *hot foil marking*.

hot-probe gate
This type of gate is used in *runnerless moulding* where material freezing in the gate or runner system could interfere with production. A pointed heated probe projects into the runner and the tip of this probe helps to keep the thermoplastics material warm in the gate region.

hot-type factice
See *brown factice*.

hour
An abbreviation used for this unit is h. A unit of time which contains 3,600 seconds.

hp
An abbreviation used for *horsepower*.

HP
An abbreviation used for hard processing (black). See *carbon black*.

HPC
An abbreviation used for hard processing channel (black). See *carbon black*.

HPHW
An abbreviation used for *high pressure hot water*.

HPIM
An abbreviation used for high pressure impingement mixing. See *reaction injection moulding*.

HPT
An abbreviation used for hexamethyl phosphoric acid triamide.

H_r
An abbreviation used for height of the rebound. See *resilience*.

HR foam
An abbreviation used for *high resilience foam*.

HRC
An abbreviation used for *Rockwell hardness* on the C scale - the scale used for metals such as *steel*.

HRH system
An abbreviation used for *resorcinol formaldehyde silica system*.

HS
An abbreviation used for high structure (black). See *carbon black*.

HS CF
An abbreviation used for *high strength carbon fibre*.

HSE
An abbreviation used for the UK-based organization called the Health and Safety Executive.

HSN
An abbreviation used for *hydrogenated nitrile rubber*. Sometimes used for *highly saturated nitrile*. See *nitrile rubber*.

HT
An abbreviation used for high temperature. For example:
  HTP = high temperature polymerization; and,
  HTV = high temperature vulcanization;

hub
That part of the *screw*, at the rear of the screw, which functions as a seal and prevents plastic from leaking back into the drive; has the same diameter as the flighted part of the screw.

hue
The obvious, or striking feature, of a *colour* and that which gives it a name, for example, blue. Determined by the wavelength of light.

humectant
A material which assists the retention of water in a compound.

hundredweight
In the UK, 112 pounds equals 1 hundredweight (cwt) - a long hundredweight with 20 cwt in a ton. In the US, 100 pounds = 1 hundredweight - a short hundredweight with 20 in a short ton. To convert from hundredweight (long) to kilograms multiply by 50·80. To convert from hundredweight (short) to kilograms multiply by 45·36. See *UK system of units* and *US Customary Measure*.

hunting
Control terminology which means that an actual value is moving around a set point: usually applied to on/off temperature control.

HV
An abbreviation used for Vickers hardness number.

HVA-2
An abbreviation used for *m-phenylenebismaleimide*.

HWM modal
An abbreviation used for *high wet modulus modal*. See *modal fibre*.

HX
Letters used in *plasticizer* abbreviations for hexyl (see *H*). For example, dihexyl phthalate (DHXP). See *phthalate*.

HXODA
An abbreviation used for *hexyl octyl decyl adipate*.

HXODP
An abbreviation used for *hexyl octyl decyl phthalate*.

hybrid fabric
See *co-woven fabric*. See *hybridized reinforcement fabric*.

hybrid yarn
A *yarn* based on more than one material. Made by stretch breaking separate materials and then spinning a combination into a staple yarn: this staple yarn may be woven into a hybrid yarn fabric.

hybrid yarn fabric
A *fabric* made from a *hybrid yarn*.

hybridized reinforcement fabric
A fabric made form more than one *yarn* to give supple drapeable fabrics with the desired stiffness. High temperature thermoplastics fibres may be combined with, for example, *carbon fibres*. May be made by stretch breaking separate materials and then spinning a combination into a staple yarn: this staple yarn may be woven into a hybrid yarn fabric. Alternatively each fibre could be processed and then combined to produce a *co-woven fabric*.

hybridized reinforcement fibre
A fibrous material based on more than one material. For example, based on a *high temperature thermoplastics fibre*, for example, polyether ether ketone (PEEK) and on *carbon fibre*. Such fibres give supple, drapeable fabrics with the desired stiffness in the moulded composite.

hydantoin-glycol fatty ester
This is a distearate ester which has a relative density of 1·03.

hydrated alumina
See *aluminium trihydrate*.

hydrated aluminium silicate
See *aluminium silicate*.

hydrated alumino silicate
See *clays*.

hydrated calcium silicate
See *calcium silicate*.

hydrated chrome oxide (green)
See *chrome oxide*.

hydrated chromium oxide
A green pigment which is also known as Guignet's green. Stable to sunlight and to chemicals but has poor hiding power and is unstable at temperatures above 260°C/500°F.

hydrated lime
See *lime*.

hydrated magnesium silicate
See *talc*.

hydrated magnesium silicate
See *chrysotile asbestos* and *asbestos*.

hydrated potassium/magnesium/aluminium silicate
See *slate*.

hydraulic actuator
A device used to obtain movement from a hydraulic system. Linear movements are obtained by feeding the hydraulic fluid to a cylinder or ram; rotary movements are obtained by feeding the hydraulic fluid to a hydraulic motor.

hydraulic balance
A term used in *hydraulics* and which means that opposing hydraulic forces are equal.

hydraulic cylinder
See *linear actuator*.

hydraulic ejection system
A system of ejection which is actuated by a hydraulic ram. Such systems are fitted to, for example, *injection moulding machines* as with such a system it is possible to adjust the speed of ejection independently of the mould opening speed. The ejection stroke can commence as soon as the mould begins to open - the ejector plate is driven forward by a small hydraulically-powered ram. Hydraulic ejector mechanisms are very useful when delicate parts are being moulded because, for example, the speed of operation of the ejector mechanism can be adjusted to suit the component.

hydraulic intensifier
Also called an intensifier or a booster. A term used in *hydraulics* and which, for example, refers to a device which is used to increase fluid pressure in a part of a circuit. Such a device, a pressure intensifier, takes in low pressure oil (large volumes) and gives high pressure output (low volumes).

A volume intensifier takes in high pressure oil (low volumes) and gives low pressure output (high volumes). The use of such a device on an *injection moulding machine*, helps explain why there has been increased usage of *direct lock machines* in recent years for machines of up to approximately 100 tonnes.

hydraulic lock
See *direct hydraulic lock*.

hydraulic motor
A term used in *hydraulics* and which refers to a device which converts fluid flow into rotational movement. It is basically a pump running in reverse and is used to obtain rotary movements, for example, of a screw or, of an unscrewing mechanism. See *hydraulic actuator*.

hydraulic press
A press operated by hydraulic power. See *hydraulic actuator*.

hydraulic pressure and flow
A moulding machine has a power unit which is an electric motor which drives a hydraulic pump or pumps. The pump circulates fluid through the system and when there is resistance to the flow of hydraulic fluid, pressure builds up. The pump tries to maintain the rate of flow as it is driven at a constant speed: that is, the pump circulates fluid and does not pump pressure. Pressure is usually measured in either bar, pounds per square inch (psi) or, in megaNewtons per square meter (MNm^{-2}).

hydraulic screw drive
A term associated with *injection moulding* and which means that the *screw* is rotated by means of an hydraulic motor. May be rotated by a (i) direct hydraulic motor or, (ii) hydraulic motor via a gearbox. The screw speed can be steplessly varied and the low torque developed during starting, helps protect the screw against breakage. However, such a motor can be noisy, inefficient and can have poor speed consistency near peak power demands. See *electric screw drive*.

hydraulics
That engineering science which deals with fluid flow and pressure. The study of the mechanical properties of liquids and their application to engineering.

hydromechanical
Another word for *toggle*.

hydroscopic
Sometimes used in place of *hygroscopic* and means that a material absorbs water from the air.

hydrazide group
May be depicted as (-NH-NH-). See *polyamide-hydrazide polymers*.

hydrazine hydrate
An *aldehyde-condensing agent*. See *constant viscosity rubber* and *storage hardening*.

hydrobenzene
See *cyclohexane*.

hydrocarbon
An organic material which contains only carbon and hydrogen atoms. See, for example, *alkane*.

hydrocarbon content
See *rubber hydrocarbon content*.

hydrocarbon resins
Relatively low molecular weight organic materials which are used as tackifying agents in rubbers and in coatings. May be solids or liquids and are either *petroleum resins* or *terpene resins*.

hydrochlorofluorocarbon
An abbreviation used for such materials is HCFC. A carbon-based compound which contains hydrogen, chlorine and fluorine atoms. Such materials have a reduced lifespan, compared to a *chlorofluorocarbon*, and so do not ferry chlorine to the upper *atmosphere* as well as a chlorofluorocarbon. They have a reduced *ODP* compared to a chlorofluorocarbon: it is approximately 10% of that of a chlorofluorocarbon. Because of factors such as the reduced ODP, and the fact that they can be used in equipment designed for chlorofluorocarbons (CFCs), such HCFC materials are being actively promoted as short-term, commercial alternatives to chlorofluorocarbons.

hydrofluoric acid acceptor
An abbreviation used for this term is HF acceptor. An additive used with, for example, a *fluororubber*. A metal oxide such as lead monoxide (which gives vulcanizates with hot acid resistance) or magnesium oxide (which gives vulcanizates with good resistance to dry heat).

hydrofluorocarbon
An abbreviation used for such materials is HFC. A carbon-based compound which contains hydrogen and fluorine atoms. Such materials have a reduced *ODP*, compared to a *chlorofluorocarbon (CFC)*, and so do not contribute to *stratospheric ozone depletion*. Used in blends with *hydrochlorofluorocarbons (HCFC)* to make refrigerant gases which are used in place of chlorofluorocarbons in, for example, humidifiers and chillers. HFC materials will replace *chlorofluorocarbons*, for example, CFC 12 in refrigerant applications.

hydroforming
A sheet-forming process which uses one mould, or one mould half, mounted on the moving platen of an up-stroking press and a rubber cushion (polyurethane) or pad, mounted on the fixed upper platen. The heated thermoplastics material in sheet form is placed on top of the mould and then the heated sheet assembly is pushed into the mould shape by the rubber pad using the press movement. Used to produce embossed key-pads from, for example, *polycarbonate* sheet.

hydrogenated natural rubber
Also known as hydrorubber. May be represented as $(C_5H_{14})_x$. Prepared from *natural rubber* by the addition of hydrogen to the unsaturated groups of isoprene units. Prepared from *natural rubber* dilute solutions in the presence of catalysts, for example, platinum black. As the hydrogenation step removes the stereospecifity, *balata* and *gutta-percha* will give a similar product. This material has a relative density (RD or SG) of approximately 0.86. A colourless material which is insoluble in acetone and alcohol. Used as an impregnating material and as a bonding agent. Because of preparation difficulties and cost, such materials are not usually commercially available.

hydrogenated nitrile rubber
An abbreviation used for this type of material is H-NBR or HNBR. This type of polymer is made by hydrogenating *nitrile rubber* (NBR) in the presence of a precious metal catalyst. Hydrogenation removes the double bonds and for this reason the abbreviation ENM has been suggested as the polymer could theoretically have been derived from ethylene and acrylonitrile. HSN has also been suggested.

The hydrogenated product is a polymethylene chain containing a controlled amount of trans-unsaturation, cyano sidegroups and a few ethyl side-groups. Can have various degrees of unsaturation, acrylonitrile content and polymer molecular weight. NBR of approximately 45% acrylonitrile content is often used.

H-NBR extends the range of use of NBR because of its improved temperature and oil resistance, high strength and ease of processing: it competes with *speciality rubbers*.

Fully saturated nitrile rubber can only be cross-linked with organic peroxides but gives the best heat and hot oil resistance: sometimes known as peroxide-cured A type. Partially saturated H-NBR has better dynamic properties than fully saturated H-NBR. 2% unsaturated nitrile rubber may be cross-linked with peroxides and/or with *sulphur*: sometimes known as type B. 5% unsaturated nitrile rubber may be cross-linked with *sulphur*: sulphur-cured H-NBR is sometimes known as sulphur cured C type. The 5% material has good fabric and fibre adhesion but a considerable reduction in heat resistance compared to type A.

hydrogenated NR
An abbreviation used for *hydrogenated natural rubber*.

hydrogenation
The introduction of hydrogen into an unsaturated material. See, for example, *hydrogenated natural rubber*.

hydrohalide
A compound which contains hydrogen and a halogen, for example, hydrogen chloride. See *rubber hydrochloride*.

hydrohalogenated natural rubber
The reaction product of a hydrohalide and natural rubber. Most usually the hydrohalide is hydrogen chloride and in this case the product is *rubber hydrochloride*.

hydrolysis prevention
See *drying agent*.

hydrolytic degradation
Decomposition, or degradation, of a polymer by water. Accelerated by acids, bases and the use of high temperatures. Polymers which are degraded by water include *cellulose* and its derivatives, *polyamides*, *polyesters* and *polyurethanes*.

hydrolytic polymerization
Polymerization initiated by water, for example, the ring opening polymerization of a lactam to give a *polyamide*.

hydromechanical
Term used in the injection moulding industry to describe a machine which has a certain type of *clamping system*. That is, a lever system (a toggle system) which is usually driven hydraulically.

hydropentafluoro propylene
An abbreviation used for this monomer is HFP. May be represented as $CHF = CFCF_3$. Such a material can be used to make fluoropolymers - see, for example, *fluororubber*.

hydroperoxide-containing gasoline
See *sour gas*.

hydroperoxide-containing polymer
Also known as a polymer hydroperoxide and is formed during *oxidation*. May be depicted as ROOH.

hydrophillic
Means that a material likes, or has an affinity, for water.

hydrophobic
Means that a material has little affinity, for water.

hydroquinone
May also be known as quinol or as 1,4-dihydroxy benzene. This material has a melting point of 170°C, a boiling point of approximately 286°C and a relative density (RD or SG) of 1·33. May be represented as $C_6H_4(OH)_2$. An inhibitor for free radical polymerization and the basis for various *antioxidants*, for example, for hydroquinone monobenzyl ether. In the presence of oxygen, hydroquinone is oxidized to *p-benzoquinone*.

hydroquinone derivatives
Such materials are, for example, *antioxidants* or polymerization inhibitors. See *hydroquinone*.

hydrorubber
See *hydrogenated natural rubber*.

hydrostatics
That engineering science which deals with the energy of confined liquids.

hydroxy-benzotriazoles
See *benzotriazoles*.

hydroxyapatite
See *calcium hydroxyapatite*.

hydroxyethyl cellulose
An abbreviation used for this material is HEC: a *cellulose ether*.

hydroxyl number
The amount of potassium hydroxide, in milligrams, needed to neutralise the free hydroxyl groups in one gram of a polymer, for example, in one gram of a polyester.

hydroxylamine hydrochloride
Used as an *aldehyde-condensing agent*. See *constant viscosity rubber* and *storage hardening*.

hydroxylamine salt
See *storage hardening*.

hydroxymethyl
See *methylol*.

hydroxymethyl phenol
See *methylol phenol*.

hydroxytoluene
See *cresol*.

hydrozincite
A naturally occurring form of zinc carbonate. See *basic zinc carbonate*.

hysteresis
The time-lag exhibited by a material in reacting to an applied stress. When rubber is deformed the difference between the energy input and output is known as hysteresis. For example, a tread compound designed to have high hysteresis (or high internal friction) will have low resilience (or bounce) and the tread will grip well on wet roads. A tread compound designed for lower hysteresis will last longer, will have higher resilience but, in general, the tread will not grip so well on wet roads.

I

i
This letter is used as an abbreviation for isotactic. See *polypropylene*.

I
This letter is used as an abbreviation for:
impact. See *high impact polystyrene*;
integral - in control terminology;
iso. For example, butyl isodecyl phthalate (BIDP);
isobutylene. See *butyl rubber*;
isophthalamide in, for example, *aramid nomenclature*; and,
isoprene. See *butyl rubber*.

I calender
A *calender* named after the letter I. A *calender* in which the rolls are stacked one directly above another: a simple example is a three roll, I calender which is also a *super-imposed calender*.

i-PP
An abbreviation used for isotactic polypropylene. See *polypropylene*.

IARC
An abbreviation used for the International Agency for Research on Cancer - a French based organization.

IB
Letters used in *plasticizer* abbreviations for isobutyl. For example, di-isobutyl adipate (BIBA). See *di-iso-butyl adipate*.

IBM
An abbreviation used for *injection blow moulding*.

ibp
An abbreviation used for initial boiling point.

IBS
An abbreviation used for *n-isopropyl-2-benzthiazole sulphenamide*.

ICR
An abbreviation used for *initial concentration rubber*.

ID
Letters used in *plasticizer* abbreviations for isodecyl. For example, butyl isodecyl phthalate (BIDP). ID is also used as an abbreviation for internal diameter.

ideal copolymer
Also known as a *random copolymer*. See *ideal copolymerization*.

ideal copolymerization
The process whereby an *ideal copolymer* is produced. That is, the growing polymer chain shows the same preference for adding either monomer, during the propagation stage and so, a random *copolymer* is produced.

ideal fluid
A rheological term: an ideal fluid is also known as a Newtonian fluid. Shear stress is directly proportional to shear rate for an ideal or Newtonian fluid. The coefficient of viscosity of such fluids is constant irrespective of the shear stresses involved and is independent of time. See *flow curve*.

identification
The identification of substances present in a material or compound is often done using *qualitative chemical analysis* and *quantitative chemical analysis*. See *material identification*.

IEC
An abbreviation used for the organization *International Electrotechnical Commission*. The English language version of Commission Electrotechnique Internationale (CEI).

IEN
An abbreviation used for *interpenetrating elastomeric network*.

IFW
An abbreviation used for *instrumented falling weight (testing)*.

Igepal
A *trade name/trademark* for an alkyl-aryl polyethylene glycol detergent (for example, Igepal CO-630). See *Bell Telephone Laboratories test method*.

ignition temperature
Also known as the ignition point. The temperature to which a material must be raised before it will burn. See *ignition test - ASTM*.

ignition test
A test used to determine to assess how readily and under what conditions a material will ignite. See ASTM D 1929-68.
Terms used in this ASTM test include *flash ignition temperature*, *self ignition temperature* and *self ignition by temporary glow*. Small specimens, for example, cut from mouldings and bound by wire, are heated in a hot air ignition furnace at a specified heating rate of 600°C/h initially and then 300°C for another specimen: the air flow rate is also specified, for example, 2.54 cm/s. Minimum *flash ignition temperatures* and *self ignition temperatures* are determined by appropriate adjustments of the furnace apparatus settings. See *burning behaviour tests*.

ihp
An abbreviation used for indicated horsepower.

IIR
An abbreviation used for *butyl rubber*.

IISRP
An abbreviation used for the International Institute of Synthetic Rubber Producers. See *regular institute numbers*.

Illinois silica
A fine white soft grade of *silica* which is classed as amorphous but which is *cryptocrystalline*.

illuminance
The luminous flux falling on a surface in unit time. See *lux*.

IM
An abbreviation used for *polyisobutylene*.

IM
An abbreviation used for *injection moulding*.

imaging
Lithography. See *photoresist*.

IMC
An abbreviation used for *in-mould coating*.

imide
A material which contains the imido group -CO·NH·CO- and which may be also known as an imido compound. In polymer technology, the term usually means a *polyimide*.

imido compound
See *imide*.

imine
A material which contains the imino group =NH and which may be also known as an imino compound. Derived from *ammonia* by replacing two hydrogen atoms with two organic radicals. See *amines*.

imino compound
See *imine*.

iminourea
See *guanidine*.

immiscible blend
A *blend* based on two polymers which are immiscible. Most homopolymers are immiscible and so most blends are immiscible blends. Commercially useful systems are still possible

and such a useful immiscible blend is classed as a *compatible blend*. One such immiscible blend is that based upon *polyphenylene oxide (PPO)* and *polyamide (PA)*: such a blend may be made compatible by modification of the PPO and the addition of a third component - a *compatibilizer*.

impact modified polypropylene
See *rubber modified polypropylene*.

impact modifier
An additive for polymer compounds which improves the *impact strength*. Strictly speaking an elastomer/a rubber, when used with a thermoplastics material is an impact modifier. However, the term is usually used in a more restricted way: for example, applied to materials used in conjunction with *unplasticized polyvinyl chloride* (UPVC). UPVC is widely used, for example, as a bottle-making material: the impact strength of such UPVC bottles can be improved by the use of some other plastics materials, for example, by the use of methacrylate-butadiene-styrene polymers (MBS). These thermoplastics materials, when used at approximately 10% addition, still give transparent bottles. However, because they are comparatively expensive, alternative cheaper solutions are now commonly used. One such solution is to introduce *biaxial orientation* into the bottle.

impact polystyrene
See *high impact polystyrene*.

impact resistant polymethyl methacrylate polymers
Thermoplastics materials with greater impact strength than *polymethyl methacrylate*. Can achieve such an increase by *copolymerization* and/or by the incorporation of a rubber such as, for example, polybutyl acrylate. See *methyl methacrylate-acrylonitrile copolymer* and *high impact methacrylate*.

impact speed
An abbreviation used for this term is V_0. See *falling weight impact strength test*.

impact strength
The ability of the plastics to resist a rapidly applied stress as measured by an *impact test*. See *falling weight impact strength test*. One of the major reasons for blending thermoplastics materials is to improve impact strength. See *blend* and *alloy*.

impact test
An impact test measures the ability of the plastics to resist a rapidly applied stress. The two main types of test are the *pendulum impact tests* and the *falling weight impact tests*.

impact test - Izod, see *pendulum impact test - Izod*.
impact test - Charpy, see *pendulum impact test - Charpy*.

impact test machine
An impact tester: a machine used to perform an *impact test*.

impact tester
A machine used to perform an *impact test*.

impeller
Also spelt impellor. The rotating member of a centrifugal pump or of a *high speed mixer*.

impellor
See *impeller*.

Imperial
Usually associated with *Imperial units*. Imperial was first used in England in connection with weights and measures in 1824. The term 'imperial' also refers to a paper size. (22 × 30 inches in the UK and 23 × 31 in the US). See *UK system of units*.

Imperial (FPS absolute)
See *foot-pound-second*.

Imperial gallon
Contains eight pints and is equal to 4·545 96 litres, 1·200 94 US gallons and contains 277·420 cubic inches.

Imperial units
The traditional *UK system of units* was based on the pound, the yard and the gallon. The metric system is slowly being introduced into the UK for general use and the SI system is now extensively used for scientific use. The pound, yard and gallon system is still widely used in North America (see *US Customary Measure*).

impingement chamber
See *mixhead*.

impingement mixer
A mixing device used for *impingement mixing* in *reaction injection moulding* (RIM). A mixing device used for RIM mixing where the polymer foaming and cross-linking reactions occur quickly at mixing temperatures. With epoxy *reaction injection moulding* (RIM) the reaction may be so slow at mixing temperatures that the use of an impingement mixer is not necessary.

impingement mixing
Also known as high pressure impingement mixing or as, high jet pressure impingement mixing. A mixing process which occurs as a result of the collision, or of the impingement, of two or more separate streams. For example, in *reaction injection moulding* (RIM), two liquid streams flow at high pressure, typically 100 to 200 bar (1500 to 3000 psi), into a mixing chamber where the streams impinge at high velocity, mix and the mixed product commences to polymerize as it flows under low pressure into the mould (this is *reaction injection moulding impingement mixing*). Solid polymer rapidly forms as a result of crosslinking or of phase separation. Impingement of the reactants brings them into close proximity: this is then followed by *micromixing* and diffusion. See *parallel stream reaction injection moulding mixhead*.

imposed pressure rheometer
A *rheometer* in which the ram speed is not fixed: it is the pressure or force which is fixed. The force is set at a certain level and the rate is measured during the test. In this case the weight of material extruded is noted. Using the melt density, the volume extruded per unit time is calculated and from this the shear rate is determined. This type of machine, and a *flow rate tester*, may be referred to as an imposed pressure type rheometer or as a shear stress rheometer. Most *high shear rate rheometers* are of the imposed rate type (imposed ram speed).

imposed ram speed
A set ram speed. See *high shear rate rheometry*.

imposed rate, laboratory capillary rheometer
See *high shear rate rheometry*.

imposed rate testing
See *high shear rate rheometry*.

imposed rate-type rheometer
A *high shear rate rheometer*. See *high shear rate rheometry*.

impregnator
A mechanized lay-up machine designed for wetting and placing woven fabric (for example, *glass fibre*) prior to moulding. The fabric feeds through rollers (from 4 to 7) which are set close together so that a pool of *unsaturated polyester resin* can pass through the fibre weave from one side only - this minimizes air trapping. The impregnated fabric is used in *hand lay-up* processes used, for example, for boat building so as to reduce styrene emissions and to decrease the resin content. The glass content, from an impregnator may be 50%: in spray up it may only be 33%.

impression
That part of a mould which imparts shape to the component and which is formed by the *cavity* and the *core*. Moulds may be specified as being single impression (i.e. with one cavity) or they may be specified as being multi-impression (i.e. with more than one cavity).

improved technology blacks
See *carbon black*.

impulse welding
Also known as impulse sealing or as, thermal impulse welding. A *contact welding process* which relies on the use of a *resistance heater*. The heater is in the form of a metal tape and it is separated from the thermoplastics film materials to be welded by a strip of PTFE-coated glass cloth. The films to be joined are clamped between the jaws of the machine and a burst, or impulse, of electricity is passed through the metal tape thus causing welding. After cooling the jaws are opened and the welded material, for example, *polyethylene* film, removed.

IMS
An abbreviation used for *industrial methylated spirits*.

in
An abbreviation for *inch*. For example:

   in H_2O = inches of water; and
   in Hg = inches of mercury.

in-house
Means that a particular operation, such as colour addition, is done within, for example, a processors establishment. For example, *in-house colouring* is colouring done within a factory or processing establishment. Thermoplastics are often coloured in-house using *natural polymer* and a *masterbatch*.

in-line
Term usually applied to processing machines and means that the melt flow path is basically straight or streamlined: the melt is not deliberately turned, or diverted, from a straight line. See *in-line screw machine*.

in-line adaptor
Extrusion terminology. The die and the extruder are on the same line, connected by the adaptor, and the product is centred on the extrusion cylinder or barrel.

in-line machine
See *in-line screw machine*.

in-line, reciprocating single-screw, injection moulding machine
See *in-line screw machine*.

in-line, reciprocating single-screw machine
See *in-line screw machine*.

in-line, reciprocating twin-screw, injection moulding machine
An early type of in-line machine which used two screws to prepare the melt. See *in-line screw machine*.

in-line screw machine
The term usually refers to a type of *injection moulding machine* which is based on an *in-line screw unit*: the most widely used type of moulding machine is an *in-line machine*. A more precise name for this type of machine is an in-line, reciprocating single-screw, injection moulding machine. Such a machine is usually a horizontal machine. A *screw* is used to prepare the melt for injection: when sufficient melt has been produced, the screw is pushed forward so as to displace the melt into the mould. After the *hold period* has lapsed, the *screw* is then rotated so as to prepare further melt for the next shot.

in-line screw unit
The *injection unit* of an *injection moulding machine*. An injection unit which uses a screw to plasticize and transport the material. Modern injection moulding machines are usually *in-line screw machines* often with *closed loop control* of the injection moulding speed and pressure.

in-mould coating
A transferrable coating applied to a mould surface before the moulding operation. Improvements in surface finish and in surface hardness can be obtained by the use of an in-mould coating (IMC). Such in-mould coatings are based on, for example, *unsaturated polyester (UP)* resins which are heavily reinforced with glass beads. May be applied to the hot mould by robot electrostatic spraying and on contact with the hot mould the coating flows and fuses. Moulding then occurs and the PMC compound bonds to the previously applied coating. Very hard scratch-resistant coatings, suitable for kitchen sinks, have been obtained in this way.

in-mould cutting
A Bekum patented process which enables a trepanning operation to take place inside a blow mould and before the mould opens. The blowing air holds the thin walled component firmly against the mould surface during the cutting (*lost dome*) or removal operation.

in-mould foiling - compression moulding
A high quality paper is printed, while flat, with the pattern or legend required. It is then impregnated with an aminoplastic resin solution, partially cured, dried and cut to the shape required. At the moulding stage it is inserted into the mould where it may be held electrostatically or by means of vacuum applied through pins. Moulding onto this printed substrate then occurs to give highly decorated products: this eliminates the problem of printing on a shaped product. By using this method of decoration, complex patterns (for example, 'willow pattern') may be moulded into/on melamine formaldehyde (MF) plates. See *in-mould transfer graphics*.

in-mould foiling - injection moulding
A moulding technique used to make changes to surface appearance. A paper or plastic foil is printed, while flat, with the pattern or legend required. It is then folded, for example, into a box-shape and inserted into the mould where it may be held electrostatically or by means of vacuum applied through pins. Moulding onto, and/or through, this printed substrate then occurs to give highly decorated products: this eliminates the problem of printing on a shaped product.

in-mould hinge
See *integral hinge*.

in-mould labelling
A system of component decoration in which labels (paper or a thermoplastic foil) are placed in, for example, a blow mould prior to collection of the next parison. The labels are held in position by vacuum pins and the labels bond to the component during moulding. As the labels may be printed flat, before moulding, this avoids the problems of printing on circular containers.

in-mould labelling - compression moulding
This is thought to have been originally done for gramophone records where two small labels were placed in the centre of the mould - one on either side of the preheated material or biscuit. Moulding then occurred and the labels bonded to the product.

in-mould transfer graphics
A component decorating system used with *glass reinforced plastics* to transfer graphics or legends to the finished surface during the moulding process. Has been used to produce, for example, signs.

in-mould trimming
The removal of un-wanted material during the moulding or forming cycle. For example, in *thermoforming* it is possible to achieve in-mould trimming by forming over a knife edge and then separating the forming from the web using a roller or pressure pad.

in-process gauging
Gauging performed during manufacture.

inactive
A term sometimes used in place of *reinforcing* in connection with fillers for rubbers is 'activity'. Inactive means no reinforcing action.

inactive copper compounds
See *copper*.

inch
An abbreviation used for this unit is " or in. Originally defined in England as being equivalent to three, dry and round, grains of barley. 12 inches make one foot and three feet make one yard. One inch is equivalent to 25·4 mm. One inch/minute is equivalent to 0·423 333 mm/s. See *UK system of units*.

inch of mercury
A unit of pressure with the abbreviation of in Hg: equals 0·033 4 atmospheres or 3,386·389 N/m^2 at 0°C.

inch of water
A unit of pressure with the abbreviation of in H$_2$O and which equals 249·073 N/m^2 at 4°C.

inch-pound-second
An abbreviation used for this term is ips. A system of measurement which uses the inch (in) as the basic unit of length and the basic unit of mass is the pound (lb). The derived unit of force is the pound force (lbf or psi). This system is sometimes known as IPS or as ips. See *UK system of units*.

incineration
The burning of a material. Incineration of *municipal solid waste* may be combined with energy generation so as to form part of a *waste-to-energy* policy. Recycling of mixed, contaminated plastics is often impractical and the plastics waste has a high energy content (higher than coal). There are however, grave concerns about the generation of *dioxins*. See *polyvinyl chloride*.

incl
Shortened form of the word 'inclusive'.

inclined mould
A term used in *blow moulding* - the mould is not vertical but is at an angle to the die so as to facilitate the production of large, thin blow mouldings.

inclined toggle
A lever system fitted to, for example, an *injection moulding machine* so as to minimize platen deflections. By using a large *anchor platen*, larger than the *moving platen*, the toggle can be inclined so that the effects of distorting forces are minimized: platens may thus be made thinner and lighter. See *locking force*.

inclined Z calender
A four roll *calender* named after the letter Z where the letter is skewed or inclined from its normal position. A type of *four roll calender* in which the rolls are offset one with respect to the other so as to minimize roll interactions. Rolls of different finishes may be fitted to the same machine: for example, the machine may be fitted with one pair of rolls polished and the other pair matt. Polished or matt sheet can then be produced by feeding either the top nip or the bottom nip. This is relatively easy to do as both nips are offset and inclined and either nip can be fed from the same side of the calender.

inclusion cellulose
Also known as occlusion cellulose. Cellulose which contains a trapped solvent: such a trapped solvent may not be lost even by heating above its boiling point. Solvent inclusion renders the *cellulose* very reactive, for example, for acetylation.

incompatible blend
A *blend*, or system, based on an *immiscible mixture* of two or more thermoplastics materials which is not considered commercially useful. Such blends may be made compatible by chemical modification, addition of third components or by varying the processing employed. Such incompatible blends may then be called polymer alloys. See *compatible blend* and *immiscible blend*.

incompatible plasticizer
A *plasticizer* which is easily lost from a system. A *plasticizer* which requires a *compatible plasticizer* to be used as well so that plasticizer loss will not occur from a material in use.

indene
A colourless liquid hydrocarbon with the formula $C_6H_4C_3H_4$. Obtained from *coal tar*, mixed with coumarone, and used to make, for example, *coumarone-indene resins*. This material has a melting point of $-2°C$ and a boiling point of 182°C.

indene resin
See *coumarone-indene resin*.

indene-coumarone resin
See *coumarone-indene resin*.

indentation test methods
See *hardness indentation test methods*.

indentor
That which is pushed into a material in a hardness test, for example, a blunt needle-shaped probe. See *hardness*.

index number
A rating in the *lighter test*. For example:

index 0 means that combustion resumes after more than three drops have fallen in the lighter test;
index 20 means that combustion resumes after three drops have fallen;
index 40 means that combustion resumes after two drops have fallen;
index 60 means that combustion resumes after one drop has fallen;
index 80 means that combustion ceases after one drop has fallen;
index 100 means that no drops fall and combustion ceases as soon as the flame is removed in the *lighter test*.

india rubber
See *natural rubber*.

India rubber
See *natural rubber*.

Indian yellow
See *potassium cobalt nitrite*.

indirect cooling system
A liquid cooling system, for example, for a mould, in which the heat transferred to the cooling liquid is removed by a heat exchanger so that the fluid may be re-circulated. Indirect systems are usually only used when the mould is required to be kept at a temperature below that of the cold-water supply at the hottest time of the year (for example, in the UK this is approximately 25°C) or if the mould temperature is required to be above approximately 90°C. A pressurised system may

be used in order to maintain the water temperature above atmospheric boiling point. For example, a heavy-duty 3 kW, low watt density, immersion heater provides rapid heating whilst cooling may be through a large surface area, coil-type heat exchanger and solenoid valve. See *chiller*.

indirect electrical heating process
A heating process in which heat transfer is from a separate, electrically heated source. See *electrical resistance element*.

indirect heating process
A heating process in which heat transfer is from a separately heated source. See *electrical resistance element*.

indirect pressure sensing
A pressure measuring system in which the *sensor* is not in direct contact with the *melt*. In injection moulding, pressure may be sensed by a force-type of transducer located behind a pin. See *cavity pressure control*.

indirect pressure transducer
An indirect sensor. A transducer which is not in direct contact with the melt. A *transducer* which gives *indirect pressure sensing*, for example, in *injection moulding*. A pressure transducer whose temperature-sensitive electronics, are located away from the hot melt.

inductance
The property of a circuit which when carrying current forms a magnetic field.

induction heating
See *high frequency heating processes*.

induction period
Also known as inhibition period. The time during which there is no observable change in, for example, a chemical reaction. At the end of the induction period there is an observable change. Often encountered in decomposition, or degradation, of polymers.

industrial methylated spirits
An abbreviation used is for this material is IMS. Methylated spirit which is free from pyridine. Approximately 95% ethanol and 5% methanol. See *methylated spirit*.

Industrial, Scientific and Medical bands
An abbreviation used for this term is ISM bands. Frequency bands allocated for industrial, scientific and medical uses by the *International Telegraphic Union*. The permitted wavelengths may differ from country to country. For example, the ISM bands in the UK are as follows.

Frequency MHz	Wavelength m	Tolerance ± MHz	Application
13.56	22.1	0.07	High frequency heating
27.12	11.0	0.163	High frequency heating
915	0.33	13	Microwave heating
2450	0.12	50	Microwave heating

Recommendations for radio interference are given by the *International Special Committee on Radio Interference (CISPR)*. See *dielectric heating*.

industrial vegetable oil
Examples of such oils include rapeseed, safflower, sunflower and soybean. See *vegetable oil*.

inelastic fluid
Also known as a purely viscous fluid. Such a material does not show any elastic effects, unlike polymer melts. See *die swell*.

inert filler
A *filler* which is mainly added to reduce costs and/or shrinkage. See *inorganic filler*.

inferior grade
See *natural rubber*.

inflammability tests
The term flammability is now used in place of inflammability. See *flammability test*.

infra-red
An abbreviation used for this term is IR. Electromagnetic radiation of wavelengths between those of visible light and radio waves. Approximately 0.8 micron to 1 mm.

infra-red pyrometer
A device which may be used to measure temperatures without contacting the product.

infusorial earth
See *diatomaceous earth*.

ingrain dye
A *dye* which is formed in-situ, by *diazotization*, on a fibre. Such dyes are frequently used with *cotton*.

inherent variation
A change which occurs even though machine set-points have not been altered. Intentional changes made to machine settings are called assignable variations. However, even when a machine is running under constant conditions, i.e. it has 'settled down' and no deliberate changes are made, a change can still occur. This is an 'inherent variation' and is so called because it cannot be attributed to a single cause: could be caused by material variations, machine variations, environment variations or operator variations. Inherent variation is also known as inherent variability or as, intrinsic variability or as, instantaneous variability.

inhibition
The delay of a free radical polymerization due the presence of an *inhibitor*.

inhibition period
See induction period.

inhibitor
An *additive* to a system, or a material such as air, which retards a chemical reaction, for example, a curing reaction. A material which retards, or prevents, a chemical reaction by reacting with free radicals. Inhibitors, unlike catalysts, are consumed during the reaction. When the inhibitor is consumed, the reaction proceeds at the usual rate. See *retarder*.

In rubber technology, inhibitors are additives, or ingredients, added to rubber compounds in order to inhibit the effect of copper. One such class of materials are the sodium salts of *ethylene diamine tetra-acetic acid*.

In *unsaturated polyester* (UP) resin technology, there are two general classes of inhibitor: these are quaternary ammonium salts and substituted phenolic derivatives. This last class includes hydroquinone, quinone and their mono and di-substituted derivatives. Salts of organic bases are usually used with these inhibitors so as to extend storage life. For example, trimethyl ammonium chlorides, bromides and acetates.

inhomogeneous melt
Melt which is non-uniform in temperature or in composition. See *screw*.

initial concentration latex
Undiluted latex which is used to produce *natural rubber*: used to produce *initial concentration rubber*.

initial concentration rubber
An abbreviation used is ICR. Natural rubber produced from undiluted latex. The resultant coagulum will contain more

initial mould shrinkage
The *shrinkage* measured within a short time of moulding, for example, within 2 h. After the specified time (for example, 2 hours) the dimensions of the moulding are measured at room temperature and the dimensions of the cavity, if not known, are also measured. The mould shrinkage (MS) is given as a percentage by $100 \times L_0 - L_1/ L_0$. Where, L_0 is the length of the cavity and L_1 is the length of the moulding.

initiation
When used in connection with polymerization, also known as chain initiation. The beginning of chain polymerization which commences when an active initiating species, obtained from an *initiator*, reacts with the monomer to give an active polymer chain: this in turn reacts with more monomer etc. See *oxidation*.

initiation radical
A primary radical formed as a result of *initiation*.

initiator
A material which causes chain polymerization. In free radical polymerization, the initiator is often a peroxide, for example, *benzoyl peroxide*. Initiators, are also known as catalysts, in the *unsaturated polyester resin industry*. Usually *organic peroxides* are used but can also use *azo compounds*. Peroxides are usually preferred as they are soluble in styrene and do not give off gaseous by-products. See *initiation*.

injection
The transfer of melt from an injection cylinder into a mould. See *injection moulding*.

injection barrel
The barrel or cylinder which contains the material which is to be injection moulded. The barrel or cylinder which is used to heat soften the material which is to be injection moulded.

injection blow moulding
A *blow moulding process* which uses an *injection moulding machine* and, in the simplest case, two moulds. A blowing stick or mandrel is placed inside the first mould - which is a heated injection mould. The thermoplastics material is then forced into the mould, from the injection unit, and flows around the mandrel thus forming a tube with a closed end. Whilst the material is still hot, the mould is opened and the mandrel is transferred to the second mould. Air is then introduced into the plastics tube, via the mandrel, and this causes the plastics tube to expand to the shape of the mould. When the product (e.g. a bottle) is rigid enough to be ejected, the mould is opened and the mandrel plus bottle is removed so that the bottle may be removed from the mandrel.

The major advantage of this process is that it produces components which require no finishing, that is, finished form production is possible. This feature is important where there must be no risk of contamination of the package or of its contents, for example, in medical applications. The major disadvantage is one of cost - high costs are involved because an injection moulding machine, and an expensive injection mould, is required.

injection blow moulding - biaxial orientation
It is relatively easy to arrange for biaxial orientation to occur in *injection blow moulding*. The mandrel is made of telescopic sections and these are extended before the air is introduced. Mandrel extension causes orientation in one direction and the blowing process causes orientation in the other. Such a stretching process can, because of the orientation, give strong products with improved resistance to gas diffusion. If a material such as *polyethylene terephthalate* (PET) is used the bottles may be suitable for the packaging of carbonated drinks, for example, lemonade.

injection blow moulding - characteristics
In the *injection blow moulding process*, the blowing pressure is relatively low: it is of the order of 7 atmospheres, 0.7 MNm^{-2} or 100 psi and this means that the blow mould costs are, in turn, relatively low. However, an injection mould is also needed and this must be constructed to withstand the forces involved in that process. The need for two moulds (one of them an expensive injection mould) and the comparatively low rate of production are the main disadvantages of this particular process. However, as the moulds used are of the *hot-runner* variety, there is little or no wastage and as a closed preformed tube is used, the product requires no finishing, i.e. finished-form production is possible. See *extrusion blow moulding*.

injection boost time
That time period for which an additional source of hydraulic fluid is available during *injection* so as to give more rapid mould filling.

injection compression moulding
An *injection moulding process*. In order to attain the mould-filling rates required, in conventional injection moulding, high injection pressures are needed (e.g. up to 245 Mn/m^2 or 35,000 psi) and this in turn results in the need for high clamping pressures - so as to stop the mould from opening during the mould filling stage. This means that injection moulding equipment is very expensive as it must be built to withstand the pressures involved.

One way of overcoming the need for high clamping pressures is to design the mould so that slight mould opening is possible without melt escaping. This means that during mould filling a precise quantity of melt is injected into the lightly-closed mould which parts or breathes. When melt transfer is complete the clamping unit is then powered so that final mould closing occurs - just like *compression moulding*. Cooling then occurs in the usual way.

The use of this process offers a number of advantages compared to conventional injection moulding. As the mould is partially open, orientation is reduced and this, in turn, gives more stable mouldings. Compression moulding of a hot melt results in mouldings with an excellent surface finish and minimizes clamping force requirements. This process is used to produce high-quality, compact discs from polycarbonate (PC).

injection line pressure
The pressure in the hydraulic line, or pipe, which is connected to the *hydraulic actuator* (a ram or piston) which causes linear screw movement during mould filling.

injection machine
See *injection moulding machine*.

injection mould
The mould used in *injection moulding*. After the material has been thermally softened in the *injection* machine barrel it is then transferred to the injection mould so that shaping followed by hardening may occur. The material is forced through the *sprue*, into the *runner system* and from there into the cavity (or cavities) via the *gate*. The component is formed between the opposing core and cavity, and the surface finish on the component is dictated by the surface finish on these two mould parts. After the hardening stage has advanced sufficiently the component is then ejected or removed from the cavity.

An injection mould is therefore rather more than a simple shaping device. It must function, for example, as a heat exchanger and an ejection device: it may also be required to separate the feed system from the mouldings. The mould must be designed and constructed so that it is capable of producing the required number of components at the desired production rate.

Where only a small number of mouldings is required it may be possible to produce injection moulds from low-cost materials or by low-cost techniques using, for example, a *prototype mould*. Such prototype moulds have been made from, for example, other plastics and dental plaster. If the moulds are required for production then the cost and complexity are many orders of magnitude higher. Mould cost is therefore influenced by the material of mould construction, but it is even more strongly influenced by part complexity and size. Some injection moulds cost more than the machines on which they run. Because of the high cost of injection moulds it is essential to ensure that both the component and the mould are well designed. The component must be designed to suit the injection moulding process and the two must be designed to produce the component at the required rate and with the minimum of supervision and repair. If the mould design is not optimised then components of the desired quality may not be produced consistently and economically.

It should be possible to design injection moulds so that high injection moulding pressures are not necessary. This may be done by using a correctly sized feed system, using high melt and mould temperatures and by using reduced *flow path:wall thickness ratios*. If the injection pressure requirements are reduced then, better components may often be produced more cheaply. Better components because, for example, the amount of shear induced brittleness is decreased for *nylon 6*: more cheaply as clamping pressure requirements will be reduced. This in turn will reduce the wear and tear on the machine and on the mould.

There are many different types of injection mould and these include, for example, *two-plate mould*, *three-plate mould* and the various types of runnerless mould (see *insulated runner mould*, *sprueless mould* and *hot runner mould*).

injection mould - cooling of

For a *thermoplastics material* the *injection mould* is usually held at a temperature below the *melt temperature*. Cooling is most commonly achieved by circulating water at a specified temperature through the *cooling channels* in the mould. It is this temperature which is often reported although it is the mould surface temperatures which are important. What is often forgotten is the rate at which the water must be circulated so as to obtain the desired cooling capacity.

To prevent large shrinkage variations and/or warpage occurring to mouldings of large surface areas, or in those of differing wall sections, the need for separate cooling circuits in the same half of the mould may be required. Each circuit should be independently controlled at a specific temperature.

For example, in the case of *high density polyethylene* (HDPE), the region of the gate should be held at a mould surface temperature of between 10 to 20°C/50 to 68°F, whereas a temperature of between 40°C to 60°C/104 to 140°F would be used at the end of the melt flow path. To obtain better cooling, particularly when moulding thick sectioned components, certain areas of the mould should be made from *beryllium-copper* in order to dissipate the heat more rapidly. The position of the water channel in relation to the mould surface is very important when designing moulds to be used for HDPE. In order to achieve economic cycle times the diameter of the water cooling channel should not be less than 8 mm and the distance from the mould surface to the edge of the cooling channel should not be greater than 1·3 times the cooling channel diameter (1·3 D). For mouldings that possess both thick and thin sections, the distance of the water channel should be closer for the thicker sections (i.e. 0·9–1·0 D) than for the thinner sections.

injection mould - damage

Mould damage commonly results from:

1. acid attack - *unplasticized polyvinyl chloride* (UPVC) is the greatest offender in this respect as the acid released during processing can cause severe damage;
2. mould burning - caused through improper venting of the moulding cavity;
3. material trapping - when the mould closes it is important to ensure that there is no trace of polymer, such as flash, which can cause mould bruising;
4. mould abuse - only tools made from soft brass or copper should be used to remove mouldings or flash from the mould; and,
5. water damage - water can cause considerable damage to moulds, particularly those which are operated at low temperatures or those which have been used to mould UPVC.

injection mould thickness

The thickness of an *injection mould* when closed. This is quoted in either inches or millimetres and is commonly quoted as a range, for example, from 250 to 600 mm for an *injection moulding machine*. The first figure is the minimum mould thickness that can be accommodated whilst still maintaining clamp pressure and the second figure is the maximum mould thickness that can be mounted on that particular machine.

injection moulding

A cyclic process in which a soft material (usually a heat softened *melt*) is injected into a mould from which it is ejected after it has set to the shape of the cavity. Injection moulding is the most common moulding process used by the plastics industry and, after extrusion, the greatest user of plastics materials.

In this process two mould halves are clamped together and a thermally softened material is injected, from a barrel, into the cavity contained in the two mould halves. To obtain the mould filling speeds required, high pressures are used: in order to stop the mould from being over-filled, when extra material is packed into the mould so as to compensate for shrinkage a reduced second-stage pressure, or *hold pressure*, may be used. The component is then set to the shape of the cavity. When the part has set to the shape of the cavity the mould is opened and this component is ejected so that the process can begin again.

The process is very extensively used for thermoplastics materials: virtually every thermoplastics material is injection moulded. For such materials the mould is kept cooler than the melt so as to achieve setting. With slight modifications the process may also be used for thermosetting plastics and for rubbers. For example, for such materials the mould is kept hotter than the melt so as to achieve setting: the barrel of the machine is kept cooler than the mould. See *rubber injection moulding*.

injection moulding - attractions of

Injection moulding is a mass production or repetitive process which is capable, when operated correctly, of producing very large numbers of identical complex mouldings at a high production rate. The mouldings may require little or no finishing although their appearance may be enhanced, if required, by a range of decorating techniques, e.g. metallising. Good dimensional accuracy even for components of complex and intricate shape is possible.

Because of the high equipment costs, injection mouldings are usually comparatively small and include gears, switches, pen and pencil barrels, plumbing fixtures, housewares, etc. Large mouldings can be made but in order to reduce the size of the clamping forces involved, various techniques or methods, have been evolved. For example, *structural foam moulding* and *injection compression moulding*.

injection moulding - automatic operation of

Only commence automatic moulding when a satisfactory melt is produced during *purging* under *hand operation*. At the end of the purging stage reduce the feed setting to a low value - below that which is needed to fill the mould. The *injection unit* is brought into contact with the *sprue bush* (i.e. brought forward) and material is forced into the mould under a low injection moulding speed and pressure. Commence moulding on an *automatic*, or *semi-automatic cycle* using pre-determined cycle times (although these may be calculated, they are usually based on experience). Adjust the *line pressure* (first stage pressure), and the injection speed, until the required rate of injection is obtained. Rapid changes often do not save any time as overshooting is often the result.

Gradually increase the *feed setting* until the mould is almost full. At this stage of the proceedings the moulding should be short but the surface finish should be that required: re-adjust the injection speed if necessary to that which is required. At the end of the injection stroke the screw/ram should *bottom*: note this position. Periodically, check to ensure that the hopper has the required amount of material (check the level of the plastics material and the colorant supply) and that material is not leaking, or weeping, from around the nozzle.

Set the final mould filling pressure (*hold pressure*) to that suggested for the particular material: set this second stage pressure to come in, to be actuated, at this point. Increase the amount of feed until the correct *screw cushion* is obtained. Increase the second stage pressure until the mouldings are free from sinks or voids; allow the machine to *equilibrate* and then weigh the moulding. Increase the second stage pressure until the moulding weight no longer increases or, until the moulding starts binding or sticking in the mould. It may prove necessary to increase the injection time if the moulding weight does not settle down to a steady value. Do not overpack the moulding: adjust the screw feed so as to maintain a constant length of screw cushion. After each adjustment allow the machine to settle down for approximately 12 shots before making another adjustment. Adjust the cooling time until the moulding can be ejected without distortion and then adjust the *screw rotational speed* to fill this time.

injection moulding - clamping pressure consistency

An injection moulding machine must be capable of applying a consistent *clamping force* to the mould if consistent mouldings are to be obtained. There is a tendency, to reduce clamping forces to the lowest value possible so as to reduce running costs and to reduce machine wear. In such cases, relatively small changes in clamping force (caused for example, by mould temperature variations) can alter the amount of mould parting on injection and, the amount of mould distortion. Direct lock machines give, in general, more consistent clamping forces, and are more maintenance free, than toggle machines. Such *direct ram machines* have an inherent, and automatic, compensation for mould size changes caused by changes in mould temperature. See *injection moulding - mould size adjustment*.

injection moulding - ease of flow

The *ease of flow* of the material should be checked during processing and before processing begins. For example, the ease of flow may be assessed using a *flow tab* or by noting the ease of mould filling. Before the material is processed it would be wise to check that it is similar to what is currently being used by measuring the flow behaviour using a *rheometer*. See *melt flow rate*.

injection moulding - feed

See *injection moulding - regrind and consistency*.

injection moulding line

The total production facility used for injection moulding, for example, moulding machines, colour dispensers, regranulators etc. The *injection moulding machine* is just one part of the total production process and must be integrated with all the other units which make up the production facility so as to provide a system which can do what is required, quickly and efficiently. This often means that all the machines are microprocessor controlled and are integrated one with the other: they are also linked and controlled by a central computer which can control, and monitor, groups of machines. For the management of a whole factory, several computers may be linked in a network so as to provide a system which can handle all aspects of production from materials ordering to component dispatch.

injection moulding machine

A machine which produces components by *injection moulding*. The power unit employed in such a machine is most commonly an electric motor which drives a hydraulic pump or pumps (see *self contained*). The pump(s) circulates fluid through the system at a certain rate or speed. Movement are obtained by the use of *actuators*. Commonly, linear movements are obtained by feeding the hydraulic fluid to a cylinder or ram; rotary movements are obtained by feeding the hydraulic fluid to a hydraulic motor.

Injection moulding machines can be grouped or classified in various ways. For example, they can be grouped according to the type of *injection unit*: this gives ram (plunger) or screw machines. Most injection moulding machines are *reciprocating screw machines*.

The machines could be classified by the system employed to power, or drive, the machine, for example, into pneumatic, hydraulic, electro-mechanical or all-electric. Most machines are hydraulically powered: if the *power unit* is built into the machine then the machine may be called self-contained. If the machine draws the hydraulic fluid from a line, then it is not self-contained but line powered or a central feed machine.

The machines may also be classified according to the type of control system employed. For example, the injection moulding machine may be classed as being a relay machine, a solid state machine or a microprocessor controlled machine. May be also be considered as being *open loop* or *closed loop*.

The machines may also be classified as being *horizontal* or *vertical*: the use of these terms indicates how the clamping system (the locking system) and the injection unit lie relative to the ground. In turn, the ways in which the clamping system can be arranged relative to the injection unit (the various permutations possible) gives rise to:
  horizontal locking and horizontal injection;
  horizontal locking and vertical injection;
  vertical locking and vertical injection; and,
  vertical locking and horizontal injection.

The horizontal locking and horizontal injection arrangement is the most common. See *in-line screw machine*.

The injection moulding process, and injection moulding machines, are constantly being modified in an effort to extend the range of use of the process and/or to produce components not possible by more conventional injection moulding. See, for example, *gas injection moulding*, *sandwich moulding* and *injection compression moulding*.

injection moulding - monitoring

See *injection moulding - product monitoring*.

injection moulding - mould size adjustments

Changes in clamping force can arise on many injection moulding machine because of changes in mould temperature. Such changes are usually associated with toggle (hydro-mechanical) machines. For example, if a 100 tonne (100 t) machine is set to give maximum clamping force at 25°C/77°F then, if the mould temperature rises to 40°C/104°F, the clamping force can easily reach 110 t. If the mould temperature rises to 65°C/149°F, the clamping force can easily reach

120 t. Direct ram machines automatically compensate for such changes in mould size. On toggle machines, it is necessary to correct for these temperature variations by, for example, changing the mould height. A sensor (transducer) is needed to measure the tie bar extension/contraction, or to measure tail-stock deflection, and the signal so generated is fed back to a mould height, adjustment motor. Such a device can lead to reductions in component variations when used with *cavity pressure control*.

injection moulding - mould temperature control
Because of the effect of changes in component dimensions, caused through mould temperature variations, it is very important to control the mould temperature precisely, for example, to within ±1°C (±2°F). Mould temperature also has a tremendous effect on component properties (see *orientation*). The mould should be preheated to the operating temperatures and then held at those temperatures during machine operation: this is often done by circulating fluid through channels machined, or cast, into the mould. See *injection mould - cooling of*.

injection moulding - non-uniform shrinkage
See *non-uniform shrinkage - injection moulding*.

injection moulding - product monitoring
The quality of many injection mouldings is often only assessed, for example, by a simple visual examination. However, it is often preferable to assess the mouldings by methods which give numerical answers: the figures so produced may then be readily subjected to statistical analysis. Usually the mouldings are either weighed or, a dimension is checked. Weight is often easier, and simpler, to measure than a dimension of a moulding: this is because of developments in modern, digital balances.

If a series of injection mouldings are produced, and both the weight and a dimension are measured then, it is often found that there is no direct correlation between the weight variations and the size variations. A heavy moulding does not necessarily give the moulding with the largest dimensions nor, does the lightest moulding give the moulding with the smallest dimensions. This could be due to the fact that the wrong dimensions are being checked as, for example, there could be a variation in cavity thickness, caused through mould opening, and length is being checked. However, to obtain minimal, dimensional variations it is best to minimize weight variations. See *automatic quality control*.

injection moulding - regrind and consistency
Many injection moulding machines run on a thermoplastics material which is a mixture of new (virgin) material and reclaimed (regrind) material. The use of regrind is only usually possible with thermoplastics materials. To obtain a desired colour, a *masterbatch* may also be used.

Surprisingly, the use of some regrind may improve the performance of an injection moulding machine i.e. its use may give more consistent mouldings as there is better feeding of the material onto the machine screw if some regrind is used. The exact amount of regrind will have to be determined experimentally and, once found, then the ratio must be held as precisely as possible if consistent mouldings are to be obtained as if the feed is not consistent, then inconsistent injection mouldings are often obtained. The differences may not be discernible to the naked eye but they are large enough to cause rejection of the product - either because the appearance is incorrect or because the size is incorrect. In many factories, it is possible to run on two levels of regrind use: 10% for critical or demanding jobs and 30% for the remainder.

injection moulding - temperature control
Because of the effect of changes in component dimensions, caused through mould temperature variations, it is very important to control the mould temperature precisely, for example, to within ±1°C (±2°F). The same degree of control must be exercised over the temperature of the hydraulic oil and, of the plastics melt. The importance of melt temperature control is more readily appreciated, as it obviously effects material colour and material flow (viscosity). Oil temperature control is often just as important, even though it may not be so obvious. If the oil temperature changes, then the way the injection moulding machine performs also changes and, component weight/dimensions also changes. So, machine consistency is vitally important if, for example, *cavity pressure control (CPC)* is to be made most effective.

injection moulding - test specimen production
Because of the speed with which test samples can be produced by injection moulding this method of production is widely employed even though the samples are *anisotropic* and the production costs are high. To save on time and materials it is also common to use a multi-impression mould. A *family mould* is often used, for example, one which produces a tensile dumb-bell, an impact bar and a flexural strength sample simultaneously. Despite its obvious attractions, injection moulding using a family mould has several disadvantages, for example:

mould and machine costs are very high;
properties are very dependent upon processing conditions;
samples are anisotropic;
it is difficult to fill and pack each separate cavity ideally; and,
compared to *compression moulding* a lot of material is required.

Because of the dramatic effect that processing can have on properties it is important to adjust the machine conditions and the moulding cycle so that samples which have 'average' properties for that material are used. To meet this requirement, and to ensure that reproducible samples are produced, the following should be observed.

1. Predry the material thoroughly if necessary (for example, for PC and PA);
2. Set the barrel temperatures so that a melt temperature approximately midway between the lowest and the highest recommended by the manufacturer is obtained;
3. Purge the machine thoroughly before sample production is attempted;
4. Use a mould with a relatively large gate - the gate should be as deep as the cavity;
5. Fill the mould relatively slowly and do not overpack;
6. Use the slowest screw-rotational speed possible;
7. If a multicavity mould is being used, concentrate on the production of one specimen at a time;
8. Record all machine settings and the production procedure;
9. Store the specimens for a predetermined time under standardized conditions;
10. Load the samples into the testing machine the same way each time.

injection mouldings
The product of an *injection moulding machine*.

injection mouldings - contamination of
Oil and/or grease contamination sometimes occurs after ejection of the mouldings (for example, it drops from the toggles, platen bushes and hydraulic cylinder seals) and should be corrected immediately; it should be kept under control by preventative maintenance (see *all-electric machine*). A common cause of grease contamination is grease leaching from around the ejector pins; this is particularly true for pastel shades of

some filled materials. When this happens, the mould should be stripped and cleaned thoroughly and then, re-assembled using a white *polytetrafluoroethylene*-based lubricant.

Many plastics and *thermoplastic elastomers (TPE)* attract dust at a tremendous rate and this can be extremely difficult, and expensive, to remove. It is far better to stop the problem occurring in the first place by, for example, keeping the mouldings covered at all times and handling them as little as possible. When they are handled, or moved, the dangers of scratching the injection mouldings should be remembered (many will even scratch each other if rubbed gently together). If touched with the bare hands then grease can be transferred to the mouldings and this, again, can be very troublesome, for example, in *electroplating*. Clean, dry gloves should be worn when injection mouldings are handled unless it has been proved that this precaution is unnecessary.

injection mouldings - impact testing of
The ability to withstand an impact, commonly referred to as toughness, is a very important requirement for many plastics mouldings and one which is not often measured satisfactorily by the standard type of test: this is usually a pendulum type of test. For this reason a standard, pendulum-type tester may be modified so that, for example, it can either be used to test small components directly or, so that it may be used to test a special moulded tab in the case of a large component. In the case of a large component, a *falling weight impact test* may be performed on the whole component.

injection mouldings - testing of
Simple tests performed on injection mouldings include visual inspection, moulding weight (density), ease of flow testing or flow tendency, oven heating and photoelastic inspection. Moisture content may be performed on the moulding material if the mouldings look unacceptable. See *fit and finish testing*.

injection mouldings - visual testing of
This is often the only test to which injection mouldings are subjected. If the machine is running semi-automatically then, the operator inspects every injection moulding for colour, colour dispersion, colour uniformity, surface finish (gloss/mattness) and, freedom from moulding defects such as stringing and silver streaking. See *fit and finish testing*.

injection moulds - uneven filling of
An *injection moulding fault*. When uneven filling occurs in a multi-cavity mould fed with a *balanced runner system*, the rate of filling in the 'slow' cavity may be conveniently increased by removing metal from the gate land, i.e. keeping the cross-section constant but reducing the gate length. Alternatively, the runner diameter may be altered so that the pressure drop at each cavity is the same.

injection port
See *tool injection port*.

injection pressure
The pressure exerted on the material in the injection cylinder by the ram or screw during injection (mould filling); also known as *first stage pressure*.

During mould filling, the injection pressure may be measured by a transducer located in he nozzle or in the hydraulic line. It does not have a constant value but increases as mould filling becomes more difficult. Once the mould is initially full then the cycle enters the *holding phase*. There is a direct relationship between *injection line pressure* and *injection pressure*.

injection speed
Sometimes called screw displacement rate and usually refers to the speed of mould filling during *injection moulding*: that is, when the screw is acting as a ram. When moulding thin sectioned components, high injection speeds are essential in order to fill the moulding before freezing occurs. For many injection moulding operations, the initial cavity filling is complete in less than one second. However, a better surface finish is obtained on mouldings with thicker sections by using a slower speed. Many moulding faults, for example jetting and air trapping, may be avoided by using a range of speeds (that is, programming the injection speed), during the mould filling stage. See *programmed injection speed*.

injection unit
Term used in *injection moulding* and refers to the screw and barrel assembly of, for example, an *in-line screw machine*.

ink embossing
See *valley printing*.

in lbf
An abbreviation used for inch pound-force.

inner liner
A *tyre* component: that which lines the inner service of a tyre so as to retain air and/or to reduce chafing of the inner tube.

inorganic accelerator
An *additive* for polymer compounds, for example, for rubber compounds. Addition of inorganic materials such as antimony sulphide, antimony pentasulphide, calcium hydroxide, lead oxide and magnesium oxide shorten the vulcanization time. That is, each of these materials can act as an inorganic vulcanization *accelerator*.

inorganic blocking agent
An inorganic additive which reduces *blocking*. See *matting agent*.

inorganic filler
An additive for polymer compounds, for example, for both plastics and rubber compounds. A *filler* based on an inorganic material, for example, china clay. See *non-reinforcing filler* and *reinforcing filler*.

inorganic pigment
A pigment, or *colorant*, which is based on an inorganic material. Such pigments generally have better heat stability and are more economical than *organic pigments*. *Titanium dioxide* is the most important inorganic pigment. Other types include iron oxides, zinc oxide and other metal oxides. Inorganic pigments may also be based on lead and other chromates, cadmium and mercury cadmiums, ultramarine blues, cobalt aluminate blues, chromium oxide greens and metal titanates. Flake pigments may be used to produce a range of metallic, pearlescent and iridescent colours. In general, however, final colours are less clear and brilliant than those obtained with an *organic pigment* system. Roughly three quarters of all pigments used are of the inorganic type. See *masterbatch*.

inorganic polymer
A polymer based on elements other than carbon. See, for example, *silicones* and *phosphonitrilic polymer*.

inorganic vulcanization accelerator
A *vulcanization* accelerator based on an inorganic compound.

inorganically bound sulphur
One of the ways the *sulphur* in rubber compounds, on analysis, may be classified. That sulphur which is chemically bound, or chemically attached, to an inorganic material such as a *filler*.

input signal
A term used in *hydraulics* and which refers to a signal (for example, an electric signal) which initiates a hydraulic action or function. See *command signal*.

insert
A piece of material placed in the mould before moulding takes place and around which, or part of which, the melt is moulded. See *insert moulding*.

insert bolster
A *bolster* which carries inserts, for example, *cavity inserts* or *core inserts*.

insert moulding
A *composite moulding technique* used, for example, to produce items such as engine mounts and handles around an *insert*, for example, a metal blade. The insert is loaded into the *injection mould*, the mould is then closed and the polymer is moulded around the insert. The insert is usually retained within the moulded component. Special moulding machines (for example, fitted with vertical clamps and equipped with *shuttle moulds*) are available for insert moulding. Both metallic and non-metallic inserts are used, for example, a glass-reinforced nylon insert imparts stiffness to a localised region of the moulding.

insert moulding machine
A moulding machine used for *insert moulding*.

insert pin
A pin used to locate an *insert* in a mould.

inside coating
See *internal coating*.

insoluble sulphur
Sulphur may be classed as soluble and insoluble. Insoluble *sulphur* may also be referred to as μ sulphur.

instantaneous variability
See *inherent variation*.

Institute of Materials
An abbreviation used for this UK-based organization is IoM. A materials organization which incorporates the Plastics and Rubber Institute.

instrumented falling weight
An abbreviation used for this test is IFW. See *falling weight impact strength test*.

instrumented pendulum impact testing
An abbreviation used for *instrumented pendulum impact* (testing) is IPI. See *pendulum impact testing*.

insulated hot-tip mould
A type of *insulated runner mould*.

insulated runner mould
A *type* of injection mould in which the material involved in the *feed system* is kept from hardening by making the runner diameter deliberately large, for example, 19 mm (0·75 in) in a *three-plate mould*. The *centre plate* of the mould is latched to the *stationary plate* during normal operation. When moulding is commenced the feed system is adjusted so that the runner system is filled. The layer of thermoplastics material in contact with the relatively cold mould hardens and because of the low thermal conductivity of plastics, this layer of hardened material insulates the material in the centre of the runner.

The large diameter of the runner, the shape of the runner (fully round) and the speed of moulding help to ensure that the material in the centre of the runner system remains hot enough to be moulded. If the material sets or hardens in the runner system then the stationary and centre plates are separated and the hardened feed system is removed. Polyolefins are commonly moulded in insulated runner moulds (such materials have very high heat contents) and the gating system is similar to that used on a three-plate mould. Initially, insulated runner moulds were equipped with insulated multiple nozzles but, because of start-up problems due to blocking, it is now more usual to equip such moulds with externally heated nozzles or to fit *heated probes* into the centres of the nozzles; nozzle temperatures are accurately controlled. Such a mould is known as an insulated hot-tip mould.

insulated sprue
A type of *extended nozzle*. The back of the *sprue bush* consists of a chamber which is filled with plastics material; the centre of the material remains mouldable (because of the thickness of material) so that when the injection pressure is applied the mould can be filled. The nozzle tip may have an undercut machined into it so that in the event of the material solidifying the carriage retraction movement will clear the blocked sprue. The Italian sprue is of this type.

insulating compound
A compound which is an electrical insulator. Materials with volume resistivities greater than 10^8 ohm/cm are classed as insulating rubbers or plastics. See *volume resistivity*.

insulating plate
A plate made of an *asbestos*-type material which fits between the injection mould and the platen. To obtain a constant mould temperature it is recommended to use insulating plates attached to the back of each *injection mould* half so as to minimize heat losses. Such plates also reduce the time period needed to heat-up the mould to the required temperature.

insulation
Also known as squeegee. A *tyre* component: a thin layer of rubber used to reduce the concentration of *stress* between the plies.

insulation strength
See *dielectric strength*.

insulation resistance
This is the total resistance between two electrodes fixed across an insulator and it is therefore a combination of both *volume* and *surface resistance* of the material. See *volume resistivity*.

integer cavity
A *cavity* formed as an integral part of the *cavity plate*.

integer core
A *core* produced by machining the form from a piece of steel.

integral
An abbreviation used for this term is I. The use of this term means that a controller, for example, a temperature controller has a circuit, or term, to eliminate the error known as *droop* or *offset*.

integral hinge
Also known as an in-mould hinge or a living hinge. A hinge which is formed during the *injection moulding* process and which is an integral part of the moulding: it is not attached subsequently to the injection moulding. The depth of the hinge section, for *polypropylene* (PP) is 0·25 to 0·6 mm (0·010 to 0·024 in). Locate all the gates on one side of the hinge, in a position which will ensure that full pressure is behind the melt as soon as it begins to flow through the hinge section.

integral ionic polymer
See *ionene* and *ionic polymer*.

integral skin moulding
A cellular (foam) moulding which has a relatively high density outer layer and a relatively low density inner core: both are produced at the same time during, for example, *reaction injection moulding*.

integrally woven honeycomb fabric
A type of fabric used, for example, in the *glass reinforced plastics* moulding industry in order to obtain woven, 3D preform shapes: such shapes give improved damage tolerance and shear strength compared to conventional weaves. Such preforms have parallel faces connected by ribs in a variety of flute configurations. When impregnated with a resin, for example, an *unsaturated polyester resin*, and cured a type of sandwich panel is formed.

integrating sphere hazemeter
See *light transmittance measurement*.

intelligent terminal
An abbreviation used for this term is IT. Part of the computer control system of, for example, an *injection moulding machine*. The interface between the operator and the control system: called intelligent because it has been designed to assist the setter/operator to make sensible decisions.

intensifier
See *hydraulic intensifier*.

intensive mixing
Melt mixing in which the ingredients are dispersed in the polymer very thoroughly. An *internal mixer* gives intensive mixing: other mixing systems are often judged against the degree of mixing possible form an *internal mixer*.

interlock
The term usually applies to guarding and means that, for example, two machine guards are linked together in such a way that machine operation is not allowed unless both guards are closed.

intermediate super abrasion furnace (black)
An abbreviation used is ISAF. See *carbon black*.

intermeshing twin screw extruder
An extruder which contains two *screws* the flights of which overlap so that the screws clean each other. The most popular type of *twin screw machine* contains screws which partially intermesh.

internal coating
The coating applied to the inside of a container or, the action of applying a coating to the inside of a container. With some *blow moulded containers* a coating or layer, may be present as either an inside layer, an outside layer, or both. Inside coating is more difficult to do on a mass production basis.

internal cooling mandrel
A *tube* sizing system which uses a cooled *mandrel* to size the inside diameter of *tubing*.

internal friction
See *hysteresis*.

internal lubricant
A *lubricant* which is primarily used to ease flow, for example, *glyceryl monostearate* (GMS) is used with *unplasticized polyvinyl chloride* (UPVC). For many UPVC products, production would not be possible without the use of relatively small amounts of lubricants as the unlubricated material will not flow sufficiently easily through the processing equipment. To act as an internal lubricant, the material must be more compatible with the polymer than a material used as an *external lubricant*.

internal mixer
A mixing or compounding machine (commonly called a Banbury mixer) in which the mix ingredients are contained in a chamber of fixed volume where they are subjected to high shear by two rotors: this results in *intensive mixing*. The sides of the chamber are formed by the heated walls, the top by a *ram* and the base of the chamber is a door (sliding door or drop door). Once mixing is complete, the mix is dumped onto a *two-roll mill* for *sheeting out, temperature conditioning* etc. See *internal mixing* and *reclaiming processes*.

internal mixing
Compound mixing performed with an *internal mixer*. An *internal mixer* should be used in conjunction with a *two-roll mill* as this often makes for safer and faster output. For rubbers, the curing system is added during the mill mixing stage, when the output from the internal mixer has cooled down as this procedure keeps the internal mixer free from curing materials and helps eliminate scorch problems. In some factories the output from the internal mixer (which is a large, intractable mass) is fed to two mills in succession. The first is used to remove the heat from the compound as quickly as possible and then the cooled material is passed to the second mill so that the curing system may be added.

internal mould release
A *mould release* which is incorporated within the moulding compound. See *external mould release*.

internal plasticization
Internal plasticization, or polymer softening, results when selected comonomers are used during polymerization. For example, vinyl acetate is copolymerized with vinyl chloride to give a *vinyl acetate-vinyl chloride copolymer* which may be referred to as internally *plasticized polyvinyl chloride* (PVC) as the material is softer than homopolymer PVC. As the effect is more permanent than *external plasticization* (as the *plasticizer* cannot be lost or extracted) then the term 'permanent plasticization' is sometimes used.

internal plasticizer
A chemical group, or copolymer, incorporated into a polymer so as to give a softening effect. See *internal plasticization*.

internal resistivity
See *volume resistivity*.

internal softening
Process used to increase the softness of a polymer by the incorporation of appropriate groups into the polymer chain. See *copolymer*.

internal stabilization
Stabilization which is incorporated within the polymer chain. May be achieved by the use of comonomers or by polymer modification. For example, the use of ethylene oxide as a comonomer, will stabilize the resultant *acetal* plastics material (polyoxymethylene). End-capping of the polyoxymethylene chains is also used.

internal undercut
See *undercut*.

internally heated, hot runner mould
A type of *hot runner mould* which has cartridge-heated rods fitted down the entire length of the feed system which are temperature controlled. Such a mould should be cheaper to operate (because of reduced heating losses) than a conventional *hot runner mould*.

internally plasticized polyvinyl chloride
A *polyvinyl chloride (PVC) copolymer*: the co-monomer is incorporated into the polymer so as to give a softening effect. See *internal plasticization*.

internally plasticized PVC
An abbreviation used for *internally plasticized polyvinyl chloride*.

international agency for standardization
See *International Standards Organisation*.

international calorie
See *international table calorie* and *specific heat*.

international candle
A unit of luminous intensity. See *candela*.

international carat
A unit of weight which is equal to 200 mg. See *carat*.

international natural rubber grades
See *standard international grades - natural rubber*.

international practical temperature scale
The international practical temperature scale uses the degree *kelvin* as the unit of temperature and has eleven fixed points on the scale. For example, the triple point of oxygen is specified as being 54·361 K and the boiling point of oxygen is specified as being 90·188 K. Interpolation between such specified temperatures is made with a platinum resistance thermometer. Above the freezing point of gold (1337·58 K) a radiation pyrometer is used.

international system of units
An English translation of *Système International d'Unité*.

international table calorie
Abbreviation used is calorie$_{IT}$. The International Table Calorie (International Calorie) is equal to 4·1868 joules. See *specific heat* and *calorie*.

International d'Unité
See *Système International d'Unité*. Abbreviation SI.

International Electrotechnical Commission
An abbreviation used for this organization is IEC. The English language version of Commission Electrotechnique Internationale (CEI). The International Electrotechnical Commission issues specifications, for example, specification 538A (1980) Procedure B, for *environmental stress cracking testing*. This procedure specifies no failures after 48 hours in a 10% solution, by volume, in water of Igepal CO-360. This solution is a more aggressive stress crack agent then 100%. See *CEI*.

International Institute of Synthetic Rubber Producers
An abbreviation used for this organization is IISRP. See *regular institute numbers*.

International Organisation for Standardization
See *International Standards Organisation*.

International Rubber Hardness Degrees
An abbreviation used for this term is IRHD. See *International Rubber Hardness testing*.

International Rubber Hardness testing
A *hardness* testing procedure which classifies vulcanizates in terms of International Rubber Hardness Degrees (IRHD). For the standard dead load test, measuring in the range 30 to 85 IRHD, the specimen needs to be a flat sheet or disc 8 to 10 mm thick which has been conditioned at the standard, or test, temperature for at least three hours immediately before testing. Tests can be made on thinner specimens by using the *Wallace Micro-hardness tester* (this instrument is essentially a scaled down version of the larger instrument). The test procedure is the same as for the larger instrument, using a 2 mm thick specimen. That procedure is the same as for *BS Softness Number* except that the result is expressed in International Rubber Hardness Degrees (IRHD). The relationship between the two scales is given in **table 3** of BS 903, 1969. However because of the very different deformation characteristics of vulcanised rubber and flexible plastics no direct comparison of the hardnesses of these two types of materials should be attempted.

International Special Committee on Radio Interference
An abbreviation used for this organization is CISPR. This committee issues recommendations for radio interference limits for *ISM bands* as industrial heating equipment, radiates energy which could interfere with communications. See *dielectric heating*.

International Standards of Quality and Packing for Natural Rubber
See the *green book*.

International Standards Organisation
An abbreviation used for this organization is ISO. Sometimes also referred to as the International Organisation for Standardization. A body which issues test standards or procedures - the ISO standards. This organization is the successor to the International Federation of the National Standardizing Association or ISA. ISA was formed in 1926 and the ISO in 1947. ISO is the international agency for standardization in all areas except those covered by the *IEC*. See *abbreviations*.

International Standards Organisation viscosity number
An abbreviation used for this term is ISO viscosity number. A measure of molecular weight widely used, for example, to characterize the molecular weight of *polyvinyl chloride* (PVC): the bigger the number, the higher is the molecular weight. Commercial polymers have values which range from approximately 65 to 140. Such values are obtained from dilute solution, viscosity measurements using cyclohexanone as the solvent at 25°C.

Defined as being $(t - t_0)/t_0)c$. Where c = the polymer concentration - usually measured at c = 0·005 gcm^{-3}: t = is the flow time of the solution and t_0 is the flow time of the solvent. See *K-value*.

International Telegraphic Union
An abbreviation used for this organization is ITU. This body allocates frequencies and tolerances, for example, for *Industrial, Scientific* and *Medical bands*.

interpenetrating elastomeric network
Also known as IEN. A blend, or alloy, formed between two elastomeric polymers and in which there are cross-links between the two polymers. May be obtained, for example, by mixing the two polymer lattices and then vulcanizing the mixture. That is, co-vulcanizing the mixture. See *interpenetrating polymer network*.

interpenetrating polymer network
Also known as IPN. A blend, or alloy, formed between two cross-linked polymers. May be obtained, for example, by adding a cross-linkable monomer to an already cross-linked polymer. If the cross-linkable monomer will swell the cross-linked polymer, and is compatible, then on polymerization/cross-linking, two separate but inter-penetrating networks will be formed. See *interpenetrating elastomeric network*.

interpolymer
A blend, or alloy, formed between two polymers and in which there are chemical links between the two polymers. May be obtained by mixing the two polymers together under high shear conditions in, for example, an *internal mixer*. May also be obtained by mixing a polymer and a monomer together under high shear conditions in, for example, an internal mixer. The product is a complex mixture of block copolymers, graft copolymers and homopolymers. Has been used for example, to prepare polymer blends from methyl methacrylate monomer and *natural rubber*. See *interpenetrating polymer network*.

interpolymerization
The method, or technique, used to produce an *interpolymer*.

intra-chain, cyclic, di-sulphide cross-link
A chemical grouping, based on a sulphidic group, and which is present, for example, in a sulphur-vulcanized *natural rubber*. The sulphur-containing group is not a cross-link as it connects regions of the same chain with a (short) *di-sulphide link*.

intra-chain, cyclic, mono-sulphide cross-link
A chemical grouping, based on a sulphidic group, and which is present, for example, in a sulphur-vulcanized *natural rubber*. The sulphur-containing groups is not a cross-link as it connects regions of the same chain with a (short) *mono-sulphide link*.

intractable
Usually means that a polymer is not capable of being melt processed: many *aromatic*, heat resistant plastics materials are intractable.

intrinsic variability
See *inherent variation*.

intumescence
The formation of an insulating char, or layer, from an *intumescent polymer compound*.

intumescent polymer compound
A compound which will decompose, or degrade, on heating to give an expanded, insulating char or layer, which protects the underlying areas of the compound from further degradation. Used as a *fire retardant system*. *Dicyanamide* is used as a char expansion agent.

inverse piezoelectric effect
See *piezoelectric effect*.

inverted L calender
A *calender* named after an upside down letter L. Also known as an F calender. A very widely used *four roll calender* as it is easy to feed such a *calender* via the off-set nip: three of the rolls form a *super-imposed calender*.

inverted temperature profile
A revers temperature profile. Conventionally the hopper-end of say, an *injection moulding machine* is cooler than the nozzle-end. With an inverted temperatures profile the reverse is true. Used, for example, for *polyhydroxybutyrate*.

inverted Z calender
A *calender* named after a reversed letter Z: sometimes called a S calender. See *four roll calender*.

IODA
An abbreviation used for *isooctyl isodecyl adipate*.

iodine
The elements of Group V11b of the Periodic table are known as halogens and consist of *fluorine*, *chlorine*, *bromine*, iodine and astatine. Iodine does not occur naturally but occurs as iodates, for example, in saltpetre deposits from which it is extracted by treatment with sodium bisulphite. It is not as reactive an element as the other halogens. It is a black lustrous solid (I_2) which is slightly soluble in water. Melts at approximately 114°C and boils at 114°C (the vapour is irritating and poisonous). The relative density (RD) is 4·94. See *iodine absorption*.

iodine absorption
A measure of the surface area of a filler. See *carbon black*.

iodine value
An abbreviation used for this term is IV. The number of grams of iodine absorbed by 100 g of oil: a measure of the unsaturation of *vegetable oils* by determination of the amount of iodine absorbed or reacted at the double bonds. Iodine absorption is determined by titrating unreacted reagent with sodium thiosulphate.

$$\text{Iodine number} = \frac{12\cdot 69 \times \text{ml of thiosulphate} \times \text{normality}}{\text{mass of sample (g)}}.$$

IODP
An abbreviation used for *isooctyl isodecyl phthalate*.

IoM
An abbreviation used for the Institute of Materials.

ion discharge rods
Rods used to treat *calendered sheet* so as to remove electrostatic charges accumulated during running. Ion discharge rods are mounted above and below the sheet and these initially cause ionisation of the surrounding air. By removing the electrostatic charges from the sheet, contamination due to dust build up can be considerably reduced.

ion exchange resin
An *ionic polymer*. This type of polymer carries electrostatic charges along its length which are neutralised by counter charges on the *counter ions*. The counter ions are free or mobile and can be exchanged for other counter ions from a surrounding medium. Ions of the same type of charge as the bound ions do not have free movement (Donnan exclusion). A cross-linked polymer with an ion content of approximately 4 equiv/kg (may also be classed as a *polyelectrolyte*) is used as an ion exchange resin. Insoluble beads, based on the cross-link polyelectrolyte, are used to, for example, de-ionize water by removing M^+ and X^- from the water and replacing these ions with H^+ and OH^- from the resin. The water contacts two types of insoluble resin: a polyacid and a polybase.

That is, $-SO_3^- H^+ + M^+ X^- \rightleftharpoons -SO_3^- M^+ + H^+ X^-$
Then, $-NR_3^+ OH^- + H^+ X^- \rightleftharpoons -NR_3^+ X^- + H_2O$.

The resins are supplied as water insoluble beads: these swell in water. Such beads can have either a dense internal structure (gel-type) or they can have a porous, multi-channelled type of structure (macroporous ion exchange resin or macroreticular ion exchange resin). Often the ionic groups are introduced by *post functionalization* of the cross-linked matrix: gel-type polystyrene resin was the first type of ion exchange resin.

ionene
An *ionic polymer* in which the bound ion is integral with the main chain, that is, it is enchained (an ionene).

ionic polymer
Also known as a polyion. A polymer which carries electrostatic charges along its length: these charges are neutralised by counter charges on the counter ions. A polymer which contains chemically bound ions along its length.

When such a polymer has relatively few ions (for example, if the ion content is 0·5 equiv/kg) then it is a melt processable thermoplastics material and may be called an *ionomer*. When such a polymer has slightly more ions (for example, if the ion content is approximately 1·5 equiv/kg) then it is an *ionomer* but used for ion-exchange purposes. When such a polymer has an ion content of approximately 4 equiv/kg then it may be classed as a *polyelectrolyte* and used as a thickener. A cross-linked polymer with an ion content of approximately 4 equiv/kg may also be classed as a *polyelectrolyte* and used as an *ion exchange resin*. When an ionic polymer has an ion content of approximately 14 equiv/kg then it may also be classed as a *polyelectrolyte* and used as a dental cement.

An ionic polymer may be classified by the type of bound ion, the position of the bound ion within the structure and the amount of bound ion. May also be classified according to the nature of the counter ions and/or the nature of the main polymeric chain. The bound ion may be pendant to the main chain (for example, as in a polysulphonate) or it can be integral with the main chain (an ionene).

Most ionic polymers are anionic and contain the bound anion (SO_3^- or, CO_2^- or, PO_3^{2-}) pendant to the main chain. Cationic ionic polymers contain bound cations (for example, $-CH_2N^+(CH_3)_3$) pendant to the main chain. Most conventional polymers can be modified to form an ionic polymer of the pendant type.

To neutralise the charges of the bound ion the ionic material contains counter ions (counterions) which may be univalent, divalent, trivalent or polyvalent. An ionic polymer with a polymeric counter ion is called a *polysalt*.

The most common ionic polymers are the *polyelectrolytes* known as polyacrylic acid and polymethacrylic acid (made by direct synthesis). Ionic polymers may be made by direct synthesis or by *post functionalization*. Ionic polymers are hydrophillic and this characteristic may be used to purify water by reverse osmosis.

ionic polymerization
A polymerization process in which the growing active centre are ions: used to produce an *ionic polymer*.

ionic vulcanization
Vulcanization induced ionically. For example, *carboxylated polybutadiene ionomer* when neutralized with, for example, *zinc oxide*, undergoes ionic cross-linking so as to produce the effect of *vulcanization*.

ionization
The process of removing electrons from an atom.

ionization energy
The energy required to remove electrons from an atom.

ionization potential
The potential difference through which an electron is moved to gain ionization energy.

ionized carboxylate
See *carboxylated latex*.

ionizing radiation
Electromagnetic, or particulate, radiation which produces ion pairs on passing through a medium. Radiation which causes ionization in gases and solids, for example, electrons and α particles are produced.

ionomer
An *ionic polymer*. For example, a thermoplastics material based on a copolymer of ethylene and methacrylic acid (up to approximately 10% acid units) which has been reacted with a metal derivative, for example, sodium methoxide (see *ethylene-acrylic acid copolymer*). The neutralization may be performed on a *two roll mill* at approximately 150°C: during milling the material changes from a soft, fluid opaque material to a stiffer, clear material which is quite rubbery. That is, the melt strength increases as the ionomer is formed.

This type of polymer carries electrostatic charges along its length which are neutralized by counter charges on the *counter ions*. The polymer chains are forced into conformations which allow the ions to associate with each other and because these ionic associations involve ions from different chains they behave as *cross-links* - they are, however, *thermally labile cross-links*. They can function as melt processable *thermoplastics* materials or as, *thermoplastic elastomers* - if the polymer backbone is elastomeric (*elastomeric ionomer*).

The ethylene ionomers have high melt strength at low shear rates (useful for *blow moulding*) but at high shear rates the melt strength is similar to that of polyethylene (PE). They are more transparent than PE as the crystallites which an ionomer contains do not aggregate into spherulites - it is these which scatter light in PE. Micro-crystallinity is enhanced by the ions but macro-crystallinity is not enhanced by the ions.

Polyethylene ionomers are tough, flexible, thermoplastics materials with good toughness at low temperatures. They have good resistance to greases and solvents. Such materials have good *environmental stress cracking resistance* and are abrasion resistant. Used in packaging and processed by *blow moulding* and *injection moulding*. Used to coat glass containers so as to reduce breakage.

ionosphere
The gaseous mantle, the atmosphere, which surrounds the planet Earth has been divided into several layers or strata. The ionosphere is the one above the *stratosphere* but below the *exosphere*. The ionosphere stretches from approximately 65 kilometres/40 miles to approximately 805 kilometres/500 miles above sea level. This layer contains the ionized layers,

or regions, (D, E and F regions) which allow long distance radio communications as such layers reflect the radio waves. Sometimes known as the heaviside layer.

IPA
An abbreviation used for *isophthalic acid*.

IPI
An abbreviation used for *instrumented pendulum impact (testing)*.

IPN
An abbreviation used for *interpenetrating polymer network*.

IPP
An abbreviation used for isotactic polypropylene. See *polypropylene*. The same letters are used for *isopentenyl pyrophosphate*.

IPPD
An abbreviation used for N-isopropyl-N'-phenyl-p-phenylenediamine. See *antiozonant* and *antioxidant*.

ips
An abbreviation used for *inch-pound-second*.

IPS
An abbreviation used for *inch-pound-second*. IPS is also used for iron pipe size and for impact polystyrene. See *high impact polystyrene*.

IR
An abbreviation used for *infra-red* and for isoprene rubber.

IRHD
An abbreviation used for *International Rubber Hardness Degrees*.

iridescent colours
Iridescent colours and pearlescent colours may be obtained with flakes of lead carbonates or of, bismuth oxychloride or of, titanium dioxide coated mica. See *metallic colours*.

iron
A *transition element* (Fe) which is a fairly soft, white, lustrous metal with a high melting point (1,530°C) and boiling point (2,750°C): the relative density is 7·7. Not usually found in nature. Ferromagnetic up to approximately 770°C: the properties of this element depend greatly upon other elements which may only be present in small amounts. When iron contains carbon (0·1 to 1·5%) in the form of cementite, it is known as *steel*. A major problem with iron and some steels is rusting: can be treated with metals such as *zinc* to reduce this problem. *Powder coating* with plastics materials is also used to decorate and protect the metal. Iron can catalyze the decomposition, or degradation, of some polymers, for example, *polyvinyl chloride* (PVC). See *stabilizer*.

iron blue
See *Prussian blue*.

iron II oxide
See *iron oxide*.

iron III oxide
See *ferric oxide* and *iron oxide*.

iron oxide
Iron II oxide is ferrous oxide FeO which is a black solid with a comparatively low melting point of approximately 370°C. Iron III oxide is ferric oxide Fe_2O_3: this is a red/brown material that occurs naturally as haematite. Iron oxide (red) has a relative density (RD or SG) of 5·14. Iron oxide (yellow) has a relative density (RD or SG) of 4·1.

iron oxide pigments
A class or type of *inorganic pigment* based on *iron oxide*. May be natural or synthetic: the natural materials are more vari-

able than the synthetic which are also much purer. Such pigments give buff, tan, red, brown, yellow and black colours. Such pigments are popular because, in general, they are low cost colouring systems with good resistance to heat, light, and chemicals such as solvents. Some iron oxides may degrade at high temperatures and the iron in the pigment (or another metal such as manganese) may cause decomposition, or degradation, of some plastics materials.

iron-based casting alloy
An alloy of iron which contains approximately 20% nickel and 2% chromium. Used for cast moulds, for example, which weigh up to 20 kg (44 lb). Used in place of *beryllium-copper* where bonding problems can occur with *natural rubber* compounds.

irons
A unit of length measurement used in the footwear industry and equal to $1/48$ of an inch (0·0208 in or 0·0528 cm).

ISAF
An abbreviation used for intermediate super abrasion furnace (black).

    ISAF-HS = intermediate super abrasion furnace (black) - high structure. See *carbon black*.

ISM bands
An abbreviation used for *Industrial, Scientific* and *Medical bands*.

iso
The inclusion of the prefix 'iso' in a chemical name means that the organic material contains a branched structure - as opposed to a normal, or straight, chain structure.

ISO
An abbreviation used for the *International Standards Organisation*.

ISO viscosity number
An abbreviation used for the *International Standards Organisation viscosity number*.

iso-octyl
See *isooctyl* as well as the following entries.

iso-octyl ester - saturated C₄ to C₆ dibasic acids
A mixed ester used as a *plasticizer* with *polyvinyl chloride* (PVC) to give good low temperature properties at an economic price. This is because it is relative cheap to produce C_4, C_5 and C_6 mixed acids during adipic acid manufacture: cheaper than producing separate acids. Such mixed acids are used to make *nylonate esters*.

iso-octyl iso-decyl adipate
Also known as *octyl decyl adipate*. An abbreviation used for this material is IODA. This material has a relative density (RD) of approximately 0·99 and a boiling point in the region of 245°C (5 mm). Adipic acid esters based on mixed isomers of branched alcohols which contain between between 8 and 10 carbon atoms. This liquid ester is used as a *plasticizer*, for example, for *polyvinyl chloride*. Also suggested as a softener for rubbers.

iso-octyl iso-decyl phthalate
Also known as isooctyl isodecyl phthalate. An abbreviation used for this material is IOD. This colourless liquid ester has a relative density (RD or SG) of 0·99 and a boiling point of approximately 235 to 250°C at 5 mm. May be used as a *plasticizer* for rubbers and vinyl polymers.

iso-propyl phenols
These synthetic alkylated phenols are feed stocks for *triaryl phosphates* and are made by reacting propylene and the phenol over a heterogeneous catalyst of *Fuller's earth*. The tri-isopropylphenol phosphates which are prepared from them, are claimed to have similar properties to *phosphates* such as TCP and TXP.

iso-rubber
Obtained when *rubber hydrochloride* is heated with pyridine. Sometimes referred to as α-iso-rubber. β-iso-rubber is obtained by reclaiming the α-iso-rubber.

isobutanol
See *isobutyl alcohol*.

isobutene-isoprene copolymer
See *butyl rubber*.

isobutene-isoprene rubber
See *butyl rubber*.

isobutyl
A prefix used in organic chemistry and used to denote the branched group $(CH_3)_2CHCH_2$-. See, for example, *isobutyl acetyl ricinoleate*.

isobutyl acetyl ricinoleate
This material is used as a softener in the rubber industry: it has a relative density (RD or SG) of 0·93, a melting point of −30°C and a flash point of approximately 216°C, an ester *plasticizer* based on *isobutyl alcohol*.

isobutyl alcohol
Also known as isobutanol or as, 2-methylpropan-1-ol or as, 2-methyl-1-propanol. May be represented as $(CH_3)_2CHCH_2$-OH. This material has a relative density (RD or SG) of 0·80 and a boiling point of approximately 108°C. Used as a solvent for, for example, synthetic rubbers and as a diluent for surface coatings. Its derivatives are used as *softeners* or *plasticizers*. See *isobutyl acetyl ricinoleate*.

isobutyl ricinoleate
An ester used as a softener in the rubber industry: it has a relative density (RD or SG) of 0·93, a melting point of −23°C and a flash point of approximately 218°C.

isobutylene-isoprene copolymer
See *butyl rubber*.

isobutylene-isoprene rubber
See *butyl rubber*.

isoceresine
See *ceresin*.

isochain polymer
See *homopolymer*.

isochronous
Meaning uniform in time. See *isochronous stress/strain curve*.

isochronous stress/strain curve
A way of expressing the results of *creep testing*. Stress is plotted against strain at constant time. A vertical line is drawn on the family of *creep curves*, parallel to the strain axis at a desired time, say at the 100 s mark and the corresponding stress and strain plotted to give the isochronous stress/strain curve. This procedure is commonly used to illustrate the effect of, for example, changes in crystallinity and water content, on creep properties.

isocyanate
A salt or ester of isocyanic acid. The inclusion of this word in a chemical term means that a compound contains the isocyanate group -N=C=O which is more commonly written -NCO: the compound may be written as RNCO. If the material contains two isocyanate groups then it is referred to as a diisocyanate: such diisocyanates are reacted with diols or

polyols to form *polyurethanes*. If 3 RNCO groups trimerize, then as isocyanurate is the result.

isocyanate trimer
See *isocyanurate*.

isocyanate-hydroxyl reaction injection moulding
A common type of *reaction injection mouding* (RIM): see *polyurethane*. A mixture of tertiary amine and tin catalysts is most effective for such RIM formulations.

isocyanic acid
An isomer of cyanic acid which forms *isocyanates*. May be represented as H-N=C=O.

isocyanurate
Also known as an isocyanate trimer. Formed when 3 RNCO groups trimerize: a cyclic trimer of an isocyanate. This trimerization reaction is promoted by tertiary amines and bases: this reaction can be used to build up highly cross-linked *polyisocyanurate* structures via *reaction injection moulding*.

isodecyl diphenyl phosphate
A *mixed alkyl aryl phosphate*.

isoindolinone pigments
A class or type of *organic pigment* sometimes referred to as isoindolinones and which give colours ranging from yellow to red. Relatively high cost systems which are lightfast, bleed-resistant and heat resistant.

isomerized rubber
See *cyclized rubber*.

isomers
See *polymer isomers*.

isometric stress/log-time curves
A way of expressing the results of *creep testing*. Stress is plotted against time at a constant strain. A line is drawn parallel to the time axis of a *creep-curve* so that it intercepts, or crosses, the family of curves. At each intercept the stress required to produce that constant strain may be noted and then used to construct the isometric stress/log-time curve by plotting the stress against the noted log time.

Such curves can provide design information; for a given component. For example, it could be that the design limitation specified is a maximum deformation or distortion. In this case the graph may be used to obtain the maximum stress which can be tolerated for a specified time.

isooctyl
Also written as *iso-octyl*. The inclusion of this word in a chemical term means that a compound contains a branched group which may be written as $-C_8H_{17}$. For example, iso-octane is $(CH_3)_2CHCH_2(CH_3)_2$.

isooctyl isodecyl adipate
See *iso-octyl iso-decyl adipate*.

isooctyl isodecyl phthalate
See *iso-octyl iso-decyl phthalate*.

isooctyl palmitate
An *ester*. This colourless liquid material has a relative density (RD or SG) of 0·86, a melting point of approximately 7°C and a boiling point of 228°C at 5 mm. May be used as a softener/processing aid for rubbers and as a *secondary plasticizer* for vinyl polymers.

ISOPA
An abbreviation used for *Association of European Isocyanate Producers*.

isopentenyl pyrophosphate
An abbreviation used is IPP. Initially formed in the biosynthesis of *natural rubber* by plants.

isophorone diisocyanate
An abbreviation used is IPDI. An isocyanate used to produce, for example, *polyurethane coatings* as the use of this material can produce systems with good resistance to yellowing.

isophthalate-based material
An *allyl moulding material* based on *diallyl isophthalate*.

isophthalic acid
Also known as benzene-1,3-dicarboxylic acid or as, m-phthalic acid or as, 1,3-benzenedicarboxylic acid. See *phthalic acids*. An abbreviation used for this material is IPA. The meta isomer of phthalic acid which may be prepared by the oxidation of m-xylene. This material has a melting point of approximately 346°C and is used, for example, to prepare *unsaturated polyester resins* which are required to have a relatively high heat resistance, a high heat distortion temperature and high modulus. Also used to prepare *plasticizers*.

isoprene
Also known as methyl butadiene or as, 2-methyl-1,3-butadiene or as, 2-methylbuta-1,3-diene. A *methyl substituted butadiene* in which a methyl group is substituted for a hydrogen atom. May be represented as $CH_2=C(CH_3)CH=CH_2$. This material has a relative density (RD or SG) of 0·68, a melting point of $-120°C$ and a boiling point of approximately 35°C. Natural rubber is a polymer of isoprene. This material is also the monomer for synthetic *isoprene rubber* by *diene polymerization*. It is obtained from the C_5 fraction in petroleum cracking processes. Upon polymerization, both cis and trans isomers can be formed. See *cis-polyisoprene*, *trans-polyisoprene*, *balata*, *gutta-percha* and *natural rubber*.

isoprene acrylonitrile rubber
An abbreviation used for this material is NIR. It is also known as isoprene acrylonitrile copolymer as it is a *copolymer* of *isoprene* and *acrylonitrile*.

isoprene rubber
An abbreviation used for this material is IR. Also known as cis-polyisoprene or as, cis-1,4-polyisoprene or as, polyisoprene or as, poly-(2-methyl-1,3-butadiene) or as, synthetic natural rubber or as, synthetic polyisoprene or as, synthetic isoprene rubber or as, synthetic polyisoprene rubber. Also suggested is, poly(1-methyl-1-butenylene).

This material has been classified as a *general purpose rubber*. This material, like *natural rubber (NR)*, is a polymer of *isoprene* and in order to differentiate it from that material it is sometimes known as synthetic polyisoprene. Like NR, polyisoprene has proved to be the only synthetic rubber which exhibits self-reinforcement at high elongations. However, in order to show these properties, the material must be made so that its chemical structure is very regular, i.e. it has a high cis content (e.g. 96% or above): because of this it is sometimes known as cis-polyisoprene or synthetic cis-polyisoprene.

The polymerization of isoprene is performed in low boiling point, aliphatic hydrocarbons (such as pentane or heptane) and using a *Ziegler-Natta catalyst*. For example, using titanium chloride ($TiCl_4$) and aluminium alkyl a grade of IR rubber is produced which may be referred to as Ti-IR. When lithium-n-butyl is used, then a grade of rubber is produced which may be referred to as Li-IR. The Li-IR is of higher molecular weight but of a narrower molecular weight distribution: it also has a lower cis-content. Because of its easier processing, higher green strength and higher cis-content the Ti-IR is often preferred (Li-IR may be considered as being similar in its processing behaviour to *styrene-butadiene rubber*.

Both straight and oil-extended varieties of IR are available and by varying parameters such as cis content, molecular weight, etc., it is possible to produce a very wide range of grades. It would be expected that lowering the molecular weight and/or the cis content would produce grades which exhibit improved flow during injection moulding. Despite its chemical similarity to natural rubber, the material does exhibit differences. For example, it cures at a slower rate than natural rubber and lightly filled compounds have poorer hot tear strength. In general, processing is possible at lower temperatures and it is usually found that there is less heat generation when fillers are mixed with this material. As IR is compatible with NR in all proportions, it can be usefully used in blends with NR where scorch (premature vulcanisation) is a problem.

isoprene rubber (natural)
See *natural rubber*.

isopropanol
See *isopropyl alcohol*.

isopropyl
The inclusion of this word in a chemical term means that a compound contains a branched group which may be written as $(CH_3)_2CH-$.

isopropyl acetate
The formula may be written as $(CH_3)_2CHOOCCH_3$. This material has a relative density (RD or SG) of 0.87 and a boiling point of 89°C. Used as a solvent in, for example, rubber-based cements: used at it reduces the viscosity of such cements.

isopropyl alcohol
Also known as isopropanol and propan-2-ol. The formula may be written as $(CH_3)_2CHOH$. This material has a relative density (RD or SG) of approximately 0.8 and a boiling point of 82°C. Used as a solvent in, for example, surface coatings: may be used in preference to *ethyl alcohol*, as it is often anhydrous.

isopropyl benzoate
The formula may be written as $(CH_3)_2CHOOCC_6H_5$. This material has a relative density (RD or SG) of 1.02 and a boiling point of 219°C. Used as a solvent for, for example, plastics materials such as *polystyrene* and *cellulosics*.

isopropylated phenol phosphate
See *tri-isopropylphenol phosphate*.

isotactic
See *isotactic polymer*.

isotactic biopolymer
See *biopolymer* and *polyhydroxybutyrate*.

isotactic index
That percentage of *polypropylene* which is insoluble in boiling *hexane*. It is a measure of the percentage of *isotactic polypropylene* present in the sample. See α *olefin*.

isotactic polymer
A stereoregular polymer: a polymer in which there is a regular, repeating structure. The repeat units, along the polymer chain, have the same configuration: that is, the groups of atoms have the same orientation in space. If the repeat unit is two carbon atoms long in a vinyl-type polymer then, the substituent group (for example, $-CH_3$) always lies on the same side of the polymer chain. See α *olefin* and *stereoregular polymer*.

isotactic polypropylene
An abbreviation used is i-PP or IPP. Polypropylene with a completely *isotactic structure*. Pure isotactic *polypropylene* is not commercially made as an *injection moulding* material - has a melting point greater than the commercial material (>170°C) and an SG of approximately 0.91.
The common, commercially used moulding and extrusion grades of *polypropylene* have a high isotactic content, for example, up to 95%. The amount of isotactic polypropylene is measured by the *isotactic index*.

isotacticity
See *isotactic polymer*.

isotopes
Atoms which have the same number of protons but a different number of neutrons. They have the same atomic number but a different mass number.

isotropic
Having properties that are the same in all directions. See *anisotropy*.

isotropic pitch-based carbon fibre
See *carbon fibre*.

IT
An abbreviation used for intelligent terminal.

Italian sprue
See *insulated sprue*.

ITU
An abbreviation used for *International Telegraphic Union*.

IV
An abbreviation used for *iodine value*.

Izod impact test
See *pendulum impact test - Izod*.

J

J
An abbreviation used for *joule*.

jack ram
Also known as an auxiliary ram: a small hydraulic ram. Used on, for example, injection moulding machines in order to save on the usage of hydraulic fluid, and to speed up cycle times. The jack rams may be used to achieve fast mould movements as they have a much smaller diameter than the main clamping ram. Once the two mould halves are together then the main ram is energized. See *direct clamp machine*.

jacket
See *blanket*.

jaune brilliant
See *cadmium yellow*.

jazz effect
Also known as a tortoiseshell or as a marbled effect. Can be produced in moulded components by using a mixture of different coloured starting materials, for example, by using a mixture of different coloured masterbatches. Such finishes can be very attractive but can only be obtained provided that intensive melt mixing does not occur during the melt preparation stage. Special *injection moulding machines* are made so that the moulded patterns can be reproduced precisely from one component to another.
Jazz effects can also be produced in moulded, or dipped rubber components, by using a mixture of different coloured starting materials

jelutong
A mixture of natural rubber and natural resins: dry rubber content approximately 22%. Obtained from countries such as

Borneo by tapping of trees of the genus Dyera Apocynaceae. Major use now appears to be as a substitute for chicle in chewing gum: can be used as a substitute for *natural rubber*.

jig
A device for holding a component, for example, while it is cooling.

jig cooling
Cooling a product after production on or, in, a *jig* so as to obtain a desired shape.

JIT
An abbreviation used for *just in time (production)*.

joint line
A *witness mark* which is formed, for example, at the junction of a pair of splits.

joule
An SI derived unit which has a special symbol, that is J. When the point of application of a force of one newton (N) is moved through a distance of one metre, in the direction of the applied force, then the work done is one joule. It is the SI derived unit of work or energy. Equivalent to 107 ergs, one watt-second or 0.74 foot pounds. Has the dimensions of kgm^2s^{-2}. See *Système International d'Unité*.

Joule-Gough effect
See *Gough-Joule effect*.

journal
The end of a roll which runs in a *bearing*: the end of the roll which is of reduced diameter and which runs in a bearing.

just in time (production)
An abbreviation used for this term is JIT. Means that components are not produced until they are required.

K

k
An abbreviation used for *kilo*. See *prefixes - SI*.

K
This letter is used as an abbreviation for:
carbazole - for example, as in polyvinylcarbazole;
consistency index - in the *power law equation*;
copolymer. A more widely used way of depicting a copolymer is to use the letters CO in conjunction with the letters used for the material, for example, PP-CO.
filler. For example, if a plastics material contains chalk or a knitted fabric;
kilobyte. It is 1024 bytes or characters and also known as Kb;
kelvin;
Mark-Houwink constant.
temperature. When used to indicate temperature, then it means degrees Kelvin (for example, 205K). See *prefixes - SI*.

Also see *K-value* of PVC. Sometimes the letter K is used to refer to the enormous plastics show (fair) periodically held in Dusseldorf in November (every three years).

K-value
Also known as Fikentscher K-value or as, Fikentscher number. A measure of molecular weight widely used, for example, to characterize the molecular weight of *polyvinyl chloride* (PVC): the bigger the number, the higher is the molecular weight. Commercial polymers have K values which range from approximately 45 to 80: increasing the K value will increase the strength and stiffness of a material but the ease of flow will become more difficult. Higher K value resins can be used for *extrusion* than for *injection moulding*.

K-values are obtained from dilute solution, viscosity measurements and are defined as being 10^3 times the value of K in the formula:

$$(1/c)\log_{10}(\eta_{rel}) = (K + 75K^2)/(1 + 1\cdot 5Kc).$$

Where c = the polymer concentration and η_{rel} is the viscosity ratio. The value of K obtained is dependent on the solvent used. See *ISO viscosity number*.

kaolin clay
Also known as kaolin. See *china clay*.

kaolinite
The major component of kaolin. See *china clay*.

kardiseed oil
See *safflower oil*.

Karl Fischer equivalent number
An abbreviation used for the Karl Fischer equivalent number is D and this is obtained by standardising *Karl Fischer reagent* with a known quantity of water. See *Karl Fischer method*.

Karl Fischer method
A method used for assessing the moisture content of polyamide materials (PA), that is, *nylon* granules.

A known quantity of polymer granules is heated in a test tube, connected to a cold trap, until the moisture in it vaporises. A stream of dry nitrogen gas is used to flush the water vapour into the cold trap where it condenses. The water collected in the cold trap is titrated with Karl Fischer reagent which reacts quantitatively with water. A Karl Fischer equivalent number is obtained by standardising the reagent with a known quantity of water.

$$\% \text{ water in granules} = \frac{C \times D \times 100}{E}$$

where C = millilitres of Karl Fischer reagent used to titrate the water in the trap
D = Karl Fischer equivalent number
E = milligrams of sample

Full details of this method can be found in ASTM D 789.

Karl Fischer reagent
An iodine-containing reagent used to determine the moisture content of a *polyamide material*. By standardising the reagent with a known quantity of water the *Karl Fischer* equivalent number (D) is obtained.

Karrer plastimeter
A parallel plate type of plastimeter. See *plastimeter*.

kcal
An abbreviation used for kilocalorie. See *calorie*.

kelvin
The basic SI unit of temperature which has the symbol K. Defined as being 1/273.16 of the temperature difference between absolute zero and the triple point of water. As a temperature interval, the kelvin is equal to the *degree Celsius* (K = °C). To obtain degrees Celsius from degrees kelvin subtract 273.15. Degrees Kelvin may also be referred to as degrees absolute. See *Système International d'Unité*.

Kelvin scale
A scale of temperature on which zero denotes the absence of heat energy. See *absolute zero*, *heat* and *kelvin*.

kerbstone rib
A moulded rib on the upper *sidewall* which is there to protect the *tyre carcass* against kerbstone abrasion.

kerosene
See *paraffin oil*.

kerosine
See *paraffin oil*.

ketone
An organic compound with the formula RR'C=O. See, for example, *acetone*.

ketone-amine condensates
Additives, *antioxidants*, widely used in rubber compound to prevent oxygen attack: may be classed as discolouring antioxidants (staining antioxidants) as they have a tendency to stain or discolour the rubber on ageing. Included in this type of material are acetone-anilines acetone-diphenylamine condensate (ADPA) and polymerized 1,2-dihydroxy-2,2,4-trimethylquinoline (TMQ). Such antioxidants are widely used in rubber compounds where heat resistance is necessary, for example, tyres and cables.

ketopropane
See *acetone*.

kettle bodying
An *oil* treatment. The refined *oil* is heat treated so as to cause polymerization. This increases viscosity and impairs brushability: it does improve wetting, colour retention, flow, gloss and drying properties.

Kevlar
A trade name/trademark of an *aramid*, or para-aramid, material in fibre form. Used, for example, to make high strength reinforced plastics or composites.

key
That part of the screw which permits the turning motion of the drive to be transmitted to the *screw*.

key coat
A *coating* applied to promote adhesion.

keying
The modification of surfaces so as to make them more receptive to inks; techniques used include *corona discharge* or *flaming*.

keying agent
See *coupling agent*.

kg
An abbreviation for *kilogram*. For example:

   kg-load = a term sometimes used for kilogram;
   kgf = kilogram-force;
   kg·m = kg-m = kilogram-metre; and,
   kg·m/s = kgms^{-1} = kilogram-metre per second.

kicker
An additive for polymer compounds, for example, for *polyvinyl chloride* (PVC) compounds. A material which is added to a formulation in order to make another additive decompose at a comparatively low temperature. Some *blowing agents* can be made to decompose at temperatures lower than normal in this way. See, for example, *dibasic lead phthalate*.

kieselguhr
See *diatomaceous earth*.

kilo
An abbreviation for *kilogram*. An abbreviation for one thousand. For example:

   kiloampere = one thousand amperes;
   kilobaud = one thousand baud;
   kilobit = one thousand binary digits;
   kilocalorie = one thousand *calories*;
   kilocycle = 1,000 cycles;
   kilohertz = one thousand hertz (one thousand cycles per second);
   kiloton = one thousand tons;
   kilovolt = one thousand volts;
   kilowatt = 1,000 watts; and,
   kilo-electron-volt = one thousand electron volts.

kilogram
The kilogram is the basic SI unit of mass and is that of the prototype cylinder (made of platinum-iridium alloy) originally used in France to define the mass of one thousand grams. It has the abbreviation of kg. One kg is equivalent to 2·204 623 pounds (pounds-mass). One *pound-mass* is equivalent to 0·453 592 kg. (Originally the kg was intended to represent the mass of a cubic *decimetre* of water at 4°C). See *metric system of units* and *Système International d'Unité*.

kilogram calorie
One thousand *calories*.

kilogram-metre
The work done by one *kilogram* of force when its point of application moves a distance of one *metre* in the direction of the force. Abbreviation kg.m and also kgm. That is, 1 m × 1 kg. A unit of work equal to approximately 7·2 foot pounds. Equal to 9·81 joules or, 9·81 × 10^7 ergs or, 9·81 watt seconds or, 2·34 calories.

kilogram-metre per second
Abbreviation kg·ms^{-1}. Also kgm/s. A unit of *power*. 75 kg·m/s is equivalent to 1 horsepower (metric).

kilogramme
An alternative way of spelling *kilogram*.

kilom
An abbreviation sometimes used for *kilometre*.

kilometre
One thousand *metres*. Equivalent to 0·621 37 miles.

kilopond
An abbreviation used for this term is kp. This term is sometimes used in place of *kilogram force*.

kilowatt-hour
Also known as the Board of Trade unit or as the B.O.T. unit. A unit of energy which has the abbreviation kWh. The energy obtained when a power of one kilowatt is maintained for one hour. Approximately equal to 1·34 horsepower per hour. 1 kWh is equivalent to 367,000 kgm or 864 kcal or 3·6 MJ.

kinematic viscosity
Sometimes referred to as kinetic viscosity and obtained by dividing the coefficient of viscosity (the *viscosity*) by the density of the fluid. The units are, for example, m^{2}s^{-1}.

kinetic coefficient of friction
An abbreviation used for this term is F_k/R. Where R is the normal force at the contact surface and F_k is the friction when steady sliding is obtained. See *coefficient of friction*.

kinetic friction
See *sliding friction*.

kinetic viscosity
See *kinematic viscosity*.

kinked fibre
A fibre which is not straight. See *aramid fibre*.

Kirksite
A trade name/trademark for a *low melting point metallic alloy* based on zinc. See *zinc-based alloy*.

kiss-polishing
Also known as nip-polishing or as, calendering. A *sheet extrusion* technique in which the extruded sheet is calendered between the hardened steel rolls of a three-roll stack. This

procedure when used, for example, for a *copolyester* requires high roll pressures of approximately 53 N/mm or 300 lb/linear inch. The extruded sheet often possesses *orientation* or *snap-back*.

kneader
A machine used for melt mixing a *compound* and which can develop high shear within that compound.

kneading pins
Also known as kneading teeth. See, for example, Buss Kno-Kneader.

knife coating
See *knife-on-blanket coating*.

knife edged die procedure
An alternative to the *Bagley correction* for correcting for end effects. By using knife-edged dies, of zero length, a positive pressure ΔP_0 is obtained at a point (equivalent to the intercept of a Bagley ΔP-vs-L/R) plot with the vertical ΔP axis.

knife line
See *sheeter line*.

knife-on-blanket coating
Also called knife-on-jacket coating. A *spread coating method* in which the substrate is supported by a continuous flexible band (the blanket) against a *doctor blade*.

knife-on-jacket coating
See *knife-on-blanket coating*.

knife-on-roller coating
A *fixed-gap spread coating method*. A *spread coating method* in which the substrate is supported by a roller against a *doctor blade*.

knit-line
Also known as *weld-line*.

knitted hose
A hose in which the reinforcing material has been applied by circular knitting.

knockout blade
See *ejector blade*.

knockout mark
See *witness mark*.

knockout pin
An ejector pin. See *pin ejection*.

knockout sleeve
See *ejector sleeve*.

Knoop hardness number
An abbreviation used for this term is HK.

knot
A measure of speed: one nautical mile per hour.

knuckle area
In reinforced plastics, the area of transition between sections of different geometry in a filament-wound part.

kp
An abbreviation used for *kilopond*.

kraftstunde
See *pferde-stärken-stunde*.

Krenckel's efficiency factor of reinforcement
See *efficiency of reinforcement*.

kV/mm
An abbreviation for kilovolts per millimetre. The units used for *dielectric strength*.

kW
An abbreviation for *kilowatt*.

kWh
An abbreviation for *kilowatt hour*.

L

l
This letter is used as an abbreviation for:
length. Also used is L;
litre. Also used is L; and,
tube diameter. See *Reynold's number*.

L
This letter is used as an abbreviation for
cellulose. If, for example, a plastics material contains a filler then the abbreviation may be modified with L;
large rotor - in a *Mooney viscosity test*.
layer. If, for example, a plastics material contains a filler in layer form;
length. For example, length of insulator through which the current flows in *volume resistivity measurements*.
linear;
linseed. For example, *epoxidized linseed oil (ELO)*.
litre;
low;
solution - for example, as in L-Br.
span width. See *flexural strength*.

L calender
A *calender* named after the letter L. A *four roll calender*, in which one roll, the first roll, is off-set with respect to the others: the remaining three rolls form a *super-imposed stack*. The calender is fed into the bottom, off-set nip. The L calenders were developed to process *unplasticized polyvinyl chloride (UPVC)* by, for example, the Luvitherm process. The main attraction of using this design was that any dry material which fell from the machine rolls dropped away from the product. The longer flow path, or *wrap*, associated with this design also gives better fusion.

L'homme and Argy balance
A purpose built instrument used to measure density of, for example, rubber compounds. See *displacement method*.

L-50
The length at *T-50*.

L-Br
An abbreviation used for solution butadiene rubber. See *butadiene rubber*.

L-malic acid
See *malic acid*.

L-SBR
An abbreviation used for solution styrene-butadiene rubber. See *styrene-butadiene rubber*.

laboratory capillary rheometer
An abbreviation used for this term is LCR. A device used to assess, or measure, the ease of flow of plastics melts. See *high shear rate rheometry*.

lac
A resinous material secreted by a tropical insect, as a protective covering, and which when refined forms the basis of *shellac* compounds.

lace cut pellets
Pellets made by chopping strands with a circular cross-section.

lacquer
A varnish based on a resinous material and which dries to a hard shiny finish. A polymer solution used as a final coating, for example, for protection of underlying layers.

lactam
Also known as a cyclic amide. Organic ring compounds in which the group -NH-CO- is part of the ring: the other groups in the ring are usually methylene groups -CH_2-. Lactams are used to produce *polyamides*, for example, for the production of *nylon 6* from caprolactam.

lactone
Also known as a cyclic ester. Organic ring compounds in which the group -O-CO- is part of the ring: the other groups in the ring are, for example, methylene groups -CH_2-. Lactones are used to produce *polyesters* from, for example, a caprolactone.

ladder polymer
A heat resistant polymer which is, in the simplest case, based upon two chains which are regularly linked so that the structure becomes a series of fused rings. This is the classical ladder, or double strand polymer, in which the two strands are cross-linked at regular intervals. If the structure is not perfect then, it may be known as a step-ladder polymer or a partial-ladder polymer. By using more than two chains, then sheet or parquet polymers result. The best known example of a ladder polymer is polyacrylonitrile or *carbon fibre*.

By linking rings (R) with double bonds (=), it is possible to produce allene ladder polymers: that is, =(R)$_n$= . By linking rings (R) with two single bonds, which in turn originate from the same ring, it is possible to produce spiro-polymers.

laddering
An extrusion defect. Defective surface finish caused by *melt fracture* at the die.

lag
A term used in *hydraulics* and which refers to a delay in response.

lake
A coloured, insoluble substance which is formed by the chemical reaction between a *dye* and a *mordant*. A lake is sometimes known as an extension or lake pigment. If the inorganic substrate (for example, aluminium hydrate) contains water, moisture problems are reduced by using a lake with the highest dye concentration. When the dye has food approval usage, such lakes are used for food packaging.

LAL
An abbreviation used for *lower action limit*.

lambert
A unit of luminance: the luminance of a light source that emits one lumen per square centimetre. It is equal to 3180 candela per square metre. See *candela*.

lamellar crystal
A flat plate-like crystal which contains folded polymer chains.

lamina
A thin sheet or layer. See *laminate*.

laminar
Means that a material has a leaf-like, or plate-like, shape.

laminar flow
Also known as streamlined flow. When a molten polymer is made to flow then this type of flow occurs because the polymer chains slip, or slide, one over the other. As the individual, plastics molecules move. one relative to the other, then this causes the molecules to change their direction or *orientation*. As the speed of a flowing fluid increases, beyond a critical speed (u_c), then the flow becomes turbulent. See *Reynold's number*.

laminate
Also known as laminated composite or as, layered composite. A composite structure formed by bonding together, layers of reinforcement with a polymeric binder. The reinforcement is often based on a fibrous material in sheet form while the polymeric binder is traditionally a synthetic, thermosetting resin. One of the best known examples is based on paper and thermosetting resins such as phenol-formaldehyde (PF) and melamine formaldehyde (MF). PF-impregnated paper is faced with MF impregnated paper and the assembly is then compression moulded in very large presses. The expensive MF is only used on the outside and the bulk of the laminate is made from the cheaper PF.

As the paper used with the transparent MF may be printed with virtually any pattern required, a wide range of patterns are available. By using fine cloth and/or textured mould plates, then a different surface may be obtained.

This type of material is most commonly used for working surfaces (for example, in kitchens) because of the attractive appearance and ease of cleaning. An advantage of this type of material, which is often over-looked, is that the surface is not only highly polished but it remains that way even after considerable exposure to the atmosphere. That is, MF-based materials have very good weathering resistance and for this reason are used for building, cladding and for signs. There is now considerable commercial interest in thermoplastic-based laminates based on *engineering thermoplastics materials*.

laminated angle section - thermosetting plastic
A moulded angle section (an L-shape) which is based on a preform formed from a resin-impregnated substrate: the assembly is bonded together under heat and pressure in an appropriately shaped cavity. The cured product may be ground to a final size.

laminated channel section - thermosetting plastic
A moulded angle section (a U-shape) which is based on a preform formed from a resin-impregnated substrate: the assembly is bonded together under heat and pressure in an appropriately shaped cavity. The cured product may be ground to a final size.

laminated composite
See *laminate*.

laminated moulded rod - thermosetting plastic
A moulded rod which is based on a preform formed by rolling a resin-impregnated substrate around a mandrel: the mandrel is then removed and the assembly is bonded together under heat and pressure in a cylindrical cavity. The cured product may be ground to a final size.

laminated moulded tube - thermosetting plastic
A moulded tube which is based on a preform formed by rolling a resin-impregnated substrate around a mandrel: the assembly is then cured under heat and pressure in a cylindrical cavity. The mandrel is subsequently removed: the cured product may be ground to a final size if required.

laminated moulding
A moulding produced from layers of polymer impregnated reinforcement and bonded together under heat and pressure.

laminated plastics panel
A *laminated plastics sheet* cut to a specified size and shape.

laminated rolled tube - thermosetting plastic
A rolled tube which is based on a preform formed by rolling a resin-impregnated substrate around a mandrel between heated pressure rollers: the assembly is then cured using a heated oven and the mandrel is subsequently removed: the cured product may be ground to a final size if required.

laminated sheet - thermosetting plastic
A moulded sheet which is based on a resin-impregnated substrate bonded together under pressure: heat may or may not be used.

laminated tee section - thermosetting plastic
A moulded angle section (a T-shape) which is based on a preform formed from a resin-impregnated substrate: the assembly is bonded together under heat and pressure in an appropriately shaped cavity. The cured product may be ground to a final size.

laminating
The process of making a *laminate*. For example, the application of resin to reinforcement and the subsequent bonding together of the assembly to make a *laminated sheet*.

laminating resin
A resin used in *laminating*.

lamination
See *plying*.

laminator
A machine used for plying film or sheet together, for example, by roller pressure.

lamp black
May also be called soot. An abbreviation used is LB. A *carbon black* of relatively low surface area compared to other carbon blacks.

lance piston reaction injection moulding machine
A type of *reaction injection moulding machine* (RIM machine) which uses a single stroke displacement pump for each of the reactants: each cylinder is fitted with a long piston or lance. The high presure seals, used for preventing the escape of reactants are fixed to one end of the cylinder. Fluid is displaced out of the cylinder by the action of the smooth walled lance. Fixed seals function better with abrasive fillers and are preferred for RIM. However, some reactant is left on the cylinder walls: this type of RIM machine contains more dead volume than a *piston reaction injection moulding machine*. Many RIM machines use lance pistons driven by separate hydraulic cylinders so as to minimize contamination from any reactants which leak.

land - compression mould
Part of a mould used to provide an obstruction to material escape during, for example, *compression moulding* of a *thermosetting plastic*. Each cavity should be surrounded by a land, which is approximately 6 mm/0.25 in wide.

land - injection mould
Part of a mould used to carry, or support, the applied pressure and to provide an obstruction to material escape during, for example, *injection moulding* of a *thermosetting plastic*. Each cavity should be surrounded by a land, which is approximately 9 mm/0.375 in wide, as this will help avoid flashing in *injection cavities*.

land length
Extrusion terminology. The land length is usually expressed as the ratio between the length of the opening in the flow direction and the die opening, for example, 10:1.

Landers test
An *environmental stress cracking test*. See, for example, ASTM D 2552-66T. It is suggested that an *annealed compression moulded sheet* is used as such sheet has a low internal stress level. Twenty small tensile dumb-bells are cut from the moulded sheet. The minimum width and thickness of each specimen is measured and used to calculate the load required for the initial stress level. This is calculated from load = stress required × cross-sectional area.

After mounting the specimens in the grips of the equipment the calculated load is applied to each specimen. They are then immersed in a bath of the stress-crack agent (Igepal CO-630) which is maintained at 50°C. The time for the complete fracture of each specimen is noted. The time for 50% failure is found by plotting the results on log-normal probability paper.

One of the main problems with this method is that some specimens may fail by yielding rather than by brittle fracture. If more than three ductile failures occur a new set of 20 specimens has to be tested at a lower initial stress.

landing pad
That part of the mould which supports the clamping force and which stops the flash lands form making contact when the mould is closed empty. or example, located in a blow mould just outside the flash pocket: must be sized to support the full clamp tonnage. See *pressure pad*.

lanolin
This is wool grease or wool fat and is a complex mixture of higher alcohol esters: it contains cholesterol. It is a pale yellow, wax-like material with a melting point of 36 to 40°C and a relative density of 0.97. Used as a softener and a filler dispersion aid in rubber compounds. Stops stearic acid bloom in *natural rubber* compounds.

large calorie
A kilogram calorie or 1,000 *calories*.

laser cutting
The cutting of components, or the cutting of sheet, using a laser, for example, a carbon dioxide laser of 1,500 W power. Such a cutting system can be extremely accurate. Used, for example, for cutting ferrous metals in sheet form. See *water-jet cutting*.

laser heating
See *surface bonding*.

laser marking
A *printing technique* used, for example, to form the letters on a keyboard moulded from *acrylonitrile-butadiene-styrene* by utilising laser light absorption and a resultant colour change. The light from a laser induces a colour change, in the desired pattern, on the moulding. The colour change may come from a colour change (pigment colour change, base thermoplastics material colour change or, from pigment extraction) or from the formation of a groove or foaming. Special grades of materials, for example, *polybutylene terephthalate (PBT)* are available for this process. The finished keyboards have good readability, abrasion resistance and sweat resistance.

laser-scan inspection
A laser-based, inspection technique for *calendered sheet* used to detect the presence of *pinholes*.

lastrile fibre
A manufactured, polymeric fibre composed of a rubber, which is based on a diene (for example, butadiene) and acrylonitrile: the acrylonitrile content must be at least 10% but not more than 50%. See *rubber fibre*.

lateral contraction ratio.
See *Poisson ratio*.

lateral mixing
See *two-roll mill*.

latex
A milky aqueous dispersion of fine polymer particles (approximately 1 μm diameter) in water: a colloidal suspension of natural or synthetic polymers. Produced synthetically by *emulsion polymerization* but latices commonly occurs

naturally, for example, as *natural rubber latex*. The latex is stabilized by the presence of an emulsifier: the solid polymer may be obtained by destroying the stabilizing action of the emulsifier. This de-stabilizing action may be performed by the action of ionic substances. See *latex processes*.

latex compounding
The compounding of *latex* with *additives*. This is necessary if solid, vulcanized, shaped products are to be obtained from the latex. The additives used include fillers (such as *clay*), curing agents (*sulphur* and an *ultra-accelerator* system), pigments (such as *zinc oxide*) and a coagulant (*acetic acid* or *calcium chloride*): a gelling agent (a heat sensitizing agent) may be used in place of a coagulant if gelation in contact with a heated surface is required. Such additives are added to the latex in the form of aqueous dispersions so as to obtain ease of mixing.

latex concentrate
Obtained from *field latex* by removing some of the water. The collected *latex* is bulked and then concentrated from approximately 33% to 60% by, for example, centrifuging. Ammonia, or a bacteriocide, is added to prevent coagulation during shipment. A by-product of centrifuging is a water-rich *skim latex*.

latex foam rubber
A cellular material produced from *latex*. See *latex processes* and *Talalay process*.

latex grade rubber
Grades of *natural rubber* produced from liquid *latex* and sold in block form. For example, *standard Malaysian rubber latex grades*.

latex - natural rubber
Natural rubber latex is called latex or rubber latex. Consists of rubber in particle form (approximately 0.15 to 3 μm diameter) in an aqueous base or serum and is obtained by *tapping* the tree *Hevea brasiliensis*. The total solids content is about 40% and 60% is water. Approximately 36% is rubber and the remaining 4% solids consists of resinous materials, proteinous materials, ash and sugars in roughly equal amounts.

latex processes
Manufacturing methods for the production of products from concentrated *latex*: such methods rely on either drying, coagulation or gelling of the latex: drying is slower than the other methods as these improve the water separation rate.

LATEX DRYING PROCESSES Processes which rely on drying the latex: such processes include *dip moulding* and *coating*. The dipping of a former into a latex with intermediate drying will produce a thin walled article, for example, a glove of the required wall thickness. Impregnation and coating of fabrics are performed by dipping, spraying or spreading techniques: the rubber being obtained from the latex by subsequent drying. By using a female, plaster of Paris mould, a film of rubber is formed by absorption on the mould surface: this is then vulcanized by heating. See *slush moulding*.

LATEX COAGULATION PROCESSES Processes which produce solid rubber from the latex by the use of a *coagulant*. In *dip moulding*, a thicker film can be obtained if the former is first dipped into a coagulant solution. To prevent dripping, and to speed up drying, the latex-coated former may also be dipped into a coagulant solution. Rubber thread may be also be produced by extrusion of the latex into a coagulant solution. By treating a female metal mould with coagulant, *slush moulding* and *rotational moulding* may be performed.

LATEX GELLING PROCESSES Processes which produce solid rubber from the latex by the use of a *gelling agent*. Such a heat sensitizing agent has an inverse solubility-temperature coefficient and so cause the latex to gel in contact with a heated mould. The latex used contains the gelling agent which causes the build up of a relatively thick layer of rubber on the mould, for example, in slush moulding.

Latex foam rubber may be produced by frothing the latex (by the beating in of air), gelling the foam by the use of a delayed action gelling agent (for example, sodium silicofluoride), filling the mould and then heating the filled mould to cause vulcanization. The cured product is washed and dried: used for cushions, mattresses etc. See *Talalay process*.

latex rubber grades
See *latex grade rubber*.

latex sprayed rubber
An abbreviation used for this material is LS rubber. See *sprayed rubber*.

latices
The plural of latex.

lauric acid
Also known as dodecanoic acid. Occurs in, for example, coconut oil as a *glyceryl ester*: has the formula $CH_3(CH_2)_{10}COOH$. The acid is a pale yellow, colourless crystalline material with a melting point of approximately 44°C and a relative density of approximately 0.89. Commercial products often have a lower melting point as they are not so pure. Used as a softener and activator for rubbers (not *butyl rubber*). Metal salts of this material are used as *lubricants* and as *stabilizers* in, for example, *unplasticized polyvinyl chloride* (UPVC).

laurolactam
See *dodecanelactam*.

lauroyl
An organic radical with the formula $CH_3(CH_2)_{10}CO$-.

lauroyl peroxide
A *diacyl peroxide* which has the formula $\{CH_3(CH_2)_{10}COO\}_2$. It is an active peroxide widely used in the polymerization of, for example, *styrene* and *vinyl chloride*.

law of mixtures
Also known as the rule of mixtures or as, the simple rule of mixtures. A relationship between the value of a property P, for the mixture, to the values of the same property for the components of the mixture: allowance is made for the volume fraction (phi or ϕ) of the components of the mixture. For a binary mixture it is usually of the form $P = \phi_1 P_1 + \phi_2 P_2$. Where ϕ_1 is the volume fraction of the first component, P_1 is the value of the property for the first component, ϕ_2 is the volume fraction of the second component and P_2 is the value of the property for the second component. Useful in predicting the value of properties (for example, the modulus) of a composite. See *logarithmic law of mixtures*.

lay
The length of twist produced by stranding filaments, such as fibres, wires or roving: or, the angle that such filaments make with the axis of the strand during a stranding operation. (The length of twist of a filament is usually measured as the distance parallel to the axis of the strand between successive turns of the filament).

lay flat film
Film produced by the *tubular film process*.

lay up
A *reinforced plastics term*: to assemble layers of resin-impregnated material ready for processing or, the assembly of layers of resin-impregnated material which are ready for processing.

lay-flat
Also spelt lay flat. The term may refer to the lay-flat extrusion process or to the output (the *film*) produced by that process. See *tubular film process*.

lay-flat extrusion process
A term used to describe the *tubular*, or bubble, process used to produce film.

lay-flat width
A measure of the size of *tubular film* and is the width of a roll of double thickness film: a tube with diameter D gives a lay-flat width of D/2 or 1·57D.

L:D ratio
An abbreviation used for *length to diameter ratio*.

layer shearing
See *melt processing*.

layered composite
See *laminate*.

layflat film
Film produced by the *tubular film process*.

lb
Abbreviation for *pound*.

LB
An abbreviation used is for *lamp black*.

lbf
An abbreviation used for *pound-force*.

lbf in^{-2}. Abbreviation for pounds-force per square inch.
lbf/s. Abbreviation for pounds-force per second.
lbf/sq in. Abbreviation for pounds-force per square inch.

LC
An abbreviation used for low colour (black). See *carbon black*. LC is also used as an abbreviation for *liquid crystal*. See *liquid crystal polymer*.

LC TP
An abbreviation used for liquid crystal thermoplastic. See *liquid crystal polymer*.

LCCM
An abbreviation used for *liquid composite compression moulding*.

LCM
An abbreviation used for *liquid curing method* or *liquid curing medium*.

LCP
An abbreviation used for *liquid crystal polymer*.

LCR
An abbreviation used for laboratory capillary rheometer. See *high shear rate rheometry*.

LDPE
An abbreviation used for *low density polyethylene*.

lead
This element (Pb) occurs in Group IVB of the Periodic table along with *carbon*, silicon, germanium and tin. It does not occur naturally. It is a soft, silvery white, lustrous metal (when first cut) which melts at approximately 327°C boils at 1,750°C and has a relative density of 11·34. It is corrosion resistant and for this reason was, at one time, widely used for roofing and plumbing. Compounds of lead are used in the polymer industry as, for example, *heat stabilizers* in *polyvinyl chloride*: such compounds are poisonous. Alloys of lead include solders (based on say, 50% tin and 50% lead), type metal (based on say, 10% tin, 75% lead and 15% antimony) and pewter (based on say, 80% tin and 20% lead). See *lead stabilizer*.

lead (II) carbonate hydroxide
See *basic lead carbonate*.

lead (phenyl amino-ethyl)-phenyl dimethyl dithiocarbamate
This grey granular material has a relative density (RD or SG) of 1·51. An accelerator for natural rubber and for styrene-butadiene rubber. See *dithiocarbamates*.

lead carbonate
Also known as normal lead (II) carbonate. May be represented as $PbCO_3$. This material is a white powder. Used as a *heat stabilizer* for *polyvinyl chloride*. See *lead stabilizer* and *basic lead carbonate*.

lead carbonate flakes
See *flake pigments*.

lead chromate
This material has a relative density of 5·70. See *chrome yellow*.

lead chromates
A class or type of *inorganic pigment*. This class includes *chrome yellows*, *chrome reds*, *molybdate oranges* and *molybdate reds*. Such pigments are popular because, in general, they are low cost colouring systems which are easily dispersed and which give bright opaque colours. The presence of lead limits their use as does a comparatively low upper-use temperature of approximately 205°C/400°F. Such materials are often strong oxidizing agents.

lead chrome green
See *Brunswick green*.

lead chromes
See *chrome yellows*.

lead compounds
Compounds of lead are used in paints and as petrol additives. In the plastics industry, they are best known as *heat stabilizers* for *polyvinyl chloride* (see *lead stabilizer*). In the rubber industry, they are best known as *accelerators*.

lead diethyl dithiocarbamate
This grey, solid has a relative density (RD or SG) of 1·87 and a melting point of 206°C. An ultra-accelerator for continuous vulcanization processes. See *dithiocarbamates*.

lead dimethyl dithiocarbamate
This white powder material has a relative density (RD or SG) of 2·38 and a melting point (with decomposition, or degradation) of approximately 320°C. An ultra-accelerator for, for example, continuous vulcanization processes: also an ultra-accelerator for *natural rubber*, *butyl rubber* and for *styrene-butadiene rubber*. See *dithiocarbamates*.

lead dithiocarbamates
See *dithiocarbamates*.

lead monoxide
See *litharge*.

lead - of a screw
The horizontal distance travelled by the material in one revolution of the *screw*.

lead oxides
There are three oxides of lead *litharge* or lead monoxide, lead dioxide and red lead (lead IV oxide). Litharge or lead monoxide, is probably the most important one for the polymer industry.

lead pentamethylene dithiocarbamate
This white powder material has a relative density (RD or SG) of 2·29 and a melting point of approximately 250°C. A delayed action accelerator with good processing safety. Used for, for example, continuous vulcanization processes: also for hot air and press vulcanization. See *dithiocarbamates*.

lead salicylate
This yellowish/white powder material has a relative density (RD or SG) of 2·36 (at approximately 45% litharge content) An activator and anti-scorch agent for *natural rubber* and synthetic rubbers. Acts as an *ultra-violet stabilizer* in vinyl plastics.

lead salts
Lead compounds which are used as heat stabilizers for, for example, *polyvinyl chloride*. See *lead stabilizer*.

lead silicate
Also known as basic lead silicate. This white powder material has a relative density (RD or SG) of approximately 5·8. An activator for *natural rubber* and for synthetic rubbers (such as *nitrile rubber* and *styrene-butadiene rubber*).

lead stabilizer
Lead stabilizers are also known as basic lead stabilizers. The term refers to any basic lead compound which has stabilizing power in *polyvinyl chloride* (PVC). Reacts with the decomposition, or degradation, products of PVC (for example, hydrochloric acid) to give white compounds and so their use minimises colour changes. A common example is *basic lead carbonate*. Other lead stabilizers include tribasic lead sulphate, dibasic lead phthalate and dibasic lead phosphite. A major problem with lead compounds is their toxicity: another is their high density, for example, lead sulphate has a relative density of 6·20.

lead stearate
This white powder material is used, for example, as a lubricant, activator and softener for *natural rubber* and *synthetic rubbers*. For example, at approximately 1 phr this material is used as an activator for *acrylic rubber*.

lead white
See *basic lead carbonate*.

lead-lag
A problem in, for example, *reaction injection moulding* (RIM), in which the flow of the various reactants is not correct when the *mixhead* is first opened. To improve mixing and homogeneity an aftermixer may be employed

leaded zinc oxide
A mixture of *zinc oxide* and *basic lead sulphate*. Used as a white *pigment for paints*.

leader
A length of material used to pull or carry an extrudate through the *haul-off system* during start-up of an *extrusion line*. For example, a length of film or paper which is pre-threaded through the system.

leading edge
Screw terminology which refers to that part of the *screw flight* which is located towards the die-end of the system.

leak testing
The testing of a container to detect an undesirable hole. Often done by noting the decay in air pressure after pressurizing a *blow moulded container* to a specified level.

leakage
See *blow moulded container*.

leakage flow
Loss in output caused through material being lost over the screw flight during *extrusion*: the material leaks through the clearance between the *screw* and the *barrel*.

leaning
Distortion of a *blow moulded container* caused, for example, by using too high a product temperature during hot filling.

leathercloth
A coated fabric which has a leather-like appearance: often based on *polyvinyl chloride* compositions.

leathery state
Also known as retarded elastic state or as, leathery region or as, transition zone or as, viscoelastic state. The state, or region, of polymeric behaviour observed at a particular temperature: the material is between the glassy and the rubbery state. The temperature of the polymer is approximately that of the *glass transition temperature*.

length of flow
See *Rossi-Peakes test*.

length to diameter ratio
A ratio which is referred to as L:D and given as, for example, 10:1 which means that the length is ten times greater than the diameter. See *screw*.

length - units of
See *UK system of units* and *US system of units*.

lens
A transparent piece of material, for example, glass or an amorphous plastics material, having two (or two main) surfaces and used for changing the convergence of radiation. Associated with changing the convergence of light rays where at least one of the lens surfaces is curved.

let down ratio
The resin-to-additive ratio. For example, the resin-to-masterbatch ratio. See *masterbatch*.

let go
This phrase is used as a noun and denotes an area in laminated glass over which an initial adhesion between interlayer and glass has been lost.

let-off
Also known as a pay-off. Part of a production line which allows the feed of a continuous length of material, under controlled tension, to a coating system, for example, a *calender* or an *extruder*.

letterpress printing
A *printing technique* in which the ink is transferred to the substrate, for example, a film, from a surface which carries the required design in relief. A raised type face applies the ink to the film surface: as the ink has a paste-like consistency, and high pressure must be applied, this process is difficult to apply to plastics films. Used originally for newspaper printing but also for stiff plastics films, for example, those based on *unplasticized polyvinyl chloride* (UPVC). Letterpress inks are based on a mineral oil, vegetable oil and resin mixture and have a viscosity of approximately 20 poise.

level gauge
A term used in *hydraulics* and which refers to a device used to indicate the amount of fluid in a reservoir.

lever type dial test indicator
A type of *dial test indicator* (DTI) which relies upon a lever and scroll system of magnification. It has only a limited range of stylus movement: little more than one complete revolution of the pointer. It is more compact than the *plunger type*.

LFRP
An abbreviation used for long fibre reinforced plastic. See *filled engineering plastics*.

Li-BR
An abbreviation used for butadiene rubber based on a lithium catalyst. See *butadiene rubber*.

Li-IR
A grade of *isoprene rubber* prepared with a lithium catalyst.

liaison block
A component of a *rubber injection moulding machine*. It contains the nozzle and links the injection unit to the extruder unit: it is heated, for example, with oil.

lidding material
A material, often a multi-layer system, which is used to seal a packaging container, for example, a yoghurt container. To assist in re-use, the lidding material needs to be compatible with the material used to make the packaging container.

lift
Also called a spray. The complete output from the moulding machine.

light
A rubber moulding defect experienced when the moulding does not completely fill out: a short which may be produced through using a stiff-flow compound. The term light is often used in connection with fillers, particularly for rubbers, and taken to mean that the filler is a synthetic material of good colour, low bulk density and of fine particle size. See *magnesium oxide*.

light ageing
Decomposition, or degradation, of polymers brought about by the action of light. When the term is applied to rubbers, then the alternative term light cracking may be applied. May be minimised in the case of rubber by the use of waxes which act as light ageing inhibitors. In the case of plastics materials, ultra-violet light must be prevented from acting on the polymer: may be done by screening out the harmful radiation (using *carbon black*) or by using *ultra-violet light absorbers*.

light ageing inhibitor
See *light ageing*.

light calcined magnesium oxide
See *magnesium oxide*.

light - composition of
Light is electromagnetic radiation with a frequency of, from approximately 4.3×10^{14} Hz (red light) to 7.5×10^{14} Hz (violet light). This corresponds to a wavelength of approximately 700 nm for red light to about 400 nm for violet light. Approximate wavelengths are:

   400 to 440 nm for violet;
   440 to 500 nm for blue;
   500 to 540 nm for green;
   540 to 600 nm for yellow;
   600 to 650 nm for orange; and,
   650 to 700 nm for red.

light cracking
See *light ageing*.

light higher alpha olefin - polyethylene
An abbreviation used for this term is light HAO-PE. Grades of *polyethylene* with a very low density. These grades of polyethylene are hardly crystalline and are rubber-like materials. See *higher alpha olefin grade*.

light magnesium carbonate
This white solid material has a relative density (RD or SG) of 2.19. Synthetic magnesium carbonate is made by precipitating, for example, the sulphate with sodium carbonate: light magnesium carbonate is made by precipitation at low temperatures and has a finer particle size than the filler obtained from natural materials. See *magnesium carbonate*.

light magnesium oxide
See *magnesium oxide*.

light oil
The fraction, obtained during the distillation of *coal tar*, and which distils at up to 170°C. From this may be obtained *benzene*, *toluene*, *xylenes* and *pyridine*.

light scattering
A method of molecular weight determination which uses a dilute solution of the polymer and gives a measure of weight average molecular weight (M_w).

light screen
A material which helps to minimize light initiated degradation by preventing the harmful radiation from reaching the polymer, for example, *carbon black*.

light stabilizer
A *stabilizer* which protects a polymeric material against the action of light, for example, against the action of natural radiation of wavelength of approximately 300 to 400 nm. A light stabilizer may be classed as an ultraviolet absorber, a quenching agent, a hydroperoxide decomposer and as a free-radical scavenger (for example, a *hindered amine light stabiliser*). Such stabilizers are used in concentrations of from 0.05% to 2%: mixed stabilizer systems are commonly used. The most commonly used, non-black, light stabilizers are those based on 2-hydroxy benzophenone, 2-hydroxyphenyl benzotriazoles, nickel (11) chelates and hindered amine light stabilizers. See *ultra-violet stabilizer*.

light transmittance
The *transparency* of a material is defined in terms of two parameters or measurements. These are its luminous, or light, transmittance, and haze. Light transmittance is a ratio obtained by measuring how much light passes through a sample of a material compared to how much light is passed without the material. Light that is not transmitted is reflected, scattered or absorbed. The upper limit for light transmittance achieved by really good quality acrylic sheet is about 92%.

Light transmittance is measured by placing a flat test specimen is placed in front of, and at right angles to, a collimated (parallel) beam of light. A photocell measures the amount of light passing directly through the material. If the specimen is then removed the amount of light reaching the photocell increases. The ratio of the two measurements, expressed as a percentage, is the light transmission.

When the specimen is thin and flexible, e.g. film rather than rigid sheet an integrating sphere hazemeter has to be used. This consists of a light source and lens which gives a collimated beam which passes through the specimen into an integrating sphere. The inside of the sphere is coated with a highly reflective white powder. The sphere can be moved so that all the light entering it either passes into a light trap, or falls on the wall of the sphere where it is repeatedly reflected until it reaches a photocell. Output from the photocell is displayed on a galvanometer.

Light transmittance measurements are made by first setting the instrument so that without a specimen in position, the galvanometer reads zero (i.e. all the light passes into the trap) and 100 when the sphere is tilted and all the available light reaches the photocell. Interposition of the specimen causes the galvanometer reading to fall to a value of less than 100, and this figure is the light transmittance percentage.

lighter test
A burning test in which the sample in the form of a clip, is burnt with a butane lighter: the lighter placed beneath the sample, is used to simulate the effect of a small flame. Such a small flame may result, for example, from faulty electrical equipment and could cause a fire.

After conditioning at 50% RH, 25 clips (95 × 2.4 × 1 mm) are burnt horizontally and 25 are burnt vertically. The combustion time is recorded as is the number of burning drops. The results are recorded as:

   index 100. No drops and combustion ceases as soon as the flame is removed.
   index 80. Combustion ceases after one drop has fallen.
   index 60. Combustion resumes after one drop has fallen.

index 40. Combustion resumes after two drops have fallen.

index 20. Combustion resumes after three drops have fallen.

index 0. Combustion resumes after more than three drops have fallen.

The percentage of specimens in each index is noted. See *Underwriters laboratory UL94 vertical burning test*.

lightfast
Not affected by light: not bleached by sunlight. See *lightfastness*.

lightfastness
The ability of a system to withstand light. Usually assessed by exposure to a carbon arc or to a xenon light. See *lightfast*.

lighting whites
A type or class of white pigments which are used to make light diffusers from plastics materials. Such white pigments will hide the illumination source but still allow the passage of most of the light. Typical pigments include *zinc oxide* and *barium sulphate*.

lightness
A characteristic of light, reflected by a surface, and which is a measure of its brightness. See *colour*.

lignin
Also known as polydehydro diconiferyl alcohol. This light brown powder has a relative density (RD or SG) of 1·30. It is a high molecular weight, natural benzene derivative which may be obtained, for example, during *cellulose pulp production*. Used as a filler and as a dispersing agent for other fillers. Gives improved ageing to rubber compounds and a smooth velvet-like finish to cured components.

ligroin
Another name for petroleum ether. A mixture of hydrocarbons, mainly paraffins, with a boiling point range of approximately 80 to 150°C. Used as a solvent for rubbers.

LIM
An abbreviation used for liquid injection moulding. See *reaction injection moulding*.

lime
Slaked lime is Ca(OH)$_2$ and calcium oxide is CaO (which is also known as quicklime). Calcium oxide is a white amorphous powder made by heating *calcium carbonate*: it has a melting point of approximately 2,600°C. Reacts vigorously with water (slaking) to form slaked lime or Ca(OH)$_2$. Calcium oxide may be used to reduce porosity in rubber compounds. An intermediate used to produce *precipitated calcium carbonate* and *calcium silicate*. Calcium hydroxide is a fine, odourless white powder. This material has a relative density (RD or SG) of approximately 2·1. Used as an *accelerator* in, for example, mixes containing *factice*.

limed rosin
A material produced by chemical modification of rosin with *lime*. The resultant resin has high gloss, good gloss retention, good adhesion and will tolerate large quantities of water when used in paints and finishes. Has poor drying time and poor resistance to water and to chemicals. A solution of limed rosin in mineral spirits is called gloss oil: used for low cost paints and varnishes.

limestone
A naturally occurring *calcium carbonate* material used as a *filler*: to produce a filler the mineral is ground and classified. Wet-ground limestone is of finer particle size and more uniform size distribution than the dry-ground limestone. Marble and calcite are hard, compact crystalline varieties of limestone.

limit switch
A switch set at the limit of movement of part of a machine and which, for example, initiates the cutting of the extrudate.

limited compatibility
See *secondary plasticizer*.

limiting oxygen index
An abbreviation used for this term is LOI. The concentration of oxygen, in an oxygen/nitrogen mixture which will just sustain burning in a, so-called, candle test: the concentration of oxygen is expressed as volume percent or as volume fraction. The test yields a number, for example, 27. This value is taken, for solid polymers, as being the self extinguishing limit for a *candle test*. For flame retardancy purposes, the higher the number, the better.

limiting oxygen index test
This is a quality control test which measures the relative flammability of plastics materials by measuring the minimum concentration of oxygen (O_2) in a slowly rising stream of oxygen (O_2) and nitrogen (N_2) that will just support combustion. It differs from the UL94 test in that the rigid specimen is ignited at the top and burns downwards in a candle-like manner. Thus burning takes place along a surface which has not been preheated by hot gases released in the burning process. As the oxygen index of a polymer can be modified by flame retardants, glass fibre reinforcement or other additives, testing of the actual compound is advisable.

Ten specimens at least, are required. They may be moulded, cast or machined from sheet. The dimensions of type 2 specimens of BS 2782 (Method 141 B) are 6·5 ± 0·5 mm wide, 3 ± 0·25 mm thick and length 80-150 mm. It is essential to remove burrs or flash from the edges; condition the specimens at 23 ± 2°C and 50 ± 5 per cent relative humidity.

A specimen is clamped, at the bottom, in an upright position inside a tall vertical glass column. A steady flow of gas, consisting of oxygen and nitrogen in known proportions, is fed into the base of the column. When all the air has been purged the specimen is ignited at the top with a propane flame. The time the specimen continues burning (or length burnt), after the igniting flame is removed is recorded. If it burns for more than 3 minutes (or 50 mm length) the oxygen content is reduced: if it burns for less than 3 minutes the oxygen content is increased.

The test is then repeated using a fresh sample surface. The test is repeated with varying proportions of oxygen until the critical or minimum oxygen concentration required for 3 minutes (or 50 mm) burning is found. When this critical concentration is expressed as a percentage it is known as the limiting oxygen index (LOI).

limiting oxygen index - typical results

Material	Oxygen Index (%)
Acrylonitrile-butadiene-styrene	19–35
Acrylic	19
Acetal	15–16
Nylon 6	21–24
Nylon 66	21–30
Nylon 11	25–32
Polyphenylene oxide (modified)	29–35
Polyester (thermoplastic)	20–37
Polycarbonate	24–37
Polypropylene	17–28
Polyvinyl chloride	23–43

The results are dependent upon the polymer grade and the additives used.

limiting value
A term applied to a *latex*. The limiting value of a latex, is a measure of the minimum quantity of electrolyte necessary for coagulation.

limiting viscosity number
An abbreviation used for this term is LVN or [η]. See *solution viscometry*.

line
A pipe, or passage, used to transmit fluid.

line fed
Term used in moulding technology. Most moulding machines are *self contained*. If, however, the machines are supplied from a central hydraulic supply, then they are said to be line fed. See *injection moulding machines*.

line pressure
Sometimes referred to as gauge (gage) pressure as it is often measured by means of a gauge which is set in an hydraulic line or pipe. It is, for example, the pressure in the main supply line from the pump to the actuator (it is not, for example, the *injection pressure*). Line pressure is usually capable of being adjusted and the line pressure should be adjusted to suit the particular moulding run.

line pressure control
Term used in injection moulding to indicate that the final mould filling part of the moulding cycle is pressure controlled by a signal which originates from a transducer located in the hydraulic line. See *velocity pressure transfer*.

line pressure transducer
A pressure sensor which is located in the hydraulic line or pipe in place of, or in addition to, a pressure gauge. What is measured is sometimes called 'injection pressure' but is really line pressure. Such transducers are more accurate than gauges and are capable of displaying rapid pressure variations. May also be used to actuate the switch from velocity to pressure control. See *velocity pressure transfer*.

linear actuator
A term used in *hydraulics* and which refers to a device which converts fluid flow into linear motion. That is, a hydraulic cylinder.

linear aliphatic polyamide
A *polyamide* material consisting of linear aliphatic sections (based on methylene groups) joined by amide groups. See *nylon*.

linear aromatic polyester
See *thermoplastic polyester*.

linear low density polyethylene
An abbreviation used for this material is LLDPE. Also known as polyethylene-linear low density which gives PE-LLD. Also known as linear low density polyethene or as, linear low.

This thermoplastics material was developed in the 1950's but it is only within the last few years that it has come into prominence. This plastic is, in fact, a copolymer and is made by polymerizing *ethylene* with a small percentage of a higher olefin, for example, octene or butene. Butene-type materials (C_4 LLDPE) can be made by any of the present available processes and are cheaper; octene-type materials (C_8 LLDPE) have slightly better mechanical properties. The comonomer introduces short chain branches into the molecular structure (more than *high density polyethylene* or HDPE) and so it is possible to produce a material, which in some respects is similar to *low density polyethylene* (LDPE), but without the need for high polymerization pressures. These short chain branches interfere with chain packing and so, crystallization is not so pronounced as for HDPE; such materials have a lower density therefore than HDPE. (The term 'linear' means that there is an absence of long chain branching. When these materials were first introduced, the lower price of them forced their use: the standard butene-based product was, in effect, being used as a filler for LDPE.

With suitable catalysts (halogen-free metal complexes) only low pressures and temperatures are required for manufacture and polymerization efficiency is high. It is possible to produce PE over the density range 0·89 to 0·955 g/cm^3 and it is also possible to produce materials which have a particular molecular weight: such materials have a narrow molecular weight distribution (a narrower molecular weight distribution than LDPE). They contain a smaller quantity of low molecular weight material and more medium molecular weight molecules; the average length of the molecular chain is longer.

For a given molecular weight therefore, LLDPE will have better mechanical properties at low (e.g. −40°C/−40°F) and high temperatures, better *environmental stress cracking* (ESC) resistance, and higher melt viscosity than LDPE. It can be stronger and tougher than LDPE and because of its narrow molecular weight distribution warp-free mouldings can be readily produced. Increasing the comonomer content increases the flexibility, the impact and tear strength, the environmental stress cracking (ESC) resistance and the ease of sealing and printing. However, the stiffness, yield strength, creep strength, hardness, Vicat softening point and the melting point decrease proportionally. Decreasing the density, or the crystallinity, also increases the flexibility, the impact and tear strength, the ESC resistance and the ease of sealing and printing. Similarly, the stiffness, yield and creep strength, the hardness, Vicat softening point and the melting point decrease proportionally.

LLDPE is less pseudoplastic than LDPE; if a LDPE plastic of a certain *melt flow rate* (index), is compared with an LLDPE plastic of the same melt flow index, then it will be found that the linear polymer will be of higher viscosity at processing shear rates. Die gaps should be larger than used for LDPE. Melt flow rate (MFR) values can reach 100 for LLDPE; with LDPE they can reach 250. (Polymers of different density but with the same melt flow index do not have the same molecular weight). Replace a LDPE grade, with an MFR of 7 to 10, with a LLDPE of MFR of 20 to 50.

The shrinkage of LLDPE is of the order of 0·015 to 0·030 mm/mm i.e. 1·5 to 3·0%. By optimizing the moulding conditions for LLDPE it is possible to obtain lower *shrinkage* values than those obtained with LDPE and HDPE. LLDPE also exhibits a much lower differential shrinkage than LDPE and so warping is considerably reduced.

There is no solvent at room temperature but it does swell in aliphatic, aromatic and chlorinated hydrocarbons. At higher temperatures (approximately 55°C/131°F) LLDPE is soluble in hydrocarbons and chlorinated hydrocarbons, for example xylene and trichloroethylene. LLDPE has better ESC resistance than LDPE and the higher molecular weight grades have the best ESC resistance. The water vapour permeability of this type of PE, as with all types of PE, is low. PE is permeable to gases and vapours (i.e. it lets them through) but LLDPE and HDPE are less permeable to gases and vapours than LDPE. Some grades of LLDPE are accepted as being suitable for containers for oil and petrol: they have been used for fuel tanks: in some cases the formed containers do not have to be chemically modified so as to make the material more impermeable to fuels.

The light resistance is similar to that of LDPE. The cheapest way of improving this is by *carbon black* incorporation, for example, addition of say 2% black can improve the weathering resistance by twelve times. Has limited resistance to oxygen at elevated temperatures; antioxidants are used for protection. Unless so protected, the electrical properties will suffer.

As the natural colour of the material is an off-white then a wide colour range is possible; this does not include transparent colours. Because of the rigidity of this material, compared to LDPE, mouldings can be made of thinner wall thickness; the higher crystallization temperature means a faster set-up time. Both considerations should result in significantly faster cycle times in processes such as injection moulding.

The surface of this material may be made more receptive to inks or to adhesives by pre-treatment using a corona discharge or by ozone. Commonly welded using techniques such as *hot plate* or *hot shoe*; when welding LLDPE it is usual to cover the plates, or shoe, with PTFE so as to prevent the melted material sticking to the hot surfaces.

Film is the major market for this material. Because of the rigidity of this material, its higher elongation at break and puncture resistance, compared to LDPE, film extrusions can be made of thinner thickness: that is, the product may be down-gauged. Similar considerations should apply to *blow moulded containers*.

Injection mouldings can have high gloss, good impact strength and good resistance to ESC. As a result of a combination of ESC resistance, dimensional stability and good torque retention, screw caps and closures, particularly in the pharmaceutical field, are made from LLDPE. Blends of LLDPE with other polyethylenes are widely used, for example, in film products. Now blends of HDPE and LLDPE are being considered for industrial paint containers. The addition of some LLDPE to high density polyethylene, improves the ESC resistance as it slightly disrupts the crystalline structure of this semi-crystalline, thermoplastics material. For a given MFR and density, LLDPE has a higher Vicat softening point (10 to 15°C/(50 to 59°F) and lower internal stresses than LDPE; this means that it can be used for components subjected to elevated temperatures. Because of the material's good low temperature properties, it may be utilized for containers used in cold storage rooms and for high quality houseware and toys (specifically snow-sledges and snowmobiles). The addition of very low density PE (VLDPE) will give the material increased elasticity and improve the low temperature impact strength.

LLDPE is becoming increasingly used for large, thin, circular and rectangular box lids - because of the flatness required (in order to decorate the lid) and because of the cheapness of the product obtained (as a result of thinner sections being used). LLDPE is now being considered for applications such as re-usable containers, for example, once an ice-cream container is emptied it becomes a quality sandwich box. The reason for such use is its attractive properties which includes the ability to be cleansed in dishwashers

linear phthalates
Also known as linear dialkyl phthalates. The term refers to a group of plasticizers which are based on phthalic anhydride and a linear, or normal, alcohol. As the ester which results has two side groups (alkyl groups) it is referred to as dialkyl.

Phthalates based on linear alcohols include, *octyl decyl phthalate (ODP), di-C(6-8-10) phthalate* and *di-C(7-9-11) phthalate*. These linear phthalates have good low temperature properties, low volatility and are now relatively cheap. Where long term retention of properties, after exposure to high and low temperatures, is a requirement then phthalates based on a linear alcohol will perform better than a phthalate based on a branched alcohol. For example, the linear phthalate, ODP will retain properties better than branched chain alcohols such as di-iso-octyl phthalate (DIOP) and di-iso-decyl phthalate (DIDP).

linear polymer
This class of polymers contains those materials which are not branched, cross-linked or of a network type. That is, the polymer chains are of unbranched chains of atoms, for example, *vinyl* polymers. Produced by *linear polymerization*.

linear polymerization
The process used to produce a *linear polymer*.

linear PU
See *thermoplastic polyurethane*.

linear shrinkage
Mould *shrinkage* quoted as a linear value, for example, 0·004 in/in or 0·004 mm/mm.

linear triblock polymer
A linear polymeric material in which two end-blocks are connected to a central block. Triblocks consist of the structure A-B-A, where A represents a block which is a glassy, or crystalline, thermoplastics material and B represents a block which is rubbery/elastomeric at room temperature. When A = PS and B = BR then, SBS is the result. When A = PS and B = isoprene rubber then, SIS is the result.

Triblock copolymers were first produced in order to see if they offered solutions to problems of excessive flow in storage, and poor green strength, of some commercial *synthetic rubbers*. To determine *green strength*, the copolymers were pressed into unvulcanized, flat sheets; when these were tested, they were found to exhibit significant, rubbery behaviour.

Triblock copolymers are phase separated systems: for example, the two phases, PS and BR, retain many of the properties of the individual homopolymers, for example, they have two glass transition temperatures (T_g). See *styrene block copolymer* and *saturated triblock polymer*.

linear velocity displacement transducer
A linear velocity displacement transducer (LVDT) is a *transducer* which is used for measuring relatively small amounts of movement in the vertical or horizontal plane. The amount of movement is detected by means of a change in an electrical signal caused by, for example, the movement of an iron core within a coil: this change is then amplified and converted into a linear measurement. Very accurate measurements of *tie bar* extensions can be obtained when using these devices. Many newer injection moulding machines have a linear displacement transducer mounted at the end of, or along, the tie bar so as to measure the *tie bar* extension when the locking force is applied. The amount of extension is displayed in digital form either on the moulding machine's visual display unit (VDU) or on a separate control panel.

liner
A long sheet of fabric used in *calendering*. Because rubber sheeting is tacky then the output from the calender cannot be reeled upon itself but the sheeting must be wound onto a liner. Such liners are not an integral part of the product but are re-used after the rubber has been removed, assembled into the product and vulcanised. The calendered sheet may also be cured in a steam pan after being wrapped in a fabric liner on a metal drum.

lining
The lining of a *barrel* with a corrosion or wear resistant alloy such as Xaloy. This wear resistant layer may be cast in during barrel manufacture or, the liner may be inserted subsequently (used, for example, to rebuild a worn machine). Such a barrel assembly may be known as a *bimetallic barrel*: used when abrasive compounds are being processed.

linoleic acid
Also known as cis-9, cis-12-octadecadienoic acid. This yellow oily liquid has a relative density (RD or SG) of 0·95, a boiling point of 229°C and a melting point of −5°C. It is an unsaturated fatty acid ($C_{17}H_{31}COOH$) with two double bonds: the double bonds are responsible for the air drying properties of oils which contain compounds of this material. Occurs in the triglycerides of plant oils such as sunflower oil and *linseed oil*. Occurs in small quantities in *natural rubber latex* with *zinc oxide* it acts as an *activator*. Once known as vitamin F.

linolenic acid
Also known as cis-9, cis-12, cis-15-octadecatrienoic acid. This oily liquid material has a melting point of −11°C. It is an un-

saturated fatty acid ($C_{17}H_{29}COOH$) with two double bonds: the double bonds are responsible for the air drying properties of oils which contain compounds of this material. Occurs in the triglycerides of plant oils such as *soya-bean oil* and *linseed oil*.

linseed oil
Derived from linum usitatissimum. Also known as flaxseed oil. A nonedible vegetable oil. This material has a relative density (RD or SG) of 0·94. A very reactive, vegetable drying oil which is obtained when flax seeds are crushed. Oxidizes on exposure to the air to form an elastic, protective film. It is used to make paints and as ingredient in *oil forming alkyds*. It is a drying oil which is composed of the glyceryl esters of *stearic, oleic, linoleic* and *linolenic acids*. Has a high linolenic acid content (approximately 50%) which gives rapid drying but also poor after-yellowing of paint films. See *factice*.

lip type - rotary shaft seal
A type of *rotary shaft seal*.

liquid
A liquid is a substance which flows freely but which is not gaseous.

liquid aromatic polyester polymer
See *liquid crystal polymer*.

liquid bath vulcanization
See *liquid curing method*.

liquid colour
Also called a liquid concentrate. A colouring system which consists of, for example, pigments) dispersed in an inert, liquid carrier. The liquid carrier (for example, a phthalate *plasticizer*) penetrates and breaks up the pigment agglomerates and then, holds the pigment particles in suspension. A concentrated, liquid colouring system or *masterbatch*: a liquid masterbatch which can be easily dispensed, or metered, during processing. Most successful applications seem to be on long runs with one colour where the precise metering, for uniform colour, can be established and held. Liquid colours require little storage space and are often easier to clean from equipment than dry colours.

liquid colouring
A method of colouring polymers, particularly thermoplastics materials. One of the cheapest ways of adding a colouring system is in liquid form (*liquid colour*). Such a colouring system may be tumble mixed (see *tumble mixer*) with the polymer granules although in practice, the preferred method of addition is at the moulding machine: this is because, as the system is liquid, automation is cheap and can give good results at low cost. A large range of colours can be produced from a small inventory of colouring systems and the colouring systems in turn, are compatible with a wide range of polymers. Although the colouring systems may be compatible with a wide range of polymers it does not follow that the same colour will be produced with differing polymers. As the liquid carrier system may affect part properties, it is important that the effect on properties be investigated and that the level of addition be kept to as low a value as possible. The melt flow behaviour of the polymer is also affected - usually it becomes easier. In some cases, for example, at high levels of addition, screw slip can sometimes become a problem.

Liquid colour is added to the polymer often within the range of 0·1–1·5%. Dispensing of the liquid is by means of a pump (e.g. a *peristaltic pump*) which operates by drawing liquid colour from a storage container through silicone rubber tubing. The colour is added to the polymer via a metal probe fitted in the machine throat, for example, of an *extruder* or of an *injection moulding machine*. As the *let down ratio* is high only a small amount of liquid is usually added. For thin-sectioned components, the amount of liquid may need to be un-acceptably high: liquid colour is not therefore usually suitable for film.

liquid composite compression moulding
An abbreviation used for this process is LCCM. A *structural liquid composite moulding process*.

liquid concentrate
See *liquid colour*. Note, a colouring system may not be the only additive that is incorporated into liquids. When *structural foam mouldings* are being produced, the *blowing agent* is commonly incorporated into the liquid colouring system as this gives ease of addition and can result in a cleaner factory environment.

liquid crystal polyester
An abbreviation used for this type of material is LCP. See *liquid crystal polymer*.

liquid crystal polymer
An abbreviation used for this type of material is LCP. Also known as liquid crystal (LC) plastic or as a, liquid crystal thermoplastic or as a, liquid crystalline polymer, or as a, liquid crystal polyester (LCP). May be referred to as a self reinforcing polymer, or as a, self reinforcing engineering thermoplastics material or as a, mesomorphic polymer or as a, thermotropic liquid crystal polyester (TLCP).

Often such a material is an aromatic polyester (for example, based on *terephthalic acid, parahydroxybenzoic acid* and *p,p'-dihydroxybiphenyl*. The presence of the p,p'dihydroxybiphenyl hinders chain alignment as it kinks the molecular chain. The polymer may be referred to as an aromatic polyester copolymer or as a, liquid aromatic polyester polymer or as a, liquid aromatic polyester copolymer or as an, aromatic polyester condensate LCP. Cheaper materials will probably be based on mixed aromatic and aliphatic types of LCP.

When an LCP is heated to melt temperatures, then the heated material retains some solid-like properties, because of the relatively rigid molecular structure, but it is still capable of being made to flow because of relatively low inter-molecular forces. The molecules are highly anisotropic and form a liquid crystalline phase on heating (a *mesophase*) because of the *mesogenic* groups which the polymer contains. A low viscosity melt develops when the melt is sheared because of molecular anisotropy.

Can have the mesogenic groups in the main chain (which gives a liquid crystalline main chain polymer) or, the mesogenic groups are in side groups attached to the main chain (which gives a liquid crystalline side chain polymer).

In general, LCPs are a class of engineering thermoplastics materials which combine high modulus, high strength, easy processing behaviour, low mould shrinkage, resistance to chemicals, resistance to solvents, low combustibility and low smoke generation on burning. The resistance of an LCP to organic and chlorinated solvents is, in general, better than that of a *sulphone polymer* or of a polyamide imide. An LCP is not attacked by acids, bases, alcohols or esters. Such materials have good *environmental stress cracking resistance* and the resistance to nuclear radiation is good. LCP-type materials have better impact resistance, easier processing and lower cost than *polyphenylene sulphide*. Such materials have high density, for example, when unfilled this is approximately 1·35 to 1·4 g/cm^3.

An LCP is a self reinforcing, thermoplastics material in which molecular alignment is frozen-in on cooling: this gives high modulus and strength. The strain to failure is low and the behaviour is *anisotropic*. However, the molecular chains do not completely re-orient when they solidify from the melt which reduces distortion on moulding. The dimensional stability of the moulded components is enhanced by the inherently low coefficient of thermal expansion of the LCP.

Such polymers usually possess little *die swell* and extrudates set up very quickly when they emerge from the die. This means that post-extrusion deformation to final size is not recommended and the gate, in injection moulding, must be carefully sized and positioned (so as to avoid *jetting*). When filled, for example, with mineral or glass fillers, LCP compounds can be thermoformed: filling is necessary so as to reduce the high level of *orientation* which develops on *extrusion* (in the *machine direction*).

An aromatic polyester LCP, such as Xydar, has a high processing temperature which necessitates injection moulding at temperatures greater than 400°C: the melting point is approximately 423°C. The melt has a high melt viscosity which means that moulding into thin sections is difficult but the material will withstand temperatures of the order of 300°C. Continuous service temperature is of the order of 240°C. Such materials have been used for cookware.

An aromatic polyester condensate LCP, such as Vectra, has a lower processing temperature which means that injection moulding and extrusion is possible at approximately 285°C to 325°C: the melting point is approximately 280°C. The melt also has a significantly lower melt viscosity. This means that moulding into thin sections is relatively easy and the moulding equipment does not require modification. Faster cycling than an aromatic polyester LCP but not so heat resistant. Used in electronic applications, for example, for printed circuit boards, because this type of material resists contact with solder, fluxes and chlorinating agents. May also be moulded into thin sectioned, dimensionally stable components very quickly and such components resist high temperatures. Water absorption is approximately 0.03%.

liquid crystal thermoplastic
An abbreviation used for this type of material is LC TP. See *liquid crystal polymer*.

liquid crystalline main chain polymer
A *liquid crystal polymer* in which the *mesogenic groups* are in the main chain.

liquid crystalline phase
See *mesophase*.

liquid crystalline polymer
See *liquid crystal polymer*.

liquid crystalline side chain polymer
A *liquid crystal polymer* in which the *mesogenic groups* are in side groups attached to the main chain.

liquid curing medium process
An abbreviation used for this term is LCM process. A *continuous vulcanisation method* which frequently uses a *molten salt bath* as the heating medium. See *liquid curing method*.

liquid curing method
An abbreviation used for this term is LCM. May also be referred to as liquid bath vulcanization or as, salt bath vulcanization. A continuous *vulcanization* process which is often used for extrudates. The phrase, salt bath vulcanization, is used because the extrudate is carried through a hot liquid which is often a salt bath so that rapid vulcanization occurs without water marking. A eutectic salt mixture is used which may be based on 53% potassium nitrate, 40% sodium nitrite and 7% sodium nitrate. Such a liquid has high density which means that the relatively low density rubber compound must be held below the liquid salt surface, for example, by a steel conveyor band. Because of the high temperatures (up to 240°C), and good heat transfer characteristics, rapid vulcanization times are obtained.

liquid filler
A liquid material which is used as an extender, or filler, in polymer compounds. See *plasticizer extender*.

liquid injection moulding
An abbreviation used for this term is LIM. See *reaction injection moulding*.

liquid NBR
A low molecular weight *nitrile rubber*.

liquid nitrile rubber
An abbreviation used for this material is liquid NBR. A low molecular weight *nitrile rubber*.

liquid nitrogen cooling
Cooling of a product, or of a machine, with liquid nitrogen. Sometimes used in, for example, *blow moulding* in order to reduce the cooling time: the product is cooled by the injection of cold nitrogen, via the *blowing pin*, from a liquid storage vessel. Liquid nitrogen has also been used for flash removal, that is, by *cryogenic* tumbling.

liquid phase adsorption methods
Liquid phase adsorption methods are used for product control and specification for materials such as *carbon black*. For example, *iodine absorption* (absorption of iodine from potassium iodide solution) is the standard ASTM method for black assessment. However, *cetyl trimethylammonium bromide (CTAB)* absorption is in some ways better as this method is not so affected by factors which limit the iodine absorption (factors such as surface oxidation and absorbed hydrocarbons).

liquid rubber
See *depolymerized rubber*.

liquid silicone rubber
An abbreviation used for this material is LSR. Liquid silicone rubbers are available which are very easy flowing and yet, when cured, they have similar properties to conventional peroxide-cured silicones. They are supplied as two component parts which when blended will cure in a few seconds (for small parts) at high mould temperatures, for example, 225°C. The components are pumped through a static mixing device and then the mix is fed to a modified *injection moulding machine*, for example, dynamic seals must be fitted to the machine so as to contain the liquid material. Low injection pressure and low clamping forces are desirable characteristics of such machines.

litharge
Also known as lead monoxide. This material has a relative density (RD or SG) of 9.3 to 9.5. A yellow to red powder used as an *accelerator* in *natural rubber* and *styrene-butadiene rubber* compounds. When used as an accelerator then the presence of acidic materials such as *stearic acid* and pine tar are desirable. When used with stearic acid, litharge reduces the tendency to scorch of thiuram-type accelerators. As black lead sulphide is formed on cure, then the material is only used in black compounds. As the presence of litharge stiffens the uncured compound, this material may be used to prevent sagging in open steam cures. Litharge (PbO) is used as *acid acceptor* for chlorine containing rubbers.

lithium salt-PEO complex
See *polyethylene oxide-salt complex*.

lithium stearate
A powdered material with the formula $C_{17}H_{35}COOLi$. Used as a lubricant for rubber compounds and a stabilizer/lubricant for *polyvinyl chloride* (PVC) compounds.

lithographic printing
A *printing technique* in which the ink is transferred to the substrate, for example, a film, from a plane surface. Lithographic

inks are grease-like and the surface of the printing plate is grease repellent except in areas (image areas) where the design has been etched to make it ink receptive. The ink is applied to the substrate either directly of from an intermediate rubber roller: this gives direct lithography or offset lithography. Of these two techniques offset lithography is the most popular as it is kinder to the plate and gives better ink transfer. The inks are based on a mineral oil, vegetable oil and resin mixture and have a viscosity of approximately 50 to 100 poise. Such inks dry by oxidation and penetration but can be by ultraviolet radiation in some cases. Widely used for paper but not so much for plastics.

lithography
Imaging or forming an image. See *photoresist*.

lithopone
A white material which consists of a mixture of zinc sulphide, barium sulphate and small amounts (approximately 1%) of zinc oxide. In the rubber industry it is used as an inert filler and as a white pigment as it has relatively good covering power: it has approximately one quarter the hiding power of *titanium dioxide*. Does not appear to affect the rate of cure or to affect ageing. Stiffens uncured stocks and so its use can be helpful in calendering and/or extrusion. For use in latex, it can be made into emulsions. Lithopone (30% zinc sulphide) has a relative density (RD or SG) of 4·15. Lithopone (40% zinc sulphide) has a relative density (RD or SG) of 4·06. The covering power of this pigment increases with the zinc sulphide content. Used in paints as a substitute for white lead: has a tendency to darken on exposure to strong light.

litre
A unit of capacity in the *metric system of units*. Sometimes abbreviated to l or to L. Formerly defined as the volume of one *kilogram* of distilled water at 4°C and at 760 mm pressure. When specified in this way then the litre is 1,000·028 cubic centimetres. However, in the SI system, it is one cubic *decimeter*. A litre is equivalent to 0·219 98 imperial gallons. See *Système International d'Unité*.

livering action
See *maturing action*.

living hinge
See *integral hinge*.

living polymer
A polymer which is still capable of further polymerization. By for example, anionic polymerization it is possible to produce a polymer with an active centre so that when another monomer is added, further *polymerization* is possible. In this way it is possible to produce block polymers. By using another chemical, it is possible to terminate the living polymer with a desired end-group. See, for example, *styrene-butadiene-styrene*.

l_1
An abbreviation used for *longer length*.

LLDPE
An abbreviation used for *linear low density polyethylene*.

lm
An abbreviation used for *lumen*. See *prefixes - SI*.

LM
An abbreviation for low modulus (black). See *carbon black*.

LMD
An abbreviation for *lead dimethyl dithiocarbamate*.

ln
An abbreviation used for natural logarithm.

l_0
An abbreviation used for *original gauge length*.

load deflection curve
A graph, or curve, of load against deflection produced during, for example, a flexural test. Most machines used for *tensile testing* will provide a load/deflection curve automatically. See *flexural properties*.

load deformation curve
A graph, or curve, of load against extension produced during, for example, a *tensile test*. Most machines used for tensile testing will provide a load/extension curve from which various results can be calculated. If necessary the load/extension curve can be converted to a stress/strain curve by dividing (i) the load values by the original cross-sectional area of the specimen and (ii) the extension values by the original length. The resulting stress/strain curve will be of the same general shape as the load/extension curve.

load/extension curve
See *load deformation curve*.

load ratio
See *flow rate ratio*.

loading
The addition of *filler* to a compound: the amount of filler added to a compound.

loading tray
A multi-impression tray with a sliding bottom used to load a multi-impression, compression mould. Each impression of the tray holds the correct amount of material for each impression of the mould. The tray is located over the *compression mould* and the base moved so that the material falls into the mould impressions.

localized overheating
Excessive heat input confined to a small area or region and often caused by high shear. See *shear heat* and *temperature uniformity*.

locating ring
See *register ring*.

location spigot
See *mould register*.

locking force
The force which opposes mould opening. The amount of force (usually measured in tonnes, tons or kilonewtons) applied to a mould so as to hold it closed during mould filling. On many injection moulding machines, for example, direct ram machines, the *clamping force* and the locking force are the same. However on a *toggle machine*, the locking force is often 10% greater than the clamping force because of the forces generated when the material flows into the injection mould. Resisting flow causes additional stretching of the *tie bars* which gives additional resistance to mould opening.

The locking, and/or clamping, force on a moulding machine may be measured by means of a strain gage assembly attached to a tie bar. Can be attached directly to a *tie bar* (for example, by welding) so that when the clamping force is applied, the change in resistance leads to a measure of the clamping force. Such systems are not often seen on production machines. When measurement is made, it is more usual to use a *dial gauge* assembly.

locking force - automatic control
Term used in injection moulding technology to indicate that a machine is equipped with devices to control the *locking force* to preset values. Automatic control, or regulation, of the clamping force is also called clamp force regulation.

For example, a sensor continuously measures the *tie bar length* and the signal so generated is used to adjust the *effective tie bar length* so as to compensate for changes introduced by, for example, changes in *mould temperatures*. For example, if a hydraulic motor is attached to the anchor platen of a toggle lock machine then, the motor may be used to move the platen assembly, in either direction, along the threaded portion of the tie bar. This means that the effective tie bar length can be altered without the intervention of moulding shop personnel during machine operation.

locking heel
Part of a mould: that part which projects from a *chase bolster* and which holds the splits, side cores and side cavities in the moulding position.

locust bean flour
See *carob seed flour*.

log
An abbreviation used for (common) logarithm.

log-log flow curve
Obtained by plotting the log of the *shear stress* at the capillary wall (τ_w) against the log of the apparent *shear rate* at the capillary wall ($\dot{\gamma}_{w,a}$). Such a plot is intended to give a straight line. See *power law equation*.

logarithmic law of mixtures
A relationship between the value of a property P, for the mixture, to the values of the same property for the components of the mixture: allowance is made for the volume fraction (phi or ϕ) of the components of the mixture. For a binary mixture it is usually of the form $\log P = \phi_1 \log P_1 + \phi_2 \log P_2$. Where ϕ_1 is the volume fraction of the first component, P_1 is the value of the property for the first component, ϕ_2 is the volume fraction of the second component and P_2 is the value of the property for the second component. Useful in predicting the value of properties of a composite. See *law of mixtures*.

logs
A form of *sheet moulding compound (SMC)*. When *injection moulding* SMC, the material can be supplied in sheet form and then rolled into 'logs' before being placed into a *stuffer unit*.

long chain branching
The presence on a main polymer chain of side chains (branches) which are of the same order of length as the main polymer chain. An example of a material which contains such branching is *low density polyethylene*.

long fibre reinforced thermoplastics material
An abbreviation used for this type of material is LFRTP. A fibre reinforced thermoplastics material may be used to produce a thermoplastic composite: that is, a composite based on a thermoplastics material and a fibre such as glass. Such a composite may be produced by drawing (pultruding) the fibres/rovings through a thermoplastics melt and chopping the coated fibres to a predetermined length. The fibre length is then equal to the pellet length.

For glass rovings the fibre content is typically 40 to 70% by weight. For carbon fibre rovings (used to obtain a very high modulus material) the fibre content is 20 to 60% by weight.

The use of such relatively long fibres (about 3 mm) gives, in turn, a relatively long fibre length in a final injection moulded component even though fibre degradation does occur during *injection moulding*. This final long length is important as the performance of a glass reinforced thermoplastic composite is enhanced, particularly the impact strength, if the fibre length is kept long. Components made of such composites are being used where great strength and stiffness are required, for example, for engineering parts for the automotive industry. However, the surface finish possible is not as good as that from the unfilled material. See *short fibre reinforced thermoplastics material*.

long fibre thermoplastic composite
A thermoplastic composite in which most of the fibrous reinforcement is based on long fibres. For example, 50 mm in length - but not less than 12 mm in length. See *fibre length*.

long glass fibre reinforced thermoplastic
A *long fibre thermoplastic composite* in which most of the fibrous reinforcement is based on long glass fibres. For example, 50 mm in length - but not less than 12 mm in length. For injection moulding, the length of the initial fibres is limited to approximately 12 mm. After processing this may drop to 2 mm. See *fibre length*.

long glass fibre thermoplastic composite
A *long fibre thermoplastic composite* in which most of the fibrous reinforcement is based on long glass fibres. For example, 50 mm in length - but not less than 12 mm in length.

long oil alkyd resin
An *alkyd resin* which contains a large amount of oil, for example, more than 70%. Such a material is soluble in aliphatic solvents to give mixtures which rapidly air dry. Used in paints. See *oil length*.

long oil length urethane oil
A *urethane oil* based on pentaerythritol or on a triol, for example, *glycerol*.

long oil resin
See *long oil alkyd resin* and *oil length*.

long term test
See *creep testing*.

long ton
A ton which contains 2,240 pounds. This is equal to 1,016·06 kg or 1·016 metric tonnes. See *UK system of units* and *US Customary Measure*.

long-reach nozzle
See *extended nozzle*.

longer length
An abbreviation used for this term is l_l. The stretched *gauge length* obtained after a *tensile test* has been performed.

longitudinal mixing
See *two-roll mill*.

longs
An undesirable feed form produced during the compounding stage when the cutter, or chopper, does not cut the thermoplastics extrudate to the required short length consistently.

loose threaded core
A hardened steel component which fits into a mould, prior to moulding, so as to form an internal moulded thread.

loss factor
Also known as dielectric loss factor or as, dielectric loss index or as, loss index. The product of the *loss tangent* (tan δ) and the *dielectric constant* (ϵ_r). A measure of the energy loss in a material: it is a measure of how well a material will be heated when placed in a high frequency, alternating electric field. Not all *dielectric materials* have the same ability to being heated by an alternating field: factors involved include the chemical composition of the material, material temperature and the frequency of the field. *Polyvinyl chloride* (PVC), which has a high loss factor, can be high frequency welded; *polyethylene* (PE), which has a low loss factor, cannot be high frequency welded. PVC is classed as a lossy material. A material with a loss factor greater than 0·02 can be consid-

ered as a candidate for *dielectric heating*. However, by pre-heating or modifying some low loss materials, successful dielectric heating may be performed. The *positive temperature coefficient* of the loss factor can result in *thermal runaway* for some materials.

Loss factors for various materials

Material	Frequency	
	10 MHz	3,000 MHz
Melamine-formaldehyde	0·23	0·23
Nylon 66	0·09	0·04
Phenol-formaldehyde	0·18	0·22
Paraffin wax	0·000 45	0·000 45
Polyvinyl chloride (PPVC) 40% plasticizer	0·4	0·1
Polyvinyl chloride (UPVC)	0·03	0·02
Polyethylene (PE)	0·000 4	0·001
Polystyrene (PS)	0·000 5	0·000 5
Polytetrafluoroethylene (PTFE)	0·000 3	0·000 3
Water	100·0	18·0

loss tangent
Also known as the dielectric loss tangent or as, damping factor or as, dissipation factor or as, the tangent to the loss angle. An abbreviation used for this term is tan δ or D. For a *dielectric material* submitted to a sinusoidal voltage, it is the ratio of the absorbed active power to the absolute value of reactive power. It is a measure of the resistance of the molecules of a material when placed in an alternating electric field: the ratio of the energy dissipated to the energy stored for each cycle. Often used in place of sin δ (power factor) as it is easier to measure and, in practice, there is often little difference for low loss dielectrics.

lossy material
A material with a high *loss factor*.

lost blowing space
Part of the *blow moulding* which is removed before filling of the product. Used in *aseptic blow moulding* where the area above the neck is sealed after blowing so as to stop the ingress of bacteria.

lost dome
Part of a blow moulded product, which is dome shaped, and which is removed after blowing. See *in mould cutting*.

low aspect ratio tyre
A tyre with a low *aspect ratio*: a wide tyre.

low cis BR
An abbreviation used for low cis-polybutadiene rubber. See *polybutadiene rubber*.

low cis-polybutadiene rubber
An abbreviation used for this material is low cis BR. See *polybutadiene rubber*.

low compression screw
See *low work screw*.

low compression, zero metering screw
A *low work screw*.

low work screw
A low compression, zero metering screw. A type of *screw* which imparts very little shear to the material being processed as the *compression ratio* of the screw is of the order of 1:1 and the screw channel is deep. See *fibre length*.

low density polyethene
See *low density polyethylene*.

low density polyethylene
An abbreviation used for this type of material is LDPE; PE-LD is also used - from polyethylene-low density. Also known as polythene, low density polythene, polyethene, poly(methylene), and high pressure polyethylene.

When ethylene is heated at high pressures (up to 3000 bar) and temperatures of 100 to 300°C/212°F to 572°F, in the presence of a free radical initiator, the plastic known as low density polyethylene is produced. This is a semi-crystalline, thermoplastics material and as such is not available as transparent mouldings. The natural colour of the material is a milky white and so a wide colour range is possible. Because of chain branching the crystallinity level is low and the material has a relatively low density, for example, 0·92 g.cm^{-3}. By varying the polymerization conditions it is possible to produce commercial materials with densities over the range 0·915 to 0·94 g.cm^{-3}. Both long chain and short chain branching is present and, because long chain, branched molecules are compact, the material flows relatively easily. LDPE has a soft, wax-like feel and is easily scratched with a knife or fingernail.

This material is tough but only has moderate tensile strength and suffers from creep; for design purposes, 5% strain may be used as the operating limit. The impact resistance, chemical resistance and electrical insulation properties are excellent. The water vapour permeability of this type of PE, as with all types of PE, is low. PE is permeable to gases and vapours (i.e. it lets them through) but *linear low density polyethylene (LLDPE)* and *high density polyethylene (HDPE)* are less permeable to gases and vapours than LDPE. Permeability for organic vapours is least for alcohols and then increases in the order shown; from acids to aldehydes and ketones, esters, ethers, hydrocarbons and halogenated hydrocarbons. (Permeability decreases with density).

LDPE is an easy flow material as rated by *melt flow rate (MFR)*: a broad range of MFR values are encountered. Low MFR materials exhibit better environmental stress cracking resistance (ESC), solvent resistance and higher impact strength. More pseudoplastic than LLDPE. because it has a broader molecular weight distribution. If a LDPE plastic of a certain melt flow rate (index), is compared with an LLDPE. plastic of the same melt flow index, then it will be found that the linear polymer will be of higher viscosity at processing shear rates.

The *shrinkage* is of the order of 0·02 to 0·05 mm/mm (i.e. 2 to 5%) when the density is 0·910 to 0·925 g.cm^{-3}. Because of LDPE's wide shrinkage range it is extremely difficult to accurately predict the necessary shrinkage value in order to achieve the desired dimensions for a particular injection moulded component. Because of this a sample cavity is manufactured and mouldings produced using typical processing conditions (i.e. temperatures, pressures and cycle time). Shrinkage values are obtained from these mouldings which are then used for the sizing of the cavity and core dimensions of the production mould.

Below 60°C/140°F, PE is insoluble in all organic solvents but it does swell in aliphatic, aromatic and chlorinated hydrocarbons: the lower the density the more it swells. At a temperature of approximately 55°C/131°F, LDPE is soluble in hydrocarbons and chlorinated hydrocarbons, for example xylene and trichloroethylene. Relatively unaffected by polar solvents, for example alcohols, phenols, esters and ketones; vegetable oils, water, alkalis, most concentrated acids at room temperature and ozone (in absence of UV light). Very low water absorption even after long immersion times, for example after one year at 20°C/68°F, the increase in weight may be less than 0·2%. The addition of carbon black, used to improve weathering, will increase the water absorption. Absorption of other liquids, for example acetone and benzene, will be greater for LDPE than for HDPE (best chemical resistance is found with HDPE and cross-linked PE).

LDPE is not resistant to fuming nitric acid and also fuming sulphuric acid; slowly attacked by halogens and chlorinating agents such as chlorosulphonic acid and phosgene. LDPE is associated with *environmental stress cracking (ESC)*; such ESC is also associated with detergents or silicone fluids although there are many other ESC agents, for example, chloroform, xylene and paraffin. The light, or UV resistance is poor; the cheapest way to improve this is by incorporation of carbon black. Has limited resistance to oxygen at elevated temperatures, antioxidants are used for protection; unless so protected the electrical properties will suffer.

This material may not be joined to itself using solvents as there is no solvent at room temperature. Because of its inert, 'non-stick' surface it also cannot be very successfully bonded using adhesives; limited success with contact or hot melt adhesives. If the surface is made polar, for example by using a flame or an electrical discharge, then this material may be bonded to metals using epoxies or nitrile-phenolic adhesives; such treatments also improve printability. Injection mouldings are commonly welded using techniques such as hot plate or hot shoe: film is commonly welded by *impulse welding*.

Because of this material's ease of moulding and low cost it has become established as a general purpose *injection moulding* material. Most of the applications do not utilize the excellent electrical insulation properties, nor the water resistance of the material. LDPE is used for example for caps and lids for containers, bottle closures and tear-off closures, bowls, beakers, pipe couplings, pots, linen baskets, bins etc. Close tolerances are difficult to mould in this material and they are difficult to hold in service because of the high coefficient of thermal expansion and the tendency to creep. The softness of this material can allow attack by insects or animals. Very widely used as a film material - the film is often produced by the *tubular film process* and welded into bags of the required size. Carbon black can reinforce LDPE and also improves light resistance.

LDPE may be cross-linked using high energy radiation or by the incorporation of peroxides.

low efficiency phthalates
Plasticizers such as dinonyl phthalate (DNP) and di-isodecyl phthalate are sometimes referred to as low efficiency phthalate plasticizers because they have a lower plasticizing efficiency than other general purpose plasticizers (materials such as DAP, DIOP and DOP). This means that more of the low efficiency phthalate *plasticizer* can be incorporated to give compounds of the same hardness as say DOP. If used, the difficulty of incorporating a larger quantity of lower solvating plasticizer should not be forgotten. DNP is also useful if low viscosity plastisols are required. See *efficiency proportion*.

low efficiency plasticizer
See *low efficiency phthalates*.

low fogging leathercloth
Leathercloth which makes little or no contribution to *automotive fogging*.

low fogging stabilizer
A *stabilizer* used to reduce *automotive fogging*. Some liquid barium/zinc stabilizers are claimed to satisfy fogging temperatures as high as 90°C.

low melting point metallic alloy
An *alloy* with a low melting point - below one that would be expected for metals. For example, Rose's metal has a melting point of 94°C: *Wood's metal* has a melting point of 71°C. Such low melting point metals are used, for example, to make fusible cores which are used to make hollow sections in injection moulded products.

Low melting point alloys are used for the construction of a *prototype mould*. Such alloys are often based on zinc and are more usually known by trade names/trademarks, for example, Zamak, Ayem or Kirksite. See *zinc-based alloy*.

low molecular weight
An abbreviation used for this term is LMW.

low nitrile NBR
A *nitrile rubber* with a low acrylonitrile content, for example, 20%.

low nitrogen natural rubber
See *deproteinated natural rubber*.

low pressure
An abbreviation used for this term is LP.

low pressure moulding
An abbreviation used for this term is LPM. A method of moulding or laminating in which the pressure applied is, for example, 1400 kPA (200 psi) or less. A *glass reinforced plastics* moulding process uses a *low pressure moulding compound*. The sheet is cut to the shape required and loaded into a heated mould (for example, at 120°C) which is mounted in a press. As the shaping pressures used are comparatively low (for example, <2 MPa), *soft tooling* may be used. Competes against, for example, *SMC moulding* as it can give very good surface finish at low cost: in such a case, matched metal moulds or nickel shell moulds would be used and the formulation would contain a low *shrinkage* additive.

low pressure moulding compound
An abbreviation used for this material is LPMC. A reinforced moulding compound which incorporates all of the materials required for the finished part, for example, resin, glass, filler and catalyst: this compound is used in sheet form. Such a compound has long shelf life and a low moulding viscosity: a wide range of formulations can be used. Competes against, for example, *SMC moulding*. See *low pressure moulding*.

low pressure moulding processes
Moulding processes which employ less than a few atmospheres shaping pressure. Included are *casting* and *powder moulding*. As there is little, or no shaping pressure, the mould costs can be extremely low. This in turn means that large components may be economically produced.

low pressure polyethylene
See *high density polyethylene*.

low pressure steam
Steam at a pressure of up to approximately 4 atmospheres. Used mainly in the rubber industry, for example, for the warming of *two-roll mills* and for the heating of extruders. See *steam heating*.

low profile grade
A type of *polyester moulding compound* which has been made to give low *shrinkage* values on moulding. A low profile grade may show slightly higher water absorption than a general purpose grade of *dough moulding compound*. Such compounds may be referred to as low profile or as, zero shrink moulding compounds.

low profile sheet moulding compound
A *sheet moulding compound*, based on a resin system plus a thermoplastic additive, which gives very little, if any, *shrinkage* on cure. This addition, to an SMC mix, reduces shrinkage and improves the surface finish. The shrinkage for general purpose SMC (GP SMC), which contains no shrinkage control system, may be 0.4% (0.004 mm/mm). If materials like *polystyrene (PS)*, or a *styrene butadiene rubber*, are added, it is possible to get *low shrinkage SMC* where the shrinkage is 0.1–0.3% (0.001–0.003 mm/mm). Adding thermoplastics such as *polyvinyl acetate* or *polymethyl methacrylate* gives even

lower shrinkage, for example, 0·02–0%. Such LP SMC compounds are also known as low profile or zero shrink sheet moulding compound.

If moulding shrinkage is reduced, then the moulding stays in contact with the highly-polished mould and products of very high gloss are obtained. This is important because of the markets that *polyester moulding compound (PMC)* is penetrating - the automotive market and the domestic appliance market.

low shear rate flow test
See *flow test*.

low shrink resin
See *low shrinkage resin* and *low profile sheet moulding compound*.

low shrinkage resin
An abbreviation used for this type of material is LS resin. A resin system used to give a *low shrinkage sheet moulding compound*.

low shrinkage sheet moulding compound
A *sheet moulding compound*, based on a resin system plus a thermoplastic additive, which gives little shrinkage on cure. This addition, to an SMC mix, reduces shrinkage and improves the surface finish. The shrinkage for general purpose SMC (GP SMC), which contains no shrinkage control system, may be 0·4% (0·004 mm/mm). If materials like polystyrene (PS) or styrene butadiene rubbers are added, it is possible to get low shrinkage SMC where the shrinkage is 0·1–0·3% (0·001–0·003 mm/mm). See *low profile sheet moulding compound*.

low styrene emission polyester resin
An abbreviation used for this type of *unsaturated polyester resin* (UP) is LSE resin. Such LSE resins are used in the *hand lay-up (HLU)* and spray-up moulding processes in an effort to reduce styrene emissions into the atmosphere: styrene is considered harmful. LSE resins contain, for example, a film forming additive which initially is homogeneously dissolved/dispersed in the liquid resin system. In use, the film forming additive rapidly forms a (styrene) vapour tight film on the liquid surface. During the HLU process, the styrene level above the mould could be reduced from say, 75 ppm when a conventional UP resin is used, to 25 ppm when a LSE resin is used.

low temperature curing tooling prepreg
See *low temperature moulding epoxy prepreg systems*.

low temperature flexibility
A test used to compare plasticizers. Test used to determine the suitability of a *plasticizer* or compound for use at low temperatures. A brittleness temperature test is usually employed. If a series of similar hardness compounds were prepared from dibutyl phthalate (DBP), dioctyl phthalate (DOP) and di-iso-octyl adipate (DIOA) then the *cold flex temperature* could be DBP −18°C, DOP −26°C and DIOA −42°C.

low temperature moulding
An abbreviation used for this term is LTM. The term is often used in connection with *prepreg moulding*, for example, with *low temperature moulding epoxy prepreg systems* and with, tooling prepreg systems.

low temperature moulding epoxy prepreg systems
Also known as LTM epoxy prepreg systems or, as frozen prepreg systems. This type of system is used to manufacture mould tools from *prepreg materials*. A wide variety of reinforcements can be used but bidirectional graphite (*carbon*) or *E-glass twill-weave fabrics* are generally used in tool construction. Highly reactive epoxide resins with low temperature curing characteristics are used in conjunction with a *vacuum bag moulding technique*. A ten ply construction would give a total laminate thickness of approximately 6 mm/0·25 in and with approximately 55% fibre (by volume) content.

low temperature plasticizer
A *plasticizer* which gives flexibility at low temperatures to a *polyvinyl chloride* (PVC) compound, for example, an *adipate plasticizer*. Industrial use of materials such as *dibutyl sebacate* is now reduced because of the availability of cheaper materials, for example, *nylonates*. Where long term retention of properties, after exposure to high and low temperatures, is a requirement then phthalates based on a linear alcohol will perform better than a *phthalate* based on a branched alcohol. For example, the linear phthalate, *octyl decyl phthalate* will retain properties better than branched chain alcohols such as di-iso-octyl phthalate (DIOP) and di-iso-decyl phthalate (DIDP). Low temperature plasticizers effect the electrical resistance characteristics of PVC compounds the least of all the *plasticizers*.

low temperature silicone rubber
A *silicone rubber* which retains flexibilty at low temperatures. PMQ and PVMQ-types are classed as silicone rubbers with excellent low temperature performance.

low viscosity moulding compound
A *polyester moulding compound*.

low viscosity rubber
An abbreviation used for this material is LV rubber. A form of *natural rubber*. Natural rubber which has a controlled low viscosity, stabilized with hydroxylamine hydrochloride.

low work screw plasticator
A type of *extruder* which is used to produce *preforms* from a thermoplastics material: these preforms may then be formed by *compression moulding*. See *thermoplastic bulk moulding compound*.

low-profile tyre
An abbreviation used for this term is LP tyre. A wide *tyre*: a tyre with an *aspect ratio* of approximately 70 to 77%. A comparatively expensive radial-ply tyre, compared to a high profile tyre, but one which gives better high speed performance, roadholding, load carrying and longer life. A low-profile cross-ply tyre gives better control at the limit of adhesion, that is, at large *slip angles*: such tyres are used, for example, for racing cars.

lower action limit
An abbreviation used for this term is LAL: a limit line which corresponds to probability points on the normal distribution curve of, say, 99·9%.

lower solvating plasticizer
A plasticizer which acts as a high temperature solvent more slowly than a *general purpose plasticizer*. In general, solvation rates are found to increase with plasticizer polarity and to decrease with the molecular weight of the *plasticizer*.

lower warning limit
An abbreviation used for this term is LWL: a limit line which corresponds to probability points on the normal distribution curve of, say, 95%.

LP
An abbreviation used for low profile (see *low profile resin*), liquid polymer (see *depolymerized rubber*) and low pressure. For example:

    LP resin = *low profile resin*.
    LP tyre = low-profile tyre;
    LPM = low pressure moulding; and,
    LPMC = *low pressure moulding compound*.

LPD
An abbreviation for *lead pentamethylene dithiocarbamate*.

LRTM
An abbreviation used for low (pressure) resin transfer moulding. See *resin transfer moulding*.

LS
An abbreviation used for low structure (black). See *carbon black*.

LS rubber
An abbreviation used for latex *sprayed rubber*.

LSE resin
An abbreviation used for *low styrene emission resin*.

LSR
An abbreviation used for *liquid silicone rubber*.

LT
An abbreviation used for low temperature. For example:
- **LTM** = low temperature moulding. See *low temperature moulding epoxy prepreg systems*;
- **LTP** = low temperature polymerization. See *styrene-butadiene rubber*.

lubricant
An additive which is added to a polymer in order to stop the polymer sticking to the processing equipment or, to ease the flow of the polymer compound. Stearic acid is an example of a lubricant which is added to, for example, *polyvinyl chloride* (PVC) in order to prevent the compound sticking to the processing equipment. Such a lubricant is an example of an *external lubricant*. Some lubricant materials are classed as *internal lubricants*.

The gloss, or surface finish, of many products is improved by the use of lubricants as the lubricated compound can be made to contact a highly finished metal surface (which imparts the gloss) more easily and it will then come away from that polished surface smoothly and easily. That is, the highly polished surface will be retained.

Easy flow grades of polymers are available which permit the production of thin-walled, high-gloss products. For example, in the case of *polystyrene* (PS), such grades are based on a comparatively low molecular weight polymer which contains an internal lubricant such as butyl stearate or liquid paraffin. Small, regularly-sized pellets are produced which are then coated with an external lubricant such as zinc stearate or a wax: such pellets feed into an *injection moulding machine* quickly and melt easily.

With some plastics materials, lubricants are added in order to reduce friction in use. For example, the addition of 2% graphite to a compound before moulding, will reduce the friction between two gears moulded from that compound. Silicone oils are used for the same purpose: it should be noted that silicone oil containing compounds are not allowed in some establishments because they can interfere with the working of electronic components. However, the use of such oils gives high gloss products which maintain this attractive appearance in use: this is because components moulded from such modified materials, for example, tape-cassette boxes, are often shipped in bulk and the presence of the oil stops the injection mouldings damaging each other, by rubbing, in transit.

lubricant bloom
See *blooming*.

lubricating size
An additive used in conjunction with *fibres* (for example, with *glass fibre*) and applied so as to minimise fibre breakage and to improve resin to glass adhesion. Such fibres are used in composites or laminates and the size is designed to suit the polymer of the future composite, The strand, to which the fibre is applies, may be converted into *chopped strand mat (CSM)*, rovings or chopped strands.

lumen
An SI derived unit which has a special symbol, that is lm. It is the SI derived unit of luminous flux and is the light emitted in a unit solid angle of one steradian by a point source having a uniform intensity of one candela. See *Système International d'Unité*.

luminescence
The emission of light from a material from any cause other than that of high temperatures. *Fluorescence* and *phosphorescence* are examples of luminescence.

luminescent effect
The effect produced when a material capable of *luminescence* is added to a polymer. Bright, vivid colours can be produced in polymer compounds by the use of *luminescent pigments* particularly when the compound is transparent.

luminescent pigment
Bright, vivid colours can be produced in polymer compounds by the use of luminescent pigments, for example, zinc sulphide (green) and zinc cadmium sulphide (orange and yellow). See *luminescent effect*.

luminous flux
The luminous flux through an area is the amount of light which passes through that area in one second. The derived SI unit is the *lumen*.

luminous intensity
The amount of light emitted in one second in unit solid angle, in a given direction, by a point source. The SI unit is the *candela*.

luminous transmittance
The ratio of the luminous flux transmitted by a body to the flux incident upon it. Parallel definitions apply to spectral and radiant transmittance. See *light transmittance*.

lump
The increased thickness produced at each end of a filament wound structure: this is caused by the need to dwell the carriage so as to prevent the filaments being pulled back down the moulding by the reversed movement. See *filament winding* and *spin winding*.

lumpiness
An extrusion defect. A roughness in the bore, or inside diameter, of a pipe or parison; usually associated with, for example, poor mixing of material along the screw in *unplasticized polyvinyl chloride* (UPVC) extrusion.

lumps
A form of *natural rubber*. Obtained from, for example, pre-coagulated natural rubber latex: may be used as *brown crepe*.

Lüpke pendulum
An instrument used to measure *resilience*.

Luvitherm process
A process developed to produce *unplasticized polyvinyl chloride* sheet using an *L calender*.

lux
An SI derived unit which has a special symbol, that is lx. It is the SI derived unit of illumination. One lux is an illumination of one lumen per square metre. See *Système International d'Unité*.

LV rubber
An abbreviation used for *low viscosity rubber*.

LVN
An abbreviation used for *limiting viscosity number*.

LWL
An abbreviation used for *lower warning limit*.

lx
An abbreviation used for *lux*. See *prefixes - SI*.

M

m
This letter is used as an abbreviation for:
  mass. See *bulk density measurement*;
  metre. For example, m/s = metres per second;
    m/s^2 = metres per second squared;
    m^2 = square metre;
    m^3 = cubic metre;
    m^3/s = cubic metres per second; and,
  milli. See *prefixes - SI*.

M
This letter is used as an abbreviation for:
  filler - for example, if a plastics material contains a filler then the letter may be used to show the presence of mineral or, metal/mass polymer;
  maleate;
  mat;
  medium;
  mega or 10^6. That is, a unit multiplied by 1,000,000. See *prefixes - SI*;
  melamine;
  mellitate;
  methyl group - for example, in silicone chemistry/technology;
  methyl;
  methoxy;
  modulus;
  molecular;
  molecular weight;
  monomer - for example, vinyl chloride monomer;
  Mooney viscosity value or, *Mooney viscosity*; and,
  moulding

$\overline{M}$
Sometimes called $\overline{M}$ bar. An abbreviation used for *molecular weight average*. $\overline{M}_n$ is an abbreviation used for number average *molecular weight*. $\overline{M}_w$ is an abbreviation used for weight average *molecular weight*. M bar is sometimes also used for the average value in statistical expressions or calculations.

$\overline{\overline{M}}$
Sometimes called M double bar. This abbreviation is sometimes used for the average of the averages in statistical expressions or calculations. μ is also used for the average of the averages in statistical expressions or calculations: one reason is that it is easier to do on word processors.

M and S tyres
An abbreviation used for *mud* and *snow tyres*.

M bar
See $\overline{M}$.

M double bar
See $\overline{M}$ and $\overline{\overline{M}}$.

m-diaminobenzene
See *m-phenylene diamine*.

m-dihydroxy benzene
See *resorcinol*.

m-methylstyrene
See *vinyltoluene*.

m-phenylene diamine
Also known as m-diaminobenzene. This white solid material has a melting point of 63°C, a boiling point of 285°C and a relative density (RD) of 1·14. An adhesion promoter: used to improve the adhesion of rubber to tyre cords. Also used as a curing agent for *epoxide resins*: the cured materials have good chemical and heat resistance.

m-phenylenebismaleimide
An abbreviation used for this material is HVA-2. Used to achieve *dynamic cross-linking* of *natural rubber* in natural rubber/polypropylene (NR/PP) blends: level of use approximately 0·5%. See *natural rubber/polyolefin blend*.

m-phenylenediamine
See *m-phenylene diamine*.

m-phthalic acid
There are three isomers of *phthalic acid*. These are, o-phthalic acid, m-phthalic acid and p-phthalic acid.

m-xylenediamine
An abbreviation used for this type of material is MXDA. A monomer for *polyaryl amide*.

M_{50} value
This figure is the mass, in grams, of the *dart* that would be expected break 50 per cent of a large number of specimens. See *falling weight impact strength (test)*.

M_{300}
An abbreviation used for *300 per cent modulus*.

mA
An abbreviation used for milliampere.

MA
An abbreviation used for *maleic anhydride*.

MABS
An abbreviation used for *methyl methacrylate-acrylonitrile-butadiene-styrene copolymer*.

machine capability
A measure of how accurately a machine, as part of a production unit, can produce components when there are no *assignable variations* (see *process capability*). It is common practice to accept that 99·99% of production will lie within a tolerance (t) of 8σ - where σ is the standard deviation.

machine direction
Term usually used in *calendering* and *extrusion* to mark a production direction. The lengthwise or longitudinal direction of an extrudate. The direction in which the product emerges from the machine. Often denoted on the product with an arrow whose head points in the direction of product removal (haul off) from the machine. See *transverse direction*. The component will have different properties, for example, strength properties, in different directions. See *orientation*.

machine guard
A safety device which may be fixed, or bolted, over a hazardous area or, it may be moveable or openable. Training and a responsible attitude towards safe working are very important safety considerations as undue reliance should not be put upon guards. See *fixed safety guard, movable interlocked safety guards, safety bars, safety stops* and *plugged braking*.

machine operation
These comments apply to extruders, blow moulding machines and injection moulding machines.

INITIAL OPERATION AND PURGING. Check that the machine is at the set temperatures. Reduce all speeds (for example, the screw speed and the injection speed) to low values and put a small amount of material/resin in the hopper. Replace the hopper lid and run a few pounds of material/resin out of the nozzle/die (see *air shot*). Check the *melt temperature* with a melt probe, check its general appearance and then dispose of this hot, sticky material in a safe way (by, for example, putting it in a bucket of cold water). Adjust the machine settings until the required melt quality is obtained. Then, if the material is feeding well, and the melt looks satisfactory, fill the *hopper* to the specified level. Check that the throat gate (the hopper slide) is open as rotation of the screw in an empty barrel may cause damage to the *screw* and to the *barrel* liner. Check that the monitoring equipment is functioning and when material/resin starts to extrude from the nozzle/die, turn on the screw cooling if fitted.

CHANGING CONDITIONS. Once operational, the appearance, colour, weight (or selected dimensions) of the product are checked and compared against preset, or specified, values. Speeds and other settings, are changed until the product conforms to standard. It is important to realize that any changes to production conditions must be well thought out in advance and must be made gradually as, for example, an increase in screw speed, on an extruder or injection moulding machine, causes not only an increase in melt production but also, an increase in melt temperature. Changes must be made one at a time, the machine must be allowed to equilibrate and the effect of that change noted. See *quality control*.

OBJECT OF PRODUCTION. The object of production is to produce components/product of the required quality, and quantity, at a specified cost within a specified time and when required.

machine-made hose
A *mandrel-made hose* which has been made by machine.

machined mould
A mould made by machining metal and usually assembled from individual components. Most moulds are machined from steel using standard toolroom equipment such as drills, lathes, milling machines and grinders. To reduce machining, *standard mould sets* are now available - by specifying their use, time and money may be saved.

macrocyclic
A polymer in which many of the polymer chains exist as large rings: this gives rise to few chain ends which in turn is thought to account for the high tensile strength of even low molecular weight material, for example, of *polyoctenamer rubber*.

macromolecular
Meaning of high molecular weight, for example, greater than approximately 10,000. See *macromolecule*.

macromolecule
A material of high molecular weight, that is a large molecule. Does not necessarily mean that the large molecule is made up of large numbers of the same small, repeating units (like conventional synthetic polymers, such as *polystyrene*). Biological macromolecules contain many different types of repeat unit. See *high polymer*.

macroporous ion exchange resin
A type of *ion exchange resin* with a porous, multi-channelled type of structure. Also referred to as a macroreticular ion exchange resin.

macroreticular ion exchange resin
See *macroporous ion exchange resin*.

Maddock mixing section
A *fluted mixer*: a *dispersive mixing section*. The fluted barrier section is a cylindrical shearing section, of approximately 2 to 3 D in length, which forms the screw tip. It has inlet passages from which melt can only reach the outlet passages via the land of the mixing section. That is, the melt must pass over a narrow gap before it can escape and this gives *dispersive mixing*.

magnesia
See *magnesium oxide*.

magnesia alba
See *magnesium carbonate*.

magnesia usta
See *magnesium oxide*.

magnesian limestone
A *limestone* which contains less than 45% magnesium carbonate. See *dolomite*.

magnesite
See *magnesium carbonate*.

magnesium
This element (Mg) is an *alkaline earth metal*. It is a lustrous white metal with a melting point of 650°C, a boiling point of 1,100°C and a relative density of 1·74. It is a good conductor of heat and electricity, burns easily and is easily machined. Used to make alloys, for example, with *aluminium*, which are strong and light: also used to make sacrificial electrodes for steel structures so as to reduce corrosion. Compounds of magnesium are used as, for example, *fillers* and as flame retardants/smoke suppressants.

Magnesium-zinc complexes may be used as *flame retardants*, for example, for *polyvinyl chloride*. Such flame retardants may be based on, for example, *magnesium hydroxide* or *magnesium carbonate* and sometimes used in conjunction with a *phosphate ester plasticizer*.

magnesium carbonate
Occurs naturally as magnesite ($MgCO_3$) and also as dolomite ($MgCO_3CaCO_3$). Magnesite occurs in the anhydrous, trihydrate and pentahydrate forms - see also *basic magnesium carbonate*. Such naturally occurring materials are generally too coarse for use as fillers although dolomite is used in *sheet moulding compounds*. For use as fillers, in the rubber industry, synthetic materials are preferred - particularly *light magnesium carbonate*. Made by a precipitation process to give a white fluffy powder with little odour and reputed to be nontoxic. Magnesium carbonate (light) has a relative density (RD or SG) of approximately 2·19: may be represented as $4MgOCO_3H_2O$. Used in the rubber industry as a white reinforcing filler, for example, for the manufacture of shoe soles and heels. Stiffens uncured stocks and reduces collapse in open steam cures. As it has a similar refractive index to natural rubber, it is possible to make translucent compounds from magnesium carbonate filled materials: can accelerate rate of cure. Magnesium carbonate is used as a *flame retardant*, for example, for *polyvinyl chloride*.

magnesium hydroxide
Also known as magnesia. A white, crystalline material which is used as a *flame retardant*, for example, for *polyvinyl chloride*.

magnesium oxide
Also known as magnesia or as, magnesia usta or as, light magnesium oxide or as, light calcined magnesium oxide. May be represented as MgO. This material has a relative density (RD or SG) of approximately 3·4 to 3·65. Light calcined magnesium oxide is a white, odourless powder with a low bulk density and a fine particle size. Light calcined magnesium oxide is sometimes used as an *accelerator* and *activator* in the rubber industry. Helps to offset the retarding effect of *factice*: can have a slight reinforcing effect. Used as a vulcanizing agent for *polychloroprene* and reduces *scorch tendency*: absorbs hydrochloric acid formed during vulcanization and reduces the tendency of polychloroprene stocks to set up on

storing. That is, it is an *acid acceptor* for chlorine containing rubbers.

magnesium silicate
The natural material is known as *talc*. Synthetic magnesium silicate is also available and this fine white material (particle size of approximately 5 μm) appears to function as a filler/lubricant in *unplasticized polyvinyl chloride* (UPVC).

magnesium stearate
This white powder material has a relative density (RD or SG) of approximately 1·04 and a melting point of 125 to 130°C. May be represented as $Mg(C_{18}H_{35}O_2)_2$. Used as a release agent/dusting agent for rubbers as it prevents adhesion of uncured compounds. An *acid acceptor* for chlorine containing rubbers.

magnesium usta
See *magnesium oxide*.

magnesium/zinc stabilizer
A polymer stabilizer based on the metals magnesium and zinc which is used as a *heat stabilizer*. Compounds of such metals are used as, for example, heat stabilizers for *polyvinyl chloride* (PVC) because of the relatively good stabilizing performance of such a *mixed metal stabilizer*. Such a stabilizer is often combined with a *metal-free organic stabilizer*.

magnetic flux
The strength of a magnetic field through an area. The unit in the cgs system is the maxwell: in the SI system it is the *weber*.

magnetic flux density
Also called magnetic induction. The *magnetic flux* passing through unit area of a magnetic field normal to the magnetic force.

magnetic material
Polymeric compounds with magnetic properties may be produced by using a filler which has magnetic properties, for example, barium ferrite. Such compounds are used because the compounds may be moulded or extruded into complex shapes. When the magnetic compound is based on a thermoplastics material then the finished magnetic component may be produced by *injection moulding*: in such a case, the magnetic properties may be enhanced by using a mould which has magnetic coils built into the mould. Immediately after mould filling, the magnetic coils are energised and this gives a more strongly magnetised component.

magnetron
Also called a magnetron electronic valve. A thermionic valve used for generating *microwave power*: such a valve requires a high tension DC electric supply as well as low voltage heaters. A 915 MHz magnetron will give a maximum power output of approximately 30 kW whereas a 2,450 MHz magnetron has an output power of between 0·5 and 5 kW.

main head
Also called the crown. Part of a moulding press: the upper part of a vertical press. On, for example, an *upstroke press* the main head carries the fixed platen.

main line pressure
Also known as pump delivery pressure. This is the pressure in the main line, or pipe, from the pump.

malachite green
A green pigment which is based on triphenylmethane dye. Malachite green, and its sulphonated derivative called brilliant green, are stable to sunlight and to chemicals.

Malayan Rubber Producers' Research Association
An abbreviation used for this organization is MRPRA. See *natural rubber*.

male forming
Forming, or *thermoforming*, which is performed using a *male mould*. See *drape forming*.

male half
That part of the mould which has a positive shape and which fits inside the female half. That part of the mould which forms the inside of a container.

male mould
A mould over which sheet is drawn in *thermoforming*. Air is removed from between the heated sheet and the mould via small holes or slots machined into, and around, the mould. The male mould initially acts as a plug, when moved relative to the sheet, and holds the heated sheet against the mould. This gives a *wall thickness distribution* which is different to that obtained from a *female mould*. This difference is exaggerated by the fact that more sheet is available to make the forming: in female forming it is only the sheet over the cavity which is available to the component. See *drape forming* and *hand lay-up*.

male plug
See *plug* and *hand lay-up*.

maleic acid
Also known as butenedioic acid or as, cis-1,2-ethylene dicarbonic acid. May be represented as HOOCCH:CHCOOH. A white solid material with a melting point of 130°C and a relative density of 1·59. A retarder for *chloroprene rubbers*: may also be used to make synthetic resins, for example, *unsaturated polyester resins* although maleic anhydride is preferred. At 120°C this material converts to fumaric acid - the more stable trans form. On strong heating (to approximately 450°C) maleic acid eliminates water to form *maleic anhydride*.

maleic anhydride
An abbreviation used for this material is MA. Also known as maleic acid anhydride. This material has a relative density (RD or SG) of 1·48, a melting point of 53°C and a boiling point of 202°C. Formed by heating maleic acid to approximately 450°C. Used to make synthetic resins, for example, *unsaturated polyester resins* as maleic anhydride has a lower melting point than maleic acid and gives less water. Isomerizes to the trans form during esterification to give fumarate units. A retarder for *natural* and *synthetic rubbers*. The incorporation of maleic anhydride (MA) or acid into a *vinyl chloride* or *vinyl acetate polymer* improves the adhesive properties of the copolymer or of the terpolymer: such materials are therefore used in adhesive applications.

malic acid
Also known as 2-hydroxybutanedioic acid or as, L-malic acid. May be represented as $HOOCCH(OH)CH_2COOH$. A white solid material with a melting point of 99°C) and a relative density of 1·60. A retarder for *natural rubber*.

MAN
An abbreviation used for *methacrylonitrile*.

man-made fibres
See *fibres*.

mandrel
A spindle or tube, often tapered, around which, for example, *glass rovings* and *unsaturated polyester resin* may be wrapped to make tube by *filament winding*. Sometimes spelt mandril. A term used to describe a sizing element either, in a die or, in a sizer which controls one dimension of the extrudate; this dimension is usually the inside diameter.

mandrel-made hose
A *rubber hose* built and then *vulcanized* on a *mandrel*.

manganese
A *transition element* (Mn) which is a hard, brittle white metal with a high melting point (1,244°C) and boiling point (2,040°C): the relative density is 7·2. Added to steel so as to stop the formation of blow holes and to increase the hardness of the *steel*. Because of the degrading effect of this metal on *natural rubber*, the maximum allowable amount is of the order of 10 ppm.

manifold
A tube machined within a die whose function is to even out the flow of material across the width of an *extrusion die*. Manifold is a term used in *hydraulics* and which refers to a type of fluid passage, or connection, which has several input ports and several output ports.

manifold die
A *sheet extrusion die* which contains a *manifold*.

manual control
A system of control which requires the operator to actuate the various stages of a control sequence.

manual flow rate testing method.
See *flow rate*.

manual over-ride
A system of control in which the operator manually actuates an automatic control device.

MAR
An abbreviation used for *maraging steel*.

maraging steel
An abbreviation used for this type of material is MAR. As delivered, this high nickel content steel (about 18% nickel) is relatively soft and easily machined: after machining the components can be hardened by heating for 3 h at approximately 490°C. This gives a very high strength steel with a hardness in the region of $R_C 52$ which is used for intricate designs.

marble
A hard, compact crystalline variety of *limestone*.

marble-effect moulding
Injection moulding of components with a surface which resembles marble. May be achieved using a *multiple injection unit machine*. See *tortoiseshell effects*.

marbled effects
See *tortoiseshell effects*.

Mark-Houwink constants
The two constants denoted as K and a in the *Mark-Houwink equation*. The Mark-Houwink constants are determined from a plot of the limiting viscosity number $[\eta]$ and the polymer molecular weight (M): a polymer of known molecular weight is used for this purpose.

Mark-Houwink equation
An empirical relationship between the limiting viscosity number $[\eta]$ and the polymer molecular weight (M) which also uses two constants (denoted as K and a - the *Mark-Houwink constants*). The equation is, limiting viscosity number $[\eta] = KM^a$.

Mark process
A rubber *reclamation* process which uses pressurized steam/naphtha mixtures.

Martens point
A measure of the heat resistance of a material. The temperature at which a specimen deflects a specified amount in a *Martens point test*.

Martens point test
A test used to measure the heat resistance of a material. The temperature at which a bar, attached to the test specimen, deflects a specified amount in flexure: this is measured and called the Martens point or the Martens temperature. See, for example, DIN 53458.

Martens temperature
The result of a *Martens test*.

martensite
A hard and brittle form of *steel* which occurs when steel is cooled so quickly that alternative forms of steel (for example, pearlite) do not form on cooling. See *hardened*.

Marzetti plastometer
A *plastimeter*. A test apparatus used to measure the extrusion speed of rubber compounds and to evaluate rod, or thread, extrusion properties. See *plasticity*.

mass
The amount of matter that a body contains: it is not the same as *weight*. Weight is the gravitational attraction of the Earth for a body. Weight varies with altitude and latitude whereas mass does not vary. Two bodies with the same mass have the same weight at a given location. The letter M is sometimes used for mass.

mass colouring techniques
Most polymeric components are coloured by mass colouring techniques whereby the colouring system is dispersed throughout the polymer. See *skin colouring*.

mass number
Also known as A and is the number of nucleons in the atomic nucleus.

mass polymerization
Also known as bulk polymerization. A *polymerization process* in which only monomer and a catalyst system are normally used: the polymerization is carried out in the absence of solvent or another aid to dispersion. Because of the low additive level, relatively pure polymer can be produced in this way although the heat removal difficulties can be severe and autoacceleration becomes a problem at high conversions.

Mass polymerization is used to produce, for example, *polystyrene* by casting the heated, partially polymerized monomer into large heated moulds: slabs, or cakes, are produced which are then pulverized to give a coarse, granular moulding material. Polystyrene may also be produced by allowing the heated, partially polymerized monomer to flow through a large heated tower which is designed as a heat exchanger: the tower terminates in an extruder which allows *pellet* production.

When mass polymerization is performed at temperatures above the polymer melting point it is known as *melt polymerization*. When mass polymerization is performed in a mould it is known as *monomer casting*. When the polymer is insoluble in the monomer then mass polymerization is known as precipitation polymerization.

masterbatch
An abbreviation used for this type of material is MB. A masterbatch consists of large concentrations of compounding additives dispersed in a carrier system. The use of masterbatches avoids the weighing of small quantities of powders and assists subsequent dispersion. Masterbatches may be solid or liquid although solid systems are more common. In such cases, the carrier system may be the parent polymer (see *dilute masterbatch*) or it may be a carrier resin which has a low melting point (*universal masterbatch*). See also *liquid colouring*.

The additive, or additives, are in a much higher concentration than in the final compound. The *let-down ratio* varies but, for example, in the *injection moulding industry* one bag of masterbatch (25 kg) may be used with every 40 bags (1 tonne) of *natural polymer*.

If masterbatches are used, in place of fully compounded material, then a number of advantages are obtained. The number of grades of polymer that the processor must purchase decreases, bulk purchase of the natural polymer gives cost savings, there are lower book keeping costs and colouring flexibility is obtained. As it is very difficult to predict exactly the quantity of material required, for a particular moulding run, it is easy to finish with unused, coloured material when fully compounded material is used.

masterbatch colour
A solid or liquid *masterbatch* which contains a large amount of a colouring system. See *dilute masterbatch*, *universal masterbatch* and *liquid colouring*.

masterbatch colouring
A method of colouring polymers, particularly thermoplastics materials. Achieved using a solid or liquid *masterbatch colour*. If masterbatches are used, in place of *fully compounded material*, then a number of advantages are obtained. The number of grades of polymer that the processor must purchase decreases, bulk purchase of the natural polymer gives cost savings, there are lower book keeping costs and colouring flexibility is obtained. As it is very difficult to predict exactly the quantity of material required, for a particular moulding run, it is easy to finish with unused, coloured material when fully compounded material is used. See *dilute masterbatch* and *universal masterbatch*.

masterbatch - rubber
In the rubber industry, a *masterbatch* is usually solid: such masterbatches are mainly used to accurately add ingredients, used in small amounts, during a mixing operation. Masterbatches are widely used in rubber processing so as to obtain consistent use of expensive additives: the improved dispersion of the *additive* which results from their use is also extremely valuable.

masterbatch - thermoplastics
In the plastics industry, a *masterbatch* may be solid or liquid and such masterbatches are mainly used to introduce colour, for example, during the injection moulding of a *thermoplastics material*. A wide colour range is possible from such blends. The solid ones are the most popular with approximately 60% of all in-house, colour addition being done by solid masterbatch. Such solid masterbatches may be based on a wax-like carrier (a universal carrier) or, on the base polymer (polymer specific concentrate). Such concentrates are the cheaper of the two. Masterbatches are also available which contain other additives, for example, UV absorbers.

Processing equipment, such as an injection moulding machine, is often fed with a mixture of thermoplastics material and *masterbatch*. The use of such a mixture can lead to significant cost savings as a compounding step may be eliminated. Most commonly, masterbatches are only used to impart colour to the finished product. The use of such a mixture can sometimes cause problems. The usual problem is one of colour shade differences between different machines; another is separation of the masterbatch from the plastics material in the hopper.

Controlled amounts of *masterbatch colour* may be added to thermoplastics materials through a tube which is positioned, for example, just above the screw of the moulding or extrusion machine. The colouring system is drawn into the flights of the machine screw, together with the material, where mixing occurs. If the colouring system is a free flowing solid, then it may be dispensed, with reasonable accuracy, by means of a rotating feed screw which is powered by a DC variable speed drive. For example. from 0 to 1000 g/min of *colour concentrate* may be dispensed in this way. See *liquid colour* and *peristaltic pump*.

masticated rubber
A rubber which has had its molecular weight reduced by the process of *mastication*. Then the masticated material (usually *natural rubber*) is compounded with the mix ingredients, shaped by *extrusion*, *calendering* or *moulding* and the shape is set by cross-linking or *vulcanization*.

mastication
The reduction in elasticity which occurs when a rubber is intensively worked. The reduction in molecular weight of a rubber, usually *natural rubber*, which occurs when the rubber is intensively mixed or sheared: that is, plasticization. The energy intensive process used to reduce the elastic nature of *natural rubber* by reducing the molecular weight of that material.

At temperatures below approximately 100°C, mastication is a *mechanochemical degradation reaction* and is more effective than at higher temperatures. At temperatures above approximately 100°C, mastication is an oxidative degradation reaction. The minimum rate is in the region of 100°C. To achieve the necessary weight reduction the material must be sheared in the presence of oxygen using, for example, a *two-roll mill*.

Both *internal mixers* and two-roll mills are commonly used for mastication. Synthetic rubbers are not commonly masticated: rubber of the correct viscosity should be chosen as many grades (for example, those which contain stabilisers) undergo very little change on mastication. However some synthetic rubbers can be masticated, for example, butyl rubber. The use of *constant viscosity natural rubber (CVNR)* can substantially reduce, and in some cases eliminate, the energy intensive mastication process.

If a two-roll mill is used for mastication, then the rolls of the mill are cooled with water so that intensive working in the presence of air (oxygen) is obtained. As a result of the changes which occur (for example, a molecular weight reduction), the rubber becomes more plastic, is capable of flow and can therefore be shaped. Mastication may be accelerated by using, for example, *peptisers*. By mastication in the presence of a monomer, block copolymers may be prepared.

mat
A fibrous material consisting of randomly oriented chopped, or swirled, filaments which are loosely held together with a binder. Used, for example, to make *fibre reinforced plastics*. Often a mat is a *glass mat*. A random arrangement of fibres which are lightly bonded together: the bonding is just sufficient to allow handling. Such mats are usually sold on the basis of weight per unit area.

mat moulding
An abbreviation used for this term is MM. A moulding process in which a continuous fibre mat, or preform, is placed in a mould: a low viscosity, reactive mixture is then injected through the mat and the assembly cured. This process may be combined with *reaction injection moulding* (RIM) to give mat moulding reaction injection moulding (MM RIM). See *structural reaction injection moulding*.

matched die moulding
Also known as matched metal moulding. A moulding technique used for reinforced plastics and which uses two mould halves (like *compression moulding*). Can be expensive because the moulds are very large and must be well made, for example, close tolerances, high finish, long lasting, pressure resistant construction etc. See *cold press moulding*, *hot press moulding*, and *resin injection moulding*.

matched metal mould
A mould used in *matched die moulding*.

matched metal moulding
See *matched die moulding*.

material changes - plastics processing
During, for example, the shaping process there should be little or no change in the plastics material. Any change is usually undesirable. Some of the changes that can occur are:

(a) water contamination - caused through the material absorbing water or by condensation;
(b) oxidation - when plastics are heated in contact with oxygen then they will oxidize or combine with the oxygen: the first sign of this is a change in colour and then a change in properties;
(c) overheating - if overheated, even if there is no air present, then plastics will suffer decomposition or thermal degradation: often gases are produced which can be dangerous;
(d) dust contamination - it is easy to generate static electricity on plastics and this attracts dust, or dirt, very quickly.

material handling
All storage and unloading areas must be kept clean and dry; to minimise the fire hazard, the store rooms should be separated from the processing area by fire resistant doors. Store the materials away from direct sunlight and on properly constructed racks. Usually the use of unheated storage areas, with natural ventilation, is sufficient. Ensure that the material does not stagnate in the stores by adopting a strict first-in, first-out (Fi-Fo) policy.

Most thermoplastics materials are supplied in paper sacks or in foil-lined sacks. It is important that the granules inside the sack be allowed to reach the temperature of the processing area, for example, the moulding shop, so as to avoid any condensation forming on the surface of the granules. It is recommended that the material be stored in the work shop for at least 8 hours prior to opening the sack. Reseal the sack immediately after use and only put sufficient material in the hopper for up to 1 hour's running. Keep the hopper covered, or wherever possible use a hopper heater.

Strict stock control is important, as within a shipment of material there could be several different batches and one of these could be faulty. By adopting a strict stock control system, a faulty batch can be readily identified and isolated thus preventing defective products being produced.

material identification
Complete identification of most polymer compounds is very difficult because of the complexity of the formulations. Often however, this is not required. What is required is the separation (or identification) of lots of material whose identity has been lost or, an indication of the type of material used to make a given moulding. Useful simple tests for plastics materials include a preliminary examination, density, melting point and behaviour on heating.

PRELIMINARY EXAMINATION. The material should be inspected before processing and the colour, particle size, particle size consistency and any obvious contamination, noted. If a component is being examined then note size, weight, colour and any clues which indicate how the component was made, for example, a gate scar or flash lines.

DENSITY. Very often the absolute *density* of a thermoplastics material is not required; what is required is an approximate value. This may be obtained by seeing if the material sinks or floats in a limited range of liquids. These may include water and saturated magnesium chloride; the former has a density of 1 gcm^{-3} and the latter has a density of 1.34 gcm^{-3}. If an accurate measurement of density is required then this may be performed by a *flotation* or a *displacement method*.

MELTING POINT. Once again a very accurate value is not required for the thermoplastics material; what is required is an approximate value. The simplest way of obtaining an approximate value is by heating a small sample of the material on a metal hot plate while measuring the temperature rise and the temperature of the hot plate just below the plastics material; a heating rate of approximately 50°C/hour or 100°F/hour, should be used. It is useful to have a glass rod so that the sample may be moved, or prodded, during heating. Note the mid-point of the obvious softening range. An *amorphous thermoplastics material* will not have a sharp *melting point* whereas a *semi-crystalline thermoplastics material* will usually have a relatively sharp melting point. See **table 10**.

BEHAVIOUR ON HEATING. Often this test may be combined with the melting point test. A glass rod is used to move the material while it is being heated. If the material does not soften on heating then it is probably a *thermoset* (some heavily filled engineering thermoplastics materials behave in the same way). If it does soften then it is a thermoplastics material. An *amorphous thermoplastics material* will have a much broader softening range (not a *melting point*) whereas a *semi-crystalline thermoplastics material* will have a melting point. The way that the material burns and the fumes evolved can indicate the type of material used. See **table 10**.

Simple identification tests are not so useful for rubber materials as they are for plastics. Most rubber compounds are black and this hinders identification. Also, for many applications a compound with the desired properties is obtained by using a blend of elastomers. Elemental analysis (i.e. to determine what elements are present) and/or infra-red spectroscopy (i.e. so as to determine what organic groups are present) is usually necessary if base polymer identification is to be successful. Fillers, being inorganic, can only be determined by traditional analytical techniques after ashing.

material modification - thermoplastics
It is generally understood in the thermoplastics industry that the plastics materials used contain small amounts of additives. Most thermoplastics materials cannot be commercially used without the addition of additives such as *heat stabilizers* and *lubricants*. The use of the term 'materials modification' usually means the addition of larger amounts of additives. Such additives may include *elastomers*, *flame retardants* and *fillers*.

material properties
Each of the many plastics and rubbers available is, in effect, a family of materials which differ in, for example, molecular weight and molecular weight distribution. What this means is that it is possible to get a wide range of properties for each material group. Any properties quoted in the literature should therefore only be used as a general guideline. The properties of, for example, plastics may also be dramatically changed by the processing conditions employed and by the use of additives. With many materials, variations of basic formulas are available with additives to provide, for example, improved heat resistance or, weatherability. Some formulations offer improved impact strength while others, which contain fillers, are used where the mouldings require greater impact strength, tensile strength and heat distortion temperature. Processing and performance modifiers can be added; these include, for example, antistatic and nucleating agents: such additives may form part of the colour *masterbatch*.

material supply - thermoplastics
Most thermoplastics materials are supplied as granules or pellets and are usually supplied in either sacks or, in bulk containers. When supplied in 50 lb/22.6 kg sacks, then the material is often supplied in 20 sack (1,000 lb) loads. If supplied in 25 kg/55 lb sacks then, the material is often supplied in 1,000 kg/2240 lb lots. Supplying the material in sacks, or

bags, is not the only way that the material can be supplied: considerable discounts may be obtained if the plastics material is purchased in bulk and, supplied in, for example, bulk containers. With some materials, this form of supply may not be acceptable as, for example, excessive water absorption may occur.

material testing - thermoplastics processing
Most plastics materials are accepted as being satisfactory until something goes wrong during, for example, the actual moulding operation. It makes sense however, to inspect the material before processing and note colour, particle size, particle size consistency and any obvious contamination. There is no point in processing 'out-of spec' material: 'out-of spec' products will be the result.

Because most methods of shaping plastics are melt processes, such as injection moulding and extrusion, the most common test done on moulding material is a *flow test*. A large number of tests have been devised to measure flow properties and of these many tests the measurement of *flow rate (FR)* is the most

The ease of flow of the material should also be checked during injection moulding using a *flow tab* or, by noting the ease of mould filling or, by the use of a *nozzle pressure transducer*. On many machines, fitted with appropriate transducers, the flow behaviour of the material may be assessed during moulding. The presence of surface splay, or silver streaking, on an injection moulding, will indicate if the moulding material, needs drying. This, for many materials, should be checked before moulding commences by, for example, *oven heating*.

matrix
The polymeric framework of an *ion exchange resin*.

matt finish
See *sharkskin*.

matting agent
Also known as a matting additive or as, a flatting additive or as, a flatting agent. An additive used to produce a matt, or silk, finish to a component. The addition is most often performed to alter the surface appearance although such materials may also act as an *anti-blocking agent* in, for example, film or sheet (see *blocking*).

Fine particle size, inorganic materials, such as *kieselguhr*, act in this way for *polyethylene*. The addition of some of the additive may reduce blocking while the addition of even more will give a more matt finish. The level of uses is approximately 0·5 to 3 phr. Precipitated silica is another example of a matting agent and anti-blocking additive: it is used for *polyvinyl chloride* (PVC) film and sheeting. Polymeric additives are also available, for use with PVC, which are based on acrylic resins. It is thought that such resins form an acrylic, resin-rich layer on the surface of the PVC which reduces surface gloss but the general clarity is not very much impaired. Such features are useful for non-glare coverings for documents.

maturation
A maturing process: wet sheets of *natural rubber* are dried at ambient temperatures (for 1 to 2 days) before being dried in the *smoke house*. Maturation changes the non-rubber constituents and gives beneficial effects during *vulcanization*. See, for example, *Michelin sheet*.

maturation agent
A thickening agent (for instance, magnesium oxide) used in *sheet moulding compound* to produce a *maturing action*.

maturing action
Also known as a livering action. A change brought about by the *maturation agent* (for instance, magnesium oxide) which results in a loss of tack and an increase in strength for *sheet moulding compound*.

maturing time
This is the time, measured in for example, hours, from the hardening of a resin (*gelation*) to the point when the resin acquires its full hardness, chemical resistance and stability. Can occur at room temperatures or more quickly by post-curing.

maximum exposure limits
An abbreviation used for this term is MEL.

maximum overall diameter
The *overall diameter* of a *new inflated tyre* with additional tolerances for growth in service and for manufacturing variations.

maximum overall width
The *overall width* of a *new inflated tyre* with additional tolerances for growth in service and for manufacturing variations.

maximum shear rate - injection moulding
In general, the gate which is the smallest part of the *feed system*, should be sized so that the maximum *shear rate* is less than $50,000 \text{ s}^{-1}$.

maximum surface stress in bending
See *cross breaking strength*.

maxwell
The unit of magnetic flux in the *cgs system*. In the SI system it is the *weber*.

Maxwell unit
In polymer science, refers to the combination of a spring (which obeys *Hooke's law*) and a dashpot (which is assumed to contain a viscous oil): the two are in series with each other. Used to describe melt flow in polymers and stress decay in rubbers. See *Voigt unit*.

MB
An abbreviation used for *masterbatch*.

MBS
An abbreviation used for *N-morpholinothiobenzothiazole-2-sulphenamide*. The same initials are also used for *methacrylate-butadiene-styrene copolymer*.

MBT
An abbreviation used for *mercaptobenzthiazole*.

MBTS
An abbreviation used for *benzthiazyl-disulphide*.

M_c
An abbreviation used for *network parameter*.

MC
An abbreviation used for methyl cellulose. A *cellulose ether*.

MCC
An abbreviation for medium colour channel (black). See *carbon black*.

MD
An abbreviation used for *machine direction*.

MDI
An abbreviation used for *4,4'-dicyclohexylmethane diisocyanate*.

MDPE
An abbreviation used for medium density polyethylene. See *polyethylene*.

mean average
Also known as the mean. A measure of central tendency i.e. usually that which is obtained by totalling the value of items in a set and dividing by the number of individuals in that set.

measured pressure
An abbreviation used for this term in, for example, *shear flow* is P. See *capillary rheometer*.

measured temperature
The actual temperature: that temperature which is measured as opposed to that which is set.

mechanical comminution
Reduction in size by mechanical means, for example, to give mechanically comminuted rubber. See *comminution process*.

mechanical ejection system
The most common type of ejector system: one which relies on a machine movement to eject components from a mould. See *pin ejection, ejector sleeve, stripper ring, stripper blade* and *stripper plate*.

mechanical press
A press operated by mechanical power.

mechanical stability test
A test used to assess the stability of rubber latex. A standard (for example, ASTM or BS) stirrer is rotated at 14,000 rpm until the *latex* is seen to flocculate. See *natural rubber*.

mechanical tests
Sometimes referred to as physical property determinations. The mechanical tests that have been devised fall into two main categories; those which are suitable for quality control and specification purposes, and those which yield data suitable for design purposes. The former tend to be called *single point tests* while the latter take into consideration the marked effect that changes in time, temperature, and environmental conditions have on the properties of polymers. See *common test* and *creep test*.

mechanically actuated side core
A *side core* which is actuated by a mechanical movement, for example, as the mould closes, the cores slide towards each other because of angled dowels.

mechanically comminuted rubber
Rubber in small pieces which have been produced by cutting or shearing. See *natural rubber* and *technically specified rubber*.

mechanically foamed plastic
A *cellular plastic* in which the cells are formed by the physical incorporation of gasses. May be done, for example, by beating or whipping a *latex*.

mechanochemical degradation
Decomposition, or degradation, introduced into a polymeric system by mechanical shear and promoted by chemical means. For example, a reduction in molecular weight can occur during extrusion of a thermoplastics material as a result of mechanochemical degradation. *Mastication* is another mechanochemical degradation reaction.

medium activity
A term sometimes used in place of *reinforcing* in connection with fillers for rubbers is 'activity'. Medium activity means that the reinforcing action lies in between no reinforcing action and a good reinforcing action.

medium cis-polybutadiene rubber
Also known as medium cis BR. See *polybutadiene rubber*.

medium density polyethylene
Also known as polyethene or, polyethylene-medium density or, polyethene-medium density. An abbreviation used for this material is MDPE or PE-MD.
Grades of *high density polyethylene (HDPE)* which have a density below 0.96 g/cm^3 are produced by using a second monomer at low levels (<1%). Strictly speaking they are therefore copolymers of PE with another alpha olefin, for example, with butene-1 or, with hexene-1. The use of the second monomer, reduces the density by introducing short, side chain branching. Such materials may be known as MDPE. HDPE/MDPE usage accounts for approximately 55% of all plastics used in blow moulding. See *polyethylene*.

medium molecular weight
An abbreviation used for this term is MMW.

medium oil resin
A type of *alkyd resin*. See *oil length*.

medium-fast accelerator
See *accelerator*.

medium-profile tyre
An abbreviation used for this term is MP tyre. A comparatively wide *tyre*: a tyre with an *aspect ratio* of approximately 88%. See *low-profile tyre*.

MEEP
An abbreviation used for *polyester substituted polyphosphazene*.

MEF
An abbreviation for medium extruding furnace (black). See *carbon black*.

mega
An abbreviation used for this term is M. For example, megabyte is Mb and is a unit of measurement of memory or, of disk storage capacity. It is 1,024,000 bytes or characters or, 1,000 K. See *prefixes - SI*.

meganewtons per square meter
A megapascal. An abbreviation used for this expression is MNm^{-2} or MPa. 1 MNm^{-2}, or 1 MPa, is approximately equal to 145 psi.

megapascal
An abbreviation used is MPa. This is equal to a *meganewton per square meter* or MNm^{-2}. That is 1,000 psi is approximately equal to 7 MPa.

MEK
An abbreviation used for *methyl ethyl ketone*.

MEKP
An abbreviation used for *methyl ethyl ketone peroxide*.

MEL
An abbreviation used for maximum exposure limits.

melamine
A cyclic product which is also known as triaminotriazine or as, 1,3,5-triamino-2,4,6-triazine or as, cyanuramide. The material, which may be represented as $C_3H_6N_6$, is produced by, for example, heating *dicyanamide*. A white crystalline material with a melting point of 354°C. *Methylol melamine* is formed during the reaction of *formaldehyde* with *melamine*, for example, when a *melamine formaldehyde (MF)* resin is being made.

melamine formaldehyde
See *melamine-formaldehyde moulding materials*.

melamine phenolic
See *melamine-phenolic moulding materials*.

melamine plastics
Plastics based on resins made by the condensation of melamine (M) and aldehydes: when formaldehyde (F) is used then the material is known as an MF. See *melamine-formaldehyde moulding materials* and *melamine-phenol-formaldehyde moulding materials*.

melamine resin
See *melamine-formaldehyde resin*.

melamine resin moulding compound
See *melamine-formaldehyde moulding materials*.

melamine resin moulding material
See *melamine-formaldehyde moulding materials*.

melamine-formaldehyde
Also known as *melamine-methanal*. An *aminoplastic*. See *melamine-formaldehyde moulding materials*.

melamine-formaldehyde moulding materials
An abbreviation used for this type of material is MF. Also known as melamine-formaldehyde or as, melamine formaldehyde or as, melamine resin moulding material or as, melamine resin moulding compound. MF, like *urea-formaldehyde* or *UF*, is an *aminoplastic* that is, the moulding powders are based on an amino-formaldehyde resin.

A *melamine resin* is combined with fillers, hardeners, pigments, etc, to make a *thermoset* moulding powder which is usually supplied as a fine powder or as a granular powder, in a very wide range of colours. As with colourless UF resins, it is possible to produce moulding powders of virtually any colour when white alpha cellulose, or a paper filler, is used.

Mouldings are bright and attractive but unfortunately they are more expensive than UF or phenol-formaldehyde (PF). They have lower water absorption than UF, are less flammable, maintain their electrical properties better in conditions of high humidity (particularly when mineral filled), are harder, and resist staining and heat to a greater degree. Like UF, they show after-shrinkage if exposed continuously to high temperatures but maximum, continuous, end-use temperature is much higher than UF at 130°C/266°F. Like PF, they are more tolerant of changes in moulding conditions than UF materials. Mouldings possess better scratch and tracking resistance than PF materials; they are not so impact resistant as PF but have good abrasion resistance. Better electrical properties are obtained from melamine-phenolics. Where mouldings are to be used alternatively in wet and dry conditions, the use of a plasticized grade is suggested: benzamide and p-toluene sulphonamide have been used as plasticizers. New improved injection moulding grades have excellent colour retention when exposed to heat and sunlight and can now be considered for use at continuous temperatures of up to 130°C/266°F, in wet and dry conditions. Such grades give high gloss, improved impact resistance and, a lower post-moulding distortion. The density of MF is about 1.5 gcm^{-3} or 24.58 g/in^3.

MF mouldings are resistant to fuels, oils, greases, common organic liquids and organic solvents - such as acetone and alcohol. Also resistant to cold dilute acids and alkalis. MF is more chemically resistant and more stain resistant than UF, that is with better resistance to weak acids, alkalis and water. UF mouldings are attacked by boiling water; fully cured MF mouldings are more resistant, being attacked only by concentrated acids and alkalis or, by hot dilute acids. MF mouldings are not resistant to strong acids and alkalis and to hot dilute acids. Continuous exposure to high temperatures can also cause problems. MF is better than UF in this respect - an MF moulding can withstand 130°C/266°F continuous maximum working temperature. A UF moulding can stand only 75°C/167°F continuous maximum working temperature. Both these aminoplastics suffer from after-shrinkage if held at high temperature for too long.

Because moulded components can be free from odour (they do not influence the flavour of food) are bright, attractive, scratch and stain resistant, they are widely used in dinnerware and kitchen utensils. Such items can be made more attractive by moulding-in foils which carry a legend or picture. Such multi-colour decorative foils (printed alpha-cellulose paper impregnated with uncured MF resin) can be introduced into *compression moulds* containing minimally-cured MF mouldings before completing the cure. These relatively cheap foils enable personalized mouldings, for a number of customers, to be produced from one tool. Although more easily produced in flat or smooth profile moulds, grades of foil are available for sharp and rectangular edges.

The excellent colour range of MF materials, also enables two-tone, compression mouldings to be made using double-punch tools; for instance, cups and mugs with a white inside and, a coloured exterior. MF is also used for handles, knobs, household appliances and electrically insulating parts - particularly where these must withstand high temperatures. It can withstand repeated cycling from 150°C/302°F into cold water. Compared to other thermosets, MF materials are expensive and are usually only selected where appearance is of prime importance.

Where improved impact strength is required, *glass fibre* is used as a filler - it gives high mechanical strength and heat resistance; the mouldings possess good arc resistance, track resistance and burn with difficulty. Examples of mouldings which utilize these properties, include base plates for circuit breakers and switches. The filling of MF with inorganic fillers, such as glass, can give hard, heat and mar resistant mouldings. Such materials can be made very fire resistant and are used in applications which demand this property, for example, on board ships.

melamine-formaldehyde resin
Melamine may be made from urea and like urea it contains amine (-NH$_2$) groups. These groups react with formaldehyde to form a resin which is used to make *melamine-formaldehyde moulding materials*. MF resins are widely used in laminates as by their use it is possible to produce a very wide range of attractive, and durable, patterned laminates - the pattern printed on the paper base shows through the colourless resin. The laminate core is made of cheaper, PF-coated paper. Such laminates have excellent light stability. MF resins are also used for textile treatments and for paint manufacture.

melamine-methanal
See *melamine-formaldehyde*.

melamine-phenol-formaldehyde
An abbreviation used for this material is MPF. See *melamine-phenolic moulding materials*.

melamine-phenolic moulding materials
An abbreviation used for this type of material is MPF. Developed to offset some of the problems associated with *melamine-formaldehyde (MF)* and *phenolic moulding materials (PF)*. For example, post moulding shrinkage can be a problem with MF and can result in component failure. The reduction of post moulding shrinkage for this class of materials is a significant advantage over MF (when *transfer moulding* of melamine phenolics, preheating is always advisable). Such materials are of lower cost than MF-type materials and a wider colour range is available compared to PF-type materials.

melt
A plastics material at its processing temperature. Term associated with thermoplastics materials and their processing. See *calendering*, *injection moulding* and *extrusion*.

melt coating
A coating method in which the substrate is coated with a fluid composition based on a thermoplastics *melt*. Melt coating has also been used for a process where a plastic film is extruded onto a preformed film; this is necessary when the processing conditions are too dissimilar for coextrusion.

melt extrusion
Extrusion performed using a plastic material in a *melt form*. See *extrusion*.

melt flow index
An abbreviation used for this term is MFI. See *melt flow rate*.

melt flow rate
A *flow test*. The term melt flow rate (MFR) was introduced to replace MFI and refers to the *flow rate* of *polyethylene* (PE) when obtained under *condition 190/2·16* (formerly known as Condition E). MFR is also called melt index (MI) or melt flow index (MFI); the terms mean the same. Melt flow rate testing is widely used as the test is easy to do and to understand.

The test uses a simple rheometer (a *melt indexer*) to test polyethylene thermoplastics materials under specified conditions. See *method A* and *method B*.

The use of terms such as MFR or MFI is not encouraged for materials other than polyethylene (PE): it is suggested, by ASTM D 1238 that the term *flow rate (FR)* be used for other plastics materials.

melt fracture
Also known as elastic turbulence or as bambooing. The distortion of the surface of an extrudate after leaving a die orifice: the effect can range from minor ripples to severe distortion. Distortion of an extrudate which occurs when extrusion is carried out at high rates. The distortion may be due either to melt fracture or to *sharkskin*.

OCCURRENCE. Melt fracture occurs when the shear rate exceeds a critical value for the polymer melt at a particular temperature. There is a corresponding critical shear stress and the point on the shear rate-shear stress diagram is known as the critical point. Melt fracture is believed to occur, or commence, in the die entry region, that is, in the region where material is being funnelled from the die reservoir into the capillary of a capillary rheometer. In a full scale extruder, this would correspond to the point where melt moves into the die parallel portion of the die. Some complicating further effects may occur at the wall of the die. Melt fracture is most likely to occur where small diameter extrudates are being extruded at high rates. The most notable example occurs with wire covering.

FACTORS AFFECTING. Experiments have shown that:

i) the *critical shear rate* increases with an increase in temperature;

ii) the product of critical shear stress and weight average molecular weight (M_w) is a constant, i.e. melt fracture will start at lower shear stresses, and hence shear rates, with high molecular weight materials than with lower molecular weight polymers;

iii) two polymers differing in their levels of branching but which have similar melt viscosities tend to have similar *critical points*;

iv) extrudate quality may be markedly improved by tapering the die entry so that externally un-distorted extrudates may be obtained at rates well above the critical point;

v) extrudate quality may also be improved by tapering the tapering the so-called die parallel (up to 1°) as this may increase the critical point;

vi) there is some evidence that increasing the L/D ratio of the die parallel also increases the critical shear rate.

The above factors have now been well known for some years so that high speed wire covering, and other operations involving high shear rates, may be operated without undue trouble from melt fracture effects.

FORM OF DISTORTION. This varies from one polymer type to another but is generally helical in nature. With materials such as *polyethylene* and *polypropylene* a distortion like a screw thread may appear; with *polystyrene* the extrudate may form a spiral while with other melts, ripples, or bamboo-like repetitive kinks, may occur. With all melts, at rates well above the *critical point*, the helical nature becomes obscured by severe distortion which look quite random.

melt index
An abbreviation used for this term is MI. See *melt flow rate*.

melt index data
The results obtained from a *melt indexer*.

melt indexer
The apparatus used to measure *melt flow rate* (*MFR* or *MFI*). A simple, weight-loaded, rheometer widely used for *flow rate*.

melt mixer
A machine used to perform *melt mixing*: may be a *batch* or a *continuous mixer*. For example, in the thermoplastics *calendering* industry, a *pre-mix* is fed to the melt mixer, for example, an *internal mixer* of the *Banbury-type*.

melt mixing
Mixing performed in a *melt mixer* when the base polymer is soft and pliable: mixing performed on a polymer *melt*. Also referred to as compounding and is usually done at relatively low temperatures in order to keep the polymer viscosity at a high value so as to increase the *shear* input. Fillers may be treated, for example, with *stearic acid*, so as to aid dispersion. Mixing may take place after the mix ingredients have been blended together (see *two-roll mill*). Melt mixing is both *dispersion mixing* and *distributive mixing*. Both *batch* and *continuous mixers* are used.

In the thermoplastics industry, melt mixing is commonly performed in a single screw extruder however, neither a conventional *screw*, or a *zero compression screw*, gives very good melt mixing; this is because of the way that the material flows, or is transported, along the screw. The material towards the centre of the flight can easily remain undisturbed and this means that, the output from the machine will not be of uniform quality. That is, it will be inhomogeneous as it has a non-uniform shear history. Even if nothing is added, the melt temperature will be non-uniform and this non-uniformity can cause, for example, product distortion and a reduction in output. Melt mixing using a *compounding extruder*, is therefore very important to the plastics industry as it gives good dispersion. It also gives a continuous output which can be in the form of rods: this output can be chopped into regularly-sized pellets and these often form the feed to moulding machines. However, in some cases a lower level of dispersion is tolerated to save on costs and to give operational flexibility (see *masterbatch*).

The level of shear encountered during the final shaping process should be less than that used in mixing so that any pigment agglomerates are not further broken down (which may cause colour streaking). See *additive*.

melt polymerization
A *mass polymerization process* which is carried out at a temperature above the melting point of the ingredients (monomers) and the products (polymers). Sometimes this is the only method that can be used to produce the required *degree of polymerization*. Used, for example, in the manufacture of polyesters and polyamides where either, or both, the *monomer* or the *polymer* are crystalline: in melt polymerization, the growing chain ends remain available for further reaction.

melt pressure
The pressure on the melt, for example, within the die or, within the mould: the pressure exerted by the *melt*. Melt pressure is measured, or sensed, by a melt *pressure transducer* which is located within the nozzle/die. Because of their location on, for example, an *injection moulding machine*, they are sometimes referred to as *nozzle pressure transducers*. Such

transducers are useful because they can improve productivity and efficiency: however, they are relatively expensive and because of their location, they can easily be damaged. For these reasons, they are often not seen in production.

melt pressure transducer
A *transducer* used to sense pressure within a polymer melt, for example, within an *extrusion die*. See *pressure transducer*.

melt processable rubber
An abbreviation used for this material is MPR. See *thermoplastic elastomer*.

melt processes
Term associated with thermoplastics processing. Most of the processes used to shape thermoplastics are melt processes. That is, the plastics material is heated until it softens and becomes a *melt*, it is then shaped and set to that shape by being cooled. See *calendering*, *injection moulding* and *extrusion*.

melt processing
Term associated with thermoplastics processing and means that the material is shaped at elevated temperatures where the polymer is a *melt*. Most of the processes used to shape thermoplastics are *melt processes* even though a plastics melt has a low thermal conductivity, a high specific heat, a high non-Newtonian melt viscosity and a limited thermal stability. Such material characteristics introduce problems.

HEAT INPUT. Because of, for example, thermal stability problems, the processing temperatures employed for polymers are limited to relatively low values. For example, those for thermoplastics material are often below 250°C (see **table 7**). Even so, thermoplastics materials require large heat inputs to raise their temperatures to those required for melt processing: the materials also differ enormously in the amounts of heat that is needed to bring them up to processing temperatures. The differences are due to the different processing temperatures employed and to the differing specific heats. When melt processing a semi-crystalline, thermoplastics material, heat must be supplied to melt the crystal structures: this heat input is not needed in the case of an amorphous, thermoplastics material. Both types of material will however, require a large amount of heat to be put into the material quickly: this causes problems as plastics are poor conductors of heat and can have limited thermal stability at the processing temperatures employed. **Table 7** gives heat contents of some thermoplastics material

HEAT REMOVAL. As plastics materials are poor thermal conductors, the removal of the large *heat input* poses severe heat removal problems if high speed production is to be maintained. Poorly designed, and poorly operated, cooling systems give the most common problems in thermoplastics processing. Also variations in the cooling rate of thermoplastics materials can have a pronounced effect on the crystalline morphology of the product and on factors such as *molecular orientation* and on *shrinkage*.

ORIENTATION INTRODUCTION. A polymer *melt* is extensively deformed during melt processing and then the hot material is cooled extremely quickly so as to achieve the high output rates demanded. The shearing of one layer of melt over another, result in the molecules taking up a deformed, or oriented, shape whereas the rapid cooling means that this deformed molecular shape is frozen-in the product. Such an *orientation* causes the product to have different properties in different directions. This effect is known as *anisotropy*. In most case such orientation is undesirable however, in some cases, orientation can be introduced or enhanced, so as to improve the properties of the product. Orientation is deliberately introduced during the manufacture of fibrillated tape and in the *extrusion blow moulding* of bottles.

THERMAL STABILITY. Thermoplastics materials differ widely in the processing temperature required and in their thermal stability. For example, *unplasticized polyvinyl chloride* (UPVC) is very unstable and even when stabilized can only be held at processing temperatures (175°C/347°F) for a few minutes: on the other hand, a *polysulphone* require melt temperatures in the region of 400°C/752°F. The thermal stability of a material is governed not only by the processing temperature, but by factors such as the *residence time* at that temperature, the atmosphere surrounding the material (oxygen or inert) and the materials in contact with the plastics material. Copper, for example, causes rapid decomposition, or degradation, of *polypropylene* (PP): copper cleaning pads should not be used therefore, to clean processing equipment used for PP. The use of too high a *shear rate* during processing can also cause degradation.

The decomposition, or degradation, products which result from heating polymers should be regarded as harmful and any gases evolved should not for example, be ingested (breathed in). Good ventilation of the working area is essential.

VISCOSITY. Because of the low processing temperatures employed for *melt processes*, polymer melt viscosities are generally high. The melt viscosities differ from one material to another and from one grade of the same material to another grade. Viscosity is affected by temperature, by the other mix ingredients, by polymer molecular weight and by molecular weight distribution. in general, viscosity goes down with a rise in temperature and also falls as the molecular weight is reduced or broadened: *lubricants* will also affect the viscosity. For this reason, strict control over both the processing conditions, and over the material fed to the machine, must be employed.

melt residence time
The time for which a *melt* is held at processing temperatures within a machine. May be practically determined by noting how long a coloured material takes to pass through the system. It is important because melt residence time is one of the factors which determines thermal stability.

melt shearing
See *melt processing*.

melt spinning
A technique used to produce a *fibre*. The fibre-forming, *thermoplastic polymer* is turned into a melt by heat. Fibres are formed by forcing the melt through a spinneret: the fibres are then hardened by cooling. See *polyamide*.

melt strength
The tensile stress needed to break a polymer *melt* during draw-down. A high melt strength is needed for processes such *blow moulding* where the melt is required to hold die shape and not sag after the melt leaves the die.

melt tearing
An extrusion defect. See *sharkskin*.

melt temperature
The temperature of the material being melt processed. It may be measured by a melt temperature thermocouple, by means of an infra-red temperature transducer or by, inserting a thermocouple into the melt, for example, by using an *air shot technique*.

If a melt temperature thermocouple is used then, the tip of the melt thermocouple must be approximately 0.25 in/6 mm away from the barrel/nozzle wall so that it does not pick up the temperature of the metal. If nozzle pressure control is used in injection moulding, and if the transducer is mounted in an adaptor, then the same adaptor may be used to house a melt temperature thermocouple. An infra-red probe has a much faster response time, and is more accurate, than a

melt temperature over-ride
An abbreviation used for this term is MTO. The amount by which the average *melt temperature* exceeds the set temperature: caused by the generation of frictional heat in processes such as *injection moulding*.

melting point
The temperature at which a solid material becomes a liquid on heating. That constant temperature at which the solid and liquid phases of a material are in equilibrium at a specified pressure (usually atmospheric). Unlike low molecular weight materials, polymers do not normally have a sharp melting point. Even a semi-crystalline, thermoplastics material will have a broad melting point of, say, approximately 20°C. An amorphous, thermoplastics material will not exhibit a melting point but on heating, a region will be entered over which softening, to a high viscosity melt, occurs. Melting is a first order transition and the melting temperature is often given the symbol T_m. For a melting point approximation see *material identification*. For a more accurate determination see, for example, *capillary tube method*.

melting temperature
The temperature at the *melting point*.

membrane osmometry
See *osmometer*.

mer
See *repeating unit*.

mercapto accelerator
An *accelerator* based on *mercaptobenzothiazole (MBT)* or its derivatives, for example, *dibenzothiazyl disulphide (MBTS)*.

mercaptobenzothiazole
Also known as 2-mercaptobenzthiazole. An abbreviation used is MBT. A thiazole *accelerator*. This material has a relative density (RD or SG) of 1·42 and a melting point of approximately 170 to 180°C. It is a fine, pale yellow powder with a characteristic smell. It has a very long curing range and its use tends to prevent *reversion* - useful in products exposed to high temperatures, for example, tyres. It is practically non-discolouring and has an *antioxidant* effect. May be employed as a peptising agent as when used in small quantities, and at temperatures above 100°C/212°F, it reduces mastication time. The product of the reaction between cyclo-hexylamine and mercaptobenzthiazole gives *cyclohexylbenzothiazyl sulphenamide*.

mercury
Group 11b of the Periodic table consists of the elements zinc, cadmium and mercury. A transition element (Hg) which is a liquid silvery-white, lustrous metal with a melting point of −39°C and a boiling point of 357°C: the relative density is 13·6. Its vapour is very poisonous. Used in thermometer, barometers and high vacuum pumps. Some of its compounds are used as additives, for example, vermilion is mercuric sulphide.

mercury (II) sulphide
See *vermilion*.

mercury cadmiums
A class or type of *inorganic pigment* which are based on mercuric sulphide, cadmium sulphides and selenides. Relatively, high priced pigments which give colours in the yellow, orange, red and maroon range. Such pigments have good heat stability at high temperatures (better than *cadmium pigments* and of lower cost) and can be lightfast, for example, in dark shades. Sensitive to a combination of light and moisture as found in outdoor usage.

mercury sulphide
See *vermilion*.

mesogenic group
A chemical group which gives molecular order to a *liquid crystal polymer*: such groups, which are part of the polymer structure, are elongated, stiff molecules which allow the polymer to form a *mesophase* on heating.

mesomorphic polymer
See *liquid crystal polymer*.

mesophase
Also known as a mesomorphic phase. A phase present in a *liquid crystal polymer* (LCP) which has some molecular order: not as ordered as a crystalline material but not as disordered as an isotropic liquid. The LCP molecules are highly anisotropic and form a liquid crystalline phase on heating - the mesophase which possesses an ordered structure.

mesophase pitch fibre
Fibre produced from *pitch* when the pitch is in a *mesophase*. The precursor for *mesophase pitch-based carbon fibre*.

mesophase pitch-based carbon fibre
A *carbon fibre* produced from a *mesophase pitch fibre*: ultra-high modulus *carbon fibre*. Pitch is treated (heated to approximately 400°C so that it forms a *mesophase*). The mesophase material is capable of *orientation* by shear and elongational forces, in a spinning process, so that highly oriented fibres are produced. These fibres are then oxidized (to cause crosslinking and thus prevent re-melting), carbonized and graphitized. The as-spun fibres (mesophase pitch fibres) convert on heating, without tension or stretching, because of thermodynamic considerations into HM CF. Such a fibrous form has a crystalline, graphitic structure and often has a high degree of structural perfection. For example, pure graphite has a relative density of 2·25, the mesophase fibre can be 2·18. Such a material can have a negative coefficient of thermal expansion and a modulus of 550 GPa.

meta diallyl phthalate
An *allyl moulding material* based on *diallyl isophthalate*.

meta-linked polymer
A long chain polymer based on aromatic groups (benzene rings) which are linked via the meta position. See *polyphenylene* and *substituted polyphenylene*. Although linking via the para position gives the highest *heat distortion temperatures*, meta linking is used so as to obtain more tractable polymers, for example, an *aromatic polyamide*.

metal complex stabilizer
Heat stabilizers used with *polyvinyl chloride* (PVC). Ill-defined mixtures of compounds (salts) based on the reaction products of short-chain organic acids (for example, phenates and octoates) and metals such as barium, cadmium, calcium and zinc. Such stabilizers are usually liquid and may be incorporated at higher levels (2 to 3·5 phr) than *metal soaps*: more compatible than a metal soap stabilizer and less prone to separation under, for example, high shear processing conditions.

metal deactivator
A *preventive antioxidant*. May also be known as a chelator.

metal detector
Usually a ferrous metal detector which is used to detect metal contamination of a polymer feed. For example, the output

from a *strainer extruder*, used in *polyvinyl chloride* (PVC) *calendering*, is in the form of a circular rod and this continuous calender feed is passed through the metal detector. Tramp, ferrous metal is detected by this device and if such metal is detected, guillotines are operated which remove the offending section of rod.

Certain iron-based pigments cannot be used in flexible PVC as their presence in large quantities upsets the operation of metal detectors. Iron compounds may also catalyze the degradation of PVC or accelerate equipment wear.

metal naphthenate
A metal compound derived from *naphthenic acid*. See *naphthenate*.

metal oxide
The reaction product of a metal and oxygen. A class or type of material widely used in the polymer industry. For example, as *inorganic pigments* - see, for example, *titanium dioxide*. Zinc oxide is used in rubber technology as part of the *activator system*. A basic material such as *lime* (calcium oxide) or magnesium oxide is also used in novolak-based, phenolic moulding powders as an *accelerator*. Metal salts are also used as *catalysts* in polymer manufacture. Also see *inorganic accelerator*.

metal replacement
A *plastics material* which replaces a metal, for example, an *engineering thermoplastics material*.

metal soap
A salt formed by a metal and a fatty acid. For example, barium/cadmium salts and zinc/calcium salts which are *heat stabilizers* for *polyvinyl chloride*. Such materials are also used as driers for paints. See *metal soap stabilizer* and *soap*.

metal soap stabilizer
A metal soap used as a *heat stabilizer* with *polyvinyl chloride* (PVC). May be simple or complex mixtures of compounds (salts) based on the reaction products of long-chain, carboxylic acids (for example, stearic or ricinoleic acid) and metals such as barium, cadmium, calcium and zinc. Such stabilizers may be solid or liquid and may also incorporate, for example, other ingredients such as *antioxidants* and *phosphites*. The properties of the resulting metal soap, stabilizer package depends on the type of metal, the metal ratio, the acid(s) used and on the other ingredients. Many packaged stabilizers can therefore be quite complex. Such stabilizers, although effective and capable of giving clear compositions, usually have a limited compatibility with the polymer: less than 1 phr may therefore be added. See *metal complex stabilizer*.

metal stearate
The reaction product of *stearic acid* and a metal. Metal stearates are widely used as *lubricants* and stabilizers in polymer compounds.

metal titanate
See *titanate pigments*.

metal-free organic
A type of *heat stabilizer* used with *polyvinyl chloride* (PVC) and which contains no metal. In general, an organic non-metal stabilizer is not a primary stabilizer for PVC although some are used as a sole stabilizer (for example, for food packaging) while others are used as synergistic co-stabilizers (for example, *epoxy compounds* and *phosphite chelators*). Such metal-free organics may be classed as aminocrotonates (aminocrotonic acid esters), urea derivatives, epoxy compounds, phosphite chelators and other organic heat stabilizers (for example, *2-phenylindole*). In general, such stabilizers appear to function by reacting with the hydrochloric acid formed when PVC degrades. See *barium/cadmium stabilizer* and *heavy metal stabilizer*.

metal-matrix composite
An abbreviation used for this type of material is MMC. A *composite material* in which the continuous phase is based on a metal or a metal alloy: reinforcement is provided by *fibres* or by *whiskers*.

metalising
See *metallising*.

metallic colour
A colouring system which imparts a metallic appearance. Produced by the incorporation of *aluminium flakes* or of, *copper flakes* or of *bronze flakes*. Dyes and/or pigments may also be used to change the colour. See *flake pigments*. Perylenes and quinacridones are also used to make metallic colours.

metallic fibre
A manufactured fibre composed of metal, metal-covered plastics material, plastics coated metal or of a material whose core is coated completely by metal.

metallic pigments
A colouring system which imparts a metallic appearance: such pigments are used to produce a *metallic colour*. See *flop*.

metallic soap
See *metal soap*.

metallized foil
A hot stamping foil, used in *hot foil marking*, to transfer a legend to a plastics substrate. Such a foil may consist of a carrier film, a release coat, a protective lacquer coat, a metallized layer and a hot melt adhesive.

metallizing
Techniques used to build up a layer of metal onto a plastics surface. Such techniques include *electroplating*, *vacuum metallizing*, *hot foil marking* and *vacuum sputtering*.

metathesis
Double decomposition or transposition as when two compounds react with each other so as to form two other compounds.

methacrylic ester of bisphenol A-epichlorhydrin
See *vinyl ester resin*.

meter
A term used in *hydraulics* and which means to regulate the amount of fluid flow in a circuit.

meter-in
A term used in *hydraulics* and which means to regulate the amount of fluid flow into an actuator: to regulate the flow in the pressure line.

meter-out
A term used in *hydraulics* and which means to regulate the amount of fluid flow from an actuator: to regulate the discharge flow in the tank or drain line.

metering zone
That section of an extruder *screw* which has a uniform flight depth and which controls the rate of flow of *melt* through the extruder into the die.

methacrylate
An ester of *methacrylic acid*: a material with the formula $CH_2=C(CH_3)COOR$. See *polymethyl methacrylate*.

methacrylate graft
An abbreviation used for this term is MG. See *polymethyl methacrylate grafted natural rubber*.

methacrylate graft rubber
See *polymethyl methacrylate grafted natural rubber*.

methacrylatochromic chloride
Also known as chrome finish. A bonding agent or *chemical coupling agent* used to improve the bond between *glass fibre* and resins, for example, *unsaturated polyester resins*. A material with the formula $CH_2=C(CH_3)COOCr_2Cl_4(OH)_2$. Bonds to the glass surface via silanol groups and to the polymer via the methacrylic groups which participate in the *cross-linking* reactions.

methacrylatosilane
Also known as γ-methacryloxypropyltrimethoxysilane. A bonding agent or *chemical coupling agent* used to improve the bond between *glass fibre* and *resin*, for example, an *unsaturated polyester resin*.

methacrylic acid
The monomer for polymethacrylic acid: a material with the formula $CH_2=C(CH_3)COOH$. An ester of this acid is *polymethyl methacrylate*. See *polymethacrylic acid ester*.

methacrylic acid ester
See *polymethacrylic acid ester*.

methacrylonitrile
An abbreviation used is MAN: a material with the formula $CH_2=C(CH_3)CN$. A *substituted acrylonitrile* which contains an additional methyl group. The monomer for polymethacrylonitrile and for various copolymers used, for example, as fibres.

methanal
See *formaldehyde*.

methane
Also known as fire-damp and or, as marsh gas. The first member of the *alkane* series of hydrocarbons. A flammable gas with the formula CH_4: the main ingredient of natural gas and an important starting material for other organic materials.

methanoic acid
See *formic acid*.

methanol
See *methyl alcohol*.

method 1/2S d/2
See *method S2 d2*.

method A
A *flow rate* (FR) testing method for polyethylene (PE): see *method B*. Method A is also known as Procedure A and is a manual cutoff operation based on times used for materials having flow rates that fall within 0.15 to 50 g/10 minutes. The piston position during the timed measurement, that is, the position of the piston foot, is between 51 and 20 mm/2.0 and 0.8 in above the die during the measurement. The test specimen can be in any form that can be introduced into the cylinder. Select the conditions (temperature and load) by consulting the standard. For example, for a PE with an expected FR of 2, the temperature would be 190°C and the load would be 2.16 kg. After packing the PE material (between 3 and 5 g) into the cylinder, or barrel, leave for 4 minutes with the piston in contact with the material then, place the weight (2.16 kg) on top of the piston. After a further 3 minutes (usually between 2 and 4 minutes) the lower reference mark on the piston must be level with the top of the barrel. Find the rate of extrusion in a specified time, for example, 3 minutes for this PE: do this 3 times and before the upper reference mark reaches the top of the barrel. The three samples are individually weighed, to the nearest milligram (0.001 g), and the average mass in grams is found. The flow rate (FR) is reported as the rate of extrusion in grams in 10 minutes: obtained in this case, by multiplying the average mass extruded (m) by 3.33. Or, FR = m × 600/t. t is the cut-off interval expressed in seconds (s).

method A to B conversion
A to B conversion may be obtained on a *melt indexer* if an appropriate computerized control system has been fitted. The operator runs a *method A* test while the machine conducts a *method B* test. Upon test completion, a display of flow rate and apparent density may be obtained: the apparent density may be used in subsequent Method B tests. It is obtained by equating two equations and solving for melt density.

By *method A*, *flow rate* FR = m × 600/t. Where m is the average mass extruded (m) within the cut-off interval (expressed in seconds). By Method B, FR = R × L × A × 600/T. Where R is the piston radius in cm, L is the effective length of the flag in cm, A is the apparent melt density in gcm^{-3} and T is the time taken for the test measurement (in seconds). If the two Frs are the same then, the equations may be solved for A.

method B
A *flow rate* testing method: see *method A*. Method B is also known as Procedure B and is an automatically timed, flow rate measurement used for materials having flow rates that fall within 0.50 to 900 g/10 minutes. To ensure reproducibility the timing device must be accurate to within ±0.1 s and the position of the piston foot at the end of the test must be 25.4 mm/1 in above the die. During the measurement, the length of timed movement must be measured to within 0.025 mm/0.001 in over a prescribed distance. This measurement may be achieved by using an opaque flag which is hung on the load (the weight) and which interferes with the passage of light to an electronic eye. If the system is computerized then, all the operator has to do is to select the test conditions (temperature and load) by, for example, consulting the standard or the memory and load the material. The test specimen can be in any form that can be introduced into the cylinder. The electronic system will automatically control the temperatures, time the measurements and change the weight during the test run if required. Once the test has been performed then the electronics system will perform flow rate, flow rate ratios, viscosity shear rate and shear stress calculations. (To do this apparent melt density must be known - see *Method A to B conversion*.) Statistical calculations for quality control purposes, SPC/SQC, may also be performed.

method S1 d4
A method of *tapping* a *natural rubber* tree which involves making a spiral cut completely around the tree: one cut is made every fourth day. A difficult tapping method which requires skilled tappers. See *natural rubber*.

method S2 d2
Also known as method 1/2S d/2. A method of *tapping* a *natural rubber* tree: a frequently used tapping method (the half spiral method or half spiral alternating daily) which allows *tapping* every other day.

methoxy compound
A compound which contains the methoxy group. This univalent organic group, which may be represented as CH_3O-, is readily introduced into organic compounds. Low molecular weight materials (such as 2-methoxyethanol are often solvents) while the higher molecular weight materials (such as methoxyethyl acetyl ricinoleate) are sometimes used as softeners or *plasticizers*.

methyl
A univalent organic group. CH_3- which forms *methyl compounds*.

methyl α-methacrylate
See *methyl methacrylate*.

methyl acetate
A colourless, liquid. This material has a relative density (RD or SG) of approximately 0·93 and a boiling point of 58°C. Sometimes used as a solvent.

methyl acrylate
An ester of *acrylic acid*: a material with the formula $CH_2=CHCOOCH_3$. Used, for example, to make *polymethyl acrylate* and acrylic copolymers.

methyl alcohol
Also known as methanol or as, wood alcohol or as, wood spirit. This poisonous liquid material has a relative density (RD or SG) of 0·79 and a boiling point of approximately 65°C. Obtained commercially from *methane* and may be represented as CH_3OH.

methyl butadiene
See *isoprene*.

methyl cellulose
An abbreviation used for this material is MC. A *cellulose ether* prepared by the reaction of methyl chloride with alkali cellulose. Some of the hydroxyl groups on the *cellulose* are replaced by CH_3O- groups. Commercially up to 2 hydroxyl groups are replaced which gives water-soluble products which are used as, for example, emulsion stabilizers, thickening agents and paper sizes.

methyl chlorosilane
Also known as methylchlorosilane. Made by the reaction of methyl chloride and *silicon*. This reaction gives mainly dimethyl chlorosilane: methyl chlorosilane and trimethyl chlorosilane are also produced. Such chlorosilanes are used to produce polyorganosiloxanes: that is, to make methyl siloxanes (silicones). See *silicone rubber*.

methyl compound
A compound which contains the methyl group. This univalent organic group CH_3- is easily and cheaply, introduced into organic compounds via, for example, *methyl alcohol*. When reacted with acids, the product (an ester) is often relatively incompatible, on a long-term basis with polymers, unless mixed esters are made or, unless the acid is a long chain acid of high molecular weight. See *plasticizer*.

methyl ethyl ketone
Also known as butan-2-one and ethyl methyl ketone. An abbreviation used is MEK. This flammable liquid has a relative density (RD or SG) of 0·83 and a boiling point of approximately 75 to 85°C. It is a good solvent for uncured *nitrile butadiene rubber (NBR)* and *chloroprene rubber (CR)*. It is a poor solvent for uncured T rubbers. This chemical causes some swelling of uncured *natural rubber (NR)* and *styrene-butadiene rubber (SBR)*. This chemical causes a large amount of swelling, or gel formation, of uncured *butyl rubber (IIR)*. A good solvent for cellulosics, polystyrene, some acrylic polymers and some vinyl polymers. It is used as a solvent for plastics materials and to make adhesives.

methyl ethyl ketone peroxide
Also known as MEK peroxide. An abbreviation used is MEKP. A *catalyst* used, for example, to set or cure *unsaturated polyester resins*. Used in conjunction with *cobalt naphthenate*, so as to obtain curing at ambient temperatures. As this material is unstable, it is used in liquid or paste form: made with solvents or with a *plasticizer* (for example, with *dimethyl phthalate*). The characteristics of this peroxide, as with most liquid peroxides, changes with time so it is important to keep track of the age of this material. Use of unsuitable grades of this catalyst can result in blisters in cured mouldings made by, for example, the *hand lay up process*.

methyl methacrylate
An ester of *methacrylic acid*: a material with the formula $CH_2=C(CH_3)COOCH_3$ and with a boiling point of approximately 100°C. The monomer for *polymethyl methacrylate*.

methyl methacrylate grafted natural rubber
See *polymethyl methacrylate grafted natural rubber*.

methyl methacrylate-acrylonitrile copolymer
An abbreviation used for this material is MMA/ACN or MMA-ACN copolymer. An impact resistant *polymethyl methacrylate polymer*. A tough, transparent thermoplastics material which is based on two monomers and used as, for example, a sheet material as it gives tough transparent products with good rigidity and surface hardness. Used as a glazing material.

methyl methacrylate-acrylonitrile-butadiene-styrene copolymer
An abbreviation used for this material is MABS. This type of material is similar to *acrylonitrile-butadiene-styrene*: some of the *acrylonitrile* is replaced by *methyl methacrylate*. A transparent thermoplastics material which is based on four monomers and used as, for example, an *impact modifier* for *polyvinyl chloride (PVC)* as it gives tough transparent products: has a similar refractive index to PVC over a limited temperature range.

methyl methacrylate-butadiene-styrene copolymer
Also known as methyl methacrylate-styrene-polybutadiene. An abbreviation used for this material is MBS. This type of material is similar to *acrylonitrile-butadiene-styrene*: the *acrylonitrile* is replaced by *methyl methacrylate*. A transparent thermoplastics material which is based on three monomers and used as, for example, an *impact modifier* for *polyvinyl chloride (PVC)* as it gives tough transparent products: has a similar refractive index to PVC over a limited temperature range.

methyl methacrylate-styrene-polybutadiene
See *methyl methacrylate-butadiene-styrene copolymer*.

methyl nadic anhydride
See *nadic methyl anhydride*.

methyl rubber
Also known as dimethylbutadiene rubber or as, dimethylbutadiene polymer or as, polydimethyl butadiene or as, poly-(2,3-dimethylbutadiene) or as, poly-2,3-dimethylbutadiene. An abbreviation used for this material is MR. A *methyl substituted butadiene rubber*: an early *synthetic rubber* produced in Germany in approximately 1917: polymerization times were several months and the product was inferior to *natural rubber*. This type of *synthetic rubber* was made in hard (methyl rubber H) and soft grades (methyl rubber W).

methyl salicylate
Also known as oil of winter-green or as, winter-green oil. This oily liquid has a relative density (RD or SG) of 1·18 and a boiling point of approximately 222°C. Used as an odourant and as a penetrating oil.

methyl silicone
See *polydimethylsiloxane*.

methyl silicone elastomer
A silicone rubber containing mainly methyl groups. See *silicone rubber*.

methyl siloxane
See *methyl chlorosilane* and *silicone rubber*.

methyl substituted butadiene
Butadiene ($CH_2=CH-CH=CH_2$) in which methyl groups are substituted for hydrogen atoms on the butadiene molecule. Such a monomer may be represented as $CH_2=C(CH_3)$-

$CH=CH_2$ for *isoprene*: as $CH_2=C(CH_3)-C(CH_3)=CH_2$ for *dimethyl butadiene*: and, as $CH_2=CH-CH=C(CH_3)$ for *piperylene*. By *diene polymerization*, synthetic rubbers may be produced from all of these materials: at the present time, those based on dimethyl butadiene and piperylene are not commercially important.

methylated spirits
A liquid used as a fuel and which consists of, by volume, approximately 90% ethanol, 9·5% methanol and 0·5% pyridine. Also contains a violet dye and a small amount of petroleum. See *industrial methylated spirits*.

methylbenzene
See *toluene*.

methylene
A divalent organic group. May be represented as $-CH_2-$.

methylene chloride
Also known as dichloromethane or as, methylene dichloride. This poisonous, chlorinated hydrocarbon (CH_2Cl_2) has a relative density (RD or SG) of 1·34 and a boiling point of approximately 40°C. A solvent for *natural* and *synthetic rubbers*: such a material will swell vulcanizates (based on say, *nitrile rubber*) considerably. A solvent used to make solutions of, for example, polyetherimide and polyarylsulphone. See *continuous fibre reinforced prepreg*.

methylene dichloride
See *methylene chloride*.

methylene-bis-(4,4'-phenyl-isocyanate)
See *4,4'-diphenylmethane diisocyanate*.

methylenedomethylenetetrahydrophthalic anhydride
See *nadic methyl anhydride*.

methylisobutyl ketone
An abbreviation used for this material is MIBK. Also known as 4,methylpentan-2-one. A material with the formula $(CH_3)_2CHCH_2COCH_3$ and with a boiling point of approximately 116°C. A solvent, for example, for natural resins, *natural rubber* and *polyvinyl chloride* (PVC).

methylmethoxy nylon
An *N-substituted nylon*: a substituted polymer formed by reacting nylon with *formaldehyde* and an alcohol (*methanol*) in solution (formic acid) in the presence of an acid catalyst. Groups such as $-CH_2OCH_3$ and $-(CH_2O)_2CH_3$ are grafted onto the polyamide so as to give a material which is softer and tougher than the parent nylon and of a lower melting point. Replacement of some of the -CONH- groups (up to approximately a third) reduces hydrogen bonding and increases the solubility of the material.

methylol
A univalent organic radical which may be represented as $HOCH_2-$ and which is also known as hydroxymethyl.

methylol melamine
A reaction product of *formaldehyde* and *melamine* which is formed, for example, when *melamine formaldehyde (MF) resin* is being made. The univalent organic radical $HOCH_2-$ replaces some, or all, of the hydrogen atoms on the melamine. Such methylol melamines will react to produce cross-linked structures during MF fabrication or they can be reacted, for example, with alcohols to produce ethers. The ether derivatives are more oil soluble and are used to produce lacquers.

methylol phenol
Also known as phenol alcohol or as, hydroxymethyl phenol. Formed during the reaction of *formaldehyde* with *phenol*, for example, when *phenol formaldehyde (PF) resin* is being made. The univalent organic radical $HOCH_2-$ replaces some of the hydrogen atoms on the benzene ring in the ortho or para positions. Such methylol phenols will react to produce cross-linked structures during PF fabrication.

methylol urea
Formed during the reaction of *formaldehyde* with *urea*, for example, when *urea formaldehyde (MF) resin* is being made. The univalent organic radical $HOCH_2-$ replaces some, or all, of the hydrogen atoms on the urea. Such methylol ureas will react to produce cross-linked structures during UF fabrication or they can be reacted, for example, with alcohols to produce ethers. The ethers are water soluble and are used as textile finishes.

methylolphenols
Chemical structures formed, for example, during *resole formation*. They are relatively very reactive and become dimethylolphenols and trimethylolphenols: the methylolphenols condense to form *polynuclear phenols*. See *methylol phenol*.

methylphenol
Alternative name for *cresol*. See *methylphenols*.

methylphenols
Also known as cresols. Aromatic chemicals obtained from *coal tar*. May be represented as $CH_3C_6H_4OH$. A mixture of three isomers is used, for example, to make cresol formaldehyde resins.

methylphenylsilicone elastomer
See *silicone rubber*.

methylphenylsilicone fluid
See *silicone fluid*.

methylphenylsilicone polymer
See *silicone resin*.

methylsilicone fluid
See *silicone fluid*.

methylstyrene
This material commercially consists of a mixture of the isomers, m-methylstyrene and p-methylstyrene. See *vinyltoluene*.

methylvinylsilicone
May also be referred to as a vinyl modified silicone: may also be known as a *polysiloxane*. A polymer based on alternating silicon and oxygen atoms in the main chain: the silicon atoms has two organic groups (R & R') attached. The most common polymer is *polydimethylsiloxane* where R = R' = CH_3. If some of the R groups are vinyl groups then methylvinylsilicones result: these are copolymers which are used as rubbers. An abbreviation used for this type of material is VMQ. See *polyorganosiloxane* and *silicone rubber*.

metre
The metre is the SI unit of length which is equal to 39·37 inches or 1·093 6 yards. Defined as being 1,650,763·73 wavelengths in vacuum of the orange-red radiation of krypton-86. It is 1,650,763·73 vacuum wavelengths of the transition radiation between specified levels, $2p_{10}$ and $2d_5$ of krypton-86. It has the abbreviation of m. See *metric system of units* and *Système International d'Unité*.

Mètre des Archives
The *metre* standard originally constructed from a flat bar of platinum and based on measurements of the Earth's polar quadrant. The metre was originally intended to be one ten-millionth part of the Earth's polar quadrant.

metre-candle
See *lux*.

metre-kilogram-second
A system of measurement which uses the metre (m) as the basic unit of length, the second (s) is the basic unit of time and the basic unit of mass is the kilogram (kg). The unit of force is the newton (N) which is the force which imparts to a mass of one kilogram an acceleration of one metre per second per second. Also known as the MKS system.

metric horsepower
See *horsepower*.

metric system of units
The metric system of units is a decimal system. That is, a system only involving multiples of ten. It was originally only intended to be based on a single natural constant which was the *metre* (see *Mètre des Archives*). The kilogram was to be the mass of a cubic *decimeter* of water at 4°C.

metric ton
A *tonne*. An abbreviation used for this unit is t. One thousand kilograms.

MF
An abbreviation used for *melamine-formaldehyde*.

MFI
An abbreviation used for melt flow index. See *flow rate*.

MFR
An abbreviation used for *melt flow rate*. See *flow rate*.

MFRP
An abbreviation used for metal fibre reinforced plastic. See *fibre reinforced plastics*.

mg
An abbreviation used for *milligram*.

MG
An abbreviation used in rubber technology for methacrylate graft. For example, MG 30 is *natural rubber* which contains 30% of *polymethyl methacrylate*. MG 40 contains 40% of *polymethyl methacrylate* and MG 49 contains 49% of *polymethyl methacrylate*. See *polymethyl methacrylate grafted natural rubber*.

MG rubber
See *MG* and *Heveaplus MG rubber*.

mho
A reciprocal ohm. A measure of electrical conductivity. See *siemens*.

mi
Abbreviation used for *mile* and/or for *miles*.

MI
An abbreviation used for melt index. See *melt flow rate*.

MIBK
An abbreviation used for *methylisobutyl ketone*.

mica
A group of minerals with plate-like particles and which are complex potassium/aluminium silicates. These complex silicate minerals, such as muscovite, are famous for their ability to be readily separated into thin sheets (approximately 25 microns in thickness). Used as a filler in polymers, for example, to improve electrical properties. Mica has excellent high temperature resistance. Has been used as a filler to improve the gas diffusion resistance, originally in *butyl rubber* but more recently in plastics, for example, for packaging produced by *blow moulding*. This material has a relative density (RD or SG) of 2·95.

mica paper
A cleaned flake form of *mica* in which the flakes are held together by Van der Waals forces. A composite may be formed by using resin impregnation so as to give resin-bonded mica paper. If a *silicone resin* is used, temperature resistance of up to 800°C may be obtained. Mica paper products offer voltage endurance and excellent corona resistance. Phlogopite mica paper has better heat resistance than muscovite mica paper.

micelle
A cluster or group of molecules. See *emulsion polymerization*.

Michelin sheet
One of the forms in which solid *natural rubber* (unvulcanized rubber) is supplied. A dark coloured form of natural rubber like *ribbed smoke sheet (RSS)*: prepared and dried like RSS but the wet sheets are subjected to *maturation* before being dried in the smoke house. See *standard international grades*.

Michler's ketone
See *4,4'bis-(dimethylamino)-benzophenone*.

micro
An abbreviation used for this term is μ. Means one millionth, that is, 10^{-6}. See *prefixes - SI*.

micro-crystalline silica
Most *silica* is considered to be amorphous but some is *crypto-crystalline*. An example of a crystalline, or micro-crystalline, form is *diatomaceous earth*.

micro-inch
An abbreviation used for this term is μ in. A millionth of an *inch*. One μm is approximately 39·370 079 μ in. See *arithmetical mean deviation*.

micro-porous polyethylene sheet
Also known as porous polyethylene sheet or as, micro-porous PE sheet. Polyethylene (PE) in sheet form which contains fine pores or holes which are inter-connecting. Made from, for example, a PE extruded sheet which is based on ultra-high molecular weight, high density PE (UHMW HDPE), hydrocarbon oil and silica. The extruder is fed with a blend of these materials and, say, a 600 mm sheet is produced: this is then ribbed and the oil extracted by trichloroethylene to create micro-porosity. The sheet is used to make battery separators.

microballoons
Small hollow particles of, for example, glass or of a plastics material. The size is in the region of approximately 0·1 mm: such materials have been used as a *filler* for plastics materials. For example, the incorporation of glass microspheres into *epoxide resin* give mixtures which are easy to apply, which cure at room temperatures and which can be readily sanded. See *syntactic foam*.

microbially produced poly-(3-hydroxyalkanoate)
See *biopolymer* and *polyhydroxybutyrate*.

microbially produced polymer
A polymer produced by the action of a *microorganism*. See *polyhydroxybutyrate*.

microcellular polyurethane
A cellular *polyurethane* which has a relative density of approximately 0·5 to 0·8 g/cm^3. See *polyurethane foam*.

microcellular rubber
A cellular rubber which has closed cells. A rubber compound, containing a *blowing agent*, is moulded at a temperatures which cause blowing agent decomposition: expansion occurs after the rubber moulding is removed from the mould. Benzene sulphonyl hydrazide (BSH) is used as a *blowing agent* as is dinitrosopentamethylene tetramine.

microgel
Very small particles of cross-linked polymer: the size of such particles is approximately 100 nm. In the rubber industry,

the term was first used in connection with *styrene-butadiene rubber*.

microlithography
The use of *photoresist* to produce fine detail, for example, in the field of electronics for microimaging in integrated circuit printing. The finer the detail required the smaller must be the wavelength of the light used, for example, deep UV radiation (wavelength below 280 nm) is used. Novolak/naphthaquinone diazide-based positive photoresists are used for features less than 3 μm in size. See *bisazide compound* and *monoazide compound*.

micrometer
A measuring device used to measure fractions of an inch or, fractions of a millimetre.

micrometre
A unit of length equal to one millionth of a *metre*. Abbreviation μm. See *micron*.

micromixing
Fine mixing achieved in a *reaction injection moulding* (RIM) process. Also known as reaction micromixing. Impingement of the reactants brings them into close proximity: this is then followed by micromixing and diffusion. A pressure at an interface between two reactants drives one reactant into the other in the form of tubules: diffusion completes the mixing process.

micron
A unit of length equal to one micrometre. It is one thousandth of a millimetre or, a millionth of a *metre*. That is, 10^{-6} m. There are approximately 25 microns in one thousandth of an inch (0·001 in). One micron equals 1/25,400 of an inch and there are 10,000 Angstroms to one micron. The symbol for a micron is μm.

micron rating
A term used in *hydraulics* and which refers to the size in microns of the particles which a *filter* will remove.

microorganism
A microscopic, or very small organism, for example, a *bacterium*.

microporous coated fabric
A *plastic coated fabric* in which the coating contains small pores. See *poromeric*.

microprocessor
A component of a microcomputer: it is the data processing part or the central processing unit (CPU).

microprocessor control system
A control system based on one or more microprocessors. Machines equipped with a microprocessor control system usually have a visual display unit (VDU) and keypad entry. Setting is accurate and reproducible. For example, digital setting, rather than analogue setting is now normal and the control system often furnishes a 'prompt' list.

The use of microprocessors (on, for example, *injection moulding machines*) has made setting and monitoring easier. Memories are available for machine and tool setting, for example, for automatic tool setting and changing. A host of additional features can be easily put onto the machine at low cost, e.g. programmed air ejection can be fitted at a comparatively low cost. Safety is another area that has benefited from the application of the microprocessor as a large number of safety checks can be incorporated. Reliability and built-in fault finding of microprocessor-based controllers are other advantages. However, to get the speed of response required, it is necessary to use more than one microprocessor as microprocessors work to a program and it takes time to work through that program. Each of these microprocessors is dedicated to one machine function (for example, core pulling, fault diagnosis, etc.) and each of these microprocessors is connected to a master computer: the master is therefore connected to slave controllers.

For example, a temperature controller is able to communicate with the control system of the injection moulding machine. The master computer is capable of, for example, setting the temperature, altering the control characteristics, or terms, of the temperature controller and then accepting the actual temperature for record purposes. There is a Euromap standard (Euromap 15) which specifies the structure and organization of a microprocessor-based controller. See *work handling device*.

microprocessor controlled machine
A type of machine, for example, an *injection moulding machine*, which has been classified according to the type of control system employed. The machine uses a microprocessor to obtain the desired sequence of movements. See *injection moulding machines*.

microprocessor-based controller
See *microprocessor control system*.

microsphere
A small particle size, hollow sphere of material used as a filler: the most common base material is glass. See *glass microspheres*.

microtome
A machine or apparatus used to cut thin sections (of plastics materials) for microscopic examination.

microwave
An abbreviation used for this term is MW. The microwave region of the electromagnetic spectrum is located between the infra-red region and the radio region. A radio wave with a frequency greater than approximately 300 MHz but below 300 GHz. The wavelength of such radiation is between 1 millimetre and one metre and is used for *dielectric heating*.

microwave heating
At microwave frequencies (UK *ISM bands* 915 and 2,450 MHz), a parallel plate capacitor (such as is used for *high frequency heating*) is not a suitable electrode arrangement as high radiation losses would occur. Conventional valves cannot generate sufficient power and devices such as magnetrons and klystrons are needed together with tubes or waveguides in place of wires: the heating may be performed in a *multimode cavity applicator* or inside the *waveguide itself*.

Microwaves are used for *pre-heating* rubber compounds before *compression moulding*. Microwaves are also used to cure rubber compounds in continuous vulcanization processes. The addition of a polar rubber (for example, *polychloroprene* or *nitrile rubber*) to a non-polar rubber mix (based on say *natural rubber* or *butyl rubber*) can improve the efficiency of microwave heating. The addition of *carbon black* has the same effect. For a material with a very high *loss factor*, attenuation of energy near the surface becomes significant at high frequencies, for example, at 2,450 MHz: differential heating between the surface and the core may result unless a lower frequency is chosen. See *penetration depth*.

middle oil
The fraction, obtained during the distillation of *coal tar*, and which distils at from 170 to 230°C. From this fraction *phenol*, *cresols* and *naphthalene* may be obtained.

mil
An abbreviation used for one thousandth of an *inch*, i.e. 0·001" or, 0·001 in.

MIL
An abbreviation used for a USA military specification.

mild steel
A tough, ductile form of *steel* with a *carbon* content of between 0.12 and 0.25%. The wear resistance of mild steel can be significantly improved by introducing more carbon into the structure. One way of doing this is by *carburising*.

mile
A unit of length equal to 1,760 *yards* in the case of the land or statute mile. Abbreviation mi.

milk - plastics from
See *casein-formaldehyde*.

mill
Usually means *two-roll mill* but see also *ball mill*.

mill mixing
Mixing performed on a mill. See *two-roll mill*.

millboard
A low cost, relatively soft low density board material which is based on *asbestos*. Has good heat resistance and modest mechanical properties. Used in plate-form, for example, as a gasketing material.

milled fibre
Fibre, for example, of *glass*, whose size has been reduced: the milled fibres are not held or bound together. See *hammer milled glass*.

milli
One thousandth part (of a unit). An abbreviation used for this term is m. For example a:

milliampere is one thousandth of an *ampere*;
millibar is one thousandth part of a bar;
milligram is one thousandth part of a *gram*;
millilitre is one thousandth part of a *litre*; and,
millisecond is one thousandth part of a *second*.

milliard
One thousand millions.

millier
French noun for 1,000 kg or one metric ton.

milligram
An abbreviation used for this unit is mg. One thousandth part of a *gram*. That is, 0.001 g which is equivalent to 0.0154 grain or 0.000 001 kg

milligramme
An alternative way of spelling *milligram*.

millilitre
One thousandth part of a *litre (l)* and which is often abbreviated to ml. That is, 0.001 l which is equivalent to 0.039 37 fluid ounces or 0.061 025 cubic inches. The term millilitre (ml) is sometimes used interchangeably with cubic centimetre even though the *litre* is not used for very accurate, scientific measurements.

millimetre
A unit of length equal to one thousandth of a *metre*, or 10^{-3} m; one millimetre equals 0.039 in.

millimetres of mercury
A unit of pressure with the abbreviation mm Hg and which is equal to 133.322 Pa.

Milori blue
See *Prussian blue*.

mineral
Chemical compounds which make up the rocks of the Earth's crust. From such minerals can extract, for example, metals.

mineral blue
See *ferric ferrocyanide*.

mineral calcium carbonate filler
A *filler* based on naturally occurring *calcium carbonate* materials: to produce a filler the mineral is ground and classified. Such materials include *limestone, dolomite, marble, whiting* and *calcite*. Wet-ground materials are of finer particle size and of more uniform size distribution than dry-ground materials.

mineral insulated element
See *electrical resistance element*.

mineral oil
Also known as crude oil and petroleum. An oily dark-coloured liquid which occurs naturally and is the basis of gasoline (petrol) and the petrochemical industry. May be separated by distillation into benzine (solvent naphtha), gasoline, lubricating oil, naphtha, paraffin oil, paraffin wax etc. This material has a relative density (RD or SG) of approximately 0.84 to 0.94: the exact value depends on the type. For example, mineral oil (aromatic) would be 1.02, mineral oil (naphthenic) would be 0.93 and mineral oil (paraffinic) would be 0.86. The boiling point ranges over approximately 40 to 350°C.

In general, mineral oil is a poor solvent for uncured high acrylonitrile *nitrile rubber (NBR)* and *thiokol rubber*. This material causes some swelling of uncured *butyl rubber (IIR)*, *NBR (low acrylonitrile NBR), natural rubber (NR)* and *styrene-butadiene rubber (SBR)*. This material causes a large amount of swelling, or gel formation, of uncured uncured *chloroprene rubber (CR)* rubbers. See *oil*.

mineral rubber
Solid bitumen-type materials such as asphalt (Gilsonite) or oxidized residues of *petroleum*. This material has a relative density (RD or SG) of 1 to 1.04. At up to 10% level it is used as a softener in dark-coloured rubber compounds but up to 100% may be used as an *extender*. Can improve flex cracking and elongation at break of rubber compounds (but not *butyl rubber*).

mineral spirit
An alternative name for *white spirit*.

mineral wax
An alternative name for *ozocerite*.

mineral-insulated metal-sheathed resistance element
See *electrical resistance element*.

minimum moulding pressure
That pressure which is the lowest that can be used successfully to produce a component. The term also refers, in the *injection moulding* industry, to a form of test for ease of flow. It is not a standard test but was fairly popular when injection moulding machines were only fitted with an injection pressure control: this valve controlled the injection line pressure and therefore indirectly, the speed of injection. The test is performed on an injection moulding machine under the production conditions and using the production mould. At a specified time, component production is stopped but the machine continues moulding the components using the set cycle: the injection line pressure is progressively reduced until the cavity, or one of the cavities, begins to short. The result is expressed, for example, as a pressure (psi or bar), obtained under the specified production conditions and is entered on the production records.

minium
See *red lead*.

MIS
An abbreviation used for management information services.

miscible blend
A *blend* based on two polymers which are miscible. Most homopolymers are immiscible and so most blends are immiscible blends. One such miscible blend is that formed between *polyphenylene oxide* and *polystyrene*. See *alloy*.

Mitsubishi Moisture Meter
This instrument measures the *moisture content* of a polymer by heating the polymer sample in an oven to vaporize the water which is then carried by dry nitrogen gas into an electrolytic cell containing *Karl Fischer* reagent. The water changes the polarisation potential within the cell causing a current to flow: this is proportional to the amount of water present. This current liberates iodine from the reagent in direct proportion to the current. This free iodine then reacts with the water and the reagent reducing the polarization potential and the current back to zero thus indicating the end of moisture evolution from the sample.

Because the total current, which flows during the reaction period is proportional to the water released from the sample, the instrument can be scaled to digitally output micrograms of water and there is no need to standardise the Karl Fischer reagent.

mixed acids
Organic acids which contain a range of carbon atoms. See *phthalates*.

mixed alcohols
Organic alcohol which contain a range of carbon atoms. Such materials are commonly used to make esters for use as *plasticizers*. See *phthalates*.

mixed alkyl aryl phosphate
A phosphate compound which contains three organic (both *alkyl* and *aryl*) groups. Such a material is used as a *flame retardant plasticizer*. Such a material usually contain one alkyl and two aryl groups and may be referred to as an alkyl diaryl phosphate. In general, such materials impart good flame retardancy (better than *trialkyl phosphates* but less than *triaryl phosphates*). Examples of such *plasticizers* include diphenyl octyl phosphate (octyl diphenyl phosphate) and isodecyl diphenyl phosphate.

mixed isomeric linking
A long chain polymer based on aromatic groups (benzene rings) which are linked via a number of different ring positions, for example, para and meta position. See *polyphenylene* and *substituted polyphenylene*. This type of linking is often present in polymers because of mixed substitution during polymer synthesis.

mixed metal stabilizer
A polymer stabilizer, for example, a *heat stabilizer*, based on two or more metals, for example, barium and cadmium. Such compounds may be physical mixtures of the components parts in which the organic groups, or parts, may be different or, they may be co-precipitates in which the organic group is the same. See *heavy metal stabilizer*, *cadmium/zinc stabilizer*, *cadmium/lead stabilizer* and *barium/cadmium/lead stabilizer*.

mixed phthalate ester
A phthalate ester made, for example, from alcohols which are a mixture of isomers. See *di-tridecyl phthalate*.

mixed triglyceride
A *triglyceride* in which there is more than one fatty acid residue: could have used 3 different acids to make the mixed triglyceride. See *fixed oil*.

mixed urea-urethane reaction injection moulding
See *urea-urethane reaction injection moulding*.

mixed urethane/sulphur curing
Curing, or vulcanization, of a diene rubber, for example, *natural rubber*, induced by the use of a *urethane cross-linking agent* plus a small amount of *sulphur*. Such a mixture is synergistic. *TMT* is used as a catalyst as *ZMDC* gives *scorch problems*. See *natural rubber - non-sulphur vulcanization*.

mixed vulcanizate
The product of *mixed vulcanization*.

mixed vulcanization
Vulcanization based on more than one type of *vulcanization* system. For example, *sulphur vulcanization* and *ionic vulcanization*. Carboxylated nitrile latex may be vulcanized with sulphur and zinc oxide so as to produce materials with high tensile strength and excellent resistance to abrasion and to oils and fuels.

mixhead
A mixing head. A place where mixing occurs: the impingement chamber of, for example, a *reaction injection moulding machine*. Such a mixhead must (i) permit recirculation of reactants near the mixing chamber, of the mixhead, so as to maintain uniform temperatures and dispersion, (ii) have rapid opening valves which control reactant stoichiometry, (iii) inlet nozzles which accelerate the jets of reactants to high velocities (critical Reynold's number of approximately 300), (iv) a mixing chamber where the streams impinge and mix uniformly throughout the shot, and, (v) a clean out piston to push all reactants out of the mixing chamber. For example, impingement of the reactants brings them into close proximity: this is then followed by micromixing and diffusion. See *high pressure metering*.

mixhead throttling device
Part of *reaction injection moulding* (RIM) *mixhead*: a mixhead which contains a restriction or throttle so as to improve mixing and homogeneity. For example, the reactants flow into a small cylinder and then into a larger cylinder at right angles to the first smaller chamber.

mixing
Usually means *melt mixing* but also see *blending*.

mixing blade
See *mixing knife*.

mixing head
See *mixhead*.

mixing hopper
Part of a moulding or extrusion machine which is used to achieve *mixing* and blending on the machine. Mixing hoppers are available which can be mounted above a machine hopper and these hopper mixers will automatically proportion, blend and feed a continuous supply of, for example, virgin material, colour concentrate and reclaimed material to the moulding machine. Hoppers are available which can handle four components and give an output of up to 4 tonnes per hour.

There are obvious attractions in feeding, for example, an *injection moulding machine* with the required ingredients and performing the initial blending operation (pre-blending) on the moulding machine followed by *melt mixing*. The adoption of such a procedure simplifies storage, materials handling, reduces the chances of contamination and lowers costs. However, in general a single screw does not give very good mixing and so use is often made of mixing elements or sections. See *distributive mixer*.

In one mixing hopper system, materials are dropped through tubes on to a metering disc. Materials are proportioned by volume onto this disc in ratios determined by the gap-height of the material metering tubes - these are mounted above the disc so that there is a gap between the tube and the disc. The higher the tube, the greater the gap and the greater the material flow. Each tube has a vernier adjustment which is set to obtain the desired mix ratio. The proportioned materials drop through the centre of the metering disc where they are mixed by a *cascade* located in the blend supply hopper.

mixing knife
A knife used to assist mixing. For example, if mixing is done on a *two-roll mill*, then a manually held knife may be used: the knife has a short (approximately 25 mm/1 in) blade. The back of the blade is straight and the blade angles at approximately 40° away from the sharp point. See *mixing knives*.

mixing knives
Also called mixing blades. A *two-roll mill* may have two such blades: one mounted above each roll and running the full length of the *roll*. As the knife is made to bear against the full length of the roll it makes a cut in the *hide*: this cut material then rolls up behind the blade and returns to the *nip* when the blade is released.

mixing pins
Pins which are protrude from the root of a *screw* and which improve *melt mixing*.

mixing section
A section of a *screw* which is designed to improve *melt mixing*. Such a mixing section should first give *dispersive mixing* and then it should give *distributive mixing*. These two mixing processes may be repeated more than once. Any mixing section should:

 (a) not cause a pressure drop and ideally, should have a positive pumping action;
 (b) not possess dead spots or material hang-up regions. That is, flow through the section should be streamlined;
 (c) completely wipe the cylinder, or barrel, surface;
 (d) be easy to strip, clean and re-assemble; and,
 (e) be cheap to purchase, install and maintain.

See *zero compression screw*.

mixing system - calendering
The mixing machines used to prepare the hot strip fed to a calender may be referred to as the mixing system. In the thermoplastics *calendering* industry, a *pre-mix*, from a *high speed mixer*, is conveyed to the *melt mixer* (for example, an *internal mixer*). Pigment may be added at this stage and the batch is mixed for the requisite time, for example, 3 minutes or until a certain temperature reached eg 180°C. The compounded material is then dumped onto a *two-roll mill* operating at say 135°C as the output from the internal mixer is not in a very convenient form. The mill converts the material to a continuous sheet or band around the front roll and by cross cutting and folding the sheet it may be made more homogenous. The temperature is therefore made more uniform and reduced to 135°C. A strip of this gelled material is continuously removed from the mill, using a double-bladed, weight loaded knife, and fed to a *strainer extruder*. See *automatic hide turning device*.

This type of mixing system is useful where changes of colour, or of formulation, are frequent: continuous compounders, for example, the Buss Kno-Kneader, are more appropriate where long runs, based on the same formulation, are usual. On short runs, cleaning could be a problem and with some continuous compounders, edge trim cannot be returned directly to the feed hopper of the machine as they can be with an *internal mixer* of the Banbury-type.

MKS system
An abbreviation used for the *metre-kilogram-second system of measurement*.

MKSA metric
See *Système International d'Unité*.

ml
An abbreviation used for *millilitre*.

mm
An abbreviation for *millimetre* or for millimetres.
 mmHg = mm Hg = millimetres of mercury.
 mmH$_2$O = mm H$_2$O = millimetres of water.

MM
An abbreviation used for *mat moulding*.

MMA
An abbreviation used for *methyl methacrylate*.

MMA/ACN
An abbreviation used for *methyl methacrylate-acrylonitrile copolymer*.

MMC
An abbreviation used for *metal-matrix composite*.

mnemonic code
Instructions written in concise, or abbreviated, language which is easy to remember. That is, a code written as a memory aid, for example, INJ for 'inject the material'.

mo
An abbreviation used for month.

MOBS
An abbreviation used for *N-morphlinyl benzthiazyl sulphenamide*.

modacrylic fibre
A manufactured, polymeric fibre composed of less than 85% by weight of acrylonitrile units -(CH$_2$-CHCN)- but more than 15% of polyacrylonitrile. See *acrylic fibre*.

modal fibre
A generic term for high tenacity, high wet strength, viscose fibre. Also known as HWM modal or, high wet modulus modal. See *viscose rayon* and *polynosic fibre*.

mode
A particular pattern of electromagnetic energy distribution within a confining structure caused by the interaction of two or more standing waves.

mode stirrer
A device which alters the *mode* within a cavity so as to obtain more uniform heating.

modified natural rubber
Natural rubber which has been chemically modified by the introduction of, for example, highly polar groups introduced by chemicals such as *ENCAF*. See *natural rubber* and *superior processing rubber*.

modified poly-(oxy-1,4-phenylene)
See *polyphenylene oxide (modified)*.

modified poly-(phenylene oxide)
See *polyphenylene oxide (modified)*.

modified polyimide
Included in this category of plastics materials are *polyamide-imide*, *polyester-imide*, *polybismaleimide* and *polyether imide*. Such materials were developed in an effort to obtain more tractable polymers while retaining the desirable heat resistance of the parent *polyimide*.

modified polyphenylene ether
See *polyphenylene oxide (modified)*.

modified polyphenylene oxide
See *polyphenylene oxide (modified)*.

modified polyphenylene oxide/polyamide blends
See *polyphenylene oxide/polyamide blends*.

modified polypropylene
See *rubber modified polypropylene*.

modified polyvinyl chloride
Term sometimes used for a *vinyl chloride copolymer* which contain a relatively small amount of the second monomer.

modifier
A material which, for example, changes a polymerization reaction. See also *impact modifier*.

modular machine
Usually means an *injection moulding machine*. An injection moulding machine made from units, for example, from an injection unit and a clamp unit and which are available in different sizes. The use of modular construction has allowed for the relatively easy development of special purpose machines.

modulus
Modulus usually means *Young's modulus (E)*: it is also referred to as tensile modulus and the modulus of elasticity in tension. It is the elastic modulus for uniaxial extension and is usually obtained from a tensile test determination. It is the slope of the stress strain plot for small extensions: that is, stress divided by strain. It has the units of force per unit area, for example, Nm^{-2} or psi. For such small extensions this elastic modulus is approximately the same as the compressive modulus. Young's modulus depends upon the chemical nature of the particular solid for conventional materials and it cannot usually be changed. To get higher modulus, a different material must be chosen. However, there appears to be an upper limit on the *specific modulus* possible from conventional engineering materials.

modulus of elasticity in tension
See *modulus*.

modulus of rigidity
See *shear modulus*.

modulus of rupture
See *cross breaking strength*.

Moh hardness
A hardness scale, from 1 to 10, which uses the hardness of minerals as a yardstick. Talc is one and diamond is ten. This is one of the oldest hardness tests. A hard object, precisely ground to a point, is drawn over the test surface to produce a scratch. Polymethyl methacrylate rates as 2 to 3 on the Moh, GP polystyrene and cellulose acetate as 1 to 2, while the average fingernail has a Moh's hardness of 2. This test is usually referred to as a scratch hardness test and although it could be used for such things as coatings it lacks precision.

Mohr's scales
A weighing machine used to measure *relative density* from the loss of weight of a test sample when immersed in water.

Mohs scale of hardness
See *Moh hardness*.

moil
A term sometimes used for the top flash on a *blow moulding*.

moiré-fringe principle
Used in *extensometers* so as to obtain very accurate measurements of displacement. See *creep testing*.

moisture absorption
See *water absorption*.

moisture content
A measure of how much water is contained in a material and often expressed as a percentage. With many plastics materials, the level of water/moisture in the material fed to the processing equipment, for example, to an *injection moulding machine*, must be kept below very small values: for example, below 0.2%. This is usually to prevent the production of mouldings with a poor surface finish.

In many moulding shops, the raw materials are dried as a matter of course so as to prevent faults in production. The use of such 'dry' material prevents the production of mouldings with a poor surface finish: with some materials, for example PC, the product can be weakened by the use of 'wet' material. Simple tests may reveal the presence of surplus moisture - see *Tomasetti's Volatile Indicator method* and *GFT test*.

The maximum permissible limit for moisture in the granules for the production of good mouldings varies from polymer to polymer. For example, 0.02% for polycarbonate (PC) and 0.4% for cellulosics (see **table 12**).

moisture content - measurement of
There are several ways of assessing the moisture content of moulding materials, that is, of granules. See, for example, *oven drying assessment, Karl Fischer method* and *Tomasetti's Volatile Indicator method*. However, the Karl Fischer method is time consuming, involves the use of pyridine, an unpleasant smelling liquid, and requires considerable technical skill. Drying to constant weight is also time consuming and often not satisfactory where very low levels of moisture need to be measured. Consequently, proprietary instruments, which require little skill and provide accurate results in most cases for batches of polymers in less than 30 minutes have been developed. For example, the Dupont 903 moisture evolution analyzer, the Brabender Aquameter and the Mitsubishi Moisture Meter. Such instruments work on different principles: the moisture content may be assessed electrically or, by reacting the moisture with calcium carbide. A separate weighing facility as well as the moisture-measuring instrument is usually required.

moisture curing silicones
See *one-pack systems*.

moisture regain
Also known as regain. For example, weight percent of moisture gained by a dry fibre (dry at about 110°C) after exposure to a 65% relative humidity atmosphere at 21°C.

mol
an abbreviation used for *mole*.

mol wt
An abbreviation used for molecular weight.

Moldflow data
Shear flow data (obtained from the use of a *laboratory capillary rheometer* which is available in the materials data base maintained by Moldflow (Europe) Ltd., Orpington, Kent, UK. Such information is used to predict how an injection mould will fill. For this purpose the effect of changing temperature at a constant *shear rate* is needed together with values which show, for example, the effect of changing shear rate at a constant temperature. By plotting *shear stress* (Nm^{-2}) against shear rate (s^{-1}) *flow curves* may be obtained; viscosity ((Nsm^{-2}) is obtained by dividing shear stress by shear rate.

mole
The basic SI unit of the amount of a substance which has the abbreviation mol. Defined as being the amount of substance that contains as many elementary units as there are atoms in 0.012 kilograms of carbon-12. See *Système International d'Unité*.

molecular orientation
The *orientation* of polymer molecules.

molecular sieve
A solid crystalline substance which is capable of trapping other molecules in small pores: formed when water of crystallisation is removed from sodium or calcium aluminium silicates. Used in chromatography and now being used to reclaim *chlorofluorocarbons* (CFC).

molecular sieve accelerator
An *accelerator* which has been adsorbed onto a *molecular sieve*. Such a chemically loaded, molecular sieve forms an *accelerator* with a delayed action. Good processing safety is obtained together with rapid vulcanization rates.

molecular weight
Also known as molar mass or as, relative molar mass or as, relative molecular mass: an abbreviation used is MW. The ratio of the mass of an individual molecule of a substance to that of $1/12$ of a carbon-12 atom. See *molecular weight distribution*.

molecular weight average
Also known as average molecular weight: an abbreviation used is $\overline{M}$ or M bar. Polymers do not have a single *molecular weight* as they are made from chains of varying length (see *molecular weight distribution*). The molecular weight average that is obtained depends on the way that the average is obtained and calculated. Can have, for example, the number average molecular weight or $\overline{M}_n$, the weight average molecular weight or $\overline{M}_w$, z average molecular weight or $\overline{M}_z$ and the z + 1 average molecular weight or $\overline{M}_{z+1}$. The number average molecular weight is more sensitive to low molecular weight species whereas the others are more sensitive to higher molecular weight species. See *molecular weight determination*.

molecular weight change
Most commercially available polymers are susceptible to *thermal* and photo-initiated *oxidation* which change the properties of the polymer. Once a polymer has been produced such a change is usually to be avoided as the change will alter the physical properties of the product. However, in some cases, the molecular weight of the polymer is changed deliberately - see *mastication* and *constant viscosity grades*.

molecular weight determination
Many methods or techniques have been evolved for determining the molecular weight of polymers. As most methods rely upon getting the polymer into solution, they can only be used to determine the *average molecular weight* of un-crosslinked materials. Vapour pressure lowering, cryoscopy, ebulliometry and osmometry give number average molecular weight. Dilute solution viscosity gives a solution viscosity, average molecular weight. The weight average molecular weight, is given by light scattering: sedimentation gives number average molecular weight, weight average molecular weight and the z average molecular weight. The formula is:

$$\overline{M} = \sum_i N_i M_i^{n+1} / \sum_i N_i M_i^n$$

When n = 1, then this is the number average molecular weight.
When n = 2, then this is the weight average molecular weight.
When n = 3, then this is the z average molecular weight.
When n = 4, then this is the z + 1 average molecular weight.

molecular weight distribution
Also known as molar mass distribution (see *molecular weight*): an abbreviation used is MWD. The distribution of molecular sizes in a polymer. Polymers do not have a single molecular weight as they are made from chains of varying length. If a polymer has a broad MWD distribution, then generally, processing is easier.

For example, such a broad MWD greatly reduces any tendency to *sharkskin* effects. However, a broad molecular weight distribution may also result in warping of injection mouldings. See, for example, *high density polyethylene* and *linear low density polyethylene*.

molecularly bound antioxidant
See *bound antioxidant*.

molten salt bath
A heated bath used to achieve *continuous vulcanization*: consists of a fused nitrate/nitrite mixture. See *molten salt bath vulcanization*.

molten salt bath vulcanization
Used for *continuous vulcanisation*. Also known as salt bath vulcanization. Uses a *molten salt bath* (*LCM process*), at say 220°C, which is heated electrically. The bath is approximately 10 m long.
This system gives good heat transfer characteristics. However, a potentially dangerous oxidising material is used, for example, any water causes a minor explosion. The rubber product must be pushed below the surface, for example, by a steel conveyor belt because of the high density of the bath: there is also a risk of marking the unvulcanized material. The product must also be washed to remove the salt and if the electricity fails then there is difficulty in re-warming the bath. However, the process is labour saving and low reject rates of visually attractive product can be achieved.

molybdate oranges
A class or type of *inorganic pigment*. See *lead chromates*.

molybdate reds
A class or type of *inorganic pigment*. See *lead chromates*.

molybdenic oxide
An oxide of molybdenum used as a *smoke suppressant*. Sometimes used as a synergistic additive, with halogen compounds, in place of *antimony oxide*.

monazo pigments
These pigments contain one *azo* group and are often yellow and red pigments. Hansa yellow pigments are examples of this class of materials as are toluidine reds and naphthol reds. See *mordant* and *azo pigment*.

mono-ester
An organic ester prepared from a *polyol* (for example, *propylene glycol*) and an organic acid (for example, *ricinoleic acid*) and in which only one of the hydroxyl groups is used in the esterification reaction. See, for example, *propylene glycol ricinoleate*.

mono-sulphide cross-link
A cross-link which is based on one sulphur atom. The introduction of short sulphur cross-links (mono-sulphide cross-links and di-sulphide cross-links) during vulcanization, improves vulcanizate properties as opposed, for example, to the formation of polysulphide cross-links which do not. To obtain more *efficient vulcanization*, use small amounts of sulphur (0.5 to 2 phr) and relatively large amounts of *accelerator* (3 to 6 phr). Even lower *efficiency parameter values (E)* may be obtained by using a thiuram efficient vulcanization system or, a cadmate efficient vulcanization system.

monoaxial orientation
Term usually associated with *film*, *tape* or *fibre*: the term refers to single direction *orientation*; this is produced by drawing, or pulling, in the *machine direction*.

monoazide compound
A compound which contains one *azide* group. See *photoresist*.

monochlorinated poly-p-xylene
See *poly-p-xylene*.

monochlorobenzene
See *chlorobenzene*.

monochloroethylene
See *vinyl chloride*.

monochromatic light
Light of mainly one colour or wavelength. In plastics technology, such light is sometimes used, in conjunction with *polarizing filters*, to detect excessive *orientation*, or *frozen-in strains*, within injection mouldings made from transparent materials.

monofil
An abbreviation used for *monofilament*.

monofilament
An abbreviation used for this term is monofil. A single strand, or filament, of a plastics material used as brush bristles, fishing line or, as rope when twisted together.

monomer
The starting material, or unit, on which a synthetic polymer is based: the starting material for *polymerization*.

monomer casting
The *casting* of a component from a *monomer*. Bulk polymerization performed in a mould so that a shaped moulding is the result. It is the problem of removing the heat of *polymerization* which restricts the application of monomer casting. Used to make cast sheet from *polymethyl methacrylate* and to produce cast *nylon 6* products from Σ-caprolactam.

monomeric plasticizer
A material, an ester, of comparatively low molecular weight (but still usually greater than 300) used as a *plasticizer*. Adipates, phosphates and phthalates are classified as monomeric plasticizers. See *permanent plasticizer*.

mononuclear phenol
Structures based on phenols which contain one aromatic nucleus. See *hindered phenol*. Mononuclear phenols are formed, for example, during *resole manufacture*.

montane wax
A wax obtained from brown coal by, for example, extraction with organic solvents. A black-brown wax, of fossil origin, with a relative density of approximately 1·0 and a melting point in the region of 75°C. Used, for example, to obtain a shine on hard rubber compounds.

Montreal protocol
55 United Nation members were originally party to this agreement which seeks, for example, to phase out the chemicals that are known to deplete the *ozone layer*. For example, *chlorofluorocarbons* which cause stratospheric ozone depletion. Carbon tetrachloride, with an *ozone depletion potential* (ODP) of 1·2, should no longer be in use by the year 2,000: methyl chloroform, with an ODP of 0·10, will be out of use shortly after the year 2,000.

Mooney
See *Mooney viscometer*.

Mooney cure time
Also known as cure time. A measure of cure behaviour assessed by a *Mooney viscometer*. The time taken for the Mooney viscosity to rise from five units to thirty five units above the minimum value. The temperature of test is that which considered appropriate for the compound and the process. See *cure index*.

Mooney scorch time
Also known as scorch time. A measure of scorch behaviour as assessed by a *Mooney viscometer*. The time taken from the beginning of the warm-up period to that at which the Mooney viscosity is five units above its minimum value. The temperature of test is that which considered appropriate for the compound and the process. See *Mooney viscosity value*.

Mooney viscometer
Also known as the Mooney: a shearing disc viscometer. A test apparatus widely used in the rubber industry. A shearing disc viscometer used to measure the plasticity or viscosity of rubbers and their compounds. Used to assess the scorch time and/or cure time of a rubber compound. The apparatus contains a closed heated cavity within which a knurled disc may be rotated: the cavity is formed in two halves which may be opened and closed. The cavity walls are knurled/grooved/serrated so as to prevent the rubber from slipping. A known amount of rubber is placed inside the preheated chamber and preheated for a specified time, for example, one minute at the test temperature of say 100°C. The rubber compound is then trapped between the disc and the walls of the cavity. At the end of that time, the rotor is driven at 2 rpm and the resistance to rotation of the rotor is measured after, for example, 4 minutes, and called the *Mooney viscosity value* or the *Mooney viscosity*.

Mooney viscometer value
See *Mooney viscosity value*.

Mooney viscosity
See *Mooney viscosity value*.

Mooney viscosity test
A test done with a *Mooney viscometer*. This apparatus is used to measure the plasticity or viscosity of rubbers and their compounds. Used to assess the scorch time and/or cure time of a rubber compound. See *Mooney viscosity value*.

Mooney viscosity value
Also known as Mooney viscosity and Mooney viscometer value. An abbreviation used for Mooney is M. The Mooney viscosity value is the result obtained from a Mooney viscosity test: that is, when a *Mooney viscometer* is used. The value may be reported as 45-ML 1 + 4 (100°C). This means that the Mooney viscosity (M) is 45 when the large rotor (L) was used: the large rotor (L) has a diameter of 38·1 mm. (The small rotor, sometimes denoted as S has a diameter of 30·5 mm). The compound warm-up time was 1 minute, at the test temperature of 100°C, and the figure 4 shows that the large rotor was rotated for 4 minutes before the torque reading was taken. See *Mooney viscometer* and *scorch time*.

Moore's efficiency parameter
See *efficiency parameter*.

mordant
A substance used to bind a *dye*. Originally developed to bind dyes to fabrics. The fabric was impregnated with the mordant before dyeing so as to form an insoluble lake. Basic metal hydroxides were used for acid dyes and acidic substances for basic dyes. See *mordant dye*.

mordant dye
A colouring system formed from a *dye* and a *mordant*. For example, a colouring system used for plastics materials is Pigment Scarlet 3B which is formed from the dye Mordant Red 9. The dye is mixed with freshly precipitated aluminium hydroxide and then barium chloride and zinc oxide are used to precipitate the dye onto this base. This gives an insoluble lake which is a *monazo-type pigment*.

morpholine
Also known as tetrahydro-1,4-oxazine or as, diethylene oximide. A six membered heterocyclic ring compound which may be represented as C_4H_9NO. This material has a boiling point of 129°C, a melting point of −5°C and a relative density (RD) of 1·0. Used as a solvent but is also a basic starting material for *antioxidants* and *accelerators*.

morpholine disulphide
Also known as 4,4'-dithiomorpholine or as, dithio dimorpholine or as, dithiodimorpholine. An abbreviation used for this material is DTDM. This grey solid material has a melting point of 122°C and a relative density (RD) of 1·29. A *sulphur donor* for rubbers which yields large amounts of sulphur at vulcanization temperatures but which is stable at processing temperatures (also yields *morpholine*).

morphology
The study of form and structure on the microscopic level, for example, the study of crystal structures in a *semi-crystalline, thermoplastics* material.

motor
A term used in *hydraulics* and which refers to a device which converts fluid flow into rotary motion. See *hydraulic motor*.

motorised roll adjustment
See *nip adjustment*.

mould
See *compression mould* and *injection mould*.

mould breathing
See *breathing*.

mould cavity
Also called the mould impression: the female part of a mould. The space in the mould which is used to form the moulded component. See *cavity dimension*.

mould changer
See *automatic mould changing*.

mould closing speed
Usually measured in seconds and is the time taken to close all parts of the mould together. Just like the mould opening speed, it should be set to as high a value as possible, as it is an 'idle' speed (that is, it produces nothing). However, the speed selected should not cause mould damage: for this reason it should be possible to program the mould closing/opening speeds so that initial mould opening, and final mould closing, are done relatively slowly.

mould cooling capacity
This is obtained by calculating the heat input to the mould. A simple formula for determining the amount of heat to be removed (Q) requires that the *enthalpy* or heat content of the shot is known and that the cycle time is known or can be estimated.
$Q = 3,600(H_m - H_e)G/T$ where H_m is the enthalpy of the material at moulding temperatures, H_e is the enthalpy of the material at ejection temperatures, G is the shot weight in kg and T is the cycle time.

mould cooling channels
Channels contained in, for example, the *injection mould*. Most of the heat that is contained within the *thermoplastics moulding material* is removed, during the setting part of the moulding cycle, by circulating a relatively cool fluid through channels contained in the mould. The size and lay-out of the channels must be such that rapid and uniform cooling of the moulding results. Rapid cooling improves process economics, whilst uniform cooling improves product quality by preventing differential shrinkage, internal stresses and mould release problems. In addition, uniform cooling ensures a shorter moulding cycle. In some factories the injection moulds are connected to the mains cold-water supply and the mould temperature is regulated by throttling the supply of water to the mould. Such a practice can result in a mould whose temperature is non-uniform and whose cooling efficiency is not very high. To minimise such problems it is necessary to ensure that the correct quantity of temperature-controlled water is circulated through the mould *cooling channels*. The difference in temperature between the inlet water and the outlet water should not be excessive (for example, above 6°C as otherwise 'hot-spots' will result and/or the mould surface temperature will be higher than expected. If the permitted temperature rise in the circulating water is 6°C then each litre of water can remove up to 6 kcal from the mould.

mould - electrical deposition of
See *electroforming*.

mould - estimate for
The companies asked for an estimate should be of the same type, that is, those who specialise in producing certain types of mould as such a specialist company will know, from experience, the quality of component that it can produce and the time it takes to produce this component. Each of the companies asked to tender for the mould manufacture should be supplied with the following details:

the component design (including critical component dimensions and the moulding material);
a tool design or specification - which should include the hardening requirements, number of impressions, position of gate, etc.;
the cycle time expected or demanded;
the moulding machine details;
the expected or demanded life of the mould - the estimated production per annum should be specified; and,
any sampling or pre-production requirements.

mould fouling
A moulding fault associated with the moulding of rubber compounds. A mould becomes fouled when deposits build up on the mould surface which eventually interfere with production. The period from initial start-up to the point where the product is unacceptable varies from compound to compound but is undoubtedly reduced by high-temperature working during *rubber injection moulding*. Both air trapping and mould fouling can be minimised by using a *vacuum injection mould*.

mould half
A part of a mould which may be identified as a *fixed mould half* or as, a *moving mould half*. A *mould plate* may comprise a mould half in some simple mould designs.

mould - heating of
The *injection moulds* used for rubbers and for thermosetting plastics materials are normally heated by *electrical resistance elements*. To reduce heat losses and to protect the operator against the danger of burning, the mould should be insulated as far as possible. To minimise heat losses to the machine platens the moulds should be insulated from the platens. If this is not done then the heat loss to the platens may cause such an expansion that the platens will no longer be able to slide on the tie-bars.

Each mould part or plate should have at least one temperature controller and the aim should be control to within ±2°C. Such control can be achieved using thermocouple-actuated instruments, for example, a *three-term controller*. Because heater bands burn out, without any obvious indication that they have done so, a useful ancillary is a meter which shows the amount of power flowing in a heating circuit when the control instrument is demanding power.

The number of installed kilowatts necessary to electrically heat a mould can be estimated if the mould weight and the heating-up time are known. Once the mould is up to temperature only a small proportion of this installed capacity is necessary to maintain the mould at the required temperature.

The number of installed kilowatts equals:

$0.277 mc(t_m - t_i)s/1{,}000t$. Where m is the mould weight in kg, c is the specific heat of steel, t_m is the required mould temperature in °C, t_i is the initial temperature, s is a safety factor (for example, 1·6) and t is the heating up time (h). In general, about 1 kW of installed power is needed for every 50 kg of mould weight.

A *compression mould* is often *steam heated*.

mould height adjustments
See *injection moulding*.

mould identification
To simplify repair, replacement and/or future installation it is important to ensure that each mould is clearly labelled with an appropriate number or code: subsequent repair by welding will be eased if each component part of the mould is identified with the material from which it was made. Each fluid inlet and outlet should be clearly identified. It may prove useful to have a wallet or bag associated with each mould which contains a fluid circulation diagram, the types and lengths of tubing used, mounting features, repair or maintenance card, etc. An acceptable moulding should also be contained within this wallet so that deteriorations in product quality, caused through damage, can be seen. See *injection mould*.

mould insert
Part of a mould which is made separately and then fitted as part of the *punch* or *mould cavity*.

mould lubricant
See *mould release agent*.

mould mark
A fault on a moulding caused by a corresponding mark on the mould.

mould - materials of construction
Most moulds are constructed from steels of various types, but because of the cost and long delivery times of such moulds there is always considerable interest in other, more easily worked, materials (see *prototype mould*). Different grades or types of steel are used to manufacture the various parts of the mould as each of these mould elements performs a different function within the total mould assembly.

Mild steel is used where there is no need for a more expensive or harder grade of material, e.g. for bolster work. For example, in *injection moulding* as the moulding is formed between the core and the cavity, it follows that the surface finish, wear resistance, etc. of these two mould elements are very important and they are commonly made from a *pre-hardened steel*. The *sprue bush* may be made from nickel chrome steel as this material resists 'pick up' or marking; the ejector pins and the *sprue-puller pin* may be made from a chrome vanadium steel as this material is relatively flexible and yet has a hard surface. See *mould steel*.

It is a mistake to allow the cost of the steel to enter into the mould cost calculations; this is relatively unimportant compared to the far greater cost incurred when machining, for example, a hard grade of steel. Before any steel is selected a number of factors should be taken into account. Amongst these are the number of components which are required, mould and part complexity and the material which is being moulded.

mould mounting
Moulds may be attached to the machine platens by means of bolts which pass through holes in the mould back plate: the bolts then engage in threaded holes in the machine *platen*. Standard tapping patterns are available, for example, for *injection moulding machines*. The moulds may also be held in position by means of wedges which engage within slots machined into the sides of the mould; such wedges are tightened by means of bolts or, in the case of some automatic systems, they are tightened hydraulically. If the mould back plates are of a standard size then significant time-saving should result during mould changing.

mould opening speed
Usually measured in seconds and is the time taken to fully open the mould, that is, as though a component had just been ejected. Just like the mould closing speed, it should be set to as high a value as possible, as it is an 'idle' speed (that is, it produces nothing). However, the speed selected should not cause mould damage: for this reason it should be possible to program the mould closing/opening speeds so that initial mould opening, and final mould closing, are done relatively slowly.

mould plate
A steel plate which either contains a *cavity* or carries a *core*. A mould plate may comprise a *mould half* in some simple mould designs.

mould plating
The plating of a mould with a corrosion resistant, or wear resistant, metal. Some plastics materials can degrade to give corrosive materials (e.g. acids) or they may contain additives which degrade to give corrosive substances. Some blowing agents, for example, fall into this last category. To resist such corrosion the cavity may be made from stainless steel or the tool steel may be plated.

Chromium is the most common plating material although other metals have been used, e.g. gold and nickel. Gold-plated moulds have been used for handling *unplasticized polyvinyl chloride* (UPVC). This is because gold has excellent resistance to acid attack and it will protect the underlying steel even when a very thin layer is used. Techniques have been developed which allow the easy application of such metals, for example, in-house metallising

mould register
Also called a location spigot. A circular disc, or ring, attached to the base plate of a mould which locates in a register hole in the *platen* of a moulding machine.

mould release
The loss of adhesion between a moulding and the mould. See *mould release agent*.

mould release agent
A material painted or sprayed onto a *mould* in order to prevent adhesion of the moulding to the mould: an *external mould release*. An *additive* used to achieve component release after a moulding operation: an *internal mould release*. See *release agent*.

mould shrinkage
Mould *shrinkage* is defined as the change in dimensions between the size of the cavity and the moulding, 24 hours after the moulding is ejected from the mould. Mould shrinkage is quoted as a percentage, for example, 0·4%, or as a linear shrinkage, for example, 0·004 in/in or 0·004 mm/mm. Usually mould shrinkage increases when increasing mould and/or melt temperatures.

mould size adjustments
See *injection moulding*.

mould sizing
See *cavity dimensions*.

mould spraying
A method of applying a *multichange* on a mould manufacturing method. Low melting point alloys (e.g. based on bismuth) may be flame-sprayed onto a model of the particular component and the shell so formed may be backed with a suitable alloy or with concrete so as to make large prototype *injection*

moulds. Because of the softness of the alloys the life expectancy of the moulds is short, e.g. 5,000 shots. The metal shell may also be produced by *electroforming*.

mould steel
Also known as tool steel and/or, as alloy steel. Several different classes, for example, *case hardening, nitriding, fully hardened, corrosion resistant, maraging, pre-hardened* and *hard material alloy steels*. The prefix used for mould steels is P in the *AISI/SAE system*.

mould temperature control
The control or regulation of the mould temperature. For example, in the case of an electrically heated mould this is performed with a *thermocouple actuated temperature controller*. In the case of a *steam heated mould* this is performed by regulating the steam pressure with a *thermocouple actuated temperature controller*. In the case of a *fluid heated mould* this is performed by regulating the fluid temperature.

In the case of thermoplastics, heating may be required in order to raise the temperature of the mould to the required operating temperature but once *injection mouldings* are being produced then heat will need to be removed. Many thermoplastics are moulded using a heated mould but even in these cases the function of the mould is to remove heat from the moulded component. Ideally, therefore, the mould should be designed as a heat exchanger so that a uniform temperature is obtained over the entire surface of each mould cavity at the beginning of each cycle. Most of the heat is removed by circulating a liquid (for example, water) through channels, machined or cast, in the mould. The temperature in the fluid system must then be reduced before it can be recirculated through the mould cooling channels. Both *direct* and *indirect cooling systems* are used.

Mould temperature obviously has a large effect on cycle time but it also has a great effect on shrinkage, surface finish, stress level and ease of cavity filling. If the temperature of the mould is not controlled then it will alter during the moulding run and components with differing properties will result.

mould thickness
Usually means the total thickness when closed. See *injection mould thickness*.

mould venting
Allowing the air to escape from a mould cavity as the material is forced into that cavity or, applying a vacuum to the mould so as to force the air from the cavity. Because of the speed of mould filling (in processes such as *blow moulding* and *injection moulding*) the mould must be vented to allow for gas escape; in most cases this gas is simply the air in the cavity although *volatiles* are emiited by some melts.

Thorough venting is necessary so as to allow the air in the cavity to escape easily so that the trapped air does not impair surface finish by stopping the material from contacting the mould surfaces. Venting is particulary important if mould filling speeds are fast: in the extreme case the plastics material can be degraded by burning as the temperature of the trapped air is rapidly raised by compression. Such venting may be via the ejection pins or via slots machined along the parting line of the mould: *vacuum venting* is also used. See *vents*.

mould-halt
A moulding technique used, for example, during *injection moulding* of a *thermosetting plastic* and useful for insert loading. After ejection the mould moves forward, to seat the ejectors, and then the mould is stopped for insert loading.

moulded hose
A *rubber hose* which is *vulcanized* inside a mould or sheath: the mould or sheath is removed after vulcanization.

moulding assessment
The quality of many mouldings is often only assessed, for example, by a simple visual examination. That is, if it looks good then, it is judged to be good. For example, if the machine is running semi-automatically then, the operator inspects every moulding for colour, colour dispersion, colour uniformity, surface finish (gloss/mattness) and freedom from moulding defects such as, in the case of injection moulding, stringing and silver streaking.

It is often preferable to assess the mouldings by methods which give numerical answers as the figures so produced may then be readily subjected to statistical analysis. In this way accurate checks on moulding quality, and predictions on future moulding quality, may be made. Usually the mouldings are either weighed and/or, a dimension is checked. Weight is often easier, and simpler, to measure for an operator than a dimension of a moulding: this is because of developments in modern, digital balances.

If density is specified instead of part or product weight as a control test, then this measurement is difficult to perform in a production environment. Probably the easiest way of performing the measurement is to use two containers each containing a solution of the appropriate density. These solutions are selected so that the moulding will sink in one and float in the other. See *density* and *product assessment*.

moulding blank
A feed form, usually a sheet, for a *compression mould*. Rubber moulding blanks may be prepared, for example, by guillotining a sheet which has been prepared on a fairly simple *calender*, for example, a two-roll calender. Where air bubbles can be tolerated, 10 mm (0·04 in) thick sheet can be calendered directly.

moulding methods
In the case of thermoplastics, it is *injection moulding* and *blow moulding* which are the major moulding methods. In the case of thermosetting plastics, it is *compression moulding*. These are *high pressure moulding processes*: *low pressure moulding processes* include *casting* and *powder moulding*. The cost and complexity of the moulding methods employed by the polymer industry, depends primarily on the amount of force which is required in order to perform the shaping process: if large forces are required in order to shape the polymer, then the machines and moulds must be designed to withstand these large forces. It is not usually the speed of shaping which limits machine output, but rather the rate at which the polymer can be set into the shape required. In the case of thermoplastics, for example, this means that it is the cooling stage which dictates the output rate from a particular process.

moulding monitoring
See *product monitoring*.

moulding pressure
The pressure applied to the material and which causes it to take the shape of the mould. See *moulding methods*.

moulding quality control
See *moulding assessment* and *automatic quality control*.

moulding set-up time - injection moulding
This is measured in seconds and is the time needed for the injection moulding to harden so that it may be ejected. May be approximately calculated by squaring the maximum part thickness and then multiplying this product by a factor. This factor differs from one material to another. This is because, of course, differing plastics materials have different cylinder temperatures, mould temperatures, specific heats, etc. Some factors for various materials are: ABS, 190; POM, 305; PA 66, 140; LDPE, 200; PP, 338; and PS, 340. Therefore, if the part thickness for a nylon 66 moulding is 0·1 cm, the calculated

mould set-up time will be 140 × 0·1 × 0·1 = 1·4 s. The addition of glass fibre to nylon 66 can change the factor from 140 to 81.

moulding shrinkage
See *shrinkage*.

movable core sections
Mechanisms used in, for example, blow moulds to provide undercut features on *blow moulded containers*. The undercut features (threads etc.) are created by shaped inserts which can be moved pneumatically, hydraulically or, electrically.

movable safety guard
A machine guard which is fitted over a hazardous area but which can be opened or moved. For example, part of the guard over the *nip* of a two-roll mill may be capable of being opened. If the gap so made is large enough to admit a limb, or part of a limb, then the guard should be electrically interlocked so that the rolls will stop immediately the guard is opened. See *fixed safety guards, safety stops, safety bars* and *plugged braking*. Training and a responsible attitude towards safe working are very important safety considerations as undue reliance should not be put upon guards.

moving head die plate
The *die plate* which is attached to the moving platen.

moving mould half
One of the halves of an injection mould and which is attached to the *moving platen*. This mould half transmits the thrust from the locking system to the *fixed mould half*. Provision is usually made to ensure that the moulding is retained in this half of the mould. This is because the moulding ejection system is associated with the moving-mould half for reasons of cost, convenience, etc. The moulding is made to stay with the moving-mould half by using a *sprue puller*. At the end of, or during, the mould opening stroke the ejector system is activated and this clears the mouldings, and their associated runner system, from the mould.

moving plate
One of the three plates of a *three-plate mould*: the plate which is attached to the moving platen of the machine.

moving platen
One of the platens of an injection moulding machine and on which the *moving mould half* is mounted. This *platen* moves as the locking system is moved to open and close the mould. See *fixed platen*.

MPa
An abbreviation used for *megapascal*.

MPC
An abbreviation used for medium processing channel (black). S301. See *carbon black*.

MPD
An abbreviation used for m-phenylene diamine. See *poly-(m-phenyleneisophthalamide)*.

MPD-I
An abbreviation used for *poly-(m-phenyleneisophthalamide)*.

MPF
An abbreviation used for *melamine-phenol-formaldehyde*.

MPR
An abbreviation used for *melt processable rubber*. See *thermoplastic elastomer*.

MPT
An abbreviation used for methylene-p-toluidine.

MPTD
An abbreviation used for *dimethyl diphenyl thiuram disulphide*.

MQ
An abbreviation used for *silicone rubber* containing methyl groups. MQ silicone rubbers are based on *polydimethylsiloxane*. Few pure MQ and PMQ-types of silicone rubber are now marketed.

MR
An abbreviation used for *methyl rubber*.

MRPRA
An abbreviation used for the Malayan Rubber Producers' Research Association.

MSW
An abbreviation used for *municipal solid waste*.

MT
An abbreviation used for medium thermal (black). N990. See *carbon black*.

MTO
An abbreviation used for *melt temperature over-ride*.

mud and snow tyres
Also known as M and S tyres: tyres with a coarse tread pattern.

Mullins' effect
The reduction in stiffness of filled vulcanizates which occurs after repeated stretching: thought to be caused by the breaking of bonds between the rubber and the filler. That is, stress softening of a filled rubber.

multi live-feed moulding
An *injection moulding* technique used to improve weld line strength. A divided melt is formed by the plastics melt flowing around a core or pin. When these melt fronts meet and join then a weld is formed. Apart from marring the surface such a weld is a source of weakness in the finished component.

Multi live-feed moulding was developed to modify the internal structure, or *orientation*, of mouldings. This is done by an auxiliary add-on device (fitted between the injection cylinder and the nozzle) and which contains computer controlled, hydraulic pistons. It is the activation of these during the hold period which controls moulding structure by forcing the material to mingle across the weld. In this way, significant improvements in weld strength can be obtained. Significant improvements in strength and stiffness can be obtained by using this technique when injection moulding composite materials.

multi-axial non-crimp fabric
A *multi-layered, non-crimp fabric* in which the unidirectional fibres of the *non-crimp fabric* lie in different directions.

multi-cavity injection moulding
See *multi-impression moulding*.

multi-chain polymer
A polymer which has more than two chains growing from one central point: a star or comb polymer. See *block copolymer*.

multi-colour injection moulding
Injection moulding of components with deliberately selected areas of the component in a different colour to the body of the injection moulding. Achieved using a *multiple injection unit machine*.

multi-daylight mould
A mould with more than one daylight or opening, for example, a *stack mould*.

multi-daylight press
Also called a multi-platen press. A press with at least one *floating platen* and therefore with more than one *daylight*.

multi-functional organosilicon cross-linking agent
Chemicals which are used to produce *room temperature curing silicones*. May be represented as R-Si-X$_3$ where for *one pack systems* X may be, for example, amine -NH-R, acetate -O-CO-CH$_3$ or oxime -N=CR$_2$. For two-pack systems X may be SiRO$_4$. The addition of such an agent turns the *silanol terminated polydimethylsiloxane* into a tetra-functional structure.

multi-impression mould
A mould with more than one impression.

multi-impression moulding
The production of components by a moulding process, for example, *injection moulding*, using a *multi-impression mould*. Where large numbers of components are required, particularly if they are small components, then the economic advantages and fast production capability of multiple-cavity injection moulding are considerable. However, it can be difficult, and/or costly to make multiple impressions, particularly those of intricate shape, which are identical. This means that it is very difficult to get identical cavity filling and then uniform material packing in such multi-impression moulds. With many moulds the problem of core/cavity concentricity arises; this applies particularly to thin-walled moulding production. Misalignment of the core to the cavity, when disposable packaging cups are being made, by more than 5% of the wall thickness will result in weld-line weaknesses, burning or, in the worst cases, *shorting* (near the lip of a container).

multi-layer blow moulding
Blow moulding performed using a *parison* which is made from more than one material: each material is in the form of a separate layer and produced by *coextrusion*.

multi-layer extrudate
See *coextrusion*.

multi-layer fabric
Also known as a pseudo, three dimensional (3D) fabric. Multi-layer fabrics are used, for example, in the *GRP* industry in order to obtain woven, 3D preform shapes: such shapes give improved damage tolerance and shear strength compared to conventional weaves. Fabrics with up to five layers can be made with cost as the only limiting factor.

multi-layered, non-crimp fabric
A *non-crimp fabric* which consists of several layers of reinforcement.

multi-material moulding
See *sandwich moulding*.

multi-mode cavity applicator
A *microwave* heating device, for example, a microwave oven. A microwave heating device in which the electric field pattern in the cavity produces many successive reflections of the microwaves within the cavity, that is many different *modes*. Industrial applicators may have an internal volume of a few metres and are usually designed for continuous processing. For example, microwaves are used to cure rubber compounds in continuous vulcanization processes.

multi-motor drive
A *unit drive system* for a *calender* in which the gearbox contains a set of gears for each roll and each set has its own motor.

multi-part mould
Also called a multi-plate mould. A mould with more than two parts or plates, for example, a three part mould.

multi-plate mould
See *multi-part mould*.

multi-platen press
See *multi-daylight press*.

multi-screw extruder
An extrusion machine, or *extruder*, which contains more than one *screw*. The screws transport, melt, mix and pump the *melt* through the *die*. The most widely used type of multi-screw *extruder* is a *twin screw extruder*. See *planetary extruder*.

multi-screw machine
Usually means a *multi-screw extruder*.

multiple injection unit machine
A machine with more than one *injection unit*, for example, two. May be arranged in a variety of positions for multi-colour injection moulding, *marble-effect moulding* or, for multi-material moulding (*sandwich moulding*).

multiple-impression mould
See *multi-impression mould*.

municipal solid waste
An abbreviation used for this term is MSW. The waste material collected by local authorities and disposed of by landfill or *incineration*. In the UK, for example, this amounts to approximately 30 million tonnes per year.

muscovite
See *glimmer*.

MW
An abbreviation used for *microwave*.

MWD
An abbreviation used for *molecular weight distribution*.

MWRP
An abbreviation used for *metal whisker reinforced plastic*. See *fibre reinforced plastics*.

MXDA
An abbreviation used for *m-xylenediamine*.

myria
A prefix sometimes used in the *metric system of units* and which means ten thousand. For example, a myriagram is 10,000 grams and a myriametre is 10,000 *metres*.

N

n
This letter is used as an abbreviation for;
flow behaviour index. See *power law equation*;
nano. See *prefixes - SI*; and,
normal - in organic chemistry.

n-alkyl methacrylate monomer
A *monomer* which on polymerization will yield a *poly-(n-alkyl methacrylate)* polymer. May be represented as CH$_2$ = C.CH$_3$.COOR where R is an n-alkyl group. The first member of the series is often considered to be *polymethyl methacrylate*.

n-alkylsulphonate
See *alkylsulphonic acid ester*.

n-heptane
An abbreviation used for normal heptane. This material has a relative density (RD or SG) of 0·75

n-hexyl n-decyl phthalate
An abbreviation used for this material is HXDP or NHDP. See *plasticizer*.

n-isopropyl-2-benzthiazole sulphenamide
An abbreviation used for this material is IBS. An additive used in rubber formulations as an *accelerator* - particularly for compounds highly loaded with *furnace black*.

n-isopropyl-n'-phenyl-p-phenylenediamine
An additive used in rubber formulations as an *antioxidant* and as an *antiozonant*. More effective if used with a protective wax and another *antioxidant*. A greyish/purplish solid material with a melting point of approximately 70°C and a relative density of 1·14.

n-octane
An *alkane* of formula $CH_3(CH_2)_6CH_3$: an isomer of *octane*.

n-octyl decyl trimellitate
An abbreviation used for this type of material is ODTM. A *trimellitate plasticizer*.

n-octyl n-decyl adipate
See *octyl decyl adipate*.

n-octyl n-decyl phthalate
See *octyl decyl phthalate*.

n-pentane
An *alkane* hydrocarbon of formula C_5H_{12}: there are three isomers of pentane. n-pentane is the straight chain isomer which has a boiling point of 36°C and a relative density of 0·62. The n-pentane is used, for example, to pre-expand, or blow, polystyrene. See *expanded polystyrene*. Also used as a propellant gas in some aerosols. See *chlorofluorocarbon* and *butane*.

n-propyl alcohol
Also known as n-propanol or as propan-1-ol. An *alcohol* which is miscible with water in all proportions. Used as a solvent and a diluent (for example, of lacquers). Has a boiling point of about 97°C.

N
This letter is used as an abbreviation for;
   filler. For example, a non-woven fabric;
   newton. See *prefixes - SI*;
   nonyl. See, for example, *butyl nonyl phthalate (BNP)*;
   normal. Sometimes used in plasticizer terminology but this practice is discouraged - see *dioctyl phthalate*;
   normal cure rate (black). See *carbon black*; and,
   novolak.

N-1,3-dimethylbutyl-N'-phenyl-p-phenylenediamine.
An abbreviation used for this material is 6PPD. An alkyl-aryl p-phenylene diamine. An additive used, for example, as an *antiozonant*. See *antioxidant*.

N-cyanosulphonamide polymer
An abbreviation used for this material is NCNS. A polymer prepared by the reaction of a primary biscyanamide with a secondary biscyanamide: this polymer may be cross-linked by heating to approximately 150°C: cross-links are formed through triazine rings. Forms clear thermosetting moulding resins which are used with fibre reinforcement and in the manufacture of printed circuit boards.

N-cyclohexyl-2-benzothiazole sulphenamide
See *cyclohexylbenzothiazyl sulphenamide*. See *accelerator*.

N-cyclohexylbenzothiazylsulphenamide
See *cyclohexylbenzothiazyl sulphenamide*.

N-cyclohexylbenzothiazole-2-sulphenamide
See *cyclohexylbenzothiazyl sulphenamide*.

N-cyclohexylbenzothiazyl-2-sulphenamide
See *cyclohexylbenzothiazyl sulphenamide*.

N-cyclohexylmaleimide copolymer
See *vinyl chloride copolymers* and *terpolymers*.

N-cyclohexylthiophthalamide
A *pre-vulcanization inhibitor for rubber compounds*.

N-cyclohexylthiophthalimide
A *pre-vulcanization inhibitor* which prevents scorch during mixing but which does not affect cross-link formation during *vulcanization*.

N-dimethyl benzthiazyl sulphenamide
An abbreviation used is DMBS. A vulcanization *accelerator* (see *benzothiazole sulphenamide accelerators*).

N-isopropyl-N'-phenyl-p-phenylenediamine
Abbreviation used IPPD. An alkyl-aryl p-phenylene diamine. An additive used, for example, as an *antiozonant*. See *antioxidant*.

N/m
An abbreviation used for newton per metre.

N-methylmethoxy nylon
See *methylmethoxy nylon*.

N-methylpyrrolidone
An abbreviation used for this material is NMP. This liquid material is used as a solvent for *aromatic polyamides*, *polyimides*, *polyetherimides* and for *polyphenylene sulphides*. It has a boiling point of 202°C.

N-morphlinyl benzthiazyl sulphenamide
An abbreviation used is MOBS. A vulcanization *accelerator* (see *benzothiazole sulphenamide accelerators*).

N-morpholinothiobenzothiazole
See *N-morpholinothiobenzothiazole-2-sulphenamide*.

N-morpholinothiobenzothiazole-2-sulphenamide
A vulcanization *accelerator* (see *benzothiazole sulphenamide accelerators*). An abbreviation used for this material is MBS. A *delayed action accelerator*. See *mercaptobenzothiazole*.

N-nitrosodiphenylamine
Also known as diphenyl nitrosamine. This brown crystalline material has a melting point of approximately 65°C and a relative density (RD) of 1·25. A vulcanization *retarder* but one which also decreases final *cross-link density*. This material can improve the processing safety of *scorchy compounds*. Used as a *reclaiming agent*, for slightly scorched compounds, by cold rolling or milling this material with the compound.

N-octyl-2-benzthiazyl sulphenamide
This white solid material has a melting point of 100°C and a relative density (RD) of 1·14. It is a *delayed action accelerator* for *natural rubber* and for *styrene-butadiene rubber*. See *benzothiazyl sulphenamide accelerators*.

N-oxydiethylbenzothiazolesulphenamide
See *N-oxydiethylbenzothiazylesulphenamide*.

N-oxydiethylbenzothiazole-2-sulphenamide
See *N-oxydiethylbenzothiazylesulphenamide*.

N-oxydiethylbenzothiazylsulphenamide
An abbreviation used for this material is NOBS. Also known as N-oxydiethylene-2-benzthiazole sulphenamide or as, 2-(4-morpholinyl-mercapto)benzthiazole or as, N-oxydiethylbenzothiazole-2-sulphenamide or as, N-oxydiethylbenzothiazolesulphenamide or as, N-oxydiethylene benzothiazole-2-sulphenamide. This brownish material has a melting point of 80 to 90°C and a relative density of approximately 1·35. It is used as a delayed action *accelerator* for rubbers such as *natural rubber* and for *styrene-butadiene rubber* when, for example, such materials are loaded with furnace blacks. The action of this material is boosted by *diphenyl guanidine*: gives very good physical properties to vulcanizates and flex cracking resistance on ageing. May be used in conjunction with 2,2-dibenzthiazyl disulphide (10%) if the delayed action require-

ments are not so stringent. When NOBS is used, about 4% *zinc oxide* is needed and stearic acid must also be present. See *benzothiazyl sulphenamide accelerators*.

N-oxydiethylene-2-benzthiazole sulphenamide
See *N-oxydiethylbenzothiazylsulphenamide*.

N-oxydiethylene benzothiazole-2-sulphenamide
See *N-oxydiethylbenzothiazylesulphenamide*.

N-oxydiethylenethiocarbamyl-N-oxydiethylene-sulphenamide
An abbreviation used for this material is OTOS. A vulcanization *accelerator*. See *benzothiazole sulphenamide accelerators*.

N-pelargonyl-p-aminophenol
A white powder with a melting point of approximately 123°C: used as an *antioxidant* for rubbers.

N-pentamethylene ammonium pentamethylene dithiocarbamate
See *piperidine pentamethylene dithiocarbamate*.

N-phenyl-N'-cyclohexyl-p-phenylene diamine
Also known as a *p-phenylene diamine* antioxidant. This gray solid material has a melting point of about 115°C and a relative density (RD) of 1.29. A *staining antioxidant* for rubbers: a very effective *antioxidant*.

N-substituted nylon
A *nylon* which has had an organic group grafted onto the -CONH- group. For example, if a precursor nylon is reacted with formaldehyde and methyl alcohol then the chemical group -CH$_2$OCH$_3$ is grafted onto the polyamide in place of the H atom in the -CONH- group to give -CON(-CH$_2$OCH$_3$)- material. This substitution reduces hydrogen bonding and increases the solubility of the material. See *methylmethoxy nylon*.

N-t-butylbenzothiazole-2-sulphenamide
See *butyl benzthiazyl sulphenamide and benzothiazyl sulphenamide accelerators*.

N-t-butylbenzothiazyl-2-sulphenamide
See *butyl benzthiazyl sulphenamide and benzothiazyl sulphenamide accelerators*.

N-tert-butyl-2-benzothiazyl sulphenamide
See *butyl benzthiazyl sulphenamide and benzothiazyl sulphenamide accelerators*.

N/tex
An abbreviation used for newton per tex.

nacreous pigments
See *pearlescent pigments*.

nadic methyl anhydride
An abbreviation used for this material is NMA. Also known as methyl nadic anhydride or as, methylenedomethylenetetrahydrophthalic anhydride. When supplied, this material is usually a mixture of isomers which are made liquid by the addition of approximately 0.1% phosphoric acid. A liquid curing agent for *epoxide resins* which can give products with a high heat distortion temperature (HDT) of approximately 200°C.

names - plastics materials
See *plastics materials, thermoplastics materials* and *thermosetting plastics materials*.

nano
An abbreviation used for this term is n (see *prefixes - SI*). Used for 10^{-9} or 1/1,000,000,000 (one thousandth millionth). For example:

    nanometre (nm) = 1/1,000,000,000 of a metre; and,
    nanosecond (ns) = 1/1,000,000,000 of a second.

naphtha
Also known as solvent naphtha and used as a solvent, for example, in the rubber industry. Obtained from either paraffin oil or *coal tar* by distillation, composed of aromatic, aliphatic and naphthenic hydrocarbons and which, for example, have a boiling point below 200°C. Wood naphtha is impure *methyl alcohol*.

naphthalen-2-ol
See *naphthol*.

naphthalene
An aromatic hydrocarbon which may be represented as $C_{10}H_8$ and which consists of two benzene rings fused together. Used, for example, in the manufacture of organic *dyes*. This material has a relative density (RD or SG) of 1.16.

naphthalene-1,5-diisocyanate
An abbreviation used for this material is NDI. Also known as 1,5-naphthalene diisocyanate. An aromatic, toxic material used to prepare cast *polyurethane rubbers* and obtained from *naphthalene*. Has a melting point of 127°C and a relative density (RD) of 1.43.

naphthenates
Compounds derived from *naphthenic acid*: metal salts which are soap-like. See, for example, *cobalt naphthenate*. Naphthenates also function as *heat stabilizers* for *polyvinyl chloride* (PVC).

naphthenic acid
A complex mixture of cyclo-aromatic acids used to prepare metal *naphthenates*, for example, *cobalt naphthenate*.

naphthenic oil
A complex mixture of materials, which contains a high proportion of naphthalene-type structures: used as a rubber plasticizer and naphthenic *processing oil*. Such a material has good heat stability and gives low heat build up under dynamic stressing conditions. See *petroleum oil plasticizer*.

naphthenic processing oil
See *naphthenic oil*.

naphthol
An aromatic material (derived from naphthalene) and used, for example, in the manufacture of organic *dyes*. There are two isomers but it is β *naphthol* which used as an antioxidant for rubbers and to prepare dyes: also known as naphthalen-2-ol. See *diazo compounds*.

naphthol reds
A type of *monazo pigment: organic pigments*.

naphthoyl
The univalent organic radical $C_{10}H_7CO-$ which is derived from naphthoic acid $C_{10}H_7COOH$.

naphthyl
The univalent organic radical $C_{10}H_7-$ which is derived from naphthalene.

naphthyl-β-mercaptan
Also known as β-thionaphthol or as, 2-naphthalene-thiol or as, 2-mercaptonapthalene or as 2-naphthalene mercaptan. This material has a boiling point of 286°C and a melting point of 81°C. A *peptizer* and a reclaiming agent for *natural* and *synthetic rubbers*.

National Bureau of Standards smoke chamber
In this smoke testing chamber a square sample is supported in a frame so that a specific surface area (4.24×10^{-3} m^2) is left exposed. The frame or sample holder is placed within a chamber ($0.91 \times 0.91 \times 0.61$ m = h $\times$ w $\times$ d) and in front of an electric radiator, or furnace, which gives an output of 2.5 watts/cm^2. The radiator voltage is capable of being ad-

justed and the heat output of the radiator is measured by means of a radiometer. When the temperature at the radiometer position stabilises at 200°F/93°C, and the walls of the test chamber are at 95°F/35°C the smoke chamber is ready for use.

Pre-dry the samples for 24 hours at 140°F (60°C), then condition them to equilibrium (constant weight) with an ambient temperature of 73°F (23°C) and a relative humidity of 50%. Cut each specimen to fit the sample holder. Each sample should be 3 inches (76 mm) square and its thickness may be up to 1 inch (25 mm) thick e.g. 0.5 inch. For example, if *plasticized polyvinyl chloride* (PVC) sheeting is being tested then several layers would be used to make one test specimen: each test specimen is of the same agreed size.

After conditioning, wrap each specimen in foil and then remove the foil from the sample face. When a flaming test is to be performed the excess foil is bent to form a spout which permits the flow of molten material.

The conditioned specimen is placed in front of the radiator and when a flaming exposure test is being performed, a six-tube burner is connected to the chamber's air and propane supply lines and centred in front of the sample. The gas flow is adjusted to give 500 cm^3/min for the air and 50 cm^3/min for the propane. Some flames are directed towards the sample and some towards the molten material which collects in a trough.

Various ways of using the NBS smoke chamber exist. For example, the specific optical density (D_s) after a specified heating time is (e.g. 1.5, 4 or 20 min) is measured. This is a dimensionless quantity independent of the dimensions of the test chamber and specimen used. D_s is assessed by measuring the reduction in the intensity of a light beam which passes vertically through the chamber. If necessary a correction is applied when fogging of the optical window results: D_s then becomes D_m corr). Six samples of the material are required: three are tested in the flaming mode and three in the non-flaming mode

National Electrical Manufacturers' Association
An abbreviation used for this USA organization is NEMA. This organization issues NEMA standards.

National Physical Laboratory
An abbreviation used for this UK-based organization is NPL.

native rubber
A term sometimes used to describe *natural rubber* which does not come from a plantation but from a smallholding: sometimes called smallholding rubber. Often an inferior grade of rubber.

natrium
A term sometimes used for sodium. See *buna rubber*.

natural
See *natural material*.

natural calcium silicate
Natural calcium silicate is derived from *wollastonite*.

natural cis-1,4-polyisoprene
See *natural rubber*.

natural fibres
See *fibres*.

natural gas
A mixture of hydrocarbons but mainly *methane*. Used as fuel and as a raw material for the preparation of, for example, monomers.

natural material
Term used in the plastics industry for material which is supplied free from colour. There is a tendency to supply thermoplastics as such natural materials which are then, for example, coloured on the machine by a *masterbatch*. Used as, for example, the base feed-stock for *in-house colouring*.

natural poly-(1-methyl-1-butenylene)
See *natural rubber*.

natural resin
An amorphous, organic material secreted by insects and/or a plant. An example of a natural resin which is used as a plastics material is *shellac*. See *synthetic resin*.

natural rubber
Sometimes referred to as Hevea rubber or as, caoutchouc or as, gum elastic or as, India rubber or as, india rubber. May also be referred to as cis-1,4-polyisoprene or as, poly-(1-methyl-1-butenylene) or as, natural poly-(1-methyl-1-butenylene) or natural cis-1,4-polyisoprene. An abbreviation used for this type of material is NR.

Natural rubber has been classified as a *general purpose rubber*. This material is rather unusual as it is one of the few commercially significant polymers which is not synthesized or man-made. It is an unsaturated aliphatic hydrocarbon of regular, chemical structure which is produced by the tree *Hevea brasiliensis* in *latex* form (see *caoutchouc*). The rubber hydrocarbon content of all grades of NR is 100% cis-1,4-polyisoprene: the molecular weight of this polymer differs only slightly when first produced.

The latex is obtained by *tapping* the tree: this latex has a *dry rubber content (DRC)* of approximately ⅓. Some latex is concentrated to 60% DRC to make transport easier - this is known as latex concentrate. Solid rubber may be obtained from diluted latex (approximately 13% dry rubber content) by *coagulation*, with acetic or formic acid, washing the product in a *washing mill* and then *smoking* and/or *drying* the resultant sheeted coagulum. Obvious imperfections may be removed by clipping and the sheet may then be graded (see *standard international grades*). The sheet may be marketed in large bales (113 kg/250 lbs) made from compressed sheet although *comminuted rubber* is available: the bales are traditionally dusted with *talc* to stop inter-bale adhesion. See, for example, *air dried sheet*, *pale crepe* and *ribbed smoked sheet*.

Sheet rubber, in such large bales, is difficult to handle and for this reason NR is often now supplied in *block* form: such blocks are made from *chemically crumbed rubber* or from mechanically comminuted rubber (see *technically specified rubber*). In the case of *latex grade rubber*, the *natural rubber latex* is bulked and blended so as to improve uniformity. It is then, for example, blended with 0.3% *castor oil* before *acid coagulation*: considerable dilution is not necessary. After washing, to remove the oil, the crumbs are dried (5 h at 110°C) cooled, baled and wrapped. That is, the NR block is made from *comminuted rubber* which has been compressed into a block or bale. Such blocks may be 33.3 kg in weight and polyethylene (PE) wrapped. PE is used to stop inter-bale adhesion and to prevent contamination. If the PE is thin gauge, low density PE (below 0.03 mm), then it need not be stripped from the bale before mixing. *Field coagulum* may be converted to *crumb* form by various size reduction devices.

NR is most commonly sold as a specified type or grade: for example, see *standard international grades*, *technically specified rubbers*, *specially prepared rubbers* and *modified types of natural rubber*. About 80% of natural rubber comes from Indonesia, Malaysia, Sri Lanka and Thailand.

Natural rubber has a relative density of approximately 0.92. The *molecular weight* of the natural rubber molecules is very high: the weight average molecular weight can reach 2.5 million, for example, 1.6 to 2.3 × 10^6: the ratio of weight average molecular weight to number average molecular weight is approximately 2.5 to 10. Such a high molecular weight material cannot often be processed as supplied and so the molecular weight, of sheet rubber, is commonly reduced by *mastication*. (By the use of *viscosity stabilized grades* this step may be avoided thus saving on mixing costs). Once the material has been masticated then it is compounded with the mix ingredients, shaped (by *extrusion*, *calendering* or *moulding*) and the shape is set by cross-linking or *vulcanization*.

If natural rubber is stretched, then crystallinity occurs and this reinforces the rubber so that it resists deformation and enhance strength. Because of this increased resistance to deformation the fillers used with natural rubber can be of the non-reinforcing variety (such fillers are cheap). Natural rubber may be a stiffer flow material (more viscous) at equal filler loadings than the synthetic rubber, *styrene butadiene rubber*, although its viscosity may be reduced by increased *mastication*: its viscosity is more temperature sensitive.

natural rubber and isoprene rubber - comparison of
Natural rubber (NR) has greater structural regularity, a faster vulcanization rate, more propensity to *scorch* and a higher green strength. Such differences have been ascribed to the non-rubber constituents present in *natural rubber*. Lightly filled compounds also have higher hot tear strength. In general, processing is possible at lower temperatures, for the synthetic material (IR), and it is usually found that there is less heat generation when fillers are mixed with this material. Lower creep or stress relaxation is often mentioned as a plus for IR. As IR is compatible with NR in all proportions, it can be usefully used in blends with NR where scorch (premature vulcanisation) is a problem.

natural rubber and styrene butadiene rubber - comparison of
Natural rubber (NR) and *styrene butadiene rubber (SBR)* are the work horses of the rubber industry and are used when a high level of resistance to oxidation, ozone and petroleum oil is not required. Both materials cost roughly the same and the maximum service temperature of moulded components is of the order of 70°C/158°F. Both materials are available in a range of grades (range of molecular weights, viscosities, etc.) and all grades of SBR are compatible with NR in all proportions: mixtures are quite often used in order to get desired properties.

There are, however, differences between them: for example, if natural rubber is stretched, then crystallinity occurs and this reinforces the rubber so that it resists deformation and enhances the strength. Because of this increased resistance to deformation the fillers used with natural rubber can be of the non-reinforcing variety (such fillers are cheap). SBR does not crystallise when stretched and so reinforcing (expensive) fillers must be used for many applications. NR has better hot tear resistance than SBR and this property is useful if there are difficulties in moulding. SBR is slower curing than natural rubber and although the cure rates can be matched, this means that more (expensive) accelerator must be used.

However, SBR has better abrasion resistance and the moulds should stay cleaner because there is sometimes less mould fouling with compounds based on this polymer - particularly at high *vulcanization* temperatures and when the NR has not been compounded for thermal stability. The ageing resistance of the SBR mouldings is also slightly better. The synthetic polymer can also be made in very high molecular weight grades (much higher than those found in NR) and these grades can then be extended with oil. Oil-extended SBR is relatively cheap and this technique helps to offset the cost of expensive reinforcing fillers (e.g. carbon black).

natural rubber and synthetic cis-polyisoprene rubber - comparison of
See *natural rubber* and *isoprene rubber - comparison of*.

natural rubber and synthetic rubber - comparison of
Despite the large number of *synthetic rubbers* (approximately 30), *natural rubber (NR)* still has a commanding position in the rubber industry. This is because of a combination of cost and the properties offered. For example, high tensile strength, high level of elasticity, very good low temperature flexibility, outstanding dynamic properties and minimum heat development under dynamic load. Chief areas of application are large area of the tyre sector, high performance conveyor belting, springs, buffers, thin-walled and high strength components such as balloons. See *natural rubber* and *isoprene rubber - comparison of*.

natural rubber - categories of
Natural rubber can be classified, or grouped, into *standard international grades*, *technically specified rubbers*, *specially prepared rubbers*, and into *modified types of natural rubber*.

natural rubber - chlorination of
See *chlorinated natural rubber*.

natural rubber - classification of
See *natural rubber - categories of*.

natural rubber coagulum
See *coagulum*.

natural rubber - copolymer graft
See *polymer modified natural rubber* and *comb-grafted natural rubber*.

natural rubber - cultivation of
Produced on *estates* and *smallholdings*. An estate, in Malaysia, would yield about 1,500 kg of *natural rubber* per hectare per year. A high yielding clone may yield about 2,500 kg of *natural rubber* per hectare per year.

natural rubber - dry rubber content
Latex has a *dry rubber content (DRC)* of approximately ⅓. See *natural rubber*. Natural rubber *latex* consists of rubber hydrocarbon and non-rubbers suspended in an aqueous base or serum.

natural rubber - grafted
See *polymethyl methacrylate grafted natural rubber*.

natural rubber - halogenation of
See *halogenated natural rubber*.

natural rubber hydrochloride
See *rubber hydrochloride*.

natural rubber - hydrogenation
See *hydrogenated natural rubber*.

natural rubber - hydrohalogenation of
See *hydrohalogenated natural rubber*.

natural rubber - inferior grades of
These include, for example, *compo crepe*, *cup lump*, *estate brown crepe* and *flat bark crepe*. Grades which are not produced by the deliberate *coagulation* of *natural rubber latex*. See *auto-coagulation*.

natural rubber latex
See *latex*.

natural rubber - molecular weight of
The *molecular weight* of the *natural rubber molecules* is very high: the weight average molecular weight can reach 2·5 million, for example, 1·6 to 2·3 $\times$ 10^6: the ratio of weight average molecular weight to number average molecular weight is approximately 2·5 to 10. Such a high molecular weight material cannot often be processed or shaped as supplied and so the molecular weight is reduced by *mastication* before processing. Once the material has been masticated then it is compounded with the mix ingredients, shaped (by *extrusion*, *calendering* or *moulding*) and the shape is set by cross-linking or *vulcanization*.

natural rubber - non-sulphur vulcanization
Accelerated sulphur systems are most widely used to *vulcanize* rubber compounds. Alternatives to *sulphur vulcanization* include *cold cure, selenium vulcanization, tellurium vulcanization, sulphur chloride vulcanization, sulphur thiocyanate vulcanization, peroxide vulcanization, radiation curing, urethane crosslinking* and *phenolic resin curing*.

natural rubber - oil extended
See *oil extended natural rubber*.

natural rubber/polyolefin blend
See *polyolefin/natural rubber blend*.

natural rubber/polypropylene blend
An abbreviation used for this type of material is NR/PP. See *polyolefin/natural rubber blend*.

natural rubber - rubber hydrocarbon content
The rubber hydrocarbon content of all grades of NR is 100% cis-1,4-polyisoprene: the molecular weight of this polymer differs only slightly when first produced. See *natural rubber*.

natural rubber - special grades of
Grades of natural rubber introduced to provide either improved processing (SP rubber), or which give improved performance in service (for example, deproteinized rubber), or which utilise by-product material (for example, skim rubber). Other grades include *initial concentrated rubber (ICR), oil extended natural rubber (OE-NR)* and *epoxidized natural rubber (ENR)*.

natural rubber - standard international grades
See *standard international grades - natural rubber*.

natural rubber - thermoplastic blend
See *polyolefin/natural rubber blend*.

natural rubber - viscosity of
Natural rubber is a stiffer flow material at equal filler loadings than the *synthetic rubbers* and its viscosity is more temperature sensitive. The *Mooney viscosity* of freshly prepared rubber varies from 50 to 90 units (ML 1 + 4 at 100°C) depending on the mixture of clonal latices used. On storage (see *storage hardening*) the viscosity increases spontaneously unless this is prevented. By bud grafting, it is possible to produce a composite tree which yields rubber of specified viscosity values.

natural tack
The tendency of a material to stick to itself. The possession of this property has hindered the development of powdered *natural rubber*.

natural tolerance limits
See *process capability*.

natural variation in production
During the manufacture of products, a natural variation in dimensions occurs due to chance errors. Before specification limits (see *tolerance limits*) are agreed, it would be advisable to check whether or not a particular production process is capable of producing components whose dimensions fall within these specification limits. See *process capability*.

nautical mile
A nautical mile (International or US) equals 1·151 land miles, 1,852 metres or 6,076·116 feet. Also known as the International nautical mile or as an air mile. The UK nautical mile was equal to 6080 feet (contained 10 cables each of 608 feet). The UK nautical mile was also known as a geographical or sea mile. See *UK system of units* and *US Customary Measure*.

NBR
An abbreviation used for *nitrile rubber*.

NBR/PVC blend
An abbreviation used for *nitrile rubber* (NBR) *polyvinyl chloride* (PVC) *blend*.

(NBR + PVC)
A *nitrile rubber polyvinyl chloride blend*.

NBR-TPV
A *thermoplastic vulcanizate* based on a crosslinked *nitrile rubber*.

NBS smoke chamber
An abbreviation used for *National Bureau of Standards smoke chamber*.

NC machine
An abbreviation used for *numerically controlled machine*.

NCNS
An abbreviation used for N-cyanosulphonamide polymer.

NCR
An abbreviation used for *chloroprene acrylonitrile copolymer* or rubber.

Nd-BR
Butadiene rubber based on a neodymium catalyst. See *butadiene rubber*.

NDI
An abbreviation used for *naphthalene-1,5-diisocyanate*.

near-field welding
See *contact ultrasonic welding*.

neck-in
A phenomenon associated with *chill roll casting*: the edge of the extruded web tends to shrink inwards towards the centre of the web and, at the same time, this edge tends to become thicker than the bulk of the film. See *neckdown*.

neckdown
Reduction of extrudate width caused by drawing down of sheet during *extrusion*, measured as the difference between the die width and the width of the extrudate.

necking
The localised reduction in cross-section which may result from the application of a tensile stress. A phenomenon associated with *tensile testing*: the rapid decrease in cross-section at one point during stretching.

needle blowing
A blowing technique used in *extrusion blow moulding* and which is preferred for some products. When the correct length of parison has been extruded, the mould closes: this closing action pinches both the top and bottom of the *parison*. Air is introduced via a needle or small pin, so as to inflate the parison.

needle burner test
A burning test in which the sample is vertically mounted and a needle burner, placed beneath the sample, is used to simulate the effect of small flames. Such small flames may result, for example, from faulty electrical equipment and could cause a fire. See CEI 695-2-2 and *Underwriters laboratory UL 94 vertical burning test*.

needle mat
A reinforcing material which is usually based on *glass fibre*: chopped strands, of approximately 50 mm length, are held together by other strands which are pushed (needled) through the original mat.

needle tearing test
A test which is used to assess the resistance to tearing of rubber products (for example, 10 mm thick sheet) when the sheet is pierced by a 1 mm thick needle.

needle-seal nozzle
See *plunger nozzle*.

needled mat
See *needle mat*.

negative coefficient of thermal expansion
On heating the material decreases in size. See *carbon fibre*.

negative image lithography
A *photoresist* process which uses a *negative photoresist*: a photoresist film is used to form an image. Exposure of the photoresist film results in areas that are less soluble than the unexposed areas during *developing*: the dissolution rate of the polymer decreases as the molecular weight increases. For example, one system is based on *cyclized rubber* with an azide-type photosensitizer.

negative photoresist
A *photoresist* which becomes less soluble in exposed areas. For example, because of increasing polymer molecular weight.

negative pressure venting
A vacuum-assisted venting technique in which cooling water is drawn through the mould with a negative pressure pump thereby preventing leakage from the mould cooling system in the event of a crack or leak developing in the mould. When this device is used leakage of water into the cavities can be prevented even when the water-cooling channels are open to the cavities. The cavities can be vented into the water-circulating lines, for example, through porous metal plugs or through flats on *ejector pins*.

negative shrinkage
The production of a component which is larger than the *mould*. For example, a *soft grade* of *polyether ester elastomer* can be compressed by excessive dwell or hold-on pressures so that a negative shrinkage may result on ejection.

Nellen tubing machine plastimeter
A *plastimeter*. A test apparatus used to measure the extrusion speed of rubber compounds and to evaluate tube extrusion properties. See *plasticity*.

NEMA
An abbreviation used for the USA organization called the National Electrical Manufacturers' Association. Issues NEMA standards.

neo-factice
A *factice-type material* which results when liquid *polybutadiene* is reacted with *sulphur* or *sulphur monochloride*.

neopentyl glycol
See *neopentylene glycol*.

neopentylene glycol
An abbreviation used for this type of material is NPG. This material is also known as neopentyl glycol or as, 2,2-dimethylpropane-1,3-diol. A *glycol* with a melting point of 128°C which is used in the manufacture of *unsaturated polyester (UP) resins* with, for example, iso-phthalic acid. The resins are used in filament winding as the cured material has a high *heat distortion temperature* and very good chemical resistance.

neoprene rubber
Neoprene is the DuPont trade name/trade mark for *chloroprene rubber*.

nepheline syenite
A complex anhydrous sodium/potassium/aluminium silicate. Used as a *filler* for *polyvinyl chloride* (PVC): because of its low opacifying power it can give semi-transparent compositions.

nerve
A term used to describe a property of unvulcanized rubber and/or unvulcanized rubber compounds: that is, the material possesses toughness, resistance to deformation and elasticity. The nerve associated with *natural rubber* may be reduced by *mastication*. Also reduced by choosing suitable compounding ingredients, for example, high structure *carbon black* and/or *factice*. The use of a proportion of *superior processing rubber* will also help to reduce *nerve*.

net
See *extruded net*.

net ton
A short *ton* of 2,000 pounds.

network parameter
An abbreviation used for this term is M_c. A measure of the *cross-link density* of a polymer.

network polymer
A highly *cross-linked polymer*.

network-bound antioxidant
Also known as a *rubber-bound antioxidant*. A *bound antioxidant* which is obtained, for example, by the addition of p-nitrosphenylamine during black mixing of *natural rubber*.

network-bound p-nitrosphenylamine
An *antioxidant* - see *network-bound antioxidant*.

neutral reclamation processes
Used to obtain *reclaimed rubber* from un-wanted vulcanized material by heating with say, 1 to 2% zinc chloride, and oils, at approximately 185°C for up to 24 h. Washing with water and drying follows. Such procedures are useful for *natural rubber* compounds, for synthetic compounds and for blends (as cyclization of synthetic rubbers is avoided). See *reclaiming processes*.

neutron
An uncharged atomic particles of approximately the same mass as a proton.

neutron beam
A form of *high energy radiation*. See *radiation*.

new candle
See *candela*.

new technology black
See *carbon black*.

new tyre
A *tyre* which has not been used or subjected to a retreading operation.

newton
An SI derived unit which has a special symbol, that is N. It has the dimensions of $kgms^{-2}$. It is the SI derived unit of force. The newton is the force required to give a mass of one kilogram an acceleration of one metre per second per second. See *Système International d'Unité*.

Newtonian fluid
A rheological term for a fluid which is also known as an *ideal fluid*. Shear stress is directly proportional to shear rate for an ideal or Newtonian fluid. The coefficient of viscosity of such fluids is constant irrespective of the shear stresses involved and is independent of time. See *non-Newtonian fluid*.

Ni-BR
A *butadiene rubber* based on a nickel catalyst.

Ni-Cr-Mn steel
A nickel-chrome-manganese steel. See *mould steel*.

Ni-Cr-Mo-V steel
A nickel-chromium-molybdenum-vanadium steel. See *pre-hardened steel*.

Ni-Mn steel
A nickel-manganese steel. See *mould steel*.

nibs
See *fish eyes*.

nickel
A *transition element* (Ni) which is a hard, malleable metal with a high melting point 1455°C and boiling point 2840°C: the relative density is 8·90. Forms many important alloys, for example, stainless steels which may contain from 8 to 25% of nickel. Nickel steels (1 to 5% Ni) retain their strength and corrosion resistance at very high temperatures. Nickel plating is an important part of most chrome plating processes: it precedes the actual chrome plating. In the case of plastics materials, the nickel plating may be sandwiched between the copper and chromium layers or, there may be only a nickel layer and a chromium layer. Eliminating the copper makes the composite more corrosion resistant. See *acrylonitrile-butadiene-styrene* and *electroplating*.

nickel chelate
Also known as nickel (11) chelate. A chemical compound which is a complex nickel compound formed by chelation between nickel and an organic material. Such compounds are used as ultra-violet light stabilizers, for example, for *polypropylene*. Such materials are used as *quenching agents* so as to minimize light initiated degradation in polypropylene. See also *hindered amine light stabilizer*.

nickel dibutyl dithiocarbamate
This green powder material has a melting point of about 88°C and a relative density (RD) of 1·29. An anti-ageing additive for rubbers - but not for *natural rubber (NR)* when used on its own. Can, however, improve the protection given by another *antioxidant* to NR. Protects *styrene-butadiene rubbers*, *nitrile rubbers* and *polychloroprene rubbers* against heat and light ageing. See *dithiocarbamates*.

nickel dimethyl dithiocarbamate
A green flake material used as an anti-ageing additive for rubbers. For example, for *chloroprene rubbers* and for *chlorosulphonated polyethylene*. An *antioxidant*. See *dithiocarbamates*.

nickel dithiocarbamates
See *dithiocarbamates*.

nickel matrix
See *bimetallic barrel*.

nickel pentamethylene dithiocarbamate
An abbreviation used for this material is NPD. A green powder material with a relative density of 1·42. Used as an anti-ageing additive for rubbers. For example, for *chloroprene rubbers* and for *chlorosulphonated polyethylene*. An *antioxidant* which improves heat ageing and discolouration caused by light. See *dithiocarbamates*.

nickel shell mould
A mould based on a shell, or skin, of nickel backed with another material and used, for example, to prepare a large or a *prototype mould*. See *electroforming*.

nicking
The insertion of a cut or nick, into a test specimen. See *Bell Telephone Laboratories test method*.

nigrometer
A test instrument, developed by Cabot Inc of the USA, and used to measure the colour depth of *carbon blacks*. See *nigrometer index*.

nigrometer index
A logarithmic scale, or index, used to measure the colour depth of *carbon blacks*. Obtained from a test instrument called a *nigrometer*.

nigrosine
A widely used *black organic pigment* which is derived from aniline. Also known as solvent black.

nip
That which is formed where two rolls meet. Also known as the bite. See *two-roll mill*.

nip adjustment
The movement of a roll relative to another. Nip adjustment on a *calender* is usually taken to mean direct adjustment. For example, on an I calender the top and bottom rolls may be moved in an upwards or downwards direction so as to directly alter the nip or gap between the rolls. Nip adjustment is achieved by rotating a screw in the machine frame and this causes the bearing, and therefore the end of the roll, to be moved. Each roll end is capable of being moved (see *roll crossing*).

Two speed, motorised roll adjustment is fitted to most calenders with the slow speed 0·0025 mm/sec (0·0001 in/sec) being used for fine adjustments during running the fast speed 0·025 mm/sec (0·001 in/sec) is used for initial setting and for emergency purposes.

nip in
See *edge tear*.

nitrogen cooling
See *liquid nitrogen cooling*.

nip interaction
A problem found in multi-nip calenders where what happens in one *nip*, effects what happens in another. See *super-imposed calender*.

nip-forming pair
Two rolls between which there is a nip, for example, on a *calender*.

nip-polishing
See *kiss-polishing*.

NIR
An abbreviation used for *acrylonitrile-isoprene rubber*.

NIRIM
An abbreviation used for the National Institute for Research in Inorganic Materials (Japan).

NIST
An abbreviation used for the National Institute for Standards - a USA based organization.

nit
A unit of luminance. It is equal to one *candela per square metre*.

nitrided
A steel which has been hardened by *nitriding*. See *nitriding steel*.

nitriding
The introduction of nitrogen into the surface of *steel* to give a hard, wear-resistant surface which is retained at elevated temperatures. This hardening process for *steel* is accomplished by heating the steel in an ammonia atmosphere for several days at approximately 550°C. Because of the low temperatures employed, only slight dimensional changes may be encountered. See *nitriding steel*.

nitriding steel
A *steel* which is capable of being *nitrided*. Can have aluminium-alloyed types and aluminium-free types of nitriding steel: the aluminium-free types have the highest core strengths.

nitril fibre
A manufactured, polymeric fibre composed of at least 85% by weight of vinylidene dinitrile units -CH$_2$-C(CN)$_2$- and where the vinylidene dinitrile content is no less than every other unit in the polymer chain.

nitrile
An organic compound which contains the group -CN. When acrylonitrile is used to make a copolymer rubber, then in the name for that rubber, acrylonitrile is often shortened to nitrile. See *nitrile rubber*.

nitrile butadiene rubber
See *nitrile rubber*.

nitrile content
Usually means acrylonitrile content. See *nitrile rubber*.

nitrile rubber
May also be known as acrylonitrile butadiene rubber or as nitrile butadiene rubber: this gives the abbreviation NBR. Also known as butadiene-acrylonitrile rubber or copolymer or as, acrylonitrile-butadiene rubber or as, poly-(butadiene-co-acrylonitrile) or as, poly-(1-butenylene-co-1-cyanoethylene). An acrylonitrile/butadiene copolymer or rubber: a butadiene-acrylonitrile copolymer diene rubber. An abbreviation formerly used for this material was BR-A or GR-A.

An oil-resistant rubber made by copolymerizing butadiene with acrylonitrile: most NBR is cold-polymerized by emulsion polymerization at temperatures between 5 and 30°C: such cold polymers have less chain branching than hot polymers. The polymer contains a double bond and also a polar acrylonitrile group. By varying the ratio of *acrylonitrile* to *butadiene*, it is possible to produce a range of grades in which improved oil resistance is given by those grades which have the highest acrylonitrile content (referred to as nitrile content). Unfortunately, high acrylonitrile content usually means more difficult processing and the finished mouldings are harder and exhibit lower resilience. It is also found that as the proportion of acrylonitrile is increased, the low-temperature flexibility of the resultant polymer is decreased progressively. This means that a compromise between oil resistance and low-temperature properties is necessary: most commonly, the materials contain approximately 34% acrylonitrile. High acrylonitrile nitrile rubber has a relative density (RD or SG) of 1·00 and low acrylonitrile, nitrile rubber has a relative density (RD or SG) of 0·98. High nitrile NBR would have a nitrile content of approximately 45% and a low nitrile NBR would have a nitrile content of 20%.

Unfilled vulcanizates have very low tensile strengths and it is therefore necessary to compound the material with a suitable reinforcing filler. For example, with a semi-reinforcing black at approximately 40 phr. Because of fears of *antioxidant* absorption by the carbon black, silica fillers are used. Sulphur donors, particularly *semi-efficient* and *efficient vulcanization systems*, give high heat resistance to NBR.

In general, NBR is compounded along lines similar to those practised with the other copolymer of butadiene, SBR. Like SBR, it is found that there is only a small change in plasticity with mastication and the polymer does not crystallise on stretching. It has a slower rate of cure (that cannot be so easily varied by the choice of accelerator) and mouldings exhibit better ageing than SBR. The material has good resistance to aliphatic-type oils and solvents and exhibits reasonable resistance to aromatic-type oils and solvents. The oil resistance of a high nitrile NBR (45%) would be better than a low nitrile NBR (20%). A high nitrile NBR would be more compatible with polar plasticizers and polar polymers (for example, a phthalate or PVC) than a low nitrile NBR. The low nitrile NBR would be more compatible with non-polar materials (for example, styrene-butadiene-rubber) than a high nitrile NBR.

Low molecular weight NBR, also known as liquid NBR, is used as a non-volatile plasticizer for high molecular weight grades. During vulcanization, such a material is chemically bound to the solid NBR and thus resists extraction. By using a small proportion of *divinyl benzene* during polymerization, prevulcanized NBR can be produced. Such a material can be blended with normal NBR to improve processing behaviour.

NBR is the most widely used oil and fuel resistant rubber. Mainly used in the automotive industry, for example, for seals and gaskets. However as the temperatures of use increases, then this material is replaced by other, more heat resistant rubbers: such as *fluorinated rubbers*, *epichlorhydrin rubbers* and *acrylic rubbers*. Compounds based on this *synthetic rubber* are also used for conveyor belts, for oil drilling equipment, for wire sheathing and for cable sheathing. Can be melt mixed with *polyvinyl chloride* (PVC): in such blends the rubber is present in rubbery domains of sub-micrometre size. Both phenol-formaldehyde resins and epoxide resins are used in blends with NBR as they reinforce the rubber. See *thermoplastic elastomers*, *polyvinyl chloride nitrile rubber blend* and *nitrile rubber polyvinyl chloride blend*.

nitrile rubber polyvinyl chloride blend
An abbreviation used for this type of material is NBR/PVC. Also known as nitrile rubber PVC blend: nitrile rubber is abbreviated to NBR. A melt compounded material based on NBR and *polyvinyl chloride* (PVC) in which the NBR is present in the largest proportion by weight. PVC is added to NBR in order to improve, for example, the ozone resistance: the products have good abrasion and oil resistance. See *thermoplastic elastomers* and *polyvinyl chloride nitrile rubber blend*.

nitrile rubber PVC blend
See *nitrile rubber polyvinyl chloride blend*.

nitrile silicone rubber
See *nitrile-silicone rubber*.

nitrile-isoprene rubber
Also known as acrylonitrile-isoprene rubber. An abbreviation used for this type of material is NIR. A rubber based on a *substituted butadiene*. See *nitrile rubber*.

nitrile-silicone rubber
Also known as a nitrile silicone rubber or as a nitrilesilicone rubber. Materials prepared by grafting nitrile groups onto a *silicone polymer backbone*. For example, by grafting β-cyanoethyl onto polydimethylsiloxane a polymer with improved resistance to solvents and with improved electrical properties, compared to silicone rubber, is obtained. Not so commercially important as the *fluorosilicone rubbers*. See *polyorganosiloxane* and *silicone rubber*.

nitrilesilicone
See *nitrile-silicone rubber*.

nitro rubber
The yellow reaction product when concentrated nitric acid reacts with *natural rubber*. Has been given various formulas, for example (C$_{10}$H$_{11}$NO)x.

nitrobenzene vulcanization
Vulcanization achieved by heating *natural rubber* with polynitrobenzenes (for example, trinitrobenzene) in the presence of lead monoxide.

nitrocellulose
See *cellulose nitrate*.

nitrocyclized rubber
The yellow reaction product when concentrated nitric acid reacts with *natural rubber* in carbon tetrachloride. May be represented as $(C_5H_7NO_2)_4$.

nitrogen
Group Va of the Periodic table consists of the elements nitrogen, phosphorus, arsenic, antimony and bismuth. About 78% of the earth's atmosphere consists of nitrogen and it is from this source (by liquefaction and fractional distillation) that nitrogen is obtained commercially. It is a colourless, odourless gas which boils at approximately −196°C and which is very inert: exists as very stable diatomic molecules (N_2). Used to make ammonia (from which is obtained *urea* and *melamine*) and sometimes as a *blowing agent* in, for example, extrusion. Phosphorus, together with nitrogen, is the basis of a group of polymeric materials called *phosphonitrilic polymers*.

nitrogen absorption method
Also known as the BET method. The Brunauer, Emmett and Teller method. Measures the amount of nitrogen absorbed onto the surface of a filler such as *carbon black* or *silica*. Defines the surface area of a filler in terms of m^2/g nitrogen. Measures total surface area and includes areas which are too small, or inaccessible, for use by the polymer (rubber). In general, fillers with BET values <10 m^2/g are non-reinforcing fillers; if the BET values range form 10 to 60 m^2/g then the filler is semi-reinforcing and if the BET values are above 60 m^2/g then the filler will be a reinforcing filler. BET values are sometimes represented as O_{BET}.

nitroso elastomer
See *nitroso rubber* and *carboxy-nitroso rubber*.

nitroso rubber
Also known as nitroso rubber and as, nitrosofluororubber and as, fluoronitrosorubber. An alternating copolymer formed from trifluoronitrosomethane and tetrafluoroethylene and which may be used as a rubber. Difficult to vulcanize but products have excellent flame and chemical resistance. By the use of a third monomer, easier curing results. See *carboxy-nitroso rubber*.

nitrosofluororubber
See *nitroso rubber*.

nitrosoperfluorobutyric acid
Also known as 4-nitroperfluorobutyric acid. One of the three monomers used to make, for example, a *carboxy-nitroso rubber*.

nitrosorubber
See *nitroso rubber*.

NMA
An abbreviation used for *nadic methyl anhydride*.

NMP
An abbreviation used for *N-methylpyrrolidone*.

NMR spectroscopy
An abbreviation used for nuclear magnetic resonance spectroscopy.

N,N-dicyclohexylbenzothiazole-2-sulphenamide
An abbreviation used for this material is DCBS. A *vulcanization accelerator*. See *benzothiazyl sulphenamide accelerators*.

N,N'-di-(3-methyl heptyl)-p-phenylene diamine
An abbreviation used for this material is DMHPPD. An *antiozonant*.

N,N'-di-naphthyl-p-phenylene diamine
See *di-naphthyl-p-phenylene diamine*.

N,N'-disec-butyl–p-phenylene
See *dibutyl-p-phenylene diamine*.

N,N'-dithiobishexahydro-2,4-azepinone
An abbreviation used for this material is DTBC. An example of a *sulphur donor vulcanization system* which is not an *accelerator* in that formulation.

N,N-dicyclohexylbenzothiazole-2-sulphenamide
An abbreviation used for this material is DCBS. A *vulcanization accelerator*. See *benzothiazyl sulphenamide accelerators*.

no-slip condition
An assumption which is made in *rheology*: that is, at the boundary between a solid surface and a flowing melt there is no relative motion. See *viscosity*.

no-twist roving
A *roving* which has been deliberately twisted during winding so that when the subsequent *package* is pulled, the twist is removed.

NOBS
An abbreviation used for *N-oxydiethylbenzothiazylsulphenamide*.

NODP
An abbreviation used for n-octyl n-decyl phthalate. See *octyl decyl phthalate*.

nodular
A material with a pseudo-spherical shape.

nodular moulding compound
An abbreviation used for this type of material is NMC. See *granular polyester moulding compound* and *polyester moulding compound*.

Nomex
A trade name/trademark of an *aramid fibre material*. See *poly-(m-phenyleneisophthalamide)*.

nominal
When used in technology this term usually means that what is being referred to, has been selected mainly because it is convenient to measure or define. See *nominal shear rate*.

nominal aspect ratio
One hundred times the ratio of the *section height* (H) to the *section width* (S) of the *tyre* on its theoretical rim.

nominal R_a value
See *roughness grade number*.

nominal shear rate
Because of the need to choose a shear rate when dealing with a *non Newtonian fluid*, the viscosity of which varies throughout the fluid, the shear rate chosen is a nominal shear rate. For example, the shear rate at the wall is chosen.

nominal strain
The strain obtained by using the original sample dimensions: sometimes referred to as engineering strain. The change in length divided by the original length in a *tensile test*.

nominal stress
Also known as engineering stress. Force divided by the original cross-sectional area of a tensile test sample. See *tensile test*.

nominal tensile strength
See *tensile strength*.

nominal tensile strength at break
See *tensile strength at break*.

nominal tensile stress
See *tensile stress*.

non-contact gauge
A gauge which does not contact the work. An example of a non-contact gauge is an open jet of, for example, an *air gauging system*.

non-contacting rotary seal
A type of *rotary shaft seal*.

non-crimp fabric
A fabric made without inter-lacing the fibres so that the fibres are unidirectional, straight and not crimped. Such a fabric is held together by stitching and is used to give selective stiffening to a plastic composite: gives a straight fibre reinforced composite. See, for example, *resin transfer moulding*.

non-drying oil
An *oil* which does not dry, or harden, on exposure to the air at an appreciable rate. An oil which contains very little *unsaturation*, for example, castor oil or cottonseed oil.

non-extended rubber
Sometimes referred to as straight rubber. Rubber for example, *styrene-butadiene rubber*, which has not been *oil extended*.

non-flaming mode
A way of smoke testing in which the sample is not ignited. See *National Bureau of Standards smoke chamber*.

non-halogenated butyl rubber
An abbreviation used for this type of material is non-halogenated IIR. Butyl rubber which has not been halogenated. See *butyl rubber* and *halogenated butyl rubber*.

non-halogenated IIR
An abbreviation used for *non-halogenated butyl rubber*.

non-ideal fluid
See *non-Newtonian fluid*.

non-intermeshing twin screw extruder
A type of twin screw machine in which the flights of the screws do not overlap as they rotate. The most popular types of *twin screw extruder* are the *intermeshing*, or the partially intermeshing, types of machine.

non-ionic surface active agent
Surface active agents are classified as being *anionic surface active agents* or as, *cationic surface active agents* or as, non-ionic surface active agents. Non-ionic surface active agents are typified by polyoxyethylene condensation products in which the active groups are oxyethylene groups (the hydrophobic part).

non-migratory plasticizer
A *plasticizer* which when used in a material does not readily pass from that material to another contacting material when it is used as the sole plasticizer.

non-Newtonian behaviour
The type of flow behaviour exhibited by a *non-Newtonian fluid*. The coefficient of viscosity of such fluids depends upon the shear stresses involved and may also be time dependent. Can have time-independent fluids, time-dependent fluids and elasticoviscous fluids.

non-Newtonian flow
A term used to describe the type of flow behaviour shown by a polymer melt: the flow is characterised by a non-linear relationship, or non-proportionality, between shear rate and shear stress.

non-Newtonian fluid
Also known, as a non-ideal fluid. A fluid which does not exhibit Newtonian flow. That is, the fluid exhibits *non-Newtonian behaviour*.

non-recoverable joint
A permanent joint: a joint which is not intended to be taken apart. For example, an adhesive joint. Joints may be classed as recoverable and as non-recoverable.

non-reinforcing filler
A *filler* which is added usually simply to reduce compound costs. May improve hardness but does not usually give any reinforcement. See *reinforcing filler*.

non-return valve
A valve which allows flow in one direction only.

non-rigid plastic
For purposes of general classification, a plastics material that has a modulus of elasticity either in flexure or in tension of not over 70 MPa (10,000 psi) at 23°C and 50% relative humidity when tested in accordance with ASTM Method D790.

non-rubber constituents
The solid material of, for example, natural rubber which is not rubber hydrocarbon. For example, proteins. See *skim rubber*.

non-staining antioxidant
An *antioxidant* which does not colour a polymer either initially or during use, for example, a hindered phenol.

non-standard abbreviation
See *abbreviations*.

non-sulphur vulcanization
See *natural rubber - non-sulphur vulcanization*.

non-toxic heat stabilizer
See *calcium/zinc stabilizer* and *metal-free organic*.

non-uniform shrinkage
Shrinkage which is non-uniform throughout a component, for example, it is different *across the flow* to *along the flow* in an *injection moulding*. Part geometry, and changes in flow path length, cause pressure differences in the mould and these in turn result in different *shrinkage values* in different directions. If the shrinkage is non-uniform then warping of the component may result. To minimise warping fill the mould as uniformly as possible and do not overpack: use uniform wall thicknesses. With *reinforced thermoplastics* replace part of the *fibre* with a particulate filler.

non-volatile solvent
A term sometimes used to describe a *plasticizer*.

non-yellowing vegetable oil
A *vegetable oil* may be classified as yellowing or non-yellowing - dependent on their behaviour in use in films. This behaviour depends, for example, on the *linolenic acid* content. For example, soybean oil can have a linolenic acid content of approximately 10% and its non-yellowing properties would not be as good as safflower and suflower oils. Semi-drying and non-drying oils are generally non-yellowing.

nonamethylene diamine
This linear aliphatic diamine may be obtained from rice bran oil. It has the formula $NH_2(CH_2)_9NH_2$: it is used as the starting material for a fibre based on the reaction with urea. That is, for a *polyurea*.

nonanoic acid
See *pelargonic acid*.

nonyl
This term is commonly seen in terms used to describe a *plasticizer*: it means that the side chain contains 9 carbon atoms.

nonyl undecyl adipate
An abbreviation used for this type of material is NUA. A higher *adipate plasticizer*.

nonyl undecyl phthalate
An abbreviation used for this type of material is NUP. A higher *phthalate plasticizer*.

norbornene
Also known as bicyclo-(2,2,1)-heptene-2. This material has a melting point of 53°C and is obtained from *ethylene* and *cyclopentadiene*. Used to prepare polynorbornene.

normal
In organic chemistry means linear: an isomer with an unbranched structure. For example, a linear alcohol is a normal alcohol. An abbreviation used is n and sometimes N. See *N*.

normal distribution curve
In the early days of statistics it was found that many groups of data could be represented by an approximately symmetrical curve which was obtained by plotting the frequency (the number of times a particular value of a dimension was recorded) against the property being measured. This curve was called the normal curve or the normal frequency distribution.

normal frequency distribution
See *normal distribution curve*.

normal lead (II) carbonate
See *lead carbonate*.

normal temperature and pressure
An abbreviation used for this term is NTP. See *standard temperature* and *pressure*.

normal temperature profile
A type of temperatures setting in which the set temperatures steadily increase from the hopper end of the machine to the die/nozzle end of a machine.

Norme Européene
See *European Standard*.

notch embrittlement
The change from ductile to brittle failure caused by notching, or scratching, a sample made from a thermoplastics material. Some materials are more *notch sensitive* than others.

notch sensitive
Having a property which is significantly reduced if a notch or cut, is present in the test sample. Some materials are more notch sensitive than others. May be assessed by *injection moulding* samples which are then notched, with notches of different radii and, for example, performing an impact test on the samples. See *notch tip radius*.

notch sensitivity
The extent to which the resistance of a material to fracture is altered by the presence of a notch or of a sudden change in cross-section. Low notch sensitivity is usually found with ductile materials and high notch sensitivity is usually found with brittle materials. See *notch tip radius*.

notch tip radius
The radius at the root of a *V-shape notch*. This is the most critical dimension of the notch as many thermoplastics materials are *notch sensitive*. Notch sensitivity of various plastics differs, for example *nylon* and *unplasticized polyvinyl chloride* (UPVC) are more notch sensitive than *high impact polystyrene* (HIPS). For HIPS, the effect of notch tip radius on Izod and Charpy impact strength (¼ in specimens with moulded in notch) at 23°C is as follows.

Izod	kJ/m^3
Notch tip radius 0·25 mm (vee-notch)	9·1
Notch tip radius 1·0 mm (vee-notch)	11·3
No notch	95
Charpy	
Notch tip radius 0·25 mm (vee-notch)	6·4
Notch tip radius 1·0 mm (vee-notch)	10·4
0·8 mm wide rectangular notch	6·1

novaculite
A form of *silica*. See *chert*.

novolac
Also spelt novolak. A phenolic-aldehyde resin which, unless a source of methylene groups is added, remains permanently thermoplastic. Used to make phenol-formaldehyde moulding powders. See *phenolic moulding materials*.

novolak
See *novolac*.

novolak/diazonaphthaquinone sulphonate ester positive photoresist
A positive *photoresist* used for features less than 5 μm in size for integrated circuit printing. The novolak binder resin is made more soluble by a photo-induced molecular re-arrangement caused by the diazonaphthaquinone sulphonate ester at near *ultraviolet wavelengths*.

novolak/naphthaquinone diazide-based positive photoresist
A positive *photoresist* used for features less than 3 μm in size for integrated circuit printing. The novolak binder resin is made more soluble by a photo-induced molecular re-arrangement caused by the naphthaquinone diazide.

novolak resin moulding compounds
See *phenolic moulding materials*.

novoloid fibre
A manufactured fibre composed of at least 85% by weight of a cross-linked *novolak*.

nozzle
Sometimes called a *die*. A part of an *injection moulding machine barrel*: the tip of the *barrel* through which the material flows to reach the mould. The melt usually flows from the nozzle into the *sprue* of an *injection mould* but, on some moulds, the nozzle forms part of the mould wall as it extends to the base of the component. There are two other, main types of nozzles, that is, open and sealing.

Open nozzles are generally used, as they are cheap and give less possibility of stagnation. Such nozzles may be used even with comparatively low viscosity melts if the injection moulding machine is equipped with *decompression*. In some cases a sealing-type nozzle must be used: such a nozzle acts like a valve and traps the material (either plastics melt or gas) in the cylinder, or barrel.

To assist in machine alignment, the nozzle of the injection moulding machine usually has a rounded tip and this fits into a corresponding recess in the *sprue bush*. For example, nozzle tip radii of 12·5 mm (0·5 in) and 19 mm (0·75 in) are commonly used for thermoplastics materials. The injection nozzle must seat correctly into the *sprue bush* and the hole in the nozzle tip should be slightly smaller than the hole in the sprue bush; this will allow the sprue to be withdrawn easily through the mould. The hole in the sprue bush should be smaller by 1 mm (0·04"); nozzle radius should be smaller than the sprue bush radius by 2 mm (0·08").

The internal nozzle diameter for *rubber injection moulding* is very important and should match a particular compound and mould - it is usually about 3 mm/0·125 in. As the diameter is decreased, heat build-up and dispersion (for example, of filler and sulphur) increases. If the compound temperature can be built-up in the nozzle then the compound temperature can be reduced so as to decrease *scorch* tendency. If the nozzle diameter is too small then it will restrict mould filling and cause scorch; if this happens increase the diameter, for example, in 0·5 mm/0·02 in steps, until *scorch* is eliminated.

nozzle filter
A *nozzle* designed to filter out contamination. See *nozzle - mixing*.

nozzle - mixing

A *nozzle* designed to give *melt mixing*. Inhomogeneities can be removed from the plastics melt, during injection moulding, by the use of extended nozzles which contain a filter element: that is, an insert with flow channels separated by narrow bands or gaps. It is these narrow bands which remove contamination and improve mixing. By extending this principle even better mixing results from the use of stationary mixers. These fit between the barrel and the nozzle and work by dividing and re-combining the melt, many times by making the melt flow through, for example, stainless steel channels.

nozzle pressure

That pressure which is measured inside the *nozzle* of an *injection moulding machine*. This is, approximately, the pressure which is causing the material to flow. It does not have a constant value but increases as mould filling becomes more difficult. There is a direct relationship between nozzle pressure, injection line pressure and injection pressure. In screw machines, the nozzle pressure is approximately 10% lower than the line pressure. Pressure losses in ram (plunger) machines are much higher and can reach 50%.

nozzle pressure control

A type of *pressure control* used in *injection moulding*. Sometimes called NPC. Nozzle pressure causes the switch from velocity to pressure control during the mould filling part of the cycle. In NPC control, a melt pressure transducer is located within the nozzle of the injection moulding machine or, within an adaptor which fits between the *nozzle* and the cylinder, or *barrel*. The same type of transducer is used in both cases. Such transducers are expensive as the pressure transducers must be built to withstand high melt pressures, and high melt temperatures, and still give consistent readings during production. Such a NPC system consistently gives good results in terms of moulding weight repeatability and in terms of dimensional repeatability: there is no need to change the transducer when the mould is changed. However, the transducer is located in an exposed position where, for example, during the clearing of a blocked sprue, it may be seriously damaged. See *cavity pressure control*.

NPC

An abbreviation used for *nozzle pressure control*.

NPD

An abbreviation used for *nickel pentamethylene dithiocarbamate*.

NPG

An abbreviation used for *neopentylene glycol*.

NPL

An abbreviation used for the UK-based organization called the National Physical Laboratory.

NR

An abbreviation used for *natural rubber*.

NR/PP

An abbreviation used for *natural rubber/polypropylene blends*.

NS

An abbreviation used for non-staining (black). For example, SRF-NS is an abbreviation used for semi-reinforcing furnace - non-staining (black). N774. See *carbon black*.

NTP

An abbreviation used for *normal temperature* and *pressure*. See *standard temperature* and *pressure*.

NUA

An abbreviation used for *nonyl undecyl adipate*.

nuclear model

A model of the atom which consists of protons and neutrons within a small nucleus at the centre of concentric electron-containing orbitals.

nuclear radiation device

See *radiation gauge*.

nucleating agent

A nucleating agent is an additive used to increase the rate of crystallization in thermoplastics materials, for example, in *polyamide* (PA) *materials*. The use of such agents improves the nucleation density. As a result of using such agents, the size of the crystal structures is reduced but more crystal structures are produced in a shorter time. The increase in the rate of crystallization can significantly improve the rate of production in processes such as *injection moulding* as faster solidification is obtained. The reduction in the size of the crystal structures, results in an improvement in the clarity of the products as the size of such structures is reduced. Metal salts of organic acids, for example, sodium benzoate, are used but many other materials will function as such agents.

nucleation

Induced crystallization in a *thermoplastics* material brought about by, for example, a solid material additive during *injection moulding* or *extrusion*. During the recirculation process, in *reaction injection moulding*, an inert gas may be injected into the liquid reactant so as to compensate for shrinkage in the final product: the introduction of the gas may also be referred to as nucleation.

nucleation density

The number of nuclei at which crystal growth is initiated. See *nucleating agent*.

nucleons

The collective name for protons and neutrons.

numeric keypad

A small key board which has keys with numerals: sometimes used for machine setting.

number average molecular weight

The total weight of a sample divided by the number of molecules. An abbreviation used for this value is $\overline{M}_n$. As it is relatively easy to determine it is often quoted. Vapour pressure lowering, cryoscopy, ebulliometry and osmometry give the number average molecular weight.

numerically controlled

An abbreviation used for this term is NC.

numerically controlled milling machine

A machine used to remove metal and used, for example, to machine the cavity of an *injection mould*: such machines are programmed. Once the program has been written, a punched paper tape may be prepared and it is via this paper tape that the milling machine receives its instructions.

NUP

An abbreviation used for *nonyl undecyl phthalate*.

nylon

A type of *thermoplastics material*, (also known as a *polyamide*), which contains many amide (CONH) groups. When these are joined together by linear aliphatic sections (based on methylene groups) the material may be known as a *linear aliphatic polyamide*. A long chain synthetic polyamide in which at least 85% of the *amide groups* are not attached directly to two aromatic rings.

The amide (CONH) groups are very polar and result in strong interchain attraction, for example, due to hydrogen bonding: this interchain bonding gives the materials toughness, rigidity and heat resistance. Because of the strong interchain attraction, and the linear aliphatic sections, a high degree of crystallization is possible.

The easiest way of differentiating between different nylon materials is by means of melting point determinations. For example, PA 6 has a lower melting point than PA 66 (PA 6

220°C/428°F) and PA 66 265°C/509°F). Nylon 6 and 66 have superior abrasion resistance and toughness to *acetal*: not as tough as *polycarbonate* but they have better stress crack and solvent resistance. See *nylon 6* and *nylon 66*.

nylon 3
Also known as nylon three or as, polyamide 3 or as, poly-β-alanine. An abbreviation used for this type of material is PA 3. Prepared from acrylamide by anionic polymerization: the polymer has a melting point of approximately 325°C. Nylons 3, 4, and 5 show a higher moisture regain and less static build-up than *nylon 6* - which is a useful property for a textile *fibre*.

nylon 4
Also known as nylon four or as, polyamide 4 or as, polypyrrolidone. An abbreviation used for this type of material is PA 4. Prepared from 2-pyrrolidone by ring-opening polymerization: the polymer has a melting point of approximately 260°C. Nylons 3, 4, and 5 show a higher moisture regain and less static build-up than *nylon 6* - which is a useful property for a textile *fibre*.

nylon 5
Also known as nylon five or as, polyamide 5 or as, polypiperidone. An abbreviation used for this type of material is PA 5. Prepared from 2-piperidone by ring-opening polymerization: the polymer has a melting point of approximately 260°C. Nylons 3, 4, and 5 show a higher moisture regain and less static build-up than *nylon 6* - which is a useful property for a textile *fibre*.

nylon 6
Also known as nylon six or as, polyamide 6 or as, poly-(ω-amino-caproamide) or as, poly-(6-aminocaproic acid) or as, poly-(ω-aminocaproic acid) or as, polycaproamide or as, polycaprolactam or as, caprolactam PA 6 or as, poly-(imino-1-oxohexamethylene). An abbreviation used for this type of material is PA 6.

PA 6 is a *linear aliphatic polyamide* or *nylon* which is prepared from *caprolactam* by a ring-opening process. A semi-crystalline, thermoplastics material which, because of the many amide (CONH) groups, is tough, rigid, heat resistant, resilient and creep resistant (to dynamic loads). It has similar physical, chemical and electrical properties to PA 66 and is often classed and handled in the same way as PA 66, that is, it is treated as a moisture-sensitive material.

PA 6 has a lower melting point than PA 66: PA 6 220°C/428°F) and PA 66 265°C/509°F. It has a wider processing temperature range, of approximately, 25°C/77°F: it is slightly lighter in colour, has a higher impact strength and slightly better low temperature properties. Has better solvent, grease and detergent resistance than PA 66 but its resistance to dilute mineral acids is poorer. PA 6 and PA 66 are reasonably good electrical insulators at low temperatures and under conditions of low humidity.

After moulding, it reversibly absorbs water (more than PA 66) which causes swelling and an increase in toughness (the toughness may be doubled). When designing PA mouldings, account must be taken of this moisture absorption as it affects dimensional tolerances and physical properties. The rate, and amount of water absorption, depends on wall thickness, relative humidity and compound composition. In the literature, many properties are shown as functions of moisture content.

PA 6 is an easy flow material - slightly stiffer than PA 66: it does not set up so sharply during *injection moulding*. At 280°C/536°F the viscosity ranges from 45 to 300 Nsm^{-2}, dependent on grade. The mechanical properties of a moulded component will be considerably reduced, when compared to a component which has been correctly processed, if it is processed too moist, too wet, too hot, or subjected to a shearing force which is too high (i.e. too fast a mould filling speed).

Shrinkage is of the order of 0.01 to 0.015 mm/mm i.e. 1.0 to 1.5%. Glass filled grades can be as low as 0.003 mm/mm, in the flow direction but as high as 0.01 mm/mm across the flow. Mineral filled grades tend to be more uniform in their shrinkage and *nucleating agents* help reduce *after-shrinkage*.

PA 6 is resistant to alcohols, aromatic hydrocarbons, esters and ketones: the resistance to chlorinated hydrocarbons is only fair. Has good resistance to oils, greases, fuels, fats, greases and, to alkalis (in say concentrations, of up to 20%). PA 6 is not resistant to acids (dilute and concentrated) and to solutions of oxidizing agents. Formic acid, concentrated sulphuric acid, dimethyl formamide, phenol and m-cresol are solvents (as for all PA materials). The density is 1.12 gcm^{-3}/0.65 oz in^{-3}. As the natural colour of the material ranges from a translucent white to translucent beige, then a wide colour range is possible; both translucent and opaque colours can be obtained.

Both PA 6 and PA 66 will absorb large amounts of water, for example, they will absorb approximately 10% of water at saturation. However, PA 6 absorbs more water than PA 66 under the same conditions. The material is supplied dry: as supplied the material is suitable for processing however, if containers are left open rapid water absorption will occur. If moisture content is $\gg 0.2\%$, flow and component properties are affected and the material must be dried. Dry in a well ventilated, hot air oven for 16 hours at 80°C.

When *injection moulding* PA 6, use only one injection moulding pressure setting as this will give more uniform crystallinity. The screw forward time (SFT) is very important and may be longer than the cooling time, for example 5 seconds per mm (0.039 in) of component wall thickness.

Because of its good mechanical strength, impact resistance, rigidity and lightness in weight, PA 6 is widely used as a light engineering material. Its excellent wear resistance makes it a useful material for bearings and other components in contact with moving parts. Its uses are similar to those of PA 66 so that, these materials compete against each other and with *acetal*.

nylon 6 reaction injection moulding
Also known as polyamide 6 reaction moulding (PA 6 RIM): a high temperature reaction injection moulding (HT RIM) process which evolved from the *monomer* casting of Σ-caprolactam used to produce cast *nylon 6* products. The anionic polymerization of Σ-caprolactam (also known as caprolactam) proceeds via a two step propagation mechanism using a metallic catalyst (for example, magnesium bromine caprolactam or sodium caprolactam) plus an acyl caprolactam as an *initiator*. 1,6 hexane diisocyanate is an initiator for PA 6 RIM.

PA 6 is a semi-crystalline, thermoplastics material which, because of the many amide (CONH) groups, is tough, rigid, heat resistant, resilient and creep resistant (to dynamic loads). Because of its good mechanical strength, impact resistance, rigidity and lightness in weight, PA 6 is widely used as a light engineering material.

The impact strength of the homopolymer is increased if elastomeric blocks are incorporated into the polymer: for example, by putting initiator groups on the ends of an elastomeric oligomer based on say, *polypropylene oxide* (PPO). A *block copolymer*, which is lightly crosslinked, is formed if a trifunctional PPO is used - the molecular weight of the PPO may be 5,000.

The mixed monomer is of low viscosity and polymerization proceeds relatively rapidly at temperatures below 220°C - the *melting point* of PA 6. The monomer melts at approximately 67°C so processing temperatures of the order of 90°C are required: that is, the RIM machine must be *heat traced*. The mould temperatures are of the order of 135°C while demould times may be 90 s: the adiabatic exotherm temperature for a typical formulation without filler can be 40°C.

The maximum crystal growth rate for PA 6 is approximately 135°C so the reaction temperatures should be approximately 100 to 160°C: rapid crystallisation is essential for rapid solidification so that quick demoulding can be achieved. The cured RIM moulding releases easily from the mould - especially if one mould half is slightly (approximately 5°C) cooler than the other. Internal mould releases are not necessary. Reaction must be taken to approximately 96% conversion if monomer release is to be avoided.

Nylon 6 block copolymers are readily made by RIM as the reactants are readily mixed and are less sensitive to mix ratio than *polyurethane* RIM. However, higher temperatures are required and demoulding times are longer.

nylon 6T

Also known as polyamide 6T or as, polyhexamethyleneterephthalamide or as, polyhexamethylene terephthalamide. An abbreviation used for this type of material is PA 6T. A partially aromatic polyamide which is only soluble in strong acids. Prepared from hexamethylene diamine and terephthalic acid: a polyamide with a high melting point (T_m) of approximately 370°C. The mechanical properties are similar to those of *nylon 66* but the mechanical properties are retained to higher temperatures.

If trimethylhexamethylene diamine and terephthalic acid are reacted together then a transparent polyamide with a high melting point is obtained (Trogamid T). See *poly-trimethylhexamethyleneterephthalamide*.

nylon 7

Also known as nylon seven or as, polyamide 7 or as, polyenantholactam or as polyheptanoamide. An abbreviation used for this type of material is PA 7. Prepared from ω-aminoenanthic acid. The polymer has a melting point of approximately 225°C. Nylon 7 is similar to *nylon 6* and is used as a textile *fibre*.

nylon 8

Also known as nylon eight or as, polycaprylamide or as, polycapryllactam or as, polyoctanoamide. An abbreviation used for this type of material is PA 8. Not thought to be a commercial *nylon* because of the expensive monomers required (capryllactam or 8-aminocaproic acid) and because the properties are not sufficiently different from other, cheaper, nylon materials.

nylon 9

Also known as nylon nine or as, polypelargonamide. The monomer may be derived from whale oil: the polymer has been used as a textile material.

nylon 11

A high carbon nylon which is a linear, aliphatic nylon. Also known as nylon eleven, or as, polyamide 11 or as, poly-(11-amino-undecanoic acid) or as, polyundecanoamide. An abbreviation used for this type of material is PA 11 or PA11.

PA 11 is a linear, semi-crystalline thermoplastics material which may be considered to be intermediate in properties between *nylon 6* (PA 6) and *polyethylene*. In general, nylon 11 and *nylon 12* have similar properties, price, and are processed similarly. PA 11 is made from aminoundecanoic acid, derived from castor oil, whereas PA 12 comes from dodecanelactam which is derived from butadiene.

Because of differences in crystal structure - caused through amide group (CONH) spacing - PA 12 has a slightly lower melting point and density than PA 11. PA 11 performs better at higher temperatures and in addition, has superior UV resistance. Both materials are reasonable electrical insulators whose electrical properties are not so sensitive to changes in humidity as other polyamides. At room temperature, the mechanical properties, of PA 11 and PA 12, are similar but PA 11 has a higher heat distortion temperature and also a better low temperature impact resistance. Both have good resistance to shock and abrasion and are resistant to many chemicals. They may be extensively modified with plasticizers (for example, sulphonamides), reinforcements etc. PA 11 components may be used continuously at 65°C/149°F with some occasional use at 100 to 130 °C/212 to 266°F. When suitably stabilized (with *antioxidants*), the temperature of continuous use can be higher, especially if there is no strain on the component. These materials can be formulated to meet non-flammable specifications. Compared to PA 6, 66 and 610, these materials have a low melting point, density and by far, the lowest moisture regain (useful, for example, in gear wheels for water meters).

PA 12 is an easier flowing material - can be processed at slightly lower temperatures than PA 11, for example, 10°C lower. For both materials, the viscosity depends on temperature, water content and residence time: water content needs to be below 0.1 percent for *injection moulding*.

Shrinkage is comparatively low and is of the order of 0.005 to 0.02 mm/mm i.e. 0.5 to 2.0%. The actual value is dependent upon grade, component thickness and processing conditions. An increase in mould temperature gives an increase in *shrinkage*.

These materials are resistant to hydrocarbons (such as petrol), most organic solvents and medium strength alkalis (such as sodium hydroxide). Also to inorganic salts (such as, zinc chloride), organic salts, esters, ethers and alcohols (except benzyl alcohol). Low molecular weight alcohols will cause swelling to an equilibrium value (methanol worse than ethanol) which is still relatively low.

They are not resistant to extremes of pH (below 4 and above 12) to strong oxidizing acids (such as nitric and chromic acids), some halogenated compounds and phenols. Raising the temperature greatly alters the chemical resistance (as for most plastics).

PA 11 is slightly denser with a density of 1.04 gcm^{-3} (PA 12 is 1.02 gcm^{-3}) and has a higher melting point of 185°C/365°F as compared to 175°C/347°F for PA 12. Both materials have similar hardnesses but the hardness depends on the grade selected. The natural colour of these materials ranges from slight milky translucent white for PA 12 and a beige-tinged, translucent white for PA 11. These materials can be separated from other types of nylons by melting point determinations or by dropping a piece into glacial acetic acid. PA 11 and PA 12 will float in glacial acetic acid while other nylons will sink.

As the natural colour of the materials is translucent white then a wide colour range is possible; this includes both translucent and opaque colours.

PA 11 and PA 12 are shear sensitive, so small gates on large components should not be used, during injection moulding, as shear-induced crystallinity may occur giving either highly stressed components or mouldings with excessive shrinkage values. Use as large a gate as possible, so as to allow the correct amount of cavity injection pressure (follow-up pressure) to be applied (so as to produce a component of the correct dimensions).

These materials are used as gear wheels (for water meters, gas meters and business machines) and as cable ties - two very important markets for these types of materials. Components maintain properties (e.g. flexibility and low coefficient of friction) over a wide range of temperatures. For these reasons PA 11 is used to make skirts for air-rifle bullets. Good chemical and shock resistance means the choice of this material as a battery casing for aircraft use. Cams, slides and bearings (for both filled and unfilled grades) are other widespread uses. In sports, used to make shuttlecocks and racket frames (carbon-fibre filled). As PA 11 has a richer drier feel than *nylon 6* or *66* it is preferred fro some textile applications, for example, underwear.

nylon 12

A high carbon nylon which is a linear, aliphatic nylon. Also known as nylon twelve or as, polyamide 12 or as, polylauroamide or as, polylauryllactam or as, polydodecanoamide. An abbreviation used for this type of material is PA 12 or PA12. In general, PA 11 and 12 have similar properties, price, and are processed similarly. PA 11 is made from aminoundecanoic acid, derived from castor oil, whereas PA 12 comes from dodecanelactam which is derived from butadiene. PA 12 has a slightly lower density and, as is usual with even-numbered nylons, a lower melting point. Used in applications where the slightly lower moisture absorption is an advantage. See *nylon 11* for a comparison of nylon 11 and 12.

nylon 46

Also known as polyamide 46 or, polytetramethyleneadipamide or, polytetramethylene adipamide. An abbreviation used for this type of material is PA 46 or PA 4·6. This polymer is used as an *injection moulding material* and has a melting point of approximately 295°C. The melting point is higher than that of *nylon 66* and so the mouldings have a higher *heat distortion temperature*.

nylon 66

Also known as nylon six-six or as, polyamide 66 or as, PA 66 nylon or as 66 nylon or as, poly-(hexamethylene adipamide) or as, polyhexamethyleneadipamide or as, polyhexamethylene adipamide or as, poly-[imino-(1,6-dioxohexamethylene)-iminohexamethylene]. An abbreviation used for this type of material is PA 66 or PA 6·6.

The homopolymer, 66 nylon, is made from *nylon 66 salt*. By increasing the number of monomers, copolymers may also be prepared and any combination of copolymers and homopolymers may be blended, filled (for example, with *glass*) and plasticized so that a wide range of grades and of properties is possible from PA 66-type materials (see *nylon - grades of*). For example, the use of molybdenum disulphide and graphite, dispersed in the original moulding material, improves moulding wear. Polymers with 50% glass fibre loading have high strength, stiffness, abrasion resistance and good dimensional stability - ease of flow is, however, reduced and cylinder wear may be a problem. Glass beads give less warpage than glass fibres.

Like PA 6, this plastics material is a semi-crystalline, thermoplastics material which is often processed by *injection moulding*. PA 66 nylon has one of the highest melting points of commercial polyamides, very high strength and stiffness and very good retention of stiffness with increasing temperature. After moulding, it absorbs water (but not as much as PA 6) which causes swelling and an increase in toughness (the toughness may be doubled). When designing PA 66 mouldings, account must be taken of this moisture absorption as it affects dimensional tolerances and physical properties - for every 1% water absorbed, the size increases by approximately 0·003 mm/mm. The rate, and amount of water absorption, depends on wall thickness, relative humidity and compound composition. In the literature, many properties are shown as functions of moisture content. At 50% relative humidity (RH) and at 23°C/73°F the moisture content is 2·5 per cent.

Moulded components are tough even in thin sections. Because of the ease of flow of this material, such thin sections can be moulded relatively easily provided the mould can be filled before the hot material sets. PA 66 should cycle faster than PA 6. High weld strengths are also possible form this class of materials. The use of nucleated grades can result in significant reductions in cycle times (for example, of 10 to 15%) and in mould ejection forces - this in turn can mean less component distortion even though smaller ejection pins may be used. Both PA 6 and PA 66, have better abrasion resistance and toughness than acetal. Super-tough PA 66 has good resistance to crack initiation and propagation and is therefore stress concentration resistant.

PA 66 is a low viscosity material which flows very easily. The viscosity is very dependent on shear rate and on temperature; the temperature, and the mould filling speed, must be controlled very precisely. For example, altering the melt temperature by 10°C can alter the viscosity by a factor of two. At 280°C/536°F the viscosity ranges from 40 to 400 Nsm^{-2} (dependent on grade).

Shrinkage is of the order of 0·010 to 0·020 mm/mm i.e. 1 to 2%. The addition of glass fibre will reduce the shrinkage to 0·2 to 1%, but differences between the transverse and longitudinal directions (i.e. along and across the flow) can be as much as five: mineral filled grades give more isotropic mouldings while the use of a *nucleating agent* helps to reduce aftershrinkage (see *nylon - shrinkage*).

PA 66 is resistant to most solvents; insoluble in esters, ketones, aromatic hydrocarbons, chlorinated hydrocarbons, alkalis, dilute acid solutions and most organic acids. Some organic solvents, for instance chlorinated hydrocarbons, cause swelling. Excellent resistance to oils, fuels, greases and fats. PA 66 is not resistant to strong acids or to oxidizing agents. Formic acid and phenol are solvents. Outdoor exposure can cause colour fading and embrittlement unless the PA is stabilized. PA 66 is more resistant to chloroform, benzyl alcohol and trichloroethylene than PA 6. PA 6 will dissolve in 4N hydrochloric acid and boiling N,N-dimethylformamide (PA 66 will not). PA 66 is reversibly swollen by water which worsens the electrical properties.

The density is 1·15 gcm^{-3}. As the natural colour of the material ranges from translucent white to translucent light brown, a wide colour range is possible; both transparent and opaque colours are available. As supplied the material is suitable for moulding however, if containers are left open, rapid water absorption will occur. Dry in a well ventilated, hot air oven for 4 to 5 hours at 85°C/185°F but be careful not to cause oxidation. When first ejected, injection mouldings are dry as they have not absorbed water: dry mouldings are more brittle than mouldings which have absorbed water and are therefore easier to reclaim.

PA 66 components may be solvent welded with aqueous phenol (12%) or a solution of PA 66 in ethanol and calcium chloride. If dimensions are very critical, then annealing, and/or moisture conditioning, may be necessary. Anneal by heating in a non-oxidizing oil for 20 minutes at 170°C/338°F. May be conditioned to equilibrium water content by immersion in water (maximum temperature 60°C/140°F): to get uniform water distribution wrap in polyethylene film after conditioning and store.

Like PA 6, this plastics material is often used in light engineering applications. Widely used in the automotive industry for fans, grilles, door handles, filters etc. Around the home, it is used in kitchen appliance housings, vacuum cleaner components and curtain fittings. Transport applications include conveyor belt components, castors, bearings and bicycle wheels. In such applications, the material is selected because of the low level of friction encountered without lubrication together with good oil and grease resistance. Often used where the components require impact resistance and impact strength. PA 6 and PA 66 are the most widely used of the so-called engineering plastics: they are also widely used as glass reinforced materials. Often used in the automotive industry, as fuel resistant, technical mouldings which must operate in the aggressive environment in, and around, the engine. That is, under the hood or bonnet. Alloys of PA are now being used for computer housings and for automotive panels: such panels may be required to withstand elevated paint shop temperatures and then, subsequent low temperatures in use.

nylon 66 salt
Also known as hexamethylene diammonium adipate. Used for the preparation of *nylon 66* as the use of the salt, produced by the reaction of hexamethylene diamine and adipic acid, gives high molecular weight polymers. This is because the use of the salt, gives the exact, equal ratios of the two monomers.

nylon 6/66
Also known as polyamide 6/66. An abbreviation used for this type of material is PA 6/66. A *nylon copolymer* which contains *nylon 6* and *nylon 66* repeat units. A softer material than the homopolymers and with a lower melt viscosity.

nylon 6/66/610
Also known as polyamide 6/66/610. An abbreviation used for this type of material is PA 6/66/610. A *nylon copolymer* which contains *nylon 6*, *nylon 66* and *nylon 610* repeat units. A softer material than the homopolymers and with a lower melt viscosity: softer than *nylon 6/66*.

nylon 610
Also known as nylon six-ten or as, polyamide 610 or as, polyhexamethylenesebacamide or as, polyhexamethylene sebacamide. An abbreviation used for this type of material is PA 610 or PA 6·10. A *polysebacamide*: a major *nylon copolymer* formed from *nylon 610 salt*. The polymer has a melting point of approximately 223°C and a relatively low water absorption: more dimensionally stable than *nylon 6* or nylon 66 and more suited for electrical insulation applications. Widely used as a fibre material.

nylon 610 salt
Also known as hexamethylene diammonium sebacate. Used for the preparation of *nylon 610* as the use of the salt, produced by the reaction of hexamethylene diamine and sebacic acid, gives high molecular weight polymers. This is because the use of the salt, gives the exact, equal ratios of the two monomers.

nylon 612
Also known as nylon six-twelve or as, polyamide 612 or as, polyhexamethylenedodecanoamide or, polyhexamethylene dodecanoamide. An abbreviation used for this type of material is PA 612 or PA 6·12. The polymer has a melting point of approximately 218°C and a relatively low water absorption: even lower than *nylon 610* and is therefore even more dimensionally stable.

nylon 69
Also known as nylon six-nine or as, polyamide 69 or as, polyhexamethylenenonamide or as, polyhexamethylene nonamide or, polyhexamethyleneazelamide. An abbreviation used for this type of material is PA 69 or PA 6·9. This material has a melting point of approximately 215°C and is similar to *nylon 610*.

nylon 77
Also known as nylon seven-seven or as, polyamide 77. An abbreviation used for this type of material is PA 77 or PA7·7. A *nylon copolymer*.

nylon 91
Also known as nylon nine-one or as, polyamide 91 or as, polynonamethyleneurea. An abbreviation used for this type of material is PA 91 or PA 9·1. A *nylon copolymer* whose properties are similar to *nylon 11*. The polymer has been used as a textile material.

nylon bandage
The fabric belt of a *steel-braced radial tyre*.

nylon belt
The fabric belt of a *steel-braced radial tyre*.

nylon copolymer
A polyamide which is based on two monomers: such a material is softer and tougher than a nylon homopolymer and of a lower melting point. See, for example, *nylon 610*.

nylon fibre
A manufactured, polymeric fibre in which less than 85% of the amide groups are attached directly to 2 aromatic rings.

nylon - grades of
A wide range of grades is available as, for example, *nylon 6* and *nylon 66* may be extensively modified with particulate or fibrous fillers, plasticizers, other plastics materials and elastomers. Glass, in one of its many forms, is a popular filler for PA as it is relatively cheap and its use can lead to significant improvements in some properties, for example, strength and stiffness. The impact resistance of PA 6 and 66 may be improved by the use of other plastics or, by the use of elastomers. For example, the impact resistance when dry, may be improved by the use of, approximately, 15% polyethylene (PE) - providing the two plastics materials are coupled together. High impact PA, results from the successful incorporation of an appropriate elastomer, for example, *ethylene-propylene (EPDM)* or *styrene-butadiene rubber (SBR)*. It is possible to get flame-retardant grades of PA, which are halogen and phosphorus free, and which are VO rated at 0·25 mm (0·010 in). A well dispersed silicone fluid, can result in a moulding material with low frictional losses. The use of a *nucleating agent*, can give fast cycling grades as such agents improve the rate of crystallization. Aluminum stearate, approximately 0·1%, eases ejection and also reduces ejection forces (for both the nucleated and the unnucleated grades) and so, minimizes ejection problems. The use of molybdenum disulphide and graphite, dispersed in the original moulding material, improves moulding wear.

nylon plastics
The collective name for the class of materials which contain the amide group used as a chain linkage. Another name for polyamide plastics. See *nylon 6*, *nylon 66* etc.

nylon salt
Also known as an AH salt. The reaction product of an aliphatic *diamine* and an aliphatic dicarboxylic acid: a 1:1 adduct. Used to produce nylons or polyamides as the use of the salt, produced for example, by the reaction of hexamethylene diamine and sebacic acid, gives high molecular weight polymers. This is because the use of the salt, gives the exact, equal ratios of the two monomers required.

nylon screw
Usually means that a screw, of an *extruder* or of an *injection moulding machine*, has been designed for *nylon* (for example, PA 6 or PA 66) that is, it is a *dedicated screw*. See *double parallel screw*.

nylon - shrinkage
Nylon 6 and nylon 66 are semi-crystalline, thermoplastics materials with relatively high *shrinkage*. The addition of *glass fibre* will reduce the shrinkage but differences between the transverse and longitudinal directions (i.e. along and across the flow) can be as much as five: mineral filled grades give more isotropic mouldings while nucleating agents help reduce after-shrinkage.

In general, thinner components exhibit less shrinkage than thick components. However, thin sections show more after-shrinkage than thick; also low mould temperatures can increase after-shrinkage. Shrinkage and after-shrinkage (*post-moulding shrinkage*) decrease component size but moisture absorption increases component size. So, after moulding, the component may increase in size due to water absorption and this, may negate the effect of shrinkage.

A lot of the after-shrinkage is due to changes in the crystallinity level. The crystallinity of thick sections (>0.2 in/5 mm) moulded in hot moulds do not alter greatly with time. The crystallinity of thinner sections (<0.2 in/5 mm) moulded in colder (<40°C) moulds can alter significantly with time - particularly if such mouldings are held at temperatures greater than 65°C. So, for mouldings which are to be used at elevated temperatures, for example, greater than 60°C/140°F, and which are to be used under conditions of low humidity, annealing may be needed if stable dimensions are to be held

High loadings of inorganic pigments can act as *nucleating agents* and decrease shrinkage. organic colorants do not usually affect shrinkage. To reduce shrinkage, cavity filling must be maximized by using a long *screw forward time*, for example, 5 seconds per mm (0.040 in) of wall thickness. Use only one pressure setting (for example, 75 MNm^{-2}) for both first and second stage injection pressures as this will give more uniform crystallinity and shrinkage. As it is difficult to accurately predict the actual shrinkage value, use a *prototype mould* for cavity sizing purposes.

nylon tyre
Also known as a tempered tyre. A tyre based on tempered nylon cords.

nylonates
Low temperature plasticizers for *polyvinyl chloride* (PVC). Mixed esters prepared from petrochemicals. Mixed acids such as *AGS acid* is reacted with alcohols (octyl, nonyl and decyl alcohol) to give the esters which are sometimes referred to as sugludates. Industrial use of such traditional materials as *dibutyl sebacate* is now reduced because of the availability of these cheaper materials.

O

o
An abbreviation used for *ortho*.

O
This letter is used as an abbreviations for:

   octyl. See, for example, *butyl octyl phthalate*;
   oil. For example, *epoxidized soya bean oil (ESO)*; and,
   oriented.

O dimension
The diameter of the small end of the *sprue*: this is made slightly larger than the hole in the tip of the *nozzle*.

o-hydroxymethyl-phenol
Also known as saligenin. A methylol phenol: a reaction product of *phenol* and *formaldehyde* which is formed during the initial stages of *phenol-formaldehyde* resin production.

o-nitrobiphenyl
An abbreviation used for this material is ONB. It is also known as 2-nitro biphenyl and as o-nitrodiphenyl. This material has a boiling point of 330°C, a melting point of 35°C and a relative density (RD) of 1.44. A *plasticizer* for cellulosics, for example, for cellulose nitrate. Also compatible with many other polymers, for example, with *polystyrene*.

o-nitrodiphenyl
See *o-nitrobiphenyl*.

o-phthalic acid
Also known as phthalic acid or as, ortho-phthalic acid or as, 1,2-benzenedicarboxylic acid. A white solid with a melting point of 207°C. When heated gives *phthalic anhydride*. See *phthalic acids*.

O-ring
A vulcanized rubber ring which fits into a groove and is used, for example, to prevent water leakage on a mould.

OB
An abbreviation used for 4,4'-oxybis-(benzenesulphonylhydrazide).

object of production
See *machine operation*.

object of testing
Over the years, many tests have been devised to enable judgements to be made on:

    the quality and consistency of raw materials;
    the quality and consistency of associated products;
    the merits, or otherwise, of new or modified materials;
or, the suitability of a particular design.

occlusion cellulose
See *inclusion cellulose*.

occupational exposure limits
An abbreviation used for this term is OEL.

occupational exposure standards
An abbreviation used for this term is OES.

ochre
A natural earth which ranges in colour from orange to red and which is based on hydrated iron oxide mixtures. For example, French ochre is a yellow pigment which is a mixture of clay and yellow hydrated iron oxide.

octa
A prefix which means eight.

octabromodiphenyl ether
An *organo-bromine* flame retardant. Such a bromine-containing compound tends to be more powerful than the equivalent chlorine-containing compound: often used in conjunction with *antimony trioxide*. For example, *high impact polystyrene* 80%, *octabromodiphenyl ether* 15% and *antimony trioxide* 5%.

octadecanoic acid
See *stearic acid*.

octadecene nitrile
An abbreviation used for this material is ODN. A *plasticizer* for nitrile rubbers and for vinyl plastics.

octadecene-1-maleic anhydride copolymer
An alternating copolymer based on octadecene-1 and maleic anhydride: a reactive material (because of the anhydride groups). Used, for example, as a cross-linking agent for *epoxy resins*.

octadecylamine
An amine which may be used, for example, to minimise dust formation in asbestos.

octamethylcyclotetrasiloxane
Obtained from *dimethylsiloxane*. May be readily purified and the pure material is used as the precursor for high molecular weight, silicone rubber (based on polydimethylsiloxane). The starting material for *silicone rubber*. The rubber is produced by the ring opening polymerization of this monomer: molecular weights may reach one million.

octamine
The reaction product of diphenylamine and diisobutylene. This wax-like material has a melting point of about 80°C and a relative density (RD) of 0.99. An *antioxidant* for rubbers.

octane
A member of the alkane series (a paraffin) and which has eight carbon atoms. This material has a relative density (RD

or SG) of 0·68 and a boiling point of approximately 126°C. May be represented as C_8H_{18}. n-octane is used as a solvent, for example, for the polymerization of hydrocarbon polymers and as a solvent for hydrocarbon rubbers. The branched chain isomer, iso-octane is known as 100-octane and is used to assess the anti-knock properties of petrol.

octanol
Also known as octyl alcohol. Mixed isomers which contain eight carbon atoms and which have the general formula $C_8H_{17}OH$. Octanol-1 is the most important isomer and is used as a solvent.

octene
See *octene-1*.

octene-1
An *alkene* which has eight carbon atoms and the formula C_8H_{16}. Has a boiling point of 121°C. Used as a comonomer for the preparation of some types of *linear low density polyethylene*.

octene-type material
See *linear low density polyethylene*.

octo
A prefix which means eight.

octyl
The organic radical which has the formula C_8H_{17}-.

octyl alcohol
See *octanol*.

octyl decyl adipate
An abbreviation used for this type of material is ODA. Also known as n-octyl n-decyl adipate. An abbreviation also used for this material is NODA or DNODA (from di-(n-octyl n-decyl) adipate or as di-n-octyl-n-decyl adipate). However, in *ISO terminology*, the letter n or N is not used in plasticizer abbreviations to indicate normal.

This material has a melting point of −60°C, a relative density (RD) of approximately 0·92 and a boiling point in the region of 220°C (4 mm). An adipic acid ester which is based on a mixture of straight chain alcohols which contain from 8 to 10 carbon atoms. This liquid material is used as a low temperature *plasticizer*, for example, for *polyvinyl chloride* and cellulose esters. Also suggested as a softener for rubbers. See *iso-octyl iso-decyl adipate*.

octyl decyl phthalate
An abbreviation used for this type of material is ODP. Also known as n-octyl n-decyl phthalate (NODP). This material is also known as di-(n-octyl-n-decyl) phthalate or as di-n-octyl-n-decyl phthalate (DNODP). However, in *ISO terminology*, the letter n or N is not used in plasticizer abbreviations to indicate normal.

This material has a melting point of −40°C, a relative density (RD) of 0·97 and a boiling point in the region of 235°C (4 mm). This liquid material is used as a *plasticizer*, for example, for *polyvinyl chloride* and as a softener for rubbers. A *linear phthalate plasticizer* based on a mixture of straight chain alcohols which should contain from 8 to 10 carbon atoms but which often contains lower esters as well as higher esters. Where long term retention of properties, after exposure to high and low temperatures, is a requirement then phthalates based on a linear alcohol will perform better than a phthalate based on a branched alcohol. For example, the linear phthalate, n-octyl n-decyl phthalate will retain properties better than branched chain alcohols such as di-iso-octyl phthalate (DIOP) and di-iso-decyl phthalate (DIDP). Has lower volatility than plasticizers such as *dioctyl phthalate*. See *n-octyl n-decyl phthalate*.

octyl decyl trimellitate
See *n-octyl decyl trimellitate*.

octyl diphenyl phosphate
See *diphenyl octyl phosphate*.

octyl phenyl salicylate
See *p-octyl phenyl salicylate*.

octylated diphenylamine
Also known as octylated diphenyl amine. An abbreviation used is ODPA. A *substituted diphenylamine* used as an *antioxidant*.

OD
An abbreviation used for outside diameter.

ODA
An abbreviation used for *octyl decyl adipate*.

ODCB
An abbreviation used for o-dichlorobenzene.

ODN
An abbreviation used for *octadecene nitrile*.

odorant
An additive used in a polymer compound to hide an undesirable odour or to impart a desirable odour. An odour control agent. An undesirable odour may be that given to the compound by, for example, a *plasticizer* or traces of a monomer: a desirable odour may be that of a natural material such as leather. Such an additive may be classed as an odorant, a reodorant or as a *deodorant*. Odorants and reodorants are often *essential oils* and perfumes which may be available as a masterbatch: vanilla is an example of an odorant used in food-grade materials.

odour control agent
See *odorant*.

ODP
An abbreviation used for *octyl decyl phthalate*. An abbreviation used for *ozone depleting potential*.

ODPA
See *octylated diphenylamine*.

ODTM
An abbreviation used for *n-octyl decyl trimellitate*.

OE
An abbreviation used for *oil extended*. For example:

OE rubber = *oil extended rubber*;
OE-BR = oil extended *butadiene rubber*;
OE-E-SBR = oil extended emulsion *styrene-butadiene rubber*.
OE-EPDM = oil extended *ethylene-propylene rubber*;
OE-L-SBR = oil extended solution *styrene-butadiene rubber*;
OE-NR = oil extended *natural rubber*;
OE-SBR = oil extended *styrene-butadiene rubber*.

Could also have the same abbreviation without the spaces or hyphens. For example, OENR is also used for *oil extended natural rubber*. OESBR is *oil extended styrene butadiene rubber*.

OEL
An abbreviation used for occupational exposure limits.

OE tyre
An abbreviation used for *original equipment tyre*.

oersted
The unit of magnetic field strength in the *cgs system*.

OES
An abbreviation used for occupational exposure standards.

off-line reciprocating screw rheometer test
An abbreviation used for this term is off-line RSR test. A test made with a reciprocating screw rheometer. The test, or measurements are made without an injection mould connected to the injection unit, for example, measurements are made while an *air-shot* is being made, The melt temperature must be carefully and accurately controlled and the melt/nozzle pressure must be capable of accurate measurement. See *on-line reciprocating screw rheometer test*.

off-line RSR test
An abbreviation used for *off-line reciprocating screw rheometer test*.

off-set calender
A *calender* in which at least one of the rolls is off-set with respect to the remainder. In order to cut down on *nip* interactions and to make feeding easier, off-set calenders were developed from *vertical superimposed calenders*. The best known example of this type is the *inverted L* in which material can be dropped in the top nip and remains there because of gravity. Nip interaction is reduced because one set of rolls is working at right angles to the pair of rolls which form the next nip. For example, variations in the feed nip (between roll 1 & 2) will cause roll no. 2 to move horizontally and because of the size of the rolls this will have only a small effect on the nip which is formed further round that roll (between rolls 2 and 3). Gauge variations could be minimised even further by fitting *pull back rams* to rolls nos 1 and 2. Roller bearings could be fitted to roll no. 3 to minimise *roll float*: gauge control could be markedly improved by fitting *cross-axis* to roll 3 and *roll bending* to roll 4.

A logical extension to the idea of roll offset is to offset all rolls with respect to one another: this gives the Z and *inclined Z* (or S) calender. With such calenders it is claimed that nip interaction is reduced to a minimum but a problem with such calenders is nip accessibility. However such calenders require less working height than other 4 roll machines and have certain attractions for tyre cord manufacture.

off-set roll
One of the rolls of an *off-set calender*. That roll which is off-set with respect to the remainder.

offset
Sometimes spelt off-set. Means that the setting of a controller is consistently different to that which is obtained: in temperature control this is minimised by using an *integral* (I) term.

offset adaptor
The die and the extruder point in the same direction but the product is not centred on the extrusion cylinder or barrel i.e. it is offset.

offset lithography
A popular *lithographic printing technique*.

offset spider arms
Part of the mandrel support system which is designed to minimize weld line formation and parison thinning at the weld point; this type of fault can be produced by the spider legs or arms. With offset spider arms, the supports do not cross from mandrel to head in one pass: they connect first to a support ring and then, another leg, displaced from the first, continues.

offset spider legs
See *offset spider arms*.

ohm
An SI derived unit which has a special symbol, that is, Ω. It is the SI derived unit of resistance and is the resistance of a conductor in which one volt produces a current of one ampere. Has the dimensions of $kgm^2s^{-3}A^{-2}$. This equals, VA^{-1}. See *Système International d'Unité*.

ohm.centimetre
The units of *volume resistivity*. The unit ohm.centimetre is sometimes abbreviated to Ω.cm. Results of *volume resistivity* measurements are sometimes published in ohm metres (Ω.m). To convert from Ω.cm to Ω.m multiply by 100.

ohmic resistance
An abbreviation used for this term is R. The ohmic resistance is proportional to the length (l) and inversely proportional to the cross-sectional area: a constant called the *volume resistivity* (ρ) is also used. That is, $R = \rho l/a$. See *surface resistivity* and *volume resistivity*.

OI
An abbreviation used for oxygen index. See *limiting oxygen index*.

oil
An oily substance which is a liquid at room temperatures is referred to as an oil: when it is a solid at room temperatures it is referred to as a *fat*. Oils may be classified according to their origin into, for example, animal oils, (obtained from animals), vegetable oils (obtained from fruits and seeds) and mineral oils (obtained from the rocks or minerals of the earth). Oils may be hardened into fats chemically by hydrogenation. See *glycerides*.

oil adsorption method
Structure levels in *carbon black*, may be determined by an oil adsorption method or by, the *DBP adsorption method*. The amount of oil which is just sufficient to form an oil or paste is determined using *linseed oil*. The oil adsorption value is expressed in ml of oil per 100 g of filler, for example, 15 ml linseed oil per 100 g of carbon black. (The oil number of the sample may be reported in $cm^3/100$ g.) Nowadays *dibutyl phthalate* (DBP) is more commonly used and the measurement is made in an adsorptiometer or in a plastograph. The larger the oil (or DBP) value is, at constant black surface, the stronger are the forces between the carbon black particles (see *24M4 test*).

oil adsorption value
See *oil adsorption method*.

oil cooling system
That which cools the oil of a moulding machine. Generally consists of a water-cooled heat exchanger. Once a moulding machine is at operating temperature then the oil cooling system must be capable of removing the heat which is being given to the oil and of holding the oil at the specified temperature. Most moulding machines are hydraulically powered using, for example, an electric motor and a pump. Even a small injection moulding machine uses a large amount of electricity, for example, a 40 tonne/ton machine will use approximately 8 to 9 kW/h. Most of this heat finds its way into the hydraulic system where it must be removed; if it is not removed then the oil will be degraded and the machine will not run properly (it could also be seriously damaged).

To remove the heat, water is passed through a coil (heat exchanger) at the base of the oil reservoir. To save on scale problems (and on water) closed loop, water recirculating systems are used; these lose the heat, in turn, by means of another heat exchanger. This heat can be recovered and used.

oil extended butadiene rubber
An abbreviation used for this material is OE-BR or OEBR. See *oil extended rubber* and *butadiene rubber*.

oil extended natural rubber
An abbreviation used for this material is OE-NR or OENR. *Natural rubber* which contains an oil added as an *extender*. A petroleum oil (up to 50 phr, for example, 25 phr of an aromatic or naphthenic oil) may be added to *natural rubber latex* before coagulation. This is done to reduce cost and/or to give an easier processing, softer compound.

oil extended rubber
A rubber which contains an *oil* added as an *extender*. Rubber which has been mixed with oil before the *latex* is coagulated. Up to 50 phr of a compatible oil may be added to rubbers (for example, NR and SBR) to reduce cost and/or to give an easier processing, softer compound. High molecular weight rubber is used and the oil is, for example, a *petroleum oil plasticizer*. Such mixtures were originally made in order to reduce cost but the oil also acts as a processing aid and reduces vulcanizate stiffness. See, for example, *natural rubber*.

oil extended styrene butadiene rubber
An abbreviation used for this material is OE-SBR or OESBR. High molecular weight *styrene butadiene rubber* which contains an oil added as an *extender*. A naphthenic oil (up to approximately 60 phr) may be added to styrene butadiene *latex* before coagulation, to reduce cost and/or to give an easy processing, soft compound. Naphthenic oils give good resistance to discolouration.

oil extender
An *oil* used as *extender*. If, during injection moulding, a compound's viscosity is too high then re-formulate (make it safer) so that it can be run at a higher temperature. Do not try and use an oil extender as this will result in lower melt temperatures.

oil forming alkyd
See *oil modified alkyd resin*.

oil furnace black
Another name for furnace black. See *carbon black*.

oil furnace process
A process used to produce furnace black. See *carbon black*.

oil length
The amount of *drying oil* in an *alkyd resin*. The amount of drying oil incorporated in an alkyd resin may be used to classify the resin as being short oil, medium oil and long oil.

oil modified alkyd resin
An *alkyd resin* modified by the incorporation of a drying oil; the *triglycerides* of a natural plant oil. The amount of drying oil incorporated in an alkyd resin may be used to classify the resin as being short oil, medium oil and long oil. The use of the oil makes the resin more soluble and imparts toughness to the final film.

oil of safflower
See *safflower oil*.

oil of winter-green
See *methyl salicylate*.

oil preheating
Raising the temperature of the hydraulic oil, for example, of an *injection moulding machine*, to its operating temperature before the moulding operation begins. This may mean heating the oil to, for example, 45°C/113°F. This may be done by either, dumping the oil back to tank or, by using a purpose-fitted preheater.

oil - relative densities of

oil - castor	0·96
oil - cottonseed	0·92
oil - mineral (aromatic)	1·02
oil - mineral (naphthenic)	0·93
oil - mineral (paraffinic)	0·86
oil - palm	0·88
oil - pine	0·93
oil - rape	0·92
oil - tall	0·95 to 1·0

oil resistance
The resistance of a polymer to an *oil*, for example, the swelling resistance. Oil resistance is commonly imparted to polymers by the incorporation of highly polar groups along the polymer chain. For example, rubbery polymers, based on *acrylic monomers*, contain polar groups such as the *nitrile group* to impart hydrocarbon resistance.

oil resistant rubber
A rubbery material which generally gives compounds with better oil resistance than a *general purpose rubber*: a rubber which is noted for its oil resistance. For example, the following materials have been classified as oil resistant rubbers, *chloroprene rubber* and *nitrile rubber*.

oil resistant silicone rubber
A *silicone rubber* which has good *oil resistance*. FVMQ-types of silicone rubber are used where high resistance to chemical attack (fuel, oil and solvent) is required. Nitrilesilicones are also oil resistant but are not so commercially important. See *polyorganosiloxane*.

oil treatment
The treatment of an oil, for example, a *vegetable oil*, so as to make such an *oil* usable or to modify its properties. Such treatments include *alkali refining*, *kettle bodying* and *blowing*.

oil-extended rubber
See *oil extended rubber*.

oilseed
A vegetable seed which yields a *vegetable oil*: such an oil is a *fixed oil*.

oiticica oil
Derived from licania rigida. A *dual purpose vegetable oil* which is usually used in heat treated form (permanently liquid oiticica oil). Surface coatings, based on heat treated oiticica oil, are more resistant to water and to alkalis than *linseed oil* but not as good as *tung oil*. Used as a substitute for tung oil when it is significantly cheaper.

olefin
An *unsaturated hydrocarbon* or alkene. See *olefins*.

olefin fibre
A manufactured, polymeric fibre composed of at least 85% by weight of ethylene, propylene or other *olefin* units (excludes materials classified as rubbers).

olefin modified styrene-acrylonitrile
Also known as olefin modified SAN as this type of material results from the incorporation of olefin rubbers into SAN: similar properties to *acrylonitrile-butadiene-styrene (ABS)* but has better weathering properties.

olefin plastics
Plastics based on the polymerisation of olefin monomers (e.g. *ethylene*); at least 50% of the plastics material must be based on an *olefin*. See, for example, *polyethylene*.

olefin thermoplastic elastomer
An abbreviation used for this type of material is OTE or TPO. See *rubber modified polypropylene*.

olefine
See *alkene*.

olefins
A series of unsaturated hydrocarbons which are also called *alkenes*. Very important *monomers* with the formula C_nH_{2n}. The systematic names end in ene: the systematic names of the first three materials are ethene, propene and butene but the systematic names are not widely used. The common names of the first three materials are ethylene, propylene and butylene. See *polyethylene*, *polypropylene* and *polybutylene*. See α olefin and *olefin plastics*.

oleic acid
Also known as red oil or as, cis-9-octadecenoic acid or as, cis-octadec-9-enoic acid or as, 9-octadecenoic acid. This unsaturated, organic acid has a relative density (RD or SG) of 0.89 and a melting point of approximately 15°C. A constituent of the *triglycerides* of most plant oils. Although it is *unsaturated* it does not air dry. Used as an *activator* and softener for rubbers and to prepare *plasticizers*.

oleum
Fuming sulphuric acid.

oligomer
A substance composed of only a few monomeric units repetitively linked to each other, such as a dimer or trimer, or their mixtures. The physical properties of an oligomer vary with the addition, or removal, of one, or a few, constitutional units from its molecules.

oligomerization
The process of converting a *monomer*, or mixture of monomers, into an *oligomer*.

on-line monitoring
See *automatic quality control*.

on-line reciprocating screw rheometer test
An abbreviation used for this term is on-line RSR test. A test made with a reciprocating screw rheometer. The test, or measurements are made with an injection mould connected to the injection unit, for example, measurements are made during mould filling, The melt temperature must be carefully and accurately controlled and the melt/nozzle pressure must be capable of accurate measurement. See *off-line reciprocating screw rheometer test*.

on-line RSR test
An abbreviation used for *on-line reciprocating screw rheometer test*.

on-line testing
Testing which is done during the moulding or production process.

on-off
A simple type of temperature controller in which the power is either fully on or, it is fully off.

ONB
An abbreviation used for *o-nitrobiphenyl*.

one pack sachet
See *one shot sachet*.

one shot process
A technique, or process, used in *polyurethane technology* to produce polyurethane foams (both flexible and rigid). All the compound ingredients are mixed and reacted in one operation so as to give a more economical process than the *prepolymer process*.

one shot sachet
Also known as a one pack sachet. A pre-weighed package of mix ingredients. Depending upon the number of ingredients used, the level of addition can be as high as 10 phr in, for example, *polyvinyl chloride* (PVC). The use of such packages enables the user, for example, a trade moulder, to easily incorporate the correct percentage of additive and eliminates the hazards associated with the handling and weighing of the individual ingredients. Because of the dust problem associated with UPVC *dry blends*, additives are now also being supplied in a granular, flake or spaghetti form. Although the dispersion characteristics of the latter forms are not as good as when the additive is a powder, they are sometimes preferred by the processors because of safety considerations.

one stage polymer
See *one stage resin*.

one stage resin
Also known as one stage polymer. A resin which will cure without the use of *cross-linking agents*, for example, a *phenol formaldehyde resole*.

one-pack systems - room temperature vulcanizing silicone rubbers
Also known as moisture curing silicones or as, single pack systems. An abbreviation used for this material is RTV-1. Based initially on a dialkylsiloxane with terminal hydroxyl groups (a silanol terminated polydimethylsiloxane) which is then, for example, turned into a tetra-functional structure by reaction with a *multi-functional organosilicon cross-linking agent* (amine, acetate or oxime). The resulting polymer then has catalysts added (for example, diarylalkyltin acrylates): such a material is stable under anhydrous conditions but when exposed to moisture/water, curing occurs as the moisture diffuses through the polymer. Such materials are widely used as sealing compounds in the building industry.

one-shot injection moulding
See *transfer moulding*.

one-shot polyester foam
A *flexible polyurethane foam* based on a *polyester polyol* and produced by a *one-shot process*. Such a foam may be produced from the *polyester polyol* by reaction with a di-isocyanate (for example, 65:35 TDI), water, a catalyst (for example, dimethylcyclohexylamine), an emulsifier (for example, sulphonated castor oil), a structure modifier (*silicone oil*) and paraffin oil (controls pore size and minimises foam splitting).

one-shot polyether foam
A *flexible polyurethane foam* based on a *polyether polyol* and produced by a *one-shot process*.

one-shot process
A process used to produce *polyurethane foam* (both flexible and rigid) in which all the ingredients are mixed and reacted in one operation. This is now a very important production process for *polyurethanes* because of the development of better, more active, catalyst systems - compared to those originally developed for the *prepolymer process*.

one-step resin
A resin, often based on a *phenol-formaldehyde* (PF) resin, which can *cross-link* without the addition of hardening additives.

OPA
An abbreviation used for ortho-phthalic acid. See *phthalic acid*.

OPC
An abbreviation used for ordinary *Portland cement*.

open aggregate
See *carbon black*.

open assembly time
The time for which coated surface are exposed to the atmosphere before being brought into contact in *adhesive bonding*.

open centre circuit
A term used in *hydraulics* and which refers to a circuit in which the output from the pump flows freely through the circuit and back to the tank when the directional control valve is in its neutral, centre position.

open cooling system
See *direct cooling system*.

open horizontal fluidised bed
A *fluid bed* through which the profile travels horizontally with respect to the ground and which is open to the atmosphere, that is, it is not pressurised.

open loop control
A type of machine control whereby a parameter is set and it is not then adjusted by feedback information.

open steam process
A rubber *reclamation* process which uses steam/oil mixtures.

open-cell cellular plastic
A *cellular plastic* in which there is a predominance of interconnected cells.

open-structured black
See *carbon black*.

OPET
An abbreviation used for oriented polyethylene terephthalate. See *orientation* and *polyethylene terephthalate*.

OPP
An abbreviation used for oriented polypropylene. See *orientation* and *polypropylene*.

OPS
An abbreviation used for oriented polystyrene (see *orientation* and *polystyrene*). OPS is also used as an abbreviation for *p-octyl phenyl salicylate*.

optical active polymer
A polymer which can rotate the plane of vibration of polarized light.

optical activity
Also known as optical rotation. A property possessed by some substances, they can rotate the plane of vibration of polarized light.

optical density
See *density*.

optical disk
A data storage device which is read by a laser. The disks are made from the clear plastics material polycarbonate (PC) by the process of *injection moulding* and must be moulded to very tight tolerances. The under-side of the disk has a coating which can be altered by heat from a laser in the optical disk drive. This layer is made from one of four types of material. These are, low melting point tellurium alloy, dye polymer, tellurium sub-oxide or platinum. Most current optical disk drives, are write once or, as they are known, write once read many (WORM) drives. In ideal applications, the data on the disk is added to, rather than changed.

optical fibre
A fibre which functions as a waveguide for light: used for communication purposes.

optical pyrometer
A device which measure a temperature change by determining the intensity of light radiation emitted from a hot object. May be based on a *platinum resistance thermometer* or on, *thermocouples*. See *radiation pyrometer*

optical rotation
See *optical activity*.

optically flat
A surface is classified as being optically flat if the surface irregularities are smaller than the wavelength of light.

optimum cure
A term used in rubber technology to describe that time of cure needed to bring a particular property to a desired value while ensuring that other properties are acceptable.

OPVC
An abbreviation used for oriented polyvinyl chloride. See *orientation* and *polyvinyl chloride*.

orange peel
An extrusion defect - see *applesauce*.

orange pigments
Inorganic orange pigments include *cadmium mercury orange*, *cadmium sulpho-selenide*, *chrome orange* and *molybdate orange*. Organic pigments are *diazo orange*, *diazo condensation orange*, *isoindolinone*, *anthraamide orange* and *pyrazolone orange*. The *organic pigments* are, in general, more soluble, have a higher tinting strength and have a lower relative density than the inorganics.

ordered copolyamide
An aromatic *polyamide copolymer*, which has a regular alternating arrangement of the aromatic units.

ordered polymer
A *copolymer* which has a regular arrangement of the units.

organic accelerator
Also known as an organic *vulcanization accelerator*. Organic accelerators are the most important type of *accelerator* for rubbers. Accelerators used, in the cross linking of rubber compounds, include dithiocarbamates, guanidines, sulphenamides, thiazoles, thiuram disulphides and xanthates. Thiazoles are generally the most commercially important organic *accelerators*.

organic blocking agent
An organic additive which reduces *blocking*. See *matting agent*.

organic filler
A *filler* based on organic material, for example, *woodflour*. See *non-reinforcing filler* and *reinforcing filler*.

organic peroxide
A peroxide which contains two organic groups, alkyl and/or aryl (R), separated by the peroxy group -O-O-. That is R-O-O-R. Benzoyl peroxide is the best known *diacyl peroxide*: another is lauryl peroxide. Such materials are widely used as free radical polymerization initiators as they may be decomposed, or degraded, by heat or UV light to give organic free radicals.

organic phosphates
See *phosphates*.

organic pigment
A pigment/colorant which is an organic compound. Two major classes of pigment are *inorganic* and organic pigments. *Carbon black* is the most important organic *pigment*. Other types of organics include azo pigments, phthalocyanines, quinacridones, perylenes, isoindolinones, anthraquinones, thioindigos, acid and basic pigments.

Generally organic pigments are aromatic compounds which contain one or more benzene-type structures: some contain several ring structures, both benzene-type structures and heterocyclic rings. Most organic pigments are made form *dyes* which have been rendered insoluble by, for example, being trapped in a polymer-type structure. That is, their molecular weight has been increased by coupling materials together with other organic groups and/or, making them insoluble with heavy metal atoms. The metal atoms may be chelated to the organic material or it may form a salt with an appropriate acid group (for example, with carboxylic or sulphonic acid groups. Organic pigments have less opacity but more tinting strength than inorganic pigments: they also produce brighter, cleaner colours. However, organics tend to have poorer heat stability and lightfastness than inorganics: they also tend to migrate more from the polymer. Phthalocyanine pigments are the most widely used chromatic (coloured) organic pigments.

Because of the importance of azo materials, organic pigments may be divided into azo and non-azo pigments. *Azo pigments* are formed by *diazotization* (of a primary aromatic amine) and *coupling* (with say, 3-hydroxy-2-naphthoic acid or BON). *Mono-azo* and *di-azo pigments* are of two types, pigment dyes and precipitated azo pigments.

organic polysulphides
See *polysulphide rubber*.

organic vulcanization accelerator
See *organic accelerator*.

organically bound sulphur
One of the ways the *sulphur* in rubber compounds, on analysis, may be classified. That sulphur which is chemically bound, or chemically attached, to organic materials such as *accelerators*. See *sulphur analysis*.

organo-bromine flame retardant
An organic *flame retardant* which contains *bromine*. For example, tribromotoluene. Such bromine-containing compounds tend to be more powerful than chlorine-containing compounds: often used in conjunction with *antimony trioxide*.

organo-chloro flame retardant
An organic *flame retardant* which contains *chlorine*. Bromine-containing compounds tend to be more powerful than chlorine-containing compounds: often used in conjunction with *antimony trioxide*.

organo-tin compound
See *organotin stabilizer*.

organodisiloxane
See *disiloxane*.

organofunctional silane
A silicon-based material: an organosilicon compound in which chemically reactive, organic groups replace one or more of the hydrogen atoms of *silanes*. For example, in vinyldichlorosilane, the vinyl group is the functional group. See *disilane* and *trilsilane*.

organoleptic
Relating to, or perceived by a sensory organ.

organoleptic response
In the packaging industry, this term usually means a change in taste caused by the uptake of materials such as oxygen or water. If a package is to provide a given shelf life for the contents then, it must be known what level of ingress of water or oxygen can be tolerated by the product before any noticeable change in taste takes place.

organophosphite
See *metal-free organic phosphite*.

organosilane
A silicon-based material: organosilicon compounds in which organic groups replace one or more of the hydrogen atoms of a *silane*. See *disilane* and *trilsilane*.

organosilanol
An organosilicon compound formed by the hydrolysis of *silicon functional silanes*, for example, by the hydrolysis of chlorosilanes. An organosilanol which contains both organic groups and hydroxyl groups: the hydroxyl groups condense to form *siloxanes* and eliminate water. See *organofunctional silane*.

organosilicon compound
An organic chemical compound based on *silicon*. See *silicone rubber* and *polydimethylsiloxane*.

organosol
A type of *polyvinyl chloride* (PVC) *paste*. A *plastisol* which contains a solvent: the solvent lowers the paste viscosity and this may ease application. The solvent is lost on gelation or fusion.

organosulphur compound
A compound based on *sulphur (S)* and on organic groups: such a compound is used, for example, as a *preventive antioxidant*. An example is a thio-bisphenol.

organotin compound
A tin compound which contains organic groups and used as a *heat stabilizer*. See *organotin stabilizer*.

organotin stabilizer
A compound based on tin (Sn) and on organic groups. Such compounds are widely used as, for example, as *heat stabilizers* for *polyvinyl chloride* (PVC) particularly where transparency is required. An organotin may also be known as a *dibutyl tin stabilizer*. The general formula for the dibutyl compounds is $(C_4H_9)_2SnX_2$ where X is an organic group - such as the laurate group (see *dibutyl tin dilaurate*). The general formula for the dioctyl compounds is $(C_8H_{17})_2SnX_2$ where X is an organic group - such as the laurate group. See *dioctyl tin stabilizer*.

organotrisiloxane
See *trisiloxane*.

orientation
The alignment of polymer chains, or portions of polymer chains, or of fibrous additives. Distortion of molecules from their randomly coiled up state by the application of external stresses. A process, for example, an extrusion process which increases the strength and stiffness of a *plastics material*, in the orientation direction, by stretching or rolling. During extrusion, *drawing down* will cause *molecular orientation*, thus increasing strength in the flow direction but decreasing it in directions transverse to the flow. This may or may not be desirable.

During *injection moulding*, orientation may also be easily introduced. This is because, when a molten polymer is made to flow then this flow occurs because the polymer chains slip, or slide, one over the other. Layers slip one over the other in what is called *laminar flow*. As the individual, plastics molecules move, one relative to the other, then this causes the molecules to change their direction or orientation. (Reasons for this change are, for example, chain entanglements and friction between layers). The chains become 'drawn out' in the direction of flow and, because of the rapid cooling employed in processes such as injection moulding, this orientation is *frozen in*. The product contains molecules which are orientated in the flow direction: the result is a grain effect - just like the one that exists in wood. Because of the grain, the wood is stronger in one direction that it is in another. The same is true of oriented mouldings.

Increased orientation may be put into a system by stretching the polymer melt just before it freezes. Orientation is deliberately introduced during the manufacture of *fibrillated tape* and in the *extrusion blow moulding* of bottles. See *melt processing*.

orientation birefringence
Birefringence caused by distortion of molecules from their randomly coiled up state by the application of external

stresses: this results in the ordering of optically anisotropic elements. The most common cause of birefringence and often used to measure *anisotropy* of thermoplastics materials such *polystyrene*. See *monoaxial orientation* and *biaxial orientation*.

orientation hardening
Increased stiffness which results from *orientation* of polymer molecules.

orientation polarization
The most significant type of *polarization* occurring in *dielectric heating*.

oriented
A product which possesses *orientation*. An abbreviation used for this term is O: this letter is placed in front of the abbreviation for the material. For example:

   oriented polyethylene terephthalate = OPET;
   oriented polypropylene = OPP;
   oriented polystyrene = OPS; and,
   oriented polyvinyl chloride = OPVC. See *biaxial orientation* and *orientation*.

oriented film fibre
See *fibrillated film fibre*.

oriented laminar filler
Plate-like filler particles become oriented as a result of processing; examples of such fillers include *mica* and *polyamide (PA)* flakes. See *permeability*.

orifice
A small flow restrictor: a hole in a hydraulic circuit of fixed length and diameter.

original equipment tyre
An abbreviation used for this term is OE tyre. The tyre fitted by the car manufacturer. Such a tyre is matched to the suspension of the car and is selected for good handling during driving, traction, comfort, low noise and price - not necessarily for durability.

original gauge length
An abbreviation used for this term is l_o. See *gauge length*.

ortho
When used in connection with aromatic organic chemical compounds, means that two groups are adjacent on the benzene ring. An abbreviation used for this term is o.

ortho diallyl phthalate
An *allyl moulding material* based on *diallyl phthalate*.

ortho substituted phosphate
The ortho isomer of an aromatic phosphate, for example, of *trixylyl phosphate* - this is more toxic than the meta and para isomers.

ortho-phthalic acid
An abbreviation used for this type of material is OPA. See *o-phthalic acid*.

ortho-tert-butyl phenol
This yellow, liquid material has a melting point of $-65°C$, a boiling point of $224°C$ and a relative density (RD) of 0.98. Has been used as a *plasticizer*. See *tert-butyl phenol*.

orthophthalic resin
An *unsaturated polyester resin* based on *phthalic anhydride*.

oscillating plate welding
A *friction welding technique* used for *thermoplastics materials* in which the necessary heat is generated by oscillating one component against another while applying pressure. The pressure is applied while the joint cools. The components need not be circular.

osmometer
An instrument or apparatus used for measuring osmotic pressure. An instrument or apparatus used for determining *number average molecular weight* by measuring *osmotic pressure* across a semi-permeable membrane.

osmometry
Also known as membrane osmometry. A *molecular weight determination* method which gives the number average molecular weight.

osmotic pressure
The pressure that must be applied to a solution in order to prevent fluid flow across a semi-permeable membrane: the membrane separates pure solvent from a polymer solution. See *osmometer*.

Ostwald de Waele equation
See *power law equation*.

Ostwald viscometer
A type of viscometer: a U-tube viscometer which gives a measure of molecular weight.

OT
A rubber which has sulphur, carbon and oxygen in the main polymer chain. A rubber with polysulphide linkages in which, for example, the polysulphide linkages are separated by organic groups (R groups) such as $-CH_2-CH_2-O-CH_2-O-CH_2-CH_2-$.

OTOS
An abbreviation used for *N-oxydiethylenethiocarbamyl-N-oxydiethylene-sulphenamide*.

ounce
A unit of weight and of mass. One ounce avoirdupois contains 437.5 grains. This is the ounce commonly used and is equivalent to $0.028\ 349\ 5$ kg. One ounce Troy contains 480 grains. This is the ounce used for precious metals and is equivalent to $0.030\ 103\ 5$ kg. There are 12 ounces in one pound (Troy). A fluid ounce contains 8 fluid drachms and is equivalent to 28.41 cubic centimetres.

out-gassing
The loss of volatile matter from a moulding during, for example, the *vacuum metallization* of plastics: such out-gassing is minimised by the use of a *base coat*.

outline mould
See *skeleton frame mould*.

outrigger
Part of a mould: an external assembly of parts, mounted on the mould, so as to provide facilities for the operation of splits, side cores and side cavities.

outsert moulding
A *composite moulding technique* used, for example, to put a plastic material (for example, *acetal*) onto a relatively large metal plate. The metal sheet is located within the mould by an appropriate pattern of pins, and corresponding holes, and such plates may extend beyond the confines of the mould. If the metal material is flexible then it may be fed from a reel into the mould. In this case the output from the mould would be a continuous strip of metal which at regular intervals has the pattern of a plastics component. This metal sheet may then be guillotined into smaller sections. The plastics material is usually keyed onto the metal plate by moulding through a hole or into an undercut. Such outsert mouldings are used, for example, in the hi-fi and audio industries to produce snap-fits and bearings.

outside coating
A coating method used in blow moulding to put an impermeable layer, for example, *polyvinylidene dichloride (PVDC)*, on the outside of a *blow moulded container*.

oven drying
Also known as convection drying. The material is heated in an oven so as to remove moisture. Oven drying is popular as it is relatively cheap to install and easy to operate. However, it is not very efficient and contamination often happens. The plastics material is placed in thin layers (below 10 mm/0·5" thick) inside a convection oven; a fan assisted oven is better. The drying conditions differ for different materials. Drying times are rated in hours and care should be taken to avoid material change by plasticizer loss or oxidation. In order to stop the air becoming saturated with water vapour, some of the air should be bled off and, to prevent contamination, the trays should be covered with fine net, or gauze. Ideally the air should be filtered and recirculated as this will remove contaminants and save on energy.

oven drying assessment
A method of determining the moisture content in a thermoplastics, material. When the permissible moisture limit for good moulding is fairly high, such as for cellulosics, a simple oven drying procedure can be used as detailed in ASTM D 817. 5 g of polymer, weighed to the nearest 0·001 g, is heated in a wide mouth bottle in an oven at 105°C for 2 hours. After stoppering the bottle it is cooled in a desiccator and then reweighed.

$$\text{The \% moisture} = \frac{\text{weight loss on heating} \times 100}{\text{grams of sample used}}$$

oven heating - preheating
Oven heating is used as a *preheating* operation. Used to raise the temperature of moulding materials before the moulding operation so as to decrease cycle times and/or to improve the uniformity of cure. Usually associated with *compression* and/or *transfer moulding* where oven heating or *high frequency heating* is used. Oven heating takes longer than high frequency heating but can give more uniform heating.

oven heating - reversion
A dimensional stability test. Used to assess the amount of *reversion* which will occur on heating a plastics product, for example, a moulding or an extrudate. Used as a measure of how well the product was produced and/or if a product will withstand use at elevated temperatures. The dimensions of the product are checked after oven heating for a specified time and at a specified temperature. As the *orientation* stresses increase in a component, then the *heat distortion temperature* reduces and the component will distort probably more, and at a lower temperature.

The upper, maximum, surface temperature likely to be encountered by the component should be ascertained and, unless otherwise specified, it is suggested that the oven temperature be set 25°C/50°F) higher than that of this upper surface temperature. The length of time for which the component is in the oven will depend upon its thickness but for most components (e.g. of less than 3 mm/0·12 in cross-section) a heating period of 30 minutes is sufficient. Alternatively, formulas used for calculating cooling times may be used to determine the heating-up time (and approximately 20 minutes added to this). The time may also be measured experimentally by inserting a thermocouple in the thickest part of the moulding and then heating the assembly while noting the time and temperature.

over-packing
Term used in *injection moulding* to indicate that a component, or cavity, has been over-packed with plastics material. This occurs, for example, if excessively high injection pressure was used for the whole of the filling part of the injection moulding cycle. Over-packing occurs because of the compressibility of plastics melts. Such over-packing results in overweight mouldings, which take a long time to cool and which are often unsatisfactory in service. If too much material is packed into the mould during injection moulding (in an effort to compensate for *shrinkage*) then most of the extra material will be concentrated and compressed in the gate area; this compressed material generates stress which can cause failure, for example, in the gate area of a *sprue gate*. See *hold pressure* and *cavity pressure control*.

over-stretching
See *tie bar*.

overall diameter
An abbreviation used for this term is D_o. The diameter of an inflated *tyre* at the outermost surface of the *tread*.

overall efficiency
The overall efficiency of a hydraulic system is the output power divided by input power.

overall width
An abbreviation used for this term is W. The linear distance between the *sidewalls* of an inflated *tyre*: includes measurements across labelling and protective features, such as bands or ribs.

overcoating
A coating process used in *extrusion coating* or laminating and which means that the plastic coating is wider than the substrate: the plastic is trimmed subsequently and re-used.

overcure
Curing a compound beyond that time which gives the *optimum cure*. A state of cure produced by exceeding, for example, cure time and/or temperature.

overgrowth
The growth of small crystals onto larger crystals.

overheating
Exceeding a desired, or set temperature, during processing. See *material changes*.

overlap gate
A *gate* used in *injection moulding* to prevent jetting. Basically the runner overlaps the cavity, at one side, so that the incoming *melt* impinges against a mould face.

overlay
An opaque coating applied to a *top coat* so as to mask part of a metallic colour. See *vacuum metallization*.

overshoot
A term usually applied to temperature control and which means that the temperature 'shoots' past what is set; often happens during warming up unless the controller is fitted with derivative (D) control action.

oversteer
A vehicle steering defect. If the front *slip angle* is less than the rear slip angle then the vehicle will oversteer. The actual course taken by the vehicle will be greater than expected. Because of, for example, the danger of spinning this is more dangerous than *understeer*.

owf
An abbreviation used for on weight of fibre.

ox-bow effect
A thickness variation found in *calendered sheet*. The sheet is not of the correct thickness between the edges and the centre when *roll crossing* is employed (to compensate for *roll separating forces*). If the sheet thickness is measured and then plotted against sheet width an ox-bow-shaped distribution results. Despite this disadvantage roll crossing is widely used as it gives a means of controlling sheet thickness without imposing additional loads on the calender and at lower cost than *roll bending*.

oxacyclobutane
See *oxetane polymer*.

oxadiazole polymer
An aromatic polymer which contains a heterocyclic five-membered ring which contain carbon, nitrogen and oxygen atoms. See, for example, *polyphenylene-1,3,4-oxadiazole*.

oxetane
An oxetane is also known as an oxacyclobutane. See *oxetane polymer*.

oxetane polymer
A polyoxetane: a polymer based on a repeat unit of 3 carbon atoms and an oxygen atom or -C-C-C-O-. Because *chlorinated polyether* is the only commercial member of this family of polymers, the term is often applied to this material.

oxidation
Also known as oxidative degradation. The decomposition, or degradation, caused by oxygen and which alters the properties of polymers. Usually the first change is one of colour and this colour change is often enough, in the case of plastics, to cause rejection. Oxidation will affect flow properties, electrical properties and mechanical properties even if the colour change cannot be seen.

Oxidation is commonly associated with diene rubbers and with thermoplastics materials (but not with *polyvinyl chloride* which is associated with, for example, *dehydrochlorination*). Addition polymers and copolymers derived from olefin monomers are often prone to oxidation: the effects are minimized by the uses of antioxidants. In particular, polyethylene (low and high density) polypropylene, high impact polystyrene and acrylonitrile-butadiene-styrene plastics, require *antioxidant* protection. Condensation polymers such as polyamides, polyesters and polyurethanes tend to be more resistant to oxidation as does general purpose polystyrene. Natural and synthetic diene rubbers can suffer from both attack by oxygen and from ozone. Non-diene rubbers with few double bonds are more ozone resistant and, in general, oxygen resistant.

oxidation - mechanism
The mechanism of oxidative decomposition, or degradation, in polymers is a chain process (an autooxidation process) which is initiated and propagated by *free radicals*. Some of the radicals and species involved are depicted as follows:

R· A polymer radical formed by the removal of hydrogen from a hydrocarbon RH.
RO· A polymer alkoxy radical formed by the decomposition, or degradation, of a hydroperoxide-containing polymer.
ROO· A polymer peroxy radical (peroxide radical) formed by the reaction of a polymer radical R· with oxygen (O_2).
ROOH A hydroperoxide-containing polymer (polymer hydroperoxide) formed by the reaction of a polymer peroxy radical (ROO·) with a hydrocarbon polymer RH.
RH A hydrocarbon material involved in the oxidation reactions which, when hydrogen is removed, forms the radical R·.

The steps involved in *oxidation* are initiation, propagation and termination (plus chain scission).
Initiation. RH giving R· + H· and, less likely, RH + O_2 giving R· + ·OOH
Propagation.
R· + O_2 giving ROO·
ROO· + RH giving ROOH + R·
ROOH giving RO· + OH·
2ROOH giving RO· + ROO· + H_2O
Termination.
2ROO· giving non-radical products
2R· giving R-R
ROO· + R· giving ROOR
Chain scission.
A variation of the termination reactions.

Initiation reactions may be brought about by mechanical shear, thermal decomposition, or degradation and UV decomposition, or degradation. Once started, it is the hydroperoxides which are responsible for the autooxidation as each cycle of the propagation stage produces more hydroperoxides and thus, more free radicals. The peroxy radicals, which form the hydroperoxides, are more likely to abstract hydrogen from a tertiary carbon atom than from primary or secondary carbon atoms (polypropylene has tertiary carbon atoms). The decomposition, or degradation, of a polymer hydroperoxide is induced by light quanta of approximately 300 nm and is catalyzed by metals. See *antioxidant* and *oxidation*.

oxidation-reduction polymer
Also known as a redox polymer and as electron charge polymer. A polymer which has chemical groups which can exchange electrons with other chemical groups.

oxidative coupling
Also known as oxidative polymerization. A polymerization process in which the polymer is formed by the monomer units being linked by an *oxidation* reaction. Polymers based on aromatic rings may be produced in this way.

oxidative degradation
Degradation caused by heating a polymer in an atmosphere of oxygen. A chain process which is initiated and propagated by free radicals. The reactions may be accelerated by, for example, copper or slowed down by *antioxidants*. See *oxidation* and *thermally-initiated oxidation*.

oxidative polymerization
See *oxidative coupling*.

oxidative surface treatment
See *surface treatment*.

oxides of lead
See *lead oxides*.

oxidized PAN fibre
See *oxidized polyacrylonitrile fibre*.

oxidized polyacrylonitrile fibre
An abbreviation used for this type of material is oxidized PAN fibre. An intermediate product in the manufacture of *carbon fibre* which is non-flammable and acid resistant.

oxidized rubber
Natural rubber which has been heated in air with an oxidation catalyst, for example, with cobalt linoleate. A comparatively low molecular weight material which has good resistance to thermal degradation.

oxiran
See *ethylene oxide*.

oxiran-chloromethyloxiran
An abbreviation used for this type of material is ECO. An epichlorhydrin copolymer rubber. See *epichlorhydrin rubber*.

oxirane
See *ethylene oxide*.

oxo
A chemical prefix which shows that the group O= is present in a compound.

OXO process
May also be called the carbonylation reaction. A process used to produce a mixture of aldehydes from olefins: water gas (carbon monoxide and hydrogen) mixed with an *olefin* is passed over a catalyst at 100 to 150°C at a pressure of 100 to 400 atmospheres. The aldehydes so produced are, in turn, used to produce alcohols by catalytic hydrogenation of the aldehydes: the mixed primary alcohols which result may then be used to make *plasticizers*.

oxyethyl nylon
A graft copolymer formed by grafting ethylene oxide on to a *polyamide*. For example, by grafting ethylene oxide (50%) onto PA 66 then oxyethyl 66 nylon results. This is softer and tougher than the parent nylon and of a lower melting point. Replacement of some of the -CONH- groups reduces hydrogen bonding and increases the solubility of the material.

oxygen
Group V1b of the Periodic table consists of the elements oxygen, *sulphur*, *selenium*, tellurium and polonium. About 21% of the earth's atmosphere consists of oxygen and it is from this source (by liquefaction and fractional distillation) that oxygen is obtained commercially. It is a colourless, odourless gas which boils at approximately $-183°C$ and which is very reactive: most of the earth's crust and oceans consists of combined oxygen. The element exists as diatomic molecules (O_2). Liquid oxygen is pale blue and strongly magnetic. Oxygen is used to make, for example, phenol and phthalic anhydride by the oxidation of hydrocarbons. *Oxidation* is a major problem with many widely used polymers. Oxygen also exists in the allotropic form which is known as *ozone*.

oxygen barrier properties
The ability of a polymer to stop oxygen diffusion. This is very important in, for example, the packaging industry when plastics are used to package food. See *permeability*.

oxygen index
An abbreviation used for this term is OI. See *limiting oxygen index*.

ozokerite
Also called earth wax, mineral wax and ozocerite. This material has a relative density (RD or SG) of approximately 0.93. A naturally occurring mixture of solid branched chain, paraffinic hydrocarbons which can be refined into paraffins and *ceresin*. See *paraffin wax*.

ozone
Three oxygen atoms combined together make one ozone molecule: an allotropic form of oxygen. Can be made by, for example, an electric discharge through air. Has a smell reminiscent of weak chlorine and is a powerful oxidizing agent. Used for bleaching and for sterilizing. Found in the Earth's atmosphere in minute quantities: most atmospheric ozone is located in the *ozone layer*. Atmospheric ozone is now being depleted by the action of *stratospheric chlorine*.

ozone cracking
Surface cracking which results when a rubber component, based on a diene rubber, is exposed to *ozone*, particularly when the component is stressed. See *ozone-induced degradation*.

ozone depleting potential
An abbreviation used for this term is ODP. Depending on the longevity and on the perceived *ozone* damage, chemicals are assigned an ODP to that chemical with *CFC 11* being used as a standard. Fully halogenated chemicals have a high ODP. For example, CFC 11 has an ODP of 1: carbon tetrachloride has an ODP of 1.2: methyl chloroform has an ODP of 0.10. *HFC* materials have an ODP of 0 and *HCFC* materials have an ODP in between 0 and 1 (the lower the ODP value the better). Halon 1301 has a value of 10.5. See *stratospheric ozone depletion*.

ozone hole
An area in the Earth's atmosphere which is deficient in ozone and which thus allows harmful *ultraviolet radiation* to reach the surface of the Earth. See *stratospheric ozone depletion*.

ozone layer
Most atmospheric *ozone* is located in the ozone layer or ozonosphere. It is a layer, or region, in the Earth's upper atmosphere between 7.5 miles/12 kilometres and 30 miles/48 kilometres above sea level. Atmospheric ozone blocks out radiation harmful to life. See *stratospheric ozone depletion*.

ozone protection wax
See *ceresin wax*.

ozone-induced degradation
Also known as *ozonolysis*. Ozone causes degradation of diene rubbers: the *ozone* adds to the double bonds (ozonide reaction) and the cyclic ozonides formed then break down: chain scission and carbonyl group formation is, for example, the result.

ozonide reaction
A chemical reaction involving diene rubbers: the ozone adds to the double bonds to form ozonides: the cyclic ozonides formed then break down and chain scission is, for example, the result.

ozonides
Cyclic compounds formed between ozone and diene rubbers. See *ozone-induced degradation*.

ozonolysis
See *ozone-induced degradation*.

ozonometer
An apparatus used to determine the ozone concentration in the atmosphere surrounding a test specimen which is being used to determine ozone resistance. See *ozone-induced degradation*.

ozonosphere
See *ozone layer*.

P

p
An abbreviation used for:

stress;
pico (see *prefixes - SI*); and,
para.

P
This letter is used as an abbreviations for :

filler - for example, for mica or paper.
force;
mould steels - in the *AISI/SAE system*;
peta. See *prefixes - SI*;
phenol;
phthalate;
phosphate;
pigment (in colour technology):
plasticized or plasticised;
pressure. See *capillary rheometer*;
propyl; and,
pyro. For example, *tetraoctyl pyromellitate (TOPM)*.

P & J plastometer
An abbreviation used for *Pusey and Jones plastometer*.

p-benzoquinone
Also known as *quinone* or as cyclohexadiene-1,4-dione. This solid yellow material has a melting point of approximately 116°C. May be represented as $O:C_6H_4:O$. A free radical polymerization *inhibitor*.

p-dinitrosodimethyl aniline
An *accelerator* similar in action to *diphenyl guanidine*. This green crystalline material has a melting point of approximately 90°C.

p-methylstyrene
See *vinyltoluene*.

p-nitroso-ENCAF-modified natural rubber
See *p-nitroso-ethyl N-phenylcarbanoylazoformate-modified natural rubber*.

p-nitroso-ethyl N-phenylcarbanoylazoformate-modified natural rubber
An abbreviation used for this material is p-nitroso-ENCAF-modified natural rubber. A chemically modified *natural rubber* with good resistance to solvent swelling: better than *ethyl N-phenylcarbanoylazoformate-modified natural rubber*.

p-nitrosomethyl aniline
An abbreviation used for this material is PNDA. This green solid material has a melting point of about 87°C and functions as a vulcanization *accelerator*: similar to *diphenyl guanidine* in its curing action.

p-octyl phenyl salicylate
May also be referred to as octyl phenyl salicylate. An abbreviation used for this material is OPS. An *ultra-violet stabilizer* for polyolefin plastics.

p-phenyl phenol
This white solid material has a melting point of about 165°C and a relative density (RD) of 1·20. A good *antioxidant* for light coloured, rubber compounds.

p-phenylene diamine
See *para-phenylene diamine*.

p-phenylene diamines
An abbreviation used for para-phenylene diamines. See *antiozonant*, *antioxidant* and *alkyl-aryl p-phenylene diamine*.

p-phthalic acid
An abbreviation used for para-phthalic acid: also known as *terephthalic acid*. There are three phthalic acids, i.e. three isomers. These are, o-phthalic acid, m-phthalic acid and p-phthalic acid.

p-quinone dioxime
Also known as p-quinonedioxime or as PQD. This solid brown material has a decomposition temperature of about 216°C. May be represented as $HO-N:C_6H_4:N-OH$. May be used for vulcanizing in the absence of sulphur if used with an oxidizing agent such as *red lead* or *lead peroxide*. Red lead (10%) and p-quinone dioxime (2%) or, lead peroxide (6%) p-quinone dioxime (2%). (Best physical properties are however obtained with compounds which contain *sulphur* and *carbon black*.) Used particularly with *butyl rubber*.

p-quinonedioxime
See *p-quinone dioxime*.

p-tert-butyl catechol
Also known as 4-tert-butyl-1,2-dihydroxybenzene. An abbreviation used for this material is TBC. This solid material has a melting point of 56°C, a boiling point of 285°C and a relative density (RD or SG) of 1·05. A strong reducing agent which is used as a polymerization inhibitor and as an *antioxidant* for rubbers.

p-tert-butyl phenol
See *para-tert-butyl phenol*.

P_0
An abbreviation used for the initial Wallace plasticity. See *plasticity retention index*.

P2
A chrome-manganese steel (AISI-type P2) is a commonly used *case hardening steel* as it can be hobbed and also used for the manufacture of relatively large injection moulds.

P20 + S steel
A chromium-manganese-molybdenum-sulphur steel. A high sulphur content *pre-hardened steel*.

P20 steel
A chromium-manganese-molybdenum steel. A low sulphur content *pre-hardened steel*.

P_{30}
An abbreviation used for the Wallace plasticity after ageing for 30 m at 140°C. See *plasticity retention index*.

P_{30}/P_0
See *plasticity retention index*.

P4MP1
An abbreviation used for *polymethyl pentene*.

Pa
An abbreviation used for *pascal*.

PA
Sometimes used as an abbreviation for *process aid*, but more commonly, used for *polyamide* (see *nylon*).
 PA 4   = polyamide 4. See *nylon 4*.
 PA 6   = polyamide 6. See *nylon 6*.
 PA 11   = polyamide 11. See *nylon 11*.
 PA 12   = polyamide 12. See *nylon 12*.
 PA 46   = polyamide 46. See *nylon 46*.
 PA 610   = polyamide 610. See *nylon 610*.
 PA 66   = polyamide 66. See *nylon 66*.
 PA 69   = polyamide 69. See *nylon 69*.
 PA 6T   = polyamide 6T. See *nylon 6T*.
 PA 91   = polyamide 91. See *nylon 91*.
 PA 6/66 = polyamide 6/66. See *nylon 6/66*.
 PA 6/66/610 = polyamide 6/66/610. See *nylon 6/66/610*.

PA 6 RIM
An abbreviation used for polyamide 6 reaction injection moulding. See *nylon 6 reaction injection moulding*.

PAA 6
An abbreviation used for an aromatic polyamide. See *polyaryl amide*.

PA resin
An abbreviation used for *pre-accelerated resin*.

PABH-T
An abbreviation used for a *polyamide-hydrazide polymer*.

PABM
An abbreviation used for *polyaminobismaleimide*.

package
A term used in the reinforced plastics industry for units of *yarn* or *rovings* which are suitable for handling, shipping and use.

packaged stabilizer
A heat stabilizing systems used with, for example, *polyvinyl chloride* (PVC) and which contain more than one ingredient. May therefore be complex mixtures of compounds. Such stabilizers may be solid or liquid and may also incorporate other ingredients such as *antioxidants* and *phosphites*. See *metal soap stabilizer*.

packing pressure
See *hold pressure*.

packing time
See *dwell time*.

pad
A local mould insert which, for example, is used to facilitate a change of engraving.

pad printing
See *printing techniques*.

PAI
An abbreviation used for *polyamide-imide*.

PAK
An abbreviation suggested for the plastics material known as a *polyacrylate*.

pale crepe
One of the forms in which *natural rubber* is supplied. A pale (light coloured) form of natural rubber from which the yellow pigments present in natural rubber have been removed. Latex low in yellow pigmentation (carotenoids) is selected and may be bleached (using tolyl mercaptan) and/or fractionally coagulated to remove pigment-rich latex. After *acid coagulation* of diluted latex (about 20%), further impurities are removed by thorough washing and mechanical working in a *crepeing battery*. The sheets so produced are *air-dried* for approximately ten days or vacuum dried at 70°C/158°F for two hours. If vacuum drying is used, the sheets must be rapidly cooled so as to prevent *oxidation* and colour formation. See *white* and *pale crepes*.

palm oil
An orange-yellow vegetable oil which is obtained from the fruit of the oil-palm (not the coconut palm). This material has a relative density (RD or SG) of 0·88. See *palmitic acid*.

palm-kernel oil
A white fat-like material obtained by crushing the kernels obtained from the fruit of the oil-palm.

Palmer process
A reclamation process for vulcanized rubber which utilizes the softening effect produced by the use of super-heated steam. See *reclamation processes*.

palmitic acid
Also known as hexadecanoic acid. May be represented as $C_{15}H_{31}COOH$. A wax-like carboxylic acid, with a melting point of 64°C which occurs in the triglycerides of plant *oils* (for example, in *cottonseed oil* and in *palm oil*) as tripalmitin.

palmitoyl
The organic radical $C_{15}H_{31}CO-$: derived from *palmitic acid*.

PA MXD6
An abbreviation used for *polyaryl amide*.

PAN
An abbreviation used for *polyacrylonitrile*.
   PAN CF   = polyacrylonitrile carbon fibre.
   PAN fibre = polyacrylonitrile fibre.
   PAN HM CF = polyacrylonitrile high modulus carbon fibre.
   PAN HS CF = polyacrylonitrile high strength carbon fibre.
         See *carbon fibre*.

PANA
An abbreviation used for *phenyl-α-naphthylamine*.

PAPA
An abbreviation used for *polyazelaic polyanhydride*.

paper
A sheet material based on *cellulose*. May be obtained from *wood pulp* from which non-cellulosic materials have been removed. The wood pulp slurry is fed to a paper making machine where it is dried and calendered. Used, for example, in the plastics industry to make *laminates* based on thermosetting materials.

PAR
An abbreviation suggested for the plastics material known as a *polyarylate*.

para
The use of this term in a chemical formula, is used to indicate the relative position of substituents on a benzene ring: the substituents are at opposite apexes of the ring. An abbreviation used for this term is p. Also see *para rubber*.

para rubber
Rubber which came from South America: a form of *natural rubber*. Para is the European name of Belem - a trading centre near the mouth of the Amazon. Produced by *coagulation* of rubber in wood smoke. A wooden paddle was dipped into the *latex* and then the rubber was dried by holding the paddle over a fire.

para-linked polymer
A long chain polymer based on aromatic groups (benzene rings) linked head to tail via the *para* position. See *polyphenylene* and *substituted polyphenylene*.

para-phenylene diamine
Also known as benzene-1,4-diamine. The basis of many anti-ageing chemicals. See *antiozonant*, *antioxidant* and *alkylaryl p-phenylene diamine*.

para-tert-butyl phenol
Sometimes referred to as 4-tert-butyl phenol. An abbreviation used for this material is PTBP. This white, solid material has a melting point of 98°C, a boiling point of 230°C and a relative density (RD) of 1·03. One of the isomers of *tert-butyl phenol* which has been used as a *plasticizer*.

paracasein
Solid casein obtained from skim milk by the addition of rennet (an enzyme) to give rennet casein. See *casein*.

paraffin
Another word for an *alkane*. See *paraffin oil* and *paraffin wax*.

paraffin oil
Also known as kerosene or as, kerosine. This liquid material has a relative density (RD or SG) of 0·82 and a boiling point of approximately 150 to 300°C. Based on *alkanes* which have chains which contain approximately 12 carbon atoms. A mixture of hydrocarbons which may be used in rubber compounds and which have good heat and light stability. Because of their lubricating action, their use gives low heat build-up. See *paraffin wax*.

paraffin wax
A wax based on the higher *alkanes*: that is, for example, on unbranched alkanes which have a chain which contains more than 17 carbon atoms. If the chains contains between 17 and 30 carbon atoms then the wax will have a comparatively low melting point (approximately 50 to 60°C). If the chain is of higher molecular weight and contains between approximately 40 and 70 carbon atoms, then the wax will have a higher melting point (of up to 105°C): such materials may be called hard paraffins. Paraffin waxes have a relative density (RD or SG) of from 0·88 to 0·91 (refined paraffin wax has a relative density (RD or SG) of 0·90). May be supplied in solid form or as an emulsion.

In general, waxes with low melting points are sometimes used as processing aids for rubbers: materials with higher melting points are used as *ozone protection waxes*. When used in concentrations of more than 1%, paraffin wax blooms to the surface of rubber compounds so as to form a protective film of wax which acts as an *antiozonant*. Also acts as a processing aid as, for example, it eases mould release and gives a sheen to vulcanizates. The use of paraffin wax improves the transparency of chlorinated rubbers.

paraffinic processing oil
A type of *processing oil* in which paraffinic-type structures predominate. See *petroleum oil plasticizer*.

paraform
See *paraformaldehyde*.

paraformaldehyde
Also known as polymethanal and as paraform. A low molecular weight, solid polymer of *formaldehyde*. May be represented as $HO-(CH_2O)_n-H$: where n equals from 10 to 100. It is readily converted into formaldehyde by heating to approximately 150°C: used in fumigation and as a disinfectant. May also be used to cross-link *phenol-formaldehyde resins*.

paraldehyde
Also known as ethanal trimer. A low molecular weight, liquid oligomer of acetaldehyde with the formula $(CH_3CHO)_3$.

parallel extruder
A *twin screw extruder* which contains two *screws* the outside diameters of which are the same along their whole length.

parallel laminate
A *laminate* in which the layers of *anisotropic reinforcement* are arranged with the strongest direction of the reinforcement layers lying in the same direction: the grain runs the same way.

parallel operation
A method of operating an *injection moulding machine* in which the screw is rotated during mould opening and/or mould closing. This allows the use of lower *screw speeds* as the amount of time available for screw rotation is increased. The use of lower screw speeds often means more uniform *melt temperatures* which, in turn, result in products with less tendency to warp - because of cooling variations.

parallel plate plastimeter
Also called a parallel plate instrument. A test apparatus used to measure *plasticity*. With this type of machine, the change in thickness of a test piece when loaded between two heated parallel plates is measured. See *plastimeter*.

parallel plate viscometer
Used to determine the *viscosity* of viscous liquids. Consists of two large plates which are capable of being rotated, so as to shear the material, one relative to the other.

parallel screw
See *constant depth screw*.

parallel stream reaction injection moulding mixhead
An abbreviation used for this term is parallel stream RIM mixhead. A type of *mixhead* used for *reaction injection moulding* (RIM) in which the *polyol* emerges from the mixhead surrounded by a stream of *isocyanate*. The runner in the mould forms the mixing chamber so that there is no *clean out piston* required as an *impingement mixer*.

parallel vicinal cross-links
Chemical groupings, based on sulphidic groups, and which are present in a sulphur-vulcanized *natural rubber*. The sulphur-containing groups are cross-links which are attached to adjacent, main chain atoms and are thought to have about the same influence as a single cross-link. See *mono-sulphide cross-link*.

paraphenylene polybenzobisoxazole
An abbreviation used for this material is PBO. An *aromatic polymer*.

Paris blue
See *Prussian blue* and *ferric ferrocyanide*.

Paris green
Also known as schweinfurt green. A double salt of copper acetate and of copper arsenate which is used as a *pigment*.

parison
The length of tube used to make a *blow moulding*.

parison length control
An abbreviation used for this term is PLC. Obtained by, for example, using an electronic feedback system which continuously monitors (via a photocell) the length of each *parison* and adjusts the extruder screw speed, as necessary, so as to maintain the length within preset limits.

parison programmer
That which alters the wall thickness of the *parison* by parison programming. This may be done by changing the position of the *pin* and the *die* using a pneumatic cylinder.

parison programming
Altering the wall thickness of the *parison* during parison production using a *parison programmer*. Ideally the changes should correspond to the degree of expansion required in the finished product.

parison sag
The thinning of a *parison* caused by its own weight as it leaves the die. Usually associated with *extrusion blow moulding* where the parison distorts under its own weight during extrusion. Also called 'draw down' as the parison draws down, or changes its thickness distribution during extrusion. May be reduced by the use of high molecular weight material or by fast production of the parison. Part of the sag may be due to an *elastic* effect and part due to *viscous flow*.

The elastic component of the sag increases as a proportion of the total as the:

i) molecular weight, and hence viscosity, increases;
ii) melt temperature decreases (increasing viscosity); and,
iii) the length of parison per unit weight increases.

This is because an elastic deformation under a standard load depends on the length of the part being stretched whereas, the viscous flow does not depend on the length (as long as the weight of the parison is constant). See *flow rate ratio*.

Parkes process
A *cold cure process* which uses *carbon disulphide* to cure thin-walled components at room temperatures. See *Peachey process*.

parquet polymer
See *ladder polymer*.

partial aromatic
A term associated with thermoplastics materials, such as with *polyimide* and with *polyamide*. Such a material is based on main chains which contain both aliphatic and aromatic groups (for example, benzene rings) linked in the main chain. For example, a partially aromatic polyamide is *polytrimethylhexamethyleneterephthalamide*. The irregular structure of this *copolymer* prevents crystallization and the material is therefore transparent. A high *heat distortion temperature* (HDT) is imparted by the p-phenylene group in the main chain as this gives chain rigidity.

partial-ladder polymer
Also known as a step ladder polymer or as, a *semi-ladder polymer*. An incomplete *ladder polymer* which consists of, for example, sequences of fused rings linked by single bonds.

partially compatible blend
Also known as a partially miscible blend. A *blend* based on two polymers which are only miscible over a restricted range of composition and/or of temperature. See *alloy*.

partially intermeshing twin screw extruder
A type of twin screw machine in which the flights of the screws partially overlap as they rotate. Intermeshing, or par-

tially intermeshing, types of machine are the most popular type of *twin screw extruder*.

partially miscible blend
See *partially compatible blend*.

partially purified crepe
An abbreviation used for this material is PPC or PP crepe. Crepe rubber produced from *latex* which was diluted to approximately 10%. See *pale crepe*.

partially saturated nitrile rubber
Nitrile rubber which has some of the unsaturation removed by *hydrogenation* to give H-NBR. See *hydrogenated nitrile rubber*.

particle recycling
A *polyester moulding compound (PMC)* recycling process, for example, for the recycling of *sheet moulding compound (SMC)*. In 1992 a car in Western Europe contained approximately 2 kg of PMC, for example, SMC or BMC. Such components are, for example, collected, sorted, shredded, metals are removed and then the material is hammer milled to the required size before drying. The PMC regrind is then used to make a new PMC compound by replacing part, or all, of the filler and/or glass fibre. Typically up to 15% of SMC regrind may be added without a significant drop in properties: as the regrind has a lower density than chalk (1.8 as opposed to 2.7 g/cm^3) the component density is lowered when a regrind filled mix is used. See *reclamation*.

particle recycling process
A recycling process for plastics materials. The plastics materials, for example, based on the interior panels of automobiles, are ground into particles, wetted with a *polyurethane* (PU) binder and then compression moulded into panels under low pressure. The self-supporting panels can then be used in a vehicle as floor panels or as cladding panels for the construction industry. See *reclamation*.

particulate calcium carbonate
See *whiting* and *calcium carbonate*.

particulate composite
A composite based on a polymer and a *particulate filler*.

particulate filler
A material, used as a *filler*, which consists of small particles. The particles are of roughly the same size.

parting line
Sometimes called a cut-off line. Such a line on a moulding is made by the junction of two mould halves and/or a loose mould section. A witness mark.

parting surface
The surface of a *mould plate*, adjacent to the *impression*, which butts against another mould plate and so prevents loss of material from a mould.

parylene polymer
A type of plastics material which is derived from di-1,4-xylene: a *poly-p-xylene*. Used as coatings in, for example, electronics equipment.

PAS
An abbreviation used for polyarylsulphone. See *sulphone polymers*.

pascal
An SI derived unit which has a special symbol, that is Pa. It is the SI derived unit of pressure and is equivalent to one newton per metre square (Nm^{-2}) or $kgm^{-1}s^{-2}$ or $kg/m^1/s^2$. See *Système International d'Unité*.

pascal second
Units used to measure *viscosity* in the SI system of units. Abbreviation Pas or Pa·s. 1 Pas = 1 Nsm^{-2} = 0.102 kgf s m^{-2} = 10P = 0.020 88 lbf s ft^{-2} = 0.000 145 lbf s in^{-2}.

passage
A term used in *hydraulics* and which refers to a pathway for *fluid* in a hydraulic component: such a passage may be machined or cast.

passive copper compounds
See *copper*.

paste
A finely-dispersed mixture of a solid polymer and a liquid, for example, *polyvinyl chloride* (PVC) paste consists mainly of *paste making polymer* and *plasticizer*: other ingredients include *stabilizer*, *filler* and a *lubricant*. Can have different types of paste: see, for example, *organosol*, *plastisol*, *plastigel* and *rigisol*.

paste making polymer
A *polyvinyl chloride* (PVC) polymer which is used for the manufacture of a *paste*. Such polymers are nearly always based on *emulsion polymers* as the particles of such materials are usually spherical and of relatively small size, for example, below 100 microns in diameter. Being spherical the particles have a relatively small surface area and so the *plasticizer* is not readily absorbed: being small the particles can remain suspended in the plasticizer for relatively long periods once dispersed: dispersion is achieved using, for example, a *three-roll mill*.

patch
Term often associated with the depletion of *stratospheric ozone* in a patch or hole, for example, over the Antarctic.

path dependent switchover
See *screw position switching*.

PAUR
An abbreviation used for a polyester polyurethane rubber. See *polyurethane rubber*.

PAUS
An abbreviation used for pale amber unsmoked sheet. See *natural rubber*.

pause delay timer
A timer which determines the *pause time*. Most injection moulding machines are fitted with a pause delay timer which is actuated when the *moving-mould half* reaches its fully open position. When the timer lapses, the moulding cycle begins again as the moving-mould half moves forward. The setting on the timer should, therefore, be just long enough to allow the mouldings to fall clear. Check the pause delay setting (in seconds) by dividing the square root of d by 22. Where d is the distance (cm) from the topmost point of the moulding as it fits in the mould to the lowest part of the mating services of the mould, or of any other obstruction on which it can be caught as the mould closes.

pause time
The time for which the mould is held open before the signal is given for mould closing. Sometimes referred to as die halt. Time must be allowed for the moulding to drop or fall clear. See *pause delay timer*.

pay-off
See *let-off*.

Pa·s
An abbreviation used for *pascal second*.

PB
An abbreviation sometimes used for *butadiene rubber* and for *polybutylene*. An abbreviation used for pigment blue: an *organic pigment*.

PBA
An abbreviation used for *thermoplastic elastomer-amide based*.

PBD
An abbreviation used for *butadiene rubber*.

PBI
An abbreviation used for *polybenzimidazole*. Also used as an abbreviation for the *Plastics Bottle Institution*.

PBN
An abbreviation used for phenyl-β-naphthylamine. See *phenyl-naphthylamines*.

PBO
An abbreviation used for *paraphenylene polybenzobisoxazole*.

PBR
An abbreviation used for *pyridine-butadiene rubber*.

PBT
An abbreviation used for *polybutylene terephthalate*.

PBT/PET blend
An abbreviation used for *polybutylene terephthalate/polyethylene terephthalate blend*.

PBTP
An abbreviation used for *polybutylene terephthalate*.

PBZ
An abbreviation used for *polybenzobisoxazole*.

PC
An abbreviation used for *polycarbonate*.
- **PC/ABS blend** = *polycarbonate/acrylonitrile-butadiene-styrene blend*.
- **PC/PBT blend** = polycarbonate/thermoplastic polyester blend.
- **PC-BPA** = polycarbonate based on *bisphenol A*.
- **PC/PE blend** = *polycarbonate/polyethylene* blend.
- **PC/PET blend** = polycarbonate/thermoplastic polyester blend.
- **PC-TMBPA** = *tetramethyl bisphenol A polycarbonate*.
- **PC-TMC** = *tetramethylcyclohexane bisphenol polycarbonate*.

Also see *polycarbonate/thermoplastic polyester blend*.

pcb
An abbreviation used for printed circuit board.

PCBs
An abbreviation used for *polychlorinated biphenyls*.

PCC
An abbreviation used for *polymer cement concrete*.

PCD
An abbreviation used for *polycarbodiimide*.

PCDDs
An abbreviation used for *polychlorinated dibenzo-para-dioxins*.

PCDFs
An abbreviation used for *polychlorinated dibenzofurans*.

pcf
An abbreviation used for pounds per cubic foot.

PCI
An abbreviation used for process capability index.

pd
An abbreviation used for potential difference.

PDCP
An abbreviation used for *polydicyclopentadiene*.

pdl
An abbreviation used for *poundal*. See *foot-poundal*.

Peachey process
A *cold cure process* which uses firstly sulphur dioxide and then hydrogen sulphide to form active sulphur which brings about cure at room temperatures: curing is, in effect, with sulphur thiocyanate at room temperatures. Although only suitable for thin components (below 1 mm), the cured products can have properties similar to hot cured components.

peak cavity pressure
The maximum pressure during mould filling in *injection moulding*. See *cavity pressure control*.

peak exotherm
The maximum temperature recorded during resin cure. See *exotherm curve*.

peaky curing
A property of a rubber compound obtained from a curing curve. The value of a particular property, for example, *tensile strength*, is plotted against time of vulcanization and if the curve shows a sharp drop after a maximum value the compound is said to exhibit peaky cure.

pearl polymerization
See *suspension polymerization*.

pearl spar
See *dolomite*.

pearlescent colour
A colour obtained with a *pearlescent pigment*. Pearlescent and iridescent colours may be obtained with, for example, flakes of lead carbonates or of, bismuth oxychlorides or of, titanium dioxide coated mica. See *metallic colours*.

pearlescent pigment
Also known as nacreous pigment. A plate-like pigment which allows some light to be transmitted and which also reflects some of the light. See *pearlescent colour*.

PEB
An abbreviation used for *polyether block amide*.

PEBA
An abbreviation used for *polyether block amide*.

PEC
An abbreviation used for *polyester-carbonate*.

peck
A unit of measurement (usually for dry measurements) and equal to 2 gallons or 8 quarts. Abbreviation pk.

PEEK
An abbreviation used for *polyether ether ketone*.

PEG
An abbreviation used for *polyethylene glycol*.

PEI
An abbreviation used for *polyether imide*.

pelargonic acid
Also known as nonanoic acid. May be represented as $CH_3(CH_2)_7COOH$. This colourless liquid material has a boiling point of 256°C and a melting point of 13°C. Used as an acid modifier in the manufacture of some *alkyd resins*.

pelletized moulding compound
An abbreviation sometimes used for this type of material is PMC. See *granular polyester moulding compound* and *polyester moulding compound*.

pellets
A raw material *feed form* for *thermoplastics materials* which are usually produced by chopping the strand, or strands, produced by an extruder. The thermoplastics material is commonly supplied in the form of small pellets approximately 0·12 in or 3 mm in length and breadth.

pencil bank
A small, uniformly thin, *bank of material* in a *nip*. In *double bank calendering*, the pencil bank is formed by simply folding the sheet over before it is fed to the second nip, and in order to ensure that accurate sheeting is produced this bank must be kept at a constant size. Such a bank is kept small in order to keep the temperature uniform.

pendant ionic polymer
An *ionic* polymer in which the bound ion is pendant to the main chain. For example, a polysulphonate. Most conventional polymers can be modified to form an ionic polymer of the pendant type.

pendant sulphidic group
A chemical grouping, based on a sulphidic group, and which is present in a sulphur-vulcanized *natural rubber*. The sulphur-containing group is not a cross-link but is pendant to the main chain and is terminated by a fragment from the *accelerator*.

pendant unsaturation
A side chain of a polymer which contains a double bond. See, for example, *alkylene sulphide rubber*.

pendulum
The striker, tup, or hammer used in an impact test. See *pendulum impact test*.

pendulum feeding device
See *wig-wag*.

pendulum impact machine
One of the most popular types of impact test machines. Commonly used to assess the impact resistance of thermoplastics materials. See *pendulum impact test*.

pendulum impact test
A test which is done on a *pendulum impact machine*: one of the most popular types of impact test. Commonly used to assess the impact resistance of *thermoplastics materials*. Usually a moulded, notched bar is used and this supported bar is struck by a swinging pendulum: the energy absorbed in breaking the bar is measured. For thermoplastics, *injection moulding* is widely used to prepare the specimens: an end-gated bar is employed. For thermosets, compression moulding is used. All samples should be stored at the test conditions for a standard time, for example, 24 hours, before testing.

CHARPY. The Charpy impact test uses a rectangular bar which is struck by a pendulum or hammer (sometimes also referred to as a tup). The sample may be notched or unnotched and the results are now usually given as the energy required to break a specified area, that is, in kJm^2. Specimen preparation is similar to that used for the *Izod* test although the test is markedly different from the Izod test. For example, the sample is supported on anvils at each end, not clamped, and is struck at the centre immediately behind the notch (the notch may be a different shape to that used in the Izod test).

The standards specify a range of specimens, the smallest of which is much smaller than the Izod impact specimen. If this size of specimen is adopted, then specimen preparation is relatively easy and in this respect the test is superior to the Izod. However, the test does not appear to be very useful for assessing the impact strength of very flexible materials as the specimens can bend and pass through the anvils on which they rest.

Charpy impact strength - of some plastics
(DIN 53454) at 23°C)

	kJ/m^2
ABS	11
PC	35
PA 6	25[a]
PA 6	4[b]

[a] When conditioned to 2·5% water absorption
[b] When tested as moulded.

IZOD. The specimen for an Izod impact test consists of a rectangular bar which can be, for example, $2·5 \times 0·5 \times 0·25$ in. A standard V-shaped notch is moulded, or more usually machined, into the ¼" side of the bar so as to obtain more reproducible results. One end of the bar is gripped in a vice. The other end of the bar is then struck on the same side as the notch, at a specified height and speed, with a swinging pendulum: the energy expended in breaking the specimen is obtained.

The results may be expressed in various ways: foot-pounds per inch of notch or, its SI equivalent, Joules per metre. In the ft/lb method the energy to break is multiplied by that number which is obtained when the notched face width is inverted, i.e. 4 if ¼ in samples were used. However, the results are now often reported in kJ/m^2 and in this system the energy to break is divided by the area fractured: if a notched sample is used the fracture area is obtained by multiplying the width of the specimen by the depth behind the notch.

As many plastics are *notch sensitive*, and the notch dimension varies from one standard to another, it is essential that the standard employed is quoted. Flexible materials may not break completely at some temperatures and in such cases a 'no break' result is quoted.

It is difficult to imagine how this test ever became a standard as the specimens are not very representative of plastics parts. For example, they are very thick, e.g. 6·35 mm and they contain notches. Both features are unusual in plastics technology. Good specimen production (e.g. free from voids) is always a problem in Izod impact testing and such considerations may explain why the test appears to have declined in popularity in recent years.

Izod impact strength - of some plastics
(ASTM D 2562A) at 23°C using ⅛ in specimens

	ft lb/in of notch	kJ/m^2
UPVC	0·4–20	2·1–105
LDPE	no break	no break
Nylon 66	3·0[a]	15·8
PC	14–16	74–78

[a] Conditioned to equilibrium at 50% relative humidity

TENSILE IMPACT. A high speed *tensile test* performed on equipment similar to that used for the Charpy impact test. One end of the dumb-bell shaped sample has a large protrusion or jig attached to it and the other end is bolted to the pendulum of the testing machine. When the pendulum swings the sample tries to pass through the jaws of two stationary anvils but is restrained by the protrusion and so broken by this blow, delivered along the longitudinal axis of the sample. The energy required to do this is recorded.

Specimen preparation is by *injection* or *compression moulding*. However, as the test is commonly used for testing thin, sheet materials the specimens are usually pressed or blanked from a larger sheet. Specimens must be kept at the test conditions prior to testing.

Results are given in kJm^{-2}; the energy required to break the specimen is divided by its minimum cross-sectional area. For example, the tensile impact strength of polyoxymethylene (Delrin 500 at 23°C is 210 kJ/m^2: (100 ft lb/in^2). The Izod impact strength of this same material when tested according to ASTM D 256 is 75 J/m (1·4 ft/lb/in) when notched: when an unnotched specimen is used it is 1280 J/m (24 ft/lb/in). 1 ft lb/in of notch = 53·36 J/m. If it is assumed that impact strength is proportional to the distance behind the notch 1 ft lb/in of notch = 5·25 kJ/m².

penetration depth
An abbreviation used for this term is $1/\alpha$. The distance into a *dielectric* at which the field strength has decayed to 0·368 of its original value. For a particular material this depends on the frequency used. For example, for *plasticized polyvinyl chloride* it is 46 m at 10 MHz and 0·54 m at 3,000 MHz (for *loss factors* of 0·4 and 0·1 respectively). Energy penetration can sometimes be a problem at *microwave* frequencies if components of large dimensions and of large loss factors are being heated.

$$1/\alpha = \frac{1}{\frac{2\pi}{\lambda}\left[\frac{\epsilon_r}{2}\sqrt{(1+\tan^2\delta-1)}\right]^{1/2}}$$

Where ϵ_r = dielectric constant, $\tan\delta$ = the loss tangent and λ = the wavelength in metres of the speed of light divided by the frequency (radio or microwave) being used.

penetrometer
A *hardness* measuring apparatus which uses standardized needles to measure hardness. The needle is pressed into the substrate under precisely specified conditions, For example, the test temperature, the load applied, the time for which the load is applied, substrate thickness and needle tip diameter are all standardized.

penta
A prefix which means five.

pentabromodiphenyl ether
An *organo-bromine flame* retardant. Such a bromine-containing compound tends to be more powerful than the equivalent chlorine-containing compound: often used in conjunction with other flame retarders, for example, *antimony trioxide*. For example, a *plasticized polyvinyl chloride* (PPVC) composition may contain 53% polymer, 30% phthalate plasticizer, 10% phosphate plasticizer, and 5% pentabromodiphenyl ether.

pentachloroethane
Also known as pentaline. May be represented as CCl_3CHCl_2. This colourless liquid material has a boiling point of 161°C, a melting point of −29°C and a relative density (RD) of 1·67. A solvent, for example, for resins.

pentachlorophenol
This white solid material has a boiling point of 310°C, a melting point of 190°C and a relative density (RD) of 1·98. A preservative for *natural* and *synthetic rubbers*.

pentachlorophenol lauryl ester
An odourless liquid which may be used to preserve vulcanized rubber components against attack by insects, bacteria and fungi.

pentachlorothiophenol
A grey solid material: an additive which will act as a peptizer for synthetic rubbers over the temperature range 100 to 160°C in, for example, an *internal mixer*.

pentaerythritol
Also known as tetra(hydroxymethyl)methane or as, tertramethylolmethane or as, 2,2-bis(hydroxymethyl)-1,3-propanediol. May be represented as $C(CH_2OH)_4$. This white, crystalline powder material has a melting point of approximately 260°C. Used in the manufacture of some *alkyd resins*, *polyurethane foams* and *pentaresin*.

pentaline
See *pentachloroethane*.

pentane
An alkane hydrocarbon of formula C_5H_{12}: there are three isomers. n-pentane has a boiling point of 36°C and a relative density of 0·62. It is used, for example, to pre-expand, or blow, polystyrene (see *expanded polystyrene*). Also used as a propellant gas in some aerosols. See *chlorofluorocarbon* and *butane*.

pentaresin
A material produced by chemical modification of rosin with *pentaerythritol*. The resultant resin has a higher melting point, has high gloss, good colour, good colour retention, adhesion properties and heat stability. Varnishes based on this resin have good adhesion, drying properties, reasonable water resistance and alkali resistance.

PEO
An abbreviation used for *polyethylene oxide*.

PEO-alkali metal salt complex
An abbreviation used for polyethylene oxide-alkali metal salt complex. See *polyethylene oxide-salt complex*.

PEO-metal salt complex
An abbreviation used for polyethylene oxide-metal salt complex. See *polyethylene oxide-salt complex*.

peptizable chloroprene copolymer rubber
A *chloroprene copolymer rubber* which is capable of being peptized, for example, by thiuram disulphides. See *chloroprene sulphur rubber*.

peptized rubber
A form of *natural rubber*. A low-viscosity rubber prepared by adding a *peptizer* to *latex* (before *coagulation*) or, by addition to dry rubber. Rubber which has been softened by the use of a *peptizing agent*. By the use of peptizing agents (0·1 to 0·5%) the amount of power needed for mastication can be significantly reduced.

peptizer
A plasticizing agent: a reclaiming agent. A material which chemically softens rubber. An additive which improves the rate of oxidation and molecular weight reduction in, for example, *natural rubber* at high temperatures. An additive which will act as a peptizer for synthetic rubbers is *pentachlorothiophenol*. See *mastication*.

peptizing agent
See *peptizer*.

per
When this prefix is used in chemical terms, it means that a particular compound has, what appears to be, an excess amount of an element, for example, as in a *peroxide*.

percentage elongation
The *elongation* expressed as a percentage of the *gauge length*. See *elongation at break*.

percentage elongation at break
The *percentage elongation* at the instant that the sample breaks. See *elongation at break*.

percentage error
Obtained by dividing the error by the true value, or the most probable value, and multiplying by 100.

percentage modulus
The stress needed to produce an elongation of a certain per cent in a rubbery material. Not a true modulus but the *tensile stress* at that percentage extension.

percentage shrinkage
Mould shrinkage expressed as a percentage, for example, 0·4%. As a linear shrinkage, the same shrinkage would be 0·004 in/in or 0·004 mm/mm.

percentage strain
One hundred times the *strain* (strain × 100).

perfluorinated elastomer
Also known as perfluorinated rubber.

perfluoro(methyl vinyl ether)
An abbreviation used for this monomer is FMVE. Also known as perfluoro(methylvinylether). May be represented as $CF_2 = CF\text{-}O\text{-}CF_3$. Such a material is used to make fluoropolymers. The *fluororubber* formed between this monomer and tetrafluoroethylene is given the abbreviation PFE.

perfluoro(methylvinylether)
See *perfluoro(methyl vinyl ether)*.

perfluoroalkoxy copolymer
Also known as perfluoroalkylvinyl ether polymer, copolymer or rubber or as, polyperfluoroalkylvinyl ether. An abbreviation used for this type of material is PFA. A *perfluorinated elastomer (FFKM)* which contains a cure-site monomer. For example, a perfluorovinyl ether which contains an aromatic alkoxy group may be used and cross-linking achieved by an amine-based system via the para-fluorine atom on the benzene ring. See *fluororubber* and *polyphosphazene rubber*.

perfluoroalkoxy rubber
See *perfluoroalkoxy copolymer* and *fluororubber*.

perfluoroalkyl copolymer
See *perfluoroalkoxy copolymer* and *fluororubber*.

perfluoroalkyl ether copolymer
See *perfluoroalkoxy copolymer* and *fluororubber*.

perfluoroalkyl-dinitrile
See *perfluoroalkylenetriazine*.

perfluoroalkylenetriazine
A highly specialized *fluororubber* which may be prepared from a perfluoroalkyl-dinitrile with ammonia. The polymers have *vinylidene fluoride* sequences linked to the three carbon atoms of a six-membered ring (based on alternate carbon and nitrogen atoms). Such materials, like a *carboxy-nitroso rubber (AFMU)* are resistant to acids and oxidizing agents but not to bases.

perfluoroalkylvinyl ether copolymer
See *perfluoroalkoxy copolymer* and *fluororubber*.

perfluoroalkylvinyl ether polymer
See *perfluoroalkoxy copolymer* and *fluororubber*.

perfluorocarboxylate
A *polytetrafluoroethylene* ionomer.

perfluorosulphonate
A *polytetrafluoroethylene* ionomer.

perforated coated fabric
A *plastic coated fabric* in which the coating, or the coated substrate, are perforated.

performable continuous strand mat
A continuous strand mat with a thermoplastics polyester binder: used in *resin transfer moulding*.

performance modifier
An *additive* used to improve, or alter, the properties of a component made from a polymeric material. A *reinforcing filler* and an *antistatic agent* are both examples of performance modifiers. Sometimes such additives may form part of the colour *masterbatch*.

peripheral drilled roll
A roll which has the heating/cooling channels close to the outside surface of the roll so that rapid, accurate temperature control is possible. Used, for example, on a *calender* where very accurate temperature control is essential if accurate thicknesses are to be held during high speed production. See *high pressure hot water*.

perishing
A general deterioration with time of a rubber component.

peristaltic pump
A pump used, for example, to dispense *liquid colour*. The liquid is held in a storage container and passes through silicone rubber tubing. It is propelled by rollers which compress the tube and thus cause the liquid to be moved along. A control unit ensures that the correct preset amount of colour is added to the polymer via flexible tubing which passes through a metal probe fitted in the machine throat. Pump operation is controlled by a contact in the screw circuit which allows material to be metered only during the screw-turn part of the cycle during, for example, *injection moulding*. See *masterbatch*.

permanent green
A green pigment which is a mixture of hydrated chromium oxide and barytes. Stable to sunlight and to chemicals but has poor hiding power and is unstable at temperatures above 260°C/500°F.

permanent plasticization
Softening of a material by, for example, the use of a *permanent plasticizer*. See *internal plasticization*.

permanent plasticizer
Also called a polymeric plasticizer. A highly permanent, non-extractable and non-migratory *plasticizer*: a material of high molecular weight which once incorporated into a material, such as *polyvinyl chloride* (PVC), is difficult to remove. Often complex, linear polyesters made by reacting dibasic acids (such as adipic, azelaic and sebacic) with a polyglycol are used. Such plasticizers give compounds which retain their flexibility, compared to *monomeric plasticizers*, for very long periods. The plasticizers are also more resistant to evaporation, migration and extraction. Such materials include polypropylene adipate and polypropylene sebacate (capped with lauric acid end groups). Polyester plasticizers have relative densities of 1·0 to 1·1.

Sometimes a rubber when blended with a *thermoplastics materials* is classed as a permanent plasticizer. For example, *nitrile rubber* (NBR) mixed with PVC. However, NBR is not strictly a plasticizer as it is not dispersed on a molecular level and it does not affect the *glass transition temperature* of the compound in the same way as a true plasticizer.

permanent set
A permanent change of shape caused by irreversible chain slippage of molecules one over the other. Residual deformation which remains in a vulcanizate after a stress has been removed. Can have compression set or tension set.

permanent set in tension
See *tension set*.

permanent yellow
A *Hansa yellow pigment* which is used for plastics materials such as *polystyrene, polypropylene* and *unplasticized polyvinyl chloride*. Has good brightness, colour strength and chemical resistance.

permeability
The passage of a gas or vapour through a solid material: means that a material will allow diffusion. This means that ordinary plastic bottles will lose gas from, for example, fizzy drinks, because they are permeable. A rubber *tyre* will lose air quickly if it is too permeable.

Of particular interest in the usage of plastics is the permeability of a gas, liquid or vapour through a plastics material or, component. Permeation is a three part process and involves solution of the small molecules in the plastics material, migration (or diffusion) through the body of the plastics material and emergence of the small molecule from the far surface. Permeability is therefore a product of solubility and *diffusion*.

Permeability may be expressed as:

P = Q/(a. t. $\delta p/\delta x$) where,
P = the proportionality constant known as permeability;
Q = the quantity of gas (either volume or mass);
a = the area of the barrier;
t = the time taken
δp = the gas pressure; and,
δx = the thickness of the barrier.

Various units of measurement are used. The SI unit is mol/Ns; however, the permeability of gases is often expressed as $(10^{-10}cm^2)/(s.cm.Hg)$. The permeability of vapours, also known as transmittance, is often expressed as $(10^{-10}g)/(cm.s.cm.Hg)$. To get a useful scale (for example, from 1 to 100,000) transmission rates may be quoted in a different way. The permeability of gases may be expressed as $cm^3/m^2.24$ h and that of vapours as $g/m^2.24$ h: stating in each case the pressure difference and the film thickness. On a 1 to 100,000 scale the permeability to nitrogen of some polymers would be PVDC = 1, PA 6 and POM = 10, SAN = 50, HDPE = 250, and LDPE approximately 1,500. On a 1 to 100,000 scale the permeability to carbon dioxide, of some polymers would be cellulose = 5, PVDC = 30, PA 6 = 180, SAN = 1,080, HDPE = 3,500, and LDPE approximately 17,000 to 35,000. See *gas permeability* and *vapour permeability*.

permeability - factors affecting
Permeation is a function of materials, design and processing method. Metal or glass containers, when sealed correctly, may be considered impermeable, i.e. nothing can move across the metal or glass barrier. Such is not the case with polymers: a polymer becomes more permeable as the temperature is raised: permeability also increases if a plastics material is plasticized. However, it is decreased by cross-linking, crystallinity and by the use of fillers, for example, by the use of oriented, laminar fillers such as mica or, polyamide (PA) flakes. It is also decreased if a polyethylene (PE) *blow moulded container* is treated with, for example, fluorine, before ejection (*fluorination*).

Permeability coefficients of plastics, vary by many orders of magnitude for a given gas and the permeability of a given plastics material varies for different gases. It is difficult to find a plastics material which is a good barrier to both oxygen and to water vapour and which is low in cost. this fact has lead to the growing use of multi-layer extrusions and of coatings (see *coextrusion blow moulding*).

The major factors in package design which influence the permeation rate are the surface-to-volume ratio and the thickness of the container walls. the lower the ratio, and the thicker the walls, the lower the rate. However, the more material that is used, the higher is the cost. The moulding method can influence the permeation rate as it influences the crystallinity, or the orientation, of the thermoplastics material. The higher the degree of either, the lower the permeability. For example, if a blow moulding parison is stretched before being blown (so as to give different molecular orientation in the blow moulded component) permeability will be reduced. Coating with a relatively impermeable material, such as coating with *polyvinylidene dichloride* (PVDC), may be another.

permeability resistance
The resistance of a film to diffusion of gases or vapours and the reciprocal of permeability.

permeable coated fabric
A *plastic coated fabric* which allows the diffusion of gases or vapours.

permittivity
See *dielectric constant*.

peroxide
A chemical compound: an oxide that gives hydrogen peroxide with an acid. Such materials are often used as polymerization catalysts. See *peroxide initiator*.

peroxide curing
Curing, or vulcanization induced by the use of a *peroxide*. See *natural rubber*.

peroxide decomposer
A *preventive antioxidant*. A peroxide destroyer. A material which promotes the decomposition, or degradation, of organic hydroperoxides (see *oxidation*) so as to form stable products rather than form active free radical species. Such peroxide decomposers include mercaptans, sulphonic acids, zinc dialkylthiophosphate and zinc dimethyldithiocarbamate. Thioester antioxidants and phosphites are also examples of peroxide decomposers: such materials are usually used in combination with primary antioxidants (*hindered phenol* and *amine antioxidants*).

peroxide destroyer
See *peroxide decomposer*.

peroxide initiator
A chemical compound: a *peroxide* that is capable of initiating a polymerization reaction by generating free radicals. A very widely used group of polymerization catalysts. See, for example, *methyl ethyl ketone peroxide*.

peroxide vulcanization
Vulcanization of a rubber compound by means of a *peroxide*. A peroxide is used which is capable of initiating a *vulcanization* reaction by generating free radicals. These, in turn, abstract hydrogen from the rubber to form polymer free radicals and it is these which react to form the cross-links. One such peroxide is *benzoyl peroxide*.

petrochemical
A chemical derived from *petroleum* or *natural gas*.

perchlorethylene
A chlorine containing solvent used as a dry cleaning solvent: has a relative density of 1·62 and a boiling point of 121°C. A solvent for hydrocarbon rubbers and for polystyrene.

perthioaccelerator
See *accelerated sulphur system*.

perthiosalt
See *accelerated sulphur system*.

perylene pigments
A class or type of *organic pigment* which have complex polycyclic structures. They give transparent colours ranging from yellow-red to maroon. Strong colouring systems, with good chemical resistance and which are also suitable for making *metallic colours*. Relatively high cost systems which are lightfast but which may bleed in some polymer materials.

PET
An abbreviation used for *polyethylene terephthalate*.

peta
An abbreviation used for this term is P: it means 10^{15}. See *prefixes - SI*.

PETG
An abbreviation used for *polyethylene terephthalate copolymer*.

PETP
An abbreviation used for *polyethylene terephthalate*.

petrol
A complex mixture of hydrocarbons. For example, hexane, heptane and octane. Known as gasoline in some countries.

petrolatum
Also called *petroleum jelly*.

petroleum
Also known as crude oil and as *mineral oil*.

petroleum ether
Also known as ligroin. A range of low boiling point hydrocarbons (approximate range 20 to 135°C) with a relative density in the range 0.6. Used as solvents and for extraction.

petroleum jelly
This material has a relative density (RD or SG) of 0.84 to 0.89 and a melting point of approximately 40°C. Refined hydrocarbons which may be white or pale yellow and which are semi-solid. Used in rubber compounds as a processing aid and in order to improve electrical properties. Used in *sponges* as it has a greater softening effect at processing temperatures than at room temperatures. See *paraffin wax*.

petroleum oil
See *rubber oil*.

petroleum oil plasticizer
See *rubber oil*.

petroleum pitch
This type of pitch is produced during the distillation of petroleum or *mineral oil*. This material is used as a processing aid and as a compound stiffener. The use of such materials confers lower resilience and higher heat build-up in the end-use of rubber compounds.

petroleum resin
A resin made from olefins which are derived from petroleum by cracking. Aliphatic petroleum resins are derived from the low molecular weight fraction (which contains mainly isoprene and piperylene). Aromatic petroleum resins are derived from the higher molecular weight fraction (which contains styrene, α-methyl styrene, indene, vinyltoluene and dicyclopentadiene). *Dicyclopentadiene resins* may be made from fractions which consist mainly of this monomer.

petroleum spirit
See *gasoline*.

PEUR
An abbreviation used for a polyether polyurethane rubber. See *polyurethane rubber*.

pewter
Low melting point alloys of tin and lead (based on say, 80% tin and 20% lead).

PF
An abbreviation used for *phenol-formaldehyde*.

PF RP = *phenol-formaldehyde reinforced plastic*.

pferde-kraft
See *pferde-stärke*.

pferde-stärke
Abbreviation PS. A metric horsepower. A unit for measuring the rate of work: a unit of power measurement. Sometimes known as cheval vapeur or pferde-kraft. Equivalent to 0.736 kilowatts or 735.499 watts. One metric horsepower is the rate of doing work of 75 m kg/s which is equivalent to 0.75 kilowatt or 550 foot-pounds per second. See *horsepower*.

pferde-stärken-stunde
One metric horsepower hour or kraftstunde.

PFP
An abbreviation used for *plastic faced plaster*.

PG
An abbreviation used for planetary gear. See *planetary gear extruder mixer*.

PG extruder mixer
An abbreviation used for *planetary gear extruder mixer*.

PGE
An abbreviation used for planetary gear extruder. See *planetary gear extruder mixer*.

phase angle
The angle between two periodic disturbances, for example, voltage and current.

phase shift
The time difference between the input and output signal of a control unit and which may be measured in degrees.

phenate
The reaction product of phenol (which acts as a weak acid) and a strong base. Sodium and potassium phenates are soluble in water and will react with carbon dioxide to give the phenol. Barium and cadmium phenates are used as liquid *heat stabilizers* for *polyvinyl chloride* (PVC) as they exhibit little *plate out*. See *metal soap stabilizer*.

phenol
Also known as carbolic acid. An abbreviation used for this material is P (in abbreviations for plastics materials). This white solid material has a melting point of 42°C and a boiling point of 182°C. May be produced in various ways but the cumene process is now the most widely used (uses benzene and propylene). Used, for example, to make *phenol-formaldehyde* and *bisphenol A* (polycarbonates). See *phenate*.

phenol alcohol
See *methylol phenol*.

phenol alkane bisphenol
A *bridge hindered phenol*. See *phenolic antioxidant* and *phenylalkane*.

phenol formaldehyde
See *phenolic moulding materials*.

phenol formaldehyde resole
See *resole*.

phenol-based resole
A *resole* which uses phenol as a starting material (rather than, say, cresol).

phenol-formaldehyde
Also known as phenol-methanal or as, phenoplast. An abbreviation used for this material is PF. See *phenolic moulding materials*.

phenol-formaldehyde reinforced plastic
Also known as a phenolic reinforced plastic. May be abbreviated to PF-RP or PF RP. A composite which is based on a *thermoset matrix* (*phenol-formaldehyde resin*) and appropriate reinforcement, for example, *glass fibre (PF-GRP)*. The resin can be formulated so that low temperature/low pressure cure is possible and such systems can be moulded into composites by processes similar to those used for *unsaturated polyester resins*.

Phenolic, glass reinforced plastics (PF GRP) are now being used in a variety of interior design applications, particularly in the mass transport sector, for example, for interior surfacing or cladding panels in trains and aircraft. Mouldings can be glossy but are opaque and dark-coloured: they have the tremendous advantage of being inherently flame retardant, low smoke systems which do not require halogen or phosphorus-based additives to give them such properties. Such composites can be formulated to be fire and thermally resistant. When combustion does occur, the *smoke* evolved is relatively clean and of low concentration. As judged by the *NBS chamber*, it can be 90% less than that of a fire retardant polyester.

These materials are stiff, hard, have low elongations and possess good creep resistance. Paints are available to enhance the appearance without sacrificing the desirable properties, for example, flame retardancy and low smoke emission.

phenol-methanal
See *phenol-formaldehyde*.

phenol-substituted phosphonitrilic polymer
See *phosphonitrilic polymer*.

phenolic antioxidant
A phenol-based *antioxidant* which acts as a chain breaker. Chain breaking antioxidants are amines and phenols. Phenolic antioxidants are widely used in polyolefin thermoplastics and to a lesser extent in rubber compounds. In rubber technology phenolic antioxidants are classed as non-staining antioxidants. There are two main classes of phenolic antioxidant: (i) simple *hindered phenols*, which are also known as mononuclear phenols or substituted phenols and, (ii) *bridged hindered phenols*, which are also known as polynuclear phenols.

phenolic glass reinforced plastics
See *phenol-formaldehyde reinforced plastic*.

phenolic laminated sheet
A *laminated* plastics sheet material based on a *phenol-formaldehyde resin* or on a homologue of phenol.

phenolic moulding materials
An abbreviation used for this type of material is PF. Also referred to as phenolics or as, phenol-formaldehyde or as, Bakelite or as, phenoplasts or as, phenolic resin moulding compounds or as, novolak resin moulding compounds.

Unreinforced *phenolic resin* is very brittle and requires extensive modification with fillers to produce useful products. A very wide range of properties can be obtained from PF materials because of their compatibility with a variety of reinforcements and fillers: this means that the properties of a PF are very dependent on the *filler* used. Woodflour gives reasonable properties at an acceptable cost and so woodflour-filled PF is regarded as a general purpose (GP) material. The use of a more fibrous, organic filler (cellulose fibres) gives improved toughness and impact strength. Modification with a mineral filler yields increased rigidity, improved dimensional stability, thermal stability, lower water absorption and lower thermal expansion. Glass fibre (GF) addition, can improve the dimensional stability and rigidity even more: the UL index of use can reach 180°C/356°F. Because of density differences, organic fillers are used in a weight ratio of 1:1 with the resin and inorganic fillers (mineral and glass) used in a ratio of 1·5:1.

Because of the formation of complex molecules (quinone methides), PF materials are naturally dark-coloured materials which darken even more in sunlight or, on warming. As the unmodified material is virtually useless without filler addition, the resin is extensively modified by the use of fillers and the fillers makes the moulding compounds opaque. The colour range is therefore very limited, with black, dark brown, dark red and dark green being the most common colours. Recently comparatively pale shades of red have been produced.

Mouldings are glossy, opaque and dark-coloured but have the tremendous advantage of being inherently flame retardant, low smoke systems which do not require halogen or phosphorus-based additives. They are stiff, hard, have low elongations and possess good creep resistance. the materials are commonly supplied in powder form and have a useful combination of low cost, ease of moulding, temperature resistance, solvent resistance, chemical resistance and good electrical insulation properties. They have better water resistance than *melamine formaldehyde* (MF) and possess a more stable melt rheology than aminoplastics, i.e. they are not so temperature dependent: their impact resistance is not very good. Electrical properties, especially tracking resistance is inferior to aminoplastics. Melamine phenolics (MPF) have superior electrical properties to phenolics and have a wider colour range: they are used in decorative and electrical application - that is, in areas which are beyond PF. The reduced post moulding *shrinkage* of this class of materials, is a significant advantage over MF.

PF materials are resistant to dilute acids (not all grades), alcohols, aromatic and chlorinated hydrocarbons, petrol, greases, fats and oils. PF mouldings are resistant to organic chemicals, oils and fats even at elevated temperatures. Mineral-filled grades may also be detergent resistant. PF mouldings retain their properties on ageing and have good dimensional stability over a wide temperature range. They are not resistant to strong mineral acids and alkalis but may withstand weak solutions. However, some grades are not so resistant and may even be attacked by dilute acids, alkalis, ketones and detergents. Those with organic fillers, and therefore higher water absorption, are more strongly attacked, on long exposures, than those grades filled with inorganic fillers. All grades should resist short term exposure to water. Sunlight will cause general darkening.

PF is used for ashtrays, saucepan handles, knobs and percolator bases. The material is chosen for these applications because it is heat-resistant, durable and cheap. Glass fibre (GF) filled PF has a high mechanical strength in thin sections when compared to other materials. Because of this it is used for parts with thin sections, such as small bobbins, thin-walled connectors and housings. Also used for electrical items such as coil bobbins, terminal blocks, relay bases and component housings.

phenolic plastics
Plastics based on resins made by the condensation of phenols such as phenol or cresol, with aldehydes. See *phenolic moulding materials*.

phenolic reaction injection moulding
An abbreviation used for this term is PF RIM. See *reaction injection moulding process*.

phenolic reinforced plastic
See *phenol-formaldehyde reinforced plastic*.

phenolic resin
Also known as phenol-formaldehyde polymer or as, phenoplast. The starting material for, for example, *phenolic moulding materials*. Commercially such a material is a comparatively low molecular weight polymer based on the resinous material formed when a phenol-type material is reacted (or condensed) with an aldehyde, such as *formaldehyde* - this is the most widely used aldehyde.

Phenol itself is widely used as the resins produced have good mechanical strength and cure quickly. Cresols may be used for more acid-resistant products and the use of phenol/cresol mixtures lowers the cost and controls the flow in processing. For example, in compression moulding, the use of 20% cresol may be employed. Where improved alkali resistance is required, xylenols may be used. The unreinforced resin is very brittle and requires extensive modification with fillers to produce useful products. See *phenolic moulding materials*.

phenolic resin compound, single-stage
A phenolic material in which the resin, because of its reactive groups, is capable of further polymerization by application of heat (see *resole*).

phenolic resin compound, two-stage
A phenolic material in which the resin is essentially not reactive at normal storage temperatures, but contains a reactive additive (often hexamethylene tetramine) which causes further polymerization upon the application of heat (see *novolak*).

phenolic resin curing
Curing, or *vulcanization*, induced by the use of a *phenolic resin* (PF), for example, dimethylol phenol resins. Used mainly with *synthetic rubbers* as the PF improves heat resistance: with *natural rubber* the PF interferes with crystallization and so worsens other properties. PF resins are used with *butyl rubber* to give vulcanizates of excellent heat resistance.

phenolic resin moulding compounds
See *phenolic moulding materials*.

phenolic sheet moulding compound
A *sheet moulding compound* based upon a *phenolic resin*. See *high performance sheet moulding compound*.

phenolics
See *phenolic moulding materials*.

phenols
A class of aromatic chemicals which have at least one hydroxyl group attached directly to a benzene ring. They will form esters, ethers and salts.

phenoplast
See *phenolic resin* and *phenol-formaldehyde*.

phenoxies
See *phenoxy resin*.

phenoxy resin
A high molecular weight, amorphous thermoplastics material based on *bisphenol A* and developed from *epoxy resin* technology. Also known as phenoxies. A linear, high molecular weight, amorphous polyester-type plastics material formed by the reaction of a dihydric phenol and *epichlorhydrin*. Commercial materials are based on *bisphenol A* and are similar in some respects to *polycarbonate* but, because of the aliphatic chain segments, this type of material has a comparatively low softening point. Such a material is acid resistant, alkali resistant and has low gas permeability. Can be melt processed (by, for example, *injection moulding*) but not used in such processes because of the superiority of the polycarbonates. Used as primer coatings because of the good adhesion between the polymer, the base and the top coat.

phenoxyethyl oleate
An ester of *oleic acid* which is a primary *plasticizer* for cellulose acetate butyrate and a secondary plasticizer for *polyvinyl chloride* (PVC).

phenyl
An aromatic organic radical with the formula C_6H_5-: a benzene ring minus one hydrogen atom. A univalent radical.

phenyl chloride
See *chlorobenzene*.

phenyl ethylene
See *styrene*.

phenyl tolyl xylyl guanidine
This solid resinous material is used as a secondary *accelerator* for rubbers. Has a relative density (RD) of 1·08 and has vulcanization properties similar to those of *di-o-tolyl-guanidine*. See *guanidines*.

phenyl-α-naphthylamine
An abbreviation used for this material is PANA. This material has a melting point of 50°C and a relative density (RD) of about 1·19. A *staining antioxidant* for rubbers. See *phenylnaphthylamines*.

phenyl-β-naphthylamine
Also known as N-phenyl-β-naphthylamine. An abbreviation used for this material is PBN: PBNA is sometimes also used. This material has a melting point of 105°C and a relative density (RD) of about 1·2. A *staining antioxidant* for rubbers which is not now widely used because of the danger of cancer (caused through the presence of β-naphthylamine). See *phenylnaphthylamines*.

phenyl-β-naphthylamine-acetone reaction product
This yellow solid material is an *antioxidant*. This material has a melting point of 120°C and a relative density (RD) of 1·17. Can cause staining in light coloured rubber compounds. See *phenylnaphthylamines*.

phenyl-o-tolyl guanidine
An abbreviation used for this material is POTG. This white solid material is used as a secondary *accelerator* for rubbers. Has a relative density (RD) of 1·10 and vulcanization properties between those of *diphenyl guanidine* and *di-o-tolyl-guanidine*. See *guanidines*.

phenylalkane
Also known as a *bisphenol* of the phenol alkane type or as, a phenol alkane bisphenol. A *bridge hindered phenol*. Obtained by linking phenolic-type structures (with methyl, ethyl etc, linkages) so as to obtain less volatile materials which have good heat stability. For example, an antioxidant which is widely used in polyolefins is bis-[2-hydroxy-5-methyl-3(1-methylcyclohexyl)phenyl] methane. Also used as an *antioxidant* is 1,1,3-tris-(4-hydroxy-2-methyl-5-t-butylphenyl)butane. See *phenolic antioxidant*.

phenylene
An aromatic organic radical with the formula -C_6H_4-: a benzene ring minus two hydrogen atoms. A bivalent, aromatic radical. When present in aromatic polymers then the ring is usually linked via the para position as this gives the best heat resistance.

phenylene polymer
An *aromatic polymer* which contains *phenylene groups* in the polymer chain. The phenylene ring is usually substituted in the para position as this gives the best heat resistance. See *polyphenylene*.

phenylethene
See *styrene*.

phenylmethylsilicone
May also be referred to as phenyl modified silicone or as methyl phenyl polysiloxane. A polymer based on alternating silicon and oxygen atoms in the main chain: the silicon atoms have two organic groups (R & R') attached. May also be known as a *polysiloxane*. The most common polymer is *polydimethylsiloxane* where R = R' = CH_3. If some of the R groups are *phenyl* groups then a phenylmethylsilicone results: such materials are copolymers which retain their low temperature flexibility to lower temperatures than the dimethyl polysiloxane. An abbreviation used for this type of material is PM. If the material also contains vinyl groups, then it may be referred to as PVM. See *polyorganosiloxane* and *silicone rubber*.

phenylnaphthylamines
Phenyl-α-naphthylamine and phenyl-β-naphthylamine. Additives, *antioxidants*, at one time widely used in rubber com-

pound. Amine antioxidants, which may be classed as *staining antioxidants*. A widely used material was phenyl-β-naphthylamine (PBN or PBNA). This material is not now widely used because of the danger of cancer (caused through the presence of β-naphthylamine).

phenylsilicone rubber
See *phenylmethylsilicone*.

phenylsulphide
Also known as a bisphenol of the phenolic sulphide type, A *bridge hindered phenol*. Obtained by linking phenolic-type structures (with linkages based on sulphur) so as to obtain materials of low volatility which have relatively good heat stability. Such a material is 4,4'-thiobis-(6-t-butyl-m-cresol). The *antioxidant* efficiency with plastics materials is improved if carbon black is also used. See *phenolic antioxidant*.

phenylurea
A *urea* derivative used as a *heat stabilizer*: a *metal-free* organic, heat stabilizer.

Phillips process
A polymerization process: a low pressure polymerization process for *high density polyethylene*.

phosgene
Also known as carbonyl chloride. This poisonous gaseous material has the formula of $COCl_2$ and a boiling point of 8°C. Used in the preparation of, for example, a *polycarbonate*.

phosphate
A chemical compound: a salt of phosphoric acid. See *phosphates*.

phosphate esters
See *phosphates*.

phosphate plasticized compound
A *polyvinyl chloride* (PVC) compound which incorporates a significant proportion of a *phosphate plasticizer*.

phosphate plasticizer
See *phosphates*.

phosphates
Also known as phosphate esters or as organic phosphates. Organic esters of phosphoric acid (H_3PO_4) which are used as plasticizers in materials such as *polyvinyl chloride* (PVC) to give fire resistance. Phosphate plasticized compounds are used to produce mine belting and in cable insulation. The use of such plasticizers also improve the speed of high frequency welding of PVC sheet, and film, but compounds containing such plasticizers have poor low temperature properties.

The classic phosphate plasticizer was the triaryl phosphate known as *tricresyl phosphate* (*TCP* or *TTP*) but, because of its high price, this material is often replaced by the cheaper *trixylyl phosphate* (*TXP*) which is petrochemical-based. Synthetic alkylated phenols are also petrochemical-based and available at a relatively stable price: such materials include the tri-isopropylphenol phosphates which are claimed to have similar properties to the older phosphates. See *iso-propyl phenols* and *mixed alkyl aryl phosphate*.

When cheap routes were developed by the petrochemical industry for the production of *phthalate plasticizers*, the importance of the phosphates declined and they are now used only when flame resistance is specified. There are also worries about the toxicity of this class of materials: the high *cold flex* temperatures which their use can impart is also a disadvantage for most applications. Mixtures of phosphates and *phthalates* are used in order to obtain desired properties.

phosphazene polymer
See *phosphonitrilic polymer*.

phosphinate polymer
A *spiro polymer* which consists of phosphinate groups -O-P(RR)-O- linked together with metal atoms: a coordination polymer. Because of the double strand structure, such materials can have good thermal stability. The metals used include zinc, iron and beryllium.

phosphite
A chemical compound which is a salt of phosphorous acid (H_3PO_3). Phosphites are used as *metal-free organic stabilizers*.

phosphite chelator
An organic *phosphite* which is typically a liquid and which is used, for example, as a *synergistic co-stabilizer* with an *epoxy compound* in a *polyvinyl chloride* (PVC) formulation. May be represented as $P(OR)_3$ where R is an organic group, for example, a long chain alkyl group and/or an aryl group. Examples include tris(nonylphenyl) phosphite, trilauryltrithiophosphite and distearyl pentaerythritol diphosphite. Such materials improve clarity when used with primary metal stabilizers by forming PVC-soluble complexes with the insoluble chlorides of the stabilizer metal. The level of use is often below 1 phr.

phosphonitrilic fluoroelastomer
See *polyphosphazene rubber*.

phosphonitrilic polymer
An inorganic polymer. Phosphorus, together with nitrogen, is the basis of a group of inorganic polymeric materials called phosphonitrilic polymers or as, polyphosphazenes or as, phosphazene polymers. The repeat unit is $-(PX_2=N-)_2-$ where X may be a halogen such as *chlorine*. In this case the polymer would be called polydichlorophosphazene. X may also be alkyl, aryl, alkoxy or aryloxy groups (or their halogenated derivatives). Such materials can make useful rubbers, see for example, *polyphosphazene rubber*. The phenol-substituted phosphonitrilic polymer has very good fire resistance.

phosphor-bronze
An alloy of copper, tin and phosphorus. Used to make bearings. Has been incorporated into thermoplastics materials, such as a *polyamide*, so that bearings can be made by *injection moulding*.

phosphorescence
Phosphorescence occurs when radiation of a short wavelength is absorbed and radiation of a longer wavelength is emitted. Fluorescence and phosphorescence are examples of *luminescence*. Unlike *fluorescence* the emission of the radiation can continue after the radiation source has been removed.

phosphorus
Group Va of the Periodic table consists of the elements nitrogen, phosphorus, arsenic, antimony and bismuth. Phosphorus does not occur naturally but phosphorus, as *phosphates*, is an essential constituent of all living matter. The element exists in several allotropic forms of which the white and red are the most common. The allotropes have different properties, for example, white has a melting point of 44°C and a relative density of 1·83 whereas the red has a melting point of approximately 600°C and a relative density of 2·2. Phosphorus, together with nitrogen, is the basis of a group of polymeric materials called *phosphonitrilic polymers*.

This element has been used with, for example, *nylon 66* so as improve flame retardancy. Red phosphorus in the presence of PA and glass fibres, is thought to form a glassy layer of polyphosphate which prevents further oxygen in-flow and stops flame propagation. Also promotes the formation of a char rather than active burning. Relatively low levels of red phosphorus provide effective flame retardancy and the properties of the matrix are hardly affected. The use of phosphorus may eliminate the need for cadmium oxide - added to stop the plate-out of oxidation products on the surfaces of injection mouldings.

photo-chemical polymerization
See *photo-polymerization*.

photo-chemical reaction
A photo-initiated chemical reaction. Light initiated reactions such as *cross-linking* that are assisted by the presence of light. See *photo-sensitizer*.

photo-crosslinking
Cross-linking induced by visible or UV light. See, for example, *cyclized rubber/azide*-type photo-sensitizer.

photo-degradation
Light initiated degradation. The light is usually *ultraviolet* as many polymers contain structures which absorb particular wavelengths in this region of the spectrum. See *photo-sensitizer*.

photo-dimerization
The joining together of two molecules using, for example, *ultraviolet radiation*. See *polyvinyl cinnamate*.

photo-elastic inspection
Inspection of transparent materials using *polarized light*. For transparent materials the easiest way of detecting moulded-in, or *frozen-in*, stresses is to examine the component through polarized light. Such light is produced when ordinary light passes through certain materials. The crystals which cause polarization are usually carried within sheets of plastic and for the purpose of moulding examination, two such sheets are required. One sheet is placed over an ordinary light source (which is polychromatic) and the second sheet is placed on top of the first. The second sheet is rotated until the light passing through the assembly is at a minimum. If the moulding is now placed between the two sheets, then a series of coloured fringes will be seen; if the number of fringes increases, from one moulding to another, then the stress level is also increasing. If monochromatic light is used, light bands on a dark background will be seen: the greater the number of bands seen, the greater is the amount of frozen-in stress or *orientation*.

photo-elastic stress analysis
See *photoelasticity*.

photo-elasticity
A property of some transparent materials: when a suitable material is stressed it becomes doubly refracting and this allows strain detection. Has been used, for example, with models made from unfilled *epoxide materials* to determine the effect of loading on components. That is photo-elastic stress *analysis* may be performed. Could also coat a model with a suitable transparent material and use that model for assessment.

photo-electric effect
The emission of electrons from a surface and which is caused by light falling on that surface. The incident light must have a frequency greater than a certain threshold frequency.

photo-induced molecular re-arrangement
A molecular re-arrangement caused by light. For example, a novolak binder resin may be made more soluble by a photo-induced molecular re-arrangment caused by naphthaquinone diazide. See *photoresist*.

photo-initiated oxidation
Also known as photo-oxidation. Light initiated *oxidation*: photo-degradation which proceeds in the presence of oxygen. See *thermally-initiated oxidation*.

photo-initiated polymerization
See *photo-polymerization*.

photo-lithography
See *photoresist*.

photo-micrograph
A photograph taken through a microscope. Used, for example, to study rubber particle deformation in *acrylonitrile-butadiene-styrene*.

photo-oxidation
See *photo-initiated oxidation*.

photo-polymerization
Also known as photochemical polymerization or as, photo-initiated polymerization. Polymerization initiated by light, usually *ultraviolet light*, which interacts with a light sensitive compound, for example, with a *photosensitizer*.

photo-response
See *photo-sensitizer*.

photo-sensitized polymerization
See *photo-sensitizer*.

photo-sensitizer
An additive which induces polymerization when exposed to light, usually *ultraviolet light*: this interacts with the light sensitive compound, for example, with benzophenone, to give free radicals. In a degradable polymer, the use of such additives can result in *photo-degradation*. Photo-sensitizers such as the nitroamines, or quinones or, aromatic amino ketones improve the photo-response of polyvinyl cinnamate for *photoresist* purposes. See *azide*.

photo-stabilizer
See *ultraviolet stabilizer*.

photogravure printing
See *gravure printing*.

photolocking
An example of a process which uses a *plasma* for electronic *photoresist* applications. A plasma degradable acrylic polymer, for example, polydichloropropyl acrylate, is mixed with a monomer such as N-vinyl carbazole and used to form a resist. *Ultraviolet* radiation is then applied to the resist through a mask which locks (photolocks) the plasma degradable acrylic polymer onto the exposed areas. The un-exposed polymer is then removed by vacuum leaving plasma sensitive areas and non-plasma sensitive areas: this allows selective development.

photon
A quantum of electromagnetic radiation.

photoresist
A polymeric material (a polymeric resist) capable of being degraded by light. A polymer, or a polymer compound which contains a polymer binder and a photo-active material, which when applied selectively to a surface undergoes a reaction when exposed to radiation (for example, visible or *ultraviolet* radiation) so as to change the solubility of the photoresist coating; the shorter the wavelength of the radiation, the better is the definition possible.

The difference in solubility of exposed and un-exposed areas allows selective removal of the exposed photoresist material so as to create an image (developing). The image is then transferred to the substrate by, for example, etching (solvent etching or plasma etching) or deposition. As such images are raised they are called relief images. Both positive image lithography and negative image lithography are used. The use of positive photoresists is increasing because of their higher resolution capability and better thermal stability for integrated circuit printing.

phr
An abbreviation used for parts per hundred of resin or, parts per hundred of rubber. A commonly used way of expressing the weight composition of a compound. Sometimes pphr is used in place of phr.

phthalate
See *phthalic acid ester* and *phthalates*.

phthalate esters
See *phthalates*.

phthalate-based material
An *allyl moulding material* based on *diallyl phthalate*.

phthalates
A group of plasticizers widely used with, for example, *polyvinyl chloride* (PVC). Sometimes they are referred to as *monomeric plasticizers*. These materials are based on the reaction of *phthalic anhydride* and an alcohol. Both branched and linear alcohols may be used as may mixtures of alcohols. Phthalates are the workhorses of the plasticizer industry.

Phthalates which contain 8 carbon atoms are the most commercially important *plasticizers* and dominate the plasticized PVC market. Included in this category are *dialphanyl phthalate (DAP)*, *dioctyl phthalate (DOP)* and *di-iso-octyl phthalate (DIOP)*. DOP is the yardstick by which the other plasticizers are judged but for economic reasons DIOP or DAP may be preferred. DIOP has less odour and is more permanent but DAP has better heat stability.

In general, this group of plasticizers may be divided into branched chain, dialkyl phthalates and into phthalates based on linear alcohols. Branched chain, dialkyl phthalates include dioctyl phthalate (DOP), dioctyl terephthalate (DOTP) di-iso-octyl phthalate (DIOP), di-iso-nonyl phthalate (DINP) and di-iso-decyl phthalate (DIDP). Phthalates based on linear alcohols include, octyl decyl phthalate (ODP), di-C(6-8-10) phthalate and di-C(7-9-11) phthalate. These linear phthalates have good low temperature properties, low volatility and are now relatively cheap.

In the phthalate series of plasticizers, it is generally found that increasing the length of the side chain will decrease volatility, water extraction, the relative density of the plasticizer, and plasticizer efficiency. Processing will also be more difficult and the ease of extraction by oils will be easier. Extraction by oils will be easier if a branched plasticizer is replaced by a linear one. So, for the best heat resistance use a higher phthalate plasticizer, for example, one with long side chains. Such a material may, however, have a higher viscosity than a shorter chain ester. One such material is di-tridecyl phthalate another is di-isodecyl phthalate. Ditridecyl phthalate has the best heat resistance but both are used in high temperature cable insulation. Because of their greater hydrocarbon nature, both have better water extraction resistance than shorter chain esters.

For the best cold flex resistance, using a phthalate plasticizer, use the linear phthalate, *octyl decyl phthalate*. Aliphatic diesters, such as dioctyl adipate, will give a much lower value but may give an unacceptable level of volatility and oil extractability. Mixtures of aliphatic diesters and phthalates are therefore used to get a desired balance of properties.

Where long term retention of properties, after exposure to high and low temperatures, is a requirement then phthalates based on a linear alcohol will perform better than a phthalate based on a branched alcohol. For example, the linear phthalate, *octyl decyl phthalate* will retain properties better than branched chain alcohols such as *di-iso-octyl phthalate (DIOP)* and *di-iso-decyl phthalate (DIDP)*.

phthalic acid
See *phthalic acids*.

phthalic acid dibutyl ester
See *dibutyl phthalate*.

phthalic acid diethyl ester
See *diethyl phthalate*.

phthalic acid ester
The reaction product of phthalic anhydride and an alcohol and which is commonly called a *phthalate*. The alcohol may be linear or branched and/or mixed alcohols may be used. A wide variety of phthalates are made for use as plasticizers for materials such as *polyvinyl chloride* (PVC). See, for example, *dioctyl phthalate*, *di-iso-octyl phthalate* and *dialphanyl phthalate*. Also see *phthalates*.

phthalic acids
Aromatic dicarboxylic acids which may be represented as $C_6H_4(COOH)_2$. There are three phthalic acids. i.e. three isomers. These are, o-phthalic acid, m-phthalic acid and p-phthalic acid.

phthalic anhydride
The anhydride of *o-phthalic acid*. This white solid material has a melting point of 131°C and a relative density (RD) of 1·52. May be prepared from, for example, anthracene. Widely used in the polymer industry to prepare, for example, *plasticizers*. Has been used as a retarder in some rubber compounds but does not function as a retarder in sulphurless curing systems.

phthalo blue
Phthalocyanine blue. See *copper phthalocyanine blue*.

phthalo blue pigments
See *phthalocyanine pigments*.

phthalo blues
See *phthalocyanine pigments*.

phthalo green
See *phthalocyanine green*.

phthalocyanine blue
See *copper phthalocyanine blue*.

phthalocyanine green
A green pigment which is obtained from phthalocyanine blue by replacing hydrogen atoms on the isoindole rings with chlorine atoms. This pigment has good hiding power and reasonable chemical resistance.

phthalocyanine pigments
Produced from *phthalic anhydride*, *urea* and *cuprous chloride*. *Copper phthalocyanine blue* is the major pigment but can also have green pigments. A class or type of *organic pigment* used, for example, to make blues or greens. Such pigments are popular because, in general, they are moderately-priced, lightfast colouring systems with good resistance to heat and chemicals (such as solvents). Have good colouring strength and brightness but may cause problems with some plastics materials. For example, may inhibit the cure of an *unsaturated polyester resin*. Can cause *environmental stress cracking* (ESC) in polyolefin plastics (such as *high density polyethylene*); may also introduce problems related to shrinkage, such as warping. Can give good transparency: if very well dispersed can get transparency even in acrylics.

phthalocyanines
See *phthalocyanine pigments*.

physical blowing agent
See *blowing agent* and *expanded polystyrene*.

physical cross-link
Also known as a virtual cross-link. A tie or restraint between polymer chains which is not a chemical cross-link. Such

cross-links only exist below the softening point, or melting point, of the polymer and are therefore *thermo-labile cross-links*. Could be formed by chain entanglements and/or crystallization. A term used to describe the effect that the hard plastics phase has on the properties of a *thermoplastic elastomer*.

physical property determinations
Sometimes referred to as mechanical tests. See *common tests*.

PI
An abbreviation used for *polyimide(s)*.

PIC
An abbreviation used for the Pultrusion Industry Council - a USA based organization.

pick-and-place automation
Also referred to as take-and-put automation. Usually applied to *injection moulding* and means that a machine-mounted take-off arm is used for moulding removal. Although such systems are referred to as 'robots' within the moulding industry, this terminology is not strictly correct as such devices are dedicated to the removal of mouldings from the mould (unlike true robots which can be programmed to do a multitude of tasks). Such devices remove the moulding from the mould and place it in another position. The articulated arm is often mounted above one of the machine platens on a rail or slide. The arm is capable of being driven across the machine-mounted rail, pivoting into the mould-open space, gripping or holding the moulded components, removing the mouldings from between the mould halves and then depositing the components onto a conveyor or into a container. A pick-and-place unit may therefore replace an operator who is running a machine on a semi-automatic cycle. The usefulness of these devices may be extended if they are also employed to cut or trim the components. Such devices are mainly pneumatically operated, or driven, as long stroke lengths can be economically obtained from industrially proven units. See *work handling device*.

pick-and-place unit
See *work handling device*.

picking up
Screw terminology and which refers to the *screw bending* under high loads approximately halfway along its length. A frequent cause of premature *barrel wear*, screw wear and screw breakage. Often caused by the use of too tight a screw clearance and therefore the generation of excessive loads: the screw whirls as it rotates. See *dragging*.

PICM
An abbreviation used for *4,4'-dicyclohexylmethane diisocyanate*. Also used is *MDI*.

pico
An abbreviation used for this term is p: it means 10^{-12}. See *prefixes - SI*.

picture frame
The centre plate of a *frame mould*: the plate which has a hole, or holes, cut into the plate.

piezoelectric effect
A property of some crystalline materials, for example, quartz. When such a material is subjected to a pressure then the crystal gets smaller: positive and negative electric charges build up on opposite faces of the crystal. If the crystal is subjected to a tensile force, the sign of these charges is reversed. If the crystal is subjected to an electric potential, then an inverse piezoelectric effect is obtained. That is, an alteration in crystal size occurs. Some polymers display this effect, for example, *polyhydroxybutyrate* and *polyvinylidene fluoride*.

piezoelectric oxide semiconductor field effect transistor
An abbreviation used for this term is POSFET. A device, based on for example, *polyvinylidene fluoride* (PVDF), in which the electric signal (See *piezoelectricity*) from the PVDF appears directly on the gate of a metal oxide semiconductor (MOS) transistor.

piezoelectric polymer
A polymer which can display a *piezoelectric effect*. See, for example, *polyvinylidene fluoride*.

piezoelectric pressure transducer
See *pressure transducer*.

piezoelectricity
Electricity developed as a result of the *piezoelectric effect*. An electric polarization that occurs in certain crystals when they are mechanically deformed: the polarization is proportional to the deformation and the polarity changes with a change in deformation. A piezoelectric material also possesses pyroelectric properties, that is, electric polarization is generated when the temperature changes.

pig
See *dolly*.

pig iron
Impure iron produced from a blast furnace.

pigment
A *colorant* which is insoluble in the polymer and so, in general, produces opaque colours: however, pigments vary widely in their *hiding power*. Two major classes of pigment are *inorganic* and *organic pigments*. Titanium dioxide is the most important inorganic pigment and *carbon black* is the most important *organic pigment*. Roughly three quarters of all pigments used are of the inorganic type. Pigments may be grouped or divided in various ways, for example, into *titanium dioxide, carbon black, zinc sulphide/lithopone*, other inorganic pigments and into *organic pigments*.

pigment dye
Colorant systems which are insoluble in the aqueous medium used for the *diazotization reaction*. Example include toluidine reds, para reds and azo yellows. In general, such pigments have poor bleed resistance but good acid and alkali resistance.

pigment green
An organic green pigment the composition of which depends upon the particular pigment under discussion: a Color Index classification.

pigment red
An organic red pigment the composition of which depends upon the particular pigment under discussion: a Color Index classification.

pigment yellow
An organic yellow pigment the composition of which depends upon the particular pigment under discussion: a Color Index classification.

pilot cavity
A sample cavity: a moulding cavity constructed before the production mould is manufactured so as to determine cavity dimensions, for example, for *shrinkage* purposes. It may also be possible to predict the process capability, or the natural tolerance limits, for that particular machine using simple statistics.

pilot control
A term used in *hydraulics* and which refers to a means of operating valves by using a small, control pressure signal.

pilot pressure
The pressure in the *pilot control* circuit: the pressure used for control.

pilot ring
A component of a *ring check valve*. An additional ring which increases the effective area of the *ring check valve* exposed to the plastics material and thus increases speed of valve closure and consistency of operation.

pilot valve
A term used in *hydraulics* and which refers to a small valve used to control another larger valve: the controlling stage of a two-stage valve.

pimple
An imperfection described as a small, protuberance of varied shape on the surface of a plastic product.

pin ejection
A system of ejection fitted to, for example, an *injection mould*. Such a system is extremely popular as it is cheap and relatively trouble-free. A common way of actuating this type of ejector, is to utilise the opening stroke of the machine. Knockout rods on the moulding machine cause the mould's ejector system to advance. This system consists of an ejector plate, on which a series of ejector pins are positioned, held in place by an ejector retainer plate. As the system moves forward, the pins contact the parts and the *feed system* at pre-designed points: this action either lifts the moulding free (so that they can be removed) or knocks them free to drop down and out of the parted mould. Many *injection moulding machines* are now fitted with a *hydraulic ejection system* - used to actuate the ejector system of the mould.

pin gate
See *pin-point gate*.

pin impression method
See *ball or pin impression method*.

pin-fibrillated polypropylene fibre
A *polypropylene fibre* produced from an extruded sheet which has a rough edge: the rough edge is used to improve the bonding to cement. Produced from tape by the action of rotating pins and used, for example, to replace *asbestos* in cement-based composites.

pin-point gate
Also known as a pin gate. A small, circular gate commonly used for *three-plate gating*. The diameter of the gate usually lies within the range 0·5 to 2 mm (0·02 to 0·08 in); above 2 mm dia the gate may be difficult to break during mould opening. The gate should be designed so that when it is broken it breaks cleanly and does not block the gate cavity. For example, a parallel gate land should not be used; the gate should taper or flare into the cavity.

pin-point submarine gate
A small diameter *submarine gate*.

pinch cutters
A sealing/cutting device which is usually mounted directly below the die head in *extrusion blow moulding*. When the desired length of parison has been produced the cutters close and, in so doing, they cut and seal the *parison*.

pinch draw rollers
A set of rollers used as pullers in the *extrusion process*: one form of puller unit.

pinch off
That part of a *blow mould* which welds the *parison* so that inflation, or trimming, may occur.

pinching jaws
See *pinch cutters*.

pine oil
See *turpentine oil*.

pine tar
Also known as soft wood tar. A *softener* used in rubber compounds: contains turpentine and is the residue of the dry distillation of wood. Can improve *tack* and also aids filler dispersion, for example, of *carbon black*. Has a relative density of 1·03 to 1·09. The level of use can reach 7 phr.

pine trees
See *crows feet*.

pine-wood oil
See *turpentine oil*.

pinene
A liquid terpene ($C_{10}H_{16}$) which is found in turpentine. Used to make *camphor* and *pinene resin*.

pinene resin
A resin made from the aromatic material *pinene* by cationic polymerization. These terpene resins are mainly used as adhesives to improve tack. α-pinene resin is made from α-pinene and β-pinene resin is made from β-pinene.

pinholes
Irregularly spaced small holes in a sheet and associated with the *calendering* of *polyvinyl chloride* (PVC). Product marking caused by, for example, poor gelation and thermal homogeneity of the PVC. May be detected by laser-scan inspection during *calendering*.

pint
An abbreviation used is pt. One eighth of a *gallon*. A unit of measurement (usually for liquid measurements but may also be for dry measurement). As an imperial gallon and a US gallon are not the same size, units derived from them are also not the same size. An *imperial pint* is equivalent to 568·26 cc. In US measure a pint may be of two values. In US Customary Liquid Measure, it is 16 fluid ounces or, 28·7875 cubic inches or, 0·473 litres. In US Customary Dry Measure, it is 33·6 cubic inches or, 0·551 litres.

piperazine
Also known as hexahydropyrazine. A heterocyclic, aromatic single-ring compound with the formula $C_4H_8(NH)_2$. This colourless material has a melting point of 110°C and a boiling point of 145°C. Has been used to make a type of *nylon* known as *piperazine polyamide*.

piperazine polyamide
A type of nylon based on *piperazine*. Heat resistant polyamides which are more water sensitive than most other *polyamides*.

piperidine
This colourless liquid material has a boiling point of 106°C. A heterocyclic ring compound with the formula of $C_5H_{10}NH$. Has been used as a solvent but will also cure *epoxide resins* with comparatively low exotherms.

piperidine pentamethylene dithiocarbamate
Also known as piperidine-1-piperidine carbodithionate or as, N-pentamethylene ammonium pentamethylene dithiocarbamate. An abbreviation used for this material is PPD. This yellowish white material has a melting point of approximately 170°C and a relative density (RD) of 1·14. An ultra-fast *accelerator* for low temperature vulcanization when used with *zinc oxide*.

piperidine-1-piperidine carbodithionate
See *piperidine pentamethylene dithiocarbamate*.

piperylene
Also known as 1,3-pentadiene. A *methyl substituted butadiene*: a methyl group is substituted for a hydrogen atom on the butadiene. See *petroleum resins*.

piston
A cylindrically shaped component which lies within a cylinder and which transmits or receives motion through a connecting, or piston, rod. The component which lies within a cylinder and on which the fluid acts to generate motion.

piston motor
A motor which uses pistons to generate motion from fluid flow.

piston pump
A pump which uses pistons to generate fluid flow.

piston reaction injection moulding machine
A type of *reaction injection moulding* machine (RIM machine) which uses a simple stroke displacement pump for each of the reactants: the high pressure seals, used for preventing the escape of reactants, are fixed to the reciprocating piston. This type of RIM machine contains less dead volume than a *lance piston* reaction injection moulding machine.

piston rod
See *piston*.

pit
An imperfection; a small crater in the surface of the plastics product and with a width of approximately the same order of magnitude as its depth.

pitch
Black solid materials that become tar-like materials on heating. Obtained by distillation of various organic substances such as *coal tar*, *palm oil* and *petroleum*. A complex mixture of hydrocarbons. Petroleum pitches increase the hardness and stiffness of rubber compounds: they also give low resilience and high heat build-up in service. See *pitch-based carbon fibre*.

pitch-based carbon fibre
This type of *carbon fibre* may be classed into isotropic pitch-based carbon fibres or, into mesophase pitch-based carbon fibres. It is the *mesophase pitch-based carbon fibres* which give fibres with very high modulus.

pitch - of a screw
The horizontal distance between corresponding points of two successive lands on a *screw*.

Pittsburgh University Test
A test used to assess thermal decomposition product toxicity. The test employs a simple pass/fail assessment which specifies that acceptable or approved products will not be more toxic than wood when tested in the same manner. The test involves the use of animals and is expensive to perform.

pk
An abbreviation used for *peck*.

PK
An abbreviation suggested for the plastics material known as a *polyketone*. See *polyether ether ketone*.

plain end
A soft hose end in which the internal diameter at the end is the same as that of the main body of the hose.

plane angle unit
See *radian* and *Système International d'Unité*.

planetary extruder
A *multi-screw extruder*. A number of small screws surround a larger central screw for part of its length.

planetary gear extruder mixer
A compounding extruder: the *torpedo* is machined with fluted sections (like a gear wheel) and these flutes engage with other pinions or gears. A modified, single-screw extruder which is probably the best all round, single screw machine for mixing (see *mixing section*). The excellent heat transfer obtained, because of the large surface areas in contact with the melt, make them well suited for processing heat sensitive materials, such as *unplasticized polyvinyl chloride* (UPVC). Both good *distributive mixing* and good *dispersion mixing* is obtained.

plantation
An *estate* whose size is greater than, for example, 40 ha. See *natural rubber*.

plantation rubber
Natural rubber which originates from trees (*Hevea brasiliensis*) which are cultivated on a *plantation*. In the latter half of the nineteenth century there was an enormous increase in the demand for rubber. This demand was met by setting up plantations in, for example, Malaya (Malaysia), Indonesia, West Africa and the Congo. The demand was due to the invention of the pneumatic tyre and the widespread use of rubber compounds for electrical insulation purposes.

plasma
A gas which contains free electrons, ions and neutral particles and which is formed through the application of external excitation. Plasma etching is a dry etching process. A plasma is used in place of *ultraviolet* or visible light because of the finer definition possible in electronic *photoresist* applications. A plasma developable resist does not involve solvents, that is, the process is dry and this reduces the problems of image swelling and resolution loss. An example of a process which uses a plasma is *photolocking*.

plasma polymerization
A polymerization technique performed within a *plasma*. Plasma polymerization may be used to coat *blow mouldings* with relatively impermeable polymers, for example, with high molecular weight, pore-free, crosslinked, *polyethylene*.

plaster of Paris
Also known as calcium sulphate hexahydrate. When mixed with water it sets and hardens. As it is a liquid when first mixed with water, it may be used to make castings or mouldings. This material has a relative density of 2.32.

plastic
See *plastics material*.

plastic coated fabric
A fabric which has been coated with a plastics material: the two are bonded together, for example, *leathercloth*.

plastic coated paper
A paper which has been coated with a plastics material: the two are bonded together.

plastic deformation
See *plastic flow*.

plastic faced plaster
An abbreviation used for this material is PFP. A type of *soft tooling* made from plaster casts which have been subsequently treated, for example, with a resin solution.

plastic film
A thin sheet material based on a *thermoplastic material*. See *tubular film process* and *flat film process*.

plastic flow
In rubber technology, another term for *plasticity*. More generally, the term refers to the deformation that occurs after yielding, for example, in a tensile test. Such post-yield deformation is wholly or partially recoverable with polymeric systems unlike metals.

plastic foam
See *cellular plastic* (the preferred terminology).

plastic material
Usually means a *plastics material*.

plastic nets
See *extruded net*.

plastic pipe
A hollow cylinder made of a plastics material in which the wall thicknesses are usually small when compared to the diameter and in which the inside and outside walls are essentially concentric.

plastic sulphur
See *polymeric sulphur*.

plastic tubing
A particular size of *plastic pipe* in which the outside diameter is essentially the same as the corresponding size of copper tubing or, small diameter flexible pipe.

plastic yield
Also called non-elastic deformation. The deformation which remains when the deforming force is removed.

plastication
Also called plastification. The process of plasticizing a material i.e. turning a plastic material into a melt. See *heat softening*.

plasticised polyvinyl chloride
See *plasticized polyvinyl chloride*.

plasticiser
See *plasticizer*.

plasticity
The susceptibility of a rubber compound to deformation and the retention of that deformation. The ability of a material to be shaped by the application of a stress and to retain that shape when the stress is removed. In rubber technology, the term is used as a measure of the viscosity of unvulcanized rubber or of unvulcanized rubber compounds. The degree of flow which occurs under given conditions of temperature and pressure. Plasticity measurements are made in a *plastimeter* which gives the *plasticity number* of a rubbery material.

plasticity number
The *plasticity number* of a rubbery material is a test result obtained from a *plastimeter*. See *plasticity retention index*.

plasticity retention index
An abbreviation used for this term is PRI. Measured by the ratio of P_{30}/P_0 where P_0 is the initial *Wallace plasticity* and P_{30} is the Wallace plasticity after ageing for 30 m at 140°C. Usually expressed as a percentage and is a measure of, for example, the resistance of raw *natural rubber to oxidation*.

plasticization
An increase in softness and flexibility. Could be brought about by the use of a *plasticizer* but, sometimes the term means that the softening occurred as a result of heating, that is, thermal plasticization. See *heat softening*.

plasticized
In extrusion technology means that a plastics material has been turned into a melt; in *polyvinyl chloride* (PVC) technology can also mean that the resin has been made softer by the incorporation of a plasticizer.

plasticized compound
A compound which has been made relatively soft by the incorporation of a *plasticizer*.

plasticized polyvinyl chloride
Also known as plasticised polyvinyl chloride or, polyvinyl chloride-plasticized or, plasticized polychloroethene. An abbreviation used for this material is PPVC or PVC-P. A *polyvinyl chloride* (PVC) compound which contains a *plasticizer*.

Plasticizers are used in PVC compounds in order to confer flexibility, softness and ease of processing. The plasticizers most commonly employed are high boiling point esters of C_{8-10} alcohols: such as phthalates, phosphates and sebacates. Examples of common plasticizers are *dioctyl phthalate (DOP)*, *di-iso-octyl phthalates (DIOP)* and *dialphanyl phthalate (DAP)*.

The discovery that plasticized PVC could be converted to an elastomeric-type material, by the addition of low molecular weight plasticizers, was made many years ago (by Dr Waldo Semon in the 1920s) when attempts were made to dissolve the polymer. Polymeric plasticizers were used approximately 10 years later and the use of rubbers, as property modifiers, was introduced shortly afterwards.

As there are many different plasticizers, which may be used in different amounts and/or in combination, then a wide range of plasticized compounds is possible. Such compounds will differ not only in flexibility and softness, but will also differ in other respects, for example, tensile strength, resilience and ease of flow. These latter properties can also be affected by the combination of different plasticizers and of different molecular weight resins.

The term 'plasticized' (also spelt 'plasticised') simply means that there is plasticizer (plasticiser) present in the compound as well as other additives - see *unplasticized polyvinyl chloride* (UPVC). PPVC is a much easier flowing material than UPVC. To realize the potential of this material it is necessary to put in just sufficient work and heat so that the blend is fully gelled, or fused, but not so much that it is degraded.

In general, the chemical resistance of PPVC is good, for example, it is resistant to water and salt solutions; dilute acids and alkalis have little effect at room temperature but, at elevated temperatures, some hydrolysis and extraction of the plasticizer may occur. Concentrated acids and alkalis hydrolyse plasticizers slowly when cold but more rapidly when heated. Most organic liquids will extract plasticizers and cause compound hardening.

As the natural colour of the plasticized material can be clear a wide colour range is possible - both transparent and opaque. With PPVC, only pigments should be used because of colour bleeding, or leaching problems, with dyestuffs.

PPVC sheet is produced with an excellent surface finish and/or with remarkable transparency, by for example, *calendering*: such sheet can be rapidly, and strongly, welded into large and complex shapes by the technique known as high, or radio, frequency welding. Typical injection moulded components include washers, grommets, electrical cable ends, footwear, heel tags, watch straps, electrical shields, plugs, automobile arm rests, knobs, metal reinforced steering wheels and components associated with the medical industry.

plasticizer
A material added to a polymer system in order to improve processing behaviour and/or to improve compound flexibility.

Camphor is best known in the plastics industry as a plasticizer for cellulose nitrate and was first used as such by John and Isaiah Hyatt in approximately 1868. Now, it is *polyvinyl chloride* (PVC) which is associated with plasticizer use although, due to the growing use of *unplasticized polyvinyl chloride* (UPVC), the relative importance of *plasticized polyvinyl chloride* (PPVC) has decreased in recent years. However, it is probable that approximately 80% of all materials produced as plasticizers are still used with PVC. A plasticizer is usually an organic non-volatile liquid with a high boiling point: such materials usually have a relatively high molecular weight (approximately 300). Esters are widely used as such materials are relatively easy to produce and are compatible with the polar PVC.

Plasticizers were first added to plastics materials in order to allow the plastics material to be processed at temperatures below the decomposition temperature of the polymer. Now, *processing aids* are used to ease processing. Most commercial plasticizers are based on *phthalates* and include *dioctyl phthalate (DOP)* and *di-iso-octyl phthalate (DIOP)*. By adding such materials to PVC, plasticized PVC (PPVC) is produced: in general, this material flows much more easily than the unplasticized material (UPVC) and is softer and more flexible. Initially the addition of small amounts of plasticizer may result in a more brittle material (see *critical concentration*) but further additions of plasticizer will produce a progressively softer material.

Plasticizers may be grouped in various ways, for example, into *adipates*, *azelates*, *chlorinated hydrocarbons*, *phosphates*, *phthalates* and *sebacates*. May also be classed as *primary plasticizers*, secondary plasticizers and as plasticizer extenders. See *general purpose plasticizer*.

plasticizer alcohol
An alcohol used to make a *plasticizer*, for example, a *phthalic acid ester*. Such an alcohol usually contains from 6 to 11 carbon atoms and these relatively cheap materials are commonly esterified with *phthalic anhydride* to make plasticizers.

plasticizer - comparison
Plasticizers are usually compared against a well known plasticizer such as *dioctyl phthalate (DOP)*. A particular property is chosen (modulus or hardness) and then the new plasticizer and DOP are used to make separate compounds which, for example, have the same value. For example, to give a modulus of 1,500 lbf/in^{-2} at 100% elongation (see *efficiency proportion*). The ratio of the plasticizer concentrations (test material to DOP) may be designated the plasticizer efficiency.

Hardness is often compared on the basis of using 50 parts per hundred of resin (phr). The test plasticizer is turned into a compound at this concentration and DOP is turned into another compound. All other ingredients are kept the same and the two compounds are then compared. Such DOP compounds typically have Shore hardnesses (A scale) of 80.

Other relevant properties are determined by appropriate tests. These may include gelation rate, plasticizer loss by volatilization (oven heating), loss by extraction (water, oil, hexane and iso-octane), loss by migration into other materials and low temperature flexibility tests.

plasticizer efficiency
See *plasticizer - comparison*.

plasticizer extender
See *extender*.

plasticizing unit
A unit of a machine (for example, an *injection moulding machine*) which produces a polymer *melt*.

plastics calender
A *calender* used for plastics materials, usually *polyvinyl chloride* (PVC) compounds.

plastics copolymer
See *copolymeral* and *abbreviations*.

plastics masterbatch
See *masterbatch*.

plastics material
A material that contains as an essential ingredient one or more high molecular weight polymers which is solid at room temperature and which, at some stage, is capable of being shaped by flow.

All plastics are polymers but not all polymers are plastics. For example, cellulose is a *polymer* but it cannot be processed like a plastics material unless it is modified. A plastics material is therefore a polymer, which is capable of being shaped or moulded under, for example, conditions of moderate temperature and pressure: distinguished from a rubber/elastomer, by having a higher stiffness/modulus and a lack of reversible elasticity. There are two main categories of plastic and these are *thermoplastics* and *thermosetting plastics (thermosets)*. In terms of tonnage thermoplastics are by far and away the most important. Some desirable properties of plastics (which are not possessed by any one polymer) include physical strength, resilience, corrosion resistance, elasticity, electrical insulation, wide colour range, thermal insulation, lightness in weight, chemical resistance and mouldability. Plastics materials are therefore versatile and their use, for example, may simplify production due to the ease with which several discrete components may be incorporated into one component by, for example, *injection moulding*. Neither the materials, the moulds, nor the machines are cheap but the products of the plastics industry are available at an economic price due to such processes.

Plastics materials are therefore widely used because they can be turned into complex components, or shapes, relatively easily at an economic price: they also have a useful combination of properties which can be amended or altered within wide limits.

Many new markets are created for plastics materials by replacing metal components with thermoplastics mouldings. This is because of the ease with which complex components may be produced from polymer compositions by processes such as *injection moulding*. Compared to metals, such materials lack strength, stiffness, temperature resistance, fire and flame resistance; not only do plastics material burn relatively easily but they evolve large quantities of smoke and fumes when they do burn. Many plastics components, such as mouldings, will creep or change their dimensions if subjected to relatively small loads for prolonged periods of time. Environmental stress cracking (ESC), changes of dimensions with humidity and sudden change from tough to brittle behaviour are other factors which should be investigated before, for example, materials such as thermoplastics are used.

plastics melt processes
See *thermoplastics melt processes*.

plastics paste
Plastics material in a *paste* form can be used to coat a substrate (for example, a metal) and/or to make mouldings. See *plastisol*.

plastics powder
A plastics material in a powder form or, a plastics compound in a powder form. A plastics compound in *powder* form can be used to coat a substrate (e.g. a metal) and/or to make mouldings (see *powder moulding*). However because of heat generation, the reduction of the compound to a powder can be difficult.

Plastics Bottle Institution
An abbreviation used for this USA-based organization is PBI: the PBI issues codes as an aid to sorting before recycling. See *container identification*.

Plastics Recycling Development
An English translation of *Entwicklungsgesellschaft für die Wiederverwertung von Kunstoffen*. An abbreviation used for this German-based organization is EWvK.

Plastics and Rubber Institute
See *Institute of Materials*.

plastification
See *plastication*.

Plastificator shear-cone compounder
See *shear-cone compounder*.

plastigel
A type of *plastisol*: a very high viscosity plastisol. A type of *polyvinyl chloride* (PVC) paste which incorporates a filler such as *fumed silica* so that a putty-like material results. The material may be hand shaped before gelling.

plastimeter
Also known as a plastometer. A test instrument used to determine the *plasticity* of a rubber compound. This type of machine is used in rubber technology to obtain a measure of the viscosity of unvulcanized rubber or of unvulcanized rubber compounds. The results are empirical.

Many different plastimeters have been developed and of these the *parallel plate type* (for example, *Wallace rapid plastimeter*) is probably the best known. With this type of machine, the change in thickness of a test piece when loaded between two heated parallel plates is measured. A test piece, a cylinder of rubber, is compressed between a plate and the foot or bed of the instrument: this causes plastic deformation. The depth or thickness, in say 0·01 mm stages, is taken as a measure of the plasticity of the material. Other plastimeters of this type include Goodrich, Hoekstra, Karrer, Scott and Williams.

As plasticity is dependent on shear rate, the results obtained in a low shear rate test may not correlate with those obtained in higher shear rate tests. For this reason other types of plastimeters have been developed. Extrusion plastimeters measure the time taken for a test sample to flow through a die under, for example, constant pressure. Examples of this type of machine include Marzetti, Behre quick plastimeter, Nellen, and Firestone extrusion plastimeter. With some machines, it is the shearing force which is measured: for example, the Mooney viscometer.

plastisol
A liquid suspension of a finely divided *polyvinyl chloride* (PVC) polymer or copolymer in a *plasticizer*. The polymer does not dissolve appreciably in the plasticizer at room temperature, but does at elevated temperatures so as to form a homogeneous plastic mass (plasticized polymer). The plastisol gels, or fuses, at higher temperatures by plasticizer adsorption.

plastomer
This term is seen in books which have been translated into English and usually means a *thermoplastics material*.

plastometer
See *plastimeter* and *Pusey and Jones plastometer*.

plate die
A simple *die* made by machining a hole in a metal plate and bolting that piece of metal onto the die holder. Such a die is not streamlined.

plate mark
Any imperfection in a pressed plastic sheet resulting from an imperfection on the surface of the pressing plate.

plate out
The transport of additives to the surface of a product by, for example, an *external lubricant*. Such a movement can occur during the high shear conditions experienced during processing, for example, during *calendering* and *extrusion*. Plate out can result in a loss of clarity in transparent products.

plate welding
Sometimes called shoe welding. See *oscillating plate welding*.

plate-like
Also called scaly and platy. Referring to a particle shape of a *filler* such as *talc*. See *plate-like filler*.

plate-like filler
A filler whose shape resembles that of a plate: consists of plate-like particles. Examples of such fillers include *mica* and *polyamide* (PA) *flakes*: the plate-like filler particles can become oriented as a result of processing. See *permeability*.

plateau effect
See *flat curing*.

plateau zone
See *rubbery state*.

platen
Part of a press: large, thick metal plates which carry the mould and which are usually capable of being heated and which can be moved one towards the other by, for example, a hydraulic ram. Sometimes the platens are also capable of being both heated and cooled.

platen dimensions
The length, breadth and thickness quoted in either inches or millimetres, for example, for an *injection moulding machine*. The height and width of the platen are normally specified, but of equal importance is the thickness. It is important because if it is too thin then platen deflection may occur during the moulding operation. The dimensions of the centring hole in the stationary platen are also important in this respect as too large a centring hole can cause unnecessary platen deflection.

platinum
A hard silvery white metal with a relative density of 21·4. Used as a catalyst, for example, for the production of certain monomers and polymers.

platinum melter
Also called a bushing. See *glass fibre*.

platinum resistance thermometer
A device which measure temperature changes by using the fact that the electrical resistance of platinum increases with an increase in temperature. Such devices may be used to construct a *pyrometer*.

platy
Meaning *plate-like*.

PLC
An abbreviation used for *parison length control*.

Pliofilm
See *rubber hydrochloride*.

plucking
Also called tracking. Associated with the *calendering* of *polyvinyl chloride* (PVC). Product marking caused by, for example, poor release of the sheet from the roll aggravated by under-lubrication of the PVC. The same term is applied to an extrusion fault which shows as intervals along the extrudate; it looks as if a long needle has been stitched along the extrudate.

plug
A male shape or mould. In *GRP processes* it means a *male mould* (see *plug - hand lay-up*). In *thermoforming processes*, it means a male shape which approximates to the cavity shape but is slightly smaller than the cavity: used for *prestretching* the heated sheet. A plug is also a metal screw used to block a *flow-way*.

plug assist forming
A female forming process. A *thermoforming process* which uses a *female mould* and which offers a way of producing components of relatively deep draw cheaply: such components can have uniform wall thickness distribution although plug marking can be a problem. The sheet is heated by an infra-red heater (mounted approximately 150 mm above the sheet) while being held in a clamping frame. At *forming tem-*

perature, the heater is removed and a (heated) *plug* deforms the sheet into the cavity of the *female mould*. A vacuum is applied to the space between the sheet and the female mould: this completes the forming operation and the plug is removed. The product is then cooled by mould contact, or by blowing air across the forming, and ejected. See *air slip forming*.

plug flow
Flow through a *die* in which the central core of the melt moves as one piece. That is, the core has zero velocity gradient. Associated with unplasticized *polyvinyl chloride* (PVC) compounds.

plug - hand lay-up
A *male mould* used to produce the (production) female mould in *hand lay-up*. A male plug is used because it is generally easier to produce from simple materials such as *plaster of Paris*, wire netting and wood. The final stages of plug production often require hand finishing, for example, sandpapering and polishing, to get the smooth finish required. The porous surface is then sealed with a resin dope, for example, based on French polish (shellac in methylated spirits), polished with a wax polish and then treated with a release agent which allows the *unsaturated polyester (UP) resin* to wet out on the plug surface. Such a release agent is *polyvinyl alcohol* - this is often applied, by brush, from a water/methylated spirits mixture so as to get rapid drying. A dye may be added to the solution so that un-treated surfaces may be readily seen.

plugged braking
An emergency machine braking system. Such a system is part of the safety control devices on, for example, a *two-roll mill*. When actuated by, for example, a *safety bar*, the drive to the rolls is thrown momentarily into reverse so as to immediately stop the rolls. Ensure that the rolls do not rotate in reverse when plugged braking is applied as this can be very dangerous due to a lack of guarding under the mill. Training and a responsible attitude towards safe working are very important safety considerations as undue reliance should not be put upon guards.

plugging
The stoppage of the *latex* flow from the tree *Hevea brasiliensis* by *auto-coagulation*.

plumbago
See *graphite*.

plunger
See *ram*.

plunger injection moulding machine
A type of *injection moulding machine*: a moulding machine which uses a ram or plunger to displace the material from the *barrel*. Despite the relative inefficiency of such moulding machines, they are still made as they are much cheaper than a *screw-type moulding machine*.

plunger nozzle
A *nozzle* used for *sprueless moulding*: a hydraulically controlled needle-seal nozzle. For example, if a single-impression mould is being used then sprueless mouldings can be produced if the needle, which forms the seal, is extended so that it protrudes beyond the tip of the nozzle and into the *sprue bush*. The advantage of such a system is that the gate size can be varied and it is possible to open and close the needle at whatever is considered to be the most advantageous time. This type of nozzle is also used on *insulated runner moulds* in order to prevent material *drooling*.

plunger type dial test indicator
A type of *dial test indicator* which uses a rack and pinion, followed by a gear train, to magnify the movement of the plunger to the main pointer. This type of instrument has a long plunger movement and is fitted with a secondary scale and pointer for indicating the number of complete revolutions made by the main pointer.

ply - of a tyre
A layer of rubber-coated parallel cords.

ply rating
An indication of the strength of the *carcass* and of load carrying capacity of a *tyre*: originally related to the number of plies.

ply-type belt
A *belt* based on more than one *ply* of fabric.

plying
Building up increased thickness from separate layers. For example, multi-layer construction is possible by laminating a series of flexible substrates together, using a *calender* to generate the conditions necessary to achieve bonding. In rubber technology, both *single-bank calendering* and *double-bank calendering* may be used to build up section thickness by plying. This is usually done for lightly-loaded stocks where the production of thick sheets, free from blisters, would be extremely difficult. Thin sheets are therefore first produced and plied until the required thickness is obtained. If the rubber sheet is applied, for example, to a previously frictional fabric then the process is known as *skim coating*.

PMC
An abbreviation sometimes used for pelletized moulding compound. See *polyester moulding compound*.

PMC regrind
An abbreviation used for *polyester moulding compound regrind*.

PMDA
An abbreviation used for *pyromellitic dianhydride*.

PMGI
An abbreviation used for *polydimethyl glutarimide*.

PMIPK
An abbreviation used for *polymethyl isopropenyl ketone*.

PMMA
An abbreviation used for *polymethyl methacrylate*.

PMMA-HI
An abbreviation used for *high impact polymethyl methacrylate*.

PMQ silicone rubber
A *silicone rubber* which is based on *polydimethylsiloxane* where some of the methyl groups have been replaced with phenyl. Few pure MQ and PMQ-types of *silicone rubber* are now marketed.

PMS
An abbreviation used for *post-moulding shrinkage*;

PMS_{168h} = post-moulding shrinkage after 168 h heat treatment; and,
PMS_{48h} = post-moulding shrinkage after 48 h heat treatment.

PNDA
An abbreviation used for p-nitrosomethyl aniline.

pneumatic
Operated by compressed air. Containing compressed air, for example, a *pneumatic tyre*.

pneumatic ejection system
A system of ejection which is actuated by a *pneumatic* ram.

pneumatic gauging
See *air gauging*.

pneumatic tyre
A *tyre* which is inflated before use. Most tyres now used are tubeless tyres as, for example, they are less liable to deflate suddenly when punctured and, in an emergency, can be sealed by plugging without being removed from the wheel. The seal between the tyre and the wheel is provided by using a soft rubber lining to the *casing*.

pneumatically actuated sweep
A system used to clear an *injection mould*; the pneumatically actuated sweep may be mounted above the mould and moves across the mould surface at the end of the opening stroke.

pocket
See *band*.

poise
The unit of *viscosity* in the *centimetre-gram-second (cgs) system* and which is measured in dyne-second per square centimetre. One poise = 1P = 10^{-1} Pa·s = 10^{-1} Nsm^{-2} = 0.000 014 5 lbf s in^{-2}. That is, one poise = 1P = 0.1 Pa·s = 0.1 Nsm^{-2}.

Poiseuille equation
Also known as the Poiseuille-Hagen equation. A rheological equation which states that the volume of liquid (V) flowing through a tube of radius R and length L, in unit time (t) is related to its viscosity (η) by the equation

$$\eta = \pi R^4 Pt/8LV.$$

The pressure difference (between the tube ends or between two points along the tube) may be written as P or as ΔP. The equation is only true when laminar flow is occurring with a *Newtonian fluid* and if the fluid adjacent to the walls of the tube is not moving. See *pressure flow*.

Poiseuille flow
Flow to which the *Poiseuille equation* applies: laminar flow of a Newtonian liquid. See *pressure flow*.

Poiseuille-Hagen equation
See *Poiseuille equation*.

Poisson ratio
Also known as lateral contraction ratio and often given the symbol v. The ratio of the tensile strain (produced by a tensile force) to the contraction in cross-sectional area produced by that tensile force. Given by the ratio d/D to l/L where d is the decrease in area, D is the original area, l is the decrease in length and L is the original length. For most rubbers v is about 0.5 which shows that the rubber is largely incompressible: for plastics materials the value is in the range of 0.3.

v = E − 2G/G where E is the elasticity modulus (Young's modulus) and G is the shear modulus.

polarimeter
Also known as a polariscope. An apparatus for measuring the rotation of the plane of vibration of polarized light. Contains two *polarising filters* which are crossed or, at right angles to each other. If a birefringent sample of a transparent plastics material is placed between these filters then a pattern of lines or fringes will be seen. Used to study, for example, the properties of transparent, injection moulded components. See *photo-elastic inspection*.

polariscope
See *polarimeter*.

polarization
Electric polarization occurs within a *dielectric* as a result of that material being placed in an electric field. Part, or all, of a molecule becomes oriented with respect to the field. When the field is a high frequency, alternating field then polar molecules, such as water, interact with the high frequency field and generate heat as a result of molecular friction. For example, in the case of a *high frequency preheating* process, the field may reverse polarity 80 million times a second and the polar molecule tries to follow this rapid reversal. The application of the high frequency field causes a rapid and comparatively uniform temperature rise within the material thus minimizing problems of conventional heat transfer. Wet areas in a product will be selectively heated and dried but dry areas will absorb little or no energy.

polarized light
Polarized light may be obtained by passing light through a crystal of, for example, *tourmaline* or through a *polarizing filter*. The rays of light which emerge from the filter, are largely confined to a single plane. See *photo-elastic inspection*.

polarizing filter
Used to produce *polarized light*. A thin sheet of transparent thermoplastics material which contains doubly refracting crystals and so produces plane-polarized light. That is, the rays of light which emerge from the filter, are largely confined to a single plane. Used to study, for example, the properties of transparent, injection moulded components in a *polarimeter*. See *photo-elastic inspection*.

polarizing microscope
A microscope which contains two *polarising filters* which are crossed or, at right angles. If a birefringent sample of a plastics material is placed between these filters then a pattern of lines or fringes will be seen. Used to study, for example, the properties of moulded components, spherulites and crystallites.

polepiece
In reinforced plastics, the supporting part of the mandrel used in *filament winding* and usually on one of the axes of rotation.

poling
Permanent polarization. Polarization induced by, for example, thermal poling or by corona poling (polarization induced by a corona discharge process). Corona poling is faster than thermal poling. A *polyvinylidene fluoride* (PVDF) film is, for example, subjected to a corona discharge from a close needle electrode while being stretched at, for example, room temperatures. In thermal poling, electrodes are evaporated onto the PVDF and an electric field of approximately 1 MV/cm is applied at about 100°C for 1 h: the PVDF is cooled while the field is maintained.

polished sheet
Seasoned sheet, based on a cellulose plastic, which has been press polished. See *slicing machine* and *cellulosics*.

pollution
A change in the environment that is unwanted: an addition to the environment that is unwanted.

polvinylbenzene
See *polystyrene*.

poly
A prefix which means many.

poly α-olefin
See *poly-(α-olefin)*.

poly-1-butene
See *polybutylene*.

poly-(α-methylstyrene)
See *poly-α-methylstyrene*.

poly-(α-olefin)
A *polyolefin* which may be represented as -(CH$_2$-CHR)-$_n$ where R is an alkyl or cycloalkyl group. If R is a methyl group then *polypropylene* is the result.

poly-(Σ-aminocaproic acid)
See *nylon 6*.

poly-(ω-amino-caproamide)
See *nylon 6*.

poly-(ω-aminocaproic acid)
See *nylon 6*.

poly-(1-butenylene)
See *butadiene rubber*.

poly-(1-butenylene-co-1-cyanoethylene)
See *nitrile rubber*.

poly-(1-butenylene-co-1-phenylethylene)
See *styrene-butadiene rubber*.

poly-(1-butenylene-g-1-phenylethylene-co-1-cyanoethylene)
See *acrylonitrile-butadiene-styrene*.

poly-(1-methyl-1-butenylene)
See *natural rubber*.

poly-(1-phenylethylene-b-1-butenylene-b-1-phenylethylene)
See *styrene block copolymer*.

poly-(1-phenylethylene)
See *polystyrene*.

poly-(1-phenylethylene-co-1-cyanoethylene)
See *styrene-acrylonitrile copolymer*.

poly-(1,4-cyclohexylenedimethylene terephthalate-co-isophthalate
May also be referred to as 1,4-cyclohexylenedimethylene terephthalate/isophthalate. This type of *amorphous thermoplastics material* is prepared from a *glycol (1,4-cyclohexanedimethanol)*, *terephthalic acid* and *isophthalic acid*. A *polyethylene terephthalate (PET)* material which can give high clarity mouldings and formings, for example, Kodar PETG: this type of material is sometimes referred to as a *copolyester*. Being irregular in structure the resultant material cannot crystallize and so, is an amorphous clear material with a glass transition temperature (T_g) of 88°C/190°F. This material offers wider processing latitude than conventional crystallizing polyesters and a useful combination of properties, for example, clarity, toughness and stiffness. PETG is resistant to dilute aqueous solutions of mineral acids, bases, salts, aliphatic hydrocarbons, alcohols and a range of oils

poly-(2,3-dimethylbutadiene)
See *methyl rubber*.

poly-(2,6-dimethyl-p-phenylene) oxide
See *polyphenylene oxide - modified*.

poly-(3,3-bis-(chloromethyl)oxatane
See *chlorinated polyether*.

poly-(3,3-bis-(chloromethyl)oxacyclobutane
See *chlorinated polyether*.

poly-(4-methylpentene-1)
See *polymethyl pentene*.

poly-(4,4'-isopropylidenediphenylene carbonate)
See *polycarbonate*.

poly-(6-aminocaproic acid)
See *nylon 6*.

poly-(11-aminoundecanoic acid)
See *nylon 11*.

poly-(amino-caproamide)
See *nylon 6*.

poly-(aminocaproic acid)
See *nylon 6*.

poly-(butadiene-co-acrylonitrile)
See *nitrile rubber*.

poly-(m-phenyleneisophthalamide)
Better known by the trade name/trademark Nomex. Sometimes referred to as MPD-I. MPD comes from m-phenylene diamine and the I from isophthalamide: synthesised from these two monomers by solution polymerization. An *aramid fibre* material which is also available in film and paper form. An *aromatic polyamide* with excellent heat resistance.

poly-(m-xylene adipamide)
See *polyaryl amide*.

poly-(methylstyrene)
See *alpha methyl styrene*.

poly-(monochloroethylene)
See *polyvinyl chloride*.

poly-(n-alkyl methacrylate)
A polymer based on an n-alkyl methacrylate monomer. If the first member of the series is considered to be *polymethyl methacrylate*, then as the side group is lengthened the *Vicat softening point* steadily falls as the inter-molecular attraction is reduced. When the side group contains approximately 12 carbon atoms then the Vicat softening point steadily increases because of side-chain crystallization. The first few members of this series are amorphous thermoplastics materials: a number of higher n-alkyl methacrylate polymers have found uses as leather finishes. For example, poly-(n-butyl methacrylate), poly-(n-octyl methacrylate) and poly-(n-nonyl methacrylate) polymer. See *polymethacrylic acid esters*.

poly-(n-butyl methacrylate)
See *poly-(n-alkyl methacrylate)*.

poly-(n-nonyl methacrylate)
See *poly-(n-alkyl methacrylate)*.

poly-(n-octyl methacrylate)
See *poly-(n-alkyl methacrylate)*.

poly-(N-vinyl-2-pyrrolidone)
See *polyvinyl pyrrolidone*.

poly-(N-vinyl carbazole)
See *polyvinyl carbazole*.

poly-(N-vinylpyrrolidone)
See *polyvinyl pyrrolidone*.

poly-(oxy-1,4-phenylene-dimethylmethylene-1,4-phenylene-oxy-carbonyl)
See *polycarbonate*.

poly-(oxytetramethylene-oxyterephthalate)
See *polybutylene terephthalate*.

poly-(p-benzamide)
Sometimes referred to as PPB. An *aramid fibre* material.

poly-(p-phenylene sulphide)
See *polyphenylene sulphide*.

poly-(p-phenylene sulphone)
The simplest *sulphone polymer*. Such a simple aromatic polysulphone could not be melt processed.

poly-(p-phenyleneterephthalamide)
Sometimes referred to as PPD-T. An *aramid fibre* material which is better known by the trade name/trademark Kevlar (DuPont). An *aromatic polyamide* which is tougher than *glass* or *carbon fibre*. Kevlar-type materials may be used for composite reinforcement and/or for tyre reinforcement.

poly-(parabanic acid)
See *polyparabanic acid*.

poly-(propylene oxide-b-ethylene oxide)
See *propylene oxide-ethylene oxide block copolymer*.

poly-(propylene)
See *polypropylene*.

poly-(styrene-b-butadiene-b-styrene)
See *styrene block copolymer*.

poly-(styrene-b-ethylene-co-butylene-b-styrene)
See *styrene block copolymer*.

poly-(styrene-b-ethylene-co-propylene-b-styrene)
See *styrene block copolymer*.

poly-(styrene-b-isoprene-b-styrene)
See *styrene block copolymer*.

poly-(styrene-co-acrylonitrile)
See *styrene-acrylonitrile copolymer*.

poly-(styrene-co-butadiene)
See *styrene-butadiene copolymer*.

poly-(styrene-co-maleic anhydride)
See *styrene-maleic anhydride*.

poly-(sulphur nitride)
See *sulphur-nitride polymer*.

poly-(tetrafluoroethylene-co-ethylene)
See *tetrafluoroethylene-ethylene copolymer*.

poly-(tetrafluoroethylene-co-hexafluoropropylene)
See *fluorinated ethylene propylene copolymer*.

poly-(tetrafluoroethylene)
See *polytetrafluoroethylene*.

poly-(tetrafluoroethylene-co-perfluoromethylvinyl ether)
See *perfluorinated elastomer*.

poly-(thio-1,4-phenylene)
See *polyphenylene sulphide*.

poly-(thiocarbonyl fluoride)
A highly specialized *fluororubber* which contains sulphur as part of its structure. May be represented as $-(-CF_2-S-)_n-$ in the case of the homopolymer. Copolymers are also made with allyl chloroformate as this reduces crystallization which occurs on standing. Such materials may be vulcanised with *zinc oxide*.

poly-(vinyl acetal)
See *polyvinyl acetal*.

poly-(vinyl acetal) polymers
See *polyvinyl acetal polymers*.

poly-(vinyl butyral)
See *polyvinyl butyral*.

poly-(vinyl chloride)
See *polyvinyl chloride*.

poly-(vinyl chloride-co-acrylonitrile)
See *vinyl chloride-acrylonitrile copolymer*.

poly-(vinyl chloride-co-butadiene)
See *vinyl chloride-butadiene copolymer*.

poly-(vinyl chloride-co-ethylene)
See *vinyl chloride-ethylene copolymer*.

poly-(vinyl chloride-co-propylene)
See *vinyl chloride-propylene copolymer*.

poly-(vinyl chloride-co-vinyl acetate)
See *vinyl chloride-vinyl acetate copolymer*.

poly-(vinyl chloride-co-vinyl propionate)
See *vinyl chloride-vinyl propionate copolymer*.

poly-(vinyl ester)
See *polyvinyl ester*.

poly-(vinyl ether)
See *polyvinyl ether*.

poly-(vinyl formal)
See *polyvinyl formal*.

poly-(vinyl ketal) polymers
See *polyvinyl ketal polymers*.

poly-(vinylethyl ether)
See *polyvinylethyl ether*.

poly-(vinylidene chloride)
See *polyvinylidene chloride*.

poly-(vinylidene chloride-co-acrylonitrile)
See *polyvinylidene chloride* and *vinylidene chloride-acrylonitrile copolymer*.

poly-(vinylidene chloride-co-vinyl chloride)
See *polyvinylidene chloride* and *vinylidene chloride-vinyl chloride copolymer*.

poly-(vinylidene fluoride-co-chlorotrifluoroethylene)
See *vinylidene fluoride-chlorotrifluoroethylene copolymer*.

poly-(vinylidene fluoride-co-hexafluoropropylene)
See *vinylidene fluoride-hexafluoropropylene copolymer*.

poly-(vinylmethyl ether)
See *polyvinylmethyl ether*.

poly-{2,2'-bis(3,4 dicarboxyphenoxy)phenylpropane-2-phenylene bisimide}
See *polyether imide*.

poly-α-methylstyrene
The polymer of α-*methylstyrene* (alpha-methylstyrene): polymerization is performed in the presence of ionic catalysts at low temperatures, for example, $-60°C$. Homopolymers of this monomer may be used as *plasticizers* for natural or synthetic rubbers: for this application low molecular weight, liquid polymer is required. Such liquids are colourless and odourless and have boiling points in the range of 150 to 300°C: the relative density lies between 1·01 and 1·04. By conversion to higher molecular weights, hard clear thermoplastics materials may be prepared. Although the *glass transition temperature* of such polymers may reach 170°C, they have not become of great commercial significance. Copolymers of α-methylstyrene and styrene (styrene-α-methyl styrene copolymers) are transparent thermoplastics materials which are marketed as they improve on the softening point of polystyrene with only a slight increase in melt viscosity.

poly-α-olefin
See *poly-(α-olefin)*.

poly-β-alanine
See *nylon 3*.

poly-1,2,4-oxadiazole
An *aromatic polymer* based on fused rings: based on an aromatic ring linked to a heterocyclic, five membered ring (which contains two carbon atoms, two nitrogen atoms and one oxygen atom): the main chain links are via the carbon atoms. The heterocyclic rings are spaced out with benzene rings. This polymer is less heat stable than *poly-1,3,4-oxadiazole* has good hydrolytic stability but is intractable: for example, it is only soluble in strong acids.

poly-1,3,4-oxadiazole
An abbreviation used for this material is POD. An *aromatic polymer* based on fused rings: based on an aromatic ring linked to a heterocyclic, five membered ring (which contains two carbon atoms, two nitrogen atoms and one oxygen atom): the main chain links are via the carbon atoms. The arrangement of the five-membered ring is different to that of *poly-1,2,4-oxadiazole*. Prepared in fibre form from a precursor as the high molecular weight polymer is intractable: such polymers have excellent heat resistance.

poly-2,3-dimethyl butadiene
See *methyl rubber*.

poly-2,3-dimethylbutadiene
See *methyl rubber*.

poly-4-hydroxy styrene
An aqueous base-soluble polymer used to make a *photoresist* with an *azide* compound.

poly-4-methylpentene-1
See *polymethyl pentene*.

poly-m-xyleneadipamide
See *polyaryl amide*.

poly-n-butyl acrylate
An amorphous polymer which together with polyethyl acrylate, forms the basis of *acrylic rubbers*. See *acrylic monomer*.

poly-p-xylene
An abbreviation used for this material is PPX. Also known as poly-para-xylene: a parylene polymer. An *aromatic polymer* with exceptional heat resistance. Used as film and as coatings in, for example, electronics equipment. The molecular weight is very high - approximately 500,000 and the polymer cannot be melt processed. A monochlorinated polymer is also commercially available.

poly-para-xylene
See *poly-p-xylene*.

poly-phenylethylene
See *polystyrene*.

poly-[1-(methoxycarbonyl)-1-methylethylene]
See *polymethyl methacrylate*.

polyacetal
See *acetal*.

polyacrylamate reaction injection moulding
An abbreviation used for this term is polyacrylamate RIM. Also known as acrylamate reaction injection moulding. A *reaction injection moulding* (RIM) process which produces highly cross-linked products: the process of producing such a moulding.

A variation of *unsaturated ester reaction injection moulding* in which reaction occurs in two steps. Two molecules of a mono-functional alcohol (based on methacrylic acid, propylene oxide and maleic anhydride) combine with *4,4'-dicyclohexylmethane diisocyanate* (MDI) to form an unsaturated polyurethane (an unsaturated ester-urethane). The heat from this reaction starts the second step in which a peroxide initiates free radical polymerization of the double bonds in the unsaturated polyurethane.

The temperature of the reactants is typically 25°C, mould temperatures are of the order of 100°C while demould times may be 60 s: the adiabatic exotherm temperature for a typical formulation without filler can be 100°C. The cross-linked polymer can have a high modulus and a heat distortion temperature which can reach approximately 130°C: impact strength is however relatively low but is good when used with glass mats. See *reaction injection moulding* process and *structural reaction injection moulding*.

polyacrylate
A polymer of an acrylic ester. An abbreviation suggested for this type of material is PAK. May be represented as $-(-CH_2-CH\cdot COOR-)_n-$ where R is an alkyl group. If R is a methyl group then polymethyl acrylate is the result. See *acrylic rubber*.

polyacrylate elastomer
See *acrylic rubber*.

polyacrylate rubber
See *acrylic rubber*.

polyacrylic acid
The polymer of acrylic acid. May be represented as $-(-CH_2-CH\cdot COOH-)_n-$. Not a thermoplastics material as it does not become a melt when heated. Very soluble in water: aqueous solutions of this polymer may be cross-linked with a metal oxide, for example, and used as dental materials. See *polymer cement*.

polyacrylic elastomer
See *acrylic rubber*.

polyacrylic rubber
See *acrylic rubber*.

polyacrylonitrile
Also known as polyvinyl cyanide. An abbreviation used for this material is PAN. May be represented as $-(-CH_2-CH\cdot CN-)_n-$. An important *fibre* material: often used as a copolymer, for example, with vinyl acetate, so as to improve dyeability. In general, PAN-type materials have very strong inter-chain attractions which means that the polymer cannot be processed in melt form as the polymer degrades before a useful melt is obtained. Fibre are produced by spinning from solution. If this material is heated to high temperatures under controlled conditions then a *ladder polymer* is formed. See *carbon fibre*.

polyacrylonitrile carbon fibre
An abbreviation used for this type of material is PAN CF. A *carbon fibre* based on *polyacrylonitrile* (PAN) fibre. The PAN fibre, is first stabilized by being heated to approximately 250°C while being drawn through the continuous process: only that tension sufficient to draw the fibre through the equipment is needed. The temperatures are then raised in subsequent furnaces so that at approximately 1,500°C the nitrogen is eliminated from the fibre: the yield of carbon is approximately 45%. By varying the heat treatment, which can reach 2,500°C, a range of fibres are produced, for example, to give high modulus (HM), high strength (HS) and high strain carbon fibre.

A PAN HS CF can have a tensile strength of approximately 3.4 GPa, a modulus of 240 GPa, a breaking strain of 1.5% and a specific gravity of 1.7. A PAN HM CF can have a tensile strength of approximately 2.4 GPa, a modulus of 400 GPa, a breaking strain of 0.7% and a specific gravity of 1.84.

polyaddition
A *polymerization process* in which no small molecule is eliminated. Used to produce, for example, *polyurethanes*, in which addition is at the double bond of the isocyanate group: this gives chain extension without the elimination of a small molecule.

polyalkenamer rubber
An alkenamer formed by the ring-opening polymerization of a cyclo-olefin by *Ziegler-Natta polymerization*: both cis and trans isomers may be possible. A polyalkenamer which may be represented as $-(-R-CH=CH-)_n-$ where R is $(CH_2)_{>2}$. When n = 2 then the important diene rubbers are included. Many of the polymer chains probably exist as large rings (macrocyclic) which gives rise to few chain ends: this in turn is thought to

account for the high tensile strength of even low molecular weight material, for example, of *polyoctenamer rubber*.

polyallomer
Also called an allomer: a propylene copolymer. An *ethylene-propylene block copolymer*. The ethylene was originally added to improve the low temperature impact strength of *polypropylene* homopolymer.

polyaluminosiloxane
Also known as polyalumosiloxane. Polymers which consist of chains of silicon, oxygen and aluminium: some of the silicon and aluminium atoms have *alkyl groups* attached. Such polymers are usually *cross-linked*.

polyaluminoxane
Polymers which consist of chains of alternating atoms of oxygen and aluminium: ideally each of the aluminium atoms has one *alkyl group* attached.

polyalumosiloxane
See *polyaluminosiloxane*.

polyamic acid
Also known as polyamide-acid. A polymer which contains both amide and carboxylic groups. Such materials are often *aromatic polymers* and are used, for example, to prepare polyimides. The polyamic acid is the precursor polymer used to prepare the intractable *polyimide* in situ.

polyamidation
A chemical reaction which results in the formation of a *polyamide*.

polyamide
A polymer which contains many amide (CONH) groups. An abbreviation used for this type of material is PA. A PA may be prepared by reacting a diamine with a dibasic acid via a *nylon salt*. Such a polyamide is identified by the number of carbon atoms in both the diamine and in the acid: the first number indicates the number of carbon atoms in the diamine. For example, nylon 66 is based on an amine with 6 carbon atoms and on an acid with 6 carbon atoms. May be referred to as PA 66 or as, PA 6.6 or as, PA 6·6 (also without the spaces between the PA and the numbers). When the PA is followed by a single number, preparation from either an ω-amino-acid or a lactam is indicated (see, for example, *nylon 6*). By an extension of this abbreviation system, PA 66, GF 35 or, PA 66 GF35, means that the material being referred to is 'polyamide 66 with 35%, by weight, of glass fibre'. See the entries under *nylon*. For example, for:

polyamide 6 see *nylon 6*;
polyamide 11 see *nylon 11*;
polyamide 66 see *nylon 66*;
polyamide 6/66 see *nylon 6/66*; and for
polyamide 6/66/610 see *nylon 6/66/610*.

polyamide 6 reaction injection moulding
See *nylon 6 reaction injection moulding*.

polyamide copolymer
See *nylon copolymer*.

polyamide elastomer
See *thermoplastic elastomer - amide based*.

polyamide/elastomer block copolymer
A category of *thermoplastic elastomer* which may be referred to as *polyether block amide*.

polyamide-acid
See *polyamic acid*.

polyamide-hydrazide
A polymer which contains both amide (-NHCO-) and hydrazide (-NH-NH-) groups. An abbreviation used for such a material is, for example, PABH-T: this is a high temperature resistant, aromatic polyamide-hydrazide fibre-forming material.

polyamide-imide
An abbreviation used for this type of material is PAI. A polymer which has both amide and imide groups in the main chain. A *modified polyimide*. Used as high temperature, insulation varnish and in film form. Polyamide-imide polymers are also available as moulding materials: based, for example, on trimellitic anhydride and m-phenylene diamine. The polymers have heat resistance properties intermediate between those of a *polyamide* and a *polyimide*. More tractable than a polyimide.

polyamides
See the entries under *nylon*.

polyaminobismaleimide
Also known as polybismaleimide or as, polymaleimide. An abbreviation used for this material is PABM. A *modified polyimide* prepared by rearrangement polymerization of a bismaleimide with a diamine: a polymer which contains aromatic-type structures. No volatiles are eliminated during rearrangement polymerization and this minimizes void formation.

polyaryl amide
An abbreviation used for this type of material is PAA 6 or PA MXD6. An injection moulding material based on the semi-aromatic polyamide known as poly-m-xyleneadipamide. This material is prepared by reacting m-xylenediamine (MXDA) with adipic acid to give a polyamide which contains phenylene aromatic groups spaced with aliphatic groups (provided by the adipic acid sections): this gives a polymer with easier processing than the fully aromatic structure. Because of the improvement that reinforcement gives, such aromatic PA materials are usually sold reinforced. They may be reinforced or filled with inorganic particulate fillers and/or fibres (carbon or glass at up to 60 percent). Used as an *injection moulding* material.

The mouldings are glossy, very stiff and yet possess good impact resistance. Thermal expansion is low (can be similar to steel) and this, coupled with a slow and low water pick-up, gives excellent dimensional stability and low warpage. This means that metal inserts and outserts can be incorporated without inducing too much stress. Water absorption (at equivalent filler content) is lower than for PA 6 and PA 66 but slightly higher than for PA 11 and PA 12. Fire resistance is good (oxygen index 27·5) and materials are rated as V.0. down to 1·57 mm (0·062 in). The material has good electrical insulation properties and some grades have excellent tracking resistance. By using high mould temperatures (around 120°C/248°F) high gloss products with a scratch resistant surface are obtained. Impact strength is better than that of 30% glass fibre filled polycarbonate (GF PC) and the material retains its impact strength at low temperatures (around −60°C/−76°F). Flame retardant and impact modified grades are available.

An easy flow material because the melt viscosity is fairly low and hot moulds are used. Shrinkage is of the order of 0·001 to 0·007 mm/mm (0·1 to 0·7%). As the material is supplied in a filled condition, the density range is 1·42 to 1·77 gcm^{-3}. The natural colour of this material is a translucent greenish white but it is usually seen as natural or black although a restricted colour range is available.

This type of material is resistant to aliphatic hydrocarbons, aromatic hydrocarbons, ketones, esters, ethers, aldehydes (except formaldehyde) and alcohols. This material has similar properties to PA 66 - resists stress cracking and is oil and petrol resistant (even at elevated temperatures). Chemical resistance is dependent on degree of crystallinity and is therefore influenced by moulding conditions as amorphous surface layers may be formed; these lay the mouldings open to easier attack and decrease flexural strength. By increasing the mould temperature the surface crystallinity increases.

This type of material is not resistant to concentrated acids and bases, dilute acids and bases, salt solutions and strong oxidizing agents. Has better acid resistance than other PA materials. Steam can cause hydrolysis.

Fast cycles possible during injection moulding because the cooling times are relatively short due to high set-up temperatures and the stiffness of this material. Various masterbatches can be added to the material to overcome processing difficulties: for example, the adhesion of material to the surface of the mould. This adhesion will cause the component to deform during ejection.

Because of the very high stiffness of this material it is used as a replacement for metal components - such as load bearing die castings based on zinc. Used for spray guns, mowing machine components and domestic appliance housings. Used in the electrical and electronics industry for plugs, sockets, TV tuner blocks, connector blocks and electrical power cable clamps. High temperature resistance, rigidity and moulding accuracy are some of the reasons for material choice. Used in applications where dimensional stability, wear resistance and stiffness are important. Leisure applications include propellers and fishing reels - apart from rigidity it has profile stability, toughness and low water pick-up which are good reasons for its choice in these applications.

polyaryl ether sulphone
See *sulphone polymers*.

polyaryl sulphone
See *sulphone polymers*.

polyarylamide
See *polyaryl amide*.

polyarylate
An *aromatic polymer*: an *aromatic polyester* derived from the reaction of a dihydric phenol with an aromatic dicarboxylic acid. An abbreviation suggested for this type of material is PAR. The homopolymers have high heat resistance but are somewhat intractable: copolymers are more tractable. For example, copolymers prepared from isophthalic acid, terephthalic acid and, bisphenol A are used as *engineering thermoplastics materials* as they have properties between *polycarbonate* and *polyether sulphone*. Such amorphous, thermoplastics materials can have service temperatures which can reach 150°C. See *polycarbonate*.

polyarylenesulphone
See *sulphone polymers*.

polyarylether ketone
See *polyether ether ketone*.

polyazelaic polyanhydride
An abbreviation used for this material is PAPA. A curing agent for *epoxide resins* which imparts flexibility to the cured products.

polybenzimidazole
An abbreviation used for this material is PBI. An *aromatic polymer* which has excellent heat resistance. The chain contains a benzene ring and a heterocyclic five-membered ring: this ring contains carbon and nitrogen. Prepared from an aromatic tetramine and a dicarboxylic acid (or a derivative). Fully aromatic materials are intractable and are initially processed as low molecular weight intermediates: they have however, excellent resistance to hydrolysis and to heat.

polybenzobisoxazole
An abbreviation used for this material is PBZ. An *aromatic polymer* which has good heat resistance. The chain contains a benzene ring and two other aromatic units each based on a benzene ring joined to a heterocyclic five-membered ring: this ring contains carbon, oxygen and nitrogen. The two other aromatic units may be linked by a methylene group. An *aromatic polymer* which has good heat resistance.

polybenzooxazole
An *aromatic polymer* which has good heat resistance. A plastics material which has as a repeat unit a benzene ring joined to a heterocyclic five-membered ring: this ring contains carbon, oxygen and nitrogen. These aromatic materials are intractable and are processed as solutions in strong acids: they have however, excellent resistance to hydrolysis and to heat. See *polybenzobisoxazole*.

polybismaleimide
A *modified polyimide*. See *polyaminobismaleimide*.

polyblend
See *polymer blend*.

polyborate
See *borate glass*.

polybut-1-ene
See *polybutylene*.

polybutadiene elastomer
See *butadiene rubber*.

polybutadiene rubber
See *butadiene rubber*.

polybutadiene tyre
A tyre which contains a high proportion of *polybutadiene*. Such tyres have a hard and bouncy feel compared to the slight tackiness of other tyres. A hard wearing tyre which is resistant to cutting, tearing and slitting: often a relatively cheap product. Such tyres are not generally suitable for high rainfall areas (Europe) as they do not perform very well on wet surfaces: they are prone to skidding and aquaplaning on wet roads.

polybutene
See *polybutylene*.

polybutene-1
See *polybutylene*.

polybutyl acrylate
A rubbery material: an *acrylate* polymer used, for example, to improve the impact resistance of some thermoplastics materials. For example, *polymethyl methacrylate* and *polyhydroxybutyrate*.

polybutyl methacrylate
An amorphous thermoplastics material. See *polymethacrylic acid esters*.

polybutylene
Also known as polybutene or as, poly-1-butene or as, polybut-1-ene or as, polyethylethylene. An abbreviation used for this material is PB. A *poly-(α olefin)* made from the α-olefin, butene (butylene) by Ziegler - Natta polymerization into a highly crystalline and isotactic form. May be represented as $(CH_2 - CHR)_n$ where R is C_2H_5 and n is a very large number as the molecular weight may be up to 3,000,000 - much higher than that of *low density polyethylene*.

Probably because of this very high molecular weight, this type of material has very good *creep resistance* and *environmental stress cracking resistance*. For these reasons the material is used to make pipes: such pipes may be solvent welded and may be of thinner cross-section than those made of, for example, *polypropylene* (PP).

When first processed, the density is low at 0.89 g/cm³ but on standing, for example, for five days, the density increases to 0.95 g/cm³: the stiffness and hardness also increase. When first processed, using conditions similar to those used for PP, the products must be handled with care as they are weak: strength improves on standing or ageing. On ageing a strong, stiff, but still flexible, product is the result. The melting point of the aged material is approximately 135°C and although inert to many common solvents, it may be dissolved in n-alkyl acetates. See *polyolefins*.

polybutylene terephthalate

An abbreviation used for this type of material is PBT or PBTP. A thermoplastic polyester which is also called polytetramethylene terephthalate (PTMT). Also known as poly-(oxytetramethylene-oxyterephthalate).

This material is a *thermoplastic polyester* (like *polyethylene terephthalate* or PET), produced by the reaction of terephthalic acid and 1,4 butane-diol. A semi-crystalline material and, in order to obtain the required level of physical properties, the crystallinity must reach a high level. With PBT this readily develops, because of the low glass transition temperature (T_g), but it is more difficult for PET. Because of chain flexibility, compared to PET, PBT has a lower melting point and heat distortion temperature (HDT) but, it is a better electrical insulator and can be moulded at lower temperatures (the polymer chains contain longer sequences of methyl groups).

PBT has a fairly high HDT and is stiff, tough and hard. Has a lower Vicat softening point (VST) than PET. It has good abrasion resistance, resists failure by fatigue (and creep) and displays good electrical insulation properties. Even thin sections have high dielectric strength and changes in humidity do not cause large changes in such properties. It has good resistance to chemical stress cracking and excellent dimensional stability. Widely used as an injection moulding material: rapid injection moulding cycles are possible because of a fast set-up rate.

Mouldings possess a low coefficient of friction, have hard surfaces and outstanding chemical resistance. The low moisture regain, means excellent dimensional stability and consistent mechanical and electrical properties. Unreinforced PBT can have a low notched impact strength but this can be improved by making a copolymer with an acid which has a longer chain than terephthalic. Blends with other polymers are also used, for example, *polycarbonate* (PC) and *polyethylene terephthalate* (PET). Blends with glass fibre (GF 20 to 50%) are extensively used as glass significantly improves physical properties. Possible to get grades which are classified as self extinguishing by UL. PBT/PET blends are often glass reinforced as the unfilled compound is of limited value; such blends can be very easy flowing and can give good surface finishes.

PBT is an easy flowing, hygroscopic material (similar to *nylon*): the viscosity of this material is considerably affected by the amount of moisture it contains when being processed. Like PET, this material is sensitive to hydrolysis at high temperatures and so, the material must be thoroughly dry before being injection moulded. Therefore it is very important to dry the material, at the correct time and temperature, as otherwise components with inferior mechanical strength and dimensional stability will result. Moisture content must be below 300 ppm (0.03%). Shrinkage is of the order of 0.015 to 0.028 mm/mm (1.5 to 2.8% but is reduced by glass fibre addition.

PBT is resistant to aliphatic hydrocarbons alcohols, ethers, high molecular weight esters and ketones, dilute aqueous solutions (acids, bases and salts), oil and petrol (has good long term resistance to petrol at temperatures up to 60°C/140°F. Resists hydrocarbon oils without stress cracking. Slightly affected by aromatic solvents at room temperature. Good resistance to concentrated and dilute mineral acids. It has excellent resistance to common household fats including those used in the kitchen (pantry). Always check chemical resistance at intended use temperature as, for example, chemical resistance at 90°C/194°F is much inferior to that at 25°C/77°F.

PBT is not resistant to strong bases and hot water (sensitive to hydrolysis at 60°C). Swollen by low molecular weight esters, low molecular weight ketones and by partially halogenated hydrocarbons. Attacked by sodium hydroxide and ethylene dichloride.

The unfilled material has a density of 1.31 gcm⁻³, a high melting point of approximately 225°C (PET 260°C) and a Vicat of 170°C/338°F. As the natural colour of the material is milky white then a wide colour range is possible.

PBT is widely used in domestic equipment, electronics, electrical, telecommunications and automotive markets, e.g. pump housings, impellers, gears etc. Glass fibre filled grades have V0 ratings in sections down to 0.7 mm - used for oven handles (in place of phenolics) as the material resists elevated temperatures (≫100°C/212°F), is non-flammable, heat-insulating and resists hydrocarbons, oils, grease and abrasive kitchen cleaners. Elastomer modified PBT is used to make car spoilers - painted on line and stoved at temperatures of up to 140°C/284°F. PBT is notch sensitive so stress concentrations caused by sharp angles must be avoided

polybutylene terephthalate/polyethylene terephthalate blend

A blend of *polybutylene terephthalate* (PBT) and *polyethylene terephthalate* (PET). A PBT/PET blend. Such blends are often glass reinforced as the unfilled compound is of limited value; such blends can be very easy flowing and can give good surface finishes. PET is often used in blends with PBT as the use of PET improves gloss and scratch resistance (PET is hard and has a low coefficient of friction). See *thermoplastic polyesters*.

polycaproamide
See *nylon 6*.

polycaprolactam
See *nylon 6*.

polycaprylamide
See *nylon 8*.

polycapryllactam
See *nylon 8*.

polycarbamate
See *polyurethane*.

polycarbodiimide

An abbreviation used for this type of material is PCD. A polymer formed by the reaction of isocyanates with each other. That is, OCN-R-NCO + OCN-R'-NCO gives -(-CN-R-NCN-R'-NC-)$_n$- plus carbon dioxide. Such polymers are of relatively little commercial significance. See *flexible polyurethane foam*.

polycarbodiimide foam

An abbreviation used for this type of material is PCD foam. A foamed material which is based on carbodiimide formation. See *flexible polyurethane foam*.

polycarbodiimide masterbatch

An additive used to improve the resistance to hydrolysis of *polyether ester elastomer*: resistance may be improved by blending with 2% polycarbodiimide masterbatch. Take care if this material is used as it can evolve irritating fumes when heated. Treat fumes as toluene diisocyanate (TDI) and provide adequate ventilation.

polycarbonate

An abbreviation used for this type of material is PC. Also known as a bisphenol-A polycarbonate. May be referred to as poly-(4,4'-isopropylidenediphenylene carbonate) or as, poly-

(oxy-1,4-phenylene-dimethylmethylene-1,4-phenylene-oxy-carbonyl).

A linear polymer of carbonic acid which is produced from the sodium salt of *bisphenol A* and phosgene: a polyester in which the carbonate ester groups are linked by aromatic groups. PC has high heat resistance but a high melt viscosity compared to other thermoplastics.

PC is strong, stiff, hard, tough, transparent and maintains its properties over a wide range of temperatures. For example, it can maintain rigidity and toughness up to 140°C. Impact resistance is good, particularly at low temperatures, and PC is thermally resistant up to 135°C/275°F. PC is rated as slow burning (flame retardant grades are V1 and can be V0). It has reasonable electrical insulation properties but is not recommended for use in the presence of an electric arc. Because of the attractive range of properties that it offers, this clear, impact resistant material forms the basis of a wide variety of blends. See *polycarbonate blends*.

Closely related to traditional PC are aromatic polycarbonates and *polyester-carbonate*. Aromatic polycarbonates are prepared by the reaction of bisphenol A with terephthalic (or isophthalic) acid derivatives. To make polyester carbonates, carbonic acid is also used. As both materials have increased aromatic content, they are more heat resistant but less easy flowing than PC: these types of material are sometimes called polyarylates.

PC has molecular weights of, approximately, from 20,000 to 35,000 for injection moulding. If the molecular weight is higher than this then the material becomes too stiff flowing. PC is available as low, medium and high molecular weight grades. Reduced molecular weight grades of PC, cycle faster as they are processed at a lower temperature and have lower viscosities. Suggested lubricants, to improve the flow of PC, include stearyl stearate, calcium stearate and montan esters. As with all lubricants, levels should be kept as low as possible, for example, well below 1%.

PC is a stiff flow, hygroscopic material which is processed at high temperatures and at these high temperatures, decomposition, or degradation, by hydrolysis can be severe so, ensure that the material is very dry before it is put into the barrel. Drying of PC materials is essential as any trace of moisture will either cause a loss of optical quality or, a loss of impact strength as the molecular weight is reduced by hydrolysis. Usually if a 'wet' plastics material is being processed then, excessive moisture is shown by the presence of bubbles in the melt. Such is not the case with PC: in some cases a loss of strength is only noticed in service. The moisture level must be reduced to below 0·02% so dry the material at 120°C/248°F, in a dehumidifying drier for approximately 3 hours and then place in a heated hopper. Wetness in the granules may be assessed by the *Thomasatti's Volatile Indicator* (TVI) test.

Shrinkage is of the order of 0·006 to 0·008 mm/mm ie 0·6 to 0·8%. The addition of 30% glass fibre will reduce the shrinkage to 0·003 to 0·005 mm/mm. PC/PBT blends will be roughly 0·008 to 0·01 mm/mm.

PC is resistant to inorganic acids and most dilute organic acids: aliphatic hydrocarbons, saturated cyclic hydrocarbons, oxidizing and reducing agents, greases, fats, oils, alcohols (but not methanol) and, detergents. Good resistance to ionizing radiation. It is not resistant to hot water; avoid contact with water at temperatures above 60°C/140°F. Not resistant to amines, alkaline solutions and to, ammonia and its solutions. Swollen by benzene, chlorobenzene, acetone and carbon tetrachloride. Soluble in solvents such as methylene chloride, ethylene chloride, chloroform, trichlorethane and metacresol. In general, not resistant to strong bases, aromatic and chlorinated hydrocarbons, esters and ketones. As PC is swollen by benzene, it is liable to stress cracking when in contact with aromatic fuels: PC/PBT blends resist this type of fuel and also methanol-based fuels.

A PC based on bisphenol A has a density of approximately 1·3 gcm/^{3} and being clear a very wide colour range is available. PC is serviceable up to temperatures of 135°C/275°F. Such materials have a very high carbon content and give very characteristic, infra-red absorption spectra.

PC materials should not be allowed to solidify onto the surface of a screw and/or barrel as the bond formed between the polymer and the steel can be very strong; strong enough to detach the surface from the remaining core. If for any reason the machine needs to be shut down, always purge out with high density polyethylene (HDPE).

PC may be fabricated with a very smooth surface but it has limited scratch and abrasion resistance. This type of plastics material can have limited resistance to notches, chemicals and UV light. It is susceptible to crazing when strained and this last point, mars an otherwise excellent resistance to creep. More sensitive to abrasion than *nylon (PA)* so, PC is not recommended for components subjected to abrasion, for example, gears. The sensitivity to notches, as measured by the Izod impact strength, is dependent on specimen thickness and on molecular weight. For a given molecular weight material As measured by MFR) the notched impact strength will remain roughly constant as the specimen thickness is steadily increased. At a certain increased thickness, the impact strength will fall dramatically (to about a quarter of its original value). High MFR materials (lower molecular weight materials) reach this critical thickness sooner than low MFR materials. So, to avoid reaching this critical thickness keep the wall thickness of components as thin as is practical and use the highest molecular weight grade possible: always use generous radii on all corners and angles.

PC is an engineering thermoplastics material which combines a high level of mechanical, optical and thermal properties. It has a Vicat softening point (VST) of 150°C/302°F and an impact range that stretches from −40 to 135°C/−40 to 275°F. However, good impact strength is only achieved in the products if they have been produced from very dry material. The main applications are in electronics, electrical engineering, medical applications, glazing and lighting engineering. Used for syringes, covers, spectacles, safety helmets, cameras, hair dryers, heated combs, iron handles. Has been used to replace stainless steel in medical applications: inert to blood and readily sterilized (autoclaving, gas and gamma radiation). Used to produce compact audio discs which are required to have high dimensional accuracy and good surface quality low internal stress levels and orientation. Used to produce impact-resistant sheet by extrusion.

polycarbonate/acrylonitrile-butadiene-styrene blend
Also known as a polycarbonate/ABS blend or as a PC/ABS blend. A thermoplastic blend based on polycarbonate (PC) and acrylonitrile-butadiene-styrene. ABS/PC blends may be made compatible because of the copolymer effect - the styrene units and the acrylonitrile units make the *immiscible system* compatible.

PC is used in blends with ABS as such blends are readily coloured, are strong, have good light fastness, high heat distortion temperatures and, are easier to mould by, for example, injection moulding than PC alone. Such materials are used in automotive applications. Where weathering resistance is required, then PC/ASA blends are preferred. Flame retardant (FR) grades of PC/ABS blends are used in the computer industry for mouldings which must have V0 ratings. Such blends use significantly reduced quantities of halogenated flame retardants and do not use antimony trioxide. See *polycarbonate blend*.

polycarbonate blend
A *thermoplastic blend* which is predominantly polycarbonate. Polycarbonate (PC) thermoplastics materials are being used to make blends or alloys so as to overcome some deficiencies

of one, or both, of the base materials. May wish, for example, to improve the notch sensitivity of the PC, the ease of processing of the PC or to extend its range of use. For example, the use of PC in the automotive industry is hampered by poor petrol resistance, low temperature behaviour and hydrolysis resistance. Rubber modified blends of PC and polybutylene terephthalate (PC/PBT) are tough at −50°C/−58°F, have high heat resistance, fuel and weather resistance and can also be painted and ultrasonically welded. Fire resistance is improved by blending with polyphosphonate (POP).

Such blends are often *immiscible blends* which may be made compatible by chemical modification. See *polycarbonate/thermoplastic polyester blend, polycarbonate/acrylonitrile-butadiene-styrene blend* and *polycarbonate/polyethylene blend*.

Blends with properties similar to those of PC/ABS, with roughly similar properties, may be produced by blending PC with SMA. Blends of PC with *thermoplastic polyurethane (TPU)* are used for tough flexible components.

polycarbonate copolymer
An abbreviation used for this type of material is PC-CO. Such a material may also be referred to as a copolycarbonate. A copolymer which is based on, for example, two different types of bisphenol. See, for example, *tetramethylcyclohexane bisphenol polycarbonate*.

polycarbonate fibre
A manufactured, polymeric fibre which contains the linkage -O-CO-O- and which is based on *polycarbonate*. Has been used to make temporary stitchings (in the form of monofilaments) and used, for example, in tailoring as the fibre is removed by dry-cleaning solvents.

polycarbonate/PBT blend
See *polycarbonate/thermoplastic polyester blend*.

polycarbonate/PE blend
See *polycarbonate/polyethylene blend*.

polycarbonate/PET blend
See *polycarbonate/thermoplastic polyester blend*.

polycarbonate/polyethylene blend
Also known as polycarbonate/PE blend or as a PC/PE blend. A thermoplastic blend based on polycarbonate (PC) and polyethylene (PE): used as the *blend* is less notch sensitive than PC. There is no inter-action between the two phases: the PE makes the PC less notch sensitive by introducing holes which hinder crack propagation.

polycarbonate/thermoplastic polyester blend
Such blends are also known as a polycarbonate/PBT blend or as, polycarbonate/PET blend or as PC/PBT blend or as, PC/PET blend. Polycarbonate (PC) thermoplastics materials are being used to make thermoplastic blends with thermoplastic polyesters such as *polyethylene terephthalate* (PET) and *polybutylene terephthalate* (PBT). Such blends are often *immiscible blends* which may be made compatible by chemical modification. With thermoplastic polyesters (such as PET and PBT) an ester inter-change reaction during processing results in the compatibilization of an otherwise incompatible system. Such materials are used in automotive applications because of their relative ease of processing together with fuel resistance. See *polycarbonate blend*.

polycarbonate-based blend
See *polycarbonate blend*.

polycarboxylate
A *pendant ionic polymer*, that is, an *ionic polymer* in which the bound anion (−CO$_2^-$) is pedant to the main chain.

polychlorinated biphenyls
An abbreviation used for this type of material is PCBs. A family of aromatic chlorinated hydrocarbons which are similar to *dioxins*.

polychlorinated dibenzo-para-dioxins
An abbreviation used for this type of material is PCDDs. See *dioxins*.

polychlorinated dibenzofurans
An abbreviation used for this type of material is PCDFs. A family of aromatic chlorinated hydrocarbons which are similar to *dioxins*.

polychloromethyloxiran
An abbreviation used for this type of material is CO. Epichlorhydrin homopolymer rubber. See *epichlorhydrin rubber*.

polychloroprene
See *chloroprene rubber*.

polychlorotrifluoroethylene
A *chlorofluorohydrocarbon* plastics material. An abbreviation used for this type of material is PCTFE. May be represented as -(-CF$_2$-CFCl-)$_n$-. In many respects similar to *polytetrafluoroethylene* (PTFE) but is easier processing as it does not crystallize so easily. Can be *melt* processed at approximately 250°C. Also used as a coating for chemical plant and may be preferred to PTFE as pin-hole free coatings may be produced. Even though the chemical resistance is slightly inferior to PTFE the chemical resistance of the coated substrate may be better because of this feature. A harder material than PTFE and with a higher tensile strength. Transparent films amy be produced from this material. Chlorotrifluoroethylene forms useful copolymers with ethylene: see *chlorotrifluoroethylene-ethylene copolymer*.

polychlorotrifluoroethylene elastomer
See *fluororubber*.

polychlorotrifluoroethylene rubber
An abbreviation used for this material is CFM. Sometimes referred to as polychlorotrifluoroethylene elastomer. A copolymer of chlorotrifluoroethylene and *vinylidene fluoride*. See *fluororubber*.

polychromatic light
Light of many colours or wavelengths. See *chromatic aberration*.

polycondensation
See *condensation polymerization*.

polycyanurate reaction injection moulding
An abbreviation used for this term is polycyanurate RIM. A *reaction injection moulding* (RIM) process which produces highly cross-linked products based on *polycyanurate*: the process of producing such a moulding. See *cyanuric acid*.

polycyclic
A material with more than one aromatic (benzene) ring: usually means that many rings are present. See *ladder polymer* and *graphite*.

polydehydro diconiferyl alcohol
See *lignin*.

polydialkylsilicone
See *polyorganosiloxane*.

polydichlorophosphazene
See *phosphonitrilic polymer*.

polydichloropropyl acrylate
Also known as poly-(polydichloropropyl acrylate). This material is a plasma degradable acrylic polymer which is used to make a *plasma resist*. See *photoresist*.

polydicyclopentadiene
An abbreviation used for this type of material is PDCP. A polymer developed specifically for *reaction injection moulding* (RIM). A cross-linked polymer rapidly formed by metathesis of the low viscosity monomer (dicyclopentadiene or DCP) initiated by a coordination-type catalyst in the absence of oxygen and/or water by, for example, *reaction injection moulding* (RIM).

polydicyclopentadiene reaction injection moulding
Also known as dicyclopentadiene reaction injection moulding. An abbreviation sometimes used for this term is PDCP RIM. A *reaction injection moulding* (RIM) process which produces highly cross-linked products: the process of producing such a moulding

The highly strained norbornene ring, of *dicyclopentadiene*, is probably opened first during polymer production but the less strained cyclopentene ring also opens to form cross-linked structures, that is *polydicyclopentadiene* is formed in the presence of a coordination-type catalyst.

The coordination-type catalyst is formed by the reaction of a tungsten (for example, WCl_6) or molybdenum compound with alkylaluminium chloride (Et_2AlCl). A two-part RIM system can be formulated by putting, for example, a soluble tungsten compound (based on WCl_6) together with DCP in one tank and DCP plus alkylaluminium chloride (Et_2AlCl) in another. Di-n-butyl ether is used to delay formation of the initiator and so reduce the polymerization rate (this can be too fast even at room temperatures).

Products have high modulus, high impact strength but a comparatively low distortion temperature. The impact strength is improved, and air-entrainment reduced during mould filling, by the use of liquid *styrene-butadiene-styrene* thermoplastic elastomer (approximately 5%). Gas addition to provide mould packing is not essential. The temperature of the reactants is typically 35°C, mould temperatures are of the order of 60°C while demould times may be 30 s: the adiabatic exotherm temperature for a typical formulation without filler can be 170°C.

After mould release, the surface skin (approximately 50 μm) oxidizes and this oxidized layer gives good paint adhesion. PDCP has good solvent resistance because of its olefinic nature: it readily releases from the mould and so mould releases are not required.

polydimethyl glutarimide
Also known as poly-(dimethyl glutarimide). An abbreviation used for this type of plastics material is PMGI. This material is used as a *positive photoresist* in place of *polymethyl methacrylate* (PMMA) as, for example, it absorbs at longer wavelengths than PMMA in *photoresist* applications.

polydimethyl siloxane
See *polydimethylsiloxane*.

polydimethylsiloxane
An abbreviation used for this type of material is MQ. Also known polydimethyl siloxane or as, dimethyl silicone or as, methyl silicone. A polymer based on alternating silicon and oxygen atoms in the main chain: the silicon atoms has two methyl (alkyl) groups (R & R' = CH_3) attached. The most common polymer which is formed by hydrolysis of *dimethyldichlorosilane* and the basis of most silicone fluids, greases, rubbers and plastics materials. A *silicone rubber* containing methyl groups.

polydisperse polymer
A polymer whose molecules are of different sizes or lengths: a polymer which has a molecular weight distribution. Most commercial synthetic polymers are of this type. See *polydispersity index*.

polydispersity index
A measure of the width of a *molecular weight* distribution. Defined as the ratio of weight average molecular weight to the number average molecular weight, that is, $\overline{M}_w/\overline{M}_n$. See *polydisperse polymer*.

polyelectrolyte
An ionic polymer with a high concentration of ionic groups: usually a water soluble polymer which acts as an electrolyte. For example, salts of acrylic and methacrylic acids are polyelectrolytes. See *ionic polymer*.

polyelectrolyte complex
See *polysalt*.

polyenantholactam
See *nylon 7*.

polyene
An organic chemical with more than one double bond. See *polybutadiene*.

polyester
A polymer in which the repeated structural unit in the chain is of the ester type. The polyester is linear and thermoplastic if derived, either actually or formally, from either monohydroxy-mono-carboxylic acids by self-esterification, or by the interaction of diols and dicarboxylic acids.

polyester alkyd
See *alkyd resin, granular polyester moulding compound* and *polyester moulding compound*.

polyester elastomer
See *polyether ester*.

polyester elastomer block copolymer
A category of *thermoplastic elastomer* which may be referred to as *polyether ester* elastomer.

polyester fibre
A manufactured, polymeric fibre composed of at least 85% by weight of an ester of a substituted, aromatic, carboxylic acid.

polyester moulding compound
An abbreviation used for this type of composite material is PMC. Such compounds are based on *unsaturated polyester* (UP) resins and a reinforcement such as *glass fibre*: other additives include catalysts, fillers, release agents, thermoplastics materials and colouring systems. There are many different types of PMC and these are nearly always referred to by abbreviations such as DMC (which stands for dough moulding compound).

Included in the polyester moulding compound category is bulk moulding compound (BMC), continuous roving moulding compound - a wound moulding compound (XMC), continuously impregnated compound (CIC), *dough moulding compound* (DMC), *granular moulding compound* (GPMC or GMC), *low viscosity moulding compound* (ZMC), *sheet moulding compound* (SMC) and *thick moulding compound* (TMC).

polyester moulding compound recycling
Polyester moulding compounds (PMC) can be can be disposed of by landfill or by incineration: such compounds may also be recycled by, for example, *particle recycling*. See *reclamation*.

polyester moulding compound regrind
An abbreviation used for this type of material is PMC regrind or PMC-R. The ground material obtained when a component made from *polyester moulding compound* is ground or reduced in size. See *particle recycling*.

polyester plasticizer
A polyester which has a relatively high molecular weight (a few thousand) but which is still liquid. Often complex, linear

polyesters made by reacting dibasic acids (such as adipic, azelaic and sebacic) with a polyglycol. See *permanent plasticizer*.

polyester plastics
Originally this term was synonymous with alkyd plastics but now the term is more usually applied to other materials based on ester linkages. See *alkyd plastics, polyester moulding compound, thermoplastic polyesters* and *unsaturated polyester*.

polyester polyol
A polyester which has a relatively high molecular weight (a few thousand) and which is hydroxyl terminated. Prepared by *polyesterification* using an excess of a *diol* and used in *polyurethane* production. Complex, linear polyesters made by reacting dibasic acids (such as adipic or phthalic acid) with a diol (such as ethylene diol or propylene diol).

polyester polyurethane rubber
An abbreviation used for this material is PAUR or AU. A type of *polyurethane rubber* which is based on polyols containing ester groups.

polyester reaction injection moulding
See *unsaturated polyester reaction injection moulding* and *reaction injection moulding process*.

polyester resin
Usually taken to mean *unsaturated polyester resin*. See also *thermoplastic polyester*.

polyester sheet moulding compound
A *sheet moulding compound* based upon an *unsaturated polyester resin*. Also see *high performance sheet moulding compound*.

polyester substituted polyphosphazene
An abbreviation used for this type of material is MEEP. A *polymeric electrolyte* based on a *polyphosphazene*. MEEP forms complexes with a large number of metal salts, for example, $LiCF_3SO_3$. MEEP-$LiCF_3SO_3$ has a conductivity of approximately 10^{-4} ohm^{-1} cm^{-1} which means that in thin film form the complex is suitable for battery use. See *polyethylene oxide-salt complex*.

polyester TPU
An abbreviation used for a polyester *thermoplastic polyurethane*.

polyester-carbonate
A copolymer polyester which is based on the reaction of *bisphenol A, phosgene* and a *dicarboxylic acid*, for example, terephthalic acid. An abbreviation used for this material is PEC. See *polycarbonate*.

polyester-imide
A polymer which contains both ester and imide groups. Such materials are usually *aromatic plastics materials* which are used for their temperature resistance in, for example, wire coatings. A *modified polyimide*.

polyesterification
The formation of a *polyester* by forming ester groups which lead to chain extension.

polyether
A polymer which contains the ether group -O-. That is $(-R-O-)_n$ where R can be a simple or complex organic group, for example, methylene to give polyoxymethylene. See *acetal*.

polyether amide
A polymer which contains the ether group and the amide group in the polymer chain. Often such materials are block copolymers. See *polyether block amide*.

polyether block amide
An abbreviation used for this type of thermoplastics material is PEBA: PEB and TPE-A are also used. Also known as polyether amide or as, elastomeric polyamide or as, polyamide elastomer or as, polyether polyamide block (sequential) copolymer or as, thermoplastic copolyether or, thermoplastic elastomer - amide based.

The synthesis of block or sequential copolymers provides a route to polymers which have interesting properties such as resilience. For example, if synthesis is carried out in the molten state by *polycondensation* between polyether (diol) blocks and polyamide (dicarboxylic) blocks then a linear thermoplastic copolymer results. The long chain molecules of this material can consist of numerous blocks; those based on *polyamide* (PA) will confer rigidity to the system, whereas those based on polyether (PE) will confer flexibility. The appropriate association of rigid blocks and flexible blocks within the long chain can lead either to a tough, flexible plastic or to an elastomer. A very wide range of materials is therefore possible as the basic ingredients may be varied as well as the ratios used, e.g. PE:PA ratio may vary from 80:20 to 20:80 (as PE increases, a more flexible, soft amorphous material results).

The PE blocks may be based on polyethylene glycol (PEG), polypropylene glycol (PPG), polytetramethylene glycol (PTMG) etc, whereas the polyamide blocks may be based on homopolymers (e.g. PA 6, PA 66, PA 11, PA 12) or copolymers (for example, PA 6/11). The nature of the PE block influences hydrophillic and antistatic properties whereas that of the PA block influences melting point, density and chemical resistance. The melting point is also influenced by the length of the PA blocks, whereas hardness and flexibility are influenced by the mass ratio of PE:PA.

In general, PEBA-type materials have good impact strength, abrasion resistance and high tensile strength. Flexibility is unaffected by quite large changes in temperature, for example, from -40 to $80°C$. They recover well even after large and rapid deformations and have a comparatively low density (1·01 to 1·15): *polyether-ester* (PEEL) ranges from 1·18 to 1·25 and *thermoplastic polyurethane* (TPU) is 1·16 to 1·38. PEBA materials cover a wide hardness range (from 65D to 60A) which means they are softer than PEEL and as hard as some plastics. Hardness is almost independent of temperature variations.

These materials can tolerate large additions of filler (resemble high *vinyl acetate* (EVA) copolymers). Correct choice of polymer and carbon black addition can achieve a surface resistivity of 104 ohms (semi-conductive). Easy to over-mould onto other materials as they are very tolerant during processing and adhere well. PEBA does not break in an impact test (even when notched) at temperatures of $20°C$ - most specimens do not break even at $-40°C$. Can be a very easy flow material. Shrinkage is in the order of 0·5 to 1%: soft grades can give *negative shrinkage* values. Vicat softening points range from 63 to $186°C$. As the materials are virtually colourless, a wide range of colours is possible - including pastel shades.

Low PE content grades similar to PA, that is, resistant to most chemicals except strong acids and alkalis. Very slightly swollen by dilute sulphuric acid, dilute sodium hydroxide and water, more swelling with alcohols (butanol, propanol, ethanol). Slight swelling with paraffin, more with petrol. High PE content materials, liable to attack and subsequent swelling by contact with aromatic hydrocarbons. At similar hardness, the chemical resistance of PEEL and PEBA is similar.

PEBA is not resistant to hydraulic fluids based on aromatic oils. Swollen by aromatic hydrocarbons. The higher the PE content the softer is the material and the lower is the chemical resistance, for example, to ethylene glycol. Softer grades may be swollen by more than 300% in contact with chlorinated solvents such as trichloroethylene.

Widely used as an injection moulding material, for example, as keyboard pads for calculators and computers. Sports footwear, such as trainer and football boot soles; parts of ski-boots. In this application, advantage is taken of the material's good over-moulding characteristics and resistance to flexing. Also used for pump membranes and bellows for automotive use. When filled with an inorganic filler, PEBA is used for watch straps. Semi-conductive materials are used for sleeving on car aerials in order to get earthing. Because of the material's sound dampening characteristics, it is used for gaskets in loudspeakers.

polyether ester
An abbreviation used for this type of material is PEEL. Also abbreviated to TPE-E or COPE or YBPO or TP-EE or TEEE or Y-BPO. Also known as polyether ester elastomer or as, thermoplastic copolyester or as, polyether ester block copolymer or as, thermoplastic polyether ester or as, block polyether ester or as, copolyether ester or as, polyester elastomer.

This type of material is a block copolymer (for instance based on butylene terephthalate and polytetramethylene glycol terephthalate) and is made from (i) dimethyl terephthalate (ii) a polyglycol (polytetramethylene glycol terephthalate) and (iii) a short chain diol (such as butanediol). Such materials are *thermoplastic elastomers* which are available in a range of hardnesses (35 to 82 Shore D). Each molecular chain of these block polymers consists of hard segments which can crystallize together (based on butylene terephthalate). These hard blocks link the chains together by thermally reversible, crystalline network structures. The other (soft) segments are joined to the hard segments by ester linkages and it is these soft segments which are responsible for elastomeric behaviour.

The structure consists of a crystalline network superimposed on an amorphous network. As the crystal structures have a high melting point these materials resist relatively high temperatures. Softer grades are more resilient but have a lower tear strength than the harder grades; they also absorb more water and give higher elongations. Impact strength is excellent for all grades even at $-40°C/-40°F$. These materials have a crystalline melting point from 145°C to 225°C/293 to 437°F - dependent on the grade selected, for example, 145 to 195°C/293 to 383°F for a 40 D grade; 220°C/338°F for a 72 D grade.

PEEL materials also have high tensile strength, tear strength, good colouring possibilities and resistance to abrasion. By the incorporation of suitable stabilizers, grades can be produced which are more heat, light and UV resistant. The material competes with both rubbers and plastics. In typical rubber applications: flexible diaphragms can be injection moulded directly in PEEL replacing the multi stage assembly of fabric reinforced, *nitrile rubber* which has to be vulcanised.

PEEL materials perform like conventional cross-linked elastomers over a wide temperature range but soften and flow like more conventional *thermoplastics* at elevated temperatures so can be processed by *ignition moulding* and *extrusion*. Generally shrinkages range from 0·004 to 0·016 mm/mm (i.e. 0·4 to 1·6%). Post moulding shrinkage (PMS) occurs and the value is dependent upon the processing conditions used (in particular the melt and mould temperatures); typical values of up to 0·2% are obtained.

PEEL is resistant to very low temperatures (softer grades down to $-55°C/-67°F$; harder grades $-30°C/-22°F$) without becoming brittle. They are resistant to many oils and solvents. Harder compounds are more resistant than soft. Serviceable in dilute acids and bases, hydrocarbons, alcohols, ketones, esters, petroleum based solvents, oils and hydraulic fluids. Resistant to hydrolysis: resistance may be improved by blending with 2% *polycarbodiimide* masterbatch. PEEL is attacked by hot, concentrated acids and bases; affected by phenols, glycols, cresols and some chlorinated solvents (such as chloroform). Some swelling of components can occur when kept immersed in solutions such as toluene, ethanol, methanol, cyclohexanone, acetone, ethyl acetate and methyl isobutyl ketone.

As the natural colour of this material ranges from cream to white then a wide colour range is possible. The density is approximately 1·12 to 1·27 gcm^{-3}: the harder the grade, the higher is the density and the melting point (190°C/374°F for a 40 Shore D and 221°C/430°F for a 74 D). The mouldings do not feel as waxy as polyethylene as their frictional characteristics are approximately half-way between those of PE-LD and a traditional *natural rubber* (NR) vulcanizate (for the softest grades). High elongations are possible - 700% for a 40 D; 350% for a 72 D grade.

All grades must be dried before use as these materials are prone to hydrolytic degradation at their processing temperatures. Dry for 3 to 4 hours at 110°C/230°F in a desiccant dryer. When drying materials in ovens without dehumidifiers, dry for 5 to 6 hours at 110°C/230°F.

These materials have good low temperature flexibility, abrasion resistance and flex-fatigue resistance. Some grades can withstand prolonged exposure to temperatures exceeding 120°C/248°F. PEEL has been used for segmented tracks of snow vehicles, seals, gaskets, gears, fasteners, connectors, sport-wear (such as ski-boot soles), shoe soles, diaphragms, thin membranes, watch straps, shock and noise isolators, seals and packings, wheels, rollers and slow speed tyres. Also used as exterior panels, bumpers, radiator grilles and body protection trim components for the automotive industry. Widely used as a flexible pipe material. When used outdoors the materials must be protected by incorporating a UV masterbatch or a black masterbatch.

polyether ester block copolymer
See *polyether ester*.

polyether ester elastomer
See *polyether ester*.

polyether ether ketone
An abbreviation used for this type of material is PEEK. Also known as polyarylether ketone or as, aromatic polyether ketone. A semi-crystalline thermoplastics material (30%) with a high sharp melting point of 334°C/634°F.

PEEK contains aromatic groups (p-phenylene rings), carbonyl groups (CO) and ether linkages. May be represented as $-(-\phi-CO-\phi-O-\phi-)-)_n-$ (where ϕ is a benzene ring). It therefore resembles a *sulphone polymer*: if the CO group was replaced by a sulphone group (SO_2) then a sulphone polymer would result. PEEK has, however, a higher maximum continuous-use temperature. The Underwriters Laboratory (UL) rating is up to 250°C/482°F whereas for polyether sulphone (PES) it is 180°C/356°F. PEEK is tough, abrasion resistant, fatigue resistant and flame resistant: it has extremely low smoke and toxic/corrosive gas emission. Retains its good electrical insulation properties to over 200°C/392°F. Resists wear and has low friction so it is therefore a good bearing material.

Granules for *injection moulding*, are sold in unreinforced and reinforced grades, such as 20 and 30% glass fibre and 30% carbon fibre. Adding *carbon fibre* raises the heat distortion temperature from 160°C/320°F to 315°C/599°F. GF PEEK is more durable at 200°C/392°F than glass fibre (GF) filled polyphenylene sulphide (GF PPS) although the PPS is stiffer. As the natural colour of PEEK is a grey/brown and because of the high processing temperature, the colour range is limited.

Although the processing temperature of PEEK is higher than for many thermoplastics the melt viscosity is similar to *polycarbonate*. The unreinforced grade has a melt viscosity of 100 Pa·s at a temperature of 380°C/716°F and at a shear rate of 1000 s^{-1}. Shrinkage is of the order of 0·013 to 0·020 mm/mm, that is, 1·3 to 2·0%. The shrinkage of fibre filled grades can be very low (for example, 0·2%). The unfilled material has a density of 1·3 gcm^{-3}.

PEEK is resistant to gamma radiation (it is twice as good as *polystyrene* which is regarded as exceptional) and *environmental stress cracking*. Insoluble in common solvents. Complex mixtures of aromatic solvents (for example, trichlorophenol/phenol) will slowly dissolve it at high temperatures. Hydrolysis resistance (to high pressure, hot water) is exceptional: can be used for thousands of hours at temperatures in excess of 250°C/482°F. Has excellent tribological properties over a wide range of conditions. PEEK is not resistant to concentrated nitric acid and liquid bromine: these will cause degradation particularly at high temperatures. Dissolves, with degradation, in concentrated sulphuric acid.

PEEK will absorb 0·5% w/w atmospheric moisture during storage and this means that drying is necessary as, the recommended level of moisture before processing by, for example, injection moulding is <0·1%. Dry in a hot air oven for 3 hours at 150°C/302°F, in a desiccant drier for 2 hours at 180°C/356°F. If PEEK is left in the machine barrel during cooling, barrel damage may result because of the strong adhesion than can develop between this plastic and metal.

Can be bonded to itself or other materials. The surfaces are usually degreased or etched by a solvent or chromic acid. An acid etched surface often increases the bond strength by more than 30% compared to a surface which has been solvent degreased. The types of adhesives used are epoxy, cyanoacrylate and silicones. Epoxy based adhesives tend to give the best bond strength. In addition to being joined by adhesives, the mouldings can be welded together by ultrasonics or friction.

A high priced engineering thermoplastics material processed by injection moulding into, for example, electrical connectors which can withstand high temperatures - during soldering and in-service. Because of the stiffness, heat resistance and hydrolysis resistance of the GF material, it is used as pump impellers and as valve components for circulating high pressure, hot water. Also used for bearings and piston linings.

polyether group
May be represented as -R-O-: where R is an organic group. If R is the methylene group then the material is *acetal*: if R is phenylene then the material is *polyphenylene oxide*.

polyether imide
Also known as polyetherimide. An abbreviation used for this type of material is PEI. Injection moulding materials are *modified polyimides*, for example, poly-{2,2'-bis(3,4 dicarboxyphenoxy)phenylpropane-2-phenylene bisimide} which may be referred to as PEI-M.

Polyimides contain the chemical group -N< and are thus related to the polyamides. Each nitrogen atom is joined to two carbonyl (CO) groups and this means that each main chain contains ring, or ladder, sections as each of the CO groups extends the chain and both are attached to the same benzene ring. Such ring or ladder sections are responsible for the high temperature resistance of polyimides. However, they also contribute towards chain stiffness and so unmodified polyimides have very high viscosities: too high to be processed by conventional means. Chain flexibility can be introduced by the incorporation of other chemical groups, for example, the ether group -O-.

Such modified materials may be prepared from bis-phenol A, 4-nitrophthalimide and m-phenylenediamine. This gives an amorphous, thermoplastics material in which the aromatic imide units are connected by more flexible ether groups. The structure gives, strength, stiffness, inherent flame resistance, good thermal stability, retention of properties over a wide temperature range and good electrical insulation properties - even without reinforcement. The stable dielectric constant and low dissipation factor make it suitable for use at frequencies in excess of 10^9, even at elevated temperatures. Has a high oxygen index (LOI is 47) and is therefore difficult to ignite; only gives off a little smoke when burnt. Has a high glass transition temperature (T_g) of 215°C/410°F and a maximum, UL, continuous use temperature, with impact, of 170°C/338°F. (Copolymers are available which have even higher heat resistance but these are approximately five times more expensive). When reinforced, for example, with glass fibre (GF), the strength characteristics (for example, the tensile yield strength) are improved up to a loading of approximately 40%.

PEI-M is a stiff flow material which exhibits similar flow characteristics to *polycarbonate* (PC); because of its thermal stability may be processed, for example, by injection moulding, over a wide temperature range. Shrinkage is of the order of 0·006 mm/mm ie 0·6% and the addition of 30% GF will reduce the shrinkage to 0·2 to 0·4% but will put the density up from 1·27 to 1·51 gcm^{-3}. Unfilled PEI-M has a natural colour of a transparent orange/brown: this means that a very wide colour range is not possible but both transparent and opaque colours are available. Has a heat deflection temperature of 200°C/392°F at 1·82 N/mm^2.

PEI-M is resistant to most hydrocarbons (including petrol and oils), freon-based cleaners and refrigerants, alcohols and fully halogenated solvents. Good resistance to mineral salt solutions, mineral acids (dilute or concentrated), but only has short term resistance to bases - even dilute bases. Hydrolytically stable to boiling water and resistant to autoclaving, UV, and ionizing (for example, gamma) radiation; such resistance is inherently good and does not depend upon the use of additives. It is not resistant to partially halogenated solvents, for example, methylene chloride and chloroform will dissolve PEI-M. Components are not recommended for use in an electrical arcing environment at temperatures above 170°C/338°F.

As PEI-M will absorb 0·25% water in 24 hours at room temperature this means that drying is always necessary before melt processing by, for example, injection moulding. Dry in a hot air oven for a minimum of 4 hours at 150°C/302°F: best dried in a desiccant dryer for 4 hours at 150°C/302°F and until the water content is less than 0·05%.

A lot of stress may be built into injection moulded components and this may be shown by immersion in benzyl alcohol or dichloromethane. Dried over-stressed components can be stress relieved by annealing at 200°C/392°F for 2 hours per mm of wall thickness. PEI-M may be joined to itself using solvents such as methylene chloride. Suggested adhesives include polyurethanes and epoxides. Commonly welded using techniques such as ultrasonic and hot plate (mirror). This material is notch sensitive and so good design practice must be followed, i.e. avoid sharp corners, abrupt changes in wall thickness etc.

PEI-M is used in applications which require some of the following properties: high temperature stability, high physical strength, inherent flame resistance, low smoke evolution, excellent electrical insulation properties (over a wide temperature and frequency range), UV stability and chemical resistance to aliphatic hydrocarbons, acids and dilute bases. Can withstand the conditions of wave and vapour phase soldering: used in connectors, printed circuit boards, switches, controls and integrated test devices. May be moulded into sections as thin as 0·25 mm/0·01 in. A useful feature of this material is the low coefficient of thermal expansion: when filled, it is similar to that of steel and lower than that of zinc or aluminium. This material is available as a multi-layer, coextrusion for microwave cookware: the three layers are PEI-M/PC/PEI-M.

polyether ketone
Also known as polyetherketone: a type of *polyether ether ketone* (PEEK). An abbreviation used for this material is PEK. A semi-crystalline (approximately 35% crystalline) thermoplastics material which contains both ether (-O-) and ketone groups (-CO-) between the aromatic (benzene) rings. May be represented as $-(-\phi-CO-\phi-O-)_n-$ (where ϕ is a benzene ring) which is a shorter repeat unit than PEEK: has a higher melt temperature (approximately 370°C) than PEEK. Has good chemical resistance, hydrolysis resistance and is of low flammability. Has high continuous use temperatures, higher than PEEK, particularly when reinforced with, for example, *carbon fibre*.

polyether ketone ketone
Also known as polyetherketone ketone. An abbreviation used for this type of material is PEKK. A semi-crystalline thermoplastics material which contains both ether (-O-) and ketone groups (-CO-) between the aromatic (benzene) rings. May be represented as $-(-\phi-CO-\phi-CO-\phi-O-)_n-$ (where ϕ is a benzene ring) which means it is a type of *polyether ether ketone* (PEEK): has a high melt temperature (approximately 385°C), good chemical resistance, hydrolysis resistance and is of low flammability. Has high continuous use temperatures particularly when reinforced with, for example, *carbon fibre*.

polyether polyamide block (sequential) copolymer
See *polyether block amide*.

polyether polyol
A *polyether* which is hydroxyl terminated: used in *polyurethane* production, with a diisocyanate, to form a polyether-polyurethane. Polyoxypropylene triols are used as the basis for *flexible polyurethane foam production*.

polyether polyurethane rubber
Also known as a polyether-polyurethane or as a, polyether polyurethane. An abbreviation used for this type of material is PEUR or EU. A type of *polyurethane rubber* which is based on a *polyol* containing ether groups.

polyether sulphone
See *sulphone polymers*.

polyether TPU
An abbreviation used for a polyether *thermoplastic polyurethane*.

polyether-polysiloxane
See *silicone-polyether block copolymer*.

polyether-polyurethane
See *polyether polyurethane rubber*.

polyetherimide
See *polyether imide*.

polyetherketone
See *polyether ketone*.

polyethyl acrylate
An amorphous polymer which together with poly-n-butyl acrylate, forms the basis of *acrylic rubbers*. See *acrylic monomer*.

polyethyl methacrylate
An amorphous thermoplastics material based on the *acrylic monomer*, ethyl methacrylate. The *Vicat softening point* of this material is approximately 40°C lower than that of the more widely used polymer, *polymethyl methacrylate*. See *polymethacrylic acid esters*.

polyethylene
A polymer prepared by the polymerization of ethylene as the sole monomer. May be represented as $-(CH_2-CH_2)_n-$: a very important semi-crystalline thermoplastics material. See *polyethylene plastics* and *ethylene plastics*. Also known as polyethene or, poly(ethylene). An abbreviation used for this type of material is PE. See *high density polyethylene* and *low density polyethylene*.

polyethylene - density measurement of
See *flotation method*.

polyethylene fibre
Fibre made from *polyethylene*. Very strong fibres are based on *ultra-high molecular weight polyethylene*: the fibres have few surface imperfections. Claimed to be ten times stronger than steel and significantly stronger than nay other fibre currently commercially available. Such fibres (for example, Dyneema and Spectra) can give very strong composites: such composites have been used to make automotive (car) armour for doors and seats. Such fibres have also been incorporated into *glass reinforced plastic* mouldings so as to reduce impact damage to, for example, boat hulls.

polyethylene glycol
An abbreviation used for this material is PEG. PEG is better referred to as polyethylene glycols as PEG consists of a range or blend of materials with different degrees of polymerization. In the following formula n may range from say, 4 to 60 dependent on grade. $HO(CH_2CH_2O)_xH$. Low molecular weight materials are viscous liquids whereas the higher molecular weight grades are hard wax-like solids. PEG forms the basis of a number of plasticizers, for example, *polyethylene glycol dibenzoate*.

polyethylene glycol alkyl ester
A *polyester* formed by the esterification reaction of *polyethylene glycol* with an organic acid. May be represented as $RCO(OCH_2CH_2)_nOOCR$ where R is an *alkyl* group.

polyethylene glycol di-2-ethylhexanoate
A *polyester* formed by the esterification reaction of *polyethylene glycol* with an organic acid. A comparatively high molecular weight *plasticizer* or softener. Gives good retention of low temperature flexibility.

polyethylene glycol dilaurate
A *polyester* formed by the esterification reaction of *polyethylene glycol* with lauric acid. A comparatively high molecular weight plasticizer or softener with a relative density of about 0·97.

polyethylene glycol dibenzoate
A *polyester* formed by the esterification reaction of *polyethylene glycol* with benzoic acid. Materials which are used as softeners (for rubbers) and plasticizers (for example, for PF resins). Has a relative density of about 1·15 and a molecular weight which can reach approximately 800 to 900. May be represented as $C_6H_5CO(OCH_2CH_2)_nOOCC_6H_5$ where n may have a value of up to about 12.

polyethylene glycol monophenyl ether
A comparatively high molecular weight *plasticizer* or softener. Used, for example, with *polyvinyl acetate*.

polyethylene - high density
See *high density polyethylene*.

polyethylene - ionomer
See *ionomer*.

polyethylene - linear low density
See *linear low density polyethylene*.

polyethylene - low density
See *low density polyethylene*.

polyethylene oxide
An abbreviation used for this material is PEO or PEOX. Polyethylene oxides consists of a range of materials with different degrees of polymerization. In the following formula n may range from say, 4 to 25,000 dependent on grade. -$(CH_2CH_2O)_n$-. Low molecular weight materials, which are hydroxyl terminated, are *polyethylene glycols*. When the degree of polymerization is very high then the materials may be processed as thermoplastics materials by calendering, extrusion etc. As the products are water soluble, they are used for example, in packaging to make water soluble films. PEO forms *polyethylene oxide-salt complexes* which are polymeric electrolytes and may be used for alkali metal rechargeable batteries.

polyethylene oxide-alkali metal salt complex
A *polyethylene oxide-salt complex* which contains *alkali metal ions*. See *polyethylene oxide-salt complex*.

polyethylene oxide-salt complex
An abbreviation used for this type of material is PEO-salt complex. The regular *polyethylene oxide helix* is filled by metal ions (M^+), for example, an alkali metal ion: the counterions (X^-) are outside the parallel strands of the regular helix. Polyethylene oxide form ion complexes with, for example, lithium complexes such as $LiClO_4$ and $LiCF_3SO_3$. Such a compound is a polymeric *electrolyte* and it may be used for alkali metal rechargeable batteries.

Compared to inorganic solid electrolytes, the *polymeric electrolyte* has a low conductivity: this can be compensated for by using thin films and working at high temperatures, for example, 120°C. A cell for a alkali metal rechargeable battery may consist of a lithium-based foil anode, an alkali metal salt-PEO complex (for example, $PEO/LiCF_3SO_3$ of thickness 40 μm) and a vanadium oxide-based cathode. Other polymeric electrolytes offer the potential of room temperature use, for example, polyester substituted polyphosphazene.

Compared to lead acid batteries, polymeric batteries may offer faster recharging, smaller volume and the ability to be shaped to a given space.

polyethylene plastics
Plastics based on polymers made with ethylene as essentially the sole monomer. In common usage for this plastic, essentially means no less than 85% ethylene and, no less than 95% total olefins. Can have, for example, *linear low density polyethylene* (LLDPE), *low density polyethylene* (LDPE) and *high density polyethylene* (HDPE). These are produced as follows:

Process	Product	Density (kg/m³)
Gas phase	LLDPE, HDPE	900 to 960
High pressure	LDPE	915 to 935
Modified high pressure	LLDPE, HDPE	880 to 940
Slurry	HDPE	935 to 960
Solution	LLDPE, HDPE	900 to 960.

When the density of *polyethylene* (PE) is low, 0·910 to 0·925 g/cm³, a PE material is sometimes known as Type I; when the density is medium at 0·926 to 0·94 g/cm³, a PE material is sometimes referred to as Type II. Any PE, for example, HDPE, with a density of 0·940 to 0·959 g/cm³, is sometimes known as Type III. If the density is higher than 0·959 g/cm³, that is, very high, then the material is known as type IV.

The property differences, compared to the one left blank, are as follows.

Property	Density		
	Low Type I	Medium Type II	High Type III
Tensile strength at rupture	Highest	Higher	
Elongation at break	Highest	Higher	
Impact strength	Highest	Higher	
Modulus	Highest	Higher	
Transparency	Highest	Higher	
Long term load bearing		Higher	Highest
ESC resistance	Highest	Higher	
Softening temperature		Higher	Highest
Melt strength		Higher	Highest
Gloss	Highest	Higher	
Resistance to shrinkage	Highest	Higher	
Resistance to warpage	Highest	Higher	
Resistance to brittleness at low temperatures	Highest	Higher	
Resistance to grease and to oil absorption		Higher	Highest
Impermeability to gases and to liquids		Higher	Highest
Freedom from (film) haze		Highest	Higher
Moulding cycle times		Higher	Highest

polyethylene terephthalate
An abbreviation used for this type of material is PET. PETP is sometimes used while PETG is used for a copolymer; OPET means oriented PET and CPET means crystalline PET. A *thermoplastic polyester* which is sometimes referred to as a linear aromatic polyester.

PET is prepared from ethylene glycol (EG) and, either terephthalic acid (TPA) or, the dimethyl ester of terephthalic acid (DMT). It is an aromatic polyester (like *polycarbonate* or PC), as it contains both carboxylic groups and benzene (aromatic) groups. Number average molecular weights range from 18,000 to 42,000. The glass transition temperature (T_g) is approximately 165°C/329°F and the material crystallizes over a temperature range, from 120 to 220°C/248 to 428°F, with a maximum rate occurring at a temperature of 190°C/374°F.

PET is best known as a fibre (Terylene, Dacron), a film material (Melinex, Mylar) and as a bottle making material. It was not originally considered as an *injection moulding* material because of high temperature moisture sensitivity, poor impact strength, excessive warpage when glass-fibre (GF) filled and slow rate of crystallinity - this slows the moulding cycle. However, it has a higher modulus, gloss and heat distortion temperature (HDT) than *polybutylene terephthalate (PBT)*: such properties can only be achieved if mouldings are crystalline.

During injection moulding crystallinity can be promoted by using a *nucleating agent*, on which the crystals can grow, plus a crystal growth promoter or accelerator. High mould temperatures (e.g. 130°C (266°F)) and fibre reinforcement are necessary with such materials if high stiffness is required (plasticized materials such as Rynite may be run at lower mould temperatures). Particulate fillers such as mica help to minimize warping (caused by GF addition). To off-set the brittleness introduced by the use of particulate fillers, a rubber (*acrylic*-type) may be added. Super tough, reinforced grades are available which have high flexural modulus and excellent impact resistance; Claimed to be 50% tougher than GF PC at room temperature.

PET is often used in blends with PBT, for injection moulding, as the use of PET improves gloss and scratch resistance (PET is hard and has a low coefficient of friction). GF PET has good load bearing characteristics and low creep; the coefficient of thermal expansion is similar to that of brass or aluminum. Low moisture pick-up means retention of good electrical insulation characteristics under hot and moist conditions. Unfilled materials can give clear mouldings if low mould temperatures are used. High clarity mouldings are also obtained if a copolymer, also known as a *copolyester*, is used (see *glycol modified polyethylene terephthalate*).

PET can have easier or better flow characteristics than PBT: this allows the use of thin-walled sections and helps to reduce warpage problems in injection moulding. When unfilled, shrinkage of the amorphous grades is of the order of 0.2 to 0.35 mm (0.2 to 0.35%): the crystalline grade is of the order of 1.8 to 2.1 mm/mm (1.8 to 2.1%). By adding glass fibre the shrinkage value is dramatically reduced: a 36% GF PET grade may be 0.2% in the flow direction but 1.8% in the transverse.

PET is resistant to many of the more usual chemicals like oils, fats, aromatic solvents, chlorinated hydrocarbons and most acids and dilute alkalis (attacked by concentrated sulphuric, nitric and acetic acids). Similar chemical resistance to PBT. Resistant to bleaching solutions and reducing agents. Degree of crystallinity affects chemical resistance. Solvents which cause swelling (e.g. esters and partially halogenated hydrocarbons) will turn amorphous components opaque by promoting crystallinity. Amorphous materials subject to environmental stress cracking (ESC by petrol, some alcohols, esters and ketones) but crystalline components are not subject to ESC. Very good stain resistance; good outdoor and UV resistance. PETG is resistant to dilute aqueous solutions of mineral acids, bases, salts, aliphatic hydrocarbons, alcohols and, a range of oils

PET is not resistant to hot water or steam; do not use continuously in water above 50°C/122°F dissolved by m-cresol, o-chlorophenol, phenol/chlorinated hydrocarbon mixtures (e.g. tetrachlorethane) and hot benzyl alcohol. Not resistant to oxidizing mineral acids and concentrated basic solutions - particularly when hot. GF PET is more prone to alkali attack than an unreinforced material. PETG is swollen, and/or dissolved, by halogenated hydrocarbons, low molecular weight ketones and aromatic hydrocarbons.

PET can exist as an amorphous, thermoplastics material, in an oriented partially-crystalline state and, in a highly ordered, crystalline state. The density of the amorphous material is approximately 1.33 g/cm^3, that of the oriented partially-crystalline state is 1.37 g/cm^3 and, that of the highly ordered, crystalline state, is 1.45 g/cm^3. (The SG of PETG is approximately 1.27). Many applications for PET (for example, film) require orientation and/or crystallization because of the improved properties that result. Crystallization in extruded film is induced by re-heating, drawing and biaxially orienting amorphous PET at temperatures of approximately 90°C followed by *annealing* at about 200°C. Since oriented PET (OPET) is a pure and regulated material meeting, for example, FDA food contact regulations, it is widely used in food packaging. As a *blow moulding* material, it is most often seen as bottles used, for example, to package carbonated drinks: the preforms on which the bottles are based may be produced by injection moulding. OPET containers are transparent, creep resistant, shatterproof and resistant to gas and vapour passage.

The structure of PET injection mouldings, may be amorphous or crystalline in both the unfilled and filled materials. Mouldings may therefore be transparent or opaque - if filled they are opaque. As the natural colour of the crystalline material is an off-white then a wide colour range is possible. PETG is brilliantly clear and mouldings have a high gloss with a scratch resistant surface. Light transmission can be high (around 90%).

The melt processing of PET is done at high temperatures, for example, 270 to 290°C/518 to 554°F and after extensive pre-drying. Being an ester, melt processing in the presence of water will cause decomposition, or degradation and this causes molecular weight reductions and embrittlement. Drying in an ordinary oven is not good enough as it is difficult to consistently dry the material to the required level of below 0.02%. Dry the material in a dehumidifying dryer at 165°C/330°F for 4 hours.

Thin complex sections (for example, for transformer bobbins) may be formed easily by injection moulding because of the materials' ease of flow even when fibre-filled. Components are very stiff and are used as metal replacements - such as for the housings and other components of toasters, coffee-machines, industrial plugs and sockets, and water meter housings. Used in heater housings on cars and trucks. Super tough grades used for car grilles and fuel filler flaps. Usage within the automotive industry is growing rapidly, for instance, for windshield wiper arms, ignition coil housings, lamp sockets, sun roof frames, window cranks, pump housings, distributor and alternator components. Amorphous grades are used mainly for bottles and medical ware. Copolymers are used in toys, face shields, components for medical use e.g. test tubes and bottles.

polyethylene terephthalate glycol
An abbreviation used for this material is PETG. See *polyethylene terephthalate* and glycol modified *polyethylene terephthalate*.

polyfluorophosphazene rubber
See *polyphosphazene rubber*.

polyheptanoamide
See *nylon 7*.

polyhydric
Means that an organic chemical has more than one hydroxyl group. See *glycerol*.

polyhydroxybutyrate
An abbreviation used for this material is PHB. Also known as poly-(3-hydroxybutyrate) or as biopolymer or as, isotactic biopolymer or as, microbially produced poly-(3-hydroxyalkanoate).

Polyhydroxybutyrate plastics materials are polyesters which are non-oil based, biodegradable, thermoplastics materials which can be produced by microorganisms from various carbon-based substrates, for example, from glucose or, methane or, gaseous mixtures of carbon dioxide and hydrogen. Marketed by Marlborough Biopolymers (an ICI subsidiary) as Biopol.

The homopolymer, poly-(3-hydroxybutyrate), is a polyester with the formula -[CH(CH$_3$)CH$_2$COO-]$_n$- and which is produced by the bacterium alcaligenes eutropus when this bacterium is grown in solutions containing carbon-based materials. Copolymers can also be produced if this bacterium is fed with acetic and propionic acid under conditions of nitrogen or phosphorous depletion. The copolymer contains 3-hydroxybutyrate (HB) and 3-hydroxypentanoate (also known as 3-hydroxyvalerate or HV). A given species of bacteria will reproducibly synthesize polymers of a fixed molecular weight under specified conditions.

PHB is soluble in a number of solvents, for example, in chloroform, trifluoroethanol and dichloreacetic acid: such solvents can be used to determine the molecular weight. Poorer solvents include propylene carbonate, butan-2-one and 1,2-dichloroethane (DCE). Gels can be produced form such solvents and the plastics material can be processed into fibres from the gel if required. The standard, high molecular weight biopolymers do not contain low molecular weight oligomers.

PHB is a relatively brittle, semi-crystalline, thermoplastics material with a melting point of approximately 175°C/347°F and with mechanical properties which resemble those of *polypropylene*. When heated it will degrade at comparatively low temperatures. It will degrade to crotonic acid at melt processing temperatures (2-pentenoic acid is also liberated from copolymers). At 190°C/374°F the decomposition, or degradation, time for a material with a molecular weight of half a million is approximately 2 minutes. A HB-HV copolymer (15% HV) is much more stable as it can be processed at 165°C/329°F. When any biopolymer is melt processed, care must be taken to minimize melt temperatures and melt residence times: the use of an *inverted temperature profile* may be justified. Standard stabilizing systems do not appear to be effective: repolymerization of the degraded plastics material with dicumyl peroxide and triallyl cyanurate can be effective. Cold rolling will improve the ductility of brittle PHB film and sheet as spherulitic cracks are healed by this process. Self seeding will also improve ductility as a higher nucleation density and a smaller spherulitic size is obtained. Drawn oriented fibres can be obtained but the extruded material must be partially crystalline before drawing.

Highly polar plasticizers are effective with PHB and its copolymers. Polybutyl acrylate appears to be an impact modifier for this type of material. Calcium hydroxyapatite filled PHB is being evaluated as a material for use as a bone substitute as it is biocompatible and biodegradable. Most immediate use for PHB, and its copolymers, could be as processing aids for *unplasticized polyvinyl chloride* (UPVC). At a 1% addition level, they reduce the gelation time, power requirements and improve the surface finish of extrudates. PHB polymers exhibit piezoelectric effects and have good oxygen barrier properties.

polyimide
An abbreviation used for this type of material is PI. These intractable polymers are used where exceptional heat resistance is required (see *modified polyimide*). The structure may be based on a fully aromatic structure (to give *aromatic polyimide*) or on a partial aromatic structure to give the so-called *aliphatic polyimide*. Both contain in the main chain two carbonyl groups (>C=O) linked to the same nitrogen atom to give imide carbonyl groups.

The fully aromatic polyimides are the best known and are produced from an aromatic dianhydride and an aromatic diamine. Because the final polymer is intractable, the materials are processed as low molecular weight precursors which polymerize and cross-link when heated - used, for example, to make polyimide sheet moulding compound. See *high performance sheet moulding compound*.

polyion
See *ionic polymer*.

polyion complex
See *polysalt*.

polyisobutene
See *polyisobutylene*.

polyisobutylene
Also known as polyisobutene. The standard abbreviation used for this material is IM although PIB is probably more commonly used. See *butyl rubber*.

polyisobutylethylene
See *polymethyl pentene*.

polyisocyanurate
An abbreviation used for this type of material is PIR. A polymer based on the six-membered isocyanurate ring. Used to prepare cellular materials which are of interest as they have better fire resistance than *rigid polyurethane foam*. As such materials are brittle, polyisocyanurate-polyurethane combinations are used.

polyisocyanurate reaction injection moulding
A *reaction injection moulding* (RIM) process which produces highly cross-linked products based on *isocyanurate*: the process of producing such a moulding. As the polyisocyanurate polymer is brittle, and large exotherms are generated, approximately 30% *polypropylene glycol* may also be used and the material reinforced with *glass mat*. The cross-linked polymer is based on urethane bonds and isocyanurate groups. See *structural reaction injection moulding*.

polyisoprene
A polymer of *isoprene* ($CH_2=C(CH_3)CH=CH_2$. Upon polymerization, both cis and trans isomers can be formed. In the cis structure, the CH_3 group and the pendant hydrogen atom lie on the same side of the main chain and are prevented from rotating, to another configuration, by the double bond. See *balata*, *gutta-percha* and *natural rubber*.

polyisoprene rubber
See *isoprene rubber*.

polyketone
An abbreviation suggested for this type of plastics material is PK. A polyketone contains aromatic groups (for example, p-phenylene rings) and carbonyl groups (CO): it may be represented as $-(-\phi-CO-)-)_n-$ (where ϕ is a benzene ring). Such a material is therefore an aromatic, heat resistant polymer. However, most commercial polyketones contain ether linkages so as to obtain chain flexibility: such polymers may then be processed on conventional equipment as used for *thermoplastics materials*. The modified polymers may be called aromatic polyether ketones and may be represented as $-(-\phi-CO-\phi-O-\phi-)-)_n-$ (where ϕ is a benzene ring). See *polyether ether ketone*.

polylauryl methacrylate
A *polymethacrylic acid* ester which is made from lauryl methacrylate: the polymer is a rubbery material which is, for example, used as a viscosity modifier in lubricating oils.

polymer
A chemical compound made up of a large number of repeating units. See *polymer - synthetic*.

polymer alkoxy radical
A radical formed during *oxidation*. May be depicted as RO·.

polymer alloy
Also simply called an *alloy* or *blend*. Incompatible blends may be made compatible by chemical modification, addition of a third component or by varying the processing employed. Such an *incompatible blend* may then be called a polymer alloy.

polymer analogous reaction
A chemical reaction which may be performed on both a polymer and on lower molecular weight material or homologues.

polymer blend
Sometimes called a polyblend. In the simplest case, a physical mixture of two or more polymeric materials. If the properties of a product are significantly better than would be expected from the use of a simple mixture then the material may be referred to as an *alloy* or as a polymer alloy.

polymer cement
A hard cross-linked material: produced by ionically cross-linking an aqueous solution of *polyacrylic acid* with a metal oxide, for example, *zinc oxide*. Used as dental materials.

polymer cement concrete
Obtained by mixing a water soluble resin with the ingredients used to make OPC (ordinary *Portland cement*). Such a mix is used to make road and bridge deck repairs.

polymer concrete
An abbreviation used for this type of material is PC. Consists of a mineral or synthetic filler (for example, quartz or granite) and resin binder. By using different resins, and fillers, the polymer concrete, can be given different characteristics. After cold mixing of the major ingredients, a hardening system is added (for example, *catalysts, hardeners and accelerators*) so that the mix will subsequently set or harden. The catalyzed mix is poured into an appropriately shaped, and sized, mould

so as to make, for example, bases for machinery and equipment. Such PC bases have good vibration and damping resistance, better than the traditional material which is cast iron. The base can be self-coloured and can be accurately cast to size so that post-moulding operations (such as painting and machining) are eliminated. PC bases are also inert and thermally stable; very important for laser-based machines. Cutting tools mounted on such bases last longer (may be 30% longer) because of the reduced, transmitted vibration.

Unsaturated polyester (UP) resins are widely used as the properties are adequate from this relatively cheap material. Epoxides give better casting tolerances whereas vinyl ester resins give better chemical resistance than UP. Methyl methacrylate, furane resins and polyurethane have also been used.

polymer crystallites
See *self seeding*.

polymer homologues
Polymeric materials which have the same structure but which have a different molecular weight. See *molecular weight distribution*.

polymer hydroperoxide
See *hydroperoxide-containing polymer and oxidation*.

polymer impregnated concrete
Made by injecting a resin into hardened cement concrete so as to give a material with properties slightly better than *Portland cement* concrete.

polymer isomers
Materials of the same chemical composition but the molecules have different molecular arrangements and thus different structures.

polymer modified natural rubber
A chemically modified *natural rubber* produced by the polymerization of a *vinyl monomer* in either solution or in *latex form*. Monomers include styrene, vinyl acetate, acrylonitrile, and methyl methacrylate: the products is a rubber copolymer graft. Peroxides or hydroperoxides are also used, that is, a redox system is employed. See *comb-grafted natural rubber*.

polymer oxidation
The reaction of a *polymer* with *oxygen*. A chain process which is initiated and propagated by free radicals. See *thermally-initiated oxidation*.

polymer peroxide radical
A radical formed during oxidation. May be depicted as ROO·. See *oxidation*.

polymer radical
A radical formed during oxidation. May be depicted as R·. See *oxidation*.

polymer - synthetic
Many commercial rubbers/elastomers and plastics are based on the element carbon and are synthesized, or made, from simple, oil-based raw materials. These starting materials are called monomers and these simple, low molecular weight materials are put together, by a process known as *polymerization* (polymerisation), so as to form polymers. This term means that the final product consists of chains which contain many identical, repeat units. Polymers may be linear, branched or cross-linked and usually a range of molecular weights are present in commercial materials.

Because the final molecular weight, or mass, is so large the material may also be referred to as a 'high polymer' or, as a 'macromolecule'. Many rubbers and plastics are based on one monomer and are known as 'homopolymers': some are based on two monomers and are known as 'copolymers'. When they are based on three monomers they may be known as 'terpolymers'. Natural polymers may also be referred to as 'biopolymers'.

All plastics are polymers but not all polymers are plastics. Cellulose is a polymer but it cannot be processed like a *plastics material* unless it is modified.

Over the past 50 years the usage of polymeric materials throughout the world has increased dramatically: it now stands (1991) at approximately 113 million tons. Of this total, plastics materials account for about 100 million tonnes. In terms of tonnage thermoplastics are by far and away the most important type of plastics material. Approximately 80% of all plastics used throughout the world are thermoplastics.

It should be noted that a large number of grades are usually available for any one material, therefore, any properties quoted in the literature should only be used as a general guideline. Also, the properties of any plastics material may also be dramatically changed by the processing conditions employed and by the use of additives. See *additives and orientation*.

polymeric 4,4-diphenyl methane diisocyanate
An abbreviation used for this material is polymeric MDI. This material is also known as crude MDI. See *4,4'-diphenyl methane diisocyanate*.

polymeric electrolyte
An *electrolyte* which is based on a polymer such as, for example, *polyethylene oxide*, polypropylene oxide or polyester substituted polyphosphazene. See *polyethylene oxide-salt complex*.

polymeric MDI
An abbreviation used for *polymeric 4,4-diphenyl methane diisocyanate*.

polymeric plasticizer
Also called a *permanent plasticizer*.

polymeric resist
See *photoresist*.

polymeric sulphur
A polymeric form of sulphur. Obtained by heating *sulphur* to temperatures greater than 150°C but below 180°C. If this material is quenched rapidly, then plastic sulphur is produced.

polymerisation
See *polymerization*.

polymerizable plasticizer
A *plasticizer* which is capable of being polymerized, for example, on heating. See *rigisol*.

polymerization
The process used to produce a *polymer*. That is, monomers are converted to polymers which may be, for example, *homopolymers* or *copolymers*. For the polymer industry, it is important to produce polymers of high molecular weight as low molecular weight materials will not possess the required properties, for example, of strength and stiffness. With some materials, which precipitate out of solution, *melt polymerization* is the only method that can be used to produce the required *degree of polymerization*. Synthetic polymers may be produced by *condensation polymerization* or *addition polymerization*.

Biopolymers are also commercially available. The best known *biopolymer*, and the one most commercially used, is *natural rubber*: another is *cellulose*. Now biopolymers, whose biosynthesis was instigated by man, are available: these are plastics materials produced by bacterial action on materials such as glucose. See *polyhydroxybutyrate, emulsion polymerization, mass polymerization, solution polymerization and suspension polymerization*.

polymerization inhibitor
A substance which retards, or inhibits, *polymerization*.

polymerization number
See *degree of polymerization*.

polymerized 1,2-dihydroxy-2,2,4-trimethylquinoline
An abbreviation used is for this type of material is TMQ. This is a strongly discolouring antioxidant. See *staining antioxidant, dihydroquinoline derivatives* and *ketone-amine condensates*.

polymetallosiloxane
A *polysiloxane* with some of the silicon atoms replaced with metal atoms: a heat resistant copolymer. See *polyorganosiloxane*.

polymethacrylic acid esters
Also known as polymethacrylates: esters of methacrylic acid. The methyl, ethyl, propyl and butyl esters of methacrylic acid are amorphous thermoplastics materials. See, for example, *polymethyl methacrylate*. If longer side chains are used then more rubbery materials result. See, for example, *polylauryl methacrylate*.

polymethanal
See *paraformaldehyde*.

polymethyl acrylate
An abbreviation used for this type of material is PMA. A tough, leathery polymer produced from the *acrylic monomer, methyl acrylate* by *free radical polymerization*. It is used as, for example, a textile size.

polymethyl isopropenyl ketone
Also known as poly-(methyl isopropenyl ketone). An abbreviation used for this type of material is PMIPK. This material is used as a *positive photoresist* in place of *polymethyl methacrylate* as it has better spectral response and reacts more quickly in *photoresist* applications.

polymethyl methacrylate
An abbreviation used for this type of material is PMMA. Also known as acrylic and sometimes as poly-[1-(methoxycarbonyl)-1-methylethylene].

When methyl methacrylate is polymerised the beautifully clear thermoplastics material PMMA is obtained. It is a water white (or crystal clear), amorphous plastic with a glass transition temperature (T_g) of approximately 100°C/212°F. At room temperature it is a hard and rigid material which maintains its clarity even in thick sections and after long exposures to outside atmospheres. What limits the clarity after outdoor exposure is surface scratching caused by airborne dust or by cleaning. The clarity can be approximately 92% and so this material is widely used as a glass replacement. It is significantly lighter than glass and is easier to shape and machine and is considered as one of the hardest thermoplastics: the hardness of PMMA is about the same as aluminium.

To obtain good clarity, it is important to maintain surface finish to a very high standard; a high standard of storage or drying is also required and/or, a vented barrel is necessary when *injection moulding*. PMMA will absorb 0.3% water in 24 hours at room temperature and, for injection moulding a moisture level of below 0.1% is necessary. If drying is necessary, dry in a hot air oven for 2-4 hours at 75°C/167°F.

Acrylic components are rigid, dimensionally stable, odourless, resistant to many common chemicals and easy to decorate. They have low moisture absorption and have tremendous UV resistance. It is also possible to obtain grades which are softer, tougher and which have better environmental stress cracking (ESC) resistance: such materials are almost as transparent as basic grades but possess more haze. The better craze resistance is useful when products come into contact with aqueous detergents or soap solutions. The *Vicat softening point* (VSP) of such materials is slightly lower than basic grades by approximately 5°C.

A stiffer flowing material than polystyrene (PS): shrinkage is of the order of 0.004 to 0.007 mm/mm (0.4 to 0.7%). The density is 1.18 gcm^{-3} and the natural colour is a very clear, water-white and so a very wide range of both clear and transparent colours is possible.

PMMA is resistant to dilute acids and alkalis; concentrated alkalis and concentrated hydrochloric acid; aqueous solutions of salts and oxidising agents, fats, oils and aliphatic hydrocarbons (for example, white spirit and paraffin), dilute alcohols and detergents. It is not resistant to concentrated, oxidising acids (such as nitric and sulphuric) and alcoholic alkalis which cause decomposition. Soluble in most aromatic and chlorinated hydrocarbons (such as toluene and chloroform), esters (ethyl acetate) and ketones. Plasticised by ester-type materials such as tritolyl phosphate and dibutyl phthalate; swollen by alcohols, phenols, ether and carbon tetrachloride. Ethyl acetate can be used to detect strain.

This material may be joined to itself using solvents such as chloroform or, by using solutions of PMMA in methylene chloride. Could also use solutions of PMMA in MMA; these cements can be set by the addition of a catalyst or hardener. PMMA mouldings are commonly welded using techniques such as ultrasonic welding, hot plate and friction.

PMMA possesses excellent light fastness and resistance to weathering; it is also hard, rigid, transparent and mouldings can have a very good gloss. These properties dictate the applications of this class of material (it is also not subject to microbiological attack). Some of its major applications are lenses, outdoor and indoor light fitting covers, housings for vending machines, dials, nameplates and control panels. Often seen as lenses/covers for rear light displays on cars; in this application multi-coloured, injection mouldings are often used. The ease of decoration of this type of material should not be forgotten as such decoration can lead to an enhanced appearance of injection mouldings. PMMA cast sheet is used for thermoforming and for signs. See *polymethacrylic acid esters*.

polymethyl methacrylate grafted natural rubber
A modified *natural rubber (NR)*. Natural rubber which has been grafted with methyl methacrylate (MMA) so as to produce a grafted natural rubber (NR-MG). The MMA is polymerized in the presence of NR *latex*: the grafted latex is *coagulated* and made into crepe.

polymethyl pentene
Also known as polymethylpentene or as, poly-4-methylpent-1-ene or as, poly-(4-methylpent-1-ene) or as, poly-4-methylpentene-1 or as, poly-(4-methylpentene-1) or as, polyisobutylethylene. A polymer made from an α olefin (methyl pentene) by *Ziegler-Natta polymerization* into isotactic polymers. The monomer has the formula $CH_2 = CHR$ where R is the branched alkyl group which may be represented as $-CH_2CH(CH_3)_2$. An abbreviation used for this type of material is P4MP1 although it is more generally known by the trade name/trademark TPX.

A stereoregular, isotactic polymer which is crystalline and yet, which is also highly transparent - this is because the density/refractive index of the amorphous phase is similar to that of the crystalline phase at approximately 0.83 g/cm^3. Has low density, good transparency, chemical inertness and a high softening point (*Vicat softening point* of approximately 179°C and a melting point of approximately 240°C). Copolymers, for example, with hex-1-ene, are even more transparent than homopolymers and so commercial materials are thought to be copolymers. Highly permeable to gases, easily oxidized at processing temperatures and prone to *environmental stress cracking* in the presence of polar solvents. Used, for example, for medical and laboratory ware where a heat resistant, clear, sterilizable material is required. See *polyolefins*.

polymethylethylene
See *polypropylene*.

polymethylpentene
See *polymethyl pentene*.

polymethylstyrene
See *polyvinyltoluene*.

polynonamethyleneurea
See *nylon 91*.

polynonanoamide
See *nylon 9*.

polynorbornene rubber
Also known as polynorbornene. An abbreviation used for this material is PNR. A polymer obtained by the ring-opening polymerization of norbornene. The repeating unit is of a five-membered ring and a double bonded section so that both cis and trans isomers are possible. May be represented as -(-R–CH=CH-)$_n$- where is R is the five-membered ring (C_5H_8). Only shows rubbery behaviour when plasticized, for example, with hydrocarbon oils. The polymer is supplied as a powder and the additives, for example, *plasticizer* and curing system, may be initially added in a *high speed mixer*. Capable of accepting large amounts of filler and of plasticizer so as to offset the high price of the base polymer. Can be cured by conventional *sulphur* systems to give products with useful damping properties.

polynosic fibre
A regenerated cellulose fibre with a high initial wet modulus and good resistance to swelling by alkalis. See *viscose rayon* and *modal fibre*.

polynuclear phenol
Structures based on phenols which contain more than one aromatic nucleus. Formed, for example, during *resole manufacture*. See *bridged hindered phenol* and *phenolic antioxidant*.

polyoctanoamide
See *nylon eight*.

polyoctenamer rubber
An abbreviation used for this material is TOR. Also known as trans-polyoctenamer rubber. An alkenamer formed by the ring-opening polymerization of cyclooctene by *Ziegler-Natta polymerization*: both cis and trans isomers are possible. Useful rubbery materials are made with a high trans content. A *polyalkenamer* which may be represented as -(-R-CH=CH-)$_n$- where is R is $(CH_2)_6$. Many of the polymer chains probably exist as large rings (macrocyclic) which gives rise to few chain ends: this in turn is thought to account for the high tensile strength of even low molecular weight material. Main use appears to be as an additive (<30 phr) for other rubbers so as to improve *green strength* and flow properties. More commercially successful than *polypentamer rubber*.

polyol
A material which contains two or more hydroxyl groups. Often taken to mean materials which are polymeric and which have hydroxyl end groups. These groups can be used for chain extension by a chemical reaction. Such polyols are therefore prepolymers and are used, for example, to make *polyurethanes* by reaction with a *diisocyanate*. Can have *polyester polyols* and *polyether polyols*.

polyolefin
An abbreviation used for this type of material is PO. A polymer prepared by the polymerization of an olefin(s) as the sole monomer(s); at least 50% of the resin must be olefin.

The polymer may be represented as -(CH$_2$CRR')$_n$- in which R, for example, is hydrogen: R' is either hydrogen or an alkyl group, such as -CH$_3$. Both homopolymers and copolymers are available. Commercially important members of the polyolefin thermoplastics family include:

very low density polyethylene or *VLDPE*;
low density polyethylene or *LDPE*;
linear low density polyethylene or *LLDPE*;
high density polyethylene or *HDPE*; and,
polypropylene or *PP*.

Such materials are also called poly-α-olefins and may be called vinyl polymers although this is not common usage.

Some rubbers are also sometimes classed as polyolefins, for example, *ethylene-propylene rubber* and *butyl rubber*. Some copolymers are also sometimes classed as polyolefins, for example, *ethylene-vinyl acetate copolymer* and *ionomer*.

polyolefin/natural rubber blend
Also known as natural rubber/polyolefin blend: NR/PO blend. A relatively hard composition which is based on *natural rubber* and one or more *polyolefins*. A physical mixture of *natural rubber*, a polyolefin, for example, polypropylene (NR/PP blend) and sometimes also with high density polyethylene. Such a blend may be used as a *thermoplastic elastomer* if there is sufficient polyolefin present as this allows the mixture to be melt processed by, for example, *injection moulding*.

The micro-crystalline regions of the PP impart stiffness and reinforcemt: the *high density polyethylene* improves the impact strength of the blend. The NR improves the low temperature impact strength of the blend while the uses of cross-linking chemicals such as *m-phenylenebismaleimide* improves the impact strength of the blend even further: see *dynamic cross-linking*. The PP is subjected to high shear in the presence of NR in an *internal mixer*: temperatures greater than the softening point of the PP are necessary, for example, 10 m at 180°C. The products are tough, flexible materials if the rubber particle size is below 1 μm and if they are well dispersed.

polyolefin rubber blend
See *rubber modified polypropylene*.

polyolefins
The collective name for the class of materials which are polymers of olefins: the *olefin* usually only has one double bond. See *polyolefin*.

polyorganosiloxane
A polymer based on alternating silicon (Si) and oxygen atoms (O) in the main chain: the silicon atoms have two organic (often alkyl) groups (R & R') attached. May be represented as -(-SiRR'-O)$_n$-. May also be known as a *polysiloxane* or as polydialkylsilicone. The most common polymer is *polydimethylsiloxane* (MQ) which is formed by hydrolysis of *dimethyldichlorosilane*. Groups other than methyl may be used so as to make a range of copolymers. For example, if some of the R groups are phenyl groups then phenylmethylsilicones (PMQ) result. If some of the R groups are vinyl groups then methylvinylsilicones (VMQ) result. If some of the R groups are β-cyanoethyl groups then nitrilesilicones result. If some of the R groups are trifluoropropyl groups then fluorosilicones result (see *silicone rubber* and *silicone plastics*). If some of the silicon atoms are replaced with metal atoms, then *polymetallosiloxanes* result.

polyoxetane
An *oxetane polymer*.

polyoxymethylene
A polymer in which the repeated structural unit in the chain is oxymethylene. May be represented as -(-CH$_2$-O-)$_n$-. Polyoxymethylene is theoretically the simplest member of the generic class of polyacetals. See *acetal*.

polyoxymethylene plastics
Acetal plastics based on polymers in which oxymethylene is essentially the sole repeated structural unit in the chains (see *acetal*).

polyoxypropylene glycol
An abbreviation used for this type of material is PPG. Also known as polypropylene glycol. A *polyether* prepared from *propylene oxide* and a monomeric *polyol*. If, for example, the diol, ethylene glycol is used then the product is the linear polymer, polyoxypropylene glycol. If, for example, a triol, glycerol is used then the product is a branched polymer.

A polyoxypropylene polyol is used in the manufacture of a *polyurethane* as prepolymers: they are reacted with *diisocyanates* to give both cellular materials and elastomers. The ends of the molecular chain are secondary hydroxyl groups and as such are not very reactive. To increase the reactivity, the materials may be capped or tipped with ethylene oxide: such a treatment gives a capped polyol.

polyoxytetramethylene glycol
Also known as polyoxytetramethylene. An abbreviation used for this material is PTMG. A *polyol* prepolymer which is a comparatively low molecular weight material made from *tetrahydrofuran*: may be referred to as polytetrahydrofuran (PTHF). A linear polymer with the formula HO-[-$(CH_2)_4$-O-]$_n$-H and which is used in *polyurethane* production.

polyparabanic acid
Also known as poly(parabanic acid) or as 2,4,5-triketoimidazolidine. An abbreviation used for this material is PPA. A heat resistant polymer used in film or lacquer form for electrical insulation purposes. The main chain contains five membered, heterocyclic rings (based on 3 oxygen atoms and 2 nitrogen atoms) separated by organic linkages.

polypelargonamide
See *nylon 9*.

polypentamer rubber
An abbreviation used for this material is TPA. Also known as trans-polypentenamer rubber. An alkenamer formed by the ring-opening polymerization of cyclopentene by *Ziegler-Natta polymerization*: both cis and trans isomers are possible. Useful rubbery materials are made with a high trans content. A *polyalkenamer* which may be represented as -(-R-CH=CH-)$_n$- where R is $(CH_2)_3$. Many of the polymer chains probably exist as *macrocyclic* structures. Can be cured by conventional *sulphur systems* to give products with useful ageing and strength properties even when very highly filled. Does not compete as a tyre rubber with more traditional rubbers because of cost and technical limitations (such as poor skid resistance) and so now, not commercially available.

polyperfluoroalkylvinyl ether
See *perfluoroalkoxy copolymer*.

polyphenol
A *bridge hindered phenol*. Obtained by linking phenolic-type structures with other ring and cyclic structures, or linkages, so as to obtain materials of very low volatility which have good heat stability. See *phenolic antioxidant*.

polyphenylene
Also known as polyphenyl. A long chain polymer based on fused benzene rings, for example, linked head to tail via the para position. Such a polymer has excellent heat stability but is intractable.

Polymers which contain aromatic groups (for example, p-phenylene) have desirable properties such as high heat distortion temperatures (HDT) and stiffness: processing is however difficult because of the high processing temperatures and because of the high pressures needed to obtain flow. Easier processing is obtained if the aromatic groups are spaced with other groups or linkages, for example, spacing with oxygen gives ether linkages. Easier polymerization, and more tractable polymers, may be obtained if the phenylene group is substituted, that is a more tractable material is obtained by using a substituted polyphenylene. Linking via the meta position gives a more tractable material: mixed isomeric linking is often present in polymers. See, for example, *aromatic polyamide*.

polyphenylene-1,3,4-oxadiazole
See *poly-1,3,4-oxadiazole*.

polyphenylene ether
An abbreviation used for this material is PPE. Usually commercial materials are based on modified *polyphenylene oxide*.

polyphenylene oxide
An *aromatic polymer* which may be represented as -(-ϕ-O-)$_n$- where ϕ is a benzene ring. In order to obtain a more tractable polymer the ring is, however, usually substituted in the 2 and 6 positions. See *polyphenylene oxide - modified*.

polyphenylene oxide - modified
Also known as modified poly-(phenylene oxide) or as, alkyl substituted polyphenylene oxide or as, polyphenylene ether (PPE) or as, poly-(2,6-dimethylphenol) or as, poly-(2,6-dimethyl-1,4-phenylene oxide). An abbreviation used for this material is PPO: also used is MPPO, PPO-M or PPE.

An *aromatic polymer* based on substituted *polyphenylene oxide* which is poly-(2,6-dimethyl-p-phenylene oxide). Confusingly, it is still referred to as polyphenylene oxide or PPO. Thermoplastics materials which are much cheaper, easier to process and which have acceptable properties for many applications, are obtained if the substituted polyphenylene oxide is blended with other thermoplastics materials, for example, with styrene-based plastics. Once again these blends are called PPO.

The term PPO therefore covers a wide range of materials as these plastics are commonly blends of modified PPO and PS or, modified PPO and TPS. It is these blends or alloys, based on substituted polyphenylene oxide, which are the materials referred to as PPO but more correctly as PPO-M. As the basic material contain ether linkages, it may also be referred to as polyphenylene ether or PPE. Now, the range of use of this type of material is being extended by making blends with other polymers, for example, with *nylon*: such materials may also be referred to as PPO or modified PPO.

In general, PPO-M is an attractive, tough, stiff material with a wide temperature range of use (for example, from $-40°$ to $130°C/-40°$ to $86°F$: at elevated temperatures it maintains its good load bearing characteristics. Because of this, and because of low water absorption, mouldings have good dimensional stability. The material also has excellent dielectric properties, a low coefficient of expansion and may be rated as self-extinguishing and non-dripping, it is an opaque material with good resistance to hydrolysis.

Viscosity is lower than polycarbonate (PC) but greater than acrylonitrile-butadiene-styrene (ABS). At $300°C/572°F$, and at $1000\ s^{-1}$, the viscosity ranges from about 90 to 365 Nsm^{-2} (dependent on grade). This amorphous material exhibits low shrinkage, e.g. 0·005 to 0·007 mm/mm (i.e. 0·5 to 0·7%) which can be considerably reduced by the addition of glass fibre (GF), for example, reduced to 0·002 mm/mm (0·2%) when 30% GF is used.

The density is approximately 1·06 gcm^{-3}. As the natural colour of the material is beige then a wide colour range is possible; usually seen as black or opaque coloured, injection mouldings. This type of material will absorb 0·07% water in 24 hours at room temperature: this means that drying is not normally necessary.

PPO-M has outstanding resistance to aqueous environments - only absorbs about 0·2% from boiling water at equilibrium. Not attacked by dilute acids, alcohols, detergents, dilute alkalis. Can withstand exposure to elevated temperatures; the heat distortion temperature ranges from 90 to 150°C/194 to 302°F - dependent on grade. It is not resistant to halogenated and aromatic hydrocarbons, ketones and low molecular weight esters. Unless adequately stabilized, UV light resistance can appear poor: unfilled grades may rapidly yellow. However, physical properties are only marginally affected: this drop is caused through microcrazing. PPO-M has good resistance to detergents.

The ability of PPO-M to be injection moulded to close tolerances has helped this material to become established for washing machine, dishwasher and pump components. It has a useful combination of properties such as good heat resistance, impact strength and flame resistance. The most important markets are in the electrical and automotive industries, for example, used for the back-plates of TV sets. Structural foam components are a major market for PPO-M, for instance for machine housings.

polyphenylene oxide/polyamide blend
Also known as modified polyphenylene oxide/polyamide blend or as, PPO/PA blend or as, PPO-M/Pa blend. An *immiscible blend* which may be made compatible by modification of the PPO and the addition of a third component - a *compatibilizer*. The PPO is modified with functional groups, for example, with hydroxyl groups so that it becomes compatible with the PA via the compatibilizer.

polyphenylene sulphide
An abbreviation used for this type of material is PPS. Also known as sulphide polymer or as, aromatic sulphide polymer or as, poly-(p-phenylene sulphide) or as, poly-(thio-1,4-phenylene).

An *aromatic* crystalline thermoplastic (see *polyphenylene*) which is based on para-phenylene groups linked by sulphur atoms: a heat resistant, engineering thermoplastics material. Because of its symmetrical structure it can easily obtain a high level of crystallinity (maximum 65%) and this rapid rate of crystallization means rapid set-up in the mould and short moulding times.

A relatively brittle material whose properties, for example, impact strength, arc and tracking resistance, are greatly enhanced by the incorporation of fillers. It is usually sold already compounded with glass fibre (40% GF), or with mixtures of glass fibres and mineral fillers for injection moulding. Such PPS composites offer good mechanical and electrical insulation properties, high temperature resistance (may withstand continuous service up to 240°C/464°F), chemical and moisture resistance, inherent flame resistance, excellent stiffness (can compare with aluminum), creep resistance, easy processing into complex precision mouldings, dimensional stability and excellent property retention at elevated temperatures. Non-flammable materials have a high oxygen index, for example, 47 - the same as *polyvinyl chloride* (PVC); glass fibre filled (GF) polybutylene terephthalate (GF PBT) could be 20 and GF polyamide (PA) 28. The notched impact strength (GF reinforced materials) is similar to PBT but lower than polyphenylene oxide (PPO) and polycarbonate (PC). PPS is now available in higher molecular weight grades which have better physical properties than the original PPS and which are more crack resistant, such as when moulded into thick sections.

PPS is rated as easy flow (easier flowing than PC or PPO: above its melting point (around 282°C/540°F) the melt becomes very fluid - gates with a diameter of 0·5 mm (0·020 in) have been used and very thin-walled moulding is possible. Shrinkage is of the order of 0·001 to 0·006 mm/mm (0·1 to 0·6%) but can be non-uniform because of fibre orientation.

PPS is insoluble in all known solvents below 200°C/392°F. Second only to PTFE in its chemical resistance. Resistant to solvents, fuel, oils, cleaning agents, most acids and bases, hydrocarbons and chlorinated hydrocarbons. Combines chemical resistance with hydraulic and dimensional stability at high temperatures. Useful properties retained at temperatures greater than 200°C/392°F, It is not resistant to hot concentrated oxidizing acids (such as sulphuric), some amines, benzaldehyde, nitromethane and some halogenated compounds.

Unreinforced PPS has a density of 1·4 gcm^{-3}: 40% GF PPS has a density of 1·6 g cm^{-3}. As the natural colour of the material ranges from a light tan to dark brown then only a restricted colour range is available. Absorbs only small amounts of water, for example, 0·05%, but this may increase if mineral fillers are used. Even if not required, drying is suggested as the heating will assist *injection moulding* because of the high heat input required.

The ability of this material to withstand high temperature without deformation has allowed it to be used in connectors, terminal blocks, sockets, coil formers, bobbins and relay components. Arc resistant grades are available which are used to mould relay bases, motor housings, lamp holders and switch components. Used as a metal replacement in the automotive industry, for carburettor parts, ignition plates, lamp sockets, flow control valves (for heating systems) and brake circuit valves. Because of the increasing use of quartz-halogen lamps, components may reach temperatures above 200°C/392°F - PPS has replaced ceramics in such applications. Replaces metals in other automotive applications: when used to make an exhaust gas emission control valve, 14 steel components were replaced by 3 made of PPS, ultrasonically welded together.

polyphenylene sulphide sulphone
An abbreviation used for this material is PPPS. An *aromatic thermoplastics* material (see *polyphenylene*) which is based on para-phenylene groups linked by sulphur atoms and by sulphone groups: a heat resistant, engineering thermoplastics material.

polyphenylene sulphone
An abbreviation used for this material is PPSU. See *sulphone polymers*.

polyphenylene-1,3,4-oxadiazole
Also known as a polyphenyloxadiazole. An abbreviation used for this material is POD. An aromatic polymer based on fused rings: based on a repeat unit of a benzene ring linked via a heterocyclic, five membered ring which contains two carbon atoms, two nitrogen atoms and one oxygen atom. Obtained by, for example, the reaction of *phthalic acid* and *hydrazine*. Such a polymer has excellent heat stability but is intractable and so is produced in, for example, film form. Used for electrical insulation. See *poly-1,2,4-oxadiazole*.

polyphenylethylene
See *polystyrene*.

polyphenyloxadiazole
An aromatic polymer based on fused rings: based on a repeat unit of a benzene ring linked via a heterocyclic, five membered ring which contains two carbon atoms, two nitrogen atoms and on oxygen atom. See *polyphenylene-1,3,4-oxadiazole*.

polyphosphazene rubber
Also known as fluorophosphonitrilc polymer or rubber or as, fluoropolyphosphazene rubber or polymer or as, polyfluorophosphazene rubber or as, fluorophosphazene rubber or as, a phosphonitrilic fluoroelastomer. An abbreviation used for this type of material is PNF. A *phosphonitrilic*-type of polymer which has good hydrolytic stability, excellent fire resistance

and solvent resistance. May be *sulphur* or *peroxide* cured to give compounds with a very low glass transition temperature and properties which are maintained over a wide range of temperatures.

Such materials are included in the category of *speciality rubber* - because of their good oil resistance and temperature resistance, for example, from −65 to 175°C. See *fluororubber*.

polyphosphonate
A *pendant ionic polymer*, that is, an *ionic polymer* in which the bound anion ($-PO_3^{2-}$) is pendant to the main chain.

polyphthalamide
A suggested abbreviation for this type of material is PPA (but see *parabanic acid*). An *aromatic* thermoplastics material based on the reaction of terephthalic acid and an amine (Amodel): an engineering thermoplastics material which resembles a crystalline *polyamide*. Such a material has good creep resistance, rigidity, mechanical strength and dimensional stability and these properties are relatively unaffected by absorbed water. When glass-reinforced the heat resistance is excellent, for example, a *heat distortion temperature* of 285°C is possible. During *injection moulding*, melt temperatures of 310°C and mould temperatures of 140°C may be required. Has a high water absorption of approximately 0.2%, a low shrinkage and a density of 1.45 g/c.cm³ when filled with 33% glass: like PPS it is extensively modified as the base polymer is brittle. When injection moulding, treat this material as glass-filled *nylon*.

polypiperidone
See *nylon 5*.

polypro
See *polypropylene*.

polypropene
See *polypropylene*.

polypropyl methacrylate
An amorphous thermoplastics material based on the *acrylic monomer*, propyl methacrylate. The *Vicat softening point* of this material is approximately 65°C lower than that of the more widely used polymer, *polymethyl methacrylate*. See *polymethacrylic acid esters* and *poly-(n-alkyl methacrylate)*.

polypropylene
An abbreviation used for this type of material is PP. Also used are PPR: PPN: PP-H: PP-K: PP-C: PP-HO: PP-CO (may also see the abbreviations without the hyphens). Also known as polypro or as, polypropene or as, poly(propylene) or as, polymethylethylene.

By the use of *Ziegler-Natta catalysts*, high molecular weight PP may be produced from the monomer propylene. Commercial PP is mainly isotactic PP, for example, 90 to 95% *isotactic*, and has a number average molecular weight of 40,000 to 60,000. May be represented as -(-CH$_2$-CH.CH$_3$-)$_n$-. The homopolymer often consists of an approximately 50:50 mixture of amorphous material and crystalline material: the amorphous material has an SG of 0.85 and the crystalline material has an SG of 0.94. The homopolymer (PP-H) is a linear, hydrocarbon plastic like *polyethylene (PE)*, but is stiffer, harder and has a higher melting point than PE. PP-H has high strength and stiffness but a low notched impact strength: becomes very brittle at about 0°C and for this reason block copolymers, with ethylene, are often preferred. Both random copolymers, with approximately 1 to 4% ethylene, and block copolymers, which can contain higher ethylene contents, are made.

It is the ethylene-propylene block copolymers which are traditionally used in place of the homopolymers. Such block copolymer materials may be known as polyallomers and referred to as either PP or, PP-K or, as PP-C or, or as PP-CO or, as PP-B (the term PP may therefore refer to either a homopolymer or to a copolymer). Copolymers have a lower heat distortion temperature (HDT), less clarity, gloss, rigidity but greater impact strength compared to PP-H. As the proportion of ethylene increases the material becomes softer and tougher. Commercial plastics materials often have 5 to 15% of ethylene blocks.

PP (both PP-H and PP-CO) can have a Rockwell hardness (R scale) of approximately 90, a density of 0.9 gcm^{-3}, a Vicat softening point of about 150°C/302°F and a heat distortion temperature of approximately 100°C/212°F. (Random copolymers have lower values than this). Mouldings can withstand boiling water and steam sterilization and do not suffer from environmental stress cracking (ESC) problems; the maximum service temperature is above 100°C, for example, 110°C/230°F. Electrical insulation properties are good: a high gloss, scratch-resistant surface is possible.

Being a semi-crystalline, thermoplastics material, PP has a relatively high *shrinkage* which is of the order of 0.018 mm/mm, that is, 1.8% but can reach 2.5% in thick sections: gives more uniform shrinkage than HDPE. Shrinkage uniformity, in *injection moulding*, can be improved by removing the high molecular weight fraction from the PP, for example, by the use of peroxides. Such treatments give easy flow materials with a reduced tendency to warp. Grades which have such a narrow molecular weight distribution may be known as controlled rheology (CR) materials. The addition of 30% glass fibre (may be known as PP GF30) can reduce the shrinkage to approximately 0.7%.

PP is resistant to a wide range of common solvents, hot water and to chemicals. Relatively unaffected by aqueous solutions, including quite strong acids and alkalis. Because of swelling problems, at room temperature, PP is not recommended for use with aromatic hydrocarbons (for example, benzene) and chlorinated hydrocarbons (for example, carbon tetrachloride). Also swollen by esters (for example DBP and DOP), ethers (for example, diethyl ether), asphalt, camphor oil and various aqueous oxidizing agents (for example, dilute nitric acid and potassium permanganate).

PP is not resistant to outside exposure, unless protected by, for example, carbon black and/or UV stabilized. Dissolved by aromatic and chlorinated hydrocarbons at elevated temperatures, for example, 85°C/185°F. Degraded by strong oxidizing agents, for example, oleum and fuming nitric acid (especially when warm). High temperatures, and contact with copper or cuprous alloys, will cause rapid decomposition; however, grades are available which contain certain types of thermal stabilizers so as to reduce this degradation problem. Not so resistant to oxidation at high temperatures as PE.

As the natural colour of the material is a translucent, ivory white, then a wide colour range is possible. Being a semi-crystalline, thermoplastics material, transparent mouldings are not usually possible by injection moulding. However, transparent extruded film is possible because of the increased cooling rate which can be applied in some processes. PP will absorb approximately 0.02% water in 24 hours at room temperature and this means that drying is not normally necessary. The surface of this material may be made more receptive to inks, paints, lacquers or to adhesives by various forms of surface pretreatment such as a *corona discharge, flame* and *chemical treatment*.

With a combination of light weight, toughness, high temperature resistance, rigidity and excellent resistance to chemical attack, PP is suitable for a wide range of components produced by injection moulding. These include automobile fascias, bumpers, bottle crates, washing machine tubs, kitchen ware, textile bobbins, tool handles and domestic waste systems.

PP is extensively modified by the addition of glass fibres, mineral fillers, thermoplastic rubbers or a combination of these. For example, the rigidity, hardness and heat distortion temperature of PP can be markedly improved by the incorporation of talc. Rubber modification improves the low temperature, impact strength (both PP-H and PP-CO) with some reduction in stiffness. PP continues to develop and to create new markets: this is because of the versatility of the material.

New polymerization technology, allows the production of directly polymerized, soft grades which compete with *rubber modified PP*. Such grades of PP have not been subject to the traditional heat history of melt compounded PP: the spherical particles (1 to 4 mm) are surface coated after production and they melt faster as they are not crystalline (as supplied). Random copolymers are also being promoted, for example, for blow moulding applications as they are clear materials with a high gloss and a Vicat softening point (VSP) of approximately, 128°C/263°F (about 20°C lower than PP-H). At the moment they are slightly more expensive than traditional materials.

polypropylene adipate
A *polyester* formed by the reaction of propylene glycol and adipic acid: used as a *polymeric plasticizer* for *polyvinyl chloride*. See *polypropylene glycol ester*.

polypropylene azelate
A *polyester* formed by the reaction of propylene glycol and azelaic acid: used as a *polymeric plasticizer* for *polyvinyl chloride*. See *polypropylene glycol ester*.

polypropylene - controlled rheology grade
An abbreviation used for controlled rheology is CR. In some cases, the molecular weight of polypropylene (PP) is changed deliberately so as to eliminate very high molecular weight material (also see *mastication*). Such a treatment may involve, compounding with a *peroxide* so as to give constant viscosity grades by chain scission. Such constant viscosity grades (PP-CR) give more uniform *shrinkage* than untreated materials.

polypropylene copolymer
An abbreviation used for this type of material is PP-CO or PP-K or PP. See *polypropylene*.

polypropylene/EP(D)M blend
An abbreviation used for polypropylene/ethylene propylene diene monomer blend: an abbreviation used for polypropylene/ethylene propylene rubber blend. See *rubber modified polypropylene*.

polypropylene ester
See *polypropylene glycol ester*.

polypropylene/ethylene propylene rubber blend
See *rubber modified polypropylene*.

polypropylene glycol
See *polyoxypropylene glycol*.

polypropylene glycol ester
A *polyester* formed by the reaction of propylene glycol and an organic acid, for example, azelaic acid. Such materials are used as *polymeric plasticizers* for *polyvinyl chloride*. They are usually capped with lauric acid end groups. Such materials are used as *plasticizers* when non-volatility and hydrocarbon resistance is required.

polypropylene homopolymer
An abbreviation used for this type of material is PP-H or PP. See *polypropylene*.

polypropylene/natural rubber blend
See *polyolefin/natural rubber blend*.

polypropylene oxide
An abbreviation used for this material is PPG when the degree of polymerization is low. Higher molecular weight materials may be referred to as PPOX. Polypropylene oxide is made from propylene oxide by polymerization in propylene glycol. See *polypropylene glycol*. Copolymers of propylene oxide are marketed as *propylene oxide rubbers*. PPO forms *polypropylene oxide-salt complexes* which are polymeric electrolytes: such compounds have been suggested for use in alkali metal rechargeable batteries.

polypropylene oxide-alkali metal salt complex
A *polypropylene oxide-salt complex* which contains *alkali metal ions*. Such compounds have been suggested for use in alkali metal rechargeable batteries.

polypropylene oxide-salt complex
An abbreviation used for this type of material is PPO-salt complex. The *polypropylene oxide helix* is filled by metal ions (M^+), for example, an alkali metal ion: the counterions (X^-) are outside the polymer chains. See *polyethylene oxide-salt complex*.

polypropylene plastics
Plastics based on polymers made with propylene as essentially the sole monomer. See *polypropylene*.

polypropylene sebacate
A *polyester* formed by the reaction of propylene glycol and sebacic acid: used as a *polymeric plasticizer* for *polyvinyl chloride*. Has a relative density of 1·06. See *polypropylene glycol ester*.

polypropylene sulphide/ethylene sulphide copolymer
See *alkylene sulphide rubber*.

polypropylene sulphide rubber
See *alkylene sulphide rubber*.

polypyrole
An aromatic material which is insoluble in common solvents and is also infusible. Has an extensive conjugated double bond type of structure which can be made electrically conductive by doping. Produced as an electrically conductive film (for example, marketed by BASF as Lutamer P 160) and also as a powder: the powder may be added to other thermoplastics as a conductive component. Polypyrole powdered polymer is used as the positive electrode in heavy-metal-free, rechargeable batteries.

polypyrrolidone
See *nylon 4*.

polysalt
An *ionic polymer* with a polymeric counter ion. Also known as a polyelectrolyte complex or as, a polyion complex or as, a simplex, or as, a coacervate. Hard, brittle materials formed as a result of mixing an anionic polyelectrolyte solution with a cationic polyelectrolyte solution. The material is cross-linked as a result of the ionic bonds formed. Used for filtration purposes and can be used to make contact lenses.

polysebacamide
See *nylon 610*.

polysilicate
A *polymer* which may be considered as being a condensed polymer of silicic acids or, which may be considered as being based on salts derived from silicic acids. Silicate polymers occur in great abundance naturally (see *silica*). Silicate polymers may be amorphous (glasses), crystalline (*wollastonite*) or *cryptocrystalline*.

polysiloxane
A polymer based on alternating silicon and oxygen atoms in the main chain: the silicon atoms has two groups attached. May also be known as a *polyorganosiloxane* when the two groups are organic groups.

polysiloxane rubber
See *silicone rubber*.

polystyrene
A polymer prepared by the polymerization of styrene as the sole monomer (see *styrene plastics*). Polystyrene, the thermoplastics material, is also known as general purpose polystyrene or as, unmodified polystyrene or as, crystal polystyrene or as, standard polystyrene or as, polyphenylethylene or as, poly-(1-phenylethylene) or as, polvinylbenzene. An abbreviation used for this type of material is PS; GPPS or PS GP is also used. Made from *styrene* and may be represented as -(-CH$_2$-CH.ϕ-)$_n$- where ϕ is based on a benzene ring and is C$_6$H$_5$.

Although it is now possible to produce *styrene plastics* which are both clear and tough, it is not possible to make tough, clear products from PS as this material is very brittle unless of high molecular weight and/or biaxially orientated. Biaxial orientation is not usually possible in conventional moulding and the use of high molecular weights results in poor melt flow properties. Normal *injection moulding* materials have excellent flow characteristics. very good thermal stability at processing temperatures and a low moulding shrinkage. PS is a hard, rigid material which, in its natural form, has a high gloss, sparkle and transparency. As PS is transparent a very, wide colour range is possible; the mouldings can also be decorated by a wide range of techniques and so attractive components can be easily produced. Low water absorption (for example, 0.04% after 24 hours immersion) mean no drying before processing and component stability. Excellent, electrical insulating characteristics are other desirable features.

However the material is brittle, burns easily and has poor weathering properties (prolonged exposure to sunlight, or fluorescent lighting, can cause yellowing of clear PS and fading of pigmented grades). PS has, however, excellent resistance to high energy radiation and micro-organisms. It is easy to process by injection moulding but this apparent ease can be misleading as unless care is taken, (to avoid overpacking and excessive *orientation*), mouldings which will fail in service as a result of stress cracking, can easily be produced. A simple test for this can easily be developed using n-heptane.

PS is relatively unaffected by water, alcohols (low molecular weight), alkalis, non-oxidising acids, solutions of inorganic salts and some aliphatic hydrocarbons. If the material contains high levels of internal stress, then *environmental stress cracking* (ESC) will result in the presence of liquids or vapours which normally have no effect, for example, paraffins. white spirit, fats and milk products. It is not resistant to high temperatures (softens in hot water) and a wide range of organic solvents. Soluble in aromatic and chlorinated hydrocarbons (for example, toluene and carbon tetrachloride) and attacked by oils, ethers, aldehydes, ketones and esters. Also attacked by some aliphatic hydrocarbons (for example, n-hexane), cyclohexane and nitrobenzene. Decomposed by prolonged contact with oxidising acids such as concentrated sulphuric or nitric acids.

The density is approximately 1.05 gcm^{-3} and as the natural colour is clear is very wide colour range is possible. PS has poor impact strength and low abrasion resistance and these facts should be taken into account when the mouldings are handled. May be bonded to itself by the use of solvents or by the use of solutions of PS in solvents. Bonded to other substrates by the use of impact adhesives. Mouldings are also easily joined by *ultrasonic welding*. Normally this material readily attracts dust and so both the moulding material and the mouldings should be kept covered at all times. However, special antistatic grades are available which reduce dust pickup to an acceptable level.

PS is used to produce injection mouldings that require a combination of colour, clarity, stiffness, low cost and good appearance but which do not require high heat, solvent and impact resistance. Typical mouldings include toys, containers, tape cassettes and disposable tumblers. When suitably stabilised against light, the material may be used for light fittings, for example, for light diffusers.

Oriented PS (PS-O) is available as monoaxially oriented filament and as biaxially oriented sheet. Both are prepared by *extrusion* followed by *hot stretching* at approximately 130°C - this gives a material which is stronger in the stretching direction but which has a lower *heat distortion* temperature. The monofilament is used as a fibre: the biaxially oriented PS sheet has been used a packaging material and as a *thermoforming* material (the sheet is heated by *contact heating*).

polystyrene/elastomer block copolymer
A category of *thermoplastic elastomer* which may be referred to as, for example, *styrene-butadiene block copolymer*.

polystyrene-polydiene block copolymer
See *styrene block copolymer*.

polysulphide cross-link
A chemical grouping, based on a sulphidic group, and which is present in a sulphur-vulcanized natural rubber. The sulphur-containing group is a cross-link which contains more than two sulphur atoms: see *mono-sulphide cross-link*.

polysulphide rubber
Polysulphide rubbers are also known as thioplasts or as, thiokol rubbers or as thiokols, or as, elastothiomers. Abbreviations used for this type of material are T or TM or TR.

This type of material has been classified as a *special purpose rubber* and is the oldest synthetic rubber still in commercial production. In these materials (which contain no carbon-to-carbon double bonds) the polymer chains consist of organic sections and sections which contain more than one sulphur atom - these groups are referred to as polysulphide and this type of structure confers some unusual properties on the final mouldings.

The materials have excellent resistance to oils and solvents and in this respect they are better than *nitrile rubber* and *chloroprene rubber*. The materials also have good chemical stability, resist degradation by oxygen and ozone, and are fairly impermeable to gases (in this respect they are better than most other rubbers but they are not as good as *butyl rubber*). However, the materials are not very strong and have poor resilience; the feature which most people remember them for is their disagreeable odour.

polysulphonate
A *pendant ionic polymer*, that is, an *ionic polymer* in which the bound anion (-SO$_3^-$) is pendant to the main chain.

polysulphone
See *sulphone polymers*.

polysulphone ionomer
An *ionomer* based on a *polysulphone*, for example, the sulphonate. See *polysulphone sulphonate*.

polysulphone sulphonate
A sulphonate *ionomer* based on a *polysulphone*. The sulphonate groups are generated by *post functionalization* from the precursor prepolymers. Because of their ionic properties, solvent resistance, chemical resistance and ease of manufacture (by casting) such materials are used as ion exchange membranes for water purification.

polyterephthalate plastics
A *thermoplastic polyester* in which the terephthalate group is a repeated structural unit in the chain, the terephthalate being in greater amount than other dicarboxylates which may be

present. See *polybutylene terephthalate* and *polyethylene terephthalate*.

polytetrafluoroethylene
Also known as poly-(tetrafluoroethylene). An abbreviation used for this type of material is PTFE. A fluoropolymer, a crystalline plastics material with an SG of 2·1 to 2·3. May be represented as -(-CF_2-CF_2-$)_n$- and is prepared from tetrafluoroethylene by free radical polymerization. Commercial materials are of very high molecular weight and of very high melt viscosity. A very inert, solvent-resistant material which is an excellent electrical insulator and which has very good thermal stability.

Since PTFE has a melt viscosity in the neighbourhood of 10^{10} poises it never really melts: above 327°C/621°F it becomes a translucent gel. It cannot therefore be moulded or extruded as a melt by conventional equipment but must be processed with techniques similar to those used in powder metallurgy: by cold compaction, sintering and controlled cooling. If both PTFE and *fluorinated ethylene propylene copolymer (FEP)* are suitable for a particular application then it must be decided whether the components are better made by specialized PTFE processing techniques or conventional processing using FEP. Either method may be followed by supplementary finishing operations: factors such as shape, size, number of components and tolerances must be taken into account.

polytetrafluoroethylene ionomer
An abbreviation used for this type of material is PTFE ionomer. An *ionomer* based on *polytetrafluoroethylene* (PTFE). Can have, for example, perfluorocarboxylates (Flemion) and perfluorosulphonates (Nafion): based on tetrafluoroethylene and a perfluorovinyl monomer. The acid groups are generated by *post functionalization* from the precursor prepolymers. Because of their ionic properties, solvent resistance and chemical resistance such materials are used as ion exchange membranes in fuel cells. See *ionomer*.

polytetrahydrofuran
Also known as polyoxytetramethylene and as *polyoxytetramethylene glycol*. A *prepolymer*.

polytetramethylene adipate
An abbreviation used for this material is PTMA. A *polyester* prepolymer: a comparatively low molecular weight material made from adipic acid and butanediol. A linear polymer with the formula -[-$(CH_2)_4$-COO-$(CH_2)_2$-COO-$]_n$- and which is used in *polyurethane* production.

polytetramethylene ether glycol
An abbreviation used for this material is PTMEG. Also known as polyoxytetramethylene and as *polyoxytetramethylene glycol*. A prepolymer which is a *polytetrahydrofuran*. This material is used in the manufacture of *polyester thermoplastic elastomers*.

polytetramethylene oxide
See *polytetrahydrofuran*.

polytetramethylene terephthalate
See *polybutylene terephthalate*.

polytetramethyleneadipamide
See *nylon 46*.

polythiazyl
See *sulphur-nitride polymer*.

polytrimethylhexamethyleneterephthalamide
Also known as poly-(trimethylhexamethyleneterephthalamide) but better known by the trade name/trademark of Trogamid T. An abbreviation used for this type of material is TMDT. Prepared from trimethylhexamethylene diamine and terephthalic acid so as to give a transparent (non-crystalline) polyamide with a high glass transition temperature (T_g) of approximately 150°C. A partially aromatic polyamide which is soluble in 80:20 mixtures of chloroform and methanol. An amorphous, transparent thermoplastics material which contains, for example, *para-phenylene* groups and amide groups. Has a high modulus, yield strength, tracking resistance and impact strength. Not so chemically resistant as the aliphatic polyamides but has a lower water absorption, a lower coefficient of thermal expansion and good light ageing properties. Used as an *engineering thermoplastics* material.

polyundecanoamide
See *nylon 11*.

polyurea
Also known as polyureylene. A polymer which contains the repeat unit -NH-CO-NH-. Such a chemical structure results during *polyurethane foam* production when a *diisocyanate* reacts with water. Polyurea fibres have been made from *nonamethylene diamine* and urea. Despite the low price of the starting material (urea), and the low relative density of the fibre (1·07), this fibre has not made a large impact as nonamethylene diamine is relatively expensive.

polyurea reaction injection moulding
Also known as urea reaction injection moulding. An abbreviation used for this term is urea RIM. A *reaction injection moulding* (RIM) process which has been formulated to give predominantly urea linkages. For *polyurea systems*, diamine terminated oligomer and diamine chain extender, as catalyst systems are required.

polyurethane
Also known as polycarbamate or as, urethane. A polymer which contains urethane groups (-NHCOO-) as part of the main polymer chain. Abbreviations used include PUR or PU. A polymer prepared by the reaction of an organic diisocyanate with compounds containing hydroxyl groups. Polyurethanes may be thermosetting, thermoplastic, rigid, flexible, cellular, or solid. See *polyurethane foam*.

polyurethane block copolymer
A *multi-block copolymer*. See *segmented copolymer*.

polyurethane elastomer
See *polyurethane rubber*.

polyurethane/elastomer block copolymer
A category of *thermoplastic elastomer* which may be referred to as *thermoplastic polyurethane elastomer*.

polyurethane foam
An abbreviation used for this type of material is PU foam. A cellular material, or expanded material, based on polyurethane structures which are usually cross-linked. Depending upon its mechanical behaviour a polyurethane foam is either classed as *flexible polyurethane foam* or as, *rigid polyurethane foam* or as a *semi-rigid polyurethane foam*. See, for example, *reaction injection moulding* and *reinforced reaction injection moulding*.

polyurethane recycling
See *recycled polyurethane*.

polyurethane rigid foam moulding
The production of a rigid moulding from a *polyurethane* material or, a rigid cellular (foam) moulding based on a polyurethane polymer. See *reaction injection moulding*.

polyurethane rubber
A *polyurethane* which exhibits rubbery properties. Also known as polyurethane elastomer or as, urethane rubber (PUR). Abbreviations used for this type of material are PAUR (also AU) and PEUR (also EU).

This group of rubbers is based on the reaction of polyols with diisocyanates; if polyols containing ether groups are used then the product will be a polyether polyurethane (EU type) and if the polyol used contains ester groups then the product will be a polyester polyurethane (AU type). PU rubber has been classified as a *special purpose rubber*.

By varying the molecular weight it is possible to produce materials which range from liquids to solids. If the molecular weight is low, then the liquid can be chain extended and crosslinked by the addition of liquid polyols or diamines; such liquids may be cast into a mould and allowed to set. Very high strengths are possible from the products, without the addition of reinforcing fillers, even for quite soft materials.

Higher molecular weight polymers are soft gums which can be mixed and moulded on conventional rubber processing equipment; the crosslinking agents may be diamines, sulphur, etc. Such materials seem to possess the disadvantages of traditional rubbers without any of the major advantages (for example, cost) as they must be vulcanised and they are expensive. It is also possible to produce high molecular weight linear molecules which are *thermoplastic elastomers* and which are available in a very wide range of grades.

In general, polyurethane components are noted for their good ageing, resistance to oils and solvents, their high strength and excellent abrasion resistance. However, their acid and alkaline resistance is not outstanding and the compression set resistance of the thermoplastic materials is poor.

polyureylene
See *polyurea*.

polyvinyl acetal
Also known as poly-(vinyl acetal). A *polyvinyl acetal polymer* formed by the reaction of acetaldehyde with polyvinyl alcohol. Polyvinyl acetal is an amorphous material which is used, together with nitro-cellulose, as a wire enamel and as a lacquer. The polymer is not widely used, for example, as a melt processable, thermoplastics material as it has no outstanding properties.

polyvinyl acetal polymer
Also known as poly-(vinyl acetal) polymer. A polymer formed by the reaction of an aldehyde with *polyvinyl alcohol*. As the aldehyde adds to adjacent chain hydroxyl groups, acetal ring formation is the result. If *formaldehyde* is used then the product is *polyvinyl formal*.

polyvinyl acetate
A *polyvinyl ester* which is also known as poly-(vinyl acetate) or as, poly(vinyl acetate). An abbreviation used for this type of material is PVAC or PVA or PVAc. May be represented as $-(-CH_2-CH\cdot OOCCH_3-)_n-$. A polymer prepared by the polymerization of vinyl acetate as the sole monomer. Often emulsion polymerization is used as the product is used in latex form for surface coatings; for example, this material often forms the basis of emulsion paints. Cannot normally be melt processed because it decomposes at approximately 100°C.

polyvinyl alcohol
Also known as poly-(vinyl alcohol). An abbreviation used for this type of material is PVAL or PVA or PVAl. May be represented as $-(-CH_2-CH\cdot OH-)_n-$: this material is prepared from *polyvinyl acetate* by alcoholysis as vinyl alcohol cannot be isolated. Because of chain degradation, the PVAL is of lower molecular weight than the acetate. Commercial polymers are offered in a range of grades which, for example, differ in the residual acetate content.

A crystalline thermoplastics material which is highly polar and thus resistant to hydrocarbons and very solvent resistant. Used to make water-soluble films for packaging: also used as a barrier layer in multi-layer co-extrusions and *blow moulding containers*. Polyvinyl alcohol is used as a *release agent* - which is often applied, by brush, from a water/methylated spirits mixture so as to get rapid drying in *hand lay-up*. Has been used to make fibres which are made insoluble by a chemical reaction so as to make a *polyvinyl acetal polymer* by, for example, reaction with *formaldehyde*.

polyvinyl butyral
Also known as poly-(vinyl butyral). A *polyvinyl acetal polymer* formed by the reaction of butyraldehyde with polyvinyl alcohol. Polyvinyl butyral is an amorphous material which is used as a safety glass interlayer: it is tough, strong, light resistant, water resistant and adheres well to glass. Also used to prepare adhesives and coatings, for example, for textiles.

polyvinyl carbazole
Also known as poly-(N-vinyl carbazole). An abbreviation used for this type of material is PVK or PVCZ. A brittle *vinyl polymer* with very good electrical insulation properties and heat resistance: used, for example, as a dielectric in capacitors. May be represented as $-(-CH_2-CH\cdot R-)_n-$ where the side group R is a 3-ringed structure: 2 benzene rings joined to a central five-membered ring which contains nitrogen. A photoconductive polymer which is used in xerography. May be melt processed by, for example, *injection moulding* although the material is more brittle than *polystyrene*.

polyvinyl chloride
Also known as polychloroethene or as, poly-(1-chloroethylene) or as, poly-(monochloroethylene) or as, poly(vinyl chloride). An abbreviation used for this type of material is PVC. May be represented as $-(-CH_2-CHCl-)_n-$ where Cl is a chlorine atom. A polymer prepared by the polymerization of vinyl chloride (VC) as the sole monomer. See *plasticized polyvinyl chloride* and *unplasticized polyvinyl chloride*.

Most of the PVC polymer produced throughout the world is homo-polymer. PVC, as made, is a substantially amorphous, thermoplastic material, which is produced in the form of a fine powder: it has a syndiotactic structure. That is, the chlorine atoms alternate on either side of the main chain. Because of the strong inter-chain attractions, which these chlorine atoms generate, the material is harder and stiffer than polyethylene (PE). Because of weak points along the chain, the material starts to decompose, or degrade, at temperatures of approximately 74°C/165°F.

VC polymers may be made by emulsion (giving PVC-E), suspension (PVC-S) and bulk, also known as mass, polymerization (PVC-M). As VC is toxic the level of monomer left in the plastic must be kept very low, for example, less than 1 part per million (<1 ppm). Plasticized polyvinyl chloride PPVC is usually based on PVC-S or PVC-M while unplasticized polyvinyl chloride is often based on PVC-E as the presence of the emulsifying agents eases processing.

In general, PVC is a relatively stiff flow material with a limited temperature processing range. The ease of flow is dependent on the molecular weight which is characterized for PVC by the K value; the bigger the number the higher is the molecular weight and the more difficult the flow. For a particular K value the ease of flow may be dramatically altered modified by the uses of additives (see *polyvinyl chloride - additives* by for). For example, by the use of a *lubricant system*.

Relatively low K value resins are now used for injection moulding. Up to fairly recently the typical (DIN) K value used was between, approximately, 55 and 62 (these figures correspond to ISO viscosity numbers of, approximately 74 and 95. The inherent viscosity (ASTM) is approximately 1/100 of the viscosity number); now that the K value range has been reduced to 50 to 60 (corresponding to a number average molecular weight of approximately 36,000 to 55,000): these lower molecular weight materials can be processed much more easily.

The PVC material may be bought in as compound (which means that it has been melt compounded with the additives) or, it may be blended, in for example, a *high speed mixer* and then fed to the *injection moulding machine* or, it may be blended, compounded and moulded in-house. For this reason the material may be known by the suppliers name of the parent plastic or, it may be known by the name of a compound. In general, the injection moulding of PVC compounds gives the best properties but the moulding of dry blends can offer cost advantages, however, the throughput of dry blend must be sufficient to justify the equipment installation and costs.

PVC is widely used in packaging: such packaging is often disposable. A common way of disposing of waste/rubbish is by burning in incinerators. PVC has for many years been under suspicion as being potentially harmful to health, and to the environment, as being a precursor in the formation of dioxins and furans. Incineration studies often do not appear to support this view and some people therefore believe that incineration is a safe and viable disposal option. Operating conditions during burning, rather than the material being burnt, appear to have a more significant effect on toxic gas emissions. Do not operate incinerators at temperatures below 750°C/1400°F as otherwise high dioxin levels result. To burn any plastic does, however, seem silly and particularly so for PVC as, hydrochloric acid (HCl) is formed. Waste PVC can be recycled and the trend towards recycling such materials is growing.

polyvinyl chloride acetate copolymer
See *vinyl chloride-vinyl acetate copolymer*.

polyvinyl chloride - additives for
Examples of additives used in *unplasticized polyvinyl chloride (UPVC)* compounds include:

(i) heat stabilizers - often based on calcium/zinc mixtures or, on tin complexes. Epoxidized soya bean oil, at low concentrations, is used as a component of some stabilizer systems. It is important to ensure that the compound contains sufficient stabilizer so that reprocessing is practical at the desired level of regrind use;

(ii) lubricants - for example, both *internal* and *external lubricants* are necessary. Calcium stearate is an example of an internal lubricant whereas synthetic waxes and fatty acid esters are used as external lubricants. Lubricant packages are formulated specifically to suit a particular application/machine combination. the level of use is often kept below 1 phr in order to reduce *plate out* problems and a loss of impact strength;

(iii) processing aids - which may be based on, for example, acrylic polymers;

(iv) impact modifiers - based on plastics such as *acrylonitrile-butadiene-styrene* (ABS) or on, *methacrylate-butadiene-styrene* (MBS). Such additives are very important in, for example, blow moulded products. They are used at high levels (approximately 12 phr), for non-oriented products, as they give the product good impact strength.

(v) fillers - these are added to extend the scope of PVC, and/or to make it cheaper, fillers (such as *china clay* and *glass*) are used.

(vi) pigments - as the natural colour of the material is clear then a wide colour range is possible; this includes both transparent and opaque colours.

(vii) other polymers - both rubber and thermoplastics materials are used. Rubbers to improve the impact strength and other plastics to improve, for example, the *heat distortion temperature* (HDT) - see *polyvinyl chloride blend*.

At some stage the polymer plus additives is melt mixed in, for example, a batch mixer or a continuous compounder - such as a *twin-screw extruder*. To realize the potential of this material it is necessary to put in just sufficient work and heat so that the blend is fully gelled, or fused, but not so much that it is degraded.

Examples of additives used in plasticized polyvinyl chloride (PPVC) compounds include the above together with relatively large quantities of *plasticizers*. Such plasticizers are added to PVC compounds in order to confer flexibility, softness and ease of processing to the resulting PVC compounds. The plasticizers most commonly employed are high boiling point esters of C_{8-10} alcohols: such as *phthalates*, *phosphates* and *sebacates*.

polyvinyl chloride blend
A *blend* based on *polyvinyl chloride* (PVC). An abbreviation used for this type of material is PVC blend or, PVC alloy. As with many other plastics materials, PVC is now being blended, or alloyed, with other plastics or rubbers. This permits the development of new materials with different properties at comparatively low cost. One such plastics blend is that between PVC and *acrylonitrile-butadiene-styrene* (ABS) which gives a flame retardant ABS-type material. When blended with appropriate grades of *nitrile rubber* (NBR) then it is possible to produce a *thermoplastic elastomer* (TPE) - the properties of some of these TPE materials can be further enhanced by ionic crosslinking.

Additives are available (for example, Monsanto sell Elix polymer-based additives) which will improve some of the properties of *unplasticized polyvinyl chloride* (UPVC), for example, the *heat distortion temperature* (HDT) may be improved by 30°C at a 20% addition level. In this way, PVC, a relatively cheap material, can compete with the more expensive engineering thermoplastics. Because of the improvements which blending brings, it is probable that PVC alloys/blends will become of increasing importance in the injection moulding industry. See *polyvinyl chloride nitrile rubber blend*.

polyvinyl chloride compound - production of
Polyvinyl chloride (PVC) compounds may be produced using a *batch* or a *continuous mixer*. Continuous compounders are available which combine the mixing and process control advantages of twin-screw extruders with the conveying advantages of single-screw machines.

In one system, widely used for pelletizing both rigid and flexible PVC formulations, the mixing-compounding operation is separated from the pumping, conveying and pelletizing function by using a two-stage machine.

A *preblend* is introduced into the feed section of a twin-screw, co-rotating, compounder: the L:D ratio is about 15:1 and there are three separate barrel zones. The high pitch screws, rotating at say 300 rpm, maximise the intake and pump the feed, in a very positive manner, through the machine where the compounding temperature is raised by externally supplied heat. The heated PVC then passes through two kneading zones. By using interchangeable screw and kneading elements, the length and width of these sections can be varied so as to maintain the correct heat levels, necessary for dispersion of additives, without degradation of either the additives (such as a *blowing agent*) or the polymer. Degradation is avoided by heating the material in this short, co-rotating twin-screw extruder

The material is then transported into the pressure-less area of a slowly rotating (say 21 rpm) single-screw with a 6:1 L:D ratio. This is the second stage. The material is conveyed through a breaker plate for die-face pelletizing. A vacuum can be applied to the transition chamber for removal of volatiles. The single-screw section of the compounder need not raise the temperature of the material, and it can be equipped with barrel and screw cooling to lower the temperature if required. Since the temperature can be accurately controlled, the two-stage machine is suited to automatic, in-line, air-cooled pelletizing, which is recommended for the production of moisture-free pellets.

polyvinyl chloride copolymer
See *vinyl chloride copolymer and terpolymer*.

polyvinyl chloride nitrile rubber blend
An abbreviation used for this type of material is PVC/NBR. Also known as PVC nitrile rubber blend: *nitrile rubber* is abbreviated to NBR. A melt compounded material based on NBR and *polyvinyl chloride* (PVC) in which the PVC is present in the largest proportion by weight. A rubber-modified PVC. By the incorporation of NBR, the toughness, flex crack resistance, low temperature flexibility and resilience of PVC is improved.

The NBR modifier may be used in liquid, crumb or powder form: crumb is usually used where *melt mixing* is employed: low, molecular weight material is used for flexible PVC compositions. Carboxylated nitrile rubber is also used: this type of material may be cross-linked ionically using zinc compounds. The ionically cross-linked compounds, exhibit higher abrasion strength, tensile strength and tear resistance.

The major attraction of PVC/NBR blends is that they can be processed just like *thermoplastics*, by processes such as *injection moulding*, and the mouldings have good abrasion and oil resistance. The NBR is regarded as a permanent plasticizer: however, NBR is not strictly a plasticizer as it is not dispersed on a molecular level and it does not affect the glass transition temperature in the same way as a true plasticizer. Such mouldings do not suffer from plasticiser loss to the same extent as *plasticised polyvinyl chloride*. See *thermoplastic elastomer* and *nitrile rubber polyvinyl chloride blend*.

polyvinyl chloride paste
Also known as PVC paste or as PVC plastisol. PVC paste is based on two major ingredients (PVC polymer and *plasticizer*) and therefore the properties of the *paste* and of the final product, can easily be varied by changing the ratio and type of these two ingredients. Such PVC pastes are used to manufacture play balls, dolls and toys. In this case the paste has the consistency of a pourable liquid (like paint): for other processes, the paste viscosity must be higher (see *aerogel*) or lower (see *organosol*).

polyvinyl chloride - plasticized
See *plasticized polyvinyl chloride*.

polyvinyl chloride - unplasticized
See *unplasticized polyvinyl chloride*.

polyvinyl cinnamate
Also known as poly(vinyl cinnamate). A *polyvinyl ester*. A polymer which is prepared by partial esterification of *polyvinyl alcohol* with cinnamyl chloride (ϕ-CH=CHCOCl) in alkaline solution. A thermoplastics material which may be represented as -(CH$_2$-CH.OOCR-)$_n$- where R is -CH-CH-ϕ and ϕ is the phenyl group.

The polymer may then be dissolved in, for example, dichloromethane and used to form a *photosensitive* coating so as to produce a *photoresist*. On exposure to, for example, UV light the polymer cross-links via the cinnamate groups on adjacent chains (photo-dimerization) and becomes insoluble forming the printing image. Photosensitizers such as the nitroamines, or quinones or, aromatic ketones improve the photo-response of polyvinyl cinnamate.

polyvinyl cyanide
See *polyacrylonitrile*.

polyvinyl ester
Also known as a poly-(vinyl ester). A polymer which is an ester of *polyvinyl alcohol*. A thermoplastics material which may be represented as -(-CH$_2$-CH·OOCR-)$_n$- where R is an alkyl group. When R is the methyl group then the polymer is *polyvinyl acetate*.

polyvinyl ether
Also known as a poly-(vinyl ether). A vinyl polymer which contains ether side groups. A thermoplastics material which may be represented as -(-CH$_2$-CH·OR-)$_n$- where R is an alkyl group. When R is the methyl group then the polymer is *polyvinylmethyl ether*. When R is the ethyl group then the polymer is *polyvinylethyl ether*. Both of these materials are vinyl alkyl ethers. Such polymers are soluble in a wide range of solvents and are used as adhesives and in surface coatings.

polyvinyl fluoride
An abbreviation used for this type of material is PVF. May be represented as -(-CH$_2$-CHF-)$_n$- where F is a fluorine atom. A crystalline thermoplastics material which is noted for its exceptional weather resistance. Has better heat resistance than *polyvinyl chloride* (PVC) but is more difficult to melt process because of a tendency to release hydrogen fluoride at processing temperatures. Marketed as a film, for example, for cladding purposes and perhaps best known by the trade name/trademark Tedlar.

polyvinyl formal
An abbreviation used for this type of material is PVFM. Also known as poly-(vinyl formal). A *polyvinyl acetal* formed by the reaction of *formaldehyde* with *polyvinyl alcohol*. Polyvinyl formal is an amorphous material which is used, together with *phenolic resin*, as a wire enamel and adhesive.

polyvinyl ketal polymer
Also known as a poly-(vinyl ketal) polymer. A polymer formed by the reaction of a ketone with *polyvinyl alcohol*. At present, such polymers are of very limited commercial significance.

polyvinyl polymer
A polymer prepared from a *vinyl monomer*. See, for example, *polyvinyl chloride*.

polyvinyl pyrrolidone
Also known as polyvinylpyrrolidone or as, poly-(N-vinyl-2-pyrrolidone) or as, poly-(N-vinylpyrrolidone). An abbreviation used for this type of material is PVP. A water soluble thermoplastics material prepared from the monomer N-vinylpyrrolidone by free radical polymerization. Originally used as a blood plasma substitute: forms complexes with many other materials, for example, dyestuffs and for this reason is used for textile dye stripping. Also used for adhesives - see *vinyl methyl ether/maleic anhydride copolymer*.

polyvinylbenzene
See *polystyrene*.

polyvinylethyl ether
A *polyvinyl ether* which is also known as poly(vinylethyl ether). This polymer is used in adhesives (for example, pressure sensitive adhesives) and in surface coatings: not soluble in water.

polyvinylidene chloride
Also known as poly-(1,1-dichloroethylene) or as poly-(vinylidene chloride). An abbreviation used for this type of material is PVDC or PVdC. The homopolymer may be represented as -(-CH$_2$-CCl$_2$-)$_n$- where Cl is a chlorine atom. When vinylidene chloride is used to make a copolymer then the product may be known as a high-vinylidene copolymer or as, polyvinylidene chloride copolymer.

The homopolymer is a solvent-resistant, hard, tough, crystalline thermoplastics material which is translucent and can have a yellowish cast. Has a high SG of approximately 1·85 because of the large amount of chlorine (about 70%) and is very difficult to melt process because of acid emissions.

To ease processing, copolymers may be produced using, for example, vinyl chloride (10 to 15%) as the second monomer:

such materials are better known by the trade name/trademark of Saran (Dow): used for film and for fibres. When melt processing such materials great care must be taken to avoid degradation, for example, caused through over-heating, contact with metals (such as iron, steel and copper), material stagnation, etc.

To obtain coatings of low moisture permeability, copolymers may be produced using, for example, acrylonitrile (10 to 15%) as the second monomer: such materials are better known by the trade name/trademark of Saran and/or Viclan. Such coatings, on other polymers and paper, are clear, tough and chemically resistant. The higher the vinylidene chloride content, the better are the barrier properties.

polyvinylidene chloride copolymer
A copolymer which contains a significant proportion of *vinylidene chloride*. See *polyvinylidene chloride*.

polyvinylidene fluoride
An abbreviation used for this type of material is PVDF: PVdF is also used. Also known as vinylidene fluoride polymer or as, poly-(1,1-difluorethylene). May be represented as -(-CH_2-CF_2-)$_n$- where F is a fluorine atom.

Of the *fluoropolymers* which can be injection moulded, PVDF has the highest dielectric constant, heat deflection temperature, flexural strength and modulus; its yield strength and creep resistance is high for a fluorinated thermoplastic. By comparison with other fluoropolymers, PVDF has the widest *processing temperature window*: for some fluoropolymers this is only a few degrees. With PVDF this is 140°C/284°F which enables conventional injection moulding machinery to be used.

As the thermal and light stability is very good, no heat stabilizers or UV stabilizers are necessary and this means that the material can be non-toxic. By keeping molecular weight relatively low, a PVDF material can be produced which develops high crystallinity on cooling (up to 65%). At its melt temperature, such a material would have a comparatively low viscosity (because of the low molecular weight). Higher molecular weight grades require higher melt temperatures so as to obtain ease of flow. Products made from such materials have a lower degree of crystallinity (for example, 40%) and a lower softening point (the Vicat softening point may be 20°C/68°F lower); they also have a lower tensile strength, elongation at break and solvent resistance. The impact strength is however, higher and there is less tendency for stress formation. PVDF builds in stress relatively easily because it is a hard material which shrinks considerably. Although both types may be used for moulding, in general, the high viscosity (high molecular weight) grades are only used for large items and it is low molecular weight (crystalline) material which is generally used for injection moulding.

PVDF is a crystalline material and behaves similarly to PP; the melt is more viscous than a PA melt. Flows easily at 225 to 245°C/437 to 473°F (additives based on PVDF are used to reduce melt instability in polyolefin (PO) extrusion). Shrinkage is of the order of 0·020 to 0·030 mm/mm (2 to 3%) but less for filled grades. PVDF has a high density of 1·78 gcm^{-3}: the Vicat softening point is approximately 145°C/293°F and the crystalline melting point (as measured by differential thermal analysis (DTA)) is approximately 177°C/351°F. The material has good fire resistance and will not propagate a flame.

PVDF has excellent resistance to a wide range of chemicals, such as halogens, salt solutions, inorganic acids and bases, aliphatic and aromatic hydrocarbons, carboxylic acids and acid chlorides, mercaptans and chlorinated hydrocarbons. Resists degradation by ultraviolet light, alpha and beta radiation. It is not resistant to oleum, fuming nitric acid, strongly basic amines, hot concentrated bases and alkali metals. Swollen by polar solvents such as acetone and ethyl acetate. Dissolves with difficulty in solvents such as dimethylformamide, dimethylsulphoxide and tetramethylurea. Some stress-cracking occurs in hot alkalis.

The natural colour of this material is a translucent milky white and, as it is a crystalline material, it is not possible to produce transparent mouldings: a wide colour range is however possible. Some pigments such as titanium dioxide and some other inorganic materials (for example, glass and boron oxide) may cause rapid decomposition of PVDF. Will absorb 0·05% water in 24 hours at room temperature: this means that drying is not normally necessary.

The chemical process industry uses this material for valves, pumps, bearings, etc, because of its outstanding chemical resistance, relative ease of processing, high strength, rigidity and abrasion resistance. Because it has a high, sharp, crystalline melting point this material maintains a great deal of its chemical resistance even at a high, continuous-use temperature of 140°C/284°F. Price is similar to that of PTFE but because processing is easier, finished articles tend to be cheaper.

PVDF can be made to have exceptional *piezoelectric* properties - this explains current interest in uses such as transducers, microphones, loud-speakers, etc. The polymer can exist in 3 morphological forms called α-PVDF, β-PVDF and γ-PVDF: it is the β-PVDF which is preferred for piezoelectric applications. This form has an all-trans conformation with dipoles normal to the molecular axis. PVDF film is largely in the non-polar α-form. In order to improve the piezoelectric properties, the material is, for example, uniaxially stretched after *extrusion*, annealed and then subjected to *poling* (polarization induced thermally or by a corona discharge process). Such processing induces the β-form of PVDF.

The pyroelectric effect is utilised when the polymer is used to make, for example, pyroelectric detectors (based on PVDF film) for energy management systems and for intrusion detection systems.

polyvinylmethyl ether
A *polyvinyl ether* which is also known as poly(vinylmethyl ether). This water-soluble polymer is used in adhesives and in surface coatings.

polyvinylpyrrolidone
See *polyvinyl pyrrolidone*.

polyvinyltoluene
Also known as polymethylstyrene. A clear glassy material with a lower softening point than *polystyrene*. Of little commercial use. See *vinyltoluene*.

POM
An abbreviation used for *acetal*.

POM-CO
An abbreviation used for *acetal* copolymer.

POM-H
An abbreviation used for *acetal* homopolymer.

POM-K
An abbreviation sometimes used for *acetal* copolymer.

poor surface finish
Undesirable dullness on the surface of a product.

popcorn
The hard, insoluble material produced as a result of *popcorn polymerization*. So-called because of the appearance of the product.

popcorn polymerization
Also known as proliferous polymerization. An undesirable polymerization as it may occur where *polymerization* is not required. The separation of polymer nodules (*popcorn*) during a free radical polymerization occurs because once a nucleus has formed, termination is low as the radicals are trapped:

such nodules are insoluble and cross-linked. May be prevented by the addition of nitrogen monoxide.

POPG
An abbreviation used for polyoxypropylene glycol. See *polypropylene glycol*.

poppet
A term used in *hydraulics* and which refers to a cylindrical valve which moves perpendicularly relative to a cone-shaped seat: fluid flow is blocked when the valve closes against the seat.

porcelain
A hard white material which is based on *china clay*. Feldspar, quartz and china clay are mixed, shaped and then fired (fused) at high temperatures.

poromeric coated fabric
A *plastic coated* fabric which is waterproof but capable of transmitting water vapour. See *poromeric material*.

poromeric material
A material which has properties similar to those of leather in some respects, for example, the material can transmit water vapour but is waterproof.

porosity
A defect, often a moulding defect, which consists of undesirable clusters of air bubbles. An extrusion defect which shows as voids; similar to blistering. In the case of *natural rubber*, porosity may be minimized by the use of *superior processing rubber*. It has been found possible to eliminate porosity in rubber extrusions by using a desiccant additive and/or using a *pressurised fluidised bed*.

porous PE sheet
An abbreviation used for porous polyethylene sheet. Usually means *micro-porous polyethylene sheet*.

port
A term used in *hydraulics* and which refers to an inlet or outlet connection of a passage used to convey fluid.

Portland cement
An hydraulic cement made by burning a mixture of limestone and clay together in a kiln. Named after the Isle of Portland, Dorset, UK. Also called OPC - standing for 'ordinary Portland cement'.

Portland stone
A type of *limestone* which is used in building and which is quarried on the Isle of Portland, Dorset, UK.

positional control
See *screw position switching*.

positional isomerism
Isomerism which arises because of the two possible ways a monomer unit $CH_2 = CHX$ may add to a growing active centre in a vinyl polymerization. If the substituted carbon atom is called the head, then the two ways are head-to-head and head-to-tail.

positive displacement
A term used in *hydraulics* and which means that there is a fixed output per revolution of, for example, a pump; the amount of fluid delivered by the pump is constant and is relatively un-affected by pressure variations.

positive image lithography
A *photoresist* process which uses a *positive* photoresist: a photoresist film is used to form an image. Exposure of the photoresist film results in areas that are more soluble than the unexposed areas during *developing*. See, for example, *novalak/naphthaquinone* diazide-based positive photoresist.

positive mould
A type of *compression mould* in which there is no provision for *flash escape*. The exact amount of material is loaded into the mould and the moulding pressure is supported by the material throughout the moulding cycle. Because of manufacturing and use difficulties such moulds are not widely used in ordinary *compression moulding*.

When a fully positive mould is used for easy-flowing, *dough moulding compound* (DMC), porosity in mouldings (caused by venting difficulties) can be a problem. Due to the material's low viscosity, the closeness of fit between punch and die must be controlled (a typical clearance is 0·08-0·18 mm or, 0·003-0·007 in).

positive photoresist
A *photoresist* which becomes more soluble in exposed areas. For example, by increasing acidity or by bond scission (degradation). The use of positive photoresists is increasing because of their higher resolution capability and better thermal stability.

positive temperature coefficient
The fractional change in any physical quantity per degree rise in temperature.

post functionalization
The chemical modification of a polymer so as to produce desired properties. For example, the sulphonation of the preformed polymer, *polystyrene*, gives the ion exchange resin, sulphonated polystyrene.

post-calender section
The equipment which is required to remove the *sheet* from the *calender*, stretch it if required, emboss it if required, cool the product, measure and control the thickness, trim the edges and wind the sheet into rolls (or cut the sheet into lengths).

post-cooling station
A term usually associated with *blow moulding*. A machine facility which is designed to increase output and, which fits between the blow mould and the punch, or finishing, station. The component is transferred from the blow mould and a long nosed, loose fitting, pin is inserted and used to pass flushing air so as to speed up cooling of the *blow mould container*.

post-curing
A curing process performed after processing or moulding. For example, *silicone rubbers* are usually cured with peroxides and the resultant decomposition products may need to be removed by heating the mouldings in an air-circulating oven at 200°C for 12 h. This post-curing treatment will often improve physical properties.

post-extrusion equipment
Once the extrudate leaves the die then it can either be set to the shape produced or, its shape may be altered and then it may be set to shape. The equipment which does this is called the post extrusion equipment, or the haul off, and in terms of size it is far larger than the extrusion machine. One reason is that plastics materials take a long time to cool; this cooling process is so long that it often determines how fast the extrusion line will operate.

post-forming
A shaping process performed after the main shaping operation. Often means the forming of cured, or partially cured, *thermosetting plastics*.

post-forming sheet - thermosetting plastic
Also known as formable sheet. A *laminated plastics sheet* which is capable of being shaped after being heated.

post-moulding shrinkage
An abbreviation used for this term is PMS. Post-moulding shrinkage occurs after the moulding has been aged and is also called environmental shrinkage. It is given as a percentage if the following formula is used:

$$PMS_{48h} = (L_1 - L_2/L_1) \times 100 \text{ or,}$$
$$PMS_{168h} = (L_1 - L_2/L_1) \times 100.$$

Where L_1 = the length in mm of the original moulding and L_2 = the length in mm, of the same dimension measured after heat treatment for 48 hours or 168 hours. The heat treatment is at 80°C for urea-formaldehyde (UF) and 110°C for other thermosets. See *shrinkage*.

post-process gauging
Gauging performed after manufacture.

post-vulcanization curing
The curing, of *sulphur* vulcanized rubbers, which occurs after curing and during storage at room temperatures. Such curing can occur because of re-arrangement of *polysulphide crosslinks*.

postcure
Also called after-bake. A heating process applied after moulding of a thermosetting material so as to advance the degree of cure. See *post-curing*.

pot
See *transfer moulding*.

pot life
The period of time during which a reacting thermosetting composition remains suitable for its intended processing after mixing with reaction-initiating agents. The working life of an *adhesive* or of a resin. The time between the preparation of an *adhesive* for use and the stage at which it becomes unusable.

potassium/aluminium silicates
See *mica*.

potassium cobalt nitrite
Also known as Fischer's yellow or as Indian yellow. A yellow pigment.

potassium stearate
A salt of stearic acid. An *acid acceptor for chlorine containing rubbers*.

POTG
An abbreviation used for *phenyl-o-tolyl guanidine*.

potting
An *embedding process* in which the mould forms part of the total assembly.

pouncing
The application of a dusting agent to rubber surfaces.

pound
A unit of weight and of mass, the actual value of which depends on the measuring system. One pound avoirdupois (pd) contains 7,000 grains or 16 ounces: this is the pound commonly used and is equivalent to 0·453 592 375 kg. One pound Troy contains 5,760 *grains*, is equivalent to 0·373 241 725 kg and contains 12 ounces (Troy).

pound-force
Also known as pound of force and abbreviated to lbf. The unit of force in the *FPS gravitational* system of measurement. May be defined as the magnitude of the force which will support a one *pound* mass at rest relative to the earth at a place where the acceleration of gravity (g) is 32·174 feet per second per second. That force which will impart to a mass of one *slug* an acceleration of one foot per second per second. One pound-force is equal to 4·448 22 newtons.

pound-mass
Also referred to as a *pound* of mass and is equivalent to 0·453 592 375 kg. Serves as a basis for defining the *pound-force*.

poundal
Abbreviated to pdl. That force which will impart to a mass of one *pound* an acceleration of one foot per second per second. One poundal is equal to 0·138 255 newtons.

pounds per square foot
Abbreviation most commonly used lbf/sq ft. 1 lbf/sq ft = 4·883 kgf/m^2 = 47·88 Pa = 0·359 Torr.

pounds per square inch
Abbreviation most commonly used psi. Also lb/sq in and lbf in^{-2}. 1 psi = 0·0703 kgf/cm^2 = 6·894 757 kPa = 6,894·757 N/m^2 = 51·715 Torr.

powder
A finely divided solid. Polymers in *powder form* are used, for example, in *powder moulding*. However, for thermoplastics materials, powdered, compounded material is more costly than pellets (used in *extrusion* or *injection moulding*) as the thermoplastics compound/material normally has to be ground to a powder. See *powdered plastics material*.

powder blend
See *dry-blend*.

powder coating
The coating of a substrate with a *polymer* in powder form or, more usually, the coating of a substrate with a powdered compound so as to form a coating which may, or may not, be transferrable. Powdered thermoplastics compounds are used, for example, to coat fabrics and metals.

powder mould coating
The coating of a mould with a powdered compound so as to form a transferrable coating: used with *polyester moulding compounds*. The powder is based on *unsaturated polyester resin* and is applied with electrostatic powder spray equipment on to the hot mould where it melts, flows, forms a film and polymerizes. After 20 seconds, moulding may begin. Such 'in mould-coating' (IMC) is used with *sheet moulding compound (SMC)*: melamine-formaldehyde (MF) resins are also used for IMC materials.

powder moulding
Low pressure moulding processes which employs a polymer in *powder* form as the feedstock. The simplest moulding technique uses a heated metal, female mould which is filled with powder (for example, *polyethylene*): after standing the excess powder is tipped out and the layer adhering to the sides is fused by further heating. After cooling it is removed. Despite the simplicity of this process it is not widely used as poor thickness distribution results. Products of more uniform wall thickness are obtained from *rotational moulding*.

powdered natural rubber
A form of *natural rubber*. Rubber in powdered form (as opposed to rubber in bale form). Spray dried rubber latex containing about 9 parts of a partitioning agent (for example, a fine particle size *silica*) and which is supplied as a free-flowing powder. Used in, for example, adhesive solutions.

Such a material may be used to prepare a powder *dry-blend*. The technique of feeding the material in the form of a powder dry blend (similar to *polyvinyl chloride* (PVC)) to injection moulding machines can give good results and saves on compounding costs.

powdered plastics material
Polymers in *powder form* are used, for example, in *powder moulding*. Examples of powdered plastics materials include *epoxide, low density polyethylene, high density polyethylene* and *nylon 11*.

powdered rubber
See *powdered natural rubber*.

power
May be defined as work done per unit of time. The rate of doing work. An abbreviation sometimes used is N. May be measured in joules per second or watts. 1 joule per second is one watt. Also used is *horsepower*. The units of power are ml^2/t^3 - see *unit dimensions*.

power factor
See *dissipation factor*.

power law equation
Also known as the Ostwald - de Waele equation. An empirical equation which describes *non-Newtonian* flow behaviour: the relationship between shear stress τ_w and the rate of shear $\dot{\gamma}$. The shear stress τ equals $K(\dot{\gamma})^n$ where K is the consistency index and n is the flow behaviour index. In logarithmic form, the equation is $n \log \tau = \log K + n \log \dot{\gamma}$ and a log-log flow curve plot should be linear. In practice the slope of the log-log flow curve is not a straight line, as n' (see *Rabinowitsch equation*) does vary slightly with shear rate. For many purposes however, it may be equated with n, the flow behaviour index, or the flow index of the power law equation.

power law fluid
A fluid which obeys a *power law equation*.

power law indices
In shear flow, the *shear stress* τ equals $K(\dot{\gamma})^n$ where K is the consistency index and n is the flow behaviour index: the rate of shear is $\dot{\gamma}$.

power pack
A term used in *hydraulics* and which refers to the assembly of electric motor, pump, tank, and control valves.

PP
An abbreviation used for *polypropylene*.

PP crepe
An abbreviation used for *partially purified crepe*.

PP/EPDM
An abbreviation used for *rubber modified polypropylene*.

PP-A
An abbreviation used for *atactic polypropylene*.

PP-C
An abbreviation used for *chlorinated polypropylene*.

PP-CO
An abbreviation used for *polypropylene copolymer*.

PP-H
An abbreviation used for *polypropylene homopolymer*.

PP-K
An abbreviation sometimes used for *polypropylene copolymer*.

PPA
An abbreviation used for *polyparabanic acid*. The same letters have also been suggested for *polyphthalamide*.

ppb
An abbreviation used for parts per billion.

PPB
An abbreviation used for *poly-(p-benzamide)*.

PPC
An abbreviation used for *partially purified crepe*. See *pale crepe*.

ppd
An abbreviation used for p-phenylene diamines. See *antiozonant, antioxidant* and *alkyl-aryl p-phenylene diamine*.

PPD
An abbreviation used for *piperidine pentamethylene dithiocarbamate*.

PPE
An abbreviation used for *polyphenylene ether*. See *polyphenylene oxide - modified*.

PPG
An abbreviation used for *polyoxypropylene glycol*.

pphr
An abbreviation used for parts per hundred of resin or, parts per hundred of rubber. See *phr*.

ppm
An abbreviation used for parts per million.

PPO
An abbreviation used for *polyphenylene oxide* and for *polyphenylene oxide - modified*.

PPO-alkali metal salt complex
An abbreviation used for *polypropylene oxide-alkali metal salt complex*.

PPO/PA blend
An abbreviation used for *polyphenylene oxide/polyamide blend*.

PPO-M
An abbreviation used for *polyphenylene oxide - modified*.

PPO-salt complex
An abbreviation used for *polypropylene oxide-salt complex*.

PPOX
An abbreviation used for *polypropylene oxide*.

PPPS
An abbreviation used for polyphenylene sulphide sulphone. See *sulphone polymers*.

PPR
An abbreviation sometimes used for *polypropylene*.

PPS
An abbreviation used for *polyphenylene sulphide*.

PPSU
An abbreviation used for *polyphenylene sulphone*. See *sulphone polymers*.

PPVC
An abbreviation used for *plasticized polyvinyl chloride*. See *polyvinyl chloride*.

PPX
An abbreviation used for *poly-p-xylene*.

p,p'-oxybisbenzene sulphonyl hydrazide
See *4,4'-oxybis-(benzenesulphonylhydrazide)*.

PQD
An abbreviation used for *p-quinone dioxime*.

pre-accelerated resin
An abbreviation used for this type of material is PA resin. An *unsaturated polyester resin* which has the *accelerator added*.

pre-blending
The mixing together of ingredients before melt mixing. Sometimes referred to as *blending*. See *compound blending and mixing*.

pre-calender section
The compounding and feed section of a *calendering line*. The equipment which is required to deliver to the *calender* a hot, uniform feed which is free from contamination and fully compounded; in the case of *polyvinyl chloride* (PVC) it is also required to be fully gelled.

pre-charge pressure
The gas pressure in a hydraulic accumulator before the fluid is put into the *accumulator*.

pre-expanded beads
Beads of expandable *polystyrene* which have been heated in steam and/or water at approximately 100°C to give puffed-up, or pre-expanded, beads which after storage are processed into blocks or mouldings. See *expanded polystyrene*.

pre-formed compatibilzing agent
See *compatibilzing agent*.

pre-hardened steel
A *mould steel* which does not require hardening once it has been machined to shape. A quenched and tempered mould, or tool, steel which may be referred to as a pre-toughened steel. For example, the core and cavity of an injection mould are commonly made from a pre-hardened steel; the hardness as measured on the Rockwell C scale is up to approximately 50. By eliminating hardening, the risk of distortion (which may ruin weeks of expensive work) is reduced. Air-hardened or oil-hardened steels are generally used and such materials are commonly called tool or mould steels; they are of high uniform quality and their production is carefully supervised during all stages of manufacture.

The term 'alloy steel' is also used to describe these materials as they contain elements such as nickel, vanadium, silicon, carbon and manganese. For example, the pre-hardened chromium-manganese-molybdenum steels contain 0·35% carbon, 1·5% chromium and 0·4% molybdenum, or 0·35% carbon, 1·5% nickel, 1% chromium and 0·3% molybdenum. A pre-hardened steel is a compromise between factors such as machinability, toughness, hardness, distortion and wear.

Can have low sulphur content pre-hardened steel (for example, P20 steel) and a higher sulphur-content pre-hardened steel (for example, P20 + S steel). The higher sulphur content gives improved machinability but gives more non-uniform structures on photo-etching. Can improve the wear resistance of such steels by chromium plating or *nitriding*.

pre-mixing
Mixing of compounding ingredients before *melt* mixing using a pre-mixer, for example, a ribbon blender.

pre-squeeze device
A term usually associated with *blow moulding*. A sealing device which is usually mounted directly below the die head in *extrusion blow moulding*. When the desired length of parison has been produced, jaws close so as to seal the parison: used in conjunction with either cold or hot knives.

pre-toughened steel
See *pre-hardened steel*.

pre-treatment process
See *fabric*.

pre-vulcanization inhibitor
An abbreviation used for this type of material is PVI. A retarder of *vulcanization*: a *scorch* preventer. For example, N-cyclohexylthiophthalamide.

preblend
A simple mix: a mixture produced as a result of a blending operation and which is fed to another mixing machine for *melt mixing*. See *blend*.

preblow
A term usually associated with *blow moulding*. Air is introduced into a parison so as to, for example, open the *parison* before the blow pin is inserted.

precipitated azo pigments
Mono-azo (*monazo*) and *diazo pigments* are of two types, *pigment dyes* and precipitated azo pigments. These contain salt forming groups, for example, sulphonic acid groups, and are precipitated when salts such as those based on barium, calcium, manganese and strontium are used. Examples of such pigments include Lithol red and Lake red C. In general, such pigments have good bleed resistance but poor acid and alkali resistance.

precipitated calcium carbonate
Also known as precipitated $CaCO_3$. Such a *filler* may be incorrectly referred to as precipitated *whiting*. A pure form of *calcium carbonate* which can have a particle size of as low as 0·1 μm. This material has a relative density (RD or SG) of 2·62.

Produced by calcining (strongly heating) *chalk* to give calcium oxide, slaking this material in water, removing impurities and then passing carbon dioxide into the mix so as to give the pure precipitated calcium carbonate. May be obtained as a by-product from water-softening plants. In both cases the particle size is small and for this reason such fillers are often coated so as to reduce agglomeration and to assist dispersion. Used widely as a filler as it is of comparatively low cost - compared to other white fillers such as *silica* and *aluminium silicate*.

precipitated silica
May also be called colloidal silica. See *precipitation process*.

precursor polymer
A polymer which is used to make another polymer. Such a procedure is employed if the second polymer is difficult to make or if it is *intractable*.

precipitated silicic acid
See *precipitation process*.

precipitated whiting
See *precipitated calcium carbonate* and *whiting*.

precipitation polymerization
A polymerization process in which the monomer is insoluble in the monomer and precipitates as the *polymerization* proceeds. *Polyvinyl chloride*, for example, is insoluble in vinyl chloride and so will normally precipitate during polymerization. See *solution polymerization*.

precipitation process
A solution process used to refine and produce, for example, fine fillers such as *silica*. Most important method of producing silica for rubbers is by the precipitation process. A mineral acid is added to an alkaline silicate solution (for example, of sodium silicate): the product is filtered, washed and dried. Still contains about 13% of water and so may be referred to as silicic acid or as precipitated silicic acid. Primary particle size of precipitated silica may be 15 to 20 μm. Hydrated calcium silicate and hydrated aluminium silicate may also be obtained by slight modifications to this process. See *silicon dioxide* and *pyrogenic process*.

precise
When this term is applied to the reproducibility of measurements then a measurement is said to be more precise than another if, the plus and minus deviations are closer to the average or smaller. The results may be precise but they need not be *accurate*.

precoagulation
The first coagulation step, in *sole crepe* production. By using about one quarter of the acid required for *coagulation*, a coloured (yellow) fraction may be removed from the rubber as it precipitates. Coagulation may then be completed.

precure
A state of cure of a *thermosetting material* usually caused by delaying the application of the moulding pressure so that the material will not flow properly: a degree of cure.

preferential absorption
See *accelerator deactivation*.

prefixes - SI
Prefixes are used to construct decimal multiples of SI units. The names and abbreviation are summarized in a separate appendix. See appendix 3.

preform
In *blow moulding*, the term refers to the shape which is first produced before the actual container is formed: associated with *injection blow moulding*.

Reinforcement which is in the approximate shape of the cavity. Preforms are used to make polyester moulded products. Chopped glass rovings are sucked or blown onto a fine metal screen of the required shape while the screen is rotated. The strands are bound together with a resinous binder and the assembly is then baked in an oven for about 2 minutes at 15°C. See *hot press moulding*.

preform moulding
See *hot press moulding*.

preform tempering
In stretch blow moulding, the *preform* is either heated or cooled, i.e. tempered, in order to bring the temperature of the preforms into the thermoelastic region prior to the next *orientation* (blowing) step.

preforming CFM
An abbreviation used for *preforming continuous filament mat*.

preforming continuous filament mat
Also known as a thermoformable, random fibre, continuous strand mat or as, a thermoformable, random fibre, continuous fibre mat. An abbreviation used for this material is preforming CFM. A type of *continuous filament mat* made with a thermoplastics *binder* which is insoluble in the resin. Such a mat may be stretched cold but if the stretching is excessive this can break the fibres. As the binder used to make the mat is thermoplastic, the mat may be *thermoformed*: for example, after being heated at 120°C for 1 minute the mat may be formed by being pressed between lightly-constructed, matched moulds which may be cooled (depends on production requirements). The shaped mat is then used in *resin transfer moulding*.

pregel
A term used in *polyvinyl chloride* (PVC) paste technology for the gelled material which initially forms on the walls of a heated former, or mould, when the heated metal first contacts the *paste*. Also used as a verb to indicate the action of forming the partially gelled material. See *slush moulding*.

preheater
Device used to preheat material before processing, for example, used to raise the material temperature before it is fed to the *extruder* or moulding machine. Could also refer to the device used to preheat wire before it is fed to an *extruder*.

preheating
The raising of the material temperature before it is fed to the processing equipment by means of the *preheater*: used to increase output, quality and/or dry the material. Usually associated with *compression* and/or *transfer* moulding where oven heating or *high frequency heating* is used. The action of bringing a temperature some way between ambient and the desired processing temperature. The term is also associated with *extrusion*, or blow moulding operations, where it refers to the heat soaking of large metal masses.

preload device
A mechanism, or device, which loads, or stresses. the system prior to working. See *roll float*.

premasticated rubber
Rubber which has been *masticated* in a separate mixing operation.

premature vulcanization
See *scorch*.

premix
A moulding composition. A mixture of resin, fillers, etc., which is prepared when required and then the appropriate amount of this shapeless mass is then fed to a mould.

prepolymer
A comparatively low molecular weight polymer which is then used to make a higher molecular weight *polymer* by chain extension and/or cross-linking. See *polyurethane foam*.

prepolymer process
A term often associated with polyurethanes. A term used, for example, to describe a process used to make *polyurethane foam* (both flexible and rigid). Polyurethane formation is done in two steps or stages. Firstly, excess diisocyanate is reacted with a polyol to form an isocyanate terminated prepolymer. Then, this prepolymer is chain extended/cross-linked in the second stage: for flexible PU foam, the reactants may include water, blowing agents and catalysts. Such processes were primarily developed to overcome the low reactivity of the original catalyst systems now, because of improved catalyst systems, the *one shot* process is more commercially important.

prepreg
A moulding composition. A reinforcement system, usually fibrous, which has been impregnated with a resin mix: the appropriate amount of this flat material is then fed to a mould. See *continuous fibre reinforced prepreg*.

preservative
A chemical used to stop *natural rubber* coagulating. Ammonia, formaldehyde or sodium sulphite may be added to the *tapping* cup to prevent premature coagulation.

pressblowing
A *blow moulding* technique which uses both injection moulding and extrusion. A neck section is first injection moulded and the *parison* is produced by extrusion: this is subsequently blown.

pressure
May be defined as *force per unit area*. The force per unit area acting on a surface. The SI unit of pressure is the *pascal* (Pa = N/m^2). See *Système International d'Unité*. Other common units include the *atmosphere*, the *bar* and *pounds per square inch*. One atmosphere = 101 325 Pa: one bar = 100 000 Pa and one pound per square inch (1 psi) = 6894·757 361 Pa. See *absolute pressure and gauge pressure*.

During processing, it is only when there is resistance to flow that, pressure exists or builds up: no resistance to flow, no pressure. See *injection pressure*.

pressure assisted thermoforming
Also known as high pressure forming or as pressure assisted forming. An abbreviation used for this process is PAT. A *thermoforming process* which offers a way of achieving a good quality product (similar to that produced by *injection moulding*) using tooling that is less expensive, and quicker to produce, than that used for injection moulding. The technique uses a combination of an air pressure and vacuum to form sheet into, usually, a female mould. The sheet is heated by *contact heating*: air applied through the female mould blows sheet up against the heating platen. At forming temperatures, vacuum is drawn through the female mould and air in introduced (up to 7 atm) through the heated platen. The product is then cooled and ejected. See *hydroforming*.

pressure bag moulding
See *bag moulding*.

pressure break
A defect in a *laminate* caused by the moulding pressure breaking one of the outer layers of reinforcement.

pressure cast beryllium-copper cavity
A casting produced by *hot hobbing*.

pressure die
A wire coating *die* in which the guide tip stops short inside the die so that the melt makes contact with the wire inside the die.

pressure difference
A term used in *hydraulics* and which refers to the difference in pressure between any two points in a circuit.

pressure flow
Also called Poiseuille flow. Flow induced by the application of a pressure to a liquid: the boundaries of the fluid are fixed as opposed to *drag flow* where the boundaries move and cause flow.

pressure - injection mould filling stage
The maximum pressure recorded during the mould filling stage: this depends upon the position of the transducer in the mould. Pressure is highest in the gate region of an injection mould and decreases steadily away from the gate. See *cavity pressure control*.

pressure intensifier
See *hydraulic intensifier*.

pressure line
The *line* which carries the *fluid* from the *pump* outlet to the *actuator*.

pressure over-ride
A pressure difference: the difference between the cracking pressure and the pressure when a valve is passing full flow.

pressure pad
A term sometimes used for *screw cushion* in *injection moulding*. May also mean a part of a mould; that is, that part of a compression mould which is designed to support the moulding force if the mould is closed empty.

pressure roll
A rubber covered roll, often water cooled by contact with another roll, which forms part of a nip roll assembly. Often used to improves bonding, for example, between a substrate and a *coating*.

pressure sizing die
See *sizing die*.

pressure spike
A saw-toothed pressure signal generated, for example, during mould filling in *injection moulding*. The formation of pressure spikes can lead to an erratic signal in *cavity pressure control*.

pressure switch
A term used in *hydraulics* and which refers to an electric switch which is actuated by pressure.

pressure switching
Term used in injection moulding to indicate that the final mould filling part of the moulding cycle is pressure controlled. A pressure transducer is used to initiate the *velocity pressure transfer* (VPT). Of the various places available for pressure transducer location, the following two are the most widely used: within the hydraulic line which feeds the hydraulic cylinder and which, in turn, pushes the screw forward and, from a signal generated within the mould (see *cavity pressure control*. In both cases the pressure transducer simply acts as a pressure switch, as a pressure (switch) value is selected by the operator: when the pressure measuring circuit 'sees' this value then, it tells the hydraulic circuit to change to the *hold pressure*. That is. this action, replaces one hydraulic valve (set at a high, pressure value) with another, set at a lower, pressure value.

No matter what system is used, for pressure switching, it is important to have a transducer which is robust, reliable and, which gives consistent results. However, even the best transducers can give poor results unless they are installed correctly and checked periodically. This usually means with *cavity pressure control* that, for example, the user must ensure that the equipment has been calibrated to suit the size of ejector pin. During use the pressure measuring system should be checked periodically so as to ensure that the preset values are being obtained.

pressure transducer
A *transducer* used to sense pressure. Two main type of electrical pressure transducer are used. The *piezo-electric* pressure transducer and the strain gauge pressure transducer. For optimum dynamic performance the piezo-electric type is usually selected: however, they are expensive, relatively insensitive and not suitable for the measurement of static pressure. Transducers based on strain gauges are most commonly found in the polymer industry and most often in the thermoplastics industry where they are used for *extrusion control* and for *cavity pressure control*.

It is usually melt pressure which such a transducer is required to measure, for example, the pressure within an *extrusion die* or, within an injection mould. Such a transducer may be classed as a *direct pressure transducer* or as an *indirect pressure transducer*. In general, direct pressure sensing is best and should be selected wherever possible: however, the transducers are often more expensive than the indirect type as they must be built to withstand melt temperatures and still give consistent readings over long production periods.

pressure-break
A defect in a laminated plastic a break apparent in one or more outer sheets of the paper, fabric, or other base visible through the surface layer of resin which covers the base.

pressure-casting
See *hot hobbing*.

pressure-compensated
A term used in *hydraulics* and which refers to a pump or control valve whose output is automatically adjusted when the system pressure varies.

pressure-reducing valve
A term used in *hydraulics* and which refers to a valve whose output is automatically adjusted so that it is at a lower pressure than the inlet pressure.

pressure-relief valve
A term used in *hydraulics* and which refers to a valve which automatically limits the maximum pressure in a circuit.

pressurised fluidised bed
A *fluidised bed* which uses relatively high working pressures, for example, for *vulcanization*. For example, the pressure could be approximately 4 bar/60 psi (still with a 1 psi pressure drop as for an *open fluidised bed*). The problem is getting the unit sealed, i.e. so that extrudate goes in but the heating gas does not escape. A production unit, which uses steam as the heating gas, may be 15 m in length.

prestretching
Stretching the sheet before *thermoforming*: this may be done by either mechanical means (using a *plug*) or, by using compressed air.

pretreatment of fabric
See *fabric*.

preventive antioxidant
An additive, an *antioxidant*, which functions by stopping the initiation of oxidative chain reactions which cause degradation. Preventive antioxidants include organosulphur compounds, for example, thio-bisphenols. In practice, there are three main classes of preventive antioxidant. These are (i) *peroxide decomposers*, (ii) *metal deactivators* and, (iii) *ultra-violet light* (UV) *absorbers*.

prevulcanization inhibitor
An abbreviation used for this type of material is PVI. An additive in a rubber compound which is used in order to improve the resistance to *scorching*. For example, N-cyclohexylthiophthalimide.

prevulcanized NBR
An abbreviation used for prevulcanized *nitrile rubber*.

PRI
An abbreviation used for *plasticity retention index*. The Plastics and Rubber Institute (of the UK) was also referred to as the PRI. This organisation is now part of the Institute of Materials (IoM).

primary accelerator
An *accelerator* which is either used alone or, which is the main component of a mixed system: the other component of a mixed system may be referred to as the secondary accelerator.

primary acetate
See *cellulose triacetate*.

primary amine
An *amine* with the formula RNH_2, where R is an organic group.

primary antioxidant
An *antioxidant* which is either used alone or, which is the main component of a mixed system: for example, a *hindered phenol* or *amine antioxidant*.

primary colour
Red, yellow and blue are the primary colours: colours which cannot be imitated by mixing together other colours. Longest wavelength in the visible spectrum is red at approximately 700 nm: shortest wavelength in the visible spectrum is violet at 400 nm. Blue has a wavelength of about 470 nm and yellow is about 570 nm.

primary particle
The smallest, most stable particle which is not readily destroyed by shear during processing. For example, pigment primary particles are aggregates of small crystallites: such aggregates in turn may *flocculate* into larger, easily broken particles.

primary plasticizer
A *plasticizer* which has very good compatibility with a polymer: a plasticizer which is compatible with a polymer in all reasonable proportions. Primary plasticizers are more compatible with the basic plastics material than secondary plasticizers which, in turn, are more compatible than the plasticizer extenders.

primary structure
See *structure* and *secondary structure*. See *carbon black*.

print embossing
See *valley printing*.

printed circuit
A term used in electronics to describe a circuit in which the wiring between components, and even some of the components themselves, are printed or etched onto a *printed circuit board*.

printed circuit board
An abbreviation used for this is pcb. A term used in electronics to describe the insulating board which has a circuit printed or etched onto its surface. The board may, for example, be an epoxy glass laminate. See *printed circuit* and *terpenes*.

printing techniques
Techniques used to apply letters or designs to plastics products, for example, to extrusions and/or to mouldings. The major printing techniques are *letterpress printing*, *lithographic printing*, *silk-screen printing*, *gravure printing* and *flexographic printing*. Other techniques include ink-jet printing, pad printing, hot transfer printing, laser marking and diffusion printing. See *photoresist* and *photo-sensitizer*.

pro-oxidant
A material which promotes *oxidation*, for example, copper in *polypropylene*.

procedure A
Also known as *method A*. See *melt flow rate*.

procedure B
Also known as *method B*. See *melt flow rate*.

process aid
An abbreviation used for this term is PA. Also called a processing aid. An additive used to alter the processing characteristics of a particular material or compound, for example, a *processing oil*. An additive which improves processing. A rubber additive used to give compounds which, on processing by *calendering* and *extrusion*, gives less *die swell*, better surface smoothness and higher throughputs. For example, PA-80 is an *SMR grade* which has a concentration of 80 parts of *vulcanized latex*. The most immediate use for *polyhydroxybutyrate* could be as processing aids for *unplasticized polyvinyl chloride* (UPVC). At a 1% addition level, it reduces the gelation time, power requirements and improves the surface finish of extrudates.

process capability
A measure of how accurately a production unit can produce components when there are no *assignable variations*. For example, statistics may be used to determine how accurately a machine/material/mould/operator combination can produce components: production capability questions can be answered and, what is more, such questions can be answered relatively easily.

It is common practice to accept that 99.73% of production will lie within the tolerance (t). That is $t = 6\sigma/2$ where σ is the standard deviation. If 6σ is less than 2t then the process is more capable than the minimum requirement.

process capability index
An abbreviation used for this term is PCI. Also called the capability index. Used to indicate the *process capability* with regard to the specified tolerance (t). That is $PCI = 2t/6\sigma$ where σ is the standard deviation. The minimum acceptable value of the index is 1.

process control chart
Also known as a quality control (QC) chart. A chart used to assess and to evaluate the manufacturing process. A process control procedure which makes use of statistical methods based upon the *normal distribution*. A statistical control procedure used to assess whether a process which has been set up to produce an acceptable product is continuing to function as

it was initially set and, if not, to detect when the process drifted out of control.

Samples of predetermined size are taken at regular intervals, measured, averaged and points are plotted on the chart: when average values are plotted the chart is called an *average chart*. A *range chart* is also plotted as it is conceivable that samples could maintain an average value but with increasing range. See *production capability*.

process printing
The reproduction of coloured images. The object is photographed through red, green and blue filters so as to produce negatives. These negatives are then used to produce plates for use with inks which reproduce the original colours. This three colour process does not reproduce black very well and so a four colour process is more widely used: this uses a black plate to strengthen the dark areas.

process timer
A timing device used to switch an electrical circuit at a pre-selected time.

process-based monitoring
A control strategy which monitors production process conditions so as to achieve product improvements. For example, in *injection moulding*, screw cushion size and plasticizing time may be monitored in place of component weight (*product-based monitoring*).

processability
The ease with which a plastics material can be processed, for example, extruded. At a given screw speed, reduced power use (amps) and reduced die pressures characterise improved processability. One of the major reasons for blending different thermoplastics materials is to improve processability. See *blend and alloy*.

processing aid
See *process aid*.

processing material changes
See *material changes*.

processing modifier
An additive used to alter the processing characteristics of a particular material or compound. Such additives include *nucleating agents* and *lubricants*. See *masterbatch*.

processing oil
A *process aid*. Such additives are usually associated with the rubber industry where they are used to alter (increase) the *plasticity* of rubber compounds and to improve *filler* dispersion. There are three main groups of such oils which are based on the chemical composition of the *oil*. The three groups are aromatic, naphthenic and paraffinic processing oils. Naphthenic and aromatic processing oils give the best processing properties: paraffinic processing oils and naphthenic processing oils give the best low temperature properties. Aromatic oils have relatively poor ageing properties whereas paraffinic processing oils have relatively good ageing properties.

processing temperature window
The temperature range between melting/softening point and decomposition temperature for a thermoplastics material.

product assessment
See *product quality*.

product identification
The identification of a component by a moulded or printed legend. Such identification could indicate when the component was produced and the material(s) employed. Such identification is being used in an effort to make re-use of materials easier. On some large mouldings, each moulding can be identified with a moulded-in legend so that product identification throughout life is possible if required. As the information generated, on the way that the component was produced, can also be logged or stored then the effect of processing on properties can be determined.

product marking
See *hot foil marking*.

product monitoring
Checks on *product quality* may now be made by the machine operator, by the control system or by a separate system such as a robot-type device. See *moulding assessment* and *automatic quality control*.

product quality
Verification of product quality at the point of manufacture, for example, during *injection moulding*, can be documented in statistical process control (SPC) records. To do this the system designer must identify what needs to be monitored and then, for example, equip the machine with appropriate transducers. The output from these transducers is monitored and if the signal from one transducer, or from a combination of several, is different from previously set limits then various *quality control* (QC) actions may be taken if selected.

In one such system, seven parameters are set and monitored, for example, injection speed and hydraulic pressure. If the set limits are exceeded, or not met, then the control system judges that the particular injection moulding is unlikely to be acceptable. various control strategies can then be implemented, for example, an alarm sounds and the injection moulding is diverted to an inspection area. If the number of rejects exceeds a certain preset, percentage then the machine will stop once that percentage is exceeded.

Where robots are used for moulding removal then, a robot measuring system (for example, based on the measurement of different dimensions of an injection moulding using a video camera and appropriate software) may be used to judge, or assess, the component. The robot arm can orientate, or position, the moulding in different positions for video camera examination.

The information that the robot measuring system generates may be transmitted to the microprocessor control system and plotted on a *process control chart*: on-line QC, based on moulding measurements, is therefore relatively easy and can be done for every moulding produced. Every tooth of a gear wheel, for example, may be assessed for size and shape using such a system as the wheel is travelling alone a conveyor to the next stage of the process. If it is judged to be incorrect then it may be diverted for re-assessment of for regranulation.

product weight
Component or moulding weight. Of all the tests which can be performed alongside a production unit, for example, alongside an *injection moulding machine*, the weighing of components is now the easiest one to perform by an operator. This is because a modern, digital electronic balance needs little, if any, setting up and component weight can be determined very easily and rapidly. Such a determination indicates how well a mould, or a particular cavity within a multicavity mould, has filled. Component weight can be readily correlated with the dimensions of a particular component and, if required, the digital balance used can be fitted with an electronic calculator so that the readings obtained may be statistically analyzed and/or used to construct a *process control chart*. See *production capability*.

product-based monitoring
A control strategy which monitors the product so as to achieve product improvements. For example, in *injection moulding*, component weight may be monitored in place of screw cushion size and plasticizing time (*process-based monitoring*).

production capability

The accuracy of production which is possible from, for example, an *injection moulding* unit. If a very large number of mouldings are produced and a particular dimension (e.g. their diameter) is measured, then a *normal distribution* curve could be constructed. If a large enough number of samples were taken, then in the ideal case, a normal bell-shaped curve would result which would be symmetrical about the average value. Approximately 68% of all the measurements would lie within ±1 *standard deviations* of the average, 95% would lie within ±2 standard deviations, and approximately 99.73% would lie within ±3 standard deviations of the average. This means that approximately 998 injection mouldings in every 1,000 produced will have dimensions which can be specified from easily obtained information. One in a thousand will have a dimension which lies below the specified value and, one in a thousand will have a dimension which lies above the specified value. So, for many practical purposes the average, plus and minus three standard deviations, gives the production capability.

production condition recording

The systematic recording of all production conditions so that it is known how a component was produced. It must never be forgotten that the object of production is to produce component/product of the required quality at a specified cost within a specified time and when required. To do this it is essential to keep accurate records. On many machines this can be done at the press of a button. Where this is not possible then, an appropriate record sheet should be completed at periodic intervals: production samples should also be retained for future reference.

production control - use of statistics

Changes in components, can be readily identified, and quantified, if the changes are capable of being statistically analyzed. Such analysis is not necessarily difficult as the use of only relatively simple statistics can yield a surprising amount of useful information. On many machines, if fitted with microprocessor control, the control system can display the results of such statistical calculations. See *production capability*.

profile

That which is seen when an extrudate is cut across, i.e. at right angles to the direction of extrusion: usually refers to a complex cross-section. See *extruded profile*.

profile control system

A system which may be adjusted so as to control the size of an emerging profile, for example, lay flat film. See *automatic profile control system*.

profile die

A *die* which produces a complex cross-section by *extrusion*, for example, a window frame section.

profile of a tyre

See *tyre numbering*.

profiled roll

See *roll cambering*.

programmed clamp force reduction

A technique used in *injection moulding*. The amount of energy used by the clamp system can be reduced by programming a reduction in the force applied as the moulding cools.

programmed injection

See *injection speed* and *programmed injection speed*.

programmed injection speed

A term used in *injection moulding* and which means that the speed of mould filling can be changed within one injection moulding cycle. Many moulding faults, for example, jetting and air trapping, may be avoided by using several speeds (that is, programming the *injection speed*), during the mould filling stage.

When moulding thin sectioned components, high injection speeds are essential in order to fill the mould before freezing occurs. For many injection moulding operations, the initial cavity filling is complete in less than one second. However, a better surface finish is obtained on mouldings with thicker sections by using a slower speed.

Now, despite the desirability of having a uniform wall thickness in the injection moulding, and an absence of holes, many injection mouldings contain such features and their presence give rise to moulding faults, for example, air trapping. By varying the speed of filling, air which could be trapped between two merging melt fronts (formed by the plastics melt being divided by flowing around a core or pin) is allowed to escape. If it was not allowed to escape then it would cause a blemish on the surface - in the extreme case it can cause discolouration by burning as it is compressed.

projected area

The area of a moulding as projected onto a plane normal to the direction of opening of the mould in which it was made. For an *injection moulding* it is the area of the mould at right angles to clamp force application.

proliferous polymerization

See *popcorn polymerization*.

promoter

See *activator*.

propagation

Also known as chain propagation. The step in a *polymerization* reaction whereby monomer units add to a growing active centre so as to form a polymer molecule. As propagation is much faster than initiation, the *rate of polymerization* and the *rate of propagation* are virtually the same.

propagation - oxidation

See *oxidation - mechanism*.

propan-1-ol

See *n-propyl alcohol*.

propan-2-ol

See *isopropyl alcohol*.

propan-2-one

See *acetone*.

propane

An *alkane* with the formula C_3H_8 and with a boiling point of $-42°C$. Obtained from crude oil or natural gas and widely used as a fuel.

propane-1,2-diol

See *propylene glycol*.

propane-1,2,3-triol

See *glycerol*.

propanoic acid

See *propionic acid*.

propanone

See *acetone*.

propellant

See *aerosol propellant*.

propenal
See *acrolein*.

propene
See *propylene*.

propenoic acid
See *acrylic acid*.

propenonitrile
See *acrylonitrile*.

property guidelines
Test results. Because of the large number of grades available any properties quoted in the literature should only be used as a general guideline. The properties of any plastics material may also be dramatically changed by the processing conditions employed and by the use of additives. With many materials, variations of basic formulas are available with additives to provide, for example, improved heat resistance or, weatherability. Some formulations offer improved impact strength while others, which contain fillers, are used where, for example, thermoplastic mouldings are required to possess greater impact strength, tensile strength and heat distortion temperature. Processing and performance modifiers can be added; these include, for example, antistatic and nucleating agents: such additives may form part of the colour *masterbatch*.

propionic acid
Also known as propanoic acid. A carboxylic acid with the formula C_2H_5COOH and with a boiling point of 151°C. A bread preservative: also used to prepare *cellulosics*, for example, cellulose acetate propionate.

proportional
A control action which is denoted by 'P'; means that within the proportional band the power supply is progressively reduced as the set point is approached.

proportional band
Sometimes shortened to 'PB' and usually means a temperature band or range (for example, 10% of set point) over which power is proportioned or, reduced as the set point is approached.

proportional control valve
A control *valve* whose output is proportional to a control signal: output flow is proportional to the input signal.

proportional flow
A term used in *hydraulics* and which refers to a method of *filter operation*: the flow which passes through the filter is proportional to the pressure drop.

proportional limit
The position on the stress strain curve beyond which Hooke's law is no longer obeyed. The greatest *stress* which a material is capable of carrying in a *tensile test*, without any deviation of proportionality of stress to strain.

proportional valve
See *proportional control valve*.

proportioning
A term often used interchangeably with proportional.

propyl
The organic radical which has the formula C_3H_7-.

propyl alcohol
An alcohol with two isomers, *n-propyl alcohol* and *isopropyl alcohol*. Both are used as, for example, solvents.

propylene
Also known as propene. An olefin with a boiling point of −48°C. Has the formula C_3H_6 and may be represented as $CH_2=CH.R$ where the side group R is CH_3. Obtained from crude oil or natural gas. An important feedstock for the polymer industry and the monomer for *polypropylene*.

propylene glycol
Also known as propane-1,2-diol or as, 1,2-propylene glycol. Prepared from propylene oxide and may be represented as $CH_3CHOHCH_2OH$. Has a boiling pint of 189°C. A *diol* widely used in the polymer industry, for example, for the preparation of *unsaturated polyester resins*: the esters of this material may be used as softeners/plasticizers. Both mono-esters and di-esters are used, for example, *propylene glycol ricinoleate* and *propylene glycol diricinoleate*.

propylene glycol diester
An organic ester prepared from *propylene glycol* and an organic acid, for example, *ricinoleic acid*, and in which both of the hydroxyl groups were used in the esterification reaction.

propylene glycol diricinoleate
An organic di-ester prepared from *propylene glycol* and the organic acid, *ricinoleic acid* and in which both of the hydroxyl groups were used in the esterification reaction. This material has a melting point of −51°C and a relative density (RD) of 0·94. A softener for rubbers.

propylene glycol ester
An organic material prepared from *propylene glycol* and an organic acid, for example, *ricinoleic acid*. Both mono-esters and di-esters are used, for example, *propylene glycol ricinoleate* and *propylene glycol diricinoleate*.

propylene glycol mono-ester
An organic ester prepared from *propylene glycol* and an organic acid, for example, *ricinoleic acid*, and in which only one of the hydroxyl groups was used in the esterification reaction.

propylene glycol mono-laurate
Also known as 1,2-propylene glycol monolaurate. An organic mono-ester prepared from *propylene glycol* and the organic acid, *lauric acid* and in which only one of the hydroxyl groups was used in the esterification reaction. This material has a melting point of about 8°C and a relative density (RD) of 0·91. A softener for rubbers.

propylene glycol mono-oleate
Also known as 1,2-propylene glycol mono-oleate. An organic mono-ester prepared from *propylene glycol* and the organic acid, *oleic acid* and in which only one of the hydroxyl groups was used in the esterification reaction. This material has a melting point of about −20°C and a relative density (RD) of 0·92. A softener for rubbers.

propylene glycol mono-stearate
Also known as 1,2-propylene mono-stearate. An organic mono-ester prepared from *propylene glycol* and the organic acid, *stearic acid* and in which only one of the hydroxyl groups was used in the esterification reaction. This material has a melting point of about 39°C and a relative density (RD) of 0·93. A softener for rubbers.

propylene glycol ricinoleate
An organic mono-ester prepared from *propylene glycol* and the organic acid, *ricinoleic acid* and in which only one of the hydroxyl groups was used in the esterification reaction. This material has a melting point of −15°C and a relative density (RD) of 0·96. A softener for rubbers.

propylene oxide
Obtained from *propylene* and used, for example, to prepare *propylene glycol*. Has a boiling point of 34°C. Contains the *epoxide* group and may therefore be used to make polymers by ring opening polymerization. Used to make a range of

polymers. For example, some copolymers of propylene oxide are also marketed as *propylene oxide rubbers*. See *polypropylene oxide* and see *propylene oxide-ethylene oxide block copolymer*.

propylene oxide rubber
Both homopolymers and copolymers can be produced from *propylene oxide*. An abbreviation used for propylene oxide (homopolymer) rubber is PO and that for propylene oxide (copolymer) rubber is GPO. The copolymer is the most important type and is produced by copolymerization of propylene oxide and the *cure-site monomer allylglycidyl ether* (approximately 10%).

Also known as propyleneoxide copolymer or as, propylene oxide-allylglycidyl ether rubber. The material can be *sulphur vulcanized* to give products which exhibit excellent flex life, high resilience, good low temperature properties, good heat resistance, good ozone resistance and moderate oil resistance.

propylene oxide-allylglycidyl ether rubber
See *propylene oxide rubber*.

propylene oxide-ethylene oxide block copolymer
Also known as poly-(propylene oxide-b-ethylene oxide). A triblock copolymer made by anionic polymerization of *propylene oxide* followed by polymerization onto the preformed polymer of ethylene oxide. Such materials are used as non-ionic detergents. Some copolymers of propylene oxide are also marketed as *propylene oxide rubbers*.

propylene plastics
Plastics based on polymers or copolymers of *propylene* with other monomers, the propylene being in the greatest amount by mass. See *polypropylene*.

propylene sulphide
Obtained from *propylene oxide* and used, for example, to prepare *polypropylene sulphide*. Has a boiling point of 76°C. Contains a cyclic thioether group and may therefore be used to make polymers by ring opening polymerization.

protecting group
Also known as a blocking group. A chemical group which is attached to a material so as to prevent an undesirable chemical reaction occurring with that material.

protective breaker
A *tyre component*: a layer of additional *ply* material between the *tread* and the *belt* and incorporated to reduce damage to the belt.

protein
A *biopolymer*: a *copolymer* based on nitrogen-containing materials (amino acids) which has an incredible regular structure. Proteins are a major component of living cells.

proton
A hydrogen nuclei with a single positive electric charge.

proton beam
A form of *high energy radiation*. See *radiation*.

prototype mould
A mould constructed and used, for example, to evaluate a design: a mould used to produce only a few mouldings. In spite of the large amount of knowledge now available on plastics materials, it is often found that the quickest way of obtaining data for design purposes is to make a *prototype moulding*. Such a moulding may also be used to demonstrate the appearance and properties of a plastics component. Such a prototype moulding will be better than one produced by machining as it will have been produced by a moulding process and so its properties should more closely resemble those of the actual components.

Such trial mouldings have been produced from non-metal moulds, e.g. epoxide resins or high melting point thermoplastics materials such as polyimides. To withstand the clamping forces involved such materials are commonly encased in a steel bolster. A more usual material for prototype moulding is a low melting point, *zinc-based alloy* and more usually known by the trade name/trademark of Zamak, Ayem or Kirksite. Such alloys may be cast into the shape required using, for example, plaster moulds. If necessary, water-cooling coils may be incorporated during the casting process and the casts may be supported within a steel bolster so that the casts are not required to support the full clamping pressure. See *hot hobbing*.

prototype moulding
The process of producing trial mouldings or, the moulding produced from a *prototype mould*.

Prussian blue
A *blue pigment* which is ferric ammonium ferrocyanide and which is produced by the reaction of ferrous sulphate, ammonium sulphate and sodium ferrocyanide. Also known as iron blue. By varying the process, can produce slightly different shades of blue called Chinese blue, Paris blue and Milori blue. Has good hiding power in high concentrations but can appear transparent in low concentrations because of a low index of refraction. Has limited heat resistance (be careful above 180°C) and for this reason its use in thermoplastics is limited, for example, to low density polyethylene. See *ferric ferrocyanide*.

PS
An abbreviation used for *polystyrene*. An abbreviation used for *pferde-stärke*.

PS-E
An abbreviation used for *expanded polystyrene*.

PS-X
An abbreviation used for *expanded polystyrene*.

PSBR
An abbreviation used for pyridine/styrene-butadiene rubber.

pseudo three dimensional fabric
A pseudo 3D fabric is a multi-layer fabric which is used, for example, in the *GRP industry*.

pseudoplastic
Term often applied to plastic melts: a *melt* which becomes less viscous (more easier flowing) if it is moved more quickly or, sheared more quickly.

pseudoplasticity
Also known as *shear thinning*. A type of *non-Newtonian flow behaviour* in which the *apparent viscosity* decreases as the *shear rate* increases. Most polymer melts are shear thinning but see *shear-induced crystallization*.

psi
An abbreviation used for *pounds per square inch*.

PSi
See *silicone rubber*.

PSU
An abbreviation used for polysulphone: see *sulphone polymers*.

pt
An abbreviation for *pint*.

PTBP
An abbreviation used for *4-tert-butyl phenol*.

PTD
An abbreviation used for *dipentamethylene thiuram disulphide*.

PTFE
An abbreviation used for *polytetrafluoroethylene*.

PTFE ionomer
An abbreviation used for *polytetrafluoroethylene* ionomer.

PTHF
An abbreviation used for *polytetrahydrofuran*.

PTM
An abbreviation used for *dipentamethylene thiuram monosulphide*.

PTMEG
An abbreviation used for *polytetramethylene ether glycol*.

PTMG
An abbreviation used for *polyoxytetramethylene glycol*.

PTMT
An abbreviation used for *polybutylene terephthalate*.

PU
An abbreviation used for *polyurethane*.

PU-RIM
An abbreviation used for *polyurethane* components produced by the *reaction injection moulding process*.

puff-up ratio
See *die swell*.

pull-back rams
Preload devices fitted to a *calender* equipped with bush bearings. Such rams pull the rolls into their correct working positions and hold them in those positions during machine operation. These rams are hydraulically operated and pull the roll into a certain position by means of a collar which passes around the roll journal. In the case of a top roll on a three roll calender, pull back rams would pull the roll against the top of its bearing by means of two rams, that is, one on each end. See *roll float*.

pulled-surface
A defect in a laminated plastic: an imperfections in the surface ranging from a slight breaking or lifting in spots, to pronounced separation of the surface from the body.

puller
Device used to remove the extrudate from the die region at a controlled rate: usually either, a pair of rollers or, two belts which trap the extrudate between them and which pull the extrudate smoothly away from the *die*.

pulp moulding
A moulding formed from a mixture of paper pulp and a *thermosetting resin*.

pulse blowing
A technique used to speed up cooling in *blow moulding*. Once the container is blown, the air pressure is reduced and fresh, cool air is introduced.

pulsed cooling
A cooling technique which uses increased flow rates of cooling medium at appropriate points in the moulding cycle.

pultrusion
A process used to make very strong composites by drawing a *fibre* assembly through catalyzed resin and then consolidating the impregnated fibre by winding onto a mandrel. Such composites have been used to make corrosion resistant products for the construction industry, for example, walk-way gratings, hand-rails and pole holders. Approximately one third of all pultruded products go into the electrical industry, for example, for use as ladders, cable trays and lighting poles. See *fibre positioning*.

pumice
Volcanic glass which when powdered is sometimes used as a filler in rubber compounds to make abrasive compounds. Powdered pumice has a relative density of 2.35.

pump
A device which changes power into fluid flow.

pump delivery pressure
Also known as main line pressure.

punch
Also called a force. The male part of a mould.

punching
The removal of squeezed flash from a *blow moulding* via a male/female sliding punch system: the flash must be virtually severed by the mould knife edges as no real shearing effect is possible on lightweight plastics components.

puncture tapping
Method used to obtain *natural rubber latex* from the tree *Hevea brasiliensis*. The trees are tapped by inserting sharp needles into the bark. May be combined with *stimulation tapping*. Stimulation tapping, and puncture tapping, may increase the yield or maintain the same yield with a reduction in tapping intensity: this conserves the tree bark. See *tapping methods*.

PUR
An abbreviation used for *polyurethane*.

pure gum compound
See *high gum compound*.

purely viscous fluid
See *inelastic fluid*.

purge compound
Also known as a purge or as, a flushing compound or as a, cleaning compound. A compound specifically designed, or used, to assist purging or machine cleaning and which may contain, for example, large amounts of a filler such as pumice. Before such purge compounds are used (in processes such as *extrusion*) it is advisable to remove the die and head assembly as many purge compounds do not melt, or flow, like ordinary thermoplastics materials. Such compounds are introduced into a cylinder or barrel when the screw has been pumped dry, or free of, an unwanted material, for example, the unstable material unplasticized *polyvinyl chloride* (PVC). The nozzle may then be thoroughly cleaned and once the purge compound is coming through, the shut down procedure may then be followed.

purge material
A material used for *purging*. Such a polymeric material is usually of high molecular weight (high viscosity) and relatively stable at processing temperatures: a natural, non-flame retardant grade of, for example, *polymethyl methacrylate* or *high density polyethylene*. See *purge compound*.

purged
Term usually associated with processes such as *injection moulding* and means that the contents of the cylinder, or barrel, have been ejected into the air so as to remove unwanted material or, for *melt temperature measurement*.

purging
The action of cleaning an extrusion cylinder or barrel by running material through the barrel. When purging *polyvinyl chloride* (PVC) and *acetal* (POM) it must be remembered that these plastics degrade relatively easily and, if heated together, then decomposition, or degradation, can be extremely rapid and even violent. If a change is being made, from or to, another plastic and if it is thought that acetal (POM) or

polyvinyl chloride (PVC) is involved, then purge with a natural, non-flame retardant grade of *polystyrene* (PS) or *polyethylene* (PE). NEVER mix POM (acetal) and PVC (vinyl) or, follow one with the other without thorough purging with an inert material such as PS or PE. See *purge compound* and *purge material*.

purging pins
Pins, or mandrels, which are inserted into a *blow moulded* product, in a post moulding station, so as to obtain more rapid cooling.

purified mineral wax
See *ceresin wax*.

purified rubber
Natural rubber which has had most of the non-rubber constituents removed. Gives a grade of *natural rubber* with a low degree of water absorption.

Pusey and Jones plastometer
A dead-weight apparatus used to measure *hardness*, for example, of rubber covered rolls. A 0.125"/3.175 mm indentor is loaded with a weight of 1 kg and the penetration is measured in hundredth's of a mm after one minute.

push pull
See *multi-live feed moulding*.

push-back pin
A return pin. A hardened steel pin which returns the *ejector* assembly to a rear position as the mould is closed.

push-back ram
A return ram. A hydraulically operated ram fitted to a *compression moulding press* in order to ensure mould opening.

pushing flight
That face, or edge, of the *screw flight* which drives the plastic towards the *die*: the face nearest the die.

Pussey-Jones plastometer
See *Pusey and Jones plastometer*.

PVA
An abbreviation which has been used for *polyvinyl acetate* and for *polyvinyl alcohol*.

PVAc
An abbreviation used for *polyvinyl acetate*.

PVAC
An abbreviation used for *polyvinyl acetate*.

PVAl
An abbreviation used for *polyvinyl alcohol*.

PVAL
An abbreviation used for *polyvinyl alcohol*.

PVB
An abbreviation used for *polyvinyl butyral*.

PVC
An abbreviation used for *polyvinyl chloride*.

PVC calender
A *calender* used for *polyvinyl chloride* (PVC) compounds.

PVC modified rubber
See *nitrile rubber polyvinyl chloride blend*.

PVC plastisol
An abbreviation used for *polyvinyl chloride* plastisol. See *paste and plastisol*.

(PVC + NBR)
A *polyvinyl chloride nitrile rubber blend*.

PVC-P
An abbreviation used for *plasticized polyvinyl chloride*.

PVC-U
An abbreviation used for *unplasticized polyvinyl chloride*.

PVCZ
An abbreviation used for *polyvinyl carbazole*.

PVdC
An abbreviation used for *polyvinylidene chloride*.

PVDC
An abbreviation used for *polyvinylidene chloride*.

PVdF
An abbreviation used for *polyvinylidene fluoride*.

PVDF
An abbreviation used for *polyvinylidene fluoride*.

PVF
An abbreviation used for *polyvinyl fluoride*.

PVF$_2$
An abbreviation used for *polyvinylidene fluoride*.

PVFM
An abbreviation used for *polyvinyl formal*.

PVI
An abbreviation used for *pre-vulcanization inhibitor*.

PVK
An abbreviation used for *polyvinyl carbazole*.

PVMQ silicone rubber
A phenylvinylmethyl silicone rubber. Such a *silicone rubber* is based on *polydimethylsiloxane*: some of the methyl groups are replaced with phenyl and vinyl groups. The *PMQ* and the PVMQ-types of rubber are classed as silicone rubbers with excellent low temperature performance.

PVP
An abbreviation used for *polyvinyl pyrrolidone*.

PVT control
An abbreviation used for pressure (P), volume (V) and temperature (T) control. If an *injection moulding* machine is equipped with appropriate transducers, and if the machine is also equipped with the right software and computer then, the operator can implement PVT control. The idea behind this is relatively simple: it is that pressure (P), volume (V) and temperature (T) are inter-related. As the moulding cools in the cavity then the *holding pressure* may be progressively reduced so as to produce components free from over-packing. In this way, injection mouldings which have the correct amount of material contained in them are produced. Such mouldings, being free from over-packing, have reduced stress levels. Because of the correct, and lower weight, more rapid and uniform cooling occurs and this in turn gives faster outputs of more uniform components. The instrumentation is also used for *quality control* purposes.

pyknometer
An apparatus used to determine, for example, the *density* of a liquid. Consists of a glass vessel which is graduated to hold a definite volume of liquid at a specified temperature. By weighing it when full of liquid at different temperatures, variations in density may be determined.

pyknometer method
A *bulk density* measurement method which uses a pyknometer.

pyrex glass
Trade name/trademark for a *borosilicate glass*.

pyridine
This colourless heterocyclic material (C_5H_5N) has a boiling point of approximately 115°C. It is a good solvent for uncured *chloroprene rubber* (CR), *nitrile rubber* (NBR) and *styrene-butadiene rubber* (SBR). This chemical causes swelling, or solvation, of uncured *natural rubber* (NR) and *thiokol rubbers* (T). It is a poor solvent for *butyl rubbers* (IIR). Used to produce *iso-rubber*.

pyridine-butadiene rubber
An abbreviation used for this type of material is PBR. Also known as vinylpyridine-butadiene rubber.

pyridine-styrene-butadiene rubber
An abbreviation used for this type of material is PSBR. Also known as vinylpyridine-styrene-butadiene rubber.

pyroelectric effect
A property of some crystalline materials, for example, quartz. This *piezoelectric* material possesses pyroelectric properties, that is, electric polarization is generated when the temperature changes.

pyroelectric polymer
A polymer which can display a *pyroelectric effect*. See, for example, *polyvinylidene fluoride*. Such a pyroelectric polymer may be used to make, for example, pyroelectric detectors for energy management systems and for intrusion detection systems.

pyrogenic process
Burning or high temperature heating process used to produce, for example, *silica*.

pyrogenic silica
Also known as thermal silica. Silica produced by a *pyrogenic process*.

pyrolysis
Burning or high temperature heating process performed in an inert atmosphere. Sometimes used, for example, to *burn clean* polymer coated, metal components.

pyromellitic dianhydride
An abbreviation used for this material is PMDA. An aromatic organic compound with a boiling point of 286°C. Being a dianhydride this monomer is tetrafunctional and as it contains a benzene ring, it is suitable for the manufacture of *aromatic polymers*. Also used as a curing agent for *epoxide resins*: gives products with a high degree of *cross-linking* and good heat resistance.

pyrometer
An instrument used to measure elevated temperatures, for example, of a polymer melt. There are several different types which may be may be based on a *platinum resistance thermometers* or on *thermocouples*. Can also have *radiation pyrometers* and *optical pyrometers*.

pyropolymer
A polymer obtained by the *pyrolysis* of a *precursor polymer*. See *carbon fibre*.

Q

Q
This letter is used as an abbreviation for:
   silicon—as a filler;
   silicone-based rubber. See *polysiloxane*;
   tetra or quadrifunctional unit in a polyorganosiloxane: and,
   volumetric output rate. See *capillary rheometer*.

QA
An abbreviation used for *quality assurance*.

QC
An abbreviation used for *quality control*. QC is also used as a prefix in British Standards, for example, BS QC. See *British Standard*.

QMC
An abbreviation used for *quick mould changing*.

QMS
An abbreviation used for *quality management system*.

qualitative chemical analysis
Analysis which is concerned with identifying what chemical substances are present in a mixture or compound. See *quantitative chemical analysis*.

quality
A characteristic property or attribute. The totality of features and characteristics of a product, or service, that bear on its ability to satisfy stated, or implied, needs.

quality assurance
The planned and systematic actions which give confidence that a product or service will satisfy stated quality requirements. Quality assurance (QA) is concerned with the ability of an organisation to be able to give a formal guarantee, or a positive declaration, that a product is of a required quality. However, it must be remembered that quality assurance testing, like any other testing, is expensive to perform and is therefore often only undertaken when necessary. It may become necessary if customers or suppliers of an organization themselves participate in a quality assurance scheme, such as *BS 5750* or *ISO 9000*, and insist that other organizations also perform to such a relevant standard before they will trade. See *Quality Assurance Handbook 22*.

quality concepts
See, for example, BS 4778 Part 2·1991 which is entitled 'Quality concepts and related definitions'.

quality control
The application of the theory of mathematical probability to production sampling so as to detect any variation of quality. The object of quality control is to help keep manufacturing tolerances between predetermined limits (see *process control charts*). For quality control purposes, numerical data is often generated so that the results can be readily subjected to statistical analysis. Such analysis can help to keep a process under control. Quality control is concerned with implementing *quality assurance*.

Quality control (QC) therefore exists as a production support function within the manufacturing enterprise, to help answer questions about the product. Three commonly asked questions are:

1. Can we make the product to specification?
2. Are we making the product to specification?
3. Have we made the product to specification?

The first question to answer, therefore, is whether or not the production facility, for example, a moulding machine, is capable of producing components whose dimensions fall within the specification limits. Once this has been decided then, a check on the production capability can be kept by means of, for example, quality control charts. See *production capability*.

quality control charts
See *process control charts*.

quality - correlation with machine settings
With components produced by *injection moulding*, it has been found that there is a close correlation between final features

of the component (such as weight and size) and production conditions such as *cushion size, injection pressure* and *injection rate*. This means that it is possible to check that injection mouldings are satisfactory, in many cases, without actually doing any measurements on the actual mouldings themselves. During each shot, selected parameters are measured and compared against set, or stored, values. Provided the measured values fall within preselected limits, the injection moulding is judged by the control system to be acceptable: If the measured values are outside the set limits then, the injection moulding will be either rejected or, if it is only just outside, it may be retained for a second opinion by a qualified person.

quality management system
An abbreviation used for this term is QMS. In Europe it is the BS 5750/ISO 9000/EN 29000 standards which form the basis of most quality management systems or policies. Such standards specify the organisational framework which is necessary for quality management and also specify requirements for product design, product development, production, installation and servicing. To be effective a quality management system must have excellent record keeping as accurate documentation helps ensure that specified procedures are being followed and that such procedures are yielding the desired results. Third party product certification confirms, by regular audit testing, that a product complies with all aspects of the relevant specification supported by a quality management system to BS 5750/ISO 9000/EN 29000.

quality vocabulary
See BS 4778 Part 1:1987 or ISO 8402-1986: these standards are identical and give international terms. Also see BS 4778 Part 2:1991 which is entitled 'Quality concepts and related definitions'.

Quality Assurance Handbook 22
Also known as the BSI Handbook 22:1992 or as, Handbook 22. The 1992 edition is in two parts. Part 1 is entitled 'Quality Assurance' and part 2 is entitled 'Reliability and maintainability'. Part 1 contains all parts of BS 5750 'Quality systems' as well as BS 7229 'Quality systems auditing': parts 1, 2 & 3; BS 6143 'Guide to the economics of quality'; BS 7000 'Guide to managing product design'; and, BS 7373 'Guide to the preparation of specifications'. Part 2 'Reliability and maintainability' contains BS 4778 'Availability, reliability and maintainability terms': part 3: section 3.1 and 3.2; BS 5760 'Reliability of systems, equipment and components': parts 1, 2, 3, 4, 5, 6 & 7; BS 6548 'Maintainability of equipment': parts 1 & 2; and, DD 198 'Assessment of reliability of systems containing software'.

quantitative chemical analysis
Analysis which is concerned with identifying the amounts of substances present in a mixture or compound by chemical procedures. See *qualitative chemical analysis*.

quart
A unit of measurement with the abbreviation qt. One quarter of a *gallon*. To convert from a US dry quart to cubic metres multiply by 1.101×10^{-3}. To convert from US liquid quart to cubic metres multiply by 9.464×10^{-4}.

quartz
A crystalline form of *silica*: quartz crystals exhibit the *piezoelectric effect*. Finely divided quartz powder has been used as a filler for silicone-based polymers. See *polysiloxane*.

quartzite
See *crystalline silica*.

quasi-prepolymer process
Also known as the semi-prepolymer process. A process used in *polyurethane technology* to produce both rigid and flexible cellular materials. A *polyol* is first reacted with a large excess of *diisocyanate* so as to produce a low viscosity prepolymer which has an excess of isocyanate: this prepolymer is then reacted with more polyol when required. This process lies between the *prepolymer process* and the *one-shot process*.

quaternary ammonium compound
Also known as a quaternary ammonium salt. Such chemical compounds have the general formula of NR_4Hal where R is an organic radical or group and Hal is a halogen. On treatment with alkali forms a quaternary ammonium hydroxide. Such salts are used, for example, for the *inhibition* of *unsaturated polyester resins (UP)*. Quaternary ammonium compounds are also used as *antistatic agents* - for example, for *polystyrene*

quenched and tempered mould steel
See *pre-hardened steel*.

quenched and tempered tool steel
See *pre-hardened steel*.

quencher
See *quenching agent*.

quenching
Heating *steel* to redness and then cooling suddenly. Quenching and *tempering* have a tremendous effect upon hardness and toughness.

quenching agent
Also known as a quencher. An additive capable of suppressing or quenching photo-induced excited states. A material which helps to minimize light initiated degradation by deactivating the light activated, polymer molecules. Examples of such materials are nickel (11) chelates and *hindered amine light stabilizers*.

quick acting coupling
Also known as a quick change coupling or as, a quick release coupling. A coupling which has been designed for rapid connection and removal. Available for water, oil and power as the use of such couplings, for example, in *injection moulding*, saves considerable amounts of time. See *quick change*.

quick change
Reducing the amount of time spent during mould and/or material changes. Quick acting water and electrical couplings are now readily available and are used on both cylinders and moulds. Mould changing, mounting and setting has been simplified in recent years with the advent of *quick acting coupling* systems. The same philosophy has been applied to injection moulding barrels. For example, quick release couplings may be used on the barrel, so that the barrel and screw may be separated from the injection unit rapidly and easily. Quick release water and electrical couplings are then disconnected so that the barrel and screw can be placed on a pre-heating station for purging and pre-heating.

quick change coupling
See *quick acting coupling*.

quick mould changing
An abbreviation used for this term is QMC. A system used for changing the moulds rapidly at the end of production, for example, in *injection moulding*. On smaller machines (below 100 t) the standard-sized mould may be lowered from the top into plates which contain accurately machined grooves. Larger machines may, because of the weight of the mould require side loading. The use of *quick acting couplings* for water, oil and power save considerable amounts of time.

quick release coupling
See *quick acting coupling*.

quicklime
Also known as calcium oxide. See *lime*.

quicksilver
Another name for mercury.

quinacridone pigments
A class or type of *organic pigment* which are transparent and are also used to make metallic colours. Used to make gold, red and violet colours. Such pigments are used even though they are of high cost because, in general, they have good resistance to heat, light, bleeding and to chemicals such as solvents. In some plastics materials (for example, polystyrene, polymethyl methacrylate and polyamides) some systems can lose some lightfastness: they may become soluble at processing temperatures. Some systems are difficult to disperse well.

quinol
See *hydroquinone*.

quinone
A di-ketone: an aromatic compound based on a six-membered carbon ring in which two carbon atoms are attached to two separate oxygen atoms. The simplest material is *p-benzoquinone* which has the formula $O:C_6H_4:O$ and is sometimes also referred to as quinone.

quinone methide
A conjugated, aromatic chemical structure which is formed during the curing of *phenol-formaldehyde materials* and is believed to help give such materials a dark colour.

quinones
Photosensitizers which improve the photo-response of *polyvinyl cinnamate* for *photoresist purposes*.

R

r
An abbreviation used for *end to end distance*.

R
This letter is used as an abbreviation for;

radius;
random copolymer;
range. An alternative representation is 'w';
recycled material, for example, HDPE-R is *recycled high density polyethylene*;
resilience;
resistance (ohmic);
reinforced;
resin;
resol; and,
ricinoleate.

R:Si ratio
The ratio of organic groups (R = methyl and phenyl groups) to silicone (Si) which may be used to classify a *silicone resin*. The ratio will be two if a dichlorosilane is used and one if a trichlorosilane is used. Commercially the ratio varies from 1·2 to 1·6.

r-f
An abbreviation used for radio frequency.

R_a
An abbreviation used for *arithmetical mean deviation*.

Rabinowitsch correction
A correction applied to the viscosity data obtained from a capillary rheometer to allow for the *non-Newtonian* character of that fluid. While in theory the magnitude of the correction could range from 1 to infinity, the maximum error in the viscosity at any shear rate has been shown to be 15% when n' is 0·23. Generally it is less than 15% when 20:1 dies, or greater, are used. See *Rabinowitsch equation*.

For comparative purposes applying the correction will not alter the comparability of flow data. For a tube the relationship between shear stress at the wall and the apparent shear rate at the wall is unique and does not depend on the size of the tube or capillary. The apparent flow curve (i.e τ_w vs $\dot{\gamma}_{w,a}$) can be used for calculating data for flow through slits with a maximum error of as little as 3%, which is within the accuracy of most measurements. For these reasons, the only occasion where it is necessary to use the correction is where it is required to know the true shear rate. For practical scale up purposes it is not important. However, a knowledge of n' is useful as a measure of the degree of non-Newtonian behaviour of the melt, the lower the value found, the more non-Newtonian the melt.

Rabinowitsch equation
Also known as the Weissenberg-Rabinowitsch-Mooney equation. For *Newtonian fluids*, the apparent shear rate at the wall ($\dot{\gamma}_{w,a}$) is equal to the true shear rate ($\dot{\gamma}_w$) but with non-Newtonian fluids, such as polymer melts, the two are not equal but are related by the equation $\dot{\gamma}_w = [(3n'+1)/4n']\dot{\gamma}_{w,a}$.

Where $n' = [dlog(R\Delta P/2L)]/[dlog(4Q/\pi R^3)]$

This equation is a form of the Rabinowitsch equation and the term in squared brackets is known as the *Rabinowitsch correction*. The use of this equation permits the calculation of the wall shear rate from three measurable quantities R, Q and P. Where R is the die radius, Q is the volumetric output rate and L is the die length.

racked rubber
Unvulcanized *natural rubber* which has been cooled quickly under tension.

rad
An abbreviation used for *radian*. This abbreviation was also used for the former unit used to measure absorbed radiation and which is equal to 0·01 gray.

radial block copolymer
Also known as a star block copolymer. A *radial polymer* in which the arms, attached to the centre, are *block copolymers*.

radial multichain
See *radial teleblock polymer*.

radial polymer
Also known as a star polymer. A polymer which has arms, which are polymer chains, attached to a common centre. See *radial block copolymer*.

radial teleblock polymer
A radially-shaped material which has three or more polymer (rubbery) chains extending from the one central hub and with a rigid block attached to the outer end of such chains. See *styrene block copolymer*.

radial tetrachain
A radially-shaped material which has four polymer chains extending from the one central hub. See *radial teleblock polymer*.

radial trichain
A radially-shaped material which has three polymer chains extending from the one central hub. See *radial teleblock polymer*.

radial tyre
A tyre in which the *plies* of the textile reinforcement are at right angles to the direction of movement: they are held in place by a belt around the tyre's circumference. The fabric-braced radial has a multi-ply textile belt beneath the tread. A

steel-braced radial has one or more belts of steel wire instead of fabric. Most automotive tyres are now radial-ply tyres as opposed to *cross-ply tyres* as such tyres give better traction, braking, cornering ability, and comfort at speeds over approximately 50 km/h or 30 mph. The lower internal friction of radials means lower rolling resistance and a higher road speed or better fuel consumption (approximately 50% more).

radial-ply tyre
See *radial tyre*.

radian
An SI supplementary unit for plane angle and which has the abbreviation rad. The radian, which is the unit of plane angle, is the angle between two radii of a circle which on the circumference cut off an arc equal in length to the radius. See *Système International d'Unité*.

radiation cross-linked polyethylene
A *cross-linked polyethylene*: produced by the action of high energy radiation which produces the required free radicals.

radiation cross-linking
Cross-linking induced by the use of radiation. Cross-linking induced by the use of *high energy radiation* (see *radiation - effect of*). The reduced chemical content of the cross-linked compound has obvious attractions but, in the case of rubber compounds, the vulcanizates have relatively poor strength.

radiation curing
See *radiation cross-linking*.

radiation degradation
Degradation of polymers caused by *high energy radiation*. See *radiation - effect of*.

radiation - effect of
High energy radiation can cause cross-linking of polymers and/or it can cause degradation. When exposed to high energy radiation most, but not all, polymers of a mono-substituted ethylene will cross-link whereas most, but not all, polymers of di-substituted ethylenes degrade. Polymers which cross-link include polyethylene, polyacrylic acid, polymethyl acrylate, natural rubber, polydimethyl siloxane polymers and styrene-acrylonitrile polymers. Polymers which do not cross-link, but which degrade by chain scission, include polyisobutylene, poly-α-methylstyrene, polymethyl methacrylate, polymethacrylic acid, polyvinylidene chloride, polychlorotrifluoroethylene, polytetrafluoroethylene, cellulose and polypropylene.

radiation gauge
A measurement device, based on the emission of nuclear radiation, and used in sheet production and profile extrusion. Most radiation devices emit either gamma, beta or X-ray radiation. Isotopes such as Americium 241 will emit gamma radiation and strontium 90 will emit beta particles. Such devices can be calibrated against weight per unit area. See *calendering*.

radiation polymerization
Also known as radiation-induced polymerization. Polymerization caused by *high energy radiation*.

radiation pressure
The pressure exerted on a surface by radiation falling on that surface.

radiation pyrometer
A device which measure temperature changes by determining the heat radiation emitted from a hot object. See *pyrometer*.

radical
A chemical grouping which cannot exist on its own but which, when part of a larger structure or molecule, maintains its identity throughout chemical reactions. See *free radical*.

radical polymerization
Also known as *free radical polymerization*.

radiation-induced polymerization
See *radiation polymerization*.

radical scavenger
Also known as a free radical scavenger or as, a free radical trap or as, a radical trap. A material, or another *free radical*, which acts to remove active free radicals from a system. Chain breaking *antioxidants* act in this way. A radical scavenger which is a free radical itself is diphenylpicryl hydrazyl.

radical trap
See *radical scavenger*.

radio frequency
An abbreviation used for this term is rf or, r-f, or RF. That portion of the electromagnetic spectrum used in radio: about 3 MHz to 300 MHz.

radio frequency heating
Also known as *high frequency heating*. Sometimes abbreviated to RF heating or to HF heating.

radio frequency preheater
Also known as a *high frequency preheater*. See *high frequency heating*.

radioactivity
The spontaneous disintegration of the nuclei of some isotopes of certain elements, with the emission of ionizing radiation.

radiolytic polymerization
See *radiation polymerization*.

radiation polymerization
Also known as radiation-induced polymerization or as, radiolytic polymerization. Polymerization caused by *high energy radiation*. Useful for *solid state polymerization* because the radiation finds it easy to penetrate a solid monomer.

ram
A single-acting cylinder with a single diameter plunger. A cylinder in which the element which moves has the same cross-sectional area as the piston rod. A plunger used to transmit force: a *linear actuator*. When used as an abbreviation, ram stands for *relative atomic mass*.

ram displacement rate
See *injection speed*.

ram extruder
An extrusion machine which uses direct plunger pressure on the material to cause *extrusion*.

ram machine
A type of *injection moulding machine* which has an *injection unit* which contains a *ram* or plunger. An early type of moulding machine which is still made as it is relatively simple and cheap.

ram mixer
An *internal mixer*.

ram position
Term used in injection moulding and often called *screw position*.

ram speed - injection moulding
The speed of movement of the ram (plunger machine) or of the screw (in-line screw, injection moulding machine) when injection, or mould filling, is being performed. Measured in, for example, mm/s. Ideally the *injection speed* is set and held at a specified value which may change throughout the stroke.

ram speed - of a rheometer
The speed of movement of the ram which is causing material displacement from a *capillary rheometer*. An abbreviation used for this term is V. Measured in, for example, mm/s.

random copolymer
See *copolymer* and *ideal copolymer*.

random polymer chain scission
See *random scission*.

random prepolymer
A *prepolymer* based on a thermosetting system in which curing has been stopped, by cooling, before *gelation* has occurred. On reheating, network formation will occur. For example, a phenol formaldehyde *resole* will cure without any further chemical addition simply as a result of re-heating.

random scission
Also known as random polymer chain scission. Chain scission which occurs, with equal probability, at any repeating unit in the polymer chain.

randomizing agent
An additive used in polymerization to give a random structure to the resulting polymer. For example, ethers or amines are used for this purpose in the *polymerization* of *solution styrene-butadiene rubber*.

range chart
A *process control chart* on which range values are plotted. Such a chart has three lines, the mean range line and a set of tramlines. The set of tramlines has two lines called the upper warning limit (UWL) and the upper action limit (UAL): there is no lower set of tramlines (see *average chart*) as a smaller range is acceptable.

Rankine temperature.
An abbreviation used for this term is °R (see *Réamur scale*). Rankine temperatures are absolute Fahrenheit temperatures. To obtain °R, add 459·67 to the Fahrenheit temperature.

rape seed oil
Also known as rapeseed oil or as, colza oil. Derived from *Brassica napus* or from *Brassica campestris*. An edible and/or an industrial *vegetable oil*. This material has a relative density of 0·92. A yellow, edible oil obtained from various Brassica plants and used, for example, in the quenching of steel and in the manufacture of *factice*. By treating this *drying oil* with less than the amount of *sulphur* (needed to make *factice*) sulphurized oil is made.

rapeseed oil
See *rape seed oil*.

rapid plasticity testing
See *Wallace rapid plasticity*.

RAPRA
An abbreviation used for the Rubber and Plastics Research Association (of the UK).

rate of cure
The time of *vulcanization* required to reach a specified level of cure by a particular compound compared to that required by a standard compound. The rate at which cross-links are formed in a thermosetting polymer compound.

rate of polymerization
The rate at which monomer is converted to polymer. A symbol used is R_p. See *rate of propagation*.

rate of propagation
A symbol used is R_p. The rate at which monomer units add to a growing active centre so as to form a polymer molecule. As propagation is much faster than initiation, the *rate of polymerization* and the rate of propagation are virtually the same.

rate of spread of flame
See *Underwriters Laboratory horizontal burning test*.

raw material feed
The materials fed to the processing equipment. See *feed*.

raw tyre
See *green tyre*.

rayon
This material is regenerated cellulose in *fibre* form: that is, it is a semi-synthetic material - it is a man-made material but produced from a natural polymer (*cellulose*). May be defined as a manufactured fibre composed of regenerated cellulose in fibre form: may also be defined as regenerated cellulose fibre which contains some hydroxyl substitution on the cellulose molecule (not more than 15%). Three materials are covered by these definitions; *viscose rayon, cuprammonium rayon* and *saponified cellulose acetate*. Viscose rayon is the most commercially important of these three materials. Cuprammonium rayon gives the finest, silk-like yarn and saponified cellulose acetate gives strong, stable fibres. Rayon is not a thermoplastics material: it does not melt or become tacky on heating.

rayon and cotton - differences
Rayon, like cotton, is based on *cellulose*. The structural differences between the two are due to the differences in the degree of polymerization (DP) and to differences in the arrangement of the molecules, and of the crystal structures, in the filament. Cellulose has a high molecular weight and is highly crystalline. For example, cellulose may contain up to 10,000 glucose units whereas rayon may only contain one tenth of this number: the crystallinity in cellulose may reach approximately 70% (over twice that of rayon). The crystal structures, present in cotton, are themselves, in turn, highly ordered This high molecular weight, highly ordered structure helps explain the differences between these materials. For example, the high strength of cotton which is retained when the material is wet.

R_c
An abbreviation used for *Rockwell hardness* on the C scale - the scale used for metals such as *steel*.

RCR
An abbreviation used for *reciprocating screw rheometer*.

rd
An abbreviation used for *relative density*.

RD
An abbreviation used for *relative density*.

re-coiling
See *elastic effects*.

Re
An abbreviation used for *Reynold's number*

re-knit line
A line on the surface of an extrudate, or on a blow moulding, which is formed by poor re-combination of the *melt* stream after it has been divided, for example, by a *spider leg*.

reaction injection moulding
A batch moulding process which is associated with polyurethanes and is commonly known as the RIM or RSG process. May also be referred to as liquid injection moulding (LIM) or as, high pressure impingement mixing (HPIM).

The use of the term RIM implies that a chemical reaction is occurring while an *injection moulding process* is taking place. This is true but the moulding operation is totally unlike that normally used for thermoplastics materials as the starting materials are usually low viscosity liquids at room temperature. Large, complex polyurethane mouldings may, however, be produced in one step or stage using, for example, a *quasi-prepolymer process* in which two liquid streams (A + B) are mixed together: metering is simplified, and accuracy improved, if A and B are of roughly similar size. The two streams are a quasi-prepolymer and the polyol (plus various reaction promoters and additives).

The ingredients are, for example, pumped into a *mixhead* (mixing head) in the form of fine streams of liquid which are atomised and mixed by *impingement mixing*: the still-liquid material is then pumped into the mould. If a *blowing agent* (for example, methylene chloride) has been incorporated then the mould is only partially filled and the reacting mixture expands to fill the moulding cavity. Foaming is used to balance shrinkage.

Because of the low viscosity of the reactants, only low clamping pressures are needed and this in turn means that relatively large products may be made cheaply. By varying the raw materials it is possible to produce either rigid or flexible components which may or may not be microcellular in nature and which may also contain reinforcing sections and/or fillers.

The low viscosity (0·1 to 1 Pa·s) of the reactants means it is difficult to eliminate gas bubbles: such bubbles can become trapped during mould filling. To prevent air entrapment the maximum filling rate, as measured by the critical Reynold's number, for both the gate and the cavity is less than 100 in both cases.

Flashing of the mould can be a problem and the *flash* can be difficult and expensive to remove. The handling of highly reactive, and in some cases highly toxic, ingredients poses problems and mould release can be difficult. Mould release can be eased by the use of an *internal mould release* and/or an *external mould release*.

RIM began with *polyurethanes* and such polymers, together with *polyureas*, constitute most RIM production. The stoichiometry must be balanced to within less than 1% so as to obtain high molecular weight products: most RIM machines use lance-type pistons driven by separate hydraulic cylinders to achieve the accuracies desired. This process is used, for example, to produce components for the automotive industry, for example, fascias. Thermosetting materials other than polyurethanes may also be used, for example, polyesters and epoxides. See *reaction injection moulding process*.

reaction injection moulding machine
A machine used for *reaction injection moulding* (RIM). See *lance piston reaction injection moulding machine* and *piston reaction injection moulding machine*.

reaction injection moulding process
An abbreviation used for this term is RIM process. The term covers the supply of *liquid reactants*, conditioning of reactants, high pressure metering, mould filling, setting (polymerization or phase separation), demoulding and finishing (flash removal, post-curing, cleaning and painting). Most RIM production is based on polyurethane polymers: other polymer systems include, for example, nylon 6, dicyclopentadiene, polyester, acrylamate, phenolic and epoxy. See *reaction injection moulding*.

reaction moulded polyurethane elastomer
A component, based on a rubbery polyurethane material, which was moulded and set in one stage or process by *reaction injection moulding*.

reactive adhesive
A category of *adhesive* that set by a chemical reaction; *polymerisation* and/or *crosslinking* occur after the adhesive, in liquid form, has been applied.

reactive compatilization
A *compatibilization* process, used to make a useful *blend*, in which functional groups in the polymers react during processing. For example, with a *thermoplastic polyester* an ester inter-change reaction with *polycarbonate* during processing, results in the compatibilization of an otherwise incompatible system.

reactive dye
A *dye* that reacts chemically with the substrate so as to form a strong covalent bond with the substrate.

reactive flame retardant
A type of *flame retardant* which enters into a chemical reaction when the polymer is made, so as to become part of the chemical structure of the polymer. Examples include chlorinated polybasic acids or alcohols (as used in *unsaturated polyester resins*) and chlorinated monomers used in *synthetic rubber* production.

reactive processing
Term used to describe a process, usually *extrusion compounding*, whereby the plastic is chemically reacted, or modified, as well as being shaped.

Réamur scale
A temperature scale on which the melting point of ice is 0 and the boiling point of water is 80. Abbreviation °R. To convert from °R to °C multiply the °R by 1.25.

rear roll
See *back roll*.

rear slip angle
See *slip angle*.

rebound elasticity
See *resilience*.

recapping
See *retreading*.

receiving half
That half of the *blow mould* which steers or guides the *parison* when the mould is inclined.

reciprocal ohm
See *mho and siemens*.

reciprocating screw machine
A type of *injection moulding machine* which has an *injection unit* which contains a screw: the screw rotates to *plasticize* the melt and then acts as a plunger to force the melt into the mould.

reciprocating screw rheometer
Also known as RSR. A *reciprocating screw machine* used to perform a *flow test*: a high shear rate flow test performed using an *injection moulding machine* as a laboratory capillary rheometer (LCR). The job of a LCR can be performed by an in-line screw, injection moulding machine provided that the machine is equipped with a suitable (a nozzle) *pressure transducer* and that the *injection speed* can be set and held at a specified value. The cylinder, or barrel, is charged with the plastics material using a slow screw rotational speed and a low back pressure; this charge of material is allowed to stand in the barrel for, say, 2 minutes. The use of such conditions will promote temperature uniformity and should give residence times similar to those found in a LCR. The melt is then purged from the barrel, into the air, and the pressure/speed/melt temperature recorded: if a melt temperature cannot be measured directly then the temperature of the purged material may be measured using a probe. The conditions are then changed and the measurements repeated. A *flow curve* may be obtained from such measurements.

recirculation process
See *conditioning tank* and *reaction injection moulding*.

reclaim
Could refer to any polymeric material which has been recovered or reclaimed but generally taken to mean *reclaimed rubber*.

reclaimator process
Also known as the dip process. See *reclamation processes*.

reclaimed rubber
The product of a *reclaiming process* or operation. The product which results when vulcanized components are treated with, for example, heat and chemicals, so that the product can be used like virgin rubber.

reclaimed rubber
Also known as reclaim. An additive for rubber compounds: a solid material recovered from *vulcanized rubber* (natural and/or synthetic). Rubber chips or crumb obtained from vulcanized rubber components. Obtained by, for example, steam heating the finely ground rubber, in an oil/caustic soda solution (may also contain *reclaiming agents* such as *anthranyl-9-mercaptan*) for several hours. That is, by heat treatment in the presence of reclaiming chemicals at approximately 200°C.

The reclaim still contains the vulcanization chemicals and the non-rubber mix ingredients which are therefore available to function in the new compound. Fast cure times are therefore often obtained from compounds which contain reclaim. The reclaim still contains particulate fillers such as *carbon black* although very long fibre may have been removed from, for example, tyre reclaim. Reclaimed rubber is used as a filler in some rubber compounds as by doing so the processing properties of the compound may be improved. For example, mixing time may be reduced as may the power consumption and the *die swell*. The use of high loadings of reclaim may result in a loss of properties such as tensile strength and elongation at break.

The following types of reclaim my be obtained: whole tyre, tyre tread, tube, mechanical, drab and floating. Reclaim may also be produced by alkaline or neutral processes. Whole tyre reclaim may contain, for example, 50% by weight of rubber hydrocarbon. See *reclaiming processes*.

reclaimed thermoplastics material
Also known as regrind: material which has been recovered from, for example, scrapped components. One of the most common additives used with thermoplastics materials, is often reclaimed material. For example, many *injection moulding machines* run on a thermoplastics feedstock which is a mixture of new (virgin) material and reclaimed (regrind) material. The use of regrind is only usually desirable with thermoplastics materials.

The reclaimed material is obtained by grinding rejected, or unwanted, components and the feed system. This reclaimed material is usually added to reduce component costs. It is most important that this additive, as with any other additive, is added at a definite, pre-selected ratio so that, for example, the flow properties of the resultant blend are consistent. If the flow properties are inconsistent then properties, such as the surface finish, will be inconsistent. The exact amount of regrind will have to be determined experimentally and once found, then the ratio must be held as precisely as possible if consistent mouldings are to be obtained.

Great care should be taken to ensure that the reclaimed material (regrind) is clean, dry and of regular particle size. If the regrind is dirty, then mould or machine damage may occur and the appearance of the product will suffer. If the material fed to the machine contains unacceptably high moisture levels, then the properties of the moulded components will be affected, for example, in the case of a clear material then the clarity will be affected and for all materials (both clear and opaque) the quality of the surface finish will be reduced. If the feed is not of consistent particle size then the material will not feed in a regular way and inconsistent product will be obtained. The differences between mouldings may not be discernible to the naked eye but they may be large enough to cause rejection of the product, for example, because the size of the moulded component is incorrect.

Care should be taken to ensure that the original material contains sufficient stabilizer so that reclamation is possible without degradation, and/or a colour change, occurring on re-use. See *reclamation*.

reclaiming agent
A material which assists in the production of *reclaimed rubber*: a material which minimises the loss in tensile properties which results during *reclamation*. From 0.5 to 20% of a material such as, anthranyl-9-mercaptan, dipentene and solvent naphtha may be used. An aromatic disulphide can also improve reclamation.

A material which assists in the production of rubber components by minimizing the effects of scorch. For example, N-nitrosodiphenylamine is used as a reclaiming agent, for slightly scorched compounds, by cold rolling or milling the affected compound with this additive.

reclaiming chemical
See *reclaiming agent*.

reclaiming processes - rubber
Reclaimed rubber is obtained from un-wanted vulcanized material. The so-called scrap is sorted according to its origins or production process - which often indicates the presence of textiles or metals. The material is then freed from metals and ground to a particulate material (powder or to a coarser crumb). The particulate material is then subjected to the reclamation process: the choice of this depends upon factors such as the type of rubber and the type and amount of textile present. *Acid, alkali* and *neutral reclamation processes* are used.

Such *reclamation processes* involve heating, often under pressure: *reclaiming chemicals* are also used. Such a heating process causes depolymerization, also called replasticization, but not devulcanization as the chemically-bound sulphur is not removed. (It should be noted that heating to high temperatures can cause cyclization (hardening) of many synthetic rubber compounds. After reclamation, the treated rubber is blended with *softeners*, homogenized by mixing using, for example, an *internal mixer* or a *two-roll mill*.

reclamation
The process of reclaiming polymer from unwanted product.

reclamation - chlorofluorocarbon
The reclamation of such materials is now becoming economically attractive as companies have cut the production of chlorofluorocarbons (CFC) and so, such materials are becoming more expensive as they become less widely available, Now known that CFC cause *stratospheric ozone depletion* so reclamation is an environmental necessity as CFC can only be destroyed by heating to temperatures above 1,200°C. Recovered by the use of *molecular sieves* from, for example, scrapped refrigerators. Refrigerators contain CFC in the chiller units and, even more, in the *rigid polyurethane foam* which is used to insulate the walls.

reclamation - plastics materials
Using materials several times over conserves valuable raw materials and protects the environment. However, reprocessed feedstocks are often of inconsistent quality and there are concerns about incorporating reclaimed material into components intended for food or for technical components.

Legislation and public concern forces the reuse of plastics materials.

Reclaiming is very much easier for thermoplastics materials than for thermosetting materials. Most thermoplastics materials can be reclaimed and re-used again. They are either reclaimed and used on their own, reclaimed and used as additives for other similar plastics compounds or, cracked into petrochemical raw materials (see *degradative extrusion*).

Reclamation is relatively easy for materials such as thermoplastics during component manufacture: regranulation and blending with virgin material on, for example, the injection moulding machine is common. Used components from a single source, such as car bumpers, have also been collected, sorted, granulated, cleaned and reprocessed into, for example, radiator grilles.

Problems of material re-use arise when components from different, and often unknown origins, are required to be re-used. This is because different, and often incompatible, materials are involved. Sometimes such mixed scrap is used as the central filling layer for a co-extruded product or, used as the backing layer for flooring. Thick sectioned mouldings or extrusions have also been formed from mixed scrap (also see *particle recycling process*). One of the most difficult problems in the recycling of plastics materials is the separation of mixed plastics into single batches on materials, for example, the separation of *polystyrene* and *polypropylene* from domestic refuse. Typically the waste is shredded, washed and then separated by flotation (density difference or fractionation) processes: the relatively pure material (for example, 99% pure) is then dried and compounded.

Components from thermosetting materials are seldom recovered and re-used as polymers, or as polymer additives, because the addition of the reclaim to virgin material, causes a severe drop in many properties (an exception is SMC). When a cured, thermosetting material is reclaimed then the reclaim is often used in a non-polymeric way, for example, as a soil additive (urea-formaldehyde) or as a filler for road surfacing compounds (components based on sheet moulding compound SMC, are used in this way). Because of the long fibre length of *polyester moulding compounds*, particularly SMC, useful plastics products can be produced which are based on waste, or reclaimed, SMC components. The parts may be, for example, collected, sorted, granulated (shredded) and added to the feedstock for new SMC raw material. See *particle recycling*.

reclamation processes - machine-based
Reclamation processes for *vulcanized rubber* which achieve the required softening by working the vulcanized rubber compound in a machine: the material is subjected to high temperatures while being sheared. For example, processing may be done in an *internal mixer* (Banbury–Lancaster process) or in a modified extruder (dip process). See *reclaiming processes*.

reclamation processes - steam processes
Reclamation processes for *vulcanized rubber* which utilize the softening effect produced by the use of super-heated steam and chemicals at elevated temperatures. In the Mark process, finely divided *pneumatic tyre scrap* is heated in pans or digesters at temperatures of approximately 190°C together with naphtha in a caustic soda solution for several hours. The caustic soda solution destroys the fabric and the naptha swells and softens the rubber. The softened rubber is then washed and refined by passing through the tightly set rolls of a refiner which produces reclaimed rubber in the form of thin sheet: this may be rolled and sheeted into slabs if required. Materials such as anthranyl-9-mercaptan and dixylyl disulphide (a dark brown oil) are also used during the reclaiming of rubber.

It should be noted that heating to high temperatures can cause cyclization (hardening) of many synthetic rubber compounds and so such processes may only be suitable for *natural rubber* and for *butyl rubber*. Pressure of up to 50 atmospheres have been used.

In the open steam process, used for reclaiming material with little or no fabric, very finely divided scrap is heated with oils and sodium hydroxide in horizontal pans: the cooked material is then washed and refined. See *reclaiming processes*.

record keeping - importance of
The importance of careful and accurate records cannot be over-emphasised. Not only is it useful to have a full and accurate record for machine re-setting but, such records are useful for product liability reasons. With date marking of mouldings now easily possible then the precise details of how a particular batch of injection moulding was produced, can be produced and kept, easily and cheaply. A microprocessor-based machine will record all relevant data, print it out if required and then, instantly reset the machine when that particular job is re-run. Such instant re-setting saves, of course, a great deal of time and gives more accurate setting.

recoverability
The ability to revert to the original shape or dimensions when the forces causing change are removed. Polymer compounds may recover their original dimensions completely as long as they are not stretched beyond the yield point.

recoverable joint
A joint which can be taken apart and then re-assembled.

recovery
The degree to which an unvulcanized rubber, or component, will return to its original dimensions after removal of an applied stress. Decrease of strain after removal of the *stress* causing *strain*.

recrystallization
Crystallization from a partial molten material, or a partially dissolved material, around the original crystal structures.

rectangular edge gate
An edge *gate* whose cross-section is that of a rectangle.

rectified spirit
A form of *ethyl alcohol* which is usually produced by fermentation followed by distillation. Contains about 95.6% of ethyl alcohol with the rest being mainly water.

rectifier
A device which only allows electricity to pass in one direction: such a device is used to convert a.c. into d.c.

recut
A type of tyre: a tyre which has had a new tread cut after the original tread has worn. May also be called a regroove. Some bus and truck tyres are built with an extra thickness of tread rubber to allow such a practice but automotive tyres are not so constructed: in general, regrooving should not be performed because of safety considerations.

recyclate
A plastics material which has been reclaimed so that it may be re-used.

recycled plastics material
A thermoplastics material prepared from discarded articles that have been cleaned and reground. An abbreviation used for this type of material is R. For example, HDPE-R is *recycled high density polyethylene*. See *reclamation*.

recycled polyurethane
An abbreviation used for this type of material is PUR-R. Polyether polyols are recovered from, and re-used in, flexible foam applications by, for example, the process known as *glycolysis*. See *particle recycling process*.

recycled sheet moulding compound
See *particle recycling*.

recycling symbol
See *green dot*.

red
A *primary colour*. The longest wavelength in the visible spectrum is red at approximately 700 nm.

red circle rubber
A *technically classified rubber* which is slow curing. See *circle rubber*

red lead
Also known as lead orthoplumbate or as, minium. May be represented as Pb_3O_4. This bright red material has a melting point of 830°C and a relative density (RD) of about 9. An activator and *pigment*. An *acid acceptor* for chlorine containing rubbers.

red oil
See *oleic acid*.

red pigments
Inorganic red pigments include cadmium mercury red, cadmium sulpho-selenide and ultramarine red. Organic red pigments are permanent red, perylene red and quinacridone red. The organic red pigments are more soluble, have a higher tinting strength and have a lower relative density than the inorganics.

redistribution
Term used in silicone chemistry to describe what happens when chlorosilanes are heated. The substituents on the silicone atom can be swapped so that, for example, heating a mixture of methylchlorosilane and trimethylchlorosilane at 200 to 400°C with aluminium chloride, gives *dimethylchlorosilane*.

redox
An abbreviation used for reduction-oxidation (oxidation-reduction).

redox initiation
Initiation of a free radical *polymerization* reaction by the *free radicals* generated by a redox reaction. Low temperature reactions (for example, –50°C) are possible. See *cobalt naphthenate*.

redox initiator
A system which will cause *redox initiation*. Peroxides and heavy metals, such as iron and cobalt, may be used. See *cobalt naphthenate*.

redox polymer
See *oxidation-reduction polymer*.

redox polymerization
Reduction-oxidation polymerization. A *free radical polymerization reaction* initiated by the free radicals generated by a *redox initiator*. Low temperature reactions (for example, –50°C) are possible. See *styrene-butadiene rubber* and *oxidation-reduction polymer*

reduced specific viscosity
See *specific viscosity*.

reduced viscosity
See *specific viscosity*.

reducibility
The ease with which a *masterbatch* or concentrate can be let down in a polymer.

reeling
The winding of flexible sheet onto reels. For example, most flexible *polyvinyl chloride* (PVC) sheet, produced by *calendering*, is sold in reels and two main systems of winding are employed: these are *centre-core winders* and *surface winders*: to simplify operation more than one winding station is employed. A turret winder may be employed if the output justifies the expense.

refined tall oil
Obtained by distilling *tall oil*: resembles *linseed oil* and is used in *alkyd resin manufacture*.

refiner
A type of *two-roll mill* used for the refining (purifying) of *natural rubber*: a mill which is capable of giving a very tight nip. The rolls rotate at high speed with a large friction ratio, for example, 1:2·5. Some impurities are crushed by this treatment; others are moved to the side of the *band* where they are removed by cutting. Straining after refining will remove most impurities.

refining
Homogenization and impurity removal from natural rubber using a *refiner*.

refractory material
A material which is not damaged by being heated to high temperatures.

regain
See *moisture regain*.

regenerated cellulose
A man-made material but produced from a natural polymer (*cellulose*). Cellulose which has been recovered (regenerated) after it has been shaped as a soluble intermediate. Regeneration makes it possible to utilise the virtually unlimited supplies of cheap cellulose which are readily available in the form of, for example, wood pulp. See *rayon*.

regenerated cellulose fibre
A man-made material but produced from a natural polymer. The polymer (see *cellulose*) is dissolved in a solvent, shaped by spinning and then the solvent is removed. See *rayon*.

regenerated fibre
See *regenerated cellulose fibre*.

regenerative circuit
A term used in *hydraulics* and which refers to a circuit arrangement whereby the discharge fluid from a cylinder (the rod-end) is combined with the fluid flow from a pump and introduced into the head-end of a cylinder so as improve efficiency of operation.

register hole
A circular hole in the *platen* of a machine which accepts the *mould register*.

register ring
Also called a locating ring. A circular ring fitted to a mould for location purposes, for example, to locate the mould with respect to the *nozzle*.

regranulated material
Thermoplastics material, in granular form. which has been obtained by grinding fabricated material, for example, mouldings or extrusions. See *feed form* and *reprocessing*.

regranulation
The process of reclaiming output which is not required. Regranulation of, for example, sprues, runners and faulty mouldings is usually achieved by feeding such scrap through a grinder which may be located by the side of the *injection moulding machine*. See *feed form* and *reprocessing*.

regrind
The name given to material which has been reclaimed by grinding. See *reclamation*.

regroove
See *recut*.

reground tyre crumb
A particulate material obtained from worn tyres: the tyres are freed from metal (for example, edge beads) and granulated. A wide variety of sizes are possible from the micron range to the centimetre range. Such a material is used, for example, as a filler for a cast *polyurethane* so as to make PU tyres.

regular institute numbers
A numbering system established by the International Institute of Synthetic Rubber Producers: it divides *synthetic rubbers* (for example, styrene-butadiene rubber) into groups or classes. For example, into black filled styrene-butadiene rubber and/or oil extended etc.

regulator
An additive to a system which controls molecular weight during *polymerization*.

reinforced plastics material
A plastics material with high strength fillers imbedded in the composition, resulting in some mechanical properties superior to those of the base resin. The reinforcing fillers are usually *fibres, fabrics*, or *mats* made of fibres.

reinforced reaction injection moulding
An abbreviation used for this process is RRIM. A *reaction injection moulding process* in which *reinforcing fillers* are incorporated into the product. Both fibres and powders may be added. When using powders, for example, reclaimed material, the *amine* used to achieve cross-linking {for example, *diethyl toluene diamine (DETDA)*} can become adsorbed by the powder: such absorption can be minimized by using three separate streams fed to the mixing head. One is the powder-charged polyol, the second is the isocyanate and the third is the DETDA which can contain catalysts and internal mould release agents. See *reaction injection moulding*.

reinforced thermoplastic
See *reinforced thermoplastics material*.

reinforced thermoplastic polyurethane
An abbreviation used for this material is R-TPU or TPU-R. A *thermoplastic polyurethane* which has been reinforced, for example, with short *glass fibres* of relatively fine diameter as these give acceptable properties at a reasonable cost: such fibres are treated to give good adhesion to the polymer. Such thermoplastics materials are used to injection mould large body panels for the automotive industry as the components so produced can have an excellent surface finish, good impact strength (even at low temperatures) and a thermal expansion similar to that of steel. The components can have a heat deflection temperature greater than 120°C and good paintability.

reinforced thermoplastics material
A *thermoplastics material* which has been filled. or reinforced, with a fibrous filler (usually *glass fibre*): the composite material may be classed as a *long fibre filled thermoplastics* or as a, *short fibre filled thermoplastics*.

reinforcement
Can refer to that component of a composite which results in the improvement of selected properties or, the term may refer to the effect of component addition. The incorporation of a *filler* into a polymer so as to improve some properties. For example, in the case of rubber the incorporation of small particle size fillers can improve tensile strength and tearing resistance. With plastics, reinforcement (for example, with *glass fibre*) usually imparts strength and stiffness to a polymer matrix. See *specific modulus*.

reinforcement angle
See *angle of helix*.

reinforcement factor
The ratio of the *Young's modulus* of a composite or laminate to that of the un-reinforced base polymer.

reinforcing
A term usually associated with fillers for rubbers - although resins can also cause reinforcement. Addition of fillers with a small particle size increases abrasion resistance, tear strength and stiffness. A term sometimes used in place of 'reinforcing' in connection with fillers for rubbers is 'activity'. Higher activity means higher reinforcing action, for example, with *carbon black*: inactive means no reinforcing action. Medium activity means that the reinforcing action lies in between no reinforcing action and a good reinforcing action.

reinforcing filler
The term is usually applied to a particulate *filler* which when used in a compound improves a property, or properties, of the final product. For example, a strength property (such as *tensile strength*) is improved. Carbon black reinforces rubber without treatment whereas other fillers may require treatment. For example, *calcium carbonate* may be made a reinforcing filler by coating the filler with an unsaturated carboxylic polymer. Reinforcing fillers, such as *carbon black*, cause greater temperature rises than non-reinforcing fillers during plasticization (thermal softening) and injection in *rubber injection moulding*.

relative atomic mass
An abbreviation used for this term is ram or r.a.m. More commonly known as atomic weight. A ratio: the average mass of an atom of an element to that of $1/12$ of the mass of a carbon-12 atom. The natural isotopic composition is assumed unless otherwise stated.

relative density
More usually known as specific gravity (SG). Relative density (rd or RD) equals the ratio of the mass of a substance to the mass of an equal volume of water at a specified temperature: for example, the substance may be at 20°C and the water is at 4°C. Specific gravity or relative density has no units as it is a ratio.

The RD may be calculated by dividing the weight in air by the difference between the weight in air and the weight in water. Special balances (for example, L'homme and Argy) are available for direct measurement of relative density (SG) based on this principle.

RD = 145/145 − B and rd = 0·5 Tw + 100/100, where B = degrees Baumé and Tw = degrees Twaddell. Some relative densities of common materials are as follows.

	Relative Density
Filler	
Carbon black	1·80
Whiting (ground calcium carbonate)	2·70
China clay	2·50
Zinc oxide	5·57
Plasticiser	
Di-iso-octyl phthalate (DIOP)	0·98
Di-ethyl hexyl phthalate (DOP)	0·98
Tricresyl phosphate	1·13
Tri-tolyl phosphate (TTP)	1·18
Polypropylene sebacate (PPS)	1·06
Di-iso octyl adipate (DOA)	0·93
Polymers	
Polypropylene	0·90
Low density polyethylene	0·91–0·925
Medium density polyethylene	0·926–0·940
High density polyethylene	0·941–0·96
Polystyrene	1·00

Nylon 11	1·04
Nylon 66	1·14
Polycarbonate	1·2
Polyvinyl chloride	1·4
Styrene butadiene rubber	0·94
Natural rubber	0·93
EPDM rubber	0·86
Butyl rubber	0·92
Nitrile rubber	1·00
Neoprene rubber	1·23
PP/EPDM blend	0·88

Liquids

n-heptane	0·75
Ethyl alcohol	0·78
Industrial methylated spirits (IMS)	0·78
Methyl alcohol	0·79
Kerosene or paraffin oil	0·82
Toluene	0·87
Water:methylated spirits 25:35 (by volume)	0·87
Dioctyl sebacate	0·91
Proof spirit - water:ethanol 43:57 (by volume) water:ethanol 50·7:49·3 (by weight)	0·92
Rape seed oil	0·92
Cotton seed oil	0·92
Pine oil	0·93
Methyl acetate	0·93
Di-iso octyl adipate (DOA)	0·93
Dibutyl sebacate	0·94
Castor oil	0·96
Di-ethyl hexyl phthalate (DOP)	0·98
Water	1·00
Dibutyl phthalate	1·04
Ethylene glycol	1·12
Tricresyl phosphate	1·13
Glycerol	1·26
o-dichlorobenzene	1·30
Magnesium chloride - saturated solution	1·34
Chloroform	1·50
Carbon tetrachloride	1·60
Pentachlorethane	1·67
Ethyl iodide	1·93
Bromoform	2·85
Acetylene tetrabromide	2·95

relative flammability
See *limiting oxygen index test*.

relative molecular mass
More commonly known as *molecular weight*: also known as relative molar mass. A ratio: the average mass of a molecule to that of $1/12$ of the mass of a carbon-12 atom.

relative permittivity
See *dielectric constant*.

relative viscosity
Also known as viscosity ratio. An abbreviation used for this term is η_{rel}. The ratio of the viscosity of a solution (η) to that of the pure solvent (η_0) or to water.

relaxation
The stress decrease with time which occurs when a rubber strip is held under tension. Very pronounced with unvulcanized rubber but not with vulcanized material. If unvulcanized rubber is extended, and kept extended, then flow will occur as the molecules move past each other. With vulcanized material, such flow is not possible as the cross-links prevent this type of molecular movement.

relay
A mechanical/electrical device in which a small electrical signal is used to initiate the switching of a much larger electrical supply. See *cavity pressure control*.

relay logic
A machine controlling system which uses the logic possible when *relays* are inter-connected.

relay machine
A type of machine, for example, an *injection moulding machine*, which uses *relay logic*.

release agent
A material painted or sprayed onto a substrate in order to prevent adhesion. An additive or coating used to achieve, for example, component release. For example, in the production of a *plug*, the sealed surface is polished with a wax polish and then treated with a release agent which allows the *unsaturated polyester (UP) resin* to wet out on the plug surface. Such a release agent is *polyvinyl alcohol*—this is often applied, by brush, from a water/methylated spirits mixture so as to obtain rapid drying. See *mould release agent*.

relief image
An image which is raised above a surface. See *photoresist*.

relief valve
A term used in *hydraulics* and which refers to a valve which controls maximum system pressure. A pressure operated valve: when the pressure exceeds a pre-set value, the valve opens and the delivery from the pump is diverted to the tank.

relieve—mould
To relieve a mould means to reduce the contact area between sealing faces of that mould so that gas or escape material can escape from the cavity.

rem
Roentgen equivalent man. The unit of ionizing radiation whose biological effect is the same as that of one roentgen of X-rays. The former unit of dose equivalent: the rem is equal to 10^{-2} Sv. See *sievert* and *Système International d'Unité*.

remilling
Re-working of *natural rubber* using a type of *two-roll mill*. Used for the blending of *natural rubber* scrap or contaminated material. A large friction ratio is used and the rubber is washed while being milled. The rubber is then sheeted out and sold after drying. See *refining*.

remote ultrasonic welding
An *ultrasonic welding* technique for thermoplastics which is sometimes called far-field welding or transmission welding. This is the *ultrasonic welding* technique used for joining moulded components, welds have been achieved with the interface at 250 mm/10" from the point of ultrasonic contact. Irregularly shaped components, such as *polystyrene* injection mouldings, can be rapidly and easily joined.

remould
A type of *tyre*: a worn tyre which has had all the rubber renewed from bead to bead. A good remould and a good retread will probably give the same milage as an *OE tyre* but are not recommended for prolonged, high speed driving because of worries about *carcass* wear and *tyre* integrity.

remoulding
A process used to replace the *tread, shoulder* and *sidewalls* of a worn *tyre*.

rennet
The dried extract of rennin: obtained from the fourth stomach of a calf. A powerful coagulant used to obtain *rennet casein*.

rennet casein
Solid casein obtained from skim milk by the addition of rennet (an enzyme). See *casein*.

rennin
An enzyme which coagulates the protein in milk: contained in *rennet* which is obtained from the fourth stomach of a calf. Used in the production of *casein*.

reodorant
See *odorant*.

repeat unit
See *repeating unit*.

repeatability
A term used in *hydraulics* and which means a motion or position can be repeated.

repeating group
See *structural unit*.

repeating unit
Sometimes referred to as a repeat unit or as, a 'mer' or as, a constitutional unit. The smallest unit of a macro-molecule

replaceable pad
Part of a mould used to carry, for example, the *gate* when *injection moulding* a *thermosetting plastics material*. Such pads are used so that they can be replaced when worn.

replasticization
Depolymerization of vulcanized rubber. See *reclaiming processes*.

replenish
A term used in *hydraulics* and which refers to the action of adding fluid to a hydraulic system so as to obtain a full system.

repolymerization
The building up of high molecular weight materials from degraded material produced, for example, during melt processing as a result of decomposition, or degradation. The joining together of degraded material so as to increase molecular weight. For example, repolymerization of the degraded plastics material *polyhydroxybutyrate (PHB)*, with dicumyl peroxide and triallyl cyanurate, can be effective.

Reppe process
A method of preparing monomers, for example, *butadiene*, from acetylene. Used to prepare the monomers for vinyl ethers by vinylation of alcohols with acetylene.

reprocessed plastic
A plastics material which has been processed more than once. See *regrind*.

reprocessing
A term usually applied to thermoplastics and which means that they can be reclaimed (by *regranulation*) so as to be used again.
When faulty mouldings, and feed systems, are scheduled for reclamation then they should be looked after very carefully and only those parts which are free from contamination and colour changes should be used. This is because it is common practice to blend virgin (new) material with reclaimed material (*regrind*) and if the reclaimed material is contaminated, then a lot more reject material is produced. If unchecked the problem can quickly get out of hand. So, ruthlessly reject any mouldings or any feed system which is suspected of being contaminated, i.e. only reclaim good quality material. Scrap any material purged from the injection cylinder.
Keep all material, which is to be reclaimed, covered. Keep the grinder, and the storage containers, spotlessly clean. Inspect the grinder blades regularly for breakage, bluntness, and wear and then, replace when necessary. Put the reclaimed material through equipment which will remove dust or fines, metal fragments and water. Then treat this material as new stock, for example, store the reclaimed material in tightly, sealed containers in a clean, dry storeroom. Blend with virgin material in a precisely agreed ratio and ensure that this ratio is adhered to during production as otherwise inconsistent mouldings will result.
It must be emphasised that on every moulding job the object must be to minimise the creation of reclaim. For example, calculations should be done on conventional, cold-runner, tool designs so as to ensure that the feed system is not over-sized. The use of *hot runner moulds*, or hot sprues, can eliminate/reduce feed system generation and therefore the need for *reclamation*.

reservoir
A term used in *hydraulics* and which refers to a container used to store hydraulic fluid.

residual crystallites.
See *self seeding*.

resilience
May also be referred to as rebound elasticity or as R. A ratio: may be defined as the ratio of recovered work to expended work. When a ball is dropped, the rebound resilience is determined from the height of the rebound. When a pendulum is dropped against a sample, a measure of the rebound resilience is given by the ratio height of the rebound (H_r) to height of fall (H_f). The result must be corrected for damping effects: the temperature of test is very important. Resilience may be measured by instruments such as the Dunlop pendulum, Dunlop tripsometer, Goodyear pendulum, Lüpke pendulum and Schob's pendulum.

resin
A solid, or pseudo-solid, organic material which is often of high molecular weight and which exhibits a tendency to flow when subjected to stress. Usually has a relatively low softening or melting range. Often the term is used to designate any polymer that is a basic material for plastics. In North America the term is often used in place of 'a plastics material'. The term is also often taken to mean a liquid plastics material, or a solution of a plastics material, which can be set or cured. See *natural resin*.

resin bonded plywood
Plywood which is bonded with a synthetic *resin*.

resin concentrate
A *masterbatch* based upon a *resin* which is often a thermoplastics material. Most commonly, resin concentrates are used to introduce colour. Could be based upon the same thermoplastics material as that being coloured (see *dilute masterbatch*) or, it could be based upon a different thermoplastics material. This different thermoplastics material is usually softer than the parent thermoplastics material so that it melts and disperses easily. For example, ethylene/vinyl acetate copolymer resin concentrates may be used with *polyethylene, polypropylene* and *polystyrene*. Because such masterbatches contain less pigment than other types of masterbatch, they take up more storage space. See *liquid colour*.

resin injection moulding
An abbreviation used for this process is RIM. A *resin transfer moulding process* used to produce large components at relatively low cost.

resin injection VA
An abbreviation used for resin injection vacuum assisted. See *vacuum assisted resin injection*.

resin injection vacuum assisted
See *vacuum assisted resin injection*.

resin modified natural rubber
An abbreviation used for this material is RMNR. See *resin modified rubber*.

resin modified rubber
An abbreviation used for this material is RMNR. Also known as resin reinforced rubber. The reaction product of *natural rubber (NR)* and a *phenol-formaldehyde (PF)* resin. A partially condensed PF is further reacted in the presence of the rubber as fully condensed PF materials are incompatible. Initially, during mixing, the resin acts as a *processing aid* and then it reinforces the final compound so as to increase the hardness: can also act as an *extender* and thus permits the use of high filler loadings. RMNR has better oil and solvent resistance than NR and gives better bonding to substrates. If sufficient resin is used then the compounds can be made to give ebonite-type materials. Such ebonite-type materials have improved impact strength and a higher softening point than traditional *ebonite* and are used to make pipe and ducting.

resin reinforced rubber
See *resin modified rubber*.

resin streak
A streak of excess resin on the surface of a laminated plastic.

resin to glass ratio
The weight of resin expressed as a ratio with respect to the weight of *glass fibre* reinforcement. In the *hand lay up* process, for example, the resin to glass ratio is approximately 1:3.

resin transfer moulding
An abbreviation used for this term is RTM. A *glass reinforced plastic moulding* (GRP moulding) technique in which a catalyzed thermosetting resin mix is transferred from the outside of a mould to the inside. Usually the mould contains reinforcement which has been preformed to the mould shape: the amount of reinforcement is often less than 40% (see *preforming continuous filament mat*). Pumps (for example, piston-type positive displacement pumps) are used to transfer the catalyzed resin through flexible hoses into an injection head (containing a motionless mixer) and then through a centrally located feed port, or point, into the mould.

Injection is via a tapered nozzle fitted to the end of the mixer outlet. This connects to the tool injection port which is generally machined of mild steel and contains a plug or check valve made of *polytetrafluoroethylene* (PTFE) or *polyethylene* (PE): the injection pressure pushes the check valve off its seat and allows resin to fill the mould.

Because of the pumping action, considerable pressures (several atmospheres) can be built up within the mould. This means that the moulds must be strong enough to withstand this pressure: because of the strengthening therefore applied, they can become too heavy to lift manually and may need to be press-mounted. The edges of the mould contains a seal, or rubber gasket, so as to minimize resin loss: air escapes from the mould through vents. A *gel coat* may be applied to the mould, so as to improve component surface finish, before the moulding process. Moulds can be heated or used at room temperatures. If vacuum is applied to the moulds, to counter the pumping pressure, moulds of lighter construction may be used (see *vacuum assisted resin injection*). RTM is usually associated with *unsaturated polyester resins* although *vinyl ester resins, polyurethanes, methacrylate resins, epoxides* and *nylons* are all used. Examples of mouldings made with RTM include automotive body panels, recreational vehicle components, truck air deflectors, chemical tanks, propellers and wind blades. The size of mouldings produced by RTM ranges from a few grams to hundreds of kilograms: resin transfer rates can be about 25 kg/minute. See *structural reaction injection moulding*.

resin-bonded mica paper
See *mica paper*.

resist
See *photoresist*.

resistance
See *ohmic resistance*.

resistance cuff heater
See *electrical resistance element*.

resistance element
See *electrical resistance element*.

resistance heated
If a machine, or part of a machine, is said to be resistance heated then this means that *electrical resistance elements* are in use. More than one element type may be used on a mould or die in an effort to get the most uniform temperature distribution. Often the heated area is divided into zones so as to obtain more accurate temperature control: each *zone* should be controlled separately using, for example, a deep seated thermocouple and a *three term controller*.

resistance heater
An electrical heating element which relies on the resistance to flow of an electric current through a wire element based, for example, on a nickel-chrome alloy. See *electrical resistance element*.

resistance heating
Heating which relies on the use of a *resistance heater*.

resistance to tracking
A test used to assess materials for their resistance to surface *tracking* when subjected to an electric stress in a wet environment. The test specimen is usually 3 to 5 mm thick and has a flat surface of 15 mm square minimum. An ionic solution (a salt solution such as 0.1% ammonium chloride) is dripped between two electrodes (brass or platinum) 5 mm wide and 4 mm apart which have a specified 50 Hz voltage between them. Drops of solution are allowed to fall at the rate of one every 30 s until *tracking* occurs or until 100 drops have fallen.

The test is performed at a number of different voltage levels and, for example, the comparative tracking index (cti) is the maximum voltage (X) at which 50 drops can be applied without tracking. The BS test uses brass electrodes whereas the DIN test uses platinum. Two salt solutions are used in the DIN test: solution A is 0.1% ammonium chloride and solution B is 0.1% ammonium chloride plus 0.5% wetting agent. See standard CEI 112, DIN 53480 and *arc resistance testing*.

resistivity
Sometimes called specific resistance: the reciprocal of conductivity. Most plastics are good insulators, that is, they do not conduct electricity very well as they have a high resistivity (a large resistance to the passage of electricity). There may well be a difference between the resistivity of the surface of the plastic and that of the bulk, or body, of the plastic. For this reason both surface and volume resistivity are quoted. In both cases the larger the number quoted, the better is the insulation. A good conductor such as gold has a volume resistivity of 10^{-6}; carbon is 10^{-3}; a conductive plastic is approximately 10^2; cellulose is 10^6; polyvinyl chloride (PVC) is 10^{14} and polystyrene (PS) is about 10^{18}. Insulation resistance is a combination of surface resistivity and volume resistivity. It is the ratio of the direct current voltage (applied to the electrodes) to the total current between them.

resit
A resit is a stage C resin, that is, a highly cross-linked structure. When heated a *resole* gives a *resitol* and then a resit.

resitol
A resitol is a stage B resin, that is, a gelled, rubbery-type of material. When heated a *resole* gives a resitol and then a *resit*.

resole
Also known as a phenol formaldehyde resole or as, a resol. Prepared by reacting *phenol* with an excess of *formaldehyde*

under alkaline conditions. If ammonia is used then a *spirit soluble resole* is obtained: if caustic soda is used then a *water soluble resole* is obtained. A resole is a thermosetting system which will cure without any further chemical addition simply as a result of re-heating. For example, it will cure if heated to approximately 150°C: neutralization with acid is often employed before heating. When heated, the resole gives a *resitol* and then a *resit*.

The phenol reacts with the formaldehyde to give methylolphenols: these, as they are very reactive, become dimethylolphenols and trimethylolphenols: the methylolphenols condense to form polynuclear phenols. For example, mononuclear phenols, dinuclear phenols and polynuclear phenols are formed.

resorcinol
Also known as 1,3-dihydroxybenzene or as, benzene-1,3-diol or as, m-dihydroxy benzene. A dihydric phenol. This material has a melting point of 110°C, a boiling point of 280°C and a relative density (RD) of 1·27. A preservative for rubbers and also used to improve the bonding of rubbers to textiles. Used in the preparation of *phenol-formaldehyde*-type resins (for example, for cold-setting adhesives) as the material is more reactive to *formaldehyde* than *phenol*. See *resorcinol-formaldehyde resin*.

resorcinol-formaldehyde adhesive
See *resorcinol-formaldehyde resin.*

resorcinol-formaldehyde-latex dip
An abbreviation used for this type of material is RFL dip. Such an aqueous dip is used to treat nylon and rayon fabrics. The *resorcinol-formaldehyde resin* solution (aqueous) is blended with a rubber latex to give the coating or dip. The latex used depends upon the polymer, for example, *styrene-butadiene rubber* (SBR) latex for rayon and styrene-vinyl pyridine-butadiene terpolymers with SBR for nylon. The molar rates of resorcinol to formaldehyde is typically 1:2 and the total solids content of the dip is about 20%.

The coating is applied by dipping and after the dip has been applied the fabric is dried and stretched in ovens whose temperature may reach 155°C for rayon and 230°C for nylon. This drying and stretching operation is used where the fabric is to be used for tyres (a heat setting or annealing operation) as it helps to prevent tyres increasing in size during service. This stretching must be carefully controlled so that the fabric cords are not reduced in size too much as weakness would result: rayon may be stretched by up to 4% and nylon by up to 10%. The treated fabric is wound up on hollow steel rolls or shells and at this stage it may be wound up without a *liner*. See *resorcinol-formaldehyde-silica system*.

resorcinol-formaldehyde resin
Also known as resorcinol-formaldehyde or as, resorcinol-formaldehyde polymer. Such resins are usually based on the reaction products of *phenol* and *resorcinol* with formaldehyde where the presence of the resorcinol improves reactivity. Used in the preparation of cold-setting adhesives as *resorcinol* is more reactive to *formaldehyde* than *phenol*. Such adhesives will harden at room temperatures and under neutral conditions to give strong, durable joints. For cold-setting adhesives for wood, the *resorcinol* is reacted with excess *formaldehyde* under alkaline conditions to give a *prepolymer* this prepolymer is then mixed when required with extra formaldehyde solution or with *paraformaldehyde*.

resorcinol-formaldehyde-silica system
An abbreviation used for this system is RFS system: sometimes referred to as RFK system or HRH system. A textile bonding system for rubbers based on the action of *resorcinol*, a *formaldehyde* donor (such as *hexamethylene tetramine*) and *silica*. The system is mixed into the rubber and can eliminate the need for textile pre-treatments. Sometimes referred to as the direct-adhesive process and used in place of a *resorcinol-formaldehyde-latex dip*.

restricted gate
A type of *gate* which seriously interferes with material flow. Where there is no serious flow obstruction, the gate may be known as an unrestricted gate.

restriction
A term used in *hydraulics* and which refers to a reduced cross-sectional area in a line or passage and which produces a pressure drop.

restrictor bar
Device used to even out flow along a *sheet die*: the bar can be bent so that, for example, the gap through which the material flows is thinner in the die centre than at the edges. Used so as to permit the production of sheet of uniform thickness. By the use of thermal bolts, and an appropriate control system, it is possible to get automatic adjustment during production.

restrictor - choke
See *choke.*

restrictor - orifice
See *orifice.*

retaining plate
Part of an injection mould: a steel plate which is part of the *ejector plate* assembly. It retains the *ejector* element.

retardation
The reduction in the rate of a chemical reaction caused by the presence of a *retarder*.

retarded elastic state
See *leathery state.*

retarder
A material which slows down a chemical reaction, for example, a polymerization (see *inhibitor*). Retarders are also used to minimise *scorch*. For example, if a mix is too scorchy during *rubber injection moulding*, a retarder may be used to give more processing safety.

reticulate chain structure
A *carbon black* structure which consists of a fused chain of particles: often abbreviated to structure.

reticulated foam
A material of reduced density which is based on a *cellular* material: the walls of the cells are crushed or partially removed so that a very open canal-like structure is formed. Used to make printing rollers, from *polyvinyl chloride* (PVC) plastisols, by heating the ink-containing *plastisol*.

RETP
An abbreviation used for reinforced engineering thermoplastic. See *filled engineering plastics*.

retracted spew
See *back rinding.*

retread
A type of *tyre*: a worn tyre which has had a new tread moulded onto an original carcass, after the tyre has been used, by *retreading*. A good remould and a good retread will probably give the same milage as an *OE tyre* but are not recommended for prolonged, high speed driving because of worries about carcass wear and tyre integrity. Such tyres do not carry an *E mark* but should be made to a standard, e.g. BS AU 144b.

retreading
Also known as recapping. The renewal of the treads and cushion of a worn *tyre*.

return line
A term used in *hydraulics* and which refers to the *line* which carries fluid from an actuator back to the sump or tank: this is usually a low pressure line as it is carrying exhaust fluid.

return pin
See *push-back pin*.

return ram
See *push-back ram*.

return spring
Part of a mould: a spring which returns the ejector assembly after ejection. Used in place of an *ejector plate* return pin.

reverse roller coating
A *roller coating* method in which the fluid composition is applied to the moving substrate by a roller: the surface of the roller moves in the opposite direction to that of the substrate.

reverse taper
Also called a counter-draft. A taper on part of a mould which is there to ensure that the moulding is retained in that part of the mould.

reverse temperature profile
A temperature profile which is higher at the hopper end of a machine than it is at the die/nozzle end of the machine. The use of such a profile on the *barrel*, for example, can help to preserve *fibre* length as heat, and not shear, is used to plasticize the material when a *reinforced thermoplastics material* is being processed.

reversing valve
A term used in *hydraulics* and which refers to a four-way valve used, for example, to reverse a *double-acting cylinder*.

reversion
To a rubber technologist, the term reversion means over-vulcanization and a loss of cross linking; this is achieved by holding a compound for too long at elevated temperatures.

To a plastic technologist, the term reversion means the shrinkage that occurs when a thermoplastics material is heated to a specified temperature. For example, if flexible *polyvinyl chloride* (PVC) was stretched at low temperatures, for example, 100°C during *calendering*, then the sheet would remember this and revert on subsequent heating: such heating could be encountered during end-use. To avoid freezing strain into the material it should therefore be stretched quickly at the highest possible temperature and then conditioned at a high temperature (see *stretching*).

For thermoplastics materials a measure of *reversion* may be obtained by placing the component on a piece of paper, marking around the component, cutting out the shape and weighing this shape. Oven heating the component then follows: the distorted component is then placed on another piece of paper similar to the first, the outline of the distorted component is marked and the shape cut out and weighed. A measure of reversion is obtained by comparing the two weights. If the product is a plastics sheet then, reversion may be assessed by marking a circle on the sheet and measuring the change in length of two diameters at right angles to each other. Usually the two dimensions are the *machine direction* and the *transverse direction*.

Revertex process
See *evaporation*.

revolutions per minute
An abbreviation used for this term is RPM or rpm. See *screw rotational speed*.

reworked plastic
See *regrind*.

Reynold's number
An abbreviation used for this term is Re. A dimensionless quantity which is used to relate density (ρ), speed of flow (u), tube diameter (l), and liquid viscosity (η).
$Re = \rho \, u \, l \, / \, \eta$.
At low speeds streamlined or laminar flow is observed: as the speed increases, beyond a critical speed (u_c), then the flow becomes turbulent.

rf
An abbreviation used for *radio frequency*.

RF
An abbreviation used for *radio frequency*. RF is also used as an abbreviation for *resorcinol-formaldehyde*.

RFK system
An abbreviation sometimes used for *resorcinol-formaldehyde-silica system*.

RFL dip
An abbreviation used for *resorcinol-formaldehyde-latex dip*.

RFS system
An abbreviation used for *resorcinol-formaldehyde-silica system*.

rh
An abbreviation used for relative humidity.

RH
A way of depicting a hydrocarbon material in oxidation reactions. On removal of hydrogen from the hydrocarbon RH, the radical R· is formed. See *antioxidant and oxidation*.

RHC
An abbreviation used for *Rockwell hardness scale*. RHC is also used as an abbreviation for *rubber hydrocarbon content*.

rhe
See *fluidity*.

rheogoniometer
A *rheometer*. A test instrument used to study the rheological and elastic properties of a sheared liquid. A well known type of machine is the *Weissenberg cone-and-plate rheogoniometer*.

rheogram
See *flow curve*.

rheology
The study of the response of materials, in *melt* form, to stresses. See *flow curve*.

rheometer
A test instrument used to study the rheological properties of liquids (see *viscometer*) or of polymer melts. Capillary rheometers are widely used to study polymer melts: cone and plate viscometers give more precise data but are only suitable for low viscosity melts. Dynamic techniques are used to study rubbers, for example, eccentric rotating disc rheometers and oscillating disc rheometers.

rheometry
The procedure or techniques used to study the rheological properties of liquids or of polymer melts. See *capillary rheometer* and *rheometer*.

rheopexy
The solidification of a thixotropic system when such a system is gently moved. More precisely, an increased rate of solidification of a thixotropic system when such a system is gently moved.

rhombic sulphur
Also known as alpha sulphur or as α sulphur. The stable form of *sulphur* at room temperatures.

RI
An abbreviation used for resin injection: as in *vacuum assisted resin injection* or VARI.

ribbed metal roller
A roller used in the *glass reinforced plastics moulding* industry for *compaction*. The roller cylinder is roughly 40 mm in diameter and 100 mm long: the dimensions are varied to suit the particular application. It has metal fins, or ribs, running along its length and is used, for example, in the *hand lay-up process* to consolidate the resin/glass mixture.

ribbed smoked sheet
An abbreviation used is RSS. One of the forms in which solid *natural rubber* (unvulcanized rubber) is supplied. A dark coloured form of natural rubber which still contains considerable amounts of non-rubber constituents.

The diluted *coagulum* produced from *natural rubber* latex is *acid coagulated* in tanks fitted with partitions so that a continuous sheet results which is approximately 3 mm thick. This long sheet is fed to a *sheeting battery*: the rolls of the final mill have a grooved pattern which is transferred to the surface of the rubber. This pattern reduces adhesion between the rubber when it is baled and helps the rubber to dry when being smoked. The rubber is smoked/dried for approximately 3 to 4 days at approximately 60°C/140°F in a smoke-house. The dried, brown sheets are then pressed into bales of less than 114 kg/250 lb.

Ribbed smoked sheet material is given a number from one to five: the lower the number the paler the material and, in general, the better the mechanical properties. Nothing but coagulated rubber sheets, properly dried and smoked, can be used in making RSS grades. See *standard international grades—natural rubber*.

ribbon blender
A machine used as a *pre-mixer* for example, for *calendering*. Consist of a trough (which can be heated) and a central shaft which can be rotated, for example, at 20 rpm: around this central shaft is wrapped a long steel blade or ribbon which rotates with the shaft.

For example, the blender is fed with the required amounts of *plasticiser* and *polymer* from the respective bulk stores by pipeline. The requisite quantities are blended at a known temperature for a minimum time, for example, 80°C for 20 minutes. The plasticiser may be measured using a pump and a flow meter and then sprayed onto the polymer to give a free flowing powder which can be easily fed to the next stage.

Such ribbon blenders are popular in factories where frequent colour or formulation changes are the rule. This is because they give operational flexibility. For example if there is a hold up further down the line then the mix can be kept in the blender. The batch in the blender may also be split, for example, part drawn off and more plasticiser added to the remainder if a more flexible compound is required.

If convenient the other additives could be added at this stage. However for health reasons, and also because of mixer contamination problems, opacifying stabilisers such as basic lead carbonate should not be used. See *high speed mixer*.

ribbon polymer
A high polymer based on double bonds: all the bonds in the main chain are either double bonds or, the polymer is based on conjugated ring structures.

ricinoleic acid
Also known as 12-hydroxy-cis-9-octadecanoic acid. This material has a melting point of 5°C and a boiling point of 228°C (10 mm). The mixture of acids obtained from *castor oil* is referred to as ricinoleic acid. An organic acid used to prepare, for example, *alkyl ricinoleates* and as the source of the monomer for *nylon 11*. See *propylene glycol*.

Riemann's green
Also known cobalt green. An *inorganic blue-green pigment* made by heating *zinc oxide* with cobalt salts.

rigid plastic
A hard, stiff material such as *polystyrene* (PS).

rigid polyurethane foam
A cellular material, or expanded material, based on a *polyurethane* and which is stiff or rigid: the material usually has a closed *cell* structure—*flexible polyurethane foam* has an open cell structure.

In this case, the polyurethane structure is formed by the reaction between a *diisocyanate* and a *polyol*: the diisocyanate is often *methylene diisocyanate* (MDI) and the polyol is usually a polyether polyol, for example, a poly-(oxypropylene) glycol. Such foams achieve their rigidity by using low molecular weight polyols which result in high cross-link densities, compared to flexible polyurethane foams, after the setting reaction is completed. Both the diisocyanate and the polyol are liquid and may therefore be dispensed relatively easily, for example, so as to fill a cavity—see *reaction injection moulding*.

Expansion or foaming during the cross-linking or setting reaction may be achieved by the use of water (which reacts with isocyanate groups to form carbon dioxide). However, for many years the use of a *chlorofluorocarbon* (CFC), sometimes in combination with water was preferred. A CFC was preferred because a CFC acts as a thinner as it is miscible with the diisocyanate and the polyol. When these two materials react, they do so exothermically and the boiling points of the CFC used is such that it vaporises, takes heat from the system and expands the reacting materials. In the final product (the rigid foam) the CFC is non-toxic, non-flammable and does not react with the foam. As the CFC has a high molecular weight it is not easily lost from the foam where it's presence reduces thermal conductivity. However, a CFC is no longer acceptable for environmental reasons: long term alternatives to CFCs, for example, to *CFC 11*, will probably be based on *HCFC* systems.

Such rigid PU foams are used as thermal insulation in the walls and doors of refrigerators because of the low thermal conductivity of this type of material and because, for example, the material may be foamed in place or, foamed in situ. This saves on assembly costs. In addition, the rigid foam is an excellent adhesive and bonds the component parts of the assembly together. It is also used as a lagging, and sometimes as structural lagging, in buildings, automobiles and hot water systems.

rigid polyvinyl chloride
An abbreviation used for this type of material is RPVC. At one time, the term meant that the PVC composition could contain a small amount of *plasticizer* but now the term is synonymous with *unplasticized polyvinyl chloride*.

rigid PU foam
See *rigid polyurethane foam*.

rigid PVC
An abbreviation used for *rigid polyvinyl chloride*.

rigidity modulus
See *shear modulus*.

rigidity of plastics
In general the rigidity of plastics materials is low and this has hindered their applications in many areas. Also, a material which is suitable for use at one temperature will not be suitable for use at another. With some rubber modified thermoplastics, excessive stiffening may occur at low temperatures and for many applications this can be considered a disadvantage. Car bumpers are one example. See *flexural properties* and *specific modulus*.

rigisol
A *plastisol* which yields a comparatively rigid product on gelation. The rigidity may be obtained by using a *plasticiser* which polymerises on heating (see *polymerizable plasticizer*) and/or, by using a range of polymer sizes to make the plastisol. That is, using a so-called *filler polymer* so as to increase the amount of polymer in the plastisol.

rim strip
A strip used with tyres which contain inner tubes: protects the inner tube against the rubbing action of the rim.

RIM
An abbreviation used for *reaction injection moulding*.

rind
Also known as *flash or spew*.

rind-back
See *backrinding*.

ring check valve
A common type of *check valve* used on an *injection moulding machine*. A steel ring, which is approximately the same outside diameter as the internal diameter of the *barrel*, seats against a shoulder on the *screw* and thus prevents material flowing back down the screw during injection. To increase speed and consistency of operation, the valve may be fitted with an extra ring called a *pilot ring*.

ring gate
A *gate* used in injection moulding for the moulding of, for example, cylindrical mouldings: as its name implies, the *runner* circles the cavity at its normal dimensions. Ring gates have been used for hollow cylindrical parts, such as pen barrels, as it is found that *air-trapping* is minimised due to the uniform material flow that is obtained. This type of *gate* may be considered to be a circular flash gate.

ring opening polymerization
Polymerization obtained when the ring of a cyclic monomer opens and the units join together. For example, *propylene oxide* contains the *epoxide group* and may be used to make polymers by ring opening polymerization. Compounds which contain a *cyclic thioether group* may also be used to make polymers by ring opening polymerization. Lactams, cyclic amines, and cyclic sulphides may also be used to make polymers by ring opening polymerization.

ringed spherulite
Also known as a banded spherulite. A *spherulite* which appears as a series of light and dark rings when viewed between crossed polarising filters under microscopic examination.

ringing
An *extrusion fault* which takes the form of concentric rings of a thicker or thinner wall section. In *polyvinyl chloride* (PVC) extrusion, caused by screw pulsations or, by slipping in the haul off or, by the extrudate sticking in the sizing box.

ripening
The process of allowing reacting materials to achieve a desired condition, for example, of uniformity and/or of molecular weight reduction. See *viscose rayon*.

ripple
An extrusion fault which takes the form of a steady undulation or wave-like appearance; sometimes found in *polyvinyl chloride* (PVC) extrusions which are non-uniform in temperature.

rise
In open steam curing, that period of time during which the temperature is raised gradually from room temperature to the actual curing temperature.

rise time
The time required for a free-rise *cellular* plastics material to achieve its ultimate expansion under controlled conditions.

risk
A risk is the likelihood that a substance will damage health in normal practice. Risk considerations take account not only of a substance's capacity to inflict harm but also of the way that it is used in a particular process, the duration and extent of the user's exposure and the way that exposure can be controlled. Poor control can result in even a low hazard substance becoming an appreciable risk.

rivet
A headed pin, or bolt, used to make a non-recoverable joint: the shank of the pin is passed through holes in the objects to be joined and then forced or hammered over to make another head.

RM
An abbreviation used for resin modified. See *resin modified rubber*.

RMNR
An abbreviation used for resin modified natural rubber. See *resin modified rubber*.

RMPP
An abbreviation used for *rubber modified polypropylene*.

robot
The Robotics Institute of America defines a robot as 'a re-programmable, multi-function manipulator designed to move material, parts, tools, or specialised devices through variable programmed motions for the performance of a variety of tasks'. By definition, therefore, a robot can do a variety of jobs provided that it is given specific instructions on how to perform the job in question. Once the job is finished then the robot may be told, by means of a program, to perform another type of job. Robots therefore differ from the parts-handling devices described in *pick-and-place* automation.

Robots may be classified according to their level of technological sophistication into low-technology robots (these are not servo-controlled), medium-technology robots (these utilise servo-mechanisms for accurate position and velocity control and have microprocessors as the basic control element) and high-technology robots (these are equipped with sensors which, in turn, provide the microprocessor control element with information about the external environment).

robot assessment
See *product assessment*.

rock
Usually an aggregate of *minerals* but some rocks consist of only one mineral, for example, pure limestone consists only of *calcite*.

Rockwell hardness
A number which indicates the hardness of a material as measured by an indentation test. The use of different loads and procedures gives several Rockwell hardness scales: scale C being used for metals, while those used for plastics include R, L and M. Two procedures are used. The first is used when there is no appreciable recovery when the indenting force is removed and the second gives a Rockwell hardness values.

For the first procedure, the test specimen is placed on a flat anvil below the indentor, and a minor load of 10 kg is applied, forcing the indentor into the material. Within 10 seconds the scale is set to zero. The major load of 60 kg, or 100 kg depending on the scale, is applied for 15 seconds and then immediately removed, but with the minor load still operating. The hardness reading is taken from the scale 15 seconds after removing the major load.

The second method is used with the R scale only and gives α Rockwell values. The indentation is noted 15 seconds after application of the major load, with the major load still applied. The α Rockwell hardness value is then obtained by subtracting the indentation from 150.

Rockwell Hardness Scale	Minor Load	Major Load	Indentor Diameter
R	10 kg	60 kg	12·70 mm
L	10 kg	60 kg	6·35 mm
M	10 kg	100 kg	6·35 mm
E	10 kg	100 kg	3·173 mm
K	10 kg	150 kg	3·173 mm

The test increases in severity down the table and the scales overlap to a certain extent.

Rockwell hardness meter
A test instrument used, for example, to obtain a *Rockwell hardness number*.

Rockwell hardness number
A number which indicates the hardness of a material and which is obtained by pressing an indentor into the test specimen and measuring the depth of penetration. See *Rockwell hardness*.

rod die
A simple die which produces an extrudate of circular cross-section. Such a die is commonly used in a *rheometer*.

rod flammability test
See *flammability*.

rod-climbing effect
See *Weissenberg effect*.

rodding
A technique used to make an *extruder* take in material; a rod of plastic (the same as that being extruded) is pushed onto the *screw* until the feedstock is gripped and feeds by itself.

roentgen
Sometimes spelt rontgen. A unit or dose of ionizing radiation. The amount of ionizing radiation that will produce one electrostatic unit of electricity in one cubic centimetre of dry air at 0°C and atmospheric pressure. To convert from roentgen to C/kg multiply by $2·58 \times 10^{-4}$. See *rem* and *Système International d'Unité*.

roll
A component part of a machine, made from steel and which heats/cools, mixes and shapes the material: more than one roll is usual. Rolls are sometimes known as bowls, for example, as in a 'four bowl calender'. Rolls are commonly heated by steam, cooled by water and driven electrically. See *two-roll mill* and *calender*.

roll bearings - calender
The ends of the rolls (the journals) run in bearings and these, for a *calender*, may be either smooth bush bearings or roller bearings. Because roller bearings work well at slow speeds and under large loads, rubber calenders are now commonly fitted with roller bearings. Calenders fitted with such bearings can give sheet of good accuracy whilst consuming less power (e.g. 10% lower than a machine fitted with bush bearings. However, if a bearing fails then it is easier to fit a bush bearing (see *roll float*). Bearings need to be lubricated and continuous bearing lubrication is of vital importance to a modern calender. The circulating oil is used, not only to lubricate, but also to maintain the bearings at a specified temperature. Bearing lubrication is by means of a closed circuit pressure system fitted with a pump and fitters. The pressure and temperature is also automatically monitored and controlled in such a system.

Synthetic lubricants are now used on calenders as such lubricants have many advantages over mineral oils. Their initially higher cost is therefore offset by advantages such as long life, temperature resistance, low coefficient of friction and good thermal conductivity. See *pull-back rams*.

roll bending
A system widely used in *calendering* to compensate for roll deflection caused by the *roll separating forces*. Extra bearings are fitted to each end of the roll and hydraulic pressure may then be applied to the roll, via these bearings, thus producing a straightening effect.

Unfortunately this compensating force works in the same direction as the roll separating forces and thus an additional load is added to the machine frame and bearings. Because of this extra load it is preferable to restrict roll bending to machines fitted with *roller bearings*. The amount of correction also produced is limited by practical consideration to about a quarter of that possible with *roll crossing*. If the hydraulic cylinders are made double acting then roll bending can be used to decrease the crown of a roll and so produce uniform thick sheet from a calender designed to produce thin film, that is, roll bending could be used to offset the effect of *roll cambering*.

roll cambering
Also known as roll contouring or as, crowning. A technique employed in *calendering* to offset the effects of the *roll separating forces* which tend to force the rolls apart. As the rolls are restricted at the ends, the rolls deflect more in the centre and if the rolls were parallel cylinders then the sheet produced would be correspondingly thicker in the centre. To avoid such thickening, the roll is crowned or profiled so that it is thicker at the centre. During running, the roll separating forces cause a deflection and uniform sheet should be produced. However, crowns only suit a very restricted range of products and therefore recourse must be had to more complex systems such as *roll bending* and *roll crossing*.

roll coating
The process of applying a thin layer of fluid material to a substrate from a roll previously coated with the fluid material.

roll configurations
See *calender and calendering*.

roll contouring
See *roll cambering*.

roll cooling
Rolls are most commonly heated by steam and is such a case the steam channels may be used to cool the roll by circulating water through the steam channel (ensure the steam is turned off first). Where steam cannot reach the operating temperatures, electrically heated systems are used: for safety reasons, air is normally used as the cooling medium for electrically heated rolls. See *two-roll mill* and *roll heating*.

roll crossing
A technique of compensating for thickness variations which is also known as roll skewing or as, cross axis adjustment or as, cross axis roll adjustment. With this technique one roll is moved relative to another about its horizontal axis. This movement, for example, about 25 mm/1" at each end on a roll of length 2 m/78", does not change the separation in the middle but increases the separation at the ends by a controlled amount. By this technique it is possible therefore to increase the thicknesses of the sheet edges until their thickness is the same as that measured at the centre. It is however found that the sheet is not of the correct thickness across the complete width of the sheet, that is, between the edges and the centre it is thinner. Despite this *ox-bow effect*, roll cross-

ing is widely used as it gives a means of controlling sheet thickness without imposing additional loads on the calender and at lower cost than *roll bending*.

Cross-axis roll adjustment may be fitted to more than one roll (for example, to numbers 1 and 4 on a *Z type machine*). It is fitted to roll no. 3 of an inverted L (as then 2 nips are effected) or to roll no. 4 of a Z type. The apparent crown increase is large, for example, up to 0.5 mm/0.020". See *roll separating forces*.

roll crowning
See *roll cambering*.

roll deflection
See *roll separating forces*.

roll float
Movement of a *roll* in its bearing. With a *bush bearing* there is a comparatively large clearance between the journal and the bearing e.g. 0.50 mm/0.020" cold and 0.25 mm/0.010" hot. The clearance is needed to ensure that seizing does not occur during heating of the rolls and also during machine operation. During the heating up period the temperature of the bearings and the roll journals are different and this difference demands a large clearance. Such a clearance is also necessary if adequate lubrication is to be provided during running. However if the pressure on the rolls changes during operation then the rolls can move or float in their bearings and affect the accuracy of the sheet. See *pull-back rams*.

roll heating
The rolls of a *calender* and of a *two-roll mill* are often required to be heated, and then held, at different temperatures. For such reasons rolls are commonly heated using steam/water mixtures for temperatures below 100°C. If the roll temperatures are greater than 100°C then steam, high pressure hot water or heated oil are circulated through a central bore or through a series of peripheral bores. The fluid heats the roll and then maintains the temperature to what is required by removing heat which comes in from high speed operation.

Where steam cannot reach the operating temperatures, electrically heated systems are sometimes used: for safety reasons, air is normally used as the cooling medium for electrically heated rolls, for example, on a two-roll mill.

roll journal
See *journal*.

roll profiling
See *roll cambering*.

roll separating forces
Forces which push the rolls apart. Such forces are caused through squeezing a viscous material through a relatively small gap at high speed. Pressures of up to 200 MNm^{-2} (14,000 psi) have been known but the actual pressures developed depends on many factors including the material viscosity, film thickness, roll speed, roll diameter etc. Many attempts have been made to relate the various factors to each other but most of the treatments are of limited practical use because of their complexity. See *Ardichvili's equation*.

Various techniques have evolved to offset the effects of the roll separating forces involved in *calendering*. See, for example, *roll float, roll cambering, roll crossing* and *roll bending*.

Roll bending is commonly fitted to the final roll, for example, roll number 4 of an inverted L; cross-axis would be fitted to roll number 3. On very wide calenders (for example, greater than 2.2 m face width) one of these devices may also be fitted to the first roll so as improve bank and spread.

Such compensating devices are commonly fitted to *polyvinyl chloride* (PVC), or tyre calenders where comparatively long runs are common. However many rubber calenders do not operate under such conditions. If the production run on any particular product is shorter than 30 minutes to one hour then it is debatable if the product would benefit from camber correction. This is because it would be difficult to get the machine in a stable running condition so as to obtain the correct camber setting. A large amount of off-gauge material would result before the camber correction was found and adjusted to the correct value.

roll skewing
See *roll crossing*.

roll-fed machine
A machine, for example, a *thermoforming machine*, which is fed with sheet material in roll form.

rolled laminated tube - thermosetting plastic
See *laminated rolled tube - thermosetting plastic*.

roller bearing
See *roll bearings - calenders*.

roller coating
A *spread coating method* in which the fluid composition is applied to the substrate by a roller.

roller knife
A knife used on a *two-roll mill* to produce a strip for *rubber injection moulding*. The knife has two circular blades set at the required distance apart. When this knife is pressed against the roll face and traversed across its width, the required strip is produced.

rolling bank
The excess material which is held in the *nip* of two rolls—see *two-roll mill*. Sometimes called the sausage.

rolls - speed difference of
The rolls of machines such as *calenders* and *two-roll mills* are often run at different speeds. (See *friction ratio*). Such a speed difference may be used to improve mixing, by increasing shear, or to raise compound temperatures. The speed is most commonly measured in rpm but roll surface speed is more relevant - particularly for a *calender* where the roll diameter can be very large. The rolls of a calender rotate more quickly as the *stack* is descended because the compound is getting thinner and wider and must be removed from the system.

Roninger method
A method or technique used to assess the dispersion of fillers in a rubber compound. To harden the rubber, it is dipped into molten sulphur (<24 h at 135°C), cooled and polished. The polished surface is then examined with a microscope at say, 600× magnification.

rontgen
See *roentgen*.

ROOH
A hydroperoxide-containing polymer (a polymer hydroperoxide) formed by the reaction of a polymer peroxy radical (ROO·) with (RH) a hydrocarbon polymer. The decomposition, or degradation, of a polymer hydroperoxide is induced by light quanta of approximately 300 nm and is catalyzed by metals. See *antioxidant* and *oxidation*.

room temperature vulcanization
Vulcanization at room temperatures, for example, 15°C.

room temperature vulcanizing silicone rubber
Also known as RTV vulcanizing silicone elastomer or as, RTV silicone elastomer. A silicone rubber which is liquid or semi-liquid at room temperatures and which may be cured, or set, without the application of heat. Often supplied in ready to use form. Used in potting, encapsulating and sealing. Such materials are also used in mould making, for example, to make flexible moulds for use in *glass reinforced plastic* mould-

ing. Can have *one-pack systems* and *two-pack systems* (RTV-1 and RTV-2). Silicone rubbers can be classified according to the technology employed for their processing into *high temperature vulcanizing silicone rubbers*, room temperature vulcanizing silicone rubbers and liquid *silicone rubbers.*

root
The root of a *screw* is the central shaft of the screw i.e. what is left after thread machining.

ROO·
A polymer peroxy radical (peroxide radical) formed by the reaction of a polymer radical R· with oxygen (O_2) during an oxidation reaction. See *antioxidant and oxidation.*

Rose's metal
A low melting point alloy (melting point 94°C) which is based on bismuth (50%), lead (25%) and tin (25%).

rosin
Also known as colophony. A brittle *natural resin* which is obtained by distillation of the resin obtained from pine trees (wood rosin is obtained by distillation of tree stumps).

This material has a relative density of 1·08 and softens at about 80°C: it consists mainly of abietic acid, its isomers, and esters of these acids. May be represented as $C_{20}H_{30}O_2$. This material is used to make varnishes but must be neutralised before it can be used. When reacted with lime it may than be known as limed rosin; when reacted with glycerol it may then be known as ester gum (or rosing ester); and, when reacted with pentaerythritol it may then be known as pentaresin.

rosin ester
See *rosin.*

Ross flexer
A continuous flexing machine used to test the resistance of an aged rubber test piece (24 h at 100°C) to crack growth: an ASTM test.

Rossi-Peakes flow tester
Used to assess melt flow behaviour. A test apparatus specified in, for example, BS and ASTM test methods and used to conduct a *Rossi-Peakes test.*

Rossi-Peakes test
A *flow test* for thermoplastics materials performed using a *Rossi-Peakes flow tester*. The polymer melt is forced upwards from a chamber through an orifice by a ram using a specified pressure (for example, 10·3 MN/m^2/1,500 psi): the length of material extruded is measured, at a temperature suitable for the material, after 2 minutes using a following, or tracking, rod and *dial gauge indicator*. Two samples are tested at each temperature and three temperature used: one which gives a reading below 25 mm and one which gives a reading above 25 mm. By plotting the results, the temperature which gives a length of flow of 25·4 mm/1.00" may be obtained: it is this temperature which is used to classify a material.

rotary blow moulding machine
A machine with more than two sets of *blow moulding platens* and moulds: each set of moulds is presented in turn to the *extrusion head* to collect parisons so as to give very high outputs. The platens may be mounted on vertical or horizontal carousels. Also referred to as 'wheel machines'.

rotary moulding
A type of moulding process in which the moulding machine uses multiple moulds mounted on a rotary table: as the table rotates, an individual mould is presented to the *plasticizing unit.*

rotary moulding machine
A machine used for *rotary moulding*: a high output machine.

rotary shaft seal
A seal which is used where there is a requirement to seal fluids in, contaminants out, or keep different fluids separated in the presence of a common rotating shaft. There are two main categories of rotary shaft seal each of which has two types. (1) Those which seal against a cylindrical surface: sealing parallel to the rotating shaft. (2) Those which seal against a face: sealing perpendicular to the rotating shaft. (1) can be subdivided into non-contacting (commonly called clearance bush seals) and contacting (commonly called lip type, rotary shaft seals). (2) can be subdivided into non-contacting (commonly called finger seals) and contacting (commonly called face seals).

rotary wheel blow moulding
See *rotary blow moulding machine.*

rotating blender
A mixing machine: such machines are usually based on either drums or conical containers. Widely employed in the thermoplastics industry to blend granules with other granules and/or with additives. A system based on a steel drum is the simplest mixer (often used as a *tumble mixer*) that can be imagined. See *compound blending.*

rotating cavity mould
An unscrewing mould in which threaded cavity inserts rotate thus unscrewing the moulding from the *cores.*

rotating core
Also called an unscrewing core. Part of an *undercut mould* used to put a *screw thread* into a moulded component.

rotating core mould
An unscrewing mould in which threaded core inserts rotate thus unscrewing the moulding.

rotating die
An extrusion *die* used in *lay flat film* extrusion to hide thickness variations and so eliminate *gauge bands.*

rotation viscometer
See *rotational viscometer.*

rotational casting
A technique used to put an internal coating within, for example, a pipe. A measured quantity of material (for example, *polyethylene*) is placed inside the metal pipe which is slowly rotated while being heated and then subsequently cooled.

rotational moulding
A moulding technique usually associated with plastics *pastes* or *powders* although rubber components are also moulded in this way. A moulding is made by putting the exact amount of material into a mould (made of, for example, cast aluminium). The mould is then closed and rotated while the material fuses, for example, by being heated in an oven at say 250°C. In order to obtain mouldings with uniform wall thickness the mould may be rotated around two axes at right angles to each other, or the mould may be rotated around one axis whilst it is being vibrated or tapped. After setting, and if necessary cooling, the mould is taken apart, or the component is collapsed, so that the product may be removed. The process is capable of giving consistent products of relatively uniform wall thickness and of consistent weight.

PLASTICS. *Polyvinyl chloride* (PVC) pastes were the original plastics materials used in this process. For such a thermoplastics material a hot mould is used and so cooling is required before the product can be removed from the mould.

The usefulness of this process was extended when it was discovered that powdered plastics material (e.g. *low and high density polyethylene*) could be used in a similar way. Such materials can be used to produce a range of tanks and containers which, because of low equipment costs, can be very large. Some of the largest thermoplastic mouldings are made in this way as, because of the low pressures involved, equipment and tooling costs can be kept fairly low. However, the raw material is more costly than that used in *extrusion* or *injection moulding* as the thermoplastic material normally has to be ground to a powder. For large mouldings, required in small numbers, this additional material cost is more than offset by the low mould costs. The moulds may be made, for example, from welded steel plate or from cast aluminium.

RUBBERS. Rotational moulding is used to produce components from rubber latex using moulds made from *plaster of Paris*. A known volume of *latex* (compounded with fillers, vulcanizing systems, etc.) is poured into the mould and then the mould is rotated so that even deposition of the rubber results. The plaster absorbs the water so that *vulcanization* can occur and the polymer sets.

rotational viscometer
Also known as a rotation viscometer or as, a rotational rheometer. A type of viscometer which, in general, is operated at comparatively low shear rates. Used, for example, for measuring the viscosity of *polyvinyl chloride* (PVC) paste and rubber *latex*. Can have a *concentric cylinder viscometer*, a cone and plate viscometer and a rotating bob rheometer.

rotocure
A continuous rotary curing machine. A machine used for the continuous vulcanization of rubber: used for the continuous vulcanization of rubber conveyor belting and for sheet. The unvulcanized material is trapped between a moving, pressurized belt and a rotating, heated drum so that continuous vulcanization is possible.

rotometer
Instrument for seeing and regulating water flow, for example, on *injection moulding machines*.

rotomill
A *continuous internal mixer*.

rotten stone
See *tripoli*.

rottenstone
A form of *silica*: a type of *tripoli* which is used as a filler.

rough
A rough surface is one which is uneven from projections or irregularities: the opposite of smooth.

rough bore hose
An *externally corrugated hose* in which the reinforcing helix is substantially exposed within the bore.

roughness
The noun which describes a characteristic of a surface, that is, a *rough surface*. The irregularities on any manufactured, for example, machined surface, occur as a result of roughness (peaks and troughs arising from the inherent action of, say, the machining process) and, *waviness* (attributable to vibration and machine deflection). Roughness is superimposed upon waviness and may be assessed by *arithmetical mean deviation*. See *surface texture*.

roughness grade number
A number which is prefixed with N and which indicates the roughness of a surface. For example, N 12 has a nominal R_a value of 50 μm or 2,000 μ in. N 6 has a nominal R_a value of 0·8 μm or 32 μ in. N 1 has a nominal R_a value of 0·25 μm or 1 μ in.

rove
See *spun yarn*.

roving
A form of fibrous reinforcement, for example, continuous strands of glass fibre (about 60) wound onto a spool with no twist. Used in *glass reinforced plastics*. See *spun roving* and *no-twist roving*.

roving package
A package of rovings.

RO·
A polymer alkoxy radical formed by the decomposition, or degradation, of a hydroperoxide-containing polymer during oxidation. See *antioxidant* and *oxidation*.

R_p
An abbreviation used for *rate of propagation* and for *rate of polymerization*.

RP
An abbreviation used for reinforced plastic.

rpm
An abbreviation used for *revolutions per minute*. See *screw surface speed*.

RPM
An abbreviation used for revolutions per minute, for example, of an extruder screw. UPM is also sometimes used.

RPVC
An abbreviation used for rigid *polyvinyl chloride* (PVC).

RRIM
An abbreviation used for *reinforced reaction injection moulding*.

RRPP
An abbreviation used for rubber reinforced polypropylene. See *rubber modified polypropylene*.

RSG process
RSG is the German equivalent to RIM. See *reaction injection moulding*.

RSR
An abbreviation used for *reciprocating screw rheometer*.

RSS
An abbreviation used for *ribbed smoked sheet*.

RT
An abbreviation used for room temperature.

RTM
An abbreviation used for *resin transfer moulding*.

RTP
An abbreviation used for *reinforced thermoplastic*. See *filled engineering plastics*.

RTV silicone elastomer
See *room temperature vulcanizing silicone rubber*.

RTV-1
See *room temperature vulcanizing silicone rubber* and *one-pack systems - room temperature vulcanizing silicone rubbers*.

RTV-2
See *room temperature vulcanizing silicone rubber* and *two-pack systems - room temperature vulcanizing silicone rubbers*.

rubber
A word sometimes used in place of *elastomer* but in general it refers to rubbery materials which have not been cross-linked. A rubber is the starting material on which an *elastomer*, or elastomeric material, is based and is often used as a generic term for materials of high reversible elasticity: instant recovery from large deformations (high elasticity) is only possible from *vulcanized* materials. The word rubber is derived from

an observation by Joseph Priestly (approximately 1770) that *caoutchouc* would rub-out (erase) pencil marks.

rubber abbreviations
See *abbreviations - rubbery materials*.

rubber accelerator
See *accelerator*.

rubber addition - thermoplastics
Rubbers are added to plastics materials in order to improve certain properties. For example, in the case of *polystyrene* (PS), the incorporation of approximately 10% rubber, during polymerization, changes PS from a very low impact strength material into *high impact polystyrene* (HIPS).

Rubber is added to *polypropylene* (PP) homopolymer as PP has a transition temperature at approximately 0°C and at this temperature the material changes from being tough to being brittle. That is, the low temperature impact strength of polypropylene (PP) can be improved by the addition of rubber up to approximately 30% of ethylene-propylene rubber (EPR) can be added to the PP and the resulting thermoplastics material still has useful rigidity combined with impact strength. Such mixtures, or blends of materials, are commonly seen as automotive fenders and as bumper guards on the sides of automobiles.

Such blends are based on rubber which has not been vulcanized or cross-linked. Now what is often required, is a material which can be processed like a thermoplastic and yet, has many of the properties of a cross-linked rubber: this is one reason for the current interest in polymer alloying and *thermoplastic elastomers*. One approach to making such thermoplastics materials is to disperse cross-linked rubber particles in a thermoplastics matrix as this approach gives thermoplastics materials with more rubbery properties (see *dynamic vulcanization*).

In general, as with fillers, the addition of a rubbery polymer to a thermoplastics material worsens the surface finish. That is the modified material has a comparatively poor gloss finish compared to the un-modified material. Because of the elasticity of the rubber, the rubber particles tend to deform during processing and then protrude through the surface as the deforming stress is relaxed.

The reduction in gloss may be minimized in various ways, for example, by laminating a layer of unmodified material onto the rubber-reinforced material during sheet extrusion. In the case of toughened polystyrene (HIPS) a layer of pure polystyrene (PS) may be bonded to the surface.

rubber calender
A *calender* used for rubber compounds.

rubber chloride
See *chlorinated rubber*.

rubber - classification
There are many different materials which can be classified as rubbers and they can be divided in various ways. For example, into *natural rubber* and into *synthetic rubbers*. Such materials could also be divided into *general purpose rubbers*, *oil resistant types*, and *special purpose (SP) rubbers*. There are approximately a dozen or so rubbers which are widely used commercially. These materials offer a very wide range of properties and it is possible to vary the characteristics of any rubber by use of a range of compounding ingredients. Compound flow behaviour and final product properties can therefore vary enormously, even if the same rubber is used, and for this reason only general guidelines on the processing and properties of *elastomeric materials* are usually given unless the processing of a specific formulation is being discussed.

rubber copolymer graft
See *polymer modified natural rubber* and *comb-grafted natural rubber*.

rubber fibre
A manufactured, polymeric *fibre* composed of *natural* or synthetic rubber.

rubber glass transition
See *glass transition*.

rubber hose
A flexible pipe based on rubber and a reinforcement, for example, textile or metallic reinforcement.

rubber hydrocarbon
The polymer in a *rubbery material*. For example, in *natural* rubber the polymer is cis-1,4-polyisoprene.

rubber hydrocarbon content
The amount of polymer in a *rubbery material*. See *natural rubber*.

rubber hydrochloride
The reaction product of hydrogen chloride and *natural rubber* (NR). A hydrohalogenated NR which may be represented as $(C_5H_9Cl)_n$. This white solid material has a softening point over the range 50 to 130°C and contains approximately 30% chlorine. Obtained by reacting hydrochloric acid with NR either in solution or in the solid under pressure. NR can absorb one molecule of acid for each isoprene unit. At one time this tear-resistant material was widely used as a clear packaging film and was more commonly known by the trade name/trademark of Pliofilm. Used in packaging as it has low permeability to gases, to water and is chemically resistant. Dechlorinated on heating and by *ultraviolet* light. See *iso-rubber*.

rubber hydrofluoride
This solid thermoplastic-type material contains approximately 25% fluorine. Obtained by reacting hydrofluoric acid with *natural rubber*. The rubber can absorb one molecule of acid for each isoprene unit. A tough material used as an adhesive for rubber-to-metal bonding.

rubber injection moulding
A product from a rubber injection moulding machine: a moulding process in which a thermally-softened rubber compound is transferred from a cylinder, or barrel, via a feed system and into a mould where *vulcanization* occurs, for example, using a two-stage rubber injection moulding machine.

BARREL HEATING. Both the injection unit and the plasticizing unit have a heating/cooling system. Around each barrel zone there is an outer jacket through which oil can be circulated; *resistance elements* (cuff heaters) are clamped around this outer jacket. If the temperature exceeds what is desired or, if the machine needs to be crash-cooled, then the oil will remove the excess heat. The *liaison block*, is also oil heated. Thermocouples are used to sense temperatures and these sensors are connected to, for example, *three term controllers*.

The set temperatures should be hot enough to give the desired speed of vulcanization but not so hot as to cause *scorch*, *reversion* and/or non-uniformly cured mouldings.

Location	Temperatures °C		Temperatures °F	
	From	To	From	To
Plasticizing unit	80	110	176	230
Injection unit	90	125	194	257
Nozzle or liaison block	90	120	194	248
Mould	170	200	338	392

COMPOUND PREPARATION. Rubber compositions are nearly always supplied to the moulding machine in the fully compounded form in which the additives (including the *carbon black*) has been fully dispersed by *melt compounding*: such compounding may be performed by either batch or continuous processes. For reasons of speed, output and economy, internal mixing operations are still the ones most widely used although the compound may be produced in the form of a suitable strip of consistent composition by *twin screw extrusion*: a *two-roll mill* may also be used to prepare the strip.

To ensure consistent flow properties the rubber must be of consistent quality and must be available in a narrow, specified viscosity range. Design the compound for scorch safety and ensure that the mixing procedure is uniform as mixing time and temperature influence compound flow. Viscosity and scorch measurements should be made on each batch of compound before any attempt is made to mould that batch. A *natural rubber compound* is a stiffer flow material at equal filler loadings than a *synthetic rubber compound* and its viscosity is more temperature sensitive.

If a compound's viscosity is too high then re-formulate (make it safer) so that it can be run at a higher temperature. That is, do not try and use an oil *extender* as this will result in lower melt temperatures. A *reinforcing filler*, such as carbon black, will cause a greater temperature rise than a *non-reinforcing filler* during plasticization (thermal softening) and injection. All compounds will cure in the barrels if left to stand for too long – as the temperature increases the barrel residence time decreases. If there is any break in moulding drop the temperatures and purge the barrel(s) clean.

CYLINDER EQUIPMENT. To stop the injection pressure being transmitted to the screw's thrust bearings on a *two stage machine*, it is important that there is a back-flow valve between the plunger and the screw. The nozzle diamater is very important and should match a particular compound and mould – it is usually about 3 mm/0·125 in. As the diameter is decreased, heat build-up and dispersion (for example, of *filler* and *sulphur*) increases. If the compound temperature can be built-up in the nozzle then the compound temperature can be reduced so as to decrease *scorch tendency*. If the nozzle diameter is too small then it will restrict mould filling and cause scorch; if this happens increase the diameter, e.g. in 0·5 mm steps, until scorch is eliminated.

On a *two stage machine*, the plunger cushion is usually as small as possible (e.g. 1 to 2 mm/0·04 to 0·08 in) so that *barrel residence time* is kept as low as possible. As little as 5% of the machine's rated capacity may be run, although it is better to match the injection unit to the mould if at all possible.

FINISHING OF COMPONENTS. Because rubbers are soft, trimming is relatively easy and many gates may be removed by simply twisting. With some mouldings, it is difficult to avoid the formation of flash and, even though this is much less than that produced by *compression moulding*, it still must be removed, e.g. by knife or scalpel. Because it is thin and flexible it may present problems if it is removed by barrelling. See *cryogenic tumbling*.

MELT TEMPERATURE. As high as possible consistent with *scorch-free* working (e.g. 100°C/212°F). Higher temperatures may be employed if a *screw delay facility* is available. If there are signs of *scorch* at a high temperature (for example, specks of overheated material in the moulding and slow or hesitant screw and/or plunger movements), reduce barrel temperatures in 5°C/9°F steps and/or reduce the screw speed until the signs of scorch disappear.

MOULD CONSIDERATIONS. The mould must be designed to take account of the material's characteristics and the operating procedures employed. For example, high operating temperatures, the fluidity of many rubber compounds, the pressure absorbing nature of the material and the tendency to evolve fumes at high temperatures.

The mould is usually heated by electrical resistance elements – these may be located in the mould itself or in the machine platens and/or in both. Thermocouples are used to sense the mould temperatures and these sensors are connected to *three term controllers*.

As the mould is hotter than the compound, long flow paths into thin sections are possible. This means that moulds must be well constructed; some compounds can penetrate 5 micron cracks. The *steel* used to make the mould must be of very good quality (to withstand, for example, the pressures involved, erosion and corrosion): types used include prehardened, hardened and stainless. Chrome plating has been used but may be eroded by some fillers. The steel must machine well and take a good polish. A well *balanced runner system* must be used and the use of long narrow runners, excessively restricted gates and abrupt changes in direction are to be avoided as these increase the risk of scorch and lengthen mould filling times. Good mould venting is essential. Such venting may be via the ejection pins: vacuum venting is also used.

Ejection may be done using pins but air assist, robot removal and hand removal are also used. Components are often removed from the mould by hand, or by *robot*, because of component flexibility and because of part complexity. Ejector pins should be hardened and may be of the valve-type to prevent flash penetrating into the gap between the pin and the mould.

MOULD FILLING SPEED. Use a safe (scorch resistant) mix, adjust the machine controls (injection speed, pressure, back pressure, etc.) so as to obtain the highest possible injection temperature and mould filling speed consistent with freedom from scorch. The temperature of the material may be raised by up to 40°C/72°F if high speeds and a small nozzle are employed.

REPROCESSSING OF MATERIAL. For non-critical applications, some companies regrind the waste and compound the cured, granulated material in with another batch. Such reprocessing is not normally employed as either the compound properties suffer or regranulation is difficult and expensive. The use of *cold runner moulds* can help to reduce the amount of scrap produced (manifold temperature 125°C/257°F).

PRESSURES USED. The clamping pressures required per unit of projected area are relatively high (compared to compression moulding) and are of the order of 2 tsi/30 MN/m^2. The actual figure used depends on, for example, the ease of flow of the component, depth of draw of the component and the amount of flash that can be tolerated.

Back pressure is used to generate a temperature rise and is also used to exclude air from the compound. On a *two stage machine*, back pressures of up to 250 bar/25 MNm^{-2}/3,620 psi may be required.

The machine should be capable of giving the following injection pressures. First stage: up to 1,500 bar/150 MNm^{-2}/21,400 psi. If possible, do not use a compound which demands the maximum pressure as any small variation will make moulding difficult and may demand compound changes. Second stage pressure up to 1,200 bar/120 MNm^{-2}/17,400 psi.

SCREW ROTATIONAL SPEED. Screw speeds of up to 200 rpm are employed. Because of the low temperatures set on the extruder barrel in two stage machines (up to 120°C/248°F), the use of a high screw speed (e.g. 150 rpm) provides a way of generating heat if required. See *screw surface speed*.

SHRINKAGE. Because of the high mould temperatures employed, shrinkage may be higher than that found in *compression moulding*. Because of the more complex flow patterns found in injection moulds, it will also be non-uniform. If this non-uniformity is excessive then it may be reduced by using a slow curing compound and/or a lower mould temperature; this should give the material more time to relax or randomise.

Allowances should be made in mould manufacture for final sizing of cavities and pins to be made, i.e. after initial moulding and measurement.

STOPPAGES. If the stoppage is going to be a long one then cool the barrel as quickly as possible (crash cool) and purge out; if the stoppage is only going to be for a short time then reduce the temperature settings. Each 10°C/18°F drop will double the resistance to scorch.

STRIP PRODUCTION. A skilful operator may, at the end of the mixing operation on a *two-roll mill* produce a strip of the required width by gradually traversing a knife across the roll face. To stop this strip sticking to itself, it is dusted with talc before being wound on a reel. The strip, necessary to feed the injection moulding machine, may also be produced by using a *roller knife*. Alternatively, the banded material on the roll may be removed as a sheet, rolled and fed to an extruder (ram or screw) so that a strip of the required dimensions may be produced by extrusion.

TYPICAL COMPONENTS. Both mouldings and *rubber-to-metal* bonded products are produced by injection moulding. The mouldings are made in this way for a number of reasons. For example, because they are cheaper than compression moulding and little or no finishing is necessary. Rubber-to-metal products benefit because fresh, clean rubber is presented to the metal surface and this results in a stronger, more consistent product (e.g. a typical product would be an engine mount).

The automotive industry uses rubber injection moulding to produce windscreen wiper blades, seals, bushes, engine mounts, tyre valves, etc. In the pharmaceutical industry this process is used to produce bottle stoppers, teats, pipettes, etc. The shoe industry uses injection machines to produce shoe-sole units and to mould the shoe-soles onto the uppers. Another large market is the manufacture of seals, for example, for oil and water pipes. In short it is difficult to find an industry that does not use injection moulded rubber components.

rubber injection moulding machine
A moulding machine used to produce components based on compounds by the injection moulding process. See *two stage, rubber injection moulding machine*.

rubber inking roller
A flexible rubber roller used in *flexographic printing*.

rubber latex
See *latex*.

rubber masterbatch
See *masterbatch - rubber*.

rubber modified material
Usually means a *thermoplastics material* which has been modified by the addition of a rubber. See *rubber addition*.

rubber modified polypropylene
An abbreviation used for this type of material is RMPP. A material which is processed like a *thermoplastics material*. Rubber modified PP may consist of a blend of the thermoplastics material polypropylene (PP) with an unvulcanized *ethylene propylene rubber*: may also consist of polypropylene (PP) with a vulcanized ethylene propylene rubber. Both types of material may contain smaller amounts of other materials, for example, *high density polyethylene (HDPE)*, fillers and *carbon black*. See *rubber modified polypropylene - unvulcanized rubber* and *dynamically crosslinked polyolefin elastomer*.

rubber modified polypropylene - unvulcanized rubber
A *rubber modified polypropylene (RMPP)* in which the rubber is not cross-linked: it is unvulcanized. Sometimes referred to as rubber reinforced polypropylene (RRPP). May be referred to as an elastomer modified thermoplastic (EMT) or, as an olefin thermoplastic elastomer (OTE) or, as impact modified polypropylene (IMPP). The material may also be known as a *thermoplastic elastomer* (TPE) or as PP/EPDM.

The major reason for adding rubber to PP plastics is to overcome the poor low temperature properties of such materials while obtaining a *thermoplastic elastomer*. Some blends are impact resistant down to sub-zero temperatures, for example, $-40°C/-40°F$. One reason for this is that the rubber stops large spherulite growth; large spherulites are stress concentration sites in PP.

Such compounded blends may be produced by intensively melt mixing the plastic and the rubber together, for example in an internal mixer, at a temperature greater than the melting point of the PP. A two phase structure should result, in which the rubber is in the form of 0·5 to 5 micron particles which are embedded in a continuous phase of PP. Highly isotactic, polypropylene homopolymers (PP-H or PP-HO) or, block copolymers of propylene and ethylene (PP-B) may be used. The best balance between, for example, stiffness (E modulus), cold temperature impact strength and ease of flow, is obtained by the use of high molecular weight, amorphous EP(D)M and PP-B of medium viscosity. The use of a copolymer, in place of a homopolymer, can reduce the rubber content by approximately 10%.

Many RMPP compounds contain 10 to 30% of EPDM and so their densities will be less than that of PP as the SG of ethylene propylene rubbers is approximately 0·86. The hardness range of this class of materials is from 60 Shore A to 60 Shore D. Hardness, modulus and tensile strength all increase as the PP content increases. To improve the *heat distortion temperature* and stiffness, fillers may be used: for example, by using approximately 25% *talc* (of particle size 3 to 15 microns) a harder, stiffer compound results. Talc filling is detrimental to cold impact behaviour and to antioxidant performance. However, by ringing the changes on the types, and quantities, of materials used, it is possible to produce a very wide range of compounds or blends.

The addition of un-vulcanized rubber worsens the chemical resistance of the PP and the chemical resistance worsens as rubber content increases. RMPP is resistant to oils, fats, alcohols and glycols; also to caustic alkali, strong and weak acids, and low to medium strength oxidising chemicals. Also resistant to most chemicals of a polar nature, for example, water. In general, *ethylene-propylene* (EP) rubbers will not swell when in contact with highly polar liquids (such as water) but swelling may occur with aliphatic, non-polar or slightly polar liquids. RMPP mouldings have relatively poor resistance to petrol and other hydrocarbon solvents.

The natural colour of the material is similar to that of PP (i.e. an off-white), and so a large colour range is possible. However, this large colour range is not often seen as these materials are often used in automotive applications (fender, rubbing strips, door guards) where good outdoor stability is essential, As this may be achieved relatively cheaply by the addition of carbon black then, many PP/EPDM materials are black. Natural grades may however, be coloured by, for example, by the use of a solid *masterbatch*.

Major limitations of this class of materials are associated with surface properties and heat resistance. Compounds usually have poor hardness, scratch resistance and limited heat resistance (particularly under load). See *dynamically crosslinked polyolefin elastomer*.

rubber - molecular structure
See *vulcanized rubber - molecular structure*.

rubber oil
Also known as petroleum oil or as, petroleum oil plasticizer. A high boiling point petroleum fraction (see *mineral oil*). A complex mixture of hydrocarbons which is used as a *plasti-

cizer or *extender* for *natural* and *synthetic rubbers* and which may contain unsaturated aromatic rings, saturated rings (naphthenic rings) and paraffinic side chains. Such a plasticizer may be classified as aromatic, naphthenic or paraffinic – dependent on which type of structure is predominant. Aromatic or naphthenic oils tend to be used for oil-extension of rubbers. The specific gravity decreases from aromatic (1·02) to naphthenic (0·93) to paraffinic (0·86). Aromatic oils also contain the highest proportion of non-hydrocarbon materials (such as nitrogen, oxygen and sulphur). See *viscosity-gravity constant*.

rubber - production of
The total world usage of all rubbers is of the order of 11·5 million tons (11 Mt in 1991). Two of the most widely used rubbers are styrene butadiene rubber (SBR) and natural rubber (NR). Between them, they account for at least two thirds of all rubber usage. The market split is roughly *natural rubber* (NR) 37% and *synthetic rubber (SR)* 63%. For example, in Western Europe total rubber consumption was approximately 3·5 Mt in 1991: 1·25 Mt of SBR, 1·04 Mt of NR and 0·27 Mt of thermoplastic elastomers. Most new rubber goes into *tyre* products: in such applications, rubbers are used because of their elastic properties rather than because of their flexibility.

rubber - raw material form
Apart from polychloroprene, most rubbers are supplied in polyethylene-wrapped bales which may weigh, for example, 25 kg (55 lbs). Such a bale may be supplied and stored on pallets but to prevent creep problems the pallets may be fitted with sides so that the weight of the stack is not supported by the rubber. With large internal mixers the bale may be fed directly into the throat of the machine. Where appropriate the size of the bale is reduced, by a *bale-cutting* or *guillotining* operation.

rubber reinforced polypropylene
See *rubber modified polypropylene*.

rubber substitute
See *factice*.

rubber thread
Thread which may be made from either vulcanized sheet or from *latex*. The vulcanized sheet is stuck to a roll with shellac and cut into strip. In the spinning process, the threads are formed by casting the latex into a coagulating bath; they are then dried and vulcanized.

rubber to metal bonding
The bonding of rubber to a metallic substrate at the same time as a moulding is being formed: the metal is an insert within a mould. When bonding rubber to a substrate such as metal or ceramic, then an intermediate layer is often necessary: this layer must bond to both the rubber and the substrate. (See for example, *brass coating*). Rubber-to-metal bonded products are components produced by a moulding process in which the metal is placed in the mould and the rubber flows around the metal and bonds to the metal surfaces. Rubber-to-metal products benefit from being produced by *rubber injection moulding* because fresh, clean rubber is presented to the metal surface and this results in a stronger, more consistent product. A typical product would be an engine mount, produced for the automotive industry.

rubber toughened polypropylene
See *rubber modified polypropylene*.

rubber toughened polystyrene
See *high impact polystyrene*.

rubber tubing
A flexible pipe based on rubber but which contains no reinforcement. Tubing which is based on a rubbery material - often a vulcanized rubber. Once the tube has been extruded, using a short length to diameter *extruder* if prewarmed *stock* is used, the material must be heated further to cause vulcanisation. In a typical batch process the tube would be dusted with *talc* (for example, by extruding into a bed of talc located near the *die*) then *vulcanised* in an autoclave. For *continuous vulcanisation* the extruded tube would, for example, be passed continuously through a heated bath.

Rubber Manufacturers Association
The publishers of the *green book* which is entitled International Standards of Quality and Packing for Natural Rubber.

rubber-bound antioxidant
See *network-bound antioxidant*.

rubber-coated fabric
See *coated fabric*.

rubber-like material
See *thermoplastic rubber-like material*.

rubber-modified PVC
See *polyvinyl chloride nitrile rubber blend*.

rubberiness
The property of high reversible elasticity.

rubberized hair
Fibres (horse hair or coconut fibres) bound together by rubber *latex* (natural or synthetic rubber latex).

rubbery behaviour
A material exhibits rubbery behaviour if it can be extended by approximately 100% and on removal of the retracting force it rapidly springs back to its original dimensions. See *natural rubber* and *vulcanized rubber*.

rubbery flow
The region of viscoelastic behaviour which follows the *rubbery state* on further heating of an amorphous thermoplastics material. In this region, viscous flow occurs.

rubbery material
A material which exhibits *rubbery behaviour*. See *elastomer*.

rubbery plateau
See *rubbery state*.

rubbery state
Also called the rubbery plateau or the plateau zone. The region of viscoelastic behaviour which follows the retarded elastic state (see *leathery state*) on further heating of an amorphous polymeric material. In this region, the material exhibits elasticity. This region lies about 30 to 80°C above the *glass transition temperature*.

rugose
Meaning rough. For example, the extrudate produced from a highly filled, *natural rubber* compound often has a rough, or rugose, finish.

rule of mixtures
See *law of mixtures*.

runner
This term describes the channel in the mould which connects the *sprue bush* to the *gate* and it also describes the set polymer which is formed in that channel. The solid runner system may be ejected when the mould opens or may remain within the mould in a semi-fluid state (see *runnerless moulding*).

In every mould the runner should be designed to fill the mould quickly, uniformly and with acceptable heat and pressure losses. It should therefore be kept as short as possible and its cross-section should be circular. That is, it should be a fully round runner as this gives the lowest pressure loss. The diameter of the runner system usually lies within the range 3–9 mm/0.125–0.375", but runners rarely need to be larger than 6 mm diameter (0.25 in). When the runner is going to be reground, so that it may be re-used. its diameter should be kept small. However, if it is made too small then this will restrict mould filling speed and cause very high pressure losses.

runner lay-out
Flow path lay-out. The runners should be laid out so that sharp corners, or sharp changes in direction, are avoided; on a multi-cavity mould the runner system should be laid out so that the flow path between each cavity and the *sprue* is of the same length. When the flow path length is identical, the system is called a balanced runner and the mould is called a *balanced runner mould*.

The first part of the flowing melt stream will be at a different temperature to the bulk of the material and in order to 'bleed-off' this thermally different material, *cold slug wells* may be incorporated in each branch of the runner system. Such cold slug wells may be blind extensions to the runner in which the thermally different material is trapped; *vents* may also be incorporated in the runner.

runner plate
See *stationary plate*.

runner pressure drop
The magnitude of the pressure drop in a *fully round runner* is proportional to the length of the runner and to its diameter. Increasing the length and decreasing the diameter will increase the pressure drop. So, to minimise the pressure lost in the runner system it is important to keep the runner length as short as possible; if this is not done then the cavity pressure may be insufficient to maintain the specified dimensional tolerances. It is important to keep the runner diameter small so as to save on material and to keep the cooling time as low as possible. However, if the diameter is made too small then again the cavity pressure may be insufficient.

The runner adjacent to the gate should be larger in diameter than the wall section of the moulding (by approximately 10%). Every time the sub-runner is joined to a common runner closer to the sprue the diameter should again be increased by approximately 10%. A full round runner may be likened to a pipe and if this analogy is correct then the pressure drop (ΔP) over a specified length of runner may be obtained by using a *flow curve* and the *Poiseulle equation*. It is thus possible to size the runner so that approximately half of the injection pressure is lost in the runner system: this loss is acceptable and gives runners of a reasonable diameter.

To reduce pressure losses in the region where the runner meets the gate a long tapered transition zone should be avoided. For a fully round runner the runner end should be hemispherical and for trapezoidally shaped runners the end taper should be similar to the side taper.

runnerless mould
A type of *injection* mould in which the material involved in the *feed system* is kept from hardening. In the case of thermoplastics materials this means keeping the material in the feed system hot, whereas in the case of thermosetting materials and rubbers this means keeping the feed material cool. In some cases the cost of a runnerless mould is not justified by the length of run envisaged.

Typical gate diameters used for runnerless moulds for thermoplastic materials are 0.75 to 0.8 mm/0.030" to 0.032", however gate diameters as low as 0.7 mm/0.028" are often used to prevent *gate vestige* occurring particularly when producing cosmetic components that require zero gate vestige. To prevent gate vestige, it is best to use hot runner nozzles that incorporate a shut-off valve or needle. Accurate temperature control of the runner manifold and nozzle assembly is also necessary for most materials. See *hot runner mould* and *sprueless moulding*.

When *thermosetting plastics* are injection moulded it is common practice to discard the feed system. This is a serious cost disadvantage particularly for small components made from expensive raw materials. As the *sprue* is often the thickest part to be moulded it dictates the minimum cure time which needs to be given. Such wastage can be minimised by maintaining part, or all, of the feed system at a lower temperature than the part of the mould which contains the cavities. The material in the feed system thus remains mouldable and is injected into the mould during the next shot. See *warm sprue mould* and *cold runner mould*.

rutile titanium dioxide
The most widely used form of *titanium dioxide* which is used as an *inorganic pigment*. One of the crystalline forms of titanium dioxide. Rutile titanium dioxide, because of its higher index of refraction (2.76) has better *hiding power* than the *anatase* form. Has a relative density of 4.20.

R_v
An abbreviation used for *volume resistance*.

R·
A way of depicting a *free radical*. A polymer radical formed by the removal of hydrogen from (RH) a hydrocarbon. See *antioxidant* and *oxidation*.

S

S
An abbreviation used for *second*.

S
This letter is used as an abbreviations for;
entropy;
sebacate. See, for example, *dialphanyl sebacate*;
siemens;
slow cure rate (black). See *carbon black*;
small rotor. Used to show that the small rotor was used in a *Mooney viscosity test*;
soya bean. For example, *epoxidized soya bean oil (ESO)*;
styrene;
suspension (polymerization); and
synthetic filler.

S calender
See *inverted Z calender*.

S-B diblock
Polymeric materials in which two blocks of two different materials are connected together: a *polystyrene* (S) hard block is connected to a softer *butadiene rubber* (B) block. See *styrene block copolymer*.

S-B-S linear triblock polymer
Polymeric materials in which two, comparatively short, end-blocks of *polystyrene (S)* are connected to a longer, central *butadiene rubber* (B) block. See *styrene block copolymer*.

S-B-S triblock
See *S-B-S linear triblock polymer* and *styrene block copolymer*.

S-EB-S
An abbreviation used for styrene-ethylene/butylene-styrene. A *saturated triblock polymer*. See *styrene block copolymer*.

S-EP-S
An abbreviation used for styrene-ethylene/propylene-styrene. A *saturated triblock polymer*. See *styrene block copolymer*.

S-glass
A type of glass which is noted for its high mechanical strength but which is relatively difficult to draw into fibres. Sometimes used as the fibrous material in *glass reinforced plastics* where the highest strength and stiffness, from a glass, is required. May be used, for example, in *dough moulding compounds* and in *sheet moulding compounds*. The percent composition by weight is approximately SiO_2 64·3%, Al_2O_3 24·8%, Fe_2O_3 0·2%, MgO 10·3%, Na_2O 0·3%. See *E-glass* and *A-glass*.

S-RIM
An abbreviation used for *structural reaction injection moulding*. Sometimes also used for structural resin injection moulding.

S-value
Also known as black value. A measure of the depth of colour of *carbon black*. A high S-value indicates a high level of light absorption.

S1 d4
See *method S1 d4* and *tapping methods*.

S2 d2
See *method S2 d2* and *tapping methods*

SA
An abbreviation used for *salicylic acid*.

SACMA
An abbreviation used for a USA-based organization called the Suppliers of Advanced Composites Materials Association.

SAE
An abbreviation used for the USA-based organization called the *Society of Automotive Engineers*.

SAF
Super abrasion furnace (black). N110. See *carbon black*.

safety bar
A bar which is part of the safety control devices on, for example, a *two-roll mill*. Usually mounted in front of the operator at knee height so that, in the event of an emergency, the bar may be quickly pressed so as to stop the machine by *plugged braking*.

safety disc
A disc of a specified shear strength used on a *two-roll mill* for roll protection.

safety stop
A stop, or button, which is part of the safety control devices on, for example, a *two-roll mill*. Usually mounted in a prominent position (say in front of the operator's working position at a convenient height) so that, in the event of an emergency, the stop may be quickly pressed so as to stop the machine by, for example, turning off the drive power. Emergency stops should also be located in sensible positions around the working area. See *plugged braking*.

safflower oil
Derived from *Carthamus tinctorius*. Also known as kardiseed oil. An *industrial vegetable oil* which is obtained from safflower seeds. A semi-drying oil, which is low in *linoleic acid triglycerides* and therefore used in *alkyd resins* as it is non-yellowing. Used in place of *linseed oil* in white finishes where colour retention is important.

Some types, for example, UC safflower oil contains approximately 80% oleic acid and is used as an edible oil as they have low iodine values.

salicylates
Ultraviolet absorbing compounds widely used to improve the *ultraviolet* (UV) resistance of polymers. See *ultraviolet absorber* and *ultraviolet stabilizer*.

salicylic acid
An abbreviation used is SA. This material has a relative density (RD or SG) of 1·44. Can act as a *scorch retarder* for rubbers when acidic *accelerators* are used. Used to make *salicylates*.

saligenin
See *o-hydroxymethyl-phenol*.

salt
In chemistry this refers to a compound formed when the hydrogen of an acid has been replaced by a metal. A salt is formed, together with water, when an acid reacts with a base.

salt bath vulcanization
A *continuous vulcanisation method*. See *molten salt bath*.

salt solution
A solution of a *salt* (for example, ammonium chloride) in a solvent such as water.

sample
A small part or portion of a material, or product, intended to be representative of the whole.

SAN
An abbreviation used for *styrene-acrylonitrile copolymer*.

sand
An impure form of *silica*. Sometimes used as a filler for *epoxide castings*.

sandwich heating
A method of heating a thermoplastics sheet material before *thermoforming:* the sheet is heated from both sides simultaneously.

sandwich laminating
An *extrusion coating process* in which the extruded thermoplastics material bonds substrates together.

sandwich moulding
Sandwich moulding is an *injection moulding technique* used to produce components based on more than one thermoplastics material: one material surrounds, or encapsulates, the other. Can have a solid material around a cellular material, a layer of pure material around a thick base of reclaim etc. The machines used to produce such mouldings are usually equipped with two injection units: one for each material.

This process extends the scope of injection moulding. For example, one of the major disadvantages of moulding a *fibre* reinforced thermoplastics material is that the surface finish is duller than that of the unmodified material (it is often a mat finish). By using two different materials, one filled and one unfilled, heavily filled mouldings with a high *heat distortion temperature* and with a very high gloss finish (based on the unfilled material on the outside) have been produced. An example of this type of moulding is automotive headlight reflectors.

saponification
The hydrolysis of an *ester* with an alkali thus forming a *salt* or a *soap*.

saran fibre
A manufactured, polymeric fibre composed of at least 85% by weight of *vinylidene chloride* units.

Saran plastics
See *vinylidene chloride* plastics.

SATRA
An abbreviation used for the Shoe and Allied Trades Research Association of the UK.

saturated
In chemistry this means that a compound does not contain double bonds.

saturated polyester
A *polyester* which does not contain double bonds. Materials of comparatively low molecular weight are used as *polymeric plasticizers* and as *precursor* polymers for *polyurethanes*. Partially aromatic polymers include *polyethylene terephthalate* (PET), *polybutylene terephthalate* (PBT), and *polycarbonate* (PC). High temperature resistant materials are obtained if fully aromatic polyesters are produced, for example, a *polyarylate*. See *unsaturated polyester*.

saturated triblock polymer
A *linear triblock polymer* consist of the structure A-B-A, where A represents a block which is a glassy, or crystalline, thermoplastics material and B represents a block which is rubbery/elastomeric at room temperature. When A = polystyrene = PS and B = butadiene rubber = BR then, SBS is the result. By modification of the BR, so as to get rid of unsaturation, S-EB-S is the result. EB is an ethylene-butylene copolymer and, as it contains little unsaturation, the new copolymer has better resistance to UV degradation and to oxidation. It has higher modulus and lower elongation than SBS and retains its tensile strength better when the temperature is raised.

Similarly, from SIS can get S-EP-S, where EP stands for an ethylene-propylene copolymer. These hydrogenated copolymers have inter-penetrating, co-continuous, polymer phases, that is, they have a thermoplastic, interpenetrating polymer network (TPIPN) structure.

saturation
Also referred to as chroma. The purity of a *colour*: the extent to which a colour departs from white.

sausage
See *rolling bank*.

saw-gin
A machine which removes the seeds from *cotton*. Saw came from the fine-toothed circular saws onto which the lint is fed: gin came from engine.

Saxton mixing section
May be referred to as a Du Pont mixing section. A *distributive mixing section* which gives good barrel wiping as it uses helical grooves: see *Dulmage mixing section*. Gives good distributive mixing with low pressure drop as the mixing section has some forward conveying action.

SB
An abbreviation used for styrene butadiene. See *styrene-butadiene block copolymer*.

SBB
An abbreviation used for *butadiene-styrene block copolymer*.

SBC
An abbreviation used for *styrene block copolymer*.

SBR
An abbreviation used for *styrene-butadiene rubber*.

SBS
An abbreviation used for *styrene-butadiene-styrene*. See *styrene block copolymer*.

SBS block copolymer
An abbreviation used for styrene-butadiene-styrene block copolymer. See *styrene block copolymer*.

SBS block polymer
See *styrene block copolymer*.

SC
An abbreviation used for super conducting (black). See *carbon black*.

scaly
Meaning *plate-like*.

scent spray pump dispenser
An old fashioned method of generating a spray which is now coming back into fashion so as to save on the use of chemicals which harm the *ozone layer*. The propelling gas (see *aerosol propellant*) is generated by hand pressure on a rubber bulb.

SCF
An abbreviation used for superconductive furnace (black). See *carbon black*.

Scheel's green
A bright green, inorganic pigment which is based on copper arsenite.

Schob's pendulum
An instrument used to measure *resilience*.

Schopper-Dalen machine
A stress/strain test apparatus used, for example, for *vulcanized rubber* (in the form of rings or dumb-bells).

Schramm apparatus
A test instrument used to determine the resistance of rubber compounds to glowing heat.

Schramm-Zebrowski apparatus
A test instrument used to determine the resistance of rubber compounds to glowing heat.

schweinfurt green
See *Paris green*.

schwerspat
See *barium sulphate*.

scorch
Premature vulcanization of a rubber compound. Signs of scorch in *rubber injection moulding* include specks of overheated material in the moulding, slow or hesitant screw and/or plunger movements. In *rubber injection moulding*, for example, the scorch resistance of a mix is controlled mainly by the accelerator system used; more than one accelerator may be used to improve speed of cure. If a mix is too scorchy, a *retarder* may be used. Blends of polymers may also help: if a natural rubber (NR) mix is too scorchy, then replace part of the NR with *isoprene rubber*.

scorch time
See *Mooney scorch time*.

scorchy
The tendency of a mix or compound to exhibit *scorch*. If a mix is too scorchy, a *retarder* may be used.

Scott plastimeter
A parallel plate type of *plastimeter*.

Scott tester
A stress/strain test apparatus used, for example, for *vulcanized rubber* (in the form of rings or dumb-bells). A Scott testing machine.

SCR
An abbreviation used for *styrene-chloroprene rubber*.

scrap
Material which cannot be re-used and must be discarded. See *reclaim*. The term is also used in the *natural rubber* industry for the rubber strips which coagulate on the tree: such

material is blended with *cup lumps* to give inferior grades of rubber - the so-called brown types. See *reclaiming processes*.

scratch hardness test
See *Moh hardness*.

screen pack
Also called a filter pack. A screen pack, supported by the *breaker plate*, is used to remove contamination from the melt and to build up pressure with a conventional, three-zone *extruder*. A simple way to improve mixing in a single screw extruder is to increase the density of the screen pack by, for example, fitting finer screens. However, the increase may not justify the loss in output, the increase in material residence time, the increase in the residence time distribution, the increase in the melt temperatures and the increase in the chances of stagnation and degradation. (There will be an increased risk of variability as the screen pack will suffer more chances of becoming blocked as the extrusion run continues). The use of screw *mixing sections*, offers a more efficient way of improving mixing. The use of longer screws gives improved mixing but perhaps, more importantly, gives scope for the use of mixing elements.

screen printing
See *silk-screen printing* and *printing techniques*.

screw
The screw consists of a round bar of steel in which is cut a helical channel (this leaves the *screw thread*): the *screw* rotates inside the extrusion *cylinder*, or barrel, and conveys the material from the *hopper* to the *die* along the *screw channel*.

screw advance
Initial screw movement during *injection moulding*. For example, on an in-line screw machine, the screw initially pushes the melt along without there being much resistance to movement: this will happen if melt decompression has been used in an effort to stop *drooling*.

screw - barrier design
A type of screw which contains a barrier to hold back unmelted material. The flight of a conventional, *single start screw* is often filled with a mixture of solid plastic and *melt*: the solid plastic floats or 'swims' in the melt and it is difficult for the screw to grip that solid resin. To improve the ability of the machine to produce melt, and to give more uniform melt, barrier design screws are used in the extrusion industry and sometimes, in the injection moulding industry. They are two start screws i.e. the screw has two separate flights which are separated by the flight land; as the resin melts it is transferred to the other flight over the narrow land. That is, the screw employs the melt pool and solid bed separation principle; this gives improved output per rpm and a lowering of melt temperature.

screw channel
The open section between the flights of the *screw*.

screw clearance
The difference in radius between the screw and the barrel: thus a clearance of 0.10 mm on a 100 mm diameter extruder will give a screw radius of 49.9 mm or, 99.8 mm diameter.

screw cooling
Means that the temperature of part, or all, of the *screw* is reduced during machine operation. Usually associated with *extrusion* as it is difficult to engineer on an *injection moulding machine*. Done by circulating water through a bore, or drilling, which comes in from the rear end of the screw. Screw cooling is done either to cure (i) a feeding problem and/or, to improve operating consistency or, (ii) to improve mixing and/or, to minimize material decomposition. In the first case it is only necessary to cool the screw in the feed section while in the second case it is necessary to bore out the whole length of the screw. That is, almost as far as the screw tip. In this case the cooling water freezes a layer of material at the screw tip and this effectively increases the *compression ratio* of the screw. Mixing efficiency is improved but the output rate goes down and the power consumption goes up. A big disadvantage in both cases is that the actual temperature of the screw is not usually known.

screw cushion
Term used in *injection moulding* when screw machines are used. Screw cushion is the amount of material which is left in the barrel, between the tip of the *screw* and the *nozzle*, after injection is completed. The amount of screw cushion is either measured in inches or in millimetres. On small injection moulding machines this cushion may be 3 mm/0.125"; on larger machines it may be 9 mm/0.375". The use of a screw cushion ensures that the *screw forward time (SFT)* is effective and a *hold pressure* is being applied.

It is very important to precisely control the amount of screw cushion as it directly affects the amount of pressure that is transmitted from the hydraulic system into the mould. If there is insufficient screw cushion, it is possible to have a *gauge pressure* and yet not have pressure on the material; this is because the screw has come up against restraints or stops. Therefore, even if the size of the screw cushion cannot be read, on the machine display, then it should be measured and recorded by the setter/operator. For example, fit a ruler to the machine and a display arrow to the injection unit: this will allow a rough measurement of screw displacement speed and screw cushion to be made. Nowadays, the screw cushion size may be controlled to within 0.1 mm/0.004".

screw - dedicated
A screw, of an *extruder* or of an *injection moulding machine*, which has been designed to suit one type of material. Common examples include screws for *nylon* (PA 66) and for *polyvinyl chloride* (PVC). Many injection moulding machines are fitted with general purpose (GP) screws and such screws are designed to suit as wide a range of plastics as possible. A GP screw will not be the ideal answer for the processing of a specific material, for example, PA 66. A screw designed for this semi-crystalline, thermoplastics material must provide a greater heat input than one designed for an amorphous, thermoplastic material. If a machine is therefore dedicated to one plastic for a long time then it is well worth considering purchasing a specially designed, or dedicated, screw.

screw delay
A feature found on some injection moulding machines whereby the screw is allowed to rest in the forward position after injection. It is only rotated at the last possible moment so as to decrease *barrel residence time*.

screw displacement rate
See *injection speed*.

screw drive
The system used to rotate the *screw*: the system used to give the desired *screw rotational speed*. The screw of an *extruder* is nearly always driven by an electric motor via a gearbox and pulleys. The screw of an *injection moulding machine* may be rotated by a (i) direct hydraulic motor, (ii) hydraulic motor via a gearbox, (iii) pole-reversing, three-phase electric motor via a gearbox and (iv), a three-phase AC electric servo motor via a gearbox. The use of the last system is claimed to save on energy usage and to permit *parallel operation*. See *electric screw drive* and *hydraulic screw drive*.

screw forward time
An abbreviation used for this term is SFT. Term used in *injection moulding*. Commences when the screw/ram starts to move forward and finishes when the *screw* starts to move back either as a result of rotation or, of *decompression*.

screw - improving mixing

A conventional screw does not necessarily mix the feed stock that well; this is because of the way that the material flows, or is transported, along the screw. The material at the centre of the flight can easily remain undisturbed, that is, the output from the machine will not be of uniform quality (it will be inhomogeneous). To improve mixing, by breaking up laminar flow, use is sometimes made of pins which protrude from the root of the screw into the plastics material or, and more usually, the tip of the screw is fitted with mixing sections or elements. The screw is in effect lengthened so that a section of rings, cams or kneading discs may be fitted. See *dispersive mixing sections* and *distributive mixing sections*.

screw machine

A type of *injection moulding machine* which has an *injection unit* which contains a *screw*. Most *injection moulding machines* are *reciprocating screw machines*.

screw mixing sections

Mixing sections fitted to a *screw* and also known as mixing elements. It is the screw and the extrusion cylinder, of a single screw machine, which interact to convey, melt and generate pressure within the plastics material. However, unless the screw is fitted with special mixing sections it is unlikely that it will do a very good job of mixing. It is essential that this is done in a controlled way as uniformly plasticized material, of constant composition, at a constant and controllable rate is required. See *dispersive mixing sections* and *distributive mixing sections*.

screw position

Term used in *injection moulding* and often called ram position. Modern injection moulding machines are usually based on *in-line screw units* and have closed loop control of the injection moulding stroke/speed. A sensor is used to track the screw position and, for example, the position of the screw with respect to time can be plotted. The information from the sensor is also fed to a controller and this asks for progressively higher injection (line) pressure so as to maintain a constant injection moulding rate. See *velocity pressure transfer*.

screw position switching

Term used in injection moulding to indicate that the final mould filling part of the moulding cycle is pressure controlled and that, the signal which initiates the change (from first stage to second stage) originates when the screw reaches a certain position. Because of its ease of incorporation, screw position switching (also known as path dependent switchover or as, screw positional control or, as positional control) is very widely used. The *screw* (sometimes referred to as the ram) is pushed forward hydraulically and this action transfers the *melt* from the barrel into the mould. When the screw reaches a certain point within the injection barrel then it trips a switch. This action replaces one hydraulic valve (set at a high, pressure value) with another, set at a lower, pressure value. Operating inconsistencies can, however, easily arise because of, for example, wear on the screw valve and/or, within the injection cylinder. By replacing the screw position switch with a pressure switch then a more accurate system is created. See *pressure switching*.

screw positional control

See *screw position switching*.

screw recovery time

Term used in *injection moulding*. Starts when the *dwell time* finishes and ends when the *screw* stops rotating during injection moulding.

screw rotation - injection moulding

On an *in-line screw injection moulding machine*, the *screw* is rotated during the cooling cycle by either an electric motor, or more commonly, by means of an hydraulic motor. This rotation causes the material to be pumped forward. The pressure developed pushes the screw back and the amount of screw travel is used to regulate the feed by making the injection assembly, for example, strike a limit switch which turns off the screw motor. By regulating the oil flow from the injection ram assembly then the *back pressure* can be increased so as to improve mixing. Keep the back pressure and the screw revolutions as low as possible so as to, for example, reduce *melt temperature* inhomogeneities.

screw rotational speed

The speed of rotation of the screw in an *injection moulding machine* or in an *extruder* and which is usually measured in revolutions per minute or rpm.

CONSISTENCY. The rate at which a screw turns controls how much material is pumped forwards. This pumping rate also controls, or affects, mixing, melt temperature and melt temperature variations. Screw speed must be therefore capable of being set accurately, it must be capable of being read accurately and it must be capable of being held to the set value. The processing machine must therefore have a display of screw speed and the screw motor must be powerful enough to keep the speed constant. On many injection moulding machines, speed variations occur because the melt temperature is too low or, because the feed form varies, or because the motor is not powerful enough.

MEASUREMENT. The speed of rotation of the screw in injection moulding machines and in extruders must be capable of being measured and displayed. May be measured by means of, for example, a *transducer* and a steel gear wheel. The gear wheel is mounted centrally on the screw so that when the screw is rotated, the teeth of the wheel pass the transducer. This creates a change in the magnetic field which is sensed by the transducer and counted.

If a machine does not have a display of screw speed then a rough indication may be obtained by marking the rear end of the screw and timing the rotation using, for example, a stopwatch. An indication of speed uniformity may be obtained with a moulding machine, by timing the speed at the beginning and at the end of the retraction stroke.

SHEAR HEAT. Some of the heat necessary to *plasticize* the plastics material (in, for example, injection moulding) is obtained as a result of rotating the screw; the faster it is rotated, the higher, in general, is the temperature. This is because the screw surface speed is increasing and as this increases so does the amount of shear. However, a high screw speed can result in uneven melt temperatures. Therefore, although high screw speeds are possible, it does not follow that a high speed should be used. It is better to adjust the rotational speed to suit the injection moulding cycle by, for example, reducing the speed of rotation to the lowest value possible. This will give more uniform temperatures, reduce wear on the machine, and reduce the residence time at the front of the injection cylinder.

Because of the importance of the *screw surface speed*, the screws on larger machines should be rotated more slowly than those on smaller machines: this will keep the shear rate down and stop localized overheating. The speed of rotation will also differ from material to material. *Unplasticized polyvinyl chloride* (UPVC) is limited to a maximum, screw surface speed of 0.25 ms^{-1} (15 m/min or, 50 ft/min), whereas with other more shear resistant materials, higher speeds can be employed.

screw size

The screws on *injection moulding machines* and on *extruders* are usually rated in terms of their diameter: measured in either inches or millimetres.

When an *injection moulding machine* is purchased, then the buyer is offered a choice of screw sizes; three are usually available and these may be referred to as A, B or C. They have different screw diameters and allow an appropriate match of *shot size* (the feed system volume plus the moulding volume) to barrel capacity. However, because they have different screw diameters, not only will the shot capacity vary but so will, for example, *injection pressure* and *plasticizing capacity*. Normally A type screws have the lowest shot capacity, and the highest injection pressure, whereas C type screws have the highest shot capacity and the lowest injection pressure.

screw speed
Usually means *screw rotational speed* which is measured in rpm. Better to use the *screw surface speed*.

screw surface speed
The speed of movement of the outside of a screw, for example, at the tips of the flights. Measured in ms^{-1} or in ft/min. The use of rpm, for suggestions for screw speed, can lead to problems as not all screws are the same size. The larger the screw, the faster will the outside surface move (when it is rotated at the same rpm as a smaller screw). This means that the screws on larger machines should be rotated more slowly than those on smaller machines: this will keep the shear rate down and stop localized overheating. To convert from rpm to ms^{-1}, multiply the rpm by the screw diameter in mm and then by 0.000 052 4. See *screw rotational speed*.

screw thread - formation of
Many injection moulded components (for example, screw caps) are required to carry a screw thread and such threads may be formed by using an unscrewing core, by moulding into the component an insert containing the thread or, by pushing a threaded insert into a moulded-in hole after moulding.

screw wear
An increase in the operating clearance of the screw/barrel system and/or wear of, for example, the screw flights. As it is easier to replace a screw than a *barrel*, the screw should be designed to wear first. Wear can be reduced by the use of *nitriding*: in the case of a nitrided screw and barrel system susceptibility to wear is approximately three times that for the screw as for the barrel.

The wear resistance of a *screw* may be improved by using the correct operating conditions and/or by nitriding. The hardness can be increased to 67 Rockwell C: this will improve wear and will also give protection against chemical attack. It will also stop plastic adhering, and then decomposing, on the screw. Ion implantation is now used as standard by some machine manufacturers because of the good wear resistance it imparts.

The screw need not be of the same composition all over as those parts which are subject to the severest use can be built from alloys which give extra protection. For example, the screw may be constructed from a central shaft of alloy steel and then it may be surfaced with a wear resistant alloy such as *Stellite*; the required flight form is then machined from this alloy.

screw whirl
See *picking up*.

screw - zero compression
See *zero compression screw*.

screw zone
A region of a *screw*. Along its length a screw amy be divided into zones or regions. Typically, general purpose plastics screws have three distinct zones: these are the feed zone, the compression (plasticating) zone and the metering (pumping) zone. The feed section starts at the rear of the hopper and has a constant depth. Then, the root diameter of the screw increases gradually from the end of the feed section to the start of the metering section where it becomes constant again. The feed section, for example, of an injection moulding screw, is approximately 50% of the length (50% L), the transition section is about 30% L and the metering section is 20% L (including the *sliding ring valve*).

scruple
A unit of measurement in *apothecaries' weight*: $1/24$ ounce Troy.

scumbing
Also called antiqueing. A printing method used to obtain a two-colour effect on the surface of an *embossed plastics product* by spreading a material of contrasting colour, onto the substrate, so that it remains in the valleys or depressions of the embossed pattern.

SDD
An abbreviation used for *sodium dimethyl dithiocarbonate*.

SDPA
An abbreviation used for *styrenated diphenylamine*.

sea mile
See *nautical mile*.

sealed end
See *capped end*.

sealing
The term sealing is sometimes used in place of *welding* when welding is accomplished using heat and pressure.

seasoned sheet
Sliced sheet which has had its solvent content reduced by storage in a warm room. See *slicing machine* and *cellulosics*.

sebacate
The reaction product of sebacic acid and an alcohol. A *sebacic acid ester*. The alcohol may be linear or branched and/or mixed alcohols may be used. A variety of sebacates may be made for use as plasticizers for materials such as *polyvinyl chloride* (PVC). See, for example, *dioctyl sebacate*.

sebacic acid
Also known as decanedoic acid. A linear carboxylic acid which is derived from *ricinoleic acid*. A precursor for *nylon 610* and its copolymers: also used to make *sebacate plasticizers*. May be represented as $HOOH(CH_2)_8COOH$. This material has a melting point of 134°C.

sebacic acid ester
The reaction product of *sebacic acid* and an alcohol and which is commonly called a *sebacate*. See *plasticizer*.

SEBS
An abbreviation used for styrene-ethylene/butylene-styrene. See *styrene block copolymer*.

sec
An abbreviation sometimes used for used for *second*. The standard abbreviation is s.

secant modulus
See *Hooke's law* and *Young's modulus*.

second
The second is the basic SI unit of time. It has the abbreviation s (although sec is sometimes used). Defined as being equal to the duration of 9,192,631,770 periods of the transition radiation between the hyperfine levels of the ground state of the caesium-133 atom. The duration of 9,192,631,770 cycles of the radiation associated with a specified transition of caesium-133. See *Système International d'Unité*.

second order transition
See *glass transition*.

second order transition temperature
See *glass transition temperature*.

second stage
See *prepolymer process*.

second stage pressure
See *hold pressure*.

second surface metallization
Part of the production process for *vacuum metallization* of plastics in which the lower, or inner surface, of injection mouldings, after treatment with a *base coat*, is coated with metal in a vacuum chamber. The reverse side of a clear moulding is coated.

secondary accelerator
An *accelerator* which is not used alone but which forms part of a mixed system. An *accelerator* which is used in combination with a *primary accelerator*: used to activate the primary accelerator or, to improve some vulcanizate properties.

secondary acetate
See *secondary cellulose acetate*.

secondary amine
An *amine* with the formula R_2NH. If R is aryl then the *amine* is a secondary aryl amine.

secondary aryl amine
A *secondary amine* where the two organic groups are aryl groups. Secondary aryl amines are the most important type of *AH antioxidant*.

secondary butyl alcohol
Also known as butan-2-ol. This material has a boiling point of 100°C and is used as a solvent. See *butyl alcohol*.

secondary cellulose acetate
Also known as secondary acetate. Cellulose which has a degree of substitution of approximately 2, that is, two of the hydroxyl groups on the *cellulose* have been reacted. More even substitution is obtained by hydrolysis of *cellulose triacetate*. At one time, this material was a major fibre (called cellulose acetate rayon or acetate) and a thermoplastics materials (cellulose acetate). See *cellulosics*.

secondary colour
A colour obtained by mixing two primary colours.

secondary flow
A term used for the circulatory flow which is common for non-Newtonian fluids: the flow lines cross the flow direction.

secondary plasticizer
A *plasticizer* which is not as compatible with a polymer as a *primary plasticizer* but is more compatible than an *extender*. Because of limited compatibility such a material may only be used in relatively small quantities as otherwise it will separate out after melt mixing. Such plasticizers are often good, *low temperature plasticizers*.

secondary structure
Agglomerates made from primary structures, or agglomerates, of a filler, such as *carbon black*. In the case of carbon black these break down on pelletization during manufacture and when a compound is prepared.

secondary transition
A transition, seen with amorphous polymers, which occurs at lower temperatures than the *glass transition temperature* and which is smaller in magnitude. May be called an a β-transition or a γ-transition: the α-transition is the *glass transition*.

section height
An abbreviation used for this term is H. Half the difference between the *overall diameter* and the nominal rim diameter of a *tyre*.

section width
An abbreviation used for this term is S. The linear distance between the outsides of the *sidewalls* of an inflated *tyre*: excludes measurements across labelling and protective features, such as bands or ribs, of a tyre.

SEDC
An abbreviation used for *selenium diethyl dithiocarbonate*.

sedimentation
The movement of particles, through a liquid, when subjected to a gravitational field. To measure the *molecular weight* of a polymer, large fields may need to be generated by centrifuging or by ultra-centrifuging. See *sedimentation equilibrium method*.

sedimentation equilibrium method
Also known as equilibrium centrifugation. A *sedimentation* method for the determination of *molecular weight* by the ultra-centrifugation of a dilute polymer solution at about 15,000 rpm. This method can give the number average molecular weight, weight average molecular weight and the z average molecular weight.

see-through clarity
The ability to see objects, or print, which are in contact with a *film* made of a thermoplastics material. A measure of the distortion and/or obscuration of an image when seen through a material. Long distance viewing is often not possible when looking through some plastics films: contact viewing is however often possible. See-through clarity may be measured by means of *Snellen charts*.

Seebeck effect
See *thermoelectric effect*.

seed crystals
See *self seeding*.

seed hair fibre
A natural fibre which is attached to the seed of a plant. See, for example, *cotton*.

seeded crystallization
See *self seeding*.

segmented copolymer
Also known as a segmented polymer. A multi-block polymer which, for example, consists of relatively long blocks of one type of unit separated by shorter blocks of the second type of unit. See *block copolymer*.

segmented polymer
See *segmented copolymer*.

segmented polyurethane
A *polyurethane* block copolymer which is a *segmented copolymer* based on alternating hard blocks and soft blocks.

segmented tyre mould
See *tyre mould*.

Sekisui process
A process used to produce *cross-linked polyethylene foam* by radiation cross-linking.

selenium
An abbreviation used for this material is Se. An element which has a melting point of 217°C, a boiling point of 688°C and a relative density (RD) of 4·8. A non-metal which resembles *sulphur* in its properties, for example, can be used as a *vulcanizing agent* (a secondary vulcanizing agent) for rubbers.

Some selenium compounds act as *accelerators* for rubber compounds as they contain available sulphur as well as selenium. See *selenium diethyl dithiocarbonate*.

selenium diethyl dithiocarbonate
An abbreviation used for this material is SEDC. This material has a melting point of 67°C and a relative density (RD) of 1·32. An ultra-fast *accelerator* for rubbers if *zinc oxide* is present.

selenium dimethyl dithiocarbonate
An abbreviation used for this material is SEMD. This solid material has a melting point of approximately 150°C and a relative density (RD) of 1·57. An ultra-fast *accelerator for rubbers*.

selenium dibutyl dithiocarbonate
A liquid material used to assist bonding in *isocyanate* bonding agents for rubbers.

selenium vulcanization
Selenium and/or tellurium could be used in place of *sulphur* for vulcanization, but, because of price, odour and toxicity considerations, such materials are not usually commercially used even though they give enhanced heat resistance.

self adhesion
The ability of a material to adhere to itself. *Natural rubber* possesses this property which is then called *tack*.

self contained
Means, in the moulding industry, that the moulding machines are driven, or powered, by a *power unit* which is located in the base of the machine. See *injection moulding machines*.

self extinguishing polymer
A polymer which will burn but only when exposed to a source of ignition. That is, burning will stop when the source is removed.

self ignition by temporary glow
A term used in the *ignition test - ASTM*. A special case of *self ignition temperature* where slow decomposition and carbonization of the test specimen results only in a glow of short duration without general ignition.

self ignition temperature
A term used in the *ignition test - ASTM*. The lowest initial temperature of air passing around the test specimen at which, in the absence of an ignition source, the self-heating properties of the specimen lead to ignition.

self reinforcement
The increase in modulus which results when *natural rubber* is stretched.

self reinforcing engineering thermoplastics material
See *liquid crystal polymer*.

self reinforcing polymer
See *liquid crystal polymer*.

self seeding
Also known as self-nucleation and seeded crystallization. Crystallization induced by residual small polymer crystallites or seed crystals. Self seeding is observed when a polymer is only partially melted before cooling and re-crystallizing. The residual crystallites in the melt act as nucleation centres and result in a higher nucleation density and a smaller spherulite size. Self seeding will improve the ductility of some thermoplastics materials (see *polyhydroxybutyrate* or PHB) as with a semi-crystalline, thermoplastics material, such as PHB, a higher nucleation density and a smaller spherulitic size is obtained. *Cold rolling* will also improve ductility.

self tuning
A term associated with control instruments. Some instruments are self tuning so that, for example, they do not overrespond or overreact to a change.

self-curing
See *self-vulcanizing*.

self-initiation polymerization
See *thermal polymerization*.

self-nucleation
See *self seeding*.

self-vulcanizing
A rubber compound, dough or cement which will vulcanize (self-cure) at room temperatures.

self-wiping
A term used in *twin screw extrusion* and which means that the screws wipe each other clean.

SEM
An abbreviation used for scanning electron microscopy.

SEMD
An abbreviation used for *selenium dimethyl dithiocarbonate*.

semi-auto
Means *semi-automatic*.

semi-automatic
This means that a machine, for example, a *blow moulding machine*, will do one cycle of operation and then it will stop until the operator starts the cycle again.

semi-conducting polymer
A polymer whose insulating properties lie in between those a conductor and of an insulator. The conductivity of such a material will be above 10^{12} Scm^{-1}. Materials with values below 10^{12} Scm^{-1} are insulators.

semi-crystalline polymer
A polymer which is only partially crystalline. Most crystalline polymer are semi-crystalline polymers. The amount of crystalline material may be expressed as a percentage.

semi-crystalline thermoplastics material
A *semi-crystalline polymer* which is also a *thermoplastics material*, for example, *polyethylene* and *polypropylene*. See *crystalline plastics material*.

semi-drying oil
An *oil* which dries more slowly than a *drying oil*, for example, *safflower oil*.

semi-embedded hose
An *externally corrugated hose* in which only the crests of the reinforcing helix are exposed within the bore.

semi-flexible polyurethane foam
A *polyurethane foam* intermediate in its mechanical properties between a *flexible polyurethane foam* and a *rigid polyurethane foam*.

semi-positive mould
A type of *compression mould* in which there is provision for *flash* escape during the *compression moulding* of *thermosetting plastics*. The flash escapes over the *flash land* which may be horizontally or vertically arranged with respect to the mouldings: this gives the sub-divisions of horizontal semi-positive mould and vertical semi-positive mould. Semi-positive moulds are now also used in *injection moulding*. See *compression-injection moulding*.

semi-prepolymer process
See *quasi-prepolymer process*.

semi-reinforcing black
Also known as semi-reinforcing furnace (black). An abbreviation used for this material is SRF. A type of *carbon black* which is moderately reinforcing. Gives rubber compounds of high elongation and low compression set.

semi-reinforcing filler
A filler which is only moderately reinforcing, for example, a *semi-reinforcing black*. See filler.

semi-reinforcing furnace black
See *semi-reinforcing black*.

semi-rigid polyurethane foam
A *polyurethane foam* whose properties (mechanical) are intermediate between those of a *flexible polyurethane foam* and a *rigid polyurethane foam*.

semi-ultra fast accelerator
See *accelerator*.

semiconductor
An electrical conductor whose resistance decreases with rising temperature and the presence of impurities. A solid material whose electrical conductivity lies between that of an insulator, for example, *polyethylene*, and a good conductor such as a metal.

semicrystalline thermoplastics material
See *semi-crystalline thermoplastics material*.

sensitisation
Part of the *electroplating* process for plastics which is also known as activation. A film of a precious metal such as palladium is deposited on the surface of the moulding from an acidic solution: the palladium is used to pull, or deposit, copper onto the plastics surface so that *electroless plating* of copper or nickel occurs.

sensitivity
A term used in *hydraulics* and which refers to the minimum input signal needed to produce a specified output signal.

separate pot mould
A type of *transfer mould* in which the transfer pot is separate from the body of the mould.

separating forces
See *roll separating forces*.

sepiolite
A type of clay which is similar to *attapulgite clay*. Both are crystalline or paracrystalline clays with chain-like structures.

SEPS
An abbreviation used for styrene-ethylene/propylene-styrene. See *styrene block copolymer*.

sequence
A term used in *hydraulics* and which refers to the order in which events occur in a system.

sequence valve
A term used in *hydraulics* and which refers to a valve which directs fluid flow in a pre-selected sequence or order.

serum
That which is obtained from *natural rubber latex* as a result of *creaming*. The serum, from creaming, contains very little rubber. See *concentrated latex*.

servo-valve
A term used in *hydraulics* and which refers to a valve whose output is automatically adjusted by a feedback signal. If the output from a pump is fed to a hydraulic actuator, via a servovalve, then the speed may be reduced to a preset value (the servovalve responds to preset electrical signals which cause oil flow restriction by moving the valve spool). See *command signal*.

servovalve
See *servo-valve*.

set
To adopt a shape: for *thermoplastics materials* this is done by cooling whereas for *thermosetting plastics materials* it is done by cross-linking. In most processing operations it is generally desirable to 'set' the polymer as soon as possible after it has been shaped. However, this can result in *frozen-in orientation*. The term set may also mean the strain remaining after complete release of the force producing the deformation in say, tensile testing. See *compression set*.

set point
That which is set on an instrument.

set temperature
That which is set on a temperature controller.

setting
That which happens when a material cures or fuses (see *cross-linked*). The process or procedure used to bring a machine into production.

setting up
See *bin curing*.

settle down
See *equilibrate*.

SFRP
An abbreviation used for *short fibre reinforced plastic*. Also used for synthetic fibre reinforced plastic. See *fibre reinforced plastics*.

SFS
An abbreviation used for *sodium formaldehyde sulphoxylate*.

SFT
An abbreviation used for *screw forward time*.

sg
An abbreviation used for *specific gravity*.

SG
An abbreviation used for *specific gravity*.

shade
The colour which results when black pigment is used with a colour or *hue*. A variety of a colour produced by changing that colour by the addition of black. See *carbon black*.

shadow graph
An optical shadow, enlargement device used to inspect the contours of blown, and injection moulded, components for quality assurance (QA) purposes. Often used to examine the thread forms on *blow moulded container* necks.

shank
Part of a screw. The shank of a *screw* is the rearward end, i.e. that part which fits into the drive and which contains the drive key.

shape modelling
The production of a model of a proposed component during the product design stage. Traditionally such models are produced by a pattern maker from, for example, wood so that the appearance and feel of a proposed design can be evaluated but it can be expensive, difficult to do well and not precisely reproducible.

Sophisticated shape-modelling systems, in the form of special computer programs, project a three-dimensional image on a cathode ray tube (CRT) display. They may be displayed from any viewpoint and can be sectioned, intersected and inter-blended with other shapes. Some of the more sophisticated programs can display the component as a multi-coloured image which resembles a photograph of the item. See *stereolithography*.

shape weaving
Term used in the *glass reinforced plastic* moulding industry to indicate that a woven preform may be produced whose shape matches that of the final component. Shape weaving has the ability to tailor width, cross-sectional thickness, fibre volumes and fibre orientations.

shaping methods - summary of
(i) Deformation of a polymer melt - either a thermoplastics or a thermosetting melt. Processes using this approach include *blow moulding, calendering, extrusion* and *injection moulding*.
(ii) Deformation of a polymer in the rubbery state - this approach is used in sheet shaping techniques such as *thermoforming* and the shaping of acrylic sheet.
(iii) Deformation of a polymer solution - either by *spreading* or by *extrusion* so as to make films and fibres.
(iv) Deformation of a polymer suspension - this approach is used in *rubber latex* technology and in *polyvinyl chloride* (PVC) plastisol technology.
(v) Deformation of a low molecular weight polymer - This approach is used in the manufacture of *acrylic sheet* and in the preparation of *glass reinforced plastics* products.
(vi) Machining operations.

In terms of tonnage those processes involving the deformation of a thermoplastic polymer melt are by far and away the most important. In turn, processes based on extrusion and injection moulding are the most common.

sharkskin
Extrudate distortion: an extrusion defect. A greater problem in industrial extrusion than *melt fracture*. The distortion consists of transverse ridges and is believed to occur as a result of the melt tearing as it exudes from the die.

CONDITIONS FAVOURING. Sharkskin appears worst when the melt is partially elastic and has the consistency of a friable cheese. Improved results may sometimes be obtained by reducing *melt temperatures* so that the melt is more strongly elastic as it emerges from the die. Alternatively some improved results have also been obtained by heating the die at the point of exit (to make the surface layers of melt more fluid) and thus making melt tearing more difficult. The severity of sharkskin may vary enormously. At one extreme, the distance between ridge and adjacent trough may be one-third of the extrudate cross-section while at the other extreme, the effect will be barely detectable to the naked eye but may show up as a matt finish or, may be felt by running a fingernail over the surface. In *blow moulding*, a rough surface on the inside of a bottle is indicative of sharkskin (the outside defect having been flattened against the wall of the blow mould).

CRITICAL RATE. Sharkskin is likely to occur above a *critical linear extrusion rate* irrespective of the die size. That is, for a particular hypothetical polymer melt, it may occur at an extrusion rate of say, 1 m/minute whatever the die size.

The critical shear rate for onset of sharkskin is inversely proportional to the die radius (R). This means that the critical shear rate is much lower with larger diameter dies. One result of this is that, although with small dies (such as those used in laboratory rheometers) melt fracture may occur at shear rates below those causing the onset of sharkskin, the reverse may be the case with typical full-scale industrial dies.

MOLECULAR FACTORS. The only molecular factor within a polymer type that appears to greatly influence sharkskin is *molecular weight distribution* (MWD). A broad distribution generally greatly reducing any tendency to sharkskin effects,

OCCURRENCE. Sharkskin occurs because within the die the melt is moving at different rates. Melt close to the wall is moving very slowly (in the case of the layer next to the wall this movement is zero). As the melt emerges, the whole extrudate moves away from the die face at a constant speed so that the outer layers are suddenly stretched and may tear.

shear
The movement of one layer with respect to another. See *shear flow* and *shear stress*.

shear flow
The most important type of flow occurring with polymer melts is shear flow. In this type of flow, one layer of melt flows over another on application of a shearing force. The process is described by the relation between two variables, the *shear rate* and the *shear stress*. The shear rate is usually designated by the Greek letter gamma plus a dot i.e. $\dot{\gamma}$. The shear stress is usually designated by the Greek letter tau (τ) and is the stress required to cause one layer to flow over another at the required rate. By dividing the shear stress by the shear rate, a viscosity (μ) value may be obtained. See *capillary rheometer*.

shear flow data
Data obtained from the use of a *capillary rheometer*. Shear flow data is used, for example, to help predict how an injection mould will fill. For this purpose the effect of changing temperature at a constant shear rate is needed together with values which show, for example, the effect of changing shear rate at a constant temperature. Shear rate is given in reciprocal seconds (that is, in s^{-1}) and the larger the number quoted the faster is the material being sheared or forced along. By plotting shear stress (Nm^{-2}) against shear rate (s^{-1}) *flow curves* may be obtained; viscosity (Nsm^{-2}) is obtained by dividing shear stress by shear rate. See *high shear rate rheometry*.

shear heat
Heat generated within a material as a result of shear or of high speed working. For example, during *injection moulding*, mould filling is often accomplished very quickly in say, less than 1 second. To do this high pressures are required and the effort that is expended is turned into heat as the material is forced or 'sheared' along. This heat is not spread uniformly throughout the material but is highest where the shear rate is the highest: that is, where the layers of material are being sheared one over the other at the highest rate. This shear heating can be so high that it can cause localized overheating, for example, in the case of *unplasticized polyvinyl chloride* (UPVC). (This is why maximum shear rates are quoted for many plastics materials.) If high *screw surface speeds* are used during extrusion, or during the plasticization stage in injection moulding, then shear heat can also be generated.

shear induced crystallization
Crystallization induced in a flowing polymer melt as a result of using a very high *shear rate*. Sometimes found in *extrusion* where increasing the extrusion rate results in a loss of output because of shear induced crystallization.

shear modulus
An abbreviation used for this term is G. Also known as rigidity modulus or as, the modulus of rigidity or as, torsion modulus. Shear modulus may be defined as the ratio of a *shear stress* to the *shear strain* so produced. Related to the *Young's modulus* (E) by $G = E/2(1 + v)$. Where v is the *Poisson ratio*.

shear rate
This is usually designated by the Greek letter gamma with a dot above the letter, that is, gamma dot or $\dot{\gamma}$. No matter what system of units is used, it is represented by 1/time (in seconds). That is, reciprocal seconds or s^{-1}. The larger the number quoted the faster is the material being sheared or

forced along. By plotting shear stress (Nm^{-2}) against shear rate (s^{-1}) *flow curves* may be obtained.

The shear rate at the wall of a die is given by:

$$\dot{\gamma}_w = \frac{(3n' + 1)}{4n'} \times \frac{4Q}{\pi R^3}$$

where, $n' = \dfrac{d\log(RP/2L)}{d\log(4Q/\pi R^3)}$.

τ_w = the shear stress at the wall of the die = PR/2L,
P = the measured pressure,
R = the die radius,
L = the die length,
$\dot{\gamma}_w$ = the shear rate at the wall of the die, and,
Q = the volumetric output rate.

See *capillary rheometer*.

shear rate relationship
For a *Newtonian fluids*, the apparent shear rate at the wall is equal to the true *shear rate* ($\dot{\gamma}_w$) but with non-Newtonian fluids, such as polymer melts, the two are not equal but are related by the equation

$$\dot{\gamma}_w = [(3n'+1)/4n']\dot{\gamma}_{w,a}.$$

Where $n' = [d\log(R\Delta P/2L)]/[d\log(4Q/\pi R^3)]$.
See *Rabinowitsch equation*.

shear rate-shear stress diagram
A *flow curve*.

shear ring
Also known as a blister ring. A *dispersive mixing section*: a region of increased thickness on a *screw* which is there to improve mixing.

shear strain
Usually denoted by γ. The change in the original right angle between two axes in a body which is caused by the application of a *shear stress*. See *shear modulus*.

shear strength
The ability of a material to withstand a *shear stress*. The stress needed to cause failure of a material in shear.

shear stress
A stress applied to a body in the direction of one of its faces. In rheology, that force which is applied to a material to cause shear flow. Usually designated by the Greek letter tau (τ) and it is the stress (i.e. the force per unit area) required to cause one layer of melt to flow over another at the required rate. The shear stress at the wall of the die is given the symbol τ_w and is equal to PR/2L, where P = the measured pressure, R = the die radius and L = the die length.

By plotting shear stress (Nm^{-2}) against shear rate (s^{-1}) flow curves may be obtained. By dividing the shear stress by the shear rate, a viscosity (μ) value (Nsm^{-2}) may be obtained. Shear stress has the units of force per unit area. 1 Nm^{-2} = 10 dyn cm^{-2} = 0·000 145 lbf in^{-2} (psi).

shear stress rheometer
See *imposed pressure rheometer*.

shear stress testing
See *imposed pressure rheometer*.

shear thickening
See *dilatancy*.

shear thinning
See *pseudoplasticity*.

shear yielding
The application of a *stress* to a material which is beyond its yield point. A distortion without a change in volume.

shear-cone compounder
A single-screw continuous compounder (a Plastificator) widely used in the *plasticised polyvinyl chloride* (PPVC) industry. The screw conveys material to an integral shear-cone which is fitted with spiral fins: this shear-cone fits inside a tapered housing and the gap between the two can be adjusted to give the required amount of *melt mixing*.

shear-deformation behaviour
The shear-deformation behaviour of an ideal fluid is derived by assuming that the *Newtonian fluid* is contained between two plates of very large area A. When a shear force (F) is applied to the top plate then the shear stress is equal to F/A. The top plate moves with a velocity u: the shear rate is equal to $\dot{\gamma}$, or gamma dot, = du/dr. Shear stress is directly proportional to shear rate for an ideal or Newtonian fluid. See *stress-deformation behaviour*.

shear-induced crystallization
See *shear induced crystallization*.

shearing disc viscometer
See *Mooney viscometer*.

shearing - effect of
See *melt processing* and *orientation*.

sheet
An individual piece of sheeting; sheet usually has a thickness greater than 0·010 in/0·25 mm.

sheet die
See *sheet extrusion die*.

sheet extrusion
The production of *sheet*, from a *thermoplastic material*, by *extrusion*. Sheet is extruded through a heated *sheet extrusion die* and, after leaving the die, it is cooled gradually, for example, by passing it over a series (a serpentine) of water-cooled rolls or tubes. With some materials, for example, *polypropylene* and *polystyrene*, it is desirable to use a series of polishing rolls to achieve sheet with a glossy surface finish. The haul-off equipment should be designed to prevent distortion of the sheet during cooling and to minimise draw-down. Rigid sheet can be cut to length with a travelling circular saw.

Corrugated sheet, with transverse corrugations, can be produced by incorporating a corrugation unit in the haul-off. A system is also available for forming longitudinal corrugations by using water-cooled forming dies.

Embossed sheet, for example, embossed acrylic sheet for lighting fittings, can be produced by passing the extruded sheet between a pair of embossing rolls mounted close to the die, and air-cooling the sheet as it leaves the die in order to retain a sharp impression.

sheet extrusion die
Also known as a sheet die. A *die* used to make plastics *sheet by sheet extrusion*. The output from the extruder may be fed into a *manifold*, contained within the die, so as to promote material flow to the edges of the die. To achieve sheet of uniform thickness the melt flow is then adjusted by means of a *restrictor bar*. Such a die may also be called a manifold die and is used, for example, to produce wide *high impact polystyrene sheet* for *thermoforming*.

sheet impact strength
See *falling weight impact strength*.

sheet measurement - dual gauge system
The use of two types of *radiation gauge* for product monitoring. For example in wire covering, by *calendering*, the sheet may be scanned by first a beta gauge and then by a gamma gauge. Dual gauges are needed because beta and gamma rays respond differently to rubber and to steel. A gamma gauge

responds primarily to wire weight but is slightly affected by rubber weight. A beta gauge is affected by the wire but responds mainly to rubber weight. Thus neither sensor alone is capable of providing an accurate selective measurement of the individual weights. However since the response equations both contain information on the two weights they can be solved simultaneously to determine the individual weights. Such equation solving is handled by a digital computer and the calculated information may be displayed and/or used to actuate a control action. for example, *nip* alteration.

sheet moulding compound
An abbreviation used for this type of material is SMC. A fibre-reinforced thermosetting compound in sheet form. The resin is usually a polyester and the fibre is usually glass. SMC can be compression moulded into complex shapes with little scrap. As this is a polyester moulding compound, then SMC is sometimes referred to as a PMC.

SMC is made from an *unsaturated polyester (UP) resin* in styrene: isophthalic, and orthophthalic acid, based resins are used. This resin solution, tailored to tight tolerances, is used to make a *polyester moulding compound (PMC)*, by mixing together the resin solution, fillers, thickening agents (also known as *maturation agents*), mould release system, inhibiting system, a thermoplastics material and a catalyst system.

The mix is spread onto two, moving, plastic sheets (PE or PA) at the same time as *glass* (usually rovings) is chopped onto the lower sheet. The two plastic sheets are then passed through the nip of two rolls which consolidates the moulding compound and makes a sandwich (a sheet) of it between the two plastic sheets. The compacted sheet is then wound on to a roll and set aside to thicken or mature at 30°C/84°F. This maturing action is brought about by the *maturation agent* (for instance, magnesium oxide) and, results in a loss of tack and an increase in strength. As a result, the covering sheets can be stripped off just before moulding (keep on as long as possible so as to prevent contamination and inter-sheet adhesion).

The fillers used are usually calcite, dolomite and aluminum trihydrate (ATH): they lower the cost, and control the rheology, so that fibre separation does not occur on moulding. A blend of fillers such as ATH and calcite may also be used so as to obtain controlled rheology and flame retardancy.

The addition of a thermoplastics material, reduces shrinkage and improves the surface finish - the use of a thermoplastics material results in materials which are called *low shrinkage sheet moulding compounds* and *low profile sheet moulding compounds*. SMC mouldings may undergo virtually no *post moulding shrinkage* and are therefore dimensionally stable: this feature also minimizes component cracking.

UP resins are polar and, because of this, they have an affinity for metal (mould) surfaces; this adhesive tendency can be counteracted by adding a *stearate* which melts just below cure temperature - magnesium *stearate* melts at 130°C/266°F. Catalysts (initiators) are essential for curing and are usually peroxides: mixed *peroxide* systems are commonly used in thick sectioned mouldings where cracking can be a problem. Glass fibres are widely used, for example, in the size range 12–50 mm/0.5 to 2" (usually 25 mm/1") and at standardized concentrations from 15–40%. Considerable improvements in strength, and other properties, result from high *glass fibre* concentrations: this is why the glass content may reach 65% - although this may only be a local area of reinforcement. Inhibitors, such as hydroquinone, are necessary to give the required storage life and to stop premature cure when the mould is being loaded.

SMC is often seen in natural colours (for example, grey and beige) as in many applications (for instance, under the bonnet) there is no demand for bright colours. SMC is turned into large area components by *compression moulding* as this process maintains fibre integrity. SMC compression mouldings are usually large components such as exterior car parts. There is great interest in injection moulding SMC as the products are mechanically stronger than DMC - due to longer residual fibre length and higher glass contents.

SMC is resistant to high temperatures and outdoor exposure. SMC resists water, alcohols, aliphatic hydrocarbons, detergents, lubricants, greases and oils. May resist weak acids, esters, benzene and boiling water. SMC has excellent resistance to arcing (compares favourably with epoxides and melamine-formaldehyde). The modulus drops as the temperature is raised and glass content has little effect. To improve temperature resistance, use heat-resistant resins and continuous glass fibres. High glass contents (up to 70%) improve stress resistance and the use of ATH can give flame retardancy to UL V-O.

SMC is not resistant to ketones and chlorinated hydrocarbons: the resistance to aromatic hydrocarbons, strong acids and alkalis is also not very good and decreases as the temperature increases. In their resistance to moisture, *polyester moulding compounds* (PMC, i.e. SMC and DMC) are often worse than some reinforced thermoplastics.

The density is approximately 1.8 gcm^{-3} and is similar to *dough moulding compound (DMC)*. SMC contains more glass fibre than DMC but less inorganic filler (other ingredients are similar): the glass fibres are much longer. Typical composition resin 100, particulate filler 150, glass fibre (approximately 25 mm/1 mm length) 80.

Components requiring bright colours are normally produced by painting the moulded components after moulding. For the car industry, SMC components require degreasing, priming, painting and force drying at elevated temperatures as the painted components must withstand arduous tests, e.g. 10 days immersion in water at 40°C/104°F.

SMC gives mouldings with excellent dimensional stability, high mechanical properties (superior to DMC/BMC), good chemical resistance and electrical insulation. Used for the volume production of high strength, compression mouldings such as lorry automotive body parts, electrical housings, chemical trays, furniture, light fittings, etc. Minimal shrink grades can produce mouldings which may be painted to automotive standards of finish after a suitable degreasing and priming operation. The use of *in mould coating* gives scratch resistant mouldings. See *high performance sheet moulding compound*.

sheet moulding compound moulding
An abbreviation used for this process is SMC moulding. See *sheet moulding compound*.

sheet moulding compound recycling
See *polyester moulding compound recycling*.

sheet moulding compound regrind
An abbreviation used for this type of material is SMC regrind or SMC-R. The ground material obtained when a component made from a *sheet moulding compound* is ground or reduced in size. See *particle recycling*.

sheet path
See *feed-path*.

sheet polymer
See *ladder polymer*.

sheet production processes
In terms of tonnage, two processes dominate thermoplastics film and sheet production: these two major processes are *calendering* and *extrusion*. Broadly speaking extrusion is used to produce the thinner sheet, based usually on *polyethylene* (PE) and *polypropylene* (PP), and calendering is used to produce the thicker sheet which is often based on *polyvinyl chloride* (PVC). Approximately 25% of all PVC passes through a

calender and as PVC is a *commodity thermoplastics,* calendering is an important process.

Where long runs and high outputs can be achieved, *calendering* is probably the most economic means of converting polymer into *sheet*. In the case of conventional rubbers (as opposed to *thermoplastic elastomers*) calendering is often the only process available for the production of wide rubber sheets.

sheet rubber
Term may be applied to uncured *natural rubber* in sheet form, for example, *ribbed smoked sheet*.

sheet-shaping
Processes used to shape flat thermoplastics sheet material into products of three dimensions. See *thermoforming* and *embossing*.

sheeter line
A fault which is also known as a knife line. Lines produced as a result of a slicing operation. See *cellulosics*.

sheeting
A term used for continuous lengths or rolls of material. A form of plastics material in which the thickness is very small in proportion to length and width and in which the plastics material is present as a continuous phase throughout - with or without filler. (See also *film*.)

sheeting battery
A production unit of say, five two-roll, *washing mills* used to produce 3 mm/0·12 in sheet from the soft gel or *coagulum* which results when *natural rubber latex* is coagulated.

sheeting rolls
A pair of large warmed rolls used to produce sheet from *dough* and used after the *hardening rolls*. The thick sheet so produced from dough is pressed in a *block press*. See *cellulosics*.

shelf ageing
Ageing which occurs during storage.

shelf life
The time for which a compound will remain useable when stored under specified conditions, for example, normal atmospheric temperatures and pressures. The useful working life of a polymeric system: the time during which a polymeric system may be turned, or moulded, into components with acceptable properties. Thermosetting moulding compositions often have a maximum permitted shelf life of, say, 6 months. For rubber compounds, the term means the time for which the unvulcanized compound may be usefully kept

shelf life improvement
See *coating of blow mould containers*.

shellac
An example of a *natural resin* which is used as a plastics material. This material has a relative density (RD or SG) of approximately 1·15 and a melting point of about 75°C. Obtained from secretions of the lac insect. A hard thermoplastics material which has good abrasion properties and when modified (with fillers such as *slate*) has low shrinkage. At one time widely used to mould gramophone records.

May be used as an additive in rubber compounds as it acts as a *processing aid* where it eases extrusion and reduces shrinkage. As it also improves electrical properties, such as dielectric properties, it is useful for cable manufacture. 10% addition will increase the hardness of vulcanizates at the expense of a slight sacrifice in tensile strength and modulus. Such shellac-based compositions are of use in the footwear industry for soles and heels. When shellac is used, the formulation will have to be adjusted: as the shellac combines with some of the sulphur (about 3% of the shellac weight) and reacts with alkaline *accelerators*.

shielding effect
See *electromagnetic interference*.

shoe sole crepe
Comparatively thick sheets of rubber produced by building up the required thickness from thinner layers of *pale crepe*. See *natural rubber*.

Shoe and Allied Trades Research Association
An abbreviation used for this organization is SATRA (UK based).

Shore A
See *Shore hardness*.

Shore D
See *Shore hardness*.

Shore durometer
A *durometer* used to test soft compounds for *hardness*. An instrument for measuring *Shore hardness*. A spring based instrument used for *hardness measurements* and which is readily portable.

Shore hardness
A measure of the hardness or softness of a material as assessed by the resistance (on a scale of 0 to 100) experienced by an indentor. 100 on the scale corresponds to a high resistance (that given by a sheet of glass) and 0 corresponds to zero resistance). Two different indentors are used: type A for soft materials and type D for hard materials. This means that there are two scales of Shore hardness, Shore A and Shore D.

Type A measurements are made on soft materials using a truncated 35° cone with a blunt tip of 0·79 mm diameter, whereas Type D are made on harder materials with a 30° steel cone rounded to 0·1 mm radius tip as indentor. The depth of penetration is inversely indicated in thousandths of an inch. The harder the surface the higher the reading.

The specimen must be conditioned prior to testing and it has to be sufficiently flat over an area of at least 6 mm radius from the indentor point to allow contact with the foot of the instrument. It must also have a thickness of at least 6 mm unless it is known that identical results can be obtained with a thinner specimen. Measurements are made at several points over the surface simply by pressing the instrument against the surface of the specimen. Readings should be taken after a standard period of time, the ISO recommends 15 seconds and the ASTM one second.

short
An imperfection in a moulded plastic part due to an incompletely filled out condition. In thermoplastics, the term 'short shot' is often used to describe this condition. In reinforced plastics, this defect may be evident either through an absence of surface film in some areas, or as lighter unfused particles of material showing through a covering surface film, accompanied possibly by thin-skinned blisters.

short chain branching
Side branches on a main molecular chain which are only a few repeat units long. See *linear low density polyethylene*.

short fibre composite
Also known as a chopped fibre composite. A composite in which short, discontinuous fibres supply the reinforcement. The most common type of reinforcement is *glass fibre* but whiskers can also be used. Both thermoplastics materials (for example, *glass reinforced nylon*) and thermosetting plastics materials (see *glass reinforced polyesters*) are available.

short fibre reinforced plastic
An abbreviation used for this material is SFRP. The term often refers to thermoplastics materials which are filled with comparatively short glass fibres, See *short fibre reinforced thermoplastics material*.

short fibre reinforced thermoplastics material
An abbreviation used for this type of material is SFRTP. Very short, discontinuous fibres supply the reinforcement: the most common type of reinforcement is *glass fibre (GF)* but whiskers can also be used. A common example is *glass reinforced nylon*.

Such a material is often produced by adding *masterbatch*, containing short glass fibres (fibre length less than 1 mm) to an unfilled material during, for example, *injection moulding* (better dispersion is obtained by extrusion compounding the GF and the *thermoplastics material* together and then chopping the extrudate into pellets). The final fibre length is much less than the original length, for example, 0.2 mm after injection moulding. If the glass content is high, at say 33%, strong stiff mouldings with a high *heat distortion temperature* may be produced: however, the impact strength of the injection mouldings may be comparatively low and the surface finish is not as good as that from the unfilled material. See *long fibre reinforced thermoplastics material*.

short glass fibres
Glass fibres which are usually less than 1 mm in length. See *short fibre reinforced thermoplastics material*.

short hundredweight
A unit of weight which is equal to 100 pounds. In the US, 100 pounds = 1 hundredweight. See *US Customary Measure*.

short oil alkyd resin
An *alkyd resin composition* which contains less than 50% oil. Used to make coatings which, on drying, are hard and glossy. See *oil length*.

short oil length urethane oil
A *urethane oil* based on a diol ether of low molecular weight or, on a *polyoxypropylene diol* of comparatively low molecular weight.

short oil resin
A type of *alkyd resin*. See *short oil alkyd resin* and *oil length*.

short stop
A *free radical* scavenger which is added to a polymerization system to stop the conversion from monomer to polymer. For example, *hydroquinone*.

short term test
A test which only gives information about test specimen behaviour over a short time period, for example, a *tensile* or *flexural test*. Such tests are not capable of giving information which could be used, for example, in the design of a continuously stressed component in a particular environment. See *creep testing*.

short ton
A unit of weight which is equal to 2,000 pounds = 1 ton (US). This is equal to 907.2 kg or 0.907 metric tonnes.

shorting
A moulding fault: incomplete filling.

shot
The total amount of polymer which is capable of being moulded at one time or, that which is being moulded.

shot weight
The total weight of polymer which is capable of being moulded at one time: includes both the feed system and the components

shoulder
The transitional area between the *sidewall* and the *tread*.

shrink mark
An imperfection, a depression in the surface of a moulded material formed where the material has retracted from the mould. Shrink mark reduction is obtained with *gas injection moulding*.

shrink wrapping
A packaging technique which utilises *orientation*; the components are placed inside a bag and heated so as to cause shrinking (tightening) of the thermal plastics *film*.

shrinkage
The reduction in volume that occurs when a moulding is ejected from a mould, and allowed to cool, is known as shrinkage or as, moulding shrinkage. It is expressed as a percentage or as a ratio (for example, mm/mm or in/in). Because of shrinkage most plastics components are smaller than the mould used to produce them.

The *shrinkage* found with an amorphous, thermoplastics material, such as *polystyrene* (PS), is totally different from that found with a semi-crystalline, thermoplastics material such as *high density polyethylene* (HDPE). For PS it is often 0.6% whereas for HDPE it may reach 4%. This is because when polymer molecules crystallize, the molecules tend to pack more efficiently than they do in the amorphous state so that higher shrinkage values result.

Shrinkage may also be different from one grade of material to another and will certainly be influenced by changing the processing conditions. It may also be significantly different in different directions, for example, *across the flow* and *along the flow*. Because of this it is usual to quote a shrinkage range for each plastics material. Often the shrinkage values quoted in the literature are based on injection moulded components. See **Table 8**.

shrinkage control additive
See *low profile sheet moulding compound*.

shrinkage - measurement of
A widely used test is ASTM D 955-51 which covers both thermoplastics and thermosets. When *compression moulding* is the moulding procedure then a positive mould which produces specimens measuring either $12.7 \times 12.7 \times 126$ mm or a disc of diameter 102 mm is used. For injection moulded specimens, an end-gated bar of thickness 3.2 mm is usually used. If a transfer mould is used the bar is $12.7 \times 12.7 \times 127$ mm and is end-gated or, gated at the top (near one end) so as to provide flow throughout the length.

The cavity length or diameter, both parallel and perpendicular to the flow direction, are measured to the nearest 0.02 mm at $23 \pm 2°C$. At least three test specimens are then moulded. After moulding the specimens are allowed to cool to room temperature; the cooling time (before initial mould shrinkage is measured) is 1 to 2 hours when the thickness is 3.2 mm and 4 to 6 hours when the thickness is 12.7 mm. The length or diameter (both along and transverse to flow for injection and transfer moulding) are measured and then the specimens are returned to storage in a standard laboratory atmosphere (23°C at 50% RH). Measurements are again made at 24 hours and 48 hours after moulding in order to obtain 24 hour shrinkage and the 48 hour (or normal mould) shrinkage.

Results are usually given as mm/mm (same as in/in) and this is obtained by subtracting the dimension of the sample from the corresponding dimension of the cavity and then dividing by the latter. The mould shrinkage (MS) is given as a percentage if the following formula is used $(L_0 - L_1 / L_0) \times 100$. Where L_0 is the length in mm of the cavity L_1 is the length in mm of the moulding (to the nearest 0.02 mm).

The values quoted (see **table 8**) should only be taken as a guide as in the case of injection moulding, for example, part thickness, cavity pressure and the time for which that pressure is applied, all markedly influence moulding shrinkage.

shrinkage - prediction of
Predicting the *shrinkage* of some materials can be very difficult because of high and non-uniform shrinkage: some

materials also experience *post-moulding shrinkage* (see *high density polyethylene* or *HDPE*). Because of the wide shrinkage range of HDPE it is extremely difficult to accurately predict the necessary shrinkage value in order to achieve the desired dimensions for a particular component. The shrinkage value is, for example, dependent upon the degree of orientation and the crystallinity that occurs in the moulded component.

Because of this uncertainty a sample cavity is often manufactured and mouldings produced using typical processing conditions (i.e. using those temperatures, pressures and cycle times used in production). The shrinkage values obtained from these mouldings can then used for the sizing of the cavity, and core dimensions, of the production mould. However, what must be taken into consideration is the degree of after-shrinkage that occurs with materials such as HDPE. What this means is that the sample mouldings should be left for several days to condition before being measured. Mould temperature has a great effect on shrinkage and must therefore be controlled very precisely.

shrinking
The controlled heating of rubber sheet produced by *calendering* so as reduce *anisotropy* and prevent subsequent distortion.

shut down procedure
The procedure involved in removing a machine from production. It is most important to adopt a sensible shut down procedure as such shut down procedures can save a great deal of time and money. If, for example, the material/resin is prevented from burning then there is not so much purging out to do on re-starts and the cost of a complete shut down, and machine clean out, may be saved. See *stoppages*.

shut-off
See *flash land*.

shut-off nozzle
A type of injection moulding machine nozzle: a *nozzle* which contains a shut-off valve. This valve is closed when the nozzle is not in contact with the mould.

shutting down
Ending production. See *shut down procedure*.

shuttle mould
A type of *injection mould* used to produce components at high rates where long loading or unloading times are involved. One half of the mould is duplicated and the system is operated so that two moulds are, in effect, used. Whilst one cavity plate is being used for moulding the other is outside the press being stripped of its mouldings. See *shuttle press*.

shuttle plate mould
A type of *three-plate mould* which has a steel frame located around the mould. The frame projects beyond the mould and carries two identical cavity plates in individually machined slots. This type of *shuttle mould* is used for *rubber injection moulding*.

shuttle press
A type of *rubber injection moulding machine* used to produce components at high rates where long loading or unloading times are involved. Such a press has two lower heated platens and two associated mould halves. See *shuttle mould*.

shuttle press mould
See *shuttle mould*.

Si
An abbreviation used for silicon. See *silicone rubber*.

SI
An abbreviation used for *silicone plastics*.

SI units
An international coherent system of units now widely used for scientific and technical work. See *Système International d'Unité*.

SiC
An abbreviation used for *silicon carbide*.

side cavity
A part of a *cavity* which can be withdrawn at right angles to the mould axis: in this way, a projection below the *parting line* can be moulded.

side cavity assembly
A *side cavity* element fitted to a *carriage*.

side cavity element
A steel member which contains the *impression*.

side core
Part of an *undercut mould* used to put a hole or slot into a moulded component. The *core* is normally mounted and operated at right angles to the mould's axis. Side cores may also be used as an alternative to a rotating core for the moulding-in of an external screw thread. The major disadvantage of this method of moulding an external screw thread is that the mould parting line is clearly visible on the component.

side core assembly
A *side core* element fitted to a *carriage*.

side core element
A steel member which contains the *impression*.

side frame
See *calender side frames*.

side gate
This type of *gate* is also known as a standard gate and is commonly used with multi-impression moulds as it allows the usage of a *two-plate tool*. The gate usually has the same type of cross-section as the runner system (round or trapezoidal) and its size is dependent upon the shape and thickness of the moulding.

sidewall
A *tyre component*: that part between *tread* and *bead*.

sidewall rubber
The rubber layer on the *sidewall* of a *tyre*.

siemens
An SI derived unit which has a special symbol, that is S. It is the SI derived unit of electrical conductance. One siemens is equal to the conductance of a circuit or element that has the resistance of one ohm. This unit was formerly called the reciprocal ohm or mho. See *Système International d'Unité*.

sievert
An SI derived unit which has a special symbol, that is Sv. It is the SI derived unit of dose equivalent. One sievert is the dose equivalent when the absorbed dose of ionizing radiation, multiplied by a stipulated dimensionless factor, gives one joule per kilogram. See *Système International d'Unité*. The former unit of dose equivalent was the rem: the rem is equal to 10^{-2} Sv.

silane
A silicon-based material with the formula SiH_4. See *silanes*.

silane coupling agent
A silicon-based material which couples, or bonds, a *filler* to a resin matrix and so improves properties such as modulus. Such a material has organic groups (R) and *functional groups* (X) attached to a central silicon (Si) atom. Typically the formula of such a material may be represented as $RSiX_3$. The X groups (for example, chlorine or acetoxy) are hydrolysed to

silanol which provide the adhesion to the filler while the R groups provide adhesion to the polymer. It is found that the R group must be selected to suit the polymer, that is, there is no universal coupling agent. For example, γ-methacryloooxypropyltrimethoxysilane is often used for *unsaturated polyester resins systems* whereas γ-aminopropyltriethoxysilane is often used for *epoxide resins*. See *accelerator deactivation*.

silanes
Silicon-based materials which consists of two, or more, *silane units joined together*. See *organosilanes, disilane* and *trisilane*.

silanol
A silicon-based material which has an hydroxyl group (OH), or hydroxyl groups, attached to a silicon atom. For example, H_3SiOH. See *organosilanol*.

silanol terminated polydimethylsiloxane
A dialkylsiloxane, or a *polyorganosiloxane*, with terminal hydroxyl groups.

silica
Also known as silicon (1V) oxide and silicon dioxide. Pure silica is *silicon dioxide* or SiO_2: the term silica is also used to mean hydrated or hydroxylated amorphous forms of silica. Silica occurs naturally in great abundance. Quartz is a pure form of silica: sand and flint are other natural sources which are less pure. *Diatomaceous earth* (diatomic earth) and *chert* are other natural forms of silica.

Many forms of silica are amorphous. *Silica gel, precipitated* and *fumed silica* are amorphous forms of silica which may be considered as being condensed polymers of *silicic acid*. Naturally occurring forms of silica (such as *diatomaceous earth, rottenstone* and *tripoli*) are micro-crystalline, or *cryptocrystalline*, forms. Crystalline forms of silica are obtained by, for example, crushing, pulverizing and purifying quartzite. The chemical inertness, and purity (> 99·6% silica) of this crushed compound, make it a useful reinforcing filler for *silicone rubber*.

Synthetic silica is a fine, white material (for example, *precipitated silica*) which is used as a reinforcing filler in rubbers but is relatively expensive (compared to materials such as *whiting*). The use of finely-divided silica in rubber compounds, often surface treated to improve dispersion and adhesion, imparts hardness, wear and tear resistance: this type of filler is particularly used in the footwear industry. More expensive than *carbon black* and is therefore used where tough, light coloured components are required. Because of the absorbed water, longer mixing times may be needed compared to carbon black and this may cause excessive heat generation. Accelerator levels may also need to be increased because of *accelerator deactivation* although surface treatments minimise this effect. Because of the fine structure of the silica, the uncured stock may also be stiffer and additional plasticizer may be required. However, considerable reinforcement is possible in rubber compounds. This material has a relative density (RD or SG) of approximately 1.95.

Synthetic silica can have a very fine, primary particle size (8 to 50 μm) but these fine, primary particles associate into larger secondary particles. (These larger secondary particles may be hollow and can give high *nitrogen absorption* readings thus indicating a large surface area: all of this large surface area is not however, available for bonding to the rubber.) The secondary particles associate in turn, into tertiary structures which may resemble chains. The secondary particles are not broken down into the primary particles during mixing or processing but the tertiary structures can be made smaller, and dispersion improved, if sufficient shear is available.

silica gel
Three dimensional networks of aggregates of *silica*. Simple removal of water from silica results in a lot of shrinkage so as to give xerogels. If, before dehydration, the water is replaced with alcohol and then the drying is done under pressure, the shrinkage is reduced and a much more bulky form, known as an aerogel, results.

silica reinforcing filler
See *crystalline silica*.

silica sol
See *colloidal silica*.

silicate
A *salt* derived from *silicic acids*. Usually means a *silicate filler*.

silicate filler
A *filler* based on a *silicate* which may be natural or synthetic. Natural silicate fillers include *china clay* and *talc*. Synthetic silicate fillers include hydrated *calcium silicate* and *hydrated aluminium silicate*: these may be prepared by a modification of the *precipitation process*.

silicate mineral filler
Fillers based on silicate minerals. A category of *filler* which includes *clays, talc* and *asbestos*.

silicate polymer
See *polysilicate*.

siliceous earth
A pale coloured material which is a *semi-reinforcing filler* for rubbers. The *quartz* to *kaolinite* ratio is 3:1 and the relative density (RD or SG) ranges from 2·2 to 2·6.

silicic acid
May be represented as $Si(OH)_4$. This material may be considered the basis for *silicate polymers*. Silica gel, precipitated and fumed silica are amorphous forms of silica which may be considered as being condensed polymers of silicic acid. See *precipitation process* and *silicic acids*.

silicic acids
Hydrated forms of *silica* obtained by the action of acids on soluble silicates. For example, can have orthosilicic acid and metasilicic acid which in turn give rise to orthosilicates and metasilicates. See *silicic acid*.

silicon
This element (Si) occurs in Group 1VB of the Periodic table along with *carbon, germanium, tin* and *lead*. It does not occur naturally but is very commonly seen as, for example, silicates. It forms an incredible number of compounds and, like carbon, this element is remarkable for its ability to form long chain compounds. It is a hard gray lustrous solid which melts at approximately 1,400°C and which has a relative density of 2·35. It is very acid resistant and conducts electricity with difficulty. Compounds of silicon are widely used in the polymer industry as fillers. See, for example, *silica*.

silicon (1V) oxide
Silicon dioxide. See *silica*.

silicon carbide
Also known as carborundum or SiC. A hard, black material commonly used as an abrasive. A silicon-carbon polymer which is available in various crystalline forms (α and β). SiC is available in *fibre* and in whisker form. SiC has high structural stability at high temperatures which makes it useful as fibre or whisker reinforcements for high temperature composite materials. Used to reinforce metal-matrix composites (MMC) and ceramic-matrix composites (CMC).

Can have substrate-based fibres and fine ceramic fibres. Substrate-based fibres are based on a 100 to 150 μm filament of tungsten (SiC/W fibre) or, on a 35 μm filament of carbon (SiC/C fibre) and are produced by a *chemical vapour deposition process*: carbon filament is cheaper than tungsten. Can use CH_3SiCl_3 as the reactant. Fine ceramic fibres are produced by the pyrolysis of a polycarbosilane precursor to give a fibre diameter of approximately 15 μm.

SiC-β whiskers have a relative density (RD or SG) of 3·15, a melting point of about 2316°C, tensile strength of 7 to 35 GPa and Young's modulus of approximately 620 GPa.

silicon controlled rectifier
Also known as a thyristor. A four-layer, silicon rectifier with a third, control electrode which prevents conduction until that third electrode is switched, or triggered by a trigger pulse.

silicon dioxide
May be represented as SiO_2. Each silicon (Si) atom is at the centre of a tetrahedron of four oxygen (O) atoms and each oxygen atom is midway between two silicon atoms. Silicon dioxide is therefore a macromolecular material whose molecular weight can be extremely high. See *silica*.

silicon functional silane
An *organosilane* in which a functional group is attached directly to silicon: such groups (for example, chlorine) are readily hydrolysed to *silanols*. See *organofunctional silane*.

silicon nitride
A silicon-nitrogen polymer (SiN): a heat resistant polymer which has been used, for example, to make turbine blades for engines. The material may be fabricated by dispersing a fine powder form in a more conventional polymer (for example, *polyethylene* or PE): the resultant compound may then be injection moulded into the shape required. The moulding is then heated to burn off the PE and the shape is then fired, at high temperatures, to give the silicon nitride moulding. SiN is also available in *whisker* form. Such whiskers have a melting point of about 1900°C, a tensile strength of 3 to 11 GPa at a Young's modulus of 380 GPa.

silicone
See *polyorganosiloxane, silicone rubber* and *silicone resin*.

silicone elastomer
See *silicone rubber*.

silicone fluid
Also known as silicone oil. Based on *polyorganosiloxanes* which are liquid or paste-like at room temperatures. The silicon atoms have two organic groups (R & R') attached: the most common polymer is *polydimethylsiloxane* where R = R' = -CH_3. This gives rise to methylsilicone fluids which have low volatility and good resistance to oxidation. For example, they can withstand heating in air at 150°C for very long periods: they can withstand heating in an inert atmospheres at 200°C for very long periods. Materials which contain phenyl groups (approximately 45% phenyl groups) are also used which are more heat resistant than the methylsilicone fluids. Silicone fluids may be dissolved in hydrocarbon solvents and in chlorinated solvents. Silicone fluids are used as lubricants, mould releases, heat transfer fluids and as pressure transfer fluids. Have a stable viscosity/temperature relationship.

silicone laminate
A *laminate* based on a *silicone resin* (often a methyl-phenylsilicone resin) and glass cloth. After cleaning, the glass cloth is impregnated with a catalyzed silicone resin solution and after solvent removal, this is partially cured by heating. The impregnated reinforcement is then plied into layers and bonded together by heat and pressure. Both high pressure (7 MPa/1,000 psi) laminates and low pressure (0·1 MPa/15 psi) can be made by the correct choice of resin. Cross-linking is achieved, by heating with a catalyst (for example, cobalt octoate) to give a cross-linked *polyorganosiloxane*: the laminates are used in the electrical industry. See *silicone polymer*.

silicone mould release agent
A mould release agent based on a *silicone polymer*. Often such a mould release agent is an aqueous emulsion of a silicone fluid. The emulsion, a 1% concentration may be used for rubber *compression moulding*, is sprayed onto the hot mould surface where it forms a thin silicone film which assists in mould release. Can also obtain *bake-on mould releases* (see *silicone resin*) and *silicone solutions*. Silicones can cause stress cracking of some thermoplastics mouldings and even minute traces, on the surface of injection mouldings, can cause problems in the micro-electronics industry.

silicone oil
See *silicone fluid*.

silicone oil containing compound
A polymer compound which contains a silicone oil. Silicone oils are added to thermoplastics materials to ease mould filling and so reduce *clamping force* requirements. It should be noted that silicone oil containing compounds are not allowed in some establishments because they can interfere with the working of electronic components. However, the use of such oils gives high gloss products which maintain this attractive appearance in use: this is because components moulded from such modified materials, for example, tape-cassette boxes, are often shipped in bulk and the presence of the oil stops the injection mouldings damaging each other, by rubbing, in transit. See *lubricant*.

silicone plastics
Plastics materials based on *silicone resins* (methyl-phenylsilicone resins), a *filler* (a heat resistant fibrous filler, for example, such as *glass fibre* and an inorganic particulate filler) and a catalyst, for example, (cobalt octoate). An abbreviation used for this type of material is SI.

Such materials are commonly moulded by *compression moulding* into components for the electrical industry which are needed to withstand high temperatures. Moulded at temperatures of approximately 160°C for 5 to 20 minutes: post-curing is usually necessary to develop the best properties.

silicone polymer
In general, such a material is noted for its high thermal stability, resistance to oxidation, resistance to chemicals and non-stick properties. See *silicone rubbers* and *silicone resins*.

silicone resin
Also known as methyl-phenylsilicone resins. Such resins are used, for example, to make silicone plastics. An abbreviation used for this type of material is SI.

Hydrolysis of dichlorosilanes and trichlorosilanes gives highly branched polymers which contain silanol end-groups: such a system may be used as a *precursor*, in solution, for the impregnation of reinforcement so as to make *silicon plastics* or a *silicone laminate*. Cross-linking, by heating with a catalyst (for example, zinc or cobalt octoate) give a cross-linked *polyorganosiloxane*. Often such resins are based on methylphenylsilicone polymers as such materials have good heat resistance. The ratio of organic groups (R = methyl and phenyl groups) to silicone (R:Si ratio) may be used to classify the resin. See *silicone polymer*.

silicone rubber
Also known as a silicone elastomer or as a polysiloxane rubber or as a silicone. There are various types of material which may be classed as a silicone rubber: all are based on high molecular weight *polyorganosiloxanes*. The silicon atoms have two organic groups (R & R') attached. The most common polymer is *polydimethylsiloxane* (dimethyl silicone elastomer) where R = R' = -CH_3: such materials are referred to as MQ. The rubber is produced by, for example, the ring opening polymerization of the monomer octamethylcyclotetrasiloxane: molecular weights may reach one million.

If some of the R groups are vinyl groups then methylvinylsilicones (VMQ) result. If the material contains phenyl, vinyl and methyl groups then this type of silicone rubber is given the abbreviation PVMQ. If the material contains fluoro, vinyl and methyl groups then this type of silicone rubber is given the abbreviation FVMQ.

VMQ-type materials may be considered as general purpose silicone rubbers. The PMQ and the PVMQ-types are classed as silicone rubbers with excellent low temperature performance. The FVMQ-types are used where high resistance to chemical attack (fuel, oil and solvent) is required.

Reinforcing fillers used with silicone rubbers are not based on carbon blacks but are usually based on *silica* which may or may not have been treated (e.g. in order to improve the storage life of the mixed compound). Traditionally silicone rubber is supplied already compounded; now a range of base rubbers and modifiers are available so that in-house compounding is easier.

Silicones are usually cured by heating with an organic *peroxide*, for example, benzoyl peroxide, and the resultant decomposition products may need to be removed by post-curing, e.g. by heating the mouldings in an air-circulating oven at 200°C for 12 h. Such a treatment will often improve physical properties, e.g. compression set. Such cross-linked materials are noted for the way they retain their properties over a wide range of temperatures (from −100 to +300°C) and for their high temperature resistance. So, even though the initial properties, for example, tensile strength is poor, the wide temperature range of use makes them useful elastomers. Compounds based on silicone polymers do not (i) appear to be affected by atmospheric exposure, (ii) show ozone cracking and (iii) taste or smell. They are chemically inert and have excellent electrical insulation properties which are maintained over a wide temperature range.

Silicone rubbers are included in the category of *speciality rubber* because of their good high and low temperature properties. Silicone rubbers can also be classified according to the technology employed for their processing into *high temperature vulcanizing silicone rubbers*, *room temperature vulcanizing silicone rubbers* and *liquid silicone rubbers*. See *silicone polymer*.

silicone rubber - chemical group classification

FLUORO, VINYL AND METHYL GROUPS. Also known as fluorinated rubber or as, fluoro silicone rubber or elastomer or as, fluorosilicone rubber or elastomer. Such a *silicone rubber* is referred to by the abbreviation FVMQ. The FVMQ-types are used where high resistance to chemical attack (fuel, oil and solvent) is required.

METHYL GROUPS. Also known as dimethylsilicone elastomer or rubber or, methyl silicone rubber or elastomer or, polydimethyl siloxane. A *silicone rubber* containing methyl groups is referred to by the abbreviation MQ.

METHYL AND PHENYL GROUPS. Also known as methylphenylsilicone elastomer or rubber or, phenylsilicone rubber or elastomer. Such a *silicone rubber* is referred to by the abbreviation PMQ. A silicone rubber with excellent low temperature performance.

METHYL AND VINYL GROUPS. Also known as methylvinylsilicone elastomer or rubber. Such a *silicone rubber* is referred to by the abbreviation VMQ. VMQ-type materials may be considered as general purpose silicone rubbers.

METHYL, PHENYL AND VINYL GROUPS. A *silicone rubber* which is referred to by the abbreviation PVMQ. A silicone rubber with excellent low temperature performance.

silicone solution

A mould release agent based on a *silicone fluid* dissolved in a solvent and applied as a liquid. See *silicone mould release agent*.

silicone-polyether block copolymer

Also known as a polyether-polysiloxane. A water soluble material based on blocks of an *ethylene oxide-propylene oxide* copolymer and on blocks of *polydimethylsiloxane*: used to give good dispersion of the mix ingredients and to stabilize the rising foam in a *one-shot process*. A foam stabiliser.

silicones

Polymeric organic siloxanes. The term could refer to *silicone resin*, *silicone fluid* or to *silicone rubber*.

silk

A natural *fibre*, based on proteins, which is, for example, produced by the silk worm. Widely used in the textile industry.

silk screen

That which is used to form the pattern, or legend, in *silk-screen printing*.

silk-screen printing

Also known as silk screening or as, screen printing. A printing process capable of giving very dense, or opaque, printing onto film or mouldings from low cost equipment. A screen is used and some pores of the screen are blocked so that the ink can only strike through in a predetermined pattern when a flexible blade is drawn over the ink-covered screen. This is basically a stencilling process which uses a fine mesh screen which is now made from a synthetic fibre such as acrylic or polyester.

Fully automatic flat-bed presses are available as are cylinder systems where the flexible blade (a squeegee) remains stationary and the substrate moves below the blade. The substrate does not have to be flat and could be a moulding or a bottle. This process is important for bottle decoration and inks can be specially formulated to suit the polymer, for example, for *unplasticized polyvinyl chloride* (UPVC) or for *polyethylene* (PE).

siloxane

A compound which contains a silicon-oxygen (Si-O) linkage. See *disiloxane* and *trisiloxane*.

silver

This element (Ag) has a relative density of 10.5 and occurs in Group 1B of the periodic table. The three metals in Group 1B of the periodic table (copper, silver and gold) are sometimes known as the coinage metals. Silver is a soft, white, lustrous metal which is renowned for its high electrical and thermal conductivity (the best conductor known). It melts at 960°C and readily forms alloys with other metals.

simple frame mould

See *frame mould*.

simple hindered phenol

See *hindered phenol* and *phenolic antioxidant*.

simple rule of mixtures

See *law of mixtures*.

simple tests - material identification

See *material identification*.

simple triglyceride

A *triglyceride* in which there is only one fatty acid residue: only one acid was used in manufacture. If there is more than one fatty acid residue then the product is a mixed triglyceride. See *fixed oil*.

simplex

See *polysalt*.

simulated end use testing

See *simulated service test*.

simulated service test

An abbreviation used for this term is SS test. A test which simulates the treatment, for example, that a component will experience in use. Used for components which must meet very critical specifications. Such a test may be a dynamic test in which, for example, the component is repeatedly flexed so as to duplicate what the product will experience in service. Such a test may, in addition, be performed at two extreme temperatures so as to obtain a better indication of component performance. See *testing of finished components*.

sin δ
See *power factor* and *loss tangent*.

Sindanyo
Originally a trade name/trademark for an *asbestos-cement product* which may be in the form of rods, tubes or plates. Non-asbestos-cement products are now available.

single acting cylinder
A *cylinder* which operates in one direction only: a cylinder which produces motion in one direction only and return is, for example, by gravity.

single bank calendering
A *calendering* process which uses only one calender nip, for example, to produce thick rubber sheeting based on highly-loaded stocks. For some applications, for example, moulding blank production, the thickness range is accurate enough and the usage of the procedure is justified by its simplicity. If a three roll calender is used, the rubber is formed into a sheet at the top roll and then passes through the open bottom nip and around a cooler bottom roll. The cooled product is removed and rolled up inside a cotton fabric liner. See *double bank calendering*.

single motor drive
A type of *calender drive* which is common for many older calenders. The electric motor is connected to a reduction gearbox and the output from the gear box is used to drive one roll: the other rolls are driven from this one by conventional gearing.

single motor unit drive
A *unit drive system* for a *calender* with fixed roll speeds having a single input shaft for the motor and multiple output shafts to match the required calender configuration. See *calender drive*.

single nip calender
A two roll *calender*.

single pack systems
See *one-pack systems - room temperature vulcanizing silicone rubbers*.

single point tests
Those tests which are carried out at a single test temperature and/or speed. See *common tests*.

single screw extruder
An *extruder* which contains one *screw* which transports, melts, mixes and pumps the *melt* through the *die*. The most widely used type of *extruder*. Single screw machines are the most popular because they are relatively simple, cheap and easily give a continuous output.

single screw extrusion
Extrusion performed using a *single screw extruder*.

single station blow moulder
A machine with one set of *blow moulding* platens: could be used with multiple extrusion heads for the high output of small *blow moulded containers*.

single-impression mould
A mould, for example, an *injection mould*, which contains one cavity or impression. Single-cavity injection moulds are smaller and simpler and may therefore be constructed more easily and cheaply than a *multiple-impression mould*. The simplicity of such moulds should give less trouble during the moulding operation and, because of this, components may be expected to be more consistent. The problem of core/cavity concentricity can be overcome by solid construction of tapered register rings on the core and the cavity but is more difficult to solve in the case of multi-impression tools. Condition setting and process control are easier to achieve when there is only one cavity to be considered. Mould and product design is also easier when there is only one cavity to be considered because the best solutions for accommodating cooling, ejection, the mould parting line, etc. can be determined most easily. The relative simplicity of single-impression moulds gives advantages such as low risk of lost production, low replacement mould costs and short repair times. In general, there will be a higher injection pressure available to fill a single-impression cavity and this, in turn, may mean that a component with a thinner wall section may be moulded.

single-screw machine
A machine which contains one screw. See *single screw extrusion*.

single-stage machine
An *injection moulding machine* which has a clamping system which is not separately powered from the injection unit: that is, it does not have a separate clamping system. The injection unit pushes the two mould halves together and then it forces melt into the mould. If the area of the component is large, relative to the ram area, then the mould will open. This is why most machines are *two-stage machines*.

sink mark
An unwanted depression on the surface of a component.

sintering
The production of components by heating: the heating is performed at temperatures which cause agglomeration. Compaction is often achieved by using shapes which have been produced by pressing. Sintering processes are used to produce components from *polytetrafluoroethylene*.

Sioplas process
A Dow process for the production of *cross-linked polyethylene*. A trialkylvinylsilane is grafted onto *polyethylene* (PE), in the presence of a *peroxide*. The PE may be cross-linked when the trialkylvinylsilane is hydrolysed: a siloxane cross-link is formed by hydrolysis of the alkoxy groups. Grafting and extrusion may be performed at the same time on an *extruder* and the extrudate cross-linked on standing in water. This type of process has been used for pipe and for wire-covering.

sipe
A small knife-like groove in the tread of a *tyre* which is there to help mop up water from a wet surface.

SIR
An abbreviation used for Standard Indonesian Rubber. A *technically specified (natural) rubber* from Indonesia.

SIRE
An abbreviation used for the Styrene Information & Research Centre - a USA based organization.

SIS
An abbreviation used for styrene-isoprene-styrene. See *styrene block copolymer*.

sisal
A natural fibre, based on *cellulose*, which is, for example, produced by the plant called agave sisalana. Widely used to make cords and twines. Sisal fibres have been used in *dough moulding compounds*.

SIX
An abbreviation used for *sodium isopropyl xanthate*.

size
Also known as dressing. A surface treatment applied to a *fibre* so as to improve the resistance of the fibres to damage caused by handling. A size (for example, dextrimized starch) is commonly applied to *glass fibre* for this reason. Dextrimized starch is removed (by burning or with solvents) before a *coupling agent* is applied. Plasticized *polyvinyl acetate* may also be used as the basis for a size: this material may be left on the fibre as it is compatible with *unsaturated polyester resins* and with *epoxides*.

size - extruder
The size of an extruder is expressed in terms of the external diameter of its screw e.g. 2 inch or 50 mm.

sizing die
A water cooled cylinder used to control the external diameter of thermoplastics tube: this sizing die is positioned close to the *extruder* die and the plastics tube may be held against the walls of this cylinder by either air pressure or a vacuum.

sizing plates
Plates used in *tube* or pipe production to control external diameter; they consist of metal plates with machined holes of the required size: the tube may be expanded, by vacuum or compressed air, until it just touches the machined holes in the plates or rings.

sizing rings
Rings used to obtain the correct external size of extruded tube or pipe. See *sizing plates*.

skeleton frame mould
A frame which outlines the desired component shape and used in, for example, *vacuum snap back thermoforming*. The use of an outline mould means that there is only minimal mould contact at the edges and therefore minimal sheet marking.

skim coating
Also known as topping. A *calendering* process used to coat fabric with a rubber compound. The application of a calendered rubber sheet to a previously frictioned fabric. If a three roll calender is used then the sheeting is applied to the fabric at the second nip: both rubber and fabric are travelling at the same speed when they come together. To improve adhesion the fabric is usually pre-treated, for example, by *frictioning* with rubber or by the use of a *bonding agent*: pressure may be applied during calendering, for example, by using an idler roll.

skim fraction
See *centrifuging*.

skim latex
Obtained when *latex concentrate* is being made from *field latex* by *centrifuging*: the by-product of latex centrifugation used to prepare the form of *natural rubber* known as *skim rubber*.

skim rubber
A form of *natural rubber* which is made from *skim latex*. Dry rubber is obtained from skim latex by *acid coagulation*, for example, using sulphuric acid at a low pH; the coagulated material is made into sheet or granulated rubber. The pH is low because of the high concentration of non-rubber ingredients, for example, proteins, which act as colloid stabilisers. With care, a light coloured material can be obtained with a rubber content of approximately 85% and which is fast curing.

skin
A relatively dense layer at the surface of a cellular polymeric material.

skin coat
A top coat: a *coating* applied to protect an underlying layer of, for example, an expanded coated fabric.

skin colouring
Imparting colour to the skin only. Most polymeric components are coloured by *mass colouring*. However, it is possible to colour only the skin, or surface of a component, in order to save on colorant, colorant costs and compounding costs: this may be done by, for example, *sandwich moulding* and *extrusion*.

skiving
A process for producing film or sheet by cutting from a block or log of material; used, for example, to make *polytetrafluoroethylene* (PTFE) tape. The cutting of a rubber section at an angle, so as to, for example, improve the strength of a joint and/or to reduce air trapping.

slab-stock process
The most common method of producing *flexible polyurethane (PU) foam*: long, continuous blocks (buns) of foam are produced and are later cut to the required size (see *one-shot process*).

Such a foam may be produced from a *polyester polyol* by reaction with a *diisocyanate* (for example, *65:35 TDI*) using a mixing head in a Henecke-type machine. The other mix ingredients, known as the *activator mixture*, are injected into the isocyanate-polyol blend and after vigourous mixing, the frothing mixture is laid evenly into a lined trough: the trough is lined so that as the mixture reacts the side walls may be raised so as decrease side wall drag and so give a bun of uniform cross-section.

Polyether foams are also produced in this way although complex catalyst mixtures may be needed to obtain the desired balance of properties at a specified rate: must obtain the correct balance of chain extension, chain branching and gas generation.

slabstock process
See *slab-stock process*.

slaked lime
Lime which has been reacted with water. This material has a relative density (RD or SG) of 2·1. See *calcium hydroxide*.

slate
A complex of muscovite mica, chlorite and quartz: a composite hydrated potassium/magnesium/aluminium silicate combined with *silica*. Can be ground to give slate flour which is used as a *filler*. For example, when *shellac* is modified with slate, compounds with low shrinkage, and high hardness, result. Slate has also been used as a low cost filler in *polyvinyl chloride* (PVC).

SLCM
An abbreviation used for *structural liquid composite moulding*.

sleeve bearing
See *bush bearing*.

sleeve ejector
A type of *ejector*: a hollow steel ejector which fits over a core or pin.

slewable machine
A type of *injection moulding machine*. See *vertical locking* and *vertical injection*.

slicing machine
A machine used to slice, or cut, thin sheets from a block. Thick sheets produced from *dough* on sheeting rolls is pressed in a block press. The *sliced sheet* is then seasoned. See *cellulosics*.

slide-type transducer
An *indirect pressure transducer* used in *cavity pressure control*.

sliding check ring
See *check ring*.

sliding friction
Also known as kinetic friction. The *friction* between bodies which are moving with respect to each other: this is less than *static friction*. The coefficient of friction in this case, is given by the ratio of the tangential force which is required to maintain motion (without acceleration) to the normal force at the contact surface

sliding punch
Also called a sliding force. The *punch*, the male part of a mould, is mounted (for example, on rails) so that it may be withdrawn from a press when required.

slip
The ability of materials to slide over each other. Slip is also a term used in *hydraulics* and which refers to the internal leakage of fluid.

slip agent
A substance added to make a sliding action easier, for example, fatty acid amides are used widely in *film extrusion*.

slip angle
The angle between the *contact patch* and the direction of travel of the wheel on which a *tyre* is mounted. Slip angle varies from one type of tyre to another. It is lowest with the radial-ply steel-braced tyre and increases from this to the radial-ply textile tyre, the radial-ply winter tyre, the cross-ply tyre and the cross-ply winter tyre. In general, the smaller the slip angle the longer the tyre will last as the *cornering force* is lower. However, when a vehicle is driven hard (for example, a racing car) a *low-profile cross-ply tyre* gives better control at the limit of adhesion, that is, at large slip angles. Such a tyre has a stiffer sidewall and a more compliant tread than a *radial-ply tyre*.

If the slip angle between the front tyres and the road (front slip angle) is greater than the slip angle between the rear tyres and the road (rear slip angle) then the vehicle will understeer. If the front slip angle is less than the rear slip angle then the vehicle will oversteer.

slip thermoforming
A *thermoforming* process in which the heated sheet is allowed to slip, or move inwards through the clamping frame, as the forming process occurs.

slip velocity
The velocity above which slip occurs at the wall of a *die*, that is, wall slippage occurs at a certain *shear rate*.

If the wall slip velocity is V_a, Q is the volumetric flow rate and R is the capillary radius then $4Q/\pi R^3 = 4V_a(1/R) + X$ (where X is a function of the shear stress). If a series of *flow curves* are produced using a set of dies of varying radius R, and then measuring $4Q/\pi R^3$ at a given value of shear stress, a plot may then be obtained of $4Q/\pi R^3$-vs-$1/R$. The slip velocity will then be one quarter of the slope of this plot. This may then be repeated at other shear stresses and this will enable a plot of slip velocity against *shear stress* to be built up.

slit film tape
See *film tape* and *fibrillated film fibre*.

slit-die extrusion
See *slot-die extrusion*.

sliver
A continuous assembly of lightly bonded *staple fibre*: the fibres are parallel.

slot die
An *extrusion die* used to produce *flat film* and for *extrusion lamination*.

slot-die extrusion
Also called slit-die extrusion. An *extrusion* process which uses a *slot die* to produce film or sheet from a thermoplastics material. See *flat film process*.

slow accelerator
See *accelerator*.

SLR
An abbreviation used for *technically specified (natural) rubber* from Sri Lanka.

slug
The unit of mass in the *FPS gravitational system of measurement*. A unit of mass which has an acceleration of 1 foot per second per second when acted upon by a force of one pound. Since a one pound mass has an acceleration of 32·1740 feet per second per second when acted upon by a force of one pound, it follows that the slug has a mass of 32·1740 pounds. That is, approximately 32·2 lbs/14·6 kg.

slush moulding
A moulding technique used for *polyvinyl chloride* (PVC) pastes. The *paste* is poured into a hot mould where some of it *pregels* on the heated metal: the excess *paste* is poured out and the mould is heated to approximately 150°C to complete fusion. After cooling the product is removed. This process has been largely superseded by *rotational casting* as this process gives more uniform wall thickness.

SMA
An abbreviation used for *styrene maleic anhydride*.

small calorie
0·001 of a kilocalorie. See *calorie*.

small holder coagulum
Natural rubber which has been coagulated either naturally or chemically by a small-scale producer who is unable to transport *field latex* to a central factory.

smallholding
A farm whose size is, for example, less than 40 ha. In Malaysia such a plantation would yield about 1,000 kg of *natural rubber* per hectare per year.

smallholding rubber
See *native rubber*.

smart fluid
A colloquial term for an *electro-rheological fluid*.

SMBT
An abbreviation used for *sodium mercaptobenzothiazole*.

SMC
An abbreviation used for *sheet moulding compound*. For example: for:

 SMC CF - see *carbon fibre sheet moulding compound*;
 SMC recycling - see *polyester moulding compound recycling*;
 SMC regrind - see *sheet moulding compound regrind*; and for
 SMC shrinkage control additive - see *low profile sheet moulding compound*.

smithsonite
A naturally occurring form of zinc carbonate. See *basic zinc carbonate*.

smoke
A suspension of fine particles of solid in air. Smoke is evolved in a *fire*. In general, fire is less life threatening than the accompanying smoke. If the toxicity of the evolved volatiles are discounted, then there still remains the problem

of the loss of vision in a fire situation which is caused by the emission of smoke. This is why it is important to reduce the density, irritancy and opacity of the evolved smoke. See *fire testing*.

smoke and gas emission tests
Smoke and toxic gases are generated in a fire and kill more people, e.g. by suffocation or poisoning, than the actual fire itself. This is why *gas testing* and *smoke testing* are regarded as being very important.

smoke house
A shed in which *smoking* of *natural rubber* is performed. Such a house is heated to about 60°C by the smoke/fire of burning wood.

smoke suppressant
An additive: a material which reduces the tendency of a polymer compound to form smoke in a fire situation. Such a material may also act as a *flame retardant*. An example of a smoke suppressant is *aluminum trihydrate* (ATH): another is molybdic oxide which can function synergistically with *antimony trioxide*. Metal borates, such as zinc and molybdenum, are suggested for use in *polyvinyl chloride* (PVC). See *smoke testing*.

smoke testing
Most plastics and rubbers are based on the element carbon and when such materials are burnt a great deal of *smoke* and soot is often produced. In a fire situation this hampers rescue and hinders escape because the smoke and soot make it difficult to see. The methods therefore used to test for smoke emission and density rely on optical measurements. One such method of measuring smoke density employs the *National Bureau of Standards* (NBS) *smoke chamber*.

In this chamber a specimen of known size. is heated by an electric radiator in a sealed chamber, or box, of standard size; that is the sample is burnt under controlled conditions. The smoke thus produced reduces the amount of light which can pass through the chamber. This reduction can be measured and used to calculate the optical density of the air in the light path: if required jets or a burner can also be used to create flaming conditions.

Various ways of using the NBS smoke chamber exist. For example, the specific optical density (D_s) after a specified heating time is (e.g. 1·5, 4 or 20 min) is measured. D_s is assessed by measuring the reduction in the intensity of a light beam which passes vertically through the chamber. If necessary a correction is applied when fogging of the optical window results: D_s then becomes D_m corr. Six samples of the material are required: three are tested in the flaming mode and three in the non-flaming mode

smoked sheet
See *ribbed smoked sheet*.

smoking
Process used to dry and preserve *natural rubber*. This process may take 4 to 7 days in a *smoke house* and utilises the preservative effects of phenolic-type materials in the smoke. See *ribbed smoked sheet*.

smooth bore hose
Also known as fully-embedded hose. An *externally corrugated* hose in which the reinforcing helix is fully embedded in the hose wall and the hose has a smooth bore.

smooth bush bearing
See *bush bearing*.

SMR
An abbreviation used for *standard Malaysian rubber*. A technically specified rubber from Malaysia.

SMR latex grades
Grades of *natural rubber* produced from liquid *latex* and sold in *block form*. See *standard Malaysian rubber latex grades*.

snap back thermoforming
See *vacuum snap back thermoforming*.

snap fit
A joining technique widely used with high strength, resilient materials such as *nylon* and *acetal*; permanent, or recoverable joints, can be produced by mating a moulded undercut on one part with a lip on another component so as to provide a strong mechanical joint.

snap-back
A term sometimes used in place of *orientation*: applied to extruded sheet of, for example, *copolyester*. Snap-back is generally desirable in film used for *thermoforming* as it minimizes webbing or bridging.

snatch
See *sprue puller pin*.

snatch pin
See *sprue puller pin*.

Snellen charts
Chartd used to assess *see-through-clarity*. Although not standardized, Snellen charts are used to assess see through clarity at a specified chart-to specimen distance. Such charts consist of sets of parallel lines which differ in, for example, line spacing. The charts are viewed with and without the specimen. The narrowest, most closely spaced lines, which can clearly be seen as lines, with and without the specimen in front of the chart is a measure of see through clarity.

soap
A soap is a metal salt of acids such as stearic, palmitic and/or oleic acid: such salts are often sodium or potassium salts although can also have *calcium soaps*.

A soap is surface active, that is, it concentrates at the water surface or interface where it modifies behaviour or properties. The soap molecules have a highly polar, water soluble end which is the ionized carboxyl group: the other end is the long hydrocarbon chain which is water insoluble and hydrophobic. At an oil and water interface, the hydrophillic anion group will associate with the water and the other end with the oil. This action reduces surface tension and emulsified oil droplets can be formed suspended in the water. Soaps are used to assist dispersion of additives in rubber compounds.

soap stone
See *talc* and *soapstone*.

soapstone
This material has a relative density (RD or SG) of 2·72. See *talc*.

Society of Automotive Engineers
An abbreviation used for this USA-based organization is SAE. The SAE issues codings for *steel* which are widely used in the plastics industry: their designation is the same as the *American Iron & Steel Institute*. The prefix used for mould steels is P.

soda-glass
See *soda-lime glass*.

soda-lime glass
Also known as soda-glass. The commonest type of *glass* which is made by fusing together sand (*silica*) sodium carbonate and lime. Has a comparatively low softening temperature of about 745°C. See *A-glass*.

sodium acetate
The sodium salt of *acetic acid*. This solid material has a melting point of 58°C and a relative density (RD or SG) of 1·45. A *retarder* (but not for butyl rubber).

sodium alginate
A thickening and creaming agent for latex. See *creaming*.

sodium aluminium silicate
See *ultramarine blue*.

sodium benzenecarboxylate
See *sodium benzoate*.

sodium benzoate
Also known as sodium benzenecarboxylate. A food preservative and antiseptic. This white powdered material has a melting point of 250°C and a relative density (RD) of 2·8. Sometimes added to emulsion paints for rust protection.

sodium bicarbonate
Also known as sodium hydrogencarbonate or as, sodium bicarb or as, bicarbonate of soda. This white solid material has a relative density (RD or SG) of 2·20. A *blowing agent* used to produce *sponge rubber*. To improve dispersion in the rubber compound, and to increase the decomposition rate, disperse the sodium bicarbonate in a light *mineral oil* (about 50% concentration). The sodium bicarbonate decomposes above 70°C to give carbon dioxide and water. The blowing agent *4,4'-oxybis-(benzenesulphonylhydrazide)*, can be used in conjunction with sodium bicarbonate to give more even cell structures.

sodium bisulphite
A sodium salt of sulphurous acid used to treat *natural rubber latex* so as to prevent enzymeaction and to obtain *white crepe* or *extra white crepe*. Approximately 0·6% of sodium bisulphite, based on the rubber content of the latex, is used so as to prevent enzyme attack.

sodium carbonate
Also known as washing soda. This white, solid material has a relative density (RD or SG) of 2·2 and a melting point of 850°C. Used in the manufacture of *glass* and of *soap*.

sodium caseinate
A soluble form of *casein*.

sodium dibutyl dithiocarbonate
An ultra-fast *accelerator* for rubber latices.

sodium diethyl dithiocarbonate
An ultra-fast *accelerator* for rubber compounds and for rubber latices.

sodium dimethyl dithiocarbonate
An abbreviation used for this material is SDD. An ultra-fast *accelerator* for rubbers. Used as a short stop in *emulsion polymerization*.

sodium formaldehyde sulphoxylate
An abbreviation used for this material is SFS. A reducing agent used in the emulsion polymerization of cold *styrene-butadiene-rubber*.

sodium hydrogencarbonate
See *sodium bicarbonate*.

sodium isopropyl xanthate
An abbreviation used for this material is SIX. This white powdered material has a relative density (RD) of 2·1. An ultra-fast *accelerator* when used with *zinc oxide*.

sodium mercaptobenzothiazole
An abbreviation used for this material is SMBT. A vulcanization *accelerator*.

sodium pentamethylene dithiocarbonate
An abbreviation used for this material is SPD. An ultra-fast *accelerator* for rubbers when used with *zinc oxide*: also used with this material is *stearic acid*. This solid material has a melting point of 280°C and a relative density (RD) of 1·4.

sodium polyacrylate
The sodium salts of *polyacrylic acid* which, when dissolved in water, give very viscous liquids: the salts are used as thickening agents for rubber *lattices* with high filler loadings. May be prepared, for example, by neutralization of *polyacrylic acid*.

sodium/potassium/aluminium silicate
See *nepheline syenite*.

sodium rubber
Polybutadiene produced using sodium as a catalyst. See *buna rubber*.

sodium salt of ethylene diamine tetra-acetic acid
See *ethylene diamine tetra-acetic acid*.

sodium salt of polyacrylic acid
See *sodium polyacrylate*.

sodium silicofluoride
A white solid material with a specific gravity of 2·68. Used as a gelling agent in the production of *foam rubber* from rubber *lattices*.

sodium stearate
The sodium salt of *stearic acid*. An *acid acceptor* for chlorine containing rubbers.

sodium vulcanizate
An *elastomeric ionomer* produced from an *ionomer* which has been neutralized with, for example, sodium hydroxide, so as to produce, for example, *carboxylated polybutadiene rubber*.

sodium xanthate cellulose
Also known as cellulose xanthate or as xanthate. Formed from *cellulose* and to used make *viscose rayon*.

sodium-catalyzed polymerization - butadiene
See *buna rubber*.

soft block
Also known as a soft segment. Part of a copolymer: that part of the copolymer structure which is relatively soft and which is responsible for the elasticity of a *thermoplastic elastomer*. See *polyurethane block copolymer*.

soft end
A hose end in which the rigid reinforcement has been deliberately omitted.

soft grade
The terms hard and soft usually refer to appropriate grades of materials such as thermoplastic elastomers. For example, with *polyether ester elastomer* if the hardness is 34-47 Shore D then the material is referred to as a soft grade: if the hardness is 55-72 Shore D then the material is referred to as a hard grade.

soft natural rubber/polyolefin blend
A *blend* with a high proportion of natural rubber, for example, above 50%. See *natural rubber/polyolefin blend*.

soft polyurethane foam
See *flexible polyurethane foam*.

soft segment
See *soft block*.

soft tooling
Mould tools which are not made from hardened metal and which therefore only have a limited life: mould tools which are made from materials such as epoxide resins and used in, for example, *resin transfer moulding* (RTM).

soft wood tar
A tar obtained from a soft wood such as pine. See *pine tar*.

softened rubber
Usually means peptized rubber. That is, chemically softened *natural rubber*: a *peptizer* is added to the latex before coagulation. See *softener*.

softener
An additive, usually a liquid, which is added to make a polymer softer; oils are used in elastomers/rubbers for this purpose. The term softener is used by the rubber industry for products such as hydrocarbon oils: in the plastics industry the term *plasticizer* would be used. Softeners are used as additives for hydrocarbon rubbers/elastomers so as to produce soft compounds which are relatively easy flowing.

softening
The process used to increase the softness of a polymer. Can have external softening and internal softening. See *softened rubber*.

softening point
An abbreviation used for this term is SP. The temperature at which a material, usually a thermoplastics material, softens by a specified amount in a *softening point test*. The amount of softening is usually measured by the indentation of a weight-loaded needle or, by the bending of a weight-loaded bar as the temperature is slowly increased. See, for example, *Vicat softening point*.

softening point test.
A test used to determine the *softening point* of a *thermoplastics material*.

softness number
A number which indicates the softness of a material. Often used to rate plasticized *polyvinyl chloride* (PVC) compounds. An indentor is pressed under a specified load into a sheet of the compound and the penetration of the indentor is measured. See *hardness*.

softwall hose
Also known as delivery hose. A hose without a reinforcing helix of, for example, wire.

sol
Part of a cross-linked polymer system or compound: that polymeric part which is not cross-linked and which can be removed by solvents. Sol rubber is that part of rubber which is soluble in benzene.

solder
Low melting point alloys of tin and lead (based on say, 50% tin and 50% lead). Such alloys melt over the range 200 to 300°C.

sole
An extra thickness of material put onto a *parison* in an *extrusion blow moulding* operation: used if large extensions, in a certain direction, to a parison are expected.

sole crepe
Crepe rubber used for shoe soles. When the rubber product is required to be very white, then two *coagulation* steps and/or bleaching may be employed. The first coagulation step may be referred to a *precoagulation*.

sole plate
A heated platen. See *contact heating*.

solenoid
A term used in *hydraulics* and which refers to an electromechanical device which converts electrical energy into mechanical motion. An electromechanical device in which the flow of an electrical current through a coil cause an iron core to move; may be used to switch a directional valve.

solid angle unit
See *steradian* and *Système International d'Unité*.

solid electrolyte
An *electrolyte* which is a solid at a specified temperature. See, for example, *polyethylene oxide-salt complex*.

solid rubber tyre
A *tyre* which is based on a solid moulding: it is not inflated with air. Such tyres are used for slow moving vehicles, for example, for large slow moving vehicles which carry heavy loads. See *polyurethane rubber*.

solid state
The type of control system which superseded relay control: it is based on electronic components which have no moving parts and yet can, for example, cause a switching action.

solid state machine
A type of machine, for example, an *injection moulding machine*, which has been classified according to the type of control system employed. The machine uses *solid state* components to obtain the desired sequence of movements.

solid state polymerization
Polymerization of a solid *monomer*, for example, a vinyl *monomer*. Radiation polymerization is useful for solid state polymerization as the radiation finds it easy to penetrate a solid monomer.

solid woven belt
A *belt* based on a carcass which consists of a complex fabric which has many layers of weft yarns but which is woven as a single web.

solidago
Also known as goldenrod or solotarnik. A perennial plant which contains rubber in the leaves.

soling
The addition of additional material to the inside or, to the outside, of a parison.

solotarnik
See *solidago*.

solubility parameter
An abbreviation used for this term is δ. The square root of the *cohesive energy density* of a material. Solubility parameter is used as a guide to material miscibility or compatibility. If two materials have similar values of solubility parameter then they are often miscible or compatible.

soluble sulphur
See *sulphur*.

solution A
One of two salt solutions used in a measurement of *resistance to tracking*. Solution A is 0·1% ammonium chloride.

solution B
One of two salt solutions used in a measurement of *resistance to tracking*. Solution B is 0·1% ammonium chloride plus 0·5% wetting agent.

solution butadiene rubber
An abbreviation used for this type of material is L-BR. See *butadiene rubber*.

solution casting
Similar to *dispersion coating* except that the polymer used is dissolved in a suitable liquid. See *solvent casting*.

solution polybutadiene
Polybutadiene rubber produced by *solution polymerization*. See *butadiene rubber*.

solution polymerization
A *polymerization* process in which the monomer is initially dissolved in a solvent: the solvent is present as a diluent. If

the resulting polymer is insoluble in the monomer, then the polymer will precipitate and the process may also be referred to as a precipitation polymerization. Although the use of a solvent minimizes some production problems (for example, heat removal) the presence of the solvent introduces others, for example, solvent recovery. Polybutadiene rubbers are produced by solution polymerization.

solution process
See *precipitation process*.

solution styrene-butadiene rubber
An abbreviation used for this type of material is L-SBR. *Styrene-butadiene rubber* (SBR) produced by *solution polymerization*: catalysts used are alkyl lithium (lithium alkyl). Block copolymers may be produced if non-polar solvents are used as the styrene only polymerizes once the butadiene has been used. To produce copolymers with a more random structure, the butadiene may be added incrementally as the polymerization proceeds. Alternatively *randomizing agents* may be used, for example, ethers and amines.

solution viscometry
Also known as dilute solution viscometry. A widely used method of *molecular weight* determination as it is relatively simple: gives the viscosity average molecular weight from the *Mark-Houwink equation*. A number of dilute polymer solutions are prepared and the times taken for these to flow through the capillary of a viscometer is measured: the flow time of the solvent is also measured. The *Mark-Houwink constants* are determined from a plot of the limiting viscosity number and the polymer molecular weight.

solvation
Swelling, gelling and dissolving of polymer by a *solvent*. In the case of a high *molecular weight linear polymer*, the solvent first diffuses into the polymer which swells; the polymer molecules then disentangle and the solvent becomes the continuous phase.

solvent
A material, usually a liquid, which can dissolve other substances. A material will be a solvent for another substance if the molecules of the two materials are compatible and if there is no tendency to separate. Solvents initially cause swelling and gelling of a polymer before a solution is formed. See *solubility parameter*. See, for example, the following entries: *acetone, benzene, carbon disulphide, carbon tetrachloride, chlorobenzene, chloroform, cyclohexanone, ether, ethyl acetate, gasoline, hexane, methyl ethyl ketone, pyridine, turpentine oil, tetrahydroxynaphthalene, toluene* and *xylene*.

solvent bonding
See *solvent welding*.

solvent casting
A solution of a polymer is cast onto a metal plate or rolls, and the solvent is evaporated; used to make ultra-thin films of, for example, *polycarbonate* (PC).

solvent cement
A solution of a *polymer* in *solvent*: commonly associated with polystyrene (PS) or *high impact polystyrene* (HIPS). Also referred to as a cement or as a bodied adhesive. See *solvent welding*.

solvent cracking
See *solvent stress cracking*.

solvent etching
A wet etching process. See *photoresist*.

solvent naphtha
See *naphtha*.

solvent polishing
A process used to improve the gloss of thermoplastics components by applying a *solvent* to the surface, so as to smooth irregularities, and then evaporating the solvent.

solvent stress cracking
A type of *environmental stress cracking* which occurs when the liquid has a solvating effect on the polymer.

solvent welding
Also known as solvent bonding. A *welding* process for thermoplastics materials and for uncured rubbers. A solvent, or solvent cement, is used to soften the surfaces to be bonded, the treated surfaces are then pressed together and held until the solvent is removed, for example, by evaporation, and/or absorption and/or, polymerization. See *welding*.

sonotrode
See *horn*.

sorbitol
A sugar substitute which is obtained from glucose. A solid polyhydric alcohol with a melting point of 110°C and which has been used to make *alkyd resins*.

sour gas
Hydroperoxide-containing gasoline. Epichlorhydrin terpolymer is sulphur vulcanizable and is more resistant to *sour gas* than the other types of *epichlorhydrin rubber*.

source-based nomenclature
See *common names*.

soya bean oil
Derived from soya max. Also known as soybean oil. A very important *vegetable oil* which is obtained from soya beans. The triglycerides of this oil contain linoleic, linolenic, oleic and stearic acid residues. The oil is used to make margarine and cooking fats. When epoxidized, a semi-drying oil, called epoxidized soya bean oil (ESBO), results. ESBO is sometimes used in, for example, alkyd resins as it has excellent colour retention.

soybean oil
See *soya bean oil*.

sp
An abbreviation used for specific. For example;

sp gr = *specific gravity*; and,
sp ht = *specific heat*.

SP
An abbreviation used for saturated polyester. See *thermoplastic polyester*. SP is also sometimes used for *special purpose* (rubber) but more usually used for *superior processing* (rubber). For example:

SP air dried sheet = superior processing air dried sheet;
SP brown crepe = superior processing brown crepe;
SP crepe = superior processing crepe (rubber); and
SP smoked sheet = superior processing smoked sheet.

spaced-strand feed
See *calendrette line*.

spandex fibre
A manufactured, polymeric fibre composed of at least 85% by weight of a *polyurethane*.

SPC
An abbreviation used for *statistical process control*.

SPD
An abbreviation used for *sodium pentamethylene dithiocarbonate*.

SPE
An abbreviation used for The Society of Plastics Engineers. (USA).

special purpose rubber
An abbreviation used is SP rubber (but see *superior processing rubber*). A rubbery material which when correctly formulated will give a significant improvement in a particular property than a *general purpose rubber*: a rubber which is noted for a particular property. Many types of special purpose rubber are possible and this sector of the market is growing rapidly. *Polyurethane rubber, silicone rubber* and *thiokol rubber*, are all classified as examples of SP rubbers. See *oil resistant rubbers*.

special synthetic rubber
Sometimes referred to as special synthetic elastomer. An abbreviation used for this type of material is special SR. A *synthetic rubber* whose exact characteristics are ill defined. See *speciality rubber*. Often included in this category are *ethylene-propylene rubber, chloroprene rubber, nitrile rubber* and *butyl rubber*.

speciality rubber
Also referred to as a speciality elastomer. A *synthetic rubber* whose exact characteristics are ill defined but which, in general, has a combination of good oil and/or temperature resistance. Often included in this category are:

silicone rubbers - because of their good high and low temperature properties;
chlorinated rubbers and *chlorosulphonated rubbers* - because of their good oil and temperature resistance;
epichlorhydrin rubbers - because of their ozone resistance;
acrylic rubbers - because of their oil resistance and their resistance to high temperatures; and,
polyfluorophosphazene rubbers - because of their good oil and temperature resistance, for example, they can withstand temperatures from −65 to 175°C.

specially prepared rubber
See *natural rubber*.

specific
In physics the use of the word specific means that unit mass is being considered. See, for example, *specific heat capacity*.

specific gravity
Abbreviations used for this term include sg, SG and sp gr. The ratio of the density of a substance to the maximum density of water. As it is a ratio it has no units. Now known as *relative density*. May be determined using Mohr's scales from the loss of weight of a test sample when immersed in water.

specific heat capacity
Also known as specific heat and sometimes abbreviated to c or to sp ht. Heat capacity divided by mass. The amount of heat which must be added to unit mass of a material in order to raise its temperature by one degree. Values are commonly expressed in J/kg K, Btu/lb °F or in cal/g °C. To convert from Btu/lb °F or cal/g °C to to J/kg K multiply by 4186·80.

Material	cal g^{-1} °C^{-1} cal/g °C	Jkg^{-1} K^{-1} J/kg K
Aluminium	0·21	887
Carbon	0·13	531
Glass	0·16	670
Iron	0·11	452
Methylated spirits	0·57	2,400
Sea water	0·93	3,900
Water	1·00	4,186

As specific heat varies with temperature it is necessary to specify the temperature employed when quoting specific heat values. With a *crystalline polymer* the heat of fusion causes a large increase in the specific heat value at the crystalline melting point: beyond that temperature the value falls again. **Table 7** gives average values for plastics materials. The high specific heats and heat contents of plastics materials should be noted together with the difference between amorphous and crystalline materials. The specific heat of filled systems, such as rubber compounds, is a cumulative property of all the compounding ingredients.

Specific heat is commonly measured by determining the heat content (enthalpy) over a temperature range (such as upper processing temperature to room temperature) by, for example, a differential scanning calorimetry (DSC) method. If such an average specific heat is known then the amount of heat which must be removed, for example, by the mould cooling system in *injection moulding* can be estimated as:

heat content = mass × specific heat × (melt temp − mould temp)

As the DSC method can take account of the latent heat of fusion at the crystalline melting point, no allowance need be made for this. However, if *enthalpy* data is available it would be best to use this directly. If the information is used for mould cooling calculations, it is suggested that it is assumed that all the heat needs to be removed from the components whereas only half the heat needs to be removed from the feed system. See *drop method*.

specific inductive capacity
See *dielectric constant*.

specific modulus
Also called specific Young's modulus. A value obtained by dividing the modulus by the relative density (the specific gravity - the SG) or, by the density. Such a modulus is important because often a designer/user is interested in how much of *modulus* (stiffness) is being obtained from a system per unit of weight. A certain amount of modulus is needed to allow the product to perform satisfactorily: in some applications, for example, aerospace, this amount is required at the lowest possible weight.

NON-PLASTICS MATERIALS

Material	Modulus 10^6 psi	SG	E/SG 10^6 psi
Metals			
Aluminium	10·3	2·7	3·8
Magnesium	6·5	1·74	3·7
Molybdenum	40·0	10·2	3·9
Steel	30	7·8	3·8
Titanium	16·5	4·5	3·7
Glass	10·0	2·6	3·9
Spruce (parallel to grain)	1·9	0·5	3·8

This table seems to show that there is an upper maximum limit for conventional materials and that, in the case of fibrous materials (such as wood), it is important to specify the grain direction.

PLASTICS COMPOSITES. If a rod is made from parallel *glass fibres* and an equal amount of *epoxide resin*, then the density will be 1·70 g/cc. The maximum modulus of the composite will be very approximately 5 × 10^6 psi/35 GPa: that is, half of the value for glass as the resin contributes little to this property. The specific modulus will be very roughly 5/1·7 = 2·9 × 10^6 psi. Reducing the amount of resin or, replacing the glass fibre with the same amount of carbon fibre will however improve the specific modulus significantly. High modulus (HM) carbon fibres have a tensile strength of approximately 2·4 GPa a modulus of 400 GPa/58 × 10^6 psi and an SG of 2·0. The specific modulus of the 50:50 composite made from such fibres will be approximately 29 × 10^6 psi/2 or 14·5 × 10^6 psi. The highest figure from conventional engineering materials is approximately 3·9 × 10^6 psi.

COMPOSITE MATERIALS

Material	Modulus 10^6 psi	SG	E/SG 10^6 psi
E glass	10	2.55	3.9
S glass	12.5	2.52	5.0
Carbon fibre - high modulus	58	2.00	29.0
Carbon fibre - high strength	29	1.70	17.0
Carbon fibre - pitch based	55	2.02	27.2
Boron fibre	61	2.65	23.0
Kevlar	19	1.45	13.1
GRP chopped mat 30% glass	1.5	1.5	1.0
GRP woven cloth 50% glass	2	1.7	1.2
GRP unidirectional 60% glass	4.4	1.6	2.8
GRP unidirectional 80% glass	7	2.0	3.5
CFRP unidirectional HS fibre	19	1.5	12.3
CFRP unidirectional HM fibre	28	1.6	17.5
Acrylonitrile butadiene styrene 20% glass fibre (GF)	0.74	1.2	20.61
Polycarbonate 30% glass	1.25	1.4	0.89
Polycarbonate 30% carbon fibre	2.15	1.3	31.62
Acetal homopolymer 20% GF	1.0	1.56	0.64
Polypropylene homopolymer 40% talc	0.58	1.27	0.46
Polyether ether ketone 30% GF	1.25	1.5	10.83

PLASTICS MATERIALS

Material	Modulus 10^6 psi	SG	E/SG 10^6 psi
Thermoplastics			
ABS Acrylonitrile butadiene styrene	0.4	1.04	0.38
PC Polycarbonate	0.35	1.2	0.29
POM Acetal homopolymer	0.52	1.42	0.37
PP Polypropylene homopolymer	0.23	0.90	0.26
PS Polystyrene	0.48	1.04	0.46
Thermosetting plastics.			
UP Unsaturated polyester	0.51	1.28	0.40
EP Epoxide or epoxy	0.43	1.2	0.36

The low stiffness, and the low specific modulus, of plastics materials should be noted. Yet, by combining such low modulus materials with fibres such as *carbon*, very stiff materials can be obtained: much stiffer than conventional engineering materials. However, it must not be forgotten that toughness, and ideally low cost, are also required.

specific resistance
See *resistivity*.

specific resistance
See *volume resistivity*.

specific viscosity
Also known as reduced specific viscosity and as reduced viscosity. An abbreviation used for this term is $\eta_s p$. It is the fractional increase in viscosity which results when a polymer is dissolved in a solvent.

specific volume
The volume, at a specified temperature and pressure, occupied by a unit mass (for example, 1 gram) of a substance, i.e. the reciprocal of the density (1/D). Units cm³/g or m³/kg or ft³/lb.

specific Young's modulus
See *specific modulus*.

specification limits
See *tolerance limits*.

specimen
A piece or portion of a sample used to make a test.

specimen blank
A part-finished test specimen. See, for example, *Bell Telephone Laboratories test method*.

specimen preparation methods
See *test specimen preparation methods*.

specks
Compounding ingredient agglomerates: poorly dispersed compounding ingredients in a rubber mix or compound.

spectrophotometer
A machine used to measure the intensity of colour. Gives a more precise measurements, or characterisation of colour, than a *colorimeter*.

specular transmittance
The preferred terminology is regular transmittance.

speed control - injection speed
See *speed programming - moulding machines*.

speed of flow
An abbreviation used for this term is u. See *Reynold's number*.

speed of testing
It is usual to specify both the speed of testing and the temperature of testing as both often affect the results of *common tests*.

speed programming - moulding machines
Most injection and compression moulding machines are *self contained*, hydraulic machines. The pump(s) circulate fluid through the system at a certain rate or speed. The greater the volume of hydraulic fluid fed to a hydraulic cylinder or motor (hydraulic actuators) the greater will be the rate of movement (the speed) obtained. Conversely if the flow is reduced, or throttled, then the slower will be the speed. Alterations in speed can be obtained in various ways. For example, by using multiple pumps, an *accumulator* or a *servovalve*.

The output from several hydraulic pumps may be combined to obtain higher speeds on a moulding machine or, the fluid stored in an accumulator may be utilized at the appropriate moment. If the output from the pump is fed to the hydraulic actuator via a *servovalve* then, the speed may be reduced to a preset value. Such a system allows, for example, the injection speed to be altered during the actual mould filling stage on an *injection moulding machine*.

On most injection moulding machines now produced, injection speed is selected initially and the hydraulic system gives the pressure necessary, up to a preselected maximum, to maintain that speed. See *cavity pressure control*.

speed ratio
See *friction ratio*.

spew
Also known as *flash* or *rind*.

spew groove
See *flash groove*.

spew line
See *flash line*.

SPF
An abbreviation used for superior processing furnace (black). See *carbon black*.

sphere filled grade
A grade of a polymer filled with *glass microspheres*. For example, with *polybutylene terephthalate* the use of such spheres improves flexural modulus, impact strength, creep resistance and also reduce shrinkage.

sphere hazemeter
See *light transmittance measurement*.

spherulite
A crystal structure which consists of *fibrils* radiating from a central point: amorphous polymer is present between the fibrils.

SPI
An abbreviation used for the Society of the Plastics Industry - a USA based organization.

spider
That part of the die assembly which supports the torpedo. See *spider legs*.

spider arms
See *spider legs*.

spider gate
A *gate* used in injection moulding for the moulding of, for example, hollow cylindrical mouldings: an alternative to a *diaphragm gate*. The material feeds into the side of the component via a number of legs or spokes and, as a result, welds are produced in the finished component. Although such welds may be of adequate strength, their presence may cause an obvious surface blemish with certain materials, e.g. metallic filled compositions.

spider legs
Also known as spider arms. That part of the *die* assembly which supports the *torpedo* and which consist of thin fins or webs.

spider lines
An extrusion fault which is caused by the plastic melt being divided by the *spider legs*.

spin welding
A *friction welding* technique used for *thermoplastics materials*, for example, *polymethyl methacrylate* (PMMA), in which the necessary heat is generated by rotating one circular component against another while applying pressure. The pressure is applied while the joint cools. See *thermoformed bottle*.

spin winding
A *filament winding* technique in which a textile spinning process is combined with a filament winding process. If a *rove* is drawn at different speeds then the *yarn* so produced will have a different cross-section. When laid on the mandrel such a yarn will, for example, reduce the size of the *lump* produced at each end of a filament wound structure.

spinneret
A device used in the manufacture of fibres. A nozzle with holes or slots of the required size. For example, fibres are formed by forcing *viscose* through a spinneret: the fibres are then hardened by immersion in a *coagulating bath*. That is, *cellulose* is regenerated.

spinning
A process used to produce continuous yarns or threads. Originally referred to the process used to produce spun yarn from staple fibre: that is, the mass of short fibres (for example, wool) is drawn into strands which are then twisted so that fibres grip each other. *Continuous filament yarns* are also made by a spinning process: a liquid fibre-forming material is forced from the holes of a *spinneret*, hardened and then twisted together. In the production of *man-made fibres*, the hardening process may be used to classify the spinning technique into *wet spinning*, *dry spinning* and *melt spinning*.

spinning process
A process used to produce *rubber thread* by casting *latex* into a coagulating bath; the threads are then dried and vulcanized.

spiral flow length
The result of a *spiral flow test*.

spiral flow test
A *flow test*: a high shear rate flow test. This test is used to assess the ease of flow of thermoplastics materials. It is not a standard test although it is widely known in the injection moulding industry. The test is performed on an *injection moulding machine* under specified conditions and using a mould which has an open-ended cavity cut into one half. The cavity consists of an Archimedean spiral which is fed from the centre via the *sprue*. As material is forced into the cavity, flow continues until the material sets or cools. After ejection, the spiral is measured: either the weight or the length is recorded. The result is expressed, for example, as a certain length (a spiral flow length) produced under specified conditions. When this test is being performed, it is important to hold the screw cushion size constant while changing another machine setting, for example, while changing the temperature or, the injection rate. The test is not a straight forward rheological test as hot material is flowing into a cooler mould.

spiral wrapping
A method of applying external pressure to a *rubber hose* during vulcanization by using a narrow strip of cloth wound helically along the hose length.

spiralled hose
A hose reinforced with layers of strands wound helically in opposite directions.

spirit soluble resole
A *resole* obtained as a result of using ammonia as a catalyst. This class of resole has good electrical insulation properties and is therefore used, for example, to prepare laminates, based on cloth or paper, for use in the electrical industry. See *water soluble resole*. Cresol-based resoles give better electrical insulation properties than phenol-based resoles.

spiro-ladder polymer
See *ladder polymer*.

spiro-polymer
See *ladder polymer*.

spirobisindan PC
An abbreviation used for *spirobisindan polycarbonate*.

spirobisindan polycarbonate
An abbreviation used for this type of material is PC-SBI or SBI-PC: also known as spirobisindan PC. A heat resistant *polycarbonate*: an aromatic polycarbonate with a high heat resistance but which is relatively brittle compared to a *polycarbonate* based on *bisphenol A*.

split core
A *core* manufactured in two or more parts so as to ease the moulding of internal undercut components.

split mould
A type of mould in which the cavity is carried in *splits*: these are held together by a chase bolster during moulding but are removed and separated for component ejection after moulding.

splits
Components of a *split mould* which carry the *impression*.

sponge rubber
A soft, elastic, cellular material the cells of which are interconnecting. Made from solid rubbers (all vulcanizable rubbers can be used) and a blowing system: either nitrogen gas or, solid blowing agents or, a mixture of both. *Sodium bicarbonate* is the preferred blowing agent and is mixed into a suitably plasticized and compounded rubber: a *fatty acid* may be used to assist in the liberation of carbon dioxide. A wide range of filler and plasticizers are used. By replacing approximately half of the filler in sponge rubber, with *aluminium trihydrate (ATH)*, a more fire retardant sponge results. The use of a *plasticizer* such as a *phosphate* also helps in this respect as does the use of a chlorine containing polymer (*chloroprene rubber*).

spool
A term used in *hydraulics* and which refers to a moving, cylindrical part of a hydraulic component which when moved, directs fluid flow through the component.

spray
The assembly of moulded components with its attendant *feed system*. Also called a lift.

spray coating
A coating process used, for example, to make a container more impermeable. After production of a *blow moulded container*, the container is spray coated with a material, such as PVDC, under humid conditions. With spray coating no drainage should be needed; the bottle is dried for 2 to 3 minutes at 65 to 75°C. Output rates can reach approximately 18,000 bottles per hour.

spray drying
Technique used to recover solid polymer from a polymer latex whereby the *latex* is sprayed (from a fast rotating disc) into a chamber and hot air in injected at the base of the chamber. Used, for example, to recover *polyvinyl chloride* (PVC) from the latex produced by *emulsion polymerization*.

spray gun
Also known as a chopper gun. A machine used to spray resin but which also may spray glass. Associated with the moulding technique, known as *spray up*, which is used with *unsaturated polyester resin*.

spray pipe
The pipe through which steam, or steam/water mixture, is introduced into, for example, a *central drilled roll* of a *two-roll mill*. Along its length, the pipe has holes through which the heating medium sprays: the pipe does not rotate with the roll.

spray up
A moulding technique commonly associated with *unsaturated polyester resin*. For example, the catalyzed resin is fed to a spray gun together with *glass rovings*. The rovings are chopped into small lengths and are then impregnated with resin during passage through the (chopper) gun. The impregnated reinforcement is then sprayed against a mould, which has been covered with *gel coat*, and then consolidated as in the *hand lay-up method*. A dry filler, such as glass microspheres, can also be introduced into the resin stream of the *spray gun* so as to minimise filler mixing problems and produce light-weight composites.

sprayed rubber
Also known as latex sprayed rubber (LS rubber) or as, whole latex rubber. A *natural rubber* which contains all the serum constituents: white flakes of rubber are produced by *spray drying* ammonia stabilized latex (Hopkinson's process). The material is compressed into blocks. The rubber so produced cures more quickly but requires more energy input during processing. Acetone and water extracts are relatively high.

spread coating
A *coating method*: a moving substrate passes beneath a spreading device, such as a *doctor blade*, which coats the moving substrate with a fluid composition.

spread of flame
See *Underwriters Laboratory horizontal burning test*.

spreader
Another word for *torpedo*.

spreader roll
A small roll used to remove wrinkles from sheeting during production. See *calendering*.

spreading
A process used to coat substrates (for example, textiles or paper) with a polymer. Originally the process was used to coat textiles with a rubber solution or with *latex*, by means of a *spreading machine*, so as to produce water-proof fabrics. Now a wide range of substrates are coated with an even wider range of polymers so as to achieve barrier properties.

spreading machine
A machine used to achieve *spreading*. Often polymer solutions are used and applied to the substrate by means of rolls or a *doctor knife*.

spring detent
A safety device used in *split* and *side core moulds*.

sprue
In a conventional *injection mould* the sprue describes the channel which joins the mould cavity to the nozzle and also the solid material which forms in that channel. A sprue is also known as a carrot.

In order to assist removal, the sprue is usually tapered (with a taper angle of from 3 to 7°) and the diameter of the small end of the sprue (sometimes known as the 'O' dimension) is made slightly larger than the hole in the tip of the nozzle. The diameter of the small end of the sprue can be as small as 1.5 mm/0.06"; however, for components with a volume of up to 300 mm^3/20 in^3 a 5 mm diameter sprue would be more usual. The diameter of the large end of the sprue should be equal to, or larger than, the diameter of the main runner system so as to ensure that the runner does not seal first and prevent the application of sufficient *hold pressure*. The sprue should blend smoothly into the runner system and it should be polished in the direction of flow as it is found that this provides less resistance to flow and gives easier ejection. To minimise pressure losses and to reduce scrap production, sprues should be kept as short as possible. See *sprue bush* and *sprue puller pin*.

sprue break
The action of withdrawing the injection unit away from the mould is referred to as sprue break. Withdrawn, for example, by 0.25 in/6 mm.

sprue bush
The part of the *injection mould* which carries the *sprue* is known as the sprue bush or the sprue bushing. It is most important that the fit between the machine *nozzle* and the sprue bush is a good one as leakage of polymer at this point increases the danger of degraded material being drawn intermittently into the mould. The radius of the sprue bush is usually slightly larger than the corresponding nozzle radius, e.g. by about 0.8 mm/0.03") so as to minimise stagnation and to ease ejection.

sprue bushing
See *sprue bush*.

sprue gate
A *sprue* joined directly to an *injection moulding*, e.g. at the base of a bowl or bucket. Such a large *gate* is commonly used for large single-impression mouldings as it is relatively easy to machine or make, gives symmetrical mould filling and the large scar (which is produced when the gate is removed) is hidden in use. However, it is often found that parts made with a sprue gate fail in service by cracking in the gate region. This is because the gate is so large that overpacking can easily occur and this *over-packing* causes stress in the gate area. To help alleviate this problem it isbeneficial to include a slight thickening of the base beneath the *sprue*; this is gradually blended back into the wall over a large diameter.

sprue puller pin
Also known as a sprue puller. A part of the *feed system* and often part of the *cold slug well*. A pin whose use ensures that the sprue stays with a specified part of the mould, for example, the *moving mould half*. The head of the pin, that part in contact with the melt, may be undercut so that the plastics material must remain with the designated part of the *mould*. The undercut is cleared on ejection as the pin is moved forward: it is attached, for example, to the ejector plate.

sprueless mould
A type of *injection mould* in which the material involved in the *sprue* is kept from hardening (see *hot runner* and *insulated runner mould*) or, the sprue is dispensed with entirely by, for example, using an *extended nozzle*.

sprueless moulding
An *injection moulding* produced without the production of a hardened *sprue* using for example, a *plunger nozzle*.

SPS
An abbreviation used for *syndiotactic polystyrene*.

spue
A term once used for *spew*.

spun roving
A strand which has been repeatedly doubled back upon itself so as to make a *roving*.

spun yarn
Also known as staple fibre yarn or as staple yarn. Traditionally manufactured from short staple fibres such as wool or cotton. Man-made staple fibres are made from continuous filaments by cutting such long fibres into short lengths: the short lengths are then collected together in a rove in which the fibres are parallel and over-lapping. In the spinning process this is elongated and rotated to make a yarn in which the separate fibres are held together by the twist. There are two main types of *yarn*; *continuous filament yarn* and *spun yarn*.

sputtering
A technique for depositing a thin film of metal under very low pressures. The metal to be deposited is made the cathode of a low pressure discharge system (below 1 mm). The object to be coated is placed between the *anode* and the *cathode* so that when a high voltage (about 20,000 volts) is applied between the anode and cathode the metal ions, which come from the cathode, coat the object.

sq
An abbreviation used for square. For example:

sq ft = square foot or square feet;
sq in = square inch; and,
sq yd = square yard.

SQC
An abbreviation used for *statistical quality control*.

square measure
Measurement of area in square units. Units of area and volume are derived from units of length (l). For example:

square centimetre = cm^2 = 0·155 000 in^2.
square foot = ft^2 = $\frac{1}{9}$ of a sq yd;
square inch = sq in = 1 in^2 = 1/144 sq ft = 6·451 6 cm^2;
square kilometre = km^2 = 0·386 102 square miles = 10 ha;
square mile = sq mi = mi^2 = 2·589 988 km^2;
square metre = m^2 = 10·763 9 ft^2 = 1·195 99 yd^2; and,
square yard = sq yd = yd^2 = 9 sq ft = 0·838 127 36 m^2.

See *UK system of units* and *US Customary Measure*.

square pitch
Screw terminology: used when the pitch of a *screw* is equal to its diameter. In this case, the screw is called a square pitched screw: such a screw has a helix angle of 17·7° and the number of turns is equal to the L:D ratio.

square pitched screw
See *square pitch*.

squeegee
See *insulation*.

squeeze bars
See *squeeze jaws*.

squeeze bottle
A soft flexible bottle initially made from *low density polyethylene* (LDPE) by *blow moulding*. PE plastics materials made the plastics bottle acceptable when, in the 1950s, 'squeeze' bottles were first used for packaging washing up liquid (detergent). Such bottles offered lightness in weight, good impact strength and ease of dispensing (by squeezing).

squeeze jaws
Also known as squeeze bars. Two bars which come together so as to weld the parison in *extrusion blow moulding* processes.

squirming
See *0° nylon belt*.

sr
An abbreviation used for *steradian*.

SR
An abbreviation used for *synthetic rubber and for* surface resistivity.

SRF
An abbreviation used for semi-reinforcing furnace (black). N770. See *carbon black*.

SRF-NS
An abbreviation used for semi-reinforcing furnace - non-staining (black). N774. See *carbon black*.

SRIM
An abbreviation used for *structural reaction injection moulding*. Also used for structural resin injection moulding.

SS test
An abbreviation used for *simulated service test*.

St
An abbreviation used for *stokes*.

St. Joe flexometer
A test apparatus: a test machine used to study the heat build up, and the resistance to flexing fatigue, of a cylindrical rubber test specimen. The cylindrical rubber test specimen (37.5 mm by 37·5 mm) is held between two plates under a load (about 200 kg) while the top plate is eccentrically driven at 875 rpm. The bottom plate has a force applied to it so that a bending strain is applied to the test specimen. When the horizontal force suddenly drops, this is taken as the start of total fatigue.

stability
A term used in *hydraulics* and which refers to the ability of a control system to maintain control when there are outside disturbances.

stabilization
Minimization of the effects of degradation caused by, for example, heat and/or light. See *stabilizer*.

stabilizer
Also known as an antidegradant or as an external stabilizer. An additive used to minimize the effects of degradation caused by, for example, heat and/or light. The type of stabilizer used depends upon the type of degradation which will degrade a particular polymer. See *heat stabilizer, antioxidant, synergism* and *stabilizer package*.

stabilizer package
A total stabilizer system designed to be added in one lot or package. See *barium/cadmium stabilizer package*.

stack
The arrangement of rolls in, for example, a *calender*.

stack mould
An *injection mould* used for *stack moulding*: a multi-daylight mould. Output is increased by employing multi-daylight moulds. Stack moulding can be used to increase the output from a moulding machine and/or to reduce the clamp pressure required for the component. The output from an existing moulding machine can be increased by 80% if two-day lights are used. Multi-daylight moulds became established only when it became possible to employ *hot runner systems* with sprueless gating of the mould cavities lying in the two mould parting surfaces. Usually the gate is always parallel to the longitudinal axis of the mould. Such moulds are used to produce symmetrically shaped components which are generally small and flat, such components usually have a large projected area relative to their weights, e.g. jar caps, cassettes, petri dishes and tape reels. Ideally there should be no need for complex ejection or side core actions with such moulds. If containers are being moulded then sequenced opening allows the tops and bottoms to be directed to separate bins.

stack moulding
An *injection moulding technique* which uses a *stack mould*.

stacking table
A device which stacks cut, extruded sheet after cooling. See *extrusion*.

stage A resin
A soluble and fusible type of material: see *novolak* and *resol*.

stage B resin
A gelled, rubbery-type of material: see *resitol*.

stage C resin
A highly cross-linked structure: see *resit*.

staining
A change of colour of a component when exposed to light or, a change of colour of a material which is in contact with, or adjacent to, for example, a rubber component.

staining antioxidant
An additive, an *antioxidant*, used in a rubber compound to prevent oxygen attack and which has a tendency to stain or discolour the rubber on ageing. See *amine antioxidant*.

stainless steel
A type of steel (iron 70 to 90%) which contains chromium (12 to 20%) and carbon (0·08 to 0·8%).

staking
The process used to push inserts into preformed holes.

stalk
See *sub-sprue*.

stand oil
A *drying oil* that has been thickened by polymerization: achieved by heating in an inert atmosphere.

standalone
Means that a system is self-contained and does not require any additional items of equipment for it to function as intended,

standard abbreviations
See *abbreviations* and **table 1**.

standard China rubber
An abbreviation used for this type of *natural rubber* is CSR. See *technically specified rubber*.

standard dead load test
See *International Rubber Hardness*.

standard deviation
Abbreviations used for this term include σ or, sd or, SD. The standard deviation may be defined as the square root of the mean of the squares of the individual deviations from the average.

standard gate
See *side gate*.

standard international grades - natural rubber
Visually inspected grades of *natural rubber*, such as *ribbed smoked sheet, white* and *pale crepes*.

standard Malaysian rubber
Also known as SMR or as, Esemar. The first, and still the most important, *technically specified rubber* (other technically specified rubbers include SIR, SLR, TTR and CSR). SMR *natural rubber* is graded for maximum dirt content, maximum ash content, maximum nitrogen content, volatile matter, *Wallace rapid plasticity* and *plasticity retention index* (PRI): for some grades other information must be supplied, for example, colour, *Mooney viscosity* and cure information. That is, the rubber must conform to a set of specifications. Such *natural rubber* is usually sold in 33·3 kg bales although a small amount of sheet is produced.

standard Malaysian rubber latex grades
Grades of *natural rubber*, which originate from Malaysia, and which are produced from liquid *latex*: such materials are sold in *block form*.

standard mould base
See *standard mould set*.

standard mould components
Components for moulds which are available in a range of specified sizes and finishes. A number of companies now sell a range of standardised mould components and standardised mould sets.

standard mould set
Also called a standard mould base. A complete mould (for example, an *injection mould*) minus the cavities, which has been assembled from *standard mould components* and which is available in a range of specified sizes and finishes. A number of companies now sell a range of standardised mould components and standardised mould sets for use mainly in *injection moulding*. These mould sets are available in pre-hardened steel if required and the cavities and cores are cut into the appropriate mould plates when required. Bolster sets are also available so that cast or machined inserts can be readily accommodated. The range of sizes available is now very wide and, because they are produced in large numbers, they can be relatively cheap. In general, the quality of such standardised units is excellent and as replacement parts are interchangeable and readily available they are very popular.

standard of measurement
A material object which, under certain specified conditions, serves to define, represent or measure the magnitude of a unit. See *units of measurement* and *Mètre des Archives*.

standard polystyrene
See *polystyrene*.

standard pressure
See *standard temperature and pressure*.

standard rubber
See *technically specified rubber*.

standard Sri Lanka rubber
An abbreviation used for this type of *natural rubber* is SLR. See *technically specified rubber*.

standard temperature
Those temperatures which have been specified by a standards organization (for example, *ASTM*) for testing purposes. The actual temperatures may differ from one material to another and may depend on the application. See *standard temperature and pressure*.

standard temperature and pressure
An abbreviation used for this term is stp or STP: has the value of 0°C and 101·3 kPa (760 mmHg). Formerly called normal temperature and pressure.

standard Thailand rubber
An abbreviation used for this type of *natural rubber* is TTR. See *technically specified rubber*.

standardisation of test methods
Whether a test is used to determine an absolute value of a property, for example, for research, or whether its purpose is routine quality control, a major requirement of the test is that it should be repeatable. People in the same laboratory, or in different laboratories, should be able to get a very similar result on the same material. To obtain repeatability, standardization of test methods is essential as there are a large number of factors which will influence the result obtained. For example, test machine design, test piece size, shape and method of production, temperature of test and the speed of testing. If a cut sample is used, then the surface finish of the cut edges must also be controlled so as to minimize errors. See *standards organizations*.

standards organisations
National and/or international standards organisations which are responsible for preparing and issuing standards on polymers. For example, the American Society for Testing and Materials (ASTM), the Association Francaise de Normalisation (AFNOR), the British Standards Institution (BSI), the Deutsches Institut Für Normung (DIN) and the International Organisation for Standardisation (ISO) and the *Comité Européen de Normalisation*. The work of the international bodies is supported by the national standards organisations who are in turn supported by trade associations, companies, government departments and local authorities. Copies of standard test methods relevant to plastics and rubbers can be obtained from these organisations.

stannic
Term is applied to tin compounds where the tin is in its +4 state, that is, tin (1V).

stannous
Term is applied to tin compounds where the tin is in its +2 state, that is, tin (11).

stannous octoate
Also known as stannous dioctoate. An organo-tin compound which causes eye irritation: skin contact should be avoided. A tin compound which is used as a catalyst, together with an amine compound, in *flexible polyurethane foam* production. The tin compound is used as a catalyst for the *cross-linking reactions* whereas the amine compounds (tertiary amines) are used as catalysts for the blowing reactions in flexible polyurethane foam production by the *one shot process*. Can balance these two types of reaction by adjusting catalyst levels.

A polyether foam could use stannous octoate, dimethylethanolamine and 1,4-diazabicyclo-2,2,2-octane as a catalyst system. The stannous octoate may be replaced by *dibutyl tin dilaurate* if there is a danger of hydrolysis in water-containing blends.

staple
Usually means staple fibre. See *spun yarn*.

staple fibre
See *spun yarn*.

staple fibre yarn
See *spun yarn*.

staple yarn
See *spun yarn*.

star block copolymer
See *radial block copolymer*.

star polymer
See *radial polymer*.

starch
Also known as amylum. A naturally occurring material which is based on polysaccharides. Extracted from plant cells and used as a *degradant* in thermoplastics materials, for example, *polyethylene*. This material has a relative density (RD or SG) of 1·5.

stark rubber
Natural rubber in which crystallization has occurred on long-term storage.

start
The number of separate threads that can be traced along the *screw*: this is usually one so most screws are single start screws.

start up
The procedures involved in getting a machine into full production. These comments apply to extruders, blow moulding machines and injection moulding machines.

PREPARATION FOR PRODUCTION. Obtain advice or, check the records, so that it is known what machine settings are needed for the job in hand. Turn the main power switches on and select, or set, the temperature specified. Ensure that the cooling water is on and check that it is flowing through the appropriate circuits, for example, through the hopper throat. (If fitted, put on the cylinder cooling and, turn off the cooling water to the screw.) Preheat all parts of the system to operating temperatures, for example, the barrel, the mould and the hydraulic oil (see *oil preheating*) Once the machine is at the set temperatures (see *warming up*) then it should be allowed to equilibrate before any material is introduced into the barrel. This may take 20 minutes so, use the time to prepare for the production run. For example, check that the nozzle/die/mould is clean and that all parts are operational. Review the production order for colour and quantity: check that all necessary tools and equipment are in position. Check that ancillary equipment, for example, the hopper equipment, is clean and is functioning.

SAFETY CONSIDERATIONS. One of the most dangerous times during processing is at start up. This is because the machine is being heated and the material may decompose and, for example, spit from the die or nozzle. The operator is involved in getting the machine running satisfactorily and this involves close contact with machinery. So, great care should be taken at start up. In particular nobody should be allowed to stand in front of the die/nozzle and the hopper lid should be firmly in place so that the screw cannot be seen (and therefore touched). No unauthorized person should be in the processing area.

starting time
The starting time of a moulding cycle. For example, with *injection moulding*, the injection cycle may be considered to commence the instant the guard begins to close.

starve fed
A feeding technique for thermoplastics processing equipment (such as a *twin screw extruder*) whereby the hopper of the machine is not filled: the material is dosed, or metered, into the *feed throat*. See *flood feeding*.

state of cure
The attainment of a certain level of a specified property in a thermosetting system. The attainment of a certain level of a specified property, for example, modulus, in a rubber compound during cure.

static
The electrical charge generated, and retained, on insulated surfaces: generated on film surfaces when they are separated.

static coefficient of friction
An abbreviation used for this term is F_s/R. Where R is the normal force at the contact surface and F_s is the friction when the body is just on the point of moving. See *coefficient of friction*.

static fatigue
Also known as creep fracture or as, stress rupture. Fracture which results as a result of long-term loading under a steady load. For example, the fracture which results as a result of loading with a weight. The greater the load, the quicker will fracture occur.

static friction
The *friction* between bodies which are motionless with respect to each other: this is greater than *sliding friction*.

static mixer blender
An in-line unit that mixes the plastics *melt* stream so as to make it uniform in temperature and composition; does this by cutting and recombining the melt stream by making it flow over many blade-type obstructions after it has been discharged by the *screw*.

static reduction
See *antistatic agent*.

station
A section, or part, of a machine. For example, the blowing station on an *injection blow moulding machine*.

stationary plate
One of the three plates of a *three-plate mould*: the plate which is attached to the *stationary* or *fixed platen*. Also known as the runner plate or as, the clamping plate

stationary platen
See *fixed platen*.

statistical process control
An abbreviation used for this term is SPC. The application of statistical techniques to the production process so as to try and keep the process under control and thus improve quality.

statistical quality control
An abbreviation used for this term is SQC. The application of statistical techniques to the production process so as to try and improve quality.

statute mile
A land *mile* of 1,760 yards. See *UK system of units* and *US Customary Measure*.

std
An abbreviation used for standard.

steam heating
The heating of processing equipment using steam. Such heating is widely used in polymer processing, particularly in the older-established *rubber* and *thermoset* industries. Used because the temperatures needed for such materials are comparatively low (below 150°C/302°F) but mainly used because this heating medium was available when the processing of such materials was first widely performed.

COMPARISON WITH ELECTRICAL HEATING. Steam will raise the temperature of the system very quickly as the heat content of the steam is very high - due to the latent heat of vaporization of water which is recovered when the steam condenses. The ability to use the steam heating channels for cooling (for example, during *mastication* or the *compression moulding* of thermoplastics) is also a tremendous advantage. Electrical heating is usually much slower but electrical heating can reach much higher temperatures. The upper temperature is, for example, 350°C/662°F.

CURING TIMES. To achieve curing of rubber compounds, in *compression moulding*, it is necessary that the mould is held closed after shaping has been performed: the cure time depends upon the mould temperature and therefore upon the steam pressure The higher the temperature, the shorter the cure time (minutes or m).

5 m at 65 psi = 8 m at 55 psi = 14 m at 45 psi = 30 m at 35 psi.
10 m at 65 psi = 15 m at 55 psi = 26 m at 45 psi = 45 m at 35 psi.
15 m at 65 psi = 24 m at 55 psi = 40 m at 45 psi = 65 m at 35 psi.
20 m at 65 psi = 29 m at 55 psi = 51 m at 45 psi = 86 m at 35 psi.

PRESSURES REQUIRED. To reach temperatures greater than 150°C/302°F, with steam, requires the use of high pressures and the equipment must then be built to withstand such pressures. Because of cost and safety considerations, steam heating is therefore usually limited to approximately 150°C/302°F.

Temperature		Steam pressure	
°C	°F	psi	atm
102	215	1	0.07
109	227	5	0.34
115	239	10	0.68
121	250	15	1.02
126	259	20	1.36
131	267	25	1.70
135	274	30	2.04
138	281	35	2.38
142	287	40	2.72
145	292	45	3.06
148	298	50	3.40
150	302	55	3.74
153	307	60	4.08
156	312	65	4.42
158	316	70	4.76
160	320	75	5.10

SAFETY CONSIDERATIONS. Because of the pressures involved when steam heating is used, the equipment must be inspected regularly to ensure that it can withstand the circuit pressure plus an adequate safety margin as rusting and corrosion can cause component weakening. Steam can cause severe burns, because of the high heat content, and also because escaping steam is initially colourless when it first escapes.

STEAM/WATER MIXTURES. To obtain low heating temperatures, for example, 40°C/104°F water may need to be mixed with steam. This should only be attempted if the system is fitted with an appropriate mixing valve as otherwise, for example, the higher pressure steam will push the lower pressure water back along the supply line.

steam-heated fluid bed
A *fluid bed* which uses to steam to aerate the bed and to provide the heat necessary for *continuous vulcanization*.

stearate
A salt or ester of *stearic acid*.

stearic acid
Also known as octadecanoic acid. May be represented as $C_{17}H_{35}COOH$ or $CH_3(CH_2)_{16}COOH$. A monobasic organic acid or *fatty acid*. This material has a relative density (RD or SG) of 0.92 and a melting point of approximately 55°C. Widely used as a *lubricant* in many polymer compositions. It is used as an *accelerator/activator* in rubber compounds and as a processing aid. In sponge mixings it aids the decomposition,

or degradation, of sodium bicarbonate and helps processing as its softening action is more pronounced at moulding temperatures. This gives good expansion but means that the compound does not suffer from cold flow: this could result from the use of liquid plasticizers. The reaction products of stearic acid, for example, metal stearates are widely used as *stabilizers*.

stearine
Also known as stearin. This white, flaky material has a relative density (RD or SG) of 0.85 and a melting point of approximately 50°C. Made by the saponification of natural fats: a mixture of *stearic* and *palmitic acids*. May be used in most cases where *stearic acid* is used. See *tristearin*.

stearoyl
The radical which is derived from *stearic acid*.

steatite
See *talc*.

steel
Iron containing from 0.1% to 1.5% carbon in the form of iron carbide (cementite). An alloy of iron with elements such as carbon, manganese, silicon and phosphorus: the last three elements may only be present in trace amounts. Alloy steels may contain appreciable amounts of elements such as chromium, cobalt, nickel, tungsten and vanadium. The properties of the steel will depend not only upon its chemical composition but also upon its heat treatment, for example, quenching and tempering have a tremendous effect upon hardness and toughness. A major problem with iron and some steels is rusting: can be treated with metals such as zinc or cadmium so as to reduce this problem. *Powder coating* with plastics materials is also used to decorate and protect the metal. This material has a relative density (RD or SG) of 7.9. The most widely used metal for *mould* and *die* manufacture. Some alternatives to steel for mould making include *beryllium-copper*, *iron-based casting alloy*, *beryllium nickel* and *aluminium*.

steel cord
A cord, based on *steel*, and used for the tread bracing layers of radial ply tyres. The metal is drawn into fine fibres (e.g. 0.1 mm/0.004 in) and a number of such fine fibres, e.g. 12) are twisted or cabled together to give a cord of high strength but which is relatively flexible: such a cord could be about 1 mm in diameter (0.04 in). The steel cord is treated to improve adhesion by *brass coating*.

steel cord conveyor belt
A *belt* based on more than one *steel cord*: a belt in which the tension member consists of a number of steel cords.

steel cord tyre
A tyre which has *steel cords* incorporated for reinforcement. The use of steel gives good resistance to mechanical damage.

steel-braced radial
Of similar construction to a *fabric-braced radial tyre* but with one or more belts of steel wire lying beneath the textile belt. Such a tyre may have a *0° nylon belt* or *bandage*: this textile belt lies parallel to the direction of movement, at right angles to the textile cord and is located beneath the tread. This belt may cover two steel belts which may lie at a slight angle to each other and to the fabric belt.

Stellite
A trade name/trade mark for a family of hard, corrosion resistant alloys. This type of material is based on cobalt (35 to 80%), chromium (15 to 40%), tungsten (10 to 25%), molybdenum (0 to 40%) and iron (0 to 5%). Has been used, for example, to surface screws which operate in aggressive environments. Has been used to surface worn nitrided screws:

the nitriding is ground away before the Stellite is applied by welding at approximately 275°C. The screw flights are made approximately 0.25 mm/0.010 in larger: the barrel is bored to remove the original nitriding and re-nitrided. The hardness of nitrided steel can be 67 Rockwell C: the hardness of Stellite, as used for plastics, may be 47 Rockwell C. Not as hard as nitriding but sufficiently hard to give a new lease of life to a worn component.

step cure
A cure performed in steps or stages. For example, when curing thick rubber articles, the temperatures may be raised in steps so as to obtain a more uniform cure.

step-growth polymer
A polymer produced by *step-growth polymerization*. A step-growth polymer may also be referred to as a *condensation polymer*.

step-growth polymerization
Also known as step-wise polymerization. A major method of producing synthetic polymers in which the long chain structure is built up by the chemical reaction which occur between two reactive, or functional, groups. For example, by reacting a diol with a diacid a *polyester* is produced. See *step-growth polymer*.

step-ladder polymer
A partial-ladder polymer. See *ladder polymer*.

step-wise polymerization
See *step-growth polymerization*.

stepped screw
See *double parallel screw*.

steradian
The steradian (sr), which is the unit of solid angle, is the solid angle which having its vortex in the centre of a sphere, cuts off on the sphere's surface an area equal to the area of a square with sides the length of the sphere's radius. See *Système International d'Unité*.

stere
One cubic *metre*. Equivalent to 1.308 cubic yards.

stereo
A plate, based on a flexible polymer such as a rubber, and used in *flexographic printing*.

stereoblock
See *stereoblock copolymer* and *stereoblock polymer*.

stereoblock copolymer
A block copolymer in which the monomer blocks have monomer sequences with the same stereochemical configuration. See *stereoblock polymer*.

stereoblock polymer
A block copolymer in which the block differences are due to differences in the spatial configuration of the repeat units within the block. Can have blocks with different tacticity which results in different crystalline structures. See *stereoblock copolymer*.

stereochemistry
That branch of chemistry which considers the arrangement in space of the atoms in a molecule.

stereolithography
A *shape modelling technique* which utilises a computer program, for example, the program developed for mould machining (CAD/CAM) and a servo-driven laser beam. A table is positioned inside a bath of acrylic monomer and the laser is moved in accordance with the dictates of the program. The projected component is sectioned and at appropriate points the laser is triggered and the light pulse causes *polymerization* to occur at those points. Another section is taken and the

table is dropped by a small amount so that polymerization over the area of another slice can be performed. By repeated sectioning complex components can be produced without machining or moulding.

stereoregular polymer
Also known as a tactic polymer or as, a stereospecific polymer. A polymer in which there is a regular, repeating structure. A polymer produced by *stereoregular polymerization*. The repeat units, along the polymer chain, have the same configuration: that is, the groups of atoms have the same orientation in space. If the repeat unit is two carbon atoms long in a vinyl-type polymer then, the substituent group (for example, CH_3) can lie on the same side of the polymer chain or it can lie on alternate sides of the polymer chain. That is, an *isotactic polymer* can be produced or, a *syndiotactic polymer* can be produced. If there is no preferred orientation, an *atactic polymer* is produced. Stereoregularity is a result of an atom in the repeat unit being asymmetric. See α *olefin*.

stereoregular polymerization
Also known as stereospecific polymerization. The process used to produce a *stereoregular polymer*. In vinyl polymerization, this is shown as *tacticity*. Ziegler-Natta polymerization is stereoregular polymerization which results in, for example, *isotactic polymer*.

stereoregular rubber
See *stereorubber and tacticity*.

stereorubber
A rubbery polymer produced by *stereoregular polymerization*: included in this category are the commercially important polymers *cis-polybutadiene* and *cis-polyisoprene*. Such a rubber is also called a stereoregular rubber and is made by *solution polymerization*. See *tacticity*.

stereospecific catalyst system
A catalyst system which allows the production of polymers with a controlled molecular architecture. See, for example, *stereospecific styrene-butadiene rubber*.

stereospecific polymer
A polymer produced by *stereoregular polymerization*. See *tacticity*.

stereospecific polymerization
See *stereoregular polymerization*.

stereospecific styrene-butadiene rubber
Also known as stereospecific SBR. *Styrene-butadiene rubber* which has been produced by *stereoregular polymerization* (solution polymerization) and which has a more regular structure than SBR produced by emulsion polymerization.

sterile air
See *container sterilization*.

stibnite
See *antimony sulphide*.

stick-slip
See *friction*.

sticking-in
Also known as jamming. An injection moulding problem caused by an incorrectly dimensioned *sprue*. For example, with *polyether ester elastomer* if the included angle of the *sprue* is less than 2·5°, then the component cannot be ejected properly as the sprue will bind in the *sprue bush*.

stiffness constant
See *elastic modulus*.

stilbene
Also known as 1,2-diphenylethene. This colourless solid material has a melting point of 124°C: it is used in *dye manufacture*.

stimulation tapping
A method used to tap a rubber tree. Ethylene gas is the active ingredient in tree stimulation treatments (See *Hevea brasiliensis*). Stimulation tapping, and *puncture tapping*, may increase the yield or maintain the same yield with a reduction in *tapping* intensity: this conserves the tree bark. See *natural rubber*.

stock
A rubber compound used as a feed-stock, for example, as a calender feed.

stock control
See *material handling*.

stokes
Units of *kinematic viscosity* in the *centimetre-gram-second system*. An abbreviation used for this term is St. Has the units of $10^{-4} m^2 s^{-1}$.

stone
A unit of weight measurement in Avoirdupois weight equal to 14 pounds.

stoppages
Breaks in production. The following comments apply to extruders, blow moulding machines and injection moulding machines running on thermoplastics materials.

AFTER PROCESSING A HEAT-STABLE MATERIAL. If a thermally stable plastic (for example, *polystyrene*) is being processed, then for an overnight stop, it is usually only necessary to close the slide at the base of the feed hopper, turn off the cylinder heaters (leave the nozzle/die heater on) and then purge the cylinder clean by pumping the screw dry. When nothing more comes from the nozzle, put any barrel cooling on maximum and when the machine is cool, turn everything off. The machine is then ready for reheating when required

AFTER HEAT SENSITIVE MATERIAL USE. Decomposition, or burning, of the polymer in a barrel, will cause colour changes which will then result in the subsequent product being rejected. When this happens a complete shut down and clean out may be necessary. To prevent this, it may be necessary to purge a heat sensitive resin with another, more heat stable, plastic as this will withstand subsequent reheating. If, for example, material oxidation is a problem (with, say polyethylene or PE) then it may be best to leave the cylinder full of the PE material rather than pumping the screw dry before switching off.

AFTER HIGH TEMPERATURE OPERATION. When high barrel temperatures are used, then the shut down procedure should be modified so as to prevent thermal decomposition of the material/resin. For example, turn off the cylinder heaters (leave the nozzle/die heater on), put any barrel cooling on maximum and then periodically pump resin through the machine while it is cooling. Then, close the slide at the base of the feed hopper and purge the cylinder clean by pumping the screw dry and/or by making air shots. When nothing more comes from the nozzle/die, and when the machine is cool, turn everything off. The machine is then ready for reheating when required.

TEMPORARY. During a temporary stoppage, periodically purge the cylinder, or barrel, by passing material through the machine and/or by making air shots. If the plastic looks discoloured then increase the frequency of this purging. During a minor repair, the heaters on the heating cylinder should be set to low values, for example 150°C, so as to minimize thermal degradation.

storage hardening
An increase in viscosity of *natural rubber*, especially *latex grade rubber*, which occurs during storage or transit: the storage hardening reaction is increased by conditions of higher temperatures and lower humidity. Aldehyde-type groups on

the rubber molecule react with the amino groups of free amino acids and proteins to give cross-links. See *aldehyde-condensing agent*.

Stormer viscometer
A test apparatus: a test machine used to study the viscosity of rubber *latex*. A rotational *viscometer*.

stp
An abbreviation used for standard temperature and pressure (0°C and 101.3 kPa).

STP
An abbreviation used for standard temperature and pressure (0°C and 101.3 kPa).

straight cross-channel barrier mixing section
A cross barrier mixer. See *dispersive mixing section*.

straight fibre reinforced composite
A composite made from a *non-crimp fabric*.

straight rubber
See *non-extended rubber*.

straight wrapping
A method of applying external pressure to a *rubber hose* during vulcanization by using a cloth laid along the hose length: the cloth is then wrapped several times around the hose.

straight-chain plasticizer
A *plasticizer* based on a normal, or straight-chain, alcohol. For example, a straight-chain trimellitate plasticizer is based on trimellitic anhydride and a straight-chain alcohol. See *trimellitate plasticizer*.

strain
An abbreviation used for this term is e or ϵ. Deformation or distortion which is caused by applying a *stress* (a load) to a body or to a material. The strain is the ratio of the dimensional change to the original dimension.

strain birefringence
Birefringence introduced by *stress*.

strain energy
The elastic energy stored in a body as a result of deforming or distorting that body by applying a *stress* (a load) to the body. For many materials, at small strains, it is fully, and instantly, recovered when the load is removed.

strain gauge
A device used to measure *strain* or dimensional changes. A simple strain gauge consists essentially of a grid of fine wires (50 μm/0.002 in diameter) bonded onto a paper or plastics foil and through which an electric current may be passed. It is found that if the resistor, the grid of wires, is strained, its resistance changes and the amount of change is proportional to the load applied. Four sets of gages or resistors are grouped together so as to obtain a strain gage assembly which gives a more sensitive, or higher output, device together with temperature compensation. When a force is applied to the assembly the output alters and this small change may be amplified so that it may be displayed and/or used to initiate a control action (see *locking force*). Some gauges use a semiconductor element in place of a wire element.

strain gauge pressure transducer
A pressure *transducer* based on a *strain gauge*.

strain hardening
Also called work hardening. An increase in the resistance to deformation caused by, for example, *orientation* and/or *crystallization*.

strain softening
Also called work softening. A decrease in the resistance to deformation caused by, a reduction in cross-sectional area during a *tensile test* or by *adiabatic heating* of the test specimen.

strain test
A test done under constant load: a rubber test specimen is held under constant load and the deformation measured. The constant load may be 5 kg so that the deformation is less than 100% to stop crystallization effects. Such a test is used to study *vulcanization behaviour*.

strainer
A term used in *hydraulics* and which refers to a coarse *filter*. See *strainer extruder*.

strainer extruder
An *extruder* which removes unwanted material from, for example, gelled *polyvinyl chloride* (PVC) by straining the hot material through a fine wire mesh. The output from the *strainer extruder* is in the form of a circular rod and this continuous *calender* feed is passed through a *metal detector*.

straining
Removal of impurities by, for example, forcing a material through fine gauze. Straining after rubber *refining* will remove most impurities.

strand
A thread-like structure which consists of many *filaments*. See *glass fibre*.

stratosphere
The gaseous mantle, the atmosphere, which surrounds the planet Earth has been divided into several layers or strata. The stratosphere is the one above the *troposphere* but below the *ionosphere*. The temperatures within this layer are relatively uniform over considerable changes in altitude. It stretches from approximately 11 kilometres/7 miles to 65 kilometres/40 miles above sea level.

stratospheric
Commonly taken to refer to all of the earth's *atmosphere* above the troposphere.

stratospheric chlorine
Chlorine which is located within the upper reaches of the Earth's atmosphere i.e. within the *stratosphere*. Because of the longevity of *chlorofluorocarbons (CFC)*, which can be up to 130 years, such materials can drift as high as the *stratosphere*. Under the influence of *ultraviolet* (UV) light, chlorine atoms are shed and it is these which attack *ozone* to give chlorine monoxide and oxygen. The chlorine monoxide breaks down to give a chlorine atom which then commences the reaction again. It is thought that one chlorine atom can in this way react with approximately 100,000 ozone molecules so as to cause depletion of the *stratospheric ozone* layer.

stratospheric ozone
Ozone which is located within the upper reaches of the Earth's atmosphere i.e. within the *stratosphere*. This ozone deflects *ultraviolet* (UV) light and prevents it reaching the surface of the Earth: such ultraviolet light, UV-B of wavelength 290 to 320 nm, is potentially dangerous to life. See *stratospheric ozone depletion*.

stratospheric ozone depletion
Reduction of depletion of *stratospheric ozone* is caused by, for example, the action of *stratospheric chlorine*: such a depletion results in increased levels of harmful UV radiation at the surface of the Earth. The increased levels of *ultraviolet light* will cause problems, for example, high levels of blindness. See *chlorofluorocarbon*.

streamlined flow
Also known as *laminar flow*. As the speed of a flowing fluid increases, beyond a critical speed (u_c), then the flow becomes turbulent. See *Reynold's number*.

strength of bonding
See *surface bonding*.

stress
An abbreviation used for this term is p or σ. It is force per unit area. See *shear stress* and *strain*.

stress concentration factor
A factor which allows the increase in *stress*, caused by a feature such as a hole or a sharp notch, to be assessed. It is a ratio which is obtained by considering the maximum stress, when the feature is present to the stress when the feature is absent.

stress corrosion cracking
See *environmental stress cracking*.

stress crazing
A type of *environmental stress cracking* which is shown in glassy polymers such as *polystyrene* and *polymethyl methacrylate*. A large number of fine surface cracks or crazes may be formed when such materials are exposed to stress. Because of the large number of fine cracks the material may appear white and so the term stress whitening is also used. The stress may be internal stress or external stress and organic liquids may, or may not, be present. White spirit causes such crazing in polystyrene mouldings if the mouldings have a large amount of moulded-in strain.

stress relaxation
The slow decay, or relaxation, of the *stress* in a polymeric material when held at a constant *strain*.

stress relieving
An *annealing* process, if an *injection moulding* material is classed as being of stiff flow, then it is easy to build into the resultant moulding high moulded-in stresses which give rise to cracks appearing or, to failure occurring in service. With some materials, for example, *polyether imide*, the stresses involved can be very high and can be released violently in some cases. For components which require, for example, improved weatherability and better dimensional stability (during subsequent decoration and welding) an *annealing*, or stress relieving process, should be carried out after moulding.

stress rupture
See *static fatigue*.

stress softening - filled rubber
See *Mullins' effect*.

stress/strain curve
See *load deformation curve*.

stress whitening
See *stress crazing*.

stress-crack
An external or internal crack in a plastics material caused by tensile stresses which are less than the short-time mechanical strength of the material. The development of such cracks is frequently accelerated by the environment to which the plastic is exposed. The stresses which cause cracking may be present internally or externally or they may be combinations of these stresses.

stress-deformation behaviour - ideal fluid
The stress-deformation behaviour of an ideal fluid is derived by assuming that the *Newtonian fluid* is contained between two plates of very large area A. When a shear force (F) is applied to the top plate then the shear stress is equal to F/A. The top plate moves with a velocity u: the shear rate is equal to du/dr. *Shear stress* is directly proportional to *shear rate* for an ideal or Newtonian fluid.

stress-strain curve
A graphical plot of *stress* against *strain*.

stretching flow
See elongational flow.

stretching- calendering
The action of reducing the thickness of *calendered sheet*. Before it is cooled the sheet may be stretched deliberately to give thinner sheet) and if such stretching is performed it should be done at elevated temperatures, e.g. 170°C. At these temperatures the sheet thickness can be reduced considerably, for example, by a factor of 3. This gives a way of producing thin sheets at high outputs with low power requirements and with acceptable amounts of *reversion*. If higher stretch ratios are required then the sheet would need to be stretched in two stages with re-heating between the first and second stages.

striker
A part of the test apparatus used in the determination of *impact strength*, for example, a dart. See *falling weight impact strength*.

strip feed
Compound in the form of a long strip and of a width suitable for a particular machine. Often prepared using a *two-roll mill* or an *extruder*. See *rubber injection moulding*.

strip heater
An *electrical resistance element*: a mineral-insulated, metal sheathed resistance element in the form of a flat strip.

strip prepreg
See *continuous fibre reinforced prepreg*.

stripper plate
A system of ejection fitted to, for example, an *injection mould* and used to eject a component where *witness marks* are objectionable or, if the *knockout pins* cause part distortion. As the mould opens, an entire plate or mould section moves forward over the core and exerts an even pressure on the moulding which causes part removal. A mould which contains a stripper plate is often referred to as a stripper plate mould.

stripper plate mould
A mould which contains a *stripper plate*.

stripper ring
A system of ejection fitted to, for example, an *injection mould*: a hollow steel disc.

stripper roll
A small, driven roll used to remove sheeting from a machine. See *calendering*.

stripping
The removal of a component from a machine, for example, by hand and/or by the use of a *pick-and-place unit*. See *ejection*.

stroke
A term used in *hydraulics* and which refers to the linear displacement of, for example, a *piston*.

strontium/zinc stabilizer
A polymer stabilizer based on the metals strontium and zinc which is used as a *heat stabilizer*. Compounds of such metals are used as, for example, heat stabilizers for *polyvinyl chloride* (PVC) because of the relatively good stabilizing performance of such a *mixed metal stabilizer*. Such a stabilizer is often combined with a *metal-free organic stabilizer*.

strontium/zinc/tin stabilizer
A polymer stabilizer based on the metals strontium, zinc and tin which is used as a *heat stabilizer*. Compounds of such

metals are used as, for example, heat stabilizers for *polyvinyl chloride* (PVC) because of the relatively good stabilizing performance of such a *mixed metal stabilizer*. Such a stabilizer is often combined with a *metal-free organic stabilizer*.

structural
In plastics technology when the term structural is applied to a product or a process it implies that the product, or the product from a particular process, has better properties (for example, stiffness) than the properties obtained when the standard or basic process is used. The improvement in stiffness (*modulus*) is obtained by making the component thicker (see *structural foam*) and/or, using long fibre reinforcement and/or, by using polymers of high modulus (see *structural liquid composite moulding*).

structural foam
A structural foam (SF) is a type of cellular plastics material in which the outer layers are more dense than the inner layers. A *structural foam moulding* may be obtained by *gas injection moulding*.

structural foam moulding
Structural foam (SF) mouldings are obtained when a gas-containing melt is injected into a mould and the gas in the outer layers escapes before the material sets to the shape of the cavity.

The gas-containing melt may be obtained by mixing a *blowing agent* into the thermoplastics material: the blowing agent decomposes at injection moulding temperatures but the gas cannot escape as it is trapped inside the injection unit. High mould filling rates are needed in this process so that the melt is transferred before the gas escapes. However, it is not found necessary to use high injection moulding pressures to obtain these rates because gas-containing melt flows easily and because the cross-section of the components is relatively large (for example, >4 mm).

The surface finish of the moulded components is relatively rough - caused by rupture of the surface as the gas escapes. However, this feature has been turned to advantage in many applications as the components are intended for wood replacements, for example, loud-speaker housings. Such units are grained and printed subsequently.

structural irregularity
A unit or group which is different from the repeating group and which forms an irregularity in the high polymer structure: this may be a source of weakness.

structural liquid composite moulding
An abbreviation used for this type of process is SLCM. Basically such processes involve the use of a liquid system which is used to impregnate a fibre-based reinforcement system in a mould: the liquid system is then polymerized within the mould. This term covers processes such as *resin transfer moulding*, liquid composite compression moulding and structural reaction injection moulding.

structural reaction injection moulding
An abbreviation used for this term is SRIM or, S-RIM or, structural RIM. Also known as mat reinforced reaction injection moulding (mat reinforced RIM) or as, fibre mat reinforced reaction injection moulding (fibre mat reinforced RIM). Also known as mat moulding reaction injection moulding (MM RIM) or as, fast resin transfer moulding (fast RTM). See *mat moulding*.

The production of a moulding (reinforced with *glass fibre preform*) from a *polymer* (usually a polyurethane material) by a *reaction injection moulding* (RIM) process or, a RIM moulding (reinforced with a fibre preform which is usually glass) which is based on, for example, a polyurethane polymer.

A *structural liquid composite* moulding process in which a fibre preform, usually based on long glass fibres, is placed inside an empty mould. The mould is then closed under pressure so as to compact the fibres - the preform is thicker than the mould cavity. A RIM *mixhead* is then used to inject the reactive monomer mixture into the mould where polymerization occurs before demoulding. Glass is the primary filler although both *carbon fibre* and *nylon fibre* have also been used.

SRIM is usually aimed at high volume production and steel moulds are commonly used: such moulds are mounted in hydraulic presses. Although similar to the resin transfer moulding (RTM) process, SRIM uses a RIM machine to fill the mould: mixing of the reactants is therefore by high pressure impingement - see *impingement mixing*. RTM typically uses slower reacting, heat activated resin formulations which are often mixed by flowing through static mixers before injection into the mould. Moulds are often made of low cost systems, like epoxy, and may be held together by manually operated clamps. Because long fibre reinforcement is used the products have high *modulus*: this feature is emphasised because of the high glass contents used. The reinforcement can be placed where it is needed so as to obtain a desired level of reinforcement.

However, the use of long glass fibres often results in the production of components with poor surface finish unless, for example, a *mould coating* is used.

structural resin injection moulding
An abbreviation used for this process is SRIM. A *resin transfer moulding process* used to produce components of high modulus. See *structural resin transfer moulding*.

structural resin transfer moulding
An abbreviation used for this process is SRTM. A *structural liquid composite moulding process* in which the properties of the products obtained from *resin transfer moulding* are improved by using high concentrations of long fibres, for example, using glass preforms.

The amount of reinforcement in conventional *resin transfer moulding* is often less than 40%: to improve on the properties of components produced by this technique, requires the use of resins with high heat resistance and improved fibre reinforcement, for example, higher fibre content and the use of higher modulus fibres. In this way, advanced structural composite elements can be produced by a moulding technique. See *specific modulus*.

structural sheet moulding compound
See *high performance sheet moulding compound*.

structural unit
The repeating group: the smallest group which repeats itself so as to make a high polymer structure.

structure
Originally known as *reticulate chain structure*. See *carbon black*.

structure level
See *DBP adsorption method*.

structure viscosity
See *apparent viscosity*.

stuffer box
A loading device used when *injection moulding, dough moulding compounds* (DMC): a device which pushes the DMC into the injection unit. An hydraulically driven piston, applies pressure only while the screw is turning so as to give *forced charging*.

styrenated alkyd resin
An *alkyd resin* which has been reacted with *styrene*. The oil component is copolymerized with styrene via unsaturation in the *oil*. Such resins are used to make coatings which have good colour retention, water resistance and alkali resistance but relatively poor solvent resistance.

styrenated diphenyl amine
An abbreviation used is SDPA. See *substituted diphenyl amines* and *antioxidant*.

styrenated diphenylamine
An abbreviation used is SDPA. See *substituted diphenyl amines* and *antioxidant*.

styrenated unsaturated polyester
See *unsaturated polyester resin*.

styrene
Also known as phenylethene or as phenyl ethylene or as, vinyl benzene or as styrene monomer. Prepared mainly from ethyl benzene which, in turn, is prepared from *benzene* and *ethylene*. A colourless aromatic liquid that may be easily polymerized to give *polystyrene*. Used to make a variety of thermoplastics and synthetic rubbers. Also used as a reactive solvent (see *unsaturated polyester resins*). This material has a relative density (RD or SG) of 0.90 and a boiling point of approximately 145°C. Styrene is a good solvent for uncured rubbers such as uncured *nitrile butadiene rubber (NBR), chloroprene rubber (CR), butyl rubber (IIR) natural rubber (NR), styrene-butadiene rubber (SBR)* and *thiokol rubber (T)*. See *high impact polystyrene, acrylonitrile-butadiene-styrene* and *styrene-butadiene rubber*.

styrene acrylonitrile
See *styrene-acrylonitrile*.

styrene block copolymer
A type of *thermoplastic rubber* which may be referred to as SBS or SBC: Other abbreviation encountered include SIS (from styrene-isoprene-styrene), SEBS (from styrene-ethylene/butylene-styrene), and SEPS (from styrene-ethylene/propylene-styrene). Alternative names include SBS block polymer and teleblock copolymer. One suggested name for a linear SBS block is poly-(1-phenylethylene-b-1-butenylene-b-1-phenylethylene). SBS could also be called poly-(styrene-b-butadiene-b-styrene). SIS could also be called poly-(styrene-b-isoprene-b-styrene). SEBS could also be called poly-(styrene-b-ethylene-co-butylene-b-styrene). SEPS could also be called poly-(styrene-b-ethylene-co-propylene-b-styrene).

SBS polymers are block copolymers in which the styrene (S) and butadiene (B) monomers have been reacted so as to give long segments, or blocks, of polybutadiene (BR) and long segments, or blocks, of polystyrene (PS). Various combinations are possible. For example, S-B (diblock), S-B-S (triblock), S-I-S (triblock) and S-EB-S (saturated triblock): radial trichain, radial tetrachain and radial multichain can also be made (see *radial teleblock polymer*).

The diblocks have relatively poor strengths unless they are vulcanized. The unvulcanized triblocks are much stronger and the radials are the strongest of all - such strengths result from the extensive networks formed as the PS blocks congregate into domains or regions. In the popular S-B-S linear triblock materials, two comparatively short end-blocks of PS are connected to a longer central BR block.

Of the various types of thermoplastic rubbers available, *linear triblock copolymers* of polystyrene and polydienes, are one of the most important in terms of the tonnages used. The rubbery behaviour of these materials is explained by the *domain theory*. To achieve rubbery properties the styrene content is kept below 40%. Can also have a *saturated triblock polymer* such as SEPS or SEBS. Because of differences in the molecular weights, between entanglements in the rubbery segments, SIS is softer than SBS. This in turn is softer than SEBS. SBS is the cheapest material, then SIS: SEBS is approximately twice the price of SBS. (SEPS is similar in its properties to SEBS).

SBS is not so fluid as *plasticized polyvinyl chloride* (PPVC) - more like a stiff grade of *low density polyethylene* (LDPE). At the same molecular weight, radial polymers will exhibit a lower viscosity. For a given viscosity the use of a radial block polymer will result in compounds with higher hot strength and greater flex strength. However, because of the molecular weights used it is commonly found that the flow of branched compounds is poorer. SBS polymers are amorphous thermoplastics which exhibit low shrinkage, for example, below 1.0%. High filler loadings will reduce shrinkage and improve hot strength.

SBS is resistant to water, alcohols, weak acids, weak bases. Addition of up to 15% of *ethylene-vinyl acetate* (EVA) addition helps to prevent ozone attack while benzophenone-type stabilisers improve *ultraviolet* resistance. The addition of plastics, such as *polypropylene* (PP) and *polyethylene* (PE), will improve the solvent resistance of SBS: solvent resistance may also be improved by adding a small amount of another rubber, for example, *nitrile rubber*. In general, SBS is not solvent resistant: a block copolymer will be soluble in those materials which are solvents for each of the respective homopolymer. It will regain its properties when the solvent is removed. Not resistant to hydrocarbons, esters and ketones; such materials dissolve or cause excessive swelling. This is useful for adhesive purposes but poor solvent resistance limits the application of SBS. An SBS material can be used without additives but additives are commonly added to reduce cost, improve processability and to achieve specific properties.

Like traditional rubbers, these materials are often blended with oils, fillers etc., so that the amount of SBS in the final compound is often between 25 and 50%. This means that the density of compounds can vary widely: usually varies over the range 0.9 to 1.2 gcm^{-3} for SBS, SIS, SEPS and SEBS materials. Compounds based on these materials consist of, for example, SBS polymer, another polymer (e.g. PS), fillers (e.g. calcium carbonate), plasticizers (e.g. naphthenic oils), stabilisers (e.g.dilauryl thiodipropionate and a hindered phenol-type antioxidant) and a colouring system. In translucent soling then the material will require a UV light absorber. Very good lightresistance can be obtained with white mineral fillers and a UV light absorber.

Components based on such materials can be translucent and a wide colour range is possible. Mouldings may be capable of large extensions even at low temperatures, but they will exhibit a very low softening point, for example, a Vicat softening point of 70°C. SBS compounds can be used in moulds designed for other plastics, for example, for those designed for PPVC.

SBS thermoplastic elastomers exhibit reasonable strength and elasticity at room temperatures - they can be compounded to give a wide range of properties or grades. In general these materials exhibit properties which are similar to those associated with vulcanized rubbers - they exhibit good grip, resilience, high elongation at break, low set values and good low temperature properties. A wide range of hardnesses is available, e.g. from 40 to 90 Shore A. At room temperature the stress-strain properties can be similar to those found for *polyisoprene rubber*, that is, high elongation at break for a low level of stress application (SBS has been used to make elastic bands). The applicational areas for these materials are footwear, injection moulded or extruded goods, adhesives, and as modifiers, for example, for asphalt and for other polymers. As the temperature is raised then elasticity is lost - do not use above 55°C.

In the area of injection moulded or extruded goods then, for outdoor applications, S-EB-S materials are usually used as these have the best ageing resistance. Such applications include gaskets and automotive filler panels. To obtain a higher temperature resistance then the compound may contain PP. SBS-type materials may be used to modify other plastics, for example, HDPE, PP, PE, PS and HIPS: such blends are easily processed on thermoplastics equipment. The SBS material is used to improve impact strength, elongation, tear strength and to reduce the tendency to warpage. The higher the rubber (butadiene) content, in the SBS, the higher will be the impact strength of the modified material. Lower molecular weight SBS is more easily dispersed but, higher molecular weight material gives better results. When melt compounding the materials together it is important to get an optimum dispersion as too good a dispersion, will give inferior properties: the rubber particle size will be too low.

styrene butadiene block copolymer
See *styrene-butadiene block copolymer*.

styrene butadiene rubber and natural rubber - comparison of
See *natural rubber and styrene butadiene rubber - comparison of*.

styrene/copolymer blends
See *acrylonitrile-butadiene-styrene*.

styrene emission
The loss of *styrene* vapour to the atmosphere during, for example, the *hand lay-up* and the spray-up moulding processes, using *unsaturated polyester resins*. Such vapour is considered harmful and so, in many countries, the level of *styrene* in the atmosphere must be kept below certain levels. The short term average exposure (15 minutes) may be specified as being no more than 20 ppm and the 8 hour average exposure may be specified as being approximately twice this at say, 40 ppm. The level differs from country to country. The styrene concentration in the atmosphere may be accurately measured, for example, with photo-ionizer equipment.

styrene ionomer
An *ionomer* which contains styrene units. Copolymers of styrene cross-linked with *divinyl benzene* (DVB) are made which are *ion exchange resins*. The ionic groups are introduced by *post functionalization* of the cross-linked matrix of the styrene-DVB copolymer. Often the beads are sulphonated to produce a *cationic ion exchange resin* (a strong acid, cation ion exchange resin): a weak acid, cation ion exchange resin may be made by polymerization of an organic acid (for example, acrylic acid) with DVB. Anionic ion exchange resin are, however, made by post functionalization so as to give, for example, quaternary amine groups. Use of a tertiary alkyl amine gives a strong base, anion ion exchange resin: use of a primary or secondary alkyl amine gives a weak base, anion ion exchange resin.

Suspension polymerization gives a *gel-type polystyrene resin*: this was the first type of ion exchange resin. A general purpose, ion exchange resin contains approximately 8 mole% of DVB (nominal DVB content). Highly porous, *macroreticular beads* can be produced by, for example, incorporating a solvent into the monomer mixture before polymerization.

styrene maleic anhydride
Also known as poly-(styrene-co-maleic anhydride) or, styrene-maleic anhydride copolymer. An abbreviation used for this type of material is SMA. An alternating copolymer of *styrene* and *maleic anhydride* which has a higher softening point than *polystyrene*. Has been used to make *acrylonitrile-butadiene-styrene*-type plastics of high softening point.

styrene monomer
See *styrene*.

styrene olefin thermoplastic elastomer
Also known as styrene-ethylene/butylene-styrene block copolymer (SEBS). See *styrene block copolymer*.

styrene phenol reaction product
A pale yellow liquid with a relative density of 1·08 and which is used as a rubber *antioxidant*.

styrene plastics
See *styrene-based plastics*.

styrene thermoplastic rubber
See *styrene block copolymer*.

styrene triblock copolymer
Usually means styrene-butadiene-styrene block copolymer but could also be, styrene-ethylene/butylene-styrene block copolymer (SEBS) or, styrene-isoprene-styrene block copolymer (SIS). See *styrene block copolymer*.

styrene-α-methyl styrene copolymer
A thermoplastics material which is a copolymer of styrene and α-methyl styrene. High molecular weight materials are transparent thermoplastics materials which are marketed as they improve on the softening point of polystyrene with only a slight increase in melt viscosity. Low molecular weight materials are liquid thermoplastics which have been used as *plasticizers* for rubbers. See *poly-α-methylstyrene*.

styrene-acrylonitrile
An abbreviation used for this type of material is SAN. Also known as styrene acrylonitrile or as, styrene acrylonitrile copolymer.

Copolymers of styrene and acrylonitrile, in the ratio of about 70:30 by weight, are rigid, transparent (refractive index 1·56) amorphous thermoplastics which in several ways resemble *polystyrene* (PS). They have however, a higher softening point (approximately 95°/202°F), they are harder, more rigid, more craze resistant (caused by temperature fluctuations), more *environmental stress cracking* (ESC) resistant and more solvent resistant; the resistance to staining by foods is good. It is not an impact resistant material (the notched impact strength of glass filled grades can be lower than equivalent PS materials).

The polymer is transparent but clear mouldings have a yellow tint which can be offset by the use of a blue dye (the higher the acrylonitrile content the worse the yellow colour). The densities of PS and SAN are similar (approximately 1·08 gcm^{-3}) but SAN has a higher resistance to creep under load, maintains its impact strength from −40 to 50°C/−40 to 122°F without change and has a higher *heat distortion temperature* (HDT). Materials with high acrylonitrile contents and higher molecular weight are tougher but more difficult to mould; they also have the best chemical resistance. SAN has better chemical resistance than PS and, like PS, it is a good electrical insulator. Maximum continuous use temperature is approximately 85°C/185°F. The addition of glass fibres (for example, 35%) improves the HDT by approximately 10°C/50°F and gives a very rigid material with a coefficient of thermal expansion similar to that of metals. Blends with *acrylonitrile-butadiene-styrene* to improve impact resistance. are also marketed.

SAN is similar in its flow behaviour to *polymethyl methacrylate* (PMMA): shrinkage is of the order of 0.004 to 0.007 mm/mm (0·4 to 0·7%) and is similar to that of PS; glass filled grades have reduced shrinkage. Transmits more than 90% of visible light but absorbs U.V. A wide colour range is possible - including transparent colours. Water absorption is similar to that of PMMA: will absorb approximately 0·25% water in 24 hours at room temperature. If SAN is correctly stored then drying is not normally necessary. If drying is necessary, then dry in a hot air oven for 3–4 hours at 70–75°C/158 to 164°F.

SAN is resistant to saturated hydrocarbons, low aromatic engine fuels, oils, vegetable fats and oils, animal fats and oils, aqueous solutions of salts and dilute acids and alkalis. It is not resistant to aromatic and chlorinated hydrocarbons; esters, ethers, ketones and various chlorinated hydrocarbons (for example, to methylene chloride, ethylene chloride and trichloroethylene); also attacked by concentrated inorganic acids. Environmental stress cracking resistance may be assessed by immersion in a mixture of olive oil and oleic acid.

Some mouldings, for example, those of low acrylonitrile content may be bonded to each other with a solvent such as toluene; those with a high acrylonitrile content will require solvents such as ethyl acetate. A useful formulation for jointing SAN would be equal proportions of methylene chloride and butyl acetate. May be welded by hot plate, spin or ultrasonic techniques.

If the UV resistance of this plastic is improved by the use of stabilisers, then this material may be used in outdoor applications, for example, in automotive applications such as rear lamp covers and in reflectors. Because of the materials better resistance to temperature fluctuations, compared to PS, it is used for high quality, household utensils; this temperature-fluctuation resistance is improved by the use of grades which have a high molecular weight and a high acrylonitrile content. The exceptional surface gloss of this plastic, together with its abrasion. chemical and thermal resistance, mean that it is used in injection mouldings for radios (dials) and TV sets (screens), refrigerators (doors and trays), record players (covers) and washing machines (trim and windows). Glass reinforced SAN (SAN-GF) is characterised by high stiffness and dimensional stability coupled with good solvent resistance. This balance of properties, together with easy processing for some grades (20% GF), makes SAN-GF suitable for large area or long mouldings; may be painted or decorated by conventional techniques.

styrene-based plastics
The family of thermoplastics materials known as styrene plastics is a large and important one as it includes not only polystyrene (PS), but materials such as *high impact polystyrene* (HIPS), *styrene-acrylonitrile* (SAN), *acrylonitrile-butadiene-styrene* (ABS) and *butadiene-styrene block copolymers* (BDS). A very wide range of types of plastics is therefore included as the range covers glassy, brittle materials and tough, ductile ones: the range included homopolymers, rubber-toughened materials and block polymers. Although it is now possible to produce styrene plastics which are both clear and tough, it is not possible to make tough, clear mouldings from PS as this material is very brittle unless of high molecular weight and/or biaxially orientated. *Biaxial orientation* is not usually possible in conventional moulding and the use of high molecular weights results in poor melt flow properties.

A common feature of all styrene plastics is their resistance to aqueous media such as salt solutions, acids of medium concentration and alkalis. Aliphatic hydrocarbons, for example, heptane and cyclohexane, readily attack PS and HIPS but do not affect SAN and ABS. Carbon tetrachloride (CCl_4) attacks SAN and ABS only slowly but quickly attacks PS and HIPS. Resistance to CCl_4 may be used to distinguish between PS and SAN. In the case of polystyrene, when it is immersed in the CCl_4 it immediately becomes sticky whilst SAN is relatively unaffected. Alternatively if a few drops of CCl_4 are put into a test tube containing either PS or SAN then, in the case of the PS, the liquid becomes milky in a short period of time whereas with SAN it remains colourless. The flexibility of BDS, separates BDS from PS and SAN.

styrene-butadiene block copolymer
An abbreviation used for this type of material is BDS. Also known as styrene butadiene block copolymer or as, butadiene-styrene copolymer or as, BDS polymer. A *styrene-butadiene copolymer*: the abbreviation SB has also been used for this material but is not recommended as the same initials have been used for HIPS.

If 75% *styrene* and 25% *butadiene* are polymerised together by sequential, anionic polymerisation in solution then a thermoplastics material results. The material contains long lengths (or blocks) of *polystyrene (PS)* and long lengths (or blocks) of *polybutadiene*; the material may contain up to, for example, six blocks. The molecules of the final plastic are star shaped as the initial multiblocks are coupled together by a polyfunctional material, for example, epoxidized linseed oil. (Because of the polymerisation technique employed, plastics with a narrow molecular weight distribution may be produced if required).

After the formation of the star-shaped blocks there is a phase separation of the polystyrene and the polybutadiene into domains or regions; the links between the two phases are similar to the links between amorphous and crystalline regions in semi-crystalline, thermoplastics materials.

Because the domains are so small, the materials are still transparent. Such materials are clear and tough but can be very notch sensitive (if assessed by a notched impact test, they appear to have similar properties to ordinary PS). However, recent developments by the material suppliers have improved the notch sensitivity problem. The notch sensitivity is dependent on the position or orientation of any crack, as well as on the molecular orientation within the sample. Tensile strength, stiffness and hardness are lower than those values found for other clear plastics, for example, the Shore D hardness ranges from 62 to 75. The heat distortion temperature and the Vicat softening point are similar to other styrene-based plastics but big advantages of these plastics is their low density, toughness and colourless transparency. May be mixed with PS, in order to reduce cost and to improve hardness, but this can lead to opacity unless done well. Also used in blends with *polycarbonate* (PC) at 5–10% BDS concentration so as to maintain impact strength in thick sections, at low temperatures and after thermal aging.

The chemical resistance of this type of material is generally similar to other *styrene-based plastics* materials. Not resistant to most organic solvents. Alcohols may affect these materials dependent on their concentration and molecular weight; for example, not affected by methanol or ethanol. Oils tend to cause softening. Although the environmental stress cracking (ESC) resistance of these plastics is better than that of PS, they should still not be used in contact with oily products. Unsaturated oils cause ESC more quickly than saturated oils. Not suitable for the packaging of butter and margarine. Unsuitable for prolonged use outdoors. This material may be joined to itself using solvents such as toluene, methyl ethyl ketone, ethyl acetate and methylene chloride; toluene is however the best. For bonding to *acrylonitrile-butadiene-styrene* (ABS) use dimethyl formamide.

This type of material has a density of approximately $1.01–1.02$ gcm^{-3} (solid, non-filled material): the higher the butadiene content the lower is the density. The natural colour of the material is sparkling clear, water-white, and so a wide colour range is possible; often seen however as glass clear, high quality mouldings which are noted for their toughness. For example, this material is used in medical applications such as centrifuge tubes, dropping bottles and mouth tubes. Used in egg packaging and for storage boxes. Has been used to make drawing instruments and ball point pens. UV masterbatches are available to improve weather resistance; such an additive can improve the Xenon light resistance by approximately four times. These materials are similar to PS with regard to yellowing. Can be flexed, or bent (i.e. exhibits a hinge effect) unlike PS.

styrene-butadiene copolymer
Also known as butadiene-styrene copolymer or, poly-(styrene-co-butadiene). An abbreviation used for a styrene-butadiene material is SB. Styrene and butadiene can be put together in various ways so as to produce a wide range of rubbers and plastics. For example, when styrene is polymerised onto polybutadiene then the graft copolymer, *high-impact polystyrene*

(HIPS) is produced. If the two monomers are polymerised together then the random copolymer known as *styrene-butadiene rubber* (SBR) is obtained; this contains approximately 75% styrene and 25% butadiene. Also see *styrene block copolymer* and *styrene-butadiene block copolymer*.

styrene-butadiene rubber
An abbreviation used for this material is SBR. The following systematic chemical name has been suggested: poly(1-butenylene-co-1-phenylethylene).

One of the major rubbers - the other being *natural rubber* (NR). Brought into large scale production during the Second World War as GR-S in the USA and Buna S in Germany. Produced mainly by emulsion polymerization (E-SBR) although solution polymerization does give some advantages.

Initially E-SBR was polymerized at approximately 30°C to give the so-called hot rubber: now, E-SBR is produced at approximately 5°C to give the so-called cold rubber. Such emulsion polymerization gives essentially random copolymers: solution polymerization can give more regular polymers (stereospecific rubbers or L-SBR) if desired.

In general, E-SBR contains approximately 23.5% styrene or 1 styrene unit per 6 butadiene units. This gives a material with a reasonable balance of properties - the glass transition temperature (T_g) is of the order of −50°C. Most of the butadiene is trans-1,4-butadiene and the polymer has a relatively high molecular weight (number average molecular weight is about 100,000) with a broad molecular weight distribution. Like NR, SBR is an unsaturated hydrocarbon polymer.

Both NR and SBR cost roughly the same and the maximum service temperature of moulded components is of the order of 70°C/158°F. Both materials are available in a range of grades (range of molecular weights, viscosities, etc.) and all grades of SBR are compatible with NR in all proportions: mixtures are quite often used in order to get desired properties.

Because of crystallinity, NR is reinforced on stretching and this means that the fillers used with NR rubber can be of the non-reinforcing variety (such fillers are cheap). SBR does not crystallise when stretched and so more expensive reinforcing fillers must be used for many applications. NR has better hot tear resistance than SBR and this property is useful if there are difficulties in moulding. SBR is slower curing than natural rubber and although the cure rates can be matched, this means that more (expensive) *accelerator* must be used. However, SBR has better abrasion resistance and the moulds should stay cleaner because there is often less mould fouling with compounds based on this polymer; the ageing resistance of the mouldings is also slightly better. SBR is not broken down by mastication - this means that it is easier to re-work stocks as there is little change on processing.

The synthetic polymer (SBR) can be made in very high molecular weight grades (much higher than those found in NR) and these grades can then be extended with oil. Oil-extended SBR is relatively cheap and this technique helps to offset the cost of expensive reinforcing fillers (e.g. *carbon black*). Oil extended SBR may be based on a naphthenic or aromatic oil at say, 37.5 phr.

styrene-butadiene-styrene block copolymer
See *styrene block copolymer*.

styrene-chloroprene rubber
An abbreviation used for this type of material is SCR. Also known as chloroprene-styrene copolymer or rubber. See *chloroprene rubber*.

styrene-divinyl benzene copolymer
An abbreviation used for this type of material is styrene-DVB copolymer or S/DVB. Ion exchange resins. Such a DVB copolymer is also known as gel-type polystyrene resin: a *styrene ionomer*.

styrene-ester polymer
See *vinyl ester resin*.

styrene-esters
See *vinyl ester resin*.

styrene-ethyl/butylene-styrene
Also known as styrene olefin thermoplastic elastomer. See *styrene block copolymer*.

styrene-ethylene/propylene-styrene
See *styrene block copolymer*.

styrene-isoprene rubber
An abbreviation used for this type of material is SIR. Also known as isoprene-styrene copolymer or rubber. See *chloroprene rubber*.

styrene-isoprene-styrene
See *styrene block copolymer*.

styrene-maleic anhydride copolymer
See *styrene-maleic anhydride*.

styrene-rubber plastics
Plastics based on styrene polymers and rubbers, the styrene polymers being in the greatest amount by mass. See, for example, *styrene block copolymer*.

styrene-vinyl pyridine-butadiene terpolymer
See *resorcinol-formaldehyde-latex dip*.

sub-plate
A term used in *hydraulics* and which refers to a *manifold base* on which surface-mounted valves are carried and through which fluid connections are made: an auxiliary plate with appropriate ports to which connections may be made.

sub-sprue
Also called a stalk or carrot. Part of the *feed system* which connect the gate (for example, a *pin-point gate*) to the main *runner* system. See *three-plate gating*.

sub-tread
Also known as a tread base. A *tyre* component: the layer of a two-component tread unit nearest to the *carcass*. It is usually of softer rubber than the *top cap*.

submarine gate
Also called a tunnel gate. A type of *gate* used in injection moulding: the material is fed below the surface of the component. This type of *gate* feeds the plastics material into the side of the component. A *runner* carries the plastics material down below the *parting line* of the mould and feeds, via a cone-shaped gate, into the cavity. The gate is usually located in the *moving half* of the mould and, when the mould opens, the gate is sheared by the ejection action. Such gates are usually only used for small components and their diameters may range from 0.5 to 2 mm/0.02 to 0.8 in; the size employed depends on the material being moulded. For example, the materials ease of flow and its strength in shear. The gate should be as short as possible: to ensure that minimal gate vestige occurs. When using a submarine gate an inclusive gate angle of 30° and a correctly positioned ejector pin (i.e. not greater than 10 mm from gate entry) are essential. A parallel land section should not be used for the gate - see *pin-point gate*. If the gate diameter is small then this type of gate is sometimes referred to as a pin-point submarine gate. A tunnel gate with a curved axis is sometimes referred to as a winkle gate.

subplate
See *sub-plate*.

substituted acrylonitrile
An organic material based on *acrylonitrile*, for example, methacrylonitrile. Such a material could be used in place of acrylonitrile to prepare copolymers in place of *butadiene* for rubbers similar to *nitrile rubber* (NBR): however, such polymers are not usually commercially available.

substituted benzophenone
A compound based on *benzophenone* and which may be called a 2-hydroxybenzophenone. Such compounds are widely used to improve the *ultraviolet* (UV) resistance of polymers so as to minimize decomposition, or degradation: they minimize degradation by acting as ultraviolet absorbers. The organic substitution is usually in the para position on the benzene ring and may be alkyl or alkoxy. By varying the structure of the substituted group, it is possible to alter the wavelength of UV absorption and to give polymer compatibility. Compounds used include;

2-hydroxy-4-methoxy-benzophenone;
2-hydroxy-4-methoxy-5-sulpho-benzophenone;
2-hydroxy-4-octoxy-benzophenone; and,
4-decyloxy-2-hydroxy-benzophenone.
Tetra-substituted compounds include:
2,2'-dihydroxy-4-methoxy-benzophenone; and,
2,2'-dihydroxy-4,4'-dimethoxy-5-sulpho-benzophenone.

substituted butadiene
An organic material based on *butadiene*, such as isoprene, piperylene, and dimethyl butadiene: such materials contain one or more additional methyl groups. They have used to prepare copolymers in place of *butadiene* such materials have been used so as to make rubbers similar to *nitrile rubber* (NBR): the use of such a material raises the *glass transition temperature* compared to NBR. Because of this, such materials are not as commercially important as the parent material. Terpolymers of butadiene, isoprene and acrylonitrile are also available.

substituted diphenylamines
Also known as alkylated diphenylamines or as substituted diphenyl amines. Additives, antioxidants, used in rubber compound to prevent oxygen attack. Amine antioxidants which may be classed as staining antioxidants, or as discolouring antioxidants, as they have a tendency to stain or discolour the rubber on ageing. They are used in rubber compounds because of the good general *antioxidant* characteristics which their use imparts together with good heat resistance.

Materials which fall into this category are octylated diphenyl amines (ODPA), styrenated diphenyl amines (SDPA) and acetonated diphenyl amines (ADPA).

Substituted diphenylamines are not as good as, for example, *ketone-amine condensates* for *natural rubber* protection but they are useful for *chloroprene rubber* protection.

substituted p-phenylene diamines
Aromatic amines. Also known as disubstituted p-phenylene diamines. Additives, *antioxidants*, used in rubber compound to prevent oxygen attack. Amine antioxidants which may be classed as *staining antioxidants* as they have a tendency to stain or discolour the rubber on ageing. Phenyl-β-naphthylamine was at one time widely used in rubber compound to prevent oxygen attack. This material is not now widely used because of the danger of cancer (caused through the presence of β-naphthylamine). Alkyl-aryl derivatives (see *alkyl-aryl p-phenylene diamine*) are good antioxidants and also function as *antiozonants*.

substituted phenol
See *hindered phenol*.

substituted phenolic derivatives
Organic compounds based on a *phenol*-type material. Includes hydroquinone, quinone and their mono and di-substituted derivatives. Used as, for example, *inhibitors* for *unsaturated polyester resins*. There are two general classes of inhibitors, quaternary ammonium salts and substituted phenolic derivatives.

substituted polyphenylene
A *polyphenylene* in which the benzene rings are separated by a linking group, for example, linear organic groups or oxygen atoms or sulphur atoms. The presence of such linkages gives chain flexibility and a tractable heat-resistant material. See, for example, *polyphenylene sulphide* and *polyphenylene oxide*.

substituted polyphenylene oxide
See *polyphenylene oxide - modified*.

substrate-based fibre
A composite *fibre* which is produced by, for example, a chemical *vapour* deposition process onto a filament. See *silicon carbide*.

subsurface gate
See *submarine gate*.

suck back
Melt decompression during the *injection moulding* process, at the end of screw rotation, the screw is pulled back hydraulically so as to decompress the melt. This action is called suck-back, or decompression, and it helps to prevent the melt from drooling from the nozzle. In this way the need for a shut-off valve, on the cylinder or barrel, may be eliminated. To prevent material leakages, *runnerless moulds* are commonly operated using this decompression facility. See *back rinding*.

suction line
A term used in *hydraulics* and which refers to the *line* which connects the pump to the fluid reservoir.

sugludates
See *nylonates*.

sulphenamide
An organic compound which contains the structure X-SNRQR". In this formula, the -SN group is connected to X, the benzthiazyl radical: R' and R" may be hydrogen or an alkyl group: both may also be cyclic groups. See *sulphenamides*.

sulphenamide accelerator
See *sulphenamides*.

sulphenamides
A class of rubber accelerators. *Delayed action accelerators* are typified by a *sulphenamide* accelerator. For example, by *cyclohexylbenzothiazyl sulphenamide* or CBS: the product of the reaction between cyclo-hexylamine and *mercaptobenzthiazole*. Other examples include N-morpholinothiobenzothiazole-2-sulphenamide (MBS), N,N-dicyclohexylbenzothiazole-2-sulphenamide (DCBS), N-oxydiethylbenzothiazole-2-sulphenamide (NOBS), N-oxydiethylenethiocarbamyl-N-oxydiethylenesulphenamide (OTOS) and N-t-butylbenzothiazole-2-sulphenamide (TBBS), See *accelerator*.

sulphide cross-link
A cross-link which is based on a sulphur atom or on sulphur atoms (see *efficiency parameter*). Typical chemical groupings, based on sulphidic groups, and which are present in a sulphur-vulcanized natural rubber include (a) mono-sulphide cross-links, (b) di-sulphide cross-links, (c) polysulphide cross-links, (d) parallel, vicinal cross-links, (e) intra-chain, cyclic, mono-sulphide cross-links, (f) intra-chain, cyclic, di-sulphide cross-links and, (h) pendant sulphidic groups.

sulphide polymer
See *polyphenylene sulphide*.

sulphidic group
A sulphur-containing group present in, for example, a sulphur-vulcanized natural rubber. See *sulphide cross-link*.

sulphonate ionomer
See *polysulphate sulphonate*.

sulphonated ethylene-propylene-diene monomer rubber
An abbreviation used for this material is sulphonated EPDM. It is also known as ethylene-propylene-ethylidenenorbonyl sulphonate or as, sulphonated EPDM rubber or as, sulphonated EPDM. An *ethylene-propylene-diene monomer rubber* (55% ethylene units) which contains sulphonated ethylidenenorbonyl units (5%). Metal ions, for example, based on zinc, are used to give ionic cross-links: such a material Is a *thermoplastic elastomer* with good heat and weather resistance. Has been used to make sheeting material and hose.

sulphone group
May be represented as (SO_2). See *polyether ether ketone*.

sulphone polymers
Polysulphone (PSU) is also known as polysulfone. Polyether sulphone (PES or PESU) is also known as polyether sulfone or as, polyaryl sulphone. The term, polyaryl ether sulphone is also sometimes used for a sulphone polymer. However, the chemical-type names used for sulphone polymers may be applied to all commercial materials.

A sulphone polymer consists of benzene groups (p-phenylene groups) linked with sulphone groups (SO_2), that is, it is an aromatic polysulphone. Such a simple aromatic polysulphone as poly-(p-phenylene sulphone) could not be melt processed. In order to make a mouldable aromatic sulphone, it is necessary to introduce ether linkages so as to obtain chain flexibility: other groups may also be introduced, for example, the isopropylidene linkage. By varying the spacing between the groups and the type introduced it is possible to produce a range of materials which helps account for the chemical-type names used.

Any plastic which contains p-phenylene groups has a high heat distortion temperature (HDT), is rigid at room temperature and requires a high processing temperature. All of the PES/PSU materials are amorphous but when the p-phenylene groups are more spaced out, in the case of Udel-type materials (PSU), they have a lower glass transition temperature (T_g) of, for example, 190°C/374°F: that of the Victrex-type materials (PES) is 230°C/446°F.

In general however, the sulphones are tough and have a wide temperature range of use. The maximum continuous use temperature may reach 180°C/356°F (UL rating) with toughness retained down to −70°C/−94°F for some grades. Because of their chemical structure, the materials are resistant to ionizing radiation and to chemical changes (such as oxidation) on heating. Electrical insulation properties are good but the tracking resistance is not outstanding. They have excellent resistance to creep, are transparent, self-extinguishing, tough but like *polycarbonate* (PC) they are notch sensitive. However, they have lower impact strengths than PC, are more expensive, possess a higher HDT, better alkali resistance and creep resistance.

The differences between the different types of sulphone polymer are not tremendous. PES (Victrex-type material) has better creep resistance, smoke emission on burning characteristics and a higher HDT than PSU (Udel-type material). PSU has a lower density (of 1·24 gcm^{-3}) compared to that of PES (1·37 gcm^{-3}): PSU absorbs less water, has a lower dielectric constant and costs less than PES.

The addition of glass fibres (GF) results in more rapid set-up in the mould; cycle time reductions of 25% are possible. About a third of sulphone use is for GF materials. Flexural modulus (stiffness) is markedly improved, with 30% GF by a factor of 3. Mould shrinkage is reduced by a factor of 3; notched impact strength and tensile strength are up by approximately a half. Replacing the GF with *carbon fibre*, reduces the mould shrinkage and coefficient of expansion even further; it also increases conductivity and flexural modulus (by a factor of 5).

These materials have a high melt viscosity: shrinkage is of the order of 0·006 to 0·007 mm/mm (0·6 to 0·7%). As the natural colour of the material is transparent amber then a relatively wide colour range is possible: this includes both transparent and opaque colours.

These materials are resistant to aqueous acids and alkalis. Heat, solvent, creep and fatigue resistance although good, is not as good as *polyether ether ketone (PEEK)*. Aliphatic hydrocarbons, alcohols, benzene, petroleum spirit, oils and fats, sterilizing solutions, cleaning and degreasing agents do not attack this type of material. Such materials are not resistant to aliphatic and aromatic hydrocarbons, chlorinated hydrocarbons, concentrated acids and bases (for example, concentrated oxidizing mineral acids such as sulphuric acid), dilute acids, dilute bases and salt solutions. Soluble in highly polar organic solvents such as ketones, some chlorinated and aromatic hydrocarbons. A large number of organic materials will therefore attack this type of material (for example, chloroform, cresols, acetone, cyclohexanone, ethyl acetate, methyl ethyl ketone). Solvents include dimethyl formamide and dimethyl acetamide. If this material is to be used outside, then it should be light stabilized (with *carbon black*).

This type of material is hygroscopic and therefore needs to be dried before moulding for example, dry in a hot air oven for 3 to 4 hours at 150°C/302°F. A moulded component will also absorb water - newly moulded specimens will increase in dimensions by about 0·2% when exposed to air or water. Moisture does not cause material hydrolysis. Because of the stiff polymer chains, and the high set-up temperatures, it is easy to get high moulded-in strains with this type of material. The use of high mould temperatures, low packing pressures and subsequent *annealing* help to reduce this and give more stable components.

In general, such materials are used when a thermoplastic is required which has good, long-term thermal stability, dimensional stability, toughness and resistance to burning. They are more heat resistant, and have better creep resistance, than PC but are much more expensive. Used in the microwave cooking field, for example, for grilles and dishes. These materials have extended the range of use of thermoplastics as, for certain applications, they may replace thermosets and metals. For example, in medical equipment, the materials are used to produce respirators, dental reamers and dialysis equipment. This is because the material is mouldable into relatively complex shapes, the use of which, often reduces the number of metal components required.

sulphonic acid mono-ester
See *alkylsulphonic acid ester*.

sulphur
Group VIb of the Periodic table consists of the elements oxygen, sulphur, selenium, tellurium and polonium. American deposits are a major source of natural sulphur as is some natural gas deposits (these contain hydrogen sulphide from which the sulphur is obtained). Also obtained from, for example, crude oil refining processes. Sulphur is marketed in the form of crystals, ground roll, precipitated and sublimed (flowers of sulphur).

Sulphur is a non-metallic element (S) which occurs in several allotropic forms: the stable form at room temperatures is rhombic sulphur (also known as alpha sulphur or as, α sulphur). This material has a relative density of 2·06, melts at 114°C and boils at 445°C. Sulphur is best known in the polymer industry as a *vulcanizing agent*. For rubber vulcanization, the sulphur must be free from sulphur dioxide, be capable of good dispersion and be at least 99·5% pure. Sulphur also tends to form catenated compounds containing sulphur-sulphur σ bonds which are analogous to the *peroxide link*. See *sulphur - vulcanization, sulphur - soluble* and *insoluble*.

sulphur (dark) factice
See *factice*.

sulphur analysis
When rubber compounds are analyzed the *sulphur* may be classified as added sulphur, *bound sulphur, extractable sulphur, free sulphur, inorganically bound sulphur, organically bound sulphur* and *total sulphur*.

sulphur bloom
Surface contamination in rubber compounds with a high *sulphur* content, super-saturated solutions of sulphur are formed on cooling: the excess sulphur then crystallizes from the compound. See *sulphur - soluble* and *insoluble*.

sulphur chloride
Also known as sulphur monochloride or as, disulphur dichloride. A formula used for this material is S_2Cl_2. This material has a relative density (RD or SG) of 1·68 to 1·71 and a boiling point of 138°C/°F. It is a clear, yellowish red, pungent, fuming liquid. It irritates the eyes, nose and throat and is toxic. Mainly used in solution in carbon disulphide or in petroleum solvents as a cold vulcanizing agent to cold cure, thin-walled articles made from *natural rubber* at room temperatures. In such cases, traces of iron may accelerate the deterioration of cold cured vulcanizates (if the iron is in a state which does not react with hydrochloric acid then this problem will not normally arise). Sulphur chloride may also be used to prepare organic sulphides.

sulphur chloride (white) factice
See *white factice*.

sulphur chloride factice
See *factice*.

sulphur chloride vulcanization
Curing induced by the use of *sulphur chloride*. See *natural rubber*.

sulphur containing polymers
Polymers based on monomers which contain sulphur atoms as part of the original monomer structure. The best known examples are rubbers: such materials are noted for their oil and solvent resistance. See *polysulphide rubber* and *alkylene sulphide rubber*.

sulphur dispersion
Also known as a sulphur premix. The dispersion of *sulphur* in rubber compounds can be difficult, particularly when added to hard rubber compounds, as it is prone to cake on the back roll of a *two-roll mill* and form undispersed flakes. This problem can be minimized by preparing a premix (100:50) of sulphur to *petroleum jelly*.

sulphur dithiocyanate
See *sulphur thiocyanates*.

sulphur donor
A sulphur compound which is capable of donating sulphur during cure. See *sulphur donor vulcanization systems*.

sulphur donor vulcanization systems
Sulphur compounds which are capable of donating sulphur during cure. That is, the rubber compound need contain no elemental *sulphur:* the *vulcanizate* has good ageing properties because of the lack of free sulphur. Sulphur donors are of two types (a) those which are *accelerators* as well as sulphur donors and (b) those which are not accelerators in that formulation. An example of (a) is *tetramethylthiuram disulphide*

(TMTD): others include tetraethylthiuram disulphide (TETD and dipentamethyl thiuram tetrasulphide (DPTT). An example of (b) is 4,4'-dithiomorpholine (DTDM): others include bis(diethylthiophosphoryl) trisulphide (ETPT) and N,N'-dithiobishexahydro-2,4-azepinone.

sulphur factice
Also known as dark *factice*.

sulphur modified polychloroprene
A *copolymer* of chloroprene and sulphur which is obtained by polymerization of chloroprene and about 1% sulphur. A softer, tackier material than the homopolymer and easier to vulcanize. See *chloroprene rubber*.

sulphur monochloride
See *sulphur chloride*.

sulphur monothiocyanate
See *sulphur thiocyanates*.

sulphur premix
See *sulphur dispersion*.

sulphur - soluble and insoluble
Sulphur occurs in several allotropic forms: the stable form at room temperatures is *rhombic sulphur*. This allotrope of sulphur is usually used for vulcanization and may be called vulcanization sulphur. It has limited solubility in rubber at room temperatures (approximately 1%) but this increases at higher temperatures (for example, to 10% at 120°C). As it is also soluble in carbon disulphide this allotrope may be called soluble sulphur.

After curing, and on cooling, it may *bloom* to the surface of the component in use. Such blooming can hinder subsequent fabrication steps if these rely on adhesion. To prevent this blooming, another allotrope of sulphur may be used which is almost insoluble in the rubber (and in carbon disulphide): this may be called insoluble sulphur or μ sulphur. The allotropic form called flowers of sulphur, is insoluble in carbon disulphide. Insoluble sulphur slowly reverts to the normal form on standing: the reversion process is accelerated by heat. This means that mixing temperatures must not exceed 120°C. See *sulphur dispersion*.

sulphur thiocyanate vulcanization
Curing induced by the use of *sulphur thiocyanates*. See *natural rubber*.

sulphur thiocyanates
The sulphur thiocyanates include sulphur monothiocyanate and sulphur dithiocyanate. These materials may be used in carbon disulphide solution to *cold cure*, natural rubber compounds. See *sulphur chloride*.

sulphur vulcanization
Sulphur is best known in the polymer industry as the material which causes the *vulcanization*, or cross-linking, of rubbery materials (unsaturated hydrocarbon rubbers) so as to allow the production of useful articles from materials such as *natural rubber (NR)* and other *unsaturated hydrocarbon rubbers*. That is, it is a *vulcanizing agent*.

The amount of *sulphur* used for this purpose varies widely. For example, for the preparation of soft rubber components 0.25 to 5 phr of *sulphur* is used: for the preparation of hard rubber components (*ebonite*) 25 to 40 phr of sulphur is used. The range from 5 to 25 phr is hardly used because, for example, the products have poor strength and are leathery.

It is possible to produce accelerator free, *natural rubber* compounds but this is unusual. The use of *accelerators* permits the uses of lower sulphur levels and gives more efficient cross-linking: *efficient vulcanization (EV)* leads to the formation of *mono* and *di-sulphidic cross-links*. Approximately 1·5 to 2·5 phr of sulphur is commonly used for soft rubber components together with approximately 0·5 to 1 phr of accelerator. When the amount of accelerator is increased the amount of sulphur is decreased so as to maintain the cross-link density at similar values. It is also necessary to use an accelerator activator such as *zinc oxide* and a fatty acid such as *stearic acid*.

The modern tendency is to use lower sulphur levels, to use *delayed action accelerators*, to use accelerator combinations (see *synergism*) and to use pre-vulcanization inhibitors. See *efficiency parameter* and *accelerated sulphur systems*.

sulphur-free cure
Curing, or vulcanization, achieved without the use of elemental *sulphur*. Such cross-linking, for example, with a peroxide, can lead to very short cross-links such as -C-C-. As the length of the cross-link influences the micro-Brownian movement of the rubber chains, the cross-link length will influence the vulcanizate properties. See *natural rubber*.

sulphur-nitride polymer
Also called a polythiazyl or poly-(sulphur nitride) polymer. A *copolymer* of sulphur and nitrogen which is a conductor of electricity: it becomes a super-conductor below 1K.

sulphurized oil
A *drying oil* (for example, *rape seed oil*) that has been treated with less than the amount of *sulphur* needed to make *factice*.

sulphurless cure
Curing, or vulcanization, achieved without the use of elemental *sulphur*. See *sulphurless vulcanization*.

sulphurless curing system
A vulcanization system which gives curing, or vulcanization, without the use of elemental *sulphur*. See *sulphur-free cure* and *sulphurless vulcanization*.

sulphurless vulcanization
Sulphur vulcanization of a diene rubber, such as *natural rubber*, but without the addition of elemental sulphur: the sulphur is derived from the *accelerator* which is based on a *thiuram sulphide* and *zinc oxide*. Mono-sulphidic cross-links are formed and this gives vulcanizates with good heat resistance. See *sulphur-free cure*.

sump
See *reservoir*.

sunflower oil
An industrial *vegetable oil* derived from *Helianthus annus* which is similar to *safflower oil*.

super high tenacity fibre
See *viscose rayon*.

super-conducting polymer
A polymer which is an excellent conductor of electricity at very low temperatures, for example, a *sulphur-nitride polymer*.

super-heated steam process
A reclamation process for vulcanized rubber which uses super-heated steam to soften *natural rubber* compounds.

super-imposed calender
A *calender* in which the rolls are stacked one directly above another, for example, an *I calender* is a super-imposed calender. The major problem with this type of construction is one of nip interaction.

supercharge
See *charge*.

superconductive furnace (black)
An abbreviation used for this material is SCF. A highly electrically conductive *carbon black*.

supercooling
Cooling a melt of a *crystalline polymer* below the melting point without crystallization occurring.

superimposed stack
An assembly of rolls which are stacked one directly above another, for example, an *I calender* is a super-imposed calender.

superior processing
An abbreviation used for this term is SP. See *superior processing rubber*.

superior processing furnace (black)
An abbreviation used for this material is SPF. See *carbon black*.

superior processing rubber
An abbreviation used is SP rubber. A rubbery material with, for example, good extrusion characteristics and which is based on *natural rubber*. A rubber which gives compounds which, on processing by *calendering* and *extrusion*, gives low *die swell*, good surface smoothness and a high throughput when lightly filled compounds are used. Produced by blending unvulcanized *latex* and *vulcanized latex* in various ratios, prior to *coagulation*. For example, PA-80 is an *SMR* grade which has a concentration of 80 parts of *vulcanized latex*. When the rubber contains more than 50 parts of vulcanized material it is sold as a *processing aid* and sold for use as a *masterbatch*.

These modified forms of *natural rubber* may be blended with both NR and SR: they are of particular value in the liquid curing medium (LCM) process where the use of high extrusion temperatures gives problems of extrudate collapse and/or, of porosity. Replacing 50% of the rubber with a PA grade minimises such problems: adjust the amount of vulcanization system to suit the reduction of unvulcanized rubber. Can have various grades of SP rubber, for example, SP smoked sheet, SP crepe, SP air dried sheet and SP brown crepe.

superplastic metal alloy
A metal *alloy* which may be extensively stretched before failing. Such alloys are now being used for the production of *prototype moulds*. One such alloy is a zinc-based material which contains 22% aluminium and small, but important, traces of copper and manganese. A blank of this material is first heated to 275°C/527°F and then quenched rapidly to room temperature: this develops the required grain structure. After quenching the blank is polished and re-heated to 260°C/500°F. The heated blank is then placed within a frame, or chase, and a metal master (e.g. made from brass) is slowly pressed into the blank using forming pressures of up to 45 MNm^{-2}/6500 psi. After cooling the forms are mounted within an appropriate bolster.

supply tank
A vessel used to store, and/or blend components, and to maintain the level in conditioning tanks on, for example, a *reaction injection moulding machine*.

support block
Part of a mould: a rectangular block which is used to support a mould plate.

surface active agent
Also known as a surfactant. Substances introduced into a liquid in order to affect (increase) spreading and wetting properties. For example, *soap* will concentrate at the water surface or interface and modify the behaviour, or properties of that interface. The surfactant molecules have two dissimilar ends. One is water soluble (highly polar or hydrophilic): the other end is water insoluble (non-polar or hydrophobic). At an oil and water interface, the hydrophillic group will

associate with the water and the other end with the oil. This action reduces surface tension and, for example, emulsified oil droplets can be formed suspended in the water. Surface active agents are classified into three main groups (dependant upon the electric charge of their active groups). They are classified as being *anionic surface active agents* or as, *cationic surface active agents* or as, *non-ionic surface active agents*.

surface bonding
The bonding between a *polymer* and the surface of a *filler*. In the case of fillers, added to polymers, good surface bonding between the polymer and the filler is required in order to improve the properties of the filled polymer compound. Similarly, good surface bonding is required between the thermoplastics material and the rubber in rubber toughened, plastics materials. (In some cases good surface bonding can be a disadvantage see *accelerator deactivation*). The strength of bonding to a filler can be measured using laser heating of a small sample. This rapid heating produces an acoustical emission which can be measured: the stronger the bond, the lower the signal. See *silane coupling agent*.

surface coat
See *gel-coat*.

surface colouring
See *skin colouring*.

surface conditioning
Part of the production process for the *electroplating* of plastics in which injection mouldings are chemical etched so as to improve subsequent metal adhesion. The thermoplastics material associated with electroplating is *acrylonitrile-butadiene-styrene* (ABS) and for this material the etching, or deglazing, stage involves removing rubber particles form the surface. For example, this may be done by immersing the injection mouldings into a strongly oxidizing acid bath which consists of a mixture of chromic, sulphuric and phosphoric acids. This treatment produces a surface which is dull without being rough as microscopic craters are formed where the rubber particles are removed. Metal is deposited in the microscopic craters, by *electroless plating* of copper or nickel, and this gives good adhesion.

surface imperfections
See *haze*.

surface resistance
If a dc voltage is applied to an insulator in such a way that the resulting small current is made to flow either through a thin surface layer or across the surface of the material then a surface resistance can be measured from which a *surface resistivity* can be calculated. However this is not really a material property as it can be markedly affected by contaminants, such as moisture, on the surface of the material.

surface resistivity
The resistance between electrodes on opposite edges of a square on the surface of a material: measured in ohms. It is found that the size of the square is immaterial as increasing the size increases the width of the conductive path but also lengthens it in proportion. See *volume resistivity*.

surface speed - of a screw
The rate of movement of the outermost surface (the tips of the flights) of the *screw*. Measured in, for example, m/s. It is this which should be quoted in machine records rather than rpm.

surface tracking resistance
See *resistance to tracking*.

surface texture
Often only *roughness* and waviness are considered to make up surface texture. Measured on profiles of plane sections taken through the surfaces, for example, using a stylus-based instrument. The stylus, tipped with a skid, is drawn over the surface and a trace (a profile graph) is produced of the stylus movement. A mean of a number of observations, made on the trace, may be taken. For example, five consecutive sections along a short length may be measured and the heights of the peaks and valleys determined so as to obtain the *arithmetical mean deviation*.

surface treatment
A treatment applied to, for example, *carbon fibre* (CF) so as to improve the bonding between the CF and a polymer. Such a surface treatment may be classed as a wet oxidative surface treatment or as a dry oxidative surface treatment. Wet oxidative surface treatments are based on the use of solutions of sodium hypochlorite or, sodium bicarbonate or, chromic acid: a dry oxidative surface treatment relies on the use of ozone. Such treatments introduce active chemical groups, or remove weak surface layers or, chemically clean the surface or, roughen the surface.

surface winder
A winding, or *reeling*, system for flexible sheet. Surface winding is also known as contact batching. Contact batching is relatively cheap to perform, but sheet tension can vary as the weight of the material roll increases. In this method of winding, commonly used for flexible materials, the roll is in contact with a wind-up drum which is capable of being rotated at the required speed. See *centre-core winder*.

surfacing agent
An additive used with *unsaturated polyester resin* to prevent air-inhibition of cure. An oily or waxy material which is present in the resin system and which rises to the surface during cure to form a protective skin.

surfacing mat
A fine *glass cloth* used with, for example, *unsaturated polyester resin* to reduce the appearance of fibres at the surface of the moulding: such a mat is applied as the last layer over the reinforcing layers and forms a protective skin.

surfactant
See *surface active agent*.

surge
A term used in *hydraulics* and which refers to an uncontrolled rise in pressure or, in flow rate.

surge pressure
A term used in *hydraulics* and which refers to the pressure generated as a result of *surge*.

surging
The unsteady flow of plastics *melt* through an *extruder*.

SUs
An abbreviation used for Saybolt Universal seconds.

suspending agent
An ingredient of a *suspension polymerization* recipe: the suspension system stops the particles sticking together when they become sticky. The suspending agent system and the agitation of the stirrer control the particle size and the particle size distribution. Examples include finely divided mineral fillers and water-soluble polymers, for example, *gelatin*.

suspension
A two-phase system which consists of small solid particles suspended in a liquid. See *suspension polymerization*.

suspension polymerization
Also known as bead or pearl polymerization. A very popular method of producing synthetic polymers as the heat liberated on *polymerization* is transferred to the water in which the

reaction occurs. The monomer is suspended in droplet form in the water by the combined action of a *suspending agent* system and a stirrer: these control the particle size and the particle size distribution.

The suspending agent system is commonly a water-soluble polymer (for example, a vinyl acetate-maleic anhydride copolymer) and a finely divided solid (for example, kaolin): the suspension system stops the particles sticking together when they become sticky. A monomer soluble initiator is used to initiate the reaction and so polymerization (*mass polymerization*) occurs within each droplet. The course of the reaction may be followed by the changes in pressure which occur during manufacture.

The solid polymer which results may be separated from the water by centrifuging before being dried in, for example, a large rotating drum drier. The resultant solid polymer contains the suspension system but as the contamination is not as much as that obtained by *emulsion polymerization*, better electrical and optical properties may result. See *polyvinyl chloride* where PVC-S means that the polymer was made by suspension polymerization.

In the case of PVC, the particles are in general larger than those produced by emulsion polymerization: the surface area of each particle is also much larger so that rapid plasticizer absorption occurs during *dry-blend* manufacture.

Sv
An abbreviation used for *sievert*.

swash-plate
An inclined plate in an axial piston pump or motor, against which the pistons slide so as to generate reciprocating piston movement as the pistons assembly is rotated. Can be used, for example, to alter the output of the pump by altering the piston stroke.

swashplate
See swash-plate.

sweat out
See *exudation*.

swelling ratio
See *die swell*.

swirl mat
See *continuous filament mat*.

swiss-roll
A compound form: a compound in sheet form which has been rolled so that on introduction into a *two-roll mill*, the mixing is improved.

symbols - for polymers
See *abbreviations*.

symmetrical diphenyl thiourea
See *thiocarbanilide*.

syndiotactic polymer
A *stereoregular polymer*: a polymer in which there is a regular, repeating structure. The repeat units, along the polymer chain, have the same configuration: that is, the groups of atoms have the same orientation in space. If the repeat unit is two carbon atoms long in a vinyl-type polymer then, the substituent groups (for example, CH_3) do not lie on the same side of the polymer chain: the substituent group points alternately in one of two directions. See α *olefin* and *isotactic polymer*.

syndiotactic polystyrene
An abbreviation used for this type of material is SPS. A *syndiotactic polymer*: a stereoregular polymer produced using a highly stereospecific catalyst system such as a Ziegler-Natta, metallocene catalyst. A styrene polymer in which there is a regular, repeating structure which allows crystallinity to develop. Such a thermoplastics material has good heat, steam and water resistance. The *heat deflection temperatures* can be 260°C for an unfilled grade: the *heat deflection temperatures* can be 480°C for a 30% glass filled grade. This material has high gloss, good dimensional stability, reasonable impact strength and stiffness.

syneresis
The contraction of a gel accompanied by the separation of a liquid. The squeezing out, or separation, of liquid from between colloid particles.

synergism
A term associated with mixture of materials and which means that the joint action of two materials increases each others effectiveness. Such mixtures give a combined action which is greater than what would be expected from a simple addition of effects. Therefore a synergistic *heat stabilizer* system would give more protection than that expected from the individual components of the mixture.

synergistic co-stabilizer
A material, used in conjunction with another primary stabilizer, and which has a *synergistic* action. Examples of *heat stabilizers* include, for example, *epoxy compounds* and *phosphite chelators*.

synergistic combination
A mixture of materials which gives a combined action which is greater than what would be expected from a simple addition of effects. Associated with *heat stabilizers* and *antioxidants*.

synergistic heat stabilizer
See *metal soap stabilizers* and *packaged stabilizers*.

synergistic mixture
A *synergistic combination*.

syntactic cellular plastics material
See *syntactic foam*.

syntactic foam
A light-weight material produced by incorporating a low density filler into a polymer. Glass microspheres are used as a filler in plastics materials to reduce the density and to produce syntactic foams. The incorporation of glass *microspheres* into *epoxide resins* give mixtures which are easy to apply, which cure at room temperatures and which can be readily sanded when cured.

synthetic
Formed by chemical reaction in a laboratory or chemical plant, for example, *synthetic rubber*.

synthetic alkylated phenols
See *iso-propyl phenols*.

synthetic calcium silicate
Made from *diatomaceous silica* and *lime*: see *calcium silicate*.

synthetic cis-polyisoprene and natural rubber - comparison of
See *natural rubber and synthetic cis-polyisoprene - comparison of*.

synthetic diene rubber
A synthetic elastomeric material made from a *conjugated diene hydrocarbon*, for example, butadiene or isoprene. See *diene polymerization*.

synthetic isoprene rubber
See *isoprene rubber*.

synthetic lapis lazuli
See *ultramarine blue*.

synthetic latex
A *latex* based on a *synthetic rubber*. For example, can have *styrene-butadiene latex, chloroprene latex, butyl latex* and

nitrile rubber latex. Such a latex may be used in place of natural rubber latex because of, for example, oil resistance and/or flame resistance.

synthetic natural rubber
See *isoprene rubber*.

synthetic polyisoprene
See *isoprene rubber*.

synthetic polymer
A man-made material: the product of a *polymerization* process initiated and controlled by man. Most commercially important thermoplastics materials are of this type. See *commercial plastics* and *synthetic rubber*.

synthetic precipitated calcium carbonate
See *whiting* and *calcium carbonate*.

synthetic resin
A man-made material (a *synthetic polymer*) which resembles a *natural resin*.

synthetic rubber
The term was originally used to denote a synthetic homopolymer of *isoprene* but is now taken to mean any rubber which is a *synthetic polymer*. An abbreviation used for this type of material is SR. There are about 30 different types of synthetic rubber: those classified as *general purpose*, account for 70% of SR consumption. *Styrene-butadiene-rubber* (SBR) is commercially the most important, followed by *butadiene rubber* (BR) followed by *isoprene rubber (IR)*. These materials have such a dominant position because 1,3 butadiene is the cheapest monomer available from crude oil.

synthetic silica
Fine *silica*, whose particle size is less than that of carbon black, may be synthesized by *precipitation* (colloidal silica) or by *pyrogenic processes* (fumed silica). All silicas and silicate fillers contain absorbed water: the water content should be controlled to obtain consistent compound properties.

system
An interrelated assembly of objects working together as a unit for a common purpose.

Système International d'Unité
An abbreviation used for Système International d'Unité is SI. The units issued by this organization are an international coherent system of units now widely used for scientific and technical work. The seven basic units are the *metre* (length), kilogram (mass), second (time), ampere (electric current), kelvin (temperature), mole (amount of substance) and the candela (luminous intensity). There are supplementary units for plane angle (radian) and solid angle (steradian) and some SI derived units which have special symbols. These SI derived units which have special symbols include, for example, the newton and the pascal. Certain other non-SI units are permitted, for example, cm and °C. SI units were derived from the *MKS system* and are sometimes occasionally known as MKSA metric.

T

t
An abbreviation used for time. An abbreviation used for *a ton* or *a tonne*.

t method
An extension of the *BET method* which is also called adsorbed layer thickness method. Distinguishes between internal and external surface area. This method and the *BET value* may rank or assess *carbon blacks* in the same order as the blacks used for rubbers are relatively non porous.

T
This letter is used as an abbreviation for:

    polysulphide rubber. A rubber which has sulphur, carbon and oxygen in the main polymer chain. For example, OT and EOT. See *polysulphide rubber*;

    talcum or cord. When a plastics material contains such fillers, then T may be used to modify the abbreviation for the plastics material;

    tera. See *prefixes - SI*;

    terephthalamide. See *poly-(p-phenyleneterephthalamide)*:

    tesla. See *prefixes - SI*;

    tetra. Used in *plasticizer abbreviations*, for example, tetraoctyl pyromellitate (TOPM);

    thermal (black). See *carbon black*;

    thermoplastic. See *thermoplastic material*;

    time. See *prefixes - SI*;

    tri. Used in *plasticizer abbreviations* for example, tricresyl phosphate (TCP); and,

    trifunctional unit. See *polyorganosiloxane*.

T-50
A temperature obtained in a *T-50 test*.

T-50 test
A test used to indicate the degree of *vulcanization* of a cured rubber compound. The rubber compound is stretched and then frozen at $-70°C$: the stretched rubber is then released from tension and slowly heated. T-50 is the temperature at which the test specimen is half the original stretched length. L-50 is the length at T-50. Stretch gum stocks and latex films by 700%, lightly loaded compounds by 500% and highly loaded compounds by 350%.

T23P
An abbreviation used for *tri-(2,3-dibromoropropyl) phosphate*.

tab gate
A *gate* used in injection moulding for the moulding of, for example, large plane areas, particularly in transparent materials where flow marks should be reduced to a minimum. Feed is through a *side gate* into a tab at the side of the cavity. A tab gate helps to obviate *jetting* by creating turbulent flow in the tab. After the moulding operation, the tab may be removed from the component; as the area adjacent to the gate is usually highly stressed, tab-gate removal can improve component reliability.

tack
The tendency of a material to stick to itself without necessarily sticking to other surfaces. Thought to be due to the mobility of molecules which allow diffusion across the interface. See *self adhesion*.

tack off sections
Also known as tack offs. Product features used in *blow moulding* to increase stiffness: two opposite *parison walls* are welded together by shaped protrusions on the mould walls.

tack offs
See *tack off sections*.

tackifier
An additive used to increase the tack of rubber compounds. Examples of tackifiers include synthetic resins and abietic acid. See *self adhesion*.

tactic polymer
See *stereoregular polymer*.

tacticity
Also known as stereoregularity. Regular spatial arrangements of atoms within the monomer repeat units. This is possible when the repeat unit contains an asymmetric carbon atom and gives rise to *stereoregular polymers*. Can have, for example, *isotactic polymer* and *syndiotactic polymer* types in common synthetic polymers.

tail
The excess material attached to the base of a *blow moulding*.

tail-to-tail link
See *tail-to-tail* structure.

tail-to-tail structure
A structure produced in a vinyl polymer molecule during polymerization and which may be represented as -CHX-CH$_2$-CH$_2$-CHX- .

take-and-put automation
See *pick-and-place automation*.

take-off
Part of a production line which allows the output of a continuously producing machine, for example, a *calender* or an *extruder*, to be conveyed away from that machine.

take-up
Part of a production line which allows the output of a continuously producing machine, for example, a *calender* or an *extruder*, to be wound on a reel.

Talalay process
A *latex foam rubber* process. Shaped mouldings may be produced by partially frothing the latex (by the beating on of air), pouring some of the mixture into a mould and then sealing and evacuating the mould so as to cause foam expansion. The mould plus contents is then chilled to $-35°C$ and carbon dioxide is admitted: this causes gelling to occur. The mould is then heated to cause vulcanization.

talc
Also known as French chalk or as, magnesium silicate or as, hydrated magnesium silicate or as, talcum, or as, steatite or as, soap stone. This crystalline silicate material has a relative density (RD or SG) of 2·72: consists of hydrated magnesium and silicon oxides. When pure, this material consists of a sheet of brucite (MgO.H$_2$O) sandwiched between two silica (SiO$_2$) sheets. These composite sheet layers are held together by weak forces which allow the composite sheets to slide past each other very easily.

An odourless, soft white powder the particles of which may be granular, *plate-like* or needle shaped. The chemical composition and the particle shape are influenced by the origin of the material: the particle shape is also influenced by the grinding process used in material preparation.

This material is widely used as an inert filler in rubber compounds. When used at high loadings it stiffens rubber stocks thus aiding calendering and extrusion processes. Gives good electrical properties and its acid resistance is useful where rubber components come into contact with acids. Used as a dusting powder on unvulcanized rubber, as a packing in open steam vulcanization and as a rubber lubricant.

One of the most popular fillers used for plastics materials. Used in thermoplastics materials, for example, in *polypropylene* as the resultant compounds can be *injection moulded* into components which have good surface finish and high heat resistance. Used in *polyvinyl chloride* (PVC) to give semi-transparent compounds and to stiffen floor-tile compositions produced by calendering.

talcum
See *talc*.

tall oil
A combination of *fatty acids* and *rosin* which behaves as a semi-drying *oil*: obtained during *wood pulp* manufacture. This material has a relative density (RD or SG) of 0·95 to 1·0. When limed, tall oil gives a low cost liquid which has high gloss but poor flexibility and poor colour retention on ageing. Tall oil fatty acids may be used in *alkyd resin* manufacture and are similar to those obtained from soya-bean oil. Tall oil may be used as a rubber softener.

tallow
Rendered animal fats: this type of material has a relative density (RD or SG) of 0·95.

tan δ
See *loss tangent*.

tangent to the loss angle
See *loss tangent*.

tank
See *reservoir*.

tape
See *film tape*.

tape laying
Sometimes called tapelaying. See *tape winding*.

tape winding
Sometimes called tapelaying or tape laying. A tape (or tapes) made from *thermoplastic impregnated fibre rovings* is heated so that the layers will adhere when the tape is wound onto a temperature controlled mandrel: the assembly is then rapidly heated (infra-red or hot air) so as to fuse the layers by melting the thermoplastic coating. Pressure is applied to the assembly, by means of a pressure roller, immediately after heating and during winding. See *thermoplastic bulk moulding compound*.

tapelaying
See *tape winding*.

taper screw
See *tapered screw*.

tapered block copolymer
See *copolymer*.

tapered extruder
An extruder which contains a tapered *screw* or screws. See *tapered twin-screw extruder*.

tapered machine
Usually means *tapered twin screw extruder*.

tapered parallel screw
An *extrusion screw* which has a gradually tapering screw root (the channel depth gradually changes) in the *compression zone* but the *feed zone* and the *metering zone* are parallel and of constant depth.

tapered screw
A screw which has a gradually tapering screw root, i.e. the channel depth gradually changes - usually the diameter of the *screw root* increases from the *hopper* end to the *die* end of an *extruder*.

tapered twin screw extruder
A type of *twin-screw extruder*. An extruder which contains two *screws* the outside diameters of which gradually reduce from the feed zone to the discharge end.

Because of the very positive feeding characteristics of twin-screw counter-rotating machines, very high forces may be generated: so high, that thrust bearing failure is a real danger. Such machines are therefore often starved fed and/or run at low screw speeds. This is because it is difficult to incorporate adequate thrust bearings in the small space available. The space available can be increased by tapering the screws and larger thrust bearings can therefore be incorporated: this permits higher outputs. Shear at the screw tips is also reduced because of the reduced surface speed of the screw at that point.

tapper
One who taps a rubber tree. For example, the tapper removes a very thin, sloping strip of bark with a curved knife (a jebong) so that *latex* is exuded and can be collected in a cup. See *tapping*.

tapping
The process used to obtain *natural rubber latex* from the tree *Hevea Brasiliensis* by an excision of the bark of the tree. The *tapper* removes, and collects, the *tree lace* and the *cup lump*. Tapping is then traditionally done by making narrow slits (the length of the cut is more important than the width of the cut for latex collection) in the bark of the tree at approximately 30° to the horizontal. The cut severs the latex vessels without affecting the normal sap circulating system: the latex is not the sap of the tree. In an effort to heal these wounds the tree exudes the latex which is collected in cups via a spout. After some time (for example, 2 hours) the latex seals the cut by *coagulation*.

Each tree will yield about 100 g of latex at each tapping. The contents of the cups are collected (*field latex*) and emptied into large tanks for further processing. The latex may be concentrated to 60% solids content or coagulated to give dry (solid) rubber. Unless processed immediately the latex is stabilized with, for example, *ammonia*.

The residual latex in the cup will coagulate (*cup lump*) as will the latex in the tapping cut (*tree lace*) by auto-coagulation: this field coagula is collected and comprises about 25% of the crop. See *natural rubber* and *tapping methods*.

tapping methods
Methods used to obtain *natural rubber* from the tree *Hevea Brasiliensis*. A frequently used tapping method is the half spiral method which allows *tapping* every other day - method S2 d2. If full spiral cuts are made, around the complete trunk, the tree is tapped every fourth day - *method S1 d4*. *Stimulation tapping* and *puncture tapping* may be used in place of such traditional methods.

target bound-styrene content
The amount of styrene contained in a copolymer based on styrene and butadiene. See *styrene-butadiene rubber*.

TBAC
An abbreviation used for *tributyl o-acetyl citrate*.

TBBS
An abbreviation used for N-t-butylbenzothiazole-2-sulphenamide. See *accelerator* and *suphenamides*.

TBC
An abbreviation used for p-tert-butyl catechol.

TBEP
An abbreviation used for *tri-(2-butoxyethyl) phosphate*.

TBLS
An abbreviation used for *tribasic lead sulphate*.

TC
An abbreviation used for *thiocarbanilide*.

TC-NR
An abbreviation used for *technically classified natural rubber*.

TCDD
An abbreviation used for *dioxin*. See *dioxins*

TCEP
An abbreviation used for *tri-(2-chloroethyl) phosphate*.

TCF
An abbreviation used for *tricresyl phosphate*.

TCP
An abbreviation used for *tricresyl phosphate*.

TCR
An abbreviation used for *technically classified rubber*.

TCT
An abbreviation used for *tricrotonylidine tetramine*.

TDBP
An abbreviation used for *tri-(2,3-dibromoropropyl) phosphate*.

TDCP
An abbreviation used for *tri-(2,3-dichloropropyl) phosphate*.

TDI
An abbreviation used for *toluene diisocyanate*. For example;
TDI index = *tolylene diisocyanate index*; and
TDI mixtures = *tolylene diisocyanate mixtures*. See *80:20 TDI* and *65:35 TDI*.

TDT test
An abbreviation used for *tensile deformation temperature test*.

TEAC
An abbreviation used for *triethyl o-acetyl citrate*.

tear resistance
Also known as tear strength. The force required to propagate a rip, or tear, in a material. The tearing energy; usually expressed as the ratio of the maximum load measured during the test to the specimen thickness. A nick, or cut, is made in the edge of a specimen and the force required to propagate the cut in a tensile-type of test is measured. The sample may be a strip of material or a specially shaped piece of material (cut or moulded): a common test sample is shaped like a pair of trousers and the use of this *test piece* gives the trouser tear test.

tear strength
See *tear resistance*.

tearing energy
See *tear resistance*.

technically classified natural rubber
An abbreviation used for this type of material is TC-NR. See *circle rubber*.

technically classified rubber
An abbreviation used for this term is TCR. A *natural rubber* whose cure properties have been assessed or graded. See *circle rubber*.

technically specified (natural) rubber
See *technically specified rubber*.

technically specified rubber
An abbreviation used for this material is TSR. A *natural rubber* product which is supplied in a convenient form (for example, a regularly-sized, compact bale or block) and which is of consistent properties (with regard to cure rate). Such rubbers have been subjected to tests which yield numerical quality control limits rather than having been graded by visual observations alone. See *standard international grades - natural rubber*.

The regularly-sized, compact bales can be produced because the natural rubber coagulum is originally in *crumb form*: the consistent cure properties result from, for example, careful control over the latex concentration before *coagulation* and the pH at which coagulation is performed: the cleanliness of the rubber is also measured by a standardized dirt test.

Technically specified (natural) rubber from China is CNR: from Indonesia it is SIR: from Malaysia it is SMR: from Sri Lanka it is SLR: from Thailand it is TTR. See *standard Malaysian rubber*.

TEDC
An abbreviation used for *tellurium diethyl dithiocarbamate*.

TEEE
See *thermoplastic elastomer - ether based*.

teleblock copolymer
See *styrene block copolymer*.

telechelic polymer
A polymer which has ben produced so that it contains specified end groups.

telescopic flow
A type of liquid flow in which the layers or laminar move relative to one another like the sleeves, or pieces, of an extending telescope. See *laminar flow*.

tellurium diethyl dithiocarbamate
An abbreviation used for this material is TEDC. An ultrafast *accelerator* for *vulcanization*. See *tellurium vulcanization*.

tellurium vulcanization
Selenium and/or tellurium could be used in place of sulphur for vulcanization, but, because of price, odour and toxicity considerations, such materials are not usually commercially used even though they give enhanced heat resistance compared to *sulphur vulcanization*.

telomer
A polymer composed of molecules having terminal groups incapable of reacting with additional monomers, under the conditions of the synthesis, so as to form larger polymer molecules of the same chemical type.

temperature
A measure of the hotness of a body which is determined by the kinetic energy of the component particles of that body. Cannot be measured directly and so temperatures are measured with respect to an arbitrary scale. For many years the system used was degrees centigrade. The centigrade scale is based on the freezing and boiling points of water with the difference divided into 100 parts or degrees. Commonly referred to as °C. It follows that temperatures below those of the freezing point of water are negative and this can be misleading. Temperature should be measured on a scale which increases from *absolute zero* as any body above absolute zero has some heat energy (see *kelvin*). Temperature determines the direction of heat flow from one body to another: heat flows from the hotter body to the colder body.

temperature coefficient of vulcanization
The ratio of the rates of vulcanization obtained at two temperatures which are 10°C apart.

temperature control
Usually means the control of some part of the processing equipment as measurement and control of melt temperatures is often difficult and expensive

Thermoplastics materials are often processed by *melt processes*. The temperature of the melt should be held to precise values as the material temperature often determines the output rate as it controls the cooling rate. Melt temperatures also determine degradation or decomposition and have a tremendous effect on *orientation*. However, because of the generation of *shear heat* in (caused by high screw rotational speeds and mould filling speeds), the melt temperature is often greater than the highest set temperature (because of practical difficulties, the actual melt temperature is often not measured directly but measured using, for example, an *air shot technique*). The problem of *melt temperature override*, in screw-base equipment, can be minimized in various ways - see, for example, *zero compression screw*.

During *calendering*, material (stock) and equipment temperatures must be accurately controlled if sheet of the required thickness is to be produced. This is because, for example, rubber mixes are never completely plastic; they retain a certain amount of elasticity and the amount, for a given mix, depends on temperature. As the mix passes through the nip it recovers and is thus thicker than the gap between the bowls. The amount of elastic recovery decreases as temperature increases.

However the temperature cannot be increased too much as otherwise the stock would become so soft that entrapped air would not be eliminated and blistered sheet would result. A lower limit on temperature is set by the appearance of V-shaped marks called *crows feet*. There is thus a range of temperatures over which satisfactory sheet can be produced. If thin gauge sheet is required then the higher end of the temperature range is used and the calender is run more slowly. This gives a longer time under pressure and reduces the elastic recovery. In general the greater the rubber content of the mix then the narrower is the optimum temperature band. As *synthetic rubbers* have lower self adhesion than *natural rubber stocks*, it is found that for these materials higher operating temperatures are needed together with lower speeds. Roll temperatures used for *natural rubber* stocks are approximately 80°C. Roll temperatures are best controlled by the use of *drilled rolls* and circulating a temperature-controlled liquid at the correct flow rate through the rolls.

temperature profile
The temperatures set on the *zones* of a machine. During melt processing it is common to have a lower temperature at the hopper-end of a machine and a higher temperature at the die-end: the temperatures of the *zones* in between alter fairly uniformly. This is a normal temperature profile: can also have an *inverse temperature profile*.

temperature reversion test
An abbreviation used for this term is TR test. A test sometimes used for the examination of the effects of crystallinity in rubbers. The temperatures measured are those at which a previously stretched and frozen test sample retracts by 10% (TR-10) and then by 70% (TR-70). As crystallization increases so does the difference between TR-10 and TR-70.

temperature - testing
It is usual to specify both the speed of testing and the temperature of testing as both often effect the results of a *common test*.

temperature transducer
Another word for *thermocouple*.

tempered and quenched tool steel
See *pre-hardened steel*.

tempered tyres
See *nylon tyres*.

tempering
Heating *steel* to approximately 250 to 300°C and then cooling slowly so as to impart a definite degree of *hardness* to the *steel*. *Quenching* and tempering have a tremendous effect upon hardness and toughness.

temporary stoppage
Usually means a halt in production of less than approximately 30 minutes. See *stoppages*.

tenacity
The tensile strength of a *fibre* expressed as the breaking load in grams per denier or decitex.

tenasco-type fibre
High tenacity, viscose fibre. See *viscose rayon*.

tensile
Means pulling. For example, a *tensile test* is a pulling test.

tensile creep modulus curve
See *creep modulus/time curve*.

tensile deformation temperature test
An abbreviation used for this term is TDT test. The tensile deformations produced as a result of applying a *tensile stress* as a function of temperature.

tensile flow
See *elongational flow*.

tensile impact strength
The impact strength of a material measured in a tensile mode: may be measured, for example, by a high speed tensile test machine or by an impact test machine. In both cases it is usual to use a *dumb-bell* specimen. See *pendulum impact test*.

tensile impact test
See *pendulum impact test*.

tensile load
The load or stress applied during a *tensile test*. See *tensile strength at break*

tensile modulus
See *modulus*.

tensile properties - typical values of (at 23°C)

	UPVC	LDPE	PA 66 (1)	PC	Rubber (2)	(3)
Tensile strength at break 10^3 psi	6–7.5	0.6–2.3	11.0	9.5	4.0	3.0
(MPa)	41–52	4–16	76	65	28	21
Elongation at break (%)	40–80	90–800	300	110	680	420
Tensile yield stress 10^3 psi	N/A	N/A	6.5	9.0	N/A	N/A
(MPa)	N/A	N/A	45	62	N/A	N/A
Tensile modulus 10^3 psi	350–600	14–38	N/A	345	0.28	0.86
(MPa)	2410–4130	96–262	N/A	2380	1.9	5.9

(1) As conditioned to equilibrium with 50% relative humidity
(2) Natural rubber gum compound (no filler)
(3) A carbon black filled natural rubber compound
Plastics tested to ASTM D 635: Rubber to BS 903 part A2.

tensile strain
An abbreviation used for this term is Σ_1. Commonly referred to as strain and may be defined as the change in length/original length of a tensile specimen when subjected to a tensile test. That is, longer length (l_1) minus original gauge length (l_o) divided by original gauge length (I_o).

tensile strength
The maximum tensile stress sustained by a material in a *tensile test*: the ultimate tensile strength. The *stress* that has to be applied to a material, in tension, so as to cause breakage. Obtained by dividing the load by the original cross-sectional area. It is usual to specify both the speed of testing and the temperature of testing as both affect the results. If a cut sample is used, then the surface finish of the cut edges must also be controlled so as to minimize errors.

tensile strength at break
Also called the ultimate tensile stress. May be obtained by dividing the force necessary to cause failure during a tensile test, by the cross-sectional area at break. As this area is difficult to measure, the tensile strength that is commonly quoted is that which is given by F/A_o. Where F is the force which causes failure and A_o the original cross-sectional area. The stress and strain obtained by using the original sample dimensions are sometimes referred to as 'nominal' or 'engineering' stress and/or strain.

tensile strength at break - nominal
The tensile strength at break when the maximum *tensile stress* occurs at break.

tensile strength machine
A machine used to perform a *tensile test*.

tensile stress
The force applied in *tensile testing*.

tensile test
A pulling test. See *tensile strength*.

tensile test machine
A machine used to perform a *tensile test*. Also known as a tensile tester. See *tensile strength*.

tensile tester
A *tensile test machine*.

tensile testing - description of
Tensile testing is probably the most widely used short-term mechanical test of all. This is because it is relatively easy to perform, gives reasonably reproducible results and yields a great deal of information. From this one test one can obtain not only tensile strength, but also elongation and modulus; the same basic machine may also be used to perform a number of other tests (for example, flexural strength) with relatively simple modifications.

Such machines are usually constructed so that as the test specimen is deformed at standard cross head speed the resistance to deformation and the amount of extension is measured. In order to ensure that breaking does not occur in the grips, but in a specified region, the sample is usually *dumb-bell* shaped or waisted

Test specimens are often cut from compression moulded sheet, or are produced by injection moulding - where the dumbbells are moulded directly. The sample dimensions in the waist region (gauge section) are first measured with a micrometer or dial gauge and then the sample is firmly gripped in the jaws of the machine. An extensometer may then be clipped to the sample. One jaw of the machine is drawn away from the other (usually by an electric motor) at the specified speed. The resistance to deformation is usually measured by a load cell which is connected to one of the jaws. A load/extension curve is usually produced automatically by the machine. Five specimens are normally tested and the average result(s) quoted together with the standard deviation.

tensile testing - terms used
The application of the tensile stress causes the specimen to stretch from its original *gauge length (l_o)* to a longer length (l_1). Tensile strain (Σ_1), commonly referred to as strain may then be defined as: change in length/original length which is longer length (l_1) minus original gauge length (l_o) divided by original length (I_o). Percentage strain = strain × 100.

If sufficient stress is applied then the sample will break. The percentage elongation at break (ΣB) is usually expressed as a percentage of the original length and this, therefore, may be expressed as longer length (l_1) minus original gauge length (l_o) divided by original gauge length (I_o) times 100.

Tensile strength at break, also called the ultimate tensile stress, may be obtained by dividing the force necessary to cause failure during a tensile test, by the cross-sectional area at break. As this area is difficult to measure, the tensile strength that is commonly quoted is that which is given by F/A_o. Where F is the force which causes failure and A_o the original cross-sectional area. Yield strength may be calculated from the load at which the specimen continues to elongate without additional load. The stress and strain obtained by using the original sample dimensions are sometimes referred to as 'nominal' or 'engineering' stress and/or strain.

The ratio of stress/strain, if stress is proportional to strain, is called the elastic modulus or the Young's modulus; this figure is commonly quoted for many materials as it gives an appreciation of the stiffness of a particular material. For materials such as steel the value of Young's modulus, obtained from the slope of a stress/strain plot (up to the yield point), does not depend markedly on the test conditions and as such has come to be regarded as a material constant. Such is not the case for polymeric materials as the results obtained depend markedly on the test conditions. For example, if the speed or temperature of testing is changed then a very different answer will be obtained.

For rubbery materials a different type of modulus is quoted, for example, the 300% modulus (M_{300}). This is the stress needed to produce an elongation of 300 per cent.

tensile viscosity
See *elongational viscosity*.

tension set
The elongation remaining after a rubber test specimen has stretched for a given time and allowed to rest for a given time. Also referred to as permanent set or as, permanent set in tension.

TEPA
An abbreviation used for *tetraethylene pentamine*.

tera
An abbreviation used for this term is T: it means 10^{12}. See *prefixes - SI*.

terephthalates
The collective name for the class of materials which are based on *terephthalic* acid and used as *plasticizers*. See, for example, *dioctyl terephthalate*.

terephthalic acid
Also known as para-phthalic acid or as, p-phthalic acid or as, 1,4-benzenedicarboxylic acid. An abbreviation used for this material is TPA. This white solid material has a melting point of 131°C and a relative density (RD) of 1.52: sublimes at 300°C. Used to prepare polyesters, for example, *polyethylene terephthalate*. See *aromatic polymer*

termination - oxidation
See *oxidation*.

termonomer
One of the three monomers used to produce a *terpolymer*.

ternary copolymer
See *terpolymer*.

terpene resin
See *pinene resin*.

terpenes
Monocyclic hydrocarbons with the formula $C_{10}H_{16}$ which occur naturally in essential or volatile oils. Terpenes have been used to make industrial cleaners, for example, which are used in the electronics industry to clean printed circuit boards. Such cleaners can be biodegradable and have been used as an alternative to *chlorofluorocarbons*.

terpolymer
Also known as a ternary copolymer or as, a *tripolymer*. A polymer based on three monomers: a polymer based on three different repeat units. See, for example, *ethylene-propylene-diene monomer*.

terpolymerization
A polymerization process used to produce a *terpolymer*.

tert-butyl phenol
May be represented as $(CH_3)_3CC_6H_5OH$. There are two isomers, *ortho-tert-butyl phenol* and *para-tert-butyl phenol*.

tertiary amine
An *amine* with the formula R_3N. Tertiary amines are used as catalysts for the blowing reactions in *flexible polyurethane foam* production by the *one shot process*. See *stannous octoate*.

tertiary colour
A colour made by blending two secondary colours, for example, brown.

tertiary structures
Particle structures based on *secondary structures*. See *silica*.

tertramethylolmethane
See *pentaerythritol*.

tervalent
Meaning trivalent or having a valency of three.

tesla
An SI derived unit which has a special symbol, that is T. It is the SI derived unit of magnetic flux density. One tesla is defined as the density of one weber of magnetic flux per square metre of circuit area. See *Système International d'Unité*.

test
Often means an examination performed so as to evaluate performance or capability: for example, of a *compound*, a *test piece* or a component. Such a test is performed as specified by a test standard which is issued by a *standards organisation*. See *common test*.

test piece
Sometimes referred to as a sample or as a, specimen or as a, test sample or as a, test specimen: test piece is preferred by *ISO*. In general, testing may be performed using finished components, test pieces cut from sheeted compounds, test pieces cut from finished components or, specially moulded test pieces. See *testing*.

test - classification of
The tests performed on plastics and rubbers can be grouped in various ways, for example, into mechanical tests, electrical tests, optical tests etc. However, look under the following entries for information on specific tests:

 i   stress-strain testing, for example, *tensile* and *flexural*;
 ii *impact testing*, for example, Charpy, Izod and instrumented falling weight;
 iii *creep testing*, that is, long term testing;
 iv flow testing, for example, *flow rate* and *high shear rheometry*;
 v   *thermal properties*, for example, *Vicat* and *heat distortion temperature*;
 vi *environmental stress cracking*;
 vii optical properties, for example, *light transmittance measurement* and *haze*;
 viii electrical properties, for example, *volume resistivity*;
 ix flammability, for example, *limiting oxygen index* and *flammability rating*;
 x   weathering, for example, natural and artificial;
 xi *testing of finished components*, for example, *unplasticized polyvinyl chloride* (UPVC) window frames.

test - repeatability of
See *standardisation*.

test results - factors influencing
There are a large number of factors which will influence the results obtained from a test. Factors such as test machine design, test piece size, shape and method of production,

conditioning of the samples, temperature of test, etc. This means that standardisation of test methods is essential. See *standards organizations*.

test sample conditioning
No matter which test method is employed it is important to ensure that the samples have received the same conditioning before testing. For many materials this conditioning is simply storing the specimens at a standard temperature, for a specified time, before testing. Despite its simplicity such physical ageing is important as post-moulding changes commonly occur. For some materials, for example, *engineering thermoplastics materials*, the specimens should be stored for a specified time under conditions of specified humidity.

test specimen
See *test piece*.

test standards
See *standards organisations*.

testing
The performance of a *test*: such testing may be performed on finished components, on test pieces cut from sheeted compounds, on test pieces cut from finished components, or on specially moulded test pieces

FINISHED COMPONENT TESTING. Testing performed using the production components as the test pieces or samples: the tests are often performed using conditions which simulate end-use conditions. Such tests give the most reliable indication of component performance but the obvious disadvantage is that the components must be available. A test widely used for this purpose is *instrumented falling weight testing*.

MOULDED TEST PIECE TESTING. Testing performed on test samples of a standardized size and shape which are either *compression moulded* or *injection moulded*. The testing of such specially moulded specimens is widely performed as test specimens of uniform thickness, and which are free from edge defects, can be prepared relatively easily: such samples give fairly reproducible results. However, it is difficult to relate the results of the tests to the performance of the finished product as, for example, orientation levels within the test piece will be different to those found in the components.

TEST PIECES/SPECIMENS CUT FROM SHEETED COMPOUND. For example, a sheet may be produced after mixing on a *two-roll mill* and test pieces may be blanked or cut from that sheet. Such testing is performed in both the rubber and the plastics industry as problems associated with sample preparation are minimized and results can be obtained quickly. In the rubber industry, *hardness* and *cure characteristics* are measured and in the plastics industry, such samples are often tested for *hardness/softness* and *flow rate*. With such tests the poor surface finish of the milled sheet is not important.

TEST PIECES/SPECIMENS CUT FROM FINISHED COMPONENTS. Testing performed on, for example, test pieces cut from moulded components: such testing can give realistic answers but it is important to ensure that the samples produced are free from *edge defects* as these could cause premature sample failures. As more than one sample is required for most tests it is important to have a number of mouldings available so that appropriate samples may be cut from identical positions within the mouldings. A template is a great help in this respect.

TESTING - RESTRICTING TESTING. Testing is expensive and it is therefore important to restrict testing as much as possible. Unnecessary testing, on materials or products, should therefore not be performed. For example, the testing of gloss, or finish, on parts which are to be hidden in service would seem to be pointless. There is not only the cost of performing the test itself but, there is also the cost of putting things right if the test results show this to be necessary. Before a test is performed it must always be decided what will be done, with the results and with the components, after testing. It is very important that test specifications are drawn up at a very early stage in the production process: the use of written specifications and procedures is vital. Once production has started, then periodic reviews of the tests are essential so as to ensure that the test programmes continue to provide what is required.

UNITS OF MEASUREMENT. One of the major problems associated with the testing of polymers is the variety of units in which results can and are reported. Although the SI system (Système International d'Unité) has been widely adopted, both the metric and imperial systems continue to be used. For this reason it is essential to be able to convert back and forth between units. Some of the most frequently used conversions are set out in a separate appendix.

testing - fit and finish
See *fit and finish testing*.

testing organizations
See *standards organisations*.

testing - raw material tests
Inspect the incoming raw material and note colour, particle size, particle size consistency and any obvious contamination. The ease of flow of the material should be checked either during processing or before processing begins. For example, during a process such as *injection moulding* the ease of flow may be assessed using a *flow tab* or by noting the ease of mould filling. Before the material is processed it would be wise to check that it is similar to what is currently being used by measuring the flow behaviour using a *rheometer*.

testing - speed of
See *speed of testing*.

TET
An abbreviation used for *triethylene tetramine*.

TETA
An abbreviation used for *triethylene tetramine*.

TETD
An abbreviation used for *tetraethylthiuram disulphide*.

tetra
A prefix which means four. The term is sometimes also used as an abbreviation for *carbon tetrachloride*.

tetra-(hydroxymethyl) methane
See *pentaerythritol*.

tetraalkyltitanate
A *coupling agent*, based on titanium and used to improve the bonding of a *filler* in a compound.

tetrabromophthalic anhydride
An aromatic anhydride which has four bromine groups attached to the benzene ring. Used to make, for example, *unsaturated polyester resins* which have greater *fire resistance* than those based on *phthalic anhydride*.

tetrabutyl thiuram disulphide
At approximately 2·5 phr this material is used as an *accelerator* for *acrylic rubber*.

tetrachloroisoindolinone pigments
A class or type of *organic pigment* and which have complex polycyclic structures. Yellow, red and orange pigments which have good resistance to heat (up to approximately 300°C): such pigments also resist light and bleeding.

tetrachloromethane
See *carbon tetrachloride*.

tetrachlorophthalic anhydride
An aromatic anhydride which has four chlorine groups attached to the benzene ring. Used to make, for example, *unsaturated polyester resins* which have greater *fire resistance* than those based on *phthalic anhydride*.

tetraethyl thiuram disulphide
See *tetraethylthiuram disulphide*.

tetraethyl thiuram monosulphide
See *tetraethylthiuram monosulphide*.

tetraethylene pentamine
An abbreviation used for this material is TEPA. A catalyst activator used in emulsion polymerization of, for example, *synthetic rubbers*. Also used in the *vulcanization* of foam rubber.

tetraethylthiuram disulphide
Also spelt tetraethyl thiuram disulphide. An abbreviation used for this material is TETD. An example of a *sulphur donor vulcanization system* which is an *accelerator* as well as being a sulphur donor. This solid material has a melting point of 65 to 72°C and a relative density (RD or SG) of about 1·2.

tetraethylthiuram monosulphide
Also spelt tetraethyl thiuram monosulphide. An example of a *sulphur donor vulcanization system* which is an *accelerator* as well as being a sulphur donor. This solid material has a melting point of 103 to 114°C and a relative density (RD or SG) of about 1·4.

tetrafluoroethylene
An abbreviation used for this monomer is TFE. May be represented as $CF_2 = CF_2$. Also known as tetrafluoroethene. This gaseous material has a boiling point of $-77°C$: a monomer which must be stored, transported and handled with great care as it can polymerize very violently to give *polytetrafluoroethylene*. Also used to prepare tetrafluoroethylene copolymers. One of the three monomers used to make, for example, a *carboxy-nitroso rubber*. See *fluororubber*.

tetrafluoroethylene-ethylene copolymer
Also known as poly-(tetrafluoroethylene-co-ethylene). An abbreviation used for this type of material is ETFE or TFE/E. An alternating copolymer of *tetrafluoroethylene* and *ethylene* with a melting point of approximately 270°C. A semi-crystalline copolymer which can be *melt processed* as a thermoplastics material. Similar properties to *fluorinated ethylene propylene copolymer* but has superior abrasion resistance. Has a very high *impact strength* and, when reinforced with *glass fibre*, high *tensile strength*. Used as electrical insulation, for example, for wire.

tetrafluoroethylene copolymer
A polymer based on two monomers one of which is *tetrafluoroethylene*. See, for example, *fluorinated ethylene propylene copolymer*.

tetrafluoroethylene-propylene copolymer
An abbreviation used for this type of material is TFE/P. An alternating copolymer of *tetrafluoroethylene* and *propylene*. A fluororubber copolymer which can resist some chemical combinations (concentrated acids) better than the more widely used FKM-type of rubber although the FKM-type of rubber resists petroleum fluids better. A highly specialized *fluororubber* which can be cross-linked with an *bisphenol-AF/amine*-based system. A *tetrafluoroethylene-propylene terpolymer* can be cross-linked with a peroxide system, as it contains a *cure-site monomer*: it is more convenient to use than a tetrafluoroethylene-propylene copolymer.

tetrafluoroethylene-perfluoromethylvinyl ether copolymer
See *perfluorinated elastomer*.

tetrafluoroethylene fluororubber
See *fluororubber*.

tetrafluoroethylene - hexafluoropropylene copolymer
See *fluorinated ethylene propylene copolymer*.

tetrafluoroethylene-propylene terpolymer
A terpolymer of *tetrafluoroethylene*, *propylene* and a *cure-site monomer* (a fluorinated vinyl material). A tetrafluoroethylene-propylene terpolymer which can be cross-linked with a peroxide system and is more convenient to use than a *tetrafluoroethylene-propylene copolymer*. Has better low temperature flexibility than the copolymer. See *fluororubber*.

tetrafluoroethylene terpolymer
A polymer based on three monomers one of which is *tetrafluoroethylene*. See, for example, *fluororubber*.

tetraglycidyl methylene dianiline epoxy resin
Also known as glycidylamine epoxy or as TGMDA. An epoxy, or epoxide, resin used in the fabrication of, for example, *carbon fibre* reinforced hardware for the aerospace industries as the use of such resins gives high temperature resistant matrices. The resultant composites are also tough and have good mechanical properties.

tetrahydro-1,4-oxazine
See *morpholine*.

tetrahydrofuran
Also known as 1,4-epoxy butane and may be represented as C_4H_8O: a five-membered ring compound. An abbreviation used for this material is THF. This aromatic liquid material is soluble in water and has a boiling point of 65°C and a relative density (RD or SG) of 0·89. Used as a solvent, particularly for *polyvinyl chloride* (PVC), and is the monomer for *polytetrahydrofuran*.

tetrahydrofurfuryl alcohol
This liquid material has a boiling point of 178°C: it is used as a solvent for *cellulosics* and for *natural resins*. This material is the basis for a range of rubber softeners as it may be reacted with organic acids (for example, with *oleic acid* and palmitic acid) to give, respectively, *tetrahydrofurfuryl oleate* and *tetrahydrofurfuryl palmitate*.

tetrahydrofurfuryl oleate
Obtained from *oleic acid* and *tetrahydrofurfuryl alcohol*. An organic liquid ester which is used as a rubber *softener*: a secondary *plasticizer* for *polyvinyl chloride* (PVC).

tetrahydrofurfuryl palmitate
Obtained from *palmitic acid* and *tetrahydrofurfuryl alcohol*. An organic liquid ester which is used as a rubber *softener*.

tetrahydronaphthalene
Also known as tetralin or as tetraline. This aromatic material has a relative density (RD or SG) of 0·98 and a boiling point of approximately 205°C. It is a good solvent for uncured *chloroprene rubber (CR), natural rubber (NR), styrene-butadiene rubber (SBR)* and *thiokol rubber (T)*. This chemical causes some swelling, or solvation, of uncured *butyl (IIR)* and *nitrile butadiene (NBR)* rubbers. At elevated temperatures it will dissolve polyethylene.

tetrahydrophthalic anhydride
An aromatic anhydride which a melting point of approximately 104°C: this anhydride, unlike *phthalic anhydride*, does not sublime. Used to prepare *unsaturated polyester resins* which will cross-link by atmospheric oxidation. Also used to produce light-coloured components from *epoxide resins* as it acts as a hardener for such materials.

tetralin
See *tetrahydronaphthalene*.

tetraline
See *tetrahydronaphthalene*.

tetramethyl bisphenol A polycarbonate
An abbreviation used for this type of material is PC-TMBPA or TMBPA-PC: also known as tetramethyl bisphenol A PC. A heat resistant *polycarbonate* (PC): a PC with a high heat resistance, low density (approximately 1·1 g/cm^3) but which is relatively brittle and has poorer environmental stress cracking resistance (ESC) than a *polycarbonate* based on *bisphenol A*.

tetramethyl bisphenol PC
An abbreviation used for *tetramethylcyclohexane bisphenol polycarbonate*.

tetramethyl thiuram disulphide
See *tetramethylthiuram disulphide*.

tetramethyl thiuram monosulphide
See *tetramethylthiuram monosulphide*.

tetramethyl thiuram tetrasulphide
See *tetramethylthiuram tetrasulphide*.

tetramethylcyclohexane bisphenol polycarbonate
An abbreviation used for this type of material is PC-TMC or TMC-PC: also known as tetramethyl bisphenol PC. A heat resistant *polycarbonate* (PC) obtained from hydrogenated isophorone: a PC with a glass transition temperatures (T_g) of approximately 238°C. The heat resistance is higher than that of a *polycarbonate* based on *bisphenol A*: the impact strength is however lower, the *ultraviolet* resistance is good but the melt viscosity although high is still reasonable. Such a material could have a melt viscosity of, for example, 380 Pa·s at 360°C and at 1000 s^{-1}. A *bisphenol A polycarbonate* could have a melt viscosity of, for example, 120 Pa·s at 360°C and at 1000 s^{-1}. As the reactivity of the starting materials (the bisphenols) is similar, a range of copolymers can be readily prepared and such materials have glass transition temperatures (T_g) of approximately 149 to 238°C dependent on composition.

tetramethylthiuram disulphide
Also spelt tetramethyl thiuram disulphide. An abbreviation used for this material is TMTD or TMT. An example of a *sulphur donor* vulcanization system which is an *accelerator* as well as being a sulphur donor. This material has a relative density (RD or SG) of about 1·3 and a melting point of approximately 150°C.

tetramethylthiuram monosulphide
Also spelt tetramethyl thiuram monosulphide. An abbreviation used for this material is TMTM or TMS. This material has a relative density (RD or SG) of 1·38. An ultra-fast *accelerator* for rubbers which is similar to *tetramethylthiuram disulphide*. A thiuram sulphide which gives safer processing than *tetramethylthiuram disulphide*.

tetramethylthiuram tetrasulphide
Also spelt tetramethyl thiuram tetrasulphide. An ultra-fast *accelerator* for rubbers which has good processing safety in the absence of sulphur: when used with sulphur, the compounds will usually scorch. This material has a relative density (RD or SG) of 1·52 and a melting point above 90°C: active at temperatures above 110°C.

tetramethylurea
An abbreviation used for this material is TMU. A liquid material with a boiling point of 175°C and which is used as a solvent for, for example, *aramid polymers*.

tetraoctyl pyromellitate
An abbreviation used for this type of material is TOPM. Also known as tri-2-ethyl hexyl pyromellitate. A high molecular weight plasticizer which resists extraction. See *trimellitate plasticizer* and *pyromellitic dianhydride*.

tetrapolymer
A polymer based on four monomers. See, for example, *fluororubber*.

tex
A unit of weight used for textiles (synthetic). Used to measure the fineness or coarseness of fibres. The weight in grams of 100 metres of the fibre or yarn. To convert to kg/m, multiply by 1·0 x 10^{-6}. See *denier*.

Texanol isobutyrate
See *2,2,4-trimethylpentane-1,3-diol di-isobutyrate* and *3,3,5-trimethylpentane-1,4-diol di-isobutyrate*.

textile back moulding
See *back injection moulding*.

textile belt
The fabric belt of a *steel-braced radial tyre*.

textile glass
The collective name for the textile class of materials which are based on *glass:* may be based on *staple fibre* and/or, on *continuous filaments*.

textile size
A material added to a *fibre* to facilitate textile operations. See, for example, *glass*.

TFE
An abbreviation used for *tetrafluoroethylene*. See *fluororubber*.

TFE/E
An abbreviation used for *tetrafluoroethylene-ethylene copolymer*.

TFE/FMVE
An abbreviation used for a tetrafluoroethylene - perfluoro-(methyl vinyl ether) polymer. See *perfluorinated elastomer*.

TFE-HFP
An abbreviation used for *fluorinated ethylene propylene*.

T_g
An abbreviation used for the *glass transition temperature*.

TGMDA
An abbreviation used for *tetraglycidyl methylene dianiline epoxy resin*.

th
An abbreviation used for thickness, for example, of a test specimen.

THBS
An abbreviation used for *trihydroxybutyrophenone*. An *antioxidant* for *polyethylene*.

therm
A heat unit. 100,000 British thermal units or 25,200,000 calories.

thermage
Proprietary system of component decoration, often in line with a *blow moulding machine*, in which labels are applied to *blow moulded components*: the labels are clear film, adhesive coated and multi-coloured.

thermal analysis
This term covers a number of techniques, or methods, used to study polymers: used to study the temperature dependence of selected properties or the way that polymers degrade, or change mass, when heated. See, for example, *thermographic analysis*.

thermal black
See *carbon black*.

thermal conductivity
Also known as heat conductivity. The rate of transfer of heat along a body by conduction. The rate of transfer of heat along a test specimen, by conduction, is measured at equilib-

rium. This may be done by clamping two circular test specimens on either side of an electrical heating element which supplies a constant supply of heat. Cooling plates are pressed against the other side of the specimens and the temperatures of the faces measured with thermocouples: this allows the temperature differences to be measured. The conductivity may be calculated from the electrical energy supplied, the temperature difference and the specimen dimensions. The units are $Js^{-1}m^{-1}K^{-1}$.

thermal cyclization
A process used to produce *cyclized rubber*.

thermal decomposition product testing
Assessment of, for example, the toxicity of the products evolved during a thermal decomposition test. Such a test is the *Pittsburgh University Test*.

thermal degradation
Degradation caused by exposure to an elevated temperature. Such degradation often starts at structural irregularities within the polymer which lower the thermal stability. When the heating is done in an inert atmosphere it is called pyrolysis. When the heating is done in an oxygen-containing atmosphere, then thermo-oxidative reactions occur due to the oxygen participating in the free radical (R·) reactions. See *stabilizer* and *thermally-initiated oxidation*.

thermal impulse welding
See *impulse welding*.

thermal plasticization
Plasticization, or softening, which occurs as a result of heating. See *heat softening*.

thermal polymerization
Also known as self-initiation *polymerization*.

thermal runaway
The dramatic rise in temperature during, for example, *dielectric heating* and caused by an increase in *loss factor* on heating.

thermal shrinkage
The *shrinkage* which occurs as a result of cooling. See *post moulding shrinkage*.

thermal silica
Fumed *silica*. Silica prepared by a *pyrogenic* process.

thermal softening
See *heat softening*.

thermal testing
See *thermal analysis*.

thermally foamed plastic
A *cellular* plastics material produced by applying heat to effect gaseous decomposition or volatilization of a constituent of the plastics compound. See *blowing agent*.

thermally-initiated oxidation
Heat initiated *oxidation*. Most commercially available polymers are susceptible to thermal and photo-initiated oxidation which change the properties of the polymer. Condensation polymers (such as *polyamides, polyesters,* and *polyurethanes*) are more resistant than addition polymers (such as polyolefins, for example, *polyethylene* and *polypropylene*). Materials, such as *acrylonitrile-butadiene-styrene* and *high impact polystyrene*, will require more protection than *polystyrene*: antioxidant use level ranges between 0·05 to 1 phr.

thermionic device
A device which relies on the emission of electrons from hot bodies, for example, a triode valve.

thermistor
A semiconductor whose resistance decreases with temperature. Used to measure temperature and/or to compensate for changes in temperature in, for example, control instruments.

thermo-junction
A junction of different metals joined for the purpose of temperature measurement. See *thermocouple*.

thermo-labile cross-links
Cross-links which only exist below the softening point, or melting point, of the polymer: they are destroyed on heating. See *physical cross-link*.

thermocouple
A type of *transducer*. A device which measure temperature changes by using the *thermoelectric effect*. Basically a thermocouple consists of two dissimilar wires which are joined at each end, that is it contains a *thermo-junction*. If one end of the assembly is made hotter that the other, a small electrical signal, or emf, will result: the emf is dependent on the choice of metals and on the temperature difference of the junctions. For a given combination, the more the junction is heated, the greater will be the electrical signal. If one end is kept at a stable temperature (reference temperature) then the thermocouple may be calibrated so that a simple and accurate measurement of temperature may be made. The measurement may be displayed in either analogue or digital form. Usually the temperatures are displayed in either °C or in °F.

A thermocouple is commonly used to feed a control instrument with information. Such a thermocouple should be sunk as deep into the *barrel* wall, for example, of an extruder or of an injection moulding machine, as safety will allow and a *three term control* action used.

thermoelectric effect
Also called the Seebeck effect. If two dissimilar wires are joined at each end then, if one end of the assembly is made hotter that the other, a small electrical signal, or emf, will result. See *thermocouple*.

thermoformable, random fibre, continuous fibre mat
See *preforming continuous filament mat*.

thermoformed bottle
A bottle made from thermoplastic sheet. Sheet can be thermoformed into two container halves (top and base) and the trimmed halves can then be spin-welded together. As the halves can be stacked one inside the other (before welding) container storage is simplified and unusual designs (e.g. two-colour bottles) are possible. See *blow moulding*.

thermoforming
This term embraces a number of techniques (or shaping methods) in which a thermally softened, thermoplastics material in sheet form is distorted to the shape required by means of a pressure difference: once the shape has been achieved the material is set into that shape by cooling. Usually the pressure difference is achieved by using a reduced air pressure on one side of the sheet which is obtained by removing the air via small holes or slots machined into a mould: both *male moulds* and *female moulds* are used. See *vacuum forming* and *hydroforming*.

Thermoforming processes are relatively simple and both machinery and mould costs are incredibly low compared to *injection moulding*: one reason is only one mould half is required. Because of the low pressures involved (usually below 1 atm), the moulds are easily and cheaply made (for example, from wood, cast epoxide resin and aluminium) and such moulds can be easily re-worked if tool modifications are required.

It is possible to make formings whose wall thicknesses are lower than those possible by injection moulding and it is also possible to make very large formings economically. Thin walled components are usually produced using a roll-fed machine: in-mould trimming is possible by, for example, forming over a knife edge and then separating the forming from the web using a roller or pressure pad.

Thermoforming is widely used for making items such as disposable cups and refrigerator liners; both of these items being manufactured from *high impact polystyrene* extruded sheet. A major disadvantage of any thermoforming process is the need to trim the forming from the web as this results in design limitations, increased labour charges and considerable material wastage.

As *amorphous, thermoplastics materials* maintain their integrity when heated to forming temperatures, it is amorphous materials which can be trimmed which are mainly used as thermoforming materials. Such materials include *acrylonitrile-butadiene-styrene (ABS), polyvinyl chloride (PVC), ABS/PVC blends, acrylics (PMMA), acrylic/PVC blends, cellulosics, high impact polystyrene (HIPS), modified polyphenylene oxide (PPO/PS) blends* and *polycarbonate (PC)*. It is not usually possible to produce a component with a wall thickness greater than the original sheet: however, this may be done if the sheet contains a *blowing agent*. See *thermoforming processes*.

thermoforming - heating
If a thermoplastics sheet material will shrink or revert on heating then it must be restrained while being heated. The sheet is therefore usually heated by an infra-red heater (mounted approximately 150 mm/6 in above the sheet) while being held in a clamping frame (for example, extruded *high impact polystyrene* (HIPS) sheet. If the thermoplastics sheet material will not shrink or revert on heating then it may be heated in an oven (for example, *polymethyl methacrylate* (PMMA) cast sheet may be heated in this way.

thermoforming processes
There are a number of techniques or shaping methods which are classed as *thermoforming*. Both *male moulds* and *female moulds* are used and the use of one rather than the other will give a different material thickness distribution to the forming (see *thickness distribution control*). The material thickness distribution can be altered, or controlled, by stretching the sheet before forming (prestretching): this may be done by either mechanical means (using a *plug*) or, by using compressed air.

From the basic technique *vacuum forming*, which uses a female mould, the techniques known as *plug assist forming* and *air slip forming* are derived. From the basic technique *drape forming*, which uses a male mould, the technique known as *billow forming* is derived. See *pressure assisted forming*.

thermoforming temperature
The temperature at which thermoforming is performed. This is not usually measured but determined experimentally. For example, the sheet is clamped and heated while being timed. Initially the sheet sags (through thermal expansion) and then it tightens (through *reversion*). At this point the sheet is deemed ready to be formed. See *forming temperature*.

thermographic analysis
Methods of *thermal analysis* in which the rate of mass loss is studied.

thermoplastic
The term may be used to indicate that a material is capable of being softened by heat. A thermoplastic is also a type of plastics material. Thermoplastic products, for example an injection moulding or an extrudate, may be softened and re-shaped whereas a thermoset product cannot. See *thermoplastics material*.

thermoplastic alloy
An *alloy* which can be processed like a thermoplastics material. See *blend*.

thermoplastic blend
See *blend*.

thermoplastic bulk moulding compound
An abbreviation used for this material is TP BMC or BMC TP. A *bulk moulding compound* which consists of fibres held together with a *thermoplastics material*. As the fibres are relatively long (for example, 50 mm) the composite may be known as a long fibre thermoplastic composite: if the fibres are long glass fibres, then the term long glass fibre thermoplastic composite may be used. The performance of a *glass reinforced thermoplastic composite* is enhanced, particularly the impact strength, if the fibre length is kept long.

Such a composite may be produced by drawing (pultruding) the *fibres/rovings* through a thermoplastics melt and chopping the coated fibres to a predetermined length (if rovings are used then the product may be called *thermoplastic impregnated fibre rovings*). The fibre length is then equal to the pellet length. Hot *preforms* may be produced, using for example, a *low work screw plasticator*, and these preforms may then be formed by *compression moulding*: this type of BMC moulding, maintains the fibre length better than injection moulding.

thermoplastic chlorinated polyethylene
A *chlorinated polyethylene* which is a *thermoplastics material*. Thermoplastic chlorinated polyethylene is used as an additive for other polymers, for example, it is used as an *impact modifier* for *polyvinyl chloride* (PVC).

thermoplastic chloroolefin elastomer
See *elastomeric alloy melt processable rubber*.

thermoplastic comb-graft
A *copolymer* which consists of side chains of one polymer grafted onto the main chains of another polymer. See *comb-grafted natural rubber*.

thermoplastic composite
A composite based on a *thermoplastics material*. See, for example, *thermoplastic bulk moulding compound*.

thermoplastic copolyester
See *polyether ester*.

thermoplastic copolyether
See *polyether block polyamide*.

thermoplastic elastomer
A rubbery type of material which can be processed like a *thermoplastics material*. An abbreviation used for this type of material is TPE. Also known as a thermoplastic rubber (TPR). Such materials exhibit some of the properties of an *elastomer* and some of the properties of a *thermoplastics material*. A TPE can be stretched fairly easily and will then retract after stretching: that is, it shows, or exhibits, significant rubbery behaviour but can also be processed like a thermoplastics material on equipment designed for thermoplastics, unlike traditional vulcanized rubbers. Such materials exhibit the speed, and ease of processing, of thermoplastics as once shaped, the products do not have to be cured like traditional rubbers: such curing can be a very time consuming process. The change from a melt to a solid, rubbery article takes place on cooling in the case of a TPE.

ATTRACTIONS OF. When components, made from a thermoplastic elastomer (TPE) replace vulcanized elastomers then, the components are often easier or cheaper to make as production is so much faster and may be more automated. Such considerations offset the often higher cost of a TPE. Such materials are a group of polymers which can offer a wide range of properties. In general, however, they occupy a position between plastics and rubbers although there can be considerable overlap in each direction. The major advantages of such materials are that:

they are thermoplastics,
they are elastomeric or rubbery,

their scrap can be re-cycled,
they have high strength at moderate temperatures,
they can be prepared, or compounded, to give a wide hardness range,
vulcanization is not required,
they have good low temperature properties,
they are compatible with a wide range of materials.

Such materials also usually have reasonable wear resistance, resilience, high strength, toughness and, resistance to chemical attack.

CROSS-LINKS. To exhibit thermoplastic behaviour, the 'cross-linked' TPE material must possess cross-links which disappear on heating (i.e. it must contain heat fugitive or, *thermally labile cross-links*). This may be achieved by:

i) ionic cross-linking;
ii) cross-linking by hydrogen bonding;
iii) cross-linking by thermally, unstable covalent cross-links which reform on cooling;
iv) linking of molecules by small crystalline structures; and,
v) the use of block copolymers.

Included in this category of materials are plastics which are modified, after polymerization, so that they exhibit increased elasticity (for example, *polyvinyl chloride* (PVC) and materials which are made, or polymerized, so that they have built-in elasticity (for example, *thermoplastic polyurethane*). Five classes of TPE materials are commercially important and these are:

i) polystyrene/elastomer block copolymers;
ii) polyurethane/elastomer block copolymers;
iii) polyester/elastomer block copolymers;
iv) polyamide/elastomer block copolymers, and;
v) hard thermoplastics/elastomer blends.

See, for example, *styrene-butadiene block copolymer, thermoplastic polyurethane, polyether ester, polyether block amide* and *rubber modified polypropylene*.

TWO PHASE STRUCTURE. Most TPE materials have two phase structures (with the exception of *elastomeric alloy - melt processable rubber*): they consist of a hard, plastic phase and a soft, rubbery (elastomeric) phase. It is the hard plastic phase which gives the material its strength and stops the soft, rubbery phase from flowing under stress and so becoming unusable. Such hard phases behave as cross-links and are sometimes called *physical cross-links*.

As most TPE materials are phase separated systems they will show many of the characteristics of the individual polymers which make up the phases. Each phase will have its own glass transition temperature (T_g) and its own crystalline melting point (T_m) if it is a semi-crystalline, thermoplastics material): these transition temperatures will, in turn, determine the temperatures at which changes in properties (transitions) will occur.

If modulus (stiffness) is measured as the temperature is changed then, three distinct regions will be seen. At very low temperatures, both phases are below their glass transition temperatures, and so the material is stiff, brittle and hard. At a higher temperature, the rubbery phase is above its (T_g) and becomes soft so that the material behaves like a traditional cured rubber. As the temperature is further increased, the modulus stays constant (the rubbery plateau) until the plastic phase softens and the material becomes a melt. (The property changes are reversed on cooling). Thus most TPE materials have two service temperatures: the lower service temperature is dictated by the glass transition temperature (T_g) of the rubber/elastomer while the upper temperature is dictated by the glass transition temperature (T_g), or the crystalline melting point T_m, of the plastics material.

At the present time, these materials do not have, in general, the high temperature resistance of vulcanized materials. For example, the styrene blocks in a *styrene-butadiene block copolymer (SBS)* soften at approximately 75°C and this limits the maximum use temperature to about 65°C. The upper service temperature of *thermoplastic polyurethane* is higher and can reach 120°C. The upper service temperature of *polyether ester* is higher still and can reach 160°C. A thermoplastic fluoro elastomer should have a very high service temperature.

OIL AND SOLVENT RESISTANCE. Most TPE materials have two phase structures: the choice of the plastics material used as the hard phase, strongly influences the oil and solvent resistance of a TPE. Even if the rubbery phase resists a particular chemical, all useful properties of the TPE, will be lost if the hard phase is swollen. Thus the *polystyrene/elastomer* block copolymers have very little resistance to organic solvents (solvent resistance is improved by blending with plastics such as *polypropylene or PP)*. With the other types of *block copolymer*, there are crystalline hard phases and these, together with the use of polar elastomer phases, gives such materials good oil and solvent resistance.

Polymer blends are usually based on oil resistant materials, for example, *polyvinyl chloride* (PVC), and so are usually oil resistant. Some of these TPE materials are more oil resistant than traditional rubbers and so, one TPE material may replace several traditional rubber compounds: this simplifies materials and component handling within the factory. However, when a material type is stated as being resistant to a certain chemical (for example, oil) then, this should only be taken as a guideline as there are usually many different grades of each material and some of them can be very different.

Nevertheless it can be said that plastics have, in general, good chemical resistance. Usually the addition of unvulcanized rubber, to a plastics material so as to get a TPE, would be expected to worsen the chemical resistance: however, because rubber addition can reduce the effects of stress, problems such as *environmental stress cracking* (ESC) can sometimes be reduced or, eliminated by rubber addition. With all plastics, and with a TPE, chemical resistance worsens as the temperature is raised and processing influences properties (see, for example, *orientation*). The chemical resistance of a TPE is therefore very dependent upon the particular application and closely linked with the time and temperature of exposure.

MATERIALS HANDLING. Thermoplastic elastomers (TPE) are relatively expensive and so the amount delivered as bulk shipments is small. Most are supplied in multi-walled, paper sacks which have a plastics layer as a moisture-barrier: each bag holds, for example, 25 kg and 40 such sacks may be shrink-wrapped onto a disposable, wooden pallet. No matter how they are delivered, the materials must be carefully looked after as such polymers are expensive and many will burn relatively easily. Adopt a strict *stock control system* and, in general, treat such materials as *thermoplastics materials*. A great deal of TPE usage is accounted for by compounded material although, as with *thermoplastics materials* colouring on the machine is possible using, for example, solid *masterbatches*.

FLOW BEHAVIOUR OF. These materials, like *thermoplastics material*, differ widely in their viscosity, or ease of flow, and the problem is made more difficult by the fact that the flow properties of a TPE is *non-Newtonian* and so there is not a linear relationship between pressure and flow. See *flow properties representation*.

INJECTION MOULDING. TPE materials are most commonly processed on *in-line, single screw machines* - this is because that type of machine is the most popular in the injection moulding industry. The materials are treated as though they are thermoplastics: that is, the barrel is run hot (at, say 200°C) and the mould is run cool (at, say 40°C). This is the exact opposite to the settings of a machine for *rubber injection moulding*. Several heating zones will also be used on the electrically heated barrel and the feed zone may well be run at temperatures above room temperature.

The material used to feed the injection moulding machine will be in the form of small pellets (as opposed to strip for traditional rubbers) which may well be hot if drying has been performed. Excessive heating will usually cause oxidation of a TPE as opposed to the setting, or curing, of traditional rubbers. As the mould is cool, and the material is hot, then TPE flow and mould filling will be different to traditional rubbers. The way the material flows will also introduce *anisotropy* and variations in moulding *shrinkage*.

COMPONENT DESIGN. The difference in behaviour between thermoplastic elastomers and traditional, vulcanized rubbers must be remembered if these materials are to be successfully used. A thermoplastic elastomer (TPE) often does not have good, long term creep resistance and products should be designed, and used, accordingly. This means that creep data must be available and used, that components should be designed to uncoil rather than stretch, that components should not operate at extreme strains and at high temperatures; it is also sensible to design bending points in the component, if appropriate.

MARKETS. Thermoplastic elastomers (TPE) are a comparatively new group of materials but already they have become significant in terms of both tonnage and in terms of value. They either create new markets, as they span the gap between rubbers and plastics, or they replace existing materials. For example, a low cost TPE, based on a *rubber modified polypropylene blend*, may replace vulcanized *styrene-butadiene rubber (SBR)*, vulcanized *ethylene-propylene rubber* and *plasticized polyvinyl chloride (PVC)*. In general, aggressive marketing, a change in attitudes and the careful use of technical information are helping to develop markets for thermoplastic elastomer (TPE) materials.

Output is usually increased when these materials are used, in place of traditional rubbers, and man-power is reduced. The ability of this category of materials to be re-processed is also now particularly important: for example, in view of concerns about the environment. Of major importance is the wide colour range possible from this type of material and the world-wide availability of this type of material. The elimination of the compounding stage, essential for traditional rubbers, is also not to be dismissed lightly.

A major reason for the selection of a TPE material is often one of global availability: that is, the same material can be obtained in a large number of countries. Traditionally, rubber compounds were manufactured (within the actual factory where they were processed) from the individual components thus leading to variability. However, it should not be forgotten that a TPE must be regarded as a thermoplastic and this means that processing influence properties (the injection mouldings will be *anisotropic*). Bonding to metals will, in general, be more difficult for a TPE than for a traditional elastomer.

Major markets for TPEs are in automotive applications and in footwear. Acoustic 'deadened' components will increasingly be made from TPE materials as greater demands are made for quieter cars which use high performance engines. Medical applications, for example, disposable items, could be a major applicational area if a TPE is used in place of vulcanized rubber as repeated sterilisation can cause embrittlement of re-usable, vulcanized materials. Gaskets and seals could be another growth area as, for example, a TPE can be injection moulded into self coloured items so as to make possible the production of, for example, machines for customized kitchens. Materials can be combined, for example, in layers, so as to make dual hardness, or multicolored mouldings.

However, at the present time, these materials do not have, in general, the high temperature resistance, the compression set resistance, the oil resistance and the solvent resistance of conventional vulcanized rubbers. The lack of such properties has influenced the applicational areas so that, these materials are used in footwear, wire insulation and tubing and not in tyres, fan belts and radiator hoses. The workhorse, or commodity TPE, is *styrene-butadiene block copolymer* followed by *rubber modified polypropylene*. The *polyether ester materials (PEEL)* have a very healthy growth potential.

thermoplastic elastomer - amide based
An abbreviation used for this type of material is TPE-A. See *polyether block polyamide*.

thermoplastic elastomer - crosslinked rubber
An abbreviation used for this type of material is TPE-OXL or TPE-XL. See *rubber modified polypropylene* and *thermoplastics vulcanizate*.

thermoplastic elastomer - ether based
An abbreviation used for this type of material is TPE-E. See *polyether ester*.

thermoplastic elastomer - ethylene-vinyl acetate elastomer
An abbreviation used for this type of material is TPE-EVA. See *thermoplastic ethylene-vinyl acetate copolymer* and *ethylene-vinyl acetate*.

thermoplastic elastomer - ionomer
See *elastomeric ionomer*.

thermoplastic elastomer - olefin based
Also known as thermoplastic polyolefin. An abbreviation used for this type of material is TPE-O. See *rubber modified polypropylene* and *thermoplastics vulcanizate*.

thermoplastic elastomer - olefin based with crosslinked rubber
An abbreviation used for this type of material is TPE-OXL. See *rubber* modified *polypropylene* and *thermoplastics vulcanizate*.

thermoplastic elastomer - styrene based
An abbreviation used for this type of material is TPE-S. See *styrene-butadiene block copolymer*.

thermoplastic elastomer - urethane based
An abbreviation used for this type of material is TPE-U. See *thermoplastic polyurethane*.

thermoplastic ether ester
See *polyether ester*.

thermoplastic ethylene-propylene elastomer
See *rubber modified polypropylene*.

thermoplastic ethylene-propylene rubber
An abbreviation used for this type of material is TPE-EPDM. See *rubber* modified polypropylene.

thermoplastic ethylene-vinyl acetate copolymer
This type of material is also known as ethylene-vinyl acetate (EVA or EVAc) copolymer or as, vinyl acetate-ethylene copolymer (VAE) or as, ethylene vinyl acetate copolymer or as, ethylene vinyl acetate.

Polymerizing vinyl acetate (VA) with ethylene, disrupts the crystal structures that are present in polyethylene (PE) and eventually gives an amorphous material. By varying the percentage of vinyl acetate (VA) in the composition, polymers with significantly different properties are produced. As the percentage of VA is increased the transparency and flexibility of the copolymers increases.

The EVA copolymer which is based on a medium proportion of VA (approximately 4 to 30%) is the type referred to as a thermoplastic ethylene-vinyl acetate copolymer and is processed as a thermoplastics material by, for example, *injection moulding*. It is not vulcanized but has some of the properties of a rubber or of *plasticized polyvinyl chloride* (PPVC) - particularly at the higher end of the range. May be filled and both filled and unfilled materials have good low temperature properties and are tough, semi-opaque thermoplastics with a comparatively low, upper working temperature, for example, 65°C/149°F. The materials with approximately 11% VA are used as hot melt adhesives.

EVA is softer, clearer and more permeable than LDPE; has better *environmental stress cracking resistance (ESC)*, is tougher and will accept fillers more readily. Some grades of EVA can retain their flexibility at temperatures as low as −70°C. Can be cross-linked by peroxides and/or, may also be chemically blown to give cellular products; the density of crosslinked, foamed EVA can be very low indeed and is far lower than that possible without crosslinking. Such crosslinked, foamed EVA can have a very fine structure - similar to micro-cellular rubber. The inherent flexibility, good processing characteristics, low odour and no plasticizer migration problems make this material an attractive alternative to PPVC, *natural rubber* and *synthetic rubbers*.

This type of material is resistant to flex cracking at low temperatures and also to ozone cracking. Disinfectants, ethylene oxide and sodium hypochlorite may be used for cleaning or disinfection of EVA products. Chemical resistance is similar in many respects to LDPE but EVA is not as chemically resistant. Low *flow rate* grades have the best resistance. EVA is not resistant to steam sterilization; use gamma irradiation if sterilization is required. Resistance to aromatic and chlorinated solvents is poor. Soluble in aromatic and chlorinated hydrocarbons at elevated temperatures.

This type of material (solid, non-filled material) has a density of 0.926 to 0.95 gcm^{-3}. The natural colour of the material is similar to that of LDPE at low VA content (i.e. an off-white material) but at higher VA contents (for example, 20%) the material is colourless. A wide colour range is therefore possible; this includes both transparent and opaque colours. The weathering behaviour of EVA is similar to that of LDPE and like LDPE, it can be significantly improved by the incorporation of 2 to 3% carbon black.

Since the coefficient of friction is much higher than other polymers, EVA is used for applications where slippage needs to be overcome, for example, such materials are used to make record player, turn-table mats. Initial uses for this material were as replacements for rubbers, LDPE and PPVC. It replaces rubber in some applications, for example, in traffic signal hoods, because of its resistance to ozone attack, flexibility, low temperature resistance and flex crack resistance. Because of the low temperature flexibility of this material, it is used in place of LDPE for ice cube trays. Other applications include barrel bungs, bicycle saddles, railway sleeper pads, disposable baby bottle teats, ear tags for animals, refrigerator gaskets, freezer door gaskets, closure wads and WC pan connectors. Also used to make base pads for staplers and for small items of electrical equipment, When crosslinked, foamed EVA can be used to make cellular shoe soles and tyres, for example, for push chairs and golf trolleys.

thermoplastic ethylene-vinyl acetate elastomer
See *thermoplastic ethylene-vinyl acetate copolymer*.

thermoplastic fluoro elastomer
An abbreviation used for this type of material is TPE-FKM. A *thermoplastic elastomer* which, for example, contains tetrafluoroethylene segments, as hard segments, in an effort to extend the upper service temperatures of thermoplastic elastomers. See *fluororubber*.

thermoplastic impregnated fibre rovings
A semi-finished product produced by, for example, drawing (pultruding) *rovings* through a thermoplastics melt or solution. Powder coating is also used to coat the fibres. For glass rovings the fibre content is typically 40 to 70% by weight: For carbon fibre rovings the fibre content is 20 to 60% by weight. See *thermoplastic bulk moulding compound*.

thermoplastic isoprene rubber
An abbreviation used for this type of material is TPE-IR. See *thermoplastic elastomer*.

thermoplastic natural rubber blend
A thermoplastics material which is based on a *blend* of *natural rubber* and one, or more, thermoplastics materials (see *natural rubber/polyolefin blend*). An abbreviation used for this type of material is TPE-NR or TP-NR or TPNR. The natural rubber (NR) may also be modified by, for example, grafting *polystyrene* side chains onto the NR molecule so as to give a *comb-grafted natural rubber*

thermoplastic nitrile-butadiene elastomer
Also known as thermoplastic nitrile-butadiene rubber. An abbreviation used for this type of material is TPE-NBR or TP-NBR or Y-NBR. See *polyvinyl chloride nitrile rubber blend*.

thermoplastic nitrile-butadiene rubber
See *thermoplastic nitrile-butadiene elastomer*.

thermoplastic olefin elastomer
See *thermoplastic elastomer - olefin based*.

thermoplastic polyester
A *polyester* which is a *thermoplastics material* (see also *unsaturated polyester resin*). Polyethylene terephthalate and polybutylene terephthalate are the two major thermoplastic polyesters. *Polybutylene terephthalate* (PBT) and *polyethylene terephthalate* (PET) compete with other engineering plastics, for example, *nylon* (PA) and *polycarbonate* (PC). They have better moisture and chemical resistance and are more dimensionally stable: they also have better fatigue endurance. Slightly cheaper than PA but PA is better known. PBT has a lower moisture absorption and a lower heat distortion temperature (HDT) than PC. Blends with other polymers are also used, for example, *polyphenylene oxide*. Blends with *glass fibre* (for example, 20 to 50%) are extensively used as glass significantly improves the physical properties of thermoplastic polyesters. Possible to get grades which are classified as self extinguishing by UL. PBT/PET blends are often glass reinforced as the unfilled compound is of limited value; such blends can be very easy flowing and, can give good surface finishes.

thermoplastic polyether ester
See *polyether ester*.

thermoplastic polyetherester
See *polyether ester*.

thermoplastic polyolefin
See *thermoplastic elastomer - olefin based*.

thermoplastic polyolefin - cross-linked rubber
See *dynamically vulcanized polyolefin rubber*.

thermoplastic polyolefin rubber
An abbreviation used for polypropylene/ethylene propylene diene monomer blend: an abbreviation used for polypropylene/ethylene propylene rubber blend. See *rubber modified polypropylene*.

thermoplastic polyurethane
Also known as thermoplastic elastomer - urethane based or as, thermoplastic urethane elastomer or as linear PU. An abbreviation used for this type of material is TPE-U or TPU.

A thermoplastics material which consists of interpenetrating domains of soft blocks (which have a low glass transition temperature) and hard blocks (which retain their rigidity almost up to the crystalline melting point). A type of *thermoplastic elastomer* which is obtained by reacting a hydroxyl-terminated polyester, or polyether, with a diisocyanate so as to give a linear polymer which contains the urethane group -NHCOO-. Such polymers are noted for their resistance to wear and abrasion; they have high strength, resilience and flexibility. The materials are suitable for use over a wide temperature range (-40° to 140°C) have high compressive strength and low compression set. Both polyester TPU and polyether TPU are available in a range of grades, the hardness of which can be widely varied from 78 Shore A to 74 Shore D. The mouldings are strong and elongations of up to 500% are possible.

At the same hardness, a polyether TPU will have better low temperature flexibility, hydrolytic stability, fungus resistance and a higher resilience than a polyester TPU, A polyester TPU will have marginally better abrasion resistance, oil resistance, superior resistance to heat ageing and will be tougher than a polyether TPU. Both types can have similar flow properties.

Such materials can be very easy flowing: grades are available which can be processed on low pressure machines developed for moulding shoe soles onto uppers. However, a TPU sold for *injection moulding* is of high molecular weight and is a relatively viscous material. Flow depends not only on the processing conditions but also on the age of the material, moisture content and storage temperature. A *lubricant* may be added in an effort to reduce viscosity, to improve mould release and to reduce shrinkage. Only small amounts should be added and minimise the risk of degradation by avoiding the use of a stearate-type lubricant.

Shrinkage depends on grade and on thickness but is in the range 0.5 to 2%. The materials absorb water and this causes dimensional changes; up to 1.5 per cent water may be absorbed as the mouldings come into equilibrium with their surroundings. In general, harder grades exhibit less shrinkage than softer grades. Shrinkage obtained is of two types - *thermal shrinkage* and *post curing shrinkage*. It is therefore difficult to predict the actual shrinkage: as a rule of thumb 80% is thermal and 20% cent is post moulding. For critically dimensioned parts, a prototype cavity is used to determine the size of cavity required. To stop shrinkage variations it is essential that the temperature of the mould is held very accurately and adequate venting should be provided.

If possible do not use components until they have been conditioned, or annealed, for example, by being heated at 100°C for 24 hours, or by being stored at room temperature for a month. Room temperature storage is not as good as high temperature conditioning (post curing) as this permits stress relaxation and allows crystalline regions to stabilize. Post cured components have a higher maximum service temperature, better creep resistance and lower compression set. In practice not many components are post cured (they are allowed to further stabilize for 1-2 months at room temperature) but where components require maximum resilience (compression set properties) post curing should take place after moulding.

Both an ester and an ether TPU have good resistance to oil, petrol and grease as they contain polar groups; they have good stress crack resistance. They have good resistance to aromatic hydrocarbons and excellent resistance to aliphatic hydrocarbons but beware of swelling effects. Strong acids, strong alkalis, alcohols, ketones and, chlorinated hydrocarbons cause swelling to both an ester and an ether TPU (acids and alkalis also cause hydrolytic attack with a polyester TPU). In general, a polyester TPU has better resistance to hydrolysis although above 80°C hydrolytic degradation will occur. A polyether TPU based on polytetramethylene ether has the best resistance to hydrolysis of the ether-type materials. Among the polyester materials, those based on polycaprolactone (also known as caproesters) are probably the best. Swelling is dependent upon the type of TPU (ether or ester) and the equilibrium level of solvation. Up to 100% swelling can occur with chlorinated hydrocarbons but paraffin-type materials such as kerosene have minimal swelling effect. A TPU is soluble in dimethyl fluoride solution.

The specific gravity is approximately 1.11-1.22 and the natural colour varies from milky white to clear, dependent upon type of PU, but is usually translucent: some grades can be very clear. A wide colour range is possible - this may be achieved by using masterbatches. Being thermoplastic these materials soften on heating but there is no sharp melting point: melting is usually completed by about 140°C. These materials are hygroscopic and are therefore supplied in foil-lined sacks or PE lined drums to minimise moisture absorption. Adopt a strict *Fi-Fo policy*, only open sacks only when required and only remove what is strictly necessary from each sack. Reseal sacks immediately and if possible use a heated hopper on the moulding machine. If more than 0.1% moisture is absorbed then the material must be dried - 3 hours at 80°C in air circulating oven or 1 hour in a vacuum oven. It is recommended that pre-drying is carried out as a routine procedure to ensure that moisture-free material is always processed as a wet TPU will suffer from degradation resulting in a loss of physical properties.

Compared to traditional rubbers these materials are expensive (a polyether is more expensive than a polyester) but as they can be handled like thermoplastics, fast production on standard machines is possible. A very wide range of grades is available: for example, the hardness can range from 78 Shore A to 74 Shore D. Linear PU can bridge the hardness gap between rubbers and plastics. PE has a hardness of approximately 65 Shore D; PP, 85; PS, 100. These thermoplastic elastomers are used where a flexible, durable material with good oil resistance is required. They are used in the form of grommets, seals, washers, wheels and rollers in industries such as automotive, mechanical hauling and mining: the relatively poor resistance of some grades to synthetic lubricants (phosphates and diesters) should not be forgotten. On outdoor exposure, fading (particularly of pale colours) may occur - use UV stabilized grades for outdoor use such as for animal ear tags. Fibre modified grades are also available, for example, for car components. Applications for the polyester types are restricted as the material in certain conditions is prone to microbiological degradation (rotting).

thermoplastic rubber
An abbreviation used for this type of material is TPR. See *thermoplastic elastomer*.

thermoplastic rubber-like blend
A blend of a rubber and a thermoplastics material which gives rubber-like materials: a *hard thermoplastics/elastomer blend*.

thermoplastic styrene-butadiene rubber
An abbreviation used for this type of material is TPE-SBR or TPE-S or Y-SBR. See *styrene-butadiene block copolymer*.

thermoplastic urethane elastomer
See *thermoplastic polyurethane*.

thermoplastic vulcanizate
An abbreviation used for this type of material is TPV. This type of material may also be referred to as *elastomeric alloy thermoplastic vulcanizate* (EA-TPV): a two-phase systems in which a crosslinked rubber phase is dispersed in a continuous plastics phase, which is usually a polyolefin. When the crosslinked rubber is an *ethylene propylene rubber* then, the material may be referred to as an EPDM-TPV. When the crosslinked rubber is a *nitrile rubber* then, the material may be referred to as an NBR-TPV. See *dynamically vulcanized polyolefin rubber*.

thermoplastic(s) injection moulding
An abbreviation used for this term is TIM. See *injection moulding*.

thermoplastics
Commercially the most important type of high polymer. See *thermoplastics material*.

thermoplastics blends
A *blend*, or mixture, of two or more polymers which may be processed like a thermoplastics material. Because of their ease of manufacture, using (for example, twin-screw compounding extruders) there is a lot of interest in blends of plastics or, in blends of plastics with rubbers: either may be modified with fillers or glass fibre. There are a number of reasons for blending materials and these include the generation of unusual plastics materials, extending the performance of existing plastics, extending this performance quickly, extending the performance cheaply and re-using plastics materials. See *blend* and *compatibilization*.

thermoplastics copolymers
A *thermoplastics material* which is a *copolymer*. Often, many so-called hompolymers are in fact copolymers (for example, *polyethylene*) but as the percentage of the second monomer is small, this fact is seldom mentioned. See *abbreviations*.

thermoplastics film
See *film*.

thermoplastics material
A thermoplastics material is one which softens on heating and hardens on cooling: this heating and cooling process can be repeated a large number of times. Commercially the most important type of high polymer. Approximately 80% of all plastics used throughout the world are thermoplastics. An abbreviation used for this type of material is TP. Many thermoplastics are based on one monomer and are known as *homopolymers*: some are based on two monomers and are known as *copolymers*.

CATEGORIES OF. Thermoplastics materials may be divided into two main categories; these are *amorphous* and *crystalline*. Thermoplastics materials can be also divided into four groups (each of which may contain *amorphous* or *crystalline* materials):
 i) commodity or bulk thermoplastics;
 ii) engineering thermoplastics;
 iii) thermoplastic elastomers or rubbers; and,
 iv) blends or alloys.

The materials in each of these categories can also be filled or reinforced. Some *common names* and alternative names of some plastics are given in **table 1**.

DEGRADATION OF. With thermoplastics materials the first sign of thermal degradation is a slight change of colour, for example, yellowing occurs during melt processing. This slight change is often unacceptable and is therefore important as, in many applications the material was used, or selected, because of its colour (or its lack of colour for a transparent material). Any colour change cannot therefore be tolerated: a *stabilizer* may be classed as good, simply because it delays colour formation and not necessarily because it stops polymer degradation. See *thermally-initiated oxidation*.

COLOURED COMPONENT PRODUCTION. There are several methods which may be used to produce coloured components, for example, injection mouldings, of the required colour. Most components are coloured, one-colour throughout and many also have a surface decoration, for example, achieved by printing. Coloured components, which are the same colour throughout, may be produced by using *fully compounded material* (for example, in granular form) or, by colouring on the machine, by *dry colouring*, *liquid colouring* and *masterbatch colouring*.

COMPOUND PRODUCTION.
Fully compounded, thermoplastics materials are commonly produced using single-screw *extruders* as such machines are relatively cheap and a continuous flow of regularly-shaped material (granules) is easily obtained. However, single-screw machines are often not very good mixers. To improve the mixing action of single-screw machines use is made of *mixing sections*. However, single-screw machines are often not suitable for heat-sensitive materials such as *polyvinyl chloride* (PVC).

CHOICE OF GRADE. Once a plastics material has been selected for a particular application, then a high viscosity grade should be selected if the components are to be subjected to severe mechanical stresses; this is because the high viscosity grades usually have the highest molecular weight and exhibit the best mechanical properties. However, in some cases this advice cannot be followed as unacceptable levels of frozen-in strains result in the products. Easy flow grades are preferred, for example, in *injection moulding*, for filling thin walled sections or for use where very smooth surfaces are specified.

FILLED AND REINFORCED. Materials modification, such as with fibres or with fillers, is extensively adopted with thermosetting plastics and now, to an increasing extent, with thermoplastics (for example with engineering plastics). It is done in order to obtain a desirable combination of properties: it is seldom carried out in order to save money as often a moulding, made from a filled compound, is the same price as one made from the unfilled plastics material. This is because of the high density of most fillers and because of high compounding costs. Many of the fillers used are fibrous fillers as the use of such materials improves properties such as modulus: a commonly used fibrous filler is *glass*. By the use of such a *filler* it is possible to lift, or move, a plastics material from one category to another. In the case of *polypropylene* (PP), a commodity plastics material, it can be changed into an engineering plastics material by materials enhancement.

HYGROSCOPIC BEHAVIOUR. If a compound contains water, or if it contains another material which has a low boiling point, then the energy input, needed for *melt processing*, could cause vapour formation. Bubbles could then form within the mass of the thermoplastics material when the pressure falls, for example, when the material emerges from the die of an *extruder*. Generally speaking, the higher the processing temperatures, the lower is the amount of water that can be tolerated. This is because the higher temperatures will generate larger quantities of steam from the same quantity of water. The water could be introduced into the system by using additives which contain excessive water or by using damp material. Usually commodity thermoplastics do not suffer from water-related problems to the same extent as the engineering thermoplastics: such materials are often *hygroscopic*.

RE-USE. Most thermoplastics materials can be re-used, for example, in melt processes such as *injection moulding* and *extrusion*. The *reclamation* of thermoplastics materials during *melt processing* is relatively straightforward as the reclaimed material is added to the *virgin polymer*. Re-use of materials, or of mixed materials, from unknown sources poses considerable problems. For example, because it is often not known what material was used to make a particular product: one approach to the problem is product identification.

DISADVANTAGES OF. In general, some basic disadvantages of thermoplastics components are low stiffness, high coefficient of thermal expansion, low heat distortion temperature and flammability.

thermoplastics melt processes
Most of the processes used to shape thermoplastics materials are melt processes. That is, the plastics material is heated until it softens and becomes a *melt*: usually this is not a liquid but a high viscosity, sticky material which does not flow quickly under its own weight. Once the material is heated to the melt state (plasticized) then it is shaped and set to that

shape by being cooled. The heating process is made more efficient if the material is stirred/sheared during heating using for example, an *extruder*. The shaping processes can either produce a continuous output (for example, *calendering* and *extrusion*) or they can produce discrete components or mouldings (for example, *injection moulding*).

Before a thermoplastics material is melt processed there are certain factors which should be considered. Such factors include the hygroscopic behaviour of the material, the feed form (granule) characteristics, the thermal properties (such as heat transfer and the thermal stability), the flow properties, crystallization behaviour, shrinkage and molecular orientation.

thermoplastics - resistance to oxidation
See *thermally-initiated oxidation*.

thermoplastics - rubber addition
Rubbers are added to plastics materials in order to improve properties such as low temperature impact strength. An example is *rubber modified polypropylene (PP/EPDM)*. Often such a material is a mixture or blend of unvulcanized rubber in a plastics material. What is required is a material which can be processed like a thermoplastic and yet, has many of the properties of a cross-linked material. This is one reason for the current interest in making *thermoplastic elastomers* by dispersing cross-linked rubber particles in a thermoplastics matrix - the resultant material displays significant high elasticity and resembles a traditional rubber in this respect. See *dynamic vulcanization*, and *high impact polystyrene*.

thermoplastics sheet
See *sheet*.

thermoplastics tube
A *tube* based on a *thermoplastics material*. Tube is extruded from an *annular die* which is usually mounted in an in-line extension: a *torpedo* forms the inner surface of the tube. A low air pressure may be applied through the torpedo so that as the tube leaves the die it is held to the required shape by internal air pressure. The hot tube proceeds to a sizing arrangement which fixes its dimensions and thence to a water-bath or to a series of air-cooling rings.

To control the external diameter of the tube, the sizing equipment is normally a water-cooled cylinder (sizing die) positioned close to the extruder die. For small-diameter tube an arrangement of *sizing plates* or rings located in the cooling bath can be used instead. The internal air pressure forces the molten tube outwards against the inner surface of the sizing die, plates or rings.

Initially the tube must be sealed within a few inches of its emergence from the sizing die by pressing the walls together while the interior of the tube is still hot: flexible tube may be folded tightly just before a length, or coil, is cut off. To avoid the need for sealing, an internal floating plug may be used, attached by a hook arrangement to the torpedo, and over which the tube passes. Alternatively, the circular profile can be maintained by applying a partial vacuum through fine holes in the inner surface of the sizing die.

After passing through the cooling bath, the tube passes to a haul-off (e.g. a unit based on solid-rubber or foam-rubber caterpillar bands or a system of moving clamps) and from there to coiling gear (for flexible tube) or to a cutting device (for rigid tube).

thermoplastics vulcanizate
See *thermoplastic vulcanizate*.

thermoprene
A *cyclized rubber* which has been made by cyclization with an acid.

thermoset
The short form, and often used form, of *thermosetting plastics material*.

thermoset heat treatment
A post-moulding treatment sometimes used in the measurement of *post-moulding shrinkage*. The heat treatment is at 80°C for urea-formaldehyde (UF) and 110°C for several other thermosets.

thermoset matrix
The resin-based part of a composite: that part of the composite which gives the product its shape. See, for example, *phenol-formaldehyde reinforced plastic*.

thermosetting plastics material
A thermosetting plastics material is also known as a thermoset or as, a thermosetting plastic. It is a compound which on heating, flows and then yields a rigid cross-linked structure: that is, it yields a cross-linked plastics material on heating. This cured material does not soften significantly if re-heated. The resin-based part of such a composite may be referred to as the *thermoset matrix*. The first synthetic plastics material, on which the modern plastics industry was built, was a by-product of another industry. Phenol was obtained from the *coal* distillation industry and in an effort to utilize this material it was reacted with aldehydes such as *formaldehyde*. The old-fashioned names of the two starting materials, or monomers, are linked to give a name for the plastics material, that is, phenol-formaldehyde (PF). In the same way the names of the two major aminoplastics are obtained, that is *urea-formaldehyde* (UF) and *melamine-formaldehyde* (MF). Another material, developed to give some of the advantages of the parents, is known as melamine-phenol-formaldehyde (MPF).

Source-based nomenclature is not, however, universally used: names which refer to the type of chemical linkage involved (such as *unsaturated polyester resin and epoxides*) are also encountered. Some common names and alternative names of some plastics are given in **table 1**.

thermosetting polymer
A low molecular weight polymer which may be cured, or cross-linked, so as to yield a cross-linked plastics material or a vulcanized rubber.

thermotropic liquid crystal polyester
An abbreviation used for this type of material is TLCP. See *liquid crystal polymer*

THF
An abbreviation used for *tetrahydrofuran*.

thiazole
A colourless to yellowish liquid with a boiling point of approximately 117°C and with a relative density of 1.20. A heterocyclic ring compound whose derivatives are used as *accelerators* for rubbers. Such a derivative may also be referred to as a thiazole or as, a benzothiazole. Thiazole has the structure of a five membered ring which contains as part of its heterocyclic ring structure a -S-CH=N- group. A thiazole derivative, a benzothiazole, has the -S-CH=N- group attached to a benzene ring: this gives a heterocyclic five membered ring. The hydrogen of the central CH group is replaced with another group (X) to give -S-CX=N-. When X is SH then the *accelerator mercaptobenzothiazole* is obtained. Also see *benzthiazyl-disulphide* (MBTS).

thiazole accelerators
An *accelerator* class of great commercial importance which can be sub-divided into *mercapto accelerators* and *benzothiazole sulphenamide accelerators*. See, for example, *mercaptobenzothiazole (MBT)* and *benzthiazyl disulphide* (MBTS).

thick moulding compound
An abbreviation used for this type of material is TMC. A type of *dough moulding compound* which results if the *glass fibres* are not degraded during compound manufacture. Such

lack of fibre degradation gives products with superior strength properties and which are often compared to *sheet moulding compound (SMC)*.

thickener
An additive which may also be known as a thickening agent: a material added to a system in order to raise the viscosity of the system so that, for example, a liquid resin will not drip. A *clay* is often used for this purpose. See *sodium polyacrylate*.

thickening agent
Also known as a *maturation agent* and used in *sheet moulding compound*. See *thickener* and *creaming*.

thickening paste
A paste-like material used to increase the viscosity of an unsaturated polyester *sheet moulding compound*. Such pastes may be based on magnesium oxide or magnesium hydroxide.

thickness control
The control of thickness during production. When processing thermoplastics materials it is nearly always the rate of cooling which limits the rate of production. Accurate thickness control saves not only on material use but also gives faster cooling as the product can be held to the bottom of the allowed tolerance band.

BLOW MOULDING. The wall thickness of a blow moulding is usually controlled by die gap adjustment (which adjusts the concentricity of the *parison*) and/or by parison programming (which adjusts the thickness along the length of the parison).

CALENDERING. Thickness variations in the sheet from a *calender* can have two basically different causes. Gauge variations parallel to the calendering direction (*along the sheet*) are mainly due to variations in roll speed, roll movement within its bearings, roll eccentricity, temperature variations and inhomogeneities of the raw material. Gauge variations perpendicular to the calendering direction (*across the sheet*) are primarily due to a non-uniform nip which is caused through *roll separating forces*.

EXTRUSION. Thickness variations in the extrudate (for example, sheet) from an *extruder* can have two basically different causes. Gauge variations parallel to the extrusion direction (along the sheet) are mainly due to variations in screw speed (caused through changes in material viscosity), haul-off speed, temperature variations and inhomogeneities of the raw material. Gauge variations perpendicular to the extrusion direction (across the sheet) are primarily due to a non-uniform die gap. The die gap and/or the die temperatures can be adjusted: if sheet is being extruded it is often preferable to leave the die gap constant and to adjust the *spreader bar*. See *automatic profile control system*.

INJECTION MOULDING. It is often not appreciated that the thickness of an injection moulding can, and does, vary. One of the reasons for this, during initial start-up, is poor, or non-existent oil temperature control. Variation are more common with the *direct thrust type* of machine particularly with components of large *projected area*. Many machines are now fitted with automatic control of the *locking force* so as to maintain product dimensions within selected limits.

THERMOFORMING. The use of a *male mould* will give a different wall thickness distribution to that obtained using a *female mould*. It is however, more usual to pre-stretch the sheet before the forming operation: that is, the *wall thickness distribution* in *thermoforming processes* can be altered, or controlled, by *pre-stretching* the sheet before forming. This may be done by either mechanical means (using a *plug*) or, by using compressed air. When the heated sheet is stretched using a plug, or a male mould, the thickness of the sheet in contact with the plug stays constant while the sheet not in contact stretches. When the heated sheet is stretched before forming using air, the sheet in the centre thins more than the sheet near the edges. It is also possible to heat the sheet to different temperatures in different regions so as to give different amounts of stretching although this is not often done commercially.

thinner
A liquid material added to a coating system in order to reduce the viscosity.

thio
When a chemical term contains thio, it means that part, or all, of the oxygen in the compound may be considered to have been replaced by sulphur so as to give sulphur analogues of oxygen compounds. See *thiocarbanilide*.

thio-bisphenol
An organosulphur compound which is used as a *preventive antioxidant*. See *bisphenol*.

thioalkyl glycidylether units
See *alkylene sulphide rubber*.

thiocarbamide
See *thiourea*.

thiocarbanilide
Also known as thio or as, diphenyl thiourea or as, symmetrical diphenyl thiourea or as, 1,3-diphenyl-2-thiourea. An abbreviation used for this material is TC or DPTH or DPTU. May be represented as $(C_6H_5NH)_2C=S$. This solid material has a melting point of about 150°C and a relative density of approximately 1.3. An early accelerator for rubber vulcanization which is now little used as there are other better systems: this material gives low curing speeds and poor mechanical properties.

thioester antioxidant
An example of a *peroxide decomposer*: an *antioxidant*. For example, dilauryl thiodipropionate is an efficient peroxide decomposing antioxidant for saturated hydrocarbon polymers. Other examples include distearyl thiodipropionate and dimyrstyl thiodipropionate

thioindigo pigments
A class or type of transparent *organic pigment* which is not widely used in plastics materials because of poor heat stability and poor resistance to bleeding. Give colours ranging from red to maroon. Relatively high cost systems which give bright colours: some red-violets are used in plastics because of their excellent lightfastness.

thiokol rubber
See *polysulphide rubber*.

thiokols
See *polysulphide rubber*.

thioplasts
See *polysulphide rubber*.

thiourea
Also known as thiocarbamide. An abbreviation used for this type of material is TU. This solid material has a melting point of 182°C and a relative density (RD or SG) of 1·40. May be represented as $H_2N.CS.NH_2$. Has been used as a rubber *accelerator* and to form the plastics materials known as thiourea-formaldehyde.

thiourea-formaldehyde
An *aminoplastic* formed by the reaction of *thiourea and formaldehyde*. Slower curing than *urea resins*: the products are more brittle and water repellent than urea resins.

thiourea-urea-formaldehyde
An *aminoplastic* formed by the reaction of *thiourea, urea* and *formaldehyde*. Slower curing than *urea resins* and the products are more water repellent: such resins have been used to make decorative laminates.

thiuram disulphides
A group of sulphur containing compounds (*thiuram sulphides*) which may be represented as $(R_2NCS)_2S_x$ where R is an organic group (such as a methyl group) and x is 2. Group of sulphur containing compounds which acts as rubber *accelerators*. Very powerful accelerators which tend to be too fast for use as the primary accelerator in dry compounds based on *natural rubber*. Used in *latex* technology and in slower-curing diene rubbers such as *ethylene-propylene rubbers*. They are related to the *dithiocarbamates*: obtained by the oxidation of soluble dithiocarbamates with hydrogen peroxide. See, for example, *dipentamethylene thiuram disulphide*.

thiuram efficient vulcanization system
See *efficiency parameter*.

thiuram polysulphides
Group of sulphur containing compounds which may be represented as $(R_2NCS)_2S_x$ where R is an organic group (such as a methyl group) and x is greater than 2, See *thiuram disulphides*.

thiuram sulphide
Could refer to a thiuram monosulphide such as *tetramethyl thiuram monosulphide* (TMTM). See *thiuram disulphides*.

thiuram sulphides
Group of sulphur containing compounds which may be represented as $(R_2NCS)_2S_x$ where R is an organic group (such as a methyl group). Group of sulphur containing compounds which can be used as rubber *accelerators*: in general, slow curing and with short scorch times. When x is 2, then the *thiuram disulphides* are obtained. When x is greater than 2, then *thiuram polysulphides* are obtained. Thiuram disulphides and thiuram polysulphides can act as curing agents in the absence of sulphur. That is, sulphur vulcanization of a diene rubber, such as *natural rubber*, may be achieved without the addition of elemental sulphur: the sulphur is derived from the *accelerator* system (which is based on a thiuram disulphide, or a thiuram polysulphide, and *zinc oxide*).

thiurams
See *thiuram polysulphides* and *dipentamethylene thiuram disulphide*.

thixotropy
A term used in rheology which means that the *viscosity* of a material decreases significantly with the time of shearing and then, increases significantly when the force inducing the flow is removed.

thread up
The procedure of starting an *extrusion line* by leading the extrudate through the downstream equipment

three bowl calender
See *three roll calender*.

three colour printing process
A type of *process printing*.

three nip calender
See *four roll calender*.

three point bending jig
The apparatus used to apply *three-point loading*.

three roll calender
Also known as a three bowl calender. A *calender* with three main rolls. That is, with two nips. The rolls may be arranged in different ways or configurations. Can have them superimposed (an *I calender*) or, in the form of a triangle (an *A calender*).

three term control
Usually associated with temperature controllers and means that the instrument has circuits, or terms, for *proportional*, *integral* and *derivative control*. That is, it is a PID controller.

three term controller
A controller, usually a temperature controller which has *three term control*.

three-plate gating
Gates, for example, *pin-point gates*, which are carried in a *three-plate mould*. For example, simple thin-walled mouldings (e.g. beakers) are made using multiple-impression three-plate moulds in which the cavities are fed via central pin-point gates in the base of the cavity. Sub-sprues (stalks or carrots) connect the gates to the main runner system. Such pin-point gates need to be fed with a generous runner system as otherwise premature freezing will occur. The major disadvantages of three-plate gating are runner ejection difficulties, runner entanglement (during conveying and regrinding), more runner to be reground (compared to the use of *edge gates*), more complex mould designs and more complex operating sequences.

three-plate injection mould
A *three-plate mould* used for *injection moulding*. The mouldings drop from one daylight and the feed system, in the case of a *cold runner type*, drops from the other. The feed system can therefore be easily segregated from the components. One of the big advantages of this type of mould is that *pin-point gates* may be used for multi-impression moulds. The three main parts or plates which make up this type of mould are the stationary plate which is attached to the stationary or fixed platen. The centre plate which is linked to the two other main plates and the moving plate, or the front cavity plate, which is attached to the moving platen of the machine.

Compared to a *two-plate mould*, this type of mould therefore has an additional plate which floats between the fixed and moving plates. The feed system is contained between the fixed plate and the centre plate whereas the mouldings are formed between the moving plate and the centre plate. The cavity is usually cut in the centre plate and each cavity is fed via a pin-point gate from secondary sprues which are also cut into the plate. In such tools it is necessary to arrange not only for a sprue puller but for an ejector device for the section of the runner leading to the pin-point gate.

three-plate mould
This type of mould has three main parts or plates, which on opening are separated one from the other to give two daylights: such moulds are usually associated with injection moulding. See *three-plate injection mould*.

three-point loading
The application of a *stress* at, say, the centre of a rectangular bar which is supported at two other points. See *flexural properties*.

three-roll mill
A mill which consists of three horizontal rolls which lie adjacent to each other. The rolls are often un-heated but the nip between the rolls can be adjusted so as to control dispersion. Used for paint dispersion and for *paste manufacture*.

three-roll stack
A set of three metal cooling rollers used to cool and size plastics *sheet* in a *sheet extrusion line*.

threshold frequency
The frequency of light above which photoelectric emission will occur.

threshold stress
The stress above which above *ozone cracking* of rubber compounds will occur.

throttle
A term used in *hydraulics* and which refers to a device which restricts fluid flow.

through hardening steel
See *fully hardening steel.*

thrust bearing
The main bearing which takes the load (or thrust) generated by the *screw*: designed to last at least 100,000 hours at the extreme thrust load of the screw. See *starve fed.*

thrust load
The load generated by the turning action of the screw; assumed to be twice the head pressure. For example, if a 3·5" (approximately 90 mm) machine is operating against a maximum head pressure of 5000 psi then the direct back force is 50,000 lbs (223 kN); the extreme thrust load is therefore twice this or, 100,000 lbs (446 kN).

thrust type machine
A term used in the injection moulding industry and means that an injection moulding machine is a *direct hydraulic-type* of machine.

THTM
An abbreviation used for *triheptyl trimellitate.*

Ti-BR
An abbreviation used for *butadiene rubber* based on a titanium catalyst.

Ti-IR
A grade of *isoprene rubber* prepared with a titanium catalyst.

tie bar
Sometimes known as a tie rod. A part of a press: that part on which the platens slide. On a conventional *injection moulding machine*, the machine is built around, for example, 4 tie bars. The platens and the injection unit are located one with respect to the other by the tie bars: movement along the tie bars occurs by sliding. See *tiebarless machine.*

tie bar clearance
The clearance between the tie bars. Such dimensions are important they determine whether or not a mould, of a given length or width, can be accommodated in a particular *injection moulding machine.*

tie bar extension
Also known as tie bar stretch. When an injection mould is closed and clamped then, because of the force applied, the *tie bars* stretch or extend. The amount of extension can be as much as 1% of the original effective length of the tie bar. The locking force on a toggle operated clamping unit of an injection moulding machine is determined by the amount of tie bar extension.

It is therefore very important to ensure that the *effective tie bar length*, on a toggle lock injection moulding machine, is capable of being measured quickly and easily as incorrect over-setting will, eventually, lead to tie bar failure. Tie bar extension measurement can be carried out in various ways and can, on newer machines, lead to automatic control or regulation, of the *clamping force* during the injection moulding cycle.

tie bar extension - bending moment measurement
The locking, or anchor, platen remains stationary but deflects, or bends backwards, in an arc, when the locking force is applied to a toggle-lock machine. The formation of the arc creates a bending moment at a point at the base of the platen. This moment may be measured using a *strain gauge assembly.*

tie bar extension-locking force graph
A graph of *tie bar extension* against *locking force* obtained when a range of locking forces is used. Most moulding machine manufacturers will supply tie bar extension values needed to obtain the correct locking force on a toggle lock machine. Such graphs are used to assist mould setting in injection moulding shops as their use enables the moulding shop personnel to set the required locking force for a particular design of mould. Correct usage of such graphs helps ensure that over-stretching of the *tie bar* does not occur.

tie bar extension value
Term used in *injection moulding technology* and refers to the amount of *tie bar extension* needed for a mould of a given projected area. See *locking force.*

tie bar failure
Failure of a *tie bar* caused by over-stretching (over-stressing): usually caused by unequal loading of the tie bar system. Regular monitoring of the extension values for each tie bar should be performed during moulding on toggle lock machines. See *locking force.*

tie bar stretch
See *tie bar extension.*

tie rod
A steel rod used to couple together parts of a mould. The term is also used in place of *tie bar.*

tie-layer
A layer of material, usually a thermoplastics material, which binds two layers together in a multi-layer product. For example, in a multi-layer *blow moulded container.* See *coextrusion blow moulding.*

tiebarless machine
A type of *injection moulding machine* which does not employ *tie bars*. On a conventional injection moulding machine, the machine is built around, for example, 4 tie bars: these stiffen the machine but can hinder mould mounting and platen accessibility. A tiebarless machine gives more mould mounting area than a conventional machine and there is more space for robot operation. Such machines are usually built around a C frame and are of relatively small size, for example, <100 tonnes: they may be horizontal or vertical machines.

tilting head press
A press used in *compression moulding* and in which, the head tilts during mould opening: used, for example, in gramophone record moulding.

TIM
An abbreviation used for thermoplastics injection moulding and for thermoplastic injection moulding. See *injection moulding.*

time measurement unit
Time is usually measured in seconds. The SI unit is the *second.*

time-dependent fluid
A fluid in which shear stress-shear rate relationships depend upon how the fluid has been sheared and upon previous shear history.

time-independent fluids
A *Newtonian fluid.* The rate of shear at a single point, in such a fluid, is some function of the shearing stress at that point and is not dependent upon anything else.

timed acceleration
A control circuit which limits the rate at which an electric motor comes up to speed so as to prevent excessive loads which may damage the *screw* or the drive.

tin
This element (Sn) occurs in Group 1VB of the Periodic table along with *carbon*, silicon, germanium and lead. It does not occur naturally. It is a soft, silvery white, lustrous metal which melts at approximately 232°C and which boils at

2,300°C. It is very corrosion resistant and for this reason is used to coat steel so as to make tin plate. Compounds of tin are used in the polymer industry as, for example, catalysts in *polyurethane foams* and *heat stabilizers* for *polyvinyl chloride* (PVC). Alloys of tin include solders (based on say, 50% tin and 50% lead), type metal (based on say, 10% tin, 75% lead and 15% antimony) and pewter (based on say, 80% tin and 20% lead). *Bronze* can be an alloy of copper and tin.

tin (11)
See *stannous*.

tin (1V)
See *stannic*.

tin compound
Compounds based on *tin* are used as *stabilizers and catalysts*: such compounds are often *organo-tin compounds*. For example, *stannous octoate* is used as a catalyst for the *cross-linking reactions* in flexible polyurethane foam production by the *one shot process*.

tinplate container
Tinplate was the original material for can manufacture. Such a container is virtually indestructible, is light in weight, impermeable and readily decorated. When used to contain liquids, the can offers long shelf life, sterilizability and, with the development of the ring-pull, ease of opening. Now, can-type containers can be made from *polyethylene terephthalate* (PET) - such containers can be clear if required.

tint
The colour which results when white pigment is used with a colour or *hue*. A variety of a colour produced by diluting that colour with white. See *titanium dioxide*.

tinting strength
A measurement of the colouring power of *carbon black* which is obtained by comparison with standard compounds.

TINTM
An abbreviation used for *tri-iso-nonyl trimellitate*.

TIOTM
An abbreviation used for *tri-iso-octyl trimellitate*.

tip printing
Also called tipping. A printing method used to obtain a two-colour effect on the surface of an *embossed plastics* product by spreading a material of contrasting colour, onto the substrate, so that it remains on the tips, or crests, of the embossed pattern.

tipped polyol
Also known as a capped polyol. A *polyol* which has been made more reactive. For example, to increase the reactivity of a *polyoxypropylene polyol*, the materials may be capped or tipped with ethylene oxide: such a treatment gives primary hydroxyl end-groups rather than less-reactive, secondary end-groups. Used to make a *polyurethane foam*.

tire
See *tyre*.

tire rubber
A form of *natural rubber*. Made from 30 parts of rubber (as latex), 30 parts of sheet rubber, 40 parts of *cup lump* and 10 parts of an aromatic process oil. Viscosity stabilized and requires no *mastication*.

titanate pigments
A class or type of *inorganic pigment*: obtained from *rutile titanium dioxide* by calcination with various other metallic compounds. By combining nickel and antimony with the *rutile titanium dioxide* then yellow pigments are obtained. By combining chromium and antimony with the rutile titanium dioxide then buff pigments are obtained. By combining nickel and cobalt with the rutile titanium dioxide then green pigments are obtained. By combining cobalt and aluminium with the rutile titanium dioxide then blue pigments are obtained. Relatively, high priced pigments which have low colouring strength and opacity. Such pigments have good heat stability at high temperatures, are chemically inert, are usually light-fast and weather-resistant.

titania
See *titanium dioxide*.

titanic acid anhydride
See *titanium dioxide*.

titanium
A transition element (Ti) which is a hard, lustrous gray metal with a high melting point (1,720°C), a high boiling point (3,280°C) and a relative density of 4·5. One of the most common elements in the earth's crust. Titanium and its alloys are noted for their corrosion resistance and high strength coupled with a low density. *Titanium dioxide* is the most important compound of titanium as it is widely used as a white pigment.

titanium (1V) oxide
See *titanium dioxide*.

titanium dioxide coated mica flakes
See *flake pigments*.

titanium trichloride
See *Ziegler-Natta catalyst system*.

titanium white
See *titanium dioxide*.

titanium dioxide
Also known as titania or as, titanium (1V) oxide or as, titanium white or as, titanic acid anhydride. The most important compound of titanium which is widely used as a white pigment. A class or type of *inorganic pigment*: the most important inorganic pigment. Has a melting point of approximately 1,850°C. Compared to white pigments based on lead, barium or zinc, it has superior opacity and covering power. It is thought to be non-toxic, is stable and relatively cheap. Titanium dioxide is often used to pigment a material a base white which is then tinted with other pigments. As titanium white is slightly yellow, blue pigments may be used to mask this colour. Titanium white is relatively low in cost and also relatively inert: cost may be reduced by extending with finely divided, white fillers such as alumina or silica. Weatherability may be improved by coating the titanium dioxide with alumina or silica.

There are two major crystalline forms, *rutile* and *anatase*. The relative density of titanium dioxide (anatase) is 3·90 and that of titanium dioxide (rutile) is 4·20. Rutile titanium dioxide is the most widely used form because of its high colouring strength and hiding power which gives it good opacity. The anatase form has a higher chalking tendency than the rutile form: such chalking and yellowing may be reduced by the addition of alumina before calcination during manufacture. When titanium dioxide is used with some *antioxidants,* compound yellowing may be seen. Titanium dioxide is also abrasive and can cause machine wear. See *titanate pigments*.

TLCP
An abbreviation used for *thermotropic liquid crystal polyester*. See *liquid crystal polymer*.

TLV
An abbreviation used for threshold limit value.

T_m
An abbreviation used for the melting point or temperature.

TM
An abbreviation used for *polysulphide rubber*. TM is also used in plasticizer abbreviations for trimellitate - see *trimellitate plasticizer*.

TMA
An abbreviations used for trimellitic anhydride and for thermomechanical analysis. See *trimellitate plasticizer*.

TMBPA-PC
An abbreviation used for a *tetramethyl bisphenol A polycarbonate*.

TMC
An abbreviation used for *thick moulding compound*: a *polyester moulding compound*.

TMC-PC
An abbreviation used for *tetramethylcyclohexane bisphenol polycarbonate*.

TMDT
An abbreviation used for *poly-trimethylhexamethyleneterephthalamide*.

TMIB
An abbreviation used for *3,3,5-trimethylpentane-1,3-diol diisobutyrate*.

TMQ
An abbreviation used for *polymerized 1,2-dihydroxy-2,2,4-trimethylquinoline*.

TMS
An abbreviation used for *tetramethylthiuram monosulphide*.

TMT
An abbreviation used for *tetramethylthiuram disulphide*.

TMTD
An abbreviation used for *tetramethylthiuram disulphide*.

TMU
An abbreviation used for *tetramethylurea*.

tn
An abbreviation sometimes used for *ton*.

TN
An abbreviation used for trade name.

TNPP
An abbreviation used for *tris-(p-nonylphenyl) phosphite*.

TODI
An abbreviation used for *3,3'-dimethyl-4,4'-diphenyl diisocyanate*.

toe cap
Also called a tread cap. That layer of a two-component *tyre tread* assembly into which the pattern is moulded.

TOF
An abbreviation used for *trioctyl phosphate*.

toggle
A lever system. See *toggle lock machine*.

toggle lock machine
Also known as a hydromechanical machine. A type of *injection moulding* machine which uses a system of levers, driven hydraulically, to achieve the clamping force required. During initial mould closing, such a system gives rapid movements but is only capable of generating low forces: as the mould halves come together the speed of movement drops and force generation increases. Clamping force generation is by *tie bar* extension the *locking force* is then generated because of the forces generated when the material flows into the injection mould.

On a conventional toggle machine (where the anchor platen and the moving platen are the same size) the fully extended toggles lie parallel to the tie bars: the use of such a lever system can generate forces which distort the platens. However by using a larger *anchor platen*, larger than the *moving platen*, the toggle can be slightly inclined so that the effect of such distorting forces are minimized: platens may thus be made thinner and lighter.

Over the size range of approximately 100 to 800 tonne, toggle machines are often preferred because of ease of manufacture and cost considerations. See *clamping system* and *locking force*.

toggle-type machine
A toggle lock machine.

tolerance limits
The limits on dimensions which will be acceptable for production purposes. In design specifications, limits on dimensions are usually set in the form of a nominal value and a plus or minus tolerance, for example 10.00 + 0.02 mm. It should be noted that the plus and minus deviations need not be equal. Such limits ensure that the items manufactured are serviceable and, for example, can be assembled together with other parts. These are the specification limits. See *injection moulding*.

toluene
Also known as methylbenzene or as toluol. An aromatic chemical which may, for example, be obtained from *coal tar*. May be represented as $CH_3C_6H_5$. This material has a relative density (RD or SG) of 0.87 and a boiling point of approximately 111°C. It is a good solvent for uncured low *nitrile butadiene rubber (NBR), chloroprene rubber (CR), butyl rubber (IIR), natural rubber (NR), styrene-butadiene rubber (SBR)* and *thiokol rubber (T)*. This chemical causes some swelling of uncured NBR (low acrylonitrile NBR). This chemical causes a large amount of swelling, or gel formation, of uncured high acrylonitrile NBR. Used as a solvent for many plastics materials: for example, for thermoplastics materials such as *polystyrene* and for thermosets such as *epoxides*.

toluene di-isocyanate
See *tolylene diisocyanate*.

toluene diisocyanate
See *tolylene diisocyanate*.

toluene-2,4-diisocyanate
See *2,4-tolylene diisocyanate*.

toluene-2,6-diisocyanate
See *2,6-tolylene diisocyanate*.

toluidine reds
Red *monazo pigments*: organic pigments.

toluol
See *toluene*.

tolyl diphenyl phosphate
See *cresol diphenyl phosphate*.

tolyl mercaptan
A bleaching agent used in *pale crepe* production.

tolylene diisocyanate
Also known as tolylene di-isocyanate or as, toluene diisocyanate. An abbreviation used for this material is TDI. A *diisocyanate* used in the manufacture of *polyurethane foams*. The two isocyanate groups (-NCO) can be attached to different positions (2, 4 or 6) around the benzene ring and so different isomers are possible. Most commonly, mixed isomers are used. For example, 80:20 TDI is a mixture of 2,4-tolylene diisocyanate with 2,6-tolylene diisocyanate in a ratio of 80:20. This is the most widely used mixture and is used particularly for *flexible polyurethane foams*. 65;35 TDI is a mixture of 2,4-tolylene diisocyanate with 2,6-tolylene diisocyanate: this has a lower reactivity than the 80:20 mixture.

These mixtures are liquids and are produced from *toluene*. The liquid 80:20 mixture has a melting point of approximately 12°C: the liquid 65:35 mixture has a melting point of approximately 4°C.

tolylene diisocyanate index
An abbreviation used for this term is TDI index. A measure of the TDI content of a polyurethane foam formulation. A TDI index of 105, a commonly used amount, means that there is a 5% excess of *isocyanate* groups: such an excess is used to take account of wasteful side reactions.

tolylene diisocyanate mixtures
An abbreviation used for this term is TDI mixtures. See *80:20 TDI* and *65:35 TDI*.

tolylene-2,4-diisocyanate
One of the possible isomers of the diisocyanate, *tolylene diisocyanate*. The two isocyanate groups (-NCO) are attached to the 2 and 4 positions on the benzene ring.

tolylene-2,6-diisocyanate
One of the possible isomers of the diisocyanate *tolylene diisocyanate*. The two isocyanate groups (-NCO) are attached to the 2 and 6 positions on the benzene ring.

Tomasetti's Volatile Indicator method
An abbreviation used for this term is TVI. A method used for assessing the moisture content of *polycarbonate* (PC) granules.

Two microscope glass slides are put on a hot plate at 270°C. Three or four granules are put on one slide and then the other slide pressed on top of the granules until they spread to a diameter of about 10-13 mm. After further heating at 270°C (for one minute) the slides are removed and cooled to ambient temperature (260°C is suggested for PC/PBT blends). The water content can be estimated from the number and size of bubbles present. If there are no bubbles the moisture content is less than 0.15% and the material can be safely processed: if more then it must be dried.

ton
A unit of *avoirdupois weight*. In the UK, 2,240 pounds = 1 ton (a long ton). In the US, 2,000 pounds = 1 ton (a short ton). A freight ton is 40 cubic feet and a register ton is 100 cubic feet.

tone
A variety of a colour. The colour which results when both black and white pigments are used with a colour or *hue*. A variety of a colour produced by changing that colour with other pigments See *tint* and *shade*.

tonne
A metric ton. A unit of weight of 1,000 *kilograms (kg)* which is roughly equal to 1 ton (UK) or a long *ton*.

tool injection port
Part of a *resin transfer moulding system*: that part of the mould through which the catalyzed resin enters the mould. It is generally machined of mild steel and contains a plug or check valve made of PTFE or PE: the injection pressure pushes the check valve off its seat and allows resin to fill the mould.

tool rule
The strip electrode used for *high frequency welding*.

tooling inserts
Replaceable parts of the mould used, for example, in *blow moulding* so as to create threads and other undercut features.

tooling prepreg systems
Systems used to manufacture mould tools from *prepreg* materials. A wide variety of reinforcements can be used but bi-directional graphite (carbon) or E-glass twill-weave fabrics are generally used in conjunction with *epoxide resins*. Can have both high temperatures curing on, for example, heat resistant models at cure temperatures in excess of 90°C and room, or low temperature, curing tooling prepregs. See *low temperature moulding epoxy prepreg systems*.

top blow machine
A type of *blow moulding machine* which produces components by top blowing. See *top blow process*.

top blow process
An *extrusion blow moulding* process. After the tube has been produced by extrusion, a mould closes around this tube and the closing action causes the tube to be sealed or pinched at the base. A knife or hot wire then travels across the top of the mould and this action separates the material in the mould from the material extruding from the die. In order to allow a continuous extrusion process, the mould then moves away from the die region and a blowing pin or mandrel is forced into the open neck. This, in effect, compression moulds the material in the neck region: the body of the bottle is formed when air is introduced through the blowing pin. As the moulding is cooling, another length of tube (parison) is being extruded. If only one mould set is being employed the extrusion rate and the cooling rate must obviously be synchronised. To speed-up production, more than one mould set may, however, be employed. See *form-and-fill*.

top blowing
See *top blow process*.

top coat
A lacquer or varnish whose application forms part of the production process for the *vacuum metallization* of plastics. The metallized surface of an injection moulding is coated with lacquer so as to protect the thin layer of aluminium against oxidation and in-service scratching. The thickness of aluminium is of the order of 0.05 μm and this is protected by an air-drying cellulose lacquer or a stoving lacquer. The use of a clear lacquer gives a bright silver appearance whereas the use of a coloured lacquer changes the appearance of the metal, for example, to gold. By the use of an *overlay*, part of the metallic colour can be masked.

top ejection
Ejection from the upper part of *compression mould*. Compression moulds may have both top and bottom ejection so as to ensure automatic operation.

top plate
Part of a mould: a steel plate which is bolted to the upper part of a mould and usually used to secure the mould to the *platen*. The plate may be cored for heating or cooling.

TOPM
An abbreviation used for *tetraoctyl pyromellitate*.

Töpfer decree
A German decree which specifies that, for example, 64% of all plastics packaging should be recycled by July 1995. See *recycled material* and *reclamation*. See *Duale System Deutschland*.

topping
See *skin coating*.

TOR
An abbreviation used for transpolyoctenamer. See *polyoctenamer*.

torque
That which produces (e.g. a force) or which tends to produce rotation. A measure of rotary force, or turning effort of a motor, and which is measured in newton meters.

torque motor
A term used in *hydraulics* and which refers to a device which can generate rotary motion and which is used to actuate a *servo-valve*.

torr
A unit of pressure. Used for measuring very low pressures. It is equivalent to 1 mm of mercury or 133·322 pascals. 1 torr = 0·0193 psi or 2·784 lb/sq ft.

torsion modulus
See *shear modulus*.

tortoiseshell effects
Also known as marbled effects. May be produced in moulded components, by using a mixture of different coloured starting materials provided that intensive mixing does not occur during the melt preparation stage, for example, by using a *plunger injection moulding machine*.

total cycle time
This is the sum of the mould open time and the mould closed time in a process such as *injection moulding*: the time needed to produce one component.

total luminous transmittance
See *light transmittance measurement*.

total quality control
An abbreviation used for this term is TQC. Means that everyone in the factory is committed to quality production.

total shrinkage
The total shrinkage, experienced by a moulding, is made up of mould *shrinkage* and *post-moulding shrinkage*. Usually mould shrinkage increases when increasing mould and/or melt temperature. However, total shrinkage often decreases as mould temperature is increased and, in practice, more dimensionally stable mouldings are obtained. If the mould temperature is very high, then post-moulding shrinkage may be neglected even for semi-crystalline, thermoplastics materials. Such high mould temperatures may be an advantage where close tolerances are required. It should be noted that, in general, it is not possible to produce components by *blow moulding* to the same tolerances as are possible by *injection moulding*.

total sulphur
One of the ways the *sulphur* in rubber compounds, on analysis, may be classified. The total sulphur content no matter what its origins. See *sulphur analysis*.

touch dry
Dry to the touch: not sticky. A state of cure often associated in the plastics industry with the curing of *unsaturated polyester resin* systems. When the resin is touch dry, for example, the next layer of resin can be added during *hand lay up*.

tough
A material is classified as tough if it is not easily broken or cut.

tough fracture
See *ductile fracture*.

tough-brittle transition
A change from tough to brittle behaviour. Caused, for example, in the case of thermoplastics materials by a lowering of temperature and/or by the incorporation of a stress-raiser such as a notch.

toughened polystyrene
See *high impact polystyrene*.

toughness
The ability to withstand fracture.

tourmaline
A mineral based on complex silicates (for example, boron and aluminium) and which can produce *polarized light*.

tow
A loose rope of many *filaments* which has, for example, been prepared for spinning.

towpreg
See *continuous fibre reinforced towpreg*.

toxic gas testing
See *gas testing*.

TP
An abbreviation used used in *plasticizer abbreviations* for terephthalate, for example, *dioctyl terephthalate (DOTP)*. An abbreviation used for thermoplastic or for a *thermoplastics material*. For example:
TP BMC = *thermoplastic bulk moulding compound*;
TP-EE = *thermoplastic elastomer - ether based*;
TP-EPDM = a *thermoplastic elastomer* based on an *ethylene-propylene rubber*. See *rubber modified polypropylene*;
TP-EVA = thermoplastic elastomer - ethylene-vinyl acetate elastomer. See *thermoplastic ethylene-vinyl acetate copolymer*;
TP-NBR = a *thermoplastic elastomer material* based on *nitrile rubber*. See *polyvinyl chloride nitrile rubber blend*; and
TP-NR = a *thermoplastic elastomer* based on *natural rubber*.

TPA
An abbreviation used for *terephthalic acid* (see *phthalic acid*). TPA is also used as an abbreviation for transpolypentenamer. See *polypentenamer*.

TPE
An abbreviation used for *thermoplastic elastomer*. For example:
TPE-A = *thermoplastic elastomer - amide based*;
TPE-E = *thermoplastic elastomer - ether based*;
TPE-EPDM = a *thermoplastic elastomer* based on *ethylene-propylene rubber*. See *rubber modified polypropylene*;
TPE-EVA = a *thermoplastic elastomer* based on *ethylene-vinyl acetate*. See *thermoplastic ethylene-vinyl acetate copolymer*;
TPE-FKM = a *thermoplastic elastomer* based on a *fluororubber*;
TPE-NBR = a *thermoplastic elastomer* based on *nitrile rubber*. See *polyvinyl chloride nitrile rubber blend*;
TPE-NR = a *thermoplastic elastomer* based on *natural rubber*;
TPE-O = a *thermoplastic elastomer - olefin based*. See *dynamic vulcanization*;
TPE-OXL = a *thermoplastic elastomer* which is polyolefin based - the rubber is crosslinked. See *dynamic vulcanization*;
TPE-S = a *thermoplastic elastomer* which is *styrene* based. See *styrene-butadiene block copolymer*;
TPE-SBR = a *thermoplastic elastomer* which is based on *styrene-butadiene rubber*. See *styrene-butadiene block copolymer*;
TPE-U = a *thermoplastic elastomer* which is urethane based. See *thermoplastic polyurethane*;
TPE-XL = *thermoplastic elastomer* which is usually polyolefin based - the rubber is crosslinked. See *dynamic vulcanization*; and,
TPE-XLV = a *thermoplastic vulcanizate*.

TPG
An abbreviation used for *triphenyl guanidine*.

TPIPN
An abbreviation used for thermoplastic, interpenetrating polymer network (structure). See *styrene block copolymer*.

TPNR
An abbreviation used for *thermoplastic natural rubber*.

TPO
An abbreviation used for *olefin thermoplastic elastomer*. See *thermoplastic elastomer*.

TPO-XL
An abbreviation used for a *thermoplastic elastomer* which is polyolefin based - the rubber is crosslinked. See *dynamic vulcanization*.

TPP
An abbreviation used for *triphenyl phosphate*.

TPR
An abbreviation used for thermoplastic rubber. See *thermoplastic elastomer*.

TPS
An abbreviation used for toughened polystyrene. See *high impact polystyrene*.

TPU
An abbreviation used for *thermoplastic polyurethane*.

TPU-R
An abbreviation used for reinforced *thermoplastic polyurethane*.

TPV
An abbreviation used for *thermoplastic vulcanizate*.

TPX
An abbreviation used for *poly-4-methylpent-1-ene*.

TQ
An abbreviation used for *tubular quench*.

TQC
An abbreviation used for *total quality control*.

TR
An abbreviation used for *polysulphide rubber*. An abbreviation used for *temperature reversion (test)*. For example:

TR-10 = the temperature which a previously stretched and frozen test sample retracts by 10% in a temperature reversion test; and,

TR-70 = the temperature which a previously stretched and frozen test sample retracts by 70% in a temperature reversion test.

track resistance index
An abbreviation used for this term is TRI. It is the maximum voltage (X) at which 50 drops of ionic solution can be applied without *tracking*. See *resistance to tracking*.

tracking
The formation of a conducting track across a polymer surface such as those based on *phenol-formaldehyde* and *polyvinyl chloride*. A conductive path may be formed along the surface of a plastic by a spark or arc which means that insulation properties are lost. In general, those plastics which degrade on heating to give volatiles (gases) are more non-tracking: they are more track-resistant than those plastics which do not give volatiles on heating. See *resistance to tracking and plucking*.

trade name
An abbreviation used for this term is TN. See **table 5**.

trademark
See **table 5**.

traditional rubber
A rubber such as *natural rubber (NR)*, which must be *vulcanized* after being shaped unlike a *thermoplastic elastomer*.

trailing edge
Also called trailing flight. The rearward part of the *screw flight*.

trailing flight
See *trailing edge*.

tramlines
Sets of two lines on a *process control chart*.

trans form
See *cis-trans isomerism* and *trans-polyisoprene*.

trans-1,4-butadiene
An isomer of *butadiene*. See *styrene-butadiene rubber*.

trans-1,4-polychloroprene
A stereo-isomer of *chloroprene rubber* (CR): most commercial CR is predominantly trans-1,4-polychloroprene.

trans-configuration
See *cis-trans isomerism*.

trans-polyisoprene
Also known as trans-1,4-polyisoprene. A polymer of *isoprene* - which may be represented as $CH_2=C(CH_3)CH=CH_2$. Upon polymerization, both cis and trans isomers can be formed. In the trans structure, the CH_3 group and the pendant hydrogen atom lie on opposite sides of the main chain and are prevented from rotating, to another configuration, by the double bond. See *balata*, *gutta-percha* and *natural rubber*.

trans-polyoctenamer rubber
See *polyoctenamer rubber*.

trans-polypentenamer rubber
See *polypentamer rubber*.

transducer
A device which changes energy from one form to another, for example, force may be changed to an electrical output as in a *pressure transducer*.

transfer chamber
Also called a transfer pot. The heated chamber used to contain the charge of moulding material in a *transfer mould*.

transfer coating
A *coating* method for a substrate. A carrier layer is first coated and then the substrate is applied to the coated carrier: at a later stage the carrier is removed.

transfer mould
A mould used in *transfer moulding*.

transfer moulding
A more descriptive name for this process is one-shot injection moulding. It is used for rubbers and for thermosetting plastics materials and is similar to *injection moulding* in that it utilises a closed mould. It is different to injection moulding in that the material is placed in a loading chamber and not in a heated barrel: the exact amount of material for a shot is placed in the loading chamber.

The mould contains a feed system (sprue, runner and gate) and this is connected to the pot or transfer chamber. A preheated charge of material is loaded into the pot and then forced, by means of a plunger, to flow through the feed system and into the cavity. Usually used with thermosetting plastics so setting is applied by further heating. It is often the closing action of the press, in which the mould is mounted, which causes the material to be displaced from the transfer pot: transfer pressures of 45 MN/m^2 (3 tsi) may be required.

Because the mould is closed and clamped before the material is admitted it is possible to produce components with good dimensional accuracy. Rapid production rates can be achieved and the production of mouldings with thin sections and delicate inserts is possible. Because all of the material flows through a narrow gate (transfer port), cross-linking or cure is more uniform than it is for compression moulded components. Thicker sectioned components can also be produced more easily by this process because of the more uniform heat input - caused, for example, by the generation of *shear heat*. However, the moulds are more elaborate than those used in *compression moulding* and are therefore more expensive. As the material cures in the feed system this cured material (sometimes called the cull) must be discarded at the end of the moulding operation.

transfer port
See *transfer moulding*.

transfer pot
See *transfer chamber*.

transition - first order
A change of state associated with crystallisation, or melting in a polymer; the melting point of a semi-crystalline, thermoplastics material (T_m).

transition metal salt
A chemical compound based on a transition metal. An example is titanium trichloride which is used with aluminium triethyl to give a *Ziegler-Natta catalyst system*.

transition metals
The term transition metals, was originally applied to *iron*, *cobalt* and *nickel* but now the term now refers to all elements between calcium and gallium in the periodic table. These elements are all hard, lustrous metals with high melting points and boiling points. Includes titanium, vanadium, chromium, manganese and nickel. Many transition elements are noted for their catalytic powers and/or for the useful alloys which they form.

transition - second order
See *glass transition temperature (T_g)*.

transition temperature
The temperature at which the physical properties of a polymer change: it is usually a temperature range rather than a specific temperature.

transition zone
One of the *zones* of a three-zone screw. The centre zone which lies between the *feed zone* and the *metering zone* and in which the material is compressed as the volume of the *flight* is decreased. Also see *leathery state*.

translucent
Semi-opaque. A material may be considered as translucent when the percentage of *haze* is greater than 30%.

transmission belt
A *belt* used to transmit power.

transmission beta ray gauge
A measuring device used in *calendering*. The gauge traverses across the sheet and sheet weight per unit area is monitored by means of such a *radiation gauge*.

transmission plate
Part of an *unscrewing mould*: a steel plate through which the *transmission system* operates.

transmission system
That which transmits the power to the operating gear of an *unscrewing mould*.

transmission welding
See *remote ultrasonic welding*.

transparent crystalline thermoplastics material
See *transparent semi-crystalline thermoplastics material*.

transparent plastics materials
A plastics material which permits the passage of light so that objects can be clearly seen through the material. A material may be considered as transparent when the percentage of *haze* is less than 30%. If all the light falling onto, for example, a piece of thermoplastics material passed through it, then the plastics material would be completely transparent. However, there are many factors which stop this happening, for example, the material may contain crystal structures or, the surface may be scratched. Production conditions can also alter transparency as the choice of incorrect conditions can lead to the formation of volatiles (gas bubbles) or frozen-in stress. Good quality acrylic sheet (PMMA) will transmit approximately 92% light. Light that is not transmitted will be reflected, scattered or absorbed.

An *amorphous*, thermoplastics material is transparent unless it contains something, for example, a filler, which interferes with the passage of light. Transparent plastics are therefore usually unfilled. Matching the refractive index of two polymers, or of an additive to a plastics material, will give a transparent compound: the transparency may, however, only match over a limited temperature range.

Two phase materials are also usually opaque as the different phases have different densities and so light travels at differing speeds through each phase. Even if the densities are the same at one temperature, they are unlikely to match as the temperature is changed. This is why some rubber toughened materials become less transparent as the temperature is altered.

Semi-crystalline, thermoplastics materials are usually opaque as the crystal structures have a different density to the amorphous regions (an exception is polymethyl pentene-1 where the the density of the crystal regions are the same as those of the amorphous regions). For a given semi-crystalline, thermoplastics material, reducing the size and number of the crystal structures, will improve the transparency, that is, make the plastics more clear. This may sometimes be achieved by quick cooling and/or by using additives such as a *nucleating agent*. By incorporating stiff, aromatic segments in the molecular structure of a polyamide (PA) chain, which discourage crystallinity, it is possible to make a transparent PA thermoplastics material.

Assessment of transparency requires the measurement of a number of different factors, for example, *light transmittance*, *haze* and *see-through clarity*.

transparent polyamide thermoplastics material
By incorporating stiff, aromatic segments in the molecular structure of a polyamide (PA) chain, which discourage crystallinity, it is possible to make a transparent PA thermoplastics material. See *nylon*.

transparent rubber
An *amorphous* material is transparent unless it contains something, for example, a filler, which interferes with the passage of light. Fillers are therefore not used in transparent compounds or, are only used in small amounts. The additives are chosen because of their solubility in the *pale crepe grade* (100 phr) of *natural rubber* and/or because they mix and disperse very easily. A fine grade of *zinc oxide* (0.5 phr) is used to activate the *accelerator* (for example, *tetramethyl thiuram monosulphide* at 0.25 phr. Other mix ingredients could be *aldehyde-amine antioxidant* 0.75 phr *stearic acid* 0.5 phr and *sulphur* 1.5 phr.

transparent semi-crystalline thermoplastics material
Above its melting point (T_m) a semi-crystalline, thermoplastics material will be a clear melt; below its T_m it will be opaque as crystallization will have occurred. However, there will still be a lot of amorphous material present that is, two phases are present. It is the difference in density, and therefore refractive index, between the two phases which causes the loss of transparency. When the density of each phase is the same, as in polymethyl pentene-1 (TPX), then a semi-crystalline, thermoplastics material is transparent.

transparent thermoplastics materials
Many plastics, and some elastomers, are inherently transparent and it is now possible to obtain plastics which are hard, rigid and brittle (for example, *polystyrene*), soft, flexible and tough (for example, *cellulose acetate*) and, hard, rigid and tough (for example *polycarbonate*). Some of the newer thermoplastics are also more heat resistant materials, for example, a *sulphone*

polymer. In general, rubber toughened thermoplastics materials are not transparent as the rubber particles are relatively large: it is possible to get some which are relatively clear, for example, *acrylonitrile-butadiene-styrene*.

transverse direction
Term usually used in *calendering* and *extrusion* to mark a production direction. In this case the direction at right angles to the *machine direction*. The component will have different properties, for example, strength properties, in different directions. See *orientation*.

transverse flow
The circulatory flow pattern generated between the barrel and the screw: the melt is assumed to circulate in the rectangular channel which is formed by the *screw root*, the *barrel wall* and the *flights* of the screw (leading and trailing).

transverse ridges
An extrusion defect. See *sharkskin*.

transverse strength
See *cross breaking strength*.

trapezoidal runner
A *runner* with a trapezoidal cross-section. The advantage of a trapezoidal-shaped runner is that it need only be cut into one mould half. Provided it is dimensioned correctly, it is almost as good as the *fully round runner*. A very generous radius should be given to he corners in the base of the runner; if possible, the base should be semi-circular. The side walls should taper by about 5° as this will ease ejection. Such runners are often used with *three-plate moulds* as with this type of mould the runner is retained on one plate and then stripped away during part ejection. The cross-sectional area of this type of runner is approximately the area of that circle which will fit inside the runner cross-section.

trapped air process
A *blow moulding process*. A technique used to produce hollow parts by utilizing the air contained, or trapped, in the *parison*.

Traube's process
See *creaming*.

travelling saw
A saw which travels with the moving, cooling extrudate and cuts that extrudate to a predetermined length when required. See *extrusion*.

tread
See *tread rubber and tyre tread*.

tread base
See *sub-tread*.

tread cap
See *toe cap*.

tread cushion
See *undertread*.

tread depth
In the EEC a tyre must have a minimum of 1·6 mm of *tread* across the central three-quarters of the *tyre tread*: if below this the tyre should be changed. More sensible to change the tyre when the tread wears below 1·6 mm at any point. See *tread wear indicator*.

tread groove
A water drainage channel contained in the tread of a *tyre*.

tread pattern
A pattern formed by the *tread grooves*.

tread radius
The radius subtended by the *tread*.

tread rubber
A shaped strip of unvulcanized rubber which subsequently forms the *tyre tread*.

tread sipe
A very narrow tread groove. See *sipe*.

tread wear indicator
A bar of rubber across the *tyre tread*: when the *tread* depth has worn down to this bar then there is 1·6 mm of tread left and the tyre should be changed.

tread width
The width of the *tread* when measured between the points where the *tread radius* abruptly diminishes.

treatment of fabric
See *fabric*.

tree lace
The thin layer of coagulated rubber in the *tapping cut* and which is formed by *auto-coagulation*. See *natural rubber*.

tree stimulation treatments
See *stimulation tapping*.

trepanning
A *blow moulding* trimming operation in which a dome, or head, is removed by the action of an epicyclic, or orbiting, cutting knife: generally used to produce a wide necked jar or canister.

tri
A term used in chemical nomenclature for three.

TRI
An abbreviation used for *track resistance index*.

tri-(2-butoxyethyl) phosphate
Also known as tri-butoxyethyl phosphate. An abbreviation used for this material is TBEP. A *halogenated alkyl phosphate*. A *flame retardant plasticizer*.

tri-(2-chloroethyl) phosphate
An abbreviation used for this material is TCEP. A *halogenated alkyl phosphate*. A *flame retardant plasticizer*.

tri-(2-ethylhexyl) phosphate
See *trioctyl phosphate*.

tri-(2,3-dibromoropropyl) phosphate
An abbreviation used for this material is TDBP or tris, or T23P. A *halogenated alkyl phosphate*. A *flame retardant plasticizer*.

tri-(2,3-dichloropropyl) phosphate
An abbreviation used for this material is TDCP. A *halogenated alkyl phosphate*. A *flame retardant plasticizer*.

tri-(monochloropropyl) phosphate
A *halogenated alkyl phosphate*. A *flame retardant plasticizer*.

tri-2-chloroethyl phosphate
See *tri-(2-chloroethyl) phosphate*.

tri-2-ethylhexyl phosphate
See *trioctyl phosphate*.

tri-2-ethylhexyl trimellitate
See *trioctyl trimellitate*.

tri-butoxyethyl phosphate
See *tri-(2-butoxyethyl) phosphate*.

tri-iso-nonyl trimellitate
An abbreviation used for this material is TINTM. A *flame retardant plasticizer*. See *trimellitate plasticizer*.

tri-iso-octyl trimellitate
Also known as triisooctyl trimellitate. An abbreviation used for this material is TIOTM. A *flame retardant plasticizer*. See *branched-chain trimellitate*.

tri-iso-propylphenol phosphate
Also referred to as an isopropylated phenol phosphate. A type of *phosphate plasticizer*. Such materials were developed to replace *triaryl phosphates*. They are claimed to have similar properties, in many respects, to TTP and TXP. See *phosphate* and *iso-propyl phenols*.

tri-meta-cresyl phosphate
Also known as tri-m-cresyl phosphate. An isomer of *tricresyl phosphate*.

tri-meta-xylyl phosphate
Also known as tri-m-xylyl phosphate. An isomer of *trixylyl phosphate*.

tri-n-butyl citrate
A *citrate plasticizer*.

tri-n-octyl n-decyl trimellitate
An abbreviation used for this material is NODTM. A *flame retardant plasticizer*. See *straight-chain trimellitate*.

tri-ortho-cresyl phosphate
Also known as tri-o-cresyl phosphate. An isomer of *tricresyl phosphate* which is more toxic than the meta and para isomers.

tri-ortho-xylyl phosphate
Also known as tri-o-xylyl phosphate. An isomer of *trixylyl phosphate* which is more toxic than the meta and para isomers.

tri-para-cresyl phosphate
Also known as tri-p-cresyl phosphate. An isomer of *trixylyl phosphate*.

tri-para-xylyl phosphate
Also known as tri-p-xylyl phosphate. An isomer of *tricresyl phosphate*.

triacetate
See *cellulose triacetate*.

triacetate fibres
Fibres made from cellulose acetate in which at least 92% of the hydroxyl groups on the original cellulose have been acetylated. See *acetylation* and *acetate fibres*.

trialkyl phosphate
A phosphate compound which contains three *alkyl* groups, for example, *trioctyl phosphate*. Such materials are used as *flame retardant plasticizers* which offer good low temperature properties. In general, such materials only impart moderate flame retardancy (less than *triaryl phosphates* and *mixed alkyl aryl phosphates*). Of limited compatibility with *polyvinyl chloride*.

triaminotriazine
See *melamine*.

triangular accelerator system
A system which contains three accelerators. See *accelerator*.

triaryl phosphate
A phosphate compound which contains three *aryl* groups. Such materials are used as *flame retardant plasticizers*. In general, such materials impart very good flame retardancy (better than *trialkyl phosphates* and *mixed alkyl aryl phosphates*), give good high frequency welding characteristics and they also resist microbial attack: triaryl phosphates can tolerate high levels of *extender* addition. See *tricresyl phosphate* and *cresol diphenyl phosphate*.

tribasic lead sulphate
Also known as tribase. An abbreviation used for this type of material is TBLS. A *lead stabilizer* which is relatively cheap and is effective as a stabilizer in *polyvinyl chloride* (PVC). Gives good electrical properties in the final compound and is not as toxic as *basic lead carbonate*. Other lead stabilizers include basic lead carbonate, dibasic lead phthalate and dibasic lead phosphite.

triblock copolymer
A polymeric material in which two end-blocks are connected to a central block based on another monomer. See *linear triblock polymer*.

triblock polymer
A polymeric material in which two end-blocks are connected to a central block. See *linear triblock polymer*.

tributyl o-acetyl citrate
An abbreviation used for this material is TBAC. A *citrate plasticizer*.

tributyl phosphate
An abbreviation used for this material is TBP. A liquid plasticizer with a boiling point of about 290°C: it is relatively volatile but imparts low flammability. Compatible with cellulosics, polyvinyl chloride and phenol-formaldehyde. See *trialkyl phosphate*.

trichloroethylene
Also known as trichlorethene or as, acetylene trichloride. An abbreviation used for this material is TCE. This liquid material has a melting point of −73°C, a boiling point of 87°C and a relative density (RD or SG) of 1·47. This chlorinated hydrocarbon is used as a solvent.

trichloromethane
See *chloroform*.

tricresyl phosphate
Also known as tritolyl phosphate. An abbreviation used for this material is TCP, TCF or TTP (tritolyl phosphate). A *triaryl phosphate* which has a relative density (RD or SG) of 1·16. A *flame retardant plasticizer* for cellulosics and for *polyvinyl chloride*. Derived from *cresols* which may be obtained from coal tar or from petroleum. The petroleum-based materials are preferred as they are less toxic: they are mainly based on a mixture of tri-meta-cresyl phosphate and tri-para-cresyl phosphate. The coal-based material can contain the more toxic ortho isomer.

tricrotonylidine tetramine
An abbreviation used is TCT. An aldehyde-amine compound which is used as an *accelerator* for hard rubber compounds.

tricyanic acid
See *cyanuric acid*.

triethyl citrate
A *citrate plasticizer*.

triethyl o-acetyl citrate
An abbreviation used for this material is TEAC. A *citrate plasticizer*.

triethyl trimethylene triamine
Also known as triethyltrimethylenetriamine. An abbreviation used for this material is TTT. This dark coloured liquid material has a relative density (RD or SG) of 1·10 and is used as a curing agent, for example, for *natural rubber* and for *styrene-butadiene rubber*.

triethylene diamine
See *1,4-diazabicyclo-2,2,2-octane*.

triethylene tetramine
Also known as triethylenetetramine. An abbreviation used for this material is TETA or TET. This liquid material has a boiling point of 277°C and a relative density (RD or SG) of 0·98. It is used as a curing agent for *epoxide resins* and in the redox polymerization of *styrene-butadiene rubber*.

triethylenetetramine
See *triethylene tetramine*.

trifluoronitrosomethane
One of the three monomers used to make, for example, a *carboxy-nitroso rubber*.

trifunctional unit
A chemical unit, or grouping, which has three reactive positions: an abbreviation used for this type of unit is T in a *polyorganosiloxane*.

triglyceride
Esters of *glycerol* and a *fatty acid* are called glycerides: when 3 fatty acids are combined with one molecule of glycerol then the ester is called a triglyceride. Such compounds are the major constituent of fats and *oils*. If there is only one fatty acid residue then the product is a simple triglyceride: if there is more than one fatty acid residue then the product is a mixed triglyceride. Most fatty acids have 18 carbon atoms in the chain. See *fixed oil*.

triheptyl trimellitate
An abbreviation used for this type of material is THTM. A *trimellitate plasticizer*.

trihydroxybutyrophenone
An abbreviation used for this material is THBS. An *antioxidant for* polyethylene.

triisooctyl trimellitate
See *tri-iso-octyl trimellitate*.

trilauryltrithiophosphite
A *phosphite antioxidant*. A *preventive antioxidant*.

trillion
A multiple of a million. In the UK it is 10^{18}, in the US it is 10^{12}.

trimellitate plasticizer
A *plasticizer* based on the aromatic anhydride, *trimellitic anhydride*, for example, *trioctyl trimellitate*. Such materials are *primary plasticizers* for *polyvinyl chloride* (PVC) which are used when compounds are required to have resistance to high temperatures and to water extraction. For high temperature applications, the plasticizer should have an *antioxidant* incorporated.

A branched-chain trimellitate plasticizer will give better electrical properties than a straight-chain trimellitate plasticizer. A branched-chain trimellitate plasticizer will give worse low temperature properties than a straight-chain trimellitate plasticizer.

trimellitate plasticizers
The collective name for the class of materials which are based on *trimellitic anhydride* and used as *plasticizers*. See, for example, *tri-iso-octyl trimellitate*.

trimellitic anhydride
An abbreviation used for this material is TMA. An aromatic anhydride used to make *trimellitate plasticizers*. When reacted with an alcohol, three acid groups are available to give products such as, for example, *trioctyl trimellitate*.

trimer
An *oligomer* with a degree of polymerization of three.

trimethyl ammonium acetates
Salts of organic bases. Used, for example, with inhibitors for *unsaturated polyester resins* so as to extend storage life. See *inhibition*.

trimethyl ammonium bromides
Salts of organic bases. Used, for example, with inhibitors for *unsaturated polyester resins* so as to extend storage life. See *inhibition*.

trimethyl ammonium chlorides
Salts of organic bases. Used, for example, with inhibitors for *unsaturated polyester resins* so as to extend storage life. See *inhibition*.

trimethyl chlorosilane
A *chlorosilane* which has three methyl groups attached to each silicone atom. See *methyl chlorosilane* and redistribution.

trimethylchlorosilane
See *trimethyl chlorosilane*.

trimming
The removal of un-wanted material, for example, *flash*. See *calendered sheet trimming*.

trioctyl mellitate
See *trioctyl trimellitate*.

trioctyl phosphate
Also known as tri-2-ethyl phosphate. An abbreviation used for this material is TOF or TOP. A *trialkyl phosphate*.

trioctyl trimellitate
Also known as trioctyl mellitate or as, tri-(2-ethylhexyl) mellitate or as, tri-2-ethylhexyl trimellitate. An abbreviation used for this material is TOTM or TOM. A *trimellitate plasticizer*.

triode valve
A thermionic valve used, for example, as the main amplifier oscillator for *high frequency heating*.

trioxane
A polymer of *formaldehyde*: the cyclic trimer of formaldehyde which is used in the preparation of *acetal* copolymers and as a source of formaldehyde for *curing* purposes, for example, for *phenol-formaldehyde* materials.

tripalmitin
A glyceride of *palmitic acid*. A wax-like material with a melting point of 66°C, which occurs as the triglycerides of *plant oils*, for example, in *cottonseed oil* and in *palm oil*. May be represented as $(C_{15}H_{31}COO)_3C_3H_5$.

triphenyl guanidine
Also known as triphenylguanidine. An abbreviation used for this material is TPG. This solid material has a relative density (RD or SG) of 1·10 and a melting point of approximately 145°C. An early vulcanization *accelerator* which because of its slow cure properties is now seldom used.

triphenyl phosphate
An abbreviation used for this material is TPP. This material has a relative density (RD or SG) of 1·19 and a melting point of 49°C. A flame retardant *plasticizer*, for example, for cellulosics which is used with another plasticizer so as to prevent the crystallization of the TPP. Also used in some *synthetic rubbers*, for example, with *styrene-butadiene rubber (SBR)*.

triphenylguanidine
See *triphenyl guanidine*.

tripoli
A naturally occurring form of *silica* which is used as a *filler*: an amorphous silica. Rottenstone is a type of tripoli.

tripoli stone
See *tripoli*.

tripolitic ore
The source of *tripoli*: a source of amorphous *silica*.

tripolymer
See *terpolymer*.

tris
An abbreviation used for *tri-(2,3-dibromoropropyl) phosphate*.

tris(nonylphenyl)phosphite
See *tris-(nonylphenyl) phosphite*.

tris-(1-aziridinyl) phosphine oxide
An abbreviation used for this type of material is APO. This material is used to impart flame retardant properties to *cellulose fibres*. APO treatment can however result in fabric yellowing.

tris-(nonylphenyl) phosphite
A *phosphite antioxidant*. A *preventive antioxidant*.

tris-(p-nonylphenyl) phosphite
An abbreviation used for this type of material is TNPP. An *ultraviolet light absorber*.

trisilane
A silicon-based material which consists of three *silane units* joined together. The formula of such a material may be represented as $H_3Si-SiH_2-SiH_3$. See *disilane*.

trisiloxane
A *siloxane* which has the structure $H_3Si-O-SiH_2-O-SiH_3$: if some of the hydrogen atoms are replaced by organic groups then the compound should be referred to as an *organosiloxane*, for example, as an organotrisiloxane.

trisnonyl phenyl phosphite
An antioxidant used, for example, as part of a stabilizing system. See *metal soap stabilizer*.

tristearin
Also known as *stearin*. A glyceride of *stearic acid*. A fatty substance with a melting point of 54°C.

tritolyl guanidine
Also known as tritolylguanidine. An abbreviation used for this material is TTG. This solid material is a vulcanization *accelerator* which gives slightly faster cure rates than *triphenyl guanidine*.

tritolyl phosphate
An abbreviation used for this material is TTP. See *tricresyl phosphate*.

tritolylguanidine
See *tritolyl guanidine*.

trixylyl phosphate
An abbreviation used for this material is TXP. This material has a relative density (RD or SG) of 1·14 and a boiling point of 420°C. A *flame retardant plasticizer* for *cellulosics* and for *polyvinyl chloride*. Derived from xylenols which are derived from coal tar or from petroleum. The petroleum-based materials are preferred as they are less toxic: they are based mainly on a mixture of the meta and para isomers. The coal-based material can contain more of the toxic ortho isomer.

Trogamid T
See *polytrimethylhexamethyleneterephthalamide*.

troposphere
The gaseous mantle, the *atmosphere*, which surrounds the planet Earth has been divided into several layers or strata. The troposphere is the one adjacent to the surface of the Earth and below the *stratosphere*: separating it from the stratosphere is a shallow layer known as the tropopause. The thickness of the troposphere varies from approximately 10 kilometres/6 miles to 20 kilometres/12 miles. It is the layer within which all cloud formations occur: there is a steady fall in temperature with an increase in altitude within this layer of approximately 2°C for every 305 m/1,000 feet.

trouser tear test
Tear strength measured using a test sample shaped like a pair of trousers. See *tear resistance*.

Troy weight
A system of weights for gems and precious metals. In the UK system, 24 grains = 1 pennyweight, 20 pennyweights = 1 ounce, 12 ounces = 1 pound (Troy), 25 pounds = 1 quarter, 4 quarters = 1 hundredweight (100 pounds Troy) and 20 hundredweight made one ton (precious metal). One pound Troy contains 5,760 grains. The grain, ounce and pound are the same as in Apothecaries' weight. The grain is the only unit that has the same value in Apothecaries' weight, in avoirdupois weight and in Troy weight. In Troy weight, 3·17 grains make 1 carat (0·205 g).

TSPP
An abbreviation used for tetra-sodium pyrophosphate.

TSR
An abbreviation used for *technically specified (natural) rubber*.

TTG
An abbreviation used for *tritolyl guanidine*.

TTR
An abbreviation used for *technically specified (natural) rubber* from Thailand (Thai tested rubber).

TTT
An abbreviation used for *triethyl trimethylene triamine*.

TTT curves
Curves used to show the hardening behaviour of *steel*. One T stands for temperature, another for transformation and the third for time. See *case hardening steel*.

TU
An abbreviation used for *thiourea*.

tube - thermoplastics
See *tubing* and *thermoplastics tube*.

tuber
An old name for an *extruder*.

tubing
Small bore pipe, for example, with an internal diameter of approximately 12 mm/0·5" or less; the term may also be applied however, to larger bore pipe if such pipe is flexible and has thin walls.

tubing die
A *die* used to produce tube: has an outer bushing and an inner pin or *torpedo*; the plastics melt flows in between the two to form the product.

tubular film process
Also called the layflat film process. The *melt* is extruded through an annular die to give a tube of controlled diameter and wall thickness. Upward vertical extrusion is generally used, the die being mounted in a right-angle crosshead. The extruded melt is partially cooled in the vicinity of the die, usually by passing the tube through a carefully controlled stream of air from a cooling ring, and the tube of film is inflated to a bubble of the required diameter by low-pressure air introduced through the torpedo. The film is hauled off between a pair of nip rolls, located about 3 m from the die, so that the inflation air is contained within the bubble formed between the nip rolls and the die. It is important to maintain a steady bubble which feeds to a constant position in the nip rolls. After passage through the nip rolls, the cold layflat film is reeled up under constant tension.

The diameter of the bubble and hence the width of the flattened tube (the lay-flat width) is determined by the amount of inflation air introduced. The thickness of the film depends on the output from the extruder, the bubble blow ratio (diameter of bubble: diameter of die) and the film haul-off rate. Thus, by variation of the blow ratio and the haul-off rate, a range of lay-flat widths and thickness may be produced from the same die.

If sufficient air-cooling time is not allowed before the tube of film is closed at the nip rolls, *blocking* will result: this is one of the most serious limitations to high-speed production. Another is achieving high speed, uniform film cooling. Both the optical and the mechanical properties of the films produced from a given polymer can be varied by varying the extrusion conditions.

tubular quench
An abbreviation used for this term is TQ. A *tubular film process* in which the bubble is quenched with water so as to improve the clarity of films made from a semi-crystalline, thermoplastics materials such as *polypropylene* (PP).

tubule
A fine tube or stream of liquid. See *micromixing*.

tumble finishing
A finishing process for mouldings, usually for *thermoset mouldings*. The mouldings are tumbled (rotated) in a barrel which contains pegs and/or particles, for example, wooden pegs which remove the *flash*.

tumble mixer
A mixer used to pre-blend a mixture (often a thermoplastics formulation) before it is fed to a melt processing machine, for example, an *injection moulding machine*. See *rotating blender*.

tumble mixing
The procedure used to produce a blend of a *thermoplastics* formulation: the ingredients are tumbled together in, for example, a *rotating blender*.

tumble polishing
A finishing process for mouldings, usually for thermoset mouldings. The mouldings are tumbled (rotated) in a barrel which contains slightly-abrasive particles which remove fine *flash* or alter the surface finish.

tumbling
See *barrelling*.

tung oil
Derived from *Aleurites fordii* or from, *Aleurites montana*. A *duel purpose vegetable oil* which is also known as Chinese wood oil or as, wood oil. Obtained from the kernel of the nuts of the tung tree and has very good drying properties as it contains conjugated double bonds. Used in place of *linseed oil* when heat treated (kettle bodied). Surface coatings, based on heat treated tung oil, can be clear dry films which are more resistant to water and to alkalis than linseed oil. Has the best drying properties and gives the most durable film of any of the common paint oils.

tungsten carbide
A very hard material which is sometimes used in areas where there is the likelihood of very high wear - such as the *gate area* in a mould used for heavily filled polymers. This material has also been used to re-build worn screws and barrels because of wear caused by, for example, the use of *linear low density polyethylene*. For small screws (<60 mm) both the flights and the screw root may be encapsulated: for larger screws only the *leading edge* of the flights and the tips may be coated (because of cost considerations). Worn barrels have also been coated with tungsten carbide. See *bimetallic barrel*.

tunnel gate
See *submarine gate*.

tup
The striker, or hammer, used in a Charpy impact test. See *pendulum impact test*.

turbo trepanning
A *blow moulding* trimming operation in which a dome, or head, is removed by the action of an epicyclic, or orbiting, cutting knife which is driven by an air motor.

turbulent flow
Flow which is no longer *laminar flow* as the critical Reynolds number has been exceeded.

turn up
Free end of a *ply*, chafer or filler strip which is folded around the *bead core*.

turpentine oil
Also known as turpentine or as, pine-wood oil or as, pine oil. A pale coloured liquid obtained from distillates of pine trees. This material has a relative density (RD or SG) of approximately 0.93 and a boiling point of approximately 170 to 210°C. Used as a solvent thinner in paints and varnishes. It is a good solvent for uncured *chloroprene rubber (CR)*, *butyl rubber (IIR)*, *natural rubber (NR)*, and *styrene-butadiene rubber (SBR)*. It is a poor solvent for uncured high acrylonitrile NBR. This chemical causes some swelling of uncured low acrylonitrile NBR and *thiokol rubbers*.

turret winder
A *reeling machine* with more than one winding position. As one core or reel becomes full then another can be swung into position.

TVI
An abbreviation used for *Tomasetti's Volatile Indicator (method)*.

Tw
An abbreviation used for degrees Twaddell. See *Twaddell scale*.

Twad
An abbreviation used for *Twaddell*. See *Twaddell scale*.

Twaddell scale
A scale for measuring the *relative density (rd)* of liquids. Degree Twaddell = 200(rd -1). rd = 145/145 - B and rd = 0.5 Tw + 100/100. Where B = degrees Baumé and Tw = degrees Twaddell.

twin screw extruder
An extruder which contains two *screws* which transport, melt, mix and pump the *melt* through the *die*. The most widely used type of *multi-screw extruder*. The screws may intermesh or they may not intermesh (non-intermeshing): the screws may be parallel or tapered. Usually the two screws are of the same length but this need not be so. The two screws may rotate in the same direction (corotation) or they may rotate in opposite directions (counter or contra-rotation). See *tapered twin-screw extruder*.

With a single screw extruder the hopper is filled and the screw takes what material it wants: such a scheme of flood feeding is often not possible with twin screw machines. This is because of the very positive feeding characteristics of twin screw machines and this in turn means that very high forces may be generated: so high, that thrust bearing failure is a real danger. Such machines are therefore often starve fed and thus the through-put is independent of screw speed. Residence time decreases with increasing output and thus the shear, and the energy input, decreases with increasing through-put. The output is also often virtually independent of the size of die fitted. Because of the mixing action obtained, twin screw machines can achieve more melting, mixing and conveying in a shorter machine length than a single screw machine; they are however, much more expensive.

Twin screw machines have always been popular for certain types of job, for example, where there is a need for a compounding step as well as for an extrusion step: this is particularly true for *unplasticized polyvinyl chloride* (UPVC). This material is often stabilized against heat degradation by the use of heavy metal compounds (e.g lead) and such stabilizers are expensive. For economic reasons the amount of these *heat stabilizers* must therefore be kept as low as possible. One way of doing this is to compound and extrude in one step as this saves a further heating stage (compared to the process whereby the material is first compounded, cooled and then re-extruded).

Twin screw machines are widely used to make PVC pipe and profiles; now they are often used to compound other plastics, or resins, with additives so as to make compounds for use in other extrusion or injection moulding operations.

twin screw extruder - contra-rotation
The two screws rotate in opposite directions (counter or contra-rotation). Both corotating and counter-rotating machines have distinct advantages which have resulted in their use in specific applications. Counter-rotating machines are widely used for the extrusion of *unplasticized polyvinyl chloride* (UPVC) because the counter-rotating machine has very positive material feed and conveying characteristics; the residence time, and temperature, of the material in the machine is also even or uniform. However, air entrapment, high pressure generation, low maximum screw speed and output are usual disadvantages.

twin screw extruder - corotation
The two screws rotate in the same direction (corotation). Both corotating and counter-rotating machines have distinct advantages which have resulted in their use in specific applications. The advantages of the corotating machine are that the screws wipe each other clean (self-wiping) and that high screw speeds and outputs are possible particularly for materials which are not very shear, or heat, sensitive (for example, *polyethylene*). Lower screw and barrel wear is also found. The output is however, dependent on die-head pressure and at high die-head pressures the residence time distribution, of the material, becomes wider (the clearances between the flanks of the screws is usually greater for this type of machine). At the much higher shear rates possible in this type of machine, this non-uniform residence time can result in decomposition for heat sensitive materials but the greater inter-channel flow results in better mixing or compounding.

twin screw extruder - counter-rotation
See *twin screw extruder - contra-rotation*.

twin screw extrusion machine
See *twin screw extruder*.

twin screw machine
Usually means a *twin screw extruder*.

twin sheet forming
A *thermoforming* process which uses two mould halves and which offers a way of producing hollow components cheaply: such a process is an alternative to *blow moulding*. Two sheets are clamped together and heated. At *forming temperatures*, the top and bottom mould halves close and clamp and weld the edge of the sheets together. Simultaneously the forming is produced by a combination of vacuum and air pressure. After cooling the moulds part and the forming is removed for secondary operations.

twin station blow moulder
A machine with two identical sets of blow moulding platens: each set of platens is presented in turn to the extrusion head to collect parisons so as to give double the output of a *single station machine*.

two axis winding
See *filament winding*.

two bowl calender
See *two roll calender*.

two nip calender
A three roll *calender*.

two roll calender
Also known as a two bowl calender. A *calender* with two main rolls. That is, with one *nip*. The rolls may be arranged horizontally or vertically. Used for rough sheeting jobs or, when two such calenders are used together, for the production of thermoplastics flooring. That is, one calender is used to feed another. Such a layout avoids nip interactions but is impractical for thinner sheets because of film transport and heating problems.

two-roll mill
A machine used to heat/cool, shape and mix polymers and their compounds. A two-roll mill consists of two horizontal, steel rolls which are of the same diameter and length and which lie parallel to each other at a convenient working height horizontal to the ground. The rolls rotate in opposite directions and towards each other. Because of the way the rolls rotate, added materials are carried towards the *nip* (see *safety*). The two rolls are identified as the *front roll* and the *rear roll*. Both rolls are usually capable of being heated and, on some machines, the rolls may also be cooled when required. The rolls may be rotated at different speeds by means of an electric motor and gear box assembly - see *friction ratio*. The effective width of the rolls can be reduced by the setting of the *cheek plates*.

CLASSIFICATION. It is the working length of the rolls which are primarily used to classify a *two-roll mill*. A usual laboratory machine may have rolls which are 300 mm/12 in in length. The *friction ratio* may also be used to describe the machine as well as the heating system.

COMPOUND PREPARATION. A two-roll mill is a *melt* mixing device as the material is mixed above the softening point of the polymer. To achieve both longitudinal and lateral mixing the material must first be banded onto the roll and the nip adjusted to give a *rolling bank* of material above the *nip*. The *band* so formed should be part-cut, first from the edge of one side and then from the other. The cut material should be folded so that the edge of the cut band is moved towards the centre of the roll and material is exchanged with the rolling bank. If required the bank is cut completely across and the material rolled (a so-called swiss roll); the roll is then passed (end-on) immediately through the nip. This 'swiss-rolling' operation may be repeated several times. Such cutting and folding actions will give *distributive mixing*: dispersion mixing is achieved mainly by the shearing action of the nip. Once established the mixing procedure should be standardized and recorded.

HEATING OF. The rolls of a *two-roll mill* may need to be heated to different temperatures. As the rolls may also need to be cooled, for example, during the *mastication* of rubber, such rolls are commonly heated by steam (or steam/water mixtures) and cooled by water circulation. The steam, or water, flows through a central channel, or bore, in the roll and is then collected for recirculation or disposal. For rubber mixing, low pressure *steam* is employed (as the temperatures do not need to exceed 100°C/212°F) whereas for plastics materials, steam at higher pressures is required. Usually the maximum temperature that can be reached, even with high pressure steam, is approximately 160°C/320°F. For temperatures higher than this electrical heating with *cartridge heaters*, is used

INTERNAL MIXER USAGE. If *natural rubber* is being mixed in an *internal mixer*, then the first stage in the mixing cycle is often *mastication*: when the viscosity of the polymer has dropped to the desired level, the compounding ingredients are added to the mixing chamber. (Not all grades of rubber require mastication, for example, most *synthetic rubbers* do not and neither do *constant viscosity grades* of NR). Mixing and dispersion occur very rapidly as the mix is held in the chamber by an air-operated ram. When the temperature reaches a preset value, mixing is judged to be complete and the batch is dumped onto a first *two-roll mill* where its temperature is reduced rapidly: it is passed through this mill several times and then transferred to a second mill where the curing system is added. At this stage the compound is banded around the front roll of the mill and the operator achieves the desired degree of mixing by cutting and folding the banded material into the nip formed by the two rolls. If required a strip may be produced for *rubber injection moulding* or the hide may be stripped from the roll in one piece.

PVC COMPOUND PREPARATION. If a *polyvinyl chloride* (PVC) *homopolymer* is the base polymer, then the roll temperatures should be first adjusted to approximately 150°C front and 130°C rear. (A *friction ratio* of approximately 1·1:1 is commonly employed). The mix ingredients are commonly roughly blended together and the resultant mix is then slowly fed to the closed *nip* of the two-roll mill. As the material forms a band or hide on the hotter roll, the nip is progressively opened until all the mix has been added: the nip is then adjusted until a small, rolling sausage is formed above the nip. The mixing process takes approximately 10 minutes (see compound preparation). If a *polyvinyl chloride* (PVC) copolymer is the base polymer, then the roll temperatures should be first adjusted to lower values, for example, approximately 100°C front and 80°C rear.

RUBBER COMPOUND PREPARATION. If *natural rubber* (NR) is the base polymer, then the material may first have to be *masticated*. The masticated rubber is banded onto one roll and the mix ingredients are added in a definite order. For example, the first materials added are those which are difficult to mix and which are used in small quantities - for example, additives such as protective agents and *accelerators*. Half of the *filler* is then added followed by *stearic acid*. If softeners are now added the band will probably split and the band will therefore have to be worked before the additional filler can be added. If no *ultra-fast accelerators* are present, then the sulphur may be blended/mixed into the compound: if ultra-fast accelerators are present, then the sulphur is blended/mixed into the compound on a warm-up mill. The mixing procedure takes approximately 10 to 15 minutes (see compound preparation). On an industrial scale, an *internal mixer* is preferred for mixing as large scale mill mixing is exhausting and time consuming.

SAFETY OF. Because of the way the rolls rotate, added materials are carried towards the *nip* of a two-roll mill. This is why these machines can be incredibly dangerous as the operator's hand, or loose clothing, may also be carried towards the nip. Once the object is in the nip then it will be crushed unless a sophisticated, fast-acting, nip opening device has been fitted. These are not usually seen industrially and so great care must be taken when these machines are used. Before the machine is used the safety hazards should be explained to the operator and appropriate safety devices fitted. Such devices may include *fixed safety guards, movable interlocked safety guards, safety bars, safety stops* and *plugged braking*. In some cases it may be possible to restrict access to the hazardous area by making that area normally inaccessible by, for example, elevating the machine. Training and a responsible attitude towards safe working are very important as undue reliance should not be put upon guards.

THERMOPLASTICS COMPOUND PREPARATION. Broadly speaking, the procedure is similar to that used for *plasticized polyvinyl chloride* (see PVC compound preparation). However, if the compound temperatures need to be higher than 160°C then an electrically heated mill will probably have to be used. Because of the temperatures employed, the mill will become difficult and dangerous to use and so machine mounted, *mixing knives* are best employed. These are mounted above each roll and when an extended lever is pulled the *hide* is cut: releasing the lever allows the material to be carried towards the nip.

THERMOSET COMPOUND PREPARATION. Broadly speaking, the procedure is similar to that used for *plasticized polyvinyl chloride* (see *PVC compound preparation*) although mixing is more difficult to do well with, for example, a *phenol-formaldehyde* material. The addition of a small amount of water can help to band the material. The compound temperatures need to be lower than those used for PVC. Before mixing is commenced, the roll temperatures should be adjusted to comparatively low values, for example, approximately 100°C front and 80°C rear. *Mixing knives* or blades can be a great help as thermoset compounds can be very difficult to work because they are stiff and hard even at processing temperatures.

USAGE. A two-roll mill is usually associated with the older established branches of the polymer industry: that is, with rubbers and with thermosetting plastics. It is used to prepare compounds from mixes based on polymers, in the form of a sheet or *hide*. The same type of machine may be used to feed another machine by cutting strip from the *band* on one of the rolls: in such cases, the mill may be used to cool the compound before feeding the next machine. For example, an *internal mixer* may dump onto a two-roll mill which is used to lower compound temperatures and to free the internal mixer for further mixing: that is, the mill is used as a storage/temperature conditioning device. Two-roll mills are also used in laboratories to make thermoplastics compounds because of the relative ease of cleaning of such machines and because of the small amount of material which can be used.

two stage machine
Usually means a two stage *injection moulding machine*. An injection moulding machine which has a clamping system which is separately powered from the injection unit: that is, the clamping system applies the clamping force before the injection unit forces melt into the mould. As most machines are now of this type (see *single-stage machine*) the term has come to be applied to machines which heat the material in one barrel from where it is transferred to a second 'shooting' barrel. See *two stage, rubber injection moulding machine.*

two stage, rubber injection moulding machine
The rubber injection moulding industry makes extensive use of the so-called two-stage injection moulding machine. These are machines in which the rubber is softened by means of a screw in one barrel and then transferred to another barrel which contains a ram or plunger; at this point, the rubber has a temperature of approximately 110°C. The rubber is then forced into the closed heated mould by the ram: the mould is held at, for example, 185°C. At these high temperatures, the setting, or vulcanization, is extremely rapid and once it has been taken to the desired degree, the rubber moulding is ejected from the machine. Such *injection moulding machines* are often vertical machines and as such take up relatively little floor space: they can therefore directly replace a compression moulding machine. See *rubber injection moulding*.

two start screw
A *screw* which has two separate flights separated by the flight land. See *barrier screw*.

two term control
Usually associated with temperature controllers and means that the instrument has circuits, or terms, for proportional and integral or, proportional and derivative control. That is, it is either a PI controller or, a PD controller.

two-pack systems - room temperature vulcanizing silicone rubbers
An abbreviation used for this type of material is RTV-2. There are two main categories which are *condensation cross-linked silicone polymers* and *addition cross-linked silicone polymers*.

two-phase material
A heterogeneous material based on, for example, an amorphous phase and a crystalline phase. For a two-phase material to be transparent the refractive indices of the two phases must be the same. The transparency of a two-phase material may change appreciably with a change of temperature as the refractive indices of the two phases is usually only matched over a limited temperature range.

two-plate injection mould
An *injection mould* which separates into two main pieces or halves on either side of the mould split line. The parting line is normal to the injection direction as this simplifies mould construction and use. At the parting line the moulds butt together and are kept together by the application of the clamping force. The two mould halves are closed and clamped before the material is injected into the cavity.

Of the many types of injection mould possible, such a simple two-plate mould is often used because of its relative simplicity and low cost. However for many moulding jobs, mould types other than the simple two-plate mould are specified because the use of such mould types may simplify, for example, component feeding and component finishing.

two-plate mould
A mould which separates into two main pieces or halves. See *two-plate injection mould*.

two-stage machine
See *two stage machine*.

two-stage polymer
See *two-stage resin*.

two-stage process
See *prepolymer process*.

two-stage resin
Also known as a two-stage polymer. A *prepolymer* which can only be cross-linked by the addition of a cross-linking agent, for example, a novolak *phenol-formaldehyde*.

two-stage stretching
Stretching performed in two steps or stages: reheating may be employed between the two stages. See *stretching*.

two-way valve
A term used in *hydraulics* and which refers to a valve which has two flow paths.

TXIB
An abbreviation used for 3,3,5-trimethylpentane-1,4-diol di-isobutyrate and for *2,2,4-trimethylpentane-1,3-diol di-isobutyrate*.

type A
The indentor used in a *Shore hardness* test for soft materials.

type D
The indentor used in a *Shore hardness* test for hard materials.

type I carbon fibre
See *high strength carbon fibre*.

type I PE
When the density of *polyethylene* (PE) is low, 0·910 to 0·925 g/cm^3, a PE material is sometimes known as Type 1.

type II carbon fibre
See *high modulus carbon fibre*.

type II PE
When the density of *polyethylene* (PE) is medium at 0·926 to 0·94 g/cm^3, a PE material is sometimes referred to as Type II.

type III PE
When the density of *polyethylene* (PE) is high with a density of 0·940 to 0·959 g/cm^3, a PE material is sometimes known as Type III.

type IV PE
If the density of a *polyethylene* (PE) is higher than 0·959 g/cm^3 that is, very high, then the material is sometimes known as type IV.

type metal
Low melting point alloys of tin, lead and antimony (based on say, 10% tin, 75% lead and 15% antimony).

tyre
The running surface of a wheel. Can have solid tyres (used for heavy, slow moving vehicles) and *pneumatic tyres* (as used for most road-using vehicles). Solid tyres are often based on cast *polyurethane* whereas pneumatic tyres are based on traditional vulcanized rubbers reinforced with fibres and/or fabric (metal or polymeric). Pneumatic tyres may be classified according to their method of construction, for example, as *cross-ply*, *radial*, *bias belted* and *concave tread*.

The earliest tyres were based on canvas reinforcement but the tyre life was short because as the tyre flexed in use, the cross-woven yarns rubbed over each over and breaks occurred. This type of construction was replaced by one in which the textile strands or cords ran in one direction only. This textile was covered with a thin layer of rubber and successive plies were super-imposed so that the cords in one layer crossed those in another layer at an angle: this construction gives a very strong casing and avoids tyre cord chafing. Such a tyre is known as a *cross-ply tyre* and although still made, this type of tyre is not as important as the *radial-ply* type.

A tyre must be flexible and yet strong enough to resist, and then cushion against, impact damage. The tyre must respond accurately to steering instructions and not be deflected by, for example, ridges in the road. The tyre must provide good grip for traction, accelerating, cornering and braking, in all weathers and on a wide range of surfaces (both wet and dry). A tyre must not overheat, must run quietly, give the standard of ride required and, ideally, last a long time at an attractive price.

Although cross-ply tyres are easier to make than radials, and are cheaper initially, the cost per tyre mile is higher. General motoring, tyre requirements are difficult to achieve in a cross-ply tyre because of the method of tyre construction: the cross-ply reinforcement runs through the sidewalls and the tread. Soft springing for comfort should be provided by means of flexible sidewalls whereas a stiff tread structure is needed for steering and for stability. These two requirements are dealt with more separately in a radial-ply tyre. A radial-ply tyre has very flexible sidewalls and an in-extensible bracing layer under the tread. This means that the *cornering force* is provided at a lower *slip angle* than that needed for the cross-ply: this results in less scrubbing of the tread when cornering and improved tyre life.

Waste tyres have been recycled by, for example and cryogenic fragmentation.

tyre casing
The rubber/textile casing forms the main body or carcass of the tyre and onto this other components such as the tread and sidewalls are built. Fabric for tyres is rather unusual in its construction and is sometimes referred to as *all-warp fabric*.

tyre mould
A steel or aluminium mould used to produce a tyre and which, in the simplest case, is of two-piece construction with the split being circumferential. The mould is capable of being heated by, for example, the circulation of steam/hot water and shaping of the *green tyre* is by means of a *curing bag* or bladder in a *tyre press*. To allow moulding using a green tyre which is close to the final tyre size (so as to avoid disturbing the tyre components) the mould may be made of segments: on closing the tyre press, the segments move inwards.

tyre numbering
The legend on the side of a *tyre*. For example, 175/70R1382T. The figure 175 is the width of the tyre in mm: the figure 70 is the *aspect ratio* (the profile of the tyre): R shows that the tyre is a *radial tyre* (C would show a *cross-ply*): 13 is the diameter of the wheel rim in inches: 82 is the load index which specifies the maximum load the tyre can carry at maximum speed: and, T is a speed symbol which indicates maximum speed (for example, T = 190 kph/118 mph: a common speed rating; H = 130 mph and ZR = > 149 mph etc). A *tread wear* indicator may also be present and an *E mark*.

tyre press
A press used to mould a *tyre*. A shaped preform (a *green tyre*) is placed inside the tyre mould (for example, a watch-case mould), the mould is closed by the tyre press and pressure applied to the preform by means of a *curing bag* or bladder inflated with steam/hot water: this curing bag or bladder may be part of the press. The pressure so applied (<30 kgf/cm^2 or 425 psi) shapes the tyre and moulds the tread: the press remains closed until the tyre is vulcanized (for example, 30 m at 150°C). The curing bag or bladder is then deflated so that the tyre can be removed.

tyre reclamation
Reclaimed rubber may be obtained from tyres by various *reclaiming processes*. Waste tyres have been recycled by, for example, *cryogenic fragmentation*. Some tyres are used again as, for example, *retreads* whereas others are burnt as fuel or used to create artificial reefs in the sea. It is the large number of used tyres, their size, construction and toughness which hampers re-use.

tyre reinforcement
The reinforcement which gives a *tyre* its strength of shape retention: the reinforcement is usually fabric but may be metallic (see *steel cord*). A great deal of the fabric reinforcement is synthetic, for example, *nylon* or *rayon*. The coating of tyre fabrics is usually performed by *calendering* and in order to get the best properties from such composites, the fabrics are usually sized or treated before *calendering* with a *bonding agent*.

Coated fabrics are used for example to make the carcass plies (which give the tyre its strength) and to make the tread bracing components. Such components are used in *radial* or belted bias tyres to increase the modulus in the tread area and thus reduce tread pattern movement and distortion. Fabric for tyres is rather unusual in its construction and is sometimes referred to as *all-warp fabric*.

tyre rubber
See *tire rubber*.

tyre tread
That part of a *tyre* in contact with the road: that part of a tyre which provides the good grip necessary for driving, braking and steering. The tread is extruded as a profile from a compound which has excellent abrasion resistance and is laid on the tyre with the sidewalls before moulding.

The complex tread patterns used are designed to give the best road grip under varying weather conditions. A smooth tyre would give the best grip on a dry road, as the *contact patch* would be large, but would be virtually useless on a wet road as there would be limited road contact. Water is removed to the rear of the tyre by means of channels and zig-zag grooves: *sipes* remove most of the rest of the water. At 60 mph a tyre may have to remove a gallon of water every second so as to ensure good grip. If the tyre has insufficient tread depth it cannot remove sufficient water and *aquaplaning* may result.

tyre tread reclaim
Reclaim obtained from *tyre treads*. See *reclaiming*.

tyre-building machine
A machine used to produce the shaped preform (a *green tyre*) which is placed inside the tyre mould (for example, a watch-case mould) on a *tyre press*. The plies of rubber coated fabric are laid on the tyre-building drum: the *beads* are attached to the assembly and stitched into place. *Chafers* are attached to the edges of the last ply followed by a central cushion layer: the *tread* and sidewall rubber extrudate is then placed in position (both may be extruded into one continuous strip). The drum is then collapsed to give a green tyre which looks like an open-ended barrel. The *breaker* strips may then be applied and the tyre shaped, into a doughnut-form, on an expanding machine before being moulded.

tyre-cord line
A production system for tyre-cord and which uses a *calender*. Synthetic tyre fabric is coated with a *bonding agent* and dried under a controlled tension. Such fabric is coated by *calendering*, that is, a thin coating of rubber is applied on both sides by a *four bowl calender*. This operation forces the hot rubber sheet between and around each cord of the fabric and the final result is an arrangement of cords which run parallel to each other (see *all-warp fabric*). Each cord is separated from its neighbours by the rubber coating and the total assembly is wound into rolls, after cooling, together with a fabric *liner*.

U

u
An abbreviation used for speed of flow. See *Reynold's number*.

U
Used to modify polymer abbreviations, for example, when used in this way stands for ultra or, unplasticized or unplasticized. U is also used in *plasticizer* abbreviations for undecyl. For example, diundecyl phthalate (DUP).

UV stabilizer
See *ultraviolet stabilizer*.

UAL
An abbreviation used for *upper action limit*.

u_c
An abbreviation used for critical speed. See *Reynold's number*.

UD
An abbreviation used for *unidirectional*. For example, UD PEEK is a unidirectionally reinforced laminate based on *polyether ether* ketone. Made from carbon fibres and polyether ether ketone. See *unidirectional thermoplastics prepreg*.

UDP
An abbreviation used for *undecyl dodecyl phthalate*.

UF
An abbreviation used for urea-formaldehyde. See *urea-formaldehyde moulding material*.

UHMW
An abbreviation used for ultra-high molecular weight. For example, UHMW HDPE is *ultra-high molecular weight, high density polyethylene*. UHMWPE is ultra high molecular weight polyethylene.

UK system of units
The United Kingdom (of Great Britain) system of units. May also be known as the Imperial (imperial) system of units.

AREA. Based on square yards. 144 square inches make one square foot and nine square feet make one square yard. 30·25 square yards make 1 square rod, pole or perch. 40 square poles make 1 rood and 4 roods equal 1 acre. 4,840 square yards make one acre and 640 acres make one square mile.

AVOIRDUPOIS. 16 drams = 1 ounce, 16 ounces = 1 pound, 14 pounds = 1 stone, 28 pounds = 1 quarter, 4 quarters = 1 hundredweight and 20 hundredweight = 1 ton. 2,240 pounds make one ton. One pound avoirdupois contains 7,000 grains.

DRY MEASURE. Based on the gallon which contains 277·420 cubic inches. 2 gallons = 1 peck, 4 pecks = 1 bushel, 8 bushels = 1 quarter and 5 quarters = 1 load. May also use pints and quarts. 2 pints make one quart. 4 quarts make one gallon and 8 quarts make one peck. One bushel equals 8 gallons and 8 bushels equal 1 quarter. 36 bushels equals 1 chaldron.

LENGTH. Based on the yard. 12 lines make one inch, 12 inches make one foot and three feet equal one yard. 5·5 yards make one rod (pole or perch). Forty rods make one furlong (220 yards) and eight furlongs make one mile. One mile (land mile) equals 1,760 yards.

LENGTH (NAUTICAL). A nautical mile (UK) equals 6080 feet and contains 10 cables: one fathom equals 6 feet.

LENGTH (SURVEYORS). Thought to be based on the chain: this equals 22 yards or 66 feet. The chain was made up of 100 wire rods called links: as there are 100 links in one chain each link equals 7·92 inches. Ten chains made one furlong and eight furlongs made one mile. Ten square chains made one acre.

VOLUME. A cubic yard contains 27 cubic feet and a cubic foot contains 1,728 cubic inches. 277·420 cubic inches make an Imperial gallon which contains 10 pounds of water. 4 gills equal 1 pint (one pint contains 20 fluid ounces), 2 pints equal 1 quart, 4 quarts equal 1 gallon. Then, for example, 9 gallons equal 1 firkin, 36 gallons equal 1 barrel, 42 gallons equal 1 tierce, 54 gallons equal 1 hogshead, 72 gallons equal 1 puncheon and 108 gallons equal 1 pipe or butt.

WEIGHT. See *avoirdupois*.

UL
An abbreviation used for Underwriters Laboratory: A USA-based organization. For example:

UL 94 horizontal test = *Underwriters Laboratory horizontal burning test*; and,

UL 94 vertical test = *Underwriters Laboratory vertical burning test*.

ULDPE
An abbreviation used for ultra low density polyethylene. See *very low density polyethylene*.

ultimate compressive strength
See *crushing strength*.

ultimate elongation
The elongation reached at the breaking point during a *tensile test*: the elongation at break.

ultimate tensile strength
See *tensile strength*.

ultra high molecular weight, high density polyethylene
An abbreviation used for this type of material is UHMW HDPE. Because the impact resistance of HDPE can be low when low molecular weight grades are used, then high molecular weight materials are preferred in many applications: such materials have low values of *melt flow rate*.

A linear polyethylene with a very high molecular weight, for example, the weight average molecular weight is in the region of 1·5 million. Because of the large size of the molecules, crystallization is difficult and the density may therefore be only 0·94 g cm^{-3}. See *high density polyethylene*.

The water vapour permeability of this type of PE, as with all types of PE, is low. PE is permeable to gases and vapours but linear low density polyethylene (LLDPE) and HDPE are less permeable to gases and vapours than LDPE. Permeability for organic vapours is least for alcohols and then increases in the order shown; from acids to aldehydes and ketones, esters, ethers, hydrocarbons and halogenated hydrocarbons. (Permeability decreases with density). Some grades of HDPE, high molecular weight grades, are accepted as being suitable for containers for oil and petrol: they have been used for fuel tanks. In some cases the *blow moulded containers* have been chemically modified, by *fluorination* or sulphonation, so as to make the material almost impermeable to fuels.

ultra high molecular weight polyethylene
An abbreviation used for this material is UHMWPE. See *ultra high molecular weight, high density polyethylene*.

ultra high molecular weight polyethylene fibre
A *fibre* based on *polyethylene* of very high molecular weight. See *polyethylene fibre*.

ultra high modulus carbon fibre
See *mesophase pitch-based carbon fibre*.

ultra low density polyethylene
An abbreviation used for this term is ULDPE. See *very low density* polyethylene.

ultra-fast accelerator
An *accelerator* which promotes rapid reaction. Usually associated with the sulphur vulcanization of rubbers. See for example, thiuram disulphides.

ultra-violet
See *ultraviolet*.

ultraaccelerator
A very fast accelerator. See *ultra-fast accelerator*.

ultrafast accelerator
A very fast accelerator. See *ultra-fast accelerator*.

ultramarine blue
Also known as synthetic lapis lazuli or as, sodium aluminium silicate. A *blue pigment*, of variable composition, which is a complex aluminosilicate or aluminium sulpho-silicate: a reddish blue pigment which is widely used in plastics. Occurs naturally as lapis-lazuli: also prepared synthetically. Has poor hiding power but good resistance to sunlight, alkalis (not acids) and to heat. Redder in shade than *phthalocyanine blue* and has much less hiding power. This material has a relative density (RD or SG) of 2·35. See *ultramarine blues*.

ultramarine blues
A class or type of *inorganic pigment*. Relatively, high priced pigments which have low colouring strength and opacity. Such pigments have good heat stability at high temperatures, are chemically inert and are usually lightfast.

ultrasonic C-scan evaluation
A technique used to evaluate composites using ultra-sound, for example, using a 10 MHz *transducer*. The C-scan is used to non-destructively test a *composite* for voids, inter-facial debonding and fibre wetting problems. Good sound transmission is desirable in a C-scan test as it means the absence of voids, good inter-facial bonding and the absence of fibre wetting problems.

ultrasonic frequency
A frequency greater than approximately 20,000 hertz. See *welding* and *ultrasonic welding*.

ultrasonic heating
Heat generated by the application of ultra-sound and used in the polymer industry for *ultrasonic welding*. See *high frequency heating processes*.

ultrasonic inserting
The bonding of a metal piece to a thermoplastics component using the heat and pressure generated by an *ultrasonic press*: a

hole in the metal piece is placed over a stud or boss moulded onto the thermoplastics component. This stud is then formed over using the heat and pressure generated by the ultrasonic press.

ultrasonic press
A welding machine used for *ultrasonic welding*, ultrasonic staking and ultrasonic inserting.

ultrasonic staking
The insertion and bonding of a metal insert into a hole in a thermoplastics component using the heat and pressure generated by an *ultrasonic press*.

ultrasonic welding
Sometimes referred to as ultrasonic sealing. A joining process for thermoplastics which uses the friction induced in components by ultra-sound. Achieved by transmitting the ultrasonic vibrations through, and into, the components to be joined via a horn. This causes rapid motion or vibration of one component against another so that frictional heat is generated very quickly, for example, in less than 1 s. Ultrasonic plastics welding can be divided into *contact ultrasonic* welding and into *remote ultrasonic welding*.

ultraviolet absorber
An additive which improve the (ultraviolet) UV resistance of polymers by adsorbing harmful UV radiation. A UV stabilizer. The main types of ultraviolet absorber are the *substituted benzophenones*, the *benzotriazoles* and *salicylates*. See also *ultraviolet stabilizer*.

ultraviolet absorbing compound
A compound which contains an *ultraviolet absorber*.

ultraviolet light absorber
See *ultraviolet absorber*.

ultraviolet radiation
An abbreviation used for this term is UV or uv. Also written as ultra-violet. That invisible radiation which lies outside the violet end of the visible spectrum. That is, the wavelength of this electromagnetic radiation is shorter than the wavelength of violet radiation. It lies in the region between the visible and X-rays and has a wavelength, over the approximate range of from 100 to 4000 Angstroms.

Near ultraviolet radiation = wavelengths of 436, 405 and 365 nm.

Mid ultraviolet radiation = wavelengths of 330 and 315 nm.

Deep ultraviolet radiation = radiation with, for example, a wavelength of below 280 nm: used for high definition *microlithography*.

ultraviolet screen
An *ultraviolet stabilizer* which is also known as an ultraviolet screening agent. An additive which stops UV induced decomposition, or degradation, by stopping the light reaching the polymer: that is by screening. Examples of ultraviolet (UV) screens include *carbon black* and the rutile form of *titanium dioxide*. Most pigments will reflect UV radiation but fine particle size, carbon black is the best screen.

ultraviolet screening agent
See *ultraviolet screen*.

ultraviolet stabilizer
Also known as a UV stabilizer or as a photostabilizer. An additive which protects against decomposition, or degradation, caused by ultraviolet (UV) light attack. UV attack is common with most polymeric materials and so, UV stabilizers are usual in products which are to be used outdoors. May be grouped, or classified, in various ways, for example, into *ultraviolet absorbers* and into *ultraviolet screens*.

un-load
A term used in *hydraulics* and which means to release the pressure in a part of a circuit (usually) back to tank.

un-smoked sheet
An abbreviation used is USS. One of the forms in which solid *natural rubber* (unvulcanized rubber) is supplied. A light coloured form of natural rubber in sheet form. See *white and pale crepes*.

uncoiling
See *elastic effects*.

undecyl
This term is commonly seen in terms used to describe a *plasticizer*: it means that the side chain contains 11 carbon atoms.

undecyl dodecyl phthalate
An abbreviation used for this material is UDP. A low volatility, *phthalate plasticizer*: may be used in blends with a *trimellitate plasticizer*. See *diundecyl phthalate*.

undercure
Curing a compound for less than the time which gives the *optimum cure*. See *overcure*.

undercut
a part of a mould which will not release because of the negative angle, or reverse draft, at that part of the mould. An external undercut is formed by a recess or projection on the outside of a cavity. An internal undercut is formed by a recess or projection on the inside of a cavity: a restriction which prevents a moulding being extracted from the core in the line of mould opening.

undercut mould
A type of *injection mould* used to produce components with an *undercut*. The presence of undercuts in the moulded component means that the moulding cannot be released in the direction of mould opening unless moulds with loose cores or inserts are used. Because of the problems associated with the usage of inserts (for example, insert loading and location), moulds containing inserts may only be used for comparatively short runs. For long production runs use may be made of a mould with side cores, a mould with wedges or, a mould with rotating cores. The cores or wedges may be operated manually, mechanically, hydraulically, pneumatically or electromechanically. Collapsible cores have been used to form internal undercuts.

undercut pin
See *cold slug well*.

undercuts
Many injection mouldings contain holes, slots or threads and such features may be loosely termed undercuts. See *undercut mould*.

underfeed mould
A type of *injection mould* in which the *feed system* feeds into the inside of a component or, onto the underside of a component.

understeer
A vehicle steering defect. If the front *slip angle* is greater than the rear slip angle then the vehicle will understeer. The actual course taken by the vehicle will be less than expected. Because of, for example, the danger of spinning this is less dangerous than *oversteer*.

undertread
Also known as tread cushion. A *tyre* component: a thin layer of rubber, intermediate in hardness between the *tread* and the adjacent rubberized reinforcement *ply*, which provides a transition from tread to *carcass*.

underwater pelletizing
Another name for face cutting; process used to produce pellets from extruded strands by cutting them off, underwater, at the die face.

Underwriters Laboratory horizontal burning test
An abbreviation used for this term is UL 94 horizontal test. A standard issued by Underwriters Laboratory Inc. (UL 94). A rectangular bar is held horizontally, and at an angle of 45°, by being clamped at one end, The burning speed of the bar when exposed to a pilot flame is measured: the gas flame is applied to the unclamped end of the specimen.

Smooth bars (moulded or cut) 127 mm in length are used: width 12.7 mm: maximum thickness 12.7 mm. If appropriate the material should be tested in a range of colours, melt flows and reinforcements. Three bars from each batch are conditioned at 23 ± 2°C and 50 ± 5% relative humidity for 48 hours.

The flame is applied until the burnt portion reaches 25 mm and then the combustion speed is measured between two marks, one at 25 mm and the other at 102 mm. For a thickness of between 3.05 mm and 12.7 mm the material is rated as HB if the combustion speed is less than 38.1 mm/min.

Underwriters Laboratory UL 94 vertical burning test
Standard issued by Underwriters Laboratory Inc. (UL 94). An abbreviation used for this term is UL 94 vertical test. A rectangular bar is held vertically and clamped at the top, The burning behaviour of the bar and its tendency to form burning drips when exposed to a methane or natural gas flame (applied to the bottom of the specimen) is noted.

Smooth bars (moulded or cut) 127 mm in length are used: maximum width 13.2 mm: maximum thickness 12.7 mm. If appropriate the material should be tested in a range of colours. melt flows and reinforcements. Five bars from each batch are conditioned at 23 ± 2°C and 50 ± 5% relative humidity for 48 hours and then for 1 hour at 70°C and finally cooled (over anhydrous calcium carbonate) for 4 hours before testing.

The test should be conducted in a ventilated chamber fitted with an observation window. The lower edge of the specimen should be 9.5 mm above the top of the Bunsen burner tube and 305 mm above the surface of a layer of surgical cotton. The burner, adjusted until the flame is 19 mm high and blue, is placed under the specimen for 10 seconds, then removed. When the specimen stops burning the flame is replaced for a further 10 seconds.

A record is made of whether burning drips ignite the cotton and if the sample burns up to the clamp and the duration of burning and glowing. Five samples are tested and the materials are rated 94 V-0, 94 V-1, or 94 V-2.

94 V-0: Specimen burns for less than 10 seconds after either application of test flame. Cotton not ignited: after second flame removal glowing combustion dies within 31 seconds. Total flaming combustion time of less than 50 seconds for 10 flame applications on five specimens.

94 V-1: Specimen burns less than 30 seconds after either test flame: cotton not ignited. Glowing combustion dies within 60 seconds of second flame removal. Total flaming combustion time of less than 250 seconds for 10 flame applications on five specimens.

94 V-2: As for 94 V-1 except that there may be some flaming particles which burn briefly but which ignite the surgical cotton.

Some typical results arc as follows.

Material	UL 94 coding
ABS/PVC alloy	V-1
ABS/PC blend	V-2
Polycarbonate	V-2
Nylon 6	V-2
Nylon 66	V-2
Nylon 66	V-0
Glass reinforced polysulphone	V-0

N.B. The UL 94 codings apply only to specific material grades: the above results were obtained on 0.12"/3 mm thick specimens.

UNEP
An abbreviation used for *United Nations Environmental Program*.

unichain polymer
See *homopolymer*.

unidirectional
An abbreviation used for this term is UD. Usually used in connection with fibrous fillers and means that the filler lies in one direction: it is oriented in one direction.

unidirectional thermoplastics prepreg
A prepreg which is based on a *unidirectional* fibrous reinforcement and a thermoplastics matrix, for example, UD PEEK is made from unidirectional *carbon fibre* and *polyether ether ketone*. Such a system is similar to an epoxide-based counterpart but gives better thermal stability, hot/wet properties and impact toughness to moulded products. The prepreg has an infinite shelf life and needs no refrigeration.

unified atomic mass unit
Also known as u. It is one twelfth of the mass of a carbon-12 atom: approximately the mass of a proton or of a neutron.

unit
A quantity used as a standard of measurement: a dimension used as a standard of measurement.

unit dimensions - mechanical quantities
Dimensions of mechanical quantities are usually given it terms of mass (m), length (l) and time (t).

Mechanical quantity	Dimensions
Acceleration	l/t^2
Area	l^2
Density	m/l^3
Energy	ml^2/t^2
Force	ml/t^2
Momentum	ml/t
Power	ml^2/t^3
Velocity	l/t
Volume	l^3

unit drive
A type of *calender drive* where all the gears are mounted in an enclosed gearbox and the drives to the roll ends are via universal couplings. This is termed a unit drive gearbox and comes in two forms. See *single motor drive* and *multi-motor drive*.

unit of measurement
A precisely defined, and ideally invariable, quantity of some particular kind in terms of which the magnitudes of all other quantities of the same kind can be stated. Units of measurement are of two kinds: *fundamental units* (basic units) and *derived units*. See *standard of measurement*.

unit volume
To obtain the unit volume (specific volume) of a substance, divide the figure one by the value of the density. The units will be m^3/Mg when density is expressed as Mg/m^3.

United Nations Environmental Program
An abbreviation used for this term is UNEP. Provides a framework for the phasing out of chemicals that are known to deplete the *ozone layer*. See *chlorofluorocarbon*.

units
ABSOLUTE ELECTRICAL. Electrical quantities measured in terms of the basic units of length, mass and time plus an electrical unit.

CGS ELECTROMAGNETIC. Also known as units - cgs emu. This system was based on the *cgs system* to which was added a definition of electric current, the *abampere*. The units are prefixed by ab.

CGS ELECTROSTATIC. Also known as units - cgs esu. This system was based on the *cgs system* to which was added a definition of electric charge, the *franklin*. Such units are prefixed by stat.

CGS GAUSSIAN. Also known as units - cgs mixed. This system was based on the *cgs system* to which was added the quantities from the cgs electrostatic system and units from the cgs electromagnetic system.

GRAVITATIONAL SYSTEM OF. Such a system uses the weight of a standard body as a force unit, for example, the pound. See *gravitational system of units*.

IMPERIAL. See *foot-pound-second*.

universal constants
Also known as fundamental constants or as fundamental physical constants.

universal masterbatch
Concentrated masterbatches which can contain up to 60% of a colouring system. Made by using carrier systems which are based on low molecular weight resins; such resins have a low melting point (e.g. 60°C) and are compatible with a very wide range of polymers. Such universal' compatibility is a tremendous advantage, although at the same addition level, the same masterbatch may produce a slightly different colour in two different polymers. Such concentrated masterbatches may be supplied in a range of granule sizes so that if, for example, powdered polymer is being run, then a fine-grain masterbatch could be selected so as to achieve improved pigment dispersion. Because of the concentrated nature of the masterbatch, the metering system must be capable of delivering a very accurately dosed feed consistently and the screw must be capable of achieving a high level of dispersion. Despite these disadvantages, the use of concentrated masterbatches is increasing as the cost is low, they are easy to handle, the colour inventory is relatively small and, as they are clean to handle, the amount of contamination produced is small. See *dilute masterbatch*.

unmelted particles
Also known as unmelt. Melt contamination caused by uneven plasticization.

unmodified polystyrene
See *polystyrene*.

unplasticized polyvinyl chloride
Also known as rigid polyvinyl chloride or as, rigid PVC or as, unplasticized polychloroethene or as, unplasticised polyvinyl chloride. An abbreviation used for this type of material is UPVC: sometimes PVC-U or RPVC is used.

A great deal of the *polyvinyl chloride* (PVC) now produced is UPVC. The term unplasticized means that there is no plasticizer present in the material but, it does not mean that there are no other additives present. Additives are essential in PVC technology as without them the plastic is useless as it cannot be processed. Some additives may also enhance the properties of the base PVC and thus permit its use in a wider range of applications. The problems associated with *additive* use, for example, toxicity, are made worse in some UPVC applications as the products are often used as containers for food or drinks.

Examples of additives used in UPVC compounds include *heat stabilizers*, *lubricants*, *processing* aids, *impact modifiers*, *fillers*, and *pigments*. When correctly formulated, transparent products, for example, injection mouldings, are possible. This means using appropriate stabilizers, for example tin stabilizers, and matching the refractive index of, for example, the impact modifier and the PVC. As the natural colour of the material can be clear then a wide colour range is possible; this includes both transparent and opaque colours.

At some stage the polymer plus additives is melt mixed in, for example, a batch mixer or a continuous compounder - such as a *twin-screw extruder*. To realize the potential of this material it is necessary to put in just sufficient work and heat so that the blend is fully gelled, or fused, but not so much that it is degraded.

The important factor is to process the material at the correct melt temperature as a major problem in UPVC processing is resin decomposition, or degradation: the first sign of this is a change in colour. Resin decomposition can be caused by over-heating or by excessive shear. Once decomposition starts then it can spread very rapidly because, for example, one of the products of decomposition (hydrochloric acid, or HCl) catalyses further degradation. Being an acid it also readily attacks metals such as steel and causes pitting and corrosion; it also promotes rust as it strips protective layers away from metals such as mild steel. The effects on human beings are also very harmful.

UPVC is a relatively stiff flow material with a limited temperature processing range. The ease of flow is dependent on the molecular weight - this is characterized for PVC by the K *value*; the bigger the number the higher is the molecular weight and the more difficult the flow. If sufficiently low K value resins are used (for example 50 to 55) then in many respects injection moulding PVC compounds, based on these low 'K' value resins, may be treated similarly to other thermoplastics materials.

UPVC is widely used as it is relatively cheap, inherently flame retardant, strong and stiff. The chemical resistance is good, for example the material has good oil resistance and yet, it is possible to join it by solvent welding. UPVC is resistant to water, salt solutions, oxidizing agents (for example, hydrogen peroxide), reducing agents, hypochlorite solutions, aliphatic hydrocarbons, detergents, non-oxidizing acids and concentrated alkalis. UPVC is resistant to most oils, fats, alcohols and petrol. Highly resistant to strong acids, for example, any concentration of hydrochloric acid at temperatures up to 60°C/140°F; any concentration, up to 90%, of sulphuric acid at temperatures up to 60°C/140°F; cold, 50% nitric acid does not attack UPVC. Chemical and weathering resistance is good, and can be made excellent, but is generally made worse by the addition of impact modifiers. Homopolymers of PVC are more resistant to chemical attack than copolymers. UPVC compounds are not generally recommended for continuous use above 60°C/140°F. However, some modifiers can now raise the heat distortion temperature (HDT) by approximately 30°C to, say, 90°C/194°F.

UPVC is not resistant to concentrated, oxidizing acids such as sulphuric, nitric and chromic acids which cause decomposition; the rate of decomposition may be accelerated in the presence of metals, for example, zinc and iron. Attacked by bromine and fluorine even at room temperature. Unsuitable for use in contact with aromatic and chlorinated hydrocarbons, ketones, nitro-compounds, esters and cyclic ethers; these can penetrate the PVC and cause marked swelling. The material can have a high impact strength but can be very notch sensitive. Methylene chloride can be used to detect an under-gelled compound. The material may be subject to environmental stress cracking (ESC) if a moulding is subject to stress when in contact with acids, alkalis and aliphatic alcohols. For example, 2-propanol and sodium hydroxide.

UPVC pipe fittings for rainwater and irrigation projects are a major outlet for injection mouldings. Other injection moulded products consist of housings for computers and televisions, water filtration pressure tanks, photocopier doors, electricity and gas meter housings, transparent printer hoods, ventilation grilles, and various components associated with the electronics industry. The use of UPVC alloys for computer housings has become very popular due to its inherent flame retardant properties, excellent physical properties (e.g. rigidity), UV stability, low price and the ability to produce large flat area components to close fitting tolerances. UPVC is also calendered into sheet and there is a very large market for UPVC pipe and profile which is produced by *extrusion*. A major disadvantage of UPVC is its high density - the density is approximately 1.4 gcm^{-3} (solid, un-filled material).

unrestricted gate
A *gate* of large cross-sectional area which does not seriously interferes with material flow. See *restricted gate*.

unsaturated compound
A chemical compound in which some of the atoms are linked by more than one valence bond, for example, a double bond. Such unsaturation helps explain rubber elasticity.

unsaturated ester reaction injection moulding
An abbreviation used for this term is unsaturated ester RIM. A *reaction injection moulding process* which uses a polymer based on an unsaturated ester. See *vinyl ester resin* and *acrylamate*.

unsaturated ester RIM
An abbreviation used for *unsaturated ester reaction injection moulding*.

unsaturated ester-styrene copolymer
See *vinyl ester resin*.

unsaturated hydrocarbon rubber
A rubber which is based on a *hydrocarbon* and which has chemical unsaturation in the main chains. See *olefin* and α *olefin*.

unsaturated polyester reaction injection moulding
An abbreviation used for this term is UP RIM. A *reaction injection moulding process* which is based on an unsaturated polyester resin. The temperature of the reactants is typically 25°C, mould temperatures are of the order of 120°C while demould times may be 60 s: the adiabatic exotherm temperature for a typical formulation without filler can be 100°C.

unsaturated polyester resin
An abbreviation used for this type of material is UP. The term 'ester' is given to the reaction product of acids and alcohols, so a polyester is a reaction product (a resin) which contains many ester groups. Such resins are often obtained by reacting together propylene glycol, maleic anhydride and, phthalic anhydride (either one, or all three, reactants may be varied). As the resulting resin contains double bonds, as well as ester groups, it is also known as an unsaturated polyester (UP) resin.

The toffee-like UP material, is dissolved in styrene: this acts as a solvent but it is capable of reacting with the resin when catalyzed or initiated. To stop premature gelation, the resin solution may be inhibited. There are two general classes of inhibitor used with unsaturated polyester resins and these are quaternary ammonium salts and substituted phenolic derivatives.

The resin solution may be catalyzed, cast inside a mould (as it is a comparatively thin liquid) and then it will set to the shape of that mould. The double bonds enter into free radical copolymerization with the *styrene* and this causes the system to *gel* and then to harden. Volatile by-products are not eliminated when the system sets by *cold curing* or by *hot curing*. However, to get high strength products, fibrous fillers such as *glass* must be added; to reduce shrinkage and cost, particulate fillers, such as calcite or dolomite, are also added. The products are tough, hard and have reasonable chemical and heat resistance. See *dough moulding compound*, *hand lay-up* and *sheet moulding compound*.

unsaturated polyurethane
Also known as an unsaturated ester-urethane. A *polyurethane* which is unsaturated. See, for example, *acrylamate reaction injection moulding*.

unsaturation
See *unsaturated compound*.

unscrewing core
See *rotating core*.

unscrewing mould
A mould in which threaded components are automatically unscrewed from the mould. See *rotating core*.

UP
An abbreviation used for *unsaturated polyester*. For example:

UP BMC = unsaturated polyester bulk moulding compound. See *bulk moulding compound*; and,
UP resin = *unsaturated polyester resin*.

UP RIM
An abbreviation used for *unsaturated polyester reaction injection moulding*.

UPM
An abbreviation sometimes used for revolutions per minute, for example, of an extruder screw.

upper action limit
An abbreviation used for this term is UAL: a limit line which corresponds to probability points on the normal distribution curve of, say, 99.9%.

upper warning limit
An abbreviation used for this term is UWL: a limit line which corresponds to probability points on the normal distribution curve of, say, 95%.

upside down mixing
Also known as the upside down process. A mixing technique used in the rubber industry. A mixing technique used with an *internal mixer* to get good dispersion and short mixing times when high filler loadings are used. The filler and oil are first added to the mixer followed by the polymer, for example, *ethylene-propylene rubber*. If the mixing chamber temperatures are below 120°C, then the *curatives* are added to the mixer: if the mixing chamber temperatures are above 120°C, then the curatives are added on a *two-roll mill*. If the rubber is added complete with the outer polyethylene (PE) wrapper, then temperatures above 120°C will be needed to achieve PE dispersion. Relatively short mixing times may be obtained if the rubber is in crumb, friable bale or pellet form (see *technically specified rubber*). Upside down mixing is recommended, for example, for *butyl rubber* mixes with large filler or high oil loadings.

upside down process
See *upside down mixing*.

upstroke press
A press in which the direction of force application is from below: from its open position, the *moving mould* half moves upwards against the *fixed mould half*.

UPVC
An abbreviation used for *unplasticized polyvinyl chloride*.

urea
Also known as carbamide. An abbreviation used for this material is U. This solid material has a melting point of 130°C and a relative density (RD or SG) of 1.31. May be represented as $CO(NH_2)_2$. This material is used to prepare *urea-formaldehyde polymers* and may be used to prepare *melamine*. See *urea-formaldehyde moulding material*.

urea derivative stabilizer
A derivative of *urea* such as phenylurea, diphenylurea and diphenylthiourea. A derivative of *urea* used as a *metal-free organic*, heat stabilizer with *plasticized polyvinyl chloride* (PPVC) it is used with a co-stabilizer such as an *epoxy compound*.

urea formaldehyde
See *urea-formaldehyde moulding material*.

urea group
May be represented as -NHCONH-. The reaction of an amine group and an isocyanate results in the formation of such a group in *polyurethane* manufacture, for example, R.NH.CO.NHR' (where R and R' are organic groups).

urea laminated sheet
A *laminated plastics* sheet material based on a *urea-formaldehyde resin*.

urea moulding material
See *urea-formaldehyde moulding material*.

urea plastics
Plastics based on polymers in which the repeated structural units in the chains are of the *urethane* type, or on copolymers in which urethane and other types of repeated structural units are present in the chains. See *urea-formaldehyde moulding material*.

urea reaction injection moulding
See *polyurea reaction injection moulding*.

urea resin moulding compound
See *urea-formaldehyde moulding material*.

urea-formaldehyde
An abbreviation used for this type of material is UF. Also known as urea-methanal. The reaction product of *urea* and of *formaldehyde* in the ratio of approximately 1:3 gives an aminoplast or aminopolymer. Used to make *urea-formaldehyde moulding materials* and for fabric treatments.

Cotton is easily creased but this disadvantage can be minimized by resin treatments using amino resins (see *cotton fabrics*). Urea resins are widely used to impart crease resistance and 'drip dry' qualities. The fabric is passed through aqueous solutions of hydroxy methyl ureas and the resin is hardened by heating (130 to 160°C/266 to 320°F) in the presence of metal salt catalysts.

Urea-formaldehyde polymers may be reacted (etherified) with an alcohol (for example, n-butanol) to give resins which are more soluble in organic solvents than the base polymer: used with *alkyd resins* to make stoving enamels.

urea-formaldehyde moulding material
An abbreviation used for this type of material is UF. Also known as urea-formaldehyde (an *aminoplastic*) or as, urea or as, urea resin moulding compound or as, urea resin moulding material.

UF materials are thermosetting materials which are based on *urea-formaldehyde resin* combined with *fillers, lubricants, hardeners*, etc. They are available in fine powder or granular form. A wide colour range is possible (because of the lack of resin colour).

When *woodflour* is used as the filler brown mouldings will result: this is masked by intense colouring so as to give black or brown shades. Such moulding powders have a well balanced range of properties and are cheap. They are used in electrical and closure applications. If bleached wood pulp (paper) is used as the filler, then a wide colour range is possible as the resultant compound can be translucent: bright, intense colours are possible - including pastel shades and white. Such colours are lightfast. The properties are similar to woodflour-filled grades and the mouldings are widely used in electrical fittings for their low cost, wide colour range, rigidity and good electrical properties.

The heat resistance of UF is lower than that of *phenol-formaldehyde (PF)* but arc resistance is higher and curing time is faster. The water absorption is also significantly higher than for either PF or *melamine-formaldehyde* (MF). In order to obtain improved water resistance the resin may be fortified with melamine: acid, alkali and heat resistance are thus improved. Flame resistance, as measured by UL 94, is excellent at V-0: oxygen index of typical mouldings is approximately 35%.

Grades are available for all types of processing - automatic/manual compression moulding, transfer and injection moulding; such materials are commonly supplied in granular form.

The cured material is resistant to solvents and household chemicals. Not affected by fats, oils, esters, ether, petrol, alcohol or acetone; nor by detergents or weak acids such as acetic. In general, the acid resistance is not good but such materials have good resistance to weak alkalis: they are not resistant to strong acids and alkalis. Not as stain resistant as MF.

Moulding powder use is only a small outlet for UF resins. They are more widely used as adhesives in, for instance, particle board and plywood. However, UF moulding materials are used extensively in the electrical industry for insulating parts in plugs, sockets, switches, connectors and lamps. In domestic applications, UF find uses as handles and knobs (furniture and kitchen utensils), buttons, buckles and closures for jars and bottles: this is because of UFs good resistance to solvents etc. and the excellent surface finish possible. They are widely used for toilet seats, because of the bright, attractive appearance possible.

urea-formaldehyde resin
A low molecular weight polymer formed by the reaction of *urea* and *formaldehyde*. The reaction product is a syrup which is combined with fillers, lubricants, hardeners, etc, to give a *urea-formaldehyde moulding material*.

urea-urethane reaction injection moulding
An abbreviation used for this term is urea-urethane RIM. Also known as mixed urea-urethane reaction injection moulding. A *reaction injection moulding* (RIM) process which has been formulated to give urea and urethane linkages. The temperature of the reactants is typically 40°C, mould temperatures are of the order of 60°C while demould times may be 30 s: the adiabatic exotherm temperature for a typical formulation without filler can be 110°C.

urethane
Also known as ethyl carbamate. This solid material has a melting point of 49°C and a relative density (RD or SG) of 1·1. May be represented as $NH_2.CO.OC_2H_5$. This material is used to prepare, or to modify, *uaminoplastics* such as *urea-formaldehyde polymers*. Sometimes the term urethane is used in place of *polyurethane rubber*.

urethane compound
A compound of the type R.NH.COOR' where R and R' are organic groups: a compound which contains the urethane group -NHCOO-. See *polyurethane*.

urethane cross-linking
Curing, or vulcanization, of a diene rubber, for example, *natural rubber*, induced by the use of a urethane cross-linking agent. For toxicity reasons, may be performed by using an adduct of an *isocyanate* and a nitroso compound (a blocked diphenylmethane diisocyanate): this adduct dissociates to give the *isocyanate* and a nitroso compound. Tautomerism produces a nitrophenol (from the nitroso compound) which attaches itself to the rubber leaving a pendant aminophenol group free which then reacts with the isocyanate. The rubber is thus cross-linked. This reaction is catalyzed by *ZMDC*: to prevent hydrolysis of the isocyanate, a drying agent based on calcium oxide is also added to the mix. Slow curing systems which give good *reversion resistance*: useful for thick sections.

urethane group
May be represented as -NHCOO-. The reaction of an alcohol and an isocyanate results in the formation of such a group in *polyurethane* manufacture, for example, R.NH.COOR' (where R and R' are organic groups).

urethane oil
The reaction product of *drying oil esters* and an *isocyanate*. The drying oil esters are mixed partial esters derived from drying oil *fatty acids* and one or more polyhydric alcohols, for example, drying oils such as linseed and soya bean are used although more yellow-resistance is given by the use of safflower and tall oils. The *isocyanate* is often *tolylene diisocyanate* although more yellow-resistance is given by the use of a cycloaliphatic diisocyanate.

Long oil length urethane oil may be made from triols, for example, *glycerol*: short oil length urethane oil may be made from a diol ether of low molecular weight or from a polyoxypropylene diol of comparatively low molecular weight.

Urethane oils are widely used as varnishes and as surface coatings. See *alkyd resin*.

urethane reaction injection moulding
An abbreviation used for this term is urethane RIM. A *reaction injection moulding* (RIM) process which has been formulated to give urethane linkages. The temperature of the reactants is typically 40°C, mould temperatures are of the order of 70°C while demould times may be 45 s: the adiabatic exotherm temperature for a typical formulation without filler can be 130°C.

urethane rubber
See *polyurethane rubber*.

urethane rubber - polyester based
A *polyurethane rubber* based on a *polyester*. Isocyanate crosslinkable are abbreviated to AU-I: peroxide crosslinkable are abbreviated to AU-P.

urethane rubber - polyether based
A *polyurethane rubber* based on a polyether.

urethanes
The collective name for the class of materials which contain the *urethane group* (see *polyurethane rubber*). The term urethanes, is applied to the esters of *carbamic acid* and in particular *to ethyl carbamate*.

US Customary Measure
US means the United States of America (the USA). This system is based on the pound, the yard and the gallon.

AREA. Based on square yards. 144 square inches make one square foot and nine square feet make one square yard. 4,840 square yards make one acre and 640 acres make one square mile.

AVOIRDUPOIS. 16 drams = 1 ounce, 16 ounces = 1 pound, 2,000 pounds = 1 ton (short) and 2,240 pounds make one long ton. One pound avoirdupois contains 7,000 grains.

CAPACITY. See *volume*.

DRY. This is based on the bushel (sometimes called the stricken or struck bushel) which contains 2,150.420 cubic inches. 2 pints make one quart. 8 quarts make one peck and 4 pecks make one bushel.

LENGTH. Based on the yard. There are 3 feet in one yard and there are also 36 inches. That is 12 inches make one foot. 16.5 ft make one rod and one mile (land mile) equals 1,760 yards. A nautical mile (international) equals 1.151 land miles.

LIQUID. Based on the gallon which contains 231 cubic inches. One pint contains 16 fluid ounces and 2 pints make one quart. 4 quarts make one gallon. From 31 to 42 gallons make one barrel (according to law or usage).

VOLUME. A cubic yard contains 27 cubic feet and a cubic foot contains 1,728 cubic inches.

WEIGHT. See avoirdupois.

USS
An abbreviation used for *un-smoked sheet*

Utermark process
See *centrifuging*.

uv
An abbreviation used for *ultraviolet (ultra-violet)*.

UV
An abbreviation used for *ultraviolet (ultra-violet)*. For example:
UV light absorber = *ultraviolet light absorber*;
UV screen = *ultraviolet screen*;
UV screening agent = ultraviolet screening agent.
    See ultraviolet screen;
UV stabilizer = *ultraviolet stabilizer*; and
UV-B = ultraviolet light of wavelength 290 to 320 nm.
See *stratospheric ozone* and *stratospheric chlorine*.

UWL
An abbreviation used for *upper warning limit*.

V

v
Symbol used for *Poisson ratio* and sometimes used for velocity.

V
An abbreviation used for *volt*. Also used to modify polymer abbreviations, for example, when used in this way stands for very. Also used for vulcanized or crosslinked. If a plastics material contains a filler then the letter may be used to show the presence of a veneer. V is also used as an abbreviation for the ram speed of a *capillary rheometer*.

V-0
See *Underwriters laboratory UL94 vertical burning test*.

V-1
See *Underwriters laboratory UL94 vertical burning test*.

V-2
See *Underwriters laboratory UL94 vertical burning test*.

V-belt
A *transmission belt* which has a cross-section in the form of an isosceles trapezium.

V-shaped marks
See *crows feet*.

V-shaped notch
A notch shaped like the letter vee. Impact test samples are given a V-shape notch so as to obtain more reproducible results. See *notch tip radius*.

VA
An abbreviation used for vacuum assist: as in *vacuum assist resin injection*. VA is also used for *vinyl acetate*.

vacuum
A space containing air/gases at very low pressure. A region of space in which the *pressure* is very low.

vacuum assisted resin injection
An abbreviation used is VARI. Also known as resin injection VA (resin injection vacuum assisted). A *resin transfer moulding* (RTM) process in which a vacuum is applied to the moulds so as to counter-act the pumping pressure. Because of the vacuum applied, lighter moulds (compared to RTM) may be used. Maximum transfer pressures available for VARI is approximately 200 kPa or 2 bar. Unfilled low viscosity resins are employed as if resins with viscosities above 0.4 Pa (300-400 cps) are used, unacceptably long filling times result. Double seals may be needed at the mould edges so as to minimize the ingress of air. This process can result in higher fibre loadings than RTM.

vacuum bag moulding
See *bag moulding* and *low temperature moulding epoxy prepreg systems*.

vacuum calibrator
Equipment used to set the dimensions of an extrudate (for example, the external diameter of pipe or tubing) by expanding the material against *sizing rings*: expansion is done by means of a *vacuum*.

vacuum deposition
Part of the production process for *vacuum metallization*, of plastics in which injection mouldings, after treatment with a *base coat*, are coated with metal in a vacuum chamber. Aluminium is carried on a tungsten filament and is vaporised when the filament is electrically heated. Under the high vacuum conditions employed, the aluminium atoms fly away from the filament and coat anything in their path: to obtain uniform coating the mouldings may be rotated. Both *first surface* and *second surface metallization* are used. The thickness of the aluminium coating is of the order of 0.05 μm.

vacuum evacuation
A moulding technique used to assist the escape of gases, for example, during injection moulding of a thermosetting plastic. The cavities are connected to a vacuum supply: this can eliminate porosity and eliminate the need for *breathing*: it permits fast cavity filling and assists the filling of deep indentations or lugs.

vacuum forming
A *thermoforming process* which is also known as female forming. A thermoforming process which uses a *female mould* and which offers a way of producing components of relatively shallow draw cheaply. The sheet is heated by an infra-red heater (mounted approximately 150 mm above the sheet) while being held in a clamping frame. At *forming temperatures*, the heater is removed and a vacuum is applied to the space between the sheet and the female mould. Atmospheric pressure forces the softened plastics material into the mould cavity. The product is then cooled by mould contact, or by blowing air across the forming, and ejected. Such a forming can have poor wall thickness distribution, for example, the corners of box-type components will be thin because the material which makes the corners has been stretched excessively. To obtain a different material distribution the sheet may be pre-stretched before the vacuum is applied. See *plug assist* forming.

vacuum injection mould
A type of *injection mould* used to produce components with a surface which has not been marked by air becoming trapped in the cavity. Both *air trapping* and *mould fouling* can be minimised by using a vacuum mould. For example, the mould is fitted inside a two-part case which becomes a fully sealed unit in the final phase of the mould closing stroke. By application of a vacuum the air within the unit is largely removed during *rubber injection moulding*.

vacuum metallization
A process used to apply a coating of metal, usually aluminium, to a plastics substrate, by evaporation in a vacuum chamber: *first surface* and *second surface metallization* are both used. Thermoplastics materials are associated with vacuum metallization and for such materials the main steps are *injection moulding* of the components, cleaning, base coat application, vacuum deposition and top coat application. See *metallizing*.

vacuum sizing
Technique used to set the dimensions of an extrudate by expanding the material against a sizing sleeve, or against sizing rings, by means of a vacuum.

vacuum sizing die
See *sizing die*.

vacuum snap back thermoforming
Also known as snap back thermoforming. Originally developed for acrylic sheet (PMMA) which at forming temperatures is elastic. The sheet is heated in an oven and transferred to the forming machine where it is clamped over a vacuum box. When the vacuum is applied the sheet is drawn down into a deep bubble: a *skeleton frame mould* is then lowered into the bubble and the vacuum released. The elasticity of the sheet causes the sheet to revert (snap back) over the frame mould. See *thermoforming*.

vacuum sputtering
A process used to apply a coating of metal (for example, chromium or stainless steel) to a plastics substrate, by evaporation of the metal in a chamber at low gas pressures. Thermoplastics materials are associated with vacuum sputtering and for such materials the main steps are *injection moulding* of the components, cleaning, base coat application, vacuum deposition and top coat application. See *sputtering and metallizing*.

vacuum-assisted venting
Also known as vacuum venting. Venting of cavities using a partial vacuum and used in *rubber injection moulding*; the mould is designed so that it incorporates a sealing ring and a vacuum groove. Once the mould has closed and sealed (against the rubber sealing ring) then the air in the cavity is removed by applying a vacuum to a groove which completely surrounds the cavity. Removal of the entrapped air allows the use of fast injection speeds and can significantly speed up the curing cycle as well as reduce the number of scrap or reject mouldings. A high vacuum is not necessary for this purpose but the pressure in the mould should be reduced to below half atmospheric pressure.

VAE
An abbreviation used for vinyl acetate-ethylene copolymer. See *ethylene-vinyl acetate copolymer*.

valley printing
Also known as ink embossing or as, print embossing. A printing method used to obtain a two-colour effect. The valleys are formed by an embossing roller which also applies a contrasting colour, to the substrate, in those valleys.

valve
A term used in *hydraulics* and which refers to a device which controls the direction of fluid flow, pressure or flow rate.

valve ejector
A valve-headed *ejector element*. See *rubber injection moulding*.

valve spool
See *spool*.

valves - injection unit
It is usual to employ a *back-flow valve* on the tip of the screw of an injection moulding machine so that material is not lost during injection: with some materials, for example *unplasticized polyvinyl chloride* (UPVC), the use of a valve may not be possible. Drooling from the nozzle may also be prevented by means of a nozzle valve. Shut off nozzles (nozzle valves) are not widely used nowadays due to material leakage, and degradation, taking place within the nozzle assembly. It is more usual to use an open nozzle on an *injection moulding machine* fitted with *decompression*.

VAMA
An abbreviation suggested for *vinyl acetate-maleic anhydride copolymer*.

VAMP
An abbreviation used for vacuum assisted moulding process. A *resin transfer moulding* process.

vanadium
A *transition element* which is a very hard, lustrous gray/white metal with a high melting point (1,710°C) and boiling point (3,380°C): the relative density is 6·1. Noted for its corrosion resistance: used to make steel alloys which are noted for their great hardness and strength even when the vanadium content is very low, for example, below 1%. The pentoxide is used as an oxidation catalyst to make, for example, *phthalic anhydride*.

vanadium pentoxide
Also known as vanadium (V) oxide. Vanadium pentoxide is used as an *oxidation catalyst* to make, for example, phthalic anhydride.

vanillin
Also known as 3-methoxy-4-hydroxybenzaldehyde or as, 4-hydroxy-3-methoxy-benzaldehyde. This solid has a melting point of 82°C, a boiling point of 285°C and a relative density (RD or SG) of 1·06. Used to deodorize and to perfume polymeric products.

vapour
A vapour may be defined as moisture (water) or other substances which are normally liquid or solid, suspended in air. For example, mist or smoke are both vapours.

vapour permeability
The permeability of a material to a *vapour*. As vapour permeability may show considerable variation with barrier thickness, it is best to show the rate of transmittance, the barrier thickness and the partial pressure across the barrier. For water, the difference in relative humidity and the temperature would also need to be specified. It is a general phenomenon that vapours (below the critical temperature) diffuse through a substrate more rapidly than gases under the same pressure gradient. There are usually specific vapours that pass through a particular plastics material more easily than others. See *gas permeability* and *permeability*.

vapour pressure lowering
A *molecular weight* determination method which gives the number average molecular weight.

vapour-liquid-solid
An abbreviation used for this term is VLS. A controllable vapour phase reaction performed at high temperatures, and used, for example, to produce *whiskers*.

VARI
An abbreviation used for *vacuum assisted resin injection*.

variable volume pump
A pump whose output is varied to suit the demand. Used on *injection moulding machines* to reduce energy consumption. The pump is only accelerated to full operating speed when required.

variance
A statistical term which may be obtained by squaring the *standard deviation*.

vario screw
A *compression-less extruder screw* which incorporates a smear, or Maddock section, towards the front of the screw followed by approximately four diameters (4D) of mixing or conveying sections. These last named can be varied to suit the material or the process.

Vaseline
Trade name/trade mark for *petrolatum or petroleum jelly*. This material has a relative density (RD or SG) of 0·86 to 0·90.

vat pigments
Organic *pigments* which are based on *anthraquinone dyes* and which have complex polycyclic structures.

VC
An abbreviation used for *vinyl chloride*. VC is often used to make *copolymers*: if VC is listed first, then it is usually taken that the VC is in the greatest amount by mass. When discussing copolymers, it is recommended (ISO) that an oblique stroke / be placed between the two monomer abbreviations. However, the oblique strokes may be omitted when common usage so dictates according to ISO 1043-1:1987 (E). The oblique stroke has often been omitted in this text as it is not in common usage.

VCA = **VC/A** = *vinyl chloride-acrylonitrile copolymer.*
VCE = **VC/E** = *vinyl chloride-ethylene copolymer.*
VCEMA = **VC/E/MA** = *vinyl chloride ethylene-methylacrylate polymer.*
VCEVA = **VC/E/VA** = *vinyl chloride ethylene-vinyl-acetate terpolymer.*
VCEVAC = **VC/E/VAC** = *vinyl chloride-ethylene-vinyl acetate terpolymer.*
VCMA = **VC/MA** = *vinyl chloride-methylacrylate copolymer.*
VCMMA = **VC/MMA** = *vinyl chloride-methyl methacrylate copolymer.*
VCOA = **VC/OA** = *vinyl chloride-octylacrylate copolymer.*
VCP = **VC/P** = *vinyl chloride-propylene copolymer.*
VCVA = **VC/VA** = *vinyl chloride-vinyl acetate copolymer.*
VCVAC = **VC/VAC** = *vinyl chloride-vinyl acetate copolymer.*
VCVAMA = **VC/VA/MA** = *vinyl chloride-vinyl acetate-maleic anhydride terpolymer.*
VCVDC = **VC/VDC** = *vinyl chloride-vinylidene chloride copolymer.* See *vinyl chloride copolymers and terpolymers.*

VCM
An abbreviation used for *vinyl chloride monomer*.

VDC
An abbreviation used for *vinylidene chloride*. (See *vinyl chloride*).

VDCA = **VDC/A** = *vinylidene chloride-acrylonitrile copolymer.* See *vinyl chloride copolymers and terpolymers.*

VDF
An abbreviation used for *vinylidene fluoride*. (See *vinyl chloride*.)

VDFTEHTP = **VDF/TFE/HTP**. A FKM rubber. See *fluororubber.*

VDMA
See *German Machinery Plant Manufacturers Association*.

VDU
An abbreviation for visual display unit. Can mean both the screen and the keyboard.

VDVA
An abbreviation used for vinylidene chloride-acrylonitrile copolymer. See *vinyl chloride copolymers and terpolymers*.

VE
An abbreviation used for *vinyl ester resin*. Also see *unsaturated polyester resin*.

vegetable fibre
A fibre obtained from a plant. For example, leaf fibres.

vegetable oil
Oil obtained from the fruit and/or seed of a plant. For example, from coconuts, cotton seeds, ground nuts (peanuts), soya beans, etc. Oil obtained from an oilseed: such an oil is a *fixed oil*.

The oil may be classed according to its film forming ability or iodine value. For example, as a non-drying oil - an oil with an *iodine* value of less than 120 (cottonseed, coconut, rapeseed and castor). As a semi-drying oil - an oil with an iodine value of 120 to 150 (safflower, sunflower and soybean). As a drying oil - an oil with an iodine value of greater than

150 (linseed, perilla and tung oil). Oils may be also classified as yellowing or non-yellowing - this depends, for example, on the linolenic acid content. Semi-drying and non-drying oils are generally non-yellowing. Oils can also be classified into edible, industrial and dual purpose.

Often such an *oil* cannot be used in the raw state but must be treated to modify its properties. Examples of such treatments include *alkali refining, kettle bodying and blowing.*

vehicle
Also known as a carrier. That which is used to carry an additive. See *masterbatch*.

velocity control
Term used in injection moulding to indicate that the initial mould filling part of the moulding cycle is speed controlled. At a certain point in the injection moulding stroke, for example, when the mould is full or when the gate freezes, the resistance to flow becomes very high and it becomes unrealistic to expect the screw to maintain the desired rate. At this point control is shifted from being velocity controlled to being pressure controlled: This is known as *velocity pressure transfer.*

velocity pressure transfer
An abbreviation used for this term is VPT. Term used in *injection moulding* to indicate that after initial mould filling (speed controlled) the final mould packing is done under a controlled pressure. See *injection speed* and *programmed injection speed*.

ACCURATE SETTING. It is very important that the VPT point is capable of being set very precisely and, that it is also very reproducible. If these conditions cannot be met then mouldings with varying properties will result. As the object of injection moulding is to produce mouldings with identical properties, any variation of component properties is obviously unwelcome. The VPT is capable of being set by the machine operator provided that he/she, is given appropriate monitoring equipment: this is necessary so that, it is possible to 'see' exactly what is being set, where the set point is located and how reproducible is the set point.

OPTIONS. If a second stage, holding pressure is required then, a signal which initiates the changeover, must be generated. Changeover at the VPT point may be set, or triggered, in the following ways:
  i) screw position- also known as path dependent switchover or as, screw positional control or as, positional control;
  ii) hydraulic pressure - also known as line pressure or as, line pressure control;
  iii) cavity pressure control - also known as CPC or as, cavity pressure changeover control;
  iv) nozzle pressure - also known as melt pressure;
  v) mould opening force; and,
  vi) mould opening position.

Of these options, (i), (ii) and (iii) are the most widely used.

SYSTEM COMPARISON. Changeover at the VPT point, is often set, or triggered, by screw position switching, or by hydraulic pressure switching or by, *cavity pressure control (CPC)*. Of these three systems, CPC consistently gives the best results in terms of moulding weight repeatability and, in terms of dimensional repeatability: it is approximately twice as good as the other two, systems (these are roughly similar although line pressure control is approximately 10% better). Nowadays, with modern position transducers, it is possible to get reasonable accurate control with screw position control but, even with modern transducers, it is still often not as good, as a system switched with a pressure transducer. *Direct pressure sensing* should be selected if at all possible.

ORIENTATION MINIMIZATION. To decrease the level of *orientation* in the injection moulding it is important that the mould is filled as quickly as possible and that the plastics melt is not sheared while it is being cooled - what is called *cold, creeping flow*. Such an undesirable situation could result if the VPT point was set at the wrong position, for example, if control was handed over, from velocity to pressure, too soon. If the gate was still open, and if the packing pressure was only high enough to give slow filling then, a high level of internal stress in the moulding would result: this is because the level of retained orientation is being maximized by cooling the moulding while filling slowly. In such a case it would be better to increase the packing pressure so as to raise the mould filling rate.

vent
A hole, or port, in the *barrel* through which material can be withdrawn, or introduced, into the plastics material. An escape channel in a mould through which *volatile material* can escape.

A large quantity of air must be displaced quickly from a mould (a blow mould or an injection mould) so that the surface finish is as specified and the moulding cools at the required rate. For this reason, vents must be placed along the mould parting line and also near the last areas to be filled. Typical vents are slots 6 to 13 mm (0.25 to 0.5 in) wide. For small mouldings, such vents are of the order of 0.01 to 0.03 mm (0.0005 to 0.001 in) deep but may reach 0.5 mm (0.02") for large mouldings. Such slots are located on the mating surface of one of the mould halves. A lot of air can also escape via the clearance of a mould ejector pin. However, this should not be relied upon and deliberate vents should always be incorporated in the mould. An ejector pin can be relieved to act as a vent.

vented extruder
An extruder which contains a *vent*: often used to extract *volatiles* from a plastics material.

vented machine
Term usually applied to an *injection moulding machine* (see also *vented extruder*). The water (moisture) and volatile content in a plastics material may be reduced to an acceptable level by the use of a vented machine. The melt is decompressed part-way along the extrusion cylinder or barrel: this is done by decreasing, or reducing, the screw root diameter part-way along its length. At this point a *vent* is located in the barrel and the vapour escapes through this. (The vent may be plugged and not used when venting is not required). The vapour-free melt is then conveyed towards the die and re-compressed by increasing the screw root diameter. To ensure that melt does not escape through the vent, a dam or torpedo section is incorporated on the screw just before the vent. Because of the way that vented machines are constructed (high L:D ratio and variable screw geometry) and operated, it has been found that vented machines give very good mixing. However, it must be remembered that the action of heating some plastics in contact with water can cause decomposition, or degradation.

venting
The process whereby volatiles escape, or are removed from, a plastics material. May also mean allowing air to escape from a mould cavity via a *vent* or *ejection pin*: *vacuum venting* is also used to stop *air trapping*.

Detection of air trapping may be shown by the occurrence of polymer burning in extreme cases. The need for venting may be shown by introducing a fine spray of a hydrocarbon, such as kerosene, into the mould. If the mould is not vented the small amount of kerosene is compressed with the air in the mould during the next cycle and ignites in the mould leaving black soot on the moulding where air is trapped. In general, approximately one-third of the perimeter of the moulding should consist of *vents* or slots through which the air in the cavity can escape.

Venturi cooling ring
A unit primarily used to cool *tubular film* by using a primary air stream to draw in additional air by the Venturi effect and thus improve cooling. It is also used to stabilize, or size, the film whilst it is cooling.

VER
An abbreviation used for *vinyl ester resin*. Also see *unsaturated polyester resin*.

VE RIM
An abbreviation used for *vinyl ester resin reaction injection moulding*.

vermiculite
A group of silicate materials, like *mica*, that expand considerably (exfoliate) when heated to temperatures of about 300°C. This gives a low density material which has been used as a *filler*, for example, for *unsaturated polyester resin*.

vermiculite coated glass fibre
Heat resistant *glass fibre*. See *asbestos*.

vermilion
The scarlet form of mercury sulphide which is used as a *pigment* and which has a relative density (RD or SG) of 8·20.

vertical burning test
A burning test in which the sample is vertically mounted. See *Underwriters laboratory UL 94 vertical burning test* and *needle burning test*.

vertical extrusion
A process in which the barrel of an *extrusion* machine is at right angles to the ground, i.e. mounted vertically. When used to produce *lay-flat film*, this arrangement gives a more streamlined system than a conventional horizontal machine. The machine may be oscillated or rotated to minimize the appearance of *piston rings*.

vertical flash mould
A type of *injection mould* in which part of one mould half fits inside the other mould half. Large mouldings and/or *structural foam mouldings* are made in such moulds. Because of the small clearance between the two mating mould halves, material finds it difficult to escape down the parting or *flash line*. For ease of operation such a mould is normally mounted on a machine which is fitted with a *hydraulic clamp*. The two mould halves are closed, but not clamped, and the required volume of material is injected into the cavity. A shut-off valve in the machine nozzle is then closed and the full 'clamp' pressure is applied; this completes the moulding operation. As a result of this method of operation moulding-clamping pressures are comparatively low, e.g. 5 MNm^{-2}/0·3 tsi and so large mouldings can be made on comparatively low-powered machines. See *injection compression moulding*.

vertical flash semi-positive compression mould
A type of *semi-positive mould* in which the flash escape vertically. Such a mould type is preferred for *dough moulding compound* as direct pressure can be applied to the material: the thickness of the moulding is controlled by pressure pads mounted at the sides of the cavities.

vertical fluidised bed
A *fluid bed* through the profile travels vertically with respect to the ground.

vertical locking and horizontal injection
A compact type of *injection moulding machine* arrangement.

vertical locking and vertical injection
A type of *injection moulding machine* arrangement. The injection unit is above the clamping unit and injection is usually into the centre of the horizontal mould (which opens vertically). This gives a compact machine which is often used for insert moulding. Some machines are built so that they may be swung or slewed: they change from being *horizontal locking* and horizontal injection to vertical locking and vertical injection.

vertical machine
Usually refers to a design of *injection moulding machine*: with such a machine the injection unit is perpendicular to the floor. See *two stage, rubber injection moulding machine* and *injection moulding machines*.

vertical semi-positive mould
See *vertical flash semi-positive compression mould*.

vertical superimposed calender
A *calender* in which the rolls are stacked one directly above another, for example, an *I calender* is a super-imposed calender. For many years vertical superimposed calenders were extremely popular and were standard in the rubber industry. Such calenders suffer however from problems associated with nip interactions, that is, as the pressure in one nip varies then the roll moves in its bearings and affects the other nip. Feeding of such vertical machines is also relatively difficult. See *off-set calender*.

Verwertungsgesellschaft für gebrauchte Kunstoffe
An abbreviation used for this German organization is VGK. Sometimes also referred to as Verwertungsgesellschaft Gebrauchter Kunstoffverpackungen. The German plastics industry's own collection recycling firm. This organization, for example, collects money needed for supporting uneconomic recycling activities. It charges more than the *Duale System Deutschland* charge for this service.

Verwertungsgesellschaft Gebrauchter Kunstoffverpackungen
An abbreviation used for this German organization is VGK. See *Verwertungsgesellschaft für gebrauchte Kunstoffe*.

very high structure
An abbreviation used for this term is VHS. See *carbon black*.

very low density polyethylene
Also known as polyethylene-very low density or, polyethene-very low density or, ultra low density polyethylene. An abbreviation used for this type of material is VLDPE or PE-VLD or ULDPE.
By copolymerization ethylene with alpha olefins it is possible to produce a range of very low density materials (VLDPE) which can have a density of 0·88 to 0·91 gcm^{-3}. Such very light *HAO* grades are hardly crystalline and are rubber-like materials which can be used as an alternative to *thermoplastic elastomers*. They can be highly filled and used like other thermoplastics materials or, they can be used to enhance certain properties (for example, crack resistance) of other polyolefins. That is, they are used for materials modification, for example, for improving the impact strength of *polypropylene (PP)*. The materials at the higher end of the density range are also used as *injection moulding materials*.

very low modulus
An abbreviation used for this term is VLM. See *carbon black*.

very low structure
An abbreviation used for this term is VLS. See *carbon black*.

Vf
An abbreviation used for *flash-over voltage*. See dielectric strength.

VF_2
An abbreviation used for *vinylidene fluoride*.

VFA number
An abbreviation used for *volatile fatty acid number*.

VGC
An abbreviation used for *viscosity-gravity constant*.

VGC copolymer
See *vinyl chloride graft copolymer*.

VGC value
An abbreviation used for *viscosity-gravity constant value*.

VGK
An abbreviation used for *Verwertungsgesellschaft für gebrauchte Kunstoffe*. An abbreviation used for *Verwertungsgesellschaft Gebrauchter Kunstoffverpackungen*.

VHS
An abbreviation used for very high structure (*carbon black*).

Vicat needle instrument
The apparatus used for a *Vicat softening point* test.

Vicat softening point
A *softening point test* which is performed using a Vicat needle instrument. This term is sometimes abbreviated to VSP and the result, the Vicat softening temperature (VST), is usually quoted in °F or °C. It is measured by applying a standard load (10 N or 49 N), via a circular indentor, onto a plastics specimen while the test assembly is heated at a constant rate of either 50°C/122°F or, at 120°C/248°F per hour. The result is reported as, say, 100°C (10 N, 50°C). This means that the indentor sank into the plastic sample at 100°C for the set distance, while carrying a load of 10 Newtons and at a heating rate of 50°C. See *heat distortion temperature* and *heat resistance*.

Vicat softening temperature
An abbreviation used for this term is VST. See *Vicat softening point*.

Vicat temperature
The penetration temperature measured in a *Vicat softening point* test.

Vickers diamond pyramid test
A *hardness test* which indents a right diamond pyramid, on a square base, into a hard plastics material. The test is performed in a similar way to a *Brinell hardness test* and gives a Vickers hardness number.

Vickers hardness number
An abbreviation used for this term is HV. The result of performing a Vickers diamond pyramid test on a hard plastics material.

viewing strip
Part of a container (produced by, for example, *blow moulding*) and used to inspect the amount of product still contained after use.

vinal fibre
A manufactured, polymeric fibre composed of at least 50% by weight of vinyl alcohol units $-(CH_2-CHOH)_n-$ and in which the total of the vinyl alcohol units, and any of various acetal units, is at least 85%.

vinyl acetate
Also known as vinyl ethanoate: a monomer. An abbreviation used for this material is VAC or VA. This liquid material has a boiling point of 72°C. See *vinyl acetate plastics* and *ethylene-vinyl acetate copolymer*.

vinyl acetate modified polyethylene
An *ethylene-vinyl acetate copolymer* which is based on a low proportion of vinyl acetate (approximately up to 4 mole %). A copolymer which is processed as a thermoplastics material - just like *low density polyethylene*. It has some of the properties of a low density polyethylene but increased gloss (for film), softness and flexibility. Generally considered as a non-toxic material.

vinyl acetate plastics
Plastics based on polymers of vinyl acetate or copolymers of vinyl acetate with other monomers, the vinyl acetate being in greatest amount by mass.

vinyl acetate-ethylene copolymer
See *ethylene-vinyl acetate copolymer*.

vinyl acetate-maleic anhydride copolymer
Also known as poly-(vinyl acetate-co-maleic anhydride). An abbreviation suggested for this type of material is VAMA. A *copolymer* of *vinyl acetate* and of *maleic anhydride* in which the vinyl acetate is in the greatest amount by mass. The incorporation of maleic anhydride (MA) or acid into a vinyl chloride or vinyl acetate polymer improves the adhesive properties of the copolymer or of the terpolymer: such materials are often therefore used in adhesive applications.

vinyl alkyl ether
A monomer for a polyvinyl alkyl ether. Such a *monomer* is used to make homopolymers and copolymers, for example, with *maleic anhydride*. See *polyvinyl ether*.

vinyl benzene
See *styrene*.

vinyl chloride
Also known as chlorethene or as, monochloroethylene. An abbreviation used for this material is VC or VCM. This gaseous material has a boiling point of −14°C and is nearly always stored and used as a liquid. The monomer for *polyvinyl chloride* and for *vinyl chloride copolymers and terpolymers*.

vinyl chloride copolymers *and* terpolymers.
Polymers based on *vinyl chloride* (see VC) in which the VC is in the greatest amount by mass. Often the comonomer is present to provide *internal plasticization* - for example, this is the case for a *vinyl ether* or for *vinyl acetate*. The most important vinyl chloride copolymer is *vinyl chloride-vinyl acetate copolymer* which typically contains approximately 20% *vinyl acetate*. A vinyl chloride-acrylonitrile copolymer, which is a useful *fibre material*, is made by copolymerizing VC with *acrylonitrile* so as to make a copolymer with approximately 60% VC. A vinyl chloride-butadiene copolymer is made by copolymerizing VC with *butadiene* so as to make a copolymer which is more soluble than the VC homopolymer and which is used for lacquers and films. The copolymer (95% VC) with *vinylidene chloride* is also more soluble than the VC homopolymer. By copolymerizing VC with N-cyclohexyl-maleimide (approximately 5%) it is possible to improve the *heat distortion temperature* of PVC compounds by about 7°C.

Terpolymers of vinyl chloride are also made. For example, terpolymers of vinyl chloride, vinyl acetate and *ethyl acrylate* are soft, flexible transparent thermoplastics materials which are used as films, coatings and fibres.

vinyl chloride graft copolymer
Also known as VGC copolymer. A polymer formed by grafting *vinyl chloride* onto another polymer. For example, the polymer formed by grafting *vinyl chloride* onto *ethylene-vinyl acetate* is an *impact modifier* - this material has a relative density (RD or SG) of 1·13.

vinyl chloride monomer
An abbreviation used for this material is VCM. See *vinyl chloride*.

vinyl chloride plastics
Plastics based on polymers of *vinyl chloride* or copolymers of vinyl chloride with other monomers, the vinyl chloride being in the greatest amount by mass.

vinyl chloride-acrylonitrile copolymer
Also known as poly-(vinyl chloride-co-acrylonitrile). An abbreviation used for this type of material is VCA. A useful

fibre material made by copolymerizing *vinyl chloride* (VC) with *acrylonitrile* so as to make a copolymer with approximately 60% VC.

vinyl chloride-butadiene copolymer
Also known as poly-(vinyl chloride-co-butadiene). An abbreviation used for this type of material is VCB. Made by copolymerizing VC with *butadiene* so as to make a copolymer which is more soluble than *polyvinyl chloride* (PVC) homopolymer and which is used in lacquers and films.

vinyl chloride-ethylene copolymer
Also known as poly-(vinyl chloride-co-ethylene). An abbreviation used for this type of material is VCE. A *vinyl chloride copolymer* which contains a relatively small amount of *ethylene*. A softer, easier flowing material than the homopolymer and which can be processed at lower melt temperatures. See *vinyl chloride copolymers and terpolymers*.

vinyl chloride-ethylene-vinyl acetate terpolymer
An abbreviation used for this type of material is VCEVA. See *vinyl chloride copolymers and terpolymers*.

vinyl chloride-maleic anhydride copolymer
A copolymer of vinyl chloride and maleic anhydride: Such a copolymer is a soft flexible material used, for example, as a *plasticizer*. See *vinyl chloride copolymers and terpolymers*.

vinyl chloride-N-cyclohexylmaleimide copolymer
A *vinyl chloride* copolymer which has improved heat resistance, compared to the homopolymer, and which is still clear. Used, for example, for hot-fillable containers. See *vinyl chloride copolymers and terpolymers*.

vinyl chloride-propylene copolymer
Also known as poly-(vinyl chloride-co-propylene). An abbreviation used for this type of material is VCP. An important *vinyl chloride* copolymer and which typically contains 3 to 10% of *propylene*. A softer, easier flowing material than the homopolymer and which can be processed at lower melt temperatures. As the polymer chains are terminated with propylene groups, such copolymers have better heat stability than the PVC and need less *heat stabilizer*. Used, for example, for flame resistant films and gramophone records and for coatings. See *internal plasticization*.

vinyl chloride-vinyl acetate copolymer
Also known as polyvinyl chloride-acetate copolymer or, poly-(vinyl chloride-co-vinyl acetate). An abbreviation used for this type of material is VCVAC or VCVA or PVCA. The most important *vinyl chloride* copolymer and which typically contains approximately 20% *vinyl acetate*. A softer, easier flowing material than the homopolymer and which can be processed at lower melt temperatures. Used, for example, for gramophone records and for coatings. See *internal plasticization*.

vinyl chloride-vinyl acetate-maleic anhydride terpolymer
An abbreviation suggested for this type of material is VC-VAMA. A terpolymer of vinyl chloride (VC), vinyl acetate (VA) and of maleic anhydride (MA) in which the VC is in the greatest amount by mass. For example, VC:VA:MA = 86%:11%:3%. The incorporation of maleic anhydride (MA) or acid into a vinyl chloride or vinyl acctate polymer improves the adhesive properties of the copolymer or of the terpolymer: such materials are often therefore used in adhesive applications. Used for the heat sealing of packaging films and to produce a *plastisol* which adheres well to substrates.

vinyl chloride-vinyl propionate copolymer
Also known as poly-(vinyl chloride-co-vinyl propionate). A *vinyl chloride* copolymer which is a softer, easier flowing material than the homopolymer and which can be processed at lower melt temperatures. Used, for example, for heat sealing (of films) and for coating sheet metal. See *internal plasticization*.

vinyl chloroacetate
A *cure site monomer* for *acrylic rubber*.

vinyl compound
A compound which contains the *vinyl group*. See *vinyl chloride plastics*.

vinyl cyanide
See *acrylonitrile*.

vinyl ester resin
An abbreviation used for this type of material is VE or VER. Also known as phenylacrylate resin. The VER may be formed from an epoxy resin prepolymer and an unsaturated carboxylic acid such as acrylic or methacrylic acid so as to give, for example, the methacrylic ester or bisphenol A-epichlorhydrin. Unsaturated end groups are formed on the epoxy resin which may be used for copolymerization with styrene and initiated by a free radical initiator: a highly cross-linked network, sometimes called an unsaturated ester-styrene copolymer, is the result.

These materials are used in the *glass reinforced plastics (GRP)* industry. Such a resin is used where, for example, very good chemical resistance is required from a *sheet moulding compound* (see *high performance sheet moulding compound*). These materials are also used in the *reaction injection moulding process*: impact strength is, however, relatively low but is good when used with glass mats. See *structural reaction injection moulding*.

vinyl ester resin reaction injection moulding
An abbreviation used for this term is VER RIM or VE RIM. A *reaction injection moulding process* which uses a polymer based on an unsaturated ester. Polymerization is based on heat activated, free radical initiated polymerization. A two part RIM system can be formulated with *initiator* in one reactant stream and *accelerator* in the other.

Free radical inhibitors are used to achieve adequate storage stability and to give gel times sufficient for mould filling. Azo compounds which decompose to give nitrogen, have been used to compensate for the large shrinkage which occurs when a vinyl monomer polymerizes. Chopped glass, glass mats and/or particulate fillers are used to modify properties such as modulus and impact strength. The *vinyl ester resin* is cross-linked and relatively brittle.

vinyl ethanoate
See *vinyl acetate*.

vinyl ethanoate copolymers
See *ethylene-vinyl acetate copolymer*.

vinyl ether
Also known as vinylethyl ether. May be represented as $CH_2=CH.OC_2H_5$. A colourless liquid with a boiling point of approximately 30°C. Used as a *monomer* to make, for example, *polyvinyl ether*. See *vinyl plastics*.

vinyl group
May be represented as $CH_2=CH-$. See *vinyl chloride plastics*.

vinyl methyl ether-maleic anhydride copolymer
A copolymer of vinyl methyl ether and maleic anhydride: Such a copolymer is a soluble and reactive material used, for example, as a tackifier for pressure sensitive adhesives. The butyl ester of such a material is said to provide an adhesive for polyethylene when compounded with *polyvinyl pyrrolidone*.

vinyl monomer
A monomer for a vinyl polymer, or polyvinyl polymer, which may be represented as $CH_2=CHX$ where X is an atom or chemical group such as chlorine, fluorine or the cyanide group. See *polyvinyl chloride*.

Strictly speaking, the term also covers materials which may be represented as CWX = CYZ where W, X, Y and Z are atoms or chemical groups. However, in many cases such di-, tri- or tetra-substituted groups often due do not polymerize. Monomers containing two double bonds, such as C=C-C=C, are also vinyl monomers but are usually considered separately as 1,3-dienes.

vinyl polybutadiene
See *1,2-polybutadiene*.

vinyl polymer
A polymer prepared from a *vinyl monomer*.

vinyl polymerization
Polymerization of a *vinyl monomer*. Such a reaction is usually highly exothermic and the *polymerization* technique must be selected so that this heat can be removed rapidly and safely during the polymerization reaction.

vinyl pyridine rubber
Also known as vinyl pyridine or as, vinyl pyridine copolymer or as, butadiene-vinyl pyridine copolymer. An abbreviation used for this type of material is VP. A *copolymer* of *butadiene* and vinyl pyridine which improves the adhesion of rubber (*styrene-butadiene rubber*) to tyre cord. See *pyridine-butadiene rubber*.

vinylation
The introduction of a vinyl group. See *Reppe process*.

vinylbenzene
See *styrene*.

vinyldichlorosilane
An *organofunctional silane* in which the *vinyl group* is the functional group.

vinylethyl ether
See *vinyl ether*.

vinylidene chloride
Also known as 1·1-dichloroethylene. An abbreviation used for this material is VDC. May be represented as $CH_2 = CCl_2$. This toxic volatile material has a boiling point of approximately 32°C: the monomer for *polyvinylidene chloride* and its copolymers.

vinylidene chloride plastics
Plastics based on polymers, or resins, made by the polymerization of *vinylidene chloride* or copolymerization of vinylidene chloride with other unsaturated compounds, the vinylidene chloride being in the greatest amount by weight. See *polyvinylidene chloride*.

vinylidene chloride-acrylonitrile copolymer
Also known as poly-(vinylidene chloride-co-acrylonitrile). An abbreviation used for this type of material is VDCA. See *polyvinylidene chloride*.

vinylidene chloride-vinyl chloride copolymer
Also known as poly-(vinylidene chloride-co-vinyl chloride). An abbreviation used for this type of material is VDCVC. The copolymer (95% VC) with *vinylidene chloride* is more soluble than *vinyl chloride (VC)* homopolymer. Increasing the amount of VC increases the solubility. See *polyvinylidene chloride*.

vinylidene fluoride
An abbreviation used for this monomer is VDF or VF_2. May be represented as $CH_2 = CF_2$. Such a monomer is used to make fluoropolymers such as *polyvinylidene fluoride and fluororubbers*.

vinylidene fluoride polymer
See *polyvinylidene fluoride*.

vinylidene fluoride-chlorotrifluoroethylene copolymer
Also known as poly-(vinylidene fluoride-co-chlorotrifluoroethylene). An abbreviation used for this type of material is CFM. See *fluororubber*.

vinylidene fluoride-hexafluoropropylene copolymer
Also known as poly-(vinylidene fluoride-co-hexafluoropropylene). An abbreviation used for this type of material is FKM. See *fluororubber*.

vinylidene monomer
A monomer for a polyvinylidene polymer. May be represented as $CH_2=CX_2$ where X may be, for example, chlorine, fluorine or the cyanide group. See *polyvinylidene chloride*.

vinylpyridine-butadiene rubber
See *pyridine-butadiene rubber*.

vinylpyridine-styrene-butadiene rubber
See *pyridine-styrene-butadiene rubber*.

vinyltoluene
Also known as methylstyrene which commercially consists of a mixture of the isomers, m-methylstyrene and p-methylstyrene. A monomer which may be used to make *polyvinyltoluene* but is more generally used as a comonomer (in *unsaturated polyester resins*) and to make *alkyd resins*. See *petroleum resins*.

vinyon fibre
A manufactured, polymeric fibre composed of at least 85% by weight of vinyl chloride units -(CH_2-CHCl)-.

virgin material
Term associated with the thermoplastics industry. Material which has never been melt processed once it has been bagged by the material supplier. The original resin, or plastics material, will probably have been melt compounded by the material supplier.

virgin polymer
See *virgin material*.

virtual cross-link
See *physical cross-link*.

viscoelastic fluid
Also called an elasticoviscous fluid. A fluid which exhibits predominantly viscous flow behaviour but which also exhibits elastic recovery when the deforming stress is removed. See *die swell*.

viscoelastic state
See *leathery state*.

viscoelasticity
Material behaviour where on application of a stress both viscous flow and elasticity occur. See *leathery state and creep*.

viscometer
A test instrument or apparatus used to measure *viscosity*.

viscose
Also known as artificial silk. The solution obtained when sodium xanthate cellulose is mixed with caustic soda. See *viscose rayon*.

viscose rayon
Three materials are covered by the definition of rayon; viscose rayon, *cuprammonium rayon* and saponified cellulose acetate. Viscose rayon is the most commercially important of these materials which are all examples of regenerated cellulose. *Cellulose* is first turned into alkali cellulose (soda cellulose) with warm caustic soda, aged and then converted to sodium xanthate cellulose by reaction with *carbon disulphide*. The sodium xanthate cellulose is mixed with caustic soda to form a solution known as viscose which is filtered repeatably while being ripened. That is, the solution is allowed to stand

(ripen) until the molecular weight of the cellulose has fallen so that spinning is possible. Fibres are formed by forcing the viscose through a spinneret: the fibres are hardened by immersion in a coagulating bath which contains a mixture of acids and salts. The chemicals in the coagulating bath (sulphuric acid, sodium sulphate and zinc sulphate) turn the sodium xanthate cellulose solution back into solid cellulose. That is, cellulose is regenerated. After hardening, the *fibres* are purified and bleached: this may be done continuously, that is, continuous spinning is possible. By, for example, controlling the reduction in molecular weight which occurs during regeneration and controlling the coagulating bath conditions, it is possible to produce fibres with significantly improved properties.

The structure and form of the rayon filament can be changed in many ways. For example, the fibre cross-section can be changed to give flatter fibres which have improved covering power compared to more circular fibres. The covering power can also be increased by incorporating gas cells in the fibre: that is, cellular fibre (or bubble filled fibre) can be produced. By incorporating pigments (for example, *titanium dioxide*) before spinning, spun-dyed fibres can be produced which are very light-fast and washing resistant. To obtained improved spinning quality in staple fibres, crimp may be introduced: for example, by producing a *bicomponent fibre* or by chemical means. The properties of regular rayon may be improved by maintaining the original degree of polymerization (DP) and by maintaining the arrangement of the molecules, and of crystal structures, in the filament. If this is done, then *high modulus rayon fibre* may be obtained. See *rayon and cotton*.

viscose rayon - high modulus fibres
See *viscose rayon - high tenacity fibre.*

viscose rayon - high tenacity fibre
May also be referred to as high modulus fibre. Rayon of increased strength and dimensional stability: approximately three times the strength of normal *viscose rayon*. Used in conveyor belts and tyres. Coagulation and stretching are controlled to give a more highly oriented structure. For example, the rate of coagulation is reduced, in the *coagulating bath*, so as to permit more orientation.

viscose rayon - super high tenacity fibre
Viscose rayon of increased strength and dimensional stability compared to high tenacity fibre.

viscosity
Resistance to flow or internal friction. The term is usually used in connection with the deformation of liquids and polymer melts. The term is sometimes used as an alternative way of saying coefficient of viscosity or dynamic viscosity. A viscosity value may be obtained by dividing the *shear stress* by the *shear rate*. The units used to measure viscosity in the SI system of units are *Pascal seconds* (Pa·s). 1 Pa·s = 1 Nsm^{-2} = 0.102 kgf s m^{-2} = 10P = 0.020 88 lbf s ft^{-2} = 0.000 145 lbf s in^{-2}. The units used to measure viscosity in the *CGS system* are dyn s cm^{-2} which are called *poise (P)*. See *apparent viscosity*, *kinematic viscosity* and *capillary rheometer*.

viscosity coefficient
The shearing stress necessary to induce a unit velocity flow gradient in a material.

viscosity - output rate
The viscosity often goes down with an increase in output rate through a given extrusion die. Thus the energy used per unit output, tends to go down as the extrusion rate is increased: see *viscosity induced crystallization*.

The figures in **table 14** were obtained when samples of *nylon 6 (PA 6)*, produced by Akzo, were *flow tested* (such results are typically obtained using a die with high L:D ratio, for example, 20:1). These figures clearly show that the three, *injection moulding* grades tested, have very different viscosities, with the first grade having the lowest viscosity that is, it is the easiest flowing grade. However, the viscosity of all three materials falls as the shear rate is increased, that is, mould filling becomes easier. Raising the melt temperature, while keeping the shear rate constant, reduces the amount of injection pressure required to maintain a certain rate of flow.

viscosity ratio
See *relative viscosity*.

viscosity stabilised natural rubber
A *natural rubber* that has been chemically treated so as to prevent unwanted hardening reactions (cross-linking reactions) occurring on storage. See *constant viscosity rubber*.

viscosity units

Property	Symbol	System of measurement			
		cgs	fps	ips	S.I.
Shear stress	τ	dyn cm^{-2}	lbf ft^{-2}	lbf in^{-2}	Nm^{-2}
Shear rate	$\dot{\gamma}$	s^{-1}	s^{-1}	s^{-1}	s^{-1}
Apparent viscosity	μ_a	dyn s cm^{-2}	lbf s ft^{-2}	lbf s in^{-2}	Pa·s

cgs - an abbreviation used for centimetre-gram-second.
fps - an abbreviation used for foot-pound-second.
ips - an abbreviation used for inch-pound-second.
SI - an abbreviation used for *Système International d'Unité*.

viscosity-gravity constant
An abbreviation used for this term is VGC. A measure of the aromaticity of *petroleum oil plasticizers* which is calculated from the equation: VGC = [10G - 1.0752log(V - 38)]/[10 - log(V - 38)]. Where, VGC = viscosity-gravity constant, G = the specific gravity at 15°C and V is the Saybolt viscosity at 38°C.

If an oil has a low VGC value then it will be more compatible with an *ethylene-propylene rubber* and with a *butyl rubber*. If an oil has a high VGC value then it will be more compatible with a *chloroprene rubber* and with a *nitrile rubber*. *Natural rubber* and *styrene-butadiene rubbers* are compatible with a wider range of such oils. If an oil has a high VGC value then carbon black dispersion will be easier and better: the vulcanizate will have higher tensile and tear strength. Cure may however be adversely affected and the vulcanizate will have a greater tendency to crack growth. Heat build-up will be greater and the resilience in dynamic applications will be worse. If an oil has a low VGC value then its low temperature properties, and its high temperature properties, will be better than that of an oil with high VGC.

viscous behaviour
Refers to a *melt* and means that the molecules of the melt can flow past each other. They possess viscous rather than elastic behaviour. Such behaviour is useful in cases where stretching is important, for example, *chill roll casting*.

viscous flow
That which occurs as the molecules of a polymer *melt* slide past each other.

visually inspected grades - natural rubber
See *standard international grades*.

vitrification
See *glass transition*.

vitrification temperature
See *glass transition temperature*.

VLDPE
An abbreviation used for *very low density polyethylene*.

VLM
An abbreviation used for very low modulus (carbon black). See *carbon black*.

VLS
An abbreviation used for *vapour-liquid-solid*.

VLS
An abbreviation used for very low structure (carbon black). See *carbon black*.

VMQ silicone rubber
A *silicone rubber* which is based on *polydimethylsiloxane* where some of the methyl groups have been replaced with vinyl groups: methylvinylsilicones are the result. A VMQ-type material may be considered as a general purpose silicone rubber.

VOC
An abbreviation used for volatile organic compound.

voids
A fault found in thick sections and which develop during cooling: they are also unwanted cavities in for example, injection mouldings. See *gas injection moulding*.

Voigt unit
In polymer science, refers to the combination of a spring (which obeys *Hooke's law*) and a dashpot (which is assumed to contain a viscous oil): the two are in parallel with each other. The dashpot retards the spring and the unit represents highly elastic behaviour. See *Maxwell unit*.

vol
An abbreviation used for volume.

volatile fatty acid number
An abbreviation used for this term is VFA number. The amount of potassium hydroxide (in g) needed to neutralize the volatile fatty acids contained in 100 g of dry solids (rubber): used for *latex* assessment.

volatiles
A general term for a substance which is a gas or vapour at processing temperatures. Such volatiles therefore include water and organic materials such as monomers and/or plasticizers.

When a plastics material is dried (e.g. in an oven) then, water is often not the only contaminant that is removed: materials which are volatile at processing temperatures are also extracted from the plastic in the form of volatiles. Organic chemicals, such as *styrene*, should be treated with great care as they can be very harmful. The processing conditions should be adjusted to minimize their formation and any, which are formed, should be disposed off in an acceptable way and not allowed to pollute the factory atmosphere by, for example, venting the drying oven into the processing area. The removal of volatiles can change the processing characteristics of a plastics material.

Volatiles are also generated during the processing of plastics, for example, they are often seen rising from the die/nozzle area of a moulding machine and are formed by the burning of plastics material: such plastics material may have 'wept' from the die/nozzle during moulding or purging. It is very important to keep the outside of the die/nozzle, and of heaters, spotlessly clean as otherwise harmful vapours are formed. The operating conditions, for example, the barrel temperatures, should be kept as low as possible so as to minimize vapour formation and any escaping volatile matter should be burnt in a catalytic burner, mounted above the machine. As some volatiles, for example, some flame retardants, cannot be burnt in this way, the factory should also be equipped with an efficient extraction system.

volt
An SI derived unit which has a special symbol, that is V. It is the SI derived unit of electric potential difference or of electromotive force. Has the dimensions of $kgm^2s^{-3}A^{-1}$ which equals $JA^{-1}s^{-1}$. When one watt is dissipated between two points of a conducting wire carrying a current of one ampere, then the difference of electric potential between the two points is one volt (see *Système International d'Unité*). It is approximately equal to 10^8 electromagnetic units on the cgs. scale. One ampere times one ohm.

voltage gradient
The difference in voltage per unit thickness of a *dielectric* or, per unit length of a conductor.

volume content - of a filler/fibre
The filler volume content in a composite compound. Because of the high SG of inorganic fillers, the filler volume content in a plastics composite is low. For example, if the resin to glass ratio is 2:1 in a mix based on an *unsaturated polyester resin (UP)* then the glass content by weight is 33%. However, the volume content is only approximately 20% by volume. At a ratio of 2·5:1 the volume content will be 17%: at a ratio of 3:1 the volume content will be 14%. For example, if 100 g of UP (density 1·28 g/cc) is mixed with 50 g of glass (density 2·55 g/cc) then the total volume will equal:

$$\frac{100}{1·28} + \frac{50}{2·55} \text{ or } 97·74 \text{ cc.}$$

Of this 97·74 cc only 19·61 cc is glass. That is, 20·06% of the total.

volume intensifier
See *hydraulic intensifier*.

volume resistance
An abbreviation used for this term is R_v. See *volume resistivity*.

volume resistivity
The property used to characterise the insulation behaviour of plastics and rubber compounds. An abbreviation used for this term is ρ or, ρ_v or, VR. Also known as internal resistivity or as, specific resistance. Materials with volume resistivities greater than 10^8 ohm.cm are classed as insulating rubbers or plastics. Materials are classed as antistatic if their VR lies in the range 10^2 to 10^8 ohm.cm. For conducting grades of polymers, VR needs to be less than 100 ohm.cm. A loading of 30% of a special carbon black in ethylene vinyl acetate can, for example, reduce the volume resistivity to 10 ohm.cm.

CALCULATION OF. Volume resistivity may be defined as the electrical resistance between opposite faces of a unit cube of the material. If the small current which flows, when a dc voltage is applied across an insulator, passes through the volume of the material it is possible to measure a volume resistance, which enables a volume resistivity (ρ_v) to be calculated. Thus if the area through which the current flows is A and the length L then volume resistance R_v, is given by ρ_v times L/A where σ_v is the volume resistivity. So, $\rho_v = R_v.A/L$.

MEASUREMENT OF. Different methods are required for antistatic, conducting and insulating compounds. The method outlined here applies to materials with volume resistivities greater than 10^7 ohm.cm, that is, to an *insulator* only.

If metal foil or conductive electrodes are to be used, then the specimens must be conditioned at 23 ± 2°C and 65 ± 15% relative humidity before fitting the electrodes. For all other types of electrode, conditioning is carried out after the electrodes are applied. Conditioning is for at least 16 hours.

The test specimen consisting of a flat sheet of the material between 1 and 3·2 mm thick has special circular contact electrodes applied to its surface. Several different techniques are permitted for applying the electrodes but the primary requirement is that there should be extremely good contact between electrode and the material under test in order to minimise contact resistance.

These thin circular electrodes are then covered by metal backing plates, so that connections can be made between the electrode assembly and the measuring circuit. 500 volts dc is applied between the inner upper electrode and the lower electrode for 1 minute, and either the current flowing between these two electrodes is measured or the volume resistance is measured directly. When current is measured, volume resistance R_v is then calculated from $R_v = 500/I$ where I is the current flowing in amps.

Volume resistivity in ohm.centimetres is then calculated from: volume resistivity = volume resistance times the area through which current flows divided by specimen thickness

$$= \frac{R_v \cdot (D_1 + 5)^2}{T \cdot 40}$$

where R_v is in ohms
T is specimen thickness in mm
D_1 is the diameter of the upper inner electrode in mm. For a 50 mm electrode = $240 \cdot R_v/T$.
For a 150 mm electrode = $1900 \cdot R_v/T$.

typical values.

plastics (without fillers)	volume resistivity at 23°C (ohm.cm)
PTFE	10^{19}
LDPE	10^{18}
Polystyrene	$>10^{18}$
Polypropylene	$>10^{17}$
Polycarbonates	10^{16}
Polyacetals	10^{14}
UPVC	10^{13}–10^{14}
Cellulose acetate	10^{10}–10^{13}
Plastics with typical fillers	
Urea formaldehyde (UF)	10^{12}–10^{15}
Phenol formaldehyde (PF)	10^{9}–10^{11}
Rubbers (no fillers)	
Natural rubber gum stock	10^{15}–10^{19}
Ebonite	10^{16}
Styrene-butadiene rubber (SBR)	10^{15}
Polychloroprene	10^{11}–10^{13}
Butadiene-acrylonitrile rubber	10^{9}–10^{11}

See *surface resistivity*.

volume to surface ratio
A ratio used to characterize a container. Important when *blow moulding containers* as a change in this ratio can lead, for example, to the easier loss of gas from a drink. Polyethylene terephthalate (PET) bottles can be used uncoated for large bottles (1 to 2 litres) but, for smaller bottles they would need a coating as the surface to volume ratio is larger.

volumetric efficiency
A percentage: the volume of melt obtained from a *screw* during one revolution expressed as a percentage of the volume of the last turn of the screw (see *extrusion*). The volumetric efficiency of a pump is the actual output divided by the theoretical, or design, output.

volumetric output rate
The output rate expressed as a volume. An abbreviation used for this term in, for example, *shear flow* is Q. See *capillary rheometer*.

VP
An abbreviation used for *vinyl pyridine rubber*.

VPE
An abbreviation used for vulcanized PE. See *crosslinked polyethylene*.

VPSi
An abbreviation used for a *silicone rubber*.

VPT
An abbreviation used for *velocity pressure transfer*.

VR
An abbreviation used for *volume resistivity*.

VST
An abbreviation used for Vicat softening temperature. For example, the VST at 50°C and 10 N is the Vicat softening temperature at a heating rate of 50°C/h and using a load of 10 N. See *Vicat softening point*.

vulcanite
Another name for *ebonite*.

vulcanizable ethylene-vinyl acetate rubber
Also known as vulcanizable EVA rubber. An *ethylene-vinyl acetate copolymer*. See *ethylene-vinyl acetate rubber*.

vulcanizable EVA rubber
An abbreviation used for *vulcanizable ethylene-vinyl acetate rubber*.

vulcanizate
A product of *vulcanization*. A vulcanized rubber compound or product.

vulcanization
Also spelt vulcanisation. The term covers both a cross-linking reaction and the process used to achieve *cross-linking*. Originally associated with the curing, or cross-linking, of *natural rubber* with *sulphur*.

An irreversible process during which a rubber compound, through a change in its chemical structure (*crosslinking*), becomes less plastic and more resistant to swelling by organic liquids: elastic properties are conferred, improved, or extended over a greater range of temperature.

Approximately 0.1 kWh/kg of rubber is needed for vulcanization.

vulcanization coefficient
The percentage of bound rubber in relation to the rubber content.

vulcanization - LCM
See *liquid curing method*.

vulcanization machine
Usually means a continuous vulcanization machine where, for example, long lengths of sheet are passed over heated rolls.

vulcanization sulphur
Rhombic sulphur. This allotrope of *sulphur* is soluble in the rubber and in carbon disulphide.

vulcanized
Cross-linked material. For example, vulcanized polyethylene is *crosslinked polyethylene*. *Vulcanized rubber* is rubber which has been *cross-linked*.

vulcanized latex
Latex which has been cured or vulcanized. Latex in which the solid rubber particles contain cross-linked molecules. Produced from *latex* by using a paste of curing agents and casein: the mixture is heated (for example, at approximately 85°C for 2 h) with steam. After cooling, excess ingredients are removed by decanting or centrifuging. May be blended with ordinary latex and the mixture *coagulated* (with formic acid) to give *superior processing rubber*. Both *natural rubber* and *synthetic rubber* can be obtained in superior processing rubber grades.

vulcanized rubber
The long chain molecules on which rubbers are based, prefer to exist in a random, or coiled, configuration. However, the same, long chain molecules are not rigid structures but can untwist (or unwind), around the carbon-to-carbon bonds, on the application of a stretching force, provided the material is free to do so. This means that the material, must be above its glass transition temperature (T_g). The chain is not being stretched, it is simply made to adopt a different configuration,

or layout, but it always remembers that it would prefer to exist in a random, or coiled, state. So, when the stretching force is removed, it returns to a random, coiled state. (A rubber molecule may be likened to a steel spring which, because of the way that it is constructed, may be easily stretched).

The cross-links, which bind the chains of vulcanized (cross-linked) rubber together, serve to stop the chains slipping past each other when they are stretched. If the cross-links were not there, then the chains would find it easier to untangle, or orientate in the direction of force application and then move past each other: this would cause the stretched material to separate into pieces.

Therefore to exhibit rubbery behaviour, a material must have certain features. It should have a low glass transition temperature (T_g), bond rotation must be possible and structural features must be present which ensure that the material will retract after stretching. With a traditional rubber, the structural features which ensure that the material will retract after stretching, are the *crosslinks* between the long chain polymer molecules. However, these same crosslinks mean that once the rubber is set, or cured/vulcanized, then, it cannot be reprocessed after heating. See *thermoplastic elastomer*.

vulcanized unsaturated oil
Another name for *factice*.

vulcanizing agent
Sometimes called a bridging agent. That which causes *vulcanization*. See *sulphur*.

W

w
Sometimes used in statistical formulas for 'range'. An alternative is R.

W
This letter is used as abbreviation for
load at fracture (in a *crushing strength test*);
section width (of a *tyre*);
watt:
weight - in polymer abbreviations; and,
wood - in polymer abbreviations.

waisted
Having a waist or region of reduced cross-sectional area. For example, in a *tensile test*, the sample is usually *dumb-bell* shaped or waisted, in order to ensure that breaking does not occur in the grips of the machine.

wall slip
Slippage of *melt* at the wall of a capillary. Flow analysis generally assumes that there is no wall slip but this assumption is not necessarily true (particularly at high shear rates). If such slip exists, then a form of plug flow is superimposed on the normal *laminar flow* pattern.

wall thickness
The thickness of the walls, or sides, of a component. See *thickness control*.

wall thickness distribution
The pattern of wall thicknesses throughout a component. See, for example, *thickness control*.

Wallace hardness
The result obtained from a *Wallace hardness meter*. A Wallace hardness meter is used to measure the depth of penetration into a sample of an indentor: a preload is applied followed by a total load. The result may be expressed in degrees, for example, in British Standard Hardness Degrees or in International Rubber Hardness Degrees. See *International Rubber Hardness*.

Wallace hardness meter
A test instrument, or apparatus, used to measure the *Wallace hardness* of rubber compounds. Uses a spherical indentor of $\frac{1}{16}"$ or 1·59 mm on a flat sheet or disc of from 8 to 10 mm thick.

Wallace micro-hardness tester
A type of *Wallace hardness* tester. Tests can be made on thinner specimens than those used on larger machines. The test procedure is the same as for the larger Wallace instrument but using a 2 mm thick specimen. The depth of penetration of an indentor under a load of up to 3·5 kg after an initial preload is measured. The result may be expressed in degrees, for example, in British Standard Hardness Degrees or in International Rubber Hardness Degrees. See *International Rubber Hardness*.

Wallace rapid plasticity
The result obtained from a *Wallace rapid plastimeter*. See *parallel plate plastimeter*.

Wallace rapid plastimeter
A *parallel plate plastimeter* in which the compression surfaces are of similar size as the sample. A test piece of 1 mm thickness is placed under the plate of the instrument: a load of 10 kg is applied to the plate for 15 s at 100°C. This compresses the test piece between the plate and the foot or bed of the instrument. The depth, in 0·01 mm stages, is taken as a measure of the plasticity of the material.

warm runner mould
An injection mould used for *warm runner moulding*.

warm runner moulding
A moulding technique used in *injection moulding* of a *thermosetting plastic*. The runner is kept colder than the body of the mould so that the material in the runner does not harden: it is not ejected with the component. Used with thermosetting materials in an effort to reduce material usage.

warm sprue mould
A type of *injection mould* used for *warm sprue moulding*.

warm sprue moulding
A moulding technique used in *injection moulding* of a *thermosetting plastic*. The sprue is kept colder than the body of the mould so that the material in the sprue does not harden: for example, for UF materials the temperature is approximately 120°C) approximately 30°C higher than that of the barrel. The sprue is not ejected with the component. Used with thermosetting materials in an effort to reduce material usage. Such a system is comparatively easy to operate and can give useful cost savings.

warm-up mill
A *two-roll mill* used to pre-warm a rubber stock before further processing.

warming mill
A *two-roll mill* used to keep a rubber compound at a temperature suitable, for example, for *calendering*.

warming up
Also known as heating up. Heating a machine so that production can begin. The actual procedure may be different if the machine, for example, an extruder or an injection moulding machine, contains polymer/resin or, if it is empty.

FULL MACHINE. Must ensure that the machine is heated in a safe way so that decomposition of the polymeric material does not occur. Decomposition can produce gases under pressure: such pressurised gases can cause serious accidents by, for example, blowing hot material from the nozzle/die.

Set all temperatures to below the melting temperature of the particular material, for example, 135°C. Allow the machine to reach, and *equilibrate*, at these temperatures. Raise the temperature of the nozzle/die to above the melting temperature of the resin then, raise the temperature of the front zone and the rear zone to above the melting temperature; working towards the centre, raise the temperatures of the other zones. Allow the machine to equilibrate at these temperatures for a short while before commencing purging and production.

EMPTY MACHINE. Programme the heat input so that *overshoot* does not occur and so that the heating up time is reasonably short. With some materials, for example, *unplasticized polyvinyl chloride* (UPVC), overshoot can be very serious as it can cause degradation even before processing commences. Once the machine is at the set temperatures then it should be allowed to equilibrate before any material is introduced into the barrel. (Keep this time as short as is reasonably possible as otherwise residual material may burn by being heated in the presence of the air which is present in the screw flights). Before starting the production run check that the set conditions are satisfactory by purging some material out of the nozzle/die; check *melt temperature* with a melt probe and also check its general appearance. Only proceed if this purged material looks satisfactory. See *machine operation*.

warping
Distortion of the product which generally occurs some time after cooling is complete: lengths of extrudate may distort because of uneven cooling or, because they were hauled off unevenly (the post extrusion equipment may not be lined up correctly).

wash mill
See *washing mill*.

washing
A technique or process used to improve the purity or colour of a polymer, for example, *natural rubber*. See washing mill.

washing mill
A machine, a *two-roll mill*, used to remove impurities from *natural rubber* coagulum by working the rubber while washing it with water. Part of a *sheeting battery*.

washing soda
See *sodium carbonate*.

washings
Rubber obtained by coagulating the *latex* obtained when containers are washed.

waste-to-energy
The *incineration* of, for example, *municipal solid waste* so as to generate energy in the form of electricity.

wasted cross-link
A *cross-link* which does not contribute to the elasticity of a network in a cured or *vulcanized rubber*.

watchcase mould
A type of *tyre mould*: a single impression mould used to produce one *tyre*.

water absorption
A measure of how much water a material will absorb under specified conditions and over a certain time: can have cold water absorption and hot water absorption. Suitably *conditioned specimens* are weighed before and after immersion in water at a specified temperature and for a specified time. Some specifications state that the specimen must be re-conditioned before weighing. For moulding materials the change in weight is expressed as a percentage of the original weight. For components, the water absorption may be expressed as a weight pick-up per surface area.

A measure of how much water a thermoplastics material will absorb is given by how much it will absorb in 24 hours at room temperature. If greater than approximately 0·2%, drying is usually necessary before injection moulding can be performed; if less, this means that drying is not normally necessary. For extrusion-based processes, the level is approximately 0·1%. See *moisture content*.

water blown
A *polyurethane foam* in which most of the foaming is brought about by the water/isocyanate reaction. See *flexible polyurethane foam*.

water contamination
See *material changes* and *moisture content*.

water content
See *moisture content*.

water jet cutting
The cutting of components, or the cutting of sheet, using a fine, high speed, water jet. The jet may travel at 500 mph/845 kph and be pressurized to 700 bar. Such a cutting system can be extremely accurate and is used, for example, for cutting non-ferrous metals in sheet form (such as copper, aluminium etc). Also used for other non-metals such as rubber, ceramics and plastics composites. This type of cutting does not change the characteristics of the material at the cut edge and does not weld layers together - *laser cutting* does. Abrasive particles can be incorporated in the jet to give abrasive water jetting.

water soluble resole
A *resole* obtained as a result of using caustic soda as a catalyst. More water soluble than a *spirit soluble resole* and of a stronger (yellow) colour. This class of resole gives laminates with good mechanical properties and is therefore used, for example, to prepare laminates, based on cloth or paper, for use where the highest strength is required.

water treeing
The breakdown of cable insulation caused by moisture in the presence of an electric field.

water-cooled screw
A *cored screw* which uses water for temperature regulation. See *extrusion*.

water-extended polyester
A cross-linked material based on an *unsaturated polyester resin* and which can contain up to 90% water as fine water droplets (about 3 μm diameter). For example, the water is added to a benzoyl peroxide catalysed, styrene-resin solution and vigorously stirred until an emulsion is formed. Setting may be achieved by adding dimethyl-p-toluidine and curing at approximately 50°C. Used as a plaster of Paris replacement but water loss in service can cause warping problems.

waterway
See *flow-way*.

watt
An SI derived unit which has a special symbol, that is W. It is the SI derived unit of power and has the dimensions of $kgm^{-2}s^{-3}$ or Js^{-1}. That is, a watt is equal to one joule per second. See *Système International d'Unité*.

watt-hour
A measure of energy. A unit of work or energy which is equal to 3,600 joules = 367 kgm = 846 calories. (One kilowatt-hour is a thousand times these values.)

watt-second
A measure of energy. A unit of work or energy which is equal to 1 *joule*. 1 watt × 1 second = 1 joule = 0·102 kg.m = 0·24 calories.

waveguide
Part of a *microwave* heating system. A circular or rectangular tube which conveys the microwaves in place of the wires used in lower frequency devices.

waveiness
A periodic surface irregularity attributable to vibration and machine deflection during component manufacture. Roughness may be superimposed upon waviness. See *surface texture*.

wax
Esters of fatty acids with monohydric alcohols: such waxes are simple lipids. The fatty acids have a higher molecular weight than those used to construct oils or fats. However, the term 'wax' is often applied to any solid which is wax-like, for example, *paraffin wax*.

Wb
An abbreviation used for *weber*.

wear
See *barrel wear* and *screw*.

wear plate
Part of an *injection mould*. A hardened steel plate fitted to the locking heel of a *split mould* (or of a *side core mould*) and which allows for adjustment when wear occurs.

weather skin
See *crazing effect* and *oxidation*.

weathering
The degradation of polymers caused by exposure to a particular climate or weather. May be assessed, for example, by exposure of standard samples to sunlight and rain in a specific location and under specified conditions: such conditions may include, for example, the size of the sample, the orientation of the sample to the light etc. A weathering apparatus may be used to assess the resistance of polymers, or their compounds, to the effects of weathering. See *artificial weathering*. For improving the weathering of glass reinforced plastics see *glass reinforced plastics*.

web
A name given to film under tension during production.

weber
An SI derived unit which has a special symbol, that is Wb. It is the SI derived unit of magnetic flux. One weber is that quantity of magnetic flux which, linking a circuit of one turn, produces an electromotive force of one volt as it reduces to zero, at a uniform rate, in one second. See *Système International d'Unité*.

wedge
Part of an *undercut mould* used to put an *undercut* into a moulded component. For example, during mould opening the central ejector simultaneously actuates both the wedges and the ejectors causing the moulded component to be both released from the *side cores* and ejected in one movement.

weighing-up room
The room or area where the compounding ingredients are measured.

weight
The gravitational attraction of the Earth for a body. The amount that something weighs. Often confused with *mass* which refers to the amount, or quantity, of matter contained in a body. Weight varies with altitude and latitude whereas mass does not vary. See *UK system of units* and *US Customary Measure*.

weight average molecular weight
An abbreviation used for this term is $\overline{M}_w$. A measure of *molecular weight* which is given, for example, by light scattering. A molecular weight average which is the mean value of the weight distribution of molecular sizes: it is an average which is sensitive to higher molecular weight species.

weir
An aperture on the end of a water tank which is used to control the water overflow into the drain; weirs are usually shaped to the extrudate.

weight loading
See *creep testing*.

weight variations - injection moulding
The natural weight variations which occur when *injection moulding*. A moulding machine based on *cartridge valves*, should be capable of producing mouldings whose weight varies by less than 0·2%. A moulding machine based on spool valves, would probably be capable of producing mouldings whose weight varies by 0·4%

weight-loaded accumulator
A type of hydraulic *accumulator* which relies on the energy stored when a weight is lifted by the output from a *pump* during idle periods.

Weissenberg cone-and-plate rheogoniometer
See *rheogoniometer*.

Weissenberg effect
Also known as the rod-climbing effect. An effect observed when an elastic fluid is sheared or rotated. The fluid climbs the rod or cylinder due to *viscoelastic* effects. Has been used as the basis for extrusion machines (*elastic melt extruder*) where one disc rotates against another and the polymer melt moves towards the centre of the disc and then out through a hole in the disc towards the die.

Weissenberg extruder
See *elastic melt extruder*.

Weissenberg rheogoniometer
A cone and plate rheometer used, for example, at low shear rates to study elastic effects in polymer melts. See *rheogoniometer*.

Weissenberg-Rabinowitsch-Mooney equation
See *Rabinowitsch equation*.

weld-line
Also called a knit-line or a weld mark. A mark on, or weakness in, a moulded plastic formed by the union of two or more streams of *melt* flowing together. See *spider lines*.

weld-mark
A visible *weld-line*.

welding
Processes used to make non-recoverable joints between components. Welding is also known as sealing, for example, heat sealing. Welding is the process of uniting softened surfaces: this is often accomplished using heat and pressure. Can also have *solvent welding*. Examples of non-solvent welding processes include *hot gas welding*, *high frequency welding*, *impulse welding*, *ultrasonic welding* and *spin welding*

wet grip
A term often applied to tyres where it means the ability of a *tyre* to grip a wet road surface. Most attempts to improve wet grip, for example, by *oil extension*, worsen rolling resistance and thus increase fuel consumption. See *epoxidized natural rubber*.

wet layup
A method of making a reinforced plastics component in which the polymer compound is applied as a liquid as the reinforcement is put in place. See *hand lay-up*.

wet material
A material which as supplied contains sufficient moisture to produce unacceptable components when melt processed by, for example, *injection moulding*. Most thermoplastics

**materials are supplied dry and ready for use although some are 'wet' and must be dried before use. See *moisture absorption*.

wet spinning
A production process for *fibres* which uses a solution of a fibre-forming polymer. For example, fibres are formed by forcing *viscose* through a spinneret: the fibres are then hardened by immersion in a *coagulating bath*. That is, *cellulose* is regenerated.

wet winding
A method of making *filament-wound reinforced plastics* in which the fibre reinforcement is coated with a polymer compound as a liquid just prior to wrapping on a mandrel.

wet-ground
Usually refers to a method of filler preparation. If water is used in the grinding process, then the product is usually of finer particle size and more uniform size distribution than the dry-ground product. For example, wet-ground limestone is of finer particle size and more uniform size distribution than dry-ground limestone. See *limestone*.

wet-out
The ability of a resin system to impregnate, or saturate, the reinforcement.

wettability
The relative tendency of a fluid, for example a plastics melt, to flow onto and cover a substrate: wettability of a plastics coating is typically dependent on surface tension, viscosity and chemical compatibility.

WF latex grades
An abbreviation used for *whole field latex grades*.

whale oil
See *nylon 9*.

Wheatstone bridge
A divided electrical circuit used for the measurement of resistance.

whisker
A short, single-crystal *fibre* of high purity: such a material has mechanical strengths equivalent to the binding forces of adjacent atoms. These small diameter fibres (say 1 to 30 μm) have very high strength and modulus – particularly when the fibre diameter is reduced as this improves internal crystal perfection (see *cobweb whisker*). Whiskers are also comparatively flexible which minimizes damage during fibre processing. Examples of whisker-forming materials include, *alumina* (sapphire whiskers) and *silicon carbide*. Such whiskers are typically produced by vapour phase reactions at high temperatures, for example, by a vapour-liquid-solid (VLS) technique.

white and pale crepes
One of the forms in which solid *natural rubber* (unvulcanized rubber) is supplied. A pale coloured form of natural rubber (see *pale crepe*) from which most impurities (non-rubber constituents) have been removed. Such grades must be produced from the fresh coagula of natural liquid latex under carefully controlled conditions and the thickness of the material must be as per standard. See *standard international grades*.

white asbestos
Chrysotile asbestos. See *asbestos*.

white crepe
A pale coloured form of natural rubber (see *pale crepe*) from which most impurities (non-rubber constituents) have been removed. See *white and pale crepes*.

white factice
May be referred to as sulphur chloride (white) factice or as, cold-type factice. See *factice and sulphurized oil*.

white lead
Also known as *basic lead carbonate*.

white lead paste
A paste of *basic lead carbonate* in a *phthalate plasticizer*.

white pigments
Pigments which impart a white colour to a product. The most important white pigment is *titanium dioxide*. Others include *basic lead carbonate*, *basic lead silicate*, *basic white lead*, *lithopone*, *zinc oxide* and *zinc sulphide*. Also see *lighting whites*.

white reinforcing filler
A white filler with a *reinforcing action* in a rubber compound. Usually *silica* or a *silicate*.

white spirit
Also known as mineral spirit. A petroleum fraction with a boiling point in the range of approximately 150 to 200°C: a mixture of *alkanes*. This material has a relative density (RD or SG) of approximately 0·8.

whiting
May be represented as $CaCO_3$. Ground chalk or particulate *calcium carbonate*. This material has a relative density (RD or SG) of approximately 2·70. An odourless, white powder widely used as a cheap white filler in, for example, rubber compounds. Synthetic, precipitated calcium carbonate has a relative density (RD or SG) of 2·62 and may also be referred to as whiting.

whole field latex grade
An abbreviation used for this term is WF latex grade. See *standard Malaysian rubber latex grades*.

whole latex rubber
See *sprayed rubber*.

whole tyre reclaim
Reclaimed rubber from *tyres*.

wide vulcanization range
Usually refers to a compound which can be cured satisfactorily over a range of processing conditions, for example, the temperatures are not critical. See *accelerator*.

wig-wag
A feeding device used to obtain uniform feeding to a *calender*. The strip feed is fed into the wig-wag which swings like a pendulum in front of the *nip* thus distributing the incoming feed. This device distributes the material across the nip and results in sheeting of improved accuracy.

Williams plastimeter
A parallel plate type of *plastimeter*.

wind up
Roller system for collecting film or wire after *extrusion*.

winder
A unit used to collect an extrudate on spools or reels.

winding station
The *system* used for *reeling*.

window
A fault: an area of transparent material in an otherwise translucent component; could be due to uneven cooling or, to unmelted granules which have passed through the machine.

winkle gate
A *submarine gate* with a curved axis. Although such gates may be useful (in the production of buttons and bobbins)

they are more difficult to make and so their usage is comparatively restricted.

winter-green oil
See *methyl salicylate.*

wire covering
An *extrusion* operation for the application of a plastics coating onto a wire; usually done by crosshead extrusion. See *extrusion wire-covering process.*

wire reinforced hose
A hose reinforced with layers of spiralled, or braided, wire.

wire test
See *hot wire test.*

wire-covering
The covering of a wire or cable with an insulator. See *extrusion wire-covering process.*

witness mark
A mark on the surface of a moulding which is often caused by an *ejector pin*: also known as a knockout mark.

wk
An abbreviation used for week.

wollastonite
A naturally-occurring silicate mineral. Natural *calcium silicate* is derived from wollastonite while synthetic calcium silicate is made from *diatomaceous silica* and *lime*. Wollastonite has acicular particles and a relative density (RD or SG) of 2·9. A fine white powder which is alkaline when dispersed in water. A calcium metasilicate used, for example, as a *filler* in floor tiles and in plastisols.

wood alcohol
See *methyl alcohol.*

wood flour
A *filler* prepared by the grinding of soft wood. This material has a relative density (RD or SG) of approximately 1·25. Originally used as a water-absorbing, reinforcing filler in thermosetting plastics such as *phenol-formaldehyde* but now also used in *thermoplastics materials* such as *polypropylene* and *polyvinyl chloride* (PVC). For example, in the case of PVC it is used to impart a resemblance to wood in the final products: such products include extruded beading and sheet and may be based on cellular materials.

wood naphtha
Impure *methyl alcohol*. See *naphtha.*

wood oil
See *tung oil.*

wood pulp
An important softwood timber product: wood reduced to pulp by mechanical and chemical treatments. A major source of *cellulose* for the chemical industry, Cellulose jelly is prepared from wood pulp by making a chemical solution which is then, for example, extruded and turned into filaments. See *regenerated cellulose fibre.*

wood rosin
See *rosin.*

Wood's metal
A low melting point alloy based on bismuth (50%), lead (25%), cadmium (12·5%) and tin (12·5%). Wood's metal has a melting point of 71°C.

wool
A natural *fibre* obtained from sheep and which is very hygroscopic and durable. This material has a relative density (RD or SG) of 1·32.

wool fat
See *lanolin.*

wool grease
See *lanolin.*

work
The product of force times the distance moved. A symbol used is A. The SI unit of work is the *joule*. Other commonly seen units include the erg, the foot-pound, the foot-poundal and the kilogram-metre (kg.m). One erg is 1 cm × 1 dyne. 10^7 ergs = 1 joule = 1 watt-second = 0·24 calories = 0·102 kg.m.

work handling device
Also known as a robot or as a pick-and-place unit (see *pick-and-place automation*). Used, for example, in the *injection moulding* industry for demoulding operations. These devices usually take the moulding out in the same direction as that adopted by the operator, i.e. normal to the mould opening stroke. To allow the injection moulding machines to be placed closer together, some robots now take the moulding(s) out and bring them along the top of the clamp, i.e pass them rearwards; the device may even pass the moulding into a clean room. Work-handling devices, not only remove the moulding but can cut off sprues, can carry out follow-on automation and then place the finished assembly on a conveyor for packing. Colour checking/moulding tolerances checking etc, using a video camera and display (plus associated software) is now also possible.

As work handling devices become more complicated then the chance of something going wrong is increased. To make fault finding easier some systems will print out the fault diagnosis in plain language thus easing fault recognition. The work handling devices now used by the injection moulding industry are capable of good greater positional accuracy: electric drive devices can give positional accuracy of less than ± 0·1 mm with high speeds and smooth movements.

work hardening
See *strain hardening.*

work softening
See *strain softening.*

working face width
The width of a roll used in production. For example, a *calender* may have up to 3 m of working face width available so that sheet of this width may be produced.

working range
The temperatures over which a polymeric product shows useful properties. For thermoplastics materials this lies between the brittle to ductile transition temperature at the lower end and the softening point at the higher end.

worn machine
In the case of a machine which contains a *screw* this term usually means an increase in the operating clearance of the screw/barrel system and/or wear of, the screw flights or of a barrel *zone*. See *barrel wear* and *screw wear.*

woven fabric
A cloth or fabric made from fibres by a weaving process: the fabrics are made by inter-weaving threads or *yarns* which have been produced by twisting together long, thin fibres. Woven fabrics are most important to the reinforced plastics industry where they are used to make *composites.*

woven glass cloth
See *glass cloth.*

woven glass fabric
See *glass cloth.*

woven hose
A hose in which the reinforcing material has been applied by circular weaving.

woven roving
A heavy fabric made by weaving rovings made of, for example, *glass fibre*. See *glass cloth*.

wrap
The length of flow path in contact with the rolls of a *calender*. See *L calender*.

wrapped ply hose
A *mandrel-made hose* reinforced with woven fabric.

wrinkle
An imperfection in reinforced plastics that has the appearance of a wave moulded into one or more plies of fabric or other reinforcing material.

wrought iron
This is almost pure pig iron: produced by oxidizing impurities in *iron* at high temperatures.

X

X
Used to modify polymer abbreviations, for example, when used in this way stands for crosslinked or crosslinkable. X is also used for xylyl and for xylenol. In rubber technology, X is used to indicate that a rubber has been modified, for example, it is used for carboxylated and, sometimes, for halogenated (see *halogenated butyl rubber*).

X rays
Electromagnetic *radiation* of extremely short wavelength, for example, 10^{-9} to 10^{-11} m. A form of *high energy radiation*.

X-CR
An abbreviation used for *chloroprene rubber* with reactive groups.

X-ray gauge
A gauge which uses X-ray radiation. A measuring device used in *calendering*. X-ray devices can be mounted further away from the roll than nuclear radiation devices and are therefore less likely to be hit in service and are also relatively insensitive to roll temperature variations.

X-rays
See *X rays*.

X 36 CrMo 17
A hardenable *corrosion resistant steel* which contains 17% chromium and 1% molybdenum.

X 45 Cr-Mn-Mo 4 steel
A chromium-manganese-molybdenum steel with a relatively high carbon content. See *fully-hardening steel*.

xanthate
See *sodium xanthate cellulose*.

xanthates
Salts or esters of xanthic acids which may be represented as ROCSSH. A class, or type, of rubber *accelerator*. Very powerful accelerators which tend to be too fast for use as the primary accelerator in dry compounds based on *natural rubber*. Used in *latex* technology and in slower-curing diene rubbers such as *ethylene-propylene rubbers*.

xanthene
Also known as dibenzo-1,4 pyran. May be represented as $C_6H_4.CH_2C_6H_4$. A yellow crystalline material with a melting point of approximately 100°C: used to make *xanthene dyes*.

xanthene dyes
A class or type of organic *dye* which is based on the organic compound, *xanthene*. Some of the linked benzene rings in such materials (say 4) are substituted. Typified by rhodamine B - although, in general, this dye is not suitable for use in thermoplastics. Some xanthene dyes are compatible with acrylics, polystyrene and unplasticized polyvinyl chloride (PVC).

XCR
An abbreviation used for *chloroprene rubber* with reactive groups.

XDI
An abbreviation used for *xylylene diisocyanate*.

xenon arc lamp
A light source used in *artificial weathering* studies as the light it emits closely resembles that of sunlight.

xenon lamp exposure test
An *artificial weathering* test performed using a *xenon arc* lamp.

xerogel
See *silica and silica gel*.

XF
An abbreviation used for *xylenol-formaldehyde resin*.

XIIR
An abbreviation used for *halogenated butyl rubber*.

XL
An abbreviation used for *crosslinked* or, cured or, vulcanized.

XMC
An abbreviation used for continuous roving moulding compound. A *polyester moulding compound*.

XNBR
An abbreviation used for *carboxylated nitrile rubber*.

XPS
An abbreviation used for *expanded polystyrene*.

XSBR
An abbreviation used for carboxylic-styrene butadiene rubber. See *carboxylated styrene-butadiene rubber*.

xylene
Also known as xylol or as, dimethyl benzene. An aromatic hydrocarbon which exists in the form of three isomers. A commercial mixture of these three isomers has a relative density (RD or SG) of 0·86 and a boiling point of approximately 138 to 142°C. May be represented as $C_6H_4(CH_3))_2$. It is a good solvent for uncured *nitrile butadiene rubber* (low acrylonitrile NBR), *chloroprene rubber (CR), butyl rubber (IIR), natural rubber* (NR), *styrene-butadiene rubber* (SBR) and *thiokol rubber* (T). This chemical causes a large amount of swelling, or gel formation, of uncured high acrylonitrile NBR.

xylenol
Also known as dimethyl phenol. An aromatic material which exists in the form of six isomers. May be represented as $C_6H_4(CH_3)_2OH$. 3,5-xylenol has a melting point of approximately 63°C and a boiling point of 225°C. As it has the three reactive positions available (o and p), it is used to make *phenol-formaldehyde*-type materials (PF) with improved chemical resistance for coating purposes: it is more oil-soluble than a PF resin. 2,6-xylenol has a melting point of approximately 49°C and a boiling point of 212°C. It is used to make the monomer for polypropylene oxide-type materials.

xylenol-formaldehyde
An abbreviation used for this type of material is XF. The reaction product of *xylenol* and *formaldehyde*: the resin formation is catalysed by the use of sulphuric acid.

xylol
See *xylene*.

xylyl mercaptan
An additive used to reduce the colour of *pale crepe* by removing β-carotene. By using *sodium bisulphite* and xylyl mercaptan *extra white crepe* may be produced.

xylylene diisocyanate
An abbreviation used for this material is XDI. A *diisocyanate* which may be represented as $C_6H_4(CH_2NCO)_2$. It is used to make *polyurethanes* which have good resistance to ageing, for example, for coatings.

Y

Y
If a plastics material contains a filler then the letter may be used to show the presence of yarn. Y is also used as a prefix in *thermoplastic elastomer abbreviations*. For example:

Y-BPO or YBPO = *polyether ester elastomer*;
Y-IR or YIR = a *thermoplastic elastomer* based on *isoprene* rubber; and,
Y-NBR or YNBR = a *thermoplastic elastomer* based on *nitrile* rubber.

yard
The basic standard of length of the imperial system of units. A yard contains 3 feet and therefore 36 inches. One yard is equivalent to 0·9144 metres. See *UK system of units and US Customary Measure*.

yardage
The length of product on a reel.

yarn
Bundles of fibres or of filaments. A thread made by twisting together long, thin fibres. Two main types of yarn; *continuous filament yarn* and *spun yarn*.

yarn package
A *package of yarn*.

yd
An abbreviation used for *yard*;
 yd^2 = square yard; and
 yd^3 = cubic yard.

yellow
A *primary colour*. See *yellow pigments*.

yellow circle rubber
A *technically classified rubber* which is medium curing. See *circle rubber*.

yellow pigments
Such pigments are often based on elements such as antimony, lead, tin chromium and/or nickel. See *cadmium sulphide, calcium plumbate, chrome yellows, French ochre, hansa yellows, lead chromate* and *zinc chromes*.

yellowing vegetable oil
A *vegetable oil* may be classified as yellowing or non-yellowing - dependent on their behaviour in use as films. This behaviour depends, for example, on the linolenic acid content. For example, *soya-bean oil* can have a linolenic acid content of approximately 10% and its non-yellowing properties would not be as good as *safflower and sunflower oils*. Semi-drying and non-drying oils are generally non-yellowing.

yield
The area of film at a given thickness produced from a given weight of polymer.

yield point
That point on a *stress-strain curve* above which deformation is not completely recoverable. Polymer compounds may recover their original dimensions completely as long as they are not stretched beyond the yield point. See *yield strength*.

yield stimulant
A chemical used in *natural rubber* production to increase the output obtained from a tree by delaying the onset of *plugging*. For example, 2-chloroethane phosphonic acid.

yield strength
May be calculated from the load, in a *tensile strength test* at which the specimen continues to elongate without additional load being applied. The stress, the stress at the yield point, at which a material under test exhibits a specified limiting deviation from the proportionality of stress to strain.

yield stress
Also known as yield value. The stress at the *yield point* on a stress-strain curve.

yield value
See *yield stress*.

yielding
The beginning of permanent deformation, that is, the material has been stretched beyond a critical value.

Young's modulus
Usually denoted by the letter E. The constant of proportionality. Also known as elasticity modulus. An abbreviation used for this term is E. Measured as the slope of the stress-strain curve obtained in a *tensile test*. If the plot is not linear, then the slope of a secant is taken thus giving a secant modulus. See *Hooke's law* and *modulus*.

yr
An abbreviation used for year.

Z

Z
This letter is used as an abbreviation for:
 atomic number;
 azelate. See, for example, *dialphanyl azelate*;
 filler. For example, a non-specified filler;
 phosphonitrilic polymer. A rubber which has nitrogen and phosphorous in the main polymer chain. For example, *FZ*.

Z average molecular weight
An abbreviation used for this term is $\overline{M_z}$. A measure of *molecular weight* which is given, for example, by gel permeation chromatography. A molecular weight average which is sensitive to higher molecular weight species: more sensitive than the weight average and therefore preferred when high molecular weight effects dominate, for example, when discussing melt viscosity.

Z calender
A *calender* named after the letter Z. An off-set, *four roll calender*.

Z pin
Also called a Z puller. A type of sprue puller widely used in *injection moulding*. The pin is commonly located opposite the *sprue* and located within the *cold slug well*. The head of the pin has an undercut formed by machining: viewed from the side, the head of the pin has the shape of the letter Z. This undercut is filled during mould filling and is cleared when the pin is pushed forward on component ejection.

Z puller
See *Z pin*.

Zamak
A trade name/trademark for a *low melting point metallic alloy* based on zinc. See *zinc-based alloy*.

ZBD
An abbreviation used for *zinc dibenzyl dithiocarbamate*.

ZBDP
An abbreviation used for *zinc dibutyl dithiophosphate*.

ZD
An abbreviation used for *zero defects*.

ZDBC
An abbreviation used for *zinc dibutyl dithiocarbamate*.

ZDC
An abbreviation used for *zinc diethyl dithiocarbamate*.

zeolite
A member of a group of silicates which have water molecules held in lattice spaces; hydrated aluminium silicates which have associated alkali metals. Both the water and the alkali metals can move through the lattice structure and may therefore be replaced. The materials exhibit exchange properties and are, for example, used to exchange or separate mixtures of materials. This type of mineral occur naturally as secondary minerals in the cavities of igneous rocks. The term may also be used to describe artificial materials which work in a similar way, for example, *ion exchange resins*.

ZEPD
An abbreviation used for *zinc ethyl phenyl dithiocarbamate*.

zero compression screw
A *screw* with no *compression ratio:* such a screw is used on a *blow moulding machine* to decrease heat generation.

Typically, for a blow moulding machine running on *high molecular weight polyethylene* (HMWPE), a zero compression screw is, in effect lengthened so that a *fluted barrier section* may be fitted, followed by a section which contains mixing pins. This is because at high output rates, the fluted barrier section (for example, a *Maddock mixing section*) often cannot homogenize high molecular weight melts satisfactorily. This arrangement also allows the melt to become thermally homogenized after it has been sheared: that is, it gives *distributive mixing*. Other arrangements are possible. For example, two mixing pin sections may sandwich a Maddock mixing section.

zero defects
An abbreviation used for this term is ZD. A manufacturing objective: that is, trying to produce perfect product all the time.

zero metering screw
A *zero compression screw*. See *low work screw*.

zero shrink sheet moulding compound
See *low profile sheet moulding compound*.

zero shrink SMC
An abbreviation used for zero shrinkage sheet moulding compound. See *low profile sheet moulding compound*.

zero start
Means that the speed control, of, for example, an *extruder*, must be turned to zero before the drive can be started.

ZI
An abbreviation used for zero inventory manufacturing. See *JIT*.

Ziegler polymerization
See *Ziegler-Natta polymerization*.

Ziegler-Natta catalyst system
A stereospecific catalyst system which allows the production of polymers with a controlled molecular architecture. Often consists of two components, for example, an aluminium alkyl and a transition metal salt. An example is aluminium triethyl and titanium trichloride, See, for example, *stereospecific styrene-butadiene rubber*.

Ziegler-Natta polymerization
Sometimes referred to as Ziegler polymerization. Polymerization performed with a *Ziegler-Natta catalyst system*. Stereoregular polymerization which results in polymers with a controlled molecular architecture, for example, mainly *isotactic polymer*. See α *olefin*.

zinc
Group 11b of the Periodic table consists of the elements zinc, cadmium and mercury. Zinc does not occur naturally. A *transition element* (Zn) which is a hard, bluish-white, lustrous metal with a melting point of 419°C and a boiling point of 907°C: the relative density is 7.13. Brittle at room temperatures but malleable over the temperature range 100 to 150°C. Used in sheet form as a corrosion resistant covering: also used to plate (galvanize) steel with a corrosion resistant covering. Used to make die castings and alloys such as brass. Some zinc compounds are used, for example, as *heat stabilizers* for *polyvinyl chloride* (PVC): others are used as *accelerators* for rubbers. See also *zinc oxide*.

zinc arc spraying
A method of putting a conductive layer of zinc onto a plastics material, for example, a solvent sensitive plastics material, so as to obtain electromagnetic shielding (see *electromagnetic radiation*). The metal is melted by an electric arc and sprayed in the form of fine droplets onto the component. This gives a hard dense coating but is expensive and only suitable for *heat resistant plastics*.

zinc borate
An additive which reduces the amount of smoke produced when a compound burns. A *smoke suppressant*. Sometimes used as a synergistic additive, with halogen compounds, in place of *antimony oxide*.

zinc butyl xanthate
An *accelerator*. See *dibenzyl amine*.

zinc/calcium salts
A type of *heat stabilizer* for *polyvinyl chloride*.

zinc carbonate
A basic carbonate of variable composition and which occurs naturally, for example, as hydrozincite. This material has a relative density (RD or SG), when precipitated, of 3.30. A filler and activator for rubber compounds. See *basic zinc carbonate*.

zinc chromate
A yellow pigment produced from zinc oxide and potassium dichromate. See *zinc chromes*.

zinc chromes
Yellow pigments such as zinc chromate and zinc tetroxychromate. Used, for example, as anticorrosive primers. Such pigments are non-toxic and sunlight resistant but have less hiding power than lead chromes.

zinc dibenzyl dithiocarbamate
Also known as zinc dibenzyldithiocarbamate. An abbreviation used for this material is ZBD. This solid material has a melting point of approximately 170°C and a relative density (RD or SG) of 1.41. An *ultraaccelerator* for the *sulphur vulcanization* of rubbers, especially for latices. A very safe processing compound.

zinc dibutyl dithiocarbamate
Also known as zinc dibutyldithiocarbamate. An abbreviation used for this material is ZDBC. This solid material has a melting point of approximately 106°C and a relative density (RD or SG) of approximately 1·26. An *ultraaccelerator* for the *sulphur vulcanization* of rubbers, especially for latices. A very safe processing compound.

zinc dibutyl dithiophosphate
An abbreviation used for this material is ZBDP. An example of a *sulphur donor vulcanization system* which is not an *accelerator* in that formulation.

zinc diethyl dithiocarbamate
Also known as zinc diethyldithiocarbamate. An abbreviation used for this material is ZDC. This solid material has a melting point of approximately 176°C and a relative density (RD or SG) of approximately 1·46. An *ultraaccelerator* for the *sulphur vulcanization* of rubbers, especially for latices.

zinc dimethyl dithiocarbamate
Also known as zinc dimethyldithiocarbamate. An abbreviation used for this material is ZMD or ZDMC. This solid material has a melting point of approximately 245°C and a relative density (RD or SG) of approximately 1·67. An *ultraaccelerator* for the *sulphur vulcanization* of rubbers, especially for latices.

zinc ethyl phenyl dithiocarbamate
Also known as zinc ethylphenyldithiocarbamate. An abbreviation used for this material is ZEPD. This solid material has a melting point of approximately 205°C and a relative density (RD or SG) of approximately 1·46. An *ultraaccelerator* for the *sulphur vulcanization* of rubbers.

zinc hydroxystannate
A *flame retardant*, for example, for *polyvinyl chloride*

zinc isopropyl xanthate
Also known as zinc isopropylxanthate. An abbreviation used for this material is ZIX. This solid material has a melting point of 145°C and a relative density (RD or SG) of 1·54. An *ultraaccelerator* for the *sulphur vulcanization* of rubbers.

zinc laurate
This *fatty acid* ester has a relative density (RD or SG) of 1·10. A *lubricant*.

zinc lupetidine dithiocarbamate
Also known as zinc lupetidinedithiocarbamate. An abbreviation used for this material is ZL or ZLD. This solid material has a melting point of approximately 90°C and a relative density (RD or SG) of approximately 1·55. An *ultraaccelerator* for the *sulphur vulcanization* of rubbers, especially for latices as it is very active.

zinc MBT
An abbreviation used for *zinc mercaptobenzothiazole*.

zinc mercaptobenzothiazole
Also known as zinc-2-mercaptobenzothiazole. An abbreviation used for this material is ZMBT or zinc MBT. This solid material has a melting point of greater than 200°C and a relative density (RD or SG) of 1·64. A vulcanization *accelerator* and an antioxidant in rubber compounds.

zinc methyl phenyl dithiocarbamate
Also known as zinc methylphenyldithiocarbamate. An abbreviation used for this material is ZMPD. This solid material has a melting point of approximately 230°C and a relative density (RD or SG) of approximately 1·53. An *ultraaccelerator* for the *sulphur vulcanization* of rubbers.

zinc octoate
A metal soap used as a synergistic stabilizer, in *metal soap* stabilizers, to improve colour. Used in *silicone resin* production to achieve cross-linking so as to give a cross-linked *polyorganosiloxane*.

zinc oxide
Also known as Chinese white or as, zinc white or as, flowers of zinc or as, lana philosophica. This acicular-nodular, white material has a relative density (RD or SG) of 5·5 to 5·8. Used as part of the *activator* system in rubber technology. For example, with a *fatty acid* such as stearic acid. Reacts with the acid/accelerator to form a zinc salt which, in turn forms a perthiosalt (sometimes called a perthioaccelerator): this in turn reacts with the rubber to form cross-links. Now used with *polypropylene* (PP) compounds to obtain mildew resistance and UV resistance. To obtain UV resistance, a *synergistic* mixture of zinc oxide and a 2-hydroxybenzophenone derivative, may be used. Zinc oxide can react with free carboxyl groups in *unsaturated polyesters* and this will increase the viscosity of, for example, paints based on such materials (*alkyd* paints). At one time zinc oxide was one of the major white pigments: *leaded zinc oxide* is also used as a white pigment for paints.

zinc oxide viscosity
An abbreviation used for this term is ZOV. Originally referred to as zinc oxide thickening or ZOT test. A test performed on *natural rubber latex*: used to determine the chemical stability of latex so as to decide, for example, whether a particular batch of *latex* is suitable for use in *foam rubber* manufacture. Zinc oxide addition increases *latex viscosity* and decreases latex stability. Either or both factors are measured.

zinc palmitate
A salt of the *fatty acid*, palmitic acid. See *zinc stearate*.

zinc pentamethylene dithiocarbamate
Also known as zinc pentamethylenedithiocarbamate. An abbreviation used for this material is ZPD. This solid material has a melting point of approximately 225°C and a relative density (RD or SG) of approximately 1·35. An *ultraaccelerator* for the *sulphur vulcanization* of rubbers.

zinc stannate
Used as a *flame retardant*, for example, for *polyvinyl chloride*.

zinc stearate
This white solid material has a melting point of approximately 128°C and a relative density (RD or SG) of 1·06. The commercial material is often not solely zinc stearate but a mixture of zinc palmitate, zinc stearate and zinc oxide. A *lubricant*, for example, for *polystyrene*. Used as an *accelerator activator* in rubber compounds in place of *stearic acid* and *zinc oxide*. Widely used as a dusting agent and as a mould release agent in the rubber industry.

zinc sulphide
A white pigment. Unlike *basic lead carbonate*, this material (ZnS) does not darken on exposure to hydrogen sulphide. A major white *pigment* until the development of *titanium dioxide*. Has good *hiding power* but a tendency to chalk. ZnS has a relative density of approximately 4·0, an index of refraction of 2·37 and a colour index of 7. Used to make *lithopone*.

zinc tetroxychromate
A yellow pigment produced from zinc oxide and chromic acid. See *zinc chromes*.

zinc vulcanizate
An *elastomeric ionomer* produced from an *ionomer* which has been neutralized with, for example, zinc oxide, so as to produce an elastomeric ionomer. See *carboxylated polybutadiene ionomer*.

zinc white
See *zinc oxide*.

zinc-2-mercaptobenzothiazole
See *zinc mercaptobenzothiazole*.

zinc-based alloy
A *low melting point metallic alloy* based on zinc and used for the construction of, for example, a *prototype mould*. Such alloys are usually known by trade names/trademarks, for example, Zamak, Ayem or Kirksite.

Cast mould cavities, based on zinc-based alloys, are used mainly for mouldings that need an intricately patterned surface and which can be taken from a simple model. The components may be modelled in wax or wood and then disposable casts (sub-masters) may be prepared from them using, for example, plaster or silicone rubber. Porous materials will need to be sealed, for example, with pattern-makers varnish, before use. A shrinkage allowance of 10 mm per metre (0.125 in per foot) should be allowed for a zinc-based alloy. The sub-master is contained in a suitable box (for example, wood surfaced with a melamine-formaldehyde laminate) and a feed system (for the polymer) incorporated. The heated casting metal is then poured into the box.

The resultant cast cavities and cores can then be assembled or contained in a steel bolster which absorbs the clamping force. At room temperatures the hardness of a zinc-based alloy (about 115 Brinell) approaches that of steel but because of its relatively low melting temperature (it is prepared at 450°C) it is not recommended for high-temperature moulds (for example, above 130°C). A high degree of fluidity allows exceptional reproduction of fine detail and another advantage is that moulds made from zinc-based alloys may be re-melted and re-cast without loss of mechanical properties.

ZIX
An abbreviation used for *zinc isopropyl xanthate*.

ZL
An abbreviation used for *zinc lupetidine dithiocarbamate*.

ZLD
An abbreviation used for *zinc lupetidine dithiocarbamate*.

ZMBT
An abbreviation used for *zinc mercaptobenzothiazole*.

ZMC
An abbreviation used for low viscosity moulding compound. A *polyester moulding compound*.

ZMD
An abbreviation used for *zinc dimethyl dithiocarbamate*.

ZMDC
An abbreviation used for *zinc dimethyl dithiocarbamate*.

ZMPD
An abbreviation used for *zinc methyl phenyl dithiocarbamate*.

ZO
An abbreviation used for *zinc oxide* (see *zinc oxide viscosity*).

zone
A part or section of a machine barrel. For ease of control, the barrels of, for example, injection moulding machines and of extruders, are divided into zones or regions. Each zone should have its own heating system. See *zoned heating*.

zoned heating
The separate heating, and temperature control, of each zone of a machine. The smallest *injection moulding machine* will usually have three zones and larger machines may have, for example, twelve. Each of these zones is controlled by a temperature sensor and associated equipment, for example, a microprocessor-based, three term (PID) controller.

Zorro mixing section
A *fluted mixer*: a *dispersive mixing section*.

ZOT test
An abbreviation for zinc oxide (thickening) test. See *zinc oxide viscosity*.

ZOV
An abbreviation used for *zinc oxide viscosity*.

ZPD
An abbreviation used for *zinc pentamethylene dithiocarbamate*.

Table 1a. *Standard (based on ISO and ASTM) abbreviations of plastics*

Abbreviation	Name
ABS	Acrylonitrile butadiene styrene.
A/B/A	Acrylonitrile/butadiene/acrylate.
A/CPE/S	Acrylonitrile/chlorinated polyethylene/styrene.
A/EPDM/S	Acrylonitrile/ethylene-propylene-diene/styrene.
A/MMA	Acrylonitrile/methyl methacrylate.
ASA	Acrylonitrile/styrene/acrylate.
CA	Cellulose acetate.
CAB	Cellulose acetate butyrate.
CAP	Cellulose acetate propionate.
CF	Cresol-formaldehyde.
CMC	Carboxymethyl cellulose.
CN	Cellulose nitrate.
CP	Cellulose propionate.
CTA	Cellulose triacetate.
EC	Ethyl cellulose.
E/EA	Ethylene/ethylene acrylate.
E/MA	Ethylene/methacrylic acid.
EP	Epoxide or epoxy.
E/P	Ethylene/propylene.
EPDM	Ethylene/propylene/diene.
E/TFE	Ethylene/tetrafluoroethylene.
EVAC	Ethylene/vinyl acetate.
EVAL	Ethylene/vinyl alcohol.
FEP	Perfluoro(ethylene/propylene): tetrafluoroethylene/hexafluoropropylene.
FF	Furane-formaldehyde.
MBS	Methacrylate/butadiene/styrene.
MC	Methylcellulose.
MF	Melamine-formaldehyde.
MPF	Melamine-phenol-formaldehyde.
PA	Polyamide.
PAI	Polyamide imide.
PAN	Polyacrylonitrile.
PAUR	Polyester urethane.
PB	Polybutene-1.
PBT	Polybutylene terephthalate.
PC	Polycarbonate.
PCTFE	Polychlorotrifluoroethylene.
PDAP	Polydiallyl phthalate.
PE	Polyethylene.
PEBA	Polyether block amide.
PEEK	Polyetheretherketone.
PEI	Polyether imide.
PEOX	Polyethylene oxide.
PES	Polyether sulphone.
PET	Polyethylene terephthalate.
PES	Polyether sulphone.
PEUR	Polyether urethane.
PF	Phenol formaldehyde.
PFA	Perfluoro alkoxyl alkane.
PI	Polyimide.
PMMA	Polymethyl methacrylate.
PMP	Poly-4-methylpentene-1.
PMS	Poly-α-methylstyrene.
POM	Polyoxymethylene or, polyacetal or, polyformaldehyde.
PP	Polypropylene.
PPE	Polyphenylene ether.
PPOX	Polypropylene oxide.
PPS	Polyphenylene sulphide.
PPSU	Polyphenylene sulphone.
PS	Polystyrene.
PSU	Polysulphone.
PTFE	Polytetrafluoroethylene.
PUR	Polyurethane.
PVAC	Polyvinyl acetate.
PVAL	Polyvinyl alcohol.
PVB	Polyvinyl butyral.
PVC	Polyvinyl chloride.
PVDC	Polyvinylidene chloride.
PVDF	Polyvinylidene fluoride.
PVF	Polyvinyl fluoride.
PVFM	Polyvinyl formal.
PVK	Polyvinylcarbazole.
PVP	Polyvinylpyrrolidone.
SAN	Styrene acrylonitrile.
S/B	Styrene/butadiene.
SI	Silicone.
SMA	Styrene maleic anhydride.
S/MS	Styrene α-methylstyrene.
SP	Saturated polyester.
UF	Urea formaldehyde.
UP	Unsaturated polyester.
VC/E	Vinyl chloride/ethylene.
VC/E/MA	Vinyl chloride/ethylene/methyl acrylate.
VC/E/VAC	Vinyl chloride/ethylene/vinyl acetate.
VC/MA	Vinyl chloride/methyl acrylate.
VC/MMA	Vinyl chloride/methyl methacrylate.
VC/OA	Vinyl chloride/octyl acrylate.
VC/VAC	Vinyl chloride/vinyl acetate.
VC/VDC	Vinyl chloride/vinylidene chloride.

The above abbreviations may be modified by the addition of up to four specified letters after the abbreviation for the polymer under discussion. For example, PS could become PS-HI when high impact polystyrene is being described (see **table 2**).

When mixtures are made from two or more polymers (blends or alloys), ISO 1043 suggests that the symbols for the basic polymers be separated by a plus (+) sign and that the symbols be placed in parentheses. For example, a mixture of *polymethyl methacrylate* and *acrylonitrile-butadiene-styrene* should be represented as (PMMA + ABS). That is, a mixture of poly(methyl methacrylate) and acrylonitrile/butadiene/styrene should be represented as (PMMA + ABS).

Table 1b. *Standard (based on ISO and ASTM) abbreviations of rubbers*

Abbreviation	Name or meaning
ABR	Acrylate-butadiene rubber.
ACM	Copolymer of ethyl acrylate, or another acrylate, and a small amount of a monomer to facilitate vulcanization. See *acrylic rubber*.
AECO	Terpolymer of allyl glycidyl ether, ethylene oxide and epichlorhydrin. See *epichlorhydrin rubber*.
AEM	Copolymer of ethyl acrylate, or another acrylate, and ethylene. See *ethylene-methyl acrylate rubber*.
AFMU	Terpolymer of tetrafluoroethylene, trifluoronitrosomethane and nitrosoperfluorobutyric acid. See *carboxy-nitroso rubber*.
ANM	Copolymer of ethyl acrylate, or another acrylate, and acrylonitrile.
AU	Polyester urethane. See *polyurethane*.
BIIR	Bromo-isobutene-isoprene rubber. See *halogenated butyl rubber*.
BR	Butadiene rubber.
CFM	Polychlorotrifluorethylene. See *vinylidene fluoride-chlorotrifluoroethylene copolymer*.
CIIR	Chloro-isobutene-isoprene rubber. See *halogenated butyl rubber*.
CM	Chloropolyethylene. See *chlorinated polyethylene*.
CO	Polychloromethyloxiran. See *epichlorhydrin rubber*.
CR	Chloroprene rubber
CSM	Chlorosulphonylpolyethylene. See *chlorosulphonated polyethylene*.
EAM	Ethylene-vinyl acetate rubber.
ECO	Ethylene oxide-chloromethyloxiran. See *epichlorhydrin rubber*.
EOT	A rubber which has sulphur, carbon and oxygen in the main polymer chain. A rubber with polysulphide linkages in which, for example, the polysulphide linkages are separated by organic groups (R groups) such as $-CH_2-CH_2-O-CH_2-O-CH_2-CH_2-$ and other R groups. For example, by $-CH_2-CH_2-$.
EPDM	Ethylene-propylene diene monomer. A terpolymer of ethylene, propylene and a diene with the residual unsaturated portion of the diene in the side chain. An *ethylene-propylene rubber*.
EPM	Ethylene-propylene monomer or copolymer. See *ethylene-propylene rubber*.
EU	A polyether urethane. See *polyurethane*.
FFKM	A perfluoro rubber of the polymethylene type having all substituents on the polymer chain either fluoro, perfluoroalkyl or perfluoroalkoxy groups. See *perfluorinated copolymer*.
FKM	A fluororubber of the polymethylene type having substituent fluoro and perfluoroalkoxy groups on the main chain. See *perfluorinated copolymer and fluororubber*.
FMQ	Silicone rubber containing fluoro and methyl substituent groups on the polymer chain.
FPM	A rubber having fluoro and fluoroalkyl, or fluoroalkoxy groups, substituent groups on the polymer chain. See *fluororubber*.
FVMQ	Silicone rubber containing fluoro, vinyl and methyl substituent groups on the polymer chain. See *silicone rubber*.
FZ	A rubber which has nitrogen and phosphorous in the main polymer chain. The $-P=N-$ chain has flouroalkoxy groups on the P atoms or aryloxy groups on the P atoms: the aryloxy groups are phenoxy and substituted phenoxy groups. See *phosphonitrilic polymer*.
GPO	Propylene oxide-allylglycidyl ether rubber. See *propylene oxide rubber*.
IIR	Isobutene-isoprene rubber. See *butyl rubber*.
IM	Polyisobutene.
IR	Isoprene rubber (synthetic).
M	A rubber which has a saturated chain of the polymethylene type.
MQ	Silicone rubber containing only methyl substituent groups on the polymer chain. See *silicone rubber*.
N	A rubber which has nitrogen in the main polymer chain.
NBR	Nitrile-butadiene rubber. See *nitrile rubber*.
NCR	Acrylonitrile-chloroprene rubber.
NIR	Acrylonitrile-isoprene rubber
NR	Natural rubber. Isoprene rubber (natural).
O	A rubber which has oxygen in the main polymer chain. Rubbers which contain an ether-group contain the letter O.
OT	A rubber which has sulphur, carbon and oxygen in the main polymer chain. A rubber with polysulphide linkages in which, for example, the polysulphide linkages are separated by organic groups (R groups) such as $-CH_2-CH_2-O-CH_2-O-CH_2-CH_2-$.
PBR	Pyridine-butadiene rubber.
PMQ	Silicone rubber containing both methyl and phenyl groups substituent groups on the polymer chain. See *silicone rubber*.
PSBR	Pyridine-styrene-butadiene rubber.
PVMQ	Silicone rubber containing methyl, phenyl and vinyl substituent groups on the polymer chain. See *silicone rubber*.
Q	A rubber which has silicon and oxygen in the main polymer chain. See *silicone rubber*.
R	A rubber which has an unsaturated carbon chain, for example, natural rubber and synthetic rubbers derived at least partly from diolefins.
SBR	Styrene-butadiene rubber.
SCR	Styrene-chloroprene rubber.
SIR	Styrene-isoprene rubber.
T	A rubber which has sulphur in the main polymer chain: a rubber which has sulphur, carbon and oxygen in the main polymer chain. See *polysulphide rubber*.
U	A rubber which has carbon, oxygen and nitrogen in the main polymer chain.
VMQ	Silicone rubber containing both methyl and vinyl groups substituent groups in the polymer chain.
X	Denotes the presence in a rubber of the reactive carboxylic (COOH) group.
XBR	Carboxylic-butadiene rubber.
XCR	Carboxylic-chloroprene rubber.
XNBR	Carboxylic-nitrile rubber. See *carboxylated nitrile rubber*.
XSBR	Carboxylic-styrene butadiene rubber.
Z	A rubber which has nitrogen and phosphorous in the main polymer chain.

Table 2a. *Letters used to modify abbreviations, or symbols, for plastics (ISO and ASTM)*

Letter	Meaning or significance
A	Acetate, acrylate, acrylonitrile, alkane, alkoxy, allyl, amide, and ester.
AC	Acetate.
AL	Alcohol.
AN	Acrylonitrile.
B	Block, butadiene, butene, butyl, butylene, butyral, and butyrate.
C	Carbonate, carboxy, cellulose, chloride, chlorinated, chloro, and cresol.
D	Density, and di.
E	Ether, ethyl, ethylene, expandable or expanded.
EP	Epoxy or epoxide.
F	Flexible, fluid, fluoride, fluoro, formaldehyde, furane, and perfluoro
FM	Formal.
H	High
I	Imide, impact, and iso.
IR	Isocyanurate.
K	Carbazole, and ketone.
L	Linear or low.
M	Medium, melamine, meth, methacryl, methacrylate, methyl, methylene, and molecular.
MA	Maleic acid, and methacrylic acid.
N	Nitrate, normal or novolak.
O	Octyl, oxide, and oxy.
OX	Oxide.
P	Pentene, per, phenol, phenylene, phthalate, plasticized, poly, polyester, propionate, propylene, and pyrrolidone.
R	Raised or resol.
S	Saturated, styrene, sulfide, and sulfone.
SI	Silicone.
SU	Sulfone.
T	Terephthalate, tera, thermoplastic, and tri.
U	Ultra, unplasticized, unsaturated, and urea.
UR	Urethane.
V	Very, and vinyl.
W	Weight.
X	Crosslinked or crosslinkable.

Table 2b. *Commonly-used letters used to modify abbreviations for plastics (i.e. in addition to table 2a)*

Letter	Meaning or significance
A	Atactic or, amorphous.
B	Block copolymer.
C	Crystalline.
CF	Carbon fibre.
E	Emulsion (polymer).
EP	Engineering thermoplastic or, engineering thermoplastics material.
F	Fibre.
FR	Flame retardant and/or fire resistant.
G	Glass.
GF	Glass fibre.
GMT	Glass mat (reinforced) thermoplastics (material).
GP	General purpose.
H	Homopolymer.
HI	High impact.
K	Copolymer.
M	Mass or bulk (polymer) or, mat.
O	Oriented.
P	Plasticised.
PMC	Polyester moulding compound.
R	Random copolymer or, reinforced.
S	Sulphide, sulphone, and suspension (polymerization).
SU	Sulphone.
TP	Thermoplastic.
U	Unplasticised.
V	Vulcanized or crosslinked
XL	Crosslinked or, cured or, vulcanized

Table 2c. *Symbols used for fillers and/or reinforcing materials*

Letter	Meaning or significance
A	Asbestos
B	Boron or, beads or, spheres or, balls
C	Carbon or, chips, or, cuttings
D	Powder
F	Fibre
G	Glass or, ground
H	Whisker
K	Chalk or, knitted fabric
L	Cellulose or, layer
M	Mineral or, metal
N	Non-woven fabric (usually thin)
P	Mica or, paper
Q	Silicon
R	Aramid or, roving
S	Synthetic, organic or, scale, flake
T	Talcum or, cord
V	Veneer
W	Wood
Y	Yarn
Z	Others.

Please note that if a letter is not being used in these tables then, it does not mean that it is not being used in another branch of the polymer industry.

Table 3. *Some commonly used abbreviations and trade names of plastics and thermoplastic elastomers*

Abbreviation	Common name	Common trade names or, trade marks
ABS	Acrylonitrile butadiene styrene	Cycolac; Lustran.
AMS	Alpha methyl styrene	Elite HH.
ASA	Acrylonitrile styrene acrylate (AAS)	Luran S
BDS	Butadiene styrene block copolymer	K resin; Styrolux.
BMC	Bulk moulding compound	Freemix; Norsomix.
CA	Cellulose acetate	Cellidor; Tenite.
CAB	Cellulose acetate butyrate	Cellidor; Tenite.
CAP	Cellulose acetate propionate	Cellidor; Tenite.
CF	Casein formaldehyde	Erinoid; Lactoid.
CN	Cellulose nitrate	Celluloid; Xylonite.
CP	Cellulose propionate (CAP)	Cellidor; Tenite.
CPE	Chlorinated polyethylene (PE-C)	Bayer CM; Tyrin CM
CPVC	Chlorinated polyvinyl chloride (PVC-C)	Lucalor.
DAP	Diallyl phthalate	(from Synres)
DAIP	Diallyl isophthalate	
DMC	Dough moulding compound	Beetle DMC; ERF DMC.
EA-MPR	Elastomer alloy melt processable rubber or, melt processable rubber	Alcryn.
EA-TPV	Elastomer alloy thermoplastic vulcanizate	Lomod; Santoprene.
EP	Epoxide or epoxy	Araldite.
ETFE	Tetrafluorethylene-ethylene copolymer	Tefzel.
EVA	Ethylene vinyl acetate copolymer (EVAC)	Evatane.
EVAL	Ethylene vinyl alcohol copolymer	Clarene; Eval.
EVOH	Ethylene vinyl alcohol copolymer	Clarene; Eval.
FEP	Fluorinated ethylene propylene (TFE-HFP)	Teflon FEP.
GPMC	Granular polyester moulding compound	Freeflo; Impel.
HDPE	High density polyethylene (PE-HD)	Lupolen HD; Rigidex.
HIPS	High impact polystyrene (TPS or IPS)	Lustrex; Polystyrol.
LCP	Liquid crystal polymer	Vectra; Xydar.
LDPE	Low density polyethylene (PE-LD)	Alathon; Hostalen LD.
MBS	Methacrylate butadiene styrene	Paraloid.
MDPE	Medium density polyethylene (PE-MD)	Fortiflex.
MF	Melamine formaldehyde	Melmex; Melopas.
MPF	Melamine phenol formaldehyde	Melmex
MPR	Melt processable rubber or, elastomer alloy melt processable rubber	Alcryn.
PA	Polyamide or nylon	
PA 6	Polyamide 6 or nylon 6	Akulon K; Ultramid.
PA 11	Polyamide 11 or nylon 11	Rilsan B.
PA 12	Polyamide 12 or nylon 12	Rilsan A; Grilamid.
PA 46	Polyamide 46 or nylon 46	Stanyl.
PA 66	Polyamide 66 or nylon 66	Maranyl; Zytel.
PA 610	Polyamide 610 or nylon 610	Brulon; Perlon N.
PAA 6	Polyaryl amide or, poly-m-xylene-adipamide (PA MXD6)	Ixef.
PAN	Polyacrylonitrile	Acrilan; Barex; Orlon.
PBI	Polybenzimidazole	Celazole
PBT	Polybutylene terephthalate	Pocan; Valox.
PC	Polycarbonate	Lexan; Makrolon.
PCTFE	Polychlorotrifluorethylene	Hostaflon C2; Kel-F
PE	Polyethylene	Alathon; Lupolen.
PEBA	Polyether block amide (TPE-A)	Pebax.
PEEK	Polyether ether ketone	Victrex PEEK;
PEEL	Polyether ester (TPE-A or YPBO)	Arnitel; Hytrel.
PE-HD	Polyethylene-high density	Lupolen HD; Rigidex HDPE.
PEI	Polyether imide	Ultem.
PEK	Polyether ketone	Hostatec.
PEKK	Polyether ketone ketone	(from DuPont)
PE-LD	Polyethylene-low density	Alathon; Lupolen.
PE-MD	Polyethylene-medium density	Fortiflex.
PE-VLD	Polyethylene-very low density	Norsoflex
PET	Polyethylene terephthalate	Arnite A; Techster E.
PES	Polyether sulphone	Victrex.
PMC	Polyester molding compound	Aropol; Norsomix
PF	Phenol formaldehyde	Bakelite; Sternite.
PI	Polyimide	Vespel.

Table 3 - *contd*

Abbreviation	Common name	Common trade names or, trade marks
PMMA	Polymethyl methacrylate (acrylic)	Diakon; Plexiglas.
PMMA-T	Toughened acrylic	
PMMI	Polymethyl methacrylimide	Pleximid.
PMP	Polymethyl pentene	TPX; Crystalor.
POM	Polyoxymethylene or, acetal or, polyformaldehyde	Delrin; Hostaform
POM-H	Acetal homopolymer	Delrin and Delrin II
POM-CO	Acetal copolymer	Hostaform; Ultraform
PP	Polypropylene	Profax; Propathene.
PPA	Polyphthalamide	
PPE	Polyphenylene ether (see PPO)	
PPO	Polyphenylene oxide-usually modified polyphenylene oxide (PPO-M)	Luranyl; Noryl.
PPS	Polyphenylene sulphide	Fortron; Ryton.
PPPS	Polyphenylene sulphide sulphone	Ryton S.
PPVC	Plasticised polyvinyl chloride (PVC-P)	Solvic; Vinnol.
PS	Polystyrene (GPPS or PS-GP)	Lustrex; Polystyrol.
PSU	Polysulphone	Udel.
PTFE	Polytetrafluoroethylene	Fluon; Teflon.
PVC	Polyvinyl chloride	Corvic; Geon.
PVDC	Polyvinylidene chloride copolymers	Saran.
PVDF	Polyvinylidene fluoride	Dyflor; Kynar; Solef.
PVF	Polyvinyl fluoride	Tedlar.
SAN	Styrene acrylonitrile copolymer	Lustran SAN; Tyril.
SMC	Sheet moulding compound	ERF SMC; Flomat.
UPVC	Unplasticized polyvinyl chloride (PVC-U)	Corvic; Geon.
RMPP	Rubber modified polypropylene (RRPP or PP/EPDM)	Uniroyal TPR; Keltan.
SBS	Styrene butadiene styrene block copolymer or, thermoplastic elastomer styrene based (TPE-S)	Cariflex TR; Solprene.
SEBS	Styrene butadiene styrene block copolymer (saturated) or, thermoplastic elastomer styrene based (TPE-S)	Cariflex.
TPE	Thermoplastic elastomer (rubber)	
TPE-A	Polyether block amide (PEBA)	Pebax.
TPE-E	Thermoplastic elastomer - ether ester or, polyether ester elastomer	Arnitel; Hytrel.
TPE-OXL	Thermoplastic elastomer - polyolefin based with crosslinked rubber	Levaflex; Santoprene.
TPE-S	Thermoplastic elastomer styrene based (usually styrene butadiene styrene block copolymer)	Cariflex TR; Solprene.
TPE-U	Thermoplastic polyurethane (TPU)	Elastollan; Estane.
TPO	Thermoplastic polyolefin (RMPP)	Propathene OTE; Vistaflex.
TPR	Thermoplastic rubber (elastomer)	
TPU	Thermoplastic polyurethane	Elastollan; Estane.
TPV	Thermoplastic vulcanizte (a TPE with crosslinked rubber)	Lomod; Santoprene.
UF	Urea formaldehyde	Beetle; Scarab.
VE	Vinyl ester resins	
VLDPE	Very low density polyethylene (ULDPE)	Norsoflex.

Table 4. *Some abbreviations and names of plastics and elastomers*
Where there is more than one entry, the italicized words shall be used to find more information in the text

1,2 BR	Vinyl polybutadiene or, 1,2-polybutadiene. See *polybutadiene rubber*.		BR-Co	See Co-BR. *Butadiene rubber* based on a cobalt catalyst.
AAS	See *ASA*.		BR-E	See E-BR. Emulsion butadiene rubber. See *butadiene rubber*.
ABS	*Acrylonitrile-butadiene-styrene* or, acrylonitrile-butadiene-styrene copolymer or, poly-(1-butenylene-g-1-phenylethylene-co-1-cyanoethylene).		BR-L	See L-BR. Solution butadiene rubber. See *butadiene rubber*.
			BR-li	See Li-BR. *Butadiene rubber* based on a lithium catalyst.
ACM	*Acrylic rubber* or, acrylate rubber or, acrylic acid ester rubber or, acrylic elastomer or, polyacrylic elastomer.		BR-Nd	See Nd-BR. *Butadiene rubber* based on a neodymium catalyst.
			BR-Ni	See Ni-BR. *Butadiene rubber* based on a nickel catalyst.
			BR-OE	See OE-BR. Oil extended butadiene rubber. See *butadiene rubber*.
ACS	*Acrylonitrile-chlorinated polyethylene-styrene copolymer* or, acrylonitrile-styrene-chlorinated polyethylene.		CA	*Cellulose acetate* or, acetylcellulose or, cellulose ethanoate.
AECO	Allyl group containing *epichlorhydrin rubber* or, epichlorhydrin-ethylene oxide-allyl glycidyl ether terpolymer or rubber. ETER or ETE.		CAB	*Cellulose acetate-butyrate* or, cellulose ethanoate-butanoate.
AEM	*Ethylene-methyl acrylate rubber* or, ethylene-acrylate copolymer or rubber or, ethylene-acrylic elastomer. EAM.		CAP	*Cellulose acetate-propionate* or, cellulose ethanoate-propanoate.
			CF	*Casein-formaldehyde*. CF is also used for cresol-formaldehyde.
AES	*Acrylonitrile-ethylene/propylene-styrene rubber* or, acrylonitrile-styrene/EPR rubber or, acrylonitrile-styrene/EPR elastomer.		CFM	*Vinylidene fluoride-chlorotrifluoroethylene copolymer* or, poly-(vinylidene fluoride-co-chlorotrifluoroethylene) or, polychlorotrifluorethylene rubber or elastomer.
AFMU	*Carboxy-nitroso rubber* or, fluoronitrosorubber or, nitrosofluororubber or, nitrosorubber or nitroso rubber.		
AMMA	Acrylonitrile-methyl methacrylate.		CHR	Allyl-group-containing epichlorhydrin rubber or, epichlorhydrin-ethylene oxide-allyl glycidyl-ether terpolymer or rubber. AECO or ETER or ETE.
AMS	*Alpha methyl styrene* or, poly-(α-methylstyrene).		CIC	Continuously impregnated compound - a *polyester moulding compound* or PMC.
APP	*Atactic polypropylene* or, atactic polypropene or, atactic poly(propylene). PP-A.		CIIR	chlorinated butyl rubber or, chloro-isobutene-isoprene rubber or, *halogenated butyl rubber*. IIR-C.
ASA	*Acrylate-styrene-acrylonitrile* or, acrylate modified styrene acrylonitrile or, acrylonitrile-styrene-acrylate copolymer. AAS.		CIR	*Coumarone-indene resins* or, indene-coumarone resins or, indene resins.
ASR	Alkylene sulphide rubber.		CM	Chloropolyethylene or, *chlorinated polyethylene* or, chloro-polyethylene. CPE or PE-C.
AU	*Polyurethane rubber* (ester based) or polyurethane elastomer or, urethane rubber. PAUR or PU or PUR.		
			CN	*Cellulose nitrate* or, nitrocellulose.
AU-I	*Polyurethane rubber*, based on polyesters, isocyanate crosslinkable.		CO	Polychloromethyloxiran or, epichlorhydrin (homopolymer) rubber. EC. See *epichlorhydrin rubber*.
AU-P	*Polyurethane rubber*, based on polyesters, peroxide crosslinkable.		
			Co-BR	*Butadiene rubber* based on a cobalt catalyst. BR-Co
BDS	Styrene-butadiene block copolymer. SBB.		COPE	See TPE-E. *Thermoplastic elastomer - ether based*.
BIIR	Brominated butyl rubber or, bromo-isobutene-isoprene rubber or, *halogenated butyl rubber*. IIR-B.		CP	*Cellulose propionate* or, cellulose propanoate.
BMC	*Bulk moulding compound*. (a *polyester moulding compound* or PMC).		CPE	See CM.
			CPVC	Chlorinated polyvinyl chloride. PVC-C.
BR	*Butadiene rubber* or, cis - polybutadiene or, cis-1,4-polybutadiene rubber or, 1,4-polybutadiene or, polybutadiene rubber or, poly-(1-butenylene). BR or PB or PBD.		CR	*Chloroprene rubber* or, polychlorobutadiene) or, poly-(2-chloro-1,3-butadiene) or, poly-(1-chloro-1-butenylene) or, polychloroprene rubber.

Table 4 - *contd*

CR-X	See *X-CR*. Chloroprene rubber with reactive groups.	EPS	*Expanded polystyrene* or foamed polystyrene. PS-E or XPS or PS-X.
CSM	Chlorosulphonylpolyethylene or, *chlorosulphonated polyethylene* or, chlorosulphonated polyethylene rubber.	ETE	See AECO. Allyl-group-containing-*epichlorhydrin rubber*.
CTA	Cellulose triacetate.	ETER	See AECO. Allyl-group-containing-*epichlorhydrin rubber*.
DAIP	Diallyl isophthalate.	ETFE	*Tetrafluoroethylene-ethylene copolymer* or, poly-(tetrafluoroethylene-co-ethylene) or, ethylene-tetrafluoroethylene copolymer. TEP.
DAP	Diallyl phthalate.		
DMC	Dough moulding compound - a *polyester moulding compound* or PMC.	EU	*Polyurethane rubber* (ether based). PEUR or PU or PUR
DP-NR	*Deproteinated natural rubber* or low nitrogen natural rubber. NR-DP or DP-NR or LN-NR	EVA	*Ethylene-vinyl acetate* or, ethylene-vinyl acetate copolymer.
E-BR	Emulsion butadiene rubber. BR-E.	EVAl	See EVAL. *Ethylene-vinyl alcohol.*
E-SBR	Emulsion styrene butadiene rubber. SBR-E.	EVAL	*Ethylene-vinyl alcohol* or, ethylene-vinyl alcohol copolymer. EVOH or EVAl.
E-SR	Emulsion synthetic rubber. SR-E.	EVM	*Ethylene-vinyl acetate rubber* or copolymer or, EVA rubber or, ethylene-vinyl acetate copolymer. EAM or EVA.
EA	Elastomeric alloy.		
EA-MPR	Elastomeric alloy melt processable rubber.		
EA-TPV	Elastomeric alloy thermoplastic vulcanizate. See TPV.	EVOH	See EVAL. *Ethylene-vinyl alcohol.*
EAA	*Ethylene-acrylic acid* or, ethylene-acrylic acid copolymer.	FEP	*Fluorinated ethylene-propylene copolymer* or, tetrafluoroethylene-hexafluoropropylene copolymer or, poly-(tetrafluoroethylene-co-hexafluoropropylene). TFE-HFP.
EAM	Used for ethylene-vinyl acetate rubber and sometimes for ethylene-methyl acrylate rubber. See *EVM* or *AEM*.	FFKM	See PFE. A *perfluorinated copolymer* elastomer or rubber.
EC	*Ethyl cellulose* or, cellulose ethyl ether (a *cellulose ether*). Also see CO.	FKM	*Vinylidene fluoride-hexafluoropropylene copolymer* or, poly-(vinylidene fluoride-co-hexafluoropropylene).
ECO	*Epichlorhydrin copolymer rubber* or, ethylene oxide-chloromethyloxiran or, oxiran-chloromethyloxiran. See *epichlorhydrin rubber*.	FPM	*Vinylidene fluoride copolymer rubber*. A rubber having fluoro and fluoroalkyl, or fluoroalkoxy groups, substituent groups on the polymer chain. See *fluororubber*.
ECTFE	*Chlorotrifluoroethylene-ethylene copolymer* or, chlorotrifluoroethylene-ethylene alternating copolymer or, poly-(chlorotrifluoroethylene-co-ethylene).	FRP	Fibre reinforced plastic.
EEA	Ethylene-ethyl acrylate copolymer.	FRTP	Fibre reinforced thermoplastic.
ENM	See H-NBR. Hydrogenated nitrile rubber.	FVMQ	*Silicone rubber containing fluoro, vinyl and methyl groups*. fluorinated rubber or, fluoro silicone rubber or elastomer or, fluoro-silicone rubber or elastomer or, fluorosilicone rubber or elastomer or, silicone rubber containing fluoro, vinyl and methyl groups.
ENR	Epoxidized natural rubber. NR-E.		
EP	*Epoxy* or, epoxide or, epoxy resin or epoxide resin or, ethoxyline resin.		
		GMT	Glass mat (reinforced) thermoplastics (material).
EP(D)M	*Ethylene-propylene diene monomer* or, ethylene-propylene terpolymer. EP(D)M or EPDM (an EPR). An *ethylene-propylene rubber*.	GPMC	Granular polyester moulding compound a *polyester moulding compound (PMC)* or, granular moulding compound (GMC).
EPDM	Ethylene-propylene diene monomer. Terpolymer of ethylene, propylene and a diene with the residual unsaturated portion of the diene in the side chain. An *ethylene-propylene rubber*.	GPO	Propylene oxide (copolymer) rubber or, propylene oxide-allylglycidyl ether rubber or copolymer. See *propylene oxide rubber*.
		GPPS	See PS. *Polystyrene.*
EPM	*Ethylene-propylene monomer* or, ethylene-propylene rubber or copolymer or elastomer.	GR-S	See SBR. *Styrene-butadiene rubber.*
EPR	*Ethylene-propylene rubber* or, ethylene-propylene copolymer or, ethylene-propylene diene monomer or, ethylene-propylene elastomer or, ethylene-propylene monomer or, ethylene-propylene terpolymer. EP(D)M, EPM or EPDM	H-NBR	*Hydrogenated nitrile rubber*. NBR-H or ENM or HSN.
		HDPE	*High density polyethylene* or, high density polyethene or, low pressure polyethylene or, polyethylene-high density. PE-HD.

Table 4. *Some abbreviations and names of plastics and elastomers - contd*

HEC	Hydroxyethyl cellulose (*a cellulose ether*).	MF	*Melamine-formaldehyde* or, melamine-methanal (an aminoplastic).
High cis BR	High cis-polybutadiene rubber.	MPF	Melamine-phenol-formaldehyde.
HIPS	*High impact polystyrene* or, impact polystyrene or, rubber toughened polystyrene or, toughened polystyrene. TPS or IPS.	MPR	Melt processable rubber.
		MQ	*Silicone rubber* containing methyl groups. dimethylsilicone elastomer or rubber or, methyl silicone rubber or elastomer or, polydimethyl siloxane or, *silicone rubber* containing methyl groups.
HSN	Highly saturated nitrile rubber or *hydrogenated nitrile rubber* See H-NBR.		
IIR	*Butyl rubber* or, isobutylene-isoprene rubber or copolymer or, isobutene-isoprene rubber or copolymer or, poly-(1,1-dimethylethylene-co-1-methyl-1-butenylene) poly-(isobutene-co-isoprene).	MR	*Methyl rubber* or elastomer or, dimethylbutadiene rubber or polymer or, polydimethyl butadiene.
		N	A rubber which has nitrogen in the main polymer chain.
IIR-B	See *BIIR*. Brominated butyl rubber.	(NBR+ PVC)	A *nitrile rubber polyvinyl chloride blend*.
IM	Polyisobutylene.		
IPS	See *HIPS*. High impact polystyrene.	NBR/ PVC	Nitrile rubber/PVC blend or, *nitrile rubber polyvinyl chloride blend*.
IR	*Isoprene rubber* or, cis-polyisoprene or, cis-1,4-polyisoprene or, polyisoprene or, poly-(2-methyl-1,3-butadiene) or, synthetic natural rubber or, synthetic polyisoprene or, synthetic isoprene rubber or, synthetic polyisoprene rubber.	NBR	Nitrile-butadiene rubber or, *nitrile rubber* or, butadiene-acrylonitrile rubber or copolymer or, acrylonitrile-butadiene rubber or, poly-(butadiene-co-acrylonitrile) or, poly-(1-butenylene-co-1-cyanoethylene).
		NBR-H	See *H-NBR*. *Hydrogenated nitrile rubber*.
L-BR	Solution butadiene rubber. BR-L.	NBR-X	See *XNBR*. *Carboxylated nitrile rubber*.
L-SBR	Solution styrene-butadiene rubber. SBR-L.	NCR	Acrylonitrile-chloroprene rubber or, chloroprene acrylonitrile copolymer or rubber.
LCP	*Liquid crystal polymer* or, LC polymer or, mesomorphic polymer.	Nd-BR	*Butadiene rubber* based on a neodymium catalyst.
		Ni-BR	*Butadiene rubber* based on a nickel catalyst.
LDPE	*Low density polyethylene* or, high pressure polyethylene or, low density polyethene or, polyethylene-low density. PE-LD.	NIR	Acrylonitrile-isoprene rubber or, isoprene acrylonitrile copolymer or rubber.
Li-BR	*Butadiene rubber* based on a lithium catalyst. BR-Li.	NR	*Natural rubber* or, caoutchouc or, gum elastic or, cis-1,4-polyisoprene or, India rubber or, india rubber or, poly-(1-methyl-1-butenylene).
LLDPE	*Linear low density polyethylene* or, linear low density polyethene or, polyethylene-linear low density. PE-LLD.		
Low cis BR	Low cis-polybutadiene rubber.	NR-DP	See *DP-NR*. Deproteinated natural rubber. See *natural rubber*.
LSR	Liquid silicone rubber.	NR-E	See *ENR*. Epoxidized natural rubber. See *natural rubber*.
M	A rubber which has a saturated chain of the polymethylene type.	NR-OE	See *OE-NR*. Oil extended natural rubber. See *natural rubber*.
		O	A rubber which has oxygen in the main polymer chain. Rubbers which contain an ether-group contain the letter O.
MABS	Methyl methacrylate-acrylonitrile-butadiene-styrene copolymer.		
MBS	*Methyl methacrylate-butadiene-styrene copolymer* or, methyl methacrylate-styrene-polybutadiene.	OE-BR	Oil extended butadiene rubber. BR-OE.
		OE-NR	Oil extended natural rubber. NR-OE.
MC	Methyl cellulose. A *cellulose ether*.	OENR	See *OE-NR*. Oil extended natural rubber.
MDPE	*Medium density polyethylene* or polyethene or, polyethylene-medium density or, polyethene-medium density. PE-MD.	OPET	Oriented polyethylene terephthalate. PET-O.
		OPP	Oriented polypropylene. PP-O.
Medium cis BR	*Medium cis-polybutadiene rubber*.	OPS	Oriented polystyrene. PS-O.

Table 4 - contd

OPVC	Oriented polyvinyl chloride. PVC-O.	PAA 6	*Polyaryl amide* or, poly-(m-xyleneadipamide or, aromatic polyamide. PAMXD6.
P4MP1	*Poly-(4-methylpentene-1)* or, polyisobutylethylene or, poly-(4-methylpent-1-ene).	PABM	*Polyaminobismaleimide* or, polybismaleimide or, polymaleimide.
PA	*Nylon* or, polyamide.	PAI	Polyamide-imide.
		PAMXD6	See *PAA 6. Polyaryl amide*.
PA 11	*Nylon 11* or, polyamide 11 or, poly-(11-amino-undecanoic acid) or, polyundecanoamide.	PAN	*Polyacrylonitrile* or, polyvinyl cyanide.
PA 12	*Nylon 12* or, polyamide 12 or, polylauroamide or, polylauryllactam or, polydodecanoamide.	PAUR	*Polyurethane rubber* (ester based).
		PB	See *BR. Butadiene rubber*.
		PB	*Polybutylene* or, polybutene or, poly-1-butene or, polybut-1-ene or, polybutene-1 or polyethylethylene.
PA 4	*Nylon 4* or, polyamide 4 or, polypyrrolidone.		
PA 46	*Nylon 46* or, polyamide 46 or, polytetramethyleneadipamide or, polytetramethylene adipamide.	PBA	See *TPE-A. Thermoplastic elastomer - amide based*.
		PBD	See *BR. Butadiene rubber*.
PA 6/66	*Nylon 6/66* or, polyamide 6/66.	PBI	Polybenzimidazole.
		PBO	Paraphenylene polybenzobisoxazole.
PA 6	*Nylon 6* or, polyamide 6 or, poly-(ω-amino-caproamide) or, poly-(6-aminocaproic acid) or, poly-(ω-aminocaproic acid) or, polycaproamide or, polycaprolactam or, poly-(imino-1-oxohexamethylene).	PBT	*Polybutylene terephthalate* or, polytetramethylene terephthalate or, poly-(oxytetramethylene-oxyterephthalate). (a thermoplastic polyester - see also PET). PTMT.
		PBZ	Polybenzobisoxazole.
PA 6/ 66/610	*Nylon 6/66/610* or, polyamide 6/66/610.	PC	*Polycarbonate* or, bisphenol A polycarbonate or, poly-(4,4'-isopropylidenediphenylene carbonate) or, poly-(oxy-1,4-phenylene-dimethylmethylene-1,4-phenylene-oxy-carbonyl).
PA 610	*Nylon 610* or, polyamide 610 or, polyhexamethylenesebacamide polyhexamethylene sebacamide.		
		PCTFE	Polychlorotrifluorethylene.
PA 612	*Nylon 612* or, polyamide 612 or, polyhexamethylenedodecanoamide or, polyhexamethylene dodecanoamide.	PE	*Polyethylene* or, polyethene or, poly(ethylene).
		PE-C	See *CM. Chlorinated polyethylene*.
PA 66	*Nylon 66* or, polyamide 66 or, polyhexamethyleneadipamide or, polyhexamethylene adipamide or, poly-[imino-(1,6-dioxohexamethylene)-iminohexamethylene].	PE-HD	See *HDPE. High density polyethylene*.
		PE-LD	See *LDPE. Low density polyethylene*.
		PE-LLD	See *LLDPE. Linear low density polyethylene*.
		PE-MD	See *MDPE. Medium density polyethylene*.
		PE-VLD	See *VLDPE. Very low density polyethylene*.
PA 69	*Nylon 69* or, polyamide 69 or, polyhexamethylenenonamide or, polyhexamethylene nonamide or, polyhexamethyleneazelamide.	PEBA	See *TPE-A. Thermoplastic elastomer - amide based*.
		PEC	Phenylene ether copolymer.
PA 6T	*Nylon 6T* or, polyamide 6T or, polyhexamethyleneterephthalamide or polyhexamethylene terephthalamide.	PEEK	*Polyether ether ketone* or, polyetherether ketone.
		PEEL	See *TPE-E. Thermoplastic elastomer - ether based*.
PA 91	Nylon 91 or, polyamide 91 or, polynonamethyleneurea.	PEI	*Polyether imide* or, polyetherimide.

Table 4. *Some abbreviations and names of plastics and elastomers - contd*

PEK	*Polyether ketone* or, polyetherketone.	PO	Propylene oxide (homopolymer) rubber. See *propylene oxide rubber*.
PEKK	*Polyether ketone ketone* or, polyetherketone ketone.	PO	Polyolefin.
PEO	See *PEOX*. Polyethylene oxide.	POD	Polyphenylene-1,3,4-oxadiazole.
PEOX	Polyethylene oxide. PEO.	POM	*Acetal* or, acetal homopolymer and/or, acetal copolymer or, polyacetal or, polyformaldehyde or, polyoxymethylene. POM-H or POM-CO.
PES	*Sulphone polymer* or, polyarylenesulphone or, polyaryl ether sulphone or, polyaryl sulphone or, polyether sulphone or, polyethersulphone or, polysulphone. PSU.		
		POM-CO	Acetal copolymer. See *acetal*.
		POM-H	Acetal homopolymer. See *acetal*.
PET	Polyethylene terephthalate, (a thermoplastic polyester). PETP.	POM-K	See *POM-CO* and *acetal*.
PET-O	See *OPET*. Oriented polyethylene terephthalate.	(PP + EPDM)	See *RMPP*. Rubber modified polypropylene.
PETG	Polyethylene terephthalate glycol.	PP/ EPDM	See *RMPP*. Rubber modified polypropylene.
PETP	See *PET*. Polyethylene terephthalate.		
PEUR	*Polyurethane rubber* (ether based).	PP	*Polypropylene* or, polymethyl ethylene or, polypropene or, poly(propylene). PP-H or PP-CO or PPR.
PF	*Phenol-formaldehyde* or, phenol-methanal or, phenoplast.		
		PP-A	See *APP*. Atactic polypropylene. See *polypropylene*.
PFA	*Perfluoroalkoxy copolymer* or, perfluoroalkylvinyl ether polymer or copolymer or, polyperfluoroalkylvinyl ether.	PP-CO	Polypropylene copolymer. See *polypropylene*.
		PP-H	Polypropylene homopolymer. PP. See *polypropylene*.
PFE	*Perfluorinated elastomer or rubber* or, tetrafluoroethylene-perfluoromethylvinyl ether copolymer or, poly-(tetrafluoroethylene-co-perfluoromethylvinyl ether). FFKM.	PP-K	See *PP-CO*. Polypropylene copolymer. See *polypropylene*.
		PP-O	See *OPP*. Oriented polypropylene.
		PPE	Polyphenylene ether. Usually polyphenylene oxide or, *polyphenylene oxide (modified)*. PPO or PPO-M.
PHB	Polyhydroxybutyrate.		
PI	Polyimide.		
PIB	*Polyisobutylene* or, polyisobutene.	PPO	*Polyphenylene oxide (modified)* or, modified poly-(phenylene oxide) or, poly-(2,6-dimethylphenol) or, poly-(2,6-dimethyl-1,4-phenylene oxide). PPO or, PPO-M or PPE.
PIR	Polyisocyanurate.		
PMA	Polymethyl acrylate.		
PMC	Polyester moulding compound. bulk moulding compound (BMC) continuous roving moulding compound - a wound moulding compound (XMC) continuously impregnated compound (CIC) dough moulding compound (DMC) granular moulding compound (GPMC or GMC) low viscosity moulding compound (ZMC) sheet moulding compound (SMC) thick moulding compound (TMC).	PPO-M	See *PPO*. Polyphenylene oxide (modified).
		PPOX	Polypropylene oxide.
		PPPS	Polyphenylene sulphide sulphone.
		PPR	See *PP*. Polypropylene.
		PPS	*Polyphenylene sulphide* or, poly-(p-phenylene sulphide) or, poly-(thio-1,4-phenylene) or, sulphide polymer.
PMMA	*Polymethyl methacrylate* or, acrylic or, poly-[1-(methoxycarbonyl)-1-methylethylene].	PPSU	Polyphenylene sulphone.
		PPVC	*Plasticized polyvinyl chloride* or, plasticised polyvinyl chloride or, polyvinyl chloride-plasticized or, plasticized polychloroethene. PVC-P.
PMQ	*Silicone rubber* containing methyl and phenyl groups. methylphenylsilicone elastomer or rubber or, phenylsilicone rubber or elastomer.		
		PPX	Poly-p-xylene.
PNF	*Polyfluorphosphazene rubber* or, fluorophosphonitrilc polymer or, fluoropolyphosphazene rubber or polymer.	PS	*Polystyrene* or, general purpose polystyrene or, polyphenylethylene or, poly-(1-phenylethylene) or, polvinylbenzene. PS-GP or GPPS.
PNR	*Polynorbornene* or, polynorbornene rubber.		

Table 4 - contd

PS-E	See *EPS*.	R	A rubber which has unsaturated carbon chain, for example, natural rubber and synthetic rubbers derived at least partly from diolefins.
PS-GP	See *PS. Polystyrene*.		
PS-O	See *OPS. Oriented polystyrene*.	RMPP	*Rubber modified polypropylene* or, rubber reinforced polypropylene or, rubber toughened polypropylene. (PP + EPDM) or RRPP or PP/EPDM.
PS-X	See *EPS. Expanded polystyrene*.		
PSU	See *PES. Sulphone polymer*.	RPVC	See *UPVC. Unplasticized polyvinyl chloride*.
PTFE	*Polytetrafluoroethylene* or, poly-(tetrafluoroethylene).	RRPP	See *RMPP. Rubber modified polypropylene*.
PTMT	See *PBT. Polybutylene terephthalate*.	SAN	*Styrene-acrylonitrile copolymer* or, poly-(1-phenylethylene-co-1-cyanoethylene) or, poly-(styrene-co-acrylonitrile).
PU	See *PUR. Polyurethane*.		
PUR	*Polyurethane* or, polycarbamate. PU.	SB	*Styrene-butadiene copolymer* or, butadiene-styrene copolymer or, poly-(styrene-co-butadiene).
PVA	See *PVAC* and also *PVAL. Polyvinyl acetate and polyvinyl alcohol*.	SBB	See *BDS. Styrene-butadiene block copolymer*.
PVAC	*Polyvinyl acetate* or, poly-(vinyl acetate). PVA or PVAc.	SBR	*Styrene-butadiene rubber* or, poly-(1-butenylene-co-phenylethylene).
PVAl	See *PVAL. Polyvinyl alcohol*.	SBR-E	See *E-SBR. Emulsion styrene butadiene rubber*.
PVAL	*Polyvinyl alcohol* or, poly-(vinyl alcohol). PVA or PVAl.	SBR-L	See *SBR-L. Solution styrene-butadiene rubber*.
PVC	*Polyvinyl chloride* or, polychloroethene or poly-(1-chloroethylene) or, poly-(monochloroethylene).	SBS	See *TPE-S*. Styrene-butadiene-styrene block copolymer or, poly-(1-phenylethylene-b-1-butenylene-b-1-phenylethylene). This is a linear block copolymer) or a, SBS block copolymer or, SBS block polymer or, styrene triblock copolymer or, teleblock copolymer or, *thermoplastic elastomer-styrene based*.
PVC-C	See *CPVC. Chlorinated polyvinyl chloride*.		
PVC-O	See *OPVC. Oriented polyvinyl chloride*.		
PVC-P	See *PPVC. Plasticized polyvinyl chloride*.		
PVC-R	Rigid polyvinyl chloride (RPVC). See *UPVC. Unplasticized polyvinyl chloride*.	SCR	Styrene-chloroprene rubber or, chloroprene-styrene copolymer or rubber. See *chloroprene rubber*.
PVC-U	See *UPVC. Unplasticized polyvinyl chloride*.	SEBS	See *TPE-S*. Styrene-ethylene/butylene-styrene block copolymer or, styrene olefin thermoplastic elastomer. See *thermoplastic elastomer-styrene based*.
PVCZ	See *PVK. Polyvinyl carbazole*.		
PVdC	See *PVDC. Polyvinylidene chloride*.	SI	*Silicone plastics or resins* or, methyl-phenylsilicone resins.
PVDC	*Polyvinylidene chloride* or, high-vinylidene copolymer or, poly-(1,1-dichloroethylene) or, polyvinylidene chloride copolymer or, poly-(vinylidene chloride). PVdC.	SIR	Styrene-isoprene rubber or, isoprene-styrene rubber
		SIS	See *TPE-S*. Styrene-isoprene-styrene block copolymer. *A thermoplastic elastomer-styrene based*.
PVDF	*Polyvinylidene fluoride* or, poly-(1,1-difluoroethylene). PVF$_2$.	SMA	*Styrene maleic anhydride* or, poly-(styrene-co-maleic anhydride) or, styrene-maleic anhydride copolymer.
PVF	Polyvinyl fluoride.		
PVF$_2$	See *PVDF. Polyvinylidene fluoride*.	SMC	Sheet moulding compound a *polyester moulding compound* or PMC.
PVK	*Polyvinyl carbazole* or, poly-(N-vinyl carbazole). PVK or PVCZ.	SR	Synthetic rubber.
PVMQ	*Silicone rubber* containing methyl, phenyl and vinyl groups.	SR-E	See *E-SR. Emulsion synthetic rubber*.
PVP	*Polyvinyl pyrrolidone* or, poly-(N-vinyl-2-pyrrolidone).	T	A rubber which has sulphur, carbon and oxygen in the main polymer chain.
		T	A thermoplastics material.
Q	*Silicone rubber* or, silicone elastomer or, polysiloxane rubber.	T	*Polysulphide rubber* or, elastothiomer or, thioplast. TM or TR.

Table 4. *Some abbreviations and names of plastics and elastomers - contd*

TC-NR	Technically classified (natural) rubber.	TPE-OXL	Thermoplastic elastomer-olefin based with crosslinked rubber. TPE-OXL or TPO-XL.
TE	*Thermoplastic elastomer.*	TPE-S	*Thermoplastic elastomer - styrene based.* A styrene triblock copolymer. TPE-S or SBS - usually styrene-butadiene-styrene block copolymer but could also be, styrene-ethylene/butylene-styrene block copolymer (TPE-S or SEBS) or styrene-isoprene-styrene block copolymer (TPE-S or SIS).
TE-EE	See *TPE-E. Thermoplastic elastomer - ether based.*		
TFE-HFP	See *FEP. Fluorinated ethylene propylene.*		
TM	See *T. Polysulphide rubber.*		
TMC	Thick moulding compound. - a polyester moulding compound or PMC.	TPE-U	*Thermoplastic elastomer - urethane based.* thermoplastic polyurethane or, thermoplastic urethane elastomer. TPU.
TOR	Transpolyoctenamer. *Polyoctenamer* rubber.	TPE-XL	Thermoplastic elastomer - crosslinked rubber.
TP-EE	See *TPE-E. Thermoplastic elastomer - ether based.*	TPE-XLV	See *TPV. Thermoplastic vulcanizate.*
TP	A thermoplastics material.	TPO	See *TPE-O. Thermoplastic elastomer - olefin based.*
TP-EPDM	See TPE-EPDM. *Thermoplastic ethylene-propylene elastomer* or rubber.	TPO-XL	See *TPE-OXL*. Thermoplastic elastomer-olefin based with crosslinked rubber.
TP-EVA	See *TPE-EVA*. Thermoplastic elastomer - ethylene-vinyl acetate elastomer.	TPR	See *TPE*.
TP-NBR	See *TPE-NBR*. Thermoplastic nitrile-butadiene elastomer.	TPS	See *HIPS. High impact polystyrene.*
TP-NR	See *TPE-NR*. Thermoplastic natural rubber.	TPU	See *TPE-U. Thermoplastic elastomer - urethane based.*
TPA	Transpolypentenamer. *Polypentenamer* rubber.	TPV	Thermoplastic vulcanizate. EA-TPV or TPE-XL.
TPE	*Thermoplastic elastomer* (or rubber).	TR	See *T. Polysulphide rubber.*
TPE-A	*Thermoplastic elastomer - amide based* or, elastomeric polyamide or, polyamide elastomer or, polyether block amide or, thermoplastic copolyether or, thermoplastic elastomer - amide based. PEBA or PBA.	TSR	Technically specified (natural) rubber. CNR - technically specified (natural) rubber from China. SIR - technically specified (natural) rubber from Indonesia. SMR - technically specified (natural) rubber from Malaysia. SLR - technically specified (natural) rubber from Sri Lanka. TTR - technically specified (natural) rubber from Thailand.
TPE-E	*Thermoplastic elastomer - ether based* or, block polyetherester or, copolyetherester or, polyether ester or, polyether ester block copolymer or, polyether ester elastomer or, polyester elastomer or, thermoplastic copolyester or, thermoplastic ether ester or, thermoplastic polyetherester or, thermoplastic polyetherester. PEEL or, COPE or, TP-EE or TEEE or Y-BPO.	U	A rubber which has carbon, oxygen and nitrogen in the main polymer chain.
		UF	*Urea-formaldehyde* or, urea-methanal. (an aminoplast).
		ULDPE	See *VLDPE. Very low density polyethylene.*
		UP	Unsaturated polyester.
TPE-EPDM	*Thermoplastic ethylene-propylene elastomer* or rubber. TPE-EPDM or TP-EPDM.	UPVC	*Unplasticized polyvinyl chloride* or, rigid polyvinyl chloride or, rigid PVC or, unplasticized polychloroethene or, unplasticised polyvinyl chloride. PVC-U or RPVC or PVC-R.
TPE-EVA	Thermoplastic elastomer - ethylene-vinyl acetate elastomer. TP-EVA.		
TPE-FKM	Thermoplastic fluoro elastomer.	VCA	*Vinyl chloride-acrylonitrile copolymer* or, poly-(vinyl chloride-co-acrylonitrile).
TPE-I	Thermoplastic isoprene rubber. TPE-S or Y-IR.	VCE	*Vinyl chloride-ethylene copolymer* or, poly-(vinyl chloride-co-ethylene).
TPE-NBR	*Thermoplastic nitrile-butadiene elastomer* or, thermoplastic nitrile-butadiene rubber. TP-NBR or Y-NBR.	VCEVA	*Vinyl chloride-ethylene-vinyl acetate terpolymer.*
		VCP	*Vinyl chloride-propylene copolymer* or, poly-(vinyl chloride-co-propylene).
TPE-NR	*Thermoplastic natural rubber. TP-NR.*		
TPE-O	*Thermoplastic elastomer - olefin based* or, thermoplastic polyolefin or, olefin thermoplastic elastomer. TPO.	VCVA	*Vinyl chloride-vinyl acetate copolymer* or, polyvinyl chloride-acetate copolymer or, poly-(vinyl chloride-co-vinyl acetate). PVCA.

Table 4 - *contd*

VDCA	*Vinylidene chloride-vinyl chloride copolymer* or poly-(vinylidene chloride-co-vinyl chloride).	XNBR	Carboxylic-nitrile rubber or, *carboxylated nitrile rubber*.
VDVA	*Vinylidene chloride-acrylonitrile copolymer* or, poly-(vinylidene chloride-co-acrylonitrile).	XBR	Carboxylic-butadiene rubber or, *carboxylated butyl rubber*. See *halogenated butyl rubber*.
VE	Vinyl ester resin.	XMC	Continuous roving moulding compound - a wound moulding compound - a *polyester moulding compound* or PMC.
VLDPE	*Very low density polyethylene* or, polyethylene-very low density or, polyethene-very low density or, ultra low density polyethylene. PE-VLD or ULDPE.	XPS	See *EPS. Expanded polystyrene*.
VMQ	*Silicone rubber* containing methyl and vinyl groups.	Y	When used as a prefix indicates a thermoplastic elastomer.
VP	*Vinyl pyridine rubber* or copolymer or, butadiene-vinyl pyridine copolymer.	YBPO	See *TPE-E. Thermoplastic elastomer - ether based*.
		YNBR	See *TPE-NBR. Thermoplastic nitrile-butadiene elastomer*.
X	Denotes the presence in a rubber of the reactive carboxylic (COOH) group.	Z	A rubber which has nitrogen and phosphorous in the main polymer chain.
XCR	Carboxylic-chloroprene rubber or, carboxylated chloroprene rubber.	ZMC	Low viscosity moulding compound - a *polyester moulding compound* or PMC.

Table 5a. *Some trade names/trade marks, abbreviations and suppliers of polymers and polymer moulding compounds (sorted by alphabetical order of trade name)*
Key for Table 5a on p. 517

Trade name/trade mark	Abbreviation	Supplier
Aclyn	ION	Allied Signal Inc.
Acrylite	PMMA	Canada Color & Chemicals
Admer	PO	Mitsui Sekka
Adpro	PP	Advanced Global Polymers
Airvol	PVAL	Air Products and Chemicals Inc.
Akulon	PA 6	DSM (Dutch State Mines)
Akulon	PA 66	DSM (Dutch State Mines)
Akuloy	PA/PP	DSM (Dutch State Mines)
Alcoryl	ABS	Rhône-Poulenc Chimie
Alcryn	TPE EA-TPV	Du Pont
Altuglas	PMMA	Elf Atochem S.A.
Altulite	PMMA	Altulor
Amoco PE-HD	PE-HD	Amoco Performance Products
Amodel	PPA	Amoco Performance Products
Ampal	UP	Ciba Geigy
Apec	PC-HT	Bayer
Apel	PO-A	Mitsui Sekka
Appryl	PP	Appryl
Appryl	PP	Elf Atochem S.A.
Apscom	TP-COM	DSM (Dutch State Mines)
Araldite	EMC	Ciba Geigy
Araldite	EP	Ciba Geigy
Arcel	PE-X	Arco
Arcomid	PA 6	Resinmec
Arcomid	PA 66	Resinmec
Arcoplen	PP	Resinmec
Ardel	Polyarylate	Amoco Performance Products
Arlen	PA	Mitsui Sekka
Arnite	PBT	DSM (Dutch State Mines)
Arnite	PET	DSM (Dutch State Mines)
Arnitel	TPE-E	DSM (Dutch State Mines)
Aropol	PMC	Ashland Chemical Co.
Arpro	PP-X	Arco
Arpylene	PP-COM	Hydro Polymers
Arrhadur	ABS	Elf Atochem S.A.
Arylon T	ABS/PSU	USS Chemicals
Ashlene	PA 6	Ashley Polymers Inc.
Ashlene	PA 612	Ashley Polymers Inc.
Ashlene	PA 66	Ashley Polymers Inc.
Ashlene	PA-amorphous	Ashley Polymers Inc.
Astryn	PP-COM	Himont
Attane	PE-VLD	Dow
Avalon	TPE-U	ICI
Azdel	GMT/PP	Azdel Inc.
Azloy	GMT/PC/PBT	Azdel Inc.
Azmet	GMT/PBT	Azdel Inc.
Bakelite	PF	Bakelite/Sterling
Bakelite Polyester Alkyd	GPMC	Sterling Moulding Materials
Bapolan	PS	Bamberger
Bapolene	PE-HD	Bamberger
Bapolene	PP	Bamberger
Barex	PAN	Sohio
Barex	PAN	Standard Oil
Basopor	UF	BASF
Bayblend	ABS/PC	Bayer
Baycomp	FF-TP	Baycomp
Bayer CM	CPE	Bayer
Baygal	PUR resins	Bayer
Baymidur	PUR resins	Bayer
Baymod A	ABS modifier	Bayer
Baymod L	EVA modifier	Bayer
Baypren	CR	Bayer
Beaulon	PB	Mitsui Sekka

Table 5a - contd

Trade name/trade mark	Abbreviation	Supplier
Beetle	MF resins	BIP Chemicals
Beetle	PC	BIP Chemicals
Beetle	POM	BIP Chemicals
Beetle	UF	BIP Chemicals
Beetle	UF resins	BIP Chemicals
Beetle	UP resins	BIP Chemicals
Beetle DMC	DMC	BIP Chemicals
Beetle nylon 6	PA 6	BIP Chemicals
Beetle nylon 66	PA 66	BIP Chemicals
Beetle PBT	PET	BIP Chemicals
Beetle PET	PET	BIP Chemicals
Benvic	PVC	Solvay
Benvic EV	PVC-GF	Solvay
Bergacell	CA	Bergmann
Bergadur	PBT	Bergmann
Bergamid A	PA 66	Bergmann
Bergamid B	PA 6	Bergmann
Bergaprop	PP	Bergmann
Bexloy	ETP	Du Pont
Bioceta	CA-BIO	Tubize Plastics
Biopol	PHB-CO	ICI
Biopol	PHB-H	ICI
BP Polystyrene	PS	BP Chemicals
Buna AP	EPDM	Hüls
Buna CB	BR	Bayer
Buna EM	SBR-E	Hüls
Buna SL	SBR-L	Hüls
Buna V	SBR-L	Hüls
Butachlor	CR	Distugil
Bynel	PO	Du Pont
Cabelec	PP-conductive	Cabot Plastics
Cadon	SMA	Monsanto
Cadon 300	SMA-impact modified	Monsanto
Calibre	PC	Dow
Caprolan	TPE-U	Elastogran/BASF
Capron	PA 6	Allied Signal Inc.
Carbaicar	UF	S A Aicar
Carbopol	Acrylic acid	BF Goodrich
Carboset	Acrylic polymer	BF Goodrich
Cariflex	SR	Shell
Cariflex	TPE-S (SBS)	Shell
Caril	PPO-M	Shell
Carinex	PS-HI	Shell
Celanese Nylon	PA 66	Hoechst/Hoechst Celanese
Celanex	PBT	Hoechst/Hoechst Celanese
Celanex	PBT-GF	Hoechst/Hoechst Celanese
Cellasto	PUR-X	BASF
Cellidor	CA	Albis
Cellidor	CAP	Albis
Celsir	UF resins	SIR (Societá Italiana Resine)
Celstran	TP-LF	Hoechst/Hoechst Celanese
Centrex	ASA/AES	Advanced Elastomer Systems
Cevian	ABS-FR	Hoechst Daicel Polymers
Clarene	EVOH	Solvay
Clearflex	PE-LLD	Enichem
Corton	PP-COM	Poly Pacific Pty
Corvic	PVC	European Vinyl Corp.
Crastine	PBT	Du Pont
Craston	PPS	Ciba Geigy
Cristamaid	PA-CO	Elf Atochem S.A.
Crystic	UP	Scott Bader
Crystic Impel	GPMC	Scott Bader
Crystic Impreg	PMC	Scott Bader
Crystic Kollerdur	PUR	Scott Bader
Crystic Kollernox	EP	Scott Bader

Table 5a. *Some trade names/trade marks, abbreviations and suppliers of polymers and polymer moulding compounds (sorted by alphabetical order of trade name) - contd*

Trade name/trade mark	Abbreviation	Supplier
Cyanacryl	ACM	Enichem
Cyanacryl	ACM	Enichem
Cycolac	ABS	General Electric Co.
Cycoloy	ABS-AL	General Electric Co.
Cymel	MF	Cyanamid
DAIP 6000	DAIP	Synres Amoco
DAP 5000	DAP	Synres Amoco
Daplen	PE-HD	Chemie Linz
Daplen	PE-LD	Chemie Linz
Daplen	PP	Chemie Linz
Daplen	PP-COM	Chemie Linz
Degadur	MMA-resins	Degussa AG
Degalan	PMMA-CO	Degussa AG
Degalan LP	PMMA-CO	Degussa AG
Degalan S	MMA-resins	Degussa AG
Degament	MMA-resins	Degussa AG
Delrin	POM-H	Du Pont
Delrin 100 ST	POM-H (HI)	Du Pont
Delrin II	POM-H	Du Pont
Delrin P	POM-H	Du Pont
DER	EP	Chemroy Canada
Derakane	VE	Dow
Desmopan	TPE-U	Bayer
Dexel	CA	Courtaulds Speciality Plastics
Dexel S	CA	Courtaulds Speciality Plastics
Dexflex	TPE-O	Dexter Plastics
Diakon	PMMA	ICI
Dialac	ASA	Misubishi
Dimension	PA/PPE	Allied Signal Inc.
Dion	UP	Fiberglass
Disco	DCF	Technical Fibre Products
Dowlex	PE-LD	Dow
Dowlex	PE-LLD	Dow
DSM Nyrim	RIM PA	DSM RIM Nylon
Du Pont 20 series	PE	Du Pont
Dunlocrumb	NR-TSR	Dunlop Plantations
Dunlocrumb S	NR-skim rubber	Dunlop Plantations
Duraflex	PIB	Shell
Dural	UPVC-COM	Dexter Plasics
Duralex	PVC/PU/NBR AL	Dexter Plasics
Duralon	PA 11	Thermoclad
Duranit	SBR	Hüls
Durapol	PMC	Isola Werke
Durethan	PA 6	Bayer
Durethan	PA 66	Bayer
Durethan	PA 66	Mobay
Durethan	PA 66-GF	Bayer
Durethan	PA 6-GF	Bayer
Durez	DAP	Occidental
Durez	PF	Occidental
Dutral	EPR	Enichem
Dyflor	PVDF	Dynamit Nobel
Dyflor	PVDF	Hüls
Dyflor	PVDF	Kay Fries
Dylile	PS-X	Arco
Dynaform	TPE-RMPP	Dynamit Nobel
Dynapor	PF-resin-X	Hüls
Dynaset	PF	Reichold
Dynat	NR-TSR	KGSB
Dynat S	NR-skim rubber	KGSB
Dytron XL	TPE-OXL	Advanced Elastomer Systems
Eccomold	EP	Emmerson and Cuming
Ecdel	TPE-E	Eastman Chemicals
Ecolyte	TP-photodegradable	Ecoplastics/Eco Chemicals

Table 5a - *contd*

Trade name/trade mark	Abbreviation	Supplier
Ecolyte II & IV	PE	Ecoplastics/Eco Chemicals
Ecolyte S	PS	Ecoplastics/Eco Chemicals
Edistir	PS	Enichem
Edistir	PS-HI	Enichem
Elastocoat	PUR-coating	BASF
Elastollan	TPE-U	BASF/Elastogran
Elastopal	PUR-casting	BASF
Elastopreg	GMT laminates	BASF/Elastogran
Elexar	TPE	Shell
Elmit	PA-COM	Mitsui Sekka
ELP	T-liquid	Morton International
Eltex	PE-HD	Solvay
Eltex P	PP-CO	Solvay
Eltex P	PP-H	Solvay
Elvaloy	PO-CO	Du Pont
Elvamide	PA	Du Pont
Elvanol	EVOH	Du Pont
Elvax	EVA	Du Pont
EMI-X	TP-EMI	LNP Engineering Plastics Inc.
Empee PP	PP	Monmouth Plastics
Envex	PI-COM	Rogers Corp
Epikote	EP resins	Shell
Epoester	EP resins	SIR (Societá Italiana Resine)
Epolan	ABS	Industrial Resistol
Epomik	EP resin	Mitsui Sekka
Epon	EP	Shell
Eponac	EP	SPREA
Eponite	EP-COM	Shell
Eposir	EP resins	SIR (Societá Italiana Resine)
Epox	EP-COM	Mitsui Sekka
Epoxyprene	NR-E	KGSB
Eraclene	PE-HD	Enichem
Eref	PP	Solvay
ERF DMC	DMC	ERF
ERF SMC	SMC	ERF
Ertalan	PA-cast	Erta
Escor EAA	EAA	Exxon Chemical
Escorene	PE-LD	Exxon Chemical
Escorene	PE-LLD	Exxon Chemical
Escorene	PE-LLD/EVA (<5%)	Exxon Chemical
Escorene Optema	EMA	Exxon Chemical
Escorene PP	PP	Exxon Chemical
Escorene Ultra	EVA	Exxon Chemical
Escorene α	PE-LLD (a HAO)	Exxon Chemical
Esrel	TPE-E	Cheil Synthetices Inc.
Estaloc	TPE-U (COM)	BF Goodrich
Estane	TPE-U	BF Goodrich
Esterform	PMC	Chromos Ro-Polimeri
ETA	TPE	Republic Plastics
Ethocel	EC	Dow
Europrene	ACM	Enichem
Europrene	BR	Enichem
Europrene	NBR	Enichem
Europrene	NBR/PVC	Enichem
Europrene	SBR	Enichem
Europrene	TPE-S (SBS)	Enichem
Europrene	TPE-S (SIS)	Enichem
Eval	EVOH	Kuraray/EVAL Co
Evatane	EVA	Elf Atochem S.A.
EVOH SF	EVOH	Elf Atochem S.A.
Evoprene	TPE	Evode Plastics Ltd.
Evoprene E	TPE	Evode Plastics Ltd.
Evoprene G	TPE	Evode Plastics Ltd.
Evoprene Super S	TPE	Evode Plastics Ltd.
Extir	PS-X	Enichem

Table 5a. *Some trade names/trade marks, abbreviations and suppliers of polymers and polymer moulding compounds (sorted by alphabetical order of trade name) - contd*

Trade name/trade mark	Abbreviation	Supplier
Exxelor	modifiers-plastics	Exxon Chemical
Exxon Bromobutyl	IIR-X	Exxon Chemical
Exxon Butyl	IIR	Exxon Chemical
Exxon Chlorobutyl	IIR-X	Exxon Chemical
FA	T-millable	Morton International
Faradex	TP-SS	DSM (Dutch State Mines)
Fenochem	PF	Chemiplastica Spa
Fenoform	PF	Chromos Ro-Polimeri
Ferrolene	PP-COM	Ferro
Ferrolene-TPE	TPE-RMPP	Ferro
Fiberloc	PVC-GF	BF Goodrich
Fibresinol	PF-GF	Raschig
Finaclear	PS/SBS	Petrofina
Finaprene	TPE-SBS	Petrofina
Finapro	PP	Petrofina
Finathene	PE-HD	Petrofina
Finathene	PE-MD	Petrofina
Flexene	PE-LLD	Enichem
Flomat	SMC	Freeman Chemicals
Flowmat	SMC	Freeman Chemicals
Fluon	PTFE	ICI
Fluromelt	PTFE-CO	LNP Engineering Plastics Inc.
Foam Kon	SF-MB	LNP Engineering Plastics Inc.
Foraflon	PVDF	Elf Atochem S.A.
Fortiflex	PE-HD	Soltex Polymer Corp
Fortiflex	PE-MD	Soltex Polymer Corp
Fortilene	PP	Soltex Polymer Corp
Fortron	PPS	Hoechst/Hoechst Celanese
Freeflow	GPMC	Freeman Chemicals
FR-TPX	TPX-GF	Mitsui Sekka
Garaprene	TP-AL	Evode Plastics Ltd.
GARY	PVC-COM	Evode Plastics Ltd.
Gedex	PS	Orkem
Gedexcel	PS-X	Elf Atochem S.A.
Geolast	PP/NBR	Advanced Elastomer Systems
Geon	PVC	BF Goodrich
Geon CIM	PVC-COM	BF Goodrich
Geon HTX	PVC-HT	BF Goodrich
Geon RX	PVC-medical	BF Goodrich
Glilax	TPE-A	Dianippon
Goodmer	TPE-O	Mitsui Sekka
Greenflex	EVA	Enichem
Grilamid	PA 12	EMS-Chemie
Grilamid TR	PA-TR	EMS-Chemie
Grilon	PA 6	EMS-Chemie
Grilon C	PA 6/PA 12	EMS-Chemie
Grilon T	PA 66	EMS-Chemie
Grilonit	EP	EMS-Chemie
Grivory	PA-AR	EMS-Chemie
Halar	ECTFE	Ausimont
Halon	PTFE	Ausimont
Haysite	PMC	Haysite Reinforced Plastics
Hercules HPR	PE-HD (HMW)	Hercules
Hetron	PMC	Ashland Chemical Corp.
Hifax	PP-CO	Himont
Higlass	PP-COM	Himont
Hipol	PP	Mitsui Sekka
Hi-zex	PE-HD	Mitsui Sekka
Hi-zex Million	PE-HD UHMW	Mitsui Sekka
Hostacom PP reinforced	PP-COM	Hoechst/Hoechst Celanese
Hostaflon FEP	ETFE	Hoechst/Hoechst Celanese
Hostaflon FEP	FEP	Hoechst/Hoechst Celanese
Hostaflon FEP	PFA	Hoechst/Hoechst Celanese
Hostaflon FEP	PTFE	Hoechst/Hoechst Celanese

Table 5a - *contd*

Trade name/trade mark	Abbreviation	Supplier
Hostaform	POM-CO	Hoechst/Hoechst Celanese
Hostalen	PE-HD	Hoechst/Hoechst Celanese
Hostalen	PE-MD	Hoechst/Hoechst Celanese
Hostalen GUR	PE-UHMW	Hoechst/Hoechst Celanese
Hostalen PP	PP	Hoechst/Hoechst Celanese
Hostalit	PVC	Hoechst/Hoechst Celanese
Hostalit Z	PVC-HI	Hoechst/Hoechst Celanese
Hostapren	PE	Hoechst/Hoechst Celanese
Hostatec	PEK	Hoechst/Hoechst Celanese
Huntsman Polystyrene	PS	Huntsman
Huntsman Polystyrene	PS-HI	Huntsman
Hyflo	NR-powdered	Golden Hope Plantations
Hypalon	CSM	Du Pont
Hyrub	NR-TSR	Golden Hope Plantations
Hytrel	TPE-E	Du Pont
Idemitsu Polycarbonate	PC	Idemitsu Petro Chem.
Illandur	PMC	Illing
Impact	PET alloy	Allied Signal Inc.
Impel	GPMC	Scott Bader
Impet	PET	Hoechst/Hoechst Celanese
Innovex	PE-LLD	BP Chemicals
Intene	BR	Enichem
Intol	SBR-E	Enichem
Iotek	ION-Na	Exxon Chemical
Iotek	ION-Zn	Exxon Chemical
Isomin	MF	Perstorp
Iupital	POM-CO	Misubishi
Ixan	PVDC	Solvay
Ixef	PAA 6	Solvay
Iztavil	PVC	Polimeros De Mexico
Jonylon	PA 6	BIP Chemicals
Jonylon	PA 66	BIP Chemicals
K F	PVDF	Kureha
Kadel	Polyketone	Amoco Performance Products
Kamax	Acrylic imides	Rohm and Haas
Kane Ace	PVC-C	Kaneka/E W Seward Ltd.
Kane Ace B	MBS	Kaneka/E W Seward Ltd.
Kane Ace PA	Proc. aid-acrylic	Kaneka/E W Seward Ltd.
Kane Ace XEL	PVC-X	Kaneka/E W Seward Ltd.
Kelburon	TPE-RMPP	DSM (Dutch State Mines)
Kelon A	PA 66	Lati
Kelon B	PA 6	Lati
Keltan TP	TPE-RMPP	DSM (Dutch State Mines)
Kematal	POM-CO	Hoechst/Hoechst Celanese
Kerimid	PI	Rhône-Poulenc Chimie
Keripol	PMC	Pheonix
Kinel	PI	Rhône-Poulenc Chimie
Koblend	ABS/PC	Enichem
Kodapek PET	PET	Eastman Chemicals
Kodar PETG	PET-CO	Eastman Chemicals
Kostil	SAN	Enichem
Kraton TR	TPE-S (SBS)	Shell
Kynar	PVDF	Elf Atochem S.A.
Kynar	PVDF	Penwalt
K-resin	BDS	Phillips Petroleum Chemicals
Lacovyl	PVC	Elf Atochem S.A.
Lacovyl	VC/VA	Elf Atochem S.A.
Lacqrene	PS	Elf Atochem S.A.
Lacqrene	PS-COM	Elf Atochem S.A.
Lacqrene	PS-HI	Elf Atochem S.A.
Lacqtene IID	PE-HD	Elf Atochem S.A.
Lacqtene HX	PE-LLD	Elf Atochem S.A.
Lacqtene LX	PE-LLD	Elf Atochem S.A.
Ladene	PE-LLD	Sabic
Larflex	TPE-EPDM	Lati

Table 5a. *Some trade names/trade marks, abbreviations and suppliers of polymers and polymer moulding compounds (sorted by alphabetical order of trade name) - contd*

Trade name/trade mark	Abbreviation	Supplier
Larton	PPS	Lati
Lastane	TPU	Lati
Lastane	TPU	Lati
Lastiflex	PVC/SR	Lati
Lastil	SAN	Lati
Lastilac	ABS	Lati
Lastilac	ABS/PC	Lati
Lastirol	PS	Lati
Lasulf	PSU	Lati
Latamid 12	PA 12	Lati
Latamid 6	PA 6	Lati
Latamid 66	PA 66	Lati
Latan	POM	Lati
Latene HD	PE-HD	Lati
Later	PBT	Lati
Latilon	PC	Lati
Latilub	TP-LUB	Lati
Latishield	TP-EMI shielding	Lati
Latistat	TP-antistatic	Lati
Legupren	UP	Bayer
Leguval	UP	Bayer
Lekutherm	EP	Bayer
Levaflex	TPO-XL	Bayer
Levapren	EVA	Bayer
Lexan	PC	General Electric Co.
Linpac Polystyrene	PS	Linpac
Lomod	TPE-E	General Electric Co.
Lotader	TP-CO	Orkem
Lotrene	PE-LD	Orkem
Lotrex	PE-LLD	Orkem
LP	T-liquid	Morton International
LP-R	T-compounds	Morton International
Lubmer	PO-LUB	Mitsui Sekka
Lubricomp	TP-LUB	LNP Engineering Plastics Inc.
Lucalen	PE-CO	BASF
Lucalor	PVC-C	Elf Atochem S.A.
Lucobit	PE-CO/bitumen	BASF
Lucolene	PVC-U	Elf Atochem S.A.
Lucorex	PVC-U	Elf Atochem S.A.
Lucryl	PMMA	BASF
Lupolen	PE-LD	BASF
Lupolen HD	PE-HD	BASF
Luran	SAN	BASF
Luran S	ASA	BASF
Luranyl	PPO-M	BASF
Lustran	ABS	Monsanto
Lustran Ultra ABS	ABS-HG	Monsanto
Magnacomp	TP-MAG	LNP Engineering Plastics Inc.
Magnum	ABS	Dow
Makroblend	PC/PBT	Bayer
Makrolon	PC	Bayer
Makrolon	PC-GF	Bayer
Maranyl	PA 6	ICI
Maranyl	PA 66	ICI
Marlex	PE-HD	Phillips Petroleum Chemicals
Marub	NR-TSR	MARDEC
Marvylex	TPE-PVC	LVM
Marvylflo	PVC	LVM
Megapoly	NR-MG	Asiatic Developments
Megolon	TP-COM	Lindsay & Williams
Melaicar	MF	S A Aicar
Melamine moulding comp	MF	Perstop Ferguson
Melinar	PET	ICI
Melmex	MF	BIP Chemicals

Table 5a - *contd*

Trade name/trade mark	Abbreviation	Supplier
Meloplas	MF	Ciba Geigy
Meloplas	MPF	Ciba Geigy
Melsprea	MF	SPREA
Menzolit	PMC	Menzolit Werke
Merlin	PC	Mobay
Methocel	MC	Dow
Metton	LMR	Shell/Hercules
Miapol	UP resin	Mia Chemical
Milastomer	TPE-O	Mitsui Sekka
Mindel	PSU-COM	Amoco Performance Products
Minlon	PA 66-COM	Du Pont
Mipelon	PO-UHMW	Mitsui Sekka
Mitsui EPT	EPDM	Mitsui Sekka
Mitsui FR-PP	PP-GF	Mitsui Sekka
Mitsui Hi-wax	PE-VLD	Mitsui Sekka
Modar	TST-acrylic resin	ICI
Moldsite	PF	SPREA
Moplen	PP	Himont
Nakan	PVC-COM	Elf Atochem S.A.
NAS	PS/PMMA	Novacor
NAS	PS/PMMA/BD	Novacor
Naycar-A	PA 66	Polymer Trading
Naycar-B	PA 6	Polymer Trading
Nealid	PP	Neste
Neocis	BR	Enichem
Neocis	BR	Enichem
Neoflon	FEP	Daikin
Neonite	EMC	Ciba Geigy
Neoprene	CR	Du Pont
Neo-zex	PE-MD	Mitsui Sekka
Nepol	PP-LF	Neste
Neste HDPE	PE-HD	Neste
Neste LD	PE-LD	Neste
Neste LLD	PE-LLD	Neste
Neste PS	PS-HI	Neste
Nestorite	PF	Perstorp
Nike	CN	Punda Inc.
Nivionplast A	PA 6-COM	Enichem
Nivionplast B	PA 6-COM	Enichem
Norchem	PE-HD	Enron/Delong Prochem
Norchem	PE-LLD	Enron/Delong Prochem
Norchem	PP	Enron/Delong Prochem
Nordel	EPR	Du Pont
Norlin	PE-LLD	Northern Petrochemicals
Norpol	UP	Jotun Polymer
Norpol	UP-resin	Jotun Polymer
Norsoflex	PE-VLD	Orkem
Norsomix	DMC	Orkem
Norsorex	SR	Elf Atochem S.A.
Nortuff	PP	Norchem Inc
Norvinyl	PVC	Norsk Hydro
Noryl	PPO-M	General Electric Co.
Noryl GTX	PPO-M/PA	General Electric Co.
Novablend	PVC	Novatec
Novalloy	ABS-AL	Hoechst Daicel Polymers
Novamid	PA	Misuibishi
Novapol HD	PE-HD	Novacor
Novapol LD	PE-LD	Novacor
Novex	PE-LD	BP Chemicals
Novodur	ABS	Bayer
Novolen	PP	BASF
Novon	TP-starch based	Warner Lambert
Nucrel	EMA	Du Pont
Nuloy	PA 6	Terlon Polimeros
Nydur	PA 6	Mobay

Table 5a. *Some trade names/trade marks, abbreviations and suppliers of polymers and polymer moulding compounds (sorted by alphabetical order of trade name) - contd*

Trade name/trade mark	Abbreviation	Supplier
Nylafil	PA-COM	Wilson Fibrefil
Nypel	PA 6-RC	Allied Signal Inc.
Ongro	CPE	Borsodchem
Oppanol	PIB	BASF
Orgalloy R	PA 66/PP	Elf Atochem S.A.
Orgalloy R	PA 6/PP	Elf Atochem S.A.
Orgamide	PA 6	Elf Atochem S.A.
Orgater	PBT	Elf Atochem S.A.
Orkot	PF composite	Orkot Engineering Plastics
Oroglas DR	PMMA-HI	Rohm and Haas
Oroglas V	PMMA	Rohm and Haas
Palapreg	DMC	BASF
Palapreg	SMC	BASF
Palatal	UP	BASF
Paraloid	Impact modifiers	Rohm and Haas
Paraloid EXL	Impact modifiers	Rohm and Haas
Parapol	PIB	Exxon Chemical
Parr	DAP	US Prolam Inc.
Paxon	PE-HD	Allied Signal Inc.
Pax-Purge	Purge compounds	Canada Color & Chemicals
Pebax	TPE-A	Elf Atochem S.A.
Pekema	PVC	Punda Inc.
Pekevic	PVC	Neste
Pemex	PE-LD	Petroleos Mexicanos
Perbunan	NBR	Bayer
Perbunan	NNBR	Bayer
Petlon	PET	Bayer
Petra	PET	Allied Signal Inc.
Petrothene	PE-HD	USI/Quantum
Petrothene	PE-LLD	USI/Quantum
Pevikon	PVC	Norsk Hydro
Pibiter	PBT	Enichem
Pibiter	PBT-HI	Enichem
Plaskon	EP	Plaskon Molding Div.
Plaskon DAP	DAP	Plaskon Molding Div.
Plastech	PP-COM	Cabot Plastics
Plasticlean	Purge compound	W S Wood Assoc.
Plenco	MPF	Plastics Engineering Co.
Plenco	PF	Plastics Engineering Co.
Plenco	UP	Plastics Engineering Co.
Plexar	PE/EVA-CO	DSM (Dutch State Mines)
Pocan	PBT	Bayer
Pocan	PBT-GF	Bayer
Pocan 7918	PBT/SR/PC	Bayer
Polloplas	UF	Dynamit Nobel
Poly DAP	DAP	US Prolam Inc.
Polychem	DAP	Budd Co.
Polyfort	PP-COM	Schulman Inc
Polyidene	PVDC	Scott Bader
Polyloy	PA 6	EMS-Chemie
Polyloy	PA 6	Illing
Polyloy	PA 66	Illing
Polymer E	PE-LD	Asia Polymer Corp.
Polyplastol	Proc. aid-rubbers	Enichem
Polystal	GMT	Bayer
Polystyrol	PS	BASF
Polystyrol	PS	Norsk Hydro
Polystyrol	PS-HI	BASF
Polystyrol	PS-HI	Norsk Hydro
Polytron	PVC-conductive	BF Goodrich
Prevex	PPO-M	General Electric Co.
Primacor	EAA	Dow
Primef	PPS	Solvay
Procom	PP-COM	ICI

Table 5a - *contd*

Trade name/trade mark	Abbreviation	Supplier
Profax	PP	Himont
Progilite	PF	Rhône-Poulenc Chimie
Propathene	PP-CO	ICI
Propathene	PP-H	ICI
Propathene OTE	TPE-RMPP	ICI
PTS Thermoflex	TPE-S (SEBS)	Plastics Technology Services
Pulse	PC/ABS	Dow
Quantum	PE-LD-HMW	Quantum Chemical Corp
Quatrex	EP	Dow
Quimcel	CN	Punda Inc.
Radel	PSU	Amoco Performance Products
Radiflam	PA-COM	Radicinovacips
Radilon	PA-COM	Radicinovacips
Radlite	GMT	Azdel Europe
Ralupol	UP	Raschig
Resarit	PMMA	Resart
Resartherm	PMC	Resart
Resilon	PVC	Canadian General-Tower
Resinol	PF	Raschig
Resinol V	CF (cresol-based)	Raschig
Rexene	PP	El Paso
Rhodester CL	LCP	Rhône-Poulenc Chimie
Riblene	PE-LD	Enichem
Rigidex	PE-HD	BP Chemicals
Rigipore	PS-X	BP Chemicals
Rilsan	PA 11	Elf Atochem S.A.
Rilsan A	PA 12	Elf Atochem S.A.
Riteflex	TPE-E	Hoechst/Hoechst Celanese
Ronfalin	ABS	DSM (Dutch State Mines)
Ronfaloy	ABS-AL	DSM (Dutch State Mines)
Ronfaloy-E	SAN/CPE/EPDM-AL	DSM (Dutch State Mines)
Rosite	BMC	Rostone
Rosite	SMC	Rostone
Rossi Lightflex	SBS	Rossi
Rovel	TP-styrene based	Dow
Royalene	EPDM	Uniroyal
Rutaform	MF	Sterling Moulding Materials
Rutaform Polyester	GPMC	Sterling Moulding Materials
Rutamid 6	PA 6	Bakelite
Rutamid 66	PA 66	Bakelite
Rynite	PET	Du Pont
Rynite PBTP	PBT	Du Pont
Ryton	PPS	Phillips Petroleum Chemicals
Santoprene	TPO-XL	Advanced Elastomer Systems
Saran	PVDC	Dow
Scarab	UF	BIP Chemicals
Scarnol	EVOH	Nippon Gobsei
Sclair	PE-LLD	Du Pont
Sclairlink	PE-crosslinkable	Du Pont
Selar OH	EVOH	Du Pont
Selar PA	PA-amorphous	Du Pont
Series 20	PE-LD	Du Pont
Series 20	PE-MD	Du Pont
Sinkral	ABS	Enichem
Sinvet	PC	Enichem
Sirester	UP resins	SIR (Societá Italiana Resine)
Sirfen	PF resins	SIR (Societá Italiana Resine)
Sirfen X	PF	SIR (Societá Italiana Resine)
Siritle	UF	SIR (Societá Italiana Resine)
SMA Resins	SMA	Sartomer
Smokeguard	TP-AL	Evode Plastics Ltd.
Smokeguard HF	TP-AL	Evode Plastics Ltd.
Smokeguard II	TP-AL	Evode Plastics Ltd.
Snialoy	TP-AL	Snia
Sniamid	PA 6	Snia

Table 5a. *Some trade names/trade marks, abbreviations and suppliers of polymers and polymer moulding compounds (sorted by alphabetical order of trade name) - contd*

Trade name/trade mark	Abbreviation	Supplier
Sniamid	PA 66	Snia
Sniasan	ABS	Snia
Sniasan	SAN	Snia
Sniatal	POM	Snia
Sniater	PBT	Snia
Sniater	PET	Snia
Soarnol	EVOH	Elf Atochem S.A.
Solef	PVDF	Solvay
Solvic	PVC	Solvay
Spheretex	TP + microspheres	Buch and Kolce
Spherilene	PP-LLD	Montecatini
ST	T-millable	Morton International
Stamylan HD	PE-HD	DSM (Dutch State Mines)
Stamylan LD	PE-LD	DSM (Dutch State Mines)
Stamylan P	PP	DSM (Dutch State Mines)
Stamylex	PE-LLD	DSM (Dutch State Mines)
Stamyroid	PP-amorphous	DSM (Dutch State Mines)
Stanyl	PA 46	DSM (Dutch State Mines)
Stapron C	ABS/PC-AL	DSM (Dutch State Mines)
Stapron M	ABS/PA-AL	DSM (Dutch State Mines)
Stapron S	SMA/SR-AL	DSM (Dutch State Mines)
Staramide	PA-GF	Ferro
Starflam	ETP-FR	Ferro
Starglas	ETP-GF	Ferro
Starpylen	PP-GF	Ferro
Star-C	TP-CF	Ferro
Star-L	TP-LUB	Ferro
Star-X	PA-HI	Ferro
Stat Kon	TP-CON	LNP Engineering Plastics Inc.
Statoil polyethylene HDPE	PE-HD	Statoil
Statoil polyethylene LDPE	PE-LD	Statoil
Statoil polypropylene PP	PP	Statoil
Stat-Rite	TP-antistatic	BF Goodrich
Sternite	PF	Sterling Moulding Materials
Sternite	PS	Sterling Moulding Materials
Sternite	PS-HI	Sterling Moulding Materials
Strippex	PE-XL	Neste
Styrocell	PS-X	Shell
Styrolux	BDS	BASF
Styron	PS	Dow
Styron	PS-HI	Dow
Styropor	PS-X	BASF
Sunlet	PP-COM	Mitsui Sekka
Sunpreme	TPE	Elf Atochem S.A.
Supec	PPS	General Electric Co.
Super Hexene	PE-LLD	Mobil
Superclean	PE-XL	Neste
Supersmooth	PO-XL	Neste
Supopoly	NR-SP	Asiatic Developments
Supraplast	BMC	Süd West Chemie
Supraplast	DAP	Süd West Chemie
Supraplast	DMC	Süd West Chemie
Supraplast	EP	Süd West Chemie
Supraplast	MF	Süd West Chemie
Supraplast	MPF	Süd West Chemie
Supraplast	PF	Süd West Chemie
Supraplast	SMC	Süd West Chemie
Supraplast	UF	Süd West Chemie
Suramin	MF resin	SIR (Societá Italiana Resine)
Suramin	UF resin	SIR (Societá Italiana Resine)
Surlyn	ION	Du Pont
Synolite	UP resins	DSM Resins
Tactix	EP	Dow
Taffen	GMT/PP	Exxon Chemical

Table 5a - *contd*

Trade name/trade mark	Abbreviation	Supplier
Tafmer	PO-CO	Mitsui Sekka
Tancin	PP	Washington Penn Plastics
Teamex	PE-VLD	DSM (Dutch State Mines)
Techmore	EP-HT	Mitsui Sekka
Technopolymer	GMT	General Electric Co.
Technoprene	PP-GF	Enichem
Technorub	NR-TSR	Hecht Heyworth & Alcan
Technyl	PA 66	Rhône-Poulenc Chimie
Techster E	PET	Rhône-Poulenc Chimie
Techster T	PBT	Rhône-Poulenc Chimie
Tecnoprene	PP-GF	Enichem
Tecolit	PF	Toshiba Chemical Products
Tedur	PPS	Bayer
Teflex	FEP	Nitechim
Teflon	FEP-COM	Du Pont
Teflon AF	FEP-amorphous	Du Pont
Teflon FEP	FEP	Du Pont
Teflon PFA	PFA	Du Pont
Teflon TFE	PTFE	Du Pont
Tefzel	ETFE	Du Pont
Telcar DVNR	NR/PP	Texnor Apex
TempRite	PVC-C	BF Goodrich
Tenite	CA	Eastman Chemicals
Tenite	CAB	Eastman Chemicals
Tenite	CAP	Eastman Chemicals
Tenite	CP	Eastman Chemicals
Tenite PET	PET	Eastman Chemicals
Tenite Polyethylene	PE-LD	Eastman Chemicals
Tenite Polypropylene	PP	Eastman Chemicals
Terblend S	ASA/PC	BASF
Terluran	ABS	BASF
Therban	HNBR	Bayer
Therban	NBR	Bayer
Thermaflo	PVC-P	Evode Plastics Ltd.
Thermocomp	TP-COM	LNP Engineering Plastics Inc.
Torlon	PAI	Amoco Performance Products
Toyobo MXDA	PAA 6	Toyobo
TPR	TPE	Advanced Elastomer Systems
TPX	PMP	Misubishi
TPX	TPX	Mitsui Sekka
Transpalene	PP-trans	Neste
Trefsin	TPE	Advanced Elastomer Systems
Triax	TP alloys	Monsanto
Triax 1000	PA/ABS alloys	Monsanto
Triax 2000	PC/ABS alloys	Monsanto
Trithene	PE-LD	Petroquimica Triunfo
Trithera	EVA	Petroquimica Triunfo
Trogamid T	PA-amorphous	Hüls
Trolon	PF-resin	Hüls
Trosiplast	PVC-COM	Hüls
Tyril	SAN	Dow
Tyrin	CPE	Dow
Udel	PSU	Amoco Performance Products
Ugikral	ABS	General Electric Co.
Ultem	PEI	General Electric Co.
Ultrablend	PBT/PC	BASF
Ultrablend S	PBT/ASA	BASF
Ultradur	PBT	BASF
Ultraform	POM-CO	BASF
Ultramid	PA 6	BASF
Ultramid A	PA 66	BASF
Ultramid C	PA-CO	BASF
Ultramid RC	PA-recycled	BASF
Ultramid S	PA 610	BASF
Ultranyl	PPO-M/PA	BASF

Table 5a. *Some trade names/trade marks, abbreviations and suppliers of polymers and polymer moulding compounds (sorted by alphabetical order of trade name) - contd*

Trade name/trade mark	Abbreviation	Supplier
Ultrapek	PEK	BASF
Ultraplas	MF	Dynamit Nobel
Ultrason E	PSU	BASF
Ultrason S	PSU	BASF
Ultrastyr	AES	Enichem
Ultzex	PE-LLD	Mitsui Sekka
Unidene	SBR-S	Enichem
Unipol	PE-LLD	Mobil
Urochem	UF	Chemiplastica Spa
Uroplast	UF	Sterling Moulding Materials
Valox	PBT	General Electric Co.
Valtec	PP	Himont
Vamac	SR-EMA	Du Pont
Vandar	PBT-HI	Hoechst/Hoechst Celanese
Vector	TPE-S	Dexco Corp
Vectra	LCP	Hoechst/Hoechst Celanese
Verton	LF-TP	ICI
Vespel	PI	Du Pont
Vestamid	PA 12	Hüls
Vestamid	PA 612	Hüls
Vestamid	PEBA	Hüls
Vestenamer	TPA	Hüls
Vestodur	PBT	Hüls
Vestolen	TPE-EPDM	Hüls
Vestolen A	PE-HD	Hüls
Vestolen P	PP	Hüls
Vestolit	PVC	Hüls
Vestolit	PVC-COM	Hüls
Vestopal	UP-resin	Hüls
Vestoplast	PO-amorphous	Hüls
Vestopren	TPE-O	Hüls
Vestoran	PPO-M	Hüls
Vestypor	PS-X	Hüls
Vestyron	PS	Hüls
Vestyron	PS	Svenska
Vibrin	UP resins	Fiberglass
Victrex PEEK	PEEK	ICI
Victrex PES	PSU	ICI
Vinex	PVAL-CO	Air Products and Chemicals Inc.
Vinidur	PVC-CO	BASF
Vinoflex	PVC	BASF
Vinuran	PVC modifiers	BASF
Vipla	PVC	European Vinyl Corp.
Vista	PVC	Vista Chemicals
Vistaflex	TPE-O	Advanced Elastomer Systems
Vistalon	EPDM	Exxon Chemical
Vistalon	EPM	Exxon Chemical
Vistanex	PIB	Exxon Chemical
Vitacom DVNR	NR/PP	Vitacom
Vitalon	PA 46 - see Stanyl	
Vitax	ASA	Hitachi Chemicals
Viton	FKM	Du Pont
Voltalef	PCTFE	Elf Atochem S.A.
Vydox	PTFE	Du Pont
Vydyne	PA	Monsanto
Vydyne R	PA 66 COM	Monsanto
Vynite	PVC/NBR	Alpha Chemical & Plastics
Vyram	TPE	Advanced Elastomer Systems
Vythene	PVC/PU	Alpha Chemical & Plastics
Wacker Polyathylen	PE-HD	Wacker-Chemie
Welite	PBT	Wellman Inc.
Wellamid	PA 6	Wellman Inc.
Wellamid	PA 66	Wellman Inc.
Welpet	PET	Wellman Inc.

Table 5a - *contd*

Trade name/trade mark	Abbreviation	Supplier
Xantar	PC	DSM (Dutch State Mines)
Xenoy	PC/PBT	General Electric Co.
Xydar	LCP	Amoco Performance Products
Zytel	PA 66	Du Pont
Zytel ST	PA 66 (HI)	Du Pont

Key

A	=	amorphous.
AL	=	alloy.
AR	=	aromatic.
COM	=	compound.
ETP	=	engineering thermoplastics material.
GF	=	glass fibre.
GMT	=	glass mat reinforced thermoplastics material.
HI	=	high impact.
HG	=	high gloss.
HT	=	high temperature.
ION	=	ionomer.
LMW	=	low molecular weight.
HMW	=	high molecular weight.
DCF	=	discontinuous fibre composite.
Encap	=	encapsulating.
FF-TP	=	fibre filled thermoplastics moulding compounds
LF-TP	=	long fibre thermoplastics moulding compounds.
LMR	=	liquid moulding resin.
Na	=	sodium (neutralised).
NR-SP	=	superior processing natural rubber.
NR-E	=	epoxidized natural rubber.
NR-MG	=	NR methacrylate graft rubber.
RC	=	recycled.
SF-TP	=	short fibre thermoplastics moulding compounds.
SF-MB	=	structural foam masterbatch.
SS	=	stainless steel (filler).
TP	=	thermoplastics material.
TP-AL	=	thermoplastic alloy.
TP-BIO	=	thermoplastic compounds which are designed to be biodegradable.
TP-COM	=	thermoplastic compounds.
TP-CON	=	thermoplastic compounds which are designed to be conductive.
TP-EMI	=	thermoplastics compounds which are EMI shielding.
TP-LUB	=	thermoplastics compounds which contain a lubricant, for example, PTFE, silicone oil, graphite etc.
TP-MAG	=	thermoplastics compounds which contain metal fillers and which are capable of being turned into magnets.
trans	=	transparent material.
TST	=	thermosetting material.
X	=	expanded or expandable.
XL	=	crosslinked or crosslinkable.
Zn	=	zinc (neutralised)

Company alternative names or abbreviations
AES - see Advanced Elastomer Systems.
Amoco Chemical - see Amoco Performance Products.
Atochem - see Elf Atochem S.A.
Dutch State Mines - see DSM.
EMS-Grilon - see EMS-Chemie.
GE Plastics - see General Electric Co.
KGSB = Kumpulan Guthrie Seridirian Berhad
   Plastiques Techniques - see Rhône-Poulenc Chimie.
RP - see Rhône-Poulenc Chimie.
SWC - see Süd West Chemie.

Table 5b. *Trade names/trade marks, abbreviations and suppliers of polymers and polymer compounds (sorted by alphabetical order of abbreviation)*
Key for Table 5b on p. 531

Abbreviation	Supplier	Trade name/trade mark
ABS	BASF	Terluran
ABS	Bayer	Novodur
ABS	Dow	Magnum
ABS	DSM (Dutch State Mines)	Ronfalin
ABS	Elf Atochem S.A.	Arrhadur
ABS	Enichem	Sinkral
ABS	General Electric Co.	Cycolac
ABS	General Electric Co.	Ugikral
ABS	Industrial Resistol	Epolan
ABS	Lati	Lastilac
ABS	Monsanto	Lustran
ABS	Rhône-Poulenc Chimie	Alcoryl
ABS	Snia	Sniasan
ABS modifier	Bayer	Baymod A
ABS-AL	DSM (Dutch State Mines)	Ronfaloy
ABS-AL	General Electric Co.	Cycoloy
ABS-AL	Hoechst Daicel Polymers	Novalloy
ABS-FR	Hoechst Daicel Polymers	Cevian
ABS-HG	Monsanto	Lustran Ultra ABS
ABS/PA-AL	DSM (Dutch State Mines)	Stapron M
ABS/PC	Bayer	Bayblend
ABS/PC	Enichem	Koblend
ABS/PC	Lati	Lastilac
ABS/PC-AL	DSM (Dutch State Mines)	Stapron C
ABS/PSU	USS Chemicals	Arylon T
ACM	Enichem	Cyanacryl
ACM	Enichem	Cyanacryl
ACM	Enichem	Europrene
Acrylic acid	BF Goodrich	Carbopol
Acrylic imide	Rohm and Haas	Kamax
Acrylic polymer	BF Goodrich	Carboset
AES	Enichem	Ultrastyr
ASA	BASF	Luran S
ASA	Hitachi Chemicals	Vitax
ASA	Misubishi	Dialac
ASA/AES	Advanced Elastomer Systems	Centrex
ASA/PC	BASF	Terblend S
BDS	BASF	Styrolux
BDS	Phillips Petroleum Chemicals	K-resin
BMC	Rostone	Rosite
BMC	Süd West Chemie	Supraplast
BR	Bayer	Buna CB
BR	Enichem	Europrene
BR	Enichem	Intene
BR	Enichem	Neocis
BR	Enichem	Neocis
CA	Albis	Cellidor
CA	Bergmann	Bergacell
CA	Courtaulds Speciality Plastics	Dexel
CA	Courtaulds Speciality Plastics	Dexel S
CA	Eastman Chemical	Tenite
CA-BIO	Tubize Plastics	Bioceta
CAB	Eastman Chemical	Tenite
CAP	Albis	Cellidor
CAP	Eastman Chemical	Tenite
CF (cresol-based)	Raschig	Resinol V
CN	Punda Inc.	Nike
CN	Punda Inc.	Quimcel
CP	Eastman Chemical	Tenite
CPE	Bayer	Bayer CM
CPE	Borsodchem	Ongro
CPE	Dow	Tyrin
CR	Bayer	Baypren

Table 5b - *contd*

Abbreviation	Supplier	Trade name/trade mark
CR	Distugil	Butachlor
CR	Du Pont	Neoprene
CSM	Du Pont	Hypalon
DAIP	Synres Amoco	DAIP 6000
DAP	Budd Co.	Polychem
DAP	Occidental	Durez
DAP	Plaskon Molding Div.	Plaskon DAP
DAP	Süd West Chemie	Supraplast
DAP	Synres Amoco	DAP 5000
DAP	US Prolam Inc.	Parr
DAP	US Prolam Inc.	Poly DAP
DCF	Technical Fibre Products	Disco
DMC	BASF	Palapreg
DMC	BIP Chemicals	Beetle DMC
DMC	ERF	ERF DMC
DMC	Orkem	Norsomix
DMC	Sd West Chemie	Supraplast
EAA	Dow	Primacor
EAA	Exxon Chemical	Escor EAA
EC	Dow	Ethocel
ECTFE	Ausimont	Halar
EMA	Du Pont	Nucrel
EMA	Exxon Chemical	Escorene Optema
EMC	Ciba Geigy	Araldite
EMC	Ciba Geigy	Neonite
EP	Bayer	Lekutherm
EP	Chemroy Canada	DER
EP	Ciba Geigy	Araldite
EP	Dow	Quatrex
EP	Dow	Tactix
EP	Emmerson and Cuming	Eccomold
EP	EMS-Chemie	Grilonit
EP	Plaskon Molding Div.	Plaskon
EP	Scott Bader	Crystic Kollernox
EP	Süd West Chemie	Supraplast
EP	Shell	Epon
EP	SPREA	Eponac
EP resin	Mitsui Sekka	Epomik
EP resins	Shell	Epikote
EP resins	SIR (Societá Italiana Resine)	Epoester
EP resins	SIR (Societá Italiana Resine)	Eposir
EPDM	Exxon Chemical	Vistalon
EPDM	Hüls	Buna AP
EPDM	Mitsui Sekka	Mitsui EPT
EPDM	Uniroyal	Royalene
EPM	Exxon Chemical	Vistalon
EPR	Du Pont	Nordel
EPR	Enichem	Dutral
EP-COM	Mitsui Sekka	Epox
EP-COM	Shell	Eponite
EP-HT	Mitsui Sekka	Techmore
ETFE	Du Pont	Tefzel
ETFE	Hoechst/Hoechst Celanese	Hostaflon FEP
ETP	Du Pont	Bexloy
ETP-FR	Ferro	Starflam
ETP-GF	Ferro	Starglas
EVA	Bayer	Levapren
EVA	Du Pont	Elvax
EVA	Elf Atochem S.A.	Evatane
EVA	Enichem	Greenflex
EVA	Exxon Chemical	Escorene Ultra
EVA	Petroquimica Triunfo	Trithera
EVA mofifier	Bayer	Baymod L
EVOH	Du Pont	Elvanol
EVOH	Du Pont	Selar OH

Table 5b. *Trade names/trade marks, abbreviations and suppliers of polymers and polymer compounds (sorted by alphabetical order of abbreviation) - contd*

Abbreviation	Supplier	Trade name/trade mark
EVOH	Elf Atochem S.A.	EVOH SF
EVOH	Elf Atochem S.A.	Soarnol
EVOH	Kuraray/EVAL Co	Eval
EVOH	Nippon Gobsei	Scarnol
EVOH	Solvay	Clarene
FEP	Daikin	Neoflon
FEP	Du Pont	Teflon FEP
FEP	Hoechst/Hoechst Celanese	Hostaflon FEP
FEP	Nitechim	Teflex
FEP-amorphous	Du Pont	Teflon AF
FEP-COM	Du Pont	Teflon
FF-TP	Baycomp	Baycomp
FKM	Du Pont	Viton
GMT	Azdel Europe	Radlite
GMT	Bayer	Polystal
GMT	General Electric Co.	Technopolymer
GMT laminates	BASF/Elastogran	Elastopreg
GMT/PBT	Azdel Inc.	Azmet
GMT/PC/PBT	Azdel Inc.	Azloy
GMT/PP	Azdel Inc.	Azdel
GMT/PP	Exxon Chemical	Taffen
GPMC	Freeman Chemicals	Freeflow
GPMC	Scott Bader	Crystic Impel
GPMC	Scott Bader	Impel
GPMC	Sterling Moulding Materials	Bakelite Polyester Alkyd
GPMC	Sterling Moulding Materials	Rutaform Polyester
HNBR	Bayer	Therban
IIR	Exxon Chemical	Exxon Butyl
IIR-X	Exxon Chemical	Exxon Bromobutyl
IIR-X	Exxon Chemical	Exxon Chlorobutyl
Impact modifiers	Rohm and Haas	Paraloid
Impact modifiers	Rohm and Haas	Paraloid EXL
ION	Allied Signal Inc.	Aclyn
ION	Du Pont	Surlyn
ION-Na	Exxon Chemical	Iotek
ION-Zn	Exxon Chemical	Iotek
LCP	Amoco Performance Products	Xydar
LCP	Hoechst/Hoechst Celanese	Vectra
LCP	Rhône-Poulenc Chimie	Rhodester CL
LF-TP	ICI	Verton
LMR	Shell/Hercules	Metton
MBS	Kaneka/E W Seward Ltd.	Kane Ace B
MC	Dow	Methocel
MF	BIP Chemicals	Melmex
MF	Ciba Geigy	Meloplas
MF	Cyanamid	Cymel
MF	Dynamit Nobel	Ultraplas
MF	Perstop Ferguson	Melamine moulding comp
MF	Perstorp	Isomin
MF	S A Aicar	Melaicar
MF	Süd West Chemie	Supraplast
MF	SPREA	Melsprea
MF	Sterling Moulding Materials	Rutaform
MF resins	SIR (Societá Italiana Resine)	Suramin
MF resins	BIP Chemicals	Beetle
MMA-resins	Degussa AG	Degadur
MMA-resins	Degussa AG	Degalan S
MMA-resins	Degussa AG	Degament
modifiers-plastics	Exxon Chemical	Exxelor
MPF	Ciba Geigy	Meloplas
MPF	Plastics Engineering Co.	Plenco
MPF	Süd West Chemie	Supraplast
NBR	Bayer	Perbunan
NBR	Bayer	Therban

Table 5b - *contd*

Abbreviation	Supplier	Trade name/trade mark
NBR	Enichem	Europrene
NBR/PVC	Enichem	Europrene
NNBR	Bayer	Perbunan
NR-E	KGSB	Epoxyprene
NR-MG	Asiatic Developments	Megapoly
NR-powdered	Golden Hope Plantations	Hyflo
NR-skim rubber	Dunlop Plantations	Dunlocrumb S
NR-skim rubber	KGSB	Dynat S
NR-SP	Asiatic Developments	Supopoly
NR-TSR	Dunlop Plantations	Dunlocrumb
NR-TSR	Golden Hope Plantations	Hyrub
NR-TSR	Hecht Heyworth & Alcan	Technorub
NR-TSR	KGSB	Dynat
NR-TSR	MARDEC	Marub
NR/PP	Texnor Apex	Telcar DVNR
NR/PP	Vitacom	Vitacom DVNR
PA	Du Pont	Elvamide
PA	Misuibishi	Novamid
PA	Mitsui Sekka	Arlen
PA	Monsanto	Vydyne
PA 11	Elf Atochem S.A.	Rilsan
PA 11	Thermoclad	Duralon
PA 12	Elf Atochem S.A.	Rilsan A
PA 12	EMS-Chemie	Grilamid
PA 12	Hüls	Vestamid
PA 12	Lati	Latamid 12
PA 46	DSM (Dutch State Mines)	Stanyl
PA 46 - see Stanyl		Vitalon
PA 6	Allied Signal Inc.	Capron
PA 6	Ashley Polymers Inc.	Ashlene
PA 6	Bakelite	Rutamid 6
PA 6	BASF	Ultramid
PA 6	Bayer	Durethan
PA 6	Bergmann	Bergamid B
PA 6	BIP Chemicals	Beetle nylon 6
PA 6	BIP Chemicals	Jonylon
PA 6	DSM (Dutch State Mines)	Akulon
PA 6	Elf Atochem S.A.	Orgamide
PA 6	EMS-Chemie	Grilon
PA 6	EMS-Chemie	Polyloy
PA 6	ICI	Maranyl
PA 6	Illing	Polyloy
PA 6	Lati	Kelon B
PA 6	Lati	Latamid 6
PA 6	Mobay	Nydur
PA 6	Polymer Trading	Naycar-B
PA 6	Resinmec	Arcomid
PA 6	Snia	Sniamid
PA 6	Terlon Polimeros	Nuloy
PA 6	Wellman Inc.	Wellamid
PA 6-COM	Enichem	Nivionplast A
PA 6-COM	Enichem	Nivionplast B
PA 6-GF	Bayer	Durethan
PA 6/PA 12	EMS-Chemie	Grilon C
PA 6/PP	Elf Atochem S.A.	Orgalloy R
PA 6-RC	Allied Signal Inc.	Nypel
PA 66	Ashley Polymers Inc.	Ashlene
PA 66	Bakelite	Rutamid 66
PA 66	BASF	Ultramid A
PA 66	Bayer	Durethan
PA 66	Bergmann	Bergamid A
PA 66	BIP Chemicals	Beetle nylon 66
PA 66	BIP Chemicals	Jonylon
PA 66	DSM (Dutch State Mines)	Akulon
PA 66	Du Pont	Zytel

Table 5b. *Trade names/trade marks, abbreviations and suppliers of polymers and polymer compounds (sorted by alphabetical order of abbreviation) - contd*

Abbreviation	Supplier	Trade name/trade mark
PA 66	EMS-Chemie	Grilon T
PA 66	Hoechst/Hoechst Celanese	Celanese Nylon
PA 66	ICI	Maranyl
PA 66	Illing	Polyloy
PA 66	Lati	Kelon A
PA 66	Lati	Latamid 66
PA 66	Mobay	Durethan
PA 66	Polymer Trading	Naycar-A
PA 66	Resinmec	Arcomid
PA 66	Rhône-Poulenc Chimie	Technyl
PA 66	Snia	Sniamid
PA 66	Wellman Inc.	Wellamid
PA 66 COM	Monsanto	Vydyne R
PA 66-COM	Du Pont	Minlon
PA 66-GF	Bayer	Durethan
PA 66-HI	Du Pont	Zytel ST
PA 66/PP	Elf Atochem S.A.	Orgalloy R
PA 610	BASF	Ultramid S
PA 612	Ashley Polymers Inc.	Ashlene
PA 612	Hüls	Vestamid
PAA 6	Solvay	Ixef
PAA 6	Toyobo	Toyobo MXDA
PAI	Amoco Performance Products	Torlon
PAN	Sohio	Barex
PAN	Standard Oil	Barex
PA-A	Ashley Polymers Inc.	Ashlene
PA-A	Du Pont	Selar PA
PA-A	Hüls	Trogamid T
PA-AR	EMS-Chemie	Grivory
PA-cast	Erta	Ertalan
PA-CO	BASF	Ultramid C
PA-CO	Elf Atochem S.A.	Cristamaid
PA-COM	Mitsui Sekka	Elmit
PA-COM	Radicinovacips	Radiflam
PA-COM	Radicinovacips	Radilon
PA-COM	Wilson Fibrefil	Nylafil
PA-GF	Ferro	Staramide
PA-HI	Ferro	Star-X
PA-recycled	BASF	Ultramid RC
PA-TR	EMS-Chemie	Grilamid TR
PA/ABS alloys	Monsanto	Triax 1000
PA/PP	DSM (Dutch State Mines)	Akuloy
PA/PPE	Allied Signal Inc.	Dimension
PB	Mitsui Sekka	Beaulon
PBT	BASF	Ultradur
PBT	Bayer	Pocan
PBT	Bergmann	Bergadur
PBT	DSM (Dutch State Mines)	Arnite
PBT	Du Pont	Crastine
PBT	Du Pont	Rynite PBTP
PBT	Elf Atochem S.A.	Orgater
PBT	Enichem	Pibiter
PBT	General Electric Co.	Valox
PBT	Hoechst/Hoechst Celanese	Celanex
PBT	Hüls	Vestodur
PBT	Lati	Later
PBT	Rhône-Poulenc Chimie	Techster T
PBT	Snia	Sniater
PBT	Wellman Inc.	Welite
PBT-GF	Bayer	Pocan
PBT-GF	Hoechst/Hoechst Celanese	Celanex
PBT-HI	Enichem	Pibiter
PBT-HI	Hoechst/Hoechst Celanese	Vandar
PBT/ASA	BASF	Ultrablend S

Table 5b - *contd*

Abbreviation	Supplier	Trade name/trade mark
PBT/PC	BASF	Ultrablend
PBT/SR/PC	Bayer	Pocan 7918
PC	Bayer	Makrolon
PC	BIP Chemicals	Beetle
PC	Dow	Calibre
PC	DSM (Dutch State Mines)	Xantar
PC	Enichem	Sinvet
PC	General Electric Co.	Lexan
PC	Idemitsu Petro Chem.	Idemitsu Polycarbonate
PC	Lati	Latilon
PC	Mobay	Merlin
PC-GF	Bayer	Makrolon
PC-HT	Bayer	Apec
PC/ABS	Dow	Pulse
PC/ABS alloys	Monsanto	Triax 2000
PC/PBT	Bayer	Makroblend
PC/PBT	General Electric Co.	Xenoy
PCTFE	Elf Atochem S.A.	Voltalef
PE	Du Pont	Du Pont 20 series
PE	Ecoplastics/Eco Chemicals	Ecolyte II & IV
PE	Hoechst/Hoechst Celanese	Hostapren
PE-X	Du Pont	Sclairlink
PE-CO	BASF	Lucalen
PE-CO/bitumen	BASF	Lucobit
PE-HD	Allied Signal Inc.	Paxon
PE-HD	Amoco Performance Products	Amoco PE-HD
PE-HD	Bamberger	Bapolene
PE-HD	BASF	Lupolen HD
PE-HD	BP Chemicals	Rigidex
PE-HD	Chemie Linz	Daplen
PE-HD	DSM (Dutch State Mines)	Stamylan HD
PE-HD	Elf Atochem S.A.	Lacqtene HD
PE-HD	Enichem	Eraclene
PE-HD	Enron/Delong Prochem	Norchem
PE-HD	Hoechst/Hoechst Celanese	Hostalen
PE-HD	Hüls	Vestolen A
PE-HD	Lati	Latene HD
PE-HD	Mitsui Sekka	Hi-zex
PE-HD	Neste	Neste HDPE
PE-HD	Novacor	Novapol HD
PE-HD	Petrofina	Finathene
PE-HD	Phillips Petroleum Chemicals	Marlex
PE-HD	Soltex Polymer Corp	Fortiflex
PE-HD	Solvay	Eltex
PE-HD	Statoil	Statoil polyethylene HDPE
PE-HD	USI/Quantum	Petrothene
PE-HD	Wacker-Chemie	Wacker Polyathylen
PE-HD UHMW	Mitsui Sekka	Hi-zex Million
PE-HD (HMW)	Hercules	Hercules HPR
PE-LD	Asia Polymer Corp.	Polymer E
PE-LD	BASF	Lupolen
PE-LD	BP Chemicals	Novex
PE-LD	Chemie Linz	Daplen
PE-LD	Dow	Dowlex
PE-LD	DSM (Dutch State Mines)	Stamylan LD
PE-LD	Du Pont	Series 20
PE-LD	Eastman Chemicals	Tenite Polyethylene
PE-LD	Enichem	Riblene
PE-LD	Exxon Chemical	Escorene
PE-LD	Neste	Neste LD
PE-LD	Novacor	Novapol LD
PE-LD	Orkem	Lotrene
PE-LD	Petroleos Mexicanos	Pemex
PE-LD	Petroquimica Triunfo	Trithene
PE-LD	Statoil	Statoil Polyethylene LDPE

Table 5b. *Trade names/trade marks, abbreviations and suppliers of polymers and polymer compounds (sorted by alphabetical order of abbreviation) - contd*

Abbreviation	Supplier	Trade name/trade mark
PE-LD-HMW	Quantum Chemical Corp	Quantum
PE-LLD	BP Chemicals	Innovex
PE-LLD	Dow	Dowlex
PE-LLD	DSM (Dutch State Mines)	Stamylex
PE-LLD	Du Pont	Sclair
PE-LLD	Elf Atochem S.A.	Lacqtene HX
PE-LLD	Elf Atochem S.A.	Lacqtene LX
PE-LLD	Enichem	Clearflex
PE-LLD	Enichem	Flexene
PE-LLD	Enron/Delong Prochem	Norchem
PE-LLD	Exxon Chemical	Escorene
PE-LLD	Exxon Chemical	Escorene α
PE-LLD	Mitsui Sekka	Ultzex
PE-LLD	Mobil	Super Hexene
PE-LLD	Mobil	Unipol
PE-LLD	Neste	Neste LLD
PE-LLD	Northern Petrochemicals	Norlin
PE-LLD	Orkem	Lotrex
PE-LLD	Sabic	Ladene
PE-LLD	USI/Quantum	Petrothene
PE-LLD/EVA (<5%)	Exxon Chemical	Escorene
PE-MD	Du Pont	Series 20
PE-MD	Hoechst/Hoechst Celanese	Hostalen
PE-MD	Mitsui Sekka	Neo-zex
PE-MD	Petrofina	Finathene
PE-MD	Soltex Polymer Corp	Fortiflex
PE-UHMW	Hoechst/Hoechst Celanese	Hostalen GUR
PE-VLD	Dow	Attane
PE-VLD	DSM (Dutch State Mines)	Teamex
PE-VLD	Mitsui Sekka	Mitsui Hi-wax
PE-VLD	Orkem	Norsoflex
PE-X	Arco	Arcel
PE-XL	Neste	Strippex
PE-XL	Neste	Superclean
PE/EVA-CO	DSM (Dutch State Mines)	Plexar
PEBA	Hüls	Vestamid
PEEK	ICI	Victrex PEEK
PEI	General Electric Co.	Ultem
PEK	BASF	Ultrapek
PEK	Hoechst/Hoechst Celanese	Hostatec
PET	Allied Signal Inc.	Petra
PET	Bayer	Petlon
PET	BIP Chemicals	Beetle PBT
PET	BIP Chemicals	Beetle PET
PET	DSM (Dutch State Mines)	Arnite
PET	Du Pont	Rynite
PET	Eastman Chemicals	Kodapek PET
PET	Eastman Chemicals	Tenite PET
PET	Hoechst/Hoechst Celanese	Impet
PET	ICI	Melinar
PET	Rhône-Poulenc Chimie	Techster E
PET	Snia	Sniater
PET	Wellman Inc.	Welpet
PET alloy	Allied Signal Inc.	Impact
PET-CO	Eastman Chemicals	Kodar PETG
PF	Bakelite/Sterling	Bakelite
PF	Chemiplastica Spa	Fenochem
PF	Chromos Ro-Polimeri	Fenoform
PF	Occidental	Durez
PF	Perstorp	Nestorite
PF	Plastics Engineering Co.	Plenco
PF	Raschig	Resinol
PF	Reichold	Dynaset
PF	Rhône-Poulenc Chimie	Progilite

Table 5b - *contd*

Abbreviation	Supplier	Trade name/trade mark
PF	Süd West Chemie	Supraplast
PF	SIR (Societá Italiana Resine)	Sirfen X
PF	SPREA	Moldsite
PF	Sterling Moulding Materials	Sternite
PF	Toshiba Chemical Products	Tecolit
PF composite	Orkot Engineering Plastics	Orkot
PF-GF	Raschig	Fibresinol
PF-resins	Hüls	Trolon
PF-resins	SIR (Societá Italiana Resine)	Sirfen
PF-resins-X	Hüls	Dynapor
PFA	Du Pont	Teflon PFA
PFA	Hoechst/Hoechst Celanese	Hostaflon FEP
PHB-CO	ICI	Biopol
PHB-H	ICI	Biopol
PI	Du Pont	Vespel
PI	Rhône-Poulenc Chimie	Kerimid
PI	Rhône-Poulenc Chimie	Kinel
PI-COM	Rogers Corp	Envex
PIB	BASF	Oppanol
PIB	Exxon Chemical	Parapol
PIB	Exxon Chemical	Vistanex
PIB	Shell	Duraflex
PMC	Ashland Chemical Corp.	Hetron
PMC	Ashland Chemical Co.	Aropol
PMC	Chromos Ro-Polimeri	Esterform
PMC	Haysite Reinforced Plastics	Haysite
PMC	Illing	Illandur
PMC	Isola Werke	Durapol
PMC	Menzolit Werke	Menzolit
PMC	Pheonix	Keripol
PMC	Resart	Resartherm
PMC	Scott Bader	Crystic Impreg
PMMA	Altulor	Altulite
PMMA	BASF	Lucryl
PMMA	Colors & Chemicals	Acrylite
PMMA	Elf Atochem S.A.	Altuglas
PMMA	ICI	Diakon
PMMA	Resart	Resarit
PMMA	Rohm and Haas	Oroglas V
PMMA-CO	Degussa AG	Degalan
PMMA-CO	Degussa AG	Degalan LP
PMMA-HI	Rohm and Haas	Oroglas DR
PMP	Misubishi	TPX
PO	Du Pont	Bynel
PO	Mitsui Sekka	Admer
PO-A	Hüls	Vestoplast
PO-A	Mitsui Sekka	Apel
PO-CO	Du Pont	Elvaloy
PO-CO	Mitsui Sekka	Tafmer
PO-LUB	Mitsui Sekka	Lubmer
PO-UHMW	Mitsui Sekka	Mipelon
PO-XL	Neste	Supersmooth
Polyarylate	Amoco Performance Products	Ardel
Polyketone	Amoco Performance Products	Kadel
POM	BIP Chemicals	Beetle
POM	Lati	Latan
POM	Snia	Sniatal
POM-CO	BASF	Ultraform
POM-CO	Hoechst/Hoechst Celanese	Hostaform
POM-CO	Hoechst/Hoechst Celanese	Kematal
POM-CO	Misubishi	Iupital
POM-H	Du Pont	Delrin
POM-H	Du Pont	Delrin II
POM-H	Du Pont	Delrin P
POM-H (HI)	Du Pont	Delrin 100 ST

Table 5b. *Trade names/trade marks, abbreviations and suppliers of polymers and polymer compounds (sorted by alphabetical order of abbreviation) - contd*

Abbreviation	Supplier	Trade name/trade mark
PP	Advanced Global Polymers	Adpro
PP	Appryl	Appryl
PP	Bamberger	Bapolene
PP	BASF	Novolen
PP	Bergmann	Bergaprop
PP	Chemie Linz	Daplen
PP	DSM (Dutch State Mines)	Stamylan P
PP	Eastman Chemicals	Tenite Polypropylene
PP	El Paso	Rexene
PP	Elf Atochem S.A.	Appryl
PP	Enron/Delong Prochem	Norchem
PP	Exxon Chemical	Escorene PP
PP	Himont	Moplen
PP	Himont	Profax
PP	Himont	Valtec
PP	Hoechst/Hoechst Celanese	Hostalen PP
PP	Hüls	Vestolen P
PP	Mitsui Sekka	Hipol
PP	Monmouth Plastics	Empee PP
PP	Neste	Nealid
PP	Norchem Inc	Nortuff
PP	Petrofina	Finapro
PP	Resinmec	Arcoplen
PP	Soltex Polymer Corp	Fortilene
PP	Solvay	Eref
PP	Statoil	Statoil polypropylene PP
PP	Washington Penn Plastics	Tancin
PP-A	DSM (Dutch State Mines)	Stamyroid
PP-CO	Himont	Hifax
PP-CO	ICI	Propathene
PP-CO	Solvay	Eltex P
PP-COM	Cabot Plastics	Plastech
PP-COM	Chemie Linz	Daplen
PP-COM	Ferro	Ferrolene
PP-COM	Himont	Astryn
PP-COM	Himont	Higlass
PP-COM	Hoechst/Hoechst Celanese	Hostacom PP reinforced
PP-COM	Hydro Polymers	Arpylene
PP-COM	ICI	Procom
PP-COM	Mitsui Sekka	Sunlet
PP-COM	Poly Pacific Pty	Corton
PP-COM	Schulman Inc	Polyfort
PP-CON	Cabot Plastics	Cabelec
PP-GF	Enichem	Technoprene
PP-GF	Enichem	Tecnoprene
PP-GF	Ferro	Starpylen
PP-GF	Mitsui Sekka	Mitsui FR-PP
PP-H	ICI	Propathene
PP-H	Solvay	Eltex P
PP-LF	Neste	Nepol
PP-LLD	Montecatini	Spherilene
PP-trans	Neste	Transpalene
PP-X	Arco	Arpro
PP/NBR	Advanced Elastomer Systems	Geolast
PPA	Amoco Performance Products	Amodel
PPO-M	BASF	Luranyl
PPO-M	General Electric Co.	Noryl
PPO-M	General Electric Co.	Prevex
PPO-M	Hüls	Vestoran
PPO-M	Shell	Caril
PPO-M/PA	BASF	Ultranyl
PPO-M/PA	General Electric Co.	Noryl GTX
PPS	Bayer	Tedur
PPS	Ciba Geigy	Craston

Table 5b - contd

Abbreviation	Supplier	Trade name/trade mark
PPS	General Electric Co.	Supec
PPS	Hoechst/Hoechst Celanese	Fortron
PPS	Lati	Larton
PPS	Phillips Petroleum Chemicals	Ryton
PPS	Solvay	Primef
Proc. aid-acrylic	Kaneka/E W Seward Ltd.	Kane Ace PA
Proc. aid-rubbers	Enichem	Polyplastol
PS	Bamberger	Bapolan
PS	BASF	Polystyrol
PS	BP Chemicals	BP Polystyrene
PS	Dow	Styron
PS	Ecoplastics/Eco Chemicals	Ecolyte S
PS	Elf Atochem S.A.	Lacqrene
PS	Enichem	Edistir
PS	Huntsman	Huntsman Polystyrene
PS	Hüls	Vestyron
PS	Lati	Lastirol
PS	Linpac	Linpac Polystyrene
PS	Norsk Hydro	Polystyrol
PS	Orkem	Gedex
PS	Sterling Moulding Materials	Sternite
PS	Svenska	Vestyron
PSU	Amoco Performance Products	Radel
PSU	Amoco Performance Products	Udel
PSU	BASF	Ultrason E
PSU	BASF	Ultrason S
PSU	ICI	Victrex PES
PSU	Lati	Lasulf
PSU-COM	Amoco Performance Products	Mindel
PS-COM	Elf Atochem S.A.	Lacqrene
PS-HI	BASF	Polystyrol
PS-HI	Dow	Styron
PS-HI	Elf Atochem S.A.	Lacqrene
PS-HI	Enichem	Edistir
PS-HI	Huntsman	Huntsman Polystyrene
PS-HI	Neste	Neste PS
PS-HI	Norsk Hydro	Polystyrol
PS-HI	Shell	Carinex
PS-HI	Sterling Moulding Materials	Sternite
PS-X	Arco	Dylile
PS-X	BASF	Styropor
PS-X	BP Chemicals	Rigipore
PS-X	Elf Atochem S.A.	Gedexcel
PS-X	Enichem	Extir
PS-X	Hüls	Vestypor
PS-X	Shell	Styrocell
PS/SBS	Petrofina	Finaclear
PS/PMMA	Novacor	NAS
PS/PMMA/BD	Novacor	NAS
PTFE	Ausimont	Halon
PTFE	Du Pont	Teflon TFE
PTFE	Du Pont	Vydox
PTFE	Hoechst/Hoechst Celanese	Hostaflon FEP
PTFE	ICI	Fluon
PTFE-CO	LNP Engineering Plastics Inc.	Fluromelt
PUR	Scott Bader	Crystic Kollerdur
PUR resins	Bayer	Baygal
PUR resins	Bayer	Baymidur
Purge compound	W S Wood Assoc.	Plasticlean
Purge compounds	Canada Color & Chemicals	Pax-Purge
PUR-casting	BASF	Elastopal
PUR-coating	BASF	Elastocoat
PUR-X	BASF	Cellasto
PVAL	Air Products and Chemicals Inc.	Airvol
PVAL-CO	Air Products and Chemicals Inc.	Vinex

Table 5b. *Trade names/trade marks, abbreviations and suppliers of polymers and polymer compounds (sorted by alphabetical order of abbreviation) - contd*

Abbreviation	Supplier	Trade name/trade mark
PVC	BASF	Vinoflex
PVC	BF Goodrich	Geon
PVC	Canadian General-Tower	Resilon
PVC	Elf Atochem S.A.	Lacovyl
PVC	European Vinyl Corp.	Corvic
PVC	European Vinyl Corp.	Vipla
PVC	Hoechst/Hoechst Celanese	Hostalit
PVC	Hüls	Vestolit
PVC	LVM	Marvylflo
PVC	Neste	Pekevic
PVC	Norsk Hydro	Norvinyl
PVC	Norsk Hydro	Pevikon
PVC	Novatec	Novablend
PVC	Polimeros De Mexico	Iztavil
PVC	Punda Inc.	Pekema
PVC	Solvay	Benvic
PVC	Solvay	Solvic
PVC	Vista Chemicals	Vista
PVC modifiers	BASF	Vinuran
PVC-C	BF Goodrich	TempRite
PVC-C	Elf Atochem S.A.	Lucalor
PVC-C	Kaneka/E W Seward Ltd.	Kane Ace
PVC-CO	BASF	Vinidur
PVC-COM	BF Goodrich	Geon CIM
PVC-COM	Elf Atochem S.A.	Nakan
PVC-COM	Evode Plastics Ltd.	GARY
PVC-COM	Hüls	Trosiplast
PVC-COM	Hüls	Vestolit
PVC-CON	BF Goodrich	Polytron
PVC-GF	BF Goodrich	Fiberloc
PVC-GF	Solvay	Benvic EV
PVC-HI	Hoechst/Hoechst Celanese	Hostalit Z
PVC-HT	BF Goodrich	Geon HTX
PVC-medical	BF Goodrich	Geon RX
PVC-P	Evode Plastics Ltd.	Thermaflo
PVC-U	Elf Atochem S.A.	Lucolene
PVC-U	Elf Atochem S.A.	Lucorex
PVC-U-COM	Dexter Plastics	Dural
PVC-X	Kaneka/E W Seward Ltd.	Kane Ace XEL
PVC/NBR	Alpha Chemical & Plastics	Vynite
PVC/PU	Alpha Chemical & Plastics	Vythene
PVC/PU/NBR AL	Dexter Plastics	Duralex
PVC/SR	Lati	Lastiflex
PVDC	Dow	Saran
PVDC	Scott Bader	Polyidene
PVDC	Solvay	Ixan
PVDF	Dynamit Nobel	Dyflor
PVDF	Elf Atochem S.A.	Foraflon
PVDF	Elf Atochem S.A.	Kynar
PVDF	Hüls	Dyflor
PVDF	Kay Fries	Dyflor
PVDF	Kureha	K F
PVDF	Penwalt	Kynar
PVDF	Solvay	Solef
RIM PA	DSM RIM Nylon	DSM Nyrim
SAN	BASF	Luran
SAN	Dow	Tyril
SAN	Enichem	Kostil
SAN	Lati	Lastil
SAN	Snia	Sniasan
SAN/CPE/EPDM-AL	DSM (Dutch State Mines)	Ronfaloy-E
SBR	Enichem	Europrene
SBR	Hüls	Duranit
SBR-E	Enichem	Intol

Table 5b - *contd*

Abbreviation	Supplier	Trade name/trade mark
SBR-E	Hüls	Buna EM
SBR-L	Hüls	Buna SL
SBR-L	Hüls	Buna V
SBR-S	Enichem	Unidene
SBS	Rossi	Rossi Lightflex
SF-MB	LNP Engineering Plastics Inc.	Foam Kon
SMA	Monsanto	Cadon
SMA	Sartomer	SMA Resins
SMA-impact modified	Monsanto	Cadon 300
SMA/SR-AL	DSM (Dutch State Mines)	Stapron S
SMC	BASF	Palapreg
SMC	ERF	ERF SMC
SMC	Freeman Chemicals	Flomat
SMC	Freeman Chemicals	Flowmat
SMC	Rostone	Rosite
SMC	Sd West Chemie	Supraplast
SR	Elf Atochem S.A.	Norsorex
SR	Shell	Cariflex
SR-EMA	Du Pont	Vamac
T-compounds	Morton International	LP-R
T-liquid	Morton International	ELP
T-liquid	Morton International	LP
T-millable	Morton International	FA
T-millable	Morton International	ST
TP + microspheres	Buch and Kolce	Spheretex
TP-AL	Evode Plastics Ltd.	Garaprene
TP-AL	Evode Plastics Ltd.	Smokeguard
TP-AL	Evode Plastics Ltd.	Smokeguard HF
TP-AL	Evode Plastics Ltd.	Smokeguard II
TP-AL	Monsanto	Triax
TP-AL	Snia	Snialoy
TP-antistatic	BF Goodrich	Stat-Rite
TP-antistatic	Lati	Latistat
TP-CF	Ferro	Star-C
TP-CO	Orkem	Lotader
TP-COM	DSM (Dutch State Mines)	Apscom
TP-COM	Lindsay & Williams	Megolon
TP-COM	LNP Engineering Plastics Inc.	Thermocomp
TP-CON	LNP Engineering Plastics Inc.	Stat Kon
TP-EMI	Lati	Latishield
TP-EMI	LNP Engineering Plastics Inc.	EMI-X
TP-LF	Hoechst/Hoechst Celanese	Celstran
TP-LUB	Ferro	Star-L
TP-LUB	Lati	Latilub
TP-LUB	LNP Engineering Plastics Inc.	Lubricomp
TP-MAG	LNP Engineering Plastics Inc.	Magnacomp
TP-photodegradable	Ecoplastics/Eco Chemicals	Ecolyte
TP-SS	DSM (Dutch State Mines)	Faradex
TP-starch based	Warner Lambert	Novon
TP-styrene based	Dow	Rovel
TPA	Hüls	Vestenamer
TPE	Advanced Elastomer Systems	TPR
TPE	Advanced Elastomer Systems	Trefsin
TPE	Advanced Elastomer Systems	Vyram
TPE	Elf Atochem S.A.	Sunprene
TPE	Evode Plastics Ltd.	Evoprene
TPE	Evode Plastics Ltd.	Evoprene E
TPE	Evode Plastics Ltd.	Evoprene G
TPE	Evode Plastics Ltd.	Evoprene Super S
TPE	Republic Plastics	ETA
TPE	Shell	Elexar
TPE EA-TPV	Du Pont	Alcryn
TPE-A	Dianippon	Glilax
TPE-A	Elf Atochem S.A.	Pebax
TPE-E	Cheil Synthetics Inc.	Esrel

Table 5b. *Trade names/trade marks, abbreviations and suppliers of polymers and polymer compounds (sorted by alphabetical order of abbreviation) - contd*

Abbreviation	Supplier	Trade name/trade mark
TPE-E	DSM (Dutch State Mines)	Arnitel
TPE-E	Du Pont	Hytrel
TPE-E	Eastman Chemicals	Ecdel
TPE-E	General Electric Co.	Lomod
TPE-E	Hoechst/Hoechst Celanese	Riteflex
TPE-EPDM	Hüls	Vestolen
TPE-EPDM	Lati	Larflex
TPE-O	Advanced Elastomer Systems	Vistaflex
TPE-O	Dexter Plastics	Dexflex
TPE-O	Hüls	Vestopren
TPE-O	Mitsui Sekka	Goodmer
TPE-O	Mitsui Sekka	Milastomer
TPE-OXL	Advanced Elastomer Systems	Dytron XL
TPE-PVC	LVM	Marvylex
TPE-RMPP	DSM (Dutch State Mines)	Kelburon
TPE-RMPP	DSM (Dutch State Mines)	Keltan TP
TPE-RMPP	Dynamit Nobel	Dynaform
TPE-RMPP	Ferro	Ferrolene-TPE
TPE-RMPP	ICI	Propathene OTE
TPE-S	Dexco Corp	Vector
TPE-S (SBS)	Enichem	Europrene
TPE-S (SBS)	Shell	Cariflex
TPE-S (SBS)	Shell	Kraton TR
TPE-S (SEBS)	Plastics Technology Services	PTS Thermoflex
TPE-S (SIS)	Enichem	Europrene
TPE-SBS	Petrofina	Finaprene
TPE-U	BASF/Elastogran	Elastollan
TPE-U	Bayer	Desmopan
TPE-U	BF Goodrich	Estane
TPE-U	Elastogran/BASF	Caprolan
TPE-U	ICI	Avalon
TPE-U (COM)	BF Goodrich	Estaloc
TPO-XL	Advanced Elastomer Systems	Santoprene
TPO-XL	Bayer	Levaflex
TPU	Lati	Lastane
TPU	Lati	Lastane
TPX	Mitsui Sekka	TPX
TPX-GF	Mitsui Sekka	FR-TPX
TST-acrylic resin	ICI	Modar
UF	BASF	Basopor
UF	BIP Chemicals	Beetle
UF	BIP Chemicals	Scarab
UF	Chemiplastica Spa	Urochem
UF	Dynamit Nobel	Polloplas
UF	S A Aicar	Carbaicar
UF	Sd West Chemie	Supraplast
UF	SIR (Societá Italiana Resine)	Siritle
UF	Sterling Moulding Materials	Uroplast
UF resins	BIP Chemicals	Beetle
UF resins	SIR (Societá Italiana Resine)	Celsir
UF resins	SIR (Societá Italiana Resine)	Suramin
UP	BASF	Palatal
UP	Bayer	Legupren
UP	Bayer	Leguval
UP	Ciba Geigy	Ampal
UP	Fiberglass	Dion
UP	Jotun Polymer	Norpol
UP	Plastics Engineering Co.	Plenco
UP	Raschig	Ralupol
UP	Scott Bader	Crystic
UP-resin	Jotun Polymer	Norpol
UP-resins	Hüls	Vestopal
UP resins	BIP Chemicals	Beetle
UP resins	DSM Resins	Synolite

Table 5b - *contd*

Abbreviation	Supplier	Trade name/trade mark
UP resins	Fiberglass	Vibrin
UP resins	Mia Chemical	Miapol
UP resins	SIR (Societá Italiana Resine)	Sirester
VC/VA	Elf Atochem S.A.	Lacovyl
VE	Dow	Derakane

Key

A	= amorphous.
AL	= alloy.
AR	= aromatic.
COM	= compound.
ETP	= engineering thermoplastics material.
GF	= glass fibre.
GMT	= glass mat reinforced thermoplastics material.
HI	= high impact.
HG	= high gloss.
HT	= high temperature.
ION	= ionomer.
LMW	= low molecular weight.
HMW	= high molecular weight.
DCF	= discontinuous fibre composite.
Encap	= encapsulating.
FF-TP	= fibre filled thermoplastics moulding compounds
LF-TP	= long fibre thermoplastics moulding compounds.
LMR	= liquid moulding resin.
Na	= sodium (neutralised).
NR-SP	= superior processing natural rubber.
NR-E	= epoxidized natural rubber.
NR-MG	= NR methacrylate graft rubber.
RC	= recycled.
SF-TP	= short fibre thermoplastics moulding compounds.
SF-MB	= structural foam masterbatch.
SS	= stainless steel (filler).
TP	= thermoplastics material.
TP-AL	= thermoplastic alloy.
TP-BIO	= thermoplastic compounds which are designed to be biodegradable.
TP-COM	= thermoplastic compounds.
TP-CON	= thermoplastic compounds which are designed to be conductive.
TP-EMI	= thermoplastics compounds which are EMI shielding.
TP-LUB	= thermoplastics compounds which contain a lubricant, for example, PTFE, silicone oil, graphite etc.
TP-MAG	= thermoplastics compounds which contain metal fillers and which are capable of being turned into magnets.
trans	= transparent material.
TST	= thermosetting material.
X	= expanded or expandable.
XL	= crosslinked or crosslinkable.
Zn	= zinc (neutralised)

Company alternative names or abbreviations
AES - see Advanced Elastomer Systems.
Amoco Chemical - see Amoco Performance Products.
Atochem - see Elf Atochem S.A.
Dutch State Mines - see DSM.
EMS-Grilon - see EMS-Chemie.
GE Plastics - see General Electric Co.
KGSB = Kumpulan Guthrie Seridirian Berhad
   Plastiques Techniques - see Rhône-Poulenc Chimie.
RP - see Rhône-Poulenc Chimie.
SWC - see Süd West Chemie.

Table 6. Drying conditions for injection moulding materials

Abbreviation	Water absorption %	Hot air drying Temp. °C	No. of hours	Temp. °C	Dessicant drying No. of hours
ASA	>0.1	80 to 85	2 to 4	90	2 to 3
ABS	0.2 to 0.35	70 to 80	2 to 4	70 to 80	2
BDS	0.08	60	1 to 1.5	60	0.5
CA	4.5 to 6.0	*55 to 85	3 to 4	85	1 to 2
CAB	2.2	*55 to 85	3 to 4	85	1 to 2
CAP	2.8	*55 to 85	3 to 4	85	1 to 2
FEP	0.01	150	2 to 4	150	2 to 3
HIPS	0.08	70	2 to 3	70	1 to 2
PA6	1.6	80	16	105	12
PA66	1.5	85	16	105	12
PA11	0.4	85	5 to 6	85	3
PA12	>0.4	85	5 to 6	85	3
PBT	0.08	120 to 150	3 to 5	120 to 150	2 to 3
PC	0.16	120	2 to 4	120	2
PEBA (hard grades)	0.5	80	4	80	3
PEBA (soft grades)	2.5	70	6	70	4
PEEL (GP grades)	1.5	120	10	120	2 to 4
PEEL (HP grades)	0.6	90	10	90	2 to 4
PEEK	0.5	150	3	150	2 to 3
PE-HD	<0.01	65	3	80	1 to 1.5
PE-LD	<0.2	65	3	80	1 to 1.5
PE-LLD	<0.2	65	3	85	1 to 1.5
PES	0.6	135 to 150	3 to 4	135	2 to 3
PET-A	0.03	**135	4	135	2
PET-C	0.03	135	4	135	2
PMMA	0.3	75	2 to 4	90	3 to 4
POM-H	0.4	110	2 to 3	110	1 to 2
POM-CO	0.22	110	2 to 3	110	1 to 2
PPO	0.1	100	2	100	2
PPS	<0.05	150	6	150	3
PP-H and PP-CO	<0.2	65	3	85	1 to 1.5
PS (GPPS)	0.08	70	2 to 3	70	1 to 2
PSU	0.3	135 to 150	3 to 4	135 to 150	2 to 4
PVDF	0.05	80	2 to 4	80	2 to 4
SAN	0.25	75 to 80	3 to 4	85 to 90	1.5
TPU/PUR	0.3	80	3	80	1
UPVC	<0.2	65	3	80	1 to 1.5

*The temperature at which the material is dried at is dependent upon the material's flow characteristics, i.e. soft flow materials 55°C to 68°C and hard flow materials 70°C to 85°C.
**Some grades cannot be dried at temperatures >60°C.
HP grades = high performance grades; A = amorphous; C = crystalline.

Table 7. *Heat contents of some injection moulding materials.*

Material Abbreviation	Temperature			Specific Heat $Jkg^{-1}K^{-1}$	Heat to be removed Jg^{-1}
	Melt °C	Mould °C	Difference °C		
FEP	350	200	150	1600	240
PES	360	150	210	1150	242
CA	210	50	160	1700	272
CAB	210	50	160	1700	272
CP	210	50	160	1700	272
PEEK	370	165	205	1340	275
PET	240	60	180	1570	283
PETP(C)	275	135	140	2180	305
PEEL	220	50	170	1800	306
POM	205	90	115	3000	345
SAN	240	60	180	1968	354
BDS	220	35	185	1968	364
PC	300	90	210	1750	368
ABS	240	60	180	2050	369
PMMA	260	60	200	1900	380
PPS	320	135	185	2080	385
PS	220	20	200	1970	394
ASA/AAS	260	60	200	2010	402
HIPS	240	20	220	1970	433
PPO	280	80	200	2120	434
PSU	360	100	260	1675	436
PETP(A)	265	20	245	1970	483
PA 11/12	260	60	200	2440	488
PA 6	250	80	170	3060	520
LDPE	210	30	180	3180	572
PA 66	280	80	200	3075	615
PP	260	20	240	2790	670
HDPE	240	20	220	3640	801

Where (A) is amorphous and (C) is crystalline.

Table 8. Shrinkage values

Abbreviation	Material	Mould shrinkage	Percentage
		in/in or mm/mm	
Thermoplastics			
ABS	Acrylonitrile–butadiene–styrene	0·004–0·007	0·4–0·7
POM	Acetal	0·020–0·035	2·0–3·5
PMMA	Acrylic	0·002–0·010	0·3–1·0
CA	Cellulose acetate	0·003–0·007	0·3–0·7
CAB	Cellulose acetate butyrate	0·002–0·005	0·2–0·5
CP	Cellulose propionate	0·002–0·005	0·2–0·5
EVA	Ethylene vinyl acetate	0·007–0·020	0·7–2·0
FEP	Fluorinated ethylene propylene	0·030–0·060	3·0–6·0
PA6	Nylon 6	0·010–0·015	1·0–1·5
PA66	Nylon 66	0·010–0·020	1·0–2·0
PBT	Polybutylene terephthalate	0·015–0·020	1·5–2·0
PBT GF 30%	Polybutylene terephthalate + 30% glass fibre	0·003–0·008	0·3–0·8
PC	Polycarbonate	0·006–0·008	0·6–0·8
PC GF 30%	Polycarbonate + 30% glass fibre	0·003–0·005	0·3–0·5
PES	Polyethersulphone	0·006–0·008	0·6–0·8
LDPE	Polyethylene (low density)	0·015–0·040	1·5–4·0
HDPE	Polyethylene (high density)	0·015–0·040	1·5–4·0
PPO	Polyphenylene oxide (modified)	0·005–0·007	0·5–0·7
PPO GF 30%	Polyphenylene oxide (modified) + 30% glass fibre	0·002	0·2
PP	Polypropylene	0·010–0·030	1·0–3·0
PS	Polystyrene(GP)	0·002–0·008	0·2–0·8
TPS	Polystyrene (toughened)	0·002–0·008	0·2–0·8
PTFE	Polytetrafluoroethylene	0·050–0·100	5·0–10·0
UPVC	Polyvinyl chloride (rigid)	0·002–0·004	0·2–0·4
PVC	Polyvinyl chloride (plasticized)	0·015–0·050	1·5–5·0
SF	Structural foam	0·006	0·6
PVF	Polyvinylidene fluoride	0·020–0·030	2·0–3·0
SAN	Styrene–acrylonitrile	0·002–0·006	0·2–0·6
Thermoplastic elastomers			
PP/EP(D)M	Rubber reinforced polypropylene	0·010–0·020	1·0–2·0
SBS	Styrene–butadiene–styrene	0·004–0·010	0·4–1·0
PEEL	Thermoplastic polyether ester	0·004–0·016	0·4–1·6
TPU	Thermoplastic polyurethane	0·005–0·020	0·5–2·0
Thermosets			
MF	Melamine formaldehyde	0·006–0·010	0·6–2·0
PF	Phenol formaldehyde	0·007–0·012	0·7–1·2
UF	Urea formaldehyde	0·006–0·010	0·6–1·0
DMC	Dough moulding compound	0·0005–0·002	0·05–0·2

Table 9. *Relative densities of some compounding ingredients and other materials*

Material	Relative density (RD) or specific gravity (SG)	Material	Relative density (RD) or specific gravity (SG)
Acetone	0.79	Chalk	
Activated calcium carbonate	2.6	crushed	1.44
Acrylonitrile butadiene styrene	1.01 to 1.07	solid	2.48
Aluminium	2.7	China clay (kaolin)	2.50
Aluminium (cast)	2.9	Chlorinated biphenyl	1.2 to 1.7
Aluminium oxide (fibre)	3.9	Chlorinated paraffin	
Aluminium silicate	2.60	44% chlorine content	1.16
Aluminium stearate	1.07	51% chlorine content	1.25
Aluminium trihydrate	2.42	Chlorinated polyethylene (CPE)	1.16
Ammonia	0.91	Chlorinated rubber	1.64
Ammonium bicarbonate	1.58	Chlorobutyl rubber	0.92
Ammonium carbonate	1.59	Chloroform	1.48
Aniline	1.02	Chlorobenzene	1.10
Antimony sulphide (without free sulphur)	3.6	Chrome oxide green	5.21
Antimony trioxide	5.4	Clays	
Artificial silk (viscose)	1.52	Kaolinite	2.60
Asbestos (chrysotile)	2.55	Calcined	2.63
Asbestos (hornblende)	2.7 to 3.6	Coal tar	1.18
Asbestos (serpentine)	2.3 to 2.8	Coal tar pitch	1.2
Asphalt	0.95 to 1.5	Copper	8.93
Balata	0.97	Cork (ground)	0.4 to 1.4
Barium sulphate (barytes)	4.4 to 4.6	Cornflour	1.5
Barium sulphate (blanc fixe)	4.25	Corundum	4.0
Barytes (ground barium sulphate)	4.45	Cotton	1.45
Barytes (barium sulphate)	4.4 to 4.6	Cotton flock	1.45
Basic zinc carbonate or zinc oxide (transparent)	3.5	Cottonseed oil	0.92
Beeswax	0.96	Coumarone-indene resin	1.11
Bentonite clay	2.50	Cresyl diphenyl phosphate (CDP)	1.21
Benzene	0.88	Cyclic oil	0.92
Benzoic acid	1.27	Cyclized rubber	0.99
Benzthiazyl-disulphide or		Cyclohexanone	0.94
dibenzothiazoledisulphide (MBTS)	1.5	Cyclohexyl-2-benzthiazyl sulphenamide (CBS)	1.28
Benzyl alcohol	1.04	Dekalin (decahydronaphthalene)	0.88
Benzyl butyl phthalate	1.10	Dialphanyl phthalate (DAP or di-C7-C9 phthalate)	1.00
Bitumen (oxidized)	1.04	Diatomaceous earth	2.15
Beryllium (fibre)	1.84	Dibenzothiazoledisulphide (MBTS)	1.5
Beryllium oxide (fibre)	1.8	Dibenzyl ether	1.04
Black — see Carbon black		Dibenzyl sebacate	1.05
Blanc fixe (barium sulphate)	4.25	Dibutoxyethoxy ethyl adipate	1.03
Boron (fibre)	2.59	Dibutoxyethyl adipate	1.00
Brass	8.4 to 8.7	Dibutoxyethyl sebacate	0.97
Brown iron oxide	5.15	Dibutyl amine	0.75
Butadiene (0°C)	0.65	Dibutyl adipate (DBA)	0.96
Butyl acetyl ricinoleate	0.93	Dibutyl phthalate (DBP)	1.05
Butyl alcohol	0.8	Dibutyl sebacate (DBS)	0.94
Butyl benzyl phthalate (BBP)	1.12	Diethyl phthalate (DEP)	1.12
Butyl benzyl sebacate (BBC)	1.02	Dicapryl phthalate (DCP)	0.97
Butyl cyclohexyl phthalate (BCHP)	1.08	Di-C7-C9 phthalate (DAP)	1.00
Butyl rubber	0.92	Diethylene glycol dibenzoate (DEGB)	1.18
Butyraldehyde-aniline product (BA)	0.96	Di-2-ethylhexyl adipate (DOA)	0.93
Cadmium oxide	8.20	Di-2-ethylhexyl phthalate (DEHP or DOP)	0.99
Cadmium red	4.4 to 5.4	Di-2-ethylhexyl sebacate (DOS)	0.92
Cadmium yellow	4.1 to 4.6	Di-iso-butyl adipate (DIBA)	0.95
Cadmium sulphide	4.40	Di-iso-butyl azeleate (DIBZ)	0.94
Calcium carbonate (activated and precipitated)	2.6	Di-iso-butyl phthalate (DIBP)	1.04
Calcium carbonate (ground whiting)	2.70	Di-iso-decyl adipate (DIDA)	0.92
Calcium hydroxide	2.28	Di-iso-decyl phthalate (DIDP)	0.97
Calcium silicate		Di-iso-octyl adipate (DIOA)	0.93
Wollastonite	2.90	Di-iso-octyl azelate (DIOZ)	0.92
Precipitated	2.23	Di-iso-octyl phthalate (DIOP)	0.98
Calcium stearate	1.04	Di-iso-octyl sebacate (DIOS)	0.92
Camphor	1.0	Di-2-methoxy phthalate	1.17
Carbon black	1.81	Dinonyl phthalate (DNP)	0.97
Carbon disulphide	1.26	Di-(o-benzamidophenyl) disulphide	1.35
Carbon tetrachloride	1.63	Dioctyl adipate	0.93
Carnauba wax	0.99	Dioctyl phthalate (DOP or	
Casein (lactic)	1.25 to 1.30	di-2-ethylhexyl phthalate DEHP)	0.99
Cast aluminium	2.9	Di-ortho tolyl guanidine (DOTG) or	
Castor oil	0.96	di-o-tolylguanidine (DOTG)	1.19
Cellulose acetate butyrate (CAB)	1.15 to 1.21	Dipentamethylene thiuram tetrasulphide (DPTT)	1.50
Cellulose acetate CA)	1.26 to 1.30	Di-t-butyl peroxide	0.79
Ceresin wax	0.93	Diphenyl guanidine (DPG)	1.19

Table 9. *Relative densities of some compounding ingredients and other materials - contd*

Material	Relative density (RD) or specific gravity (SG)	Material	Relative density (RD) or specific gravity (SG)
Ditridecyl phthalate (DTDP)	0.95	Kaolin (kaolin clay or china clay)	2.50
Dioxane	1.04	Kerosene	0.82
Diphenylguanidene (DPG)	1.19	Kieselguhr	2.20
Ditridecyl phthalate	0.95	Lanolin (wool grease)	0.97
Dolomite	2.34	Lauric acid	0.90
Duraluminium	2.8	Lead	11.37
Earth wax (ozocerite or ozokerite or mineral wax)	0.9	Lead chromate	5.70
Ebonite dust	1.15 to 1.2	Lead monoxide (litharge)	9.3 to 9.5
Emery	3.7 to 4.0	Lead powder	1.34
Epoxidized soya bean oil (epoxidized soybean oil)	0.99	Lead sulphate	6.20
Ether	0.72	Lead	11.34
Ethyl acetate	0.9	Leather	0.9 to 1.0
Ethyl alcohol	0.78	Light magnesium carbonate	2.19
Ethylene bis-stearamide (EBS)	0.97	Lignin	1.30
Ethylene chloride	1.26	Lime (hydrated or slaked)	2.10
Ethylene diamine	0.90	Linseed oil	0.94
Ethylene glycol	1.12	Litharge (lead monoxide)	9.3 to 9.5
Ethylene propylene rubbers	0.87	Lithopone (30% zinc sulphide)	4.15
Ethylene thiourea	1.28	Lithopone (40% zinc sulphide)	4.06
Ethylene vinyl acetate	0.95	Lycopodium	1.6
Ethyl iodide	1.93	Magnesia (heavy calcined)	3.20
Factice — brown	1.0 to 1.1	Magnesia (light calcined)	3.20
Factice — white	1.06 to 1.15	Magnesium	1.74
Ferric oxide	5.14	Magnesium carbonate (light)	2.19
Fluoroelastomers	1.72 to 1.86	Magnesium oxide	3.60
Formaldehyde (30%)	1.09	Magnesium silicate	2.72
Fossil flour	2.15	Mercaptobenzthiazole (MBT) or 2-mercaptobenzthiazole	1.42
French chalk	2.72	Mercury	13.55
Fuller's earth	2.15	Methacrylate butadiene styrene (MBS)	0.99
Gelatin	1.27	Methanol (methyl alcohol or wood alcohol)	0.79
Gilsonite	1.10	Methyl acetate	0.9
Glass (E type fibre)	2.55	Methyl alcohol (methanol or wood alcohol)	0.79
Glass (S type fibre)	2.49	Methylene chloride	1.34
Glue (bone)	1.27	Methylene iodide	3.32
Glycerine — see Glycerol		Methyl ethyl ketone (MEK)	0.83
Glycerol	1.27	Mica (muscovite)	2.75
Glycerol diacetate	1.19	Mineral oil (aromatic)	1.02
Glycerol monoacetate	1.19	Mineral oil (naphthenic)	0.93
Glycerol monolaurate	0.97	Mineral oil (paraffinic)	0.86
Glycerol mono-oleate	0.95	Mineral rubber	1.04
Glycerol monoricinoleate	0.98	Mineral spirit (white spirit)	0.8
Glycerol monostearate (GMS)	0.97	Mineral wax	0.9
Glycerol triacetate	1.16	Montan wax	0.81
Glycerol tributyrate	1.04	Myristic acid	0.86
Gold	19.25	Naphthalene	1.16
Granite	2.56	Natural rubber	0.93
Graphite	2.2 to 2.6	N-cyclohexylbenzthiazole-2-sulphenamide (CBS)	1.30
Graphite flake	2.25	Nepheline syenite	2.60
Ground glass	2.4 to 2.6	n-Heptane	0.75
Ground slate	2.8	Nibran wax	1.6 to 1.7
Ground whiting (calcium carbonate)	2.70	Nitrile rubber (high acrylonitrile)	1.00
Guayule	0.96	Nitrile rubber (low acrylonitrile)	0.98
Gutta percha	0.98	Nitrobenzene	1.21
Hard rubber dust	1.17 to 1.20	N-nitroso-diphenylamine	1.27
Hexa or hexamethylenetetramine	1.02	n-octyl n-decyl adipate (NODP)	0.92
Hexamethylenetetramine or hexa	1.02	n-octyl n-decyl phthalate (NODP)	0.98
Hexalin	0.9	Nylon 66	1.14
Hornblende (asbestos)	2.7 to 3.6	o-Dichlorobenzene	1.30
Hydantoin-glycol fatty ester (a distearate ester)	1.03	Oil — castor	0.96
Hydrated alumina	2.42	Oil — cottonseed	0.92
Hydrated chrome oxide (green)	3.4	Oil — mineral (aromatic)	1.02
Hydrated lime	2.1	Oil — mineral (naphthenic)	0.93
Hydrated magnesium silicate or talc or french chalk	2.72	Oil — mineral (paraffinic)	0.86
Hydrogenated tallow glyceride (HTG)	0.96	Oil — palm	0.88
Infusorial earth	2.15	Oil — pine	0.93
Iron	7.5 to 7.9	Oil — rape	0.92
Iron oxide (red)	5.14	Oil — tall	0.95 to 1.0
Iron oxide (yellow)	4.1	Oleic acid or red oil	0.89
Isooctyl isodecyl phthalate (IODA)	0.99	Ozokerite (ozocerite or earth wax or mineral wax)	0.9
Isooctyl palmitate	0.86	Palm oil	0.88
Isoprene	0.68	Paraffin oil	0.86
Isopropyl alcohol	0.8	Paraffin wax	0.88 to 0.91

Table 9. - *contd*

Material	Relative density (RD) or specific gravity (SG)	Material	Relative density (RD) or specific gravity (SG)
Paraffin wax — refined	0.90	Talc	2.72
Pentachloroethane	1.67	Tall oil	0.95 to 1.0
Perchlorethylene	1.62	Tallow	0.95
Petrolatum	0.84	Turpentine oil	0.87
Petroleum ether	0.6	Tetra ethyl thiuram disulphide (TETD)	1.26
Petroleum jelly	0.84 to 0.89	Tetra methyl thiuram disulphide (TMTD)	1.29
Petroleum spirit	0.68 to 0.75	Tetra methyl thiuram monosulphide (TETD)	1.38
Phenolic resin	1.27	Tetrahydronaphthalene (tetralin)	0.98
Phenyl-α-naphthylamine	1.17 to 1.22	Tetralin (tetrahydronaphthalene)	0.98
Phenyl-β-naphthylamine	1.18 to 1.24	Tetramethylthiuram disulphide (TMTD)	1.42
Phthalic anhydride	1.52	Tetramethylthiuram monosulphide (TMTM)	1.38
Pine oil	0.93	Thiocarbanilide	1.30
Pine pitch	1.11	Tin	7.28
Pine tar (soft wood tar)	1.03 to 1.09	Titanium dioxide (anatase)	3.90
Plaster of Paris	2.32	Titanium dioxide (rutile)	4.20
Platinum	21.4	Toluene	0.87
Polyacrylic rubber	1.05 to 1.15	Tri-butoxyethyl phosphate (TBEP)	1.02
Polybutadiene rubber	0.94	Tributyl phosphate (TBP)	0.98
Polyester plasticizers	1.0 to 1.1	Trichloroethylene	1.47
Polyethylene — high density	0.94 to 0.97	Tricresyl phosphate (TCP)	1.16
Polyethylene — low density	0.92 to 0.94	Tri-2-ethylhexyl phosphate (TOF)	0.93
Polyisoprene rubber	0.92	Triphenylguanidine	1.10
Polymethylpentene	0.83	Triphenyl phosphate (TPP)	1.19
Polypropylene	0.90	Tritolyl phosphate (TTP)	1.17
Polypropylene sebacate	1.06	Trixylyl phosphate (TXP)	1.14
Polystyrene	1.05	Ultramarine blue	2.35
Polytetrafluoroethylene (PTFE)	2.1 to 2.3	Urea	1.34
Precipitated whiting (calcium carbonate)	2.62	Vaseline	0.86 to 0.90
Pumice — powdered	2.35	Vermilion	8.20
Rape oil	0.92	Vinyl chloride graft (VGC) copolymer PVC onto EVA	1.13
Red oil or oleic acid	0.89	Viscose (artificial silk)	1.52
Red lead	8.7	Water	
Rosin	1.08	fresh	1.00
Rosin oil	0.99	salt	1.03
Salicylic acid	1.44	White lead	6.27
Selenium	4.8	White spirit (mineral spirit)	0.8
Shellac	1.15	Whiting (ground calcium carbonate)	2.70
Silica	1.95	Whiting (precipitated calcium carbonate)	2.62
Silica — colloidal	2.1	Whole tyre reclaim	1.20
Siliceous earth	2.2 to 2.6	Wollastonite	2.9
Silicon carbide	3.17	Wood alcohol (methyl alcohol or methanol)	0.79
Silicon nitride	3.2	Wood flour	1.25
Silver	10.5	Wool	1.32
Slaked lime	2.1	Wool grease (lanolin)	0.97
Slate powder	2.8	Xylene	0.86
Soapstone	2.72	Zinc	7.14
Sodium acetate	1.45	Zinc dibutyl dithiocarbamate	1.21
Sodium carbonate	2.2	Zinc diethyl dithiocarbamate	1.50
Sodium benzoate	2.8	Zinc dimethyl dithiocarbamate	1.75
Sodium bicarbonate	2.20	Zinc carbonate (precipitated)	3.30
Soft wood tar (pine tar)	1.04	Zinc ethyl phenyl dithiocarbamate	1.50
Starch	1.5	Zinc isopropylxanthate (ZIX)	1.54
Stearic acid	0.92	Zinc laurate	1.10
Stearine (pale flake)	0.85	Zinc mercaptobenzothiazole (ZMBT)	1.64
Steel	7.9	Zinc oxide	5.57
Styrene butadiene rubber	0.93	Zinc stearate	1.06
Sulphur	2.04	Zinc sulphide	3.9 to 4.2
Sulphur monochloride	1.68 to 1.71		

Table 10. *Plastics identification chart*

Abbn	S.G.	Softening Temp.°C	Ease of ignition	Self ext. property	Colour of flame	Odour on burning	Behaviour on burning
Opaque Materials							
ABS	1·02–1·06	104	Quite easily	No	Yellow and smoky.	Sweet and rubbery smell.	Softens, blackens, and bubbles
DAP/DAIP	2·0–2·2	110	Quite easily	No	Yellow and smoky.	Pungent smell.	Discolours chars and cracks.
EPOXY	1·11–1·40	149 to 260	Easily	No	Yellow and smoky.	Like burnt flour.	Blackens and softens.
HIPS	1·05	73 to 78	Easily	No	Yellow with black	Distinctly styrenic (flowery).	Melts and bubbles.
MF	1·48	120	Difficult	Yes	Yellow.	Formaldehyde.	Discolours chars and cracks.
PA6	1·12–1·13	220	Easily	Yes	Blue with yellow edge.	Burnt hair.	Melts and froths.
PA66	1·13	256	Easily	Yes	Blue with yellow edge.	Like celery.	Melts and froths.
PBT	1·31	225	Easily	Yes	Mostly white and smoky		Melts forming burning droplets.
PE–LD	0·90–0·92	104	Easily	No	Blue with a yellow top.	Candle wax	Melts with burning droplets.
PE–HD	0·94–0·96	120	Easily	No	Blue with a yellow top.	Candle wax.	Melts as burning droplets.
PF	1·15–2·08	120	Quite easily	Yes	Mostly yellow with black smoke.	Phenol.	Chars and cracks.
PETPc	1·4	260	Easily	Yes	Yellow.	Little odour.	Blackens and swells.
PP	0·90	79 to 116	Easily	No	Blue with yellow top.	Candle wax.	Melts forming burning droplets.
POM	1·41	104	Not so easily	No	Pale blue	Pungent smell.	Melts forming droplets.
PPO	1·06	80 to 102	Easily	No	Yellow and smoky.	Sweet, floral, faintly phenolic.	Sooty deposits.
PPS	1·4	282	Difficult	Yes	No flame	Sulphur.	Chars and blisters.
PPVC	1·3–1·4	66 to 88	Not so easily	Yes	Yellow and smoky	Biting acidic.	Melts and forms burning droplets.
PTFE	2·13	290	Does not ignite	Yes	No flame	Odourless.	Becomes clear and putty like.
UF	1·5	77	Difficult	Yes	Yellow with	Fishy.	Swells, cracks, discolours.
UPVC	1·4–1·5	66 to 92	Not so easily	Yes	Yellow	Biting acidic.	Softens and blackens.
Transparent Materials							
BDS	1·01–1·02		Easily	No	Yellow and sooty.	Styrenic smell.	Blackens and bubbles.
CA	1·23–1·34	60 to 100	Easily	No	Yellow with black smoke	Sharp vinegary acidic smell.	Melts forming burning droplets.
CAB	1·15–1·22	60 to 100	Easily	No	Dark yellow, blue edges.	Rancid cheese or butter.	Melts forming burning droplets.
EVA	0·93–0·97	88 to 93	Easily	No	Blue with a hint of yellow.	Faintly vinegary.	Melts forming burning droplets.
GPPS	1·04	78 to	Easily	No	Yellow and black smoke.	Flowery with distinct smell of styrene.	Melts and bubbles.
PEI	1·27	200	Easily	Yes	Yellow.	Faintly phenolic.	Blackens and bubbles.
PC	1·20	135	Not so easily	Yes	Yellow and smoky.	Faintly phenolic.	Softens, bubbles and carbonises.
PES/ PSU	1·37 1·24		Not easily	No	White.	Sulphur	Chars.
PETPa	1·38	230	Easily	Yes	Bright yellow.	Bitter-sweet smell	Blackens and forms burning droplets.
PMMA	1·19	60 to 88	Easily	No	Blue with yellow top.	Fruity and smoky.	Melts and bubbles.
PUR/TPU	1·11–1·22	80 to 100	Easily	No	Bright		Melts candle-like forming flame burning droplets.
SAN	1·08	60 to 96	Easily	No	Yellow and smoky.	Sweet, floral.	Blackens and bubbles.

a = amorphous; c = crystalline.

Table 11. *Suggested (based on BS and ISO) conditions for MFR tests*

Abbreviation	Common name	Test temp. °C	Ref time s	Nominal force N
ABS	Acrylonitrile butadiene styrene	220	600	98.0
PE	Polyethylene	190	600	21.2
PE	Polyethylene	190	150	49.0
PE-HMW	Polyethylene–high molecular weight	190	600	212.0
PP	Polypropylene–powder	190	600	98.0
PP	Polypropylene	230	600	21.2
PS	Polystyrene (GPPS)	200	600	49.0
SAN	Styrene acrylonitrile copolymer	220	600	98.0
SAN	Styrene acrylonitrile copolymer	230	600	37.3

Table 12. *Moisture content limit for good injection mouldings*

Material	Moisture content limit %
Acrylic	0.05
Acrylonitrile–butadiene–styrene*	0.02
Cellulosics	0.40
Nylon 6 and 66	0.25
Nylon 11 and Nylon 12	0.01–0.10
Polycarbonate	0.02
Polystyrene	0.10
Polyvinyl chloride	0.08
Styrene-acrylonitrile	0.10
Thermoplastic polyester*	0.01

*The actual moisture level may depend on the application: for some ABS mouldings the moisture may be above that given in the table but for plating grades very low levels are recommended

Table 13. *Suggested temperature settings for high shear rate rheometry*

Abbreviation	Temperature setting °C/°F	Range °C/°F
ABS	240/464	230–270/446–518
ASA	260/500	250–280/482–536
BDS	220/428	190–230/374–446
EVA	180/356	140–225/284–437
FEP	350/662	300–380/572–716
HDPE	240/464	205–280/401–536
HIPS	240/464	200–270/392–518
LDPE	210/410	180–280/350–536
LLDPE	210/410	160–280/350–536
PA	See PA 6, PA 11, PA 12 and PA 66	
PA 6	250/482	230–280/446–536
PA 11	255/491	240–300/464–572
PA 12	255/491	240–300/464–572
PA 66	280/536	260–290/500–554
PBT	250/482	220–260/428/500
PC	300/572	280–320/536–608
PE	See HDPE, LDPE and LLDPE	
PEEK	370/698	360–380/680–716
PEI	380/716	340–425/640–800
PET	275/527	260–300/518–572
PES	360/680	330–380/626–716
PMMA	240/464	210–270/410–518
POM–CO	205/401	190–210/374–410
POM–H	215/419	190–230/374–446
PP	240/464	220–275/428–527
PPE	See PPO	
PPO	280/536	260–300/500–572
PPS	320/608	290–360/554–680
PPVC	180/356	175–200/347–392
PS	220/428	200–250/392–482
PSU	360/680	330–380/626–716
PVC	See PPVC and UPVC	
PVDF	225/437	220–250/428–482
SAN	240/464	200–270/392–518
UPVC	195/383	185–205/364–401
RMPP	240/464	220–275/428–527
TPE	See TPE–A, TPE–E, TPE–S, TPE–U and TPE–OXL	
TPE–A	200/392	185–240/364–464
TPE–E	220/428	195–255/383–491
TPE–S	170/338	150–200/302–394
TPE–U	200/392	180–230/356–446
TPE–OXL	190/374	180–200/356–392

Table 14. Moldflow data for PA 6

Grade	Viscosity (Nsm^{-2}) at 1,000s^{-1} at the following temperatures			
	°C/°F	°C/°F	°C/°F	°C/°F
Akzo 'Akulon'	*240/464*	*260/500*	*280/536*	*300/572*
Easy flow grade	98	78	63	50
Medium flow grade	149	115	89	69
Stiff flow grade	394	319	259	210

Grade	Viscosity (Nsm^{-2}) at 280°C/536°F.			
	100s^{-1}	1,000s^{-1}	10,000s^{-1}	100,000s^{-1}
Easy flow grade	197	63	20	6
Medium flow grade	323	89	25	7
Stiff flow grade	1220	259	55	12

Table 15. Carbon black classification

Iodine absorption g/kg	Particle diameter nm	Old Code	Type of black	ASTM No.
145	11 to 19	SAF	Super abrasion furnace	N110
121	20 to 25I	SAF	Intermediate super abrasion furnace	N220
121	—	ISAF–HS	Intermediate super furnace - high structure	N2242
82	—	HAF–LS	High abrasion furnace - low structure	N326
82	26 to 30	HAF	High abrasion furnace	N330
90	—	HAF–HS	High abrasion furnace - high structure	N346
—	26 to 30	EPC	Easy processing channel	S300
—	26 to 30	MPC	Medium processing channel	S301
90	—	HAF–HS(NT)	High abrasion furnace - high structure (new technology)	N339
—	31 to 39	FF	Fine furnace	N440
43	40 to 48	FEF	Fast extrusion furnace	N550
36	49 to 60	GPF	General purpose furnace	N660
—	49 to 60	HMF	High modulus furnace	N601
—	61 to 100	SRF	Semi–reinforcing furnace	N770
29	—	SRF–NS	Semi–reinforcing furnace - non–staining	N774
—	101 to 200	FT	Fine thermal	N880
—	201 to 500	MT	Medium thermal	N990

Appendix A: SI units - advice on use

SI is an abbreviation used for *Système International d'Unité*.

Singular and plural forms of SI unit abbreviations or symbols, are the same. That is, do not put the letter s after the unit abbreviation, or symbol, if specifying more than one of a particular unit. SI symbols are always in roman type.

A period (full stop) is not used with the unit abbreviations or symbols, except at the end of the sentence.

A space is left between the number and the unit abbreviation or symbol, except when the temperature is specified in degrees Celsius (centigrade).

When the temperature is specified in degrees Celsius (centigrade) then Celsius begins with a capital C (upper case) and the unit abbreviation or symbol is also written with a capital C. The capital C (upper case C) is prefixed with a small zero written level with the top of the C. That is, °C. The small zero is associated with the numerical value of the temperature rather than with the capital C (upper case C).

Unit abbreviations or symbols, are written in lower case letters except when the unit abbreviation or symbol, is derived from a proper name. The full name of the unit abbreviation or symbol, is written in lower case letters even when it is derived from a proper name. So, the units named after Pascal would be written as pascal and abbreviated to Pa. One pascal would be abbreviated to 1 Pa and 14 pascals would be abbreviated to 14 Pa and not to 14 Pas.

Plurals of unit names.

Compound units formed by multiplication are written in a number of ways. For example, newton metres may be written as Nm or as, N.m or as, N m with a large dot between the two letters and level with the top of the lower case letter. That is, for example, as N·m. Although this last suggestion is preferred it is difficult to do well on many typewriters or word processors.

Compound units formed by division are written in a number of ways. For example, newtons per square metre may be written as N/m² or as, Nm⁻². The suggestion N/m², is preferred. That is, the unit abbreviations or symbols, are separated by a solidus or oblique stroke.

Compound prefixes are not used. That is, one million metres would be written as 1 Mm and not 1 kkm.

Common fractions, are not used. That is, one half of a kilogram would be written as 0·5 kg and not 1/2 kg.

Prefixes are not used in the denominator of a compound unit - except for kilograms as a kilogram is a base unit of the SI system. That is, one million newtons per square metre would be written as 1 MN/m² and not as 1 N/mm². (One newton per square millimetre is the same as one million newtons per square metre.)

It is suggested that for simplicity, when calculations are being performed, that prefixes are changed so that powers of ten are used. That is, decimal multiples are used.

It is suggested that when decimal multiples are used, that the prefix used should be 10 raised to a power that is a multiple of 3.

It is suggested for ease of understanding that when density is discussed, that the units are Mg/m³ rather than kg/m³. This gives values which have the same numerical values as the well established g/cm³ values or SG values.

To avoid misunderstandings avoid the use of the word billion; use a prefix such as giga (G) (Note, prefixes E P T G & N are capitalized).

Do not use a comma to separate groups of digits as it is used as a decimal marker in some countries.

Appendix B: SI Prefixes

Please note that the prefixes hecto, deca, deci and centi are non-preferred.

Prefix	Symbol	Value	Multiply unit by
Exa	E	10^{18}	1 000 000 000 000 000 000·
Peta	P	10^{15}	1 000 000 000 000 000·
Tera	T	10^{12}	1 000 000 000 000·
Giga	G	10^{9}	1 000 000 000·
Mega	M	10^{6}	1 000 000·
Kilo	k	10^{3}	1 000·
Hecto	h	10^{2}	100·
Deca	da	10^{1}	10·
Deci	d	10^{-1}	0·1
Centi	c	10^{-2}	0·01
Milli	m	10^{-3}	0·001
Micro	μ	10^{-6}	0·000 001
Nano	n	10^{-9}	0·000 000 001
Pico	p	10^{-12}	0·000 000 000 001
Fento	f	10^{-15}	0·000 000 000 000 001
Atto	a	10^{-18}	0·000 000 000 000 000 001

Symbol	Prefix	Value	Multiply unit by
E	exa	10^{18}	1 000 000 000 000 000 000·
P	peta	10^{15}	1 000 000 000 000 000·
T	tera	10^{12}	1 000 000 000 000·
G	giga	10^{9}	1 000 000 000·
M	mega	10^{6}	1 000 000·
k	kilo	10^{3}	1 000·
h	hecto	10^{2}	100·
da	deca	10^{1}	10·
d	deci	10^{-1}	0·1
c	centi	10^{-2}	0·01
m	milli	10^{-3}	0·001
μ	micro	10^{-6}	0·000 001
n	nano	10^{-9}	0·000 000 001
p	pico	10^{-12}	0·000 000 000 001
f	fento	10^{-15}	0·000 000 000 000 001
a	atto	10^{-18}	0·000 000 000 000 000 001

Value	Prefix	Symbol	Multiply unit by
10^{18}	Exa	E	1 000 000 000 000 000 000·
10^{15}	Peta	P	1 000 000 000 000 000·
10^{12}	tera	T	1 000 000 000 000·
10^{9}	giga	G	1 000 000 000·
10^{6}	mega	M	1 000 000·
10^{3}	kilo	k	1 000·
10^{2}	hecto	h	100·
10^{1}	deca	da	10·
10^{-1}	deci	d	0·1
10^{-2}	centi	c	0·01
10^{-3}	milli	m	0·001
10^{-6}	micro	μ	0·000 001
10^{-9}	nano	n	0·000 000 001
10^{-12}	pico	p	0·000 000 000 001
10^{-15}	fento	f	0·000 000 000 000 001
10^{-18}	atto	a	0·000 000 000 000 000 001

Appendix C: Unit conversion

In order to make recognition easier, the figures have been divided by spaces where appropriate. Spaces have been used, for example, as a thousand marker to the left of the decimal point; to the right of the decimal point, a space divides the digits into groups of three. In some cases a back slash, or /, has been used in place of the word 'per'. The use of an asterisk * indicates an exact number.

Knowing	Multiply by	To get
Acres	0·404 686	hectares
Acres	4 046·856 421	square metres
Acres	4 840·0 *	square yards
Angstroms	0·000 000 1	millimetres
Angstroms	0·000 000 000 1	metres
Angstroms	0·000 10	microns
Angstroms	0·100	millimicrons
Ares	0·010	hectares
Ares	100·0 *	square metres
Ares	0·000 1	square kilometres
Ares	119·599 005	square yards
Astronomical unit (AU)	$1·496 \times 10^{11}$	metres
Atmospheres - means standard atmospheres unless otherwise stated.		
Atmospheres	1·013 250	bars
Atmospheres	75·999 989	centimetres of mercury
Atmospheres	33·900 579	feet of water (at 4°C)
Atmospheres	29·921 256	inches of mercury (at 0°C)
Atmospheres	1·033 228	kilograms square centimetre
Atmospheres	101·325 0	kilonewtons/sq metre
Atmospheres	101 325·0	pascals
Atmospheres	14·695 949	pounds per square inch
Atmospheres	1·058 108	tons (short) per sq foot
Atmospheres	759·999 892	mm of mercury (at 0°C)
Atmospheres (metric)	1·0	kilograms force/sq cm
Avdp = avoirdupois.		
Bar	0·986 923	atmospheres
Bar	100 000·0 *	newtons per square metre
Bar	1 000 000·0	dynes per square centimetre
Bar	750·061 576	millimetres of mercury
Bar	14·503 774	pounds per square inch
Barn	$1·0 \times 10^{-27}$	square metres
Barrels (UK)	0·163 6	cubic metres
Barrels (UK)	36·0 *	gallons (UK)
Barrels (UK)	163·659 24	litres
Barrels of oil (US)	0·158 987	cubic metres
Barrels of oil (US)	42·0 *	gallons (US)
Barrels of oil (US)	158·987 3	litres
Barrels (US liquid)	31·5 *	gallons (US)
Baryes	0·000 001	bars
Baryes	1·0	dynes per square centimetre
Blots	10	joules/tesla
Bohr magneton	$9·274 08 \times 10^{-24}$	amperes metre squared
Bohr radius	$5·291 67 \times 10^{-11}$	metres
Btu = British thermal units.		
Btu	251·995 764	calories (gram calories)
Btu	778·169 270	foot-pounds (force)
Btu	1 055·055 863	joules (newton metres)
Btu	0·251 996	kilocalorie
Btu	1·055 058	kilojoules
Btu	0·000 293	kilowatt hours
Btu/cubic foot	8·899 146	kilocalories/cubic metre
Btu/cubic foot/°F	67·066 1	kilojoules/metre³ Kelvin
Btu/cubic foot	37·258 948	kilojoules/cubic metre
Btu/hour	0·069 998	gram-calories/second
Btu/hour	0·293 071	watts
Btu/hour foot squared	3·154 591	joules/second metre squared
Btu/hour foot squared	0·003 155	kilowatts/metre squared
Btu/hour foot squared °F	5·678 263	joules/second metre² K
Btu/hour foot squared °F	5·678 263	watts/metre² K
Btu/pound °C	2·326 0	joules/gram °C
Btu/pound °F	4 186·80	joules/kilogram °C

Appendix C: Unit conversion - *contd*

Knowing	*Multiply by*	*To get*
Bushels	0·036 369	cubic metres
Calories - large	1 000·0 *	calories
Calories$_{15\ degree}$	4·185 50	joules (newton metres)
Calories$_{IT}$	0·003 968 321	British thermal units
Calories$_{IT}$	0·000 001 163	kilowatt-hours
Calories$_{IT}$	4·186 80	joules (newton metres)
Calories/gram °C	4 186·80	joules/kilogram °C
Calories/second cm^2	41 868·0	joules/second metre squared
Calories/second cm^2	41·868 0	kilowatts/metre squared
Cals/second cm^2 °C	41·868 0	watts/metre^2 K
Cals/second cm^2 °C	41·868 0	joules/second metre^2 K
Candle power (spherical)	12·566 371	lumens
Candles (International)	1·0	lumens (Int)/steradian
Carats - metric	200·0	milligrams
Cental	45·359 238	kilograms
Centares (centiares)	1·0 *	square metres
Centigrade heat unit = Chu		
Centimetres	0·1 *	decimetres
Centimetres	0·032 808	feet
Centimetres	0·393 701	inches
Centimetres	0·000 010	kilometres
Centimetres	0·010 *	metres
Centimetres	10 000·0	microns
Centimetres	10·0	millimetres
Centimetres	10 000 000·0	millimicrons
Centimetres/second	1·968 504	feet/minute
Centimetres/second	0·032 808	feet/second
Centimetres/second	0·036 0	kilometres/hour
Centimetres/second	0·60	metres/minute
Centimetres squared - see square centimetres.		
Centipoises	1·0 *	dyne second/sq centimetre
Centipoises	0·010 *	grams/centimetre second
Centipoises	0·010	poises
Centipoises	1·0 *	meganewtons second/sq metre
Centipoises	0·001 0	pascal-seconds
Centistokes	0·000 001	square metres per second
Centistokes	1·0 *	sq millimetres/second
Centistokes	0·01 *	stokes
Chains	20·116 80	metres
Chains	22·0 *	yards
Chains - engineers	100·0 *	feet
Chains - engineers	30·480	metres
Cheval vapeur	1·0 *	horsepower (metric)
Cheval vapeur	0·735 499	kilowatts
Cheval vapeur	735·499	watts
Chu - centigrade heat unit.		
Chu/hour foot squared	5·678 264	joules/second metre squared
Chu/hour foot squared	0·005 678	kilowatts/metre squared
Chu/pound	1·0	calories/gram
Chu/pound °F	4 186·80	joules/kilogram °C
Coulombs	1·0 *	ampere seconds
Coulombs/second	1·0 *	amperes
Cubic centimetres	0·000 035 315	cubic feet
Cubic centimetres	0·061 023 744	cubic inches
Cubic centimetres	0·000 001	cubic metres
Cubic centimetres	0·000 219 969	gallons UK
Cubic centimetres	0·000 264 172	gallons US (liquid)
Cubic centimetres	0·001 0	litres
Cubic centimetres	0·000 227 021	US gallons (dry)
Cubic decimetres	1·0	litres
Cubic feet	28 316·846 59	cubic centimetres
Cubic feet	1728·0	cubic inches
Cubic feet	0·028 316 847	cubic metres
Cubic feet	6·228 835	gallons UK
Cubic feet	28·316 866	litres

APPENDIX C

Appendix C: Unit conversion - *contd*

Knowing	Multiply by	To get
Cubic feet/minute	1·0	cumins
Cubic feet/minute	0·124 675	gallons US/second
Cubic feet/minute	0·471 947	litres/second
Cubic feet/pound	0·062 428	cubic metres/kilogram
Cubic feet/second	1·0	cusecs
Cubic feet/second	0·028 317	cubic metres/second
Cubic feet/second	28·316 877	litres/second
Cubic inches	16·387 064	cubic centimetres
Cubic inches	0·000 578 704	cubic feet
Cubic inches	0·000 016 387	cubic metres
Cubic inches	16 387·064 00	cubic millimetres
Cubic inches	0·003 604 650	gallons (UK)
Cubic inches	0·016 317	litres
Cubic inches	0·576 744	ounces (fluid UK)
Cubic metres	1 000 000·0	cubic centimetres
Cubic metres	61 023·744 09	cubic inches
Cubic metres	35·314 666	cubic feet
Cubic metres	1·307 950	cubic yards
Cubic metres	10 000·0	decilitres
Cubic metres	219·969 248	gallons (UK)
Cubic metres	264·172 05	gallons US (liquid)
Cubic metres	10·0	hectolitres
Cubic metres	1 000·0 *	litres
Cubic yards	0·764 555	cubic metres
Cubits	1·5 *	feet
Cumins	1·0	cubic feet/minute
Curie	3.70×10^{10}	becquerel
Cusecs	1·0	cubic feet/second
Cycles per second	1	hertz
Deciare	10·0	square metres
Decilitres	0·000 1	cubic metres
Decilitres	0·1	litres
Decimetres	10	centimetres
Decimetres	3·937 008	inches
Decimetres	0·1	metres
Decimetres	100	millimetres
Degrees (of angle)	0·017 453	radians
Dekalitres	10·0 *	litres
Dekametres	10·0 *	metres
Denier (international)	1.111×10^{-7}	kilograms per metre
Drachm (UK)	3·887 93	grams
Drachm (fluid UK)	3 551·63	cubic millimetres
Dram (avdp)	1·771 845	grams
Dram (avdp)	0·062 50	ounces
Dram (fluid US)	0·003 697	kilograms
Dram (US)	0·003 888	kilograms
Dynes	0·001 020	grams
Dynes	0·000 000 1	joules/centimetre
Dynes	0·000 001 02	kilograms-force
Dynes	0·000 01 *	newtons (joules/metre)
Dynes	0·000 072 330	poundals
Dynes	0·000 002 248	pounds-force
Dyne centimetre	1.0×10^{-7}	joules (newton metres)
Dynes/centimetre	0·001 0	newtons per metre
Dynes/square centimetre	0·10	newtons/square metre
Dynes/square centimetre	0·000 014 504	pounds force/square inch
Dynes/square centimetre	0·000 014 504	pounds per square inch
Dynes/square centimetre	0·002 088	pounds per square foot
Electron volt	$1.602\ 19 \times 10^{19}$	joules
Ell	45·0 *	inches
Ergs	0·000 000 1 *	joules (newton metres)
Fathoms	6·0 *	feet
Fathoms	1·828 80	metres

Appendix C: Unit conversion - *contd*

Knowing	Multiply by	To get
Feet	12.0 *	inches
Feet	0.304 80 *	metres
Feet	0.000 304 80 *	kilometres
Feet cubed - see cubic feet		
Feet of water (17°C)	0.029 460	atmospheres
Feet of water (17°C)	0.881 507	inches of mercury
Feet of water (17°C)	0.030 440	kilograms/square centimetre
Feet of water (17°C)	2 985.126 553	newtons per square metre
Feet of water (17°C)	0.432 955	pounds/square inch
Feet per minute	0.508 0	centimetres per second
Feet per minute	0.005 08	metres per second
Feet per minute	0.304 80 *	metres per minute
Feet per second	30.480	centimetres per second
Feet per second	1.097 280	kilometres per hour
Feet per second	18.880	metres per minute
Feet squared per second	0.092 903	metres squared per second
Fluid ounces - see ounces (fluid).		
Foot cubed - see cubic feet.		
Foot-candles	1.0 *	lumens/square foot
Foot-candles	10.763 910	lux
Foot-lambert	3.426 26	candelas per square metre
Foot poundals	0.042 140 160	joules
Foot pounds	0.001 286	British thermal units
Foot pounds	0.323 836	gram-calories
Foot pounds	32.174 049	foot poundals
Foot pounds	5.050×10^{-7}	horsepower-hour (UK)
Foot pounds	1.355 818	joules
Foot pounds	0.138 255	kilogram (force) metres
Foot pounds per second	1.355 818	watts
Foot pounds per minute	0.001 286	Btus/minute
Foot pounds per minute	$3.030 3 \times 10^{-5}$	horsepower (UK)
Foot pounds per minute	0.000 324	kilogram calories/minute
Foot pounds per minute	$2.259 7 \times 10^{-5}$	kilowatts
Furlongs	0.201 168	kilometres
Furlongs	201.168 0	metres
Furlongs	220.0 *	yards
Galileo	1.0	centimetres/second squared
Galileo	0.01	metres per second squared
Gallons UK	1.200 950	gallons US (liquid)
Gallons UK	4 546.090	cubic centimetres
Gallons UK	0.160 544	cubic feet
Gallons UK	277.419 433	cubic inches
Gallons UK	0.004 546 090	cubic metres
Gallons UK	4.546 090 *	litres
Gallons US means gallons US - liquid, unless otherwise stated.		
Gallons US - dry	0.004 405	cubic metres
Gallons US	3 785.411 785	cubic centimetres
Gallons US	231.0 *	cubic inches
Gallons US	0.133 681	cubic feet
Gallons US	0.832 674	gallons UK
Gallons US	3.785 411	litres
Gallons US (of water)	8.335 882	pounds of water
Gallons US per minute	0.002 228	cubic feet/second
Gallons US per minute	3.785 415	litres/minute
Gallons US per minute	0.063 090	litres/second
Gallons US per minute	0.003 782	cubic metres/minute
Gallons US per minute	0.227 125	cubic metres/hour
Gallons US per square foot	40.743	litres per square metre
Gauge	0.254	microns
Gauss	0.000 1	tesla
Gilbert	0.795 8	ampere
Gill - US	118.294 118	cubic centimetres
Grade	0.015 708	radian
Grains	0.064 799	grams

Appendix C: Unit conversion - *contd*

Knowing	Multiply by	To get
Grains	64·798 9	milligrams
Grains	0·002 286	ounces (avdp)
Gram calories - see calories.		
Grams	0·001 0	kilograms
Grams	0·035 274	ounces (avdp)
Grams	0·002 205	pounds (avdp)
Grams per centimetre	0·005 600	pounds per inch
Grams per denier	0·088 26	newtons per tex
Grams per cubic centimetre	1 000·0	kilograms/cubic metre
Grams per cubic centimetre	0·578 038	ounces per cubic inch
Grams per cubic centimetre	62·427 960	pounds per cubic foot
Grams per cubic centimetre	0·036 127	pounds per cubic inch
Grams (force) centimetres	$1·0 \times 10^{-5}$	kgf metres
Grams (force) centimetres	$9·806\ 65 \times 10^{-5}$	newton metres
Grams (force)/sq centimetre	10·0	kgf per square metre
Grams (force)/sq centimetre	0·014 223	pounds per square inch
Grams (force)/sq centimetre	2·048 161	pounds per square foot
Hectares	2·471 054	acres
Hectares	100·0	ares
Hectares	10 000·0	square metres
Hectares	0·010	square kilometres
Hectolitres	0·10	cubic metres
Hectolitres	1 000·0	decilitres
Hectolitres	21 996·925	gallons (UK)
Hectolitres	100·0	litres
Hectometres	100·0	metres
Horsepower (550 ft lbf/s)	745·700	watts
Horsepower (boiler)	981·0	watts
Horsepower (electric)	746·0 *	watts
Horsepower (UK)	42·407 226	Btu/minute
Horsepower (UK)	550·0 *	foot-pounds per second
Horsepower (UK)	0·745 700	kilowatts
Horsepower hour (UK)	0·745 700	kilowatts hour
Horsepower hour (UK)	2·684 52	megajoules
Horsepower (metric)	0·735 499	kilowatts
Horsepower (metric)	75·0 *	kilogram metres/second
Hundredweight (UK or long)	50·802 346	kilograms
Hundredweight (short or US)	45·359 238	kilograms
Imperial gallons - see gallons UK.		
Inches	0·025 40 *	metres
Inches	25·40	millimetres
Inches	1 000·0 *	mils
Inches cubed - see cubic inches.		
Inches of mercury (0°C)	0·033 42	atmospheres
Inches of mercury (0°C)	0·034 53	kilograms/sq centimetre
Inches of mercury (0°C)	3 386·39	newtons per square metre
Inches of mercury (0°C)	0·491 154	pounds per square inch
Inches of water (4°C)	0·002 458	atmospheres
Inches of water (4°C)	0·073 552	inches of mercury (°C)
Inches of water (4°C)	0·002 540	kgf/square centimetre
Inches of water (4°C)	0·036 125	pounds per square inch
Inches of water (4°C)	249·073 066	newtons per square metre
Inches per minute	0·423 333	millimetres per second
Inches ounces	0·007 062	newton metres
Inches ounces	0·062 50	inch pounds
Inch pounds	16·0	inch ounces
Inch pounds	0·112 985	newton metres
Inches squared - see square inches.		
International (int) nautical miles - see miles - nautical.		
Irons	0·529 167	millimetres
Joules (newton metres)	0·000 947 81	Btu
Joules (newton metres)	0·238 846	calories $_{IT}$

Appendix C: Unit conversion - *contd*

Knowing	Multiply by	To get
Joules (newton metres)	1.0×10^7	ergs
Joules (newton metres)	0.737 560	foot-pounds
Joules (newton metres)	1.0 *	newton metres
Joules (newton metres)	1.0	watt-second
Joules per centimetre	10 197.162 13	grams
Joules per centimetre	100.0	newtons (joules/metre)
Joules per centimetre	22.480 894	pounds
Joules/second	1.0	watts
Joules/second metre^2 K	1.0	watts/metre squared K
Kgf = kilogram force.		
Kgf centimetres	0.072 330	foot-pounds
Kgf centimetres	0.010	kgf metres
Kgf centimetres	0.098 067	newton metres
Kgf metres	7.233 014	foot-pounds
Kgf metres	100.0	kgf centimetres
Kgf metres	9.806 650	newton metres
Kgf/metres second squared	1.0 *	mPa seconds/sq metre
Kgf seconds/square metre	9.806 650	pascal-seconds
Kgf/sq centimetre	28.959 021	inches of mercury
Kgf/sq centimetre	0.967 841	atmospheres
Kgf/sq centimetre	1.0 *	atmospheres (metric)
Kgf/sq centimetre	98 066.50	pascals
Kgf/sq centimetre	32.810 368	feet of water (4°C)
Kgf/sq centimetre	98.066 50	kilopascals
Kgf/sq centimetre	14.223 343	pounds force/square inch
Kgf/square metre	9.806 650	pascals
Kgf/square metre	0.204 816	pounds force/square foot
Kgf/square millimetre	9.806 65	megapascals
Kgf/square millimetre	0.711 167	tons (US) force/sq inch
Kgf/square millimetre	1.0×10^6	kgf/square metre
Kilobar	100.0	mega pascals/m^2
Kilocalories	1 000.0	calories (gram calories)
Kilocalories	3.968 320	British thermal units
Kilocalories	4 186.80	joules
Kilocalories	4.186 80	kilojoules
Kilocalories/hours	0.001 163	kilowatts
Kilocalories/gram °C	4 186.80	joules/kilogram °C
Kilocalories/hours foot^2	12.518 428	joules/second metre squared
Kilocalories/hours foot^2	0.012 518	kilowatts/metre squared
Kilocalories/hours metre^2	1.163 0	joules/second metre squared
Kilogram calorie - see kilocalories.		
Kilograms	1 000.0	grams
Kilograms	2.204 62	pounds
Kilograms	0.001 0	tonnes (metric)
Kilograms	0.000 984 207	tons (long)
Kilograms	980 665.0	dynes
Kilograms	9.806 650	newtons (joules/metre)
Kilograms	70.931 6	poundals
Kilograms	2.204 623	pounds-force
Kilograms/centimetre	9.806 65	newtons per centimetre
Kilograms/centimetre	5.599 741	pounds per inch
Kilograms/cubic metre	0.001 0	grams/cubic centimetre
Kilograms/cubic metre	0.062 428	pounds per cubic foot
Kilograms/metre	9.806 65	newtons per metre
Kilograms/metre	0.671 969	pounds per foot
Kilograms/metre	2.015 907	pounds per yard
Kilograms/metre	0.000 336	tons (US) force per foot
Kilograms-force - see Kgf.		
Kilograms/cubic metre	$3.612\ 7 \times 10^{-5}$	pounds per cubic inch
Kilograms/cubic metre	1.685 553	pounds per cubic yard
Kilograms/litre	1.0	megagrams per cubic metre
Kilograms/litre	8.344 5	pounds per gallon (US)
Kilograms/sq centimetre	14.223 343	pounds per square inch
Kilograms/sq metre	0.001 422	pounds per square inch

Appendix C: Unit conversion - *contd*

Knowing	Multiply by	To get
Kilojoules	0·947 817	British thermal units
Kilojoules	1 000·0	joules (newton metres)
Kilometres	100 000·0	centimetres
Kilometres	3 280·839 895	feet
Kilometres	1 000 000·0	millimetres
Kilometres	1 000·0	metres
Kilometres	0·621 371	miles
Kilometres per hour	0·911 344	feet per second
Kilometres per hour	0·277 778	metres per second
Kilometres per hour	0·621 371	miles per hour
Kiloponds (kp or kps)	1·0	kilograms (kgf)
Kilopounds (kips)	1 000·0	pounds
Kilowatt hours	3 412·141 600	British thermal units
Kilowatt hours	859 845·227 9	calories
Kilowatt hours	1·341 022	horsepower hours (UK)
Kilowatt hours	3 600 000·0	joules
Kilowatt hours	859·845 228	kilocalories
Kilowatt hours	3·600	megajoules
Kilowatts	56·869 027	Btu/minute
Kilowatts	737·562 150	foot-pounds per second
Kilowatts	1·341 022	horsepower (UK)
Kilowatts	14·330 754	kilocalories/minute
Kilowatts/metre squared	1 000·0	joules/second metre squared
kips = kilopounds = 1,000 pounds.		
kips	4 448·221 659	newtons
kips/square inch	1 000·0	pounds per square inch
Knots (international)	1·852 0 *	kilometres per hour
Knots (international)	1·0 *	miles (nautical)/hour
Knots (international)	0·514 444	metres per second
Knots (UK)	1·853 184	kilometres per hour
kp = kps kiloponds = kgf		
kp/square centimetre	1·0	kgf/square centimetre
Lambert	0·318 310	candles/sq centimetre
lbf = pounds force		
League	3·0 *	miles
League (nautical Int)	5·556 0	kilometres
League (nautical UK)	18 240·0 *	feet
Light year	9·460 55 × 10^{15}	metres
Links (US survey)	0·660 *	feet
Links (US survey)	0·201 168	metres
Litre atmospheres	101·325 0 *	joules (newton metres)
Litres	1·0	cubic decimetres
Litres	0·035 315	cubic feet
Litres	61·022 744	cubic inches
Litres	0·001 0	cubic metres
Litres	0·219 969	gallons (UK)
Litres	0·264 172	gallons (US liquid)
Litres	0·010	hectolitres
Litres	1·759 753	pints (UK)
Litres	2·113 376	pints (US liquid)
Litres	1·056 688	quarts (US liquid)
Litres per minute	0·035 315	cubic feet/minute
Litres per minute	0·004 403	gallons (US) per second
Long tons. See tons (long).		
Lumens	0·001 471	watts
Lumens per square foot	10·763 910	lux
Lusecs	133·332	micronewtons-metre/second
Lux	0·092 9	foot-candles
Lux	1·0 *	lumens/square metre
Maxwell	0·000 000 01	weber
Maxwell/sq centimetre	1·0 *	gausses
Megabars	100·0	giganewtons/square metre
Megapascals	10·0 × 10^6	bar
Megapascals	1·0	meganewtons/square metre

Appendix C: Unit conversion - *contd*

Knowing	*Multiply by*	*To get*
Megapascals	145·038 736	pounds per square inch
Metres	100·0	centimetres
Metres	10·0	decimetres
Metres	3·280 840	feet
Metres	0·001	kilometres
Metres	1 000·0	millimetres
Metres	1·093 613	yards
Metres per second	196·850 394	feet per minute
Metres per second	3·280 84	feet per second
Metres per second	3·60	kilometres per hour
Metres per second	2·236 94	miles per hour
Metres per minute	1·666 667	centimetres per second
Metres per minute	3·280 840	feet per minute
Metres per minute	0·037 28	miles per hour
Metric horsepower - see horsepower.		
Metric tonnes	1 000·0	kilograms
Metric tonnes	2 204·622 60	pounds
Metric tonnes	0·984 207	tons (long)
Metric tonnes	1·102 311	tons (short)
Metric tonnes force	1 000·0	kgf
Metric tonnes force/m^2	9·806 650	kilopascals
Metric tonnes force/m^2	1·422 334	pounds force/square inch
Metric tonnes force/m^2	0·102 408	tons (US) force/square foot
Metric tonnes/cubic metre	1·0	megagrams/cubic metre
Metric tonnes/cubic metre	0·843	tons (US)/cubic yard
Microbar	0·10	newtons per square metre
Microinch	0·000 001	inches
Micrometres of mercury	0·133 322	newtons per square metre
Microns	3·937	gauge
Microns	10 000·0 *	Angstroms
Microns	1·0 × 10^{-6}	metres
Microns	0·001	millimetres
Microns	1 000·0	millimicrons
Mil - see mils.		
Miles	5 280·0 *	feet
Miles	1·609 344	kilometres
Miles	1 760·0 *	yards
Miles per hour	44·704 0	centimetres per second
Miles per hour	88·0	feet per minute
Miles per hour	1·466 667	feet per second
Miles per hour	1·609 344	kilometres per hour
Miles per hour	26·822 40	metres per minute
Miles per hour	0·447 04	metres per second
Miles per minute	2 682·240	centimetres per second
Miles - nautical (mi n), see Miles - n.		
Miles - n (International)	1·852 0 *	kilometres
Miles - n (telegraphic)	6087·0	feet
Miles - n (telegraphic)	1·855 32	kilometres
Miles - n (UK)	1·853 184	kilometres
Miles - n (UK)	6 080·0	feet
Millibar	100·0	newtons per square metre
Milligalileo	0·001 0	galileo
Millimetres	10 000 000·0	Angstroms
Millimetres	0·10	centimetres
Millimetres	0·010	decimetres
Millimetres	0·003 281	feet
Millimetres	0·039 370	inches
Millimetres	0·000 001	kilometres
Millimetres	0·001 0	metres
Millimetres	1 000·0	microns
Millimetres	1 000 000·0	millimicrons
Millimetres	39·370 079	mils
Millimetres of mercury (0°C)	133·322 387	newtons per square metre
Millimetres of water (0°C)	9·804 684	newtons per square metre
Millimicrons	0·001 0	microns
Millimicrons	10·0	Angstroms

APPENDIX C

Appendix C: Unit conversion - *contd*

Knowing	Multiply by	To get
Millitorr	0.133 322	newtons per square metre
Mils	0.002 54	centimetres
Mils	0.001	inches
Mils	0.000 025 4	metres
Mils	25.4	microns
Mils	0.025 4	millimetres
Minims (UK)	59.193 905	cubic millimetres
Minutes (angular)	0.000 290 9	radians
Myriagrams	10.0	kilograms
Myriametres (US)	10.0 *	kilometres
Nautical miles (n mi) - see Miles - n.		
New candles	60.0 *	candles/square centimetre
Newton centimetres	0.007 376	foot-pounds
Newton centimetres	0.101 972	kgf metres
Newton centimetres	0.010	newton metres
Newton metres	0.737 562	foot-pounds
Newton metres	10.197 162	kgf metres
Newton metres	100.0	newton centimetres
Newtons (joules/metre)	100 000.0	dynes
Newtons (joules/metre)	0.101 972	kilograms
Newtons (joules/metre)	7.233 014	poundals
Newtons (joules/metre)	0.224 809	pounds
Newtons per square metre	0.000 009 869	atmospheres
Newtons per square metre	1.0 *	pascals
Newtons per square metre	10.0 *	dynes/sq centimetre
Newtons per square metre	0.000 010 972	kilograms/sq centimetre
Newtons per square metre	0.020 88	pounds per square foot
Newtons per square metre	0.000 145 038	pounds per square inch
Nits	1	candelas per square metre
Oersteds	79.58	amperes per metre
Ounce-force inches	7 061.55	micronewton metres
Ounces (avdp)	16.0 *	drams (avdp)
Ounces (avdp)	28.349 523	grams
Ounces (avdp)	0.028 350	kilograms
Ounces (avdp)	0.278 014	newtons (joules/metre)
Ounces (avdp)	0.062 50	pounds (avdp)
Ounces (fluid UK)	28.413 063	cubic centimetres
Ounces (fluid UK)	0.028 414	litres
Ounces (fluid US)	29.573 530	cubic centimetres
Ounces (fluid US)	1.804 688	cubic inches
Ounces per cubic inch	1.733	grams per cubic centimetre
Ounces (apothecary/troy)	8.0 *	drams (apothecary/troy)
Ounces (apothecary/troy)	480.0 *	grains
Ounces (apothecary/troy)	31.103 5	grams
Ounces (apothecary/troy)	20.0 *	pennyweights (troy)
Ounces (apothecary/troy)	0.083 333	pounds (troy)
Paces	2.5 *	feet
Pascals	1.0	newtons per square metre
Pascal-seconds	10.0	dyne-seconds/sq centimetre
Pascal-seconds	0.102	kgf seconds/sq metre
Pascal-seconds	1.0	newtons-seconds/sq metre
Pascal-seconds	10.0	poise
Pascal-seconds	0.020 88	lbf seconds/square foot
Pascal-seconds	0.000 145	lbf seconds/square inch
Parsecs	$3.085\ 7 \times 10^{13}$	kilometres
Pecks (UK)	9.092 180	litres
Pecks (US)	8.809 769	litres
Pennyweight	1.555 17	grams
Pennyweights (troy)	24.0 *	grains
Perch (rod or pole)	5.029 20	metres
Perch (rod or pole)	5.5 *	yards
Pferde-stärke (metric horsepower) - see horsepower.		

Appendix C: Unit conversion - *contd*

Knowing	Multiply by	To get
Phots	1·0	lumens/sq centimetre
Pints (UK)	0·568 261	litres
Pints (UK)	20·0 *	ounces (fluid UK)
Pints (UK)	1·200 950	pints (US liquid)
Pints (US dry)	0·550 610	litres
Pints (US liquid)	0·473 2	litres
Pints (US liquid)	16·0 *	ounces (fluid US)
Pints (US liquid)	0·832 674	pints (UK)
Pole (rod or perch)	5·029 20	metres
Pole (rod or perch)	5·5 *	yards
Poncelets	100·0	kgf metres/second
Poncelets	980·665	watts
Poise	100·0	centipoises
Poise	1·0 *	dyne second/sq centimetre
Poise	1·0 *	gram/centimetre second
Poise	0·102	kgf seconds/sq metre
Poise	0·000 015	lbf seconds/square inch
Poise	0·10	pascal-seconds
Poise	0·067 197	pounds/foot second
Pound (lb) - see pounds.		
Poundals	13 825·495 57	dynes
Poundals	0·014 098	kilograms-force
Poundals	0·138 255	newtons (joules/metre)
Poundals	0·031 081	pounds-force
Pounds (avdp)	7 000·0 *	grains
Pounds (avdp)	453·592 375	grams
Pounds (avdp)	0·453 592 375	kilograms
Pounds (avdp)	16·0 *	ounces (avdp)
Pounds (avdp)	32·174 049	poundals
Pounds (avdp)	0·000 446 429	tons (long)
Pounds (apothecary/troy)	5 760·0 *	grains
Pounds (apothecary/troy)	0·373 242	kilograms
Pounds-feet	0·138 3	metres-kilogram
Pounds force	0·453 592	kilograms-force
Pounds force	4·448 22	newtons (joules/metre)
Pounds force	32·174	poundals
Pounds force foot	1·355 82	newton metres
Pounds force inch	0·112 985	newtons/metre
Pounds force per square inch. See pounds per square inch (psi).		
Pounds of water	0·016 02	cubic feet
Pounds of water	0·276 8	cubic inches
Pounds of water	0·119 8	gallons US
Pounds of water/minute (17°C)	0·000 267	cubic feet/second
Pounds of water/minute (17°C)	0·119 985	gallons (US liquid)/minute
Pounds per cubic foot	0·016 019	gram per cubic centimetre
Pounds per cubic foot	16·018 463	kilograms per cubic metre
Pounds per cubic foot	0·000 598 704	pounds per cubic inch
Pounds per cubic inch	27·679 904 98	grams per cubic centimetre
Pounds per cubic inch	27 679·9	kilograms per cubic metre
Pounds per cubic inch	1728·0	pounds per cubic foot
Pounds per square inch	0·068 046	atmospheres
Pounds per square inch	2·309 704	feet of water (17°C)
Pounds per square inch	2·036 021	inches of mercury (0°C)
Pounds per square inch	703·069 59	kilograms per square metre
Pounds per square inch	0·070 306 979	kilograms/sq centimetre
Pounds per square inch	0·006 894 76	megapascals
Pounds per square inch	6 894·757 361	newtons per square metre
Pounds per inch	178·579 675	grams per centimetre
Quarts (UK)	1·136 533	litres
Quarts (US dry)	1·101 122	litres
Quarts (US liquid)	946·352 946	cubic centimetres
Quarts (US liquid)	57·750 000	cubic inches
Quarts (US liquid)	0·946 352	litres
Quarters (UK)	12·700 6	kilograms
Quarters (UK)	28·0	pounds

APPENDIX C

Appendix C: Unit conversion - *contd*

Knowing	*Multiply by*	*To get*
Quintals (metric)	100·0 *	kilograms
Quintals (UK or long)	112·0 *	pounds
Quintals (US or short)	100·0 *	pounds
Rad	100	gray
Radians	57·295 780	degrees
Relative density - see specific gravity.		
Rods (perch or pole)	5·029 20	metres
Rods (perch or pole)	5·50 *	yards
Roentgens	0·000 258	coulombs per kilogram
Roods	0·250	acres
Roods	101 171·411 05	square metres
Second (angle)	0·000 004 848	radian
Section	0·000 002 590	square metre
Scruples (UK)	1·295 98	grams
Short tons - see tons (short).		
Slugs	14·593 9	kilograms
Small calories - see calories.		
Specific gravity	62·427 960	pounds per cubic foot
Specific gravity	16·387 064	grams per cubic inch
Specific gravity	0·578 038	ounces per cubic inch
Specific gravity	0·036 127	pounds per cubic inch
Square centimetres	197 352·524 1	circular mils
Square centimetres	127·323 955	circular millimetres
Square centimetres	0·010	square decimetres
Square centimetres	0·001 076	square feet
Square centimetres	0·155 000	square inches
Square centimetres	0·000 10	square metres
Square centimetres	100·0	square millimetres
Square chains	404·685 642	square metres
Square decimetres	100·0	square centimetres
Square decimetres	10 000·0	square millimetres
Square decimetres	0·010	square metres
Square feet	929·030 400	square centimetres
Square feet	0·092 903	square metres
Square feet (US survey)	1·000 004	square feet
Square inches	6·451 60	square centimetres
Square inches	645·160	square millimetres
Square kilometres	10 000·0	ares
Square kilometres	100·0	hectares
Square kilometres	1 000 000·0	square metres
Square kilometres	0·386 102	square miles
Square metres	0·010	ares
Square metres	0·000 10	hectares
Square metres	10 000·0	square centimetres
Square metres	100·0	square decimetres
Square metres	0·000 001	square kilometres
Square metres	1 000 000·0	square millimetres
Square metres	10·763 910	square feet
Square metres	1 550·003 100	square inches
Square metres	1·195 990	square yards
Square miles	2·589 988	square kilometres
Square millimetres	0·010	square centimetres
Square millimetres	0·000 1	square decimetres
square millimetres	0·001 550	square inches
Square millimetres	0·000 001	square metres
Square yards	8 361·273 60	square centimetres
Square yards	0·836 127	square metres
Stere	1·0 *	cubic metres
Stilbs	1·0	candle/sq centimetre
Stokes	1·0 *	square centimetre/second
Stones	6·350 293	kilograms
Stones	14·0 *	pounds (avdp)

Appendix C: Unit conversion - *contd*

Knowing	Multiply by	To get
Teslas	10 000·0 *	gausses
Teslas	1·0 *	webers/sq metre
Tex	1·0	grams per metre
Tex	0·000 001	kilogram per metre
Thermies	4·185 50	megajoules
Therms	100 000·0	Btu
Therms	105·506	megajoules
Thou. See mils.		
Thousandths of an inch. See mils.		
Tons (imperial) = tons (long).		
Tons (long)	1 016·046 919	kilograms
Tons (long)	1·016 047	metric tonnes
Tons (long)	2 240·0 *	pounds (avdp)
Tons (long)	1·120	tons (short)
Tons (long) force	2 240·0	pounds force
Tons (long) force/sq inch	152·422 81	atmospheres
Tons (long) force/sq inch	1·574 876	kgf/sq millimetre
Tons (long) force/sq inch	$1·544\ 43 \times 10^7$	pascals
Tons (refrigeration)	3 516·90 *	watts
Tons (UK). See tons (long).		
Tonnes (metric). See metric tonnes.		
Tons (metric). See metric tonnes.		
Tons short. See tons (US).		
Tons (US)	0·907 185	metric tonnes
Tons (US)	2 000·0 *	pounds (avdp)
Tons (US)	0·892 857	tons (long)
Tons (US) force	2 000·0	pounds force
Tons (US) force/foot	2 976·328	kilograms force/metre
Tons (US) force/yard	992·109	kilograms force/metre
Tons (US) force/sq inch	136·091 929	atmospheres
Tons (US) force/sq inch	1·406 139	kgf/sq millimetre
Tons (US) force/sq inch	13·789 514	megapascals
Torr	0·001 316	atmospheres
Torr	1 000·0	mm mercury (0°C)
Torr	133·322 368	newtons per square metre
US gallons - see gallons US.		
Unit poles	0·000 000 126	weber
Watt-hours	3·412 142	British thermal units
Watt-hours	859·845	gram-calories
Watt-hours	0·001 341	horsepower-hours (UK)
Watt-hours	367·097 837	kilogram (force) metres
Watts	3·412 142	British thermal units/hour
Watts	10 000 000·0	ergs per second
Watts	44·253 729	foot-pounds per minute
Watts	0·001 341	horsepower (UK)
Watts	1·0 *	joules per second
Watts	0·014 331	kilogram-calories/minute
Watts/centimetre squared	3 169·983 276	Btu/hour foot^2
Watts/metre squared	1·0	joules/second metre^2
Watts/metre squared	0·001 0	kilowatts/metre^2
Yards	3·0 *	feet
Yards	91·440 0	centimetres
Yards	0·914 40	metres

Appendix D: Temperature Conversion

In the temperature conversion table the centre column contains the numbers that you wish to convert. If you have a temperature in °C that you would like in °F then, read from the centre to the right. For example, if you like to convert 200°C into °F, then find 200 in the centre column and then look immediately right: you should see 392°F. Check this by dividing by 5, multiplying by 9 and adding on 32.

$$\frac{200 \times 9}{5} = 360$$

Then 360 + 32 = 392 (or 200 × 1·8 then + 32).

To go from °F to °C then enter the figure you have in the centre column and look immediately left 200°F becomes 93°C. Check this by taking away 32 from 200, then dividing by 9 and then multiplying by 5.

200 − 32 = 168.

Then $\frac{168 \times 5}{9} = 93$ (or 168 × 0·56)

TEMPERATURE CONVERSION TABLE

Centigrade	Starting Value	Fahrenheit
−18	0	32
−12	10	50
−7	20	68
−1	30	86
4	40	104
10	50	122
16	60	140
21	70	158
27	80	176
32	90	194
38	100	212
43	110	230
49	120	248
54	130	266
60	140	284
66	150	302
71	160	320
77	170	338
82	180	356
88	190	374
93	200	392
99	210	410
104	220	428
110	230	446
116	240	464
121	250	482
127	260	500
132	270	518
138	280	536
143	290	554
149	300	572
154	310	590
160	320	608
166	330	626
171	340	644
177	350	662
182	360	680
188	370	698
193	380	716
199	390	734
204	400	752
210	410	770
216	420	788
221	430	806
227	440	824
232	450	842
238	460	860
243	470	878
249	480	896
254	490	914
260	500	932
266	510	950
271	520	968
277	530	986
282	540	1004
288	550	1022
293	560	1040
299	570	1058
304	580	1076
310	590	1094
316	600	1112
321	610	1130
327	620	1148
332	630	1166
338	640	1184
343	650	1202
349	660	1220
354	670	1238
360	690	1256
366	690	1274
371	700	1292
377	710	1310
382	720	1328
388	730	1346
393	740	1364
399	750	1382
404	760	1400
410	770	1418
416	780	1436
421	790	1454

Appendix E: Relative Atomic Masses (Atomic Weights)

Element	Symbol	Relative atomic mass
Aluminium	Al	26.98
Antimony	Sb	121.8
Arsenic	As	74.92
Barium	Ba	137.3
Bismuth	Bi	209.0
Boron	B	10.81
Bromine	Br	79.90
Cadmium	Cd	112.4
Calcium	Ca	40.08
Carbon	C	12.01
Chlorine	Cl	35.45
Chromium	Cr	52.00
Cobalt	Co	58.93
Copper	Cu	63.55
Fluorine	F	19.00
Gold	Au	197.0
Hydrogen	H	1.008
Iodine	I	126.9
Iron	Fe	55.85
Lead	Pb	207.2
Lithium	Li	6.941
Magnesium	Mg	24.31
Manganese	Mn	54.94
Mercury	Hg	200.6
Molybdenum	Mo	95.94
Nickel	Ni	58.71
Nitrogen	N	14.01
Osmium	Os	190.2
Oxygen	O	15.99
Palladium	Pd	106.4
Phosphorus	P	30.97
Platinum	Pt	195.1
Potassium	K	39.10
Rhodium	Rh	102.9
Silicon	Si	28.09
Silver	Ag	107.9
Sodium	Na	22.99
Sulphur	S	32.06
Tin	Sn	118.7
Titanium	Ti	47.90
Tungsten	W	183.9
Vanadium	V	50.94
Zinc	Zn	65.37

Appendix F: The Greek Alphabet

alpha	α	A	nu	ν	N
beta	β	B	xi	ξ	Ξ
gamma	γ	Γ	omicron	o	O
delta	δ	Δ	pi	π	Π
epsilon	ϵ	E	rho	ρ	P
zeta	ζ	Z	sigma	σ	Σ
eta	η	H	tau	τ	T
theta	θ	Θ	upsilon	υ	Ψ
iota	ι	I	phi	ϕ	Φ
kappa	κ	K	chi	χ	X
lambda	λ	Λ	psi	ψ	Ψ
mu	μ	M	omega	ω	Ω